ENCYCLOPEDIC HANDBOOK OF BIOMATERIALS AND BIOENGINEERING

Part A: Materials

Volume 2

ENCYCLOPEDIC HANDBOOK OF BIOMATERIALS AND BIOENGINEERING

Part A: Materials

Volume 2

edited by

Donald L. Wise
Northeastern University
Boston, Massachusetts

Debra J. Trantolo
Cambridge Scientific, Inc.
Belmont, Massachusetts

David E. Altobelli
Harvard School of Dental Medicine
Boston, Massachusetts

Michael J. Yaszemski
United States Air Force
Lackland Air Force Base, Texas

Joseph D. Gresser
Cambridge Scientific, Inc.
Belmont, Massachusetts

Edith R. Schwartz
National Institute of Standards and Technology
Gaithersburg, Maryland

MARCEL DEKKER, INC.   NEW YORK · BASEL · HONG KONG

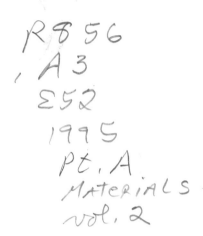

Library of Congress Cataloging-in-Publication Data

Encyclopedic handbook of biomaterials and bioengineering / edited by Donald L. Wise
... [et al.].
    p. cm.
    Contents: Pt. A., v. 1-2. Materials -- Pt. B., v. 1-2. Applications.
    ISBN 0-8247-9593-8 (v. 1 : alk. paper) — ISBN 0-8247-9594-6 (v. 2 : alk. paper)
— ISBN 0-8247-9595-4 (v. 1 : alk. paper) — ISBN 0-8247-9596-2 (v. 2 : alk. paper).
    1. Biomedical engineering -- Encyclopedias. 2. Biomedical materials--
Encyclopedias.   I. Wise, Donald L. (Donald Lee)

R856.A3E52   1995
610'.28--dc20
                                        95-21232
                                            CIP

The publisher offers discounts on this book when ordered in bulk quantities. For more information, write to Special Sales/Professional Marketing at the address below.

This book is printed on acid-free paper.

Marcel Dekker, Inc.
270 Madison Avenue, New York, New York 10016

Current printing (last digit):
10 9 8 7 6 5 4 3

PRINTED IN THE UNITED STATES OF AMERICA

Preface

The medical device and drug industry is consistently one of the strongest performers. Materials are a key ingredient in this industry. Development of these materials is in a constant state of activity, with the burdens of old materials not withstanding the tests of time and new materials needs coming to the forefront of modern applications. This handbook focuses on materials used in or on the human body—materials that define the world of "biomaterials."

The *Encyclopedic Handbook of Biomaterials and Bioengineering* covers the range of biomaterials from polymers to metals to ceramics. The depth of the field necessitated careful integration of basic science, engineering, and practical medical experience in a variety of applied disciplines. As a result, scientists, engineers, and physicians are among the chapter authors, as well as the editors. The handbook provides a detailed accounting of the state of the art in the rapidly growing biomaterials arena. Its organization reflects the diversity of the field.

The encyclopedia is a four-volume reference: Part A, "Materials," in two volumes and Part B, "Applications," in two volumes. In the Materials texts, the focus is on materials development and characterization. Volume 1 deals first with issues in the selection of a proper biomaterial from biocompatibility to biostability to structure/function relationships. Volume 2 then focuses on the use of specific biomaterials based on their physicochemical and mechanical characterizations. Integral to these chapters are discussions of standards in analytical methodology and quality control.

The users of this encyclopedia will represent a broad base of backgrounds ranging from the basic sciences (e.g., polymer chemistry and biochemistry) to the more applied disciplines (e.g., orthopedics and pharmaceutics). To meet varied needs, each chapter provides clear and fully detailed discussions. This in-depth coverage should also assist recent inductees to the biomaterials circle. The editors trust that this handbook conveys the intensity of this fast-moving field in an enthusiastic presentation.

iii

The editors are grateful for the cooperation of many friends and colleagues in their support of this work. Our appreciation extends to each of the contributors for suggestions and comments as the project developed. Their interest and enthusiasm in pulling together a comprehensive reference for all our associates in the biomaterials area have been most gratifying. The editors are especially thankful to Ms. Wanda O'Connell for her patience and competence in dealing with manuscripts from more than 100 authors.

Donald L. Wise, Debra J. Trantolo, David E. Altobelli,
Michael J. Yaszemski, Joseph D. Gresser, and Edith R. Schwartz

Contents of Part A: Materials

Part II. Biocompatibility and Tissue Response

Part III. Biodegradability, Resorption, and Stability

Part IX. Ceramic Materials

Part X. Other Materials

Contents of Part B: Applications

ENCYCLOPEDIC HANDBOOK OF BIOMATERIALS AND BIOENGINEERING

Part A: Materials

Volume 2

V
Materials Based on PLGA

27

Synthetic Resorbable Polymers Based on Glycolide, Lactides, and Similar Monomers

B. Amecke, D. Bendix, and G. Entenmann
Boehringer Ingelheim KG
Ingelheim, Germany

I. SYNTHESIS

A. General Remarks

Homopolymers and copolymers based on L-lactic acid, D,L-lactic acid, and glycolic acid have received interest in the medical and pharmaceutical field because of their degradability and toxicological safety. The application of these polymers in devices for wound closure, orthopedics, and controlled drug release, and their degradation in the physiological environment have been reviewed recently [1-9].

Low molecular weight polymers can be prepared by condensation of lactic acid and glycolic acid. The reaction may be performed without any catalyst [10] or in presence of catalysts like ion exchange resins or acid clay [10], organic or inorganic acids, acid anhydrides, metals (e.g., lead), metal oxides (e.g., litharge or antimony oxide), or Lewis acids (e.g., zinc chloride, stannous chloride, or antimony fluoride). The oligomers show some potential application in drug delivery systems. However, mainly the oligomers are used to produce lactide (from oligo lactic acid) or glycolide (from oligo glycolic acid). These cyclic dimers are starting materials for high molecular weight polymers and copolymers, which are formed by ring-opening polymerization in the presence of suitable polymerization catalysts.

Besides glycolide and lactides, other monomers like dioxanone and trimethylene carbonate are of increasing importance.

L-Lactide   D-Lactide   meso-Lactide

D,L-Lactide

Glycolide   Trimethylene carbonate   Dioxanone   Caprolactone

Starting from commercially available glycolic acid, L(+)-lactic acid, racemic D,L-lactic acid, dioxanone, or trimethylene carbonate, a variety of polymers can be derived. The properties of the polymers can vary considerably depending on the molecular weight of the polymer, the comonomer ratio and the sequence distribution in copolymers, which are controlled by the polymerization conditions.

The methods applied to produce high molecular weight polymers are melt polymerization, bulk polymerization, solution polymerization, and emulsion polymerization. These methods, their advantages and disadvantages, have been discussed in an overview [11]. A polymerization catalyst is needed, for example, metal oxides or metal salts like stannous chloride or the more common stannous "octoate" (stannous 2-ethylhexanoate). In many cases, it is useful to add substances with free hydroxylic groups (water, hydroxy acids, esters of hydroxy acids, alcohols) as cocatalysts. With increasing amounts of the catalyst, the molecular weight of the resulting polymer passes through a maximum, whereas the cocatalyst acts as a stopper for the growing polymer chain and thus lowers the molecular weight with increasing concentration. Many research groups have studied the influence of the reaction conditions (reaction time, temperature, catalyst and cocatalyst concentration) in detail, in some cases applying the response surface method [12,13].

It has been shown that impurities in the monomer feed have a tremendous effect on the molecular weight of the resulting polymer [11]. Therefore, the monomers must be purified thoroughly, for example, by distillation and/or crystallization, and the polymerization has to be run under dry inert gas. Further, the drying procedure or other pretreatment of the equipment as well as the application of vacuum during the polymerization are important factors affecting the polymer molecular weight [14]. Taking into account these experimental difficulties, it is not surprising that the results from different research groups frequently are not comparable.

After the polymerization process, the crude polymer contains unreacted monomer in amounts of about 1% or more and contains the catalyst as well. These impurities can be removed by reprecipitating the polymer, by an extraction process, by treating with vacuum, by contacting the dissolved polymer with adsorbent, or in special cases by recrystallization. The different methods are discussed below under the heading of the particular polymers.

B. Poly(L-lactide)

Poly(L-lactide) is a chiral crystalline polymer with a melting point of greater than 180°C. Due to its high initial strength and good strength retention, this polymer is preferred in the production of surgical implants for internal fixation [6]. The key data of the polymer in this application are the molecular weight, optical purity, and crystallinity.

The thermodynamics of L-lactide polymerization in solution with tin octoate as catalyst has been investigated [15] and compared with the experimental data of other authors. Kinetics and mechanism of the reaction in different solvents have been studied with catalysts like potassium t-butoxide/18-crown 6 [16], aluminum isopropoxide [17,18], n-butyllithium [18], triethylaluminum [19], diethylzinc [19], stannous tetrachloride [19,20], tetraphenyltin [20], or stannous octoate [20]. Independent of temperature and reaction medium, partial racemization was observed with basic catalysts [18]. The resulting polymers are reported to have inherent viscosities of about 0.1–0.5 deciliter per gram (dl/g). Under special conditions, molecular weights up to 222,000 (see Ref. 11; no experimental details are given) or with inherent viscosities of greater than 5 dl/g (Ref. 21; stannous octoate as catalyst, toluene dried with molecular sieve as solvent) are available by solution polymerization.

For the production of poly(L-lactide), bulk polymerization is the preferred procedure. Kinetics [22,23], mechanism [24–26] and the influence of the polymerization conditions (reaction time, temperature, amount of catalyst; see Refs. 22, 23, and 26–30) have been investigated. Numerous catalysts like metals, metal oxides, metal alkyls, metal alkoxides, or metal salts have been studied in detail and compared with respect to their catalytic activity [14,19,20,23,31–33]. The preferred catalysts are tin salts, especially stannous octoate, as high molecular weight polymers with inherent viscosities up to about 10 dl/g can be achieved without racemization. Recently, it has been found that the use of zinc-bis(2,2-dimethyl-3,5-heptanedionato-O,O′) as catalyst results in poly(L-lactide) with significantly higher crystallinity than other catalysts [31].

Poly(L-lactide) as polymerized shows a microporous structure [26]. Under physiological conditions, this material is degraded much faster than nonporous specimens prepared by injection molding [34]. Further polylactide as polymerized always contains some unreacted monomer. The residual monomer has an influence on thermal degradation during processing the polymer by injection molding and on biological degradation as well [13]. To find reproducible results, it is desirable to produce polymers with low monomer content by adjusting the polymerization conditions [13,29] and by purifying the crude polymer.

Poly(L-lactide) may be purified by recrystallization from toluene [21] and by reprecipitation from chloroform/methanol [35], dichloromethane/methanol [22], dioxane/water [36], or toluene/hexane [21]. Further purification methods are extraction with ethyl acetate, with acetone/ethanol mixtures [37], or with supercritical carbon dioxide [38]. Residual monomer is removed to a level below 0.5%, but the catalyst content is not affected, at least in the case of stannous octoate [22]. However, the tin content can be reduced significantly when the polymer is dissolved (e.g., in dichloromethane or similar solvents), washed with aqueous hydrochloric acid, and precipitated with methanol [22,39], or by treatment of the solid polymer with a weak organic acid like acetic acid [40].

C. Poly(D,L-lactide)

Poly(D,L-lactide) is prepared by polymerization of a racemic mixture of L-lactide and D-lactide. The polymer is optically inactive and completely amorphous. Strength and modulus are significantly lower, elongation is higher, and degradation is faster than in the case of poly(L-lactide) [41]. The polymer finds application in surgery with lower requirements for strength and strength retention and in the field of drug delivery.

The kinetics and mechanism of the polymerization in solution with basic catalysts have been investigated [17], but the standard procedure is bulk polymerization. Thermodynamics [42], kinetics [43–46], and mechanism [14,44,45,47,48] in the polymerization of molded poly(D,L-lactide) have been discussed in detail. It has been found that, in cationic polymerization in the presence of $SnCl_2$ dihydrate, the addition of OH-containing compounds like cetyl alcohol, glycolic acid, or water increased the reaction rate and led to polymers with lower molecular weight [44,49]. The influence of catalyst concentration, polymerization time, and temperature on monomer conversion and molecular weight is well known [12,14,27,30,43,47–50]. One of the key factors is the monomer quality; fresh D,L-lactide gave much higher molecular weight than old D,L-lactide [49].

Different catalysts like zinc chloride [45,46], zinc acetylacetonate [46], aluminumacetylacetonate [46], stannous chloride [45], stannous laurate [43], or tintetraphenyl [49] are used in the bulk polymerization of D,L-lactide. Stannous octoate, however, is the most usual one [14,30,43,48]. Zincdiethyl alone or in mixture with aluminumtriethyl is suitable as well [46], but aluminumtriethyl alone did not react [46]. Tributyltinmethoxide results in polymers with very low inherent viscosities [48]. Titanalkoxides, zirconiumalkoxides, or bisalkyltinoxides did not show catalytic activity [43]. As racemization is not of relevance to poly(D,L-lactide), basic catalysts like aluminumisopropylate may be used [46].

For the purification of poly(D,L-lactide), reprecipitation (i.e., from acetone/water) is preferred, as extraction is difficult because of swelling under the influence of most solvents. Recently, the removal of volatile impurities by degassing the polymer melt has been claimed [51]. Concerning the residual catalyst, the remarks above on poly(L-lactide) are valid for poly(D,L-lactide) as well.

D. Poly(glycolide)

Poly(glycolide) is an important raw material for manufacturing surgical sutures [1–3,52] or devices for fixation of bone fractures [53]. Because of its insolubility in most organic solvents except hexafluoroacetone sesquihydrate or hexafluoroisopropanol, the polymer has only limited application in drug delivery systems [54–57].

The polymer can by prepared from glycolide in solution with basic substances as catalysts [58]. However, for the preparation of commercial amounts of high molecular weight material, melt polymerization is preferred. The thermodynamic properties of monomer, polymer, and reaction have been determined [59], as well as kinetics and mechanism of the polymerization [60,61]. It is noteworthy that glycolide exists in two modifications [52,59] with different stability and sensitivity against moisture [52]. Suitable polymerization catalysts are antimonous fluoride, zinc chloride, stannous chloride dihydrate, or stannous octoate. Hydroxylic additives like water, acids, esters of hydroxyacids, and especially long-chain alcohols are accelerating the reaction [60] and controlling the molecular weight [61–63].

Due to its insolubility in common solvents, it is difficult to purify poly(glycolide) by reprecipitation or recrystallization [64]. Unreacted monomer can be removed by extraction with ethyl acetate [61] or acetone [62], by heating the polymer melt under vacuum in the second half of the polymerization [51], by heating the solid polymer under vacuum, or contacting the polymer particles with a flowing stream of inert gas [65]. Heating in vacuum is also applied to remove volatile impurities from polyglycolide in a shaped form like a suture [66].

E. Poly(Lactide-Co-Glycolide)

Copolymers of L-lactide and glycolide are amorphous in the range of 25% to 70% glycolide, whereas copolymers of D,L-lactide and glycolide are amorphous in the range of 0% to 70% glycolide. The other compositions exhibit crystallinities between zero and the value of the pure homopolymer [61]. Copolymers with high excess of one monomer are used for surgical sutures such as poly(glycolide-co-L-lactide) 90 : 10 (VICRYL™) [2,3], or are suitable in surgical implants such as poly(L-lactide-co-glycolide) with 70–85 % lactide [67]. The other copolymers, especially poly(D,L-lactide-co-glycolide) 50 : 50, find broad application in drug delivery systems [5].

For polymers, which are used for microencapsulation of drugs, solubility in chloroform, dichloromethane, or acetone is an important property. In the case of poly(D,L-lactide-co-glycolide), with a molar excess of lactide, no solubility problems are reported. Poly(D,L-lactide-co-glycolide) 50 : 50, however, represents a borderline polymer in respect to solubility because glycolide is more reactive than lactide. Hence, the polymer, which results from bulk polymerization of a 1 : 1 mixture of both monomers, exhibits a more or less blocky structure demonstrated by ^{13}C nuclear magnetic resonance (NMR) [68–70]. The products with a higher average length of glycolide blocks, and in particular the polymers with higher inherent viscosities, tend to form gels in solution or are insoluble in many organic solvents [68,70].

The effect of the polymerization parameters has been studied for poly(L-lactide-co-glycolide) [71,72] and for poly(D,L-lactide-co-glycolide) [68,69,73]. To reduce the average block length by transesterification during the polymerization, it is recommended to prepare the polymers at an elevated temperature [68]. The most preferred catalyst system is stannous octoate/dodecanol, but the catalyst mentioned in Sections I.B, I.C, or I.D are suitable as well. The standard procedure is bulk polymerization. Replacement of dodecanol by glycolic acid as cocatalyst results in a more hydrophilic polymer with accelerated degradation [73]. Branched copolymers are received when multifunctional molecules like cyclodextrine, glucose, mannitol, pentaerythrol, sorbitol, xylitol, or fructose are used instead of dodecanol [74].

The purification of the copolymers is performed with the same procedures as described above; highly crystalline insoluble polymers are purified by solvent extraction or devolatilization (similar to polyglycolide), the others by reprecipitation like poly(D,L-lactide).

F. Poly(dioxanone) and Copolymers

In distinction from poly(glycolide) or poly(glycolide-co-L-lactide), which can be processed only to multifilaments, poly(dioxanone) can be spun to monofilament sutures with adequate properties [8,75]. Polymers and copolymers based on dioxanone are gaining increasing attention as raw materials for surgical implants.

Low molecular weight poly(dioxanone) was received by simple heating of the monomer to 150°C for several hours [76]. High molecular weight polymers are available in emulsion polymerization with trialkylaluminum as catalyst [77], but melt polymerization is preferred. Numerous catalysts like Zr-, Fe- or Ti-acetylacetonates [78,79], dialkylzinc [79–81], dibutyltindilaurate [81,82], dibutyltinoxide, or dibutyltindi-2-ethylhexanoate [82], alkoxytitanates [79,83], or tin(II) salts [84–87] have been used. In recent developments, stannous octoate is preferred [85–88]. Under similar conditions (melt polymerization, stannous octoate as catalyst), random copolymers are prepared from mixtures of dioxanone with other reactive lactones. Addition of dioxanone to reactive prepolymers or addition of reactive lactones to prepolymerized dioxanone (sequential polymerization) leads to block polymers with various structures and properties (see Table 1) (85,88–92).

Low molecular weight hydroxylate polyesterethers as starting materials for polyurethanes have been prepared from dioxanone, adipic acid, and multifunctional alcohols like glycol or pentaerythrite with or without catalyst [93].

Polymers based on dioxanone can be purified with the same methods as described for polylactides, that is, reprecipitation [80], extraction [86], or devolatilization under vacuum.

G. Poly(trimethylene carbonate) and Copolymers

A block copolymer of the ABA type, where A is polyglycolide and B poly(trimethylene carbonate-co-glycolide), is used for a monofilament suture [8,75]. Poly(trimethylene

Table 1   Block Copolymers and Random Copolymers of Dioxanone with Different Monomers

Comonomer	Copolymer structure	Reference no.
Glycolide	Block	85, 89, 90
L-Lactide	Block	88
	Sequential	88
Hexamethylene carbonate	Random	91
Morpholinedione	Random	92

carbonate) and copolymers based on trimethylene carbonate are of increasing interest because of their enhanced flexibility in comparison with polymers made from lactide and glycolide.

Poly(trimethylene carbonate) is formed as a side product in the synthesis of trimethylene carbonate from oxetane and carbon dioxide, catalyzed by organotin iodine complexes with phosphine or phosphine oxides, or by polymerization of trimethylene carbonate with the same catalysts under pressurized carbon dioxide [94]. The monomer, which is usually prepared from propanediol and dialkylcarbonate, can be polymerized in solvents like methylene chloride [95], chloroform, dichloroethane, or nitrobenzene [96], or in toluene [95]. Numerous catalysts have been investigated like methyl triflate [96], triethyloxoniumfluoroborate [96], borontrifluoride/ether [95], aluminum chloride [95], dibutyltinoxide [95], potassium acetate [95], or sodium hydride [95].

Melt polymerization with potassium carbonate as catalyst [95,97,98] results in low molecular weight polymers [95,97]. Products with high molecular weights are received with tributyltin methoxide or acetate [99], dibutyltin dibromide [99], bismuth(III) 2-ethylhexanoate [99], zinc stearate [96], stannous octoate [96], or diethylzinc [100].

The reaction mechanism for the formation of identified end groups has been discussed for the polymerization in solution [96] and in the melt [99]. Especially in solution, decarboxylation and the formation of ether groups in the polymer is observed as a side reaction [95,96].

Copolymers of trimethylene carbonate with L-lactide [101,102], D,L-lactide [103], or glycolide [104,105] are available by melt polymerization in the presence of stannous octoate [101–103] or stannous chloride [104,105]. Dodecanol [104] or diethyleneglycol [104,105] are used as cocatalyst. The sequence of poly(L-lactide-co-trimethylene carbonate) has been investigated by means of NMR [102]. If monomers are added subsequently to the reaction mixture, block copolymers are formed [8,75,104,105]. Block polymers with one block being poly(ethylene glycol) and the other comprised of randomized glycolide and trimethylene carbonate can be prepared as well [106].

The polymers can be purified as described above, such as by reprecipitation from methylenechloride/methanol [106], chloroform/methanol [100], or acetone/methanol [97].

H. Epsilon-Caprolactone Copolymers

Properties and use of polymers and copolymers made from caprolactone, L-lactide, D,L-lactide, and glycolide have been reviewed [8]. They can be prepared by heating a mixture of the monomers in the presence of catalyst/cocatalyst systems in solution or in the melt. However, the large difference in reactivity of caprolactone and lactide or glycolide [107] leads to the formation of copolymers with blocky structures [108,109]. The average sequence length has a strong impact on the properties of the polymers and is controlled by polymerization conditions [109]. Polymers with blocky structures may be randomized to a certain degree by transesterification with suitable catalysts [110]. Addition of further monomers to a melted prepolymer results in diblock and triblock copolymers. Investigations concerning the copolymerization of caprolactone with other lactones are summarized in Table 2 [81,83,108–117].

Depending on the solubility of the copolymers, the products can be purified by reprecipitation from acetone/hexane [111], toluene/heptane [110,113], toluene/methanol [113], chloroform/methanol [114], by extraction with acetone [115], or by devolatilization in vacuum at elevated temperature [116,118].

Table 2   Copolymerization of Caprolactone with Other Lactones

Comonomer	Polymerization conditions	Copolymer structure	Reference no.
L-Lactide	Melt, stannous octoate solution, aluminium isopropylate	Random[1]	108, 111, 112
		Random[1]	113
		Diblock	113
D,L-Lactide	Melt, stannous octoate + dodecanol		114
	Solution, aluminium isopropylate	Random[1]	110
		Diblock	113
Glycolide	Solution, "Teyssie-cat." [(n-C$_4$H$_9$O)$_2$AlO]$_2$Zn	Triblock	115
	Solution, "Teyssie-cat." [(n-C$_4$H$_9$O)$_2$AlO]$_2$Zn	Triblock	115
	Melt, stannous octoate + 1,6-hexandiol	Random[1]	116
		Diblock	116
Dioxanone	Melt or solution, numerous catalysts	Random[1]	109
	Melt, stannous octoate + diethylene glycol	Random[1]	117
		Diblock	117
		Triblock	117
	Melt, dibutyltindilaurate	Random[1]	81
	Melt, alkoxytitanate	Random[1]	83

[1]Prepared by polymerization of a monomer mixture; depending on the polymerization conditions, the polymer structure is more or less blocky.

II.   ANALYTICAL CHARACTERIZATION AND QUALITY CONTROL

A.   General Remarks

Unlike a normal organic or inorganic compound, a polymer is a mixture of molecules with different molecular weights and, in the case of copolymers, even with different chemical composition. A polymer chain can be linear, branched, or cross-linked. Polymers often tend to build up networks and superstructures in the melt or in solution and can exist in different phases. They may be crystalline or amorphous, stiff or rubberlike, solid or liquid. All polymer properties also are changing gradually with the number of chain units and with the variation in the chemical composition or other structural attributes. The tacticity, that is, the sterical orientation of substituents along the polymer chain, also has a strong influence on the polymer properties.

As there are different molecular species present in a polymer sample, most of the properties of a given polymer can be described only by averaged values or ranges. Therefore, to characterize a polymer sample fully, various sophisticated and expensive techniques have to be applied. Besides the determination of the polymer properties, the identification and quantification of those substances that have been added to the polymer either intentionally during polymer processing (e.g., stabilizers, plasticizers, fillers) or unintentionally during the chemical reaction and/or purification (e.g., residual monomers, catalysts, solvents, etc.) is important.

In industrial practice, only those polymer properties will be determined that are necessary to guarantee the identity and the quality of a given polymer. In addition, those methods may be applied that check an essential property of the final application, (e.g., the release pattern in a drug application or the mechanical properties in a medical device).

B. Analytical Methods

1. Molecular Structure

Identity.   A technique often used to check the identity of a compound is infrared (IR) spectroscopy. The spectrum of the substance under investigation has to match a reference spectrum. This method, often used in the pharmaceutical industry, however, is not feasible for most of the resorbables based on lactic and glycolic acid. The IR spectrum of the highly crystalline, stereoregular poly(L-lactide), superimposed on that of the totally amorphous poly(D,L-lactide) shows practically no difference (Fig. 1) and is therefore not suitable to check the identity of the polymer.

In the case of the copolymers of D,L-lactide and glycolide, mainly used in controlled-release applications, IR is not capable of discriminating the 50 : 50, 75 : 25, and 85 : 15 mol% polymers. Only the nearly totally insoluble homopolymer poly(glycolide) may be characterized by IR (Fig. 2), but for all other polymers proton NMR spectroscopy is the method of choice. The stereoregular poly(L-lactide) (e.g., with its equivalent protons) shows a simple first-order spectrum, whereas the mainly atactical poly(D,L-lactide) develops a complex spectrum with multiple lines (Fig. 3). In the poly(D,L-lactide-co-glycolide) spectra, the glycolide and lactide parts are clearly separated and the polymer can be identified by the ratio of the height of the corresponding integrals (Fig. 4). The spectrum also gives directly the average molar comonomer composition. In addition, a high-resolution NMR spectrum will give first hints of the presence of additives or impurities in a polymer.

Average Chemical Composition.   In the case of copolymers, often the average chemical composition has to be determined. As mentioned above for lactide/glycolide copolymers,

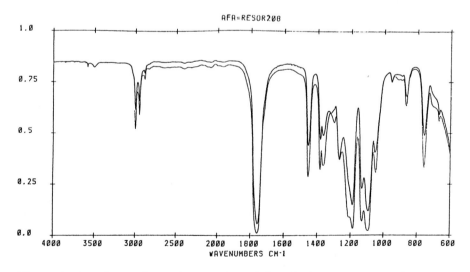

Figure 1 Superimposed Fourier transform infrared (FT-IR) spectra of poly(L-lactide) and poly(D,L-lactide) (sample preparation: film, cast out of a chloroform solution).

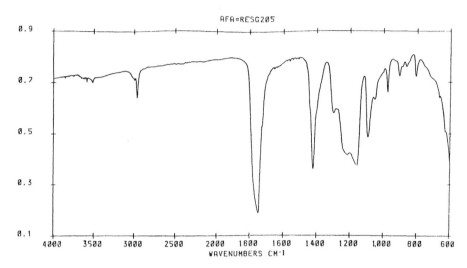

Figure 2 Fourier transform infrared (FT-IR) spectrum of poly(glycolide) (sample preparation: film, cast out of a HFIP solution).

this can be done by high-resolution proton NMR (frequency $\geq$ 250 MHz). The method is also applicable for copolymers containing the trimethylenecarbonate (TMC) moiety.

With copolymers made out of L- and D,L-lactide, the NMR method naturally does not work. Here, the comonomer composition can be determined by optical rotation. Whereas the racemic poly(D,L-lactide) is optically inactive, the pure poly(L-lactide) shows a strong rotation. It can be shown that there is a simple linear relation between the comonomer composition and the measured optical rotation (Fig. 5) and that the method can be calibrated with mixtures of poly(L-lactide) and poly(D,L-lactide).

Molecular Weight Average.   With only a few exceptions, a synthetic polymer is always a mixture of molecules with different molecular weights in different amounts and distributions. Therefore, there is no well-defined molecular weight and only averaged values can be given. Most important are the number and the weight average of the molecular weight, M_n and M_w, respectively. The methods to determine the absolute values of these averages are in general very sophisticated, need special equipment, and are therefore not used in the routine. An exception is the determination of M_n by end-group titration of low molecular weight cocondensates.

In practice, the molecular weight average is described by so-called characteristic numbers, which are directly related to the molecular weight. For this purpose, often viscosity numbers are used. For a linear, unbranched, and not cross-linked polymer of the same chemical constitution, the viscosity of a dilute polymer solution is directly related to the molecular weight average M_{vis}, which is approximately M_w. By determining the relative viscosity, that is, the ratio of the viscosity of a diluted polymer solution and the pure solvent, several viscosity numbers can be calculated (Fig. 6).

The most important numbers are the inherent viscosity η_{inh}, determined by a single-point measurement, and the intrinsic viscosity $[\eta]$, evaluated in a dissolution series. It can be shown mathematically that, by extrapolating to zero concentration, both numbers

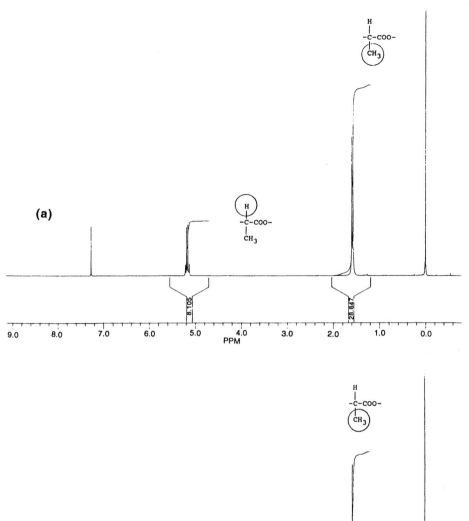

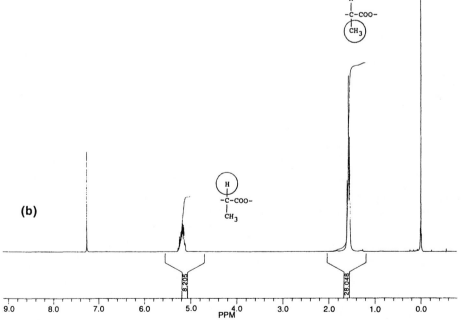

Figure 3 The 250-megahertz (MHz) ^{1}H nuclear magnetic resonance (NMR) spectra of poly(L-lactide) and poly(D,L-lactide) (sample preparation: 5–10 mg polymer in about 1 ml deuterochloroform; reference: i-TMS).

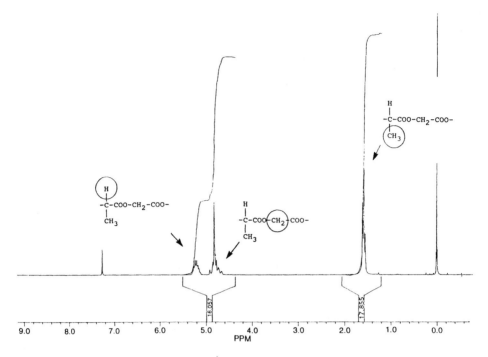

Figure 4 The 250-megahertz (MHz) ¹H nuclear magnetic resonance (NMR) spectrum of poly-(D,L-lactide-co-glycolide) 50 : 50 (sample preparation: 5–10 mg polymer in about 1 ml deutero-chloroform; reference: i-TMS).

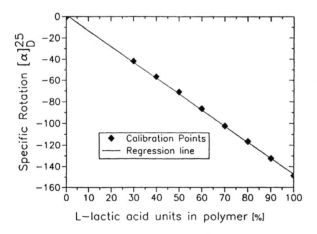

Figure 5 Determination of the comonomer composition of poly(L-lactide-co-D,L-lactides) by measuring of the specific optical rotation $[a]_D^{25}$ (sample preparation: 1.000 g polymer in 200.0 ml methylenechloride; path length: 1 dm).

Relative viscosity: $\eta_{rel} = t_{solution}/t_{solvent}$ (t = efflux time)

Specific viscosity: $\eta_{spec} = \eta_{rel} - 1$

Reduced viscosity: $\eta_{red}' = \eta_{spec} / c$

Inherent viscosity: $\eta_{inh} = \ln(\eta_{rel}) / c$ (c = concentration in g/dl)

Intrinsic viscosity: $[\eta] = \lim_{c \to 0} \eta_{spec} / c$

Solomon—Ciuta—approximation:

$$[\eta] = [2(\eta_{spec} - \ln\eta_{rel})]^{1/2} / c$$

Mark—Houwink—Equation (MHE):

$$[\eta] = K M_{vis}^{a} \quad \text{or}$$

$$M_{vis} = ([\eta] / K)^{1/a}$$

MHE—Parameter for poly(L—lactide): K = 1.29 E-4, a = 0.82

Figure 6 Viscosity numbers to characterize the molecular weight average of a polymer sample (method: viscosimetry of diluted polymer solution in a capillary viscometer [Ubbelohde, Cannon-Fenske]).

become identical. Besides out of a dissolution series, $[\eta]$ can be approximated also out of a single-point measurement with sufficient accuracy [118].

In Europe, mainly the Ubbelohde type of viscometer with DIN capillaries is used, whereas in America the Canon-Fenske type with American Society for Testing and Materials (ASTM) capillaries is quite common. Because polymer solutions are not Newtonian in their behavior, different capillaries will yield slightly different numbers. To compare data, one has to use also the same solvent at exactly the same temperature and the weight must lay in a very narrow range [119].

Out of $[\eta]$, the viscosity average of the molecular weight can be calculated by using the so-called Mark-Houwink equation (MHE) and appropriate MHE parameters.

A second number for the molecular weight average is the melt flow index (MFI), used especially in technical applications. To determine this number, a polymer sample is melted under well-defined conditions (e.g., DIN 53 735) and then pressed by a piston through a small orifice. The weight of the extruded polymer per time unit is measured. Background of this measurement is the fact that for small shear velocities the melt viscosity is related to the molecular weight average.

Average Block or Sequence Length. In copolymers, the single comonomer units can be distributed in different ways along the polymer chain. The border cases are either a totally random distribution or the building of long blocks. If the pure homopolymers are very different, then some properties of the copolymers may also vary significantly depending on the average block length. In the 50 : 50 copolymers of D,L-lactide and glycolide, for example, we find a pronounced dependence of the solubility on the average block length [70]. Whereas the more random copolymers are totally amorphous and freely soluble even in bad solvents like acetone, the copolymers with a block structure are insoluble even in normally good solvents like methylenechloride or chloroform. The reason is the development of small crystalline domains of glycolide units, which, analogous to the pure poly(glycolide), are extremely poorly soluble and tend to build up microgels.

An often-used technique to determine the sequence of a copolymer is the high-resolution NMR technique [72]. In the case of the poly(D,L-lactide-co-glycolide), one can use quantitative ^{13}C in dimethyl sulphoxide (DMSO) to determine the frequency of the GA*-GA and the G*-LA dyads [70] (Fig. 7). An indication for a pronounced buildup of blocks can be found also by thermal analysis like DSC, especially if one of the homopolymers forms a semicrystalline polymer. Another, more quantitative way to check for

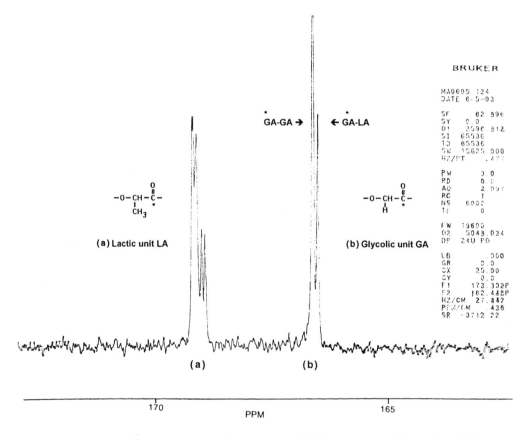

Figure 7   A 63-MHz ^{13}C nuclear magnetic resonance (NMR) spectrum of the carbonyl C atoms: (a), of the lactic acid LA; (b), glycolic acid GA units in a poly(D,L-lactide-co-glycolide) 50 : 50. The spectrum in d_6-DMSO shows a sensitivity for the G*A-GA and G*A-LA dyads (sample preparation: about 60 mg in 0.6 ml d_6-DMSO; reference: i-TMS; average block length $L = [(\text{integral G*A-GA})/(\text{integral G*A-LA})]^{+1}$).

blocky sequences is the cloud point titration, which makes use of the different solubility of random and blocky copolymers. It has been used in the case of poly(D,L-lactide-co-glycolide) 50 : 50 to show the temperature dependence of the block formation during melt polymerization [68].

Distributions.   Besides the average value of a property, often the underlying distribution is important for the characterization of the polymer quality, mainly because totally different distributions can give the same average value. During injection molding, for example, two polymer batches with an equal MFI but different molecular weight distributions may behave totally different and may lead to big production problems.

Molecular weight distribution.   By far the most important distribution of a polymer sample is its molecular weight distribution (MWD). Nearly every property of a polymer, and in particular the mechanical properties, depends on the MWD. Normally, the normalized portion of a molecular weight fraction is plotted against the logarithm of its molecular weight. The resulting differential distribution curve is then described in terms like symmetrical, broad, multimodal, and the like. A more quantitative way to describe an MWD is the calculation of the polydispersity D, defined as the quotient of M_w and M_n.

On a routine basis currently gel permeation chromatography, GPC, is used exclusively to determine the MWD. The GPC is often also called SEC (size-exclusion chromatography) because GPC is an exclusion chromatography on columns with a defined pore-size distribution. It is a relative method; it does not separate according to the molecular weight, but according to the hydrodynamic volume. This is the size of a swollen polymer coil and its volume depends on the particular solvent and the shear stress during the passage through the GPC columns. During separation, the molecules with the highest molecular weight are eluted first because only a few pores with corresponding sizes are accessible for them. The longest retention time or retention volume is for the solvent itself because all pores are accessible to its molecules. Important for the separation are the so-called exclusion limits. The upper limit is reached when a macromolecule shows practically no retention because it is so big that it cannot intrude into any of the pores of a given column combination. The lower exclusion limit is normally given by the solvent itself.

The apparatus used for GPC is the normal high-performance liquid chromatography (HPLC) equipment. Because normally the retention volume is not measured directly, the retention time is used instead. Therefore, the HPLC pump must show an extremely good flow constancy to guarantee a constant relation between flow and volume. Because resorbable polymers based on lactic or glycolic acid have no chromophores, a refractive index (RI) detector must be used. In principle, also an evaporative light-scattering detector (ELSD) can be used.

Because GPC is a relative method, it has to be calibrated with polymer standards. In general, there are three well-established methods [120]:

1. *Calibration with narrow molecular weight standards.* The retention times of different polymer standards are evaluated and in a kind of point-to-point calibration fitted to a straight line or a polynomial of higher degree.
2. *Broad standard calibration.* A standard with a broad molecular weight distribution is characterized by absolute methods. Then, with an iterative mathematical procedure, the parameters of an equation that describes the molecular weight as a function of the retention time are fitted to this equation until the right averages are calculated.

The drawback of this method is the availability of well-defined broad standards for most of the polymers.

3. *Universal calibration.* It has been shown that for all polymers the retention volume is a function of the intrinsic viscosity. If now the parameters for the Mark–Houwink Equation (MHE), which correlates the viscosity number with a molecular weight (e.g., narrow molecular weight distribution polystyrene standards and the polymer under investigation) are known, then the molecular weights of the sample can be calculated. The problem with this method is the fact that reliable MHE parameters often are not known for all polymers and MHE parameters are temperature and solvent dependent.

For resorbable polymers, there are no polymer standards, either broad or narrow, or reliable MHE parameters available for all polymers. Therefore, nearly everybody calibrates with sets of narrowly distributed polystyrene standards, which are easily available from different suppliers. Doing this, one has to keep in mind that, in solution, a given PS molecule with its pure C-C linkages is tighter folded than a polyester of the same molecular weight because of its more rigid chain. And, because GPC separates only according to the volume of the coil and not to the absolute molecular weight, this leads to totally wrong absolute numbers. For the pure poly(L-lactide), for which reliable MHE parameters are known [22,121], the GPC data are two- to threefold too high (Fig. 8). In addition, there is a strong solvent dependence of the GPC data. Another common problem in GPC, especially with polymer peaks with tailing into the low molecular weight range, is the setting of the integration marks. M_n values are extremely sensitive against slight changes in the setting of the end point of the polymer peak in the chromatogram.

So, care has to be taken if GPC data are compared. On the other hand, GPC data are useful during such things as degradation studies *in vitro* or *in vivo* (Fig. 9) [122].

Other distributions. Besides the MWD, additional distributions are present in a

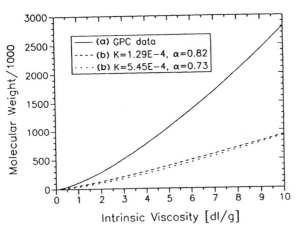

Figure 8 Comparison of molecular weight averages of poly(L-lactides) plotted against their intrinsic viscosity: (a), averages determined by gel permeation chromatography (GPC) using narrow molecular weight distribution polystyrene standards for calibration; (b), averages calculated by using the Mark–Houwink equation (MHE) and reliable MHE parameters (the GPC data have been fitted by a MHE type of equation to the data points).

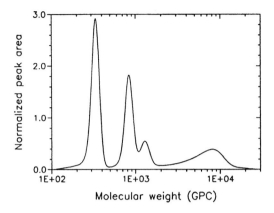

Figure 9   Molecular weight distribution of a retrieved poly(L-lactide) implant after three years of degradation *in vivo* (polymer matter aspirated in a clinical study).

polymer sample, especially in copolymers. With different reactivities of the comonomers or with special catalyst systems, there will be a chemical composition distribution (CCD) overlaying the MWD. Also, the above-mentioned average sequence length will normally not be dispersed equally over the whole sample, but will show a distribution depending on the MWD and/or CCD. The analysis of such complex systems can be performed only in special cases and requires the development of special methods.

2.   Supermolecular Structure

Of the multitude of structures and phases that may be present in a polymer, in the following we deal exclusively with two important phase transitions, glass transition and melting and crystallization, of the solid polymer. These are commonly used to characterize a polymer. Both utilize thermal analysis methods such as differential scanning calorimetry (DSC).

Glass Transition.   By rapidly cooling down the melt, a polymer solidifies to an amorphous, glassy mass in which the single polymer chains cannot glide along each other. On heating up the sample, there is a distinct temperature range in which the polymer chains become mobile again. This changing is called *glass transition* and the accompanying temperature the glass temperature T_g.

This transition now gives rise to a large number of changes in the behavior of the polymer. Below the glass temperature, a polymer tends to be more stiff and brittle, whereas above this temperature range polymers are more flexible and elastic. This is also the effect of a plasticizer, which simply lowers the glass transition. Above T_g, solvents and water can easily penetrate the material and dimensional stability is sometimes affected. In principle, polymers can be extruded only above their glass transition temperature.

The T_g can be determined by different methods, by which the resulting value is dependent on the method. Routinely, a thermal analysis such as DSC is used. At the T_g, the heat capacity of the polymer is changed, which shows up in the DSC trace by lowering the baseline. Without a thermal pretreatment of the sample, the T_g is often superimposed by an enthalpy relaxation, which makes it difficult to determine the T_g. A sample pretreatment eliminates this effect, either by annealing above the T_g or by melting and rapid cooling (Fig. 10).

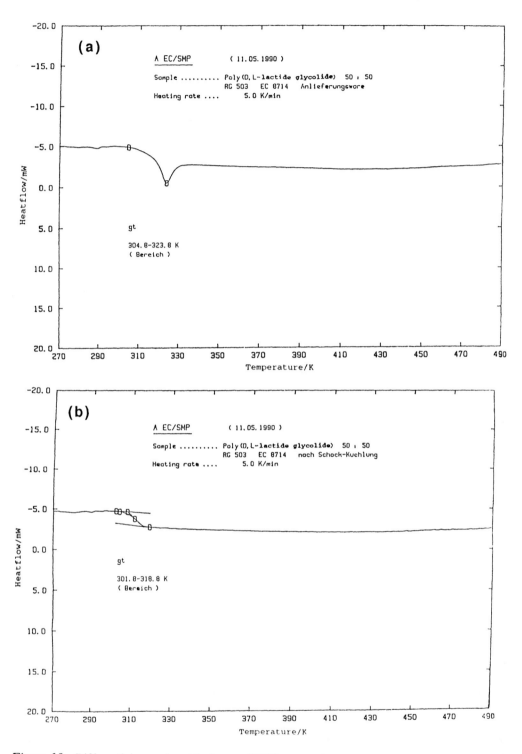

Figure 10   Differential scanning calorimetry (DSC) trace of an amorphous poly(D,L-lactide-co-glycolide) 50 : 50: (a), DSC of the material as stored with a significant enthalpy relaxation; (b), DSC after melting and quenching with liquid nitrogen (heating rate: 5°K/min).

Melting and Crystallization.   Many polymers exist only as amorphous, glassy solids. The structure of the polymer chain hinders every orientation and no melting or crystallizing can be observed. Polymers with a regular, linear structure and a high steric uniformity (tacticity), on the other hand, can form crystalline structures by joining polymer chains to small crystalline domains. These semicrystalline polymers show a more or less broad melting range, not a melting point, because they do not have a uniform crystalline structure.

Again, DSC is normally used to characterize the thermal behavior of a polymer. According to the thermal history of the sample or the possibility of different crystal structures being present in a sample, the DSC trace shows one or more melting peaks. Out of the melting enthalpy, one can calculate the crystallinity of the sample by comparing this value with that of a theoretically 100% crystalline polymer. If the melt is cooled down rapidly, even a semicrystalline polymer solidifies to a glassy mass, which on heating up shows in the DSC a T_g and an exothermic recrystallization peak (Fig. 11). This recrystallization is often delayed, but can be induced by annealing or in threads by stretching and expanding. Recrystallization is often linked with dimensional changes by shrinkage and the accuracy of size in injection-molded parts may be affected.

3. Nonpolymer Constituents

A technical polymer normally contains some nonpolymer constituents. These are either added with purpose, like additives, fillers, stabilizers, plasticizers, and the like, or are unintentionally present in the polymer, like residual monomers, catalysts, solvents, heavy metals, and so on. The characterization of the nonpolymer constituents is mostly done with classical analytical methods. In the following, some analytical methods are described for those unintentional impurities that may be present in resorbable polymers.

Residual Monomers.   A first indication if monomers are present in a polymer sample is a proton NMR spectrum. At 250 MHz, the monomeric lactide and/or glycolide signals are well separated from the polymer peaks, but the accuracy and the detection limit is not very good. Only with monomer contents well above 1% can reliable quantitative values be determined. Gas chromatography (GC) is the best and most accurate method to detect and quantify the lactide, glycolide, and also TMC monomer. Both GC systems, packed and capillary columns, can be used for this purpose. Commercially available lactones like γ-decalacton or caprolacton can be used as internal standards. Using a chiral column, it is even possible to separate the different stereoisomers of lactide. To analyze pure poly(glycolide) or to achieve a very low detection limit, hexafluoroisopropanol should be used as a solvent.

Residual Catalyst.   Nowadays, resorbable polymers exclusively are polymerized with tin salts. The most common catalysts are tin chloride and tin octoate. The tin content in the polymer is determined by atomic absorption spectroscopy (AAS). A problem in the past was the extraction of the tin into the AAS solution. Classical decomposition methods using fuming acids led to significant losses of tin during the degradation of the polymer matrix. The best way is to hydrolyze the polyesters by means of a 20% tri-n-butylammonium-hydroxide (TBA) solution. The measurements are done in a graphite furnace using an EDL as the source of radiation.

Residual Solvents.   Solvents may be present in a final polymer originating from the polymerization or purification steps. They are also sometimes added intentionally to induce a plasticizer effect or to enhance degradation. Large amounts can be detected by NMR, but gas chromatography head space has to be used to detect solvents in the

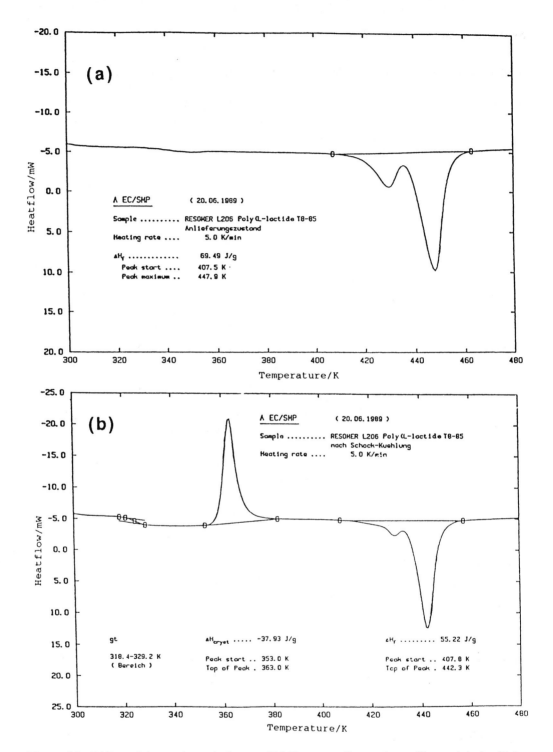

Figure 11 Differential scanning calorimetry (DSC) traces of a semicrystalline poly(L-lactide): (a), DSC of the material as stored with only an endothermic melting peak (double peak); (b), DSC after melting and quenching with liquid nitrogen with a glass transition T_g, an exothermic recrystallization peak, and an endothermic melting peak (double peak) (heating rate: 5°K/min).

996

parts-per-million (ppm) level. Dioxane at 80°–90°C has been found to be a good solvent for performing this analysis.

Water.   Water disturbs in particular all thermal processes, such as injection molding and extruding, by degrading the polymer. It may also affect the stability of the polyesters during storage. To determine the water content of a polymer, typically Karl–Fischer methods are used. The problem here is to find a good solvent/buffer/titrant system. A dioxane/methanol mixture 40 : 20 has been found to be a good solvent system for all resorbable polyesters. Polymers that are not directly soluble in this mixture such as the crystalline poly(L-lactides) are first dissolved in chloroform. The adding of buffer prior to titration depends on the amount of free carboxylic acid groups, but can normally be neglected.

Additional Tests.   For polymers used in the medical device industry or as a pharmaceutical excipients, additional tests are normally performed. These tests are typically described in different pharmacopoeias and include

Appearance
Odor
Heavy metals
Sulfated ash
Particulate

These tests have no significance for the polymer performance, but are necessary to fulfill the requirements of different authorities.

III. COMMERCIAL ASPECTS

A. Applications

Commercial interest in these polymers started in 1970 with the introduction of synthetic resorbable sutures by Davis and Geck (Danbury, CT). The production of sutures still is the dominating outlet for resorbable polyesters. In the meantime, however, many additional applications have been developed and introduced as commercial products (or reached the clinical testing stage).

The main advantage of this polymer group is the many years of *in vivo* experience, in particular with sutures. Most results show that the implants are safe. Therefore, resorbable polyesters are considered over other polymers when selecting the raw material for new developments. On the other hand, they are not foolproof as complications in some orthopedic applications have been shown. But, there is no other group of polymers as well documented. Resorbable polyesters are preferred also over natural polymers because of greater control over uniformity and purity and better mechanical properties.

Resorbable implants can be used for treating patients in many aspects. In the four sections that follow, the applications shown in Table 3 are described.

Table 3   Areas of Application

Wound closure
Orthopedic applications
Drug delivery systems
Other applications

1. Wound Closure

The most important products made from resorbable polyesters still are resorbable sutures for wound closure. Although monofilaments are considered superior for surgery, the first products reaching the market were multifilaments. The reason was that the first polymers selected (polyglycolide homopolymer or glycolide-L-lactide copolymer) were too stiff as monofilaments. Coatings such as stearates overcame the most important drawbacks and achieved smooth passage through tissue and less tendency to harbor pathogenic bacteria.

Subsequently, flexible polymers based on dioxanone and copolymers with trimethyl-enecarbonate were developed suitable for monofilament sutures. Multifilament sutures are still used extensively, but their market share is decreasing.

The total worldwide suture market is estimated to exceed $1 billion/year. In developed countries, synthetic resorbable sutures are the leading products, having reached up to 80% market share. The former resorbable workhorse, catgut, has almost disappeared. Certain market segments are covered by nonresorbable sutures required for some types of operation (e.g., cardiovascular). Only in less-developed countries is catgut still used extensively because of price reasons and to protect local production against imports.

Ethicon (Somerville, NJ) is by far the largest supplier worldwide. A list of manufacturers of synthetic resorbable sutures is given in Table 4.

Resorbable implants have also been developed for other wound closure applications. Most important are ligating clips based on polydioxanone (Absolok™R from Ethicon) and Lactomer™ staples and clips (U.S. Surgical Corp., Norwalk, CT) based on poly-(glycolide-co-lactide). Also, there are meshes available based on Vicryl™ (Ethicon) and Dexon™ (Davis and Geck). These are used mainly for soft tissue operations.

Although wound closure is the most mature market for resorbable polymers, it still offers significant growth potential, in particular for clips and staples, but also for new products such as resorbable bone wax.

2. Orthopedic Applications

The first orthopedic applications were already proposed in the early 1970s, but the first products reached the commercial stage only in the 1980s. The main reasons were larger implant weights (making the toxicology question more important) and demanding mechanical properties. Besides, the registration proved to be particularly complicated in some countries such as the United States, United Kingdom, and Germany—an obstacle for the medical device industry not used to this type of requirement.

On the other hand, resorbable polymers appear to be particularly suited for osteo-

Table 4   Commercially Available Sutures

Bondek™	Deknatel	MU	Polyglycolide
Dexon™	Davis and Geck	MU	Polyglycolide
Maxon™	Davis and Geck	MO	Poly(glycolide-co-TMC)
Medifit™	Japan Medical Supplies	MU	Polyglycolide
Monacryl™	Ethicon	MO	Poly(glycolide-co-caprolactone)
Opepolix II™	Nippon Shoji	MU	Polyglycolide
PDS™	Ethicon	MO	Polydioxanone
Vicryl™	Ethicon	MU	Poly(glycolide-co-lactide)

MU, multifilament; MO, monofilament

Table 5   Orthopedic Implants from Resorbable Polyesters

Orbital floor	(Phusis, France)
Rods and pins	Biofix™ (Bioscience, Finland)
	Orthosorb™ (Ethicon, United States)
	Polypin™ (Biovision, Germany)
Screws	Biofix™ (Bioscience, Finland)
Intramedullary plugs	(Howmedica, Ireland)
	(Zimmer, United States)
Anchoring devices	(Instrument Makar, United States)
	Suretac™ (Acufex, United States)
	(Arthrex, Germany)
	(Phusis, France)
Ligament reconstruction	(Ethicon, United States)

synthesis applications. Its main advantages are mechanical properties closer to the properties of bone than metals (promising to result in faster healing as there would be no stress protection and no need for a second operation to remove the implant).

The first products were an orbital floor for maxillofacial applications and a pin (Orthosorb™), both made from polydioxanone (from Ethicon). The market reception, however was disappointing. Significant inroads were made by the Biofix™ rod from polyglycolide (Bioscience, Tampere, Finland) used for a range of osteosynthesis applications. Again, the product did not meet completely the needs of the surgeons as the fixation is not firm. Plaster still is needed for additional fixation. Besides, the number of inflammations (sinus formation) appear to be unacceptable to many surgeons.

The introduction was also slowed down by adverse reactions of certain types of polylactide in clinical tests (sinus formation, bone resorption). The main problem is that the reasons are not well understood.

A list of commercial products is provided in Table 5. It is important to note that most of these products are available in some markets only at present. Therefore, total sales still are small. The need for more clinical experience is one reason. Another is the need for registration in many countries, a time-consuming and expensive process. It is expected that the European legislation for medical devices will help speed up registration.

A rapid market expansion is expected, however, because a new generation of implants is in clinical testing—implants better matching medical requirements with polymer properties. As the implants are small and the requirements for mechanical strength are limited, fewer failures and side effects can be expected.

3.   Drug Delivery Systems

The controlled release of peptides is one of the success stories of resorbable polyesters. The first product was a gonadorelin analog used against prostate cancer, breast cancer, endometriosis, uterine fibroma, and other indications. For effective treatment, the patient needs continuous dosage. This could only be achieved by daily injections, a rather unpleasant proposition. Therefore, the pharmaceutical industry was looking for a controlled-release device giving a four-week protection.

This could be achieved with a controlled-release formulation based on poly(D,L-lactide-co-glycolide). As the polymer is degraded, the peptide is released at a constant rate. Luteinizing hormone–releasing hormone (LHRH) was particularly suited for this type of formulation as it has a low molecular weight (fast diffusion rate) and a high activity (3.6 mg for 28 days). Therefore, small implants (or injections) were possible.

A list of commercial drug delivery systems using resorbable polyesters is given in Table 6.

4. Other Applications

Resorbable implants are suitable for other applications as well, although there are not many products commercially available yet. An important product is the anastomosis ring for colon operations developed by Davis and Geck (Valtrak™). It is made from polyglycolide with bariumsulfate as an additive to achieve radioopacity.

Also, there are two products for dental applications on the market (both based on polylactide): Drylac™ (Osmed, Duluth, MN) for the prevention of "dry sockets" and membranes for guided tissue regeneration in the treatment of periodontal disease (by Guidor, Huddinge, Sweden).

B. The Polymer Market

The market for resorbable polyesters is rather fragmented. It is characterized by many small applications (except for sutures) with different requirements concerning degradation, mechanical strength, crystallinity, and so on. The technology (production, characterization, and application) is highly sophisticated. Therefore, entry barriers are high. This market is a small niche as compared with other polymers, even in medical applications.

Unfortunately, there are no reliable data available for estimating the market size accurately. It should still be smaller than 100 metric tons per year. On the other hand, it is easy to estimate that close to 90% of this polymer quantity is ending up in wound closure applications.

Many more projects are in the pipeline. Nevertheless, the polymer consumption is expected to increase to a few hundred tons annually at best.

1. Pricing

Although the raw materials used for the polymers discussed in this chapter (e.g., lactic acid, glycolic acid, ethylene glycol, and others) are comparatively cheap (generally less than $5 per kg), the final polymers as used for implants often are costing more than $1000 per kg. What are the reasons? Which aspects have to be considered?

Above all, volumes are very small so that the contribution of labor costs is high.

Table 6 Commercial Controlled-Release Products

Trademark	Company	Type of formulation
Decapeptyl™	Beaufour-Ipsen (France) Ferring (Sweden)	Microparticle
Lupron Depot™	TAP Pharmaceutical (a joint venture of Abbott and Takeda)	Microparticle
Parlodel LA™	Sandoz (Switzerland)	Microparticle
Prostap SR™	Lederle (United States)	Microparticle
Suprefact™	Hoechst (Germany)	Implant
Zoladex™	Zeneca (United Kingdom)	Implant

Making small batches as compared to highly automated continuous polymerization plants (for technical applications) is extremely expensive. Important other reasons are

- The process is complicated as there are many steps from monomer preparation to polymerization and purification.
- As the polymer is used for medical applications, the production facility must meet the requirements in this industry. This means meeting GMP (Good Manufacturing Procedures, as set by the U.S. Food and Drug Administration) standards in the whole process. Besides, clean-room facilities are required for certain processing steps.
- Technical plastics are sophisticated blends of base polymer and additives to achieve optimum performance. The selection of raw materials for producing medical plastics is very limited. Only raw materials as acceptable as the polymer itself may be used.
- Different from nondegradable implants that remain intact within body liquid, resorbable implants are degraded completely. Each impurity will eventually be exposed. Consequently, resorbable polymers have to be purified and dried thoroughly before they may be used for human or animal applications.
- At last each batch of polymer must be accompanied by a complete certificate of analysis.

Therefore, these polymers will always remain expensive even if volumes should go up considerably and production technology is improved.

Even so there are differences among various products. Prices for monomers range from \$50 to \$250 per pound (lb), with L-lactide being at the low end. This is also true for the polymers made of them: certain poly(L-lactides) are the cheapest polymers at less than \$500 per lb. Sophisticated copolymers, on the other hand, may cost up to \$2000 per lb or more. But, even at these prices, their cost contribution to the final price of the implant is small in most applications.

2. Polymer Producers

Presently, there are two types of companies producing resorbable polyesters:

1. Implant manufacturers producing polymers and (in some cases) monomers for captive consumption. Important companies are Davis and Geck, Ethicon, Medisorb (Cincinnati, OH), Mitsui Toatsu (Tokyo, Japan), Phusis (Le Versoud, France), Southern Biosystems (Birmingham, AL), and U.S. Surgical. Some of these companies also sell polymers to other implant manufacturers.
2. Polymer (and monomer) producers just active in trade sales. Important suppliers are Boehringer Ingelheim (Ingelheim, Germany), Purac (Gorinchem, Holland), and Wako (Osaka, Japan).

C. Outlook

Originally, the industry was optimistic that all metal implants serving a temporary function in surgery could be replaced by resorbable implants (such as the fixation of long bones with plate and screw systems or intramedullary nails). In the meantime, a more modest approach can be observed. The key to developing a successful implant is the careful selection of suitable applications. Today, mainly small implants with limited mechanical requirements are developed.

The most important projects are

Anchoring devices
Small plate and screw systems (for maxillofacial and similar applications)
Stents (for cardiovascular and urological applications)
Guided tissue regeneration

Additional applications will be possible with the advent of stronger polymers and improved processing technology. Also, the success of the first controlled-release formulations will lead to many additional products, in particular for highly active drugs developed through genetic engineering.

Changes are expected on the production side as well. There will be a further concentration of manufacturers, not so much to reduce costs, but to improve reliability and to meet requirements for registration better. It will be very difficult, however, to introduce new chemical entities as polymers because determining the information required for registration is very expensive.

REFERENCES

1. Chu, C. C., Recent advancements in suture fibers for wound closure, *ACS Symp. Ser.*, 457: 167–211 (1991).
2. Benicewicz, B. C., and Hopper, P. K., Polymers for absorbable surgical sutures, Part I, *J. Bioact. Compat. Polymers*, 5:453–472 (1990).
3. Benicewicz, B. C., and Hopper, P. K., Polymers for absorbable surgical sutures, Part II, *J. Bioact. Compat. Polymers*, 6:64–94 (1991).
4. Lewis, D. H., Controlled release of bioactive agents from lactide/glycolide polymers, *Drugs Pharm. Sci.*, 45 (Biodegr. Polym. Drug. Delivery Syst.):1–41 (1990).
5. Lewis, D. H., Biodegradable polymers as drug delivery systems, *Pharm. Manufacturing Intern.*, 1993:99–105 (1993).
6. Gogolewski, S., Resorbable polymers for internal fixation, *Clinical Materials*, 10:13–20 (1992).
7. Vert, M., Li, S. M., Spenlehauer, G., and Guerin, P., Bioresorbability and biocompatibility of aliphatic polyesters, *J. Mater. Sci.: Mater. in Medicine*, 3:432–446 (1992).
8. Barrows, T. H., Synthetic bioabsorbable polymers, in *High Performance Biomaterials* (M. Szycher, ed.), Technomic Publishing Company, Lancaster, PA (1991), 243–257.
9. Amecke, B., Bendix, D., and Entenmannn, G., Resorbable polyesters: Composition, properties, applications, *Clinical Materials*, 10:47–50 (1992).
10. Yamomoto, M., Okada, H., Ogawa, Y., and Miyagawa, T. (Takeda Chemical Industries, Ltd., and Wako Pure Chemical Industries, Ltd.), Polymer, production and use thereof. European Patent (EP) 0202065 (1986), U.S. Patent (US) 4728721 (1986).
11. Nieuwenhuis, J., Synthesis of polylactides, polyglycolides and their copolymers, *Clinical Materials*, 10:59–67 (1992).
12. Munguia, O., Delgado, A., Farina, J., Evora, C., and Llabres, M., Optimization of DL-PLA molecular weight via the response surface method, *Intern. J. Pharm.*, 86:107–111 (1992).
13. Tunc, D. C., Rohovsky, M. W., Jadhav, B., Lehman, W. B., Strongwater, A., and Kummer, F., Body absorbable osteosynthesis devices, *Polymer Sci. Technol. (Plenum)*, 35:87–99 (1987).
14. Zhang, X., Wyss, U. P., Pichora, D., and Goosen, F. A., An investigation of the synthesis and thermal stability of poly(DL-lactide), *Polymer Bulletin*, 27:623–629 (1992).
15. Duda, A., and Penczek, S., Thermodynamics of L-lactide polymerization. Equilibrium monomer concentration, *Macromolecules*, 23:1636–1639 (1990).
16. Sipos, L., Zsuga, M., and Kelen, T., Living ring-opening polymerization of L,L-lactide

initiated with potassium t-butoxide and its 18-crown-6 complex, *Polymer Bulletin*, 27:495–502 (1992).

17. Dubois, P., Jacobs, C., Jerome, R., and Teyssie, P., Macromolecular engineering of polylactones and polylactides. Mechanism and kinetics of lactide homopolymerization by aluminium isopropoxide, *Macromolecules*, 24:2266–2270 (1991).

18. Kricheldorf, H. R., and Kreiser-Saunders, I., Polylactons 19, anionic polymerization of L-lactide in solution, *Macromol. Chem.*, 191:1057–1066 (1990).

19. Dittrich, W., and Schulz, R. C., Kinetik und Mechanismus der ringöffnenden Polymerisation von L(-)-lactid, *Die Angewandte Makromolekulare Chemie*, 15:109–126 (1971).

20. Kohn, F. E., Van Ommen, J. G., and Feijen, J., The mechanism of the ring-opening polymerization of lactide and glycolide, *Eur. Polym. J.*, 19:1081–1088 (1983).

21. Mueller, J., *Internal Reports 1985*, Boehringer Ingelheim KG, Ingelheim, FRG.

22. Eenink, M. J. D., Synthesis of biodegradable polymers and development of biodegradable hollow fibers for controlled release of drugs, Ph.D. thesis, Twente University, Enschede, Netherlands, 1987.

23. Nijenhuis, A. J., Grijpma, D. W., and Pennings, A. J., Lewis acid catalyzed polymerization of L-lactide. Kinetics and mechanism of the bulk polymerization, *Macromolecules*, 20:6419–6424 (1992).

24. Kricheldorf, H. R., and Sumbel, M., Polylactones 18, polymerization of L,L-lactide with Sn(II) and Sn(IV) halogenides, *Eur. Polymer J.*, 25:585–591 (1989).

25. Kricheldorf, H. R., Berl, M., and Scharnagl, N., Poly(lactones) 9. Polymerization mechanism of metal alkoxide initiated polymerizations of lactide and various lactones, *Macromolecules*, 21:286–293 (1988).

26. Leenslag, J. W., and Pennings, A. J., Synthesis of high-molecular-weight poly(L-lactide) initiated with tin 2-ethylhexanoate, *Makromol. Chem.*, 188:1809–1814 (1987).

27. Deasy, P. B., Finan, M. P., and Meegan, M. J., Preparation and characterization of lactic/glycolic acid polymers and copolymers, *J. Microencaps.*, 6:369–378 (1989).

28. Tsuji, H., Hyon, S.-H., and Ikada, Y., Stereo complex formation between enantiomeric poly(lactic acid)s. 5. Caloric and morphological studies on the stereocomplex formed in acetonitril solution, *Macromolecules*, 25:2940–2946 (1992).

29. Tunc, D. C. (Johnson & Johnson), Absorbable bone fixation device, EP 0108635 (1983).

30. Jamshidi, K., and Eberhard, R. C., Characterization of polylactide synthesis, *Polym. Prep. (Am. Chem. Soc. Div. Polym. Chem.)*, 28:236–237 (1987).

31. Nijenhuis, A. J., Grijpma, D. W., and Pennings, A. J., Highly crystalline as-polymerized poly(L-lactide), *Polymer Bulletin*, 26:71–77 (1991).

32. Dunsing, R., and Kricheldorf, H. R., Polylactones 5. Polymerization of L,L-lactide by means of magnesium salts, *Polymer Bulletin*, 14:491–495 (1985).

33. Kricheldorf, H. R., and Serra, A., Polylactones 6. Influence of various metal salts on the optical purity of poly(L-lactide), *Polymer Bulletin*, 14:497–502 (1985).

34. Bendix, D., and Entenmann, G., Molecular weight distributions of reimplanted poly(L-lactides), *World Congress on Implantology*, Paris, March 1989.

35. Tsakala, M., Gillard, J., Roland, M., Chabot, F., and Vert, M., Pyrimethamine sustained release systems based on bioresorbable polyesters for chemoprophylaxis of rodent malaria, *J. Contr. Rel.*, 1988:233–242.

36. Marcotte, N., and Goosen, M. F. A., Delayed release of water-soluble macromolecules from polylactide pellets, *J. Contr. Rel.*, 1989:75–85.

37. Leenslag, J. W., Pennings, A. J., Bos, R. R., Rozema, F. R., and Boering, G., Resorbable materials of poly(L-lactide) VII. *In vivo* and *in vitro* degradation, *Biomaterials*, 8:311–314 (1987).

38. Bendix, D., Buchholz, B., and Entenmann, G. (Boehringer Ingelheim KG), Verfahren zur Herstellung von loesungsmittel- und restmonomerenfreien und in Spritzgiess-, Extrusions- und Schmelzspinnverfahren einsetzbaren Polymeren, EP 0456246 (1991).

39. Bendix, D., and Entenmann, G. (Boehringer Ingelheim KG), Katalysatorfreie resorbierbare Homopolymere und Copolymere, EP 0270987 (1987).

40. Buchholz, B., and Liedtke, H. (Boehringer Ingelheim KG), Verbessertes Verfahren zur Reinigung resorbierbarer Polyester, German Patent Application (DE) 4218268.

41. Vert, M., Chabot, F., Leray, J., and Christel, P., Stereoregular bioresorbable polyesters for orthopaedic surgery, *Makromol. Chem.*, Suppl., 5:30–41 (1991).

42. Kulagina, T. G., Lebedev, B. V., Kiriparisova, E. G., Lyudvig, E. B., and Barskaja, I. G., Thermodynamics of DL-lactide, polylactide and the polymerization in the range of 0–430 K, *Vysokomol. Soedin. Ser.*, A24:1496–1501 (1981).

43. Dahlmann, J., Rafler, G., Fechner, K., and Mehlis, B., Synthesis and properties of biodegradable aliphatic polyesters, *British Polymer J.*, 23:235–240 (1990).

44. Barskaya, I. G., Lyudvig, E. B., and Izyumnikov, A. L., Role of hydroxyl-containing compounds in the cationic polymerization of DL-lactide, *Vysokomol. Soedin. Ser.*, A25: 1544–1548 (1983); Canadian Patent (CA) 99 88638.

45. Barskaya, I. G., Lyudvig, E. B., Shifrina, R. R., and Izyumnikov, A. L., Cationic polymerization of DL-lactide, *Vysokomol. Soedin. Ser.*, A25:1283–1288 (1983); CA 99 54204.

46. Bero, M., Kasperczyk, J., and Jedlinski, Z. J., Coordination polymerization of lactides, 1. Structure determination of obtained polymers, *Makromol. Chem.*, 191:2287–2296 (1990).

47. Rafler, G., and Dahlmann, J., Biologisch abbaubare Polymere. 2. Mitt.: Zur Homo- und Copolymerisation von D,L-Lactid, *Acta Polymerica*, 41:611–617 (1990).

48. Kricheldorf, H. R., Boettcher, C., and Tönnes, K.-U., Polylactones, 23. Polymerization of racemic and meso D,L-lactide with various organotin catalysts — Stereochemical aspects, *Polymer*, 33:2817–2824 (1992).

49. Kohn, F. E., Van den Berg, J. W. A., and Van de Ridder, G., The ring-opening polymerization of D,L-lactide in the melt initiated with tetraphenyltin, *J. Appl. Polymer. Sci.*, 29:4265–4277 (1984).

50. Marcotte, N., and Goosen, M. F. A., Delayed release of water-soluble macromolecules from polylactide pellets, *J. Contr. Rel.*, 9:75–85 (1989).

51. Shinoda, H., and Ohtaguro, A. (Mitsui Toatsu Chemicals, Inc.), Preparation process for bioabsorbable polyester, EP 0368571 (1989).

52. Frazza, E. J., and Schmitt, E. E., A new absorbable suture, *J. Biomed. Mat. Res. Symp.*, 1: 43–58 (1971).

53. Toermaelae, P., Vasenius, J., Vainionpaeae, S., Laiho, J., Pohjonen, T., and Rokkanen, P., Ultra-high-strength absorbable self-reinforced polyglycolide (SR-PGA) composite rods for internal fixation of bone fractures, *In vitro* and *in vivo* study, *J. Biomed. Mat. Res.*, 25: 1–22 (1991).

54. Hickey, A. J., Redmon, M. P., and Marson, L., A polymeric controlled delivery system for chronic administration of nicotine, *Proc. Intern. Symp. Cont. Rel. Bioact. Mater.*, 16:495–496 (1989).

55. Lee, K. C., Soltis, E. E., Newman, P. S., Burton, K. W., Mehta, R. C., and DeLuca, P. P., *In vivo* assessment of salmon calcitonin sustained release from biodegradable microspheres, *J. Contr. Rel.*, 17:199–206 (1991).

56. Debenedetti, P. G., Tom, J. W., Yeo, S. D., and Lim, G. B., Application of supercritical fluids for the production of sustained delivery systems, *J. Contr. Rel.*, 24:27–44 (1993).

57. Rivera, R., Alvarado, G., Aldaba, C. F. S., and Hernandez, A., Norethinestrone contraceptive microspheres, *J. Steroid Biochem.*, 27:1003–1007 (1987).

58. Braun, D., and Kohl, P. R., Anionische Loesungspolymerisation von Glycolid, *Die Angewandte Makromolekulare Chemie*, 139:191–200 (1986).

59. Lebedev, B. V., Yevstropov, A. A., Kiparisova, Y. G., and Belov, V. I., The thermodynamics of glycolide, polyglycolide and of polymerization of glycolide in the temperature range of 0–550°K, *Polymer. Sci. U.S.S.R.*, 20:32–43 (1978).

60. Lyudvig, Y. E., Belen'kaya, B. G., Barskaya, I. G., Khomyakov, A. K., and Bogomolova,

T. B., Cationic polymerization of lactones in presence of hydroxyl-containing compounds, *Acta Polymerica*, 34:754–761 (1983).

61. Gilding, D. K., and Reed, A. M., Biodegradable polymers for use in surgery – Polyglycolic/poly(lactic acid) homo- and copolymers, 1, *Polymer*, 20:1459–1464 (1979).

62. Casey, D. J., and Epstein, M. (American Cyanamid Co.), Process for polymerization a substantially pure glycolide, US 3912692 (1974).

63. Glick, A., and Chirgwin, L. D., Jr. (American Cyanamid Co.), Gefärbtes Polyglycolsäure-Nahtmaterial, DE 2232298 (1972).

64. Fortunatov, O. G., Trostenyuk, N. V., Selezneva, V. E., and Savin, V. A., Purification of polyglycolide, Soviet Union Patent (SU) 569600, Canadian Patent (CA) 88 153265.

65. Deprospero, D. A. (American Cyanamid Co.), Extrudable and stretchable polyglycolic acid and process for preparing some, US 3565869 (1971).

66. Glick, D., and McPherson, J. B. (American Cyanamid Co.), Method for improving the *in-vivo* strength of polyglycolic acid, US 3772420 (1968).

67. Kaplan, D. S., Muth, R. R., and Kennedy, J. J. (U.S. Surgical Corp.), Polymers for injection molding of absorbable surgical devices, US 436056 (1982), EP 01 07591 (1983).

68. Dunn, R. L., English, J. P., Strobel, J. D., Cowsar, D. R., and Tice, T. R., Preparation and evaluation of lactide/glycolide copolymers for drug delivery, in *Polymers in Medicine III* (C. Migliaresi et al., eds.), Elsevier Science Publ. B. V., Amsterdam (1988), 149–160.

69. Avgoustakis, K., and Nixon, J. R., Biodegradable controlled release tablets: 1. Preparative variables affecting the properties of poly(lactide-co-glycolide) copolymers as matrix forming material, *Intern. J. Pharm.*, 70:77–85 (1991).

70. Bendix, D., Analytical studies on the solubility problem of poly(D,L-lactide-co-glycolide) 50 : 50, *Proceed. Intern. Symp. Control. Rel. Bioact. Mater.*, 17:248–249 (1990).

71. Grijpma, D. W., Nijenhuis, A. J., and Pennings, A. J., Synthesis and hydrolytic degradation behaviour of high-molecular weight L-lactide and glycolide copolymers, *Polymer*, 31:2201–2206 (1990).

72. Kricheldorf, H. R., Jonté, J. M., and Berl, M., Polylactones 3. Copolymerization of glycolide with L,L-lactide and other lactones, *Makromol. Chem. Suppl.*, 12:25–38 (1985).

73. Brizzolara, N., Chabrand, I., Miller, D., Rosati, L., Casey, D., and Lawter, J. R., Hydrolytic degradation kinetics of poly(glycolide-co-D,L-lactide), *Proceed. Intern. Symp. Contr. Rel. Bioact. Mater.*, 17:246–247 (1990).

74. Kissel, T., Brich, Z., Bantle, S., Lancranjan, I., Nimmerfall, F., and Vit, P., Parenteral depot-systems on the basis of biodegradable polyesters, *J. Contr. Rel.*, 16:27–42 (1991).

75. Singhal, J. P., Singh, H., and Ray, A. R., Absorbable suture materials: Preparation and properties, *JMS – Rev. Macromol. Chem. Phys.*, C28:475–502 (1988).

76. Carothers, W. H., Dorough, G. L., and Van Natta, F. J., Studies of polymerization and ring formation. X. The reversible polymerization of six-membered cyclic esters, *J. Amer. Chem. Soc.*, 54:761–767 (1932).

77. Lundberg, R. D., Koleske, J. V., Pollart, D. F., and Smarook, W. H. (Union Carbide), Adhesive composition poly(vinyl alkyl ether) and a cyclic ester polymer, US 3641204 (1972).

78. Baggett, J. M., Horvath, J. W., and Wilson, B. W. (Dow Chemical Co.), Polymerization of para-dioxanone and derivates, US 3391126 (1968).

79. Doddi, N., Versfelt, C., and Wassermann, D. (Ethicon, Inc.), Synthetic absorbable surgical devices of poly-dioxanone, US 4052988 (1977).

80. Walter, E. R., and Koleske, J. V. (Union Carbide), Kristalline Polymermischungen, DE 2015262 (1970).

81. Union Carbide, Procédé catalytique de production de polymères linéaires solide d'esters cycliques, French Patent (FR) 2026274 (1969).

82. Snapp, T. C., and Blood, A. E. (Eastman Kodak Co.), Method of preparing 2-p-dioxanone polymers, US 3645941 (1970).

83. Union Carbide, Procédé de production catalytique de polymères linéaires solides de haut poids moléculaire d'esters cycliques, FR 2026275 (1969).
84. Hinsch, B., and Walther, C. (Ethicon Inc.), Implant, EP 0274898 (1987).
85. Jamiolkowski, D. D., Shalaby, S. W., Bezwada, R. S., and Newman, H. D. (Ethicon, Inc.), Glycolide/p-dioxanone block copolymers, US 4838267 (1989).
86. Mattei, F. V., and Doddi, N. (Ethicon, Inc.), Synthetic absorbable hemostatic composition, US 4440789 (1984).
87. Koelmel, D. F., Shalaby, S. W., and Jamiolkowski, D. D. (Ethicon, Inc.), Compatible blends of poly(p-dioxanone) and poly(alkylene phenylene-bis-oxyacetate) and absorbable devices made therefrom, EP 0185467 (1985).
88. Bezwada, R. S., Shalaby, S. W., Newman, H. D., and Kafrawy, A., Bioabsorbable copolymers of p-dioxanone and lactide for surgical devices, *Trans. Ann. Meeting Soc. Biomat.*, 1990:194.
89. Jamiolkowski, D. D., Shalaby, S. W., Bezwada, R. S., and Newman, H. D., Bioabsorbable block copolymers of p-dioxanone and glycolide for surgical sutures, *Trans. Ann. Meeting Soc. Biomat.*, 1990:193.
90. Bezwada, R. S., Shalaby, S. W., and Newman, H. D., Bioabsorbable fibers of p-dioxanone copolymers, *ACS Symp. Ser.*, 433:167–173 (1990).
91. Bezwada, R. S., Shalaby, S. W., and Hunter, A. W. (Ethicon, Inc.), Bioabsorbable copolymers of polyalkylene carbonate/rho-dioxanone for sutures and coatings, US 5037950 (1991).
92. Shalaby, S. W., and Koelmel, D. F. (Ethicon Inc.), Copolymers of p-dioxanone and 2,5-morpholinedione and surgical devices formed therefrom having accelerated absorption characteristics, EP 0086613 (1983).
93. Snapp, T. C., Blood, A. E., and Johnson, S. H. (Eastman Kodak Co.), Polymeric hydroxylated polyesterethers and polyurethane flexible foam prepared therefrom, US 3970619 (1976).
94. Baba, A., Kashiwagi, H., and Matsuda, H., Reaction of carbon dioxide with oxetane catalyzed by organitin halide complexes: control of reaction by ligands, *Organometallics*, 6:137–140 (1987).
95. Albertsson, A.-C., and Sjoling, M., Homopolymerization of dioxan-2-one to high molecular weight poly(trimethylenecarbonate), *J.M.S. – Pure Appl. Chem.*, A29(1):43–54 (1992).
96. Kricheldorf, H. R., and Jenssen, J., Polylactones. 16. Cationic polymerization of trimethylene carbonate and other cyclic carbonates, *J. Macromol. Sci. – Chem.*, A26(4):631–644 (1989).
97. McNeill, I. C., and Rincon, A., Degradation studies of some polyesters and polycarbonates: Part 5. Poly(trimethylenecarbonate), *Polymer Degradation and Stability*, 24:59–72 (1989).
98. Carothers, W. H., and Van Natta, F. J., Studies on polymerization and ring formation: III. Glycol esters of carbonic acid, *J. Amer. Chem. Soc.*, 52:314–326 (1930).
99. Kricheldorf, H. R., Jenssen, J., and Kreiser-Saunders, I., Polymers of carbonic acids, 6. Polymerization of trimethylene carbonate (1,3-dioxan-2-one) with complexation catalysts, *Makromol. Chem.*, 192:2391–2399 (1991).
100. Zhu, K. J., Hendren, R. W., Jensen, K., and Pitt, C. G., Synthesis, properties, and biodegradation of poly(1,3-trimethylene carbonate), *Macromolecules*, 24:1736–1740 (1991).
101. Grijpma, D. W., Nijenhuis, A. J., van Wijk, P. G. T., and Pennings, A. J., High impact strength as-polymerized PLLA, *Polymer Bulletin*, 29:571–578 (1992).
102. Draney, D. R., and Jarrett, P. K., Sequence assignments and block length of poly(L-lactide-co-trimethylene carbonate), *Polymer Preprints*, 31:137–138 (1990).
103. Buchholz, B., Analysis and characterization of resorbable DL-lactide-trimethylene carbonate copolyesters, *J. Mat. Sci. – Mater. in Medicine*, 4:381–388 (1993).
104. Brinen, J. S., Greenhouse, S., and Jarrett, P. K., XPS and SIMS studies of biodegradable suture materials, *Surface and Interface Analysis*, 17:259–266 (1991).

105. Roby, M. S., Casey, D. J., and Cody, R. D., Absorbable sutures based on glycolide/ trimethylene carbonate copolymers, *Trans. Soc. Biomat.*, 8:216 (1985).
106. Rosati, L., and Casey, D. J., Degradable thermoplastic hydrogels, *Polym. Mater Sci. Engn.*, 59:516–520 (1988).
107. Sanina, G. S., and Lyudvig, E. B., Basicity and relative reactivity of lactones, *Dokl. Akad. Nauk SSSR*, 230:153–165 (1976); CA 86 55741.
108. Grijpma, D. W., and Pennings, A. J., Polymerization temperature effects on the properties of L-lactide and epsilon-caprolactone, *Polymer Bulletin*, 25:335–341 (1991).
109. Kricheldorf, H. R., Mang, T., and Jonté, J. M., Polylactones. 1. Copolymerization of glycolide and epsilon-caprolactone, *Macromolecules*, 17:2173–2181 (1984).
110. Vanhoorne, P., Dubois, P., Jerome, R., and Teyssie, P., Macromolecular engineering of polylactones and polylactides, 7. Structural analysis of copolyesters of epsilon-caprolactone and L- or D,L-lactide initiated by Al(O^{i}Pr)$_3$, *Macromolecules*, 25:37–44 (1992).
111. Grijpma, D. W., Zondervan, G. J., and Pennings, A. J., High molecular weight copolymers of L-lactide and epsilon-caprolactone as biodegradable elastomeric implant materials, *Polymer Bulletin*, 25:327–333 (1991).
112. Sinclair, R. G., Biodegradable thermoplastic from lactic acid — Some processing and biomaterial applications, *ANTEC Plastics*, 87:1214–1219 (1987).
113. Jacobs, C., Dubois, P., Jerome, R., and Teyssie, P., Macromolecular engineering of polylactones and polylactides, 5. Synthesis and characterization of diblock copolymers based on poly-epsilon-caprolactone and poly(L,L or D,L)lactide by aluminium alkoxides, *Macromolecules*, 24:3027–3034 (1991).
114. Nakamura, T., Hitomi, S., Shimamoto, T., Hyon, S. H., Ikada, Y., Watanabe, S., and Shimuzu, Y., Surgical application of biodegradable films prepared from lactide-epsilon-caprolactone, in *Biomaterials and Clinical Applications* (A. Pizzoferrato et al., eds.), Elsevier Science Publ. B. V., Amsterdam (1987), 759–764.
115. Gu, Z., Ye, W., Yang, J., Li, Y., Chen, X., Zhong, G., and Feng, X., Biodegradable block copolymer matrices for long acting contraceptives with constant release, *J. Contr. Rel.*, 22:3–14 (1992).
116. Shalaby, S. W., and Jamiolkowski, D. D., Synthesis and intrinsic properties of crystalline copolymers of epsilon-caprolactone and glycolide, *Amer. Chem. Soc., Div. Polymer Chem. Polymer Prepr.*, 26:190 (1985).
117. Bezwada, R. S., Shalaby, S. W., and Erneta, M. (Ethicon, Inc.), Crystalline copolymers of p-dioxanone and epsilon-caprolactone, EP 0440416 (1991).
118. Solomon, O. F., and Ciuta, I. Z., Determination de la viscosite intrinseque de solution de polymers par une simple determination de la viscosite, *J. Appl. Pol. Sci.*, 6(24):683–686 (1962).
119. Bendix, D., Some problems in the molecular weight determination of resorbable polymers by GPC and viscosimetry, *Symposium on Characterization of Macromolecules Used as Pharmaceutical Excipients*, Gothenburg, Sweden, March 7–9, 1990.
120. Yau, W. W., Kirkland, J. J., and Bly, D. D., *Modern Size-Exclusion Chromatography*, John Wiley & Sons, New York, 1979.
121. Fischer, E. W., Sterzel, H. J., and Wegner, G., Investigation of the structure of solution grown crystals of lactide copolymers by means of chemical reaction, *Kolloid-Z. u. Z. Polymere*, 251:980–990 (1973).
122. Pistner, H., Bendix, D. R., Muehling, J., and Reuther, J. F., Poly(L-lactide): A long-term degradation study *in vivo*, *Biomaterials*, 14(4):291–298 (1983).

28
Lactide/ε-Caprolactone Copolymers

Steve T. Lin
Zimmer, Incorporated
Warsaw, Indiana

I. INTRODUCTION

During the last four decades, biomaterial implants have been widely used to treat a variety of bodily functions. The majority of the implant devices has been shown to be safe and effective in clinical uses, and they have greatly improved the well-being of humankind. Some implant devices are intended for permanent applications such as total hip, and knee joint replacements and heart valves. Others are intended for temporary applications such as fracture fixation devices and controlled drug delivery devices. The fixation devices provide a rigid fixation and promote bone healing. The drug delivery devices are designed to deliver an optimum amount of therapeutic agents over an extended time period. To avoid long-term complications, these devices are usually removed by surgical methods after they fulfill the intended clinical uses. This second surgery can often cause patients discomfort and also increase the risk and cost. The need to use degradable materials for the design of the temporary implant devices is obvious.

Extensive efforts have been devoted to the development of various biodegradable polymers for different medical applications. Among them, poly(alpha-hydroxy acids) have attracted the most attention due to their excellent biocompatibility, high mechanical strength, and good processability. Bioresorbable pins made of poly(glycolic acid) (PGA) have been used clinically to treat interfragmental bone fractures [1]. PGA has a high tensile strength and modulus, but it degrades very fast when compared with poly(L-lactic acid) (PLLA). PLLA is also a high-strength and high-modulus polymer, but degrades much slower. Resorbable screws made of PLLA have been investigated in a clinical study and were shown to be effective in anchoring acetabular components [2]. Because of the ability to retain their strength over a long time period [3], the PLLA screws were able to fix the implant in the acetabulum, allowing bony ingrowth to the porous surface for a permanent biological fixation. PLLA has also been shown to be a better material than polyethylene for the design of intramedullary cement plugs because of its higher modulus and ability to resorb [4]. PGA and PLLA are rigid thermoplastics with a relatively small elongation to break. They are limited to the applications that do not require a large deformation.

Unlike PGA and PLLA, poly(ϵ-caprolactone) (PCL) is a thermoplastic elastomer. It is strong, tough, and can degrade slowly. Copolymers of L-lactide (LA) and ϵ-caprolactone (CP) with the L-lactide as the major constituent have been synthesized with a ring-opening polymerization [5]. The LA/CP copolymers behave like a rigid thermoplastic solid or a thermoplastic elastomer depending on the ϵ-caprolactone content. When the ϵ-caprolactone is the major constituent, LA/CP copolymers can be synthesized to form thermoplastic elastomers that are tough, strong, and elastic [6]. The elastic copolymers have a low melting temperature, which makes them easy to process to films, fibers, tubes, and other shaped solids.

The purpose of this chapter is to report the method of copolymer synthesis and the physical, thermal, and mechanical properties of the copolymers. The results of the degradation study and the biocompatibility testing of the copolymers are also included.

II. MATERIALS AND METHODS

A. Purification of Monomers

To obtain high molecular weight polymers, monomers need to be purified before polymer synthesis. Optically active L-lactide was purified by recrystallization from ethyl acetate. The lactide was first dissolved in hot ethyl acetate at a ratio of 1 : 1 (weight to volume, w/v). As the solution gradually cooled to room temperature, the lactide recrystallized. The purified lactide crystals were separated from the solution by filtration and dried under vacuum to remove the residual solvent completely. The differential scanning calorimeter (DSC) was used to measure the melting temperature. The pure L-lactide crystals showed a sharp melting peak at 98°C.

The ϵ-caprolactone was purified by vacuum distillation. The pure portion was collected from the liquid boiling at 92°–94°C at 2-mm Hg. The pure ϵ-caprolactone was water clear with a single gas chromatography peak.

B. Copolymerization

The pure lactide and ϵ-caprolactone were charged into a three-neck glass reactor equipped with a mechanical stirrer. The reactants were blanketed by dry nitrogen passing through the reactor. Stannous octoate was used as a catalyst. The concentration of the catalyst was 0.02–0.03 wt% (weight percent) based on the total weight of the reactants. Copolymerization was carried out at atmospheric pressure for at least 24 hours at 140°–160°C.

After the synthesis was complete, the copolymer was dissolved in methylene chloride to form a 15 weight percent (wt%) solution. To the vigorously stirred polymer solution, a 4x isopropyl alcohol was then added to precipitate the copolymer. The majority of unreacted monomers and catalyst was removed from the copolymer mass. The copolymer mass was chopped, in a blender, to a fine powder at low temperature. The copolymer powder was extracted with cold isopropyl alcohol to remove further catalyst and unreacted monomers. The copolymer was white, elastic, and tough after the solvent was evaporated under high vacuum at room temperature.

Table 1 Physical Properties of Lactide/ε-Caprolactone Copolymers

LA/CP ratio (w/w)	5/95	25/75	40/60
$\overline{M}w$	419,000	364,000	323,000
$\overline{M}n$	230,000	172,000	180,000
$T_m(°C)$	54	47	42
$T_g(°C)$	-60	-32	-16
Hardness (Shore A)	100	84	75
Density (gm/ml)	1.12	1.13	1.13
Degradation rate constant (day^{-1})	1.08×10^{-2}	2.19×10^{-2}	3.04×10^{-2}

$\overline{M}w$ = weight average molecular weight; $\overline{M}n$ = number average molecular weight

Three random copolymers were prepared with the lactide contents of 5 wt%, 25 wt%, and 40 wt% (see Table 1).

C. Physicomechanical Characterization

The molecular weights (MWs) of the copolymers were determined by gel permeation chromatography (GPC) using polystyrene as standards. The dilute polymer solution in methylene chloride, about 0.1% (w/v), was injected into a microbead-packed column maintained at 25°C. A refractive index detector was used to measure the polymer peaks.

Differential scanning calorimetry was used to determine the glass transition temperature T_g and the melting temperature T_m of the copolymers. To measure the subzero glass transition temperatures, liquid nitrogen was used to chill the polymer samples. The heating rate was maintained at 15°C/minute.

The tensile properties of the copolymers were determined according to American Society for Testing and Materials (ASTM) 882. The copolymer film was prepared by casting a 10% (w/v) copolymer solution in methylene chloride on a smooth glass surface. After the solvent was removed, the copolymer film was cut to the tensile test specimens with nominal dimensions of 150 × 12.7 × 1.0 mm. A crosshead speed of 500 millimeters per minute (mm/min) was used. The ultimate tensile strength, elongation to break, and elastic modulus were calculated from the load-displacement curves. The density and hardness of the copolymers were also determined.

D. *In Vitro* Degradation

To study the hydrolytic degradation of the copolymers, the film specimens were immersed in a 0.9% saline solution at 37°C. The specimens were removed regularly and the molecular weights were determined by GPC. The weight average molecular weights were plotted against time in a semilog graph. A straight line was constructed and the first-order degradation rate constant was calculated from the slope.

E. Biocompatibility Testing

To evaluate the biocompatibility of the lactide/ε-caprolactone copolymers, the United States Pharmacopeia (USP) Class VI test for plastics was employed [7]. In addition, an Ames mutagenicity test was also conducted to determine if a saline extract of the copolymers would cause mutagenic changes in histidine-dependent mutant strains of *Salmonella typhimurium* [8]. The trace metals of iron, lead, and tin in the copolymers were also determined by atomic absorption spectrophotometry.

III. RESULTS AND DISCUSSION

Both poly(ϵ-caprolactone) and poly(L-lactide) are semicrystalline homopolymers. Poly-(ϵ-caprolactone) is a rubbery, tough polymer with an extremely low T_g of $-65\,°C$. It also has a rather low T_m of $60\,°C$. It can crystallize from the polymer melt very easily, with an average crystallinity close to 50%. Compared with poly(ϵ-caprolactone), poly(L-lactide) has a high T_g of $55\,°C$ and also a high T_m of $180\,°C$. It can crystallize at $100\,°C$ to achieve about 40% crystallinity. The physicomechanical properties of the poly(ϵ-caprolactone) were greatly modified when copolymerizing with poly(L-lactide). As the L-lactide content increased, the T_g of the copolymers tended to increase while the T_m tended to decrease (see Table 1). The T_g increased from $-65\,°C$ to $-60\,°C$ and the T_m decreased from $60\,°C$ to $54\,°C$ when copolymerizing ϵ-caprolactone (95 wt%) with 5 wt% of lactide. The T_g increased to $-32\,°C$ and the T_m decreased to $47\,°C$ for the 25/75 (LA/CP) copolymer. The T_g increased to $-16\,°C$ and the T_m decreased to $42\,°C$ for the 40/60 (LA/CP) copolymer.

The weight average molecular weights ($\overline{M}w$) of the lactide/ϵ-caprolactone copolymers as determined by the GPC were high, over 300,000. When immersed in water, the lactide/ϵ-caprolactone copolymers underwent ester linkage hydrolysis catalyzed by the carboxylic end groups. Chain scission, which is a random process, resulted in a decrease in molecular weight. The decrease in molecular weight followed the first-order degradation kinetics, which is similar to their respective homopolymers. From the plots of $\log_e$ (MW) versus time, straight lines could be constructed.

The degradation rate constants were calculated from the slopes and are shown in Table 1. The poly(ϵ-caprolactone) homopolymer degrades very slowly, with a degradation rate constant of $3-6 \times 10^{-3}\,day^{-1}$. The degradation rate increased when copolymerized with the lactide. The higher the lactide content was, the faster the copolymers degraded. The 40/60 (LA/CP) copolymer degraded three times faster and the 25/75 (LA/CP) copolymer degraded two times faster than the 5/95 (LA/CP) copolymer. The degradation rate constants for the 5/95, 25/75, and 40/60 (LA/CP) were 1.08×10^{-2}, 2.19×10^{-2}, and $3.04 \times 10^{-2}\,(day^{-1})$, respectively.

The hardness of the lactide/ϵ-caprolactone copolymers decreased with the increase of the lactide content. The copolymers were soft polymers with a hardness below 100 Shore A. The densities of the copolymers were about 1.1 grams per milliliter (gm/ml) (Table 1).

Because of high MW, the copolymers exhibited good tensile properties. The ultimate tensile strength and the elastic modulus of the 5/95 (LA/CP) copolymer were measured

Table 2   Tensile Properties of Lactide/ϵ-Caprolactone Copolymers

LA/CP ratio (w/w)	5/95	25/75	40/60
Ultimate tensile strength (MPa)	51	12	11
Initial elastic modulus (MPa)	35	8	5
Elongation to break (%)	>2000	>2000	>2000

Table 3 Biocompatibility Testing Results

Test method	Result
Cytotoxicity (polymer)	Nontoxic
Cytotoxicity (polymer extract)	Nontoxic
Acute systemic toxicity	Pass
Intracutaneous toxicity	Pass
Implantation test	Not significant
Ames mutagenicity test	Nonmutagenic
Trace metals	
Iron	< 1 ppm
Lead	< 1 ppm
Tin	3 ppm

to be 51 MPa (megapascal) and 35 MPa, respectively (Table 2). The tensile strength and modulus decreased when the lactide content increased. The 25/75 (LA/CP) copolymer had a tensile strength of 12 MPa and an elastic modulus of 8 MPa. For the 40/60 (LA/CP) copolymer, the elastic modulus decreased to 5 MPa, but the tensile strength was about the same at 11 MPa. The elasticity of the lactide/ε-caprolactone copolymers was very high; the copolymers exhibited an elongation greater than 2000%.

The biocompatibility testing results showed that the copolymers of lactide/ε-caprolactone were nontoxic and nonmutagenic. The trace metals of iron, lead, and tin in the copolymers were also very low (Table 3). The biodegradation products of the copolymers are a combined degradation product of the two homopolymers. Poly(L-lactide) biodegrades to lactic acid, which is metabolized to water and CO_2. Poly(ε-caprolactone) biodegrades to ε-hydroxycaproic acid, which is metabolized to water and CO_2. These biodegradation products are nontoxic.

IV. CONCLUSIONS

High molecular weight random copolymers of LA/CP were synthesized by ring-opening polymerizations using stannous octoate as a catalyst. These polymers were true copolymers because only a single T_g was detected by the DSC. These high molecular weight copolymers exhibited the characteristics of thermoplastic elastomers: tough, strong, and elastic. The copolymers also had good adhesive properties. The 25/75 (LA/CP) copolymer had been used as a biodegradable coating on a carbon-fiber artificial ligament. The high-modulus carbon fibers are brittle and easy to break. The highly elastic copolymer coating effectively protected the carbon fibers from breakage. It also gave the ligament a high degree of flexibility and made it easy to handle in surgeries. As the copolymer coating biodegraded, the carbon-fiber ligament acted like a scaffold, allowing the host tissue to regenerate [9].

Because a wide range of degradation rates can be tailored, the lactide/ε-caprolactone copolymers also have the utility for the design of other implantable devices (e.g., for nerve tissue regeneration, sustained drug release, urinary stent, etc.). The adhesive property also makes the copolymers a good binding agent for bioceramic particles as a bone graft material [10].

REFERENCES

1. Bostman, O., et al., Biodegradable Internal Fixation for Malleolar Fractures, *J. Bone Joint Surg.*, 69B(4):615–619 (1987).
2. Malchau, H., et al., Bioresorbable versus Titanium Screws in Acetabular Cup Fixation: A Prospective Randomized Evaluation Using Stereoradiography, *Transactions of the 40th Annual Meeting of Orthopaedic Research Society*, New Orleans, LA, February 1994.
3. Lin, S., et al., Poly(L-Lactic Acid) Orthopaedic Fixation Devices, *Transactions of the 17th Annual Meeting of the Society for Biomaterials*, 185 (May 1–5, 1991).
4. Lin, S., et al., Characterization of Bioresorbable Intramedullary Plug, *Transactions of the 4th World Biomaterials Congress*, 16 (April 24–28, 1992).
5. Sinclair, R., Copolymerization of L-Lactide and ε-Caprolactone, U.S. Patent 4,057,637 (1987).
6. Lin, S., Lactide/Caprolactone Polymer, Method of Making the Same, Composites Thereof, and Prostheses Produced Therefrom, U.S. Patent 4,643,734 (1987).
7. *The United States Pharmacopeia, XXII*, United States Pharmacopeial Convention, Rockville, MD, 1990.
8. Maron, D. M., and Ames, B. M., Revised Methods for *Salmonella* Mutagenicity Test, *Mutation Research*, 113:173–215 (1983).
9. Parsons, J. R., Resorbable Material and Composite: New Concept in Orthopaedic Biomaterials, *Orthopaedics*, 8:907–915 (1985).
10. Lin, S., Moldable Bone-Implant Material, U.S. Patent 4,645,503 (1987).

29
Synthesis and Properties of Biodegradable Lactic/Glycolic Acid Polymers

Xue Shen Wu*
Medisorb Technologies International
Cincinnati, Ohio

I. INTRODUCTION

Lactic/glycolic acid polymers are homo- and copolymers of lactic acid and glycolic acid. These polymers are generally classified as polyesters [1–3], aliphatic polyesters [4,5], poly(α-hydroxy acids) [6,7], or polylactones [8,9]. The lactic/glycolic acid polymers are usually named poly(lactic acid), poly(glycolic acid), and poly(lactic-co-glycolic acid) under the rules of polymer nomenclature [10], in which the monomer (or the presumptive monomer) is placed in parenthesis and prefixed by poly. With frequent usage, the parentheses are often dropped. They are also called polylactate, polyglycolate, and poly(lactate-co-glycolate) [11]. The lactic/glycolic acid polymers having high molecular weight are often named poly(lactide), poly(glycolide), and poly(lactide-co-glycolide) because they are usually synthesized from the cyclic diesters of lactic acid and/or glycolic acid, lactide and/or glycolide. Over the past two decades, the lactic/glycolic acid polymers have been extensively investigated for biomedical applications [12–15], in particular in the area of controlled release of biologically active agents [16–18].

The first discovery of the lactic/glycolic acid polymers was probably made by Bischoff and Walden a hundred years ago [19–21]. Carothers also did some pioneering work on these polymers [22]. Actually, the polymers synthesized by these pioneers were oligomers. The molecular weight of these polymers was only a few thousand daltons [23]. By 1954, Du Pont scientists found a way to achieve high molecular weights [24]. However, the lactic/glycolic acid polymers were not processed for commercial use until the late 1960s.

In 1967, Davis and Geck, a subsidiary of American Cyanamid Company, filed a patent for using polyglycolides to make suture fibers [25]. The fibers made from these

---

Current affiliation: Long Island University, Brooklyn, New York polyesters have satisfactory mechanical properties, a low immunogenicity, and an extremely low toxicity, in addition to biodegradability, biocompatibility, and bioabsorbability. Davis and Geck's sutures were first marketed in 1970 under the trade name of Dexon™ [26]. In the early 1970s, Ethicon, a Johnson and Johnson company, also developed a suture using the copolymer of glycolide and lactide; it is sold as Vicryl™ [27]. During the 1970s, a wealth of literature on lactic/glycolic acid sutures was published.

This early work clearly demonstrated the nontoxic nature of the polymers and provided degradation data for the polymers. The acceptance of lactic/glycolic acid polymers as sutures has made them attractive for a variety of medical and pharmaceutical applications, including wound closure [28], dental repairs [29], fracture fixation (bone plates, screws, pins, and splints) [30–33], ligament reconstruction [34,35], vascular grafts [36], tracheal replacement [37], ventral herniorrhaphy [38], nerve repairs [39], and drug delivery [16,17].

Since the lactic/glycolic acid polymers are important biodegradable materials and have a wide range of applications, it is important to clarify the synthesis, properties, and biodegradation of these polymers. It is also pertinent to consider experimental approaches that can be used to characterize these polymers. Therefore, this chapter discusses the synthesis, properties, and characterization of the lactic/glycolic acid polymers. The polymer properties that are covered include physical and chemical properties (e.g., chemical composition, molecular weight, molecular weight distribution, crystallinity, polymer chain flexibility, melting point, glass transition temperature, mechanical strength, and solubility). Then, the degradation process and mechanisms, including biodegradation and thermal degradation, are described. The storage and handling of the lactic/glycolic acid polymers are also addressed.

II. SYNTHESIS OF LACTIC/GLYCOLIC ACID POLYMERS

Using the property of molecular weight, lactic/glycolic acid polymers can be categorized into low molecular weight polymers or oligomers and high molecular weight polymers. The low molecular weight lactic/glycolic acid polymers have a molecular weight range of a few thousand daltons, whereas the molecular weight of the high molecular weight polymers is usually defined as over 10,000 daltons. It is relatively easy to synthesize the low molecular weight lactic/glycolic acid polymers. They can be synthesized directly from lactic acid and glycolic acid by condensation polymerization. On the other hand, the high molecular weight ones are usually synthesized from lactide and glycolide by ring-opening polymerization.

A. Synthesis of Low Molecular Weight Polymers

As stated above, the low molecular weight lactic/glycolic acid polymers can be prepared by direct condensation of lactic acid and/or glycolic acid [40,41]. Figure 1 schemes the chemical reactions for the synthesis of the low molecular weight lactic/glycolic acid polymers by this method.

A high temperature of around 130°–190°C is usually required for the synthesis of the low molecular weight lactic/glycolic acid polymers. The water generated by the condensation of the free acids can be removed by using either vacuum or purging with nitrogen [42,43]. Sometimes a catalyst such as antimony oxide is used in the preparation of low molecular weight lactic/glycolic acid polymers [44].

The physicochemical properties of the low molecular weight polymers limit their biomedical applications. For example, their mechanical strength is quite low and their

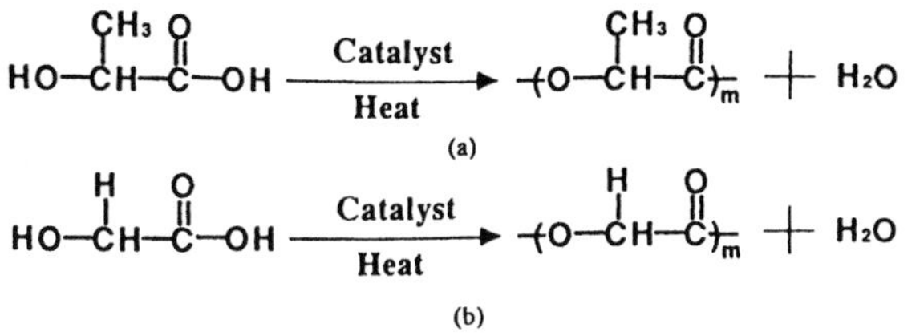

Figure 1 Formation of lactic/glycolic acid polymers: (a), from lactic acid; (b), from glycolic acid.

degradation is very fast. Therefore, the low molecular weight lactic/glycolic acid polymers are not suitable for applications in which high mechanical strength is required for the whole duration of usage. More specifically, the low molecular weight lactic/glycolic acid polymers cannot be used for suture and orthopedic devices. They are mainly used in drug delivery systems for which they serve as drug matrices that do not have a high mechanical strength requirement [45–48].

B.   Synthesis of High Molecular Weight Polymers

The simple condensation of lactic acid and/or glycolic acid can only yield low molecular weight lactic/glycolic acid polymers using current technology. The preferred method of synthesis for producing these polymers, especially high molecular weight polymers, is the ring-opening polymerization of the cyclic diester of lactic acid and/or glycolic acid (lactide and/or glycolide) (see Fig. 2) [49,50].

The advantages of the ring-opening polymerization method are many. First, there is no water removal or dehydration method needed in the polymerization system. Second, the cyclized monomer(s) and linear form of the polymers produced are sufficiently different in physical properties to allow ready purification. In addition, the stoichiometry of the monomer(s) to repeating units of the polymer is guaranteed.

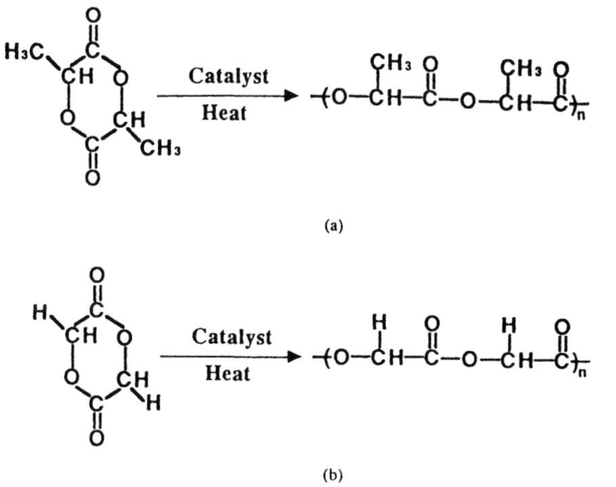

Figure 2 Formation of lactic/glycolic acid polymers: (a), from lactide; (b), from glycolide.

1. *Polymerization Mechanisms*

The general ring-opening polymerization reactions of lactide and glycolide to form the high molecular weight poly(lactic acid) and poly(glycolic acid) are shown schematically in Fig. 2. This ring-opening polymerization is usually initiated or catalyzed by a Lewis acid such as stannous octoate.

The mechanism for the ring-opening polymerization of lactic/glycolic acid polymers has frequently been discussed [51–53]. In general, it has been proposed that the polymerization mechanism involves an electrophilic attack of one of the oxygens of the ester group of a lactide or glycolide [54].

Vion et al. [54], Kumar [55], and Kissel et al. [56] have proposed a specific mechanism by which the lactic acid and/or glycolic acid diester rings are opened via a selective cleavage of the acyl oxygen bond of the ring monomers. For example, when aluminum isopropoxide is used as a catalyst in the synthesis of lactic/glycolic acid polymers, the polymerization mechanism may be postulated as shown in Fig. 3a [54]. This mechanism is supported by the apparent retention of the optical configuration of both asymmetric carbons in the diester ring of L-lactide during the polymerization. This optical configuration retention should be impossible if the ring opening took place at either the oxygen-carbon (sp^3) bond or the carbon-carbon (sp^3) bond. Additional evidence supporting the mechanism represented in Fig. 3a is the formation of hydroxy end groups when aluminum isopropoxide is used to initiate the polymerization of L-lactide (see Fig 3b) [54].

An alternative mechanism proposed by Kricheldorf and Dunsing is that the alkyl oxygen bond is cleaved to open the lactide diester ring as shown in Fig. 4 [57]. They used methyl triflate as a catalyst to initiate the ring-opening polymerization. Then, they examined the end groups of the polymer chain using ^{1}H nuclear magnetic resonance (NMR). Their finding that methyl end groups were present supports their proposed mechanism of alkyl oxygen bond cleavage [57].

The mechanism of the polymerization reaction of lactic/glycolic acid polymers ap-

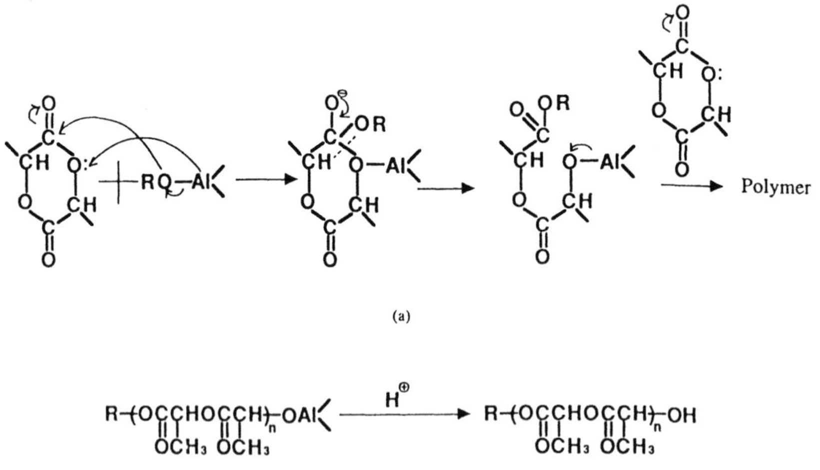

(a)

(b)

Figure 3   Hypothetical mechanism for ring-opening polymerization of (a) lactide and (b) glycolide via a cleavage of an acyl oxygen bond of the monomers.

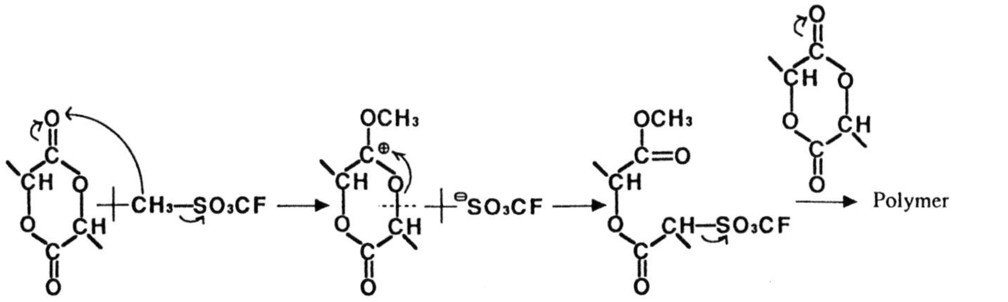

Figure 4 Alternative hypothetical mechanism for ring-opening polymerization of lactide and glycolide via a cleavage of an alkyl oxygen bond of the monomers. (Modified from Ref. 57.)

pears to depend on the polymerization conditions, such as the type of catalyst used. Kricheldorf et al. tested many different types of catalyst while investigating the polymerization mechanism for a copolymer of glycolide and ϵ-caprolactone [9]. Their conclusion was that different types of catalysts initiate different types of polymerization mechanisms, which favors the formation of polymers rich in either glycolide or caprolactone.

2. Polymerization Kinetics

It has been proposed that the polymerization kinetics for the polyesters can be described by Eq. 1 [54].

$$\frac{-d[M]}{dt} = k_p[M][C] \tag{1}$$

where $-d[M]/dt$ is the monomer conversion rate or polymerization speed, k_p is a reaction rate constant, and $[M]$ and $[C]$ are monomer and catalyst concentrations, respectively. Upon rearrangement,

$$\frac{-d[M]}{[M]} = k_p[C]dt \tag{2}$$

For a given polymerization system, the catalyst concentration is constant during the reaction. At $t = 0$, $[M] = [M]_0$, where $[M]_0$ is the initial concentration of the monomer. Integration of Eq. 2 then leads to

$$\text{Ln}([M]_0/[M]) = k_p[C]t \tag{3}$$

Equation 3 suggests that a straight-line relationship should hold between $\text{Ln}([M]_0/[M])$ and polymerization time for a given polymerization system. To illustrate, Vion et al. polymerized L-lactide using aluminum isopropoxide as a catalyst in toluene [54]. The results are shown in Fig. 5.

Gilding and Reed have studied the conversion kinetics of glycolide in the formation of glycolide homopolymer [51]. They found that 80% monomer conversion takes place within the first 30 minutes of the polymerization when the reaction is carried out at 220°C and stannous octoate (0.03%) and lauryl alcohol (0.01%) are used as catalysts. A maximum conversion of 96% is achieved in 4 hours (see Fig. 6).

Leenslag and Pennings have polymerized L-lactide using stannous octoate as the catalyst [58]; the reaction kinetics are shown in Fig. 7. At the beginning of the reaction,

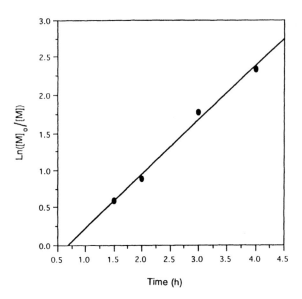

Figure 5 Semilog plot showing the decrease of monomer concentration as a function of time for polymerization of L-lactide catalyzed by aluminum isopropoxide at 70°C in toluene. (Modified from Ref. 54.)

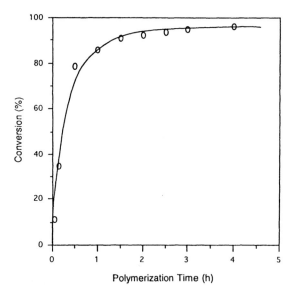

Figure 6 Conversion as a function of polymerization time for the polymerization of glycolide catalyzed by stannous octoate at 220°C. (Reproduced from Ref. 51 with permission of the copyright owner, Butterworth-Heinemann Ltd.)

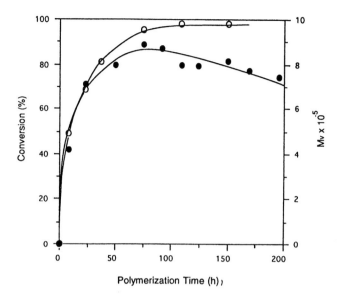

Figure 7　Conversion, -○-, and M_v, -●-, as a function of polymerization time for the polymerization of L-lactide catalyzed by stannous octoate at 100°C. (Modified from Ref. 58.)

both the monomer conversion and the increase in molecular weight proceed rapidly. After about 70 hours, the monomer conversion reaches equilibrium and the molecular weight reaches a maximum value. Then, the molecular weight levels off, probably due to thermal degradation of the polymer. Kohn et al. have investigated the reaction kinetics of D,L-lactide using tetraphenyl tin as an initiator [59]. Their results show that within the first 30 minutes there was almost no polymer formed. After this short induction period, polymerization started, and during the first 8–9 hours, 80–90% of monomer (D,L-lactide) was converted to polymer. This conversion was comparatively slow during the first 3 hours, but within 10–15 hours conversion of the monomer had reached a constant level of approximately 95%.

The glycolide monomer has a higher reactivity than the lactide monomer. The reported reactivity is 2.8 for glycolide and 0.2 for lactide. Thus, the polymerization rate of glycolide is higher than that of lactide. Glycolide has a 3-to-1 preference to react with another glycolide unit than with a lactide unit, and lactide has a 5-to-1 preference to react with a glycolide unit rather than another lactide [51]. Based on these data, when lactide is copolymerized with glycolide at a high glycolide concentration, there is an increased possibility that blocks of glycolide will be formed rather than a random copolymer structure of glycolide and lactide units. At the beginning of polymerization, blocks of glycolide will be separated by single lactide units, while lactide will be incorporated to ever-increasing extents as the glycolide is depleted. The particular monomer sequence in a copolymer of lactide and glycolide should have significant effects on its solubility in solvents, its water absorption, its biodegradation rate, and its drug-release characteristics [60].

3.　Polymerization Methods

There are two main types of polymerization methods used for the synthesis of lactic/glycolic acid polymers from lactide and/or glycolide. The preferred method is the dry or melt method (bulk polymerization). In this method, the monomers are polymerized at a temperature above their melting points. No solvent is needed. The other method is the wet or solvent method (solution polymerization). In this case, the polymerization occurs in a solvent for both the monomers and the polymer product.

Bulk Polymerization. The preferred bulk polymerization process proceeds optimally when the reaction temperature is set to slightly above the melting point of the polymer product to be formed when semicrystalline polymers (i.e., glycolide homopolymers, L-lactide homopolymer, and some L-lactide-glycolide copolymers) are to be synthesized. For the glassy polymers (i.e., D,L-lactide homopolymer and D,L-lactide-glycolide copolymers), the reaction temperature is set above the melting point(s) of monomer(s). The polymers formed are soluble in the melted monomer(s) and, as the polymerization proceeds, the viscosity of the polymerization system increases. The increase in viscosity hampers mixing and removal of product from the reaction vessel, both of which reduce yield. An example of bulk polymerization is given below. An advantage of this method is that pure polymers can be synthesized. However, heat dissipation from the bulk of the polymer mass is difficult because of the high viscosity in the later stage of polymerization and because polymers are good insulators. Consequently, this method is limited to small-scale polymerizations unless specialized equipment is used.

Example 1.   Bulk Polymerization of Lactic/Glycolic Acid Polymers [4,61]

Parameter	*Value*
Lactide and/or glycolide monomer(s):catalyst	100:0.001 to 1.0 (by weight)
Polymerization temperature	160°–250°C

Purified monomer(s) and catalyst are placed in a predried reactor (three-necked flasks, Erlenmeyer flasks, or tubes) and the polymerization system is closed under a nitrogen atmosphere or by evacuation. The reactor is then immersed in an oil bath with a thermostat set to a desired polymerization temperature and heated for six hours while being stirred and kept under an inert atmosphere or vacuum. After polymerization, the system is immersed in ice water. The solidified mass of polymer formed is dissolved in methylene chloride, chloroform, or dimethyl formamide and the solution is filtered into excess cold methanol or methanol containing 10% water (by volume) to precipitate out the polymer. After this last purification step, which removes catalyst, residual monomers, and chain-length-controlling agent if present, the polymer is vacuum dried for at least a day.

Glycolide homopolymers and the copolymers of lactide and glycolide with 50% or greater glycolide content cannot be purified or treated with the solvents because these polymers do not dissolve in common, inexpensive organic solvents [4]. When L-lactide is used in the polymerization of a homo- or copolymer, the polymerization temperature should be well controlled and kept as low as possible because elevated temperatures can cause racemization [10]. Sometimes, a monohydric, aliphatic, straight chain alcohol containing from 10 to 18 carbon atoms (e.g., lauryl alcohol) in a concentration ranging from 0.01 to 0.22 mol.% of monomer is added to the polymerization system to activate the catalyst and to aid in controlling chain length or molecular weight [51].

Solution Polymerization.   In this method, the monomer(s) is (are) dissolved in a suitable solvent and then polymerized. Both the monomer(s) and the polymer should be soluble in the chosen solvent. The concentration of the monomer(s) in the solvent should be controlled to avoid an excessively viscous final solution of polymer. Because the end product is a polymer solution, heat transfer is not a problem in comparison with the bulk polymerization method. This is an advantage of solution polymerization. The polymer that is formed can be isolated by solvent evaporation, but mostly the polymer is isolated by precipitation into a large excess of nonsolvent. Residual solvent can then be removed from the polymer by vacuum drying.

Example 2.   Solution Polymerization of Lactic/Glycolic Acid Polymers [57]

Parameter	*Value*
Lactide and/or glycolide monomer(s):catalyst	100:0.03 to 1.0 (by mole)
Lactide and/or glycolide monomer(s):solvent	50 mM:25–50 ml
Polymerization temperature	50°C

Purified monomer(s) is (are) placed in a predried reactor (three-necked flasks, Erlenmeyer flasks, or tubes). Solvent (freshly distilled and dried by phosphorous pentoxide) is added to the reactor and the reactor is closed under a nitrogen atmosphere. After the monomer(s) is (are) dissolved, a catalyst is added and the reactor is reclosed. Then, the reactor is completely immersed in an oil bath with a thermostat set to the desired polymerization temperature and heated for six hours. After polymerization, the product is poured into excess cold methanol or methanol containing 10% water (by volume) to precipitate out the polymer. After filtration, the polymer is vacuum dried at 60°–80°C for at least a day.

Toluene is often used as the solvent in solution polymerization [54,62]. Other solvents such as benzene, nitrobenzene, chloroform, dichloroethane, and acetonitrile are also used in the solution polymerization [9,57,63].

C.   Catalysts

Various kinds of compounds have been tested and used as catalysts for the polymerization of lactic/glycolic acid polymers. Table 1 lists some of these catalysts. The catalysts listed in Table 1 are not equally effective in catalyzing the polymerization of lactide and glycolide. For example, Kricheldorf and Serra evaluated different metal-salt catalysts in polymerizing L-lactide by a bulk polymerization method [10]. They found that titanium, germanium, and tin(IV) oxides were not effective. In contrast, tin(II) oxide, lead(II) oxide, tin(II) octoate, antimony octoate, and bismuth octoate were good catalysts and can be ranked among the most effective.

Although many kinds of catalysts have been tested by investigators using the ring-opening method of polymerizing lactic/glycolic acid polymers, most of these catalysts have potential problems due to their toxicity. If they cannot be thoroughly removed from the polymer after polymerization, their use will be viewed unfavorably by regulatory agencies. Concern about the fate of residual catalysts trapped inside the polymer matrix spurred a switch to nontoxic or less-toxic catalysts. For example, stannous octoate is now widely used as the catalyst because stannous salts have been approved for pharmaceutical applications [64].

D.   Effect of Reaction Parameters

1.   Effect of Monomer Purity

A basic prerequisite in achieving a high degree of polymerization is the use of very pure lactide and/or glycolide monomers. Monomers should not contain any impurities that initiate additional chains or retard the build up of chain length by forming nonreacting

Table 1   Some Catalysts Tested and Used for Polymerization of Lactic/Glycolic Acid Polymers

Stannous
  Stannous octoate
  Stannous acetate
  Stannous chloride
  Stannous fluoride
  Dibutyl stannic dilaurate
  Stannous formate
  Stannous propionate
  Stannous butyrate
  Stannous valerate
  Stannous caproate
  Stannous caprylate
  Stannous pelargonate
  Stannous caprate
  Stannous laurate
  Stannous myristate
  Stannous palmitate
  Stannous stearate
  Stannous benzoate
  Stannic oxide
  Stannous oxide
  Stannic tetrachloride
  Tetraphenyl tin

Antimony
  Antimony oxide
  Antimony octoate
  Antimony trifluoride
  Antimony trioxide

Aluminum
  Aluminum oxide
  Aluminum isopropoxide
  Diethyl aluminum chloride
  Triethoxy aluminum

Lead
  Lead oxide

Lead stearate
Lead carbonate

Titanium
  Tetraisopropyl titanate
  Titanium oxide
  Titanium tetrachloride

Other metals
  Bismuth nitrate dihydrate
  Bismuth octoate
  Calcium carbonate
  Calcium oxide
  Calcium stearate
  Ferric oxide
  Germanium oxide
  Lithium carbonate
  Magnesium oxide
  Magnesium stearate
  Sodium acetate
  Sodium carbonate
  Sodium octoate
  Zinc oxide
  Zinc stearate

Nonmetal compounds
  Boron trifluoride
  Boron trifluoride dimethyl etherate
  Chlorosulfonic acid
  Methyl triflate
  Tetraethylammonium bromide
  Triethyl amine
  Tributyl arsine
  Trifluoromethanesulfonic acid
  Tributyl amine
  Tributyl phosphine

end groups [56]. As a rule, the content of free carboxylic groups in the monomer should not exceed a value of 0.08 milliequivalents per gram (meq/g) [4].

Kohn et al. have studied the effect of monomer purity on molecular weight and molecular weight distribution of poly(D,L-lactic acid) [59]. Table 2 shows their results, which indicate that moisture, the solvent used for monomer purification (e.g., ethyl acetate), and impurities formed during storage of the monomer all have influence. High molecular weight polymers cannot be obtained when the impurities are present in the polymerization system.

Table 2 Effect of D,L-Lactide Purity on Molecular Weight and
Polydispersity

Monomer type	M_n	M_w	M_w/M_n
Fresh monomer	43,000	116,000	2.7
Monomer + ethyl acetate	9000	51,000	5.7
Monomer + water	4500	28,000	6.2
Old monomer	3700	11,500	3.1

Source: Modified from Ref. 59

2. Effect of Polymerization Temperature

The polymerization temperature has a great influence on the reaction rate, percentage of conversion of monomers, molecular weight of the polymers formed, color of the products, and even the purity of the polymers. For semicrystalline polymers such as poly(glycolic acid) and poly(L-lactic acid), the polymerization temperature should be kept above the melting point of the polymer. On the other hand, the temperature should not be too high or the color of the polymer formed will be darkened and sometimes decomposition of the polymers may even occur.

For amorphous polymers such as poly(D,L-lactide) and copolymers of lactide and glycolide, there is more latitude in selecting the polymerization temperature because these polymers do not have melting points. However, the temperature should not be too low or too high and should be optimized for the particular product desired. As shown in Fig. 8, the course of monomer conversion with time at different temperatures indicates that an increase in temperature results in acceleration of polymerization. Figure 8 also shows that a nearly complete conversion of the monomer occurs only at a reaction temperature

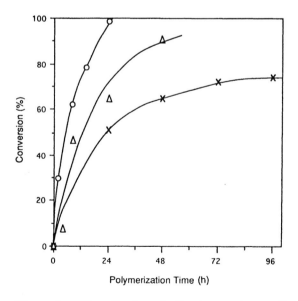

Figure 8 Effect of polymerization temperature on the conversion of D,L-lactide to polymer when catalyzed by stannous octoate at -x- 130°C, -△- 150°C, and -○- 180°C. (Modified from Ref. 4.)

above 150°C when stannous octoate is used as a catalyst at a concentration of 0.005 mol.%. The effect of polymerization temperature on the molecular weight of the polymers is discussed in Section III.

3. Effect of Catalyst Concentration

Catalyst concentration is also an important parameter in the formation of lactic/glycolic acid polymers. The catalyst concentration can affect the reaction rate, percentage of the conversion of monomers, and the nature of the products. As shown in Table 3, an acceleration of the polymerization of D,L-lactide caused by increasing the catalyst concentration leads to the formation of higher molecular weight polymers. At an optimal temperature, increasing the catalyst concentration always results in a faster reaction and a higher conversion of monomer. However, increasing the catalyst concentration will not always yield a higher molecular weight polymer [4].

E. Polymer Purification

Lactic/glycolic acid polymers can contain a significant amount of residual impurities, including unreacted monomer(s), molecular weight regulating agents, polymerization catalysts, and other volatile impurities such as linear and cyclic low molecular weight oligomers formed as by-products during polymerization. These impurities can cause problems in fabricating medical and pharmaceutical products from the polymers. Therefore, several processes have been developed to remove these residual impurities.

Example 3. Purification of Lactic/Glycolic Acid Polymers by Postpolymerization Evacuation. When bulk polymerization is used, the polymerizing system can be evacuated for a few hours at the polymerization temperature after the reaction is completed or the desired intrinsic viscosity is achieved. An example of postevacuation conditions is given below.

Parameter	Value
Vacuum	< 5 mm Hg
Temperature	The temperature used for polymerization
Time	From one hour to several hours

This method has been used to reduce the monomer residues of lactic/glycolic acid polymers to less than 1% [65].

Table 3   Effect of Stannous Octoate Concentration on Monomer Conversion and Molecular Weight in the Polymerization of D,L-Lactide

Catalyst concentration (%)	Reaction time (h)	Monomer conversion (mol.%)	M_n ($\times 10^{-3}$)
0.005	96	73.8	8.5
0.010	24	44.3	14.8
0.010	48	81.4	15.0
0.030	4	78.9	109.5

Source: Tabulated from Ref. 4
Polymerization temperature: 130°C

Example 4. Purification of Lactic/Glycolic Acid Polymers using Solvents. To achieve a pure product, and especially to remove catalyst residues, the polymer can be washed with a suitable solvent. The polymer is dissolved in a solvent (e.g., methylene chloride) or a solvent mixture. The polymer solution is then poured into a nonsolvent such as methanol to precipitate the polymer. The precipitated polymer can be recovered by filtration and vacuum drying (e.g., at 3 mm Hg for 24 hours at room temperature [65]).

The solvents frequently used for washing lactic/glycolic acid polymers are acetone, 1,4-dioxane, dimethylacetamide, tetrahydrofuran, toluene, dimethyl formamide, dimethyl sulphoxide, chloroform, methylene chloride, and other solvents that dissolve the lactic/glycolic acid polymers. The precipitation agents may be methanol, ethanol, freon®, hexane, cyclohexane, heptane, petroleum ether, and other organics that are nonsolvents for lactic/glycolic acid polymers [66]. Even water containing various agents has been used for purification. For instance, water with a small amount of an inorganic or organic acid or base, an agent that changes surface tension, or a complexing agent has been used to achieve particular, additional purifying effects [66].

III. PROPERTIES AND CHARACTERIZATION OF LACTIC/ GLYCOLIC ACID POLYMERS

The lactic/glycolic acid polymers offer great versatility in polymer properties and performance characteristics. Consequently, the polymer properties and performance characteristics affect biological and medical applications of these polymers. For example, the polymer properties determine the release rates of drugs when the polymers are used as drug delivery matrices.

A. Polymer Composition and Determination of Ratio of Lactic Acid to Glycolic Acid in a Polymer

Unlike glycolic acid, which contains only symmetric carbon atoms, lactic acid has one asymmetric carbon atom and is thus a chiral molecule that exists as two optical isomers or enantiomers, D-lactic acid and L-lactic acid (which are called R and S enantiomers according to the nomenclature of the International Union of Pure and Applied Chemistry [IUPAC]). When two lactic acid molecules combine to form the cyclic dimer lactide, three possible stereoisomers can be formed: D-lactide, which is composed of two D-lactic acid moieties; L-lactide, consisting of two L-lactic acid moieties; and meso-lactide, made of a D-lactic acid moiety and an L-lactic acid moiety. In addition, an equimolar mixture of D-lactide and L-lactide forms D,L-lactide (see Fig. 9) [67]. The lactide isomers give rise to four morphologically distinct polymers: poly(D-lactic acid) and poly(L-lactic acid), which are the two stereoregular polymers; and poly(meso-lactic acid) and poly(D,L-lactic acid), which are racemic polymers. The polymers derived solely from the optically active D-lactide or from the L-lactide are semicrystalline, while those derived from the optically inactive meso-lactide or D,L-lactide are amorphous. Poly(L-lactic acid) is much more frequently used than poly(D-lactic acid) because its biodegradation product, L-lactic acid, is a naturally occurring stereoisomer and is easily metabolized by the body [68,69].

Poly(D-lactic acid), poly(L-lactic acid), and poly(glycolic acid) are semicrystalline polymers due to the high regularity of their polymer chains. Of the three, poly(glycolic acid) has the greatest crystallinity because it does not have the methyl side groups of the poly(lactic acids). Crystallinity is lost in the racemic poly(lactic acid) homopolymers and

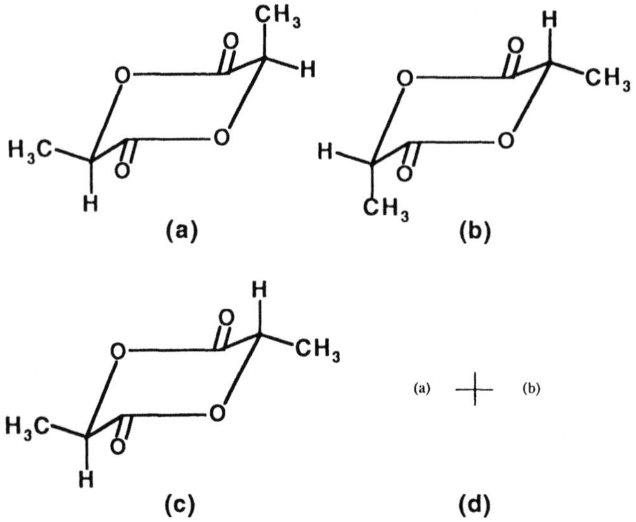

Figure 9 Stereostructure of lactide isomers: (a), D-lactide; (b), L-lactide; (c) meso-lactide; (d), D,L-lactide.

copolymers of lactic acid and glycolic acid because of the irregularities in the polymer chains. Increased irregularities in polymer chains lead to an increase in the rate of polymer hydration and hydrolysis [51,70].

Lactic acid is more hydrophobic than glycolic acid due to the presence of an extra methyl group. Therefore, lactic-acid-rich copolymers are more hydrophobic, absorb less water, are more resistant to hydrolytic attack, and degrade more slowly. The homopolymers of lactic acid and glycolic acid are the two extreme examples. Thin films of poly(L-lactic acid) take up less than 0.5% water, while those made of poly(glycolic acid) take up approximately 30% water at equilibrium, as shown in Table 4 [51]. However, the lactic acid polymers are more soluble in organic solvents than glycolic acid polymers because of their greater hydrophobicity and lower crystallinity.

Table 4 Water Uptake of Some L-Lactic/
Glycolic Acid Polymers

Polymer composition (L-lactic acid:glycolic acid)	Water uptake (%)
100:0	0.5
90:10	2.1
70:30	21.0
50:50	27.9
30:70	32.9
10:90	31.4

Source: Tabulated from Ref. 51
Film samples (250 μ thick) were incubated in 0.2 M
phosphate buffer (pH 7) for 3–4 days.

The determination of the ratio of lactic acid moieties to glycolic acid moieties in lactic/glycolic acid polymers is based on the distinct behavior of the involved hydrogens in a magnetic field (Fig. 10). Therefore, these two moieties are distinguished by nuclear magnetic resonance (NMR) and the mole fraction of lactic acid moieties and glycolic acid moieties is calculated from the integrated peaks of (b, δ = 4.8 ppm) and (c, δ = 5.2 ppm) in Fig. 10. Peak (a, δ = 1.6 ppm) is not used because it often interferes with residual lactic acid and aliphatic alcohols added into the polymerization system to control molecular weight. Peaks (c) and (b), on the other hand, are unique. Experimentally, the polymer sample is dissolved in $CDCl_3$ and tetramethylsilane is used as an internal reference [71].

Assigning A to the peak area of proton (c) and B to one-half of the peak area of protons (b), the following relationships exist:

Molar fraction of lactic acid = $A/(A + B)$
Molar fraction of glycolic acid = $B/(A + B)$
Molar ratio of lactic acid to glycolic acid = A/B

B. Molecular Weight and Polydispersity

Molecular weight is essential in characterizing lactic/glycolic acid polymers. It is particularly important in determining mechanical strength as well as a polymer's capacity to be shaped and molded. The molecular weight of a polymer usually does not directly affect drug permeability, but it does affect its biodegradation rate [72].

The molecular weights of lactic/glycolic acid polymers are affected by monomer purity, polymerization temperature, polymerization time (Fig. 7), catalyst concentration, and addition of chain transfer agents. Figure 11 shows the effect of polymerization temperature on the molecular weight of poly(L-lactic acid). The highest molecular weights are obtained at 100°C when the catalyst concentration was kept constant at 0.015 wt%. When polymerization temperature is elevated, the molecular weight of the polymer decreases and at temperatures near 275°C, no polymer is formed. Leenslag and Pennings considered 275°C as the ceiling temperature T_c of poly(L-lactic acid) (58). At T_c, the Gibbs free energy ΔG is zero. If the polymerization temperature is higher than the T_c, ΔG becomes positive and thus polymer formation is thermodynamically unfavorable. If the polymerization temperature is lower than the T_c, the molecular weight of the polymer will increase as long as time allowed for polymerization is long enough to reach the maximum molecular weight at that particular temperature.

Leenslag and Pennings also investigated the effect of catalyst concentration on molecular weight of poly(L-lactic acid). They concluded that, above a minimum level required for starting the polymerization, increasing the concentration of stannous octoate decreases the molecular weight of the polymer [58]. Using the same catalyst, Deasy et al.

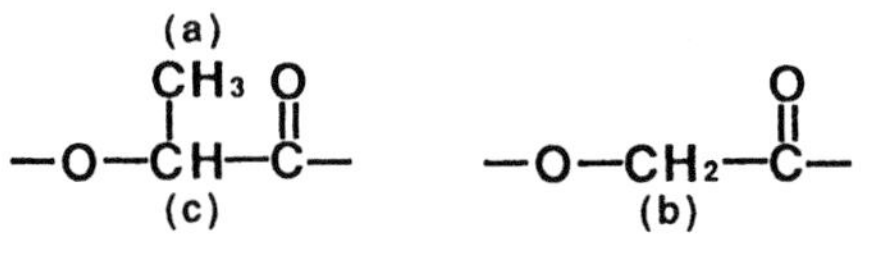

Figure 10 Chemical structures and proton assignment of (I) lactic acid moiety and (II) glycolic acid moiety in a polymer chain.

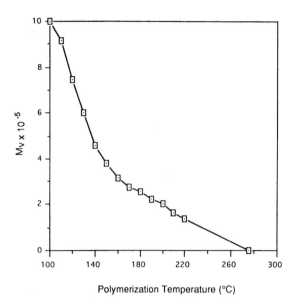

Figure 11   Effect of polymerization temperature on the molecular weight of poly(L-lactic acid) (stannous octoate concentration: 0.015 wt%; polymerization time: varied to reach maximum M_v at different temperatures). (Modified from Ref. 58.)

also found that, above a certain level, the higher the concentration of the catalyst, the lower the intrinsic viscosity of the polymer (see Fig. 12) [61]. The explanation for these results is that with higher catalyst concentration a greater number of chain reactions are initiated at the same time, resulting in a larger number of low molecular weight polymer chains being formed.

The molecular weight of lactic/glycolic acid polymers is determined directly by gel permeation chromatography (GPC). The commercially available lactic/glycolic acid polymers are commonly characterized in terms of intrinsic viscosity, which is directly related to molecular weight. End-group analysis is another method used to determine the molecular weight of lactic/glycolic acid polymers because the carboxylic end groups of the polymers can be quantified by titration with an alkali solution. These methods of characterizing molecular weight of lactic/glycolic acid polymers are now discussed in more detail.

In the GPC determination of the molecular weight and molecular weight distribution of lactic/glycolic acid polymers, monodispersed polystyrenes (molecular weight from approximately 400 to 3×10^6) are often used as the molecular weight standards. The molecular weight is expressed as the number average M_n and weight average M_w molecular weight, and the ratio of M_w to M_n is a useful marker of molecular weight distribution or polydispersity. This index of polydispersity can be used to monitor the rate of polymer hydrolysis.

Intrinsic viscosity is another common property used to characterize lactic/glycolic acid polymers. The intrinsic viscosity is related to molecular weight through the Mark–Houwink equation:

$$[\eta] = KM_v^{\alpha} \tag{4}$$

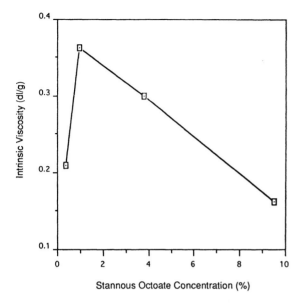

Figure 12 Effect of catalyst concentration on the intrinsic viscosity of poly(L-lactic acid) (bulk polymerization, polymerization temperature: 160°C; polymerization time: 5 hours). (Data are from Ref. 61.)

where $[\eta]$ is the intrinsic viscosity, M_v is the viscosity average molecular weight, and K and α are constants. Experiments show that lactic/glycolic acid polymers follow the Mark–Houwink equation quite well. Plots of $\log[\eta]$ against $\log[M]$ for the lactic/glycolic acid polymers give a straight line with $\log K$ as the intercept and α as the slope. Examples are shown in Fig. 13 and Table 5 lists K and α for some lactic/glycolic acid polymers.

The molecular weight M_n of lactic/glycolic acid polymers can also be determined by carboxylic end-group analysis. Since the carboxylic group can be neutralized by alkali, the carboxylic end group of the polymers can be titrated with alkali in a nonaqueous solution. For example, potassium hydroxide dissolved in benzyl alcohol has been used to titrate lactic/glycolic acid polymers dissolved in a 1 : 1 solution of benzyl alcohol and chloroform [76,78]. Phenolphthalein can be used as an indicator. The M_n can then be calculated from the results of the titration by Eq. (5) [79]:

$$M_n = \frac{W}{C(V - V_o)} \tag{5}$$

where W is the weight (g) of the polymer titrated, C is the molar or equivalent concentration of the alkali (KOH) solution used, V is the volume of alkali solution used to titrate the polymer sample, and V_0 is the volume of alkali solution used to titrate the blank.

C. Crystallinity and Melting Point

Another important property in the characterization of lactic/glycolic acid polymers is crystallinity, which is an important indicator of the physical properties of the polymers. The percentage of crystallinity that reflects the extent of formation of crystalline regions within a polymer matrix affects the mechanical strength of a polymer and can determine

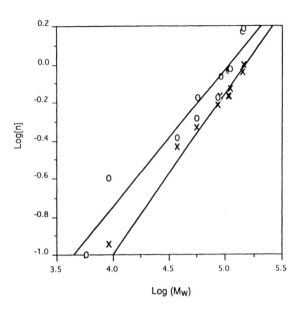

Figure 13   Viscosity/molecular weight relationships for poly(D,L-lactic acid) (-○-in chloroform; -x- in ethyl acetate). (Modified from Ref. 49.)

the end applications of the polymer [51]. For example, semicrystalline lactic/glycolic acid polymers are most useful when mechanical strength is required (i.e., in sutures and orthopedic applications for which the polymer is under tension throughout its useful lifetime). The degree of crystallinity also affects the swelling behavior of a polymer and its capability to undergo hydrolysis and, consequently, its rate of biodegradation.

X-ray diffraction and differential scanning calorimetry (DSC) are the tools used most frequently to determine the extent of crystallinity within the lactic/glycolic acid polymers. The melting point T_m of semicrystalline polymers and the glass transition temperature T_g of both semicrystalline and amorphous polymers are often determined by DSC.

The formation of crystals in lactic/glycolic acid polymers is mainly dominated by the monomer(s) used. Polymers having a chemical structure with sufficient geometric regularity can form ordered crystalline structures. For example, homopolymers of L-lactic acid or glycolic acid are semicrystalline polymers (see Table 6). The homopolymer of D,L-lactic acid and copolymers of D,L-lactic acid and glycolic acid are all amorphous polymers, while copolymers of L-lactic acid and glycolic acid are semicrystalline polymers within certain ratios of L-lactic acid to glycolic acid (see Fig. 14); the melting point also changes with the change of the monomer ratio of L-lactic acid to glycolic acid (Fig. 15).

Apart from the intrinsic susceptibility of a given polymer to have crystalline regions, the processing history of a polymer has a significant effect on the actual crystallinity value. For example, long annealing times or slow cooling or heating at temperatures between the glass transition temperature and melting point tend to increase the degree of crystallinity [14]. Figure 16 gives an example of the effect of the cooling rate on the degree of crystallinity of poly(glycolic acid).

The molecular weight of a polymer also affects the crystallinity [80]. Cohn et al. found that increasing the molecular weight of poly(L-lactic acid) increased the degree of

Table 5 *K* and α Values for Some Lactic/Glycolic Acid Polymers

Polymer	Solvent	$K \times 100$	α	Reference no.
Poly(L-lactic acid)	CHCl₃	0.0545	0.73	73–75
	CHCl₃	0.00246[a]	1.49[a]	61
	CHCl₃	0.00302[b]	0.94[b]	61
	Benzene	0.0572	0.72	75
Poly(D,L-lactic acid)	CHCl₃	0.0660[a]	0.67[a]	49
	CHCl₃	0.0606[b]	0.64[b]	49
	Ethyl acetate	0.0158[a]	0.78[a]	49
	Ethyl acetate	0.0163[b]	0.73[b]	49
	CHCl₃	0.0221	0.77	75,76
	CHCl₃	13.2[c]	0.58[c]	4
	DMF[d]	11.2[c]	0.56[c]	4
	CHCl₃	0.74[e]	0.87[e]	4
	DMF	0.64[e]	0.85[e]	4
	Benzene	0.0227	0.75	75
Poly(D,L-lactic-co-glycolic acid) 70:30	CHCl₃	0.0996	0.66	76
Poly(D,L-lactic-co-glycolic acid) 50:50	DMF	0.50[c]	0.88[c]	4
	DMF	0.50[e]	0.88[e]	4
Polycaprolactone	Toluene	0.014	0.78	77

[a]For M_n determination
[b]For M_w determination
[c]Nonprecipitated polymer
[d]Dimethylformamide
[e]Precipitated polymer
The difference of *K* and α value for precipitated and nonprecipitated poly(D,L-lactic acid) may be due to the removal of monomer residues and low molecular weight oligomers from the precipitated polymer. The small monomer and oligomer content decrease the polymer solution viscosity to a lesser degree.

crystallinity of the polymer [78]. The melting point of the polymer also increased with increasing molecular weight until a certain constant value was reached [81].

D. Chain Mobility and Glass Transition Temperature

Both homo- and copolymers of lactic acid and glycolic acid normally have glass transition temperatures T_g above physiological temperature (37°C) (see Table 6). Polymers that have been used for drug delivery applications can be classified as glassy polymers ($T_g > 37$°C) and rubbery polymers ($T_g < 37$°C) [82]. In general, the lower the glass transition temperature is, the higher the permeability of a given type of polymer will be [72]. On this basis, rubbery polymers would be more permeable than glassy polymers. However, rubbery polymers have low mechanical strength and sometimes they must be cross-linked to be used as a drug matrix. Because the lactic/glycolic acid polymers are not rubbery polymers and their polymer chains are rigid at both room temperature and physiological temperature, they all have significant mechanical strength to be used as drug delivery matrices without additional treatment.

Table 6 T_g, T_m, and Crystallinity of Some Lactic/Glycolic Acid Polymers

Polymer	T_g(°C)	T_m(°C)	Crystallinity (%)
Poly(glycolic acid)	35–40	224–230	50–52
Poly(L-lactic acid)	60–67	172–180	37
Poly(D,L-lactic acid)	55–60	–	–
Poly(D,L-lactic-co-glycolic acid) 85:15	50–55	–	–
Poly(D,L-lactic-co-glycolic acid) 75:25	50–55	–	–
Poly(D,L-lactic-co-glycolic acid) 65:35	45–50	–	–
Poly(D,L-lactic-co-glycolic acid) 50:50	45–50	–	–
Poly(D,L-lactic-co-L-lactic acid) 25:75	60	–	–
Polycaprolactone	−60–70	58–63	?

The glass transition temperature of lactic/glycolic acid polymers increases with the increase of lactic acid content in the polymer (Table 6) and with an increase in molecular weight (Fig. 17) until a certain value is reached, such as 100,000 daltons for poly(L-lactic acid) [81].

E.   Mechanical Strength

The mechanical strength of biodegradable polymers is important in implantable macrodevices and surgical sutures because these objects are often subject to considerable physical stress. Table 7 shows tensile strength and fracture strain data (at 25°C) for some lactic/

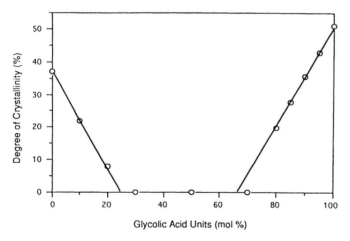

Figure 14   Degree of crystallinity as a function of composition of poly(L-lactic-co-glycolic acid) (determined by x-ray diffraction and differential scanning calorimetry). (Reproduced from Ref. 51 with permission of the copyright owner, Butterworth-Heinemann Ltd.)

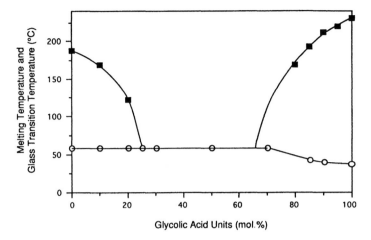

Figure 15   Melting points -■- and glass transition temperatures -○- as a function of composition for poly(L-lactic-co-glycolic acid) (measured by differential scanning calorimetry). (Reproduced from Ref. 51 with permission of the copyright owner, Butterworth-Heinemann Ltd.)

glycolic acid polymers. Polymer composition, polymer molecular weight, geometric regularity of the polymer chain, crystallinity, and orientation of the polymer all have a profound effect on the mechanical strength of a polymer.

The ratio of lactic acid to glycolic acid has an effect on the mechanical strength of polymeric devices. The polymer composition, in particular, affects strength retention during implantation because polymers of different compositions have different biodegra-

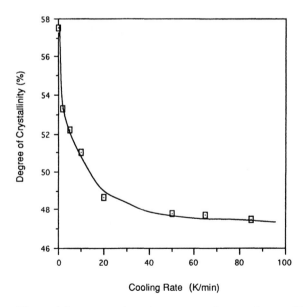

Figure 16   Effect of cooling rate on degree of crystallinity of poly(glycolic acid) after melting. (Data are from Ref. 80.)

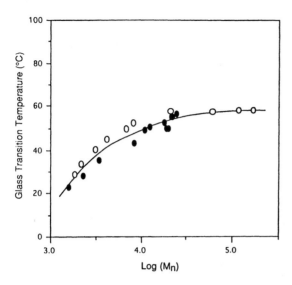

Figure 17　Effect of molecular weight on the glass transition temperature of poly(L-lactic acid) (-○-) and poly(D,L-lactic acid) (-●-). (Reproduced from Ref. 81 with permission of the copyright owner, Butterworth-Heinemam Ltd.)

dation rates. Table 8 summarizes a study on strength retention conducted by Kaplan and Muth [83]. These data were acquired by implanting injection-molded staples in lumbar muscle tissue of rats.

Vert and Chabot have studied the effect of the geometric regularity of polymer chains on the mechanical strength of a copolymer of L-lactic acid and D,L-lactic acid [84]. They found that the tensile strength of molded test bars implanted in the rat for two months decreased with increasing D,L-lactic acid content. More specifically, the D,L-lactic acid homopolymer had no strength after two months' implantation whereas the L-lactic acid homopolymer lost no strength. Poly(L-lactic acid) has greater geometric regularity than

Table 7　Mechanical Properties of Some Lactic/Glycolic Acid Polymers

Polymer	Stress at failure (Kpsi)	Strain at failure (%)	Modulus (Kpsi)
Poly(glycolic acid)	10–11.2	3.9–20	483–1000
Poly(L-lactic acid)	8.0–12	3.3–10	314–400
Poly(D,L-lactic acid)	4–6.6	2.6–10	200–400
Poly(D,L-lactic-co-glycolic acid) 85:15	6–8	2.3–10	200–400
Poly(D,L-lactic-co-glycolic acid) 75:25	6–8	3–10	200–400
Poly(D,L-lactic-co-glycolic acid) 65:35	6–8	3–10	200–400
Poly(D,L-lactic-co-glycolic acid) 50:50	6–8	2.6–10	200–400

Table 8 Strength Retention of Poly(lactic-co-glycolic acid) Staples Implanted
Intramuscularly in Rats

Polymer composition (lactic acid:glycolic acid)	Tensile strength after implantation (% of initial)			
	1 week	2 weeks	3 weeks	4 weeks
80:20	>100	>100	>100	>100
75:25	99	92	28	–
70:30	77	56	32	8
65:35	115	28	8	5

Source: Tabulated from Ref. 83

poly(D,L-lactic acid). Consequently, the higher the L-lactic acid content in a copolymer of L-lactic acid and D,L-lactic acid, the greater the geometric regularity and the mechanical strength will be.

Even though the lactic/glycolic acid polymers may initially have the proper mechanical strength for a particular application, the biodegradation process can significantly alter this property. Therefore, the mechanical properties that a polymeric device will exhibit throughout its entire lifetime must be considered when the polymer is selected.

F. Solubility

Lactic/glycolic acid polymers are linear, hydrophobic polyesters. They can dissolve in many organic solvents. Ethyl acetate and methylene chloride are currently the solvents of first choice because they are less toxic [82].

The solubility of lactic/glycolic acid polymers is affected by many factors. Polymer composition and crystallinity play an important role in the solubility of the polymers. The crystalline homopolymers of glycolic acid will only dissolve in a very strong solvent such as hexafluoroisopropanol. The crystalline homopolymers of lactic acid also do not have good solubility in most organic solvents. On the other hand, amorphous polymers of D,L-lactic acid and copolymers of lactic acid and glycolic acid that have a low glycolic acid content are soluble in many organic solvents. The structural randomness of the monomeric units in a polymer also has influence on the solubility of the polymer. Bendix found that large blocks of glycolic acid in lactic acid–glycolic acid copolymers lead to insolubility in some solvents, such as chloroform. In addition, molecular weight and temperature affect solubility [85].

The solubility parameter is a means of predicting whether or not a given compound will dissolve in a given solvent [11]. It can also be used to search for solvents for a given compound. The solubility parameters of lactic/glycolic acid polymers can be determined experimentally or calculated according to "group molar attraction constants" [11,86]. Table 9 lists the solubility parameters for some lactic/glycolic acid polymers. The values listed in Table 9 are of the same magnitude and quite close to each other. This is probably due to the similarity in the hydrophobic character of these polymers.

According to the solubility parameter concept or the principle of "dissolving the similar," the solvents with solubility parameter values close to that of a lactic/glycolic acid polymer will be suitable solvents for the polymer. Ethyl acetate and methylene chloride have solubility parameter values similar to the lactic/glycolic acid polymers (Table 10), so they are good solvents for most of the lactic/glycolic acid polymers.

Table 9   Solubility Parameters of Some Lactic/Glycolic Acid Polymers

| | $\delta \times 10^{-3} ([J/m^3]^{0.5})$ | | |
Polymer	Determined experimentally	Calculated[a]	Calculated[b]
Poly(L-lactic acid)	16.8	20.5	20.2
Poly(D,L-lactic acid)	16.4	20.2	19.9
Poly(D,L-lactic-co-glycolic acid) 58:42	16.8	18.7	18.3
Poly(D,L-lactic-co-ε-caprolactone) 90:10	16.2	20.0	19.8

Source: Retabulated from Ref. 86
[a]Calculated according to Small's data [11]
[b]Calculated according to Hoy's data [11]

Table 10   Solubility Parameters of Some Common Solvents and Nonsolvents Used with Lactic/Glycolic Acid Polymers

Compound	$\delta \times 10^{-3}$ $([J/m^3]^{0.5})$
Solvents	
Ethyl acetate	18.6
Tetrahydrofuran	18.6
Benzene	18.8
Chloroform	19.0
Methylene chloride	19.8
Ethylene dichloride	20.1
Acetone	20.3
Dioxane	20.5
Hexafluoroisopropanol	~
Hexafluroacetone sesquihydrate	~
Acetonitrile	24.3
Dimethyl sulfoxide	24.6
Dimethyl formamide	24.8
Nonsolvent	
Water	47.9
Ethylene glycol	29.9
Methanol	29.7
Petroleum ether	~
Heptane	15.1
Hexane	14.9
Toluene	18.2
Xylene	18.0

Source: Tabulated from Ref. 11

In addition to these two solvents, lactic/glycolic acid polymers are soluble in the following solvents at ambient temperatures: hexafluoroisopropanol, chloroform, dioxane, tetrahydrofuran, ethylene dichloride, hexafluoroacetone sesquihydrate, acetonitrile, dimethyl formamide, dimethylsulfoxide, and acetone (only D,L-lactic acid homo- and copolymers) [9,57,63]. These same polymers are slightly soluble or insoluble in water, methanol, petroleum ether, ethylene glycol, heptane, and hexane [66]. They are soluble in toluene and xylene at high temperatures.

C. Effect of Residual Impurities on Polymer Properties and the Determination of Residual Monomers

Lactic/glycolic acid polymers usually contain some residual impurities. These impurities are mainly unreacted monomers and volatile low molecular weight oligomers formed as by-products during polymerization. Large amounts of unreacted monomers and by-products can lower the mechanical strength of devices prepared from high molecular weight lactic/glycolic acid polymers. In the case of suture fibers, the residual impurities evaporate and generate bubbles in the polymer filament as it is extruded. Consequently, end breakage due to the bubbles frequently occurs in the spinning step. The filament obtained is also of low quality because the impurities cause fluctuations in strength and hydrolyzability.

When impure polymers are used as matrices for the sustained release of drugs, the impurities may affect the release characteristics. They may especially cause a burst phenomenon in which a large amount of drug is released in the initial period. Impure polymers may also experience more deterioration during processing and less storage stability. Unreacted lactide, glycolide, or low molecular weight volatile substances remaining in these biodegradable polymers also might cause irritation when they are used in humans or animals.

Various problems are thus caused by unreacted monomers and low molecular weight volatile substances remaining in lactic/glycolic acid polymers. Therefore, it is desirable, and even essential, to control quantitatively and minimize the amount of the volatile impurities present. Several methods have been developed to remove these residual impurities (see Section II). The monomer residues and the other volatile impurities are usually assayed by gas chromatography.

D. Heavy Metals

Because the lactic/glycolic acid polymers are mainly used for medical, pharmaceutical, and veterinary applications, the amount of heavy metals in the polymers is strictly controlled. Heavy metals that may be present include lead, mercury, arsenic, cadmium, antimony, and others. U.S. Pharmacopeia (USP) methods based on the calorimetric observation of heavy metal sulfides are used for the determination of heavy metals in the polymers [64].

IV. DEGRADATION OF LACTIC/GLYCOLIC ACID POLYMERS

A. Biodegradation and Bioelimination

Lactic/glycolic acid polymers undergo degradation in an aqueous environment. This "hydrodegradation" occurs through cleavage of the ester linkages. The hydrodegradation is usually referred to as biodegradation, differing from such other degradations as thermal degradation. The hydro- or biodegradation occurs both *in vitro* and *in vivo* [87].

1. Biodegradation Mechanisms

Biodegradable polymers can be categorized into two groups on the basis of the mechanism or process by which they degrade. These processes are bulk degradation and surface degradation. In the case of polymers that degrade in bulk, the rate of water penetration into the matrix is faster than the rate of polymer degradation. The process is a homogeneous one in which degradation occurs at a uniform rate throughout the polymer matrix. In contrast, for polymers that undergo surface degradation, the rate of water penetration into the matrix is slower than the rate of polymer degradation [88]. This process, therefore, is heterogeneous with degradation confined to a thin surface layer of polymer. The lactic/glycolic acid polymers belong to the category of polymers that undergo bulk degradation. Studies of the process in these polymers reveal that not only is there degradation at the surface of a device, but the polymer matrix also swells and degrades at the same time [7]. The biodegradation of lactic/glycolic acid polymers *in vitro* and *in vivo* is through random hydrolytic chain scissions of the swollen polymer [89]. Proposed hydrolysis reactions are shown in Fig. 18 [90,91].

The biodegradation of lactic/glycolic acid polymers has been characterized in terms of change in molecular weight and molecular weight distribution, weight loss, water uptake, and change in morphology of the hydrated and degraded polymers. Events responsible for the biodegradation of lactic/glycolic acid polymers have been viewed in terms of the four steps shown in Fig. 19 [55,92]. In the first step, a polymer absorbs water and undergoes little swelling [91]. The water penetrates into the amorphous region of the polymer and disrupts secondary and tertiary structures that had been stabilized by van der Waals' forces and hydrogen bonds. During this step, some covalent bonds in the backbone may also be cleaved, but there is almost no weight loss.

In the second step of biodegradation, cleavage of the covalent bonds in the polymer backbone by hydrolysis begins. As hydrolysis proceeds, more and more carboxylic end groups are generated. These free carboxylic groups may autocatalyze the hydrolysis reaction, which will accelerate biodegradation. A loss of mechanical strength in polymeric devices is observed in this stage.

The third step of biodegradation is characterized by the massive cleavage of the backbone covalent bonds. And, at some critical value of molecular weight, significant weight loss begins to occur. These observations have been interpreted as evidence that an autoaccelerated, random chain cleavage process occurs throughout the bulk of the polymer matrix. The polymer is first degraded to a lower molecular weight below which the mass integrity of a polymer device is lost. The molecular weight reduction necessary to reach this point depends on many factors, including glass transition temperature,

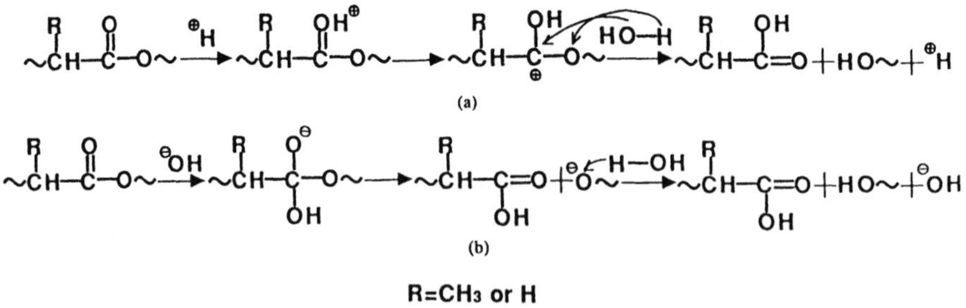

R=CH₃ or H

Figure 18 Hypothetical mechanisms of hydrolysis of lactic/glycolic acid polymers: (a), at pH < 7; (b), at pH > 7.

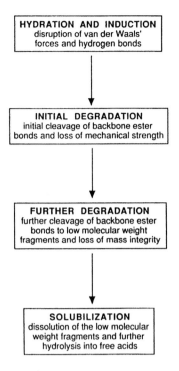

HYDRATION AND INDUCTION
disruption of van der Waals'
forces and hydrogen bonds

INITIAL DEGRADATION
initial cleavage of backbone ester
bonds and loss of mechanical strength

FURTHER DEGRADATION
further cleavage of backbone ester
bonds to low molecular weight
fragments and loss of mass integrity

SOLUBILIZATION
dissolution of the low molecular
weight fragments and further
hydrolysis into free acids

Figure 19 Scheme of the four steps proposed for the biodegradation of lactic/glycolic acid polymers.

crystallinity, and polymer conformation. Then, following the loss of physical and mechanical integrity, mass begins to get smaller.

In the fourth step, the polymer loses mass. This may occur simply by the solubilization of oligomers into the surrounding medium. As solubilization proceeds, the density of the mass decreases until the polymer disappears. Any polymeric fragments that remain are further hydrolyzed into the free acids.

Figure 20 illustrates the four steps of polymer degradation described above. Figure 20a shows that water uptake starts upon immersion of the polymer in an aqueous medium and then increases linearly. After the loss of polymer integrity, the uptake of water increases dramatically. The decrease in molecular weight also starts upon immersion of the polymer (Fig. 20b). Weight loss, however, could only be detected after 6 weeks (Fig. 20c) and corresponds to the third step of degradation. Polymer dissolution with hydrolysis into the free acids, L-lactic acid and D-lactic acid, is found after 8 weeks (Fig. 20d). This corresponds to the fourth step of biodegradation. As backbone cleavage starts and molecular weight reduction begins, the number of carboxylic end groups increases, which should result in a decrease of medium pH. However, an obvious change in pH only occurred after 8 weeks of immersion, and was most likely due to the final appearance of large amounts of the free acids (Fig. 20e).

Finally, it should be noted that the question of whether enzymes are involved in the degradation of lactic/glycolic acid polymers has not been resolved. It is still in dispute. Williams and Mort have studied the *in vitro* degradation of poly(glycolic acid) in the

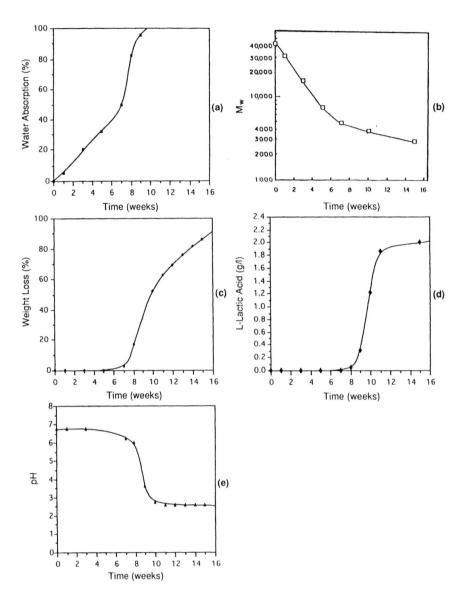

Figure 20 Degradation of poly(D,L-lactic acid) in saline at 37°C or *in vivo*: (a), water uptake; (b), molecular weight change; (c), weight loss; (d), change of L-lactic acid concentration; (e), pH change. (Figs. 20a, 20c, 20d, and 20e are modified from Ref. 7 and Fig. 20b is modified from Ref. 71.)

presence of many kinds of enzymes [93]. They reported that some enzymes, especially those with esterase activity, most likely have an effect on the degradation of the polymer. Shakenraad et al. have reported different results [94]. They could not find differences between the *in vitro* and *in vivo* breakdown rates of poly(lactic acid).

2. Biodegradation Kinetics

It has been demonstrated that the biodegradation of lactic/glycolic acid polymers is autocatalyzed by carboxylic end groups [5,71,77,95–97]. Carboxylic end groups are present in the polymer matrix before biodegradation starts because each polymeric molecule has one carboxylic group. The number of carboxylic groups increases during the biodegradation process as the polymer chains are cleaved. Therefore, the reaction kinetics of the biodegradation or hydrolysis of the lactic/glycolic acid polyesters (described in Fig. 18) can be represented by Eq. 6.

$$\frac{d[\sim COOH]}{dt} = k_h[P][H_2O][\sim COOH] \tag{6}$$

where $d[\sim COOH]/dt$ is the rate of hydrolysis, k_h is the hydrolysis rate constant, and $[P]$, $[H_2O]$, and $[\sim COOH]$ are the concentrations of the polymer, water, and carboxylic end groups, respectively. At the beginning of hydrolysis, the concentration of the polymer and water can be assumed to be constant. Thus, Eq. 6 can be simplified to

$$\frac{d[\sim COOH]}{dt} = k_h'[\sim COOH] \tag{7}$$

where $k_h' = k_h[P][H_2O]$.
Upon rearrangement,

$$\frac{d[\sim COOH]}{[\sim COOH]} = dk_h't \tag{8}$$

At $t = 0$, $[\sim COOH] = [\sim COOH]_0$, where $[\sim COOH]_0$ is the initial number or concentration of carboxylic end groups of the polymer. Integration of Eq. 8 then leads to

$$\ln[\sim COOH] = \ln[\sim COOH]_0 + k_h't \tag{9}$$

Because the number of the carboxylic end groups of a polymer is proportional to the reciprocal of the number average molecular weight of the polymer (i.e., $[\sim COOH] \propto 1/M_n$), Eq. 9 can be represented as

$$\ln M_n = \ln M_{no} - k_h't \tag{10}$$

Pitt et al. have reported that the biodegradation kinetics of a lactic/glycolic acid polymer follow Eq. 10 quite well in the initial biodegradation period (Fig. 21) [98]. Other researchers have obtained similar results [77,99].

3. Factors Affecting Biodegradation
The biodegradation rate of a polymer mainly depends on the intrinsic properties of the polymer [100], including the wettability (hydrophilicity versus hydrophobicity), the molecular weight, the degree of crystallinity, and the glass transition temperature.

Polymer Composition.   As described above, lactic acid and glycolic acid are chemically different. The methyl group makes the lactic acid moieties in polymers more hydrophobic and provides steric hindrance to attack of water molecules, and thus the lactic acid moieties are less hydrolytically labile than glycolic acid moieties. Therefore, glycolic acid polymers degrade faster than lactic acid polymers. For example, the glycolic acid homopolymer, such as poly(glycolic acid), degrades faster than the semicrystalline lactic acid homopolymer, such as poly(L-lactic acid). Incorporating two chemically different monomers into the same polymer chain is a well-established means of modifying polymer properties. This approach has been used to obtain lactic/glycolic acid polymers with properties between those of lactic acid and glycolic acid homopolymers. Moreover, varying the ratio of lactic acid and glycolic acid moieties in these polymers results in different

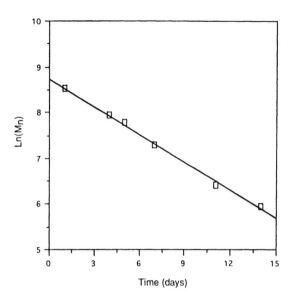

Figure 21   Semilog plot showing the decrease of molecular weight of poly(lactic-co-glycolic acid) (50 : 50) as a function of time during biodegradation (biodegradation medium: H_2O, 37°C). (Modified from Ref. 98.)

biodegradation rates (Fig. 22 [101] and Table 11 [17]). In general, copolymers with a low lactic acid content degrade more rapidly.

Molecular Weight.   Molecular weight and molecular weight distribution have an effect on the biodegradation rate of the lactic/glycolic acid polymers. Biodegradability increases as molecular weight decreases (see Fig. 23). One reason is that the low molecular weight polymers are more hydrophilic than the corresponding high molecular weight polymers because they have a higher percentage of end groups (including carboxylic end groups and hydroxy end groups). This increased hydrophilicity permits a more rapid influx of water into the polymer matrix and leads to a higher water content in the matrix, which speeds biodegradation. Another contributing factor may be that low molecular weight polymers possess more carboxylic end groups that catalyze the biodegradation. A polymer having a broad molecular weight distribution degrades faster than one with a narrow molecular weight distribution [102].

Crystallinity.   Since the degradation of the lactic/glycolic acid polymers is water dependent, the penetration of water into the polymer or the permeability of the polymer to water greatly determines the rate of biodegradation. Water will readily penetrate amorphous regions, but the semicrystalline nature of a polymer will tend to limit water accessibility. Therefore, semicrystalline polymers degrade more slowly than amorphous polymers. For example, semicrystalline poly(L-lactic acid) degrades more slowly than the amorphous poly(D,L-lactic acid) or copolymers of lactic and glycolic acid.

Chu has proposed that the degradation of the semicrystalline polymers in the lactic/glycolic acid class of polymers proceeds through two stages [103]. In the first stage, degradation occurs in the amorphous region as follows. Upon immersion of a polymer in an aqueous solution, water molecules will penetrate into amorphous regions, starting the hydrolysis of polymer chains in these regions. Then, when all the amorphous regions have

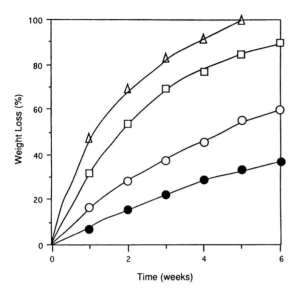

Figure 22 Effect of polymer composition on the biodegradation rate of lactic/glycolic acid polymers (molar ratio of L-lactic acid to glycolic acid: -△-, 70 : 30; -□-, 50 : 50; -○-, 30 : 70; -●-, 15 : 85; polymeric needles were implanted in Wistar rats). (Modified from Ref. 101.)

been removed by hydrolysis, the second stage of degradation begins. In this stage, the crystalline lattice will be slowly destroyed owing to the existence of imperfections in the crystalline regions. Since the crystalline region is highly packed, water penetration, and consequently the hydrolysis of the polymer chains in the crystalline regions, are much slower than those in the amorphous regions. Therefore, a high degree of crystallinity in a polymer results in slow polymer degradation.

The two degradation stages Chu has proposed are not sharply divided. During the degradation of amorphous regions, a few portions of the crystalline regions could also be destroyed. Also, during the first stage, some undegraded polymer chain segments in the amorphous regions may reorganize into a more ordered state and result in an increase in crystallinity.

Table 11 Biodegradation Rates of Some Lactic/Glycolic Acid Polymers

Polymer	Approximate time for complete degradation (months)
Poly(L-lactic acid)	18–24
Poly(D,L-lactic acid)	12–16
Poly(D,L-lactic-co-glycolic acid) 85:15	5
Poly(L-lactic-co-glycolic acid) 50:50	2
Poly(D,L-lactic-co-caprolactone) 90:10	2
Poly(glycolic acid)	2–4

Source: Retabulated from Ref. 17

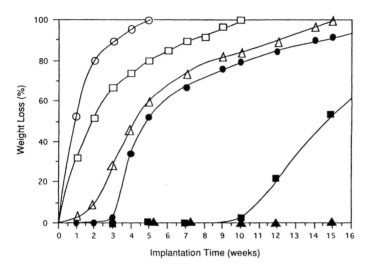

Figure 23 Effect of molecular weight on the biodegradation rate of poly(D,L-lactic acid) (M_n: -○-, 1400; -□-, 1600; -△-, 2000; -●-, 4300; -■-, 11500; -▲-, 16900; polymeric cylinders were implanted in Wistar rats). (Reproduced from Ref. 76 with permission of the copyright owner, Harwood Academic Publishers GmbH.)

Glass Transition Temperature.   The glass transition temperature reflects polymer chain mobility, which determines the ease of water diffusion and chemical attack. If the glass transition temperature of a polymer is lower than the biodegradation temperature (e.g., body temperature), the individual molecules possess more mobility to conform with chemical and/or possibly enzymatic attack, which results in faster biodegradation. On the other hand, if the glass transition temperature of a polymer is higher than the biodegradation temperature (e.g., body temperature), the individual molecules possess less mobility and are less likely to conform with chemical and/or possibly enzymatic attack, which results in slower biodegradation.

Other Parameters.   Additives, impurities, a variety of other agents, the nature and amount of drug incorporated into the polymer matrix, and even intermediate degradation products influence the biodegradation of lactic/glycolic acid polymers [104–106]. The type of initiator used for polymer synthesis has a significant effect on the degradation rate. Brizzolara et al. found that lactic/glycolic acid polymers, synthesized by using glycolic acid as an initiator, degrade faster than those synthesized by using lauryl alcohol as an initiator [102]. Kishida et al. have studied the effect of drugs on the biodegradation rate of poly(L-lactic acid) [107]. They found that addition of a basic drug to a film of the polymer greatly enhanced the biodegradation rate, and the more drug that was added, the faster the poly (L-lactic acid) degraded (Fig. 24).

Surface quality of a device composed of lactic/glycolic acid polymers affects the degradation rate. The rate of degradation of the polymers also depends on environmental conditions. Many studies have been conducted to examine the influence of temperature [69,102], pH [108], salts [91,109], enzymes [96,110], microorganisms [110], and irradiation [111] on the degradation of lactic/glycolic acid polymers. All of these investigations have provided a wealth of data; however, the findings are often in conflict with each other.

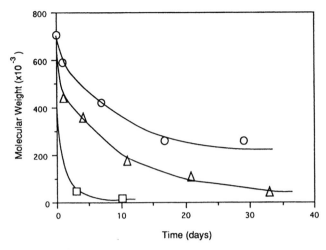

Figure 24 Effect of a basic drug (thioridazine) on the biodegradation rate of poly(L-lactic acid) *in vitro* (loading of the thioridazine: -○-, 0; -△-, 5%; □-, 30%; polymeric films were incubated in phosphate buffer, pH 7.4). (Reproduced from Ref. 107 with permission of the copyright owner, Pharmaceutical Society of Japan.)

4. Bioelimination of Biodegradation Products

Lactic acid and glycolic acid are the biodegradation products of the lactic/glycolic acid polymers. Figure 25 depicts a suggested scheme for the *in vivo* fate of these products [112]. Lactic acid is a natural metabolic product in all higher animals. It enters the tricarboxylic acid cycle, and is metabolized and eliminated from the body as carbon dioxide and water. Brady et al. have conducted a bioelimination study using ^{14}C-labeled poly(lactic acid) implants [113]. They showed that 36.8% of the radioactivity had been lost from the implants after 168 days. Of this 36.8%, only 4.6% was recovered in the urine, 2.8% in the feces, and less than 0.3% was found in tissue. They suggested that the elimination of lactic acid was mainly via respiration as carbon dioxide.

Glycolic acid is a component of several neutral mucopolysacchrides in animal tissues. It can be excreted unchanged by the kidneys or enter the tricarboxylic acid cycle and be eliminated as carbon dioxide and water [114].

B. Thermal Stability and Thermal Degradation

When heated above 200°C for a prolonged time under vacuum or a nitrogen atmosphere, lactic/glycolic acid polymers degrade into the cyclic dimer, lactide and/or glycolide [51]. At lower temperatures, the thermal degradation that leads to these products is a function of time and temperature. Thermal degradation at lower temperatures mainly results in reduction of the molecular weight or intrinsic viscosity of the polymer [18]. Figure 26 shows the effect of temperature on molecular weight of poly(L-lactic acid) under a nitrogen atmosphere [81].

From their results, Jamshidi et al. suggested that the thermal degradation of lactic/glycolic acid polymers may be due to the following: (1) polymer hydrolysis caused by a trace amount of water contained in the polymer; (2) depolymerization, catalyzed by residual polymerization catalyst; (3) oxidative, random-chain scission caused by a trace

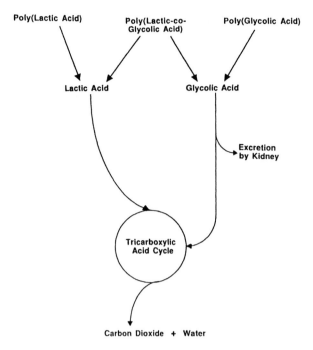

Figure 25 Metabolism and excretion of biodegradation products of lactic/glycolic acid polymers *in vivo*.

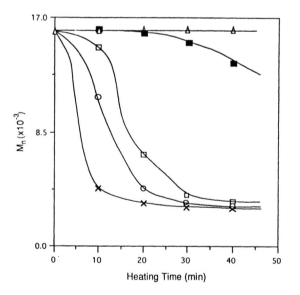

Figure 26 Thermal degradation of poly(L-lactic acid) at different temperatures as a function of heating time (-△-, 180°C; -■-, 190°C; -□-, 200°C; -○-, 210°C; -x-, 220°C). (Reproduced from Ref. 81 with permission of the copyright owner, Butterworth-Heinemann Ltd.)

amount of oxygen; and (4) inter- and intramolecular transesterification with the formation of monomers and oligomers [81].

C. Storage and Handling

Because lactic/glycolic acid polymers are hydrolytically labile, to store lactic/glycolic acid polymers properly, moisture must be excluded. It has been reported that the storage of commercial poly(lactic acid) at 40°C under a high relative humidity of 74.7% decreases the molecular weight M_w of the polymer from 139,000 to 131,370 in one week, and to 55,880 in four weeks [61]. For proper handling of the lactic/glycolic acid polymers, they should at least be vacuum sealed in a container and stored at a temperature below 0°C. Before use, the container should be brought to room temperature. After the polymer has been warmed to room temperature, the container can be opened.

V. SUMMARY

The lactic/glycolic acid polymers represent a very important class of biomaterials. These polymers have been the subject of extensive animal and clinical tests. When properly manufactured under cGMP (current good manufacturing practice) conditions from purified monomers, these polymers exhibit very little evidence of inflammatory response or any other harmful side effects. The chemical, physical, and solvent properties of these polymers can be controlled by adjusting the chemical composition (i.e., content of L-lactic acid, D,L-lactic acid, and/or glycolic acid) and molecular weight of the polymer. The biodegradation rate of these polymers can be tailored to a specific need by controlling the chemical and physical properties of the polymer. These polymers are thermoplastic and can be processed by many methods, including fiber spinning, extrusion, injection molding, and micro- and nanoencapsulations. Consequently, the lactic/glycolic acid polymers can be fabricated into a variety of wound closures items, implantable devices, and drug delivery systems, which include microspheres, fibers, films, rods, and others. There is no doubt that the lactic/glycolic acid polymers will become one of the most useful and versatile class of biodegradable polymers for biological and medical applications, particularly for the controlled release of biologically active agents.

ACKNOWLEDGMENT

The author would like to thank Dr. Janna Strobel for her editorial assistance.

REFERENCES

1. Wang, H. T., E. Schmitt, D. R. Flanagan, and R. J. Linhardt, Influence of formulation methods on the *in vitro* controlled release of protein from poly(ester) microspheres, *J. Controlled Rel.*, 17:23–32 (1991).
2. Linhardt, R. J., D. R. Flanagan, E. Schmitt, and H. T. Wang, Quantitative analysis of the monomer products formed on the hydrolysis of poly(esters) and poly(anhydrides), *Polym. Prepr.*, 30(1):464–465 (1989).
3. Linhardt, R. J., D. R. Flanagan, E. Schmitt, and H. T. Wang, Biodegradable poly(esters) and the delivery of bioactive agents, *Polym. Prepr.*, 30(1):249–250 (1990).
4. Dahlmann, J., G. Rafler, K. Fechner, and B. Mehlis, Synthesis and properties of biodegradable aliphatic polyesters, *British Polym. J.*, 23:235–240 (1990).

5. Pitt, C. G., M. M. Gratzl, G. L. Kimmel, J. Surles, and A. Schindler, Aliphatic polyester. II. The degradation of poly(DL-lactide), poly(ε-caprolactone), and their copolymers *in vivo*, *Biomaterials*, 2:215–220 (1981).

6. Li, S. M., H. Garreau, and M. Vert, Structure-property relationships in the case of the degradation of massive aliphatic poly-(α-hydroxy acids) in aqueous media. Part 1: Poly(DL-lactic acid), *Biomaterials*, 2:215–220 (1981).

7. Li, S. M., H. Garreau, and M. Vert, Structure-property relationships in the case of the degradation of massive aliphatic poly-(α-hydroxy acids) in aqueous media. Part 1: Poly(DL-lactic acid), *J. Mater. Sci.: Mater. Med.*, 1:123–130 (1990).

8. Coombes, A. G. A., and J. D. Heckman, Gel casting of resorbable polymers. 1. Processing and application, *Biomaterials*, 13:217–224 (1992).

9. Kricheldorf, H. R., T. Mang, and J. M. Jonte, Polylactones. 1. Copolymerization of glycolide and ε-caprolactone, *Macromol.*, 17:2173–2181 (1984).

10. Kricheldorf, H. R., and A. Serra, Polylactones. 6. Influence of various metal salts on the optical purity of poly(L-lactide), *Polym. Bull.*, 14:497–502 (1985).

11. Brandrup, J., and E. H. Immergut, eds., *Polymer Handbook*, 3rd ed., John Wiley and Sons, New York, 1989.

12. Arshady, R., Preparation of biodegradable microspheres and microcapsules: 2. Polylactides and related polyesters, *J. Controlled Rel.*, 17:1–22 (1991).

13. Barrrows, T. H., Degradable implant materials: A review of synthetic absorbable polymers and their applications, *Clin. Mater.*, 1:233–257 (1986).

14. Holland, S. J., and B. J. Tighe, Biodegradable polymers. In *Advances in Pharmaceutical Sciences*, Vol. 6 (D. Ganderton and T. J. Jones, eds.), Academic Press, London, 1992, pp. 101–164.

15. Rai, B. N., V. Kumar, and S. N. Upadhyay, Polymers in medicine, *CEW*, 27(2):23–30 (1992).

16. Jalil, R., and J. P. Nixon, Biodegradable poly(lactic acid) and poly(lactide-co-glycolide) microcapsules: Problems associated with preparative techniques and release properties, *J. Microencapsulation*, 7:297–325 (1990).

17. Lewis, D. H., Controlled release of bioactive agents from lactic/glycolic acid polymers. In *Biodegradable Polymers as Drug Delivery Systems* (M. Chasin and R. S. Langer, eds.), Marcel Dekker, New York, 1990, pp. 1–41.

18. Conti, B., F. Pavanetto, and I. Genta, Use of polylactic acid for the preparation of microparticulate drug delivery systems, *J. Microencapsulation*, 9:153–166 (1992).

19. Bischoff, C. A., and P. Walden, Ueber das glycolid und seine homologen, *Ber. Deut. Chem. Ges.*, 26:262–265 (1893).

20. Bischoff, C. A., and P. Walden, I. Ueber derivate der glycolsaure, *Ann. Chem., Justus Liebigs*, 279:45–70 (1894).

21. Bischoff, C. A., and P. Walden, II. Ueber derivate der milchsaure, *Ann. Chem., Justus Liebigs*, 279:71–99 (1894).

22. Carothers, W. H., G. L. Dorough, and F. J. van Natta, Studies of polymerization and ring formation. X. The reversible polymerization of six-membered cyclic esters, *J. Am. Chem. Soc.*, 54:761–772 (1932).

23. Mark, H., and G. S. Whitby, eds., *Collected Papers of Wallace H. Carothers on High Polymeric Substances*, Interscience Publishers, New York, 1940.

24. Lowe, C. E., Preparation of high molecular weight polyhydroxyacetic ester, U.S. Patent 2,668,162 (1954).

25. Schmitt, E. E., and R. A. Polistina, Surgical sutures, U.S. Patent 3,297,033 (1967).

26. First synthetic absorbable suture, *Medical World News*, February 12, 1971, p. 28G.

27. Schneider, A. K., Polylactide sutures, U.S. Patent 3,636,956 (1972).

28. Brekke, J. H., M. Bresner, and M. J. Reitman, Polylactic acid surgical dressing material. Postoperative therapy for dental extraction wounds, *Can. Dent. Assoc. J.*, 52:599 (1986).

29. Brekke, J. H., M. Bresner, and M. J. Reitman, Effect of surgical trauma and polylactate

cubes and granules on the incidence of alveolar osteitis in mandibular third molar extraction wounds, *Can. Dent. Assoc. J.*, 52:315 (1986).

30. Rokkanen, P., O. Bostman, S. Vainionpaa, K. Vihtonen, P. Tormala, J. Laiho, J. Kilpikari, and M. Tamminmaki, Biodegradable implants in fracture fixation: Early results of treatment of fractures of ankle, *Lancet*, 1:1422–1424 (1985).

31. Rudolf, R. M., G. Boering, F. Rosema, and J. W. Leenslag, Resorbable poly(L-lactide) plates and screws for the fixation of zygomatic fractures, *J. Oral Maxillofac. Surg.*, 45:751–753 (1987).

32. Wehrenberg, R. H., II, Lactic acid polymers: Strong, degradable thermoplastics, *Mater. Eng.*, 94:63 (September 1981).

33. Vert, M., P. Cristel, and F. Chobot, Bioresorbable plastic materials for bone surgery. In *Macromolecular Biomaterials* (C. W. Hastings and P. Ducheyne, eds.), CRC Press, Boca Raton, FL, 1984, pp. 120–142.

34. Bercovy, M., D. Goutallier, M. C. Voisin, D. Geiger, D. Geiger, D. Blanquaert, A. Gaudichet, and D. Patte, Carbon-PGLA prostheses for ligament reconstruction. Experimental basis and short-term results in man, *Clin. Orthop. Related Res.*, 196:159–168 (1985).

35. King, J. B., and C. Bulstrode, Polylactate-coated carbon fiber in extra-articular reconstruction of the instable knee, *Clin. Orthop.*, 196:139–142 (1985).

36. Lauritzen, C., Experimental studies on absorbable vascular grafts for microsurgery, *Scand. J. Plast. Reconstr. Surg.*, 17:133–135 (1983).

37. Mendak, S. H., Jr., R. J. Jensik, M. F. Haklin, and D. L. Roseman, The evaluation of various bioabsorbable materials on the titanium fiber metal tracheal prothesis, *Ann. Thorac. Surg.*, 38:488–493 (1984).

38. Greenstein, S. M., T. F. Murphy, B. F. Rush, Jr., and H. Alexander, The experimental evaluation of a carbon-polylactic acid mesh for a ventral herniorrhaphy, *Curr. Surg.*, 41:358–362 (1984).

39. Henry, E. W., T. H. Chiu, E. Nyilas, T. M. Brushart, P. Dikkes, and R. L. Sidman, Nerve regeneration through biodegradable polyester tubes, *Exp. Neurol.*, 90:652–676 (1985).

40. Kulkarni, R. K., E. G. Moore, A. F. Hegyeli, and F. Leonard, Biodegradable poly(lactic acid) polymers, *J. Biomed. Mater. Res.*, 5:169–181 (1971).

41. Hutchinson, F. G., and B. J. A. Furr, Biodegradable polymers for the sustained release of peptides, *Biochem. Soc. Transactions*, 13:520–523 (1985).

42. Tabata, Y., Y. Takebayashi, T. Ueda, and Y. Ikada, A formulation method using D,L-lactic acid oligomer for protein release with reduced initial burst, *J. Controlled Rel.*, 23:55–64 (1993).

43. Tabata, Y., and Y. Ikada, Macrophage phagocytosis of biodegradable microspheres composed of L-lactic acid/glycolic acid homo- and copolymers, *J. Biomed. Mater. Res.*, 22:837–858 (1988).

44. Avgoustakis, K., and J. R. Nixon, Biodegradable controlled release tablets. 1: Preparative variables affecting the properties of poly(lactide-co-glycolide) copolymers as matrix forming material, *Intern. J. Pharm.*, 70:77–85 (1991).

45. Sakakura, C., T. Takahashi, A. Hagiwara, M. Itoh, T. Sasabe, M. Lee, and S. Shobayashi, Controlled release of cisplatin from lactic acid oligomer microspheres incorporating cisplatin: *In vitro* studies, *J. Controlled Rel.*, 22:69–74 (1992).

46. Yoshikawa, H., Y. Nakao, K. Takada, S. Muranishi, R. Wada, Y. Tabata, S. H. Hyon, and Y. Ikada, Targeted and sustained delivery of aclarubicin to the lymphatics by lactic acid oligomer microspheres in rat, *Chem. Pharm. Bull.*, 37:802–804 (1988).

47. Wada, R., S. Hyon, and Y. Ikada, Salt formation of lactic acid oligomers as matrix for sustained release of drugs, *J. Pharm. Pharmacol.*, 43:605–608 (1991).

48. Wada, R., S. H. Hyon, Y. Ikada, Y. Nakao, H. Yoshikawa, and S. Muranishi, Lactic acid oligomer microspheres containing an anticancer agent for selective lymphatic delivery: 1. *In vitro* studies, *J. Bioact. Compatib. Polym.*, 3:126–136 (1988).

49. Rak, J., J. L. Ford, C. Rostron, and V. Walters, The preparation and characterization of

poly(D,L-lactic acid) for use as a biodegradable drug carrier, *Pharm. Acta Helvetiae*, 60: 162–169 (1985).

50. Marcotte, N., and M. F. A. Goosen, Delayed release of water-soluble macromolecules from polylactide pellets, *J. Controlled Rel.*, 9:75–85 (1989).

51. Gilding, D. K., and A. M. Reed, Biodegradable polymers for use in surgery – Polyglycolic/ polylactic acid homo- and copolymers: 1, *Polymer*, 20:1459–1464 (1979).

52. Lundberg, R. D., J. V. Koleske, and K. B. Wischmann, Lactone polymers. III. Polymerization of ε-caprolactone, *J. Polym. Sci.*, Part A-1, 7:2915–2930 (1969).

53. Ouhadi, T., C. Stevens, and P. Teyssie, Mechanism of ε-caprolactone polymerization by aluminum alkoxides, *Makromol Chem.*, Suppl., 1:191–201 (1975).

54. Vion, J. M., R. Jerome, and P. Teyssie, Synthesis, characterization, and miscibility of caprolactone random copolymers, *Macromol.*, 19:1828–1838 (1986).

55. Kumar, G. S., *Biodegradable Polymers: Prospects and Progress*, Marcel Dekker, New York, 1987, pp. 44–45.

56. Kissel, T., Z. Brich, S. Bantle, I. Lancranjan, F. Nimmerfall, and P. Vit, Parenteral depot-systems on the basis of biodegradable polyesters, *J. Controlled Rel.*, 16:27–42 (1991).

57. Kricheldorf, H. R., and R. Dunsing, Polylactones. 8. Mechanism of the cationic polymerization of L,L-dilactide, *Makromol. Chem.*, 187:1611–1625 (1986).

58. Leenslag, J. W., and A. J. Pennings, Synthesis of high-molecular weight poly(L-lactide) initiated with tin 2-ethylhexanoate, *Makromol. Chem.*, 188:1809–1814 (1987).

59. Kohn, F. E., J. W. A. Van Den Berg, G. Van Den Berg, and J. Feijen, The ring-opening polymerization of D,L-lactide in the melt initiated with tetraphenyltin, *J. Appl. Polym. Sci.*, 29:4265–4277 (1984).

60. Dunn, R. L., J. P. English, J. D. Strobel, D. R. Cowsar, and T. R. Tice, Preparation and evaluation of lactic/glycolic acid copolymers for drug delivery. In *Polymers in Medicine*, Vol. 3 (C. Migliaresi et al., eds.), Elsevier Science Publishers, Amsterdam, 1988, pp. 149–160.

61. Deasy, P. B., M. P. Finan, and M. J. Meegan, Preparation and characterization of lactic/ glycolic acid polymers and copolymers, *J. Microencapsulation*, 6:369–378 (1989).

62. Demirdere, A., T. Kissel, U. Siemann, and H. Sucker, Permeability and release properties of biodegradable polyesters. Part 1: Feasibility of reservoir systems, *Eur. J. Pharm. Biopharm.*, 37:42–48 (1991).

63. Kricheldorf, H. R., and I. Kreiser, Polylactones, 11. Cationic copolymerization of glycolide with L,L-dilactide, *Makromol. Chem.*, 188:1861–1873 (1987).

64. *The United States Pharmacopeia*, 23rd ed., United States Pharmacopeial Convention, Rockville, MD, 1995, p. 2038 and pp. 1727–1731.

65. Shinoda, H., and M. Ohtaguro, Preparation process for bioabsorbable polyesters, European Patent EP 0,368,571 A2 (1990).

66. Bendix, D., D. Reichert, and M. Scharfe, Process for purifying resorbable polyesters, U.S. Patent 4,810,775 (1989).

67. Benicewicz, B. C., and P. K. Hopper, Polymers for absorbable surgical suture. Part II, *J. Bioact. Compatib. Polym.*, 6:64–94 (1991).

68. Holland, S. J., B. J. Tighe, and P. L. Gould, Polymers for biodegradable medical devices. 1. The potential of polyesters as controlled macromolecule release systems, *J. Controlled Rel.*, 4:155–180 (1986).

69. Smith, K. L., M. E. Schimpf, and K. E. Thompson, Bioerodible polymers for delivery of macromolecules, *Advanced Drug Delivery Reviews*, 4:343–357 (1990).

70. Reed, A. M., and D. K. Gilding, Biodegradable polymers for use in surgery – Poly(glycolic)/ Poly(lactic acid) homo- and copolymers. 2. *In vitro* degradation, *Polymer*, 22:494–498 (1981).

71. Fukuzaki, H., M. Yoshida, M. Asano, M. Kumakura, T. Mashimo, H. Yuasa, K. Imai, and H. Yamanaka, *In vivo* characteristics of high molecular weight copoly(l-lactide/glycolide) with S-type degradation pattern for application in drug delivery systems, *Biomaterials*, 12: 433–437 (1991).

72. Pitt, C. G., and A. Schindler, The design of controlled drug delivery systems based on biodegradable polymers. In *Progress in Contraceptive Delivery Systems*, Vol. 1, *Biodegradables and Delivery Systems for Contraception* (E. S. E. Hafez and W. A. A. van Os, eds.), MTP Press, Lancaster, England, 1980, pp. 17–46.

73. Eling, B., S. Gogolewski, and A. J. Pennings, Biodegradable materials of poly(l-lactic acid): 1. Melt-spun and solution-spun fibers, *Polymer*, 23:1587–1593 (1982).

74. Gogolewski, S., and A. J. Pennings, Resorbable materials of poly(L-lactide). II. Fibers spun from solutions of poly(L-lactide) in good solvents, *J. Appl. Polym. Sci.*, 28:1045–1061 (1983).

75. Schindler, A., and D. Harper, Polylactide II. Viscosity-molecular weight relationships and unperturbed chain dimensions, *J. Polym. Sci.: Polym. Chem. Ed.*, 17:2593–2599 (1979).

76. Asano, M., H. Fukuzaki, M. Yoshida, M. Kumakura, T. Mashimo, H. Yuasa, K. Imai, and H. Yamanaka, Application of poly DL-lactic acids of varying molecular weight in drug delivery systems, *Drug Design and Delivery*, 5:301–320 (1990).

77. Pitt, C. G., and Z. W. Gu, Modification of the rates of chain cleavage of poly (ε-caprolactone) and related polyesters in the solid state, *J. Controlled Rel.*, 4:283–292 (1987).

78. Cohn, D., H. Younes, and G. Marom, Amorphous and crystalline morphologies in glycolic acid and lactic acid polymers, *Polymer*, 28:18–22 (1987).

79. Asano, M., M. Yoshida, I. Kaetsu, K. Imai, T. Mashimo, H. Yuasa, H. Yamanaka, K. Suzuki, and I. Yamazaki, Biodegradability of a hot-pressed poly(lactic acid) formulation with controlled release of LH-RH agonist and its pharmacological influence on rat prostate, *Makromol. Chem.: Rapid Commun.*, 6:509–513 (1985).

80. Moll, F., and R. Ries, Biodegradable microtablets made of low molecular weight polyglycolic acid, *Arch. Pharm. (Weinheim)*, 324:939–940 (1991).

81. Jamshidi, K., S. H. Hyon, and Y. Ikada, Thermal characterization of polylactides, *Polymer*, 29:2229–2234 (1988).

82. Linhardt, R. J., Biodegradable polymers for controlled release of drugs. In *Controlled Release of Drugs: Polymers and Aggregate Systems* (M. Rosoff, ed.), VCH Publishers, New York, 1989, pp. 53–95.

83. Kaplan, D. S., and R. R. Muth, Polymers for injection moulding of absorbable surgical devices, U.S. Patent 4,523,591 (1985).

84. Vert, M., and F. Chabot, Stereoregular bioresorbable polyesters for orthopaedic surgery, *Makromol. Chem.*, 5(Suppl.):30–41 (1981).

85. Bendix, D., Analytical studies on the solubility problem of poly(d,l-lactide-co-glycolide) 50 : 50, *Proceed. Intern. Symp. Control. Rel. Bioact. Mater.*, 17:248–249 (1990).

86. Siemann, U., Densitometric determination of the solubility parameters of biodegradable polyesters, *Proceed. Intern. Symp. Control Rel. Bioact. Mater.*, 12:53–54 (1985).

87. Spenlehauer, G., M. Vert, J. P. Benoit, and A. Boddaert, *In vitro* and *in vivo* degradation of poly(d,l-lactide/glycolide) type microspheres made by solvent evaporation method, *Biomaterials*, 10:557–563 (1989).

88. Heller, J., Controlled drug release from poly(ortho ester) — A surface eroding polymer, *J. Controlled Rel.*, 2:167–177 (1985).

89. Thies, C., and M. C. Bissery, Biodegradable microspheres for parenteral administration. In *Biomedical Applications of Microencapsulation* (F. Lim. ed.), CRC Press, Boca Raton, FL, 1984, pp. 53–74.

90. Schnabel, W., *Polymer Degradation — Principles and Practical Applications*, Hanser International, Germany, 1981, pp. 179–181.

91. Ginde, R. M., and R. K. Gupta, *In vitro* chemical degradation of poly(glycolic acid) pellets and fibers, *J. Appl. Polym. Sci.*, 33:2411–2429 (1987).

92. Kronenthal, R. L., Biodegradable polymers in medicine and surgery. In *Polymer Science and Technology*, Vol. 8 *Polymers in Medicine and Surgery* (R. L. Kronenthal, Z. Oser, and E. Martin, eds.), Plenum Press, New York, 1975, pp. 119–137.

93. Williams, D. F., and E. Mort, Enzyme-accelerated hydrogels of polyglycolic acid, *J. Bioengineering*, 1:231–238 (1977).

94. Schakenraad, J. M., P. Nieuwenhuis, I. Molenaar, J. Helder, P. J. Dijkstra, and J. Feijen, *In vivo* and *in vitro* degradation of glycine/D,L-lactic acid copolymers, *J. Biomed. Mater. Res.*, 23:1271–1288 (1989).

95. Pitt, C. G., F. I. Chasalow, Y. M. Hibionada, D. M. Kilmas, and A. Schindler, Aliphatic polyesters. I. The degradation of poly(ϵ-caprolactone) *in vivo*, *J. Appl. Polym. Sci.*, 26: 3779–3781 (1981).

96. Pitt, C. G., and A. Schindler, Biodegradation of polymers. In *Controlled Drug Delivery*, Vol. 1, *Basic Concepts* (S. D. Bruck, ed.), CRC Press, Boca Raton, FL, 1983, pp. 53–80.

97. Huffman, K. R., and D. J. Casey, Effect of carboxyl end groups on hydrolysis of polyglycolic acid, *J. Polym. Sci.: Polym. Chem. Ed.*, 23:1939–1954 (1985).

98. Pitt, C. G., Y. Cha, S. S. Shah, and K. J. Zhu, Blends of PVA and PGLA: Control of the permeability of hydrogels by blending, *J. Controlled Rel.*, 19:189–200 (1992).

99. Fildes, F. J. T., F. G. Hutchinson, and B. J. A. Furr, The development of "Zoladex" – A case history. In *Polypeptide and Protein Drugs* (R. C. Hider and D. Barlow, eds.), Horwood, London, 1991, pp. 228–250.

100. Zhu, J. H., Z. R. Shen, L. T. Wu, and S. L. Yang, *In vitro* degradation of polylactide and poly(lactide-co-glycolide) microspheres, *J. Appl. Polym. Sci.*, 43:2099–2106 (1991).

101. Kaetsu, I., M. Yoshida, M. Asano, H. Yamanaka, K. Imai, H. Yuasa, T. Mashimo, K. Suzuki, R. Katakai, and M. Oya, Biodegradable implant composites for local therapy, *J. Controlled Rel.*, 6:249–263 (1987).

102. Brizzolara, N., I. Chabrand, D. Miller, L. Rosati, D. Casey, and J. R. Lawter, Hydrolytic degradation kinetics of poly(glycolide-co-DL-lactide), *Proceed. Intern. Symp. Control. Rel. Bioact. Mater.*, 17:246–247 (1990).

103. Chu, C. C., Hydrolytic degradation of polyglycolic acid: Tensile strength and crystallinity study, *J. Appl. Polym. Sci.*, 26:1727–1734 (1981).

104. Coffin, M. D., and J. W. McGinity, Biodegradable pseudolatexes: The chemical stability of poly(D,L-lactic acid) and poly(ϵ-caprolactone) nanoparticles in aqueous media, *Pharm. Res.*, 9:200–205 (1992).

105. Vert, M., S. Li, and H. Garreau, More about the degradation of LA/GA-derived matrices in aqueous media, *J. Controlled Rel.*, 16:15–26 (1991).

106. Maulding, H. V., T. R. Tice, D. R. Cowsar, J. W. Fong, J. E. Pearson, and J. P. Nazareno, Biodegradable microcapsules: Acceleration of polymeric excipient hydrolytic rate by incorporation of a basic medicament, *J. Controlled Rel.*, 3:103–117 (1986).

107. Kishida, A., S. Yoshioka, Y. Takeda, and M. Uchiyama, Formulation-assisted biodegradable polymer matrices, *Chem. Pharm. Bull.*, 37:1954–1956 (1989).

108. Chu, C. C., The *in-vitro* degradation of poly(glycolic acid) sutures – Effect of pH, *J. Biomed. Mater. Res.*, 15:795–804 (1981).

109. Chu, C. C., An *in-vitro* study of the effect of buffer on the degradation of poly(glycolic acid) sutures, *J. Biomed. Mater. Res.*, 15:17–27 (1981).

110. Williams, D. F., The effect of bacteria on absorbable sutures, *J. Biomed. Mater. Res.*, 14: 329–338 (1980).

111. Chu, C. C., and N. D. Campbell, Scanning electron microscopic study of the hydrolytic degradation of poly(glycolic acid) suture, *J. Biomed. Mater. Res.*, 16:417–430 (1982).

112. Hollinger, J. O., and G. C. Battistone, Biodegradable bone repair materials: Synthetic polymers and ceramics, *Clin. Orthop. Related Res.*, 207:290–305 (1986).

113. Brady, J. M., D. E. Cutright, R. A. Miller, and G. C. Battistone, Resorption rate, route of elimination and ultrastructure of the implant site of poly(lactic acid) in the abdominal wall of the rat, *J. Biomed. Mater. Res.*, 7:155–166 (1973).

114. Hockaday, T. D. R., E. W. Frederick, J. E. Clayton, and L. H. Smith, Jr., Studies on primary hyperoxaluria. II. Urinary oxalate, glycolate, and glyoxylate measurement by isotope dilution methods, *J. Lab. Clin. Med.*, 65:677–687 (1965).

30
The Use of PLA-PGA Polymers in Orthopedics

C. M. Agrawal, G. G. Niederauer, D. M. Micallef, and K. A. Athanasiou
University of Texas Health Science Center
San Antonio, Texas

I. INTRODUCTION

The properties of polylactic acid (PLA) and polyglycolic acid (PGA) have been studied since the 1950s [1,2]. However, it was the work of Kulkarni and his colleagues in the 1960s that first generated interest in the use of these materials in the medical arena [3,4]. They described the use of PLA as biodegradable sutures and also as rods for repairing mandibular fractures in dogs.

In recent years there has been tremendous interest in the development of biodegradable PLA or PGA devices for orthopedic applications [5-9]. Several studies in Europe have used biodegradable fracture fixation devices in both humans and animals with varying degrees of success [6,10,11]. Biodegradable bone fixation devices are particularly attractive because, unlike metals, they transfer load gradually to the newly formed bone, thus preventing "stress shielding" or stress protection atrophy of the bone.

PLA, PGA, and their copolymers have also been used to fabricate bone graft substitutes and delivery systems for osteogenic proteins or drugs [12-14]. As these polymers biodegrade within the body, they serve not only as excipients for protein delivery but also as scaffolds for the growth of neotissue. Their biocompatibility, desirable mechanical and physical properties, ease of fabrication, predictable degradation kinetics, and regulatory approval as suture materials have contributed toward their popularity in orthopedic applications [15].

A. Structure

Both PLA and PGA are bioabsorbable polyesters belonging to the group of poly alpha-hydroxy acids (Fig. 1). It is possible to synthesize these materials by the polycondensation of lactic and glycolic acids, but the resulting polymers have a low molecular weight and inferior mechanical properties [16]. High molecular weight polymers can be prepared by

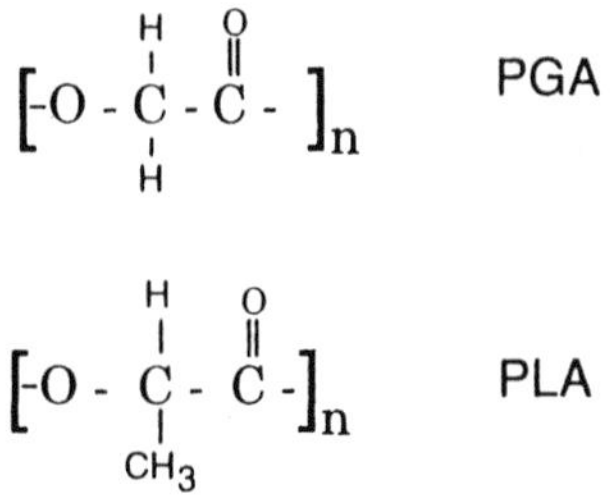

Figure 1   Chemical structure of polyglycolic acid and polylactic acid.

ring-opening melt condensation of lactide and glycolide dimers using tin, antimony, or zinc catalysts [16–18]. Stannous octoate is the catalyst most commonly used and temperatures of approximately 175°C are required for a period of 2–6 hours for polymerization (Fig. 2) [15].

A technique, involving direct polycondensation in the absence of catalysts, has been developed to produce a biodegradable copolymer of low molecular weight glycolic acid and lactones [19,20]. For this procedure, aqueous solutions of glycolic acid and lactones were placed in an ampule with bubbling nitrogen gas at a temperature of 200°C, which produced the copolyesters without further purification. The resulting copolyesters were designed to be used as drug delivery systems.

Lactic acid is a chiral compound and exists as two enantiomeric forms: L and D. Due to the asymmetric molecular structure of lactic acid, PLA can exist in forms of D or L, or the racemic DL [15,21]. The racemic DL-PLA is less crystalline than D-PLA or L-PLA, and also has a lower melting point [15].

PGA, on the other hand, is the one of the simplest linear polyesters and exists in only one form. Furthermore, copolymers of PLA and PGA are relatively easy to synthesize and exist in various ratios. Vicryl®, which is a copolymer containing 90% glycolic acid and 10% lactic acid, is widely used as a suture material [22]. The 50%:50% DL-PLA:PGA copolymer has been used in several studies as a delivery system for drugs or proteins, primarily because of its fast rate of biodegradation [5,23,24]. The properties and degradation kinetics of PLA-PGA copolymers are a function of the monomer ratio, molecular linearity, and molecular weight [15]. Variation of these parameters can yield copolymers suitable for specific applications.

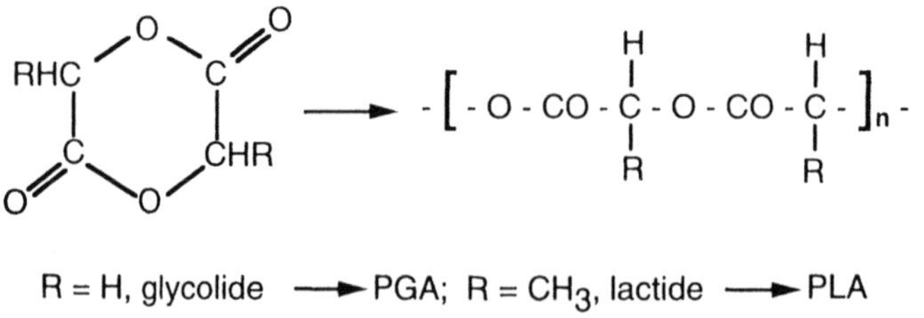

R = H, glycolide ⟶ PGA; R = CH₃, lactide ⟶ PLA

Figure 2   Polymerization of polylactic acid (PLA) and polyglycolic acid (PGA) from dimers.

B. Degradation Pathways

PLA, PGA, and their copolymers biodegrade by nonspecific hydrolytic scission of their ester bonds. Upon hydrolysis, PLA yields lactic acid, which is a normal product of muscular contraction in animals and humans. This acid then enters the tricarboxylic acid cycle and is excreted by the body as water and carbon dioxide (Fig. 3). A study using carbon-labeled PLA reported no significant amounts of accumulation of degradation products in any of the vital organs [3]. Very little radioactivity was detected in feces or urine, indicating that most of the degradation products are released through respiration.

There is evidence that, in addition to hydrolytic scission, PGA is also broken down by certain enzymes, especially those with esterase activity [25]. Monomeric units of glycolic acid produced during the biodegradation process can be excreted in urine. The glycolic acid can also react to form glycine, which then can be used to synthesize serine. After transformation to pyruvic acid, serine can enter the tricarboxylic cycle (Fig. 3) [21]. Thus, after breakdown *in vivo*, both PLA and PGA are excreted as carbon dioxide and water, even though they follow slightly different metabolic pathways.

II. CLINICAL APPLICATIONS

Internal fixation devices for reduction and stability of musculoskeletal fractures have traditionally been made of metal. Once successful tissue repair has been achieved, metallic implants are usually not needed and have to be removed at the expense of an additional surgical procedure. Material-processing advancement has allowed the development of slowly degradable polymeric fixation systems that, theoretically, can progressively transfer stresses to the bone while simultaneously providing sufficient mechanical fixation during the repair process. Furthermore, these systems obviate the need for surgical removal and thus offer clinical and financial advantages for the fixation of fractures. Their use, which has traditionally been concentrated mostly in Europe, but lately in the United States as well, has resulted in varying degrees of success.

One of the first clinical applications of PLA-PGA for fracture fixation was report-

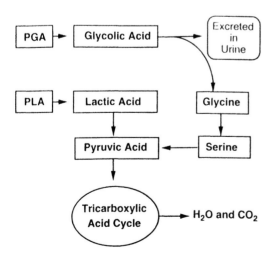

Figure 3   Degradation of polyglycolic acid and polylactic acid in the body. (From Ref. 21.)

edly begun in Finland in 1984 to treat malleolar fractures of the ankle [21], and results were published in 1985 and 1987 [7,26]. In this first application, PLA-PGA copolymeric rods were used and compared to internal fixation with ASIF screws and plates in 56 patients who had either a displaced fracture of the lateral malleolus or a bimalleolar fracture. After a one-year follow-up, it was concluded that no significant differences could be found between the two treatment groups.

In a subsequent prospective study by the same group, PLA-PGA and PGA fixation rods were used in 102 patients with malleolar fractures [10]. Sterile sinus formation was noted in 5.9% of the patients, but no second surgeries were needed. Severe ankle fractures in 41 patients were treated with biodegradable, self-reinforced PGA rods, and although sinus formation developed in 3 of these patients, satisfactory functional results were reported [27].

The clinical manifestation of an inflammatory foreign-body reaction in response to treatment with these implants has been consistently observed in the European studies, although this complication does not appear to interfere with recovery from injury or the functional and radiographic results of the treatment. A review study of 516 patients treated with PLA-PGA or PGA fixation rods indicated that in 7.9% of the cases surgical drainage due to an inflammatory tissue response was required [28]. This inflammatory response, which is clinically manifested about 12 weeks postimplantation, sometimes results in a sinus. However, bacterial cultures of the drainage from the sinus are negative [28,29]. Usually, osteolytic foci develop at the sites where a foreign-body reaction is seen, but after one year the normal structure of the bone is restored [30]. More recently, it was suggested that self-reinforced PGA pins may be successfully used to treat a variety of displaced fractures in children [31].

In the United States, most studies involving the orthopedic use of PLA-PGA devices have come from the field of podiatry. In a study using self-reinforced PGA rods to treat 10 bunionectomies, successful stabilization was achieved in 9 patients, and the absence of any unusual inflammation and hypertrophic granular formation was reported [32]. In a subsequent study, 23 bunionectomies were fixed with self-reinforced PGA rods [33]. In 4% of these (1 patient), sterile sinus discharge was detected. There was also a 30% incidence of short-lived osteolytic changes. In a recent study, 10.2% of patients (5 out of 49) treated for bunions with self-reinforced PGA pins had to undergo a secondary surgical intervention related to severe edema and erythema [34].

PGA rods have also been used in different types of fractures in the upper extremity, such as elbow intra-articular fractures in 29 patients, 15 involving the radial head, 11 in the olecranon, and 3 in the capitellum [35]. Even though successful mechanical fixation was provided by these biodegradable rods, a 13.8% incidence of late noninfectious inflammation was observed. Similarly, in a study involving 40 patients with displaced fractures of the distal radius fixed with PGA pins, inflammation was noted between 47 and 145 days postimplantation and debridement of inflamed soft tissues was needed in 22.5% of the patients [36]. Owing to complications from foreign-body reactions, the fixation of wrist fractures with PGA rods was considered unsatisfactory when compared with Kirschner wires, although no significant functional differences between the groups were seen at one-year follow-up [37].

For maxillofacial applications, the use of PLA plates and screws appears to provide effective stabilization of unstable zygomatic fractures [38] without side-effects [39], although nonspecific foreign-body reactions were observed in some patients after several years [40].

Based on this review of clinical studies performed mostly in Europe but also in North

America, it appears that several factors may influence the incidence of clinically adverse symptoms. First, local tissue tolerance may influence tissue biological response to the chemical irritation provided by degradation products. This suggests that the degree of irritation may be proportional to the amount of the biomaterial implanted and/or the anatomical site of implantation. Exceedingly essential also is the full analytical characterization of the chemical composition and purity of these implants [41]. Also, the fact that different inflammation rates are observed, for example, between the distal radius and the ankle, suggests that the *in situ* stress field the implant is exposed to may influence this biology-biomaterial interaction. In addition, the body fluid dynamics of the flow fields at and adjacent to the implantation site are expected to control the tissue's ability to eliminate the degradation products. Issues such as tissue and implant permeability and degree of vascularization seem to play an important role in by-product evacuation.

It should be emphasized that complications associated with the orthopedic use of PLA or PGA implants are only at a rate of around 5% to 10% and, although significant, they seem to resolve with time. This allows for unabated optimism that the problems of soft tissue irritation and osteopenia, associated with metallic plates and other fixation devices, may be adequately addressed with PLA-PGA or other biodegradable implants.

III. MATERIAL PROPERTIES

A. Physical Properties

1. *Molecular Weight*

PLA, PGA, and their family of copolymers are available from various suppliers in a wide range of molecular weights. Engelberg and Kohn have reported on the molecular weights of various biodegradable polymers as provided by the suppliers or measured by gel permeation chromotagraphy [42]. The starting molecular weights are often altered during fabrication of the implants due to mechanical stresses and changes in temperature and pressure. For instance, Matsusue et al. reported an initial viscosity-average molecular weight M_v of 400 kilodaltons (kDa) for L-PLA, and 220 kDa after the material was extruded and drawn into cylindrical rods [43]. It is important to account for this reduction while designing implants because the molecular weight of a polymer influences its mechanical properties and degradation characteristics.

Weight-average molecular weight M_w values have been reported in the range of 50 to 300 kDa for L-PLA and 21 to 550 kDa for DL-PLA [42]. Number-average molecular weight M_n values range from 19.6 to 150 kDa for L-PLA and 13.4 to 163 kDa for DL-PLA. Another indication of the average size of molecular chains in a polymer is its intrinsic viscosity [η]. For some commercially available polymers, viscosity has been reported to range from 0.61 to 8.2 dL/g for L-PLA and from 0.25 to 2.01 dL/g for DL-PLA [18,42].

Molecular weight measurements for PGA are relatively uncommon because the polymer is insoluble in most common organic solvents. PGA used for fiber extrusion is soluble in hexafluoroisopropanol and has an intrinsic viscosity of 0.6–1.6 dL/g at 0.5% solution. This corresponds to a M_w range of 20–145 kDa [44].

2. *Crystallinity*

Crystallinity within a polymer can be described as an arrangement of molecular chains that results in an ordered structure. Most polymers display little crystallinity and are either amorphous or semicrystalline. Crystallinity in a polymer never reaches 100%. Semicrystalline polymers contain crystalline regions as well as disordered amorphous regions (Fig. 4). Molecules in the crystalline region are packed closely together, and thus mostly crystalline polymers tend to have greater densities compared with amorphous polymers. The density of PGA has been reported as 1.50–1.64 g/cm^3 [44], while densities of 1.29 g/cm^3 and 1.248 g/cm^3 have been reported for the crystalline and amorphous phases of L-PLA, respectively [18].

A measure of crystallinity can be obtained from density measurements according to

$$\% \text{ crystallinity} = \frac{\rho_c(\rho_s - \rho_a)}{\rho_s(\rho_c - \rho_a)} \times 100$$

where ρ_s is the density of a specimen for which the percent crystallinity is to be determined, ρ_c is the density of the perfectly crystalline polymer, and ρ_a is the density of the totally amorphous polymer. The values for ρ_c and ρ_a must be obtained independently.

Parameters that influence the crystallinity of a polymer are those that provide polymeric molecular chains with the flexibility and mobility to reorganize themselves into a more-ordered, and thereby lower, energy state. Thus, the degree of crystallinity, depends upon the molecular chain structure, molecular chemistry, temperature, and the rate of cooling during solidification from a melt. Molecular structure is important because side branches and cross-linking hinder the mobility of chains. Molecular chemistry plays a significant role because monomer units containing large or complex chemical species render crystalline order very difficult. Elevated temperatures and a slow rate of cooling enable the chains to be mobile and realign themselves in a more-ordered solid structure. Thus, the crystallinity of PLA-PGA polymers can be altered as a result of fabrication processes for which heat is used. For instance, Suuronen et al. reported 74% crystallinity for L-PLA screws fabricated from a 65% crystalline batch of L-PLA [45].

The percent crystallinity of a polymer sample can be estimated by measuring its heat of fusion using a differential scanning calorimeter. This measured value is then compared with the heat of fusion for a fully crystallized sample to obtain the percent crystallinity for the test polymer. PGA sutures typically are 46–52% crystalline [16], while the crystallinity of L-PLA has been reported to range from 15% to 69% [16,42,45]. DL-PLA is predominantly amorphous. Copolymers of L-PLA and PGA are usually amorphous when

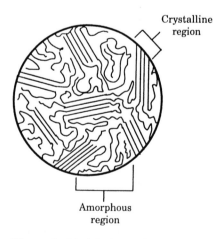

Figure 4   Model of a semicrystalline polymer showing crystalline and amorphous regions.

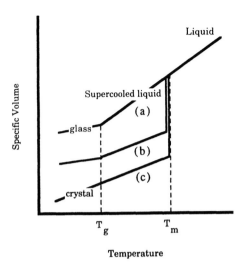

Figure 5 Trend of specific volume versus temperature: typical response of an amorphous polymer (a), semicrystalline polymer (b), and a crystalline polymer (c) upon cooling from a liquid.

the PGA content is in the range of 25% to 70%. The same is true for DL-PLA and PGA copolymers when PGA content is 0–70% [16].

3. Melting and Glass Transition Points

The semicrystalline structure of most polymers results in a unique response of specific volume (volume per unit mass) to changes in temperature. Upon cooling, glassy or amorphous polymers do not solidify in a manner similar to crystalline polymers as they do not exhibit a distinct melting temperature. The slope of the specific volume-temperature curve decreases slightly at a point called the *glass transition temperature T_g*, at which the polymer becomes a glass. For a crystalline polymer, initial cooling results in a slow, steady decrease in specific volume. At the melting temperature T_m, the heat of fusion is liberated, specific volume decreases suddenly, and a solid is formed. T_m and T_g are thus properties of the crystalline and amorphous phases of the polymer, respectively.

Generally, T_g is approximately 0.6 T_m in absolute units of temperature. For a semicrystalline polymer, the behavior is intermediate and both melting and glass transition phenomena are observed. Figure 5 shows the general trend for specific volume as a function of temperature for crystalline, amorphous, and semicrystalline polymers.

Increases in temperature cause a crystalline polymer to change from an ordered structure to a disordered viscous liquid. In the solid phase, molecules vibrate with small amplitudes and secondary bonds form between adjacent chains. At the melting temperature, however, these vibrations increase in amplitude to cause chain translations and break secondary bonds. Melting temperatures are thus a function of the ability of polymeric molecules to form secondary bonds such as van der Waals and hydrogen bonds. Other factors that affect the relative motion of the molecules also influence T_m. For instance, increasing the degree of branching and decreasing the molecular weight will result in lower melting temperatures.

For an amorphous solid, increases in temperature cause the polymer to transform from a rigid solid to one that is rubbery and soft. Chain flexibility has a significant influence on the magnitude of the glass transition temperature. A more flexible chain is more likely to undergo high vibrational amplitudes and rotations, and therefore have lower glass transition temperatures than more-rigid chains. Chain mobility is also lower for polymers with large functional groups, cross-links, and high molecular weights.

Ranges of T_g and T_m reported in the literature are summarized in Table 1. Engelberg and Kohn reported a slight tendency for T_g to increase with molecular weight for both L-PLA and DL-PLA [42]. These researchers did not find any significant relationship between molecular weight, melting point, or degree of crystallinity for L-PLA or DL-PLA.

B. Mechanical Properties

Mechanical properties of biodegradable devices in orthopedics are of critical importance because these devices act as scaffolds for tissue growth or have a structural role in the fracture healing process. Furthermore, since these properties are reduced with biodegradation, it is important to know the rate at which they diminish *in vivo*. The mechanical properties of PLA-PGA polymers are a function of their molecular weight, crystallinity, and molecular orientation. These issues are discussed below.

1. Mechanical Strength

The mechanical strength of PLA-PGA polymers is dependent on both their molecular weight and degree of crystallinity. An increase in these parameters usually leads to an increase in the tensile strength. If the crystallinity is significant, the tie molecules extend between crystalline domains, which then work cooperatively and increase the yield strength of the polymer.

The mechanical strength of L-PLA is significantly improved with increasing molecular weight and crystallinity [18,42]. For orthopedic applications of L-PLA, the use of a polymer with a molecular weight of 100 kDa or more was recommended to achieve good mechanical properties [42]. Typical tensile strength values for L-PLA have been reported to be 11.4–82.7 MPa, while flexural strengths range from 45 to 145 MPa [8,42,46].

For reinforced L-PLA composites, strengths can be greatly increased. In a recent review paper, Daniels et al. reported tensile strengths as high as 200 MPa and flexural strengths from 89.6 to 412 MPa [8]. Since L-PLA is hydrophobic and crystalline, its rate of biodegradation is relatively slower and its strength retention longer. For instance, it has been reported that the bending strength of drawn L-PLA rods exceeded the bending strength of human cortical bone for 8 weeks in the medullary canal of rabbits [43]. Also, it has been shown that the greater the diameter of the implant is, the longer the strength is maintained because the surface area/mass ratio of thick rods is lower than that of thin rods [43,47] (Fig. 6).

Table 1   Glass Transition Temperature Ranges for Various Biodegradable Polymers

Polymer	$T_g(^\circ C)$	$T_m(^\circ C)$	Reference no.
L-PLA	54–59	159–178	16, 18, 42, 56
DL-PLA	50–53	–	42
PGA	36	210–226	16, 42, 44
PGA-PLA copolymers	37–55	–	16, 46

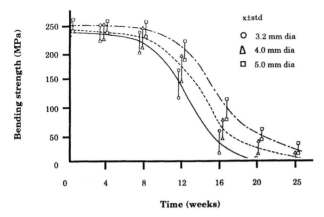

Figure 6 *In vitro* degradation of drawn L-polylactic acid rods in phosphate-buffered saline at 37°C. (From Ref. 43.)

The tensile strength of PGA in bulk form has been reported to be 57–69 MPa [46,48]. However, it was very brittle and degraded very rapidly [48]. Researchers have developed self-reinforced polyglycolic acid composite rods fabricated by sintering PGA sutures together at elevated temperature and pressure. With rods between 1.5 and 4.5 mm in diameter, initial bending strengths of 220–405 MPa and shear strengths of 165–255 MPa have been reported [47,49,50]. Table 2 shows the initial measured mechanical properties of self-reinforced (SR) PGA rods versus injection-molded PGA rods as a function of specimen diameter, and Fig. 7 shows the bending stress-strain relationship for 2.0 mm diameter SR-PGA rods and injection-molded PGA rods [47].

Usually, PLA-PGA copolymer implants have a lower initial flexural strength when compared with PGA and L-PLA implants. Tensile strengths of raw material have been shown to be in the range of 41.4 to 55.2 MPa (Table 3). Self-reinforced PLA-PGA rods fabricated from Vicryl sutures have produced initial flexural strengths of 265 MPa, compared with 150 MPa for nonreinforced specimens [51].

Table 2 Mechanical Properties of Self-Reinforced and Injection-Molded Polyglycolic Acid Rods as a Function of Rod Diameter

Sample	Shear strength (MPa)	Bending modulus (GPa)	Bending strength (MPa)
Sintered			
SR-PGA (1.5 mm diameter)	225 ± 10	12 ± 3	365 ± 40
SR-PGA (2.0 mm diameter)	190 ± 10	11 ± 3	260 ± 40
SR-PGA (3.2 mm diameter)	220 ± 35	12 ± 2	360 ± 70
SR-PGA (4.5 mm diameter)	180 ± 15	10 ± 1	350 ± 10
Injection molded			
PGA (2.0 mm diameter)	95 ± 5	7 ± 0.5	220 ± 15
PGA (3.2 mm diameter)	95 ± 15	7 ± 0.5	220 ± 10

Source: From Ref. 47

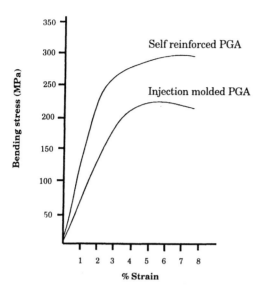

Figure 7 Bending stress-strain relationships of self-reinforced polyglycolic acid rod and injection-molded polyglycolic acid rod (2.0 mm diameter). (From Ref. 47.)

2. Elastic Modulus

The stiffness of PLA-PGA polymers is a function of crystallinity and secondary bond strength. The tensile and bending moduli for L-PLA have been reported to be 1.2–4.1 GPa and 1.4–3.2 GPa, respectively [42,46]. However, other studies report that, for nonreinforced L-PLA specimens, tensile moduli range from 0.6 to 5.0 GPa and bending moduli range from 2.4 to 10.0 GPa. For reinforced L-PLA specimens, the tensile and flexural moduli range from 6 to 29.9 GPa and 4 to 124.4 GPa, respectively [8,47].

The tensile modulus of PGA has been reported to be 6.5 GPa [48], and for self-reinforced PGA rods, bending moduli have ranged from 8 to 15 GPa [47]. Tensile moduli for PLA-PGA copolymers are lower compared with the homopolymers and have been shown to be in the range of 1.4 to 2.7 GPa (Table 3).

3. Ductility

Even though ductility is not commonly reported in the literature, it is a very important property to consider. Ductility, which may be expressed as percent elongation, is a measure of a specimen's ability to undergo plastic deformation prior to fracture. Although no values were specified, Törmälä et al. reported ductile behavior of self-reinforced PLA-PGA rods [51], while melt-molded PLA-PGA rods exhibited brittle characteristics. Load-bearing implants should be capable of some degree of plastic deformation as a safety mechanism so that catastrophic failure does not occur.

Percent elongation at break and percent elongation at yield for L-PLA from the manufacturer have been reported to be between 2.0% and 6.0% and 1.8% and 3.7%, respectively [42]. There is a clear trend toward decreasing ductility with increased molecular weight. A disadvantage of L-PLA, regardless of molecular weight, is its brittleness. Figure 8 shows the brittle behavior of L-PLA compared with the more ductile behavior of DL-PLA [42]. Fabrication techniques affect the ductility of PLA. For example, hot

Table 3 Physical and Mechanical Properties of Standard Polymers

	50/50 DL-PLG	65/35 DL-PLG	75/25 DL-PLG	85/15 DL-PLG	DL-PLA	L-PLA	PGA
Chemical formula	$(C_3H_4O_2)_x$ $(C_2H_2O_2)_y$	$(C_3H_4O_2)_x$ $(C_2H_2O_2)_y$	$(C_3H_4O_2)_x$ $(C_2H_2O_2)_y$	$(C_3H_4O_2)_x$ $(C_2H_2O_2)_y$	$(C_3H_4O_2)_x$	$(C_3H_4O_2)_x$	$(C_2H_2O_2)_y$
Inherent viscosity[a]	0.6–0.7[b]	0.6–0.7[b]	0.6–0.7[c]	0.6–0.7[c]	0.6–0.7[c]	0.9–1.0[c]	1.1–1.4[b]
Melting point (°C)	Amorphous	Amorphous	Amorphous	Amorphous	Amorphous	173–178	225–230
Glass transition (°C)	45–50	45–50	50–55	50–55	55–60	60–65	35–40
Color	White to light gold	White to light gold	White to light gold	White to light gold	White	White	Light tan
Solubility[d]	MeCl₂, THF, EtOAc, C₃H₆O, CHCl₃, HFIP	MeCl₂, THF, EtOAc, C₃H₆O, CHCl₃, HFIP	MeCl₂, THF, EtOAc, C₃H₆O, CHCl₃, HFIP	MeCl₂, THF, EtOAc, C₃H₆O, CHCl₃, HFIP	MeCl₂, THF, EtOAc, C₃H₆O, CHCl₃, HFIP	MeCl₂, CHCl₃, HFIP	HFIP, HFASH
Specific gravity	1.34	1.30	1.30	1.27	1.25	1.24	1.53
Tensile strength (MPa)	41.4–55.2	41.4–55.2	41.4–55.2	41.4–55.2	27.5–41.4	55.2–82.7	69.0+
% elongation	3–10	3–10	3–10	3–10	3–10	5–10	15–20
Elastic modulus (GPa)	1.4–2.7	1.4–2.7	1.4–2.7	1.4–2.7	1.4–2.7	2.7–4.1	6.9+

Source: From Ref. 46
[a]Inherent viscosity of material as tested
[b]HFIP, hexafluoroisopropanol
[c]CHCl₃, chloroform
[d]Partial listing only: MeCl₂, methylene chloride; THF, tetrahydrofuran, EtOAc, ethyl acetate; HFIP, hexafluoroisopropanol; HFASH, hexafluoroacetone sesquihydrate; C₃H₆O, acetone; PLG, PLA–PGA copolymer

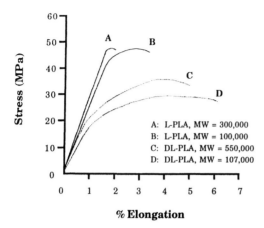

Figure 8   Stress-strain curves for various samples of L-polylactic acid and DL-polylactic acid. (From Ref. 42.)

drawn melt-spun and solution-spun fibers have shown elongations at break from 12% to 26% [18]. Braided melt-spun L-PLA fibers for ligament repair have been reported to have initial elongations of approximately 15–35% with significant decreases as a function of degradation time [52].

PGA in bulk form has been shown to be very brittle, with 0.7% elongation [48]. Similar to other polymers, the brittle nature of PGA can be manipulated through processing techniques. After being drawn into high-strength fibers of 15–25 micrometers (μm) diameter, PGA can attain elongations of 15–35% [44]. PGA sutures sintered together to form self-reinforced composite rods have been reported to have 7–8% elongation [47].

C.   Sterilization Effects

Medical devices used in surgery require some form of sterilization prior to surgery to reduce the possibility of infection. Three kinds of sterilization techniques exist: steam sterilization, gas sterilization, and radiation sterilization. Biodegradable polymeric implants fabricated from PLA-PGA require special attention because of their sensitivity to high temperatures and hydrolysis. Currently gas sterilization by ethylene oxide is not preferred because of its tendency to leave mutagenic and carcinogenic residues within and on the surface of polymers [53].

The effects of gamma radiation have been studied by various researchers. Radiation can cause chain scission in biodegradable polymers, which leads to decreases in molecular weight [16,54] and strength [49,52]. Figure 9 shows the relationship between gamma radiation dosage and molecular weight [16]. In one study, the initial flexural strength of self-reinforced PGA rods decreased from 370 MPa to 300 MPa after a 2.5 Mrad dose of gamma irradiation [49].

A new program of steam sterilization has been reported by Rozema et al. that resulted in slight decreases in molecular weight and minimal changes in the mechanical properties of L-PLA when compared with unsterilized samples and other steam-sterilized samples subjected to different procedures [55].

IV. FABRICATION TECHNIQUES

Polymers are categorized into two basic types: (1) thermosetting and (2) thermoplastic. Thermosetting polymers do not soften when heated, but instead cure by forming cross-links. Thermoplastic polymers, which include PLA and PGA, soften and melt on heating and can be formed using several techniques, such as injection molding, compression molding, and extrusion. For orthopedic applications, PLA-PGA implants are manufactured as screws, plates, rods and scaffolds; the different methods to produce these devices are described below.

A. Melting, Extrusion, and Dry Spinning of Fibers

Once the synthesis of the polymers has been accomplished, melting, extrusion, dry spinning, hot drawing, or a combination of these processes can be applied to form fibers. Melt-spun L-PLA fibers, for example, are the starting materials for the construction of implants of various shapes and textures [56]. For this process, the polymer is melt spun in air, extruded, and hot drawn at various temperatures, followed by heat treatment. Melt spinning of acetylated L-PLA produces fibers with tensile strengths comparable to conventional polymers and with high degradation resistance [56]. Fibers of L-PLA can also be produced by hot drawing of the melt-spun and solution-spun fibers [18].

Higher molecular weight (300–500 kDa) L-PLA fibers can be produced by extrusion of the polymer solution at 110°C through a 1-mm diameter capillary. Lower molecular weight (below 300 kDa) L-PLA fibers can be melt extruded at 180°C.

Solution-spun fibers exhibit superior tensile properties compared with melt-spun fibers [18]. Furthermore, the tensile strength of fibers has been found to be strongly dependent on the drawing temperature.

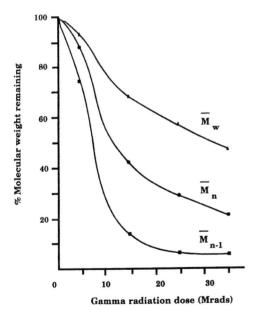

Figure 9 Effect of gamma sterilization on the molecular weight distribution of Dexon® (polyglycolic acid) sutures. (From Ref. 16.)

B. Sutures

Starting in the 1960s, one of the first medical applications of PLA and PGA polymer materials was for biodegradable sutures. In order to manufacture sutures, fibers are first produced. After synthesizing the PLA-PGA materials, the polymer is converted into fibers by melt extrusion, stretching, and heat setting [44]. For example, Vicryl, a copolymer of 90% PGA and 10% PLA, is produced by melting and extruding the polymer into fibers under specific temperature and pressure. The fibers are then braided to make the suture [57].

C. Solution Casting

For applications in which mechanical strength is not very critical, the implant device can be made by solution casting or by precipitation from solution. Schmitz and Hollinger fabricated implants for calvarial defects by solubilizing a PLA-PGA copolymer in chloroform, precipitating it with methanol, and combining with freeze-dried bone [24]. The composite was molded into the required shape and cured for 24 hours at 45°–48°C. After sterilization with ethylene oxide at room temperature, the implants were lyophilized to remove ethylene oxide residuals.

Along the same guidelines, biodegradable implants designed to assist in the repair of defects in cartilage were constructed using a 50:50 PLA-PGA copolymer. This polymer was dissolved in acetone and precipitated in ethanol. The resulting gummy precipitate was packed into Teflon molds and placed under 25 mTorr vacuum at room temperature [5]. At the end of the curing period, the implants were removed from the mold and stored in a lyophilizer until needed.

D. Emulsion-Solvent Technique

An emulsion-solvent evaporation technique can be used to prepare DL-PLA microspheres [58]. In this technique, the polymer is first dissolved in acetone and emulsified in polyvinyl alcohol solution. Microsphere formation is then achieved through a phase-separation process due to evaporation of the solvent. This process allows control of the size distribution of the microspheres, therefore making them more effective in reaching distal locations of the body.

E. Solvent-Casting Particulate Leaching

Three-dimensional scaffolds and foams have been fabricated using fiber bonding and solvent-casting particulate-leaching techniques [59,60]. As a first step in the novel fiber-bonding method, a solution of L-PLA in methylene chloride was poured in a dish containing PGA fibers as a nonwoven mesh. The solvent was allowed to precipitate, after which the composite was heated to 195°C, welding the fibers together at contact points. The L-PLA was then dissolved away to yield a three-dimensional scaffold [59].

To prepare the L-PLA foams, granular sodium chloride particles (100–500 μm) were added to a solution of L-PLA in chloroform, with the salt to polymer ratio at 9:1. The vortexed dispersion was cast in a petri dish, loosely covered, and the chloroform allowed to evaporate for 48 hours. Vacuum drying was used to remove any residual amounts of chloroform. The salt was leached out by immersing the scaffold in distilled water for 72 hours, during which time the water was changed every 6 hours. The scaffold was air dried for 24 hours, vacuum dried for 48 hours, and stored in a desiccator under vacuum until use [60].

Three-dimensional biodegradable polymer foams can also be constructed using a membrane lamination technique [59] that allows the replication of anatomical shapes and also the production of highly porous materials. This solvent-casting particulate-leaching technique allows the design and manufacture of biodegradable foams for applications in organ and tissue regeneration and reconstruction.

F. Gel Casting

Coombes and Heckman have described a procedure for making solid and microporous materials from resorbable, lactide-glycolide polymers using a gel-casting technique [61]. The basic steps involved in this gel-casting technique are (1) dissolving the polymer in a solvent; (2) casting the solution in a mold; (3) forming a gel *in situ*; (4) removing the shaped gel from the mold; (5) using solvent-exchange techniques to precipitate the polymer in shape; and (6) drying the polymer to obtain relatively thick, solid material sections (Fig. 10). This method enables production of thick-section solid and microporous materials, and blending of polymers and particulate fillers.

A similar gel-casting procedure was used to manufacture implants from a 50:50 PLA-PGA copolymer by Agrawal et al. [23,62]. Implants were prepared by dissolving the copolymer in acetone heated to 45°C. The polymer solution was then poured into a mold and allowed to stand covered for 24 hours, allowing it to gel. After the first hour of gelling, the implant was covered with acetone to prevent excess evaporation. After 24 hours, the implant was removed from the mold to start a four-day solvent-exchange process using mixtures of varying ratios of acetone, methanol, and water. This process precipitated the polymer and extracted the acetone from the implant.

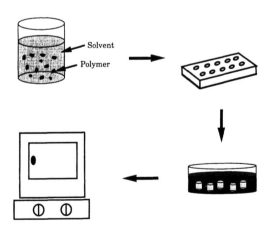

Figure 10 Schematic flowchart showing the basic steps involved in the gel-casting method: the polymer is dissolved in a solvent; the polymer solution is poured into the mold for casting and allowed to gel; implants are removed from the mold and the solvent-exchange process is started; after all solvents are extracted, polymer implants are dried in air and under ambient conditions.

H. Direct Machining, Compression Molding, Injection Molding, and Direct Compression

Biodegradable PLA-PGA implants can be directly machined from polymer blocks. For example, to fabricate plates and screws for internal fracture fixation, devices can be machined from high molecular weight L-PLA [63]. Compression molding also has been used to form L-PLA into plates and screws for internal fracture fixation [63]. The process of compression molding uses pressure and heat to fuse polymer materials into shape. Direct compression in the absence of heat or solvents has been utilized to produce biodegradable pellets using low molecular weight poly (DL-lactide) or high molecular weight L-PLA as a drug delivery system [64]. It was postulated that the energy imparted during compression was sufficient to cause fusion of the low molecular weight PLA.

Injection molding, on the other hand, forces melted polymer into a cooled mold under pressure and allows the fabrication of intricate shapes. Injection molding is often used to form polymers into detailed designs of rods, screws, plates, or other implant shapes for orthopedic applications. The properties of injection-molded products, such as the degree of polymer chain orientation and crystalline/amorphous ratio, depend on the precise details of the molding operation as well as any thermal treatments used after molding [65].

I. Composites/Reinforcing

PLA and PGA polymers and copolymers alone may not provide enough strength for high-load-bearing applications such as fracture fixation devices, so many researchers have attempted to reinforce the polymers with fibers. These fibers include carbon, PLA, and PGA [48,51,66]. For example, biodegradable PGA fibers were embedded within a L-PLA matrix to produce internal fixation plates, which showed promising mechanical characteristics [48]. Fabrication of a composite material, for which a PGA-reinforcing fabric was embedded in a L-PLA matrix, is possible by compression molding because of the difference in melting points between PGA (228°C) and L-PLA (178°C) [66]. Vainionpää et al. reinforced a PGA-PLA copolymer with 7% carbon parallel fibers by using compression molding [67]. Some of these composite plates were sputter coated with a thin layer of gold to decrease their rate of biodegradation. An ultra-high-strength self-reinforced (SR) absorbable PGA composite, in which the polymeric reinforcing elements and the binding matrix have the same chemical element composition, has also been developed for fracture fixation (Fig. 11) [47,68]. These composites have a high degree of molecular orientation that renders them both stiff and strong in the longitudinal direction.

PLA-PGA fibers have also been used to reinforce polymers other than PLA or PGA. In a study by de Groot et al. [69], L-PLA fibers were used to reinforce biodegradable poly(urethane) implants for meniscal reconstruction. L-PLA fibers, which were prepared by dry spinning and hot drawing of the filaments, were layered in the polyurethane matrices. This methodology produced homogeneous porous materials that exhibited a reproducible morphology.

V. DEGRADATION

As stated above, PLA, PGA, and their copolymers undergo bulk degradation by random hydrolytic scission of their ester bonds. The rate of degradation, however, is determined by several factors, including configurational structure, copolymer ratio, crystallinity,

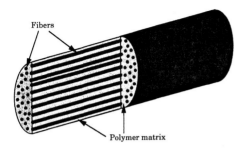

Figure 11   Diagram of a rod showing the self-reinforcing technique. Reinforcing fibers and polymeric matrix are both of the same chemical composition. (From Ref. 68.)

molecular weight, morphology, stresses, amount of residual monomer, and site of implantation. At the most fundamental level, the degradation is controlled by the degree of access that water molecules have to ester linkages in the polymeric chains. Parameters that influence this access also influence the degradation kinetics. For instance, Gilding and Reed have shown that for a PLA-PGA copolymer the water uptake increases with an increase in the PGA component [16]. This is because PLA is more hydrophobic than PGA due to the presence of a methyl group. As a result, PLA degrades at a slower rate.

The degree of crystallinity also influences the water uptake and affects degradation kinetics. Amorphous regions provide better access and mobility to the water molecules and as a result degrade at a faster rate. In general, the racemic mixture DL-PLA is less crystalline than L-PLA and degrades at a higher rate [70]. The degradation rate is also known to increase with an increase in the D-PLA isomer [71]. Gilding and Reed have shown that PLA-PGA copolymers with compositions containing the L-LA isomer with 25% to 70% of GA (glycolic acid) are amorphous in nature [16]. If the isomer used is DL-LA, then the copolymers are amorphous for the range 0% to 70% of GA.

An overall increase in crystallinity often accompanies the degradation process in semicrystalline PLA-PGA polymers [72,73]. This increase may be due to the fact that the amorphous segments are preferentially degraded at a faster rate, leaving behind a predominance of crystalline domains.

The role of monomers in the degradation process has been investigated by Nakamura et al. in a rabbit model [22]. Increasing the monomer content in various PLA-PGA implants increased their rate of degradation. The authors attributed this to the good solubility of the lactide monomer in water. They postulated that the monomer leaches out of the implants, thereby increasing their porosity and the surface in contact with water, and leading to an increase in degradation rate.

The ratio of PLA and PGA in a copolymer is another determinant of the degradation process. In an extensive study on the biodegradation of PLA, PGA, and their copolymers using carbon-14 and tritium-labeled polymers in rats, Miller et al. investigated the *in vivo* degradation of pellets of pure PLA and PGA, and PGA-PLA copolymers with ratios 75:25, 50:50, and 25:50 for periods up to 11 months [74]. They determined that the copolymers degraded at a faster rate than the hompolymers and that the 50:50 PGA:PLA copolymer degraded the fastest. Similar findings have been reported by others [15,22]. However, the degradation rate of PLA-PGA materials during any time period was a strong function of the starting molecular weight. In a recent *in vitro* study of 50:50

PLA-PGA implants (molecular weight 71 kDa), the molecular weight of the implants decreased 34% in 20 days [62].

Hydrolytic scission of the molecular chains of PLA-PGA polymers starts upon contact with water. This is clearly reflected in the change in molecular weight of the polymer [20]. However, the initial degradation products are too large to diffuse freely through the bulk material. Only after an extended period of time are they sufficiently reduced in size by hydrolysis to diffuse out as oligomers and result in a significant mass loss [75].

Such behavior has been reported by several studies [76–78]. For example, films of DL-PLA were tested subdermally in New Zealand rabbits by Pitt et al. [76]. They determined that the degradation of this polymer proceeded in two stages: first, there was only a decrease in molecular weight; subsequently, the polymer experienced weight loss with an increase in the rate of chain scission. It has been shown by Li and coworkers that the loss of material weight of DL-PLA implants does not begin until after five weeks of implant immersion in unbuffered saline solution [77].

These studies suggest that either the initial physical integrity of the implant is sufficient to prevent the degradation products of hydrolysis to exit the confines of the implant, or that the number of oligomers and monomers produced initially is very small. Therefore, although there is a rapid decrease in molecular weight [20], mass does not decline as quickly. Mass loss as a function of time typically follows a S-shaped pattern [20]. Schakenraad et al. have reported that an exponential increase in degradation occurs once the molecular weight of L-PLA decreases below 5000 daltons [79]. The time period necessary for significant mass loss is a function of the starting molecular weight of the polymer and the porosity of the specimen.

The inability of large-size degradation products to diffuse freely from within the bulk polymer leads to the interesting phenomenon of a degradation differential between the surface and interior of PLA-PGA specimens. This was first reported by Vert and his colleagues [73,77]. In a study on three different copolymers of PLA and PGA, they implanted compression-molded specimens intramuscularly in rats. It was determined that the degradation proceeds heterogenously and occurs faster in the center of the specimen and slower at the surface. Initially, degradation does occur rapidly at the surface because of the greater availability of water. However, the degradation products formed at the surface are easily dissolved in the surrounding fluid and leave the bulk polymer. On the other hand, the degradation products located within the specimen take longer to diffuse out because of their high molecular weight and large size. As a result, the concentration of carboxylic end groups increases in the implant center, creating an acidic environment that catalyzes ester breakdown and accelerates the hydrolytic scission. This phenomenon was also confirmed in an *in vitro* study [77]. This finding implies that larger implants may be subject to greater degradation differentials and may not behave like homogeneous materials.

The anatomical site of PLA-PGA implants also plays a role in determining the degradation kinetics [43,80]. For example, in a study reported by Matsusue et al. [43], L-PLA rods were surgically placed in the dorsal subcutaneous tissue and in the femoral medullary cavities of rabbits. The molecular weight was reduced to the greatest extent in the medullary cavity. The authors attributed this to a combination of the presence of stresses and enzymes.

Implants placed at sites where they are exposed to stresses degrade at a faster rate [6,45]. It is possible that under the action of stresses the implants develop cracks, which increases the effective surface exposed to water and results in enhanced degradation.

Also, stress cracks reduce the effective strength of the implants. The degradation kinetics may also be affected by fatigue loading, which might result in mechanochemically accelerated degradation of the polymer. In an *in vitro* study, Agrawal et al. have shown that implants fabricated from a 50:50 PLA-PGA copolymer biodegrade at an accelerated rate if subjected to ultrasonic irradiation [81]. It is possible that this increase is caused by mechanochemistry, although mechanochemical effects have not been decisively demonstrated in the case of PLA and PGA copolymers.

The degradation of PLA-PGA polymers is catalyzed by an acidic environment. This issue can be of concern when PLA-PGA devices are implanted at sites where the body is unable to flush away the degradation products at a steady rate, leading to a build up of carboxylic end groups. Such a scenario will result in an accelerated degradation process, which would further add to the concentration of these groups. A significant change in the pH of the surrounding media can possibly result in adverse tissue reactions [78].

VI. TOXICITY/BIOCOMPATIBILITY

Successful implant devices, permanent or biodegradable, should not cause any significant systemic or local reactions. Furthermore, biodegradable implants should degrade completely without inducing adverse effects on tissue healing. Local tissue response to biodegradable polymers depends on the rate of degradation, biocompatibility of the material components, and biocompatibility of the degradation products of the polymer. In order to evaluate toxicity and biocompatibility of PLA and PGA, both *in vivo* and *in vitro* tests have been utilized. Table 4 chronologically lists the biocompatibility/toxicity tests performed on PGA/PLA since 1966, and includes the test site locations and the general results observed.

A.   *In Vivo* Evaluation

1.   *Bone and Cartilage*

The advancing development of biodegradable polymeric implants and their minimal tissue reaction as sutures encouraged the use of these biomaterials for bone and cartilage replacement and fixation. As bulk polymers, PLA and L-PLA have shown good biocompatibility with bone [43,71]. However, once degradation is in full progress, small degraded L-PLA particles have been found in lymph nodes [82] and biocompatibility decreased [78]. To increase cement concentration in total hip replacement through intramedullary plugging, PLA was used as a plug in the femoral canal of sheep and satisfactory tissue biocompatibility was observed [71]. Klompmaker et al. used L-PLA to repair artificial meniscal lesions [83]. Even though the implant was found to guide the vascular repair tissues into the defect, tissue reaction included the presence of macrophages, fibroblasts, giant cells, and lymphocytes. This occurrence was described as chronic inflammation.

In contrast, no inflammatory or foreign-body reaction was observed histologically in the medullary cavity of rabbit femurs in response to ultra-high-strength L-PLA rods for up to one year [43]. Verheyen et al. compared various implant materials placed in the diaphyseal area of a goat femur [82]. The L-PLA polymer appeared to encourage macrophage-like cells to be formed in the lymph nodes. Crystal-like particles from the L-PLA were found in the nodes. To test biocompatibility as a function of biodegradation of L-PLA copolymers, chambers containing L-PLA were implanted in canine femurs and

Table 4 Biocompatibility/Toxicity Testing of Polyglycolic Acid and Polylactic Acid in Chronological Order

Year	Reference no.	Material	Location	Application	Results
1966	3	PLA	Guinea pigs and rats	Suture	Nontoxic and nontissue reactive
1970	118	PGA	Rabbits	Suture	Less reaction than catgut, silk, or Dacron
1971	44	PGA	Rabbits	Suture	Less inflammation than catgut
1971	119	PLA	Rat muscle	Suture	Degraded suture induced giant cell reaction
1971	120	PLA, PGA	Rabbits	Suture	Acceptable soft tissue reactions
1971	121	PLA	Monkeys	Suture	Minimal inflammatory response
1971	4	DL-PLA	Monkeys	Suture	Tissue response similar to controls
1973	122	PLA, PGA	Rat abdomen	Soft tissue	High degree of biocompatibility
1974	57	PGA, Vicryl	Humans	Suture	Vicryl tissue reaction not appreciable
1976	103	PLA, PGA	Mice and *in vitro*	Soft tissue	No foreign body reaction
1977	88	PLA, PGA	Rat tibia	Bone repair	Very tissue tolerant, little foreign-body reaction
1977	74	PLA, PGA	Rats	Soft tissue	High degree of biocompatibility
1978	123	Vicryl	Humans	Suture	Vicryl response similar to silk
1981	124	PGA	Dogs	Suture	Initial reaction intense, chronically mild
1981	125	PGA	Pigs	Suture	Negligible inflammation
1982	48	L-PLA, PGA	Rat tibia	Fracture fixation	Promising results
1983	71	PLA	Sheep femur	Bone repair	Satisfactory tissue compatibility
1983	89	PLA, PGA	Rat tibia	Bone repair	No adverse tissue host responses
1983	126	PGA, Vicryl	Rat muscle	Suture	Mild reaction
1986	117	PLA, HA	Rats	Bone repair	HA encouraged new bone formation

Year	Ref.	Material	Subject	Application	Result
1986	97	PLA, PGA	Rat muscle	Soft tissue	Slight reaction after 480 days
1987	63	L-PLA	Dogs, sheep, and in vitro	Fracture fixation	Well tolerated, increased cellular activity
1988	96	L-PLA	Rat soft tissue	Drug release	Very moderate foreign-body tissue reaction
1988	24	PLA, PGA	Rabbit calvarium	Bone repair	No adverse host-tissue responses
1990	84	PGA	Human	Fracture repair	Immunologically inert bimaterial
1990	79	L-PLA	Rat soft tissue	Drug release	L-PLA is tissue compatible
1990	100	PLA	In vitro cell culture		No reduction in cell proliferation
1991	30	PGA	Human	Fracture fixation	Foreign-body osteolytic reaction
1991	107	PGA, PLA	Human	Dermis	No inflammation
1991	99	PGA	Rats	Abdomen wall	No intrinsic bacteriocidal or bacteriostatic activity
1991	98	PGA	Human	Intestine	Well tolerated
1991	94	PLA	Rats	Soft tissue	Sufficient biocompatibility, well tolerated
1991	93	PGA	Human	Fracture fixation	No infection or foreign-body reaction
1991	83	L-PLA	Human	Meniscal repair	Chronic inflammation
1991	70	L-PLA, DL-PLA	Rat	Bone fixation	No inflammation or foreign-body reaction
1991	127	PGA	Human	Suture	PGA less infection than nylon
1991	127	PLA	Rabbit	Articular defects	Well tolerated, minimal inflammatory response
1992	12	PLA, PGA	Rabbit	Articular defects	Good long-term compatibility
1992	91	PGA	Rabbit femur	Fracture fixation	No contraindications for clinical application PGA
1992	11	PGA	Human	Fracture fixation	Variable tissue response
1992	8	PLA, PGA	In vitro		Can produce toxic solutions
1992	87	PGA	Human	Fracture fixation	Osteolysis present, no foreign-body reaction
1992	65	PLA, PGA	In vivo		No adverse response after complete degradation
1992	128	PLA, PGA	Rabbit cornea	Soft tissue	PLA nontoxic and safe, PGA some toxicity
1992	43	L-PLA	Rabbit femur and in vitro	Bone repair	No inflammatory or foreign body reaction

Table 4 Continued

Year	Reference no.	Material	Location	Application	Results
1992	55	L-PLA	Rats	Soft tissue	Some cellular reaction
1992	101	PLA	In vitro cell culture		Satisfactory biocompatibility
1993	5	PLA, PGA	Rabbits	Articular defects	No infection or inflammatory cells
1993	31	PGA	Human	Fracture fixation	No adverse clinical effects
1993	95	L-PLA	Mice	Soft tissue	L-PLA particles cause cell damage and lesion
1993	109	PGA	In vitro	Cell carrier	Hepatocytes attach to PGA mesh
1993	92	PLA, PGA, L-PLA	Rabbit femur	Fracture fixation	Insignificant inflammatory response
1993	129	L-PLA, PLA	Rat abdomen	Soft tissue	Excellent biocompatibility of PLA, larger reaction of PGA
1993	78	L-PLA	Dog femur	Bone repair	L-PLA particles induce foreign-body reaction
1993	82	L-PLA	Goat femur	Bone repair	L-PLA debris found in lymph nodes

compared to stainless steel coupons [78]. It was found that once degradation was in full progress, biocompatibility decreased, probably due to released small L-PLA particles (less than 2 μm) that induced a foreign-body inflammatory reaction and bone resorption.

Toxicity/biocompatibility testing of PGA implants for bone repair has shown some adverse effects, but, overall, long-term biocompatibility has been satisfactory. For example, to evaluate the human immunobiological response to the PGA implants, Santavirta et al. performed cytological analysis of materials aspirated from malleolar fracture repair effusions developed around a PGA implant [84]. Even though inflammatory monocytes were observed, overall it was concluded that PGA is an immunologically inert biomaterial.

Over the last few years, Böstman and coworkers have reported foreign-body reactions to PGA rods for fracture fixation. Absorbable polyglycolide rods used to treat malleolar fractures were noted initially to elicit a discharging foreign-body osteolytic reaction that lasted for approximately 10 weeks [30,85]. After one year, the bone structure was restored, and the noted foreign-body reactions in response to the PGA rods were at least partially attributed to the use of an aromatic quinone dye [86].

To study degradation in cancellous bone, this research group also used absorbable PGA screws to fix fractures of transverse distal femoral osteotomies in rabbits. Tissue response to the implant and implant replacement with bone were variable and resulted in relatively large defects in the bone [11]. In one of their most recent studies, they demonstrated a successful use of PGA pins for internal fixation of pediatric fractures without any adverse clinical effects [31]. It was thus speculated that other physicochemical factors were responsible for the inflammation observed in this group's clinical studies. Similar to reports by Böstman and coworkers [30,85,86], osteolysis around PGA pins used for pediatric fracture fixation have been observed [87]. However, subcutaneous foreign-body reaction or a discharging sinus were not found.

In order to match the degradation rates of PLA and PGA with the strength requirements and support needed during the bone healing period, PLA-PGA copolymers are often used. With regard to toxicity/biocompatibility, PLA-PGA copolymer implants for the repair of bony wounds in rat tibia have been found to be extremely tissue tolerant with little inflammatory or foreign-body reaction [88], and therefore are described as having a high degree of biocompatibility as bone implants with no adverse host tissue responses [89]. In addition, PLA-PGA copolymer implants placed in osseous defects in rat tibias were found to display an accelerated rate of healing in the polymer-treated defects compared with control bony wounds [89].

To repair articular cartilage defects in rabbits, a PLA matrix was used and was found to be well tolerated by all animals with minimal or no inflammatory response; it also showed a high biological acceptability rate and successful growth of neocartilage [90]. Similarly, for reconstruction of rabbit knee articular defects, PLA-PGA copolymer implants resulted in an absence of infection after complete degradation, and no inflammatory cells were found to be present [12].

Self-reinforced PGA screws have been used to fix a transverse transcondylar osteotomy of the distal femur. It was observed that giant cell count was highest at 80 days postimplantation, but overall no contraindications for the clinical application of PGA implants was concluded [91]. Similarly, inflammatory response to either self-reinforced PGA or L-PLA screws for fixation of transverse distal femoral osteotomies in rabbits was observed to be insignificant [92]. Self-reinforced PGA rods have also showed no evidence of infection or symptomatic foreign-body reaction [93].

2. Soft Tissue and Muscle

The biocompatibility and toxicity of PLA-PGA materials have also been evaluated in numerous studies in soft tissues and muscle. Biodegradable PLA membranes, produced by the solution-casting method, were placed transcutaneously in rats [94]. The PLA material exhibited sufficient biocompatibility and was well tolerated. In order to test the effect of L-PLA degradation on biocompatibility, Rozema et al. predegraded L-PLA and subcutaneously implanted this material in rats [55]. Histological results showed fibrous encapsulation, with macrophages and giant cells covering the smaller particles. The giant cells showing L-PLA particles internally had several swollen mitochondrial matrices, which could suggest active digestion of lactates.

Similar histological observations were reported by Lam et al. [95], who injected L-PLA particles intraperitoneally in mice. After a 7-day observation period, it was found that phagocytized L-PLA particles cause cell damage and sometimes lead to cell death. For drug-release applications, several authors have found that L-PLA causes slight foreign-body reactions, but is tissue compatible [79,96,97].

In vivo testing of PGA was evaluated by Devereux et al. [98], who implanted PGA into the peritoneal cavity of rats and observed that, although this material had no intrinsic bacteriocidal or bacteriostatic activity, it appeared to induce functional activation of leukocytes and thus appeared to stimulate inflammatory response. On the contrary, another study by Devereux and coworkers showed that a PGA mesh used in an intestinal procedure was well tolerated and demonstrated the absence of infections [99]. Gibbons examined various PLA-PGA materials and found that, as material strength decreases, small particles break off from the polymer are subsequently phagocytized by macrophages and multinucleated giant cells [65]. When all the materials and particulate have resorbed, the macrophage and multinucleated giant cells disappear. It was concluded that no adverse biological responses occur in response to these biomaterials, especially once complete degradation has been achieved. In general, if the volume of polymer material is small, the biological response appears to be mild and results in no long-lasting effects.

B. *In Vitro* Evaluation

As an alternative to *in vivo* testing of the toxicity and biocompatibility of biomaterials, *in vitro* testing has been applied. Overall, PLAs have shown satisfactory biocompatiblity, even though some reduction of cell proliferation has been observed. van Sliedregt et al. showed that the molecular weight of PLA does not seem to influence cell growth on the polymer's surface, although a reduction in cell proliferation was observed [100]. Later, van Sliedregt et al. tested various polylactides *in vitro* using cell cultures of epithelial cells, fibroblasts, and osteosarcoma cells by culturing them on films and by culturing cells with artificial aging media (based on the polymer degradation products) [101]. It was concluded that the PLAs exhibited satisfactory biocompatibility.

Daniels et al. tested the toxicity of absorbable polymers proposed for fracture fixation devices by incubating them at 37°C in buffer for various time periods and evaluating their solution toxicity using a bacterial bioluminescence toxicity assay [102]. Results showed that PGA and PLA can produce toxic solutions, probably due to accumulation of the acid being produced.

Mikos et al. prepared PGA-bonded fiber structures as a membrane for use as a transplantation device [59]. Hepatocytes were cultured on the mesh membrane and found to attach to the PGA meshes primarily as individual isolated cells with a high degree of interaction between hepatocytes and fibers observed 18 hours after plating (Fig. 12).

C. *In Vivo* and *In Vitro* Comparison

Even though *in vitro* experiments are designed to emulate *in vivo* conditions, the issue of the comparability of *in vivo* and *in vitro* results often arises. To address this, rods and beads made of lactide and glycolide polymers intended for use as drug delivery systems were evaluated *in vivo* subcutaneously in mice for up to 6 months and *in vitro* by measuring the release rate of a chemical [103]. No foreign-body reaction to the implants was observed in the animals and a one-to-one correlation between *in vitro* and *in vivo* performance was observed. Leenslag et al. examined both *in vivo* and *in vitro* degradation [104]. Up to 39 weeks, L-PLA implants were very well tolerated by the body with no chronic inflammation. A thin fibrous layer close to the L-PLA implant and increased cellular activity near the site were observed.

VII. CURRENT APPLICATIONS AND FUTURE DIRECTIONS IN ORTHOPEDICS

A plethora of experimental applications of PLA-PGA biomaterials have been examining the suitability of using these biodegradable devices in solving difficult orthopedic problems such as osteotomy fixation, articular cartilage and meniscal repair, ligament and tendon reconstructions, and substitutes of autologous bone fills. Furthermore, a biode-

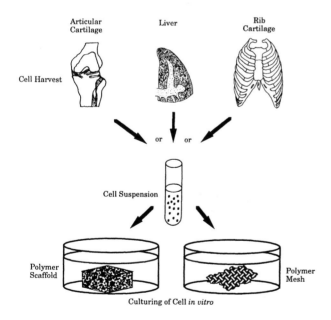

Figure 12 *In vitro* testing of polymer biocompatibility/toxicity. Cells are harvested from various tissues (e.g., articular cartilage, liver, rib cartilage). They are then isolated, suspended in media, plated, and cultured on the polymer implant devices, such as 3-D (three-dimensional) scaffolds or woven meshes.

gradable implant system that provides a steady, controlled release of drugs or bioactive factors can be an attractive delivery vehicle for substances that can enhance repair processes in the musculoskeletal system.

Biological repair of osteochondral defects remains one of the most intricate orthopedic problems, despite several techniques and theories attempting to produce healing of articular cartilage. Self-reinforced PGA rods were used in 1991 to regenerate cartilage in rabbit perichondrium, but foreign-body reaction was seen [105]. To achieve repair of large articular cartilage defects, PLA scaffolds with and without periosteal grafts were used in the rabbit knee with satisfactory results [90]. It was postulated that PLA was a suitable carrier of periosteal grafts.

In an effort to solve the elusive problem of osteochondral repair, the use of PLA-PGA copolymeric implants as carriers of growth factors has been investigated [5,12]. It was shown that the PLA-PGA biomaterial could be designed into a novel biocompatible scaffold that can be used as a carrier of either demineralized bone matrix powder [12] or transforming growth factor beta [5] and promote effective osteochondral healing. The objective of using PLA-PGA copolymers is to provide a scaffolding system that allows for new tissue ingrowth while exhibiting initial physical characteristics compatible with adjacent tissues [106]. The implant/carrier is furthermore expected to degrade and release its contents in a time-dependent fashion. Morphological and scanning electron microscopic examinations have demonstrated that the implant degrades in a gradual, extended fashion such that by 10 weeks it is completely dissolved.

PLA-PGA biodegradable polymers can also be used as graft devices on which cells are cultured. The devices plus cells can then be implanted at different sites, such as dermal tissue, bone, or soft tissue. The earliest and most frequent application of this concept has been in skin regeneration. For example, human fibroblasts were cultured on PGA meshes, grown to confluency for two weeks, and placed on mice to recreate the dermis; vascularization and epithelialization of these grafts was noted [107]. Hansbrough et al. incorporated confluent cultured fibroblasts in PGA meshes to be used as a living dermal replacement [108]. They found that after implantation in rats the grafts became vascularized and resulted in formation of organized tissue that resembled dermis. Mikos et al. used a PLA composite as a porous membrane on which hepatocytes were grown [109]. They demonstrated that it is possible to grow cells *in vitro* on these scaffolds prior to implantation.

This innovative concept of seeding cells onto PLA polymers has been applied in the treatment of osteochondral defects. Freed et al. studied neocartilage formation *in vivo* and *in vitro* using scaffolds made of PGA and porous L-PLA manufactured as described above [60]. *In vitro*, neocartilage formed on the polymer scaffold and cell growth rate was twice as high on the PGA as on L-PLA. Furthermore, cells grown on PGA produced sulfated glycosaminoglycan at a good rate. *In vivo* neocartilage was formed and the scaffold was maintained from 1–6 months. Overall, the materials were found to be suitable for regenerating neocartilage.

In meniscal reconstruction and repair, a porous L-PLA scaffold was used by Klompmaker et al. to repair meniscal lesions [83]. It was found that the biodegradable implant, which exhibited a microporous and macroporous (size 100–300 μm) structure, could provide guidance for vascular ingrowth and fibrocartilaginous tissue into the defect.

Parsons et al. reported the use of a PLA composite with filamentous carbon fibers to repair and replace tendons and ligaments [110]. They showed that when this composite was used as a patellar tendon replacement and a medial collateral ligament replacement in dogs and as an Achilles tendon replacement in rabbits, it acted as a mechanically degrading scaffold for new tissue ingrowth.

As indicated above, biodegradable fixation using PLA-PGA copolymers has been utilized in clinical studies to treat osteotomies, mostly in Europe. In the early years of research and development, successful fixation of long-bone fractures was not consistently achieved through the use of biodegradable polymers [48,111]. However, in recent years various permutations of PLA-PGA implants have been attempted in experimental animal studies. Vainionpää et al. used PLA-PGA copolymers reinforced with carbon fibers and overlaid in gold to treat bone fractures in rabbits [67]. Rapid polymer degradation, however, resulted in nonunions and fixation was deemed unsatisfactory. Hollinger and Schmitz used a PLA-PGA copolymer in combination with a proteolipid to treat bone discontinuities in canine mandibles [112]; histomorphometric evaluations revealed increased bone repair. In a pilot study, Bos et al. made bone plates and screws from high molecular weight L-PLA for fixation of mandibular fractures in sheep and demonstrated that L-PLA plates and screws provided stability and enabled normal fracture healing [6]. Antikainen et al. combined PLA and PGA with hydroxyapatite as a bone graft for rabbit skulls [113]. After 12 months, the implant became attached to the surrounding bone and no adverse effects or harmful reactions in adjacent tissues were observed.

To stimulate bone growth, researchers have also incorporated growth factors into biodegradable polymers. Turk et al. used a copolymer of PLA-PGA to deliver a calf-bone osteoinductive protein for the repair of large cranial defects in rabbits [114]. They demonstrated that protein-treated implants exhibited significantly greater amounts of ingrowth compared to untreated groups. Implant disks made of poly DL-lactide-co-glycolide with lyophilized bone matrix extracts were studied *in vitro* [115]. Most of the biological activity was found to occur during the first week of culture. It was also shown that less than half of the growth factor could retain biological activity.

Langer has reviewed various new drug delivery approaches and their methods of releasing drugs into the body [116]. One method that incorporates drugs with polymers is microparticulates/vesicles containing various drugs. For example, degradable PLA-PGA copolymers can be used for controlled release of insulin, growth factors, and angiogenesis inhibitors. Agrawal et al. have developed a microporous implant to deliver osteogenic proteins to sites of fracture non-unions [23]. In another study, a composite of DL-PLA and oligomer dideoxykanamycin B was used as a biodegradable delivery system of antibiotics to treat osteomyelitis in the rabbit femur [14]. No systemic side effects were shown and successful repair was observed after 9 weeks of implantation.

Frequently in orthopedics it is necessary to fill bony defects caused by pathologic or traumatic episodes. The golden standard is to use autogenous grafts to fill these defects. It has been proposed that biodegradable materials can be used that can eventually be replaced by newly formed bone. Higashi et al. utilized a biodegradable implant made of PLA and hydroxyapatite to replace cancellous bone in the rat [117]. They observed that histologically this implant was resorbed and replaced by new bone.

This review of *in vivo* and *in vitro* orthopedic applications of PLA-PGA suggests that these biodegradable biomaterials appear to be promising for future uses for which output of bioactive agents can be controlled and appropriately delivered in response to the needs of the body and the repair process. Thus, repair problems associated with pathophysiology or trauma of musculoskeletal tissues may be more manageable. In particular, these devices may be appropriately designed and used in tendino-ligamentous reconstruction and thus render tractable these perennially complex orthopedic problems. They can also continue to be used in bone fracture fixation, although the problems associated with noninfectious inflammatory responses persist. It is conceivable that these adverse responses may be better controlled or eliminated through improved device designs, better quality control of polymerizations so that impurities and free monomer acidity are kept below minimum levels, as well as regulation of the pH environment around the implant. The extensive experimental use of PLA-PGA materials suggests a promising future for their use as fixation devices, but also as delivery vehicles for pharmaceuticals, vaccines, growth factors, and other bioactive agents.

ACKNOWLEDGMENTS

The authors would like to thank Dave Stevens for his help with the manuscript.

REFERENCES

1. Lowe, C. E., U.S. Patent 2668162, 1954.
2. Schneider, A. K., Polymers of high melting lactide, U.S. Patent 2703316, 1955.
3. Kulkarni, R. K., K. C. Pani, C. Neuman, and F. Leonard, Polylactic acid for surgical implants, *Arch. Surg.*, 93:839–843 (November 1966).
4. Kulkarni, R. K., G. Moore, A. F. Hegyeli, and F. Leonard, Biodegradable poly(lactic acid) polymers, *Journal of Biomedical Material Research*, (5):169–181 (1971).
5. Athanasiou, K. A., R. C. Schenck, G. Constantinides, V. Sylvia, T. Aufdemorte, and B. D. Boyan, Biodegradable carriers of TGF-β in rabbit osteochondral defects, *Transactions of the Orthopaedic Research Society*, 18(1):288 (1993).
6. Bos, R. R. M., F. R. Rozema, G. Boering, A. J. Nijenhuis, A. J. Pennings, and H. W. B. Jansen, Bone-plates and screws of bioabsorbable poly (L-lactide) – An animal pilot study, *British Journal of Oral and Maxillofacial Surgery*, 27:467–476 (1989).
7. Böstman, O., S. Vainionpää, E. Hirvensalo, A. Mäkelä, K. Vihtonen, P. Törmälä, and P. Rokkanen, Biodegradable internal fixation for malleolar fractures, *Journal of Bone and Joint Surgery*, 69-B(4):615–619 (1987).
8. Daniels, A. U., M. K. O. Chang, and K. P. Andriano, Mechanical properties of biodegradable polymers and composites proposed for internal fixation of bone, *Journal of Applied Biomaterials*, (1):57–78 (1990).
9. Litsky, A. S., Clinical reviews: Bioabsorbable implants for orthopaedic fracture fixation, *Journal of Applied Biomaterials*, (4):109–111 (1993).
10. Böstman, O., E. Hirvensalo, S. Vainionpää, A. Mäkelä, K. Vihtonen, P. Törmälä, and R. Rokkanen, Ankle fractures treated using biodegradable internal fixation, *Clinical Orthopaedics and Related Research*, 238(1):195–203 (1989).
11. Böstman, O., U. Päivärinta, E. Partio, J. Vasenius, M. Manniner, and P. Rokkanen, Degradation and tissue replacement of an absorbable polyglycolide screw in the fixation of rabbit femoral osteotomies, *Journal of Bone and Joint Surgery*, 74-A(7):1021–1031 (August 1992).
12. Athanasiou, K. A., J. P. Schmitz, R. C. Schenck, M. Clem, T. Aufdemorte, and B. D. Boyan, The use of biodegradable implants for repairing large articular cartilage defects in the rabbit, *Transactions of the Orthopaedic Research Society*, 17(1):172 (1992).
13. Coombes, A. G. A., and J. D. Heckman, Gel casting of resorbable polymers 2. *In vitro* degradation of bone graft substitutes, *Biomaterials*, 13(5):297–307 (1992).
14. Wei, G., Y. Kotoura, M. Oka, T. Yamamuro, R. Wada, S.-H. Hyon, and Y. Ikado, A bioabsorbable delivery system for antibiotic treatment of osteomyelitis, *Journal of Bone and Joint Surgery*, 73-B(2):246–252 (1991).

15. Lewis, D. H., Controlled release of bioactive agents from lactide/glycolide polymers. In *Biodegradable Polymers as Drug Delivery Systems* (M. Chasin and R. Langer, eds.), Marcel Dekker, New York, 1-41 (1990).
16. Gilding, D. K., and A. M. Reed, Biodegradable polymers for use in surgery—Polyglycolic/poly (lactic acid) homo- and copolymers: 1, *Polymer*, (20):1459-1464 (1979).
17. Li, S., H. Garreau, and M. Vert, Structure-property relationship in the case of the degradation of massive poly (x-hydroxy acids) in aqueous media, Part 3: Influence of the morphology of poly (L-lactic acid), *Journal of Materials Science: Materials in Medicine*, (1):198-206 (1990).
18. Eling, B., S. Gogolewski, and A. J. Pennings, Biodegradable materials of poly(L-lactic acid): 1. Melt-spun and solution spun fibres, *Polymer*, (23):1587-1593 (1982).
19. Fukuzaki, H., et al., A new biodegradable copolymer of glycolic acid and lactones with relatively low molecular weight prepared by direct copolycondensation in the absense of catalysts, *Journal of Biomedical Materials Research*, 25:315-328 (1991).
20. Fukuzaki, H., M. Yoshida, M. Asano, and M. Kumakura, *In vivo* characteristics of high molecular weight copoly(L-lactide/glycolide) with S-type degradation pattern for application in drug delivery systems, *Biomaterials*, 12:433-437 (May 1991).
21. Böstman, O., Current concepts review: Absorbable implants for fixation of fractures, *Journal of Bone and Joint Surgery*, 73-A(1): 148-153 (1991).
22. Nakamura, T., S. Hitomi, S. Waranabe, Y. Shimizu, K. Jamshidi, S. H. Hyon, and Y. Ikada, Bioabsorption of polylactides with different molecular properties, *Journal of Biomedical Materials Research*, 23:1115-1130 (1989).
23. Agrawal, C. M., J. Best, B. D. Boyan, and J. D. Heckman, *In vitro* protein release characteristics of a bioabsorbable carrier of BMP for fracture non-unions. In *40th Annual Meeting of the Orthopaedic Research Society*, New Orleans, Louisiana, 1994.
24. Schmitz, J. P., and J. O. Hollinger, A preliminary study of the osteogenic potential of a biodegradable alloplastic-osteoinductive alloimplant, *Clinical Orthopaedics and Related Research*, 237:245-255 (1988).
25. Williams, D. F., and E. Mort, Enzyme-accelerated hydrolysis of polyglycolic acid, *Journal of Bioengineering*, (1):231-238 (1977).
26. Rokkanen, P., O. Böstman, S. Vainionpää, K. Vihtonen, P. Törmälä, J. Laiho, J. Kilpikari, and M. Tamminmäki, Biodegradable Implants in Fracture Fixation: Early Results of Treatment of Fractures of the Ankle, *Lancet*, (1):1422-1424 (1985).
27. Hirvensalo, E., Fracture fixation with biodegradable rods, *Acta. Orthop. Scand.*, 60(5): 601-606 (1989).
28. Böstman, O., E. Hirvensalo, J. Mäkinen, and P. Rokkanen, Foreign-body reactions to fracture fixation implants of biodegradable synthetic polymers, *Journal of Bone and Joint Surgery*, 72-B: 592-96 (1990).
29. Böstman, O., E. A. Mäkelä, P. Törmälä, and R. Rokkanen, Transphyseal fracture fixation using biodegradable pins, *Journal of Bone and Joint Surgery*, 71-B:706-707 (1989).
30. Böstman, O. M., Osteolytic changes accompanying degradation of absorbable fracture fixation implants, *Journal of Bone and Joint Surgery*, 73-B:679-682 (1991).
31. Böstman, O., E. A. Mäkelä, J. Södergard, E. Hirvensalo, P. Törmälä, and R. Rokkanen, Absorbable polyglycolide pins in internal fixation of fractures in children, *Journal of Pediatric Orthopaedics*, (13):242-245 (1993).
32. Yen, R. G., J. A. Giacopelli, D. P. Granoff, and R. J. Steinbroner, The biofix absorbable rod: A preliminary report, *Journal of the American Podiatric Medical Association*, 81(2): 62-66 (1991).
33. Gerbert, J., Effectiveness of absorbable fixation devices in Austin bunionectomies, *Journal of the American Podiatric Medical Association*, 82(4):189-195 (1992).
34. Parks, R. M., and G. Nelson, Complications with the use of bioabsorbable pins in the foot, *Journal of Foot and Ankle Surgery*, 32(2):153-161 (1993).

35. Hirvensalo, E., O. Böstman, S. Vainionpää, P. Törmälä, and P. Rokkanen, Biodegradable fixation in intraarticulate fractures of the elbow joint, *Acta Orthop. Scandinavica Supplementum*, 227:78–79 (1988).

36. Hoffmann, R., C. Krettek, N. Haas, and H. Tscherne, Die distale Radiusfraktur. Frakturstabilisierung mit biodegradablen Osteosynthes Stiften (Biofix), *Experimentelle Untersuchungen und erste klinische Erfahrungen*, 92:430–434 (1989).

37. Casteleyn, P. P., F. Handelberg, and P. Haentjens, Biodegradable rods versus Kirschner wire fixation of wrist fractures, *Journal of Bone and Joint Surgery*, 74B:858–861 (1992).

38. Bos, R. R. M., G. Boering, F. R. Rozema, and J. W. Leenslag, Resorbable poly(L-lactide) plates and screws for the fixation of zygomatic fractures, *Journal of Oral and Maxillofacial Surgery*, 45:751–753 (1987).

39. Gerlach, K. L., Treatment of zygomatic fractures with biodegradable poly(L-lactide) plates and screws. Clinical implant materials. In *Advances in Biomaterials* (G. Heimke, U. Soltesz, and A. C. J. Lee, eds.), Elsevier, Amsterdam, pp. 573–578 (1990).

40. Bergsma, E. J., F. R. Rozema, R. R. M. Bos, and W. C. deBruijn, Foreign body reactions to resorbable poly(L-lactide) bone plates and screws used for the fixation of unstable zygomatic fractures, *Journal of Oral and Maxillofacial Surgery*, 51:666–670 (1993).

41. Shalaby, S. W., Bioabsorbable polymers update, *Journal of Applied Biomaterials*, (3):73–74 (1992).

42. Engelberg, I., and J. Kohn, Physio-mechanical properties of degradable polymers used in medical applications: A comparative study, *Biomaterials*, 12(4):292–304 (1991).

43. Matsusue, Y., T. Yamamuro, M. Oka, Y. Shikinami, S.-H. Hyon, and Y. Ikada, *In vitro* and *in vivo* studies on bioabsorbable ultra-high-strength poly (L-lactide) rods, *Journal of Biomedical Materials Research*, 26:1553–1567 (1992).

44. Frazza, E. J., and E. E. Schmitt, A new absorbable suture, *Journal of Biomedical Materials Research Symposium*, (1):43–58 (1971).

45. Suuronen, R., T. Pohjonen, R. Taurio, P. Törmälä, L. Wessman, K. Rönkkö, and S. Vainionpää, Strength retention of self-reinforced poly-L-lactide screws and plates: An *in vivo* and *in vitro* study, *Journal of Materials Science: Materials in Medicine*, (3):426–431 (1992).

46. Birmingham Polymers, *Properties of Biodegradable Polymers*, Birmingham Polymers, Birmingham, AL, 1993.

47. Törmälä, P., J. Vasenius, S. Vainionpää, T. Pohjonen, P. Rokkanen, and J. Laiho, Ultra high strength absorbable self-reinforced polyglycolide (SR-PGA) composite rods for internal fixation of bone fractures: *In vitro* and *in vivo* study, *Journal of Biomedical Materials Research*, 25:1–22 (1991).

48. Christel, P., F. Chabot, J. L. Leray, C. Morin, and M. Vert, Biodegradable composites for internal fixation. In *Biomaterials 1980* (G. L. Winters, D. F. Gibbons, and H. Plenk, eds.), John Wiley and Sons, pp. 271–280 (1982).

49. Vainionpää, S., J. Kilpikari, J. Laiho, P. Helevirta, P. Rokkanen, and P. Törmälä, Strength and strength retention in vitro, of absorbable, self-reinforced polyglycolide (PGA) rods for fracture fixation, *Biomaterials*, 8:46–48 (January 1987).

50. Vasenius, J., S. Vainionpää, K. Vihtonen, A. Mäkelä, P. Rokkanen, M. Mero, and P. Törmälä, Comparison of *in vitro* hydrolysis, subcutaneous and intramedullary implantation to evaluate the strength retention of absorbable osteosynthesis implants, *Journal of Biomaterials*, 11(9):501–504 (1990).

51. Törmälä, P., S. Vainionpää, J. Kilpikari, and P. Rokkanen, The effects of fibre reinforcement and gold plating on the flexural and tensile strength of PGA/PLA copolymer materials *in vitro*, *Biomaterials*, 8(1):42–45 (1987).

52. Laitinen, O., P. Törmälä, R. Taurio, K. Skutnabb, K. Saarelainen, T. Iivonen, and S. Vainionpää, Mechanical properties of biodegradable ligament augmentation device of poly(L,-lactide) *in vitro* and *in vivo*, *Biomaterials*, 13(14):1012–1016 (1992).

53. Rozema, F. R., R. R. M. Bos, G. Boering, J. A. A. M. V. Asten, A. J. Nijenhuis, and A. J.

Pennings, The effects of different steam-sterilization programs on material properties of poly (L-lactide), *Journal of Applied Biomaterials*, 2:23-28 (1991).
54. Gupta, M. C., and V. G. Deshmukh, Radiation effects on poly(lactic acid), *Polymer*, 24: 827-830 (1983).
55. Rozema, F. R., R. R. M. Bos, G. Boering, A. J. Nijenhuis, A. J. Pennings, H. W. B. Jansen, and W. C. deBruijn, Tissue response to predegraded poly (L-lactide). In *Degradation Phenomena on Polymeric Biomaterials* (H. Plank, M. Dauner, and M. Renardy, eds.), Springer, New York, pp. 123-131 (1992).
56. Hyon, S.-H., K. Jamshidi, and Y. Ikada, Melt spinning of poly-L-lactide and hydrolysis of the fiber *in vitro*. In *Polymers as Biomaterials* (S. W. Shalaby, A. Hoffman, B. Ratner, and T. Horbett, eds.), Plenum Press, New York (1984).
57. Horton, C. E., J. E. Adamson, R. A. Mladick, and J. H. Carraway, Vicryl synthetic absorbable sutures, *American Surgeon*, 729-731 (December 1974).
58. Grandfils, C., R. Flandroy, W. Nihant, S. Barbette, R. Jerome, P. Teyssie, and A. Thibaut, Preparation of poly (D,L) lactide microspheres by emulsion-solvent evaporation, and their clinical applications as a convenient embolic properties, *Journal of Biomedical Materials Research*, 26:467-479 (1992).
59. Mikos, A. G., G. Sarakinos, S. M. Leite, J. P. Vacanti, and R. Langer, Laminated three-dimensional biodegradable foams for use in tissue engineering, *Biomaterials*, 14(5):323-330 (1993).
60. Freed, L. E., J. C. Marquis, A. Nohria, J. Emmanual, A. G. Mikos, and R. Langer, Neocartilage formation *in vitro* and *in vivo* using cells cultured on synthetic biodegradable polymers, *Journal of Biomedical Materials Research*, 27:11-23 (1993).
61. Coombes, A. G. A., and J. D. Heckman, Gel casting of resorbable polymers 1. Processing and applications, *Biomaterials*, 13(4):217-224 (1992).
62. Agrawal, C. M., and M. B. Kennedy, The effects of ultrasound irradiation on a biodegradable delivery system, *Transactions of the Society for Biomaterials*, 19:292 (1993).
63. Leenslag, J. W., A. J. Pennings, R. R. M. Bos, F. R. Rozema, and G. Boering, Resorbable materials of poly(L-lactide): VI. Plates and screws for internal fracture fixation, *Biomaterials*, 8(1):70-73 (1987).
64. Bodmeier, R., and H. Chan, Evaluation of biodegradable poly(lactide) pellets prepared by direct compression, *Journal of Pharmaceutical Sciences*, 78(10):819-822 (1989).
65. Gibbons, D. F., Tissue response to resorbable synthetic polymers In *Degradation Phenomena on Polymeric Biomaterials* (H. Plank, M. Dauner, and M. Renardy, eds.), Springer Verlag, New York, pp. 97-104 (1992).
66. Vert, M., F. Chabot, J. Leray, and P. Christel, Nouvelles pieces d'osteosynthese; leur preparation et leur application, French Patent appl. no. 78 29878, (1978).
67. Vainionpää, S., K. Vihtonen, M. Mero, H. Patiala, P. Rokkanen, J. Kilpikari, and P. Törmälä, Biodegradable fixation of rabbit osteotomies, *Acta. Orthopaedic Scand.*, 57:237-239 (1986).
68. Suuronen, R., Biodegradable fracture-fixation devices in maxillofacial surgery, *International Journal of Oral Maxillofacial Surgery*, 22:50-57 (1993).
69. deGroot, J. H., A. J. Nijenhuis, P. Bruin, A. J. Pennings, R. P. H. Veth, J. Klopmaker, and H. W. B. Jansen, Use of porous biodegradable polymer implants in meniscus reconstruction. 1) Preparation of porous biodegradable polyurethanes for the reconstruction of meniscus lesions, *Colloid and Polymer Science*, 268:1073-1081 (1990).
70. Majola, A., S. Vainionpää, K. Vihtonen, M. Mero, J. Vasenius, P. Törmälä, and P. Rokkanen, Absorption, biocompatibility and fixation properties of polylactic acid in bone tissue: An experimental study in rats, *Clinical Orthopaedics and Related Research*, 268(7):260-269 (1991).
71. Christel, P. S., M. Vert, F. Chabot, Y. Abols, and J. L. Leary, Polylactic acid for intramedullary plugging. In *Biomaterials and Biomechanics* (P. Ducheyne, ed.), Elsevier Science Publishers, Amsterdam (1983).

72. Pistner, H., D. R. Bendix, J. Muehling, and J. F. Reuther, Poly(L-lactide): A long term degradation study *in vivo*. Part III. Analytical characterization, *Biomaterials*, 14(4):291–298 (1993).

73. Therin, M., P. Christel, S. Li, H. Garreau, and M. Vert, *In vivo* degradation of massive poly(x-hydroxy acids): Validation of *in vitro* findings, *Biomaterials*, 13(9):594–600 (1992)

74. Miller, R. A., J. M. Brady, and D. E. Cutright, Degradation rates of oral resorbable implants (polylactates and polyglycolates): Rate modification with changes in PLA/PGA copolymer ratios, *Journal of Biomedical Materials Research*, 11:711–719 (1977).

75. Buckholz, B., Accelerated degradation test on resorbable polymers. In *Degradation Phenomena on Polymeric Biomaterials* (H. Plank, M. Dauner, and M. Renardy eds.), Springer-Verlag, New York, pp. 67–76 (1992).

76. Pitt, C. G., M. M. Gratzl, G. L. Kimmel, J. Surles, and A. Schindler, Aliphatic polyesters II: The degradation of poly (DL-lactide), poly (E-caprolactone), and their copolymers *in vivo*, *Journal of Biomaterials*, 2:215–220 (October 1981).

77. Li, S. M., H. Garreau, and M. Vert, Structure-property relationships in the case of the degradation of massive aliphatic poly-(alpha-hydroxy acids) in aqueous media; Part 1: Poly-(DL lactic acid), *Journal of Material Science: Materials in Medicine*, (1):123–130 (1990).

78. Suganuma, J., and H. Alexander, Biological response of intramedullay bone to poly-L-lactic acid, *Journal of Applied Biomaterials*, 4:13–27 (1993).

79. Schakenraad, J. M., M. J. Hardonk, J. Feijen, I. Molenaar, and P. Nieuwenhuis, Enzymatic activity toward poly (L-lactic acid) implants, *Journal of Biomedical Materials Research*, 24:529–545 (1990).

80. Scherer, M. A., H. J. Frueh, R. Ascherl, H. Mau, W. Siebels, and G. Bluemel, Kinetics of resorption of different suture materials depending on the implantation site and the species. In *Degradation Phenomena on Polymeric Biomaterials* (H. Plank, M. Dauner, and M. Renardy, eds.), Springer, New York, pp. 77–96 (1992).

81. Agrawal, C. M., M. E. Kennedy, and D. Micallef, The effects of ultrasound irradiation on a biodegradable copolymer of polylactic and polyglycolic acids, *Journal of Biomedical Materials Research*, 28:851–859 (1994).

82. Verheyen, C. C. P. M., J. R. deWijn, C. A. VanBlitterswijk, P. M. Rozing, and K. deGroot, Examination of efferent lymph nodes after 2 years of transcortical implantation of poly (L-lactide) containing plugs: A case report, *Journal of Biomedical Materials Research*, 27:1115–1118 (1993).

83. Klompmaker, J., H. W. B. Jansen, R. P. H. Veth, J. H. deGroot, A. J. Nijenhuis, and A. J. Pennings, Porous polymer implant for repair of meniscal lesions: A preliminary study in dogs, *Biomaterials*, 12(11):810–816 (1991).

84. Santavirta, S., Y. T. Konttinen, T. Saito, M. Gronblad, E. Partio, P. Kemppinen, and P. Rokkanen, Immune response to polyglycolic acid implants, *Journal of Bone and Joint Surgery*, 72-B:597–600 (1990).

85. Boštman, O. M., Intense granulomatous inflammatory lesions associated with absorbable internal fixation devices made of polyglycolide in ankle fractures, *Clinical Orthopaedics and Related Research*, 278:193–199 (May 1992).

86. Böstman, O., E. Partio, E. Hirvensalo, and P. Rokkanen, Foreign-body reactions to poly-glycolide screws, *Acta Orthop. Scand.*, 63(2):173–176 (1992).

87. Fraser, R. K., and W. G. Cole, Osteolysis after biodegradable pin fixation of fractures in children, *Journal of Bone and Joint Surgery*, 74-B:929–930 (November 1992).

88. Nelson, J. F., H. G. Stanford, and D. E. Cutright, Evaluation and comparisons of biodegradable substances as osteogenic agents, *Oral Surgery*, 43(6):836–843 (June 1977).

89. Hollinger, J. O., Preliminary report on the osteogenic potential of a biodegradable copolymer of polylactide (PLA) and polyglycolide (PGA), *Journal of Biomedical Materials Research*, 17:71–82 (1983).

90. vonSchroeder, H. P., M. Kwan, D. Amiel, and R. D. Coutts, The use of polylactic acid

matrix and periosteal grafts for the reconstruction of rabbit knee articular defects, *Journal of Biomedical Materials Research*, 25:329–339 (1991).

91. Böstman, O., U. Päivärinta, E. Partio, M. Manninen, J. Vasenius, A. Majola, and P. Rokkanen, The tissue-implant interface during degradation of absorbable polyglycolide fracture fixation screws in the rabbit femur, *Clinical Orthopaedics and Related Research*, 285: 263–272 (1992).

92. Päivärinta, U., O. Böstman, A. Majola, T. Toivonen, P. Törmälä, and P. Rokkanen, Intraosseous cellular response to biodegradable fracture fixation screws made of polyglycolide polylactide, *Archives of Orthopaedic and Trauma Surgery*, 112:71–74 (1993).

93. Hope, P. G., D. M. Williamson, C. J. Coates, and W. G. Cole, Biodegradable pin fixation of elbow fractures in children, *Journal of Bone and Joint Surgery*, 73:965–968 (November 1991).

94. Galgut, P., R. Pitrola, I. Waite, C. Doyle, and R. Smith, Histological evaluation of biodegradable and non-biodegradable membranes placed transcutaneously in rats, *Journal of Clinical Periodontol.*, 18:581–586 (1991).

95. Lam, K. H., J. M. Schakenraad, H. Esselbrugge, J. Feijen, and P. Nieuwenhuis, The effect of phagocytosis of poly (L-lactic acid) fragments on cellular morphology and viability, *Journal of Biomedical Materials Research*, 27:1569–1577 (1993).

96. Schakenraad, J. M., J. A. Oosterbaan, P. Nieuwenhuis, J. Molenaar, J. Olijslager, W. Potman, M. J. D. Eenink, and J. Feijen, Biodegradable hollow fibres for the controlled release of drugs, *Biomaterials*, 9:116–120 (January 1988).

97. Visscher, G. E., R. L. Robison, H. V. Maulding, J. W. Fong, J. E. Pearson, and G. J. Argentieri, Biodegradation of and tissue reaction to poly (DL-lactide) microcapsules, *Journal of Biomedical Materials Research*, 20:667–676 (1986).

98. Devereux, D. F., S. M. O'Connell, D. A. Spain, and F. M. Robertson, Peritoneal leukocyte response following placement of polyglycolide acid intestinal sling in patients with rectal carcinoma, *Dis. Colon Rectum*, 34(8):670–674 (1991).

99. Devereux, D. F., S. M. O'Connell, J. B. Liesch, M. Weinstein, and F. M. Robertson, Induction of leukocyte activation by meshes surgically implanted in the peritoneal cavity, *American Journal of Surgery*, 162:243–246 (September 1991).

100. van Sliedregt, A., C. A. van Blitterswijk, S. C. Hesseling, J. J. Grote, and K. deGroot, The effect of the molecular weight of polylactic acid on *in vitro* biocompatibility, *Advances in Biomaterials*, 9:207–212 (1990).

101. van Sliedregt, A., A. M. Radder, K. deGroot, and C. A. van Blitterswijk, *In vitro* biocompatibility testing of polylactides Part I: Proliferation of different cell types, *Journal of Materials Science: Materials in Medicine*, 3:365–370 (1992).

102. Daniels, A. U., M. S. Taylor, K. P. Andriano, and J. Heller, Toxicity of absorbable polymers proposed for fracture fixation devices, *38th Annual Meeting of the Orthopaedic Research Society*, February 17–20, Washington, DC, 1992.

103. Schwope, A. D., D. L. Wise, and J. F. Howes, Development of polylactic/glycolic acid delivery systems for use in treatment of narcotic addiction, *National Inst. of Drug Abuse Res. Monogr. Ser.*, (4):13–18 (January 1976).

104. Leenslag, J. W., A. J. Pennings, R. R. M. Bos, F. R. Rozema, and G. Boering, Resorbable materials of poly (L-lactide): VII. *In vivo* and *in vitro* degradation, *Biomaterials*, 8:311–314 (1987).

105. Ruuskanen, M. M., M. J. Kallioinen, O. I. Kaarela, J. A. Laiko, P. O. Törmälä, and T. J. Waris, The role of polyglycolic acid rods in the regeneration of cartilage from perichondrium in rabbits, *Scandinavian Journal of Plastic and Reconstructive Hand Surgery*, 25:15–18 (1991).

106. Athanasiou, K. A., A. R. Singhal, C. M. Agrawal, and B. D. Boyan, *In vitro* degradation and release characteristics of biodegradable implants containing trypsin inhibitor, *Clinical Orthopaedics and Related Research*, in review, (1994).

107. Cooper, M. L., J. F. Hansbrough, R. L. Spielvogel, R. Cohen, R. L. Bartel, and G. Naughton, *In vivo* optimization of a living dermal substitute employing cultured human fibroblasts on a biodegradable polyglycolic acid or polylactin mesh, *Biomaterials*, 12(3): 243–248 (1991).

108. Hansbrough, J. F., M. L. Cooper, R. Cohen, R. Spielvogel, G. Greenleaf, R. L. Bartel, and G. Naughton, Evaluation of a biodegradable matrix containing cultured human fibroblasts as a dermal replacement beneath meshed skin grafts on athymic mice, *Surgery*, 111:438–446 (1992).

109. Mikos, A. G., Y. Bao, L. G. Cima, D. E. Ingber, J. P. Vacanti, and R. Langer, Preparation of poly(glycolic acid) bonded fiber structures for cell attachment and transplantation, *Journal of Biomedical Materials Research*, 27:183–189 (1993).

110. Parsons, J. R., H. Alexander, and A. B. Weiss, Absorbable polymer-filamentous carbon composites: A new concept in orthopaedic biomaterials. In *Biocompatible Polymers, Metals and Composites* (M. Szycher, ed.), Technomic, Lancaster, PA, pp. 873–906 (1983).

111. Alexander, H., S. Corcoran, J. R. Parsons, and A. B. Weiss, Internal fracture fixation with partially degradable plates, *9th Bioengineering Conference*, Elmsford, NY, 1981.

112. Hollinger, J. O., and J. P. Schmitz, Restoration of bone discontinuities in dogs using a biodegradable implant, *Journal of Oral and Maxillofacial Surgery*, 45:594–600 (1987).

113. Antikainen, T., M. Ruuskanen, R. Taurio, R. Kallioinen, W. Serlo, P. Törmälä, and T. Waris, Polylactide and polyglycolic acid-reinforced coralline hydroxy-apatite for the reconstruction of cranial bone defects in the rabbit, *Acta. Neurochirurgica*, 117:59–62 (1992).

114. Turk, A. E., K. Ishida, J. A. Jensen, J. S. Wollman, and T. A. Miller, Enhanced healing of large cranial defects by an osteoinductive protein in rabbits, *Plastic and Reconstructive Surgery*, 92(4):593–600 (September 1993).

115. Meikle, M. C., W.-Y. Mak, S. Papaioannou, E. H. Davies, N. Mordan, and J. J. Reynolds, Bone-derived growth factor release from poly(-hydroxy acid) implants *in vitro*, *Biomaterials*, 14(3):177–183 (1993).

116. Langer, R., New methods of drug delivery, *Science*, 249:1527–1533 (September 1990).

117. Higashi, S., T. Yamamuro, T. Nakamura, Y. Ikada, S.-H. Hyon, and K. Jamshidi, Polymer-hydroxyapatite composites for biodegradable bone fillers, *Biomaterials*, 7(5):183–187 (1986).

118. Postlethwait, R. W., Polyglycolic acid surgical suture, *Arch. Surg.*, 101:489–494 (October 1970).

119. Cutright, D. E., and E. E. Hunsuck, Tissue reaction to the biodegradable polylactic acid suture, *Journal of Oral Surgery*, 31(1):134–139 (1971).

120. Cutright, D. E., J. D. Beasley, and B. Perez, Histologic comparison of polylactic and polyglycolic acid sutures, *Journal of Oral Surgery*, 32(1):165–173 (1971).

121. Cutright, D. E., E. E. Hunsuck, and J. D. Beasley, Fracture reduction using a biodegradable material, polylactic acid, *Journal of Oral Surgery*, 29:393–397 (1971).

122. Brady, J. M., D. E. Cutright, R. A. Miller, and G. C. Battistone, Resorption rate, route of elimination, and ultrastructure of the implant site of polylactic acid in the abdominal wall of the rat, *Journal of Biomedical Materials Research*, 7:155–166 (1973).

123. Racey, G. L., W. R. Wallace, J. Cavalaris, and J. V. Marguard, Comparison of a polyglycolic-polylactic acid suture to black silk and plain catgut in human oral tissues, *Journal of Oral Surgery*, 36:766–770 (October 1978).

124. Varma, S., W. V. Lumb, L. W. Johnson, and H. L. Ferguson, Further studies with polyglycolic acid (Dexon) and other sutures in infected experimental wounds, *American Journal of Veterinary Research*, 4:571–574 (1981).

125. Walter, E. P., J. A. Waldhausen, A. Prophet, and W. S. Pierce, Primary vascular anastomosis in growing pigs: Comparison of polypropylene and polyglycolic acid sutures, *Journal of Thoracic Cardiovascular Surgery*, 81:921–927 (1981).

126. Salthouse, T. N., Tissue response to sutures. In *Biomaterials in Reconstructive Surgery* (L. R. Rubin, ed.), Mosby, St. Louis, pp. 131–142 (1983).

127. Wetter, L. A., M. D. Dinneen, M. D. Levitt, and R. W. Motson, Controlled trial of polyglycolic acid versus catgut and nylon for appendectomy wound closure, *British Journal of Surgery*, 78(8):985–987 (1991).
128. Kobayashi, H., K. Shiraki, and Y. Ikada, Toxicity test of biodegradable polymers by implantation in rabbit cornea, *Journal of Biomedical Materials Research*, 26:1463–1476 (1992).
129. Robert, P., J. Mauduit, R. M. Frank, and M. Vert, Biocompatibility and resorbability of a polylactic acid membrane for periodontal guided tissue regeneration, *Biomaterials*, 14(5): 353–358 (1993).

VI
CONTROLLED RELEASE

31
Experimental Delivery Systems for Bone Morphogenetic Protein

Marshall R. Urist
University of California
Los Angeles, California

I. INTRODUCTION

A basic assumption in contemporary research on skeletal system development is that the connective tissue stroma of almost every tissue in the body incorporates a noncollagenous protein glycoprotein named *bone morphogenetic protein* (BMP). The dissemination of BMP begins very early in embryonic life at the time of the differentiation of ectomesoderm from ectoderm. BMP is transient in every tissue throughout embryonic, postfetal, and adult life. BMP is stored in large quantities in the organic matrix of dentin and bone musculoskeletal tumors, both benign and malignant.

BMP is classified as a *morphogen*, an embryonic molecule that, in effective concentration gradients, induces bone development. When the gradient is high and the mesenchymal-type perivascular connective tissue cell proliferates, bone morphogenesis follows. A conclusive identification of BMP depends on the induced bone development in a heterotopic site in an adult animal. Now that purified BMP is available from endogenous and exogenous sources in both native and recombinant forms, research on delivery systems is progressing in several directions. Native BMP is an integral part of the densely packed structure of extracellular bone matrix, noncollagenous proteins. The proteins are either soluble, slightly soluble, or very insoluble in aqueous media. Under these conditions, complete purification of BMP has been one of the most difficult problems in connective tissue biochemistry. The problem requires separation of BMP from transforming growth factor-β (TGF-β) osteopontin, osteonectin, gamma-carboxyglutamic-acid-rich proteins, and various other proteoglycans and lipoproteins [1–3].

The delivery of the desired amount of BMP to a specific site with an optimum concentration gradient is an unsolved problem. Many attempts to control release of BMP in bone defects are reported in the literature. The problem is to isolate cell membrane receptors of BMP and measure the rate of the diffusion of BMP through the bone marrow stroma of the host bed. Delivery systems for BMP are classifiable as (1) endogenous or exogenous substances derived from bone matrix, (2) biological or nonbiological substances, (3) biodegradable and nonbiodegradable materials.

Implantation of the delivery system including BMP in a bone defect in close contact with bone marrow stroma on all sides ensures activation of preexisting osteoprogenitor cells and recruitment of mesenchymal-type connective tissue cell targets for BMP. The delivery system for BMP may be xenogeneic, allogenic, or autogenic in character. Calf skin atelocollagen is a commonly used xenogeneic biodegradable delivery system with extension peptides retained in levels too low to incite either cytotoxic antibodies or a delayed hypersensitivity response. Allogeneic substances, including BMP, can induce the cell-mediated, macrophage, plasma cell, lymphocyte barrier to BMP-induced bone development.

Now that the BMP gene has been cloned and recombinant human BMP (rhBMP) is available, the delivery system assumes new importance. The development of a BMP delivery system capable of precisely and predictably releasing rhBMP in effective concentration gradients for augmentation of bone repair is absolutely essential for implantation of rhBMP in patients. New developments in the field of polymer chemistry, physical chemistry, molecular biology, and bioengineering provide a broad selection of substances for a delivery system. The physicochemical characteristics of both the delivery substance and the BMP must be taken into account by multidisciplinary research groups. Toxicology studies, immunohistocompatibility research, and metabolic studies on excretion of exogenous by-products create formidable challenges for research in academia, industry, and the interaction of teams representing both fields of endeavor.

To find the ideal delivery system, BMP in bone matrix powders, crude extracts, chromatographic fractions, and purified rhBMP have been implanted in composites of various biologic and nonbiologic or synthetic substances. The composites have been implanted in heterotopic and orthotopic sites. The BMP has been endogenous and exogenous in origin, or autogenic, isogeneic, or xenogeneic in nature.

In this chapter, the literature on delivery systems for exogenous native and recombinant BMP is summarized to illustrate efforts to secure sustained release of BMP with an effective concentration gradient and the minimum inflammatory tissue barrier or delayed immune response.

II. COMPOSITES OF HUMAN FIBRIN AND BONE MORPHOGENETIC PROTEIN

Fibrin has been used as an adhesive in cardiovascular surgery, nerve surgery, abdominal surgery, and orthopedic surgery. Because of its biologic and physiologic properties, fibrin seals wounds, adheres to living tissues, and provides hemostasis. Extensive research on the fibrin sealant shows many possible applications [4].

Kawamura and Urist investigated a physiologic delivery system as a composite of BMP/NCP (noncollagenous protein) and a fibrin sealant clot in the thigh of a mouse [5]. Comparable amounts of BMP/NCP alone and fibrin alone were implanted as controls. New bone yields were measured by random-point analysis using a video camera and computer system, and by bone ash weight measurements. Histologic examinations, beginning as early as three to seven days postimplantation, showed more extensive proliferation of mesenchymal cells in response to the composite than to lyophilized BMP alone. Correlated microradiographic and histologic observations showed that the composite also produced larger volumes of new bone than the BMP control group. The composite produced approximately three times more bone than BMP alone, and the difference was consistent and statistically significant. Lyophilized pellets of a compacted composite of fibrin and BMP/NCP, prepared as volume control implants, also showed more new bone formation than the BMP alone, but the differences were less than with the wet clot composites. The relatively high yields of new bone from both wet clot and lyophilized composites suggest that the fibrin clot may contain a favorable BMP concentration gradient. However, composites of fibrin, hydroxyapatite (HA) and rabbit BMP/NCP implanted in bone defects in allogeneic recipients produced higher yields than fibrin alone or HA alone [6].

III. INDUCED BONE DEVELOPMENT BY PHYSICAL AND CHEMICAL MODIFICATIONS AND RECALCIFICATION IN THE IMPLANTS OF DEMINERALIZED BONE MATRIX

Chemical extraction of soluble noncollagenous proteins with 2 molar (M) $CaCl_2$, 0.5M EDTA (ethylene diaminetetraacetate), and 8M LiCl in sequence at 20°C removes about 20% of the total dry weight without denaturing of BMP of the residual gelatinized bone matrix (GBM). Although the GBM was rapidly resorbed, the specific activity (ash weight) of induced bone deposits per gram of implanted GBM was the same as the original demineralized bone matrix (DBM) (Fig. 1).

GBM was also digested with collagenase (gelatinases); the undigested noncollagenous protein was 20% higher in specific activity of BMP than the GBM [7]. Yamashita et al. demonstrated recalcification of GBM in electron micrographs [8,9]; the recalcified GBM was resorbed by multinucleated cells, and preceded induced bone development. X-ray microanalysis and infrared spectroscopy showed amorphous $CaHPO_4$ and carbonate hydroxyapatite in recalcified GBM. Previously, Linden observed recalcification of DBM beneath interfaces between old matrix and new bone [10].

Based on experiments on subcutaneous implants of bone matrix with percutaneous teflon tubes, Sela et al. proposed that acid pH in the microenvironment caused by local infections retarded matrix-induced bone development, while alkaline pH enhanced in-

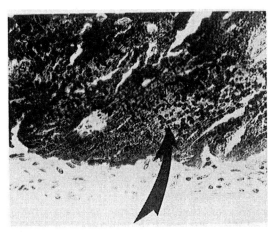

Figure 1 Photomicrograph of recalcified gelatinized bone matrix gelatin (von Kossa Hematoxylin and eosin stain, ×200). Note punctate deposits of mineral in old osteocyte lacunae.

duced bone development in rats with implants of DBM [11]. Inflammation associated with calcium hydroxide and DBM retarded bone formation in subcutaneous sites [12].

In implants in patients with periodontal disease, composites of allogeneic demineralized bone matrix and partially purified human BMP stimulated bone repair. Comparable patients treated with a bovine collagen (Collaplug, American Biomaterials, Plainsboro, NJ) with or without BMP showed significantly less bone regeneration [13]. The components of the organic matrix enhanced the effects of BMP on bone regeneration. In fact, a minimum quantity of 10 milligrams (mg) of DBM is essential for transfer of bovine BMP-3 in allogeneic rats; 10 to 25 mg produced dose-dependent increase in the yield of new bone [14]. The addition of rat tail tendon collagen Type I had no effect.

DBM fluoridated *in vitro* and implanted in the anterior abdominal muscles in rats did not cause localized fluorosis or increase the quantity of BMP-induced new bone formation [15]. In osteotomy sites in sheep, the yield of new bone from DBM was also considerably less than with callus extension [16]. In rats, demineralized dentin matrix induced larger quantities of new bone than DBM [17]. However, the viability of the host bed is the critical factor in the volume of induced bone development; implants of DBM in rats with local irradiation of doses of 20 Gy (gray) showed only small scattered areas of bone with hardly appreciable elements of bone marrow [18].

Nilsson et al. implanted DBM prelabeled with ^{3}H-proline in allogeneic rats and correlated induced heterotopic bone formation with progressive release of the isotope [19]. Injections of indomethacin inhibited uptake of ^{45}Ca and corresponding release of ^{3}H-proline. Thus, induced bone development was associated with progressive resorption of the implanted DBM.

To increase the surface area of exposure of cells to endogenous BMP and to facilitate ingrowth of host bed cells, Gendler implanted rat DBM perforated with 0.35-millimeter (mm) holes with a high-speed, water-cooled drill [20]. Within 10 days, the perforations were filled with cartilage and were invaded by vascularized new bone tissues within 14 to 21 days. By mixing demineralized allogeneic bone and dentin, Pinholt et al. also increased the surface area of matrix-containing BMP, and successfully reconstructed alveolar ridge defects in rats [21].

To prove that induced heterotopic ossification occurs in nonhuman primates the same as in rodents, Hosny and Sharawy implanted pulverized allogeneic demineralized bone matrix subcutaneously in adult rhesus monkeys [22]. Some of the particles remineralized. Chondroid, woven bone, and lamellar bone formation occurred at intervals of 21 to 72 days after implantation.

A consistently reported observation on diminished (but not totally absent) BMP in bone matrix was correlated with the process of aging [23–26]. Reduction in matrix gamma carboxy glutamic acid rich protein (gla protein) was also correlated with age-related loss of BMP. Decline in BMP activity in bone matrix also occurred in men as well as in women [27]. Autoclaving rabbit bone completely destroyed all BMP activity in allogeneic implants in orthotopic sites [28]. Rats reared on a diet containing β-amino proprionitril developed bone deformities caused by lathyrism; in this condition, the enzyme lysyl oxidase is deficient and the cross-linked structure of the bone collagen is defective. For reasons unknown, the BMP activity of lathrytic matrix in rats is low or missing [29].

After ascertaining the osteoinductive activity in pulverized human gelatinized bone matrix (hGBM) in the hindquarter muscles of mice, Xiaobo et al. implanted the hGBM in 24 patients with delayed union or nonunion of femur or tibia. The fractures healed within two to six months [30].

IV. CHEMOSTERILIZED ANTIGEN-EXTRACTED SURFACE-DEMINERALIZED AUTOLYSED ALLOGENEIC BONE

When surface-demineralized, antigen-extracted autolysed allogeneic (AAA) bone is available in adequate quantities and the operation is technically performed so that the donor tissue is in tight contact with the host bed on three sides, the implant is incorporated as completely as an autogenic bone graft. The conditions for AAA bone are ideal for arthrodesis operation in young patients with extensively injured or previously infected joints. Arthrodesis is performed less frequently now than 15 years ago, that is, before the era of endoprosthetic total joint arthroplasties.

For joints other than hips and knees, a standard procedure is to supplement the AAA bone with transplants of autogenic cancellous bone from the iliac crest. Although autogenic bone is more readily available and is generally superior to allogeneic bone, the additional operation on the iliac crest is often unjustified because (1) at least seven different complications have been reported from removal of large blocks of iliac crest bone and (2) for many years after the operation, patients frequently complain of more pain in the iliac crest donor site than in the recipient bed. Table 1 summarizes a current protocol for the preparation of AAA bone and the rationale in each step of the procedure.

The important advantages of AAA bone compared with whole raw bone, either frozen or freeze dried, are as follows. Implants in athymic mice show that AAA bone retains endogenous BMP. Because the preparation eliminates the need for either absolute asepsis or uncertainties about postmortem bacteria, AAA is more practical than aseptically collected raw, freeze-dried bone. The bone is collected under the most aseptic conditions possible, but can be used to prepare AAA bone, if cultures should prove to be positive for contaminating organisms.

The reason the bone can be used if it is contaminated during collection is that chemosterilization is applied in each of the first steps shown in Table 1. In Step 3, the bone is surface demineralized in 0.6 N HCl at 20°C to increase the rate of resorption of donor tissue and at the same time retains sterility. In Step 4, transplantation antigens are extracted and the BMP is stabilized by incubation in an inhibitor of sulfhydryl-type enzymes such as iodoacetic acid (IAA), iodoacetamide (IAM), or N-ethyl maleimide (NEM) that also autolyse cells and sterilize the bone. In Step 5, transplantation antigens in cell membranes are extracted and microbes and spores are devitalized in 1:1 chloroform methanol. In Step 7, the end-product derivative of bone is freeze dried and sealed in double plastic envelopes for storage. No detectable loss of sterility or biologic activity occurred for as long as one year.

AAA bone is reconstituted under aseptic conditions in the operating room on a separate table immediately before use in a solution of antibiotics and cut into the desired form (i.e., slabs, chips, pegs) by an oscillating saw. An antibiotic solution is also used to cool the saw and prevent heat denaturation of the BMP. Cultures for aerobic and anaerobic organisms have been taken in the operating room of the saw dust of AAA bone and found to be negative in over 100 consecutive standard orthopedic operations.

Bioassay by implantation in muscle pouches of athymic mice demonstrates that human AAA bone transmits BMP activity. The costs of AAA bone are greater than those of freeze-dried whole raw bone because of the extra expense of the biochemical extraction and pharmacologic process. However, investigations on a statistically significant series of cases are necessary to determine the exact cost-benefit ratio of AAA relative to autologous bone. It seems self-evident that, in a patient population in which the morbidity of

Table 1 Preparation of Chemosterilized Antigen-Extracted Autolysed Allogeneic Bone

Procedures	Purpose
Step 1	
Collect human diaphyyseal cortical bone under aseptic conditions within 8 h after death of subjects not over 40 years of age.	Secure clean bone with optimum BMP activity.
Step 2	
Remove all soft parts; thoroughly wash in 10 mmol/L NaN₃; remove marrow with water pick; cut in 10-cm lengths with solution-cooled oscillating saw.	Grossly remove as much antigenic material as possible.
Step 3	
Surface demineralize in 0.6 N HCl, 2°C, 24 h.	Open up the densely packed structure of cortical bone to facilitate resorption.
Step 4	
Incubate for 48 h in phosphate buffer (pH 7.4) containing NaN₃ (5 mmol/L), N-ethyl malemide (3 mmol/L), and p-chloromercuribenzoate (3 mmol/L).	Autodigest bone cells, DNA, RNA, and wash out as much other intracellular antigenic material as possible; stabilize BMP.
Step 5	
Defat in 1:1 chloroform methanol.	Extract lipid and cell membrane lipoproteins.
Step 6	
Wash in NaN₃, 5 mmol/L, 3 times.	Remove retained chemicals and other water-soluble substances.
Step 7	
Lyophilize and pack diaphyseal segments in sterile double-walled plastic, inside airtight or vacuum-sealed container.	Storage and maintenance of sterility.
Step 8	
Unwrap. Culture for anaerobic and aerobic organisms, reconstitute in double antibiotic solution in operating room at time of surgery.	Ensure sterility.
Step 9	
Before the incision is made, cut segments into strips or pieces of desired size and shape with an oscillating saw cooled with antibiotic solution.	Reduce operating time and blood loss to a minimum.
Step 10	
Implant in tight contact with a host bed of bleeding bone (AAA bone has hemostatic properties).	Ensure concurrent incorporation resorption of the AAA bone.

removing large blocks of bone from the iliac crest or the tibia is high, AAA bone is an invaluable adjunctive measure for all kinds of reconstructive surgical procedures.

In adults, AAA bone is contraindicated for replacement of large total diaphyseal bone defects. In children with high proliferative potential, AAA can induce regeneration of even large bone defects. In adults, it is advisable to use a composite of autologous cancellous bone or bone marrow with AAA bone, or possibly even a microvascular anastomosed autograft. In aged adults, AAA bone is resorbed and replaced with new bone two to three times more slowly than in young adults and should always be used as a composite graft. Research on aged patients and senile animals demonstrates that aged and osteoporotic bone are BMP deficient. Consequently, the transfer of BMP from donor to aged host is both low in quantity and slow in rate.

Observations of experimental animals suggest that a composite of autologous marrow and allogeneic bone matrix enhances regeneration, but in ordinary clinical circumstances the procedure is impractical. In open operations, bone marrow transplants are difficult to retain in place and are easily lost after installation of hemovac suction drainage. Until recombinant human BMP is available, composites of native human BMP are under investigation [31–35].

Figure 2 demonstrates induced bone development including bone marrow in a composite of AAA bone and hBMP/NCP in an athymic rat. When the composite is implanted in an allogeneic patient, water-soluble hBMP/NCP is infused in the structure of the AAA bone to supplement endogenous BMP with 100 to 1000 times more exogenous morphogen than normally present in an adult human implant. Hollinger et al. demonstrated regeneration of 15-mm cranial bone defects within six months in cynomolgus monkeys with implants of a composite of human AAA bone and bovine BMP [36].

V. SURFACE VERSUS TOTAL DEMINERALIZATION

The repair of craniofacial bone defects is especially rapid in infants and young growing children and relatively slow or even incomplete in adults. In either the child or the adult, to obtain the homostructural function of the donor, 70% to 80% of the mass of the graft

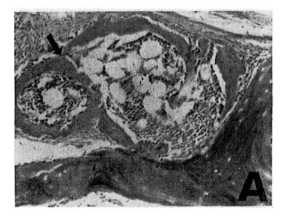

Figure 2   Photomicrograph of new bone including bone marrow (arrow) in a composite of human bone morphogenetic protein (hBMP)/NCP (noncollagenous protein) and antigen-extracted auto-lysed aloogenic (AAA) bone (A) in an athymic rat on Day 28 (Hematoxylin eosin and azure II, ×200).

is unresorbed when the donor tissue becomes completely incorporated by appositional bone formation in the host bed. Accordingly, for preparation of AAA bone, about 20% to 30% demineralization of the internal and external surfaces is sufficient to enhance resorption and induce bone formation without loss of such homostructural functions as are advantageous for arthrodesis and other orthopedic procedures.

AAA bone is a practical application of animal research on matrix-induced bone formation. The first clinical trials in 1968 were designed to compare undemineralized, totally demineralized, and surfaced demineralized bone in 35 patients. Of these, 16 patients with diaphyseal nonunion, fibrous dysplasia, aseptic necrosis of the scaphoid, juxta-articular cysts, old infected fractures, and lumbar spinal arthrodesis were treated with demineralized bone implants. Also, 9 patients were treated with undemineralized bone and 10 with surface demineralized bone for comparable conditions. Although the number of cases was small, and despite the fact that the nature of the BMP activity and the importance of neutral protease inhibitors were not known in 1968, there were enough two-year follow-up results to suggest that either total demineralization or surface demineralization induced new bone formation. In 1981, 38 additional cases of posterolateral lumbar fusion with surface demineralized AAA bone were reported.

Surface demineralization had the advantage of performing the homostructural function of cortical bone and at the same time provided the internal fixation that was necessary for arthrodesis operations. For rapid repair of rigid-walled bone cavities, pulverized total and surface demineralized bone preparations were resorbed and replaced by new bone, but it was difficult to ascertain how much was attributable to osteoinduction and how much to osteoconduction. Further experiments on animals suggested that transplantation antigens had to be reduced in quantity to obtain the optimum osteoinductive responses; this was the objective in the preparation of AAA bone. Glowacki et al. totally demineralized human bone in 0.5 N HCl at room temperature, implanted chips and powders in patients with craniofacial defects, and reported repair by induced bone formation [37]. The matrix was not resorbed, and instead "induced transformation of fibroblasts directly into osteoblasts." Although some investigators postulate transformation of fibroblasts, matrix surfaces are covered with perivascular stellate mesenchymal-type cells, which, unlike fibroblasts, lack the coarse endoplasmic reticulum. Eventually, the matrix is resorbed by macrophages and matrix-clasts.

Ordinary HCl-demineralized whole human bone matrix transfers only a fraction of the total BMP activity, and does not release enough BMP to account for bone regeneration in large bone defects. As far back as the 1890s, reports appeared in the literature to refute Senn on the value of demineralized bone matrix as a substitute for autologous bone [38]. Some of these refutations may be explained by the use of ethylene diaminetetraacetate (EDTA) and other solutions in which BMP activity is degraded. Others may be attributed to the low level of BMP activity of the bones of aged donors.

Despite the limitations of clinical experiments, animal experiments provide irrefutable evidence of matrix-induced bone formation in muscle within three weeks to affirm the validity of the response to BMP. In dogs and some other long-lived animals, including humans, induced bone formation occurs only after a lag phase, matrix resorption, and the yield is smaller than in rodents. In any case, clinical problems of osteoinduction can be resolved by quantitative analysis of systems of induced bone development. Until BMP is completely characterized and available in unlimited quantities, present efforts to enhance bone repair with crude preparations of demineralized bone matrix in orthotopic sites are bound to produce inconsistent and less-than-optimal results.

Saraf et al. implanted autoclaved, surface-demineralized bone matrix in 2.0-cm ulnar defects in allogeneic rabbits and claimed successful bridging with new bone [28]. The unautoclaved controls healed in 66% less time. Autoclaved bone is not uncommonly implanted when bone is contaminated accidentally during an operation, to fill a large bone defect temporarily.

Pike and Boyne repaired extensive mandibular defects in rhesus monkey by implanting a composite of freeze-dried surface-demineralized allogeneic bone with autogenic particulate bone grafts [39]; regeneration occurred in 12 weeks. In contrast, De Pablos et al. obtained unpredictable results in lambs with 5-cm defects in the femur with implants of massive segmental allogeneic demineralized bone matrix [40]; undemineralized control alloimplants were incorporated as well as those osteoinductive in action. More conclusive osteoinductive activity was obtained by supplementing a demineralized bone matrix with a BMP/NCP or BMP-3 [14]. However, bioassay work on recombinant BMP that relied on recombination with guanidine-HCl-extracted rat bone matrix implanted in rats was unreliable, and therefore unacceptable by careful investigators (Fig. 3).

The most accurate quantitative bioassay of native BMP was in the mouse hindquarter muscle pouch [41], which did not employ bone matrix having remnants of BMP or other undisclosed components (Figs. 4 and 5).

VI. BONE FORMATION INDUCED BY COMPOSITE OF BMP/NCP IN A POLYMETHYLMETHACRYLATE DELIVERY SYSTEM

Polymethylmethacrylate (PMMA) is commonly used for fixation of endoprostheses in orthopedic and craniofacial and nonneurosurgical operations. PMMA beads are also used as a carrier for local implantation of antibiotics. PMMA is supplied in the form of a prepolymerized p-homopolymer powder. Immediately before use, the powder is mixed with methyl methacrylate monomer liquid. The powder may be supplemented with 6% to 12% barium sulfate for radiopacity. Gentamyicin [42,43], silver sulfate [44], and other antibiotics are added for both local and systemic treatment. The additives diffuse out of the PMMA in predictably measurable quantities.

Downes et al. measured the release of human growth hormone (HGH) *in vitro* from polymethylmethacrylate bone cement in rabbits using a specific enzyme-linked immunosorbent assay (ELISA). HGH was released from the bone cement in the surrounding tissues. HGH-loaded PMMA was also inserted into one distal femur. Plain cement was inserted into the contralateral distal femur as a control, and the rabbits were sacrificed at time intervals of one, two, and four months after surgery and the distal femurs embedded whole for histology. Quantitative histomorphimetry indicated that there was a greater percentage of osteoid present at the HGH-loaded cement interface than at the unloaded cement surface ($p < 0.01$) one month after surgery. HGH released from bone cement stimulated osteogenesis and new bone formation, and improved the strength of the bone-cement interface.

PMMA cement is self-curing and expands into the host bone prior to hardening. In the process of repair, a fibrous membrane of variable thickness develops at the bone cement interface. The thickness of the membrane is determined by the density and surface area of the host bone bed. The membrane may gradually increase with time or loosening of the implant. The larger the area of host bone surface is, the more firm the fixation

BIOASSAY OF BMP

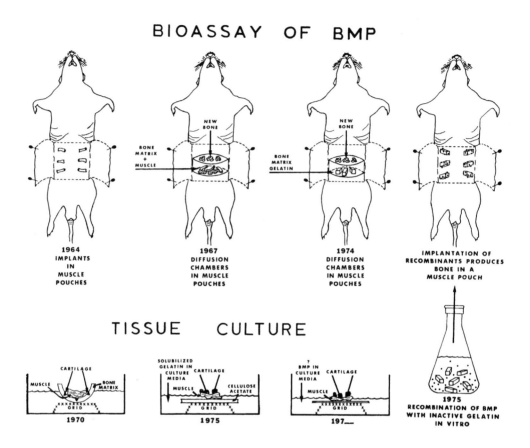

TISSUE CULTURE

Figure 3 Methods of bioassay and the year of introduction in the field of bone morphogenetic protein (BMP) research. From 1964 to 1974, investigations centered on implants of preparations of hydrochloric-acid-demineralized bone matrix and bone matrix with bone collagen converted to bone gelatin to increase the rate of resorption. From 1970 to 1989, parallel investigations were performed with bone matrix preparations and purified BMP in tissue culture. In 1975, composites of BMP and inactivated bone matrix were prepared *in vitro* for implantation *in vivo* to induce bone formation in rats. (From M. R. Urist, Ed., *Fundamental and Clinical Bone Physiology*, J. B. Lippincott, Philadelphia, 1980.)

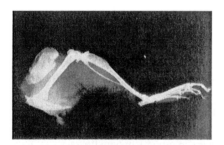

Figure 4 Roentgenogram of bone morphogenetic protein (BMP)/NCP-induced bone development in the hindquarter muscles of an adult mouse, 21 days postimplantation.

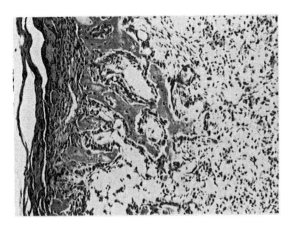

Figure 5   Photomicrograph of the ossicle shown in Fig. 4. Note bone and bone marrow (left), mesenchymal type cells (right).

will be. Loosening occurs with excessive stress, strain, and torsion, but in time internal remodeling of the host bone may account for progressive instability.

Acrylic cement is remarkably histocompatible and is even permeable to body fluids. Unless there is a microfracture or loosening secondary to host bone resorption, the cement seems almost inert. Foreign-body and local tissue reactions are generally insignificant. In some countries, antibiotics are routinely added to the cement for prevention of sepsis. In the United States, the requirements for efficacy of the antibiotic in cement have not yet been met for Food and Drug Administration (FDA) approval. *In vitro*, the antibiotic diffuses out of the cement into physiologic media [45,46]. *In vivo*, the antibiotic diffuses out of the cement into surrounding host tissue and either is excreted or degraded [42,43,45–52] (Figs. 6–10).

We investigated a BMP-PMMA composite for a delivery system for morphogenetic protein (BMP) including other bone matrix insoluble proteins (iNCP) for the purpose of locally stimulating proliferation of the host bone and inducing supplementary new bone formation [53–61]. The BMP/NCP/PMMA composite induced development of new bone by the host bed surrounding the implant. The composite induces formation of larger quantities of new bone from smaller quantities of BMP than implants of BMP without PMMA. This observation suggests slow absorption and a locally sustained high concen-

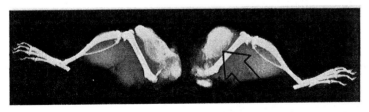

Figure 6   Roentgenogram of an implant of bone morphogenetic protein/polymethylmethacrylate (BMP/PMMA) (10 mg per 100 mg of PMMA) 21 days after implantation in the mouse hindquarter muscles. Note barium-sulfated PMMA surrounding the deposit of new bone (arrow).

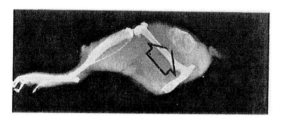

Figure 7   Roentgenogram of implant of bone morphogenetic protein (BMP) (10 mg per 100 mg of barium-free polymethylmethacrylate [PMMA]) on Day 21, after implantation in the hindquarter muscles of an adult mouse.

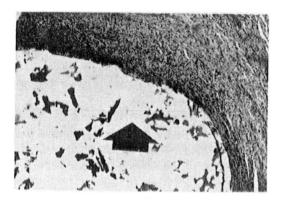

Figure 8   Photomicrograph of deposit of fibrous connective tissue (top) surrounding a control implant of polymethylmethacrylate (PMMA) only. The arrows indicate the space previously occupied by PMMA containing remnants of undemineralized barium sulfate.

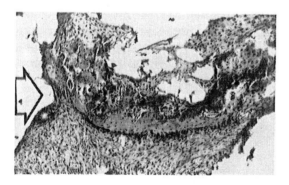

Figure 9   Photomicrograph in low-power magnification showing the clear space previously occupied by a multilobulated implant of (bovine bone morphogenetic protein) bBMP/PMMA. Note chondroosteoid (bottom), new bone (arrow), and fibrous tissue envelope (top).

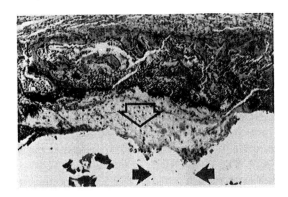

Figure 10 Photomicrograph of a deposit of new bone (top) and chondroosteoid (transparent arrow) formed in response to an implant of bovine bone morphogenetic protein (bBMP)/polymethylmethacrylate (PMMA) (black arrows indicate the space previously occupied by PMMA), including remnants of undemineralized barium sulfate additive.

tration gradient of BMP may enhance the bone morphogenetic response. In general, release of other proteins from PMMA is a sustained and long-term process.

In the above-described implants of a BMP/PMMA composite, the evidence for the BMP delivery system was a highly localized response. In the mouse assay system, the implantation of a small dose of BMP/NCP/PMMA was equivalent to the same or a greater dose of BMP/NCP alone. Denaturation of BMP by the heat of polymerization of PMMA seems not to have occurred on larger masses of cement. PMMA dissipated heat too fast to reach BMP heat denaturation temperatures [i.e., 70°C to 80°C (Table 2).]

Seligson and Henry edited 14 reports of patients with wound infections treated by local implantation of PMMA beds incorporating gentamicin [62]. The PMMA system delivered more effective concentrations of the antibiotic than intravenous infusion alone.

Since bone matrix is the delivery system for native BMP, Henrich et al. incorporated demineralized bone matrix powders in PMMA for implantation in bone defects in dogs [63]. The shear strength of the PMMA in a push-out test was improved by the DBM.

Table 2   Bone Morphogenetic Protein Transmission through Polymethylmethacrylate

BMP (iNCP) (mg)	New bone (mm^3)	mg BMP (iNCP)/16 mg PMMA*	New bone (mm^3)
0.5	0	0	0
1	0	0.1	1.0 ± 0.1
2	0.5 ± 0.1	0.2	4.0 ± 0.3
5	40.0 ± 0.9	0.5	20.0 ± 1.1
10	44.0 ± 2.5	1.0	30.0 ± 2.1

*Physical mixture of prepolymerized homopolymer powder, including weight of catalyst, and 6% barium sulfate

VII.   SINTERED BIODEGRADABLE β-TRICALCIUM PHOSPHATE

A composite of BMP/NCP and a biodegradable form of β-tricalcium phosphate (TCP) (whitlockite) was prepared as follows. Water-soluble BMP/NCP was prepared from human cortical bone by differential precipitation, gel filtration, and by separating BMP from a carboxyglutamic-acid-rich protein (M_r 14 kDa) by ultrafiltration through a 10-k pore-size membrane (Amicon). Two grams of TCP (Synthos, Miller Corporation, Columbus, OH) were immersed in an aqueous solution of BMP, 1 mg/ml, in sterile distilled water for 12 hours at 280°C. The air was removed from the pores of TCP in a partial vacuum. After slow release of the vacuum and freeze drying, the BMP was permanently entrapped within the pore structure of the TCP. The BMP/NCP/TCP (19 parts TCP to one part BMP/NCP) aggregate induced bone formation in muscle pouches in the thighs of Swiss Webster mice [64]. The quantity of new bone measured by the random-point histomorphometric analysis and expressed in terms of percent of the implants occupied by new bone tissue was 12 times greater than BMP/NCP alone. However, about half of the TCP was unabsorbed and remained as foreign material in the host bed of new bone. Although the repair was complete in skull trephine defects (1.4 cm) in dogs, the quantity of residual TCP was about the same [65].

Wu et al. implanted a composite of mouse osteosarcoma BMP and sintered β-tricalcium phosphate in mice and induced cartilage and bone development earlier than observed in implants of BMP alone [66]. Tricalcium phosphate, which has a β-whitlockite lattice after sintering, is rapidly biodegradable but less conductive to porous ingrowth of new bone [67].

To measure the effect of β-tricalcium phosphate contaminants of preparations of hydroxyapatite, Frayssinet et al. prepared ceramics composed of 25% β-tricalcium phosphate and 75% hydroxyapatite and 50:50 β-tricalcium phosphate and hydroxyapatite [68]. The mixtures were prepared by sintering to give 100% crystalline calcium phosphate. Porosity was 75%. The ceramics containing β-tricalcium phosphate produced better osteoconduction than pure hydroxyapatite ceramics. The interaction between calcium phosphate products and cells is complex and the effects of the β-tricalcium phosphate phase on the microenvironment of mesenchymal cells differentiating under the influence of BMP require further investigation.

Phagocytosis and collection of particles of β-whitlockite occurs in regional lymph nodes of rabbits with implants in bone defects in dogs but not in heterotopic implants in muscle. Implants of ^{45}Ca-labeled β-whitlockite showed a precipitous decrease in radioactivity; the hydroxyapatite lost relatively little radioactivity. There was little or no local transfer of ^{45}Ca from the donor host bed [69].

VIII.   COMPOSITES OF BMP AND β-TRICALCIUM PHOSPHATE DELIVERY SYSTEMS IN MICE

Metsger et al. reviewed the early history of research on resorbable calcium phosphates [70]. Beginning in the 1970s, a ceramic form of β-tricalcium phosphate (β-TCP) was designed for resorption and replacement by bone over a period of 6 to 18 months. β-TCP is a crystalline, porous ceramic tricalcium phosphate, $β$-$Ca_3(PO_4)_2$, that should not be confused with tribasic calcium phosphate, $Ca_{10}(PO_4)_6(OH)_2$, bearing the chemical formula for hydroxyapatite. Implants of ceramic β-TCP in rabbits, dogs, and primates were effective in repairing many types of bone defects. β-TCP has been used also in human clinical studies to repair periapical periodontal defects as well as for apexification and treatment of miscellaneous alveolar defects. No adverse reactions have been attributable to β-TCP.

Wu et al. compared pellets of mouse partially purified osteosarcoma BMP and pellets of a composite of the same preparation mixed with β-tricalcium phosphate in 1:1 mixtures of 6 mg of each component pressed into a pellet [66]. The BMP pellets were resorbed more rapidly than the BMP β-tricalcium phosphate pellet, or the β-tricalcium phosphate pellets without BMP. The tricalcium phosphate was absorbed by two types of multinucleated giant cells. One was osteoclastic as identified by calcitonin receptors, numerous mitochondria, and ruffled boarders. The other multinucleated giant cell had no osteoclast markers. The BMP pellets induced few or no multinucleated giant cells. Composite pellets induced earlier deposits of cartilage and bone formation than the pellets of BMP alone. The new bone was deposited on the surfaces of the tricalcium phosphate coated with a cement line.

IX. COMPOSITES OF BONE MORPHOGENETIC PROTEIN AND SYNTHETIC HYDROXYAPATITE

Partially purified rabbit BMP precipitated into pores of synthetic hydroxyapatite induced cartilage development in outgrowths of connective tissue minced rat muscle *in vitro*. The same composite induced cartilage and bone development in the hindquarter muscle pouches of mice. The same composite augmented bone regeneration in 4.5-mm drill holes in the distal femoral condyles of mature rabbits. These observations *in vitro* and *in vivo* demonstrated earlier ingrowth of cells and bone generation and regeneration in hydroxyapatite with pore sizes of 90 to 200 microns, rather than pore sizes of 400 or larger. Bone development occurred at the openings rather than in the interior of the pores. No ingrowth of cells or bone development occurred in pore sizes of 75 microns or less [71].

To improve the histocompatibility, hydroxyapatite must be sintered at 900°C or higher temperatures. Unsintered hydroxyapatite produces an inflammatory reaction, including a blockade of multinucleated giant cells, and does not provide the surfaces for fibronectin and other cell-adhesion molecules [72]. Osteogenesis occurs inside macroporous 60:40 HA-TCP ceramic implanted in bone and remodeling of the newly formed cell begins almost immediately by the action of mononucleated and multinucleated osteoclastlike cells [73]. The new bone was of the woven type at two months and reconstructed to the lamellar type within six months; biodegradation of 29% of the original implant was demonstrable when 88% of the HA-TCP was incorporated by osteoconduction [74].

Yamazaki implanted porous hydroxyapatite alone in 20-mm mandibular defects and observed ingrowth of woven bone at three months, with remodeling into lamellar bone at six months [75]. The bone marrow changed from red to yellow with no further increase in the quantity of bone in the interior of the implants (porosity 70%) during the interval from 6 to 24 months.

X. BMP/NCP-ATELOCOLLAGEN-HYDROXYAPATITE β-TRICALCIUM PHOSPHATE CERAMIC COMPOSITE IN SPINAL FUSION IN DOGS

Muschler et al. examined 13 dogs with multiple two-level spinal fusions [76]. Operations with autogenic bone grafts produce a high incidence of arthrodesis. Composites that included BMP/NCP produced larger volumes of paraspinal new bone than the ceramic, collagen, or composites of the two, but a lower incidence of arthrodesis.

XI. HYDROXYAPATITE-ADSORBED BMP

The affinity of BMP for hydroxyapatite (HA) observed in HA-affinity chromatographic fractionation of noncollagenous bone matrix proteins has been used in a delivery system. HA has been prepared from the calcium carbonate exoskeleton of the scleractinian reef-building corals of both genera, *Porites* and *Goniopora* [77–79]. The coral structure consists of a uniform network of interconnected channels and pores [74,77–80]. The porous microstructure of coral is maintained even after conversion of the calcium carbonate exoskeleton to HA at high temperatures in phosphate buffer solutions [79,81–83].

Coral-derived HA (CDHA) exoskeletons of the genus *Porites* (200-μm [micrometers] pore size) have been implanted extraskeletally in mice, rats, and dogs [82,84]. CDHA has been implanted also in baboons [85]. Holmes and Hagler have recently examined the 200-μm porous hydroxyapatite in skull trephine defects in dogs [86].

CDHA has osteoinductive properties in orthotopic sites in craniofacial defects and is found to compare favorably with autogenous bone grafts [71,74,86–88]. In heterotopic sites, fibrovascular tissue grew into the pores of CDHA, but no bone formation ensued. CDHA failed to induce bone formation [84]. Bone growth into the porous spaces was dependent on contact compression of the implant to viable bone interfaces [74,82,84].

Relatively little research has been done on the 600-μm pore size CDHA [89] prepared from genus *Goniopora* (Interpore 500) [80]. Morphometric data on bone ingrowth into 600-μm CDHA resembled iliac crest spongiosa [82,88].

Ripamonti et al. implanted a composite of BMP-3 and CDHA in baboons and induced bone development in both heterotopic and orthotopic skull trephine defects in baboons [85]. Long-term observations at three, six, and nine months after the operation revealed bone in the 600-μm pores of the CDHA controls. The quantity of heterotopic bone was proportional to the length of time in the muscle. The important question raised by these heterotopic deposits is whether the CDHA adsorbed BMP from the fluids of the body during an extended three-to-nine-month period of time. To prove that hydroxyapatite chromatographically adsorbs BMP, Ripamonti et al. fractionated GuHCl extracts of baboon bone matrix onto columns of CDHA [85]. Bound fractions induced bone development in rats, while unbound fractions failed to induce bone formation.

Piecuch implanted the same CDHA subcutaneously in dogs [84], and Alper et al. observed CDHA implants with and without bone matrix in rats [90]. Ohgushi et al. suspended bone marrow stroma cells in CDHA and described bone formation in subcutaneous transplants within three weeks in adult rats [91]. There was no induced bone formation either in dogs or rats, only fibrous connective tissue ingrowth. In fact, the CDHA appeared to impede bone development. In baboons, a resorbable preparation of CDHA (prepared with a ratio of 10% calcium carbonate to 90% hydroxyapatite) also failed to induce bone development, even with inclusions of BMP-3 (osteogenin).

Implants of CDHA with ratios of 10% tricalcium phosphate and 90% hydroxyapatite (600-μm pore size) after three to nine months induced bone development with or without BMP; they attributed the positive results to the three-dimensional porous spaces of the inorganic substratum. Whether BMP circulating in blood plasma and body fluids [1] was also adsorbed by the nonresorbable preparations of CDHA by mechanisms comparable to separation of BMP by hydroxyapatite-affinity chromatography [64,92,93], or whether synthesis of BMP occurred locally by mesenchymal type cells is not known.

An apatite-wollastonite-containing glass ceramic (A-W·GC) was tightly bonded to bone in orthotopic sites in dogs; microradiographic and electron probe microanalysis showed increased density of calcium phosphate across the A-W·GC interface with time [94]. *In vitro*, implants of porous hydroxyapatite ceramic were used as a substratum for cultivation of human fibroblasts. Transmission electron microscopy showed endocytosis of ceramic granules. Deoxyribonucleic acid (DNA) synthesis and translocation of protein kinase C from cytoplasm to cell membrane were correlated with cell-mediated resorption of the ceramic [95].

Miller et al. infused porous coralline hydroxyapatite with BMP/NCP for implantation in 15 × 15-mm cranial bone defects in rabbits, and observed repair within 3 to 4 months [96].

XII. COMPOSITE OF DEMINERALIZED BONE POWDER AND COMPOUND CERAMICS

A bone powder prepared from HCl-demineralized rabbit bone was mixed with a compound apatite-wollastonite ceramic and packed into defects in the medullary cavity of the proximal tibia of allogeneic recipients. The bone matrix powders significantly accelerated the process of bone growth into the surfaces of the ceramic [97]. Alper et al. implanted a composite of rat demineralized bone matrix (DBM) in a phospholipid-coral-derived hydroxyapatite paste in orthotopic sites in allogeneic rats and found a decrease in the quantity of ^{45}Ca-labeled new bone [90]; control implants without the mineral or the phospholipid paste produced 80% more new bone than the composite.

XIII. HYDROXYAPATITE COMPLEXES FOR DELIVERY OF BONE MORPHOGENETIC PROTEIN

Sato et al. prepared a composite of rabbit BMP, human fibrin, and granulated hydroxyapatite (HA) (particle size 0.15 to 0.3 mm^2) [6]. The composite induced heterotopic bone in the mouse and rabbit hindquarter muscles as well as in 3.5-mm drill holes in the femoral condyles. The tripartite composites produced larger volumes of bone than the bipartite fibrin and BMP or than HA and BMP composites.

Ono et al. observed parosteal ossification in response to composites of sintered, synthetic, porous (71.8 ± 0.7%) hydroxyapatite bovine atelocollagen and rabbit BMP [98]. Marginal bone deposits developed in response to HA alone or HA-atelocollagen. Significantly larger bone deposits developed in response to BMP-HA-atelocollagen. Comparable results were reported in rats with bovine BMP and HA without atelocollagen [99]. In rabbits, apatite-wollastonite-containing glass ceramics became bonded to bone through a calcium-phosphorus-rich layer of apatite; the process was remarkably accelerated by the addition of demineralized bone matrix powders [97].

Blocks of HA ($Ca_{10}(PO_4)_6(OH)_2$) of mean pore size 90 to 200 μm (sintered at 900°C) infused with 1.5 to 2.0 mg of rabbit BMP/NCP induced cartilage development *in vitro* and cartilage and bone deposits in the hindquarter muscles of mice [72].

XIV. CALCIUM SULFATE (PLASTER-OF-PARIS) BMP/NCP COMPOSITES

For over a century of the history of surgery, plaster of paris (calcium sulfate dehydrate, $CaSO_4 \cdot 2H_2O$) has been implanted in large bone cavities in patients with osteomyelitis, bone tumors, and comminuted fractures, and extensive literature has been reviewed along with modern experimental and clinical investigations by Peltier and Jones [100]. Calcium sulfate is biodegradable, slightly soluble in body fluids, and incites a relatively benign inflammatory response. In 26 patients with unicameral bone cysts treated by curettage and packing with plaster-of-paris pellets, resorption of the calcium sulfate and replacement by host bone over a period of about three years occurred. The recurrence rate of the tumor was 8%, which is approximately the same as with treatment by autogenic bone grafts or by filling with polymethylmethacrylate bone cement [100]. Unicameral and aneurysmal cysts and other very large bone tumors have been filled in recent years with bone cement in preference to other methods in many clinics, possibly because of the readily available supply. Plaster-of-paris pellets are commercially prepared and approved for implantation in orthotopic sites [101].

Composites of premixed anhydrous calcium sulfate and BMP/NCP were implanted in mice in heterotopic sites for evaluation as a delivery system. The BMP diffused out of the plaster and induced a shell of bone, including bone marrow, that encased the calcium sulfate. The BMP/NCP was delivered in the calcium sulfate with doses of 2, 11, and 10 mg per 25 mg, respectively, and hydrated with sterile distilled water. The volume of the envelope of new bone developing within a period of four weeks was proportional to the dose of the BMP/NCP [102].

XV. bBMP POLYLACTIC ACID POLYMER COMPOSITES

Table 3 summarizes observations on implants of BMP with polylactic acid (PLA) incorporated. Implants of 0.1 to 0.5 mg of BMP without the PLA delivery system were absorbed without inducing visible deposits of new bone; 0.7 to 1.0 mg produced deposits of new bone hardly visible in roentgenograms. Increasing the quantities of bBMP from 2.0 to 10 mg increased the quantity of bone with the yields reaching a plateau of about 5 mg.

Table 3   bBMP-PLA-Induced Bone Formation in Mouse Thigh

bBMP (mg)	New bone (mm^3)	mg of BMP/20 mg PLA	New bone (mm^3)
0.1	0.0	0.1	0.4
0.2	0.0	0.2	1.0
0.5	0.0	0.5	4.5
0.7	0.2	0.7	6.0
1.0	1.0 ± 0.1	1.0 ± 0.1	12.0 ± 0.4
2.0	3.0 ± 0.1	2.0 ± 0.2	32.0 ± 0.1
5.0	20.0 ± 0.2	5.0 ± 0.1	30.0 ± 0.1
10.0	22.0 ± 0.4	10.0 ± 0.3	30.0 ± 0.2

The quality of new bone was measured on histological sections underlying a grid with 12,600/m^3 by random-point histomorphometric method.

In the PLA composite delivery system, as little as 0.1 mg of BMP produced grossly visible deposits; 0.7 mg increased the yields of new bone to 30 times greater than from quantities from 0.2 mg without PLA. From 1.0 mg, the yield was increased 12 times; from 3.0 mg it was over 10 times greater than from BMP without the PLA delivery system. Doses above 5.0 mg produced somewhat more with than without PLA, while 10 mg produced less, rather than more, new bone. Thus, below the limits or the capacity of the thigh either to produce or contain new bone, roentgenograms showed that a BMP-PLA delivery system produced significantly higher yields of new bone than BMP without the PLA. Control implants of comparable quantities of PLA without BMP or albumin without BMP did not induce bone formation (Fig. 11).

Histological sections prepared at two- and four-day intervals from 2 to 21 days after the operation showed progressive dissolution and absorption of the implants. The residual PLA consisted of highly refractile birefringent fibers. The inflammatory reaction of the tissue to surgical injury subsided within 7 to 14 days. The PLA intrafibrillary spaces were initially filled with small round cells with densely stained nuclei, macrophages, and fusiform connective tissue cells. Later, occasionally multinucleated giant cell enveloped the free ends of a PLA fibril. By Day 21, PLA fibers on the exterior of the implants were covered with deposits of new bone tissue, including hematopoietic bone marrow. In the interior areas of the implants, the remnants of unresorbed fibers were in islands of cartilage or chondroosteoid.

Chemical derivatives and analogs of PLA are commonly used in the manufacture of absorbable suture material. PLA undergoes hydrolytic deesterfication into lactic acid, a normal body tissue metabolite [103,104]. PLA is nontoxic, slowly degraded, entirely

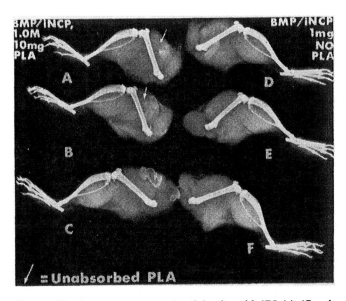

Figure 11 Roentgenogram of polylactic acid (PLA) (Southern Research Institute) with bone morphogenetic protein (BMP)/(NCP) noncollageneous protein, 1 mg per 10 mg of PLA, and 1 mg without the carrier. Without the carrier, the BMP was absorbed too rapidly for a population of responding cells to gather. With the carrier, visible heterotopic bone developed. About half of the PLA was unabsorbed. Arrow shows radiolucent PLA core.

metabolized, and possibly excreted by the respiratory system. The kinetics of the tissue reactions to PLA were comparable to stainless steel and Dacron. The degradation of PLA is generally attended by a negligible giant cell reaction. Depending upon the mass and surface area of the implant, fragments of PLA gradually dissolve over a period of 70 days or longer. Implants of biodegradable 50:50 polylactide (PLA) and polyglycolide (PGA) were resorbed more rapidly than PLA alone in bone defects in rats and healed in less time than control defects [105].

In the absence of BMP, PLA does not induce bone formation. In the presence of BMP in the form of a BMP-PLA delivery system, new bone formation is induced *de novo* in either extraskeletal or intraskeletal sites. Morphologically, the process resembles the demineralized bone matrix-induced bone development. A rapidly biodegradable BMP-PLA/PGA delivery system that is spun into sutures or fashioned into plates, screws, slabs, films, and fabrics would be useful in orthopedic, plastic, maxillofacial, neurosurgical, and dental reconstructive operations.

XVI. 1:1 PLA/PGA IN DOGS

Oklund et al. implanted a 1:1 PLA/PGA wafer containing purified BMP (50 mg) in an 18-mm cranial defect in five adult male dogs [106]. A second 18-mm defect was filled with a wafer without BMP. A third defect was not repaired. Two and three times more new bone was associated with BMP wafers than with wafers without BMP and ungrafted defects, respectively. However, differences were not statistically significant due to the variability among animals. The 1:1 PLA/PGA did not resorb as expected. More new bone grew under (next to dura mater) than over the BMP-containing wafers.

XVII. COMPOSITE OF BMP/NCP AND POLY(ϵ-CAPROLACTONE)

Poly(ϵ-caprolactone) (PCL) is another polymer frequently used in drug-release systems. Its rate of degradation in the physiological environment is lower than that of poly-D,L-lactic acid. Hydrolysis plays a decisive role in the degradation of this polymer. A detailed study of hydrolysis of PCL *in vitro* has demonstrated that the rate of degradation depends on:

The molecular weight of the polymer and its distribution
The degree of crystallinity, because hydrolysis proceeds in amorphous regions of the polymer
The process being an autocatalytic one
Within pH 4–7, the rate of hydrolysis is almost unchanged
Copolymerization with δ-valerolactone, ϵ-caprolactone, or by the incorporation of residues of glycolic and lactic acids

The higher rate of hydrolysis of copolymers compared with the homopolymer was attributed to morphological changes leading to a decrease in the crystallinity.

Composites of BMP/NCP-PCL were rapidly resorbed and replaced by a spherical ossicle of lamellar bone filled with bone marrow. The volume of the ossicle was proportional to the dose of BMP/NCP, ranging from 2 and 6 to 15 mg of BMP/NCP per 30 mg of PCL. Control implants of albumin or dextran produced results comparable to PCL. The data suggested that either rapid hydrolysis or immediate dissolution of the albumin or dextran were generally favorable for induced bone development.

XVIII. BIODEGRADABLE POLYESTER POLYMERS

Polylactic acid (PLA), polyglycolic acid (PGA), and polyparadioxanone are three samples of polyesters that have been manufactured for biodegradable sutures. Hollinger and Battistone reviewed the literature on polymeric materials that have been investigated for bone repair in rats, rabbits, and monkeys [107]. The lactic acid generated by degradation of PLA was incorporated in the tricarboxylic acid cycle, and excreted in the lungs as carbon dioxide and water.

PGA, in addition to being degraded hydrolytically, also is broken down by nonspecific esterases and carboxypeptidases that produce glycolic acid monomers. As the surface area of the polymer increases per unit volume, the susceptibility for hydrolytic degradation increases. A porous block therefore degrades more rapidly than a densely compressed block. A large volume of polymer implanted in a large bone defect is therefore degraded more slowly than a thin, threaded implant.

Dipolar microspheres or packets of polymer and BMP with and without osteoprogenitor cells are now under investigation with the use of new formulations of PLA, PGA, and PLA:PGA. Suganuma and Alexander implanted spherical crystals (spherulites) in bone defects in dogs [108]. The biocompatibility was excellent only in the early stages of degradation. Once degradation commenced, biocompatibility decreased dramatically. The smaller the particles of PLA were, the greater the foreign body inflammatory reaction was. To evaluate the process in the absence of local inflammatory cells, Meikle et al. observed degradation of PLA:PGA 50:50 *in vitro* and observed that 60% to 70% of the BMP was released from the polymer during the first week of culture (Table 3) [109].

The relationship between attachment kinetics and differentiation of osteoprogenitor cells is obscure. Meyer et al. measured the kinetics of bone cell attachments and observed the conditions most favorable for differentiation of osteoprogenitor cells [110]. Cell attachments appear to be due to different surface wettabilities and could be expressed by a linear equation. Quantitative analysis of noncollagenous protein matrix production and DNA content showed better cell differentiation on surfaces of PLA than on tricalcium phosphate or ion-leachable glass.

XIX. LOW MOLECULAR WEIGHT BIODEGRADABLE POLYMERS

Miyamoto et al. incorporated recombinant BMP-4 into two types of copolymers of poly-D,L-lactic acid and polyethylene glycol (PLA-PEG) [111]. One consisted of a poly-D,L-lactic acid (PLA) segment with a molecular weight (MW) of 650 daltons (Da) and a polyethylene glycol (PEG) segment with a MW of 200 Da (PLA-PEG 650-200). The other consisted of a PLA segment with a MW of 650 Da and a PEG segment with a MW of 600 Da (PLA-PEG 650-600). Both the block copolymers were prepared by condensation of poly-D,L-lactic acid oligomer with polyethylene glycol. After completion of these reactions, both PLA-PEG 650-200 and PLA-PEG 650-600 had a viscous and semiliquid nature at room temperature. Poly-D,L-lactic acid with a MW of 650 Da (PLA 650) and polyethylene glycol with a MW of 200 Da (PEG 200) or 600 Da (PEG 600) were also prepared and donated by Taki Chemical. PLA 650 appeared waxy, whereas PEG 200 and PEG 600 were liquid at room temperature.

The BMP-4 PLA-PEG polymers, implanted in the dorsal muscles in mice, were completely resorbed and replaced by bone within three weeks. The quantity of new bone, including bone marrow, was twice as much as induced by BMP alone. The polymers alone did not induce bone development.

PLA-PEG 650-200 was the most effective carrier for BMP of the two low MW biodegradable synthetic polymers. Composites of PLA-PEG 650-200/BMP and hydroxyapatite powder (HAP) also induced ectopic bone formation. Because PLA-PEG 650-200/BMP composite is viscous and semiliquid and PLA-PEG 650-200/BMP/HAP composite is doughy and plastic, the former can be used as an injectable osteoinductive material, and the latter as a grouting agent.

XX. COMPOSITES OF POLY-D[LACTIDE COGLYCOLIDE (PLA-PGA, 50:50)] AND DEMINERALIZED BONE MATRIX AND/OR CERAMIC HYDROXYAPATITE

Schmitz and Hollinger implanted a powder of a copolymer of polylactide-polyglycolide and demineralized bone matrix in 15-mm calvarial defects in rabbits [112]. The composite of powder consisted of 50:50 poly-D-lactide-coglycolide (PLA-PGA copolymer). Bone regeneration occurred within 20 weeks. There were no adverse local or systemic reactions to the composite. The polyester was biodegraded by nonspecific hydrolytic scission.

The rate at which PLA-PGA implants degrade is dependent on the molar ratio of the constituents, the porosity of the implants, the vascularity of the recipient site, the degree of crystallinity of the constituents, and the average molecular weight. The greater the molecular weight is, the slower the biodegradation is. The 50:50 PLA-PGA molar ratio had a half-life that was commensurate with the normal rate of bone regeneration, four to six weeks.

In 15-mm skull trephine defects in rabbits, Kleinschmidt et al. also implanted composites of two biodegradable systems, one synthetic and one (biological) allogeneic demineralized bone matrix and observed repairs in 12 weeks [113]. The quantity of new bone was significantly greater in the composite than in the PLA-PGA alone. Elgendy et al. added ceramic hydroxyapatite (CHA) to the PLA-PGA copolymer to construct a substratum delivery system for osteoinduction and osteoconduction [114]. In a system *in vitro*, the MC3T3-E1 osteogenic cell line synthesized cell adhesion and cell attachment substances slightly more visibly with than without the CHA substrate.

XXI. HIGH MOLECULAR WEIGHT POLYLACTIC ACID HOMOPOLYMERS

Miyamoto et al. implanted composites of mouse osteosarcoma BMP and PLA homopolymers with molecular weights of 105,000, 21,000, 3300, 650, and 160 Da (PLA105000, PLA21000, PLA3300, PLA650, and PLA160, respectively) in allogeneic mice [111]. Approximately 4 mg of water-soluble, semipurified osteosarcoma BMP and 100 mg of one of the PLA homopolymers were implanted into the dorsal muscles in mice. PLA105000/BMP, PLA21000/BMP, PLA3300/BMP, and PLA160/BMP composites failed to induce bone development. PLA105000, PLA21000, and PLA3300 elicited foreign-body reactions or chronic inflammation (or both). PLA160 produced tissue necrosis. PLA650/BMP induced cartilage formation within one week and induced bone with hematopoietic marrow at three weeks postimplantation. PLA650/BMP composites were completely absorbed and replaced by new bone. Of the PLA homopolymers, PLA650 was the preferred polymer for sustained delivery of BMP.

Hollinger and Battistone implanted composites of 50:50 poly-D,L-lactide-coglycolide and bovine BMP/NCP in various proportions in the mouse hindquarter muscle pouch

[107]. There was considerable variability in the polymer degradation rates. The volume of the induced bone deposits was greatest with 100% BMP/NCP, and the quantity of unabsorbed polymer may adversely interfere with formation of the optimum concentration gradient of morphogen. Miyamoto et al. implanted a rapidly degraded composite of a semisolid polymer and BMP-4 and demonstrated relatively large deposits of bone in mice [111].

XXII. COMPOSITES OF POLY-2-HYDROXYETHYL METHACRYLATE SPONGE AND ENDOGENOUS NATIVE BMP

A polymer of 2-hydroxyethyl methacrylate (HEMA) (poly-HEMA sponge) calcifies when implanted subcutis in rats and rabbits. The sponges are 80% air and 20% polymer with pore sizes of 60 to 150 microns. Only the macroporous sponges calcify consistently, and even then only at the margins [115]. In young pigs, there were sites of fibrous tissue and calcification within one month; some months later, the implants were filled with fibrous tissue and also bone [116]. When the polymer was implanted in the dermal layer of the skin, amorphous deposits of apatite mineral were noted as early as 22 days [117]. The polymer was penetrated by small blood vessels, giant cells, and loose connective tissue; within 72 days, bone developed adjacent to areas of calcification. A network of trabecular bone formed throughout the subsurface structure. The bone was enveloped in a capsule of dense fibrous connective tissue. Solid hydron sheets or particles calcified but did not show bone development. Poly-HEMA-methacrylic acid copolymer did not calcify but showed small bone deposits beginning one year after implantation.

Poly-HEMA sponges manufactured with ammonium persulfate also calcified and produced small bone deposits; cross-linking of the diester produced the same results. Polyurethane sponges produced a foreign-body giant cell reaction without any calcification or ossification. Formalinized polyvinyl alcohol sponges calcified without any associated bone development in implants observed as long as 18 months. The same material compressed to produce porous channels with average diameters of 50 μm produced small deposits of bone within 3 to 4 months.

Winter also implanted the poly-HEMA sponge in dermal layers of various parts of the body, including skeletal muscle of pigs, and observed first calcification and then ossification everywhere [117]. Species differences were critical in implants over periods as long as 18 months: mice did not form bone; rabbits and guinea pigs developed only scanty deposits. In pigs, the most consistent factor with the most effective polymers was dystrophic calcification prior to the development of bone. The spatial effect of pore size and ingrowth of blood, vessels, and connective tissue cells, (i.e., 50 μm in diameter) was remarkable. The preliminary calcification is suggestive of the mechanism as observed in intramuscular implants of porous coral-derived hydroxyapatite (CDHA), with bone deposits developing in the baboon as late as three, six, and nine months after implantation [119].

If endogenous BMP is involved in bone development induced by CDHA and poly-HEMA sponge, it is important to speculate on two possible sources. The first is BMP in circulating plasma and body fluids in nanogram or femtogram quantities [1], which could account for a lag phase in binding and accumulating in sites of calcified poly-HEMA sponge. The second possibility is localized adhesion of mesenchymal-type cells to the reactive surfaces of CDHA in the process of bonding to plasma membranes. At bonding surfaces, a residual connective tissue BMP could be converted from a latent to an active form.

XXIII. COMPOSITES OF AUTOGENIC BONE IN POLYESTER POLYMER MESH

For experimental reconstructive operations in the craniofacial regions in dogs, Kinoshita et al. implanted subcutaneous poly(L-lactide) (PLLA) mesh sheets, monofilaments, and mesh cylinders filled with fresh autogenic *particulate cancellous bone and marrow* (PCBM) [119]. Polypropylene (PP) was used as a control. The inflammatory response to PLLA mesh sheets was slight. Histiocytes and multinucleate giant cells appeared on the surface of the monofilaments of PLLA mesh sheets and gradually increased in number as the monofilaments were degraded and absorbed. Almost no inflammatory cellular infiltration was seen in the tissue around PLLA mesh sheets or between the monofilaments. A PLLA mesh cylinder filled with PCBM developed bone within one month. Vascular tissue and bone formation was also observed along the inner wall of the PP cylinder. The mechanical strength of the test material (PLLA monofilaments) did not change for almost two months after implantation; 80% of its initial strength was retained for three months. The PCBM graft supported by PLLA mesh may be very effective in the reconstruction of maxillofacial bones. Tissue reactions associated with biodegradation of the PLLA mesh were slight. PLLA mesh does not interfere with PCBM bone formation.

The relationship between attachment kinetics and differentiation of osteoprogenitor cells is obscure. Meyer et al. measured the kinetics of bone cell attachments and observed the conditions most favorable for differentiation of osteoprogenitor cells [119]. Cell attachments appear to be due to different surface wettabilities and could be expressed by a linear equation. Quantitative analysis of noncollagenous protein matrix production and DNA content show that the conditions for cell differentiation were better on surfaces of PLLA than on tricalcium phosphate or ion-leachable glass.

XXIV. COMPOSITES OF BIODEGRADABLE POLYESTER POLYMER AND ENDOGENOUS BMP IN DEMINERALIZED BONE MATRIX

Schmitz and Hollinger implanted composites of a biodegradable copolymer of polylactide-poly-glycolide combined with allogeneic demineralized freeze-dried bone (DFDB) into 15-mm diameter defects in the calvaria of 26 New Zealand White rabbits [112]. Similar defects were created in the calvaria of another 26 rabbits. Both the implants and the controls were evaluated clinically, roentgenographically, and histomorphometrically using a Zeiss Image Analysis System (Videoplan, version 4.1).

Both controls and implants were evaluated in groups at 4, 8, 12, 16, 20, and 24 weeks. When compared with the control defects, the copolymer-DFDB implants displayed a significantly greater volume of trabecular bone. Three of the 15-mm diameter defects displayed evidence of complete osseous bridging at eight weeks. No adverse host-tissue responses were observed histologically in any of the specimens.

XXV. COMPOSITES OF bBMP/NCP-PLA IN SPINAL FUSIONS IN DOGS WITH AND WITHOUT AUTOGENIC BONE GRAFTS

To investigate the action of xenogeneic bBMP/NCP in experimental spinal fusions, the lower thoracic vertebra were treated in 13 mature mongrel dogs. Four different fusion methods were used at single intervertebral levels within each dog. Three levels in each dog were used as controls for the BMP level and examined by radiohistomorphometric methods at 3, 6, and 12 weeks; the BMP level was shown to have two to three times more new bone than control levels. Fusion occurred only in five of seven of the BMP levels compared with zero of seven, one of seven, and two of seven of the control levels. The BMP level exhibited an increased number and volume of areas of *de novo* cartilage and woven bone formation at all time intervals compared with all control levels. The PLA was incompletely resorbed and retained in the fusion site [120]. These experiments should be repeated with allogeneic dog (dBMP/NCP) as suggested by Heckman et al. in implants in sites of nonunion [121].

The PLA delivery system is unsatisfactory because it is incompletely absorbed as late as six months after the operation, even in the dogs with fused segments. The residual PLA may occupy a space that ideally should contain bone. A better delivery system (i.e., 50:50 PLA/PGA or atelocollagen) may be a useful adjunct to autogenic bone grafts in inducing spinal fusion. Further investigation with a more rapidly absorbed delivery system in a double-blind, randomized, connective series of animals are essential. However, Muschler et al. obtained improved scores for union of spinal fusions in dogs by means of implants of partially purified bovine BMP and granules of ceramic tricalcium phosphate hydroxyapatite (40:60 composition by percentage).

XXVI. COMPOSITES OF BONE WAX AND BMP/NCP

A mixture of beeswax, almond, oil, and salicylic acid is the standard surgical hemostatic agent known as bone wax. To control bleeding, bone wax is packed with a spatula into the space between the diploe and the bleeding bone marrow sinusoids in neurosurgical operations and occasionally also in orthopedic procedures. In rats with two drill holes in the proximal tibia, bone wax completely inhibits bone repair in one hole while normal regeneration occurs in the adjacent hole in the same bone [122]. The effects of the bone wax were localized because there were no effects on the healing process of a drill hole less than 5 mm away in the same bone. In patients with skull trephine defects, bone wax is a bone regeneration inhibitor. Collagen sponge should be used in place of beeswax in sites of implantation of BMP/NCP.

XXVII. COMPOSITES OF POLYSULFONE AND DEMINERALIZED BONE MATRIX POWDERS IN RATS

Porous polysulfone (PPSF) was fabricated and particles of demineralized bone matrix were incorporated in a solution of 70% alcohol. The composite was encapsulated in fibrous connective tissue. Inside the capsule, the particles of demineralized bone matrix were resorbed and replaced by islands of cartilage and bone within 10 to 20 days after implantation in subcutaneous pouches. The porous polysulfone did not inhibit the process of matrix-induced bone development [123].

XXVIII.   COMPOSITES OF BMP AND COLLAGEN INCLUDING BONE MATRIX

Bone powders of HCl-demineralized rat bone matrix consistently induce bone development only if the quantity of powdered matrix is 10 to 25 mg. Less than 10 mg fails to induce bone formation. However, the addition of acid-soluble Type I collagen plus chondroitin-6-sulfate promoted induced bone development. Bone matrix deactivated by extraction of guanidine hydrochloride was inactive, but the activity was restored by the addition of partially purified BMP-3 or recombinant BMP-2. Thus, bone development required a combination of the BMP and an insoluble collagenous substratum including chondroitin-6-sulfate [14,125–126].

The same preparation of partially purified bovine BMP-3 was inactive when implanted subcutaneously without an insoluble collagenous matrix but, when combined with collagen Type IV and laminin prepared from a mouse tumor BMP-3, the composite induced bone development. To a lesser extent, collagen Type I and Type IV fulfilled the requirement of the insoluble substratum. The BMP-3 was bound to a collagen affinity chromatography column and eluted with 6.0 M urea and 1 M sodium chloride at pH 7.4. The binding of BMP-3 to collagen Type IV was not reduced by either laminin or fibronectin [127].

Bioassays of BMPs, either native or recombinant, have been most reproducible when recombined with bone matrix deactivated by extraction of BMP with guanidine HCl. To prove that this matrix need not contain residual traces of BMP, Katz et al. deactivated bone matrix by irradiation with 60 cobalt in doses of 1 to 3 megarads (Mrad) [128]. A composite consisting of irradiated rat bone matrix and partially purified bovine bone induced bone development as consistently as nonirradiated control matrix. Despite assurances that such composites are BMP free and therefore a valid bioassay system, some recent investigations have omitted the rat bone matrix and substituted Type I collagen or atelocollagen and ceramic calcium phosphate granules [129–131]. However, in the mouse an implant of BMP/NCP at or above threshold quantities for sustaining an active concentration gradient will induce bone development without any exogenous bone matrix, collagen, or mineral. Under these conditions, matrix gla protein may be a physiologic BMP carrier [41,64].

Gerhart et al. demonstrated enhancement of bone repair by composites of inactive allogeneic bone matrix and rhBMP-2 in femoral defects (2.5 cm) with bone plates in sheep [132]. No anti-BMP antibodies were detected by immunoblot. The healing process was completed in three months, the same time as with autogenic bone grafts.

Human placental Type I collagen, reconstituted with a crude extract of BMP/NCP and pressed into compact disks of various sizes and shapes, was implanted in rats; the final shape of the induced bone deposits had predictable dimensions [133]. Calf skin collagen or gelatin was an effective carrier for osteosarcoma-derived BMP-4 implanted in muscle or subperiosteal sites in mice [134].

These reports demonstrate the trend away from deactivated bone matrix as either a carrier or bioassay system for BMP for the following reasons: (1) the geometry of the bone originally considered essential is a result, not the causal factor, in induced bone development; (2) several other delivery systems are effective substitutes; (3) the matrix is not reconstituted by the matrix assay but more accurately functions as a cell surface adhesive.

XXIX. COMPOSITE OF PURE TITANIUM AND ANTIGEN-EXTRACTED AUTOLYSED ALLOGENEIC, DEMINERALIZED BONE MATRIX, OR BMP/NCP

Tubes of fiber titanium packed with AAA bone or demineralized bone matrix and isogeneic bone marrow produced substantially more bone in heterotopic sites than marrow free-implants [135]. A titanium sponge, infiltrated with a solution of bovine BMP/NCP in 4% gelatin and lyophilized, produced a BMP-Ti composite. In the mouse hindquarter muscle assay, BMP-Ti neither promoted nor inhibited the process of induced bone. In electron micrographs, cartilage and bone developed in direct contact with titanium surfaces. Bone formed inside the pores of the titanium sponge [136].

Shen et al. packed deep mineralized bone matrix powders into the pores of titanium alloy for implantation in 2.5-mm rods press-fit to defects in adult rabbits [137]. The titanium alloy delivery system for the BMP employs biomechanical measurements of push out force separation of the implants. Over a period of 12 weeks, the demineralized bone matrix provided fixation stability of the implants comparable with that of an autogenic bone graft. By substituting autogenic bone grafts for BMP and fixation with intramedullary rods, Andersson et al. replaced 3.5-cm segments of the midshaft of long bones with implants of cylinders of sintered titanium fiber in adult baboons [138].

XXX. COMPOSITES OF POLYORTHOESTER AND EITHER DEMINERALIZED BONE MATRIX POWDERS OR BMP/NCP

Because fine powders of bone matrix are difficult to manipulate owing to the lack of cohesion of the particles and dispersal in surrounding tissues, a bioerodible polyorthoester (POE) has been compacted in the solid state to construct a sustained drug-release system [139]. POE is a poly(2,2-diox-cis, trans 1, 4 cyclohexane diamethylene tetrahydrofuran). They packed 11-mm long diaphyseal radial defects in adult rats with the bone powder polyorthoester composite and noted 93.6 ± 44% regeneration compared with 76.6 ± 30% (mean and standard deviation) regeneration of comparable bone defects packed with demineralized bone alone.

POE does not inhibit induced bone development and causes only slight inflammation that subsides within three weeks and is mostly resolved within four weeks. A pseudarthrosis formed between the area of induced new bone formation and the distal bone end of the radial defect. POE is a local drug-releasing system and has been effective for local delivery of indomethacin for inhibition of heterotopic induced bone development in rats.

Composites of POE and BMP/NCP have been tested in heterotopic and orthotopic sites for a delivery system in mice, rats, and rabbits [Texiera and Urist, unpublished observations, 1992]. The BMP is released rapidly. The POE is bioeroded too slowly to be replaced entirely by new bone. In rats, the inflammatory response to the POE is more pronounced than in mice and may dissipate the earliest phase of the morphogenetic process of bone development. The fabrication of synthetic nonbiological, nonimmunogenic biodegradable substances designed to release BMP rapidly within hours and slowly over a 28-day period is an important objective of further research. Appel et al. fabricated POE disks loaded with PGE_2 and observed stimulation of localized bone formation in rats [140].

XXXI. POLYDIOXANONE

A polymer prepared by polymerizing the monomer paradioxanone is named polydioxanone (PDX) (Orthosort, Johnson and Johnson, New Brunswick, NJ). PDX is degraded by hydrolysis in the body. Polydioxanone is available as monofilament sutures and absorbable rods. Papagelopoulos et al. repaired small bone fractures in rabbits, and noted biodegradation of the PDX within 24 weeks after bone consolidation was complete [141]. New bone formation enveloped the PDX as early as 3 weeks. The rapidity of biodegradation of PDX suggests that composites of PDX and BMP/NCP warrant investigation.

XXXII. COMPOSITES OF POLYPHOSPHAZENES AND BMP

High molecular weight polymers with a backbone of alternating phosphorous and nitrogen atoms are classified as polyphosphazenes (PPZs). The polymer degrades to generate nontoxic products (i.e., ammonia phosphate) and the organic side group without inciting grossly visible inflammation. Observations by Laurencin et al. *in vitro* on the MC3T3-E1 osteogenic cell line suggests that PPZ may prove to be useful as a biodegradable BMP polymer composite for repair of bone defects, but further experiments are required to prove efficacy in heterotopic and orthotopic sites *in vivo* [142].

XXXIII. COMPOSITE OF BMP/NCP AND ISOBUTYL 2-CYANOACRYLATE

Harper and Ralston implanted isobutyl 2-cyanoacrylate (ICA) in osteochondral defects in the femurs of dogs [143]. The ICA was nontoxic and did not retard fracture healing. ICA is an industrial adhesive with the capacity to form strong bonds rapidly by undergoing anionic polymerization in the presence of hydroxyl or amino groups such as in protein. ICA does not bond to the collagen of bone. Mixed with BMP/NCP, ICA produced only a low level of heat of polymerization. The composite induced development of a shell of bone in a heterotopic site within three weeks in mice.

XXXIV. SESAME OIL BMP/NCP COMPOSITE

A mixture of olein (75%), stearin, palmitin, myristin, linolein, and sessamin is known as sesame oil. As a solvent for fat-soluble vitamins, hormones, and other biological molecules, sesame oil is a commonly used delivery system. In mice, BMP/NCP suspended in sesame oil (10 mg/ml) and injected in muscle induces bone development in the same time intervals as BMP/NCP alone. The tissue reaction is remarkably benign and the release of BMP is rapid in the first 24 hours, and slowly sustained in the subsequent 21 days [M. R. Urist et al., unpublished experiments].

XXXV. SQUALENE, A PROSPECTIVE CARRIER OF BMP

Native bBMP has been suspended in an unsaturated terpene hydrocarbon, obtained by hydrogenation of shark oil squalene, and implanted subcutaneously in rats. The composite induced markedly larger deposits than the bBMP alone. Squalene incited minimal inflammatory reaction unassociated with the bone deposits. Squalene is nonrancid, easily emulsified, and in wide use as a base for cosmetics [144].

XXXVI. CALCIUM GLYCEROPHOSPHORIC BMP/NCP COMPOSITE

In a delivery system for BMP, calcium glycerophosphate has unique properties. Glycerophosphate is very difficult to hydrolyze into glycerol and phosphoric acid. It is not decomposed by heat under alkaline conditions and it is hydrolyzed very slowly under physiologic conditions. It is efficiently hydrolyzed by phosphatase, an enzyme synthesized by chondroblasts and osteoblasts. Calcium glycerophosphate is even more insoluble in body fluids than phosphoric acid. Calcium glycerophosphate is added to culture media to promote calcification of cell cultures derived from bone marrow stroma and embryonic anlagen of chick bones. Composites composed of equal quantities of 3, 5, and/or 7 mg of calcium glycerophosphate and BMP/NCP induced development of large deposits of heterotopic bone in the hindquarter muscle pouch of adult mice. Quantity of bone is 10 times greater than produced by implants of BMP alone. Neither calcification nor bone development occurs with implants of calcium glycerophosphate alone [M. R. Urist et al., unpublished experiments].

XXXVII. SUSPENSIONS OF BMP/NCP IN SOLUTIONS OF DEXTRAN

BMP/NCP suspended in a 20% solution of dextran induces development of a spherical mass of new bone. The solutions are prepared in concentrations of 3, 5, and 7 mg of BMP/NCP in 0.15, 0.25, and 0.33 ml of dextran solution. The solution is loaded in a Number 5 gelatin capsule and implanted in the hindquarter muscle pouch of adult mice. Quantity of bone is proportional to the mass of the implanted BMP/NCP (Lietze et al., unpublished experiments).

XXXVIII. COMPOSITE OF POROUS COLLAGEN/CERAMIC CARRIER COMPOSED OF SOLUBLE BOVINE DERMOCOLLAGEN AND HYDROXYAPATITE/ TRICALCIUM PHOSPHATE CERAMIC AND RABBIT BMP/NCP

Toriumi et al. designed a facial augmentation system in rabbits to determine the effects of soluble bovine dermocollagen and compared with porous collagen ceramic hydroxyapatite composite [145]. When the composite was placed over the bony anterior plate of the maxillary sinuses, a bony mass developed within 21 days after the operation. The addition of TGF-β to the composite may or may not have increased the bone formation by stimulating cellular activity. The facial bony mass persists longer in the tissues than control implants of the carrier without BMP. Toriumi et al. also implanted demineralized bone matrix alone in facial augmentation or reconstruction operations in 53 patients [146]. Overall, 49% of the implants were resorbed within an average period of 14.3 months. Demineralized bone seemed to be incorporated only when implanted in direct contact with viable bone on all sides.

XXXIX. COMPOSITES OF BONE MATRIX POWDER AND BMP IN MANDIBULAR DEFECTS

In 3-cm full-thickness segmental defects in the mandible of 26 dogs, a composite of demineralized dog bone matrix powder and human recombinant BMP-2 induced growth of mineralized new bone in 68% of the operations. Biomechanical strength of defects increased significantly in the interval from three to six months and was correlated with the degree of mineralization and the quantity of bone bridging the defects [129]. Rat bone matrix and BMP-3 cast in silicone rubber molds transposed in flaps of adductor muscles produced a shell of bone within 10 days [147]. The overall effects on induced bone development were demonstrated by Ma et al. [148], who measured alkaline phosphatase activity in subcutaneous implants in rats. In the presence of exogenous BMP-3, bone matrix was 180 times more effective than glass, hydroxyapatite, β-tricalcium phosphate, or polymethylmethacrylate. Bovine BMP-3 (osteogenin) was avidly bound to both α and α_2 chains of collagen Type IV and to a lesser extent also to Types I and IX; neither laminin nor fibronectin interfered with binding [127].

XL. STERILIZATION OF ENDOGENOUS AND EXOGENOUS BMP

Although others have reported ethylene oxide sterilization adversely affected BMP activity (its major reaction products are ethylene glycol and ethylene chlorhydrin), this is now subject to qualification. One cycle in a sterilization chamber reduced only less than half of the BMP activity. However, irradiation sterilization in doses of 16 to 19 Gy penetrates bone tissues far better than ethylene oxide. Levels of 25 to 35 Gy, the levels recommended by the World Health Organization for sterilization of contaminated bone, eliminate all BMP activity [149,150].

Bone banks dispense only bone collected under aseptic conditions from donors extensively screened for human immunodeficiency virus (HIV) and hepatitis, then cultured for other microbial contaminants for many weeks. In any case, preservation of noncollagenous bioactive molecules should continue to be individually investigated in considerable detail.

Prolo et al. have extensively analyzed ethylene oxide (EO) gas-sterilized, allogenic freeze-dried bone, dura, and fascia for augmentation of spinal fusions or repair of cranial defects and other reconstructive procedures [151,152]. Alkylation is the probable mechanism by which EO is bactericidal. Provided that it is removed by prolonged aeration, EO is a safe, inexpensive sterilant of freeze-dried bone and soft tissues. The requirements of the AAA bone procedure, as distinguished from that for EO bone, are (1) the stabilization of BMP by the use of neutral protease inhibitors, (2) surface demineralization to facilitate rapid replacement by host bone, (3) chemosterilization of AAA bone before infusion of BMP sterilized by EO or sodium azide. Only compact cortical bone is used for preparation of AAA bone. AgCl (10^{-4} M) solutions had antibacterial effects and diminishing BMP activity in bone matrix [153].

XLI. COMPOSITES OF BMP-4 AND POLYLACTIC ACID HOMOPOLYMERS AS CARRIERS FOR BONE MORPHOGENETIC PROTEIN

Of composites consisting of 4 mg of water-soluble, partially purified BMP-4 and 100 mg of one of the PLA homopolymers with molecular weights varying from 105 to 160 daltons in mice, only the PLA 650 composites induced cartilage formation within one week, and bone including bone marrow within three weeks [154].

XLII. COMPOSITE OF BMP AND COLLAGEN TYPE I DELIVERY SYSTEM

To avoid the use of the uncertainties of incompletely deactivated demineralized rat bone matrix, Takaoka et al. bioassayed BMP-4 in pellets of pure pig skin Type I atelocollagen without any other carrier [131].

XLIII. CARBON

The remarkable histocompatibility of carbon has been investigated in twisted carbon fibers and in rabbits implanted with it for repair of tendons [155,156]. Carbon particles have been tested as a delivery system for BMP in the form of a physical mixture and compressed into the form of tablets 3 mm in diameter and 1 mm thickness. Dose responses to 0.1, 0.5, 1.0, and 5.0 mg BMP per 10 mg of carbon particles implanted in the hindquarter muscle pouch in mice were measured. Induced bone development occurred within 10, 15, and 28 days. The volumes of new bone were 10% to 50% less than control implants of BMP/NCP. Carbon incites a moderate foreign-body inflammatory cell, but the reaction culminates in fibrous connective tissue with occasional foreign-body multinucleated giant cells.

To investigate the effects of surface topography on orientation and migration of differentiating bone cells, Gomi and Davies cultured rat bone marrow stroma cells on smooth and roughened silicone carbide surfaces [157]. Rugosity influenced differentiation and/or fusion of osteoclastlike cells. Both quantity and spatial distribution of bone were modified by the substratum surface roughness. Physical mixtures of carbon fibers and BMP/NCP in composites of 1 part BMP/NCP to 10 parts carbon fibers implanted in mice did not enhance or increase the quantity of bone formation [Lietze et al., unpublished observations].

XLIV. EXPANDED POLYTETRAFLUOROETHYLENE BARRIER MEMBRANES

Expanded polytetrafluoroethylene (ePTFE) membranes have been implanted in extensive investigations in animals and patients with periodontal defects. ePTFE is a relatively inert, nonresorbable, remarkably histocompatible polymer. ePTFE has been used for many years in vascular surgery or replacement of segments of arteries in patients with aneurysms.

For the past decade, the rationale of ePTFE membrane implantation has been based on the concept of guided regeneration. This concept assumes that specific cells and the migration potential of these cells are responsible for healing following periodontal reconstructive therapy. Periodontal ligament cells with the potential for periodontal regeneration are expected to form a new root cementum with inserting collagen fibers. Guided regeneration assumes that periodontal regeneration may only take place if gingival epithelial and connective tissue cells are excluded from the root surface. The barrier membrane of ePTFE is expected to provide the optimum conditions for periodontal regeneration.

Haney et al. contended that cementum regeneration was not promoted by guided tissue regeneration with ePTFE membranes [158]. Increased root resorption by gingival connective tissue was not promoted by the membrane barrier. To oppose the guided regeneration hypothesis, Haney reviewed the sequence of events in the healing of periodontal defects: (1) connective tissue repair to the root of the surface is a function of wound stability; (2) bone regeneration is dependent on space provision; (3) exclusion of gingival connective tissue from the root surface does not prevent root resorption; (4) provisions for guided regeneration do not promote cementum regeneration; (5) complete gingival coverage of the barrier membrane appears to be critical for optimal healing. For optimum repair and control of infection, the membrane must be completely covered at wound closure with space provided adjacent to the root surface. However, periodontal defects, filled with decalcified freeze-dried allogeneic bone implants alone, with and without interposed ePTFE membranes, showed no differences in outcome of the treatment in groups of 15 patients [159]. Uretzky et al. concluded that ePTFE had inhibitory effects on DBM-induced bone development when observed in a percutaneous tube in rats [160].

XLV. METHYL PYRROLIDINONE CHITOSAN

Implants of methyl pyrrolidinone chitosan (MPC) induce bone regeneration in human tooth extraction sockets and in rabbit subperiosteal pouches. The MPC can be fabricated with growth factors in freeze-dried sponge [161].

XLVI. ETHYLENE-VINYL ACETATE COPOLYMERS IN METHYLENE CHLORIDE (ELVAX 40)

Beginning with research on neovascularization [162], ethylene vinyl acetate copolymer (ELVAX 40) has been used for delivery systems for growth factors and enzymes. Implants of BMP incorporated in ELVAX, under investigation in mice and rats, demonstrated remarkable histocompatibility of the copolymer.

XLVII. POLY(ETHYLENE OXIDE HYDANTOIN) POLY(BUTYLENE TEREPHTHALATE) SEGMENTED COPOLYMER POLYACTIVE

Van Blitterswijk et al. implanted poly(ethylene oxide hydantoin) (HPEO) and poly(butylene terephthalate) (PBT) subcutis in rats and observed dystrophic calcification and biodegradation of the copolymer. With increasing PEO content, dystrophic calcification was more prominent than biodegradation. No calcification occurred with 30/70, and maximum calcification occurred with 70/30. Biodegradation of the 70/30 was almost complete at 12 months. Calcification of PEO/PBT is a well-known phenomenon, and is associated with calcium adsorption to the PEO. Why calcification begins in the interior of PEO/PBT blocks (70/30) is not known. The deposits consist of needle-shaped hydroxyapatitelike calcium phosphates. The calcification is also correlated with bonding of the copolymer to bone, when PEO is implanted in trephine defects; 30/70 PEO/PBT did not bond to bone.

XLVIII. AN ENDOGENOUS LIPID DELIVERY SYSTEM FOR BMP

One delivery system with the potential to advance knowledge of the mechanism of delivery of BMP is an endogenous lipid found in cortical bone. The lipid is closely associated with the noncollagenous proteins, tightly bound to BMP, and quantitatively separable by extraction with cold acetone. The lipid is sudanophilic, unrelated to phospholipids, and predominantly triglyceride as revealed by thin-layer chromatography. The BMP/NCP/LIPID is inactivated by extraction of acetone-soluble lipids and reactivated by replacement of this lipid.

The biochemical character of the BMP/NCP/LIPID is presently under intensive investigation. When implanted as a composite with bovine BMP with a molecular mass of approximately 19 kDa and a 14-kDa matrix gla protein, it induces heterotopic bone development. Recombinant BMP induces only about 10% of the volume of bone induced by native BMP/NCP. In rats, as a composite with lipid, the native bovine BMP may induce 10 to 20 times greater volumes of heterotopic bone than recombinant BMP. The BMP/NCP/LIPID aggregate does not require deactivated bone matrix or any other carrier. Experiments now in progress are designed to determine whether bone matrix lipids associated with noncollagenous proteins normally function as the carrier of endogenous BMP activity (preliminary reports submitted for publication elsewhere).

REFERENCES

1. Urist, M. R., and Hudak, R. T. (1984). Radioimmunoassay of bone morphogenetic protein in serum: A tissue-specific parameter of bone metabolism. *Proc. Soc. Exp. Biol. Med.* 176: 472–475.
2. Bessho, K., Tagawa, T., and Murata, M. (1990). Purification of rabbit bone morphogenetic protein derived from bone, dentin, and wound tissue after tooth extraction. *J. Oral Maxillofac. Surg.* 48(2):162–169.
3. Nogami, H., Ono, Y., and Oohira, A. (1990). Bioassay of chondrocyte differentiation by bone morphogenetic protein. *Clin. Orthop.* 258:295–299.
4. Schlag, G., and Redl, H. (1988). Fibrin sealant in orthopedic surgery. *Clin. Orthop.* 227: 269–285.
5. Kawamura, M., and Urist, M. R. (1988). Human fibrin is a physiologic delivery system for bone morphogenetic protein. *Clin. Orthop.* 235:302–310.
6. Sato, T., Kawamura, M., Sato, K., Iwata, H., and Miura, T. (1991). Bone morphogenesis of rabbit bone morphogenetic protein-bound hydroxyapatite-fibrin composite. *Clin. Orthop.* 263:254–262.
7. Urist, M. R., Iwata, H., Ceccotti, P. L., Dorfman, R. L., Boyd, S. D., McDowell, R. M., and Chien, C. (1973). Bone morphogenesis in implants of insoluble bone gelatin. *Proc. Natl. Acad. Sci. USA* 70(12):3511–3515.
8. Yamashita, K., Horisaka, Y., Okamoto, Y., Yoshimura, Y., Matsumoto, N., Kawada, J., and Takagi, T. (1991). Effect of bupivacaine on muscle tissues and new bone formation induced by demineralized bone matrix gelatin. *Acta Anat.* 141(1):1–7.
9. Yamashita, K., Horisaka, Y., Satomura, K., and Takagi, T. (1991). Analysis of minerals on initial calcification induced by bone matrix gelatin. *Jap. J. Oral Biol.* 33:166–173.
10. Linden, G. J. (1975). Bone induction in implants of decalcified bone and dentin. *J. Anat.* 119(2):359–367.
11. Sela, J., Applebaum, J., and Uretzky, G. (1986). Osteogenesis induced by bone matrix is inhibited by inflammation. *Biomat. Med. Dev. Artif. Organs* 14(3&4):227–237.
12. Hirschfeld, Z., Gozd, D., Bab, I., Ulmansky, M., and Sala, J. (1984). Comparative study of the effect of calcium hydroxide and demineralized matrix on subcutaneous tissue in the rat. *J. Dental Sci.* 1:227–237.
13. Bowers, G., Felton, F., Middleton, C., Glynn, D., Sharp, S., Mellonig, J., Corio, R., Emerson, J., Park, S., Suzuki, J., Ma, S., Romberg, E., and Reddi, A. H. (1991). Histologic comparison of regeneration in human intrabony defects when osteogenin is combined

with demineralized freeze-dried bone allograft and with purified bovine collagen. *J. Periodontol.* 62(11):690–702.

14. Muthukumaran, N., Ma, S., and Reddi, A. H. (1988). Dose-dependence of and threshold for optimal bone induction by collagenous bone matrix and osteogenin-enriched fraction. *Collagen Rel. Res.* 8:433–441.

15. Saraf, S. K., and Tuli, S. M. (1991). Osteoinductive property of fluoride impregnated decalcified allogeneic bone matrix. *Indian J. Exp. Biol.* 29:159–161.

16. Ehrnberg, A., De Pablos, J., Martinez-Lotti, G., Kreicbergs, A., and Nilsson, O. (1993). Comparision of demineralized allogeneic bone matrix grafting (the Urist procedure) and the Ilizarov procedure in large diaphyseal defects in sheep. *J. Orthop. Res.* 11(3):438–447.

17. Katz, R. W., Hollinger, J. O., and Reddi, A. H. (1993). The functional equivalence of demineralized bone and tooth matrices in ectopic bone induction. *J. Biomed. Mater. Res.* 27(2):239–245.

18. Jergesen, H. E., Chua, J., Kaban, L., Kao, R., Fu, K., Chan, A., and Juster, R. (1993). Delayed implantation of demineralized bone powder after local irradiation in rats. *Clin. Orthop.* 294:325–332.

19. Nilsson, O. S., Persson, P.-E., and Ekelund, A. (1990). Heterotopic new bone formation causes resorption of the inductive bone matrix. *Clin. Orthop.* 257:280–285.

20. Gendler, E. (1985). Osteogenesis induced by perforated bone matrix. In *Current Advances in Skeletogenesis* (A. Ornoy, A. Harell, and J. Sela, Eds.), Elsevier Science Publishers, Amsterdam, pp. 21–26.

21. Pinholt, E. M., Bang, G., and Haanaes, H. R. (1990). Alveolar ridge augmentation by osteoinduction in rats. *Scand. J. Dent. Res.* 98:434–441.

22. Hosny, M., and Sharawy, M. (1985). Osteoinduction in rhesus monkeys using demineralized bone powder allografts. *J. Oral Maxillofac. Surg.* 43(11):837–844.

23. Urist, M. R. (1971). The substratum for bone morphogenesis. *Dev. Biol.* (Suppl.) 4:125–163.

24. Irving, J., LeBolt, S. A., and Schneider, E. L. (1981). Ectopic bone formation and aging. *Clin. Orthop.* 154:249–253.

25. Syftestad, G. T., and Urist, M. R. (1982). Bone aging. *Clin. Orthop.* 162:288–297.

26. Nishimoto, S. K., Chang, C.-H., Gendler, E., Stryker, W. F., and Nimni, M. E. (1985). The effect of aging on bone formation in rats: Biochemical and histological evidence for decreased bone formation capacity. *Calcif. Tissue Int.* 37:617–624.

27. Urist, M. R., Nilsson, O. S., Hudak, R., Huo, Y. K., Rasmussen, J., Hirota, W., and Lietze, A. (1985). Immunologic evidence of a bone morphogenetic protein in the *milieu interieor*. *Ann. Biol. Clin.* 43:755–766.

28. Saraf, S. K., Agarwal, K., Tuli, S. M., and Khanna, S. (1991). Autoclaved partially decalcified bone as osteogenic substances — An experimental study. *Indian J. Exp. Biol.* 29:39–42.

29. Strates, B. S., and Urist, M. R. (1969). Origin of the inductive signal in implants of normal and lathyritic bone matrix. *Clin. Orthop.* 66:226–240.

30. Xiaobo, H., Lunlong, Y., Chuanzin, L., Shucheng, W., and Yankun, C. (1993). Experimental and clinical investigations of human insoluble bone matrix gelatin: A report of 24 cases. *Clin. Orthop.* 293:360–365.

31. Urist, M. R., Kovacs, S., and Yates, K. A. (1986). Regeneration of an enchondroma defect under the influence of an implant of human bone morphogenetic protein. *J. Hand Surg.* 11A:417–419.

32. Johnson, E. E., and Urist, M. R. (1989). Bone morphogenetic protein augmentation grafting of segmental defects of the lower extremity. In *Symposium on Bone Transplantation: Update on Osteochondral and Allograft Surgery* (M. Aebi and P. Regazzoni, Eds.), Springer-Verlag, Berlin, pp. 314–315.

33. Johnson, E. E., Urist, M. R., and Finerman, G. A. M. (1988). Repair of segmental defects

of the tibia with cancellous bone grafts augmented with human bone morphogenetic protein: A preliminary report. *Clin. Orthop.* 236:249–257.

34. Johnson, E. E., Urist, M. R., and Finerman, G. A. M. (1990). Composite of human BMP (hBMP/AAA) implantation grafting for resistant non-unions and segmental defects of long bones. *Proc. Soc. Intern. Orthop. and Trauma* 1:585.

35. Johnson, E. E., Urist, M. R., and Finerman, G. A. M. (1990). Distal metaphyseal tibial nonunion: Deformity and bone loss treated by open reduction, internal fixation, and human bone morphogenetic protein (hBMP). *Clin. Orthop.* 250:234–240.

36. Hollinger, J. O., Schmitz, J. P., Mark, D. E., and Seyfer, A. E. (1990). Osseous wound healing with xenogeneic bone implants with a biodegradable carrier. *Surgery* 107(1):50–54.

37. Glowacki, J., Kaban, L. B., Murray, J. E., Folkman, J., and Mulliken, J. B. (1981). Application of the biological principle of induced osteogenesis for craniofacial defects. *Lancet* 1(8227):959–962.

38. Senn, N. (1889). On healing of aseptic bone cavities by implantation of antiseptic decalcified bone. *Am. J. Med. Sci.* 98:219–243.

39. Pike, R. L., and Boyne, P. J. (1974). Use of surface-decalcified allogeneic bone and autogenous marrow in extensive mandibular defects. *J. Oral Surg.* 32:177–182.

40. De Pablos, J., Alfaro, C., Martinez-Lotti, G., Nilsson, O., and Barrios, C. (1992). Valor de la desmineralización de aloinjertos en el tratamiento de grandes defectos óseos. *Rev. Ortop. Traum.* 36(4):495–503.

41. Kawai, T., and Urist, M. R. (1988). Quantitative computation of induced heterotopic bone formation by an image analysis system. *Clin. Orthop.* 233:262–267.

42. Buchholz, H. W., and Engelbrecht, E. (1970). Uber die depotwirkung einiger antibiotika bie vermischung mit dem kunstharz palacos. *Chirurgie* 41:511–515.

43. Dueland, R., Spadaro, J. A., and Rahn, B. A. (1982). Silver antibacterial bone cement: Comparison with gentamicin in experimental osteomyelitis. *Clin. Orthop.* 169:264–268.

44. Downes, S., Wood, D. J., Malcolm, A. J., and Ali, S. Y. (1990). Growth hormone in polymethylmethacrylate cement. *Clin. Orthop.* 252:294–298.

45. Schurman, D. J., Trindade, C., Hirshman, H. P., Moser, K., Kajiyama, G., and Stevens, P. (1978). Antibiotic-acrylic bone cement composites. *J. Bone Joint Surg.* 60A(7):978–984.

46. Törholm, C., Lidgren, L., Lindberg, L., and Kahlmeter, G. (1983). Total hip joint arthroplasty with gentamicin-impregnated cement: A clinical study of gentamicin excretion kinetics. *Clin. Orthop.* 181:99–106.

47. Bayston, R., and Milner, R. D. G. (1982). The sustained release of antimicrobial drugs from bone cement. *J. Bone Joint Surg.* 64B(4):460–464.

48. Carlsson, A. S., Josefsson, G., and Lindberg, L. (1978). Revision with gentamicin-impregnated cement for deep infections in total hip arthroplasties. *J. Bone Joint Surg.* 60A(8):1059–1064.

49. Dingeldein, E., Bergmann, R., Wahlig, H., Metallinos, A., Simane, Z., and Hermanek, P. (1980). Pharmacokinetics and tolerance of gentamicin-polymethylmethacrylate-beads in beagle dogs. In *Biomaterials*, (G. D. Winter, D. F. Gibbons, and H. Plenk, Jr., Eds.), John Wiley and Sons, New York, pp. 315–320.

50. Hoff, S. F., Fitzgerald, R. H., and Kelly, P. J. (1981). The depot administration of penicillin G and gentamicin in acrylic bone cement. *J. Bone Joint Surg.* 63A(5):798–804.

51. Josefsson, G., Lindberg, L., and Wiklander, B. (1981). Systemic antibiotics and gentamicin-containing bone cement in the prophylaxis of postoperative infections in total hip arthroplasty. *Clin. Orthop.* 159:194–200.

52. Mizutani, H., and Urist, M. R. (1982). The nature of bone morphogenetic protein (BMP) fractions derived from bovine bone matrix gelatin. *Clin. Orthop.* 171:213–223.

53. Takagi, K., and Urist, M. R. (1982). The reaction of the dura to bone morphogenetic protein (BMP) in repair of skull defects. *Ann. Surg.* 196(1):100–109.

54. Takagi, K., and Urist, M. R. (1982). The role of bone marrow in bone morphogenetic protein-induced repair of femoral massive diaphyseal defects. *Clin. Orthop.* 171:224–231.

55. Urist, M. R. (1981). New bone formation induced in postfetal life by bone morphogenetic protein. In *Mechanisms of Growth Control* (R. O. Becker, Ed.), Charles C. Thomas, Springfield, IL, pp. 406–434.

56. Urist, M. R., Conover, M. A., Lietze, A., Triffitt, J. T., and DeLange, R. (1981). Partial purification and characterization of bone morphogenetic protein. In *Hormonal Control of Calcium Metabolism* (D. Cohn, R. Talmage, and J. L. Matthews, Eds.), Excerpta Medica, Amsterdam, The Netherlands, pp. 307–314.

57. Urist, M. R., Lietze, A., Mizutani, H., Takagi, K., Triffitt, J. T., Amstutz, J., DeLange, R., Termine, J., and Finerman, G. A. M. (1982). A bovine low molecular weight bone morphogenetic protein (BMP) fraction. *Clin. Orthop.* 162:219–232.

58. Urist, M. R., Mizutani, H., Conover, M. A., Lietze, A., and Finerman, G. A. M. (1983). Dentin, bone and osteosarcoma tissue bone morphogenetic proteins. In *Factors and Mechanisms Influencing Bone Growth* (A. D. Dixon and B.G. Sarnat, Eds.), Allen R. Liss, New York, pp. 61–81.

59. Wahlig, H., and Dingeldein, E. (1980). Antibiotics and bone cements. Experimental and clinical long-term observations. *Acta Orthop. Scand.* 51(1):49–56.

60. Wahlig, H., Dingeldein, E., Bergmann, R., and Reuss, K. (1978). The release of gentamicin from polymethylmethacrylate beads. *J. Bone Joint Surg.* 60B(2):270–275.

61. Weinstein, A. M. (1980). Polymers in orthopaedic surgery. *Orthopaedic Surgery: A Weekly Update* 1(15):3–5.

62. Seligson, D., and Henry, S. L. (1993). Newest knowledge of treatment for bone infection: Antibiotic-impregnated beads. *Clin. Orthop.* 295:2–118.

63. Henrich, D. E., Cram, A. E., Park, J. B., Liu, Y. K., and Reddi, H. (1993). Inorganic bone and demineralized bone matrix impregnated bone cement: A preliminary *in vivo* study. *J. Biomed. Mater. Res.* 27(2):277–280.

64. Urist, M. R., Huo, Y. K., Brownell, A. G., Hohl, W. M., Buyske, J., Lietze, A., Tempst, P., Hunkapillar, M., and DeLange, R. J. (1984). Purification of bovine bone morphogenetic protein by hydroxyapatite chromatography. *Proc. Natl. Acad. Sci. USA* 81(2):371–375.

65. Urist, M. R., Nilsson, O. S., Rasmussen, J., Hirota, W., Lovell, T., Schmalzreid, T., and Finerman, G. A. M. (1987). Bone regeneration under the influence of a bone morphogenetic protein (BMP) beta-tricalcium phosphate (TCP) composite in skull trephine defects in dogs. *Clin. Orthop.* 214:295–304.

66. Wu, C.-H., Hara, K., and Ozawa, H. (1992). Enhanced osteoinduction by intramuscular grafting of BMP-β-TCP compound pellets into murine models. *Arch. Histol. Cytol.* 55(1):97–112.

67. Klein, C. P. A. T., Driessen, A. A., de Groot, K., and van den Hooff, A. (1983). Biodegradation behavior of various calcium phosphate materials in bone tissue. *J. Biomed. Mater. Res.* 17(5):769–784.

68. Frayssinet, P., Trouillet, J. L., Rouquet, N., Azimus, E., and Autefage, A. (1993). Osseointegration of macroporous calcium phosphate ceramics having a different chemical composition. *Biomaterials* 14(6):423–429.

69. den Hollander, W., Patka, P., Klein, C. P. A. T., and Heidendal, G. A. K. (1991). Macroporous calcium phosphate ceramics for bone substitution: A tracer study on biodegradation with ^{45}Ca tracer. *Biomaterials* 12(6):569–573.

70. Metsger, D. S., Driskell, T. D., and Paulsrud, J. R. (1982). Tricalcium phosphate ceramic—A resorbable bone implant: Review and current status. *J. Am. Dent. Assoc.* 105(6):1035–1038.

71. Holmes, R. E., Bucholz, R. W., and Mooney, V. (1987). Porous hydroxyapatite as a bone graft substitute in diaphyseal defects: A histometric study. *J. Orthop. Res.* 5(1):114–121.

72. Kawamura, M., Iwata, H., Sato, K., and Miura, T. (1987). Chondroosteogenetic response to crude bone matrix proteins bound to hydroxyapatite. *Clin. Orthop.* 217:281–292.

73. Baslé, M. F., Chappard, D., Grizon, F., Filmon, R., Delecrin, J., Daculsi, G., and Rebel, A. (1993). Osteoclastic resorption of Ca-P biomaterials implanted in rabbit bone. *Calcif. Tissue Int.* 53:348–356.

74. Holmes, R. E. (1979). Bone regeneration within a coralline hydroxyapatite implant. *Plast. Reconstr. Surg.* 63(5):626–633.

75. Yamazaki, Y. (1984). Experimantal study on porous apatite as artificial bone. Experimentation on implantation in the mandible. *J. Stromatol. Soc. Japan* 51(2):184–218.

76. Muschler, G. F., Huber, B., Ullman, T., Barth, R., Easley, K., Otis, J. O., and Lane, J. M. (1993). Evaluation of bone-grafting materials in a new canine segmental spinal fusion model. *J. Orthop. Res.* 11(4):514–524.

77. Chiroff, R. T., White, E. W., Weber, J. N., and Roy, D. M. (1975). Tissue ingrowth of replamineform implants. *J. Biomed. Mater. Res.* 9(4):29–45.

78. Roy, D. M., and Linnehan, S. K. (1974). Hydroxyapatite formed from coral skeletal carbonate by hydrothermal exchange. *Nature* 247(5438):220–222.

79. White, E. W., Weber, J. N., Roy, D. M., Owen, E. L., Chiroff, R. T., and White, R. A. (1975). Replamineform porous biomaterials for hard tissue implant applications. *J. Biomed. Mater. Res.* 9(4):23–27.

80. Wells, J. W. (1956). Scleractinia. In *Treatise on Intervertebrate Paleontology*, Part F (R. C. Moore, Ed.), University of Kansas Press, Kansas City, pp. 328–444.

81. Tencer, A. F., Shors, E. C., Woodard, P. L., and Holmes, R. E. (1990). Mechanical and biological properties of a porous polymer-coated coralline ceramic. In *Handbook of Bioactive Ceramics*, Vol. 2, *Calcium Phosphate and Hydroxyapatite Ceramics* (T. Yamamuro, L. L. Hench, and J. W. Hench Eds.), CRC Press, Boca Raton, FL, pp. 209–221.

82. White, E., and Shors, E. C. (1986). Biomaterial aspects of interpore-200 porous hydroxyapatite. *Dent. Clin. North Am.* 30(1):49–67.

83. Shimazaki, K., and Mooney, V. (1985). Comparative study of porous hydroxyapatite and tricalcium phospate as bone substitute. *J. Orthop. Res.* 3(3):301–310.

84. Piecuch, J. F. (1982). Extraskeletal implantation of a porous hydroxyapatite ceramic. *J. Dent. Res.* 61(12):1458–1460.

85. Ripamonti, U., Yeates, L., and van den Heever, B. (1993). Initiation of heterotopic osteogenesis in primates after chromatographic adsorption of osteogenin, a bone morphogenetic protein, onto porous hydroxyapatite. *Biochem. Biophys. Res. Comm.* 193(2):509–517.

86. Holmes, R. E., and Hagler, H. K. (1988). Porous hydroxyapatite as a bone graft substitute in cranial reconstruction: A histometric study. *Plast. Reconstr. Surg.* 81(5):662–671.

87. Holmes, R. E., and Hagler, H. K. (1987). Porous hydroxyapatite as a bone graft substitute in mandibular contour augmentation: A histometric study. *J. Oral Maxillofac. Surg.* 45(3): 421–429.

88. Piecuch, J. F., Topazian, R. G., Skoly, S., and Wolfe, S. (1983). Experimental ridge augmentation with porous hydroxyapatite implants. *J. Dent. Res.* 62(2):148–154.

89. Holmes, R. E., Bucholz, R. W., and Mooney, V. (1986). Porous hydroxyapatite as a bone-graft substitute in metaphyseal defects. *J. Bone Joint Surg.* 68A(6):904–911.

90. Alper, G., Bernick, S., Yazdi, M., and Nimni, M. E. (1989). Osteogenesis in bone defects in rats: The effects of hydroxyapatite and demineralized bone matrix. *Am. J. Med. Sci.* 298(6): 371–376.

91. Ohgushi, H., Dohi, Y., Tamai, S., and Tabata, S. (1993). Osteogenic differentiation of marrow stromal stem cells in porous hydroxyapatite ceramics. *J. Biomed. Mater. Res.* 27(11):1401–1407.

92. Takaoka, K., Nakahara, H., Yoshikawa, H., Masuhara, K., Tsuda, T., and Ono, K. (1988). Ectopic bone induction on and in porous hydroxyapatite combined with collagen and bone morphogenetic protein. *Clin. Orthop.* 234:250–254.

93. Ko, L., Ma, G.-X., and Gao, H.-L. (1990). Purification and chemical modification of porcine bone morphogenetic protein. *Clin. Orthop.* 256:229–237.

94. Nishimura, N., Taguchi, Y., Yamamuro, T., Nakamura, T., Kokubo, T., and Yoshihara, S. (1993). A study of the bioactive bone cement-bone interface: Quantitative and histological evaluation. *J. Appl. Biomat.* 4:29–38.

95. Cheung, H. S., and Tofe, A. J. (1993). Mechanism of cell growth on calcium phosphate particles: Role of cell-mediated dissolution of calcium phosphate matrix. *S.T.P. Pharma. Sciences* 3(1):51–55.

96. Miller, T. A., Ishida, K., Kobayashi, M., Wollman, J. S., Turk, A. E., and Holmes, R. E. (1991). The induction of bone by an osteogenic protein and the conduction of bone by porous hydroxyapatite: A laboratory study in the rabbit. *Plast. Reconstr. Surg.* 87(1):87–95.

97. Kotani, S., Yamamuro, T., Nakamura, T., Kitsugi, T., Fujita, Y., Kawanabe, K., and Kokubo, T. (1992). Enhancement of bone bonding to bioactive ceramics by demineralized bone powder. *Clin. Orthop.* 278:226–234.

98. Ono, I., Ohura, T., Murata, M., Yamaguchi, H., Ohnuma, Y., and Kuboki, Y. (1992). A study on bone induction in hydroxyapatite combined with bone morphogenetic protein. *Plast. Reconstr. Surg.* 90(5):870–879.

99. Horisaka, Y., Okamoto, Y., Matsumoto, N., Yoshimura, Y., Kawada, J., Yamashita, K., and Takagi, T. (1991). Subperiosteal implantation of bone morphogenetic protein adsorbed to hydroxyapatite. *Clin. Orthop.* 268:303–312.

100. Peltier, L. F., and Jones, R. H. (1978). Treatment of unicameral bone cysts by curettage and packing with plaster-of-paris pellets. *J. Bone Joint Surg.* 60A(6):820–822.

101. Coetzee, A. S. (1980). Regeneration of bone in the presence of calcium sulfate. *Arch. Otolaryngol.* 106:405–409.

102. Yamazaki, Y., Oida, S., Akimoto, Y., and Shioda, S. (1988). Response of the mouse femoral muscle to an implant of a composite of bone morphogentic protein and plaster of paris. *Clin. Orthop.* 234:240–249.

103. Kulkarni, R. K., Pani, K. C., Neuman, C., and Leonard, F. (1966). Polylactic acid for surgical implants. *Arch. Surg.* 93(5):839–843.

104. Kulkarni, R. K., Moore, E. G., Hegyeli, A. F., Leonard, F. (1971). Biodegradable poly(lactic acid) polymers. *J. Biomed. Mater. Res.* 5(3):169–182.

105. Kopecek, J., and Ulbrich, K. (1983). Biodegradation and biomedical polymers. *Prog. Polym. Sci.* 9:1–58.

106. Oklund, S. A., Prolo, D. J., Urist, M. R., and King, S. E. (1987). Repair of canine skull defects implanted with polylactic acid copolymer (PLA) wafers containing BMP. *Transactions of the 11th Annual Meeting of the American Association of Tissue Banks.* Washington, DC, September 28–30.

107. Hollinger, J. O., and Battistone, G. C. (1986). Biodegradable bone repair materials: Synthetic polymers and ceramics. *Clin. Orthop.* 207:290–305.

108. Suganuma, J., and Alexander, H. (1993). Biological response of intramedullary bone to poly-L-lactic acid. *J. Appl. Biomater.* 4(1):13–27.

109. Meikle, M. C., Mak, W.-Y., Papaioannou, S., Davies, E. H., Mordan, N., and Reynolds, J. J. (1993). Bone-derived growth factor release from poly(α-hydroxy acid) implants *in vitro*. *Biomaterials* 14(3):177–183.

110. Meyer, U., Szulczewski, D. H., Moller, K., Heide, H., and Jones, D. B. (1993). Attachment kinetics and differentiation of osteoblasts on different biomaterials. *Cells and Mater.* 3(2):129–140.

111. Miyamoto, S., Takaoka, K., Okada, T., Yoshikawa, H., Hashimoto, J., Suzuki, S., and Ono, K. (1993). Polylactic acid-polyethylene glycol block copolymer. A new biodegradable synthetic carrier for bone morphogenetic protein. *Clin. Orthop.* 294:333–343.

112. Schmitz, J. P., and Hollinger, J. O. (1988). A preliminary study of the osteogenic potential of a biodegradable alloplastic-osteoinductive alloimplant. *Clin. Orthop.* 237:245–255.

113. Kleinschmidt, J. C., Marden, L. J., Kent, D., Quigley, N., and Hollinger, J. O. (1993). A

multiphase system bone implant for regenerating the calvaria. *Plast. Reconstr. Surg.* 91(4): 581–588.

114. Elgendy, H. M., Norman, M. E., Keaton, A. R., and Laurencin, C. T. (1993). Osteoblast-like cell (MC3T3-E1) proliferation on bioerodible polymers: An approach towards the development of a bone-bioerodible polymer composite material. *Biomaterials* 14(4):263–269.

115. Kronman, J. H., Green, R. E., Goldman, M., and Hauschka, B. (1979). Poly-HEMA sponge: A biocompatible calcification implant. *Biomater. Med. Dev. Art. Org.* 7(2):299–305.

116. Winter, G. D., and Simpson, B. J. (1969). Heterotopic bone formed in a synthetic sponge in the skin of young pigs. *Nature* 223(5201):88–90.

117. Winter, G. D. (1973). Studies, using sponge implants, on the mechanism of osteogenesis. In *Biology of Fibroblast* (E. Kulonen and J. Pikkarainen, Eds.), Academic Press, London, pp. 103–125.

118. Ripamonti, U. (1993). The generation of bone in nonhuman primates experimental studies in the baboon (*Papio ursinus*). Ph. D. thesis, University of Witwatersrand, Johannesburg, South Africa.

119. Kinoshita, Y., Kirigakubo, M., Kobayashi, M., Tabata, T., Shimura, K., and Ikada, Y. (1993). Study on the efficacy of biodegradable poly(L-lactide) mesh for supporting transplanted particulate cancellous bone and marrow: Experiment involving subcutaneous implantation in dogs. *Biomaterials* 14(10):729–736.

120. Lovell, T. P., Dawson, E. G., Nilsson, O. S., and Urist, M. R. (1989). Augmentation of spinal fusion with bone morphogenetic protein in dogs. *Clin. Orthop.* 243:266–274.

121. Heckman, J. D., Boyan, B. D., Aufdemorte, T. B., and Abbot, J. T. (1991). The use of bone morphogenetic protein in the treatment of non-union in a canine model. *J. Bone Joint Surg.* 73A(5):750–764.

122. Howard, T. C., and Kelley, R. R. (1969). The effect of bone wax on the healing of experimental rat tibial lesions. *Clin. Orthop.* 63:226–232.

123. Vandersteenhoven, J. J., and Spector, M. (1983). Osteoinduction within porous polysulfone implants at extraosseous sites using demineralized allogeneic bone matrix. *J. Biomed. Mater. Res.* 17(5):793–806.

124. Hammonds, R. G., Jr., Schwall, R., Dudley, A., Berkemeier, L., Lai, C., Lee, J., Cunningham, N., Reddi, A. H., Wood, W. I., and Mason, A. J. (1991). Bone-inducing activity of mature BMP-2b produced from a hybrid BMP-2a/2b precursor. *Mol. Endocrinol.* 5(1):149–155.

125. Ripamonti, U., Ma, S., and Reddi, A. H. (1992). The critical role of geometry of porous hydroxyapatite delivery system in induction of bone by osteogenin, a bone morphogenetic protein. *Matrix* 12:202–212.

126. Ripamonti, U., Ma, S.-S., and Reddi, A. H. (1992). Induction of bone incomposites of osteogenin and porous hydroxyapatite in baboons. *Plast. Reconstr. Surg.* 89(4):731–739.

127. Paralkar, V. M., Nandedkar, A. K. N., Pointer, R. H., Kleinman, H. K., and Reddi, A. H. (1990). Interaction of osteogenin, a heparin binding bone morphogenetic protein, with Type IV collagen. *J. Biol. Chem.* 265(28):17281–17284.

128. Katz, R. W., Felthousen, G. C., and Reddi, A. H. (1990). Radiation-sterilized insoluble collagenous bone matrix is a functional carrier of osteogenin for bone induction. *Calcif. Tissue Int.* 47:183–185.

129. Toriumi, D. M., Kotler, H. S., Luxenberg, D. P., Holtrop, M. E., and Wang, E. A. (1991). Mandibular reconstruction with a recombinant bone-inducing factor: Functional, histologic, and biomechanical evaluation. *Arch. Otolaryngol. Head Neck Surg.* 117:1101–1112.

130. Yoshimura, Y., Hirano, A., Nishida, M., Kawada, J., Horisaka, Y., Okamoto, Y., Matsumoto, N., Yamashita, K., and Takagi, T. (1993). Purification of water-soluble bone-inductive protein from bovine demineralized bone matrix. *Biol. Pharm. Bull.* 16(5):444–447.

131. Takaoka, K., Yoshikawa, H., Hashimoto, J., Miyamoto, S., Masuhara, K., Nakahara, H.,

Matsui, M., and Ono, K. (1993). Purification and characterization of a bone-inducing protein from a murine osteosarcoma (Dunn type). *Clin. Orthop.* 292:329–336.

132. Gerhart, T. N., Kirker-Head, C. A., Kriz, M. J., Holtrop, M. E., Hennig, G. E., Hipp, J., Schelling, S. H., and Wang, E. (1993). Healing segmental femoral defects in sheep using recombinant human bone morphogenetic protein. *Clin. Orthop.* 293:317–326.

133. Deatherage, J. R., and Miller, E. J. (1987). Packaging and delivery of bone induction factors in a collagenous implant. *Collagen Rel. Res.* 7:225–231.

134. Nakahara, H., Takaoka, K., Koezuka, M., Sugamoto, K., Tsuda, T., and Ono, K. (1989). Periosteal bone formation elicited by partially purified bone morphogenetic protein. *Clin. Orthop.* 239:299–305.

135. Rønningen, H., Solheim, L. F., and Langeland, N. (1985). Bone formation enhanced by induction. Bone growth in titanium implants in rats. *Acta Orthop. Scand.* 56(1):67–71.

136. Kawai, T., Mieki, A., Ohno, Y., Umemura, M., Kataoka, H., Kurita, S., Koie, M., Jinde, T., Hasegawa, J., and Urist, M. R. (1993). Osteoinductive activity of composites of bone morphogenetic protein and pure titanium. *Clin. Orthop.* 290:296–305.

137. Shen, W.-J., Chung, K.-C., Wang, G.-J., Balian, G., and McLaughlin, R. E. (1993). Demineralized bone matrix in the stabilization of porous-coated implants in bone defects in rabbits. *Clin. Orthop.* 293:346–352.

138. Andersson, G. B. J., Gaechter, A., Galante, J. O., and Rostoker, W. (1978). Segmental replacement of long bones in baboons using a fiber titanium implant. *J. Bone Joint Surg.* 60A(1):31–40.

139. Solheim, E., Pinholt, E. M., Andersen, R., Bang, G., and Sudmann, E. (1992). The effect of a composite of polyorthoester and demineralized bone on the healing of large segmental defects of the radius in rats. *J. Bone Joint Surg.* 74A(10):1456–1463.

140. Appel, L. E., Balena, R., Cortese, M., Opas, E., Rodan, G., Seedor, G., and Zentner, G. M. (1993). *In vitro* characterization and *in vivo* efficacy of a prostaglandin E_2/poly (ortho ester) implant for bone growth promotion. *J. Controlled Release* 26:77–85.

141. Papagelopoulos, P. J., Giannarakos, D. G., and Lyritis, G. P. (1993). Suitability of biodegradable polydioxanone materials for the internal fixation of fractures. *Orthop. Rev.* May, 585–593.

142. Laurencin, C. T., Norman, M. E., Elgendy, H. M., El-Amin, S. F., Allcock, H. R., Pucher, S. R., and Ambrosia, A. A. (1993). Use of polyphosphazenes for skeletal tissue regeneration. *J. Biomed. Mater. Res.* 27(7):963–973.

143. Harper, M. C., and Ralston, M. (1983). Isobutyl 2-cyanoacrylate as an osseous adhesive in the repair of osteochondral fractures. *J. Biomed. Mater. Res.* 17(1):167–177.

144. Kawakami, T., Uji, H., Antoh, M., Hasegawa, H., Kise, T., and Eda, S. (1993). Squalane as a possible carrier of bone morphogenetic protein. *Biomaterials* 14(8):575–577.

145. Toriumi, D. M., East, C. A., and Larrabee, W. F. (1991). Osteoinductive biomaterials for medical implantation. *J. Long Term Effects of Medical Implants* 1(1):53–77.

146. Toriumi, D. M., Larrabee, W. F., Walike, J. W., Millay, D. J., and Eisele, D. W. (1990). Demineralized bone: Implant resorption with long-term follow-up. *Arch. Otolaryngol. Head Neck Surg.* 116:676–680.

147. Khouri, R. K., Koudsi, B., and Reddi, H. (1991). Tissue transformation into bone *in vivo*: A potential practical application. *JAMA* 266(14):1953–1955.

148. Ma, S., Chen, G., and Reddi, A. H. (1990). Collaboration between collagenous matrix and osteogenin is required for bone induction. *Ann. N.Y. Acad. Sci.* 580:524–525.

149. Buring, K., and Urist, M. R. (1967). Effects of ionizing radiation on the bone induction principle in the matrix of bone implants. *Clin. Orthop.* 55:225–234.

150. Urist, M. R., and Hernandez, A. (1974). Excitation transfer in bone. Deleterious effects of cobalt 60 radiation-sterilization of bank bone. *Arch. Surg.* 109(4):486–493.

151. Prolo, D. J., Pedrotti, P. W., Burres, K. P., and Oklund, S. (1982). Superior osteogenesis

in transplanted allogeneic canine skull following chemical sterilization. *Clin. Orthop.* 168: 230–242.

152. Prolo, D. J., and Rodrigo, J. J. (1985). Contemporary bone graft physiology and surgery. *Clin. Orthop.* 200:322–342.

153. Kramer, S. J., Spadaro, J. A., and Webster, D. A. (1981). Antibacterial and osteoinductive properties of demineralized bone matrix treated with silver. *Clin. Orthop.* 161:154–162.

154. Miyamoto, S., Takaoka, K., Okada, T., Yoshikawa, H., Hashimoto, J., Suzuki, S., and Ono, K. (1992). Evaluation of polylactic acid homopolymers as carriers for bone morphogenetic protein. *Clin. Orthop.* 278:274–285.

155. Forster, I. W., Ralis, Z. A., McKibbin, B., and Jenkins, D. H. R. (1978). Biological reaction to carbon fiber implants: The formation and structure of a carbon-induced "neotendon." *Clin. Orthop.* 131:299–307.

156. Aragona, J., Parsons, J. R., Alexander, H., and Weiss, A. B. (1981). Soft tissue attachment of a filamentous carbon-absorbable polymer tendon and ligament replacement. *Clin. Orthop.* 160:268–278.

157. Gomi, K., and Davies, J. E. (1993). Guided bone tissue elaboration by osteogenic cells *in vitro. J. Biomed. Mater. Res.* 27(4):429–431.

158. Haney, J. M., Nilveus, R. E., McMillan, P. J., and Wikesjo, U. M. E. (1993). Periodontal repair in dogs: Expanded polytetrafluoroethylene barrier membranes support wound stabilization and enhance bone regeneration. *J. Periodontol.* 64(9):883–890.

159. Guillemin, M. R., Mellonig, J. T., Brunsvold, M. A., and Steffensen, B. (1993). Healing in periodontal defects treated by decalcified freeze-dried bone allografts in combination with ePTFE membranes. Assessment by computerized densitometric analysis. *J. Clin. Periodontol.* 20(7):520–527.

160. Uretzky, G., Appelbaum, J., and Sela, J. (1988). Inhibition of the inductive activity of demineralized bone matrix by different percutaneous implants. *Biomaterials* 9(2):1–3.

161. Muzzarelli, R. A. A., Zucchini, C., Ilari, P., Pugnaloni, A., Mattioli Belmonte, M., Biagini, G., and Castaldini, C. (1993). Osteoconductive properties of methyl-pyrrolidinone chitosan in an animal model. *Biomaterials* 14(12):925–929.

162. Langer, R., Brem, H., Falterman, K., Klein, M., and Folkman, J. (1976). Isolation of a cartilage factor that inhibits tumor neovascularization. *Science* 193(4247):70–72.

163. van Blitterswijk, C. A., Brink, J., Leenders, H., and Bakker, D. (1993). The effect of PEO ratio on degradation, calcification and bone bonding of PEO/PBT copolymer (polyactive). *Cells and Mater.* 3(1):23–26.

32
Growth Hormone Release from Biomaterials

Sandra Downes
Institute of Orthopedics
Middlesex, England

I. INTRODUCTION

There has been an increase in the need for revision after joint replacement, suggesting that radical improvements in the materials used to fix prostheses to bone are necessary [1–3]. Aseptic loosening of the femoral component tends to occur at the bone-cement interface [4,5]. In the last decade, improved cementing techniques and better prosthesis design have led to fewer fractures of prostheses and less bone resorption [6,7]. Any improvement in the quality of the bone-material bond would be a further distinct advantage.

It was shown by Willert et al. that the anchorage of cement to bone is stable immediately after surgery but postoperative changes such as bone necrosis can lead to a compromise in the strength of the bone-cement bond [8]. Bone remodeling will eventually take place, but any stimulus or improvement in this initial bonding will improve the long-term stability of the implant.

One way to improve biomaterials for such use is by the incorporation of growth hormone (GH) for delivery at the implant interface. In this chapter, the effects of growth hormone on bone both *in vitro* and *in vivo* are discussed. Methods to incorporate and release growth hormone from a variety of biomaterials are presented, with investigations into their role in improved bone remodeling.

II. THE EFFECT OF GROWTH HORMONE ON BONE

Growth hormone is a large polypeptide composed of 191 amino-acids, with a molecular weight of 22,000. Several structural features of the molecule contribute to its specific biological activity. Growth hormone is water soluble, stable with time and storage, and has low immunogenicity. These properties, coupled with the fact that it is now readily available as a result of recombinant technology as authentic 22,000 hGH (human growth hormone), have allowed it to be more widely used in many clinical applications.

Bone development and bone mass are controlled by close interactions between bone formation by osteoblasts and bone resorption by osteoclasts, which are regulated by both systemic and local mechanisms [9–11]. It is well established that polypeptide growth factors play an important role in the development and growth of osseous tissues. Bone contains an abundance of cells that sequester growth factors and modulate their biological actions through complex modes of release and presentation to responding cells. Growth factors can directly stimulate specific target cells to produce secondary factors by autocrine and paracrine actions. Systemic treatment of patients with GH shows evidence of increased bone formation [12,13] and direct stimulation of bone growth by GH and other growth factors has been described [14,15]. Growth hormone is the only hormone known to stimulate longitudinal bone growth in a dose-dependent manner [16–19]. There is also evidence that GH has a direct effect on bone cells in culture [20,21].

A.   The Effect of Growth Hormone on Bone Cells *In Vitro*

Bone is comprised of various cell types, including osteoprogenitor cells, osteoclasts, osteocytes, and osteoblasts. *Osteoprogenitor cells* are those cells that give rise to the bone-forming cells found in abundance on the endosteal and periosteal surfaces of bone; these cells resemble fibroblasts in morphology. *Osteoclasts* are a heterogeneous group of multinuclear cells with the role of bone resorption. *Osteoblasts* are responsible for the bone matrix synthesis, and appear on the surfaces of bone that are undergoing growth and development. *Osteocytes* are derived from osteoblasts and are found buried deep within mineralized bone matrix and are connected to one another and to osteoblasts on the bone surface by extensive projections or canaliculi. There is strong evidence that the osteoblast is both directly and indirectly stimulated by growth hormone.

Experiments *in vitro* with cultured osteoblasts have shown that GH significantly increases both alkaline phosphatase and IGF-I (insulin-like growth factor I) production. Stracke et al. demonstrated an increase in IGF-I and alkaline phosphatase in the medium of a GH-stimulated bone organ culture [20]. Ernst and Froesch demonstrated that GH alone was able to stimulate cell proliferation of cultured rat osteoblasts [22]. Further evidence that GH action is mediated by IGF-I was provided by Schmid et al. [23]. They showed that GH was able to stimulate osteoblasts, but only in the absence of IGF-I antibodies, and that IGF-I needs to be present in a biologically active form for GH to stimulate osteoblast proliferation.

B.   Growth Hormone Effects on Bone *In Vivo*

GH has characteristic intermediary effects on adipocytes and skeletal muscle in addition to its minor effects on skeletal growth. GH is anabolic and promotes protein synthesis in muscle cells. Conversely, it has a catabolic effect on fat and carbohydrate metabolism, inducing lipolysis in adipocytes [24].

Although GH is considered to be a systemic hormone, there is evidence to suggest that it can have both direct and indirect effects on osteoblasts, mediated by IGF-I. GH has been well characterized as a therapeutic agent for actively stimulating growth [20,22,23] and it has been well documented that systemic treatment with human GH causes an increase in bone formation and direct stimulation of chondrocytes and cartilage growth. In 1982, it was reported that unilateral injections of human GH into tibial epiphysis stimulated unilateral bone growth [25]; this was subsequently confirmed by Russell and Spencer using human GH derived from recombinant deoxyribonucleic acid (DNA). These observations support the hypothesis that GH acts locally to stimulate bone growth. Nilsson et al. demonstrated a direct effect of GH on IGF-I-producing cells in the rat growth plate [27].

C. Clinical Applications

The growth-stimulating effect of GH in man is best shown in GH-deficient patients who have been treated with GH [12,13]. There is clinical evidence in adult human volunteers that shows that GH may activate osteoblasts and bone remodeling and enhance bone mass. IGFs and GH have been implicated in the regulation of bone cell function. IGFs have been identified in bone matrix and in medium conditioned by bone organ and osteoblastic cell cultures [28,29].

There is now mounting evidence to suggest that the relevance of GH and IGFs to skeletal physiology goes beyond the well-established control of longitudinal growth, to the level of bone turnover and metabolism in the adult. Hypophysectomy results in decreased mesenchymal cell proliferation, decreased and delayed chondrogenesis, delayed and reduced vascular invasion, and impaired bone formation. Administration of GH in these subjects results in corrected mesenchymal cell proliferation, restored $^{35}SO_4$ incorporation into cartilage proteoglycans, and restored ^{45}Ca incorporation in tibial metaphyses, but not in matrix-induced osteogenic plaques [30]. Recent clinical data have confirmed that GH therapy may activate osteoblasts and bone remodeling and enhance bone mass in human volunteers and elderly males [28,31].

III. BIOMATERIALS

Biomaterials such as cements, ceramics, polymers, and metals are widely used in orthopedic surgery. The use of biomaterials to deliver biologically active agents is an attractive concept because local administration of certain therapeutic agents is often the most effective method of treatment. One example of such therapy is the local administration of antibiotics from bone cements for the prevention of deep-wound sepsis. Since biomaterials are generally used to reconstruct or replace tissues and joints, they are usually placed in a wound-healing environment in the body. In most cases, the most favorable biological response is rapid tissue repair. Since we are aware that GH is a potent stimulator of bone repair, the concept of incorporating GH in biomaterials for local delivery in the site of the wound is attractive.

The development of controllable, long-term, effective release systems for the delivery of growth hormone and other growth factors may improve wound healing and tissue repair. There are a number of ways that this can be achieved and a variety of biomaterials that can be used as vehicles for GH. These techniques are reviewed in the remainder of this chapter.

A. Bone Cement

The most common procedure is cemented joint replacement. There are problems associated with this surgery because eventually many of the implants become loose.

Polymethylmethacrylate (PMMA) bone cement shrinks during polymerization, which leaves a compromised bone-cement interface. It is possible to get remodeling of the bone at the cement interface around a stable implant, but this does not always occur. One of the early responses is often the invasion of fibroblasts and the formation of fibrous tissue around the implant. It is generally believed that early osseointegration of biomaterials is advantageous for the long-term stability of the implant. It would be a distinct advantage if bone remodeling around implants could be stimulated postoperatively.

PMMA can be used to deliver growth hormone with a rapid release, followed by a slower continuous release (Fig. 1). In a study in 1989, the release of GH from polymethylmethacrylate was investigated [32]. The most rapid release occurred during the first hour, with a decreasing release over the following 60 hours. Long-term release of growth hormone was observed for up to 40 days *in vitro*. It was noted that when bone cement was fractured *in vitro,* a further rapid release of growth hormone was observed, indicating that the release is primarily from the cement surface.

A rapid release of growth hormone in the immediate postoperative period is desirable in the application of growth hormone to stimulate osteointegration. The reduction in the rate of growth hormone release after several days is also highly desirable, since a continuous release of high concentrations of growth hormone may have unfavorable long-term effects. Having established that PMMA can be used to deliver growth hormone, an animal study (using rabbits) was constructed in order to examine any *in vivo* response to the growth hormone.

In an *in vivo* study involving a rabbit model, growth-hormone-loaded cement was inserted into the femur, with plain cement in the contralateral femur as a control. Examination of histology sections revealed that viable bone was growing in direct contact with PMMA. All sections were composed of some mineralized bone, osteoid, or cells (fat and marrow cells). One month after surgery, there was a greater percentage of osteoid bone present at the cement interface in the knees containing unloaded cement ($p < 0.01$, *t*-test). After two and four months, the difference was less significant [33].

The results of this trial indicated there was an early response of the osteoid cells to growth hormone. Stimulation of bone cell proliferation at the cement surface may in-

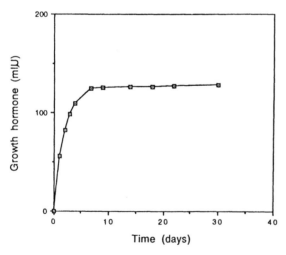

Figure 1 The release of growth hormone from polymethylmethacrylate (PMMA) bone cement into phosphate buffered saline.

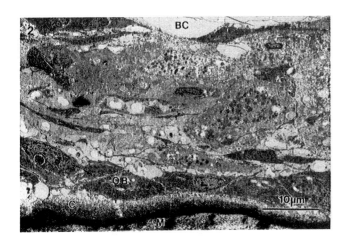

Figure 2 An electronmicrograph of growth-hormone-loaded bone cement in a rabbit model one month after surgery (OB, osteoblast; BC, bone cement; C, collagen; M, mineralized bone).

crease the strength of the bone-cement interface and thus improve stabilization of the implant. In further studies, the *in vivo* response to growth-hormone-loaded cement was compared with that of plain cement using light microscopy, transmission electron microscopy, and scanning electron microscopy in a rabbit model [34]. The results indicate that growth hormone released at the bone-cement interface stimulated osteoid formation and the reorganization of the tissue components. An advancing mineral front was observed in the direction of the bone cement, with new bone formed in direct apposition to cement.

Electron microscopy of growth-hormone-loaded bone cement in a rabbit showed active osteoblasts lining the mineralized bone and adjacent to the bone cement (Fig. 2). This was compared with the interface with plain PMMA cement, which showed little organization of the tissue components and spaces between the bone and the cement-containing areas of fibrous tissue (Fig. 3).

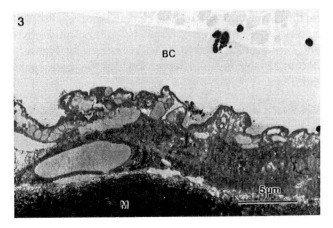

Figure 3 An electronmicrograph of plain bone cement in a rabbit model one month after surgery (BC, bone cement; M, mineralized bone). Note the bone dust and poorly organized interface.

Thus, it has been shown that PMMA can be used as a delivery agent for human growth hormone, which may directly act on target cells at the bone-cement interface. Stimulation of new bone formation in the crucial early postoperative period may improve the strength of the bond between bone and cement.

B.   Calcium Phosphate Ceramics

A wide range of calcium phosphate ceramics have been developed as potential candidates for bone substitutes. These include hydroxyapatite, fluoropatite, tricalcium phosphate, magnesium whitlockite, and tetracalcium phosphate. The so-called bioactive properties of calcium phosphate ceramics have been utilized in both dental and orthopedic applications. In orthopedics, ceramic can be used to fill bony defects or ceramic coatings can be applied to metal prostheses to improve osseointegration [35,36].

The tissue responses to tricalcium phosphate and hydroxyapatite ceramics have been examined experimentally in both bone and marrow [37,38]. At the bone-ceramic interface, there was formation of osteoid (by osteoblasts), which progressively mineralized up to the ceramic surface, allowing merger of the bone with the ceramic. At the marrow-ceramic interface, the formation of the bony covering on the ceramic surface may occur through a combination of three different mechanisms. Osteoblast precursor cells differentiate at the ceramic surface, bridging the endosteum into the marrow cavity to reach the ceramic and by an osteogenic stimulus from fragments of reamed bone within the marrow. Hence, the calcium phosphate ceramics enhance normal bone renewal mechanisms both at the reamed bone surface and in the marrow cavity. This process is initiated in the first week after surgery, and is not complete until six weeks after surgery.

1.   *Release of Growth Hormone from Dense Ceramics*

It has been shown that dense ceramics can be used to deliver GH and thus may be used to further stimulate bone remodeling around a ceramic implant. In the case of ceramics, GH must be adsorbed onto the ceramic surface. In a comparison of hydroxyapatite (HA) and tricalcium phosphate (TCP) ceramics as delivery systems [39], it was shown that both HA and TCP are good delivery systems for GH. The TCP released more GH than the HA ceramics. Figure 4 shows the release of GH during the first 24 hours of elution. It is thought that GH is released more rapidly from the more soluble ceramics and, as the TCP is more soluble than the HA, these results agree with this concept. It has been shown that GH release from ceramics is initially very rapid, followed by a slower release. Figure 5 shows the release of GH from TCP and HA for up to 25 days; note that the most rapid release occurs within the first 24 hours.

2.   *Release of Growth Hormone from Ceramic Coatings*

A conventional metal prosthesis can be coated with ceramic by plasma spraying [40]; the coating can be applied to porous [41] or nonporous surfaces [42]. In a study to compare the release of GH from ceramic coating on titanium [43], HA, heat-treated hydroxyapatite (HAH), and fluoroapatite (FA) coatings on titanium were loaded with human growth hormone and the subsequent release was monitored *in vitro*. GH elution followed the same pattern in all the materials studied, with a rapid release in the first 25 hours of the elution period, followed by a much slower continuous release for the remainder of the experimental period. The ceramic coatings tested in this study were all capable of growth hormone delivery in a dose-dependent manner. There was slightly more growth hormone

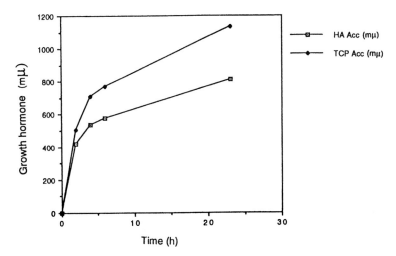

Figure 4   Release of growth hormone from ceramics during the first 24 hours of elution.

released from the FA coatings as compared with the HA coating, but the difference was not significant. However, the amount of growth hormone released from the HAH coating was six times greater than the HA and FA after two weeks elution (Fig. 6).

A further heat treatment after plasma spraying creates a coating that can deliver significantly more growth hormone than an untreated coating. Since the loading conditions were the same for each coating, it is reasonable to assume that the heat treated coating adsorbed more GH and thus was able to deliver more GH during the elution *in*

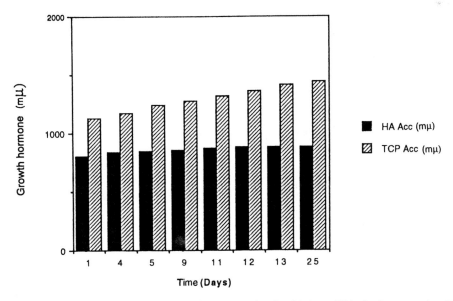

Figure 5   Growth hormone release from ceramics for 25 days (HA, hydroxyapatite; TCP, tricalcium phosphate).

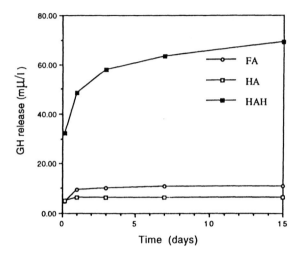

Figure 6   Cumulative release of growth hormone (GH) from hydroxyapatite (HA), fluoroapatite (FA), and heat-treated hydroxyapatite (HAH) coatings *in vitro*.

vitro. The release from the FA and HA was dose dependent; there was a linear relationship between the GH concentration in the loading solution and the amount of GH released *in vitro* (Fig. 7). Interestingly, here was more GH released from the HAH loaded with 2.0 and 4.0 international units per milliliter (IU/ml) than would have been expected in a linear relationship (Fig. 8).

Post-plasma-spraying heat treatment of HA coating has been shown to enhance the chemical bonding at the metal-ceramic interface, thus improving the mechanical properties of the coatings [44,45]. Filiaggi [45] reported a loss of the lamellar appearance and porosity of the HA after heat treatment. In addition, cracking was observed along the

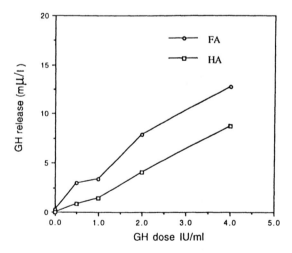

Figure 7   Relationship between growth hormone (GH) release from fluoroapatite (FA) and hydroxyapatite (HA) coatings and the growth hormone concentration in loading solution.

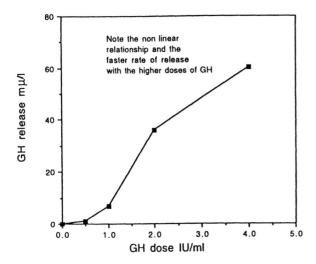

Note the non linear relationship and the faster rate of release with the higher doses of GH

Figure 8 Relationship between growth hormone (GH) release from heat-treated hydroxyapatite (HAH) coatings and growth hormone concentration in loading solution.

coating-substrate junction in the heat-treated samples. The chemical changes in HA coatings after heat treatment may result in either increased adsorption or increased release of GH from the coatings. The increased calcium release from the HAH coating indicates that the chemical composition of the HA coating may have been affected by the additional heat treatment.

C. Microspheres

There have been various drug delivery systems developed that involve the use of degradable microspheres. Microspheres may provide a useful delivery system for growth factors. There are various biomaterials that can be used to produce the carrier systems in the spheres, including polyglycolic acid, polylactic acid, gelatin, and collagen. Such systems may have important clinical uses because these biomaterials eventually degrade. In addition, microspheres provide a system that is controllable, can be moderated, and can sustain drug release over an extended period of time.

The work of Di Silvio et al. has shown that gelatin microspheres can be used to as a vehicle for human growth hormone [46,47]. A monolithic degradable drug delivery system was developed for the release of GH in a controllable manner. The results indicated that the release of GH from the microspheres was diffusion controlled. The GH detected was greater in phosphate buffered saline (PBS) than serum, which was probably due to differences in water uptake by the spheres in the different media.

Ultrasonication was used as a method for externally controlling the release of GH. Ultrasonication of the microspheres in PBS caused the release of 23% of the total GH incorporated, compared with 10% released from the spheres that had not been exposed to the ultrasonication. Figure 9 shows the total amount of GH released in PBS for microspheres subjected to ultrasonication prior to sampling over a period of 10 days.

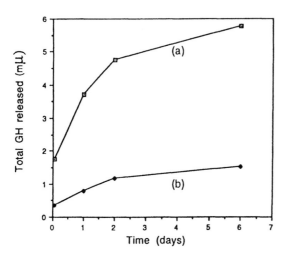

Figure 9   The release of growth hormone (GH) from microspheres: (a), after exposure to ultra-sonication; (b), without ultrasonication.

D.   Polymer Systems

In addition to polymethylmethacrylate bone cement, other polymer systems can be used to deliver therapeutic agents. Generally, polymer systems for drug delivery allow water diffusion. Transport of water within the polymer enables the transfer of the water-soluble agent from the polymer into the surrounding environment.

1.   *Polyethylmethacrylate-Based Polymers*

One of the most successful polymer systems selected for growth hormone release has been polyethylmethacrylate gelled with either tetrahydrofurfuryl methacrylate or hydroxy-ethylmethacrylate. These polymers are room-temperature polymerizing systems that exhibit low shrinkage. It has been demonstrated that these polymer systems can be modified to increase or decrease the hydrophilicity in a controlled manner. Growth hormone can be released from these polymer systems; the pattern of release is similar to that of bone cement, that is, an initial rapid release followed by a slower, sustained release.

Figure 10 shows the release of GH from polyethylmethacrylate gelled with tetrahy-drofurfurylmethacrylate (PEM/THFMA). Unlike bone cements, the porosity of the polymers affected the release of GH. The lowest release of GH occurred from the polymers with the lowest internal porosity, that is, those polymers that had undergone centrifugation or added pressure after mixing [47].

2.   *Blends of Synthetic and Natural Polymers*

In order to overcome the biological deficiencies of synthetic polymers and to enhance the mechanical characteristics of natural polymers, it is possible to blend synthetic polymers with biological polymers. Improvements in the characteristics of synthetic biomaterials could be achieved by the addition of biological macromolecules such as fibrin, collagen (C), elastin, and glycosaminoglycans. The resulting materials can combine the appropriate mechanical properties of the synthetic component with the biocompatibility of the biological component [48–50].

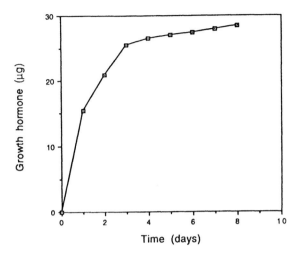

Figure 10   Growth hormone release from polyethylmethacrylate/tetrahydrofurfurylmethacrylate (PEM/THFM) (20 blocks).

Two different water-soluble synthetic polymers, poly(vinyl alcohol) (PVA) and poly-(acrylic acid),(PAA) have been blended with two different biological materials, collagen and hyaluronic acid [51–53]. These blends were used to prepare films, sponges, and hydrogels that were loaded with growth hormone (GH) to investigate their potential as drug delivery systems. The GH release was monitored *in vitro* using a specific enzyme-linked immunosorbent assay (ELISA). The results showed that the rate of GH release from HA/PAA sponges and from HA/PVA and C/PVA hydrogels is linear [54]. Figure 11 shows the release curves for GH from sponges with the same hyaluronic acid content but differing GH concentrations. There was a direct relationship between the amount of

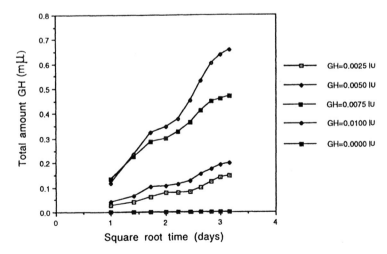

Figure 11   Release of growth hormone from hydroxyapatite/poly(acrylic acid) sponges (60/40) with different initial concentrations.

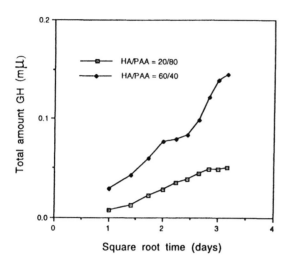

Figure 12   Release of growth hormone from hydroxyapatite/poly(acrylic acid) sponges with 20/80 and 60/40 ratios containing 0.0025 international units of growth hormone.

GH loaded and the quantity of GH released. In addition, those sponges with higher hyaluronic acid content but the same GH content released more GH (Fig. 12). These results indicated that the GH release was related to the hyaluronic acid content in the polymer blends.

IV. SUMMARY

The use of biomaterials to deliver biologically active agents is an attractive concept because local administration of certain therapeutic agents is an effective method of treatment. One example of such therapy is the local administration of antibiotics from bone cements for the prevention of deep-wound sepsis. Other examples include the delivery of growth hormone (GH) from bone cements, ceramics, degradable microspheres, and polymers. The development of controllable, long-term, effective release systems for the delivery of growth hormone and other growth factors may improve wound healing and tissue repair. There are local applications of bone-growth-promoting factors to target bone cells; these may have important clinical uses in the stabilization of implants and in the treatment of nonunited fractures and other pathological conditions.

REFERENCES

1. Salvatti, E. A., Wilson, P. D., Jolly, M. N., Valiki, F., Aglietti, P. K., and Brown, G. C. (1981). A 10-year follow-up study of our first hundred consecutive Charnley total hip replacements. *J. Bone Jt. Surg. (Am.)* 63:753–767.
2. Stauffer, R. N. (1982). Ten-year follow-up study of total hip replacement. With particular reference to roentgenographic loosening of the components. *J. Bone Jt. Surg. (Am.)* 64:983–990.
3. Sutherland, C. J., Wilder, A. H., Borden, L. S., and Marks, K. E. (1982). A ten year

follow-up of one hundred consecutive Muller curved stem total hip-replacement arthroplasties. *J. Bone Jt. Surg. (Am.)* 64:970–982.

4. Harris, W. H. (1980). Advances in surgical technique for total hip replacement: Without and with osteotomy of the great trochanter. *Clin. Orthop.* 146:188–204.

5. Oh, I., Carlson, C. E., Tomford, W. W., and Harris, W. H. (1978). Improved fixation of the femoral component after total hip replacement using methacrylate intramedullary plug. *J. Bone Jt. Surg. (Am.)* 60:608–613.

6. Crowinshield, R. D., Brand, R. A., Johnston, R. C., and Milroy, J. C. (1980). The effect of femoral stem cross sectional geometry on cement stresses in total hip reconstruction. *Clin. Orthop.* 146:71–77.

7. Huiskes, R. (1980). Some fundamental aspects of human joint replacement. *Acta Orthop. Scand.* (Suppl.) 185:109–200.

8. Willert, H. G., Ludwig, J., and Semlitsh, M. (1974). Reaction of bone methacrylate after hip arthroplasty. *J. Bone Jt. Surg.* 56A:1368–1372.

9. Mohan, S., Linkhart, T. A., Farley, J., and Baylink, D. (1984). Bone-derived factors active on bone cells. *Calc. Tiss. Int.* 36:S, 139.

10. Canalis, E., McCarthy, T., and Centrella, M. (1988). Growth factors and the regulation of tone remodelling. *J. Coni. Invest.* 81:277.

11. Raisz, L. G. (1988). Local and systemic factors in the pathogenesis of osteoporosis. *New Eng. J. Med.* 318:818.

12. Sartori, A., Conti, A., Guzzaloni, G., and Faglia, G. (1991). Serum osteocalcin levels in patients with GH deficiency before and during GH treatment. *Acta Paediatr. Scand.* 80:100–102.

13. Neilsen, H. K., Jorgensen, O. L., Brixen, K., Moller, N., Charles, P., and Christensen, J. S. (1991). Twenty-four-hour profile of serum osteocalcin in growth hormone deficient patients with and without GH treatment. *Growth Regulations* 1:153–159.

14. Schevan, B. A., and Hamilton, N. J. (1991). Longitudinal bone growth *in vitro*: Effects of insulin-like growth factor I and growth hormone. *Acta Endocrin.* 124(5):602–507.

15. McCarthy, T. L., Centrella, M., and Canalis, E. (1989). Insulin-like growth factor (IGF) and bone. *Connective Tissue Research* 20(1–4):277–282.

16. Harris, W. H., Ludwig, J., and Semlitsh, M. (1974). Growth hormone: The effect on skeletal renewal in the adult dog. (I) Morphometric studies. *Calcif. Tissue Res.* 10:1–13.

17. Henay, R. P., Harris, W. H., Cockin, J., and Weinberg, E. H. (1972). Growth hormone: The effect on skeletal renewal in the adult dog. (II) Mineral kinetic studies. *Calci. Tissue Res.* 10:14–22.

18. Thorgren, K. G., Hansson, L. I., Menander-Sellman, K., and Stenstrom, A. (1973). Effect of dose administration period of GH on longitudinal bone growth in the hypophysectomised rat. *Acta Endocrinol. (Copenh.)* 74:1–23.

19. Cheek, D. B., and Hill, D. B. (1974). In *Handbook of Physiology*, edited by E. Knobil and W. H. Sawyer, American Physiological Society, Bethesda, p. 59.

20. Stracke, H., Schulz, A., Moeller, D., Rossol, S., and Schatz, H. (1984). Effect of growth hormone on osteoblasts and demonstration of Somatomedin–C/IGF-I in bone organ culture. *Acta Endocrin.* 107:16–24.

21. Schevan, B. A., Hamilton, N. J., Fakkeldij, T. M., and Duursma, S. A. (1991). Effects of recombinant human insulin-like growth factor-I and II and growth hormone on the growth of normal adult human osteoblast-like cells and human osteogenic sarcoma cells. *Growth Regulation* 1:160–167.

22. Erst, M., and Froesch, E. R. (1988). Growth hormone dependent stimulation of osteoblast-like cells in serum-free cultures via local synthesis of insulin-like growth factor-I. *Biochemical and Biophysical Research Communications* 151(1):142–147.

23. Schmid, C., Steiner, T., and Froesch, E. R. (1984). Insulin-like growth factor-I supports differentiation of cultured osteoblast-like cells. *FEBS Letters* 1631, 173(1):48–52.

24. Pecile, A., and Mueller, E. E. (eds.) (1971). *Growth and Growth Hormone*, American Elsevier, New York.

25. Isgaard, J., Nilsson, A., Lindahl, A., et al. (1986). Effects of local administration of GH and IGF-I on longitudinal bone growth in rats. *Am. J. Physiol.* 250:E362–372.

26. Russell, S. M., and Spencer, E. M. (1985). Local injections of human or rat growth hormone or of purified human somatomedin-C stimulate unilateral tibial epiphyseal growth in hypophysectomised rats. *Endocrinology* 116:2563.

27. Nilsson, A., Isgaard, J., Lindahl, A., Dahlstrom, A., Skottner, A., and Isaksson, O. (1986). Regulation by growth hormone of chondrocytes containing IGF-I in rat growth plate. *Science* 233:571.

28. Brixen, K., Nielsen, H. K., Mosekilde, L. K., and Flyvberg, A. (1990). A short course in recombinant human growth hormone treatment stimulates osteoblasts and activates bone remodelling in normal volunteers. *J. Bone Min. Res.* 5:609–618.

29. Hock, J. M., Centrella, M., and Canalis, E. (1988). Insulin-like growth factor-I has independent effects on bone matrix formation and cell replication. *Endocrinology* 122(1):254–260.

30. Reddi, A. H., and Sullivan, N. S. (1980). Matrix induced endochondrial bone differentiation: Influence of hypophysectomy, growth hormone and thyroid stimulating hormone. *Endocrinology* 107:1291–1299.

31. Strollo, F., Bollanti, L. K., Ciamatori, A. K., More, M., Strollo, G., and Riodino, G. (1992). Effects of short-term treatment with human recombinant growth hormone in the elderly. *J. Endocrinol. Invest.* 15 (Suppl.4), Abstract No. 120.

32. Downes, S., Wood, D. J., Malcolm, A. J., and Ali, S. Y. (1989). Improvement of the bone-cement interface using growth hormone loaded cement. *J. Bone Jt. Surg.* 71B(4):727.

33. Downes, S., Wood, D. J., Malcolm, A. J., and Ali, S. Y. (1990). Growth hormone in polymethylmethacrylate. *Clin. Orthop. Rel. Res.* 252:294–298.

34. Downes, S., Kayser, M. V., Blunn, G., and Ali, S. Y. (1991). An electron microscopical study on the interaction of bone with growth hormone loaded bone cement. *Cells and Materials* 12:171–176.

35. Cook, S. D., Thomas, K. A., Kay, J. F., and Jarcho, M. (1988). Hydroxyapatite-coated porous titanium for use as an orthopaedic biologic attachment system. *Clin. Orthop. Rel. Res.* 230:303–312.

36. Osborn, J. F. (1989). Bonding osteogenesis under loaded conditions – The histological evaluation of a human autopsy specimen of hydroxyapatite coated stem of a titanium hip prosthesis. In *Bioceramics*, edited by H. Oonishi, Y. Aoki, and K. Sawai, Ishiyaku, Tokyo, p. 375.

37. Archer, R. S., Downes, S., Kayser, M. V., and Ali, S. Y. (1992). The tissue responses to dense tricalcium phosphate ceramics in bone and marrow. *Cells and Materials* (2):113–118.

38. Archer, R. S., and Downes, S. (1991). A histological and ultrastructural assessment of the tissue response to calcium phosphate ceramics. In *Bioceramics*, Volume 4, edited by W. Bonfield, G. W. Hastings, and K. E. Tanner, Butterworth Heinemann, Oxford, UK, pp. 179–185.

39. Downes, S., Di Silvio, L., Klein, C. P. A. T., and Kayser, M. V. (1991). Growth-hormone loaded bioactive ceramics. *Materials in Medicine* 2:176–180.

40. De Groot, K., Geesink, R., Klein, C. P. A. T., and Serekian, P. (1987). Plasma sprayed coatings of hydroxyapatite. *J. Biomed. Mater. Res.* 21:1375.

41. Rivero, D. P., Fox, J., Skipor, A. K., Urban, R. M., and Galante, J. O. (1988). Calcium phosphate coated porous titanium implants for enhanced skeletal fixation. *J. Biomed. Mater. Res.* 22:191–197.

42. Geesink, R. G. T. (1989). Experimental and clinical experience with hydroxyapatite-coated hip implants. *Orthopaedics* 12:1239.

43. Downes, S., Clifford, C. C., Scotchford, C. S., and Klein, C. P. A. T. (1994). A comparison of the release of growth hormone from hydroxyapatite, heat-treated hydroxyapatite and fluoroapatite coatings on titanium. *J. Biomed. Mater. Res.* (in press).

44. Filiaggi, M. J., Coombs, N. A., and Pilliar, R. M. (1991). Characterization of the interface in the plasma-sprayed HA coating/Ti-6A1-4V implant system. *J. Biomed. Mater. Res.* 25: 1211–1229.
45. Filiaggi, M. J., Pilliar, R. M., and Coombs, N. A. (1993). Post-plasma-spraying heat treatment of the HA coating/Ti-6A1-4V implant system. *J. Biomed. Mater. Res.* 27:191–198.
46. Di Silvio, L., Gurav, N., Kayser, M. V., Braden, M., and Downes, S. (1994). Biodegradable microspheres: A new delivery system for growth hormone. *Biomaterials* 15:931–936.
47. Di Silvio, L., Kayser, M. V., and Downes, S. (1994). Validation and optimization of a polymer system for potential use as a controlled drug delivery system. *Clin. Mat.* 16:91–99.
48. Giusti, P., Soldani, G., Mercoliano, R., Pracella, M., and Barbani, N. (1989). Applications of advanced polymers in medicine: Bioartificial polymeric materials. *Proc. 1st Mediterranean School on Science and Technology of Advanced Polymer-Based Materials*, Vico Equense (Naples), Italy, September 11–22, p. 715.
49. Giusti, P., Lazzerti, L., and Lelli, L. (1993). Bioartificial polymeric materials: A new method to design biomaterials by using both biological and synthetic polymers. *TRIP* 9(9):261–267.
50. Giusti, P., Lazzeri, L., Barbani, N., Lelli, L., De Petris, S., and Cascone, M. G. (1994). Blends of natural and synthetic polymers: A new route to novel biomaterials. *Macromol. Symp.* 78:285–297.
51. Barbani, N., Lazzeri, L., and Lelli, L. (1992). Nuovi materiali polimerici bioartificiali costituiti da collagene e alcool polivinilico. *Biomateriali* 1/2:59–66.
52. Lazzeri, L., Giusti, P., Barbani, N., Guerra, G. D., Lelli, L., Palla, M., and Domenici, C. (1992). Collagen-poly(vinyl-alcohol) based bioartificial materials. *Proceedings of the Fourth World Biomaterials Congress*, Berlin, Federal Republic of Germany, April 24–28, p. 462.
53. Lazzeri, L., Barbani, N., Lelli, L., Bonaretti, A., Centonze, P., and Giusti, P. (1993). Sponge-like materials based on hyaluronic acid. *Proceedings of the 10th European Conference on Biomaterials*, Davos, Switzerland, September 8–11, p. 78.
54. Gascone, M. G., DiSilvo, L., Sim, B., and Downes, S. (1994). Collagen and hyaluronic acid based polymeric blends as drug delivery systems for the release of physiological concentrations of growth hormone. *J. Mat. Sci.: Mat. in Med.* 5:770–775.

33

Preparation, Characterization, and Drug Delivery Applications of Microspheres Based on Biodegradable Lactic/Glycolic Acid Polymers

Xue Shen Wu*

Medisorb Technologies International
Cincinnati, Ohio

I. INTRODUCTION

In the past two decades, a variety of synthetic polymers have been reported to degrade in mammals. The degradation products of some of these polymers can be eliminated from the body by either metabolization or kidney filtration. Polymers that undergo degradation in a living environment, through either simple chemical reactions or enzyme-catalyzed reactions, are designated as biodegradable polymers [1].

Biodegradable polymers have become increasingly important in the development of drug delivery systems. These polymers can either function as a matrix to control diffusion of the drug, followed by polymer biodegradation and elimination of the degradation products from the body, or they can participate in and control the rate of drug release by polymer hydration and degradation [2]. A great number of biodegradable polymers have been developed and evaluated for drug delivery usage, but the most widely investigated biodegradable polymers and systems achieving the most success at this time are lactic/glycolic acid polymers. The major reasons are their reputation as safe suture materials and their degradation to products that are normal body metabolites.

Lactic/glycolic acid polymers are biodegradable, biocompatible, and bioabsorbable. Their biodegradation products are nontoxic, noncarcinogenic, and nonteratogenic. In general, they do not cause inflammatory reactions in tissues. These polymers are easily processed, have good mechanical properties, and are sterilizable. Rates of biodegradation and mechanical properties can be custom-tailored by changing polymer composition (i.e., the copolymerization ratio of lactic acid to glycolic acid) and/or molecular weight. In addition to these advantages, the monomers used for the synthesis of these polymers can

---

Current affiliation: Long Island University, Brooklyn, New York be derived from many sources, including renewable plant sources, petroleum, coal, and natural gas [3].

Yolles et al. first applied biodegradable lactic/glycolic acid polymers to drug delivery [4]. Boswell and Scribner [5], Sinclair [6], Wise et al. [7], and Beck et al. [8] all did early research in drug delivery using lactic/glycolic acid polymers. These research teams were seeking new systems to deliver conventional drugs such as narcotic antagonists and contraceptive hormones. Since then, the application of lactic/glycolic acid polymers to drug delivery has been extensively investigated. Jalil and Nixon [9], Lewis [10], Arshidy [11, 12], Conti et al. [13], and Fong [14] have published general reviews on these subjects.

Many types of drug delivery devices based on lactic/glycolic acid polymers have been investigated. These drug delivery devices include microspheres [9,11], fibers [15,16], films [17-20], implantable rods [21-23], tablets [24,25], pellets [26], beads [27], nanoparticles [28], and others. Of all these, the injectable microspheres are the major focus of research. Microspheres have some advantages over other devices. Ease of fabrication with good reproducibility and ease of administration are examples. Therefore, this chapter focuses on the preparation and characterization of microspheres composed of lactic/glycolic acid polymers and their application to drug delivery.

Microspheres are polymeric particles having a size range from a few micrometers to thousands of micrometers and can be administered by injection via intramuscular (IM), subcutaneous (SC), intravenous (IV), or intraperitoneal (IP) routes. Thus far, the targeted routes have been IM and SC as they have good potential for sustained-release applications. The desired size range for IM and SC injections is from several micrometers to several hundred micrometers. For IV administration, a microsphere size of 3-8 micrometers is desirable to prevent capillary clogging and phagocytosis [29].

II. PREPARATION OF MICROSPHERES USING LACTIC/GLYCOLIC ACID POLYMERS

There are several methods that can be used to make microspheres from lactic/glycolic acid polymers. Solvent evaporation or extraction and phase separation are the two main processes used to prepare microspheres. The solvent evaporation/extraction procedure is an aqueous system process, while the phase separation procedure is a nonaqueous process. It follows that the solvent evaporation/extraction method is suitable for water-insoluble drugs, whereas the phase separation method is preferable for water-soluble drugs. Therefore, these two processes are complementary in that the limitation of one method may be circumvented by using the other. In addition to these two methods, the (water-in-oil)-in-water multiple emulsion method is often used to encapsulate water-soluble drugs in lactic/glycolic acid polymers. Other methods, such as spray drying, are also used to microencapsulate some drugs.

When one designs a drug delivery system and selects a microencapsulation method, the following requirements should be met [14]:

1. The yield of microspheres with the targeted size range should be high.
2. The encapsulation efficiency for the drug should be high.
3. The biological activity of the drug should be maintained during the encapsulation process and in the microspheres.
4. The batch-to-batch reproducibility in terms of microsphere quality and drug-release profile should be within specified limits.
5. The drug-release profile should be adjustable by controlling composition and process variables.

6. The microspheres should not aggregate, that is, they should be a free-flowing powder.

A.   Solvent Evaporation/Extraction Method

The solvent evaporation/extraction method is essentially an oil-in-water (o/w) dispersion or emulsion process. An organic phase containing a lactic/glycolic acid polymer and a drug is mixed in an aqueous phase containing an emulsifier. The emulsified organic droplets containing the polymer and the drug are then hardened into microspheres by removing the organic solvent.

Operations involved in a solvent evaporation/extraction method are shown schematically in Fig. 1. In the first step, polymer is dissolved in an organic solvent that should be volatile and immiscible with the suspending medium, which is usually water. The drug to be incorporated into the microspheres is then added to the polymer solution to produce either a solution or a dispersion of solid drug particles in the polymer solution. Sometimes, heat is applied to keep the drug soluble in the organic solvent. If a drug does not dissolve, but instead is dispersed in the polymer solution, the drug must be previously pulverized to a fine powder with a particle size of about 20 micrometers (μm) or less [30].

In the second step, the polymer-drug-solvent system is mixed with an aqueous medium comprised of water and an emulsifier. The volume of the aqueous phase is usually much larger than that of organic phase, and an o/w emulsion is formed with controlled stirring. The oil droplets of the emulsion consist of the polymer, drug, and solvent. The droplet size, and thus the size distribution of the microsphere product, is altered through adjustment of the stir rate during the emulsion process. The temperature is also usually controlled during this step.

In the third step, the solvent is removed from the oil droplets and the droplets are converted to hardened microspheres. If the solvent evaporation process is used, stir rate is reduced and the volatile solvent is allowed to evaporate. If the solvent extraction process is used, the emulsion is slowly transferred to a large amount of water and the oil droplets eventually lose their solvent to the water or other quench medium by diffusion. Simultaneously, the emulsifier is washed away. In both cases, evaporation and extraction, the oil droplets are converted to the corresponding solid microspheres. The solvent can also be removed by freeze drying or spray drying.

In the fourth step, the microspheres are washed, are collected by sieving, filtration,

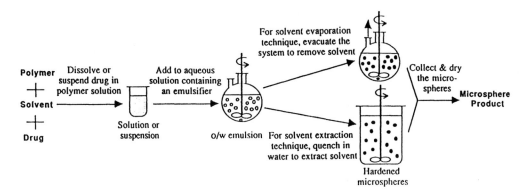

Figure 1   Schematic of a solvent evaporation/extraction process for preparing microspheres.

or centrifugation, and then are dried at ambient conditions or under vacuum or are lyophilized.

During the solvent evaporation process, the emulsion system is maintained under either reduced pressure or atmospheric pressure at the desired temperature [31,32]. The rate of solvent removal depends on pressure, temperature, and the solubility parameters of the solvent, the polymer, and the suspension medium [11]. The rate of solvent evaporation strongly affects the characteristics of the microspheres, and the optimum rate must be established empirically. Even in the solvent evaporation process, the solvent must first diffuse into the aqueous suspension medium before it can be removed from the emulsion system by evaporation. In the solvent extraction process, solvent removal is considerably more rapid than removal by evaporation. For this reason, microspheres made by solvent extraction are generally more porous in comparison with those produced by solvent evaporation under otherwise similar conditions. The rate of solvent removal in the solvent extraction method depends on temperature of quench water, volume ratio of emulsion to quench water, and solubility parameters of the solvent, the polymer, and the suspension medium [11].

This o/w solvent evaporation/extraction method is widely used to encapsulate lipid-soluble drugs. To encapsulate water-soluble drugs, a modified solvent evaporation method of oil-in-oil (o/o) is often used [27]. In this method, an organic liquid such as mineral oil [33,34], light paraffin oil [35], castor oil [27,36], cottonseed oil [37], or glycerine [38] is used as the continuous phase [33,39]. And, an oil-soluble surfactant such as lecithin, Span, or Brij is usually used as an emulsifier [27,34,40].

Tsai et al. used the o/o emulsion method to encapsulate mitomycin-C in poly(lactic acid) microspheres [39]. They dissolved the polymer and drug in acetonitrile and then emulsified the polymer-drug solution in paraffin oil containing Span 65 as an emulsifier. Acetonitrile was removed at 55°C by evaporation, and monolithic microspheres containing the drug were obtained. Jalil and Nixon used a similar o/o emulsion system to encapsulate phenobarbitone in microspheres of lactic/glycolic acid polymers [41,42]. They found that the molecular weight of the polymer, the polymer concentration, the nature and concentration of the emulsifier, and the stir rate are among the preparation conditions that affect the properties of the microspheres. For example, high molecular weight poly(L-lactic acid) generated porous microspheres, while low molecular weight poly(L-lactic acid) formed smooth, nonporous microspheres.

Although the solvent evaporation/extraction process for forming microspheres is conceptually simple, in fact a variety of variables influence the formation of the microspheres. These variables include the polymer composition and molecular weight, the nature and solubility of the drug being encapsulated, the organic solvent used, the temperature of the emulsion, the volume of the organic phase per unit volume of aqueous phase, and the physicochemical properties and amount of emulsifier used [43]. The effects of some of these variables on microsphere characteristics are summarized in Table 1 [42,44–46], and some considerations regarding the preparation of microspheres of lactic/glycolic acid polymers using the solvent evaporation/extraction process are described below.

1. Emulsifiers

The emulsifier provides a thin protective layer around the oil droplets of polymer and drug, and hence decreases the extent of droplet coalescence and coagulation and stabilizes the emulsion system. The agglomeration of the oil droplets during the fabrication process is frequently the initial difficulty encountered in developing a procedure for microencapsulation. As the solvent is being removed, the emulsifier continues to maintain the oil

Table 1   Effect of Some Formulation Parameters on Characteristics of Microspheres
Produced by Solvent Evaporation Method

Parameter	MS size	Encapsulation efficiency	Release rate
Temperature ↑	↓	↓	↑
Polymer concentration ↑	↑	↑	↑
Polymer molecular weight ↑	↑	No obvious effect	↑
Ratio of drug to polymer ↑	↑	↑	↑

Source: Tabulated from Refs. 42 and 44–46.
MS, microsphere.

droplets in their spherical configuration and prevents them from aggregating until the
solvent is completely removed and the microspheres are hardened as discrete particles.

The emulsifiers commonly employed in the solvent evaporation/extraction process
are hydrophilic polymeric colloids and anionic or nonionic surfactants. Examples are
poly(vinyl alcohol) [30,47], poly(vinyl pyrrolidone) [12], alginate [12], gelatin [48],
methyl cellulose [49], hydroxyalkyl cellulose [50], polysorbate [51,52], Span [39], Brij
[40], sodium dodecyl sulfate [53], cetyltrimethyl ammonium bromide [53], lecithin [28],
and fatty acid salts [54]. Poly(vinyl alcohol) is the most frequently used emulsifier in the
o/w method. The efficiency (required concentration) and the effectiveness (stability of
the emulsion) of each emulsifier is different and the best emulsifier for a particular
application is determined experimentally.

The physicochemical properties, structural characteristics, and concentration of an
emulsifier influence microsphere characteristics [28,49,52,55,56]. Generally speaking, for
a given emulsifier, the higher its concentration is, the smaller the resulting microspheres
are. However, there appears to be a limiting concentration above which the emulsifier has
no further effect. This is probably because the optimal packing concentration for the

Table 2   Effect of Type and Concentration of Emulsifier on Microsphere Size

Emulsifier	Concentration (wt%)	Volume mean diameter (μm)
Gelatin	1.00	50.8
Methyl cellulose	0.05	52.4
	0.50	28.9
	1.00	16.2
Sodium dodecyl sulfate	1.00	4.1
	5.00	2.4
Polyvinyl alcohol	1.00	2.5
	5.00	2.2
	10.00	1.0
Cetyltrimethyl ammonium bromide	1.00	0.5
	5.00	0.5

Source: Modified from Ref. 53
Preparation conditions: poly(lactic-co-glycolic acid) 50:50, 22,000 MW; 150 mg polymer.
in 2.5 ml methylene chloride; 10 ml emulsifier; 9800 rpm stir speed.

emulsion has already been achieved [40]. Table 2 lists the effects of some emulsifiers on the size of microspheres composed of a lactic/glycolic acid polymer.

Jalil and Nixon have studied the effect of the chemical structure of Span and of Brij emulsifiers on the size of the microspheres made by an o/o emulsion method [40]. They used acetonitrile as the solvent for poly(L-lactic acid) and paraffin oil as the continuous medium. The results, which are shown in Fig. 2, indicate that more hydrophilic emulsifiers produce smaller microspheres.

Combinations of emulsifiers have also been used in the solvent evaporation/extraction method. Spenlehauer et al. investigated the effect of emulsifier combinations and reported that microspheres of poly(lactic acid) collapsed when methyl cellulose-400 was used alone at a concentration of 1% in the aqueous phase [49]. However, spherical microspheres were formed and no aggregation occurred when a combination of 0.05% methyl cellulose-400 and 1% to 4% poly(vinyl alcohol) was used in the aqueous phase.

Cavalier et al. have reported similar results [56]. They found that irregular, aggregated microspheres were formed when 0.1% poly(vinyl alcohol) was used as an emulsifier for the preparation of 60% drug-loaded microspheres of poly(D,L-lactic acid). They also found that distorted microspheres were produced when 0.3% methyl cellulose was used as an emulsifier. However, when a combination of 0.27% poly(vinyl alcohol) and 0.05% methyl cellulose-400 was used, both microsphere sphericity and drug-loading efficiency were maximized.

2. Solvents

When the solvent evaporation/extraction process is used, the solvent for the lactic/glycolic acid polymers must be immiscible or only slightly soluble in the suspension medium (e.g., water). Bodmeier and McGinity evaluated the effects of many solvents on poly-(D,L-lactic acid) microsphere formation [57]. They found that water-miscible solvents

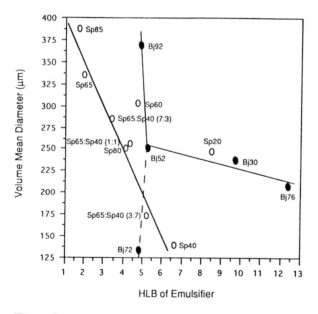

Figure 2   Influence of hydrophilic-lipophilic-balance (HLB) of emulsifiers of Span (Sp) series, -○-, and Brij (Bj) series, -●-, on the size of the microspheres of poly(L-lactic acid) (MW 41,300) prepared by solvent evaporation method at 50°C. (Modified from Ref. 40.)

such as acetone and dimethyl sulfoxide do not lead to the formation of microspheres upon emulsification. Instead, irregular agglomerates are formed because of rapid solvent exchange. Furthermore, the boiling point of the solvent must be lower than that of the suspension medium (e.g., water) if the solvent is to be removed by an evaporation process. The most commonly used solvents are ethyl acetate and methylene chloride because of their lower toxicity, ease of removal, and excellent ability to dissolve the polymers. Other solvents that have been used include chloroform and acetonitrile. If dissolution of the drug in the polymer solution is desired, then the ability of the solvent to dissolve the drug needs to be considered also.

Solvent mixtures can also be used to dissolve the polymer and the drug [37]. Such a solvent mixture usually contains a water-immiscible solvent like methylene chloride and a water-miscible solvent like methanol, ethanol, or propylene glycol [58,59]. The use of the water-miscible solvent results in rapid solvent removal and rapid polymer precipitation.

3. Drugs

Because the o/w solvent evaporation/extraction method involves an aqueous emulsion, the drugs being encapsulated in the microspheres are limited to those that have a low solubility in water. Otherwise, some amount of the water-soluble drug will partition from the organic phase into the aqueous phase. This drug loss will result in a poor encapsulation efficiency [49,51]. In other words, water-soluble drugs such as salicylic acid cannot be encapsulated in microspheres of lactic/glycolic acid polymers by using the o/w solvent evaporation/extraction method. In contrast, lipid-soluble drugs such as the steroids can be successfully encapsulated in microspheres by using this method [60].

To minimize the loss of ionizable drugs such as quinidine sulphate into the aqueous phase during an o/w microencapsulation process, the pH of the aqueous phase can be adjusted to suppress the ionization of the drug, and consequently to reduce the water solubility of the drug. Wakiyama et al. prepared poly(lactic acid) microspheres containing butamben, dibucaine, or tetracaine [55]. They used a basic aqueous phase that contained 1% alkali-treated gelatin and found that the encapsulation efficiency for butamben or dibucaine was higher than that of tetracaine. The reason was that butamben and dibucaine were unionized in the basic aqueous phase, while tetracaine was ionized. In another study, these investigators examined the effect of pH on the encapsulation efficiency of dibucaine, which has pKa of 1.6 and 8.3 [48]. With a targeted drug loading of 30 wt%, microspheres prepared using an acidic aqueous phase (pH 3.8) contained only 9.2 wt% drug. When a slightly basic aqueous phase (pH 7.5) was used, the dibucaine in the microspheres increased to 24.4 wt%. The drug content was further increased to 29 wt% when the pH of the aqueous phase was increased to 8.6.

The loss of drug to the aqueous phase can also be reduced by prior saturation of the aqueous phase with the same drug. Wakiyama et al. have shown that the tetracaine content in microspheres increased when the aqueous phase was presaturated with the drug [55]. Bodmeier and McGinity have also reported that the quinidine sulphate content in microspheres increased when the drug was added to the aqueous phase prior to forming the emulsion [51].

If a drug being incorporated into microspheres is insoluble in the solvent used for dissolving the polymer, it can be pulverized or micronized to give a homogenous distribution of discrete particles throughout the emulsion and resulting microspheres. If a drug is not completely soluble in the organic solvent used, it may crystallize inside the microspheres at certain concentrations. In this case, drug crystals may concentrate in some regions of the microspheres, such as the surface, giving a heterogenous drug distribution that will alter the drug-release profile [49,61]. Even when a drug being incorporated into microspheres is completely soluble in the organic solvent, it may also crystallize as the solvent is removed and thus form discrete drug-rich domains scattered throughout the microsphere. Drug crystals can also form on the surface of microspheres, which leads to an initial burst in the amount of drug released. However, many drugs do remain molecularly dispersed throughout the microspheres and do not crystallize during the microsphere-formation process [43].

B. Phase Separation Method

The phase separation method is another means of preparing drug-loaded microspheres from lactic/glycolic acid polymers. While the o/w solvent-evaporation/extraction method is only suitable for water-insoluble drugs, the phase separation method, being a nonaqueous method, is suitable for both water-soluble and water-insoluble drugs. However, the phase separation process is mainly used to encapsulate compounds having a low solubility in the organic solvent used to dissolve the polymer from which the microspheres are formed. That is, it is mainly used for encapsulating water-soluble compounds such as proteins, peptides, and vaccines.

Figure 3 depicts the operations used in the phase separation method. In the first step, the polymer is dissolved in an organic solvent. For lipid-soluble drugs, the drug is then either dissolved or dispersed in the polymer solution. For water-soluble drugs such as proteins and peptides, the drug is usually dissolved in an aqueous solution first. Then, a microfine emulsion is made by adding the aqueous drug solution to a small portion of the polymer solution. A microfine water-in-oil emulsion is made by gentle mixing, sonication, homogenizing, or microfluidizing. This microfine emulsion is then combined with the rest of the polymer solution with stirring.

In the second step, an organic nonsolvent, the first nonsolvent for the polymer, is slowly added to the polymer-drug-solvent system with stirring. The solvent for the polymer is slowly extracted into the first nonsolvent and the solubility of the polymer decreases. The polymer is then induced to phase separate and form coacervate droplets that contain the drug. The coacervate droplet size is controlled by the adjustment of stir speed. At this point in the process, the coacervate droplets are usually too soft to be collected. Therefore, in the third step, the emulsion system is slowly transferred with stirring to a large body of a second organic nonsolvent to harden the microdroplets, which now may be called microspheres.

In the fourth step, after the microspheres have been hardened in the second nonsolvent, the microspheres are washed with more of the second nonsolvent, are collected by

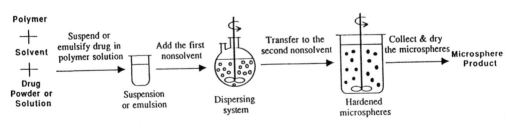

Figure 3   Schematic of a phase separation process for preparing microspheres.

sieving, filtration, or centrifugation, and are dried under ambient conditions or under vacuum. The first nonsolvent used to induce phase separation can be replaced by the addition of another polymer that is incompatible with the microsphere-forming polymer or by lowering the temperature of the system to decrease the solubility of the microsphere-forming polymer in the solvent [14].

The addition of the first organic nonsolvent should be controlled to allow the polymer solvent to be extracted slowly. Slow solvent extraction gives the wall-forming polymer sufficient time to deposit and spread evenly on the surface of the drug particles during phase separation. The polymer concentration is also important in phase separation. When the polymer concentration is too high, phase separation is too rapid and this may produce an unsatisfactory coating of polymer on the drug particles. Agglomeration is frequently encountered in the phase separation process because there is no emulsion stabilizer like there is in the solvent evaporation/extraction process. The main reason for the agglomeration is that the coacervate droplets are sticky and adhere to each other before the solvent is completely removed or before the droplets are hardened. Adjusting the stir rate and temperature [62] and the addition of an additive [63,64] to the system can prevent the agglomeration of the coacervate droplets during the phase separation process.

Many solvents for the polymer can be used in phase separation. In comparison with the solvent evaporation/extraction method, requirements for the polymer solvent are less stringent. The solvent does not have to be immiscible with water and the boiling point can be higher than that of water. However, the solvent should dissolve only the lactic/glycolic acid polymer but not the drug because it is preferred that the drug be dispersed rather than dissolved in the polymer solution. Methylene chloride, ethyl acetate, acetonitrile, and toluene all are examples of solvents for the polymer that have been used in phase separation.

The nonsolvents have an effect on both the phase separation of the lactic/glycolic acid polymer and on the hardening of the newly formed microspheres. Therefore, care should be taken in their selection. The nonsolvent should not dissolve lactic/glycolic acid polymers or the drug, but it should be miscible with the polymer solvent. The first nonsolvent should also be easily removed by washing with the second nonsolvent. The second nonsolvent should be relatively volatile. Examples of the first nonsolvents are viscous liquids such as low-molecular-weight liquid polybutadiene, low-molecular-weight liquid methacrylic polymers, silicone oil, vegetable oils, and light liquid paraffin oil [14,65,66]. Aliphatic hydrocarbons such as heptane, hexane, and petroleum ether have been commonly used as the second nonsolvent [67].

Fong has patented a unique phase separation technique that utilizes low temperatures [62]. He microencapsulated mellaril palmoate and thioridazine in poly(D,L-lactic acid), using toluene as the polymer solvent and isopropanol as the only nonsolvent. He also suggested the use of other nonsolvents such as normal propanol and isopropanol mixture, heptane, or isobutanol. In this technique, phase separation was elicited by dropwise addition of the nonsolvent at $-65\,^{\circ}\mathrm{C}$.

Ruiz et al. have studied the phase separation process used to form poly(lactic-co-glycolic acid) microspheres that contained the drug triptoreline [68]. Methylene chloride was used as the solvent for the polymer and silicone oil was used as the nonsolvent. They found that effective phase separation and coacervate formation was a function of the polymer hydrophobicity or molecular chain length. As polymer hydrophobicity increased, more silicone oil was needed to induce phase separation or to desolvate the polymer. Also, if the polymer solubility was increased such as by using a mixture of methylene chloride and methanol, more silicone oil was needed to induce phase separation. They further reported that the viscosity of the silicone oil had an effect on formation of coacervate droplets and their solubility. The silicone oil used to induce phase separation of lactic/glycolic acid polymers usually has a viscosity of approximately 350 centipoises (cp). Low-viscosity silicone oil (e.g., 20 cp) could not induce phase separation in their system. Moreover, the ratio of silicone oil to polymer solvent affected microsphere size. They concluded that the physicochemical nature (hydrophobicity or molecular chain length) of the polymer, the concentration of the polymer, viscosity of the silicone oil, and the ratio of polymer solvent to silicone oil all affect the stability of the emulsion system in phase separation and, consequently, the formation and quality of the microspheres.

C. Water-in-Oil-in-Water Multiple Emulsion Method

The (water-in-oil)-in-water (w/o/w) multiple emulsion method is a modified solvent-evaporation/extraction method. It is used for the encapsulation of water-soluble drugs, especially those that are only available in an aqueous solution (e.g., some vaccines). In comparison with the phase separation method, the w/o/w method does not need organic nonsolvents for the polymer, which can reduce the cost of production and minimize environmental pollution. When compared with the o/w solvent-evaporation/extraction method, the w/o/w method is more suitable for water-soluble drugs.

Figure 4 depicts the operations involved in the w/o/w method. In the first step, the drug is dissolved in an aqueous solution and the polymer is dissolved in an organic solvent. The drug solution is usually buffered to a desired pH and a nonionic, amphophilic polymer such as gelatin [69,70] is sometimes added to adjust the viscosity. High viscosity of the drug solution can prevent, or at least reduce, the diffusion of the encapsulate into the second aqueous solution. Sometimes an elevated temperature is needed to dissolve the drug in the aqueous solution.

In the second step, the aqueous drug solution is gradually added to the organic polymer solution with vigorous stirring. A microfine w/o emulsion is made by using sonication, homogenization, or vigorous agitation. The viscosity of both the aqueous phase and the organic phase can be increased by cooling the system after the w/o emulsion is formed.

In the third step, the w/o emulsion is slowly dropped into an aqueous medium

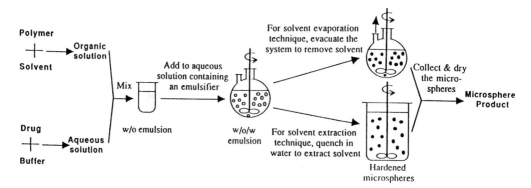

Figure 4 Schematic of a water-in-oil-in-water (w/o/w) process for preparing microspheres.

composed of water and an emulsifier such as poly(vinyl alcohol) with stirring. The volume of this second aqueous phase is larger than that of w/o emulsion, and a w/o/w emulsion is formed. The droplet size, and thus the size distribution, of the microsphere product is controlled, in part, through adjustment of the stir rate during the emulsification process.

In the fourth step, the organic solvent used to dissolve the polymer is removed from the emulsion by an evaporation or extraction technique. Evaporation can occur at either atmospheric or reduced pressure. The stir rate is usually reduced during the evaporation process. If a solvent-extraction technique is used, the w/o/w emulsion is slowly transferred to a large body of cold water with stirring.

In the fifth step, after the organic solvent has been removed, the hardened microspheres are washed, collected by sieving, filtration, or centrifugation, and then dried at atmospheric pressure or under vacuum or are lyophilized.

Ogawa et al. used this w/o/w method to encapsulate leuprolide acetate, an analog of luteinizing-hormone-releasing hormone, in a lactic/glycolic acid polymer [69,71–73]. They dissolved the drug (500 milligrams [mg]) and gelatin (80 mg) in 1 ml of distilled water to form an initial aqueous phase, which was emulsified in 5.5 ml of the polymer solution prepared with methylene chloride. This w/o emulsion was added to an outer aqueous phase with stirring to form oil droplets containing aqueous drug domains. The oil droplets were then hardened by the evaporation of methylene chloride, and the microsphere product was dried by lyophilization. They also investigated the formulation factors affecting the drug-loading efficiency and the microsphere size, and they concluded that the viscosity of the inner aqueous phase influences the drug-loading efficiency; the higher the viscosity is, the higher the drug-loading efficiency is under their encapsulation conditions. By adjusting the viscosity of the inner aqueous phase with either gelatin or poly(vinyl alcohol), they found that a higher stir speed and lower polymer concentration in the organic phase both generate smaller microspheres.

Jeffery et al. prepared ovalbumin microspheres of poly(lactic-co-glycolic acid) using a w/o/w method and evaluated the effect of formulation parameters on the microsphere characteristics [74]. They found that the ratio of ovalbumin to polymer and the volume and viscosity of both the inner and outer aqueous phases influence microsphere size and encapsulation efficiency. Table 3 summarizes the results of these investigations.

Table 3   Effect of Preparation Conditions on Microsphere Size and
Drug Encapsulation Efficiency

Variable	MS size	Encapsulation efficiency
Ratio of drug to polymer ↑	↑	↑
Gelatin concentration in inner aqueous phase ↑	↑	↑
Inner aqueous phase volume ↑	↑	↑
Outer aqueous phase volume ↑	↑	↑
PVA concentration in outer aqueous phase ↑	↓	↓

Source: Tabulated from Ref. 74.
PVA, poly(vinyl alcohol).

D. Other Methods

Spray drying [75–78], spray coating [79], the use of melts [80], supercritical fluid extraction/expansion [81,82], and other methods [83] have all been investigated for the preparation of microspheres of lactic/glycolic acid polymers. Conti et al. have published a review that summarizes most of the methods that have been used to make drug-loaded microspheres from lactic/glycolic acid polymers [13]. Wise et al. have reported the preparation of poly(L-lactic-co-glycolic acid) microspheres containing an antimalarial drug by a spray-drying technique [7]. The solution to be spray dried contained 4 grams (g) of polymer, 0.8 g or 2.0 g of drug, 30 ml of hexafluoropropanol, and 56 ml of benzene, and the microspheres generated were smaller than 125 μm in diameter.

Bodmeier and Chen have also reported the preparation of poly(D,L-lactic acid) microspheres containing progesterone or theophylline by a spray-drying method [84]. They dissolved the polymer in methylene chloride. Progesterone was codissolved in the polymer solution, while theophylline was suspended in the polymer solution. The solution or the suspension was spray dried to form microspheres having diameters of less than 5 μm. The advantage of the spray-drying method is that both water-soluble and water-insoluble drugs can be microencapsulated with high efficiency in lactic/glycolic acid polymers. However, the yield of microspheres is quite low when the batch size is small and nonspherical products are usually generated.

Shell [85] and Michaels [86] have used a spray-coating method to microencapsulate an ophthalmic drug. They coated chloramphenicol particles with a solution of poly(lactic acid), which were then used in an ophthalmic suspension.

III. CHARACTERIZATION AND STERILIZATION OF DRUG-LOADED MICROSPHERES

Once formulated, a microsphere drug delivery system must be characterized and evaluated. Characterization may include the determination of microsphere size, size distribution, surface quality, appearance, actual drug loading, *in vitro* and *in vivo* drug-release profiles, and residual solvent(s).

A. Microscopic Characteristics

Scanning electron microscopy is a basic tool used to study microspheres of lactic/glycolic acid polymers. This tool gives information on microsphere size and size distribution, shape, and aggregation. Scanning electron microscopy also reveals the surface quality or porosity of microspheres. Surface porosity of the microspheres can be of particular importance in assessing drug-release characteristics.

The combination of scanning electron microscopy and techniques for fracturing microspheres can give information about the interior quality of the microspheres and drug crystallization. The distribution of drug crystals within the microspheres can also be obtained from this combination of techniques (Fig. 5).

Many factors can affect the quality of lactic/glycolic acid microspheres. Jalil and Nixon have studied the effect of the molecular weight of poly(L-lactic acid) on microsphere size, morphology, and surface characteristics [46]. They concluded that high molecular weight polymers tend to produce larger microspheres when all other conditions are constant. Larger microspheres were tackier than smaller ones and tend to aggregate. Also, the surface of the microspheres made from high molecular weight polymers was

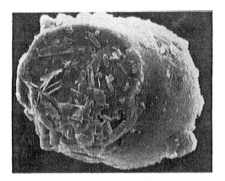

Figure 5   Scanning electron micrograph of freeze-fractured microspheres containing phenobarbitone with poly(L-lactic acid) with a molecular weight of 2,400 and a drug content of 20%. (Reprinted from Ref. 42 with permission of the copyright owner, Taylor & Francis Limited, and courtesy of Dr. Nixon.)

porous, while the microspheres made from low molecular weight polymers had a smooth and nonporous surface.

B.   Size Distribution

Scanning electron microscopy can only provide a rough estimation of the size distribution of microspheres. A quantitative value for the size distribution of microspheres is usually obtained by using particle counters.

To determine the size distribution of microspheres of the lactic/glycolic acid polymers, samples are usually suspended in an injection vehicle such as aqueous dextran to achieve a relatively stable suspension. The injection vehicle is first prepared by passing it through a 0.2-micron filter to avoid particle contamination. The microspheres are then suspended in the injection vehicle by sonication. Figure 6 is a plot showing the size distribution of a typical batch of poly(D,L-lactic acid) microspheres prepared by a solvent evaporation method.

Many factors can affect the size and size distribution of microspheres. Microsphere-formation method, stir speed during emulsification, polymer concentration, emulsifier concentration, solvents used, volume ratio of the polymer-containing phase to nonpolymer-containing phase(s), viscosity of each phase, and other parameters all have an effect. Wakiyama et al. have reported that the viscosity of the aqueous phase has an effect on microsphere size when the solvent-evaporation/extraction method is used [55]. For example, large microspheres were formed when the organic solvent was evaporated at 40°C from an aqueous phase that contained 1% gelatin as the emulsifier. In contrast, small microspheres were produced with the same formulation when the organic solvent was removed at room temperature because the viscosity of the aqueous phase at 40°C was lower than that at room temperature.

C.   Drug Encapsulation Efficiency

The drug dose required, the desired length of the drug-release period, the physicochemical properties of the drug, and the cost of manufacturing the drug-loaded microspheres all

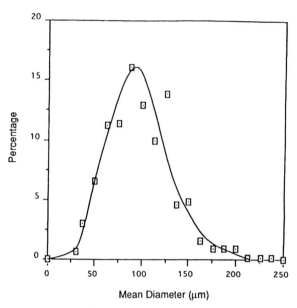

Figure 6   Size distribution of a batch of poly(D,L-lactic acid) microspheres containing norethisterone prepared by a solvent evaporation method. (Modified from Ref. 30.)

play a role in deciding on an optimal drug loading. Normally, the drug loading should be as high as possible because the quantity of microspheres administered cannot be too high. It has been reported that intramuscular and subcutaneous injections of more than 2 ml often cause patient discomfort [29]. For conventional drugs (e.g., steroids), loading of 50 wt% in microspheres of lactic/glycolic acid polymers is common, while loadings of water-soluble protein and peptide drugs have usually not been greater than 10 wt% [87]. The physical properties of the polymer matrix can be affected when the drug loading gets very high. Schwope et al. reported that high loadings of naltrexone base in polymers of D,L-lactic acid functioned as a plasticizer for the polymers, causing difficulty in the fabrication of the delivery devices of cylindrical rods and spherical beads [88].

During the process of forming drug-loaded microspheres of lactic/glycolic acid polymers, the drug is often lost to some extent, and a 100% encapsulation efficiency of any drug is unlikely [89]. With the microsphere-formation methods described above, both the yield of microspheres and the encapsulation efficiency are often in the range of 70–90% [90]. The encapsulation efficiency can be influenced by the solubility of a drug in the organic and aqueous phases, the partition coefficient of the drug between polymer-containing and nonpolymer-containing phases, the viscosity of the different phases, the microsphere size, and the nature and concentration of emulsifiers and additives [14]. Furthermore, the amount and type of drug incorporated into microspheres have an influence on the encapsulation efficiency. Jalil and Nixon reported that higher loadings of phenobarbitone resulted in higher encapsulation efficiencies when the solvent evaporation method was used (Fig. 7) [46]. On the other hand, Heya et al. reported that high loadings of thyrotropin-releasing hormone resulted in lower encapsulation efficiencies using a w/o/w method for encapsulation [91]. The results of these two groups point out differences that should be anticipated in the production of microspheres with different drugs by different encapsulation methods.

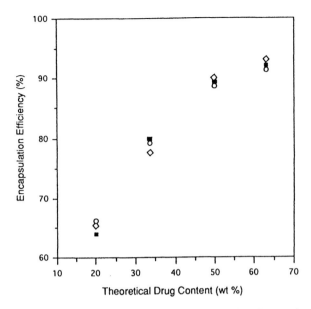

Figure 7   Effect of drug (phenobarbitone) loading on the encapsulation efficiency in microspheres prepared from poly(L-lactic acid) of different molecular weights by a solvent evaporation method (MW:-■-, 61,300; -◇-, 43,200; -○-, 2400). (Modified from Ref. 46.)

The method for determining the drug content or encapsulation efficiency of microspheres made from lactic/glycolic acid polymers must be developed for each drug. Protein and peptide drugs are usually separated from lactic/glycolic acid polymers by a two-phase extraction. That is, after dissolving the microspheres in a water-immiscible organic solvent, water (which may be buffered) is added to the organic solution to extract the protein or peptide. Then, the aqueous protein solution is analyzed to determine the quantity of protein or peptide present. Sometimes techniques such as amino acid analysis or elemental analysis are used to determine the concentration of proteins or peptides in microspheres [89].

For conventional drugs, most of which are hydrophobic, the amount of drug in microspheres can often be analyzed without the separation of lactic/glycolic acid polymer and drug. For steroids, the drug content or encapsulation efficiency is usually determined by dissolving the microspheres in an organic solvent such as methylene chloride. A spectrophotometric method can then be used to quantify the drugs by applying Beer's law. The drug encapsulation efficiency is calculated using the measured drug content and the theoretical drug content (i.e., amount of drug added in the fabrication process).

D.   *In Vitro* and *In Vivo* Drug Release

Generally, the *in vitro* release of a drug from a delivery device should be performed in a physiological medium to simulate *in vivo* conditions. Infinite sink conditions should also be maintained.

In vitro release studies with drug-loaded microspheres are usually conducted using one of the following two methods. The real-time method consists of incubating drug-loaded microspheres in the selected medium and temperature. The amount of drug released at specific time intervals is assayed. This method can be tedious because drug release from microspheres of lactic/glycolic acid polymers can take a few months to even years. The second method consists of accelerating drug release [92]. In this method, a compound is added to the "dissolution" medium to increase the rate of drug release. For example, an organic solvent (e.g., ethanol) may be added to the dissolution medium to enhance the swelling of the polymer and to increase the solubility of the drug in the medium, thereby increasing the drug diffusion rate. The accelerated release method is a useful tool for the development process and for quality control. Obviously, this method does not provide a good correlation with *in vivo* results. However, it can demonstrate the effect of changes in microsphere characteristics on the release of a drug.

Even with best imitation of the *in vivo* milieu that knowledge permits, data obtained from the *in vitro* experiments are often different from those obtained *in vivo*. Wakiyama et al. [48], for instance, have reported that the *in vivo* release of dibucaine-laden microspheres was faster than the *in vitro* release.

The *in vivo* release profiles of drugs from microspheres are often obtained from animal models and clinical trials. It is often difficult to measure polymer biodegradation rate and drug-release profiles *in vivo* because of interference from body tissue and fluids. Radiolabelling of drugs and polymers is one way of circumventing interference and measuring drug release and polymer biodegradation in animal models [93].

The selection of an injection site is important in conducting *in vivo* experiments and for obtaining optimal efficiency in testing a microsphere product. Some of the factors that should be considered are (1) possible microsphere migration from the injection site, (2) unwanted entrapment of the microspheres in a capillary bed (e.g., liver, lung, and kidney), and (3) the local blood supply [94]. One common observation has been the growth of fibrous tissue around microspheres designed for long-term drug release [95–97]. This encapsulation of the microspheres by tissue may alter the drug-release pattern through the formation of a boundary layer that is capable of controlling the drug-release rate [98].

E. Sterilization of Drug-Loaded Microspheres

Sterilization is an important consideration in the production of drug-loaded microspheres based on the lactic/glycolic acid polymers since these products are administered parenterally. The methods commonly used for producing a sterile pharmaceutical product are gamma irradiation, heat sterilization, ethylene oxide vapor sterilization, or aseptic production.

Gamma irradiation is a useful technique for the sterilization of polymeric implants. However, when this technique is used for the sterilization of drug-loaded microspheres of lactic/glycolic acid polymers, the drugs and/or the polymer matrix may be damaged due to decomposition and cross-linking. Consequently, the physical and chemical properties, such as molecular weight and tensile strength, may be changed. There may also be an alteration in the color of the microspheres [99]. Evidence has shown that gamma irradiation can break down polymer chains, decrease the molecular weight, lower the mechanical strength, increase the biodegradation rate, and change the drug-release profile (Figs. 8 and 9) [100–103].

Heat sterilization, which consists of exposing objects to a high-temperature steam environment for a prolonged period of time, is not appropriate for lactic/glycolic acid polymers because these polymers are heat- and moisture-labile polymers. The heat and

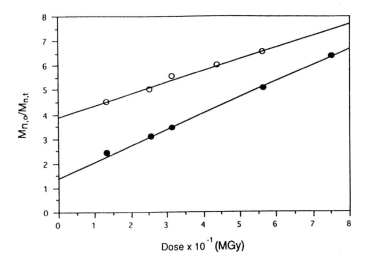

Figure 8 Effect of gamma irradiation on the molecular weight of poly(lactic acid) (M_{no}, number-average molecular weight before irradiation; M_{nt}, number-average molecular weight after irradiation; determined by end-group analysis -○-, in air, -●-, in nitrogen). (Reproduced from Ref. 102 with permission of the copyright owner, Butterworth-Heinemann Limited.)

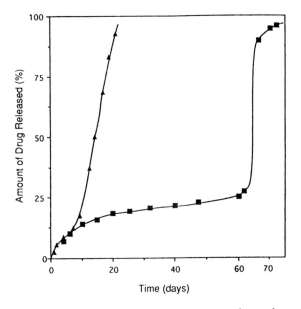

Figure 9 Effect of gamma irradiation on drug release from the microspheres of poly(D,L-lactic-co-glycolic acid) with 29 wt% drug (cisplatin) loading (-▲-, 37.7 KGy; -■-, 0 KGy). (Reproduced from Ref. 103 with permission of the copyright owner, Butterworth-Heinemann Limited.)

moisture can cause chemical and physical changes in the drug-loaded polymer products. It has been reported that the molecular weight of the lactic/glycolic acid polymers decreases after steam sterilization [99].

Ethylene oxide sterilization is an alternative method if the drug and the polymer are not reactive with the vapor of this agent [104]. However, ethylene oxide was found to soften or plasticize some lactic acid polymers. Residuals of this agent left in drug delivery devices can also be hazardous to the patient, particularly if the polymer is absorbed with time. It has been shown that ethylene oxide is mutagenic and carcinogenic [99]. Dermatologic and allergic reactions and hemolysis caused by ethylene oxide have also been found [99,105].

Finally, aseptic production is a safe but more expensive alternative. Since the lactic/ glycolic acid polymers are soluble in organic solvents, polymer solutions can be filter sterilized. Drug solutions can also be filter sterilized. Then, the drug delivery system (i.e., the microspheres) can be formulated in a clean-room environment according to current good manufacturing practice (cGMP) regulations [10].

IV. DELIVERY OF CONVENTIONAL DRUGS USING LACTIC/GLYCOLIC ACID POLYMERS

Since the first report of the application of lactic/glycolic acid polymers to drug delivery made by Yolles et al. in 1971 [4], a large number of drugs have been encapsulated in microspheres prepared with lactic/glycolic acid polymers. Microspheres that release a drug in a controlled manner can provide added value to drugs having a short biological half-life, instability, degradability in the gastrointestinal tract, and high toxicity. The drugs incorporated into microspheres can be classified as conventional drugs and protein (and peptide) drugs. Conventional drugs are relatively low molecular weight substances that are usually obtained by organic synthesis. They can be categorized as contraceptives and other steroid drugs, narcotic antagonists, anesthetics, anticancer agents, antibiotics, antiinflammatory agents, antimalarial drugs, antiasthmatic drugs, neurotropic agents, and other drugs.

From another point of view, conventional drugs can be further classified into two groups: lipid-soluble drugs (such as the steroids) and water-soluble drugs (such as some antibiotics). For the encapsulation of lipid-soluble drugs, the solvent evaporation/ extraction method is usually employed because both the polymer and the drug can be dissolved in an organic solvent. The phase separation method is primarily used for the encapsulation of those drugs that are water soluble. Other methods of encapsulation, such as w/o/w method, are also used to encapsulate water-soluble drugs in microspheres of lactic/glycolic acid polymers [106]. Table 4 lists some of the conventional drugs that have been investigated in delivery systems of lactic/glycolic acid polymers [107–167].

Conventional drugs incorporated into microspheres prepared from lactic/glycolic acid polymers by the above-described methods are either homogeneously distributed in the polymer matrix in a dissolved or dispersed form or are heterogeneously distributed in the polymer matrix in a crystalline state. If a drug is heterogeneously distributed in the polymer matrix as crystals, the drug concentration close to the surface of the microspheres is usually higher, and some drug crystals may even be bound to or embedded in the surface of the microspheres (see Fig. 10).

Table 4 Some Conventional Drugs Formulated with Lactic/Glycolic Acid Polymers

Drug	Polymer	Reference no.
Contraceptive drugs and other steroid drugs		
Norethindrone	D,L-PLA, PLGA	108, 109
Progesterone	D,L-PLA, L-PLA, PLGA	20, 105, 110, 111
Norgestrel	D,L-PLA, L-PLA, PLAC	112, 113
Levonorgestrel	L-PLA, PLGA, PLAC	94, 114–117
Norgestimate	D,L-PLA, L-PLA	118
Estradiol	D,L-PLA	20
Testosterone	PLAC	119
Dihydrotestosterone	PLA	120
Narcotic antagonists		
Cyclazocine	D,L-PLA, L-PLA	19, 20, 121
Naloxone	D,L-PLA	20
Naltrexone	L-PLA, PLGA, PLAC	20, 105, 122–124
Naltrexone pamoate	L-PLA	
L-methadone	L-PLA, PLGA, PLAC	79
Anesthetics		
Lidocaine	D,L-PLA, PLAC	125
Dibucaine	D,L-PLA	48
Butamben	PLA	48, 55
Tetracaine	PLA	48, 55
Anticancer agents		
Doxorubicin	D,L-PLA	20
Cyclophosphamide	D,L-PLA	20
Cisplatin	D,L-PLA, PLGA	20, 27, 49, 61, 126–128
Mitomycin		39
Adriamycin	PLA	129–131
1-(2-chloroethyl)-3-cyclo hydroxy nitrosourea	D,L-PLA	132, 133
Aclarubicin	D,L-PLA, L-PLA	134–136
Cytosine arabinoside	PLGA	137
5-Fluorouracil	D,L-PLA	138
Neutron-activated holmium-166	L-PLA	139
Buserelin	PLGA	140
Vanorelbine	PLGA	141
Lomustine	D,L-PLA	142
Antibiotics		
Ampicillin	PLGA	143–146
Chloramphenicol	PLGA	144, 145
Gentamicin	L-PLA, PLGA	23, 144–146
Polymyxin B	PLGA	144, 145
Tetracycline	PLGA	17, 146, 147
Clindamycin	PLGA	146
Dideoxykanamycin	D,L-PLA	148
Isoniazid	PLGA	149

Table 4   Continued

Drug	Polymer	Reference no.
Antiinflammatory agents		
Dexamethasone	D,L-PLA	20
Methylprednisolone	PLGA	150, 151
Hydrocortisone	PLA	33, 56
Indomethacin	D,L-PLA	152, 153
Triamcinolone acetonide	PLA	154
Antimalarial drugs		
Quinazoline	L-PLA, PLGA	7, 155
Sulfadrazine	PLGA	156
Pyrimethamine	PLGA	157
Antiasthmatics and cardiac		
depressants (stimulants)		
Ketotifen	PLA	158
Quinidine	D,L-PLA, L-PLA	51, 52, 159
Theophylline	PGA, D,L-PLA, L-PLA	24, 84, 159–161
Propranolol	D,L-PLA, L-PLA	159
Neurotropic agents		
Vinpocetine	D,L-PLA	80
Dopamine	PLGA	162
Bromocriptine	PLGA	162, 163
Other drugs and agents		
Thioridazine	D,L-PLA	54, 158
Acetylsalicylic acid	PLA	164
Pseudoephedrine	D,L-PLA	59
Caffeine	D,L-PLA	84
Nicotinic acid	PGA	24
Chlorine dioxide (biocide)	PLGA	165
Sucrose (drug model)	L-PLA	166
Mannitol (drug model)	L-PLA	166
Rhodamine (dye)	D,L-PLA, L-PLA, PLGA	167

D,L-PLA, poly(D,L-lactic acid); L-PLA, poly(L-lactic acid); PLGA, poly(lactic-co-glycolic acid); PLAC, poly-(lactic acid-co-caprolactone); PLA, poly(lactic acid); PGA, poly(glycolic acid).

The heterogenous distribution of a drug in microspheres is due to the difficulty of solubilizing the drug in the polymer-solvent solution, and subsequently in the polymer matrix. Because of this, some of the drug molecules crystallize during the microsphere hardening process. For example, during the encapsulation of progesterone in poly(D,L-lactic acid) microspheres, crystallization of progesterone occurs when the drug loading is increased above a certain value; the higher the loading of the drug is, the more crystals that are formed (see Fig. 11) [84]. For progesterone-containing microspheres with a 90: 10 weight ratio of poly(D,L-lactic acid) to drug, no melting peak of progesterone crystals (at 395°K and 405°K) could be detected by differential scanning calorimetry (DSC). But, with an 80:20 weight ratio, a small melt peak was observable by DSC, indicating that some drug molecules had crystallized in the microspheres. When the weight ratio of poly(D,L-lactic acid) to drug was equal (50:50), two distinct peaks (termed α-form and

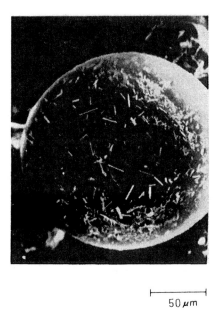

50 μm

Figure 10 Scanning electron micrograph showing cisplatin crystals on the surface of a microsphere of poly(D,L-lactic acid). (Reprinted from Ref. 36 with permission of the copyright owner, Elsevier Science Publishers, and courtesy of Dr. Sakakura.)

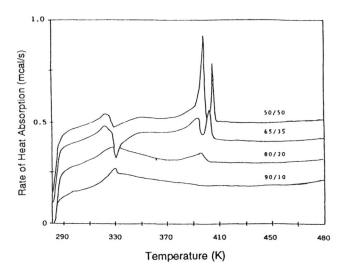

Figure 11 Differential scanning calorimetry thermograms of progesterone-containing microspheres of poly(D,L-lactic acid) prepared by spray drying. (Reprinted from Ref. 84 with permission of the copyright owner, the Royal Pharmaceutical Society of Great Britain.)

β-form) were detected, meaning that a large amount of crystal had formed. Because the hardening of microspheres proceeds from the surface to the center, drug crystallization also starts on the surface or near the surface of the microspheres. This can give rise to a heterogenous or nonuniform distribution of large drug islands in the polymer matrix, presumably on or close to the surface of a microsphere, which ultimately could result in a burst of release upon immersion of the microspheres in a release medium.

The release of conventional drugs from microspheres of lactic/glycolic acid and polymer generally occurs via diffusion through the polymer matrix as well as through the porous voids of the polymer structure. However, biodegradation of the polymer matrix and dissolution of the degraded polymer continuously changes the microsphere geometry and the texture of the polymer matrix. As a result, the drug-release pattern is a combination of diffusion and biodegradation [12].

Because the biodegradation of lactic/glycolic acid polymers generally involves bulk erosion, microspheres take up water prior to the start of matrix degradation and dissolution. After hydration of the polymer matrix has occurred, the encapsulated drug molecules begin to dissolve in the aqueous medium and diffuse out of the polymer matrix. Therefore, the mechanism of drug release may be envisioned as involving the following three phases [9,12]: (1) an initial release of drug that is either loosely bound to the surface or embedded in a superficial region of the microspheres; (2) diffusional release of drug through the polymer matrix and through the aqueous pores during the random chain scission stage of the matrix degradation; and (3) erosional release of the drugs by the polymer matrix disintegration and dissolution after the matrix loses its integrity and the polymer chains are degraded small enough to be soluble. The diffusion phase depends on polymer structure, hydrophilicity, degree of swelling, and drug solubility in the polymer matrix [9]. During this diffusion release, the polymer is hydrolyzing. But, the reduction of molecular weight has not reached a point at which the degradation products are sufficiently small to be significantly soluble. Therefore, there is no polymer matrix disintegration and no concomitant release of drug during this release phase. All of these phases may play a part in the release process, depending on the nature of the encapsulated drug, the physicochemical properties of the polymer, and the structure of the microsphere.

The release profile of a drug from microspheres depends, in part, on the distribution of the drug. If the drug is heterogeneously distributed in the polymer matrix, the release curve may possess a triphasic pattern. If the drug is homogeneously distributed in the polymer matrix, the release curve should possess a biphasic pattern. A report by Beck et al. illustrates the effect of drug distribution on the release profile [114]. They encapsulated levonorgestrel in poly(lactic-co-glycolic acid) microspheres with different drug distribution patterns.

Figure 12a shows a scanning electron micrograph of microspheres having a heterogeneous drug distribution pattern (i.e., some drug crystals were bound to the surface of the microspheres); Fig. 12b shows one of the microspheres without visible drug crystals on the surface. The microspheres were administered to baboons and serum levonorgestrel was monitored. The batch of microspheres shown in Fig. 12a exhibited a triphasic release pattern (see Fig. 13). The first peak or initial burst of release was generated by drug molecules bound to the surface. The second peak represents diffusional release through the polymer matrix and pores before occurrence of polymer matrix erosion. The third peak should be due to drug release during the disintegration and dissolution of the polymer matrix. The batch of microspheres shown in Fig. 12b, on the other hand, exhib-

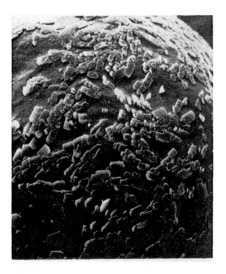

(a)

(b)

Figure 12   Scanning electron micrographs of levonorgestrel microspheres of poly(L-lactic-co-glycolic acid): (a), some levonorgestrel crystals are present on the surface of the microsphere; (b), no levonorgestrel crystals are present on the surface of the microspheres. (Reprinted from Ref. 114 with permission of the copyright owner, Kluwer Academic Publishers, and courtesy of Dr. Beck.)

ited only two peaks of drug release (Fig. 14). The first peak is the drug release from the microspheres by diffusion while integrity of the polymer matrix remains intact, and the second peak is the drug released as the polymer matrix disintegrates and dissolves. There is no distinct phase that can be attributed to superficial drug release.

The hypothetical model depicted in Fig. 15 gives an explanation for the triphasic release pattern shown in Fig. 13. As shown in Fig. 15, drug crystals bound to and/or

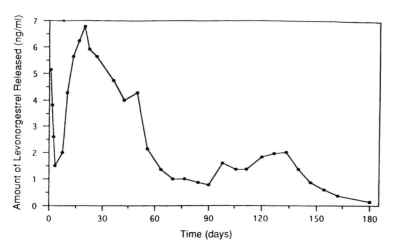

Figure 13 An example of a triphasic pattern of *in vivo* drug (levonorgestrel) release from micro-spheres (shown in Fig. 12a) with drug crystals on the surface. (Reproduced from Ref. 114 with permission of the copyright owner, Kluwer Academic Publishers.)

embedded in the surface of a microsphere are washed away upon immersion of the microspheres in a release medium. This results in an initial burst of release. The micro-sphere absorbs water and swells. The amount of swelling depends on the composition, molecular weight, and crystallinity of the polymer used. The absorbed water dissolves the drug inside the microsphere and the drug molecules diffuse out of the microsphere, leaving channels or pores inside the microsphere. Simultaneously, random hydrolysis of the polymer chains begins. As the molecular weight of the polymer decreases, the perme-ability of the matrix increases and the diffusion rate increases, providing the second peak of the drug release. This phase represents a period during which there is no disintegration

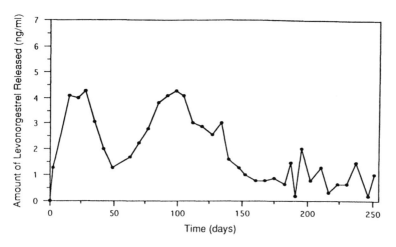

Figure 14 An example of a biphasic pattern of *in vivo* drug (levonorgestrel) release from micro-spheres (shown in Fig. 12b) with no visible drug crystals on or near the surface. (Reproduced from Ref. 114 with permission of the copyright owner, Kluwer Academic Publishers.)

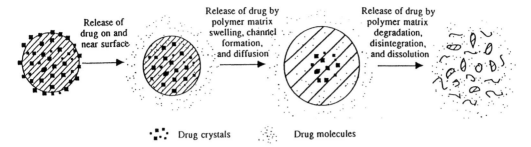

Figure 15 Hypothetical models of drug release from microspheres with some drug crystals bound to and/or embedded in the surface.

of the polymer matrix and no gross dissolution of the polymer. The duration of this phase is determined by the degradation characteristics of a particular polymer. As the amount of drug inside the microsphere decreases and the path length of diffusion through the channels increases, the drug release rate declines. Some of the drug molecules at the center of a microsphere may never be released until disintegration and dissolution of the polymer matrix occurs. When the polymer has diminished sufficiently in molecular weight to lose integrity and to become water soluble, the polymer matrix of the microsphere breaks apart and then dissolution of the polymer fragments occurs. At this point, the third phase, with a high release rate of drug, commences [168].

The initial burst of drug release can be eliminated by briefly washing the microspheres to remove the drug crystals on and/or near the surface. Mason et al. used ethanol to wash progesterone crystals from the surface of microspheres of poly(D,L-lactic acid) to eliminate a release burst [38]. Figure 16 shows the results of washing microspheres. After two hours, approximately 10% of the progesterone was released from nonwashed microspheres, while only about 2% of the drug was released from the washed microspheres.

In contrast to a burst of drug release, Suzuki and Price reported a lag-time prior to the release of drug from microspheres of lactic/glycolic acid polymers [169]. They pre-

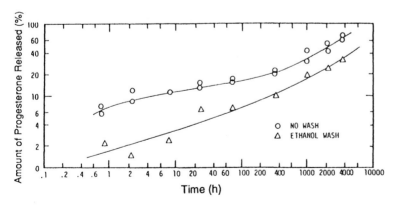

Figure 16 Release of progesterone from ethanol-washed and nonwashed microspheres of poly-(D,L-lactic acid) (drug loading, 25%; size, 180–300 μm). (Reproduced from Ref. 38 with permission of the copyright owner, CRC Press, Incorporated, Boca Raton, FL.)

pared chlorpromazine microspheres from polymers with high molecular weights of 16,300 and 20,500. There was a 24-hour lag period before the release of chlorpromazine began. The drug-release rate then increased rapidly until 40% of the drug had been released. Subsequently, the rate decreased and followed a zero-order pattern. Bodmeier and McGinity also found a lag time in the release of quinidine and quinidine sulphate from microspheres of poly(D,L-lactic acid) [170]. The length of the lag time increased with an increase in the ionic strength of the dissolution medium. Because the characteristics of the polymer matrix and possible polymer-drug interactions were largely unknown, no generally satisfactory explanation for the drug release pattern could be given.

V. DELIVERY OF PROTEIN AND PEPTIDE DRUGS USING LACTIC/GLYCOLIC ACID POLYMERS

The delivery of protein and peptide drugs has become an important research area because a wide range of genetically engineered proteins and peptides are now being investigated for therapeutic applications. The high pharmacological potency of these molecules is counterbalanced by their poor penetration through the physiological barriers, their fragility, and their very short biological half-life, which limit their usefulness [171].

Generally, protein and peptide drugs cannot be administered orally because they are hydrolyzed or denatured in the gastrointestinal tract. Conventional administration by injection is also problematic since proteins and peptides are usually quickly metabolized and eliminated [172]. In order to achieve desirable therapy, multiple injections have to be administered. Typical pharmacokinetic half-lives for proteins and peptides range from 2 to 30 minutes [173]. Due to the inadequacy of conventional dosage strategies to administer protein and peptide drugs effectively, it is necessary to develop controlled-release dosage forms. These systems should be capable of releasing proteins and peptides at a controlled rate with minimum denaturation over an extended period of time. To extend the biological half-life and achieve sustained levels of protein and peptide drugs in the body, encapsulation in microspheres composed of lactic/glycolic acid polymers has become a viable option for the delivery of these drugs [107]. Table 5 lists some protein and peptide delivery systems based on lactic/glycolic acid polymers.

Protein and peptide drugs do not dissolve in the organic solutions of lactic/glycolic acid polymers during microencapsulation processes. They do not dissolve in a solid, polymeric matrix of microspheres, either. Consequently, they are usually encapsulated in microspheres of lactic/glycolic acid polymers by either a phase separation method or a w/o/w method. In these two methods, the proteins and peptides are either dispersed in lactic/glycolic acid polymer solution or, preferably, dissolved in an aqueous solution that is then emulsified in the polymer solution. Thus, these drugs are encapsulated in the microspheres as islands or drug domains, which results in a heterogeneous drug distribution in the microspheres.

In comparison with the release of conventional drugs, the release of proteins and peptides does not involve the diffusion of the drugs through a solid polymer matrix since the drugs are not soluble in the polymer. It has been shown that the release of proteins and peptides from microspheres composed of lactic/glycolic acid polymers or other hydrophobic polymers occurs by diffusion through aqueous pores or channels [201,202]. These aqueous channels facilitate water-soluble drug release.

It has been proposed that release occurs via two different mechanisms [201]. The first mechanism involves the leaching of drug molecules from the water-soluble drug

Table 5   Some Proteins and Peptides Formulated with Lactic/Glycolic Acid Polymers

Protein/Peptide	Polymer	Reference no.
Insulin	D,L-PLA	31, 174–176
Interleukin-2	PLGA	177
γ-Globulin	D,L-PLA	174
Bovine somatotropin	PLA, PLGA	178
Lysozyme	D,L-PLA	174
Trypsin	D,L-PLA	174
Horse radish peroxidase	PLGA	179
Atriopeptin III	PLGA	180
Human serum albumin	PLGA	181
Bovine serum albumin	D,L-PLA	79, 89, 174, 181–183
Ovalbumin	PLGA	74, 174, 184–186
Dextran (protein model)	L-PLA	168
LHRH and its agonists	D,L-PLA, PLGA	187–191
Nafarelin (LHRH analog)	PLAC	192, 193
Leuprorelin (LHRH analog)	D,L-PLA, PLGA	69, 70, 73, 194, 195
Triptoreline (LHRH analog)	PLGA	196
Gonadotropin-releasing hormone and its agonist	PLGA	197, 198
Thyrotropin-releasing hormone	PLGA	91, 199, 200
Growth-hormone-releasing hormone	PLA, PLGA	178
Goserelin	PLGA	201

LHRH, luteinizing-hormone-releasing hormone

domains that exist on and near the surface of the microspheres. After the drug molecules are released from these domains, aqueous channels or pores are generated. For water-soluble drugs, and in particular for small peptides, the aqueous channels generated by this drug-leaching mechanism only facilitates the release of drug molecules on and near the surface of a microsphere. The drug molecules within the body of the microspheres, especially at the center of microspheres, cannot be released until the second mechanism becomes operative. These "inside" drug domains are isolated and are not continuous or contiguous with the surface. The second mechanism of release involves degradation of the polymer, which is associated with the generation of microporosity, enhanced water uptake, and ultimate disintegration of the polymer matrix.

These two separate aqueous-channel-generating mechanisms can lead to two different drug-release stages and often result in a biphasic drug-release profile. The first phase corresponds to the release of drug from domains on and near the surface of a microsphere by leaching. The second phase corresponds to the release of the drug from domains at the center of a microsphere by biodegradation and disintegration of the polymer matrix.

Typical parameters controlling the first phase of release include drug loading, morphology, and geometry. For example, increasing the drug loading can extend the first release phase via increased channeling. The second phase is related to the degradation properties of the polymer. Polymer composition, molecular weight, and molecular weight distribution, and consequently the water uptake and swelling ratio of the polymer matrix, all affect the time of initiation and duration of the second phase. When these two phases of release do not overlap, discontinuous release is observed (Fig. 17a). However, by selecting a polymer with the appropriate properties and by adjusting the parameters of

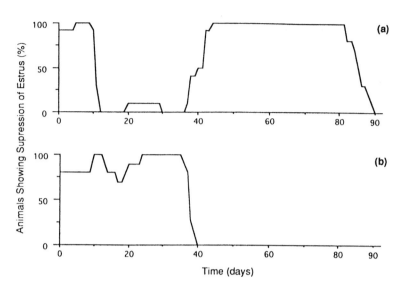

Figure 17 Biological effect (estrus suppression) of nafarelin acetate released from microspheres: (a), discontinuous release from 69:31 poly(lactic-co-glycolic acid) (IV, 0.97 dl/g; drug loading, 1.1 wt%; size, 63–125 μm); (b) continuous release from 50:50 poly(lactic-co-glycolic acid) (IV, 0.52 dl/g; drug loading, 1.05 wt%; size, 90–150 μm). (Modified from Ref. 203.)

the microencapsulation process, the first release phase can be made to overlap with the second release phase to give a continuous release profile (see Fig. 17b) [201,203]. Figure 17 shows release of an analog of luteinizing-hormone-releasing hormone, nafarelin acetate, from microspheres prepared by Kent et al. [203]. They prepared microsphere formulations using two different types of polymers, 69:31 poly(lactic-co-glycolic acid) and 50:50 poly(lactic-co-glycolic acid). The release profiles of the peptide and consequently the biological effect of the drug on animals are different for the two formulations. The formulation composed of a slowly degrading polymer having a high intrinsic viscosity, 69:31 poly(lactic-co-glycolic acid), exhibited a discontinuous release profile (Fig. 17a). There was an initial burst of drug as a result of drug release from the superficial region of the microspheres. This initial release was the first phase of drug release. After the first phase was complete, no drug was further released until the polymer began to degrade and the microspheres started to disintegrate. However, using a more rapidly degrading polymer having a lower intrinsic viscosity, 50:50 poly(lactic-co-glycolic acid), Kent and his coworkers were able to overlap these two phases (see Fig. 17a). The typical initial burst of release was followed immediately by the second phase of drug release, caused by polymer degradation and microsphere disintegration.

The release of most proteins differs from that of small peptides in that the first release phase, that is, the diffusion of drugs from aqueous pores, dominates. In this case, drugs are released as a burst. Wang et al. studied the release and biodegradation of lactic/glycolic acid polymer microspheres containing bovine serum albumin [67]. Figure 18 shows the release of bovine serum albumin from microspheres of poly(D,L-lactic-co-glycolic acid) prepared by a phase separation method and then incubated in phosphate buffered saline at 37°C. They concluded that most of the protein was released by a diffusional mechanism before the loss of polymer mass.

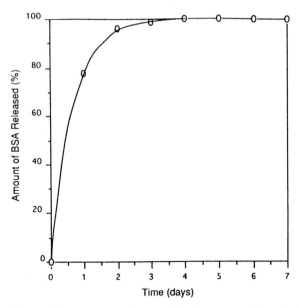

Figure 18 *In vitro* release of bovine serum albumin from microspheres of poly(D,L-lactic-co-glycolic acid). (Modified from Ref. 67.)

One explanation for the rapid release of proteins from microspheres of lactic/glycolic acid polymers follows. After the polymer matrix takes up water and swells, the protein domains are connected by aqueous micropores in the polymer matrix. These aqueous micropores form tortuous narrow channels that connect protein domains to the surface of a microsphere. Protein molecules can then diffuse out of a microsphere through these connecting channels by a random walk mechanism. After protein molecules diffuse away

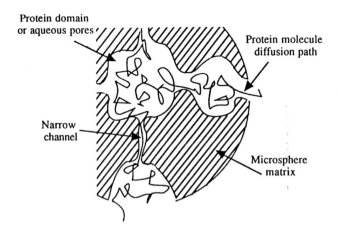

Figure 19 Hypothetical model illustrating the release of macromolecules or proteins from microspheres of lactic/glycolic acid polymers by a random walk through micropores and connecting channels.

from the domains, the protein domains themselves become aqueous pores in the polymer matrix, which further facilitates the release of protein molecules. Figure 19 depicts protein domains, connecting channels, and the random walk path of a diffusing protein molecule [204].

The retention of biological activity of the protein should be considered in the delivery of protein drugs by lactic/glycolic acid polymer microspheres. The hydrophobic character of the lactic/glycolic acid polymers and the presence of free carboxylic end groups in the polymer matrix provide a hydrophobic and possibly acidic microenvironment for the protein molecules. This microenvironment may result in the loss of biological activity of some encapsulated proteins. Using atriopeptin III as a model protein, Johnson et al. reported that proteins may be denatured if placed in contact with lactic/glycolic acid polymers [180].

Tabata et al. examined the effect of a w/o emulsion method on the biological activity of proteins [174]. They used insulin and trypsin as model proteins. Aqueous solutions of protein were emulsified in a polymer solution of chloroform by probe sonication. After emulsification, the w/o systems were freeze dried. They found that 71% of the biological activity of insulin and 40% of the bioactivity of trypsin were lost after emulsification and freeze drying. Further evidence was presented, indicating that the freeze-drying process did not influence protein bioactivity and that chloroform caused the loss of bioactivity.

In contrast, Cohen et al. have investigated the retention of enzymatic activity of horseradish peroxidase in microspheres of lactic/glycolic acid polymers [179]. These investigators reported that the microspheres stabilized the enzyme against the loss of biological activity.

The potential for a burst of release and the loss of biological activity may limit the delivery of some protein drugs using microspheres of lactic/glycolic acid polymers. Therefore, these possibilities must be considered when one designs and develops a con-trolled-release system for delivering protein drugs.

VI. DELIVERY OF VACCINES USING LACTIC/GLYCOLIC ACID POLYMERS

Effective immunization by many vaccines requires a primary dose and several boosters. Multiple visits to a clinic are required, and patient compliance is often difficult. Vaccine delivery systems that provide an effective single-injection immunization and eliminate the need for booster shots would be highly desirable. Furthermore, the majority of vaccines require the addition of adjuvants to be effective. To this end, polymeric particles with antigen adsorbed on their surface have been shown to act as adjuvants for humoral immune responses when they are injected and, to a limited extent, when they are given orally [205]. However, the oral route subjects the adsorbed antigen to degradation in the gastrointestinal tract and possibly to desorption from the polymer particles by bile salts or proteins. The microencapsulation of vaccines in lactic/glycolic acid polymers may possibly protect the antigen against degradation, may provide means of releasing antigen over an extended period of time, and may eliminate the need for adjuvants. Table 6 shows some vaccines that have been encapsulated in lactic/glycolic acid polymers.

When microspheres of lactic/glycolic acid polymers are used for vaccine delivery, they do not only function as a carrier, but they also enhance antibody production against the antigen delivered. That is, they act as adjuvants. Almeida et al. found that the association of cholera toxin B with poly(lactic acid) microspheres induced a threefold

Table 6 Some Vaccines Formulated with Lactic/Glycolic Acid Polymers

Vaccine	Polymer	Reference no.
Tetanus toxoid	PLGA	90, 206, 207
Staphylococcal enterotoxin B	D,L-PLA, L-PLA, PLGA	90, 208–210
Diphtheria toxoid	D,L-PLA	90
E. coli bacterin	PLGA	78
Panleukopenia virus	PLGA	78
Bacillus subtilis	PLA	211
Cholera toxin B	PLA	212
Hepatitis B	PGA	35
Malaria vaccine	D,L-PLA, L-PLA, PLGA	213
Contraceptive vaccine	D,L-PLA, PLGA	214
Simian immunodeficiency virus vaccine	D,L-PLA	215
Salmonella enteritidis	D,L-PLA, PLGA	216

increase in serum immunoglobulin G (IgG) titer in comparison with free antigen, which has a dose 40 times higher than the antigen associated with poly(lactic acid) microspheres [212].

Eldridge et al. have encapsulated staphylococcal enterotoxin B in poly(D,L-lactic acid) microspheres [208,209]. They found that the response of antitoxin antibody was 64 times greater than that induced by an optimal dose of nonencapsulated enterotoxin. Furthermore, they demonstrated that a mixture of two microsphere formulations, which released the vaccine at different times, provided discrete primary and booster responses after a single injection. These researchers concluded that microspheres of lactic/glycolic acid polymers are a promising adjuvant that has the added ability of providing a sustained release of antigen that can be controlled and programmed.

Hazrati et al. have shown that a tetanus toxoid vaccine formulated by using microspheres of lactic/glycolic acid polymers can provide prolonged antibody production (Fig. 20) [206]. This sustained release of antigen can eliminate many follow-up injections, which can improve patient compliance.

Nellore et al. have used poly(glycolic acid) microspheres with a size of 1–60 μm to deliver hepatitis B antigen [35]. They found that smaller microspheres (1–10 μm) elicited an earlier antibody response with an average antibody titer greater than the positive control of alum-adsorbed antigen. Their results suggested that small microspheres may be phagocytized by the macrophages. Consequently, the macrophages, which are antigen-presenting cells, contain relatively large amounts of antigen and a high degree of immuno-potentiation is observed. In contrast to the smaller microspheres, the larger microspheres (20–60 μm) provided a delayed response and a longer duration of high-level antibody production. It was suggested that the larger size of these microspheres may prevent direct phagocytic engulfment. The larger microspheres would then remain at the site of injection, where they would undergo biodegradation to form smaller-size fragments that could then be phagocytized.

Eldridge et al. have shown that the microsphere size is also important for oral vaccine delivery [210]. They found that the microspheres smaller than 5 μm are absorbed into Peyer's patches and then are carried by macrophages to the mesenteric lymph nodes and spleen, which induce a systemic antibody response, whereas microspheres larger than 10

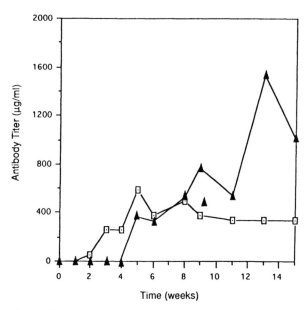

Figure 20 Mouse antibody titers following a single injection (0 week) of tetanus toxoid microspheres of poly(D,L-lactic-co-glycolic acid) (-☐-) and multiple injections (0, 4, and 12 weeks) of free tetanus toxoid (-▲-). (Modified from Ref. 206.)

μm are not absorbed at any point in gastrointestinal tract. Microspheres in the mid-size range of 5–10 μm are targeted to Peyer's patches and remain there for an extended period of time, which induces a sustained mucosal antibody response.

VII. FACTORS AFFECTING DRUG RELEASE FROM MICROSPHERES OF LACTIC/GLYCOLIC ACID POLYMERS

There are many parameters that influence drug release from microspheres of lactic/glycolic acid polymers. Physicochemical properties of the polymer used, nature of the drug and drug content in the microspheres, nature and amount of excipients added, and microencapsulation conditions all affect drug release. Discussed here are polymer composition, molecular weight of the polymer, drug content, microsphere size, microsphere porosity, and other factors.

A. Effect of Polymer Composition

Lactic acid and glycolic acid homo- and copolymers are semicrystalline to amorphous thermoplastic polymers. Poly(glycolic acid) and copolymers of high glycolic acid content do not dissolve in common organic solvents, which can limit their suitability for use in the fabrication process for drug formulations. However, the homopolymer of D,L-lactic acid and copolymers of high lactic acid content have good solubility in many commonly used organic solvents. Therefore, these are the main polymeric excipients used in drug formulations. The selection of a specific polymer is based on the requirements needed to deliver the specific drug safely and effectively. Depending on the potency of the drug and the degradation profile of the polymer used, the duration of drug release can range from a few days to even a couple of years [10].

The composition (comonomer ratio) of a polymer influences the pattern of drug release from microspheres. This is due to a number of polymer properties, including hydrophobicity, water uptake ability, swelling ratio, crystallinity, glass transition temperature, biodegradation rate, and permeability to the drug [217]. Sanders et al. have studied the effect of lactic acid content in lactic-glycolic acid copolymers on the release of an analog of luteinizing-hormone-releasing hormone (nafarelin) from subcutaneous implants in rats [192]. As shown in Fig. 21, increasing the lactic acid content caused an almost linear increase in the duration of the biological effect of nafarelin.

B. Effect of Molecular Weight of the Polymer

The molecular weight of lactic/glycolic acid polymers has a great effect on drug release from delivery systems (Fig. 22) [218].

Wakiyama et al. have studied the release of some local anesthetics (butamben, dibucaine, and tetracaine) from poly(D,L-lactic acid) microspheres [219]. They used three different polymers with molecular weights of 9100, 17,000, and 25,000, and found that the release rate increased as the polymer's molecular weight decreased. The fast release of drugs from low molecular weight polymers has been attributed to a greater water uptake, higher swelling ratio, and faster biodegradation rate than that of higher molecular weight polymers (Fig. 23) [45,220].

Jalil and Nixon found that molecular weight can have the opposite effect on the drug-release rate [42]. They microencapsulated phenobarbitone in L-lactic acid polymers with three different molecular weights: 61,300, 43,200, and 2400. In this study, the high molecular weight polymers released phenobarbitone faster than the low molecular weight

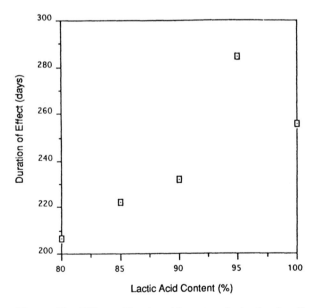

Figure 21   Effect of lactic acid content in lactic-glycolic acid copolymer on the duration of the biological effect of nafarelin released from microspheres. (Modified Ref. 192.)

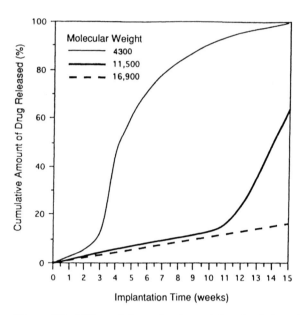

Figure 22 Effect of the molecular weight of poly(D,L-lactic acid) on the release of a luteinizing-hormone-releasing hormone agonist from microspheres. (Reproduced from Ref. 218 with permission of the copyright owner, Harwood Academic Publisher.)

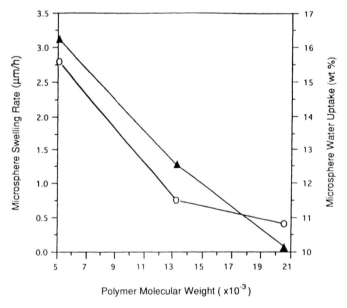

Figure 23 Effect of molecular weight of poly(D,L-lactic acid) on the microsphere swelling rate -O- and the amount of water uptake -▲- during *in vitro* dissolution (drug [phenobarbitone]: polymer = 1:2; dissolution conditions: buffer, pH 2; stir rate, 100 rpm; temperature, 37°C). (Modified from Ref. 45.)

polymers. Based on the microscopic observation of the morphology of the microspheres, they proposed that the low molecular weight polymers formed denser, more uniform microspheres than the high molecular weight polymers, and consequently released less drug than their more porous counterparts.

C. Effect of Drug Content

The drug concentration in a microsphere formulation has a considerable effect on the rate and duration of drug release. Jalil and Nixon have investigated the effect of drug loading on the release behavior of microspheres [42]. They encapsulated phenobarbitone in microspheres of poly(L-lactic acid) using an oil-in-oil (acetonitrile-in-light-liquid-paraffin) emulsion method that was followed by solvent evaporation. *In vitro* release studies were conducted at 37°C in a buffer of pH 2 with a stir rate of 100 revolutions per minute (rpm), and the results indicated that the drug-release rate increased linearly with the increase of drug content (Fig. 24).

D. Effect of Microsphere Size

Because most of the drugs encapsulated in microspheres are released prior to microsphere disintegration and polymer dissolution, the drug-release rate is proportional to the surface area of the microspheres. Smaller microspheres provide more surface area for release than larger microspheres; this effect is shown in Fig. 25 [30]. From this figure, it is evident that the *in vitro* release of progesterone from three different fractions of the same batch of microspheres decreases with increasing microsphere size.

Cowsar et al. have reported similar results [221]. They prepared norethindrone-loaded microspheres, also by a solvent-evaporation method. The microspheres were di-

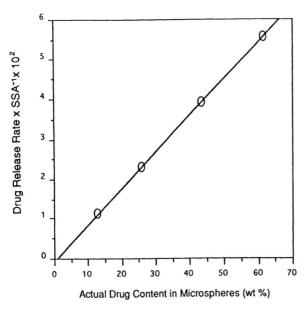

Figure 24 Effect of drug (phenobarbitone) content on release rate of microspheres of poly(L-lactic acid) (SSA, specific surface area). (Modified from Ref. 42.)

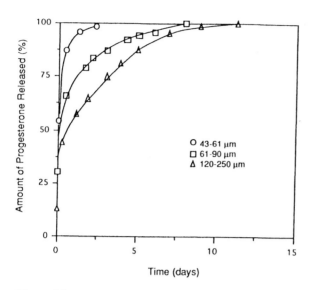

Figure 25 *In vitro* release of progesterone from different particle size fractions of poly(D,L-lactic acid) microspheres. (Reproduced from Ref. 30 with permission of the copyright owner, Kluwer Academic Publishers.)

vided into two fractions having sizes of 45–63 μm and 90–125 μm. The amount of norethindrone released from these two fractions corresponded to the amount of surface area exposed to the *in vitro* release medium.

E. Effect of Microsphere Porosity

The surface quality and porosity of microspheres can affect the drug-release rate. If other parameters are equal, the rougher the surface and the more porous the microsphere is, the faster the release rate will be because more surface area is exposed to release medium.

The porosity of a microsphere can be influenced by many factors. The concentration of the polymer solution used to prepare the microspheres, the solvent used to dissolve the polymer, the solvent-removal rate, and the method of microsphere formation all affect the porosity of a microsphere. An example of the effect of polymer concentration on porosity and thus on drug release is shown in Fig. 26, which presents the *in vitro* release profiles of methylene-blue-loaded microspheres prepared by a solvent-evaporation method [222]. Microspheres prepared from a solution with a low concentration of polymer (2.5 wt%) were depleted of methylene blue in approximately two days, while microspheres prepared from a solution with a high concentration of polymer (10 wt%) released methylene blue for eight days. These results suggest that more porous microspheres were formed with the organic solutions that contained less polymer; consequently, these more porous microspheres released laden agents more rapidly.

Juni et al. have examined the effect of fatty acid ester additives on the release rate of bleomycin from microspheres of poly(lactic acid) [223]. The addition of fatty acid esters (e.g., ethyl myristate, butyl myristate, and isopropyl palmitate) changed the morphology of the microspheres and thus changed the drug-release rate. Their results showed that increasing the concentration of fatty acid esters increased the drug-release rate and that

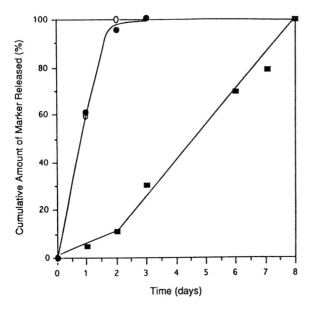

Figure 26 *In vitro* release of methylene blue from microspheres prepared from solutions with different poly(glycolic acid) concentrations (-○- and -●-, 2.5%; -■-, 10%). (Modified from Ref. 222.)

this increase was attributable to a tendency to form more porous microspheres. They also reported that the drug-release rate increased with the increase in alkyl chain length of the fatty acid esters.

F. Effect of Other Factors

The pattern of drug release also depends on many other factors, some of which are (1) the physicochemical properties of the drug, including molecular weight, solubility in the polymer matrix, and particle size; (2) the affinity of the polymer and the drug for one another; (3) the stability of the drug in the polymer matrix before and after administration; (4) the polymer's molecular weight distribution; (5) the biodegradation rate of the polymer matrix; and (6) additives. The conditions and length of storage have an effect on the release of drug from microspheres [146]. The route of administration and even the specific site of injection may also affect the release of drug from microspheres and consequently the drug's duration of action. For example, Maulding has shown differences in the duration of luteinizing-hormone-releasing hormone released from microspheres of lactic/glycolic acid polymers that are injected intramuscularly and subcutaneously [97].

VIII. SUMMARY

Microspheres are an important class of drug delivery systems. Easy fabrication, good reproducibility, and convenient administration all give microspheres broad applications in the formulation of both veterinary and human drugs. Microspheres based on the lactic/glycolic acid polymers degrade both *in vitro* and *in vivo* to nontoxic products.

When manufactured under CGMP, the microspheres of lactic/glycolic acid polymers do not exhibit toxicity or any harmful side effects.

One of the primary objectives with microsphere delivery systems is to obtain a prolonged release of a drug for periods ranging from a few days up to a year or more. Great efforts have been made over the past decades in tailoring microsphere delivery systems to release specific drugs over a required time period while the polymer matrices degrade to innocuous compounds.

Microspheres of lactic/glycolic acid polymers have been investigated for the delivery of conventional drugs, protein and peptide drugs, and vaccines. The release of drugs from microspheres of lactic/glycolic acid polymers can be controlled by adjusting polymer composition, molecular weight of the polymer, drug content in the microspheres, microsphere size, surface quality and porosity of the microspheres, and other parameters. With good control of the hydration and biodegradation properties of the polymers and good control of drug release, there is no doubt that microspheres based on the lactic/glycolic acid polymers will become one of the most useful and versatile materials for drug delivery.

Several types of veterinary drugs and human drugs incorporated into microspheres of lactic/glycolic acid polymers are currently undergoing field trials and clinical trials. It may be anticipated that some of these microsphere drug delivery products will be marketed in the near future.

ACKNOWLEDGMENT

The author would like to thank Dr. Janna Strobel for her editorial assistance.

REFERENCES

1. Vert, M., S. Li, and H. Garreau, More about the degradation of LA/GA-derived matrices in aqueous media, *J. Controlled Release*, 16:15–26 (1991).
2. Dunn, R. L., J. P. English, J. D. Strobel, D. R. Cowsar, and T. R. Tice, Preparation and evaluation of lactide/glycolide copolymers for drug delivery, in *Polymers in Medicine*, Vol. 3, C. Migliaresi et al., eds., Elsevier Science Publishers, Amsterdam, The Netherlands, 1988, pp. 149–160.
3. Lipinsky, E. S., and R. G. Sinclair, Is lactic acid a commodity chemical? *Chemical Engineering Progress*, 26–32 (August 1986).
4. Yolles, S., J. E. Eldridge, and J. H. R. Woodland, Sustained release of drugs from polymer/drug mixtures, *Polym. News*, 1:9–15 (1971).
5. Boswell, G. A., and R. M. Scribner, Polylactide-drug mixtures, U.S. Patent 3,773,919 (1973).
6. Sinclair, R. G., Slow-release pesticide system: Polymers of lactic and glycolic acids as ecologically beneficial, cost-effective encapsulating materials, *Environ. Sci. Technol.*, 7:955–956 (1973).
7. Wise, D. L., G. J. McCormick, G. P. Willet, and L. C. Anderson, Sustained release of an antimalarial drug using a copolymer of glycolic/lactic acid, *Life Sci.*, 19:293–304 (1976).
8. Beck, L. R., D. R. Cowsar, D. H. Lewis, J. W. Gibson, and C. E. Flowers, Jr., New long-acting injectable microcapsule contraceptive system, *Am. J. Obstet. Gynecol.*, 135:419–426 (1979).
9. Jalil, R., and J. R. Nixon, Biodegradable poly(lactic acid) and poly(lactide-co-glycolide) microcapsules: Problems associated with preparative techniques and release properties, *J. Microencapsulation*, 7:297–325 (1990).

10. Lewis, D. H., Controlled release of bioactive agents from lactide/glycolide polymers, in *Biodegradable Polymers as Drug Delivery Systems*, M. Chasin and R. Langer, eds., Marcel Dekker, New York, 1990, pp. 1-41.

11. Arshady, R., Preparation of biodegradable microspheres and microcapsules: 2. Polyactides and related polyesters, *J. Controlled Rel.*, 17:1-22 (1991).

12. Arshady, R., Biodegradable microcapsular drug delivery systems: Manufacturing methodology, release control and targeting prospects, *J. Bioact. Compatible Polym.*, 5:315-342 (1990).

13. Conti, B., F. Pavanetto, and I. Genta, Use of polylactic acid for the preparation of microparticulate drug delivery systems, *J. Microencapsulation*, 9:153-166 (1992).

14. Fong, J. W., Microencapsulation by solvent evaporation and organic phase separation processes, in *Controlled Release Systems: Fabrication Technology*, Vol. 1, D. Hsieh, ed., CRC Press, Boca Raton, FL, 1988, pp. 81-108.

15. Dunn, R. L., D. H. Lewis, and L. R. Beck, Fibrous polymers for the delivery of contraceptive steroids to the female reproductive tract, in *Controlled Release of Pesticides and Pharmaceuticals*, D. H. Lewis, ed., Plenum, New York, 1981, pp. 125-146.

16. Dunn, R. L., D. H. Lewis, and J. M. Goodson, Monolithic fibers for controlled delivery of tetracycline, *Proceed. Intern. Symp. Control. Rel. Bioact. Mater.*, 9:157-159 (1982).

17. Agarwal, R. K., D. H. Robinson, G. I. Maze, and R. A. Reinhardt, Development and characterization of tetracycline-poly(lactide/glycolide) films for the treatment of periodontitis, *J. Controlled Rel.*, 23:137-147 (1993).

18. Spilizewski, K. L., R. E. Marchant, C. P. Hamlin, J. M. Anderson, T. R. Tice, T. O. Dappert, and W. E. Meyer, The effect of hydrocortisone acetate loaded poly(D,L-lactide) films on the inflammatory response, in *Advances in Drug Delivery Systems*, Vol. 3, J. M. Anderson and S. W. Kim, eds., Elsevier Scientific Publishing, Amsterdam, The Netherlands, 1987, pp. 197-203.

19. Woodland, J. H. R., S. Yolles, D. A. Blake, M. Helrich, and F. J. Meyer, Long-acting delivery systems for narcotic antagonists. 1, *J. Med Chem.*, 16:897-901 (1973).

20. Yolles, S., and M. F. Sartori, Degradable polymers for sustained drug release, in *Drug Delivery Systems: Characteristics and Biomedical Applications*, R. L. Juliano, ed., Oxford University Press, New York, 1980, pp. 84-111.

21. Gresser, J. A., D. L. Wise, L. R. Beck, and J. F. Howes, Biodegradable cylindrical implants for fertility regulation, in *Progress in Contraceptive Delivery Systems, Vol. 1: Biodegradables and Delivery Systems for Contraception*, E. S. E. Hafez and W. A. A. van Os, eds., MTP Press, Lancaster, England, 1980, pp. 83-95.

22. Fukuzaki, H., M. Yoshida, M. Asano, M. Kumakura, T. Mashimo, H. Yuasa, K. Imai, and H. Yamanaka, *In vivo* characteristics of high molecular weight copoly(L-lactide/glycolide) with S-type degradation pattern for application in drug delivery systems, *Biomaterials*, 12: 433-437 (1991).

23. Sampath, S. S., K. Garvin, and D. H. Robinson, Preparation and characterization of biodegradable poly(L-lactic acid) gentamicin delivery systems, *Intern. J. Pharm.*, 78:165-174 (1992).

24. Moll, F., and R. Ries, Biodegradable microtablets made of low molecular weight polyglycolic acid, *Arch. Pharm. (Weinheim)*, 324:939-940 (1991).

25. Omelczuk, M., and J. W. McGinity, The influence of thermal treatment on the physical-mechanical and dissolution properties of tablets containing poly(D,L-lactic acid), *Pharm. Res.*, 10:542-548 (1993).

26. Marcotte, N., and M. F. A. Goosen, Delayed release of water-soluble macromolecules from polylactide pellets, *J. Controlled Rel.*, 9:75-85 (1989).

27. Ike, O., Y. Shimizu, R. Wada, S. H. Hyon, and Y. Ikada, Controlled cisplatin delivery system using poly(D,L-lactic acid), *Biomaterials*, 13:230-234 (1992).

28. Julienne, M. C., M. J. Alonso, J. L. Gomez Amoza, and J. P. Benoit, Preparation of

poly(D,L-lactide/glycolide) nanoparticles of controlled particle size distribution: Application of experimental designs, *Drug Develop. Industr. Pharm.*, 18:1063–1077 (1992).

29. Gardner, D. L., and D. J. Fink, Biodegradable microcapsules and microparticles for contraceptions, in *Progress in Contraceptive Delivery Systems, Vol. 1: Biodegradables and Delivery Systems for Contraception*, E. S. E. Hafez and W. A. A. van Os, eds., MTP Press, Lancaster, England, 1980, pp. 47–62.

30. Beck, L. R., D. R. Cowsar, and D. H. Lewis, Systemic and local delivery of contraceptive steroids using biodegradable microcapsules, in *Progress in Contraceptive Delivery Systems, Vol. 1: Biodegradables and Delivery Systems for Contraception*, E. S. E. Hafez and W. A. A. van Os, eds., MTP Press, Lancaster, England, 1980, pp. 63–81.

31. Kwong, A. K., S. Chou, A. M. Sun, M. V. Sefton, and M. F. A. Goosen, *In vitro* and *in vivo* release of insulin from poly(lactic acid) microbeads and pellets, *J. Controlled Rel.*, 4: 47–62 (1986).

32. Kishida, A., J. B. Dressman, S. Yoshioka, Y. Aso, and Takeda, Some determinants of morphology and release rate from poly(L)lactic acid microspheres, *J. Controlled Rel.*, 13: 83–89 (1990).

33. Leelarassama, N., S. A. Howard, C. J. Malanga, L. A. Cuzzi, T. F. Hogan, S. J. Kandzari, and J. K. Ma, Kinetics of drug release from polylactic acid-hydrocortisone microcapsules, *J. Microencapsulation*, 3:171 (1986).

34. Spenlehauer, G., C. F. Spenlehauer-Bonthonneau, and C. Thies, Biodegradable microparticles for delivery of polypeptides and proteins, in *Progress in Clinical and Biological Research, Vol. 292: Biological and Synthetic Membranes*, D. A. Butterfield, ed., Alan R. Liss, New York, 1989, pp. 283–291.

35. Nellore, R. V., P. G. Pande, D. Young, and H. R. Bhagat, Evaluation of biodegradable microspheres as vaccine adjuvant for hepatitis B surface antigen, *J. Parenteral Sci. Technol.*, 46:176–179 (1992).

36. Sakakura, C., T. Takahashi, A. Hagiwara, M. Itoh, T. Sasabe, M. Lee, and S. Shobayashi, Controlled release of cisplatin from lactic acid oligomer microspheres incorporating cisplatin: *In vitro* studies, *J. Controlled Rel.*, 22:69–74 (1992).

37. Herrmann, J., and R. Bodmeier, Peptide-containing biodegradable microspheres prepared by modified solvent evaporation methods, *Proceed. Intern. Symp. Control. Rel. Bioact. Mater.*, 20:258–259 (1993).

38. Mason, N. S., D. V. S. Gupta, D. W. Keller, R. S. Youngquist, and R. E. Sparks, Microencapsulation of progesterone for contraception by intracervical injection, in *Biomedical Applications of Microencapsulation*, F. Lim, ed., CRC Press, Boca Raton, FL, 1984, pp. 75–84.

39. Tsai, D. C., S. A. Howard, T. F. Hogan, C. J. Malangu, and J. K. Ma, Preparation and *in vitro* evaluation of polylactic acid-mitomycin C microcapsules, *J. Microencapsulation*, 3: 181–193 (1986).

40. Jalil, R., and J. R. Nixon, Microencapsulation using poly(L-lactic acid). II. Preparative variables affecting microcapsule properties, *J. Microencapsulation*, 7:25–39 (1990).

41. Jalil, R., and J. R. Nixon, Microencapsulation using poly(L-lactic acid). I. Microcapsule properties affected by the preparative technique, *J. Microencapsulation*, 6:473–494 (1989).

42. Jalil, P., and J. R. Nixon, Microencapsulation using poly(L-lactic acid). IV. Release properties of microcapsule containing phenobarbitone, *J. Microencapsulation*, 7:53–66 (1990).

43. Thies, C., and M. C. Bissery, Biodegradable microspheres for parenteral administration, in *Biomedical Applications of Microencapsulation*, F. Lim, ed., CRC Press, Boca Raton, FL, 1984, pp. 53–74.

44. Jalil, R., and J. R. Nixon, Microencapsulation using poly(D,L-lactic acid). I. Effect of preparative variables on the microcapsule characteristics and release kinetics, *J. Microencapsulation*, 7:229–244 (1990).

45. Jalil, R., and J. R. Nixon, Microencapsulation using poly(D,L-lactic acid). II. Effect of

polymer molecular weight on the microcapsule properties, *J. Microencapsulation*, 7:245–254 (1990).

46. Jalil, R., and J. R. Nixon, Microencapsulation using poly(L-lactic acid). III. Effect of polymer molecular weight on the microcapsule properties, *J. Microencapsulation*, 7:41–52 (1990).

47. Beck, L. R., D. R. Cowsar, D. H. Lewis, R. J. Cosgrove, Jr., C. T. Riddle, S. L. Lowry, and T. Epperly, A new long-acting injectable microcapsule system for the administration of progesterone, *Am. Fertil. Society*, 31:545–551 (1979).

48. Wakiyama, N., K. Juni, and M. Nakano, Preparation and evaluation *in vitro* and *in vivo* of polylactic acid microspheres containing dibucaine, *Chem. Pharm. Bull.*, 30:3719–3727 (1982).

49. Spenlehauer, G., J. P. Benoit, and M. Veillard, Formation and characterization of poly(D,L-lactide) microspheres for chemoemobilization, *J. Pharm. Sci.*, 75:750–755 (1986).

50. Fong, J. W., H. V. Maulding, G. E. Visscher, J. P. Nazareno, and J. E. Pearson, Enhancing drug release from polylactide microspheres by using base in the microencapsulation process, in *Controlled-Release Technology: Pharmaceutical Applications*, ACS Symposium Series 348, P. I. Lee and W. R. Good, eds., American Chemical Society, Washington, DC, 1987, pp. 214–230.

51. Bodmeier, R., and J. W. McGinity, Poly(lactic acid) microcapsules containing quinidine base and quinidine sulfate prepared by solvent evaporation technique, *J. Microencapsulation*, 4:279–288 (1987).

52. Bodmeier, R., and J. W. McGinity, Poly(lactic acid) microspheres containing quinidine base and quinidine sulphate prepared by the solvent evaporation technique. II. Some process parameters influencing the preparation and properties of microspheres, *J. Microencapsulation*, 4:289–297 (1987).

53. Jeffery, H., S. S. Davis, and D. T. O'Hagan, The preparation and characterization of poly(lactide-co-glycolide) microparticles. I. Oil-in-water emulsion solvent evaporation, *Intern. J. Pharm.*, 77:169–175 (1991).

54. Fong, J. W., J. P. Nazareno, H. J. E. Pearson, and H. V. Maulding, Evaluation of microspheres prepared by a solvent evaporation process using sodium oleate as emulsifier, *J. Controlled Rel.*, 3:119–130 (1986).

55. Wakiyama, N., K. Juni, and M. Nakano, Preparation and evaluation *in vitro* of poly(lactic acid) microspheres containing local anesthetics, *Chem. Pharm. Bull.*, 29:3363–3368 (1981).

56. Cavalier, M., J. P. Benoid, and C. Thies, The formation and characterization of hydrocortisone loaded poly(D,L-lactide) microspheres, *J. Pharm. Pharmocol.*, 38:249–253 (1986).

57. Bodmeier, R., and J. W. McGinity, Solvent selection in the preparation of poly(D,L-lactide) microspheres prepared by the solvent evaporation method, *Intern. J. Pharm.*, 43:179–186 (1988).

58. Leelarasamee, N., S. A. Howard, C. J. Malanga, and J. K. H. Ma, A method for the preparation of polylactic acid microcapsules of controlled particle size and drug loading, *J. Microencapsulation*, 5:147–157 (1988).

59. Bodmeier, R., H. Chen, P. Tyle, and P. Jarosz, Pseudoephedrine HCl microspheres formulated into an oral suspension dosage form, *J. Controlled Rel.*, 15:65–77 (1991).

60. Bodmeier, R., and J. W. McGinity, The preparation and evaluation of drug-containing poly(D,L-lactide) microspheres formed by the solvent evaporation, *Pharm. Res.*, 4:465–471 (1987).

61. Spenlehauer, G., M. Vert, J. P. Benoit, F. Chabot, and M. Veillard, Biodegradable cisplatin microspheres prepared by the solvent evaporation method: Morphology and release characteristics, *J. Controlled Rel.*, 7:217–229 (1988).

62. Fong, J. W., Process for preparation of microspheres, U.S. Patent 4,166,800 (1979).

63. Vandegaer, J. E., and F. C. Meier, Encapsulation process, U.S. Patent 3,575,882 (1971).

64. Kondo, S., and M. Nakano, Microcapsules, U.S. Patent 4,102,806 (1978).

65. Tice, T. R., W. E. Meyers, D. H. Lewis, and D. R. Cowsar, Controlled release of ampicillin and gentamicin from biodegradable microcapsules, *Proceed Intern. Symp. Control. Rel. Bioact. Mater.*, 8:108–111 (1981).

66. Kent, J. S., L. M. Sanders, D. H. Lewis, and T. R. Tice, Microencapsulation of water soluble polypeptides, European Patent Application 52,510 (1982).

67. Wang, H. T., H. Palmer, R. J. Linhardt, D. R. Glanagan, and E. Schmitt, Degradation of poly(ester) microspheres, *Biomaterials*, 11:679–685 (1990).

68. Ruiz, J. M., B. Tissier, and J. P. Benoit, Microencapsulation of peptide: A study of the phase separation of poly(D,L-lactic acid-co-glycolic acid) copolymers 50/50 by silicone oil, *Intern. J. Pharm.*, 49:69–77 (1989).

69. Ogawa, Y., M. Yamamoto, H. Okada, T. Yashiki, and T. Shimamoto, A new technique to efficiently entrap leuprolide acetate into microcapsules of polylactic acid or copoly(lactic/glycolic) acid, *Chem. Pharm. Bull.*, 36:1095–1103 (1988).

70. Okada, H., One-month release injectable microspheres of leuprolide acetate, a superactive agonist of LH-RH, *Proceed. Intern. Control. Rel. Bioact. Mater.*, 16:12–13 (1989).

71. Ogawa, Y., M. Yamamoto, S. Takada, and T. Shimamoto, Controlled release of leuprolide acetate from polylactic acid or copoly(lactic/glycolic acid) microcapsules: Influence of molecular weight and copolymer ratio of polymer, *Chem. Pharm. Bull.*, 36:1502–1507 (1988).

72. Ogawa, Y., H. Okada, M. Yamamoto, and T. Shimamoto, *In vivo* release profiles of leuprolide acetate from microcapsules prepared with polylactic acid or copoly(lactic/glycolic) acids and *in vivo* degradation of these polymers, *Chem. Pharm. Bull.*, 36:2567–2581 (1988).

73. Ogawa, Y., H. Okada, T. Heya, and T. Shimamoto, Controlled release of LHRH agonist, leuprolide acetate, from microspheres: Serum drug level profiles and pharmacological effects in animals, *J. Pharm. Pharmacol.*, 41:439–444 (1989).

74. Jeffery, H., S. S. Davis, and D. T. O'Hagan, The preparation and characterization of poly(lactide-co-glycolide) microparticles. II. The entrapment of a model protein using a (water-in-oil)-in-water emulsion solvent evaporation technique, *Pharm. Res.*, 10:362–368 (1993).

75. Conti, B., F. Pavanetto, I. Genta, and P. Giunchedi, Solvent evaporation, solvent extraction, and spray drying for polylactide microsphere preparation, *Proceed. 10th Pharm. Technol. Conf.*, Bologna, April 16–18, 1991, pp. 16–29.

76. Linhardt, R. J., D. R. Flanagan, E. Schmitt, and H. T. Wang, Quantitative analysis of the monomer products formed on the hydrolysis of poly(esters) and poly(anhydrides), *Polym. Prepr.*, 30(1):464–465 (1989).

77. Linhardt, R. J., D. R. Flanagan, E. Schmitt, and H. T. Wang, Biodegradable poly(esters) and the delivery of bioactive agents, *Polym. Prepr.*, 31(1):249–250 (1990).

78. Wang, H. T., H. Palmer, R. J. Linhardt, D. R. Flanagan, and E. Schmitt, Controlled release of protein and vaccines from poly(ester) microspheres *in vitro*, in *Cosmetic and Pharmaceutical Applications of Polymers*, C. G. Gebelein et al., eds., Plenum Press, New York, 1991, pp. 239–253.

79. Cha, Y., and C. G. Pitt, A one-week subdermal delivery system for L-methadone based on biodegradable microspheres, *J. Controlled Rel.*, 7:69–78 (1988).

80. Wichert, B., and P. Rohdewald, A new method for the preparation of drug containing polylactic acid microparticles without using organic solvents, *J. Controlled Rel.*, 14:269–283 (1990).

81. Bleich, J., B. W. Muller, and W. Wabmus, Aerosol solvent extraction system: A new microparticle production technique, *Intern. J. Pharm.*, 97:111–117 (1993).

82. Tom, J. W., and P. G. Debendetti, Formation of bioerodible polymeric microspheres and microparticles by rapid expansion of supercritical solutions, *Biotechnol. Prog.*, 7:403–411 (1991).

83. Makino, K., M. Arakawa, and T. Kondo, Preparation and *in vitro* degradation properties of polylactide microcapsules, *Chem. Pharm. Bull.*, 33:1195–1201 (1985).

84. Bodmeier, R., and H. Chen, Preparation of biodegradble poly(±)lactide microparticles using a spray drying technique, *J. Pharm. Pharmcol.*, 40:754–757 (1988).

85. Shell, J. W., Ocular systems made of biodegradable ester having linear ether, U.S. Patent 4,115,544 (1978).

86. Michaels, A. S., Bioerodible occular drug delivery systems, U.S. Patent 3,962,414 (1976).

87. Maulding, H. V., Prolonged delivery of peptides by microcapsules, *J. Controlled Rel.*, 6: 167–176 (1987).

88. Schwope, A. D., D. L. Wise, and J. F. Howes, Lactic/glycolic acid polymers as narcotic antagonist delivery systems, *Life Sci.*, 17:1877–1886 (1975).

89. Wang, H. T., E. Schmitt, D. R. Flanagan, and R. J. Linhardt, Influence of formulation methods on the *in vitro* controlled release of protein from poly(ester) microspheres, *J. Controlled Rel.*, 17:23–32 (1991).

90. Aguado, M. F., and P. H. Lambert, Controlled-release vaccines – biodegradable polylactide/polyglycolide (PL/PG) microspheres as antigen vehicles, *Immunobiol.*, 184:113–125 (1992).

91. Heya, T., H. Okada, Y. Tanigawara, Y. Ogawa, and H. Toguchi, Effect of counteranion of TRH and loading amount on control of TRH release from copoly(D,L-lactic/glycolic acid) microspheres prepared by an in-water drying method, *Intern. J. Pharm.*, 69:69–75 (1991).

92. Atkins, T. J., M. E. Rickey, M. L. Carter, S. Bhasin, D. H. Lewis, and J. M. Hotz, An injectable, 90-day testosterone system, *Proceed. Intern. Control. Rel. Bioact. Mater.*, 19: 319–320 (1992).

93. Linhardt, R. J., Biodegradable polymers for controlled release of drugs, in *Controlled Release of Drugs: Polymers and Aggregate Systems*, M. Rosoff, ed., VCH Publishers, New York, 1989, pp. 53–95.

94. Wise, D. L., H. Rosenkrantz, J. B. Gregory, and H. J. Esber, Long-term controlled delivery of levonorgestrel in rats by means of small biodegradable cylinders, *J. Pharm. Pharmcol.*, 32:399–403 (1980).

95. Visscher, G. E., R. L. Robison, H. V. Maulding, J. W. Fong, J. E. Pearson, and G. J. Argentieri, Biodegradation and tissue reaction to 50:50 poly(D,L-lactide-co-glycolide) microcapsules, *J. Biomed. Mater. Res.*, 19:349–365 (1985).

96. Visscher, G. E., R. L. Robison, H. V. Maulding, J. W. Fong, J. E. Pearson, and G. J. Argentieri, Biodegradation and tissue reaction to poly(D,L-lactide) microcapsules, *J. Biomed. Mater. Res.*, 20:667–676 (1986).

97. Maulding, H. V., Prolonged delivery of peptides by microcapsules, in *Advances in Drug Delivery Systems*, Vol. 3, J. M. Anderson and S. W. Kim, eds., Elsevier Scientific Publishing, Amsterdam, The Netherlands, 1987, pp. 167–176.

98. Benagiano, G., and H. L. Gabelnick, Biodegradable system for the sustained release of fertility-regulating agents, *J. Steroid Biochem.*, 11:449–455 (1979).

99. Rozema, F. R., R. R. M. Bos, G. Boering, J. A. A. M. van Asten, A. J. Nijenhuis, and A. J. Pennings, The effects of different steam-sterilization program on material properties of poly(L-lactide), *J. Appl. Biomater.*, 2:23–28 (1991).

100. Gilding, D. K., and A. M. Reed, Biodegradable polymers for use in surgery – Polyglycolic/polylactic acid homo- and copolymers: 1, *Polymer*, 20:1459–1464 (1979).

101. Tice, T. R., D. H. Lewis, R. L. Dunn, W. E. Meyers, R. A. Casper, and D. R. Cowsar, Biodegradation of microcapsules and biomedical devices prepared with resorbable polyesters, *Proceed. Intern. Symp. Control. Rel. Bioact. Mater.*, 9:21–25 (1982).

102. Gupta, M. C., and V. G. Deshmukh, Radiation effects on poly(lactic acid), *Polymer*, 24: 827–830 (1983).

103. Spenlehauer, G., M. Vert, J. P. Benoit, and A. Boddaert, *In vitro* and *in vivo* degradation

of poly(D,L-lactide/glycolide) type microspheres made by solvent evaporation method, *Biomaterials*, 10:557–563 (1989).

104. Yolles, S., T. Leafe, L. Ward, and F. Boettner, Controlled release of biologically active drugs, *Bull. Parent. Drug Assoc.*, 30:306–312 (1976).

105. Landrigan, P. J., T. J. Meinhardt, J. Gordon, J. A. Lipscomb, J. R. Berg, L. F. Mazzuckelli, T. R. Lewis, and R. A. Lernen, Ethylene oxide: An overview of toxilogic and epidermiologic research, *Am. J. Industr. Med.*, 6:103–115 (1984).

106. Iwata, M., and J. W. McGinity, Preparation of multi-phase microspheres of poly(D,L-lactic acid) and poly(D,L-lactic-co-glycolic acid) containing a W/O emulsion by a multiple emulsion solvent evaporation technique, *J. Microencapsulation*, 9:201–214 (1992).

107. Anderson, L. C., D. L. Wise, and J. F. Howes, An injectable sustained release fertility control system, *Contraception*, 13:375–384 (1976).

108. Cong, H., and L. R. Beck, Preparation and pharmacokinetic evaluation of a modified long-acting injectable norethindrone microsphere, *Adv. Contraception*, 7:251–256 (1991).

109. Zhou, Z. F., M. X. Zhou, S. H. Wang, F. Lin, and W. Z. Shan, Preparation and evaluation *in vitro* and *in vivo* of copoly(lactic/glycolic) acid microspheres containing norethindrone, *Biomat. Art. Cells Immob. Biotech.*, 21:71–84 (1993).

110. Skiens, W. E., F. G. Burton, and G. W. Duncan, Biodegradable delivery systems, in *Progress in Contraceptive Delivery Systems, Vol. 1: Biodegradables and Delivery Systems for Contraception*, E. S. E. Hafez and W. A. A. van Os, eds., MTP Press, Lancaster, England, 1980, pp. 3–15.

111. Benoit, J. P., V. Rosilio, M. Deyme, C. Thies, and G. Madelmont, A physicochemical study of the morphology of progesterone-loaded 85:15 poly(D,L-lactide-co-glycolide) microspheres, *Proceed. Intern. Symp. Control. Rel. Bioact. Mater.*, 13:171–172 (1986).

112. Jackanicz, T. M., H. A. Nash, D. L. Wise, and J. B. Gregory, Poly(lactic acid) as a biodegradable carrier for contraceptive steroids, *Contraception*, 8:227–234 (1973).

113. Feng, X. D., C. X. Song, and W. Y. Chen, Synthesis and evaluation of biodegradable block copolymers of ε-caprolactone and D,L-lactide, *J. Polym. Sci., Polymer Letters Ed.*, 21:593–600 (1983).

114. Beck, L. R., V. Z. Pope, T. R. Tice, and R. M. Gilley, Long-acting injectable microsphere formulation for the parenteral administration of levonorgestrel, *Adv. Contracept.*, 1:19–129 (1985).

115. Pitt, C. G., and A. Schindler, Biodegradation of polymers, in *Controlled Drug Delivery*, Vol. 1, S. D. Bruck, ed., CRC Press, Boca Raton, FL, 1983, pp. 53–80.

116. Gu, Z. W., W. P. Ye, J. Y. Yang, Y. X. Li, X. L. Chen, G. W. Zhong, and X. D. Feng, Biodegradable block copolymer matrices for long-acting contraceptives with constant release, *J. Controlled Rel.*, 22:3–14 (1992).

117. Schakenraad, J. M., J. A. Oosterbaan, P. Nieuwenhuis, I. Molenaar, J. Olijslager, W. Potman, M. J. D. Eenink, and J. Feijen, Biodegradable hollow fibers for the controlled release of drugs, *Biomaterials*, 9:116–120 (1988).

118. Gu, Z. W., W. P. Ye, J. Y. Yang, X. L. Zhao, X. L. Chen, and X. D. Feng, New block copolymer matrices based on L-lactide for controlled-release of norgestimate, *Intern. Symp. Biomat. and Fine Polym.*, October 3–7, 1991, Guilin, China, pp. 80–81.

119. Bhasin, S., R. S. Swerdloff, B. Steiner, M. A. Peterson, T. Meridores, M. Galmirini, M. R. Pandian, R. Goldberg, and N. Berman, A biodegradable testosterone microcapsule formulation provides uniform eugonadal levels of testosterone for 10–11 weeks in hypogonadal men, *J. Clin. Endocrinol. Metab.*, 74:75–83 (1992).

120. Benghuzzi, H. A., B. G. England, and P. K. Bajpai, The effects of long-term sustained delivery of dihydrotestosterone by poly(lactic acid) impregnated and noncoated biodegradable ceramic devices in male rodents, *J. Bioact. Compatib. Polym.*, 6:115–128 (1991).

121. Mason, N., C. Theis, and T. J. Cicero, *In vivo* and *in vitro* evaluation of a microencapsulated narcotic antagonist, *J. Pharm. Sci.*, 65:847–850 (1976).

122. Yolles, S., T. D. Leafe, J. H. R. Woodland, and F. J. Meyer, Long acting delivery systems for narcotic antagonists II: Release rates of naltrexone from poly(lactic acid) composites, *J. Pharm. Sci.*, 64:348–349 (1975).

123. Chiang, C. N., L. E. Hollister, A. Kishimoto, and G. Barnett, Kinetics of a naltrexone sustained-release preparation, *Clin. Pharmacol. Ther.*, 36:704–708 (1984).

124. Nuwayser, E. S., M. H. Gay, D. J. DeRoo, and P. D. Blaskovich, Sustained release inject-able naltrexone microcapsules, *Proceed. Intern. Symp. Control. Rel. Bioact. Mater.*, 15:201–202 (1988).

125. Williams, D. L., E. S. Nuwayser, D. E. Greeden, and M. H. Gay, Microencapsulated local anesthetics, *Proceed Intern. Symp. Control. Rel. Bioact. Mater.*, 11:69–70 (1984).

126. Hecquet, B., F. Chabot, J. C. D. Gonzalez, C. Fournier, S. Hilali, L. Cambier, G. Depadt, and M. Vert, *In vivo* sustained release of cisplatin from bioresorbable implants in mice, *Anticanc. Res.*, 6:1251–1256 (1986).

127. Puisieux, F., G. Spenlehauer, J. P. Benoit, M. Veillard, and C. Fizames, Preparation, *in vitro* and *in vivo* evaluation of cisplatin microspheres, *Proceed. Intern. Symp. Control. Rel. Bioact. Mater.*, 13:167–168 (1986).

128. Verrijk, R., I. J. H. Smolders, J. G. McVie, and A. C. Begg, Polymer-coated albumin microspheres as carriers for intravascular tumour targeting of cisplatin, *Cancer Chemother. Pharmacol.*, 29:117–121 (1991).

129. Juni, K., J. Ogata, M. Nakano, T. Ichihara, K. Mori, and M. Akagi, Preparation and evaluation *in vitro* and *in vivo* of polylactic acid microspheres containing doxorubicin, *Chem. Pharm. Bull.*, 33:313–318 (1985).

130. Lin, S. Y., L. F. Cheng, W. Y. Lui, C. F. Chen, and S. H. Han, Tumoricidal effect of controlled-release polymeric needle devices containing adriamycin HCl in tumor-bearing mice, *Biomat. Art. Cells Art. Org.*, 17:189–203 (1989).

131. Ike, O., Y. Shimizu, Y. Ikada, S. Watanabe, T. Natsume, R. Wada, S. Hyon, and S. Hitomi, Biodegradation and antitumour effect of adriamycin-containing poly(lactic acid) microspheres, *Biomaterials*, 12:757–762 (1991).

132. Bissery, M. C., F. Valeriote, and C. Thies, Fate and effect of CCNU-loaded microspheres made of poly(D,L) lactide (PLA) or poly β-hydroxybutyrate (PHB) in mice, *Proceed Intern. Symp. Control. Rel. Bioact. Mater.*, 12:181–182 (1985).

133. Bissery, M. C., F. Valeriote, and C. Thies, *In vitro* lomustine release from small poly-(β-hydroxybutyrate) and poly(D,L-lactide) microspheres, *Proceed Intern. Symp. Control. Rel. Bioact. Mater.*, 11:25–26 (1984).

134. Wada, R., S. H. Hyon, Y. Ikada, Y. Nakao, H. Yoshikawa, and S. Muranishi, Lactic acid oligomer microspheres containing an anticancer agent for selective lymphatic delivery: I. *In vitro* studies, *J. Bioact. Compatib. Polym.*, 3:126–136 (1988).

135. Yoshikawa, H., Y. Nakao, K. Takada, S. Muranishi, R. Wada, Y. Tabata, S. H. Hyon, and Y. Ikada, Targeted and sustained delivery of aclarubicin to the lymphatics by lactic acid oligomer microspheres in rat, *Chem. Pharm. Bull.*, 37:802–804 (1988).

136. Wada, F., S. Hyon, and Y. Ikada, Salt formation of lactic acid oligomers as matrix for sustained release of drugs, *J. Pharm. Pharmacol.*, 43:605–608 (1991).

137. Strobel, J. D., T. J. Laughlin, F. Ostroy, M. B. Lilly, B. H. Perkins, and R. L. Dunn, Controlled-release systems for anticancer agents, *Proceed. Intern. Symp. Control. Rel. Bioact. Mater.*, 14:261–262 (1987).

138. Wu, L. T., J. H. Zhu, Z. R. Shen, and J. L. Cen, Preparation of poly(D,L-lactic acid) microspheres containing 5-fluorouracil for tumor arterial chememobilization, *Intern. Symp. Biomater. Fine Polym.*, October 3–7, 1991, Guilin, China, pp. 54–55.

139. Mumper, P. J., U. Y. Ryo, and M. Jay, Neutron-activated holmium-166-poly(L-lactic acid) microspheres: A potential agent for the internal radiation therapy of hepatic tumors, *J. Nucl. Med.*, 32:2139–2143 (1991).

140. Sandow, J., W. von Rechenberg, H. Seidel, and M. Keil, Experimental studies on tissue

tolerance and on biodegradation of polylactide/glycolide-buserelin implants in rats, *Klin. Exp. Urol.*, 20:157-166 (1989).

141. Fournier, C., B. Hecquet, P. Bouffard, M. Vert, A. Caty, M. O. Vilalin, L. Vanseymortier, S. Merle, A. Krikorian, J. L. Lefebvre, A. Delobelle, and L. Adenis, Experimental studies and preliminary clinical trial of vinorelbine-loaded polymeric bioresorbable implants for the local treatment of solid tumors, *Cancer Res.*, 51:5384-5391 (1991).

142. Benita, S., J. P. Benoit, F. Puisieux, and C. Thies, Characterization of drug-loaded poly-(D,L-lactide microspheres, *J. Pharm. Sci.*, 73:1721-1724 (1984).

143. Tice, T. R., W. E. Meyers, and J. A. Setterstrom, Controlled release of antibiotics, *Proceed. Intern. Symp. Control. Rel. Bioact. Mater.*, 10:85-90 (1983).

144. Lewis, D. H., T. O. Dappert, W. E. Meyers, G. Pritchett, and W. J. Suling, Sustained release of antibiotics from biodegradable microcapsules, *Proceed Intern. Symp. Control. Rel. Bioart. Mater.*, 7:129-131 (1980).

145. Tice, T. R., C. E. Rowe, and J. A. Setterstrom, Development of microencapsulated antibiotics for topical administration to wounds, *Proceed. Intern. Symp. Control. Rel. Bioart. Mater.*, 11:61-62 (1984).

146. Tice, T. R., C. E. Rowe, R. M. Gilley, J. A. Setterstrom, and D. D. Mirth, Development of microencapsulated antibiotics for topical administration, *Proceed. Intern. Symp. Control. Rel. Bioact. Mater.*, 13:169-170 (1986).

147. Baker, R. W., E. A. Krisko, F. Kochinke, M. Grassi, G. Armitage, and P. Robertson, A controlled release drug delivery system for the periodontal pocket, *Proceed. Intern. Symp. Control. Rel. Bioact. Mater.*, 15:238-239 (1988).

148. Ikada, Y., S. H. Hyon, K. Jamshidi, S. Higashi, T. Yamamuro, Y. Katutani, and T. Kitsugi, Release of antibiotics from composites of hydroxyapatite and poly(lactic acid), in *Advances in Drug Delivery Systems*, Vol. 3, J. M. Anderson and S. W. Kim, eds., Elsevier Scientific Publishing, Amsterdam, The Netherlands, 1987, pp. 179-186.

149. Gangadharam, P. R. J., D. R. Ashtekar, D. C. Farhi, and D. L. Wise, Sustained release of isoniazid *in vivo* from a single implant of a biodegradable polymer, *Tubercle*, 72:115-122 (1991).

150. Tice, T. R., D. H. Lewis, D. R. Cowsar, and L. R. Beck, Injectable, long-acting microparticle formulation for the delivery of antiinflammatory agents, U.S. Patent 4,542,025 (1985).

151. Ratcliffe, J. H., I. M. Hunneyball, A. Smith, C. G. Wilson, and S. S. Davis, Preparation and evaluation of biodegradable polymeric systems for the intra-articular delivery of drugs, *J. Pharm. Pharmacol.*, 36:431-436 (1984).

152. Ammoury, N., H. Fessi, J. P. Devissaguet, M. Allix, M. Plotkine, and R. G. Boulu, Effect on cerebral blood flow of orally administered indomethacin-loaded poly(isobutylcyanoacrylate) and poly(D,L-lactide) nanocapsules, *J. Pharm. Pharmacol.*, 42:558-561 (1990).

153. Ammoury, N., H. Fessi, J. P. Devissaguet, M. Dubrasquet, and S. Benita, Jejunal absorption, pharmacological activity, and pharmacokinetic evaluation of indomethacin-loaded poly(D,L-lactide) and poly(isobutyl-cyanoacrylate) nanocapsules in rats, *Pharm. Res.*, 8: 101-105 (1991).

154. Krause, H. J., A. Schwartz, and P. Rohdewald, Polylactic acid nanoparticles, a colloidal drug delivery system for lipophilic drugs, sustained release of sulphadiazine, *Intern. J. Pharm.*, 27:145-155 (1985).

155. Wise, D. L., G. L. McCormick, G. P. Willet, L. C. Anderson, and J. F. Howes, Sustained release of sulphadiazine, *J. Pharm. Pharmacol.*, 30:686-689 (1978).

156. Wise, D. L., J. D. Gresser, and G. L. McCormick, Sustained release of a dual antimalarial system, *J. Pharm. Pharmacol.*, 31:201-204 (1979).

157. Tsakala, M., J. Gillard, M. Roland, F. Chabot, and M. Vert, Pyrimethamine sustained release systems based on bioresorbable polyesters for chemoprophylaxis of rodent malaria, *J. Controlled. Rel.*, 5:233-242 (1988).

158. Maulding, H. V., T. R. Tice, D. R. Cowsar, J. W. Fong, J. E. Pearson, and J. P. Nazareno,

Biodegradable microcapsules: Acceleration of polymeric excipient hydrolytic rate by incorporation of a basic medicament, *J. Controlled Rel.*, 3:103–117 (1986).

159. Bodmeier, R., and H. Chen, Evaluation of biodegradable poly(lactide) pellet prepared by direct compression, *J. Pharm. Sci.*, 78:819–822 (1989).

160. Moll, F., and G. Koller, Biodegradable tablets having a matrix of low molecular weight poly-L-lactic acid and poly-D,L-lactic acid, *Arch. Pharm. (Weinheim)*, 323:887–888 (1990).

161. Lally, J. K., and N. N. Chugh, Biodegradable polymer – Polylactic acid. Part II: Evaluation as a matrix material for controlled release and as a microcapsulating agent, *Indian Drugs*, 27:523–529 (1990).

162. McRae, A., S. Hjorth, A. Dahlstrom, L. Dillon, D. Mason, and T. Tice, Dopamine fiber growth induction by implantation of synthetic dopamine-containing microspheres in rats with experimental hemi-Parkinsonism, *Molecular Chem. Neuropathol.*, 16:123–141 (1992).

163. Kissel, T., Z. Brich, S. Bantle, I. Lancranjan, F. Nimmerfall, and P. Vit, Parenteral depot-systems on the basis of biodegradable polyesters, *J. Controlled Rel.*, 16:27–42 (1991).

164. Juretic, D., I. Cepelak, V. Jalsenjak, T. Zanic-Grubisic, K. Lipovac, and I. Jalsenjak, Effect of microencapsulated acetylsalicylic acid on glycosylation of human serum proteins *in vitro*, *Intern. J. Pharm.*, 61:219–223 (1990).

165. Gilley, R. M., K. L. Pledger, and T. R. Tice, Development of a long-acting biocidal formulation, *Proceed. Intern. Symp. Control. Rel. Bioact. Mater.*, 14:232–233 (1987).

166. Supersaxo, A., J. H. Kou, P. Teitelbaum, and R. Maskiewicz, Preformed porous microspheres for controlled and pulsed release of macromolecules, *J. Controlled Rel.*, 23:157–164 (1993).

167. Tabata, Y., and Y. Ikada, Macrophage phagocytosis of biodegradable microspheres composed of L-lactic acid/glycolic acid homo- and copolymers, *J. Biomed. Mater. Res.*, 22:837–858 (1988).

168. Sanders, L. M., B. A. Kell, G. I. McRae, and G. W. Whitehead, Prolonged controlled-release of nafarelin, a luteinizing hormone-release hormone analogue, from biodegradable polymeric implants: Influence of composition and molecular weight of polymer, *J. Pharm. Sci.*, 75:356–360 (1986).

169. Suzuki, K., and J. C. Price, Microencapsulation and dissolution properties of neuroleptic in a biodegradable polymer, poly(D,L-lactide), *J. Pharm. Sci.*, 74:21–24 (1985).

170. Bodmeier, R., and J. W. McGinity, Poly(lactic acid) microspheres containing quinidine base and quinidine sulphate prepared by solvent evaporation method. III. Morphology of the microspheres during dissolution studies, *J. Microencapsulation*, 5:325–330 (1988).

171. Lee, V. H. L., Peptide and protein drug delivery: opportunities and challenges, *Pharm. Intern.*, 7:208–212 (1986).

172. Siegel, R. A., J. Kost, and R. S. Langer, Mechanistic studies of macromolecular drug release from macroporous polymers. I. Experiments and preliminary theory concerning completeness of drug release, *J. Controlled Rel.*, 8:223–236 (1989).

173. Metcalf, D., The granulocyte-macrophage colony-stimulating factors, *Science*, 229:16 (1985).

174. Tabata, Y., Y. Takebayashi, T. Ueda, and Y. Ikada, A formulation method using D,L-lactic acid oligomer for protein release with reduced initial burst, *J. Controlled Rel.*, 23:55–64 (1993).

175. Lin, S. Y., L. T. Ho, and H. L. Chiou, Insulin controlled release microcapsules to prolong the hypoglycemic effect in diabetic rats, *Biomat. Art. Cells Art. Org.*, 16:815–828 (1988).

176. Yamakawa, I., S. Watamabe, Y. Matsuno, and M. Kuzuzya, Controlled release of insulin from plasma-irradiated sandwich device using poly(D,L-lactic acid), *Biol. Pharm. Bull.*, 16:182–187 (1993).

177. Hora, M. S., R. K. Rana, T. A. Taforo, J. H. Nunberg, T. R. Tice, R. M. Gilley, and M. E. Hudson, Development of a controlled release microsphere formulation of interleukin-2, *Proceed. Intern. Symp. Control. Rel. Bioart. Mater.*, 16:509–510 (1989).

178. Foster, T. P., Drug and vaccine delivery of proteins and peptides using poly(ester) microspheres, *Proceed. 1st Intern. Conf. Pharm. Sci. Technol.*, Chicago, IL, August 24–28, 1993, p. 76.

179. Cohen, S., T. Yoshioka, M. Lucarelli, L. H. Hwang, and P. Langer, Controlled delivery system for proteins based on poly(lactic/glycolic acid) microspheres, *Pharm. Res.*, 8:713–720 (1991).

180. Johnson, R. E., L. A. Lanaski, V. Gupta, M. J. Griffin, H. T. Gaud, T. Needham, and H. Zia, Stability of atriopeptin III encapsulated in poly(D,L-lactide-co-glycolide) microcapsules, *Pharm. Res.*, 7:S-181 (1990).

181. Marcotte, N., A. Polk, and M. F. A. Goosen, Kinetics of protein diffusion from a poly(D,L-lactide) reservoir, *J. Pharm. Sci.*, 79:407–410 (1990).

182. Park, T. G., S. Cohen, and F. Langer, Poly(L-lactic acid) pluronic blends: Characterization of phase separation behavior, degradation, and morphology and use as protein-releasing matrices, *Macromol.*, 25:116–122 (1992).

183. Park, T. G., S. Cohen, and R. Langer, Controlled protein release from polyethyleneimine-coated poly(L-lactic acid)/pluronic blends matrices, *Pharm. Res.*, 9:37–39 (1992).

184. Takagi, I., J. Nishimura, H. Itoh, S. Baba, M. Kunimatsu, and M. Sasaki, Poly(lactic/glycolic acid) microspheres containing antigen as a novel and potential agent of immunotherapy for allergic disorders, *Jpn. J. Allergol.*, 41:1388–1397 (1992).

185. O'Hagan, D. T., D. Rahman, J. P. McGee, H. Jeffery, M. C. Davies, P. Williams, S. S. Davis, and S. J. Challacombe, Biodegradable microspheres as controlled release antigen delivery systems, *Immunol.*, 73:239–242 (1991).

186. Challacombe, S. J., D. Rahman, H. Jeffery, S. S. Davis, and D. T. O'Hagan, Enhanced secretory IgA and systemic IgG antibody responses after oral immunization with biodegradable microparticles containing antigen, *Immunol.*, 76:164–168 (1992).

187. Tice, T. R., R. M. Gilley, D. F. Love, F. Labrie, and D. W. Mason, Three-month and six-month delivery of peptides (LHRH) from injectable, poly(lactide-co-glycolide) microspheres, *Proceed. Intern. Symp. Control. Rel. Bioact. Mater.*, 18:467–468 (1991).

188. Lacoste, D., F. Labrie, D. Dube, A. Belanger, T. Tice, R. M. Gilley, and K. L. Pledger, Reversible inhibition of testicular androgen secretion by 3-, 5- and 6-month controlled-release microsphere formulations of the LH-RH agonist [D-Trp[6], des-Gly-NH$_2$[10]] LH-RH ethylamide in the dog, *J. Steroid Biochem.*, 33:1007–1011 (1989).

189. Fukuzaki, H., M. Yoshida, M. Asano, and M. Kumakura, A new biodegradable copolymer of glycolic acid and lactones with relatively low molecular weight prepared by direct copolycondensation in the absence of catalysts, *J. Biomed. Mater. Res.*, 25:315–328 (1991).

190. Asano, M., H. Fukuzaki, M. Yoshida, M. Kumakura, T. Mashimo, H. Yuasa, K. Imai, and H. Yamanaka, *In vivo* characteristics of low molecular weight copoly(D,L-lactic acid) formulations with controlled release of LH-RH agonist, *Biomaterials*, 10:569–573 (1989).

191. Csernus, V. J., B. Sxende, and A. V. Schally, Release of peptides from sustained delivery systems (microcapsules and microparticles) *in vivo*, *Intern. J. Peptide Protein Res.*, 35:557–565 (1990).

192. Sanders, L. M., B. A. Kell, G. I. McRae, and G. W. Whitehead, Poly(lactic-co-glycolic) acid: Properties and performance in controlled release delivery systems of LHRH analogues, *Proceed. Intern. Symp. Control. Rel. Bioact. Mater.*, 12:177–178 (1985).

193. Burns, R. A., Jr., K. K. Vitale, and L. M. Sanders, Nafarelin controlled release injectable: Theoretical clinical plasma profiles from multiple dosing and from mixtures of microspheres containing 2%, 4% and 7% nafarelin, *J. Microencapsulation*, 7:397–413 (1990).

194. Toguchi, H., Pharmaceutical manipulation of leuprorelin acetate to improve clinical performance, *J. Intern. Med. Res.*, 18:35–41 (1990).

195. Okada, H., Y. Inoue, T. Heya, H. Ueno, Y. Ogawa, and H. Toguchi, Pharmacokinetics of once-a-month injectable microspheres of leuprolide acetate, *Pharm. Res.*, 8:787–791 (1991).

196. Ruiz, J. M., and J. P. Benoit, *In vivo* peptide release from poly(D,L-lactic acid-co-glycolic acid) copolymer 50/50 microspheres, *J. Controlled Rel.*, 16:177–186 (1991).

197. Heinrich, N., K. Fechner, H. Berger, D. Lorenz, E. Albrecht, G. Pafler, H. Schafer, and B. Mehlis, *In-vivo* release of a GnRH agonist from a slow-release poly(lactide-glycolide) copolymer preparation: Comparison in rat, rabbit and guinea-pig, *J. Pharm. Pharmcol.*, 43: 762–765 (1991).

198. Sudo, K., K. Shiota, T. Masaki, and T. Fujita, Effects of TAP-144-SR, a sustained release formulation of a potent GnRH agonist, on experimental endometriosis in the rat, *Endocrinol. Japan*, 38:39–45 (1991).

199. Toguchi, H., Drug delivery using biodegradable microspheres, *Abstracts 6th Intern. Sym. Recent Adv. Drug Del. Sys.*, Feb. 22–25, 1993, Salt Lake City, UT, pp. 29–30.

200. Heyi, T., H. Okada, Y. Ogawa, and H. Toguchi, Factors influencing the profiles of TRH release from copoly(D,L-lactic/glycolic acid) microspheres, *Intern. J. Pharm.*, 72:199–205 (1991).

201. Hutchinson, F. G., B. J. A. Furr, and J. R. Churchill, Biodegradable polymers for the delivery of polypeptides and proteins, *Ziekenhuisfarmacie*, 4:54–56 (1988).

202. Bawa, R., R. A. Siegel, B. Marasca, M. Karel, and R. S. Langer, An explanation for the controlled release of macromolecules from polymers, *J. Controlled Rel.*, 1:259–267 (1985).

203. Kent, J. S., L. M. Sanders, T. R. Tice, and D. H. Lewis, Microencapsulation of the peptide nafarelin acetate for controlled release, in *Long-Acting Contraceptive Delivery Systems*, G. I. Zatuchni, A. Goldsmith, J. D. Shelton, and J. J. Sciarra, eds., Harper and Row, Philadelphia, 1984, pp. 169–180.

204. Siegel, R. A., and R. S. Langer, Controlled release of polypeptides and other macromolecules, *Pharm. Res.*, 1:1–10 (1983).

205. O'Hagen, D. T., K. J. Palin, and S. S. Davies, Poly(butyl-2–cyanoacrylate) particles as adjuvants for oral immunization, *Vaccine*, 7:213–216 (1990).

206. Hazrati, A. M., D. H. Lewis, T. J. Atkins, R. C. Stohrer, and L. Meyer, *In vivo* studies of controlled-release tetanus vaccine, *Proceed. Intern. Symp. Control. Rel. Bioact. Mater.*, 19: 114–115 (1992).

207. Gander, B., C. Thomasin, H. P. Merkle, Y. Men, and G. Corradin, Pulsed tetanus toxoid release from PLGA-microspheres and its relevance for immmunogenicity in mice, *Proceed. Intern. Symp. Control. Rel. Bioact. Mater.*, 20:65–66 (1993).

208. Eldridge, J. H., J. K. Staas, J. A. Meulbroek, T. R. Tice, and R. M. Gilley, Biodegradable and biocompatible poly(D,L-lactide-co-glycolide) microspheres as an adjuvant for staphylococcal enterotoxin B toxoid which enhances the level of toxin-neutralizing antibodies, *Infection and Immunity*, 59:2978–2986 (1991).

209. Eldridge, J. H., C. J. Hammond, J. A. Meulbroek, J. K. Staas, R. M. Gilley, and T. R.Tice, Controlled vaccine release in the gut-associated lymphoid tissues. I. Orally administered biodegradable microspheres target the Peyer's patches, *J. Controlled Rel.*, 11:205–214 (1990).

210. Eldridge, J. H., J. K. Staas, J. A. Meulbroek, J. R. McGhee, T. R. Tice, and R. M. Gelley, Biodegradable microspheres as a vaccine delivery system, *Molecular Immunol.*, 28:287–294 (1991).

211. Pepeljnjak, S., D. Penovski, and V. Jalsenjak, *Bacillus subtilis* microcapsules, *Pharmazie*, 43:728–729 (1988).

212. Almeida, A. J., H. O. Alpar, D. Williamson, and M. R. W. Brown, Poly(lactic acid) microspheres as immunological adjuvants for orally delivered cholera toxin B subunit, *Biochem. Soc. Transactions*, 20:316S (1992).

213. Bathurst, I. C., P. J. Barr, D. C. Kaslow, D. H. Lewis, T. J. Atkins, and M. E. Rickey, Development of a single injection transmission-blocking malaria vaccine using biodegradable microspheres, *Proceed. Intern. Symp. Control. Rel. Bioact. Mater.*, 19:120–121 (1992).

214. Stevens, V. C., J. E. Powell, A. E. Lee, P. T. P. Kaumaya, D. H. Lewis, M. E. Ricky, and T. J. Atkins, Development of a delivery system for a birth control vaccine using biodegradable microspheres, *Proceed. Intern. Symp. Control. Rel. Bioact. Mater.*, 19:112–113 (1992).

215. Marx, P. A., R. W. Compans, A. Gettie, J. K. Staas, R. M. Gilley, M. J. Mulligan, G. V. Yamshchikov, D. Chen, and J. H. Eldridge, Protection against vaginal SIV transmission with microencapsulated vaccine, *Science*, 260:1323–1326 (1993).

216. Hazrati, A. M., D. H. Lewis, T. J. Atkins, R. C. Stohrer, C. A. McPhillips, and J. E. Little, *Salmonella enteritidis* vaccine utilizing biodegradable microspheres, *Proceed. Intern. Symp. Control. Rel. Bioact. Mater.*, 20:101–102 (1993).

217. Pitt, C. G., M. M. Gratzl, A. R. Jeffcoat, R. Zweidinger, and A. Schindler, Sustained drug delivery systems. II. Factors affecting release rates from poly(ϵ-caprolactone) and related biodegradable polyesters, *J. Pharm. Sci.*, 68:1534–1538 (1979).

218. Asano, M., H. Fukuzaki, M. Yoshida, M. Kumakura, T. Mashimo, H. Yuasa, K. Imai, and H. Yamanaka, Application of poly D,L-lactic acids of varying molecular weight in drug delivery systems, *Drug Design and Delivery*, 5:301–320 (1990).

219. Wakiyama, N., K. Juni, and M. Nakano, Influence of physicochemical properties of poly-(lactic acid) on the characteristics and *in vitro* release patterns of poly(lactic acid) microspheres containing local anesthetics, *Chem. Pharm. Bull.*, 30:2621–2628 (1982).

220. Kaetsu, I., M. Yoshida, M. Asano, H. Yamanaka, K. Imai, H. Yuasa, T. Mashimo, K. Suzuki, R. Katakai, and M. Oya, Biodegradable implant composites for local therapy, *J. Controlled Rel.*, 6:249–263 (1987).

221. Cowsar, D. R., T. R. Tice, R. M. Gilley, and J. P. English, Poly(lactide-co-glycolide) microcapsules for controlled release of steroids, *Method. Enzymol.*, 112:101–116 (1985).

222. Sato, T., M. Kanke, H. G. Schroeder, and P. P. Deluca, Porous biodegradable microspheres for controlled drug delivery. I. Assessment of processing conditions and solvent removal techniques, *Pharm. Res.*, 5:21–30 (1988).

223. Juni, K., J. Ogata, N. Matsui, M. Kubota, and M. Nakano, Control of release rate of bleomycin from poly(lactic acid) microspheres by additives, *Chem. Pharm. Bull.*, 33:1609–1614 (1985).

34
Biomaterials in Drug Delivery Systems

Joseph Kost
Ben-Gurion University
Beer-Sheva, Israel

I. INTRODUCTION

While newer and more powerful drugs continue to be developed, increasing attention is being given to the methods by which these active substances are administered [1]. Controlled drug delivery has evolved from the need for prolonged and better control of drug administration. In conventional drug delivery modes (such as a spray, an injection, or the taking of a pill), each time a person takes medicine, the drug concentration in the blood rises when the drug is taken, then peaks and declines. Since a drug may have a plasma level above which it is toxic and below which it is ineffective, the plasma drug concentration in a patient at a particular time depends on compliance with the prescribed routine. This is particularly problematic if the toxic and minimum effective levels are close together.

The goals of the controlled-release devices, which are already available commercially [1], are to maintain the drug in the desired therapeutic range with just a single dose, localize delivery of the drug to a particular body compartment (which lowers the systemic drug level), reduce the need for follow-up care, preserve medications that are rapidly destroyed by the body, and increase patient comfort and/or improve compliance. In general, release rates are determined by the design of the system and are nearly independent of environmental conditions.

A convenient classification of controlled-release systems is based on the mechanism that controls the release of the substance in question.

A. Diffusion-Controlled Systems

The most common mechanism is diffusion. Two types of diffusion-controlled systems have been developed; the first is a reservoir device in which the bioactive agent (drug) forms a core surrounded by an inert diffusion barrier. These systems include membranes, capsules, microcapsules, liposomes, and hollow fibers. The second type is a monolithic device in which the active agent is dispersed or dissolved in an inert matrix. As in reservoir systems, drug diffusion through the matrix is the rate-limiting step, and release rates are determined by the choice of the matrix and its consequent effect on the diffusion and partition coefficient of the drug to be released [1,2].

B. Chemically Controlled Systems

Chemical control can be achieved using bioerodible or pendant chain systems. The rationale for using bioerodible (or biodegradable) systems is that the bioerodible devices are eventually absorbed by the body and thus need not be removed surgically. *Bioerosion* can be defined as the conversion of a material that is insoluble in water into one that is water soluble. In a bioerodible system, the drug is ideally distributed uniformly throughout a matrix in the same way as in monolithic systems. As the matrix surrounding the drug is eroded, the drug escapes. In a pendant chain system, the drug is covalently bound to insoluble molecules and is released by bond scission due to water or enzymes [3,4].

C. Solvent-Activated Controlled Systems

In solvent-activated controlled systems, the active agent is dissolved or dispersed within a matrix and is not able to diffuse through that matrix. In one type of solvent-controlled system, as the environmental fluid (e.g., water) penetrates the polymeric matrix, the polymer swells and its glass transition temperature is lowered below the environmental (host) temperature [5]. Therefore, the swollen polymer is in a rubbery state and allows the drug contained in it to diffuse through the polymer.

A different solvent-controlling mechanism was developed by ALZA Corporation [6], in which a core of solid drug is coated with a semipermeable polymer membrane that admits water at a controlled rate due to osmosis. The water dissolves the drug and gradually forces the drug solution out of a single aperture in the membrane.

II. BIOMATERIALS

There are a number of important requirements specific to biomaterials used in controlled-release systems. Among these are biocompatibility, processability, and mechanical properties. The biomaterials employed for drug delivery systems can be broadly classified as liposomes, polymers, and ceramics. Ceramics are discussed elsewhere in this encyclopedia and are not covered here.

A. Liposomes

Liposomes are closed lamellar structures that are formed when certain amphipathic phospholipids are exposed to an aqueous environment. The phospholipids orient themselves in bilayers with their head groups directed toward the water molecules and the acyl chains isolated from the water phase. Liposomes entrap hydrophobic materials in the lipid bilayer, and encapsulate hydrophilic materials in the aqueous inner phase.

Liposomes can consist of one (unilamellar) or a number of parallel concentric (multilamellar) bilayers and are conventionally classified into three groups by their morphology: multilamellar vesicles (MLVs), small unilamellar vesicles (SUVs), and large unilamellar vesicles (LUVs). The diameter of liposomes can range from 25 nanometers (nm) to 20 micrometers (μm). Size and homogeneity of liposomes are important in their use as drug carriers since these characteristics affect the pharmacokinetics and distribution of liposomes *in vivo*. Liposomes can be formulated with a variety of lipid compositions and structures and are potentially nontoxic, degradable, and nonimmunogenic.

The most commonly used phospholipid for the preparation of liposomes is phosphatidylcholine, a Zwitterionic phospholipid, the structure of which is shown below.

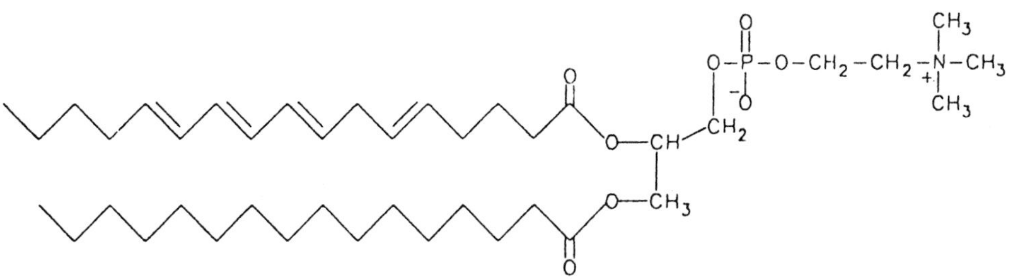

Phosphatidylcholine

The net surface charge of liposomes can be modified by the incorporation of charged molecules such as stearlyamine for positively charged liposomes and dicetylphosphate, phosphatidylglycerol, or phosphatidylserine for negatively charged liposomes.

Multilamellar vesicles (MLVs) are prepared by hydration of lipids in an aqueous solution with mechanical agitation. If the lipids are hydrated without agitation, giant oligo- or unilamellar vesicles are produced. Once liposomes are produced by hydration and then mechanically broken, they reform closed vesicles. Small unilamellar vesicles (20–100 nm) are produced by mechanical breakdown of MLVs using sonication or French press. Homogeneous-size liposomes are produced by extrusion through a polycarbonate membrane with pores of defined sizes. Liposomes can be produced by hydration of lipids from the organic phase, and by detergent-removal methods [7]. A lipid bilayer is formed in an aqueous solution of lipids dissolved in organic solvent when the solvent is removed or diluted. This technique is called the *reverse-phase evaporation method.*

The fate of injected liposomes depends on the administration route, dose, size, lipid composition, surface modification, and the encapsulated drug. Liposomes tend to remain at the injection site when they are administered intramuscularly or subcutaneously. If multilamellar vesicles are administered by intravenous (IV) injection, they tend to be taken up by reticuloendothelial systems (RESs) such as the liver and spleen. Small unilamellar vesicles tend to be retained in the blood for longer periods when administered IV [7].

The use of liposomes for site-specific delivery (targeting) of cytotoxic drugs in cancer chemotherapy has attracted an immense amount of attention. In clinical studies, liposomal doxorubicin reduces side effects such as alopecia and nausea associated with the administration of the free drug, yet permits a higher maximal tolerated dose and a reduction in cardiac toxicity of 86% [8]. Liposomal amphotericin B is more effective than the free drug in treating immunocompromised cancer patients with fungal infections [9].

Liposomes can be targeted either passively or actively. Passive targeting involves the natural uptake by cells that scavenge foreign microparticulates such as reticuloendothelial cells, which are concentrated in tissues such as the liver or spleen, or circulating monocytes. Active targeting generally involves placing a charge or recognition sequence onto the liposome such that it is more rapidly taken up by certain cell types than others.

The effectiveness of liposomes in potentiating the immune response has been widely documented [10,11]. Most of these studies monitored the antibody response to repeatedly administered haptens or proteins encapsulated in liposomes.

B. Polymers

The polymers used in controlled-release systems can be classified as hydrophobic, hydrophilic, biodegradable, and nondegradable. The decision to apply a polymer from one of these groups depends upon the route of administration, the type and amount of drug required, and the duration of release required. Oral, nasal, ocular, buccal, pulmonary, transdermal (through the skin), and rectal are common routes of drug administration. Controlled-release systems also can be administered parenterally to the body by intravenous, intramuscular, intraperitoneal, or subcutaneous injections.

The route of administration is a major consideration in the type of drug selected for the controlled-delivery system. For instance, if the drug is unstable in the presence of gastric or intestinal fluids, as are many proteins, or if it is quickly metabolized during passage through the liver, then it will have to be delivered parenterally or transdermally.

The route of administration is also a major consideration in the type of polymer selected for the controlled-delivery system. For instance, nonbiodegradable polymers can be applied for ocular drug administration as the device is removed when the drug is depleted, while biodegradable polymers are preferred for intramuscular and intraperitoneal implants for which retrieval of the nonbiodegradable polymer is less feasible.

The nature of the drug being delivered also influences the selection of the polymer used. The polymer has to be compatible with the drug and not lead to a reaction that will alter either the drug or the polymer. Also, if the drug is unstable under an aqueous environment, a more hydrophobic polymer will be preferred.

1. Nondegradable Hydrophobic Polymers

One of the most common nondegradable hydrophobic polymers is silicone rubber or poly(dimethylsiloxane), the structure of which is shown below

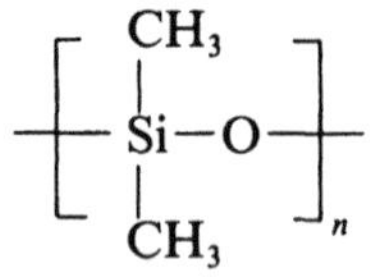

$$\left[\begin{array}{c} CH_3 \\ | \\ Si-O \\ | \\ CH_3 \end{array}\right]_n$$

Poly(dimethylsiloxane)

It can be prepared by polymerization of linear silanols and can be cross-linked by benzoyl peroxide or other agents. Silicones were one of the first polymers used for drug delivery [12]. Silicones are excellent materials for drug delivery because of their biocompatibility, their ease of fabrication, and high permeability to many drugs. As a result, a number of drug delivery products employing silicones have been commercialized, including implants for contraceptive steroids and transdermal patches. NORPLANT, for example, is an effective, long-lasting contraceptive device inserted under the skin of a woman's upper

Table 1   Nondegradable Hydrophobic Polymers
Applied in Drug-Delivery Systems

Silicones
Poly(ethylene vinyl acetate)
Polyamides
Polyacrylates
Polyethylene
Polyurethanes
Polyisobutylene
Cellulosics
Polyphosphazenes

arm that provides protection for five years. The NORPLANT device consists of tiny cylindrical reservoirs made of silicone rubber containing levonogestrel.

Another widely used nondegradable polymer is ethylene-vinyl acetate copolymer.

$$(C-C)_n-(C-C)_m$$
$$\underset{O \; Ac}{|}$$

Ethylene–vinyl acetate copolymer

This copolymer also displays excellent biocompatibility, physical stability, biological inertness, and processability. In a drug delivery application, these copolymers usually contain 30–50 wt% vinyl acetate. Ocusert is a controlled-delivery system of pilocarpine for glaucoma treatment placed in the conjuctival cul-de-sac to release pilocarpine. The pilocarpine reservoir is dispersed in alginic acid, which acts as a carrier. An ethylene-vinyl acetate copolymer membrane acts as the rate-limiting barrier for the diffusion of the drug. By controlling the thickness of the membranes, release rates of 20 or 40 micrograms/hour (μg/h) of pilocarpine can be achieved for a 1-week period [13].

Table 1 gives a list of the most commonly used nondegradable hydrophobic polymers.

2. Biodegradable Polymers

Biodegradable polymers in controlled-delivery systems have an advantage over other systems in obviating the need to surgically remove the drug-depleted device. In many cases, however, the release is augmented by diffusion through the matrix, rendering the process difficult to control, particularly if the matrix is hydrophilic and thereby absorbs water, promoting degradation in the interior of the matrix. To maximize control over the release process, it is desirable to have a polymeric system that degrades only from the surface and defers the permeation of the drug molecules. Achieving such a heterogeneous degradation requires the rate of hydrolytic degradation on the surface to be much faster than the rate of water penetration into the bulk. Highlighted below are characteristics of some of the more widely used families of synthetic biodegradable polymers used in controlled-release systems [1,14–18].

Polyesters.   The poly α-esters (ester polymers prepared from α-hydroxy acids), especially poly(lactic acid) (PLA), poly(glycolic acid) (PGA), and their copolymers poly(lactic glycolic acid) (PLGA) are the most widely used implantable degradable polymers. They are usually produced by ring-opening polymerization of glycolide and lactide and have the following structure:

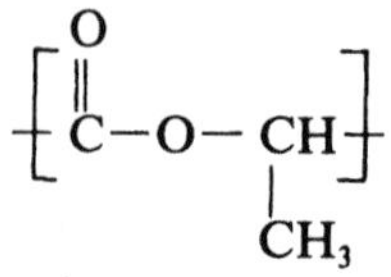

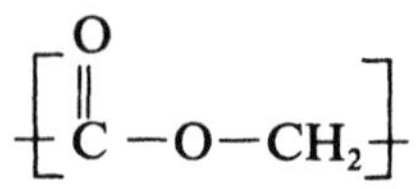

Poly(lactic acid) Poly(glycolic acid)

PGA is a highly crystalline hydrophilic polymer. PLA is more hydrophobic than PGA and therefore degrades at a slower rate. In both polymers, the degradation mechanism is largely due to hydrolysis, and the eventual degradation products are water and carbon dioxide. By varying the ratio of the monomers in the PLGA copolymer, different degradation rates can be achieved. Thus, the $t_{1/2}$ of implanted pellets (about 5 mg in rats) quickly drops from 5 months for PGA and 6 months for PLA to about 1 week for the copolymer PLGA 50:50 [19].

The first PLGA product approved by the Food and Drug Administration (FDA) to release an leutinizing hormone-releasing hormone (LHRH) synthetic analog (leuprolide acetate) for the treatment of prostatic cancer is the Lupron-Depot from Takeda-Abbot [20]. The conventional treatment required once-daily injection of the drug over a long period. The controlled-release PLGA 75:25 microspheres require injection only once a month. Similar products are currently in use or are in clinical trials at Imperial Chemical Industries in England (Zoladex) [21], at Syntex in California (Nafarelin) [22], and at Ibsen-Biotech in France (D-Trp-6-LH) [23].

Another polyester, poly(ε-caprolactone), hydrolyzes to 6-hydroxyhexanoic acid [24]. The homopolymer is less biodegradable than PLA and hence is more suitable for long-term delivery of drugs. However, its *in vivo* biodegradation can be accelerated by copolymerization with other esters. The polymer is permeable to lipophilic drugs and is therefore useful for the release of steroid drugs. This high permeability of poly(ε-caprolactone) has permitted the development of biodegradable reservoir devices that degrade completely after the drug is depleted. Devices made of this polymer were capable of subdermal delivery of the contraceptive levonogestrel at 50 μg/day for at least 1 year [25]. Poly(ε-caprolactone) has the following structure:

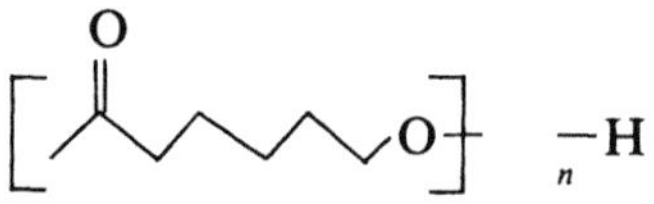

Poly(ε-caprolactone)

Poly(ortho esters) were initially developed by ALZA Corporation under the trade names Chronomer and Alzamer [26]. Chronomer is made of 2,2-dialkyloxytetrahydrofurane, 1,6-hexanediol, and 1,4-cyclohexanedimethanol to yield the following structure:

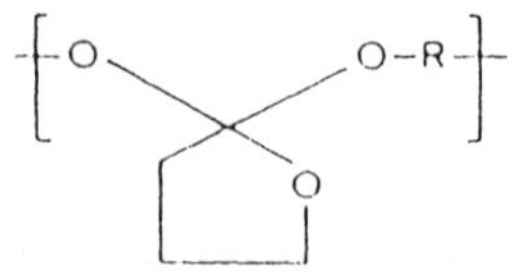

Chronomers

The poly(ortho esters) were developed to obtain surface-degrading systems. Such systems would result in zero-order (constant) drug-release rates.

Surface-degrading systems display material loss from the outside to the inside of the polymeric device. In these systems, the degradation rate is dependent on the surface area rather than the volume of the polymer matrix. Therefore, surface-degrading systems offer the potential advantage of achieving constant degradation rates if the polymer system possesses a shape with a surface area that does not change during degradation.

The poly(ortho esters) contain an ortho ester linkage that is more susceptible to hydrolysis under low pH conditions than basic or neutral conditions. Control of the degradation rate of these polymers has been achieved by means of acidic or basic excipients physically incorporated into the matrix [3,27]. These excipients maintain the matrix bulk at a different pH than the matrix surface, enabling surface erosion to be obtained if the excipient does not diffuse out prior to polymer erosion. If long-term delivery of a drug is required, then a basic salt such as magnesium hydroxide may be incorporated into the matrix. Similarly, if fast degradation of the matrix is needed, then acidic excipients such as 9,10-dihydroxy stearic acid or phthalic anhydryde that can hydrolyze to give an acidic excipient can be used.

An entirely different class of poly(ortho esters) can be prepared by the reaction of diketene acetals and polyols:

$$\left[O-\underset{\underset{CH_3}{|}}{\overset{\overset{O-R}{|}}{C}}-O-R'-O-\underset{\underset{CH_3}{|}}{\overset{\overset{O-R}{|}}{C}}-O-R''\right]$$

Because the ketene acetal-terminated prepolymer is a viscous liquid at room temperature, therapeutic agents and triol can be mixed into the prepolymer at room temperature and the mixture cross-linked at temperatures as low as 40°C. This allows incorporation of heat-sensitive therapeutic agents into a solid polymer under very mild conditions of thermal stress. However, because the prepolymer contains reactive ketene acetal groups, any hydroxyl groups present in the therapeutic agent will result in the covalent attachment of the therapeutic agent to the matrix via ortho ester bonds [3].

Another family of poly(ortho esters) was prepared by a transesterification reaction between a triol and trialkyl ortho ester [28]. When the transesterification reaction is carried out with flexible triols such as 1,2,6-hexanetriol, ointmentlike materials are obtained even though the molecular weight can be in excess of 50,000. Such materials are useful for the entrapment of therapeutic agents as the incorporation of materials into the polymer can be performed at room temperature without the use of solvents.

The poly(ortho esters) were evaluated in the development of bioerodibile implants for the release of numerous biological agents, such as contraceptives and cytotoxic and narcotic antagonists [28].

Polyanhydrides.   Polyanhydrides were first prepared as an alternative to polyester fibers for the garment industry [29]. However, due to their hydrolytic instability, polyanhydrides were unsuitable as textile fibers and useless for 50 years. In the 1980s, Langer and coworkers suggested that hydrophobic polyanhydrides might be a promising class of polymers due to the lability of the anhydride bond [30]. These polymers are generally synthesized by polycondensation.

Among the most widely used polyanhydrides are copolymers of sebacic acid (SA) and 1,3-bis(carboxyphenoxy)propane (CPP), the structure of which is shown below:

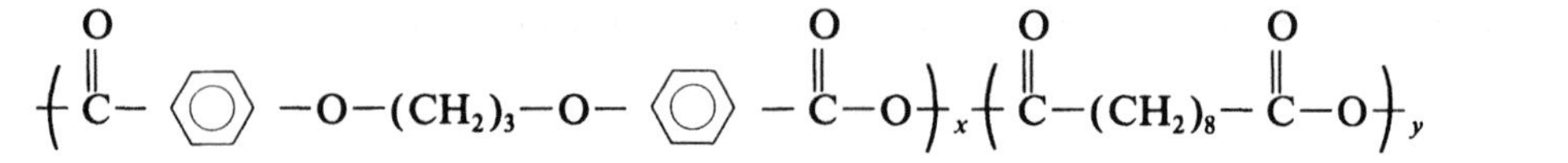

$$\left(\begin{matrix}O\\\parallel\\C-\end{matrix}\bigcirc -O-(CH_2)_3-O-\bigcirc-\begin{matrix}O\\\parallel\\C-O\end{matrix}\right)_x\left(\begin{matrix}O\\\parallel\\C-(CH_2)_8-\end{matrix}\begin{matrix}O\\\parallel\\C-O\end{matrix}\right)_y$$

By varying the ratio of the hydrophobic moiety (CPP) and the more hydrophilic SA, controlled degradation rates from days to years have been achieved for millimeter-thick disks [31,32]. To provide more linear release, polyanhydrides were prepared from aliphatic-aromatic dicarboxylic acid monomers-(carboxyphenoxy)alkanoic acids. These monomers possess an aromatic acid on one portion of the monomer and an aliphatic part on the other portion connected by a nonlabile bond. With this monomer construction, it is not possible for the polymer to be enriched in aromatic or aliphatic groups as erosion progresses; thus, erosion is completely linear [33]. Polyanhydrides with unsaturated linkages, which offer the possibility of cross-linking [34], and polyanhydrides coamides, which display excellent mechanical and thermal qualities, have also been studied [35].

Polyphosphazenes.   Polyphosphazenes consist of a chain of alternating phosphorous and nitrogen atoms, with two side groups attached to each phosphorous atom. The general molecular structure is shown below:

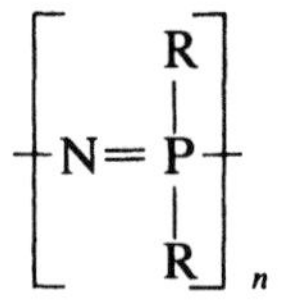

$$\left[\begin{matrix}&R\\&\vert\\N=&P\\&\vert\\&R\end{matrix}\right]_n$$

Polyphosphazenes

The physical and chemical properties of these polymers can be controlled by changing the side groups attached. These include hydrophobic groups that protect the backbone against hydrolysis, through groups that generate water solubility together with hydrolytic stability, to side groups that provide hydrolytic degradation. One of the major potentials of this class of polymers is to prepare a polymeric drug or polymeric prodrugs by attaching drugs as side chains to the polymer backbone [36].

Pseudopoly(amino acids).   The rationale for synthesizing polymers based on amino acids was that, when degraded, these would result in naturally occurring metabolites. L amino acids or dipeptides were polymerized through nonamino (e.g., ester) iminocarbonate bonds, enabling the creation of poly(amino acids) with tailor-made physical and pharmacological properties [37]. The chemical structure of poly(N-acylhydroxyproline ester) is shown below.

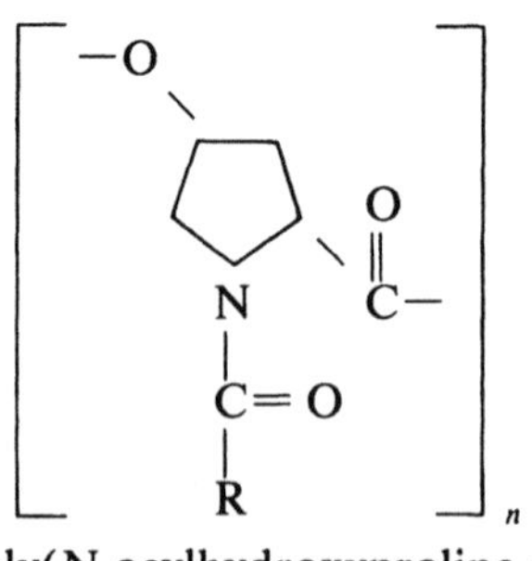

$$\left[\begin{matrix}-O\\\diagdown\\\underset{N}{\pentagon}\diagdown\begin{matrix}O\\\parallel\\C-\end{matrix}\\\vert\\C=O\\\vert\\R\end{matrix}\right]_n$$

Poly(N-acylhydroxyproline ester)

This approach permits the synthesis of biomaterials that are derived from nontoxic substances, which also have other desirable properties. For example, the incorporation of an anhydride bond into the polymer backbone results in rapid degradation, while an imide or iminocarbonate bond improves mechanical strength [35].

Hydrolysis rates of polypeptides are generally slower than those of polyesters or polyanhydrides. The degradation rate can be increased by copolymerization with hydroxy acids to form polydepsipeptide [38]. A number of polypeptides and polydepsipeptides have been studied for drug delivery, primarily for contraceptive steroids and anticancer drugs. Polyglutamic acid is one of the most studied of these [39–41].

3. Hydrogels

Hydrogels can be defined as materials that exhibit the ability to swell in water and retain a significant fraction of water within their structure. Their ability to absorb water is due to the presence of hydrophilic groups such as $-OH$, $-CONH-$, $-CONH_2$, $-COOH-$, and SO_3H [42].

Hydrogels have a special advantage over other synthetic polymers, particularly when the hydrogel and biological surface are in close contact. In many physical properties, they resemble biological tissues, having a high water content, being porous and flexible, thus reducing possible injury to the tissue. The nature of the hydrogel-water interface is such that the interfacial tension is low and protein adsorption and unfold onto hydrogel surface is minimized. (Minimal protein interaction is important for the biological acceptance of foreign materials as the denaturation of proteins by surfaces may serve as a trigger mechanism for the initiation of biological rejection mechanism). These properties were the incentive for the frequent use of hydrogels as reservoir and matrix type diffusion-controlled delivery systems [43–50]. The parameters that affect the permeability of drug and hence its release rate are polymer composition, type of releasing agent, water content in hydrogel, and crosslinking density. Each of these is discussed further below.

Polymer Composition. Poly(2-hydroxyethyl methacrylate) (PHEMA) has been the most widely studied and used hydrogel for medical applications.

$$\left[\begin{array}{cc} H & CH_3 \\ | & | \\ C - & C - \\ | & | \\ H & C=O \end{array} \right]_n$$

$$OCH_2CH_2OH$$

Poly(2-hydroxyethyl methacrylate)

It is hydrolytically stable, and its permeability and hydrophilicity depend on the monomers and cross-linking agents used [51]. Other monomers and polymers in this family that have been studied for controlled-release applications include hydroxyethoxyethyl methacrylate (HEEMA), hydroxydiethoxyethyl methacrylate (HDEEMA) [52], (HEMA-co-methyl methacrylate) copolymers [53], (HEMA-co-butyl methacrylate) copolymer [54], (HEMA-polymethacrylic acid) copolymer [55], (HEMA-N-N-dimethylaminoethyl methacrylate) (NNDMAEM) copolymer [56], and (HEMA-methyl acrylate) copolymer [57].

Other hydrogels with potential for controlled-release application include cross-linked polyvinyl alcohol (PVA) [58–60] and its copolymers with N-vinyl-2-pyrrolidone (NVP) [61], NVP-methyl methacrylate-ethyl acrylate terpolymer [62], NVP-divinyl benzene/ethyl benzene copolymer [63], unsaturated polyesters cross-linked by copolymerization with NVP [64], polyacrylamide [65,66], poly(ethylene oxide) [67], poly(ethylene oxide)/

propylene oxide block copolymer and alginic acid [68], polyamino acids [69], polyure-thanes [70], and natural polymers such as cross-linked proteins, starches, or cellulosic derivatives [70–72].

Releasing Agent.   Despite the high water content (10–100%) of the hydrogels, the sys-tems may be used for the release of both hydrophilic and hydrophobic drugs [45]. Solute permeation through hydrogels is by diffusion in the water-filled pores or partition and diffusion of the dissolved drug through the polymer. If permeation occurs via a pore mechanism, the molecular size will have an important effect on the diffusivity. An in-crease in molecular size yields lower diffusion coefficients, but it may also lead to higher solubility of the solutes, which in turn may lead to concentration dependence of the diffusivity [45]. In contrast, for the partition mechanism, the permeation is generally less dependent on molecular size [45]. Within a narrow range of temperatures (usually from 0°C to 50°C), diffusivity depends on temperature, following an Arhenius expression. At higher temperatures, significant deviations from this dependence have been report-ed [73].

Water Content.   Dehydrated hydrogels placed in aqueous solution will swell to a certain equilibrium value. The driving force for water to enter the polymer is an osmotic pressure. Strong interactions between chemical structures on the polymer, and the waterlike hydro-gen bonding, will further increase the driving force for swelling. The swelling process is counteracted by the elastic contractility of the stretched polymer network. The equilib-rium degree of swelling of a hydrogel is determined by the hydrophilicity of the polymer, the amount and nature of the cross-linking agent, the degree of dilution prior to polymeri-zation, and, to some extent, by the stereoregularity of the polymer chain [44]. Yasuda et al. found a linear dependence of the diffusive permeability coefficient, log P, on the swelling factor, as predicted by the free volume theory [74,75].

Cross-Linking Density.   Increasing the cross-link density of hydrogels should cause a lower degree of equilibrium swelling and result in reduced water content and diffusion coefficient [76–79]. In contrast, the degree of swelling of cross-linked PHEMA gels in water shows little variation as a function of the degree of cross-linking, as expressed by the concentration of cross-linking agent used. It has been proposed that there exists a secondary noncovalent network structure, consisting of hydrogen bonds superimposed upon the covalent network [80]. This secondary structure is responsible for the very high degree of cross-linking evident in the polymer and controls the swelling behavior of the hydrogel [44].

4. Dynamically Swelling Hydrogels

Recently, controlled-release systems based on swellable polymers have become popular in a variety of pharmaceutical applications [44,81]. These delivery systems, in which the drug is released during the swelling process, are based on hydrophilic polymers such as polyvinyl alcohol, PHEMA, poly(acrylic acid) and various cellulose derivatives. These polymers are generally glassy in the dehydrated state, but swell to become elastic gels upon water penetration. The entrapped drug within the swelling matrix concomitantly dissolves and diffuses through the swollen network. A slow macromolecular relaxation process at the swelling front often provides another mechanism in addition to diffusion to alter the release kinetics.

Various controlled-release systems have been developed using cross-linked PHEMA and its copolymers [44,81]. For example, such systems have been used to release tripelennamine-HCl, oxprenolol-HCl, or thiamine-HCl at a nearly constant rate for up to 24 hours. Microparticles, microspheres, and disks have been prepared.

5. Biodegradable Hydrogels

A majority of biodegradable hydrogels are based on natural polymers. However, a number of systems based on synthetic polymers have also been employed.

Based on Synthetic Polymers.   Torchilin et al. proposed hydrogels based on N-vinyl pyrrolidone or acrylamide cross-linked with N,N-methylene bisacrylamide [82]. The hydrolysis of the cross-linker is extremely slow and only hydrogels containing less than 1% cross-linker can be regarded as erodible. This limits their application because, at such low cross-link density, rapid diffusional release of the incorporated drug takes place.

A similar approach was proposed by Edman et al., who studied dextran derivatized with glycidyl methacrylate and then cross-linked with N,N-methylene bisacrylamide [83].

A more desirable hydrogel delivery system with a high cross-link density so that diffusional release is inhibited and rate of release is mainly controlled by hydrolysis of the hydrogel was proposed by Heller et al. [64], who synthesized hydrogels by cross-linking a water-soluble polyester from fumaric acid and poly(ethylene glycol) with N-vinyl pyrrolidone. Chain cleavage takes place by hydrolysis of ester links with consequent generation of poly(ethylene glycol) and poly(N-vinyl pyrrolidone) modified by vicinal carboxylic acid functions.

Rihova et al. examined water-soluble hydrogels based on copolymers of N-(2-hydroxypropyl) methacrylamide for use in site-specific therapy [84]. The three-dimensional structure of these hydrogels consists of the hydrophilic polymer chains connected together via $-$COONHCO$-$ linkages, that are stable in acidic aqueous solutions (below pH 6) but sensitive to hydrolytic cleavage at a physiologic pH of 7.4.

Based on Natural Polymers.   Natural polymers remain attractive primarily because they are natural products of living organisms, readily available, relatively inexpensive, and capable of a multitude of chemical modifications [85]. A majority of the investigations on natural polymers as matrices in drug delivery systems have centered on proteins (e.g., collagen, gelatin, and albumin) and polysaccharides (e.g., starch, amylose, dextran, inulin, cellulose, chitin, and hyaluronic acid).

The preparation of polypeptide hydrogels can be classified into (1) cross-linking of linear polypeptides, (2) graft copolymerization, and (3) copolymerization of hydrophilic and hydrophobic α-amino acids. The most widely used bioerodible hydrogel is gelatin that has been insolubilized by heat treatment, aldehyde treatment, or chromic acid treatment [86]. Other widely investigated systems are: collagen cross-linked with glutaraldehyde and albumin cross-linked by thermal denaturation or glutaraldehyde. The degradation of polypeptide hydrogels is attributed to proteolytic enzymes in the gastrointestinal tract, when given orally, or released during the acute and chronic stages of the inflammatory response when implanted.

Copolymers of hydrophilic and hydrophobic α-amino acids were prepared from hydrophilic monomer L-aspartic acid and hydrophobic monomers L-leucine, β-methyl-L-aspartate, and β-benzyl-L-aspartate [87]. The biodegradation rate of these copolymers was controlled by the ratio of the hydrophilic to hydrophobic monomers. The hydrophobic polymers showed no degradation when implanted subcutaneously in rats for up to 12 weeks, while the hydrophilic hydrogels degraded within 24 hours.

Polysaccharide hydrogels widely employed in drug delivery systems are based on modified chitin, dextran, amylose, and starch. Unmodified chitin is highly hydrophobic, but its hydrophilicity can be substantially increased by deacetylation of its N-acetylglucosamine units in strong alkali. Chitin hydrogels have been formed by the reaction between free amino groups of partially deacetylated chitin and glutaraldehyde [88]. Chitin hydrogels are degraded by a lysosome that hydrolyzes the 1-4 glycosidic linkages of chitin [89].

Dextran hydrogels have been prepared by the reaction of dextran hydroxyl groups with a glycidyl acrylate. The derivatized dextran is then cross-linked by N,N-methylenebisacrylamide [83].

Starch polysaccharides are composed of two polysaccharides, amylose and amylopectin. Amylose is essentially a linear polymer having a molecular weight of 100,000–500,000. Conversely, amylopectin is a highly branched polymer with a molecular weight in the millions. Starch hydrogels have been prepared by physical modification, heat [90] or chemical modifications, cross-linking with calcium, CS_2, or epichlorohydrin [91,92]. Amylases are endogenic enzymes capable of catalyzing the hydrolysis of $\alpha D(1\text{-}4)$ linkages in starch to yield D-glucose. Thus, starch matrices can be targeted to release the incorporated drug in organs having amylase activity such as: saliva, intestine and blood.

Kost and Shefer investigated starch, amylose, amylopectin, and dextran as bioerodible matrices for oral and injectable enzymatically controlled drug delivery [92]. They found that cross-linked high molecular weight polysaccharides can be used for entrapment and controlled release of bioactive molecules. The release rate of large molecules is degradation dependent due to amylase activity, which may target the release to the intestine when the particles are taken orally. Small molecules such as salicylic acid can be released by diffusion in addition to degradation. Enzymatic degradation of particles prepared from different ratios of amylose to amylopectin increased with the content of amylopectin. Drug release and enzymatic degradation of the studied polysaccharides can be controlled by the cross-linking density or enzyme concentration. The polysaccharide particles can also be used to temporarily alter blood flow and improve tissue uptake of drugs when injected into the bloodstream.

6. Environmentally Sensitive Hydrogels

A new class of hydrogels with swelling behavior that is sensitive to changes in environmental conditions has been of great recent interest. Several research groups have been developing drug delivery responsive systems that more closely resemble the normal physiological process [93]. In these devices, drug delivery is regulated by means of an interaction with the surrounding environment (feedback information) without any external intervention. The responsive systems utilize several approaches as rate-controlling mechanisms, such as pH-sensitive polymers, enzyme substrate reactions, pH-sensitive drug solubility, competitive binding, antibody interactions, and metal-concentration-dependent hydrolysis [94].

Temperature-Sensitive Systems.   The most commonly studied hydrogels having environmental sensitivity are either pH or temperature sensitive. Temperature-sensitive polymers can be classified into two groups based on the origin of the thermosensitivity in aqueous media. The first is based on polymer-water interactions, especially specific hydrophobic/hydrophilic balancing effects and the configuration of side groups. The other is based on polymer-polymer interactions in addition to polymer-water interactions. When polymer networks swell in a solvent, there is usually a negligible or small positive enthalpy of mixing or dilution. Although a positive enthalpy change opposes the process, the large gain in the entropy drives it.

In aqueous polymer solutions, the opposite is often observed. This unusual behavior is associated with a phenomenon of polymer phase separation as the temperature is raised to a critical value, known as the lower critical solution temperature or LCST. N-alkyl acrylamide homopolymers and their copolymers, including acidic or basic comonomers, show this LCST [95,96]. Gels characterized by LCST usually shrink as the temperature is increased through the LCST. Lowering the temperature below the LCST results in the swelling of the gels.

Bioactive agents such as drugs, enzymes, and antibodies may be immobilized on or within the temperature-sensitive hydrogels. A responsive drug-release pattern regulated by temperature changes has been recently demonstrated by several groups [95–103].

Systems Sensitive to pH.   The pH range of fluids in various segments of the gastrointestinal tract may provide environmental stimuli for responsive drug release. Studies by several research groups have been performed on polymers containing weakly acidic or basic groups in the polymeric backbone [104–116]. The charge density of the polymers depends on pH and ionic composition of the outer solution (the solution into which the polymer is exposed). Altering the pH of the solution will cause swelling or deswelling of the polymer. Thus, drug release from devices made from these polymers will display release rates that are pH dependent.

Polyacidic polymers will be unswollen at low pH since the acidic groups will be protonated and hence un-ionized. With increasing pH, a polyacid polymer will swell. The opposite holds for polybasic polymers, since the ionization of the basic groups will increase with decreasing pH. Seigel et al. found the swelling properties of the polybasic gels are influenced also by buffer composition (concentration and pKa) [110]. A practical consequence proposed is that these gels may not reliably mediate pH-sensitive swelling controlled release in oral applications since the levels of buffer acids in the stomach (where swelling and release are expected to occur) generally cannot be controlled. However, the gels may be useful as mediators of pH-triggered release when precise rate control is of secondary importance.

Recently, Annaka and Tanaka reported that more than two phases (swollen and collapsed) can be found in gels consisting of copolymers of randomly distributed positively and negatively charged groups [104]. In these gels, polymer segments interact with each other through attractive or repulsive electrostatic interactions and through hydrogen bonding. The combination of these forces seems to result in the existence of several phases, each characterized by a distinct degree of swelling, with abrupt jumps between them. The existence of these phases presumably reflects the ability of a macromolecular system to adopt different stable conformations in response to changes in environmental conditions. For copolymer gels prepared from acrylic acid (the anionic constituent) and methacryl-amido-propyl-trimethyl ammonium chloride (460 mmol/240 mmol) the largest number of phases was seven.

Electrically Regulated Systems.   Electrically controlled systems provide drug release by the action of an applied electric field on a rate-limiting membrane and/or directly on the solute, thus controlling its transport across it. The electrophoretic migration of a charged macrosolute within a hydrated membrane results from the combined response to the electrical forces on the solute and its associated counterions in the adjacent electrolyte solution [117]. Grimshaw et al. reported four different mechanisms for the transport of proteins and neutral solutes across hydrogel membranes [118]: (1) electrically and chemically induced swelling of a membrane to alter the effective pore size and permeability, (2) electrophoretic augmentation of solute flux within a membrane, (3) electrosmotic augmentation of solute flux within the membrane, and (4) electrostatic partitioning of charged solutes into charged membranes. Electrically controlled membrane permeability is also of current interest in the field of electrically controlled or enhanced transdermal drug delivery (e.g., iontophoresis) [119].

D'Emanuele and Staniforth proposed a drug delivery device that consists of a polymer reservoir with a pair of electrodes placed across the rate-limiting membrane [120]. By altering the magnitude of the electric field between the electrodes, the authors proposed to modulate the drug-release rates in a controlled and predictable manner. A linear relationship was found between current and propanolol HCl permeability through poly(2-hydroxyethyl methacrylate) (PHEMA) membranes cross-linked with ethylene glycol dimethacrylate(1% volume/volume [v/v]). Buffer ionic strength, drug reservoir concentration, as well as electrode polarity were found to have a significant effect on drug permeability [121].

Kwon et al. studied the effect of electric current on solute release from cross-linked poly(2-acrylamido-2-methylpropane sulfonic acid-co-n-butylmethacrylate) [122]. Edrophonium chloride, a positively charged solute, was released in an on-off pattern from a matrix (monolithic) device with an electric field. The mechanism was explained as an ion exchange between positive solute and hydroxonium ion, followed by fast release of the charged solute from the hydrogel. The fast release was attributed to the electrostatic force, squeezing effect, and the electro-osmosis of the gel. However, the release of neutral solute was controlled by diffusion affected by swelling and deswelling of the gel.

A different approach to electrochemically controlled release is based on polymers that bind and release bioactive compounds in response to an electric signal [123]. The polymer has two redox states, only one of which is suitable for ion binding. Drug ions are bound in one redox state and released from the other. The attached electrodes serve to switch the redox states and the amount of current passed can control the amount of ions released.

Photo-Responsive Systems.   Photoresponsive gels change their physical or chemical properties reversibly upon photoradiation. A photoresponsive polymer consists of a photoreceptor, usually a photochromic chromophore and a functional part. The optical signal is captured by the photochromic molecules and then the isomerization of the chromophores in the photoreceptor converts it to a chemical signal. Photoinduced phase transition of gels was reported by Mamada et al. [124]. Copolymer gels of N-isopropylacrylamide and a photosensitive molecule, bis(4-dimethylamino)phenyl(4-vinylphenyl)methyl leucocyanide, upon ultraviolet irradiation showed a discontinuous volume phase transition caused by osmotic pressure of cyanide ions created by ultraviolet irradiation.

Recently, Suzuki and Tanaka reported on phase transition in polymer gels induced by visible light, in which the transition mechanism is due only to the direct heating of the network polymer by light [125]. Yui et al. reported on photoresponsive degradation of heterogeneous hydrogels for temporal drug delivery comprised of cross-linked hyaluronic acid and lipid microspheres using methylene blue as a photosensitizer [126].

Inflammation-Responsive Systems. Yui et al. proposed a inflammation-responsive drug delivery system based on biodegradable hydrogels of cross-linked hyaluronic acid [127]. Hyaluronic acid is specifically degraded by hydroxyl radicals, which are produced locally at inflammatory sites by phagocytic cells such as leukocytes and macrophages. In their approach, drug-loaded lipid microspheres were dispersed into degradable matrices of cross-linked hyaluronic acid.

Saccharide-Sensitive Hydrogels. Kokufata et al. reported on a gel system that swells and shrinks in response to specific saccharides [128]. The gel consists of a covalently cross-linked polymer network of N-isopropylacrylamide in which the lectin, concavalin A, is immobilized. Concavalin A displays selective binding affinities for certain saccharides. For example, when the saccharide dextran sulphate is added to the gel, it produces an excess ionic osmotic pressure in the gel that raises the volume phase transition temperature. Consequently, the gel swells to a volume up to five times greater. Replacing dextran sulphate with nonionic saccharide α-methyl-D-mannopyranoside brings about collapse of the gel almost back to its native volume because the neutral α-methyl-D-mannopyranoside exerts no ionic pressure. The process is reversible and repeatable.

The basic principle of competitive binding and its application to controlled drug delivery was first presented by Brownlee and Cerami [129], who suggested the preparation of glycosylated insulins that are complementary to the major combining site of carbohydrate-binding proteins such as concavalin A (Con A). Con A is immobilized on sepharose beads. The glycosylated insulin, which is biologically active, is displaced from the Con A by glucose in response to, and proportional to, the amount of glucose present, which competes for the same binding sites. By encapsulating the glycosylated insulin-bound Con A with a suitable polymer that is permeable to both glucose and insulin, the glucose influx and insulin efflux would be controlled by the encapsulation membrane. The functionality of an intraperitoneally implanted device was tested in pancreatecto-mized dogs [130]. Recently, Makino et al. proposed a new system based on hydrophilic nylon microcapsules containing Con A and succinil-amidophenyl-glucopyranoside insulin [131].

Kitano et al. proposed a glucose-sensitive insulin-release system based on a sol-gel transition [132]. A phenylboronic acid (PBA) moiety was incorporated in poly(N-vinyl-2-pyrrolidone) by the radical copolymerization of N-vinyl-2-pyrrolidone with m-acryl-amidophenylboronic acid, poly(NVP-co-PBA). Insulin was incorporated into a polymer gel formed by a complex of poly(vinyl alcohol) with poly(NVP-co-PBA). PBA can form reversible covalent complexes with molecules having diol units, such as glucose or PVA. With the addition of glucose, PVA in the PVA-boronate complex is replaced by glucose. This leads to a transformation of the system from the gel to the sol state, which facilitates the release of insulin from the polymeric complex.

7. Systems Utilizing Enzymes

Heller and Trescony were the first to attempt using immobilized enzymes to alter local pH and thus cause changes in polymer erosion rates [133]. The proposed system was based on the conversion of urea to NH_4HCO_3 and NH_4OH by the action of urease. As this reaction causes a pH increase, a polymer that is subjected to increased erosion at high pH is required.

Glucose-Responsive Insulin Delivery. The systems for glucose-responsive insulin delivery consist of immobilized glucose oxidase in a pH-responsive polymeric hydrogel, enclosing a saturated insulin solution or incorporated with insulin [134–141]. As glucose diffuses into the hydrogel, glucose oxidase catalyzes its conversion to gluconic acid, thereby lowering the pH in the microenvironment of the hydrogel and causing swelling. Since insulin should permeate the swelled hydrogel more rapidly, faster delivery of insulin in the presence of glucose is anticipated. As the glucose concentration decreases in response to the released insulin, the hydrogel should contract and decrease the rate of insulin delivery (Fig. 1). As can be seen in Figs. 2 and 3, swelling and release kinetics are responsive to changes in glucose concentration. Swelling and release rates increase after a

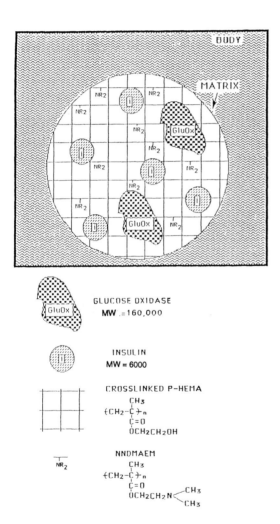

Figure 1 Schematic presentation of a glucose-responsive insulin delivery system based on a copolymer of hydroxyethyl methacrylate (HEMA) and N,N-dimethylaminoethyl methacrylate (NNDMAEM), cross-linked by tetraethylene glycol dimethacrylate, in which insulin and the enzyme glucose oxidase are uniformly distributed throughout the polymer. As glucose diffuses into the hydrogel, glucose oxidase catalyzes its conversion to gluconic acid, thereby lowering the pH in the microenvironment of the hydrogel and causing swelling, and therefore enhanced insulin release. (From Ref. 142.)

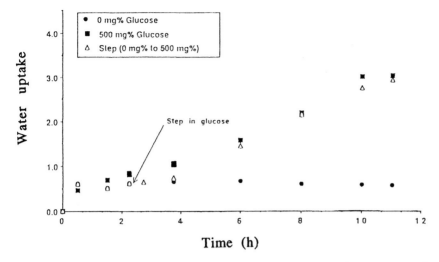

Figure 2 Swelling kinetics of glucose-sensitive hydrogels exposed to a constant glucose concentration of 0 mg% and 500 mg% and hydrogels exposed to a step change in glucose concentration from 0% to 500 mg%; water uptake is equal to (wet weight − initial weight)/dry weight. (From Ref. 142.)

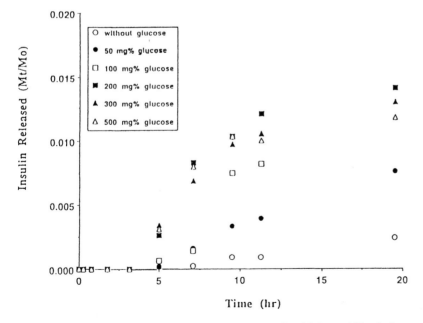

Figure 3 Fraction of insulin released from hydrogels with immobilized glucose oxidase at several glucose concentrations. M_t is the weight of insulin released at a given time, M_0 is the initial weight of insulin in the matrix. (From Ref. 142.)

step in glucose concentration from 0% to 500% (500 mg/100 ml), and reach the value of specimens that were exposed continuously to 500 mg% glucose [136].

Heller et al. suggested a system in which insulin is immobilized in a pH-sensitive bioerodible polymer prepared from 3,9-bios(ethylidene 2,4,8,10-tetraoxaspirol(5,5)-undecane and N-methyldiethanolamine, which is surrounded by a hydrogel containing immobilized glucose oxidase [141]. When glucose diffuses into the hydrogel and is oxidized to gluconic acid, the resultant lowered pH triggers enhanced polymer degradation and release of insulin from the polymer in proportion to the concentration of glucose. Initial experiments performed with this system showed very high sensitivity *in vitro* of insulin release as a function of external pulses of decreasing pH values [142]. However, when the *in vitro* studies were repeated in physiologic buffer, response of the device was only minimal, even at very low pH pulses. A more detailed study showed that the desired behavior could only be achieved in a citrate buffer [142].

Additional glucose-sensitive devices for insulin release that are not based on hydrogels were recently reviewed by Kost and Langer [94] and Heller [142].

Morphine-Triggered Naltrexone Delivery System.   Recently, Heller and coworkers have been developing a naltrexone drug delivery system that would be passive until drug release is initiated by the appearance of morphine external to the device [143,144]. Naltrexone is a long-acting opiate antagonist that blocks opiate-induced euphoria; thus, the intended use of this device is in the treatment of heroin addiction. The initial attempts have used an enzyme-degradable acidic hydrogel that surrounds a polymer stable at low pH but that erodes at a desired rate at physiological pH. Triggering of the device is accomplished by enzyme activation and consequent removal of the acidic hydrogel. Release of naltrexone then takes place when the environment surrounding the polymer changes from a low pH value to pH 7.4. As this approach was not successful, the authors have now shifted to the use of hydrophobic protective coatings [145].

III. CONCLUSIONS

During the last two decades, controlled drug delivery has become an important area of research and development. In this short time, a number of systems displaying constant or decreasing release rates have progressed from the laboratory to the clinic and, in some cases, to commercial products. Natural and synthetic materials have been used in controlled-release systems to deliver contraceptives, ophthalmics, antiarrhythmics, hormones, enzymes, anticancer agents, anticoagulants, antibodies, and antagonists. This technology is not limited to medicine. Controlled-release systems have been developed for flea collars, pesticides, antifouling agents, fertilizers, and fragrances. Liposomes are used in cosmetics [1].

Although these controlled delivery systems are advantageous compared with the conventional methods of drug administration, they are insensitive to the changing metabolic state. In order to control the physiological requirements of the specific drugs more closely, responsive mechanisms must be provided. The approaches discussed in this chapter represent attempts conducted over the past decade to achieve pulsatile release. It should be pointed out that these drug delivery systems are still in the early developmental stage and much research will have to be conducted for such systems to become practical clinical alternatives.

Critical considerations are the biocompatibility and toxicology of these multicomponent polymer-based systems, the response times of these systems to stimuli, the ability to provide practical levels of the desired drug, and addressing necessary formulation issues in dosage or design (e.g., shelf life, sterilization, reproducibility). Key issues in the practical utilization of such responsive systems as those containing enzymes or antibodies are the stability and/or potential leakage and possible immunogenicity of these bioactive agents.

For biotechnology to have a significant impact on the pharmaceutical industry, it is critical for the development of new drugs to be coupled with methodologies for delivering them appropriately. The development of approaches in genetic engineering may enable the creation of new molecules with increased ability to achieve site-specific (targeted) delivery.

Finally, advances in biomaterials science and engineering should result in improved, better characterized and processed polymers, lipids, and ceramics to be used effectively in drug delivery systems.

REFERENCES

1. Langer, R. 1990. New Methods of Drug Delivery. *Science.* 249:1527–1533.
2. Kost, J., and R. Langer. 1984. Controlled Release of Bioactive Agents. *Trends in Biotechnology.* 2:47–51.
3. Heller, J., R. V. Sparer, and G. M. Zenter (eds.). 1990. Poly(ortho Esters), in M. Chasin and R. Langer, eds., *Biodegradable Polymers as Drug Delivery Systems*, Marcel Dekker, New York, pp. 120–161.
4. Ron, E., and R. Langer. 1992. Erodible Systems, in A. Kydonieus, ed., *Treatise on Controlled Drug Delivery*, Marcel Dekker, New York, pp. 199–224.
5. Hopfenberg, H., and K. C. Hsu. 1981. Swelling Controlled, Constant Rate Delivery Systems. *Polymer Eng. Sci.* 18:18.
6. Theeuwes, F. 1987. Elementary Osmotic Pump. *J. Pharm. Sci.* 64:1975.
7. Oku, N. 1991. Liposomes in Polymeric Drugs and Drug Delivery Systems, in R. Dunn, and R. Ottenbrite, eds., *Polymeric Drugs and Drug Delivery Systems*, American Chemical Society Symposium Series 469, American Chemical Society, Washington, DC, pp. 24–33.
8. Treat, J., R. Greenspan, and A. Rahman. 1989. Liposomes Encapsulated Doxorubicin Preliminary Results of Phase I and Phase II Trials, in G. Lopez-Berestein and I. J. Fidler, eds., *Liposomes in the Therapy of Infectious Diseases and Cancer*, Liss, New York, pp. 353–365.
9. Lopez-Berestein, G. 1989. Treatment of Systemic Fungal Infections with Liposomal-Amphotericin B, in G. Lopez-Berestain and I. J. Fidler, eds., *Liposomes in the Therapy of Infectious Diseases and Cancer*, Liss, New York, pp. 317–327.
10. Alving, C. 1991. Liposomes as Carriers of Antigens and Adjuvants. *J. Immunol. Meth.* 140: 1–13.
11. Gregoriadis, G. 1990. Immunological Adjuvants: A Role for Liposomes. *Immuno. Today.* 11:89–97.
12. Folkman, J., and D. M. Long. 1964. The Use of Silicone Rubber as a Carrier for Prolonged Drug Therapy. *J. Surg. Res.* 4:139–142.
13. Baker, R., and H. Lonsdale. 1975. *Chem. Technol.* 5:668.
14. Domb, A., S. Amselem, J. Shah, and M. Maniar. 1993. Degradable Polymers for Site-Specific Drug Delivery. *Polymers for Advanced Technol.* 3:279–292.
15. Smith, K. L., M. E. Schimpf, and K. E. Thompson. 1990. Bioerodible Polymers for Delivery of Macromolecules. *Advanced Drug Deliv. Rev.* 4:343–357.
16. Dunn, R. L. 1991. Polymeric Matrices, in R. Dunn and R. Ottenbrite, eds., *Polymeric Drugs and Drug Delivery Systems*, American Chemical Society Symposium Series 469, American Chemical Society, Washington, DC, pp. 11–23.

17. Leong, K. W., and R. Langer. 1987. Polymeric Controlled Drug Delivery. *Advanced Drug Deliv. Rev.* 1:199–233.

18. Ron, E., and R. Langer. 1992. Erodible Systems, in A. Kydonieus, ed., *Treatise on Controlled Drug Delivery*, Marcel Dekker, New York.

19. Miller, R., J. M. Brady, and D. E. Cutight. 1977. Degradation Rates of Oral Resorbable Implants (Polyacetals and Polyglycolides): Rate Modification with Change in PLA/PGA Copolymer Ratios. *J. Biomed. Mater. Res.* 11:711–719.

20. Okada, H. 1989. One Month Release Injectable Microspheres of Leuprolide Acetate, a Superactive Agonit of LH-RH. *Proc. Int. Symp. Control. Rel. Bioactive Mater.* 16:12–13.

21. Hutchinson, F. G. 1988. U.S. Patent 4,767,628.

22. Sanders, L. M., B. A. Kell, G. I. McRae, and G. W. Whitehead. 1986. Prolonged Controlled-Release of Nafarelin, a LHRH Analogue, from Biodegradable Polymeric Implants: Influence of Composition and Molecular Weight of Polymer. *J. Pharm. Sci.* 75:356–360.

23. Nahlou, J. 1987. *J. Clin. Endo. Met.* 65:946.

24. Pitt, C. G., M. M. Gratzl, G. L. Kimmel, J. Surles, and A. Schindler. 1981. Aliphatic Polyesters. II. The Degradation of Poly(DL-Lactide), Poly(ε-caprolactone), and Their Copolymers *In Vivo. Biomaterials.* 2:215.

25. Pitt, C. G. 1990. Poly-ε-Caprolactone and Its Copolymers, in M. Chasin and R. Langer, eds., *Biodegradable Polymers as Drug Delivery Systems*, Marcel Dekker, New York, pp. 71–120.

26. Choi, N. S., and J. Heller. U.S. Patents 4,093,709 (1978); 4,131,648 (1978); 4,138,344 (1979); 4,180,646 (1979).

27. Heller, J., D. W. H. Penhale, B. K. Fritzinger, and S. Ng. 1987. The Effect of Copolymerized 9,10-Dihydroxystearicacid on Erosion Rates of Poly(ortho Esters). *J. Control. Rel.* 5:173–177.

28. Heller, J., Y. F. Maa, P. Wutrhrich, S. Y. Ng, and R. Duncan. 1991. Recent Developments in the Synthesis and Utilization of Poly(ortho Esters). *J. Control. Rel.* 16:3–14.

29. Hill, J., and W. H. Carothers. 1932. Studies of Polymerization and Ring Formation. XIV. A Linear Superpolyanhydride and a Cyclic Dimeric Anhydride from Sebacic Acid. *J. Am. Chem. Soc.* 4:1569.

30. Rosen, H. B., J. Chang, G. E. Wnek, R. J. Linhardt, and R. Langer. 1983. Bioerodible Polyanhydrides for Controlled Drug Delivery. *Biomaterials.* 4:131–133.

31. Leong, K. W., B. C. Brott, and R. Langer. 1985. Bioerodible Polyanhydrides as Drug-Carrier Matrices. I. Characterization, Degradation and Release Characteristics. *J. Biomed. Mater. Res.* 19:941–955.

32. Leong, K. W., V. Simonte, and R. Langer. 1987. Synthesis of Polyanhydrides: Melt-Polycondensation, Dehydrochlorination, and Dehydrative Coupling. *Macromolecules.* 20:705–712.

33. Domb, A., C. Gallardo, and R. Langer. 1989. Poly(anhydrides) 3. Poly(anhydrides) Based on Aliphatic-Aromatic Diacids. *Macromolecules.* 22:3200–3204.

34. Domb, A., C. Laurencin, O. Israeli, R. Gerhart, and R. Langer. 1990. The Step Polymerization of Poly(Propylene Fumarate) for Use in Bioerodible Bone Cement Composites. *Polym. Sci.* 28:973–985.

35. Staubli, A., E. Ron, and R. Langer. 1990. Hydrolytically Degradable Amino Acid Containing Polymers. *J. Am. Chem. Soc.* 112:4419–4424.

36. Allcock, H. R. 1990. Polyphosphazenes as New Biomedical and Bioactive Materials, in M. Chasin and R. Langer eds., *Biodegradable Polymers as Drug Delivery Systems*, Marcel Dekker, New York, 163–193.

37. Kohn, J., and R. Langer. 1987. Polymerization Reactions Involving the Side Chains of X-L-Amino Acids. *J. Am. Chem. Soc.* 109:817–820.

38. Kaetsu, I., M. Yoshida, M. Asano, H. Yamanaka, K. Imai, H. Yuasa, T. Mashimo, K. Suzuki, R. Katakai, and M. Oya. 1987. Biodegradable Implant Composites for Local Therapy. *J. Contr. Rel.* 6:249–263.

39. Sidman, K. B., A. D. Schwope, W. D. Steber, S. E. Rudolph, and S. B. Poulin. 1980. Biodegradable, Implantable Sustained Release Systems Based on Glutamic Acid Copolymers. *J. Membr. Sci.* 7:277–291.

40. Asano, M., M. Yoshida, I. Kaetsu, K. Nakai, H. Yamanaka, H. Yuasa, and M. Oya. 1983. Biodegradable Random Copolypeptides of Beta-Benzyl–Aspartate and Gamma-Methyl-L-Glutamate for the Controlled Release of Testosterone. *Makrom. Chem.* 184:1761–1770.

41. Sidman, K. B., W. D. Steber, A. D. Schwope, and Schnaper. 1983. Controlled Release of Macromolecules and Pharmaceuticals from Synthetic Polypeptides Based on Gluamic Acid. *Biopolymers.* 22:547–556.

42. Flory, P. J. 1953. *Principles of Polymer Chemistry*, Cornell University Press, Ithaca, NY.

43. Peppas, N. A., and A. R. Khare. 1993. Preparation, Structure and Diffusion Behavior of Hydrogels in Controlled Release. *Advanced Drug Deliv. Rev.* 11:1–35.

44. Peppas, N. A., and R. W. Korsmeyer. 1987. Dynamically Swelling Hydrogels in Controlled Release Applications, in N. A. Peppas, ed., *Hydrogels in Medicine and Pharmacy*, CRC Press, Boca Raton, FL, pp. 109–135.

45. Kost, J., and R. Langer. 1987. Equilibrium Swollen Hydrogels in Controlled Release Applications, in N. A. Peppas, ed., *Hydrogels in Medicine and Pharmacy*, CRC Press, Boca Raton, FL, pp. 95–108.

46. Gehrke, S. H., and P. L. Lee. 1990. Hydrogels for Drug Delivery Systems, in P. Tyle, ed., *Specialized Drug Delivery System*, Marcel Dekker, New York, pp. 333–392.

47. Colombo, P. 1993. Swelling-Controlled Release in Hydrogel Matrices for Oral Route. *Advanced Drug Deliv. Rev.* 11:37–58.

48. Kamath, K. R., and K. Park. 1993. Biodegradable Hydrogels in Drug Delivery. *Advanced Drug Deliv. Rev.* 11:59–84.

49. Yoshida, R., K. Sakai, T. Okano, and Y. Sakurai. 1993. Pulsatile Drug Delivery Systems Using Hydrogels. *Advanced Drug Deliv. Rev.* 11:85–108.

50. Bae, Y. H., and S. W. Kim. 1993. Hydrogel Delivery Systems Based on Polymer Blends, Block Co-Polymers or Interpenetrating Networks. *Advanced Drug Deliv. Rev.* 11:109–135.

51. Peppas, N. A., and H. J. Moynihan. 1987. Structure and Physical Properties of Poly(2-Hydroxyethyl Methacrylate) Hydrogels, in N. A. Peppas, ed., *Hydrogels in Medicine and Pharmacy*, CRC Press, Boca Raton, FL, pp. 49–64.

52. Gregonis, D. E., C. M. Chen, and J. D. Andrade. 1967. The Chemistry of Some Selected Methacrylate Hydrogels, in J. D. Andrade, ed., *Hydrogels for Medical Applications*, American Chemical Society, Washington, DC, pp. 88–103.

53. Cowsar, D. R., O. R. Tarwatar, and A. C. Tanquary. 1976. Controlled Release of Fluoride from Hydrogels for Dental Applications, in J. D. Andrade, ed., *Hydrogels for Medical and Related Applications*, American Chemical Society, Washington, DC, pp. 180–197.

54. Drobnik, J., P. Spacek, and O. Wichterle. 1974. Diffusion of Anti-Tumor Drugs through Membranes from Hydrophilic Methacrylate Gels. *J. Biomed. Mater. Res.* 8:45.

55. Abrahams, R. A., and S. H. Ronel. 1975. Biocompatible Implants for Sustained Zero Order Release of Narcotic Antagonists. *J. Biomed. Mater. Res.* 9:355.

56. Kost, J., T. A. Horbett, B. D. Ratner, and M. Singh. 1984. Glucose Sensitive Membranes Containing Glucose Oxidase: Activity, Swelling, and Permeability Studies. *J. Biomed. Mater. Res.* 19:1117–1133.

57. Colter, K. D., A. T. Bell, and M. Shen. 1977. Control of the Pilocarpine Release Rate through Hydrogels by Plasma Treatment. *Med. Dev. Artif. Organs.* 5:1.

58. Basmadjian, D., and M. V. Sefton. 1983. Relationship between Release Rate and Surface Concentration for Heparinized Materials. *J. Biomed. Mater. Res.* 17:509.

59. Goosen, M. F. A., and M. V. Sefton. 1983. Properties of a Heparin-Poly(Vinyl-Alcohol) Hydrogel Coating. *J. Biomed. Mater. Res.* 17:359.

60. Yamauchi, A., Y. Matsuzawa, Y. Hara, M. Saishin, K. Nishioka, S. Nakao, and S. Kamiya. 1979. The Use of Poly(Vinyl Alcohol) Hydrogels as Drug Carriers. *Polymer Prepr.* 20:575.

61. Peppas, N. A., and T. W. B. Gehr. 1979. Statistical Analysis of Nitrogen-Containing Vinyl Copolymers: Radiation-Induced Copolymerization of Vinyl Acetate and N-Vinyl-2-Pyrrolidone. *J. Appl. Polym. Sci.* 24:2105.

62. Hosaka, S., H. Ozawa, and H. Tanzawa. 1979. Controlled Release of Drugs from Hydrogel Matrices. *J. Appl. Polymer Sci.* 23:2089.

63. Hasirci, V. N. 1982. PVNO-DVB Hydrogels: Synthesis and Characterization. *J. Appl. Polymer. Sci.* 27:33.

64. Heller, J., R. F. Helwing, R. W. Baker, and M. E. Tuttle. 1983. Controlled Release of Water-Soluble Macromolecules from Bioerodible Hydrogels. *Biomaterials.* 4:262–266.

65. Rosiak, J., K. Burchak, and W. Pekala. 1983. Polyacrylamide Hydrogels as Sustained-Release Drug Delivery Dressing Materials. *Radiat. Phys. Chem.* 22:907.

66. Davis, B. K., L. Noske, and M. C. Chang. 1972. Reproductive Performance of Hamsters with Polyacrylamide Implants Containing Ethinylestradiol. *Acta Endocrinol.* 70:385.

67. Graham, N. B., and M. E. McNeil. 1984. Hydrogels for Controlled Drug Delivery. *Biomaterials.* 5:27.

68. Gregor, H. P., H. T. Hsia, S. Palevsky, R. S. Neuwirth, and R. M. Richard, eds. 1976. Fallopian Tube Cauterization by Silver-Ion Polymer Gels, in D. R. Paul and F. W. Harris, eds., *Controlled Release Polymeric Formulations*, American Chemical Society, Washington, DC, p. 147.

69. Dickinson, H., T. Sugie, J. M. Anderson, and A. Hiltner. 1979. New Poly Alpha-Amino Acids Have Been Prepared. *Polymer Prep.* 20:39.

70. Graham, N. B., M. E. McNeill, and M. Zulfigar. 1980. Hydrogels for the Controlled Release of Prostaglandin E2. *Polymer Prep.* 2:104.

71. Spang-Brunner, B. H., and P. P. Speiser. 1976. Release of a Drug from Homogeneous Ointments Containing the Drug in Solution. *J. Pharm. Pharmacol.* 28:23.

72. Harris, A. S., P. Stenberg, U. Ulmsten, and A. B. Perstrop. 1983. A Study of the *In Vivo* Release Profile of a New Prostaglandin e2, Delivery System for Local Administration in Obstetrics. *Acta Obster. Gynecol. Scand.* 113:59.

73. Stannett, V. T., W. J. Koros, D. R. Paul, H. K. Lonsdale, and R. W. Baker. 1979. Recent Advances in Membrane Science and Technology. *Adv. Polymer. Sci.* 32:69.

74. Yasuda, H., C. E. Lamaze, and L. D. Ikenberry. 1969. Permeability of Solutes through Hydrated Polymer Membranes. I. Diffusion of Sodium Chloride. *Makromol. Chem.* 118:19.

75. Yasuda, H., L. D. Ikenberry, and C. E. Lamaze. 1969. Permeability of Solutes through Hydrated Polymer Membranes. I. Permeability of Water Soluble Organic Solutes. *Makromol. Chem.* 125:108.

76. Peppas, N. A., and B. D. Bar-Howell. 1987. Characterization of the Crosslinked Structure of Hydrogels, in N. A. Peppas, ed., *Hydrogels in Medicine and Pharmacy*, CRC Press, Boca Raton, FL, pp. 27–56.

77. Reinhart, C. T., R. W. Korsmeyer, and N. A. Peppas. 1981. Macromolecular Network Structure and Its Effects on Drug and Protein Diffusion. *Int. J. Pharm. Technol. Prod. Mfr.* 2:8.

78. Lee, K. H., J. G. Jee, M. S. Jhon, and T. Ree. 1978. Solute Transport through Crosslinked Poly(2-Hydroxyethyl Methacrylate) Membrane. *J. Bioeng.* 2:269.

79. Anderson, J. M., T. Koinis, T. Nelson, M. Horst, and D. S. Love. 1976. The Slow Release of Hydrocortisone Sodium Succinate from Poly(2-Hydroxyethyl Methacrylate) Membranes, in J. D. Andrade, ed., *Hydrogels for Medical and Related Applications*, American Chemical Society, Washington, DC, pp. 167–179.

80. Refojo, M. 1967. Hydrophobic Interactions in Poly(2-Hydroxyethyl Methacrylate) Homogeneous Hydrogels. *J. Polym. Sci.* 5:3103.

81. Kim, C.-J., and P. I. Lee. 1992. Composite Poly(Vinyl Alcohol) Beads for Controlled Drug Delivery. *Pharm. Res.* 9:10–16.

82. Torchilin, V. P., E. G. Tischenko, V. N. Smirnov, and E. I. Chazov. 1977. Immobilization of Enzymes on Slowly Soluble Carriers. *J. Biomed. Mater. Res.* 11:223–235.

83. Edman, P., B. Ekman, and I. Sjoholm. 1980. Immobilization of Proteins in Microspheres of Biodegradable Polyacryldextran. *J. Pharm. Sci.* 69:838–842.

84. Rihova, B., A. Jegorov, J. Strohalm, V. Matha, P. Rossman, L. Fornusek, and K. Ulbrich. 1992. Antibody-Targeted Cyclosporin A. *J. Controlled Release.* 19:25–40.

85. Bogdansky, S. 1990. Natural Polymers as Drug Delivery Systems, in M. Chasin and R. Langer, eds., *Biodegradable Polymers as Drug Delivery Systems*, Marcel Dekker, New York, pp. 231–259.

86. Heller, J. 1987. Bioerodible Hydrogels, in N. A. Peppas, ed., *Hydrogels in Medicine and Pharmacy*, CRC Press, Boca Raton, FL, pp. 137–149.

87. Marck, K. W., C. H. Wildevuur, W. L. Sederel, A. Bantjes, and J. Feijen. 1977. Biodegradability and Tissue Reaction of Random Copolymers of L-Leucine, L-Aspartic acid, and L-Aspartic Acid Esters. *J. Biomed. Mater. Res.* 11:405.

88. Pangburn, S. H., P. V. Trescony, and J. Heller. 1982. Lysozyme Degradation of Partially Deacetylated Chitin, Its Films and Hydrogels. *Biomaterials.* 3:105.

89. Berger, L. R., and R. S. Weiser. 1957. The β-Glucoamidase Activity of Egg-White Lysozyme. *Biochem. Biophys. Acta.* 26:517.

90. Herman, J., J. P. Remon, and J. Vilder. 1989. Modified Starches as Hydrophilic Matrices for Controlled Oral Delivery. *Int. J. Pharm.* 56:5.

91. Artursson, P., P. Edman, T. Laasko, and I. Sjoholm. 1984. Characterization of Polyacryl Starch Microparticles as Carriers for Proteins and Drugs. *J. Pharm. Sci.* 73:1507.

92. Kost, J., and S. Shefer. 1990. Chemically Modified Polysaccharides for Enzymatically Controlled Oral Drug Delivery. *Biomaterials.* 11:695–698.

93. Kost, J. 1990. *Pulsed and Self-Regulated Drug Delivery*, CRC Press, Boca Raton, FL.

94. Kost, J., and R. Langer. 1991. Responsive Polymeric Delivery Systems. *Adv. Drug Deliv. Rev.* 6:19–50.

95. Hoffman, A. S. 1987. Applications of Thermally Reversible Polymers and Hydrogels in Therapeutics and Diagnostics. *J. Controlled Release.* 6:297–305.

96. Tanaka, T. 1985. Gels, in H. F. Mark and J. I. Kroschwitz, eds., *Encyclopedia of Polymer Science and Technology*, Wiley, New York, pp. 514–531.

97. Ueda, T., K. Ishihara, and N. Nakabayashi. 1990. Thermally Responsive Release of 5-Fluorouracil from a Biocompatible Hydrogel Membrane with Phospholipid Structure. *Macromol. Chem. Rapid Commun.* 11:345.

98. Urry, D. W., B. Haynes, H. Zhang, R. D. Harris, and K. U. Prasad. 1988. Mechanochemical Coupling in Synthetic Polypeptides by Modulation of Inverse Temperature Transition. *Proc. Natl. Acad. Sci. USA.* 85:3407–3411.

99. Yoshida, M., M. Asano, M. Kumakura, R. Kataki, T. Mashimo, H. Yuasa, and H. Yamanaka. 1991. Thermo-Responsive Hydrogels Based on Acrylo-l-Proline Methyl Ester and Their Use in Long-Acting Testosterone Delivery Systems. *Drug Design and Deliv.* 7:159–174.

100. Palasis, M., and H. Gehrke. 1992. Permeability of Responsive Poly(N-Isopropylacrylamide) Gel to Solutes. *J. Controlled Release.* 18:1–12.

101. Okano, T., Y. H. Bae, and S. W. Kim. 1990. Temperature Responsive Controlled Drug Delivery, in J. Kost, ed., *Pulsed and Self-Regulated Drug Delivery*, CRC Press, Boca Raton, FL, pp. 17–46.

102. Okahata, Y., H. Noguchi, and T. Seki. 1986. Thermo-selective Permeation from a Polymer-Grafted Capsule Membrane. *Macromolecules.* 19:493–494.

103. Katano, H., A. Maruyama, K. Sanui, N. Ogata, T. Okano, and Y. Sakurai. 1991. Thermoresponsive Swelling and Drug Release Switching of Interpenetrating Polymer Networks Composed of Poly(Acrylamide-co-Butyl Methacrylate) and Poly(Acrylic Acid). *J. Controlled Release.* 16:215–228.

104. Annaka, M., and T. Tanaka. 1992. Multiple Phases of Polymer Gels. *Nature.* 355:430–432.

105. Brannon-Peppas, L., and N. A. Peppas. 1989. Solute and Penetrant Diffusion in Swellable Polymers. IX. The Mechanism of Drug Release from pH Sensitive Swelling-Controlled Systems. *J. Controlled Release.* 8:267–274.

106. Firestone, B. A., and R. A. Siegel. 1988. Dynamic pH-Dependent Swelling Properties of Hydrophobic Polyelectrolyte Gel. *Polym. Commun.* 29:204–208.

107. Dong, L.-C., and A. S. Hoffman. 1990. Controlled Enteric Release of Macromolecules from pH Sensitive, Macroporous Heterogels. *Proc. Int. Symp. Control. Bioact. Mater.* 17:325–326.

108. Kou, J. H., D. Fleisher, and G. Amidon. 1990. Modeling Drug Release from Dynamically Swelling Poly(Hydroxyethyl Methacrylate-co-Methacrylate Acid) Hydrogels. *J. Controlled Release.* 12:241–250.

109. Pradny, M., and J. Kopecek. 1990. Hydrogels for Site-Specific Oral Delivery. Poly(Acrylic Acid)-co-(Butyl Acrylate) Crosslinked with 4,4′-Bis(Methacryloamino)Azobenzene. *Makromol. Chem.* 191:1887–1897.

110. Siegel, R. A., I. Johannes, A. Hunt, and B. A. Firestone. 1992. Buffer Effects on Swelling Kinetics in Polybasic Gels. *Pharm. Res.* 9:76–81.

111. Kono, K., F. Tabata, and T. Takagishi. 1993. pH-Responsive Permeability of Poly(Acrylic Acid)-Poly(Ethylenimine) Complex Capsule Membrane. *J. Membrane Sci.* 76:233–243.

112. Dong, L.-C., and A. S. Hoffman. 1991. A Novel Approach for Preparation of pH-Sensitive Hydrogels for Enteric Drug Delivery. *J. Controlled Release.* 15:141–152.

113. Hariharan, D., and N. A. Peppas. 1993. Modelling of Water Transport and Solute Release in Physiologically Sensitive Gels. *J. Controlled Release.* 23:123–136.

114. Kou, J. H., G. L. Amidon, and P. I. Lee. 1988. pH-Dependent Swelling and Solute Diffusion Characteristics of Poly(Hydroxyethyl Methacrylate-co-Methacrylic Acid) Hydrogels. *Pharm Res.* 5:592–597.

115. Siegel, R. A., and B. A. Firestone. 1988. pH-Dependent Equilibrium Swelling of Hydrophobic Polyelectrolyte Copolymer Gels. *Macromolecules.* 21:3254–3259.

116. Siegel, R. A., M. Falmarzian, B. A. Firestone, and B. C. Moxley. 1988. pH-Controlled Release from Hydrophobic/Polyelectrolyte Hydrogels. *J. Controlled Release.* 8:179–182.

117. Grodzinsky, A. J., and P. E. Grimshaw, 1990. Electrically and Chemically Controlled Hydrogels for Drug Delivery, in J. Kost, ed., *Pulsed and Self-Regulated Drug Delivery*, CRC Press, Boca Raton, FL, pp. 47–64.

118. Grimshaw, P. E., A. J. Grodzinsky, M. L. Yarmush, and D. M. Yarmush. 1989. Dynamic Membranes for Protein Transport: Chemical and Electrical Control. *Chem. Eng. Sci.* 104:827–840.

119. Rolf, D. 1988. Chemical and Physical Methods of Enhancing Transdermal Drug Delivery. *Pharm. Technol.* 12:130–140.

120. D'Emanuele, A., and J. N. Staniforth. 1991. An Electrically Modulated Drug Delivery Device. *Pharm. Res.* 8:913–918.

121. D'Emanuele, A., and J. N. Staniforth. 1992. An Electrically Modulated Drug Delivery Device. II. Effect of Ionic Strength, Drug Concentration and Temperature. *Pharm. Res.* 9:215–219.

122. Kwon, I. C., Y. H. Bae, O. T. Okano, and S. W. Kim. 1991. Drug Release from Electric Current Sensitive Polymers. *J. Controlled Release.* 17:149–156.

123. Miller, L. L., G. A. Smith, A. Chang, and Q. Zhou. 1987. Electrochemically Controlled Release. *J. Controlled Release.* 6:293–296.

124. Mamada, A., T. Tanaka, D. Kugwatchkakun, and M. Irie. 1990. Photoinduced Phase Transition of Gels. *Macromolecules.* 23:1517–1519.

125. Suzuki, A., and T. Tanaka. 1990. Phase Transition in Polymer Gels Induced by Visible Light. *Nature.* 346:345–347.

126. Yui, N., T. Okano, and Y. Sakurai. 1993. Photo-Responsive Degradation of Heterogeneous Hydrogels Comprising Crosslinked Hyaluronic Acid and Lipid Microspheres for Temporal Drug Delivery. *J. Controlled Release.* 26:141–145.

127. Yui, N., T. Okano, and Y. Sakurai. 1992. Inflammation Responsive Degradation of Crosslinked Hyaluronic Acid Gel. *J. Controlled Release.* 22:105–116.

128. Kokufata, E., Y.-Q. Zhang, and T. Tanaka. 1991. Saccharide-Sensitive Phase Transition of a Lectin-Loaded Gel. *Nature.* 351:302–304.

129. Brownlee, M., and A. Cerami. 1979. A Glucose-Controlled Insulin-Delivery System: Semisynthetic Insulin Bound to Lectin. *Science.* 26:1190–1191.

130. Kim, S. W., C. M. Pai, L. A. Makino, L. A. Seminoff, D. L. Holmberg, J. M. Gleeson, D. E. Wilson, and E. J. Mack. 1990. Self-Regulated Glycosalated Insulin Delivery. *J. Controlled Release*. 11:193–201.

131. Makino, K., E. J. Mack, T. Okano, and S. W. Kim. 1990. A Microcapsule Self-Regulating Delivery System for Insulin. *J. Controlled Release*. 12:235–239.

132. Kitano, S., Y. Koyama, K. Kataoka, T. Okano, and Y. Sakurai. 1992. A Novel Drug Delivery System Utilizing a Glucose Responsive Polymer Complex between Poly(Vinyl Alcohol) and Poly(N-Vinyl-2-Pyrrolidone) with Phenylboronic Acid Moiety. *J. Controlled Release*. 19:161–170.

133. Heller, J., and P. V. Trescony. 1979. Controlled Drug Release by Polymer Dissolution. II. An Enzyme Mediated Delivery System. *J. Pharm. Sci.* 68:919–921.

134. Albin, G., T. A. Horbett, and B. D. Ratner. 1985. Glucose Sensitive Membranes for Controlled Delivery of Insulin: Insulin Transport Studies. *J. Controlled Release*. 3:153–164.

135. Albin, G., T. A. Horbett, S. R. Miller, and N. L. Ricker. 1987. Theoretical and Experimental Studies of Glucose Sensitive Membranes. *J. Controlled Release*. 6:267–291.

136. Goldraich, M., and J. Kost. 1993. Glucose-Sensitive Polymeric Matrices for Controlled Delivery of Therapeutics. *Clinical Materials*. 13:135–142.

137. Ishihara, K., M. Kobayashi, N. Ishimaru, and I. Shinohara. 1984. Glucose Induced Permeation Control of Insulin Through a Complex Membrane Consisting of Immobilized Glucose Oxidase and a Polyamine. *Polym. J.* 16:625–631.

138. Siegel, R. A., and B. A. Firestone. 1990. Mechanochemical Approaches to Self-Regulating Insulin Pump Design. *J. Controlled Release*. 11:181–192.

139. Klumb, L. A., and T. A. Horbett. 1993. The Effect of Hydronium Ion on the Transient Behavior of Glucose Sensitive Membranes. *J. Controlled Release*. 27:95–114.

140. Klumb, L. A., and T. A. Horbett. 1992. Design of Insulin Delivery Devices Based on Glucose Sensitive Membranes. *J. Controlled Release*. 18:59–80.

141. Heller, J., A. C. Chang, G. Rodd, and G. M. Grodsky. 1990. Release of Insulin from a pH-Sensitive Poly(Ortho Ester). *J. Controlled Release*. 14:295–304.

142. Heller, J. 1993. Polymers for Controlled Parenteral Delivery of Peptides and Proteins. *Adv. Drug Deliv. Rev.* 10:163–204.

143. Heller, J. 1990. Use of Enzymes and Bioerodible Polymers in Self-Regulated and Triggered Drug Delivery, in J. Kost, ed., *Pulsed and Self-Regulated Drug Delivery*, CRC Press, Boca Raton, FL, pp. 93–108.

144. Heller, J., S. H. Pangburn, P. V. Trescony, and K. V. Roskos. 1990. Development of Enzymatically Degradable Protective Coatings for Use in Triggered Drug Delivery Systems. *Biomaterials*. 11:345–350.

145. Roskos, K. V., J. A. Tefft, B. K. Fritzinger, and J. Heller. 1992. Development of a Morphine-Triggered Naltrexone Delivery System. *J. Controlled Release*. 19:145–160.

VII
COLLAGEN-BASED MATERIALS

35

Collagen: Molecular Structure and Biomaterial Properties

Marcel E. Nimni

University of Southern California School of Medicine/Children's Hospital
Los Angeles, California

I. INTRODUCTION

Collagen, in the form of fibers, represents the single most abundant animal protein in mammals, accounting for over 30% of all proteins [1,2]. The collagen molecules, after being secreted by the cells, assemble into very characteristic fibers, and as such are responsible for the functional integrity of tissues such as bone, cartilage, skin, and tendon (Figs. 1–3). They also contribute a structural framework to other tissues such as blood vessels and most organs. Cross-links between adjacent molecules are a prerequisite for the collagen fibers to withstand the physical stresses to which they are exposed. Native cross-links are generated by a rather complicated series of intracellular and extracellular events, and significant progress has been made toward understanding the functional groups involved in their formation, nature, and location along the molecules [3].

In human tissues, there are more than 15 different collagen types, most of which have been well characterized, and several others are being studied [4–6]. We know how collagen molecules in cartilage differ from those in cornea, tendon, bone matrix, dermis, the parenchyma of organs, periodontal ligaments, and many other locations within the organism. Some tissues contain almost exclusively one type of collagen (i.e., bone Type I, hyaline cartilage Type II), while others represent mixtures of collagens (skin is 80% Type I and 20% Type III, blood vessels are 60% Type I and 40% type III) (Figs. 4, 5). These properties can vary with age, location of the tissue, and disease. Whereas in certain tissues collagen fibers become permanent, almost irreplaceable fixtures, in others, (e.g., bone, which undergoes constant remodeling) collagen turns over rather significantly even in adult individuals.

Why is cornea transparent, tendon tough and inelastic and able to sustain significant stresses, and cartilage resilient and viscoelastic? What kind of uniqueness is there in the chemical structure of collagen that enables these tissues to perform such diverse biologic

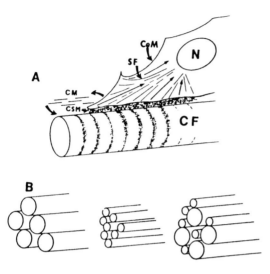

Figure 1 (A), A fibroblast secreting collagen molecules (CMs). Biomechanical and chemical factors outside the cell are sensed by the collagen fibrils (CFs), cell surface macromolecules (CSMs), receptors on the cell membrane, the cytoskeleton, and resulting stress fibers (SFs). A sequential process of message transduction eventually modulates the quantity and quality of collagen synthesized at the gene level within the necleus (N) of the cell. (B), First small collagen fibrils are deposited around the cell (lower center) that grow in size, revealing a mixed-diameter population of fibrils (lower right), and eventually give rise to the mostly mature large fibers seen in adult tissues (lower left).

functions? What regulates fiber diameter, orientation, concentration of fibers in a particular volume of tissue, and packing of the fibers into larger bundles? How can we understand the problem of aging of connective tissues and changes related to pathology? These are some of the fundamental questions being asked by biochemists, biologists, and ultrastructural anatomists that are leading to a better understanding of the large number of degenerative diseases involving connective tissues.

Degenerative arthritis, which affects the hyaline cartilage, atherosclerosis involving the vascular matrix, diabetes involving vascular and renal collagenous basement membranes, abnormal scar formation during wound healing, corneal and ocular abnormalities, impaired fracture healing, and a large number of heritable disorders of the connective tissue cause functional disabilities as a result of changes in the types, quantity, and distribution of collagen. Studies geared to understanding the etiology, pathogenesis and treatment of these diseases have led to a tremendous increase in our knowledge of the molecular structure, physical properties, biosynthesis, and metabolism of the collagen fibers. In addition, this information is starting to filter through into another area of growing interest: the use of various forms of collagen as bioprosthesis materials.

Figure 2 Cross-section of collagen fibers present in an extensor tendon of the foot of cows of different ages: (a), newborn; (b), 6 months; (c), 6 years. All photos are equally magnified. The larger fibers are approximately 300 nm in diameter (3000 Å).

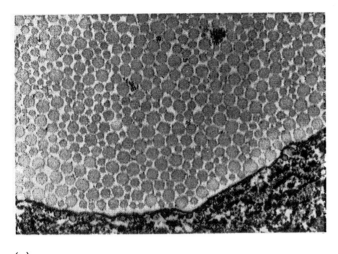

(a)

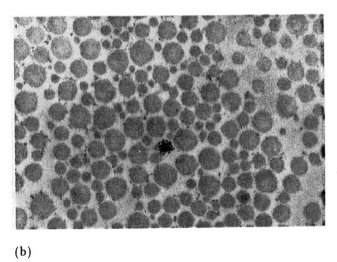

(b)

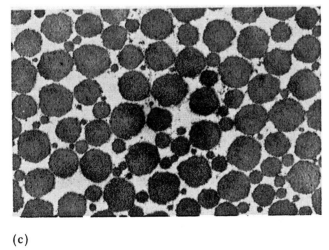

(c)

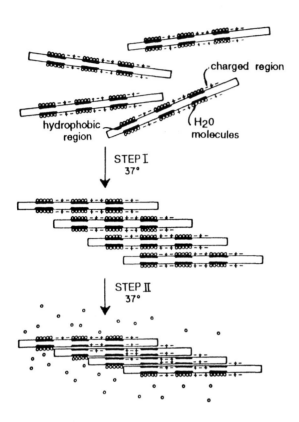

Figure 3 The nature of the forces associated with the early interactions of the helical regions of collagen via a mechanism of endothermic polymerization. This occurs *in vivo* and *in vitro* when the molecules still retain remnants of their telopeptides (i.e., nonhelical extensions), but can also occur *in vitro* with molecules devoid of such extensions. Soluble collagen can be extracted from most tissues by cold neutral salt solutions. If these solutions are warmed to 37°C, the collagen molecules reassemble into native fibers. The upper part of the drawing represents molecules that aligned in a quarter-staggered overlap. The alignment is primarily due to interactions of opposing electrostatic charges, depicted by + and −. As the temperature approaches 37°C, the hydrogen-bonded water molecules (○), clustered around the hydrophobic regions of collagen, "melt" and expose these nonpolar surfaces. Exclusion of water allows these surfaces to interact with each other, giving rise to hydrophobic bonds that greatly enhance the stability of the fiber. The driving force results from an increase of entropy of the system since the release of "organized" water from initial sites and its transformation into "random" water increase the disorder of the system. Part of the aging process that leads to a gradual insolubilization of collagen can be associated with this phenomenon, which continues to operate through the life span of an individual.

II. COLLAGEN AS A BIOMATERIAL

One of the earliest chemical modifications of collagen was that associated with the process of leather tanning [3]. This technology has evolved over the ages. This is true also for the by-products of native collagen fibers, which, in the form of denatured molecules, are used as glues and gelatins. During the last 20 years, increased interest has emerged in the use of collagen and collagen-containing tissues in the manufacturing of medical devices [7].

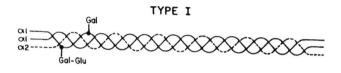

TYPE I

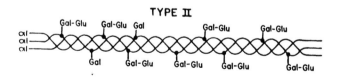

TYPE II

TYPE III

Figure 4 Diagram of the three interstitial types of collagen. Type I is present in skin, bone, tendon, and the like; Type II is present in cartilage; and Type III is present in blood vessels and developing tissues and as a minor component in skin and other tissues. There are differences in the chain composition and degrees of glycosylation. Disulfide cross-links are only seen in Type III collagen.

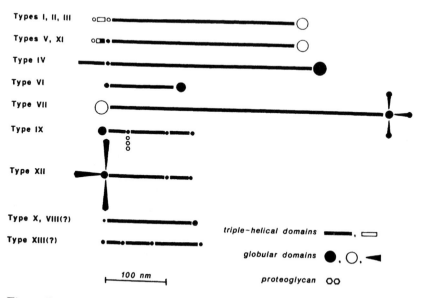

Figure 5 Collagens contain triple-helical (solid and open rods) and globular domains (open and filled circles). Portions of the initially synthesized molecules are removed prior to their incorporation into insoluble matrices (open rods and circles) and the rest of the molecule remains intact in the matrix (closed circles and rods). The domains and their distributions are drawn approximately to scale. (Courtesy of Dr. Robert Burgeson.)

There are basically two fundamental approaches that are followed in this connection. One involves the use of collagen-rich tissues, usually structural in nature, that are treated chemically in order to transform them into implantable prostheses. Examples of these are heart valves, tendons, ligaments, blood vessels, and pericardium. We briefly elaborate on the handling and preservation of tissues while discussing examples that have resulted in devices of significant value.

Another approach involves the use of purified collagen, obtained from animal tissues (mostly from bovine skin), processed in a variety of ways to generate a large number of products that not only have applications in the medical field, but also in the manufacturing of cosmetics. Collagen is used in the form of native soluble collagen, enzymatically processed native collagen, soluble collagen reconstituted into fibers, microfibrillar collagen, collagen-derived peptides, and so on. Products manufactured from these sources are used as hemostatic powders and dressings, dermal implants, implantable drug delivery vehicles, occluders of vascular flow, collagen threads, tapes, sponges, tubes, and sutures. We discuss some of these items, their advantages and disadvantages over synthetic polymers, and compare the use of prostheses made from purified collagen with those derived from natural animal tissues cross-linked and stabilized to reduce their antigenicity and preserve their native mechanical properties.

Finally, we speculate into other areas in which collagen may not have been used yet but in which significant potential exists for the development of useful implantable devices.

III. BIOPROSTHESES USING CROSS-LINKED NATURAL TISSUES

Primary examples of bioprostheses using cross-linked natural tissues are porcine heart valves, heart valves made out of bovine pericardium, bovine tendons used as prosthetic ligaments, and bovine pericardium tissue patches. These tissues are used as such with very little, if any, prefixation treatment. They are removed from animals at the slaughterhouse, cleaned from adherent tissue, and rapidly transported in cold saline to the manufacturing site. Cross-linking is usually performed using dilute solutions of buffered glutaraldehyde (0.2% to 1%).

A. Porcine Heart Valves

Among the major advances in surgery during this century has been our ability to replace diseased heart valves or blood vessels with prosthetic devices or tissue grafts. The earliest heart valves available for commercial use were mechanical ones, essentially a ball in a metal cage. After initial studies with homograft transplants, toward the latter part of the 1960s experimentation began with chemically tanned porcine heart valves. In 1969, it was found simultaneously by Alain Carpentier in Paris and Warren Hancock and myself in Southern California that porcine heart valves could be fixed with the dialdehyde glutaraldehyde, and that the cross-links introduced were chemically and biologically stable and could give rise to an essentially nonimmunogenic graft (Fig. 6).

Carpentier and his associates combined glutaraldehyde with sodium metaperiodate to destroy the interfibrillar ground substance (proteoglycans) and also to generate new sugar-derived aldehydes with potential cross-linking ability. Peridate treatment was subsequently abandoned since it seemed to contribute to instability of the cross-linked matrix.

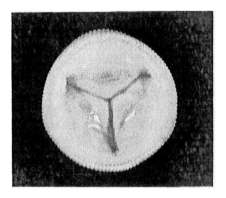 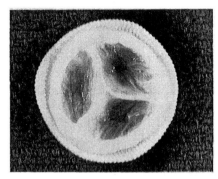

Figure 6   Inflow and outflow aspects of a porcine heart valve, cross-linked with glutaraldehyde and mounted on a polypropylene stent covered with a Dacron sewing ring.

In our studies, we used either glutaraldehyde alone (which I had shown earlier could restore cross-links to collagen devoid of cross-links synthesized in the presence of aldehyde-trapping agents) or glutaraldehyde followed by sodium borohydride to stabilize the newly formed synthetic cross-links. The last step proved unnecessary and possibly damaging since small bubbles of nascent hydrogen could disrupt the structure of the connective tissues, and was therefore abandoned. Later, other tissues (e.g., dura matter and pericardium) were used to reconstruct heart valves.

An advantage of the pericardial reconstituted valves over the porcine valves seemed to be a better hydrodynamic performance due to the improved configuration of the implant, since the porcine valve retains muscular tissue around its annulus, which prevents it from opening completely when used as a heterograft. On the other hand, the advantages of the porcine valve seem to reside on the conformation of the native cusps, which are supported by collagenous bundles or struts that allow them to retain their natural shape more readily. The pericardial tissue has to be "molded" and fixed by glutaraldehyde. It is also thicker, and therefore more resistant to initial tears, and there is some uncertainty about its ultimate durability and performance.

Nevertheless, both the porcine and pericardial reconstituted valves provide a useful prosthesis for hundreds of thousands of recipients. About half of the valves implanted throughout the world are made from natural tissues cross-linked by glutaraldehyde, while the remainder are mechanical in nature and made out of synthetic materials (floating ball, tilting disks, etc.).

Selection of a particular valve depends on many factors, but younger patients are more likely to receive mechanical valves, while older individuals are usually implanted with tissue valves. The major advantage of the mechanical valve is its proven durability. The advantages of the tissue valves are that the patients do not require anticoagulation therapy, and the fact that the problems that tend to show up are gradual and allow for reoperation. Calcification around collagen is a recurrent problem in natural tissue valves, and is more prevalent in younger people. The major problem of the mechanical valve is sudden failure due to thrombosis; therefore, in developing countries or in areas where anticoagulation monitoring is not feasible, tissue valves become the choice.

A significant effort is ongoing to devise the perfect valve, which may very well turn out to be a modified tissue valve. Improvement in flow characteristics, increasing the flexibility of the tissue through long-range cross-links, introducing of "lubricating macro-molecules" into the matrix of the cusp, increased durability by alternate modalities of cross-linking, generation of antithrombogenic surfaces, and introduction of calcification inhibitors into the connective tissue matrix are being actively investigated [8]. It is of interest that collagen in living adult tissues has a relatively very low rate of turnover. Since what cross-linking essentially does is further reduce it (it is hoped to zero), adequate preservation of the collagen in the graft should yield a very durable implant. Other experimental approaches involve the use of synthetic polymer-collagen grafted compos-ites as well as valves manufactured entirely from reconstituted collagen, hydrogels, poly-urethanes, or their extruded polymers.

B. Vascular Grafts

There is a significant need to replace damaged or degenerated blood vessels. Dacron™ grafts are satisfactory to replace larger-diameter blood vessels. Autologous veins, on the other hand, substitute very well for nonfunctional coronary vessels. In this case, the graft adapts to its new environment and responds to pulsation and the increased hydrostatic pressure by "arteriolizing." This can be evidenced histologically and biochemically. Dr. L. Robert [3] in Paris showed that the smooth muscle cells of the viable transplant alter their biosynthetic pattern and begin to produce more elastin, a characteristic of arterial cells. With the large, woven Dacron vessel substitutes, initial leakage occurs that requires the graft to be preclotted before insertion. Binding of collagen and other macromolecules to the Dacron has been attempted to make this pretreatment unnecessary. It is reasonable to say that today we have adequate synthetic substitutes for the larger vessels. In the case of the small-diameter veins and arteries (less than 5 millimeters [mm]), we have to rely on autologous grafts but, if these fail or are unavailable, the patient has no other recourse available.

Attempts have also been made to extrude collagen, reconstitute it into tubular forms, and use animal blood vessels as vascular grafts. Because of the antigenicity of xenografts, these are cross-linked with a variety of bifunctional reagents (dialdehyde starch, glutaral-dehyde, etc.) to try to make them inert. Attempts are also made to select animal vessels of suitable diameter and sequentially solubilize the noncollagenous components, leaving behind the collagen framework. Examples are the bovine aorta, digested with the protease ficin, which removes the elastin and noncollagenous proteins, leaving behind essentially a collagen tube that is then cross-linked. Some problems with these grafts are primarily related to the thrombogenicity of collagen. Even if this is overcome by adding a non-thrombogenic intima, a problem will persist at the anastomosing site.

It is always difficult to generate a suitable continuum where a synthetic material (or a nonviable collagenous matrix in this case) meets with living tissue. Surfaces that facilitate cellular ingrowth may be desirable. These could be made of biodegradable collagen com-bined with permanent cross-linked collagen or a polyurethane scaffolding. In particular, the anastomosis region will have to be further developed so as to allow for ingrowth and attachments in order to generate a continuous interphase between host and graft. This is of major concern in the small-diameter graft since overgrowth in this instance causes narrowing of the lumen, intravascular thrombosis, and finally leads to occlusion. Combi-nations of collagens of various degrees of cross-linking with cell-attachment proteins such as fibronectin and laminin may prove to be of value, and are being explored.

C. Tendons and Ligaments

Tendons and ligaments are essentially all collagen. They have a sparcity of cells, relatively few blood vessels, and small amounts of elastin. The collagen fibers are aligned parallel to the long axis in order to withstand mechanical loading, which at times can be quite significant (Fig. 7). Many attempts have been made to find replacements for these structures. Various synthetic polymers have been tested and, in the last few years, composites of carbon fibers and other materials have been explored.

Ligament-assist devices (LADs) have been designed that are supposed to act as transient scaffolds for the host to regenerate its own tissue. Sometimes, these are seeded with host-derived fibroblasts or coated with fascia from the recipient or various proteins, including collagen, or with resorbable materials such as polylactic or polyglycolic acids. Unfortunately, the granulation tissue formed under these circumstances is not organized to withstand force. It seems as if, in the presence of a device that already supports a load, the invading cells are unwilling to produce an organized, mechanically stable structure. Deposition of collagen fibers parallel to the direction of stress seems to require intermittent forces directly borne by the extracellular matrix and enclosed cells. So, when the

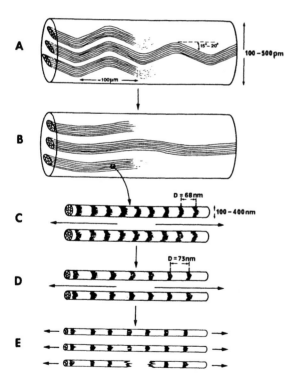

Figure 7 (A), Tendon fascicle showing bundles of collagen fibrils (fibers), arranged in a zig-zag fashion, that is responsible for the typical birefringence pattern generated by tendons; (B), stretching to the end of the toe region of the stress-strain curve eliminates the crimp; (C) and (D), further strain is accompanied by an elongation of the collagen fibers, which causes an increase in the normal periodicity from 68 nm for the fresh tissue to 73 nm in that stretched just prior to failure; (E), failure results from fiber slipping and breaking.

device fails, there is nothing to really take its place. Reconstituted collagen has also been used, but is probably not as good as the native fibrillar structures present in tendons.

D. Collagen Powder

Native insoluble collagen obtained from tissues rich in this protein, such as cowhide corium or reconstituted unsoluble collagen fibers, can be disintegrated into small particles, yielding a white, dry powder used in many medical, pharmaceutical, nutritional, and industrial areas. In general, cow skins, split below the hair follicles and above the flesh line, are used as a source of this material. The tissues are washed and cut into small pieces, allowed to swell in a variety of solvents, and exposed to an acid pH.

Under these circumstances and aided by mechanical blending, electrostatic repulsion and hydration causes the fibers to swell and disintegrate into small microfibrils that retain many of the physicochemical characteristics of native collagen fibers. This material can be lyophilized and reconstituted into a variety of forms with different textures and shapes. When suspended in physiological solutions, it is capable of forming a highly viscous gel that still retains many of the surface characteristics of native collagen fibers.

In the form of a fluffy, white powder or in the form of sheets, it has found many applications. It can be used as a hemostat and tissue adhesive when the powder is placed dry on exposed or wounded biological surfaces. If held in place by compression, hemostasis occurs, mediated by the known ability of collagen to aggregate platelets and initiate the clotting process. The degree of compacting of the collagenous material, the mode of application, the pressure applied, the degree of vascularization, and blood flow at the site of application are all important factors that relate to the effectiveness of the hemostatic effects.

Collagen powders prepared in this way have a variety of other applications, many of which have been explored with varying degrees of success. It has been mixed with hydroxyapatite to form synthetic "bonelike" materials, used as an excipient in preparing creams, pastes, and emulsions, and so on. Not always have the potential modes of application been designed in the context of what we now know about the biological functions of cells and tissues. As discussed in Section III.F, the bone matrix, although it is almost all collagen, is not the same as purified skin collagen, and therefore results differ when these two materials are used. Collagen applied directly over the skin can exhibit some characteristic physicochemical properties such as the ability to retain surface moisture or accumulate ions or act as a "glue," but cannot blend with the dermal collagen of the host since it is separated from it by an epidermal layer containing keratin, cells, and a basement membrane. Other approaches have to be used to generate a more lasting physiological effect. These should include new techniques to enhance transdermal penetration of collagen, its precursors, or agents that will stimulate the dermal cells to manufacture their own collagen.

E. Dermal Collagen Implants

The material obtained as described above is not suitable for permanent implantation because of its antigenicity, which would make it rejectable by the host. Even though collagen exhibits a relatively low degree of antigenicity because of marked similarities in amino acid composition and sequence among species, enough difference does exist to make the molecule antigenic. Various approaches have been used to decrease this antigenicity; these include enzymatic removal of the nonhelical extensions (which only account for 2–3% of the molecule) and insolubilization by the addition of synthetic cross-links.

The microfibrillar (or microcrystalline, as it is sometimes called) collagen described above retains these nonhelical extensions and is therefore antigenic; since after the hemostatic effect is displayed and tissue healing begins its presence is no longer required, this low level of antigenicity does not seem to be a major consequence.

When collagen is used as an implant to eliminate surface defects of the skin (such as seen following acne, certain age-related wrinkles, or other irregularities, depressed skin grafts, etc.), the effect derived should be permanent and rejection would be detrimental. For this reason, enzymatically degraded native collagen is used. Bovine skin collagen can be solubilized by limited digestion with pepsin or other proteolytic enzymes under conditions that preserve the helical domain of collagen (approximately 97% of the molecule) and removes only the nonhelical extensions present at both ends of the triple helix. These regions contain the more immunogenic sequences as well as most of the cross-links that stabilize the fiber. This renders collagen soluble and less antigenic. Such a collagen solution contains primarily collagen monomers, in contrast to the microcrystalline or microfibrillar preparations described above, which can be readily seen with the electron microscope.

After being purified by repeated reprecipitations at acid and neutral pH, they can be reconstituted into fibers. These fibers now lack the more immunogenic terminal extensions, but are not able to form native cross-links. Nevertheless, they can be cross-linked using bifunctional reagents. Such nativelike, extension-free, collagen molecules can be used as such or in the fibrillar or cross-linked fashion as an intradermal implant. Their use has gained popularity among dermatologists and plastic surgeons, but major reservations regarding residual immunogenicity and development of related autoimmune diseases cloud the picture.

F. Collagenous Bone Matrix as a Bone Filler

There are occasions when bone has to be replaced and the host cannot cope effectively, such as with alveolar bone loss following periodontal disease, gaps left by excised bone tumors, fractures associated with bone destruction, and accidental crushing of bone. When a large gap exists between viable bone (e.g., nonunion fractures), repair does not always proceed at a normal pace and may never occur.

Significant efforts are being made to enhance such healing using decalcified bone matrix as an inducer [9–11]. Through the pioneering work of Charles Huggins and his associates in Chicago, Marshall Urist in Los Angeles, and H. Reddi at the National Institutes of Health (NIH), we have learned that decalcified bone matrix, which is mostly collagen with small amounts of growth- and differentiation-inducing molecules, is able to stimulate bone formation, even in areas where bone did not previously exist (subcutaneously, intramuscularly, etc.). Only bone-derived matrix works in this connection since skin or other soft tissue collagens are devoid of such bone-inductive powers. Significant work is ongoing in a variety of institutions, which has led to the isolation and characterization of several bone-promoting factors associated with bone collagen matrix. Some of these have now been synthesized through recombinant ribonucleic acid (RNA) technology and are presently being tested clinically. Various forms of collagen are being used as carriers for these factors as their kinetics of release are very important in terms of their function.

G. Collagen as an Artificial Skin

In the mid-1970s, Dr. I. Yannas at the Massachusetts Institute of Technology began to experiment with fabrics made out of cross-linked gelatins and later developed composites of collagen and chondroitin sulfate, obtained from shark cartilage, which had interesting physicochemical and biological properties. Together with Dr. J. Burke at the Burn Units of the Massachusetts General Hospital and at the Shriners Institute in Boston, this concept was applied to the development of an artificial skin [12].

Many materials have been explored as a skin substitute, particularly that needed to cover extensively burned areas. Skin is a complex organ; therefore, many difficulties arise when trying to design a suitable substitute. Essentially, the composite prepared by Yannas and Burke consists of a "dermislike" layer made out of collagen and chondroitin sulfate, coprecipitated together at an acidic pH. The pore structure, chondroitin sulfate content, and cross-linked density are controlled to provide the required tensile strength, elasticity, tear quality, cellular ingrowth, and rate of biodegradation. This dermis, which has to be replaced by the natural dermis synthesized by the host, is covered with a synthetic "epidermis" made out of a silastic polymer. This silastic epidermal layer is temporary and designed to be removed later (usually after a few weeks or sooner), based on the patient's individual needs. At this time, it can be seeded or replaced by the patient's own epidermal cells.

The information learned from these experiences is of significant value since it has helped to understand how collagen is replaced, how cells migrate into the composite, how epidermal and mesenchymal cells relate to each other during the process of reconstituting the host skin, how newly synthesized collagen and proteoglycans are deposited, what controls such activities, and so on. Since many of the bioprostheses to be developed in the future will rely on tissue ingrowth, this skin model is helping us to understand how this mechanism is regulated and which are the most suitable substrates.

H. Problems Associated with Collagen Implants

Since collagen used for manufacturing devices is most often derived from animal tissues (bovine, porcine, etc.), the possibility of an antigenic reaction comes to mind. Nevertheless, this is rarely manifested with the devices so far manufactured, primarily because tissues or the derived products are cross-linked or treated enzymatically to remove the more antigenic parts of the collagen molecule. As discussed above, this problem is most significant in connection with the intradermal implant of collagen used for skin augmentation purposes. Encapsulation is a problem with most implanted materials, and is not particularly more severe with those made out of collagen. Collagen is a known nucleation surface for calcium and phosphate ions, and can contribute to the deposition of hydroxyapatite around the implanted fibers. Calcification has become a problem of some concern, particularly in younger individuals receiving collagen-derived bioprostheses.

The ultimate durability of implants has not been ascertained in some instances because of the fact that they are of recent development. Examples of problems that recur with some frequency are calcification and degeneration of the collagenous matrix of tissue heart valves, hypersensitivity to dermal implants, thrombosis of vascular grafts, and others. Most of these problems are likely to be solved, or at least considerably decreased, as our knowledge continues to increase.

IV. FUTURE DIRECTIONS

The framework of all our supporting tissues is collagen. Articular cartilage, which "wears and tears," can sometimes become extensively degenerated and require replacement. An example is osteoarthritis of the hip joint, which is now readily replaced with a metal prosthesis (ball-socket type). Other joints, such as elbows, shoulders, and knees, are not as accessible to correction, although knee prostheses are becoming more prevalent now. Collagen composites or adequately treated and molded animal cartilages may prove to be of value in the future. Blood vessels, nerve and tendon sheaths, tympanic membranes, corneas, ear and nose cartilage, and the trachea are all collagenous tissues that require replacement as a result of disease or trauma. Many of the interesting and sometimes specific properties of collagen, such as the potential for cell-attachment platelet aggregation, and the ability to initiate calcification, to generate chemotactic peptides to act as an inert or biodegradable support for chemically, enzymatically, or immunochemically active molecules are being further explored with the idea of generating new medical devices.

Scaffoldings for organ regeneration made out of nonimmunogenic collagens will be used. We are beginning to think in terms of "homograft equivalent matrices" that the host will repopulate with his or her own cells or that can be repopulated before implantation. Since the extracellular matrix contains, in addition to collagen, glycosaminoglycans (GAG) (negatively charged polysaccharide structures), in our laboratory we are currently exploring the use of such composite matrices (Fig. 8). We have recently synthesized a collagen-hyaluronan matrix in which the GAG is covalently coupled to the surface of the

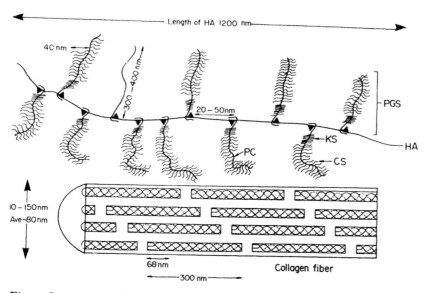

Figure 8   Collagen fibers do not exist in a vacuum. They are usually closely associated with the proteoglycans of the ground substance. In cartilage, for instance, a tissue rich in collagen and proteoglycans, one can see a collagen fiber adjacent to a proteoglycan aggregate containing hyaluronic acid (HA), proteoglycan subunits (PGSs), and link proteins (▲), which help to stabilize the structure. The PGSs consist of a protein core (PC) from which the negatively charged glycosaminoglycan chains of chondroitin surface (CS) and keratin sulfate (KS) radiate.

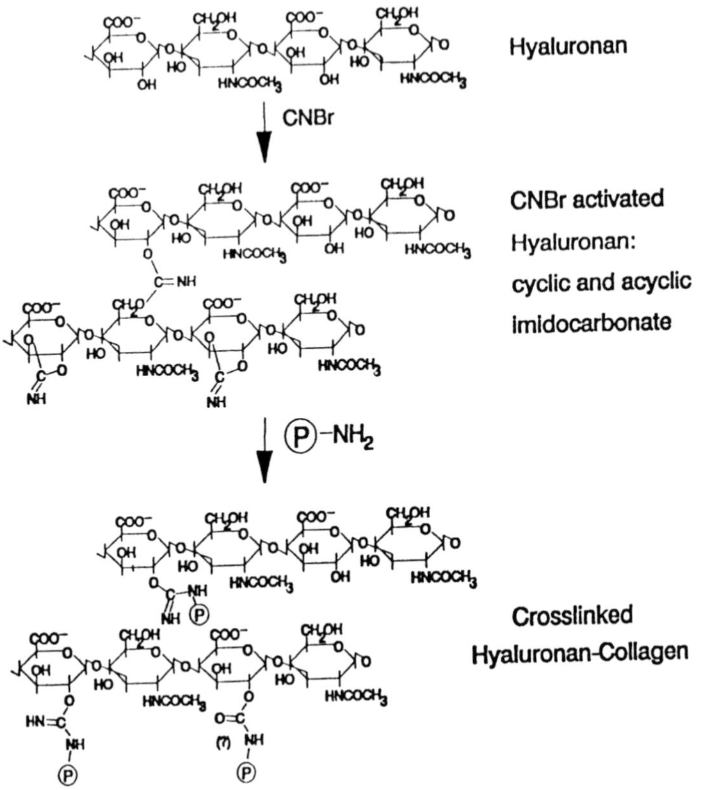

Figure 9 Illustration of the cross-linking reactions that occur between collagen and CNBr-(cyanogen bromide) activated hyaluronan (P = collagen polypeptide).

collagen fibrils [13–15] (Fig. 9). These matrices encourage fibroblast growth, but are not contracted by fibroblasts growing into their interstices. Such matrices made with nonantigenic collagens may prove of significant value for tissue repair and organ regeneration.

REFERENCES

1. Nimni, M., Collagen: Structure, function, and metabolism in normal and fibrotic tissues, *Seminars in Arthritis and Rheumatism*, 13:1 (August 1983).
2. Nimni, M. E., Collagen, structure and function. In *Encyclopedia of Human Biology*, Vol. 2, Academic Press, pp. 559–574 (1991).
3. Nimni, M. E. (Editor), *Collagen: Biochemistry, Biomechanics and Biotechnology*, Vols. 1–3, CRC Press, Boca Raton, FL (1988).
4. Nimni, M. E., and B. Olsen (Editors), *Collagen: Molecular Biology*, Vol. 4, CRC Press, Boca Raton, FL (1989).
5. Nimni, M. E., and A. Kang (Editors), *Collagen: Pathobiochemistry*, Vol. 5, CRC Press, Boca Raton, FL (1991).
6. Burgeson, R. E., and M. E. Nimni, Collagen types: Molecular structure and tissue distribution, *Clin. Ortho. Rel. Res.*, 282:250–272 (1992).
7. Nimni, M. E., D. Cheung, B. Strates, M. Kodama, and K. Sheikh, Chemically modified

collagen: A natural biomaterial for tissue replacement, *J. Biomed. Mat. Res.*, 21:741–771 (1987).

8. Nimni, M. E., D. Ertl, J. Villaneuva, and B. S. Nimni, Inhibition of ectopic calcification of glutaraldehyde crosslinked collagen and collagenous tissues by a covalently bound disphosphonate (APD), *Amer. J. Cardiovascular Pathology*, 3:237–245 (1990).

9. Nimni, M. E., S. Bernick, D. Ertl, S. K. Nishimoto, W. Paule, B. S. Strates, and J. Villanueva, Ectopic bone formation is enhanced in senescent animals implanted with embryonic cells, *Clin. Orthop. & Rel. Res.*, 234:255–266 (1988).

10. Urist, M. R., Bone morphogenetic protein. In *Bone Grafts and Bone Substitutes*, Edited by M. B. Habal and A. H. Reddi, W. B. Saunders, Philadelphia, PA, pp. 70–83 (1992).

11. Reddi, A. H., Symbiosis of biotechnology and biomaterials: Applications in tissue engineering of bone and cartilage, *J. Cellular Biochem.*, 56:192–195 (1994).

12. Yannas, I. V., and J. F. Burke, Design of an artificial skin. Basic design principles, *J. Biomed. Mat. Res.*, 14:65–81 (1980).

13. Huang-Lee, L. L. H., and M. Nimni, Fibroblast contraction of collagen matrices with and without covalently bound hyaluronan, *J. Biomater. Sci. Polymer Edn.*, 5:99–109 (1993).

14. Huang-Lee, L. L. H., and M. Nimni, Crosslinked CNBr-activated hyaluronan-collagen matrices: Effects on fibroblast contraction, *Matrix Biology*, 14:147–157 (1994).

15. Huang-Lee, L. L. H., J. H. Wu, and M. E. Nimni, Effects of hyaluronan on collagen fibrillar matrix contraction by fibroblasts, *J. Biomed. Mater. Res.*, 28:123–132 (1994).

36
Collagenous Biomaterials as Models of Tissue-Inducing Implants

Frederick H. Silver, George D. Pins, and Ming-Che Wang
University of Medicine and Dentistry of New Jersey, Piscataway, New Jersey

David Christiansen
University of Pennsylvania, Philadelphia, Pennsylvania

I. INTRODUCTION

Collagenous proteins form the structural framework of all mammalian extracellular matrices. Presently, 16 different molecular forms of collagen have been defined [1]. Fibrillar collagens (Types I, II, III, V, and XI) form structural elements 20 to several hundred nanometers (nm) in diameter that contain 1 or more collagen types [2]. The structural hierarchy and mechanical properties of collagen vary from tissue to tissue [3]. In dense, regular connective tissue such as tendon and ligament, parallel arrays of collagen fibrils are separated by cells and extracellular matrix into structural units, including fibril bundles and fascicles [3]. In dense, irregular connective tissue such as skin and dura mater, arrays of collagen fibrils or fibers form sheetlike structures that can be modeled as either woven or composite plied materials.

The ongoing objective of our research is to produce fibrous scaffolds *in vitro* from self-assembled collagen fibers with hierarchical structures and mechanical properties comparable to native connective tissue. These tissue analogs may be used as implant materials to facilitate ingrowth and repair of a variety of tissues or as model systems for evaluating cellular and noncellular events involved in the wound healing response.

To date, our laboratory has developed processes for formulating reconstituted analogs of dense regular and irregular connective tissue for replacement of soft-tissue structures, including skin, dura mater, tendons, ligaments, and peripheral nerves (Fig. 1). In addition, we have begun to develop mineralized forms of collagen to replace hard tissue. Below we review our progress in creating tissue-inducing biomaterials for each of these applications. Initial results suggest that the rate of tissue induction by collagen scaffolds is optimized if the scaffold degradation rate parallels the rate of wound healing in a particular anatomic site [4].

The Preparation of

Soluble and Insoluble Type I Collagen

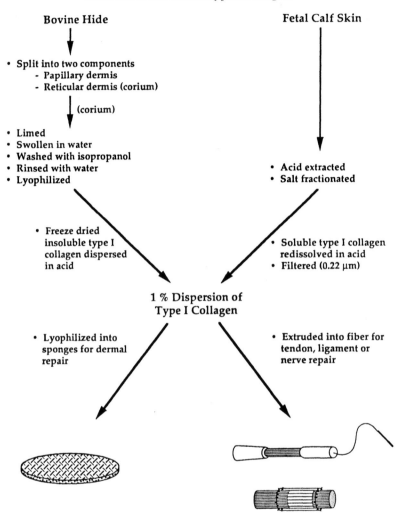

Figure 1   A flowchart describing the procedure for extracting and purifying a 1% dispersion of Type I collagen from bovine hide (insoluble collagen) and fetal calf skin (soluble collagen). The 1% collagen dispersion is subsequently engineered into a number of geometries and used in different tissue-inducing devices.

II.   SCAFFOLDS FOR SKIN REPAIR

A.   Background

Skin ulceration caused by external pressure, vascular insufficiency, and diabetes affects almost 8 million people in the United States alone [5]. In addition, replacement of skin damaged as a result of thermal or mechanical trauma is a health problem that, if left untreated, may result in systemic infection and mortality. Therefore, it is important to develop methods to enhance the rate of healing of chronic wounds or burns.

Advances in tissue engineering of skin have been a consequence of the development of collagenous matrices that induce cellular ingrowth and new extracellular matrix deposition. This work is well documented and has recently been reviewed [6]. Experimental results, reported in the literature, suggest that cell differentiation and orientation are enhanced in the presence of collagen. Extensive progress in cultivation of autogenous and allogeneic cells *in vitro* on collagen has led to advances in transplantation of cultured materials for treatment of skin loss. Collagenous matrices have also been used to promote the healing of chronic skin ulcers [7–9]. Doillon et al. demonstrated that fibrillar collagen sponges, when used to treat full-thickness wounds, facilitated fibroblast migration and acted as a template that allowed for the organized spatial deposition of newly synthesized collagen fibers [10]. These observations support the concept that a collagenous scaffold with a hierarchical structure comparable to native dermis will enhance dermal repair.

B. Hierarchical Structures and Mechanical Properties of Skin

Skin is a dense, irregular connective tissue that acts to protect underlying tissues and organs from mechanical and electrical injury; it acts as a barrier to foreign chemicals and bacteria and prevents excessive loss of heat and moisture. Morphologically, skin is regarded as a bilayered material.

The superficial or epidermal layer of the skin forms an uninterrupted layer that limits diffusion into and out of the dermis. It is composed primarily of keratinocytes (epithelial cells). These cells begin as cuboidal epithelium that differentiate and take on features and characteristics of cells with specific functions as they move up toward the epidermis-air interface. At the air-skin interface, these cells are sloughed off. The epidermal layer comprises about 10% of the skin thickness and contributes little to its mechanical properties.

The underlying layer, the dermis, contains collagen fibers in the papillary (upper dermis) and reticular (lower dermis) layers. These collagen networks align with the direction(s) of tensile loading, allowing skin to bear loads in any direction within its plane [3]. In the papillary layer, small-diameter collagen fibrils of Types I, III, and VII form a nonwoven structure almost parallel to the surface of the skin. The papillary dermis also contains Type VII collagen fibrils that connect the dermis to the epidermis. In the reticular dermis, the collagen fibrils and fibers (Types I and III) have large diameters. Table 1 contains compiled experimental results that describe the distribution of collagen fibers in skin. Small quantities of elastic fibers are also found in this layer.

The mechanical properties of skin have been recently reviewed [3]. The ultimate stress, modulus, and strain at failure range from 2 to 15 MPa, 6 to 35 MPa, and 50% to 200%, respectively [3]. These material properties should be matched by a tissue replacement in order to insure that stress concentration does not occur at the interface between the implant and host.

C. Collagenous Scaffold Research

Research on collagen scaffolds to enhance dermal repair has been directed at development of matrices to promote granulation tissue deposition in order to facilitate skin graft acceptance (large wound areas) or reepithelialization (small wound areas).

Many researchers have investigated the effects of collagenous biomaterials on dermal

Table 1 Structural Hierarchy of Skin

Component	Size
Skin	1.0–1.3 mm (thickness)
	0.8–1.0 mm (thickness)
Epidermis	38.8 μm (cheek) (thickness)
	429 μm (palm) (thickness)
Dermis	0.20–0.80 nm (thickness)
Elastic fiber	10–20 μm (diameter)
Microfibril	10–12 nm (diameter)
Collagen fiber	
Papillary	0.3–3.0 μm (diameter)
Reticular	10–40 μm (diameter)
Collagen fibril	20–40 nm (diameter)
Papillary	60–100 nm (diameter)
Reticular	89–75 nm (diameter)

Source: From Ref. 3

repair and wound healing (see Ref. 6 for a recent review). Oliver et al. reported results of cutaneous implantation studies on trypsin-purified dermal collagen [11]. They found that the implanted collagen was progressively lysed and replaced by granulation tissue. Remnants of the implant persisted up to Day 35 but had disappeared by Day 50. Oliver et al. later pretreated dermal collagen with various concentrations of glutaraldehyde to preserve collagen bundle architecture and to limit collagenolysis [11]. Glutaraldehyde cross-linked collagen was later shown to be cytotoxic to fibroblasts [12]. Yannas et al. reported the use of a collagen bilayer polymeric material to provide closure of full-thickness skin wounds [13]. The membrane was composed of a top layer of silicone elastomer and a bottom layer of a porous cross-linked network of collagen and glycosaminoglycan. They reported that a functional extension of skin over the entire wound area was formed in about four weeks.

Our laboratory has focused on the development of nonaldehyde cross-linked collagen matrices that promote rapid formation of wound tissue possessing structural and mechanical properties similar to normal dermis [10,14,15] (Fig. 2). Collagen matrices prepared by freeze drying consisted of arrays of fibers that were almost parallel to the surface and were separated by pores that had their largest dimension almost parallel to the surface [14]. Optimum conditions for collagen matrix formation included collagen dispersion preparation at pH 3.0, a dispersion freezing temperature of −30°C, and good heat transfer between the freezing tray, the dispersion, and the freezing solution. The product of such a process is characterized by an average pore size of 100 ± 50 micrometers (μm) containing channels between pores [14].

Results of other studies show that tissue ingrowth and rapid fibroblast migration occurred when the pore sizes in the collagen sponges were between 50 and 250 μm [10]. Nonaldehyde cross-linking of collagen matrices was achieved by severe dehydration at elevated temperatures combined with exposure to cyanamide, a water-soluble coupling agent [16]. Severe dehydration and cyanamide exposure results in cross-links involving formation of synthetic peptide bonds that support tissue ingrowth in an animal model [16].

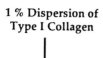

1 % Dispersion of
Type I Collagen

- Glycoproteins (e.g. HA, FN)
  mixed with collagen

- Mixture poured into Petri dishes

- Lyophilized and then crosslinked
  into collagen sponge

- Growth factors (e.g. bFGF)
  and cells (e.g. fibroblasts,
  epidermal cells) added to sponge

- Sponges sutured into full
  thickness wounds

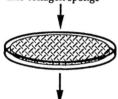

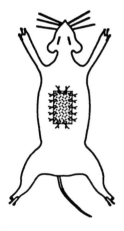

Figure 2   A flowchart describing the production of collagenous matrices used for the repair of dermal injury. Type I collagen mixed with glycoproteins is freeze dried into sponges, then infiltrated with growth factors and cells. The matrix is then trimmed to conform with the geometry of the wound bed and then sutured in place.

When collagen sponges were grafted onto full-thickness, excised dermal wounds in a guinea pig model, the matrices promoted cellular infiltration within the pores of the sponge (parallel to the surface) (Fig. 3). The collagenous matrix also encouraged the deposition of new granulation tissue that was oriented parallel to the pores of the sponge and lacked the wavy pattern of collagen seen in scar tissue below the sponge. These results suggested that the matrix acts as a template that allows for the organized spatial deposition of newly synthesized collagen [17].

Additional studies indicated that incorporation of hyaluronic acid and fibronectin into collagen matrices resulted in enhanced cellular proliferation and migration as well as an increased deposition of newly synthesized collagen into the matrices [10,15,17,18]. When collagen sponges were coated with basic fibroblast growth factor (bFGF), epider-

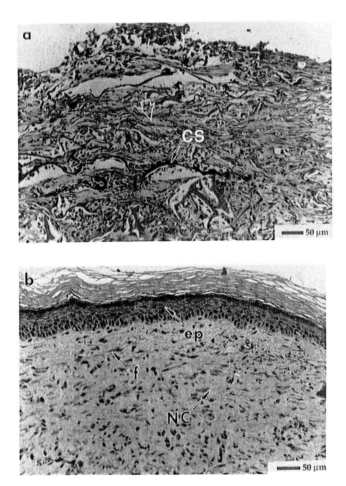

Figure 3   Light micrographs of full-thickness wounds treated with a collagen sponge: (a), 15 days postimplantation; and (b), 60 days postimplantation. Remnants of the collagen sponge (CS) are visible at 15 days as well as newly synthesized collagen (arrow) laid down in the pores of the sponge. At 60 days, the wound is covered by new collagen (NC), fibroblasts (f), and a regenerated epidermis (ep). Note the lack of hair follicles and other skin appendages.

mal cell attachment, proliferation, and differentiation into stratified layers were noted [6,19].

In the absence of a collagen sponge, the tensile strength of large-diameter excised dermal wounds is roughly proportional to the diameter of the collagen fibers in the wound tissue [20]. An early study showed that, as the fiber diameter increased from about 2 μm at 15 days to 4 μm at 30 days postwounding, the tensile strength of the corresponding wound tissue grew from 0 to 0.5 MPa [20]. A later study demonstrated that open wounds treated with collagen matrices showed increases in ultimate tensile strength from about 0.2 to 0.35 MPa 15 days postwounding and the collagen fiber diameter in the upper dermis increased from about 1.2 μm to 1.8 μm [19]. These results are consistent with the conclusion that the presence of a collagen matrix promotes the deposition of the newly synthesized collagen fibers with resultant increases in fiber diameter and tensile strength of the wound tissue. Incorporation of bFGF into the collagen sponge increased wound tensile strength 15 days postwounding without affecting collagen fiber diameter [19]. This finding may be attributed to increased levels of collagen deposition within the matrix in the presence of fibroblast growth factor.

After results of animal studies demonstrated that collagen matrices promoted deposition of large-diameter collagen fibers and facilitated closure of small open wounds, these materials were used clinically to promote granulation of pressure sores [7]. Results of clinical studies suggested that daily treatment of Stage II and III pressure sores with Type I collagen resulted in a 40% to 50% wound area reduction after 6 weeks in 85% of patients [7].

For large-area wounds such as burns, cell-seeded collagenous biomaterials promote increased wound tensile strength and improved healing by stimulating the deposition of mature collagen networks. Collagenous matrices seeded with fibroblasts and epidermal cells appear to be the long-term solution to regaining normal physiological function and appearance for patients with large open wounds or burns of an extensive surface area [6]. When implanted in large open wounds in guinea pigs, fibroblast-seeded collagen sponges show increased tensile strength 15 days postwounding and more mature collagen network formation [19]. In addition, these fibroblast-seeded collagen matrices promoted improved reepithelialization when used to treat full-thickness wounds [19].

In addition to wound healing studies, our laboratory has used collagen matrices to evaluate cellular and noncellular wound healing responses *in vitro*. Doillon et al. cultured embryonic chick fibroblasts on collagen sponges in order to examine cell-matrix interaction and new collagen deposition [17]. These studies demonstrated that, in the presence of a fibrillar matrix, fibroblasts adhered, attached, and oriented along the matrix fibrils. It was also noted that the matrix acted as a template that allowed for the organized spatial deposition of newly synthesized collagen.

In vitro cell culture studied conducted on collagen sponges containing fibronectin (FN) and hyaluronic acid (HA) indicated that matrices with these glycoproteins further enhanced fibroblast infiltration and alignment as well as collagen deposition when compared with cells cultured on sponges alone or plastic [15,17,21]. When guinea pig epidermal cells were cultured on collagenous matrices, they proliferated and differentiated into multiple stratified layers [7,21,22]. In addition, it was noted that the collagen matrix enhanced epidermal cell attachment and spreading [7,21,22].

Finally, in researching the production of an artificial skin, Doillon et al. determined that epidermal cell replication was enhanced in the presence of collagen sponge seeded with fibroblasts [21]. The results of these investigations corroborate the findings of parallel *in vivo* studies discussed above and demonstrate that the collagenous matrices provide an effective mechanism for predicting implantation responses.

D. Summary

Collagen matrices synthesized from Type I collagen can be used to facilitate rapid repair of small and large skin defects. Regardless of the defect size, the collagen matrix promotes rapid fibroblast migration and acts as a template that allows for the organized spatial deposition of newly synthesized collagen, increased fiber diameters of newly synthesized collagen, and increased wound tensile strengths 15 days postwounding. Closure of small wounds such as skin pressure sores can be achieved using Type I collagen to promote wound granulation that is followed by spontaneous reepithelialization.

For large-area wounds such as burns, collagen matrices seeded with fibroblasts and epidermal cells result in increased wound tensile strength and improved healing by promoting the deposition of mature collagen networks. In addition, implantation studies using animal models show that collagen sponges treated with matrix glycoproteins (fibronectin and hyaluronic acid) and growth factors (fibroblast growth factor) also enhance wound healing by promoting rapid cell migration and new collagen deposition and reepithelialization.

Finally, the results of numerous studies suggest that collagenous matrices may be used to evaluate cellular and noncellular wound healing responses *in vitro*, thus providing an effective mechanism for predicting implantation responses.

III. SCAFFOLD FOR TENDON/LIGAMENT REPAIR

A. Background

The anterior cruciate ligament (ACL), the major intra-articular element in the knee, is responsible for maintaining joint stability by limiting extension and rotation of the tibia with respect to the femur. Since the knee joint is among the most commonly injured during athletic activity, the ACL is a frequently compromised ligament. Many orthopedic surgeons feel that compensation for ACL deficiencies can be accomplished by strengthening the muscles that control the joint, but research has shown that untreated ACL injuries can result in chronic pain, limited athletic participation, long-term joint misalignment, and an increased risk of meniscus damage and osteoarthritis [23]. Consequently, there is great need for a treatment to repair injured ligaments.

Repair of the damaged ACL has been accomplished using a number of well-documented techniques [24]. Briefly, replacement of the injured ACL with autologous tissue such as the illiotibial band, semitendinosus and gracilis tendons, and patellar tendon and meniscus has led to restoration of normal joint function but necessitates that other tissue constructs be sacrificed. Allograft and xenograft prostheses have also been employed successfully, but foreign-body tissue responses and disease transmission present continued problems. Finally, a number of synthetic implants, including the Integraft (Osteonics Biomaterials, Livermore, CA), Leads–Keio ligament, Gore-Tex™ ligament (W. L. Gore and Assoc., Flagstaff, AZ), and Kennedy LAD™ (ligament-augmentation device) (3M Corp., Minneapolis, MN), have either not been approved or have received provisional acceptance from the Food and Drug Administration (FDA) for use in salvage cases. Consequently, there is still a demand for an "off-the-shelf" tendon/ligament replacement.

Recently, collagen fibers were reconstituted from mature, corium-derived, insoluble

Type I collagen and subsequently processed into fibrous, collagenous biocomposites as scaffolds for the repair of tendon and ligament injuries [4,25]. When used as Achilles tendon or ACL replacements, these prostheses enhanced tissue remodeling in addition to promoting rapid ingrowth of neotendon or neoligament with fibrillar substructures similar to native tissue [26–28].

B. Hierarchical Structure and Mechanical Properties of Tendon and Ligament

Tendons and ligaments are multicomponent, cablelike elements that cyclically transmit force in the absence of permanent dimensional changes [3]. Models vary for different tendons and ligaments; however, the generalized structural hierarchy for tendons and ligaments is described with compiled experimental results in Table 2.

Tendons and ligaments are composed primarily of structural units termed *fascicles* containing crimped, aligned, collagen fibril bundles or fibers that are surrounded by a paratenon or endotendineum membrane. The membrane sheath is derived from Types I, III, and IV collagen as well as elastin and proteoglycans [29]. The individuals fascicles are bound by a collagenous membrane called an *epitenon* that allows free sliding between the fascicular units. Within the fascicles, collagen fibrils, deposited as discontinuous segmented structures, laterally associate to form fibril bundles [30], which are recognized as fibers at the light microscopic level [31]. Although the fibrils appear almost parallel, they have been observed to branch and rotate [32]. Diameter distributions of collagen fibrils vary as a function of age [33] and may be affected by the presence of dermatan sulfate proteoglycan found in orthogonal arrays around collagen fibrils and attached to

Table 2 Structural Hierarchy of Tendons and Ligaments

Component	Size
Tendon/ligament	
Tail tendon (rat)	200–350 μm (diameter)
Extensor tendon (horse)	5.1 mm (diameter)
Anterior cruciate ligament	
Human, ages 48–86	57.5 mm^2 (area)
Human, ages 16–26	44.4 mm^2 (area)
Fascicle	>250 μm (diameter)
	20–400 μm (diameter)
	80–320 μm (diameter)
Collagen fiber (fibril bundles)	Up to 300 μm (diameter)
	1–20 μm (diameter)
Collagen fibril	20–50 nm (diameter)
	150–250 nm (diameter)
	0.05–0.5 μm (diameter)
	318 nm (diameter)
	189 nm (diameter)
	40 nm and 280 nm (diameters)
	185 nm (diameter)
Crystalline domain	3.8 nm (diameter)

Source: From Ref. 3

the d band [34]. The ACL is twisted 90° from the femoral to tibial attachment, providing an additional level of structural complexity.

The mechanical properties of tendons and ligaments can be estimated from uniaxial stress-strain measurements. Stress-strain profiles of these tissues are almost linear after an initial nonlinear toe region. Tendons and ligaments have been characterized by ultimate tensile strengths and strains ranging from approximately 11 to 100 MPa and 5% to 71%, respectively [3].

C. Collagen Fiber Prostheses for Tendon/Ligament Repair

Our laboratory has developed a technique for the production of high-strength collagen fibers [35] under optimum conditions for collagen self-assembly [36]. By extruding insoluble Type I collagen into neutral salt buffer maintained at 37°C, collagen fibers are formed that exhibit D periodic banding characteristic of native collagen. These fibers can then be stabilized with glutaraldehyde vapor or by subjecting them to severe dehydration and carbodiimide exposure as described in Section II.C [16]. Approximately 230 of these cross-linked collagen fibers were aligned longitudinally and coated with un-cross-linked collagen to produce implants for Achilles tendon and ACL replacement [26–28].

Fibrous collagen implants were initially investigated for repair of tendons in white New Zealand rabbits [26]. These prostheses were prepared as described above from collagen fibers that were cross-linked either with glutaraldehyde or with dehydrothermal treatment followed by exposure to carbodiimide. In conducting the study, the Achilles tendon in one leg of the rabbit was replaced with a collagen prosthetic and the same tendon in the contralateral leg of the animal was excised, devascularized, and anastomosed as an autogenous graft (Fig. 4).

Histological evaluations of the collagenous prostheses and the autografts were made 3, 10, and 20 weeks after implantation. Autogenous grafts showed slow revascularization within 3 weeks after implantation, and at 10 weeks cellular and vascular invasion from surrounding tissue had begun. Postoperatively, the autograft had been partially replaced at 20 weeks by repair tissue and exhibited reduced fibroblast cellularity. All collagenous implants were invaded with inflammatory cells and new connective tissue possessing abundant fibroblasts and capillaries 3 weeks after implantation. At 10 weeks, only remnants of the carbodiimide cross-linked implant remained. It had been replaced by repair tissue containing aligned, crimped collagen-fiber bundles. In contrast, the glutaraldehyde implant had changed little by 10 weeks. Inflammatory cells surrounded it and an increase in new connective tissue and a thin fibrous capsule were observed.

The carbodiimide implant was completely resorbed and replaced with a mature neotendon 20 weeks postoperatively (Fig. 5). The neotendon contained crimpled collagen fibers, its cellularity had decreased to that of a normal tendon, and the inflammation associated with the implant had disappeared. The glutaraldehyde implant showed little reorganization of the newly dispersed collagen after 20 weeks. Glutaraldehyde-treated implants appeared to generate a foreign-body reaction.

When the implants were examined at intervals after the operation, their mechanical properties approached those of fresh tendon. The initial strength of the carbodiimide-treated implants was lower than that of the fresh autogenous grafts. The strength and modulus of the carbodiimide-treated implants approached those of fresh tendon 20 weeks after implantation. At the same time, the autogenous tendon grafts were significantly weaker and more extensible than the carbodiimide-treated implants. The initial strength

A.

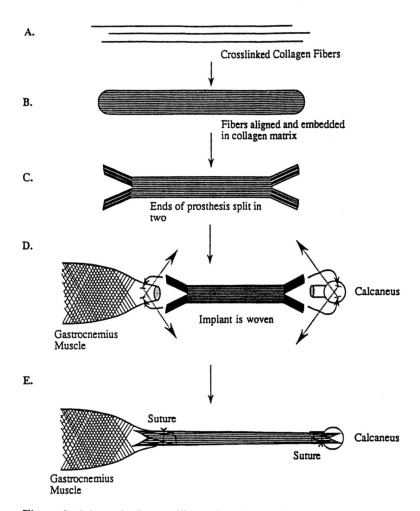

Crosslinked Collagen Fibers

B.

Fibers aligned and embedded
in collagen matrix

C.

Ends of prosthesis split in
two

D.

Calcaneus

Implant is woven

Gastrocnemius
Muscle

E.

Suture

Calcaneus

Suture

Gastrocnemius
Muscle

Figure 4 Schematic diagram illustrating the experimental steps used in preparing and implanting tendon prostheses. Cross-linked collagen fibers (A) are aligned and embedded in a non-cross-linked collagen matrix (B). After air drying, the ends are split (C). They are woven with sutures into the gastrocnemius muscle at one end and are wrapped around the calcaneous at the other end (D and E). (From Ref. 27.)

and modulus of the glutaraldehyde-treated implants were higher than those of fresh tendon, but 20 weeks after implantation the properties of the glutaraldehyde-treated implants were similar to those of devascularized autogenous tendon graft.

In summary, preliminary results indicated that, by 10 weeks after implantation, carbodiimide cross-linked implants had been replaced by neotendon in a manner that was similar to the autogenous tendon control grafts. Also by 10 weeks after implantation, glutaraldehyde cross-linked collagen implants were encapsulated and appeared to have caused a chronic inflammatory response. Finally, the histological and mechanical results together suggest that the rate at which strength and modulus of the neotendon increase correlates with the rate of biodegradation of the implant [26].

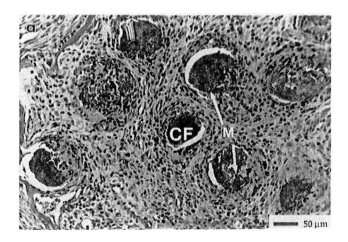

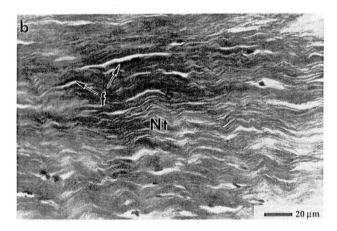

Figure 5   Light micrographs of neotendon laid down on a carbodiimide cross-linked collagenous tendon replacement of Achilles tendon: (a), 3 weeks postimplantation; (b), 52 weeks postimplantation. At 3 weeks, collagen fibers (CFs) of the prostheses are observed to be partially digested by mononuclear (M) cells. At 52 weeks, crimped neotendon (Nt) with a few scattered fibroblasts (f) are observed to replace the prostheses.

In a follow-up study, carbodiimide and glutaraldehyde cross-linked collagen implants as well as autogenous grafts that served as controls were implanted for 52 weeks as replacements for a 3-centimeter (cm) section of Achilles tendons in rabbits [27]. The absence of a crimp in a cross-linked implant and the presence of a crimp in normal tendon and in neotendon that formed after an implant had been resorbed made it possible to distinguish between a cross-linked implant and new host tendon. New collagen that had replaced the implant and autogenous (control) tendon graft were compared with normal Achilles tendon with respect to the angle and length of the crimp. The autogenous grafts and the carbodiimide cross-linked collagen implants had been completely resorbed and replaced by neotendon. The neotendon that was present 52 weeks after implantation was morphologically similar, but not identical, to normal tendon. In contrast, the glutaraldehyde cross-linked implant had not been resorbed, and was surrounded by a fibrous capsule with a relatively high cell density. The neotendon in the capsule was also similar, but not identical, to normal tendon. The fibrous capsule and relatively high cell density suggest that the glutaraldehyde cross-linked implant induced a chronic foreign-body response.

The results of this study indicated that rapid repair and tissue remodeling were achieved with a carbodiimide cross-linked collagenous implant that had structure and mechanical properties similar to those of an autogenous tendon graft and biodegraded at a comparable rate. Prolonged biodegradation of a glutaraldehyde cross-linked collagenous implant resulted in formation of a fibrous capsule, limited formation of neotendon, and caused a chronic foreign-body reaction.

Recently, a composite collagenous ACL replacement device combining both the attributes of synthetic materials (high strength, ease of fabrication and storage) and biological grafts (biocompatibility, tissue ingrowth) was designed and evaluated in a rabbit model [28]. Collagenous anterior cruciate ligament prostheses were made by embedding 225 reconstituted, cross-linked Type I collagen fibers in a Type I collagen matrix, and placing polymethylmethacrylate bone-fixation plugs on the ends (Fig. 6). The collagenous prosthesis was used to replace the anterior cruciate ligament of 31 mature rabbits.

At 4 and 20 weeks postimplantation, histologic and mechanical studies were performed on the developing neoligament tissue and compared with values for the contralateral sham-operated control. At 4 weeks, the collagen fibers of the prostheses were surrounded by inflammatory cells and infiltrated with neoligament tissue. The tibial bone tunnel attachment sites contained new bone and fibrous tissue that approached the prostheses. The glutaraldehyde-treated prosthetic fibers appeared intact, while the carbodiimide-treated prosthetic fibers began to resorb. The ultimate load and ultimate tensile strength of femur-neoligament-tibia complexes had decreased. At 20 weeks, glutaraldehyde-treated fibers appeared partially intact and encapsulated by fibrous tissue and inflammatory cells. In contrast, the carbodiimide-treated prostheses appeared to be completely degraded and were replaced by organized, crimped neoligament tissue. The ultimate tensile strength and ultimate load increased substantially due to deposition and remodeling of neoligament tissue. The neoligament ultimate load was 2 to 4 times the initial load value of the prosthesis.

From this study, it was concluded that implantation of a resorbable, composite collagenous anterior cruciate ligament prosthesis encourages the development of functional neoligament tissue parallel to the implanted fibers.

D. Summary

Reconstituted, cross-linked collagen fibers formed into parallel fibrous scaffolds have been used to replace Achilles tendons and ACLs in rabbits. In both animal models, neotendon and neoligament replaced the collagen scaffold. When the implant degraded by 20 weeks postimplantation, it enhanced tissue remodeling and promoted the rapid ingrowth of neotendon or neoligament with a fibrillar substructure and mechanical properties similar to those observed with an autograft (Achilles repair only). When the implant remained partially intact at 20 weeks, it was treated as a foreign body, encapsulated, and served to stimulate a chronic foreign-body response.

Based on these studies, it can be concluded that the resorption rate of the implant was proportional to the rate at which the strength of the host tissue regenerated and that, consequently, optimum repair of tendons/ligaments is achieved using a parallel fiber prosthesis that biodegrades by 20 weeks postimplantation.

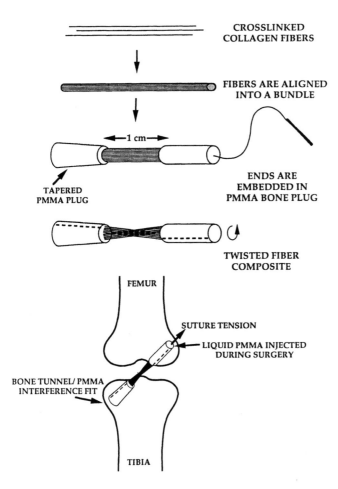

Figure 6 Schematic diagram illustrating the procedures used in the production and implantation of a fibrous anterior cruciate ligament (ACL) replacement. Cross-linked collagen fibers are aligned into a bundle, cut into 3-cm lengths, and the ends embedded in bone cement to form a nontwisted prosthesis. One end has a suture with a needle attached to it to guide the prosthesis through the bone tunnels during implantation. To form a twisted-fiber composite structure, one end was secured while the other end was twisted 1, 2, or 4 turns. (Adapted from Ref. 25.)

IV. PERIPHERAL NERVE REPAIR

A. Background

Peripheral nerve injury resulting from crush or severe compression can result in axonal discontinuity. The ensuing repair or regeneration process occurs by axonal sprouting from the proximal nerve stumps and subsequent elongation, ensheathing, differentiation, and reestablishment of contact with the target organs [37]. In most crush or compression nerve injuries, complete restoration of structure and function are believed to be a result of the preservation of Schwann cell basement membrane tubes that align and direct regenerating axons to their original target sites [37].

Severe peripheral nerve injuries such as transection produce discontinuities of both the axons and basement membrane tubes. Axonal sprouting occurs from the proximal nerve stump, across the site of injury, and into the distal stump. Since the continuity between the proximal and distal basement membrane tubes has been lost, the ability of the sprouting axons to couple with their original tubes located distally diminishes. Misdirection of the sprouting axons results in the failure to reestablish contact with the original target organs and thus impairs functional recovery of the peripheral nerve [38].

A variety of polymeric tubes consisting of silicone rubber or biodegradable polyesters have been used to bridge excised areas of peripheral nerve. These tubes align axonal growth toward the distal stump, preventing axonal misdirection and fibrosis at the suture site. Silicone tubes provide a short-term, biocompatible microenvironment by which gaps in rat sciatic nerve can repair [39], but full functional recovery does not occur when silicone rubber tubes are used alone [40].

In cell culture, collagen has long been regarded as a favorable substrate for neuritic outgrowth. Retinal neurons from chick embryo [41] and sympathetic neurons from rats [42] extend neurites over a collagen substrate. Successful growth of neurites on air-dried collagen suggests that it may be possible to direct nerve regeneration in the presence of an oriented collagen substrate. Rizvi et al. showed that silicone tubes filled with longitudinally aligned, small-diameter collagen fibers enhanced nerve regeneration by providing a scaffold over which fibroblasts and Schwann cells could migrate and axons subsequently regenerate [37].

B. Location of Collagen in Nervous Tissue

Collagen is an important component of neural connective tissue; however, its role in the peripheral nervous system and central nervous system (CNS) differs. In mammalian peripheral nerve, Type I collagen is found in the epinerium, perineurium, and endoneurium; Type III is more apparent in perineurium and endoneurium; and Types IV and V collagen are concentrated in the endoneurium [43]. Transection of rat sciatic nerve reveals collagen fibrils around the bands of Bungner [44]. Endoneurial collagen has been postulated to protect peripheral nerve fibers against compressive forces [45], while resistance to tensile stretching has been attributed to the epineurium [46].

C. Effects of Collagen Fibers on Peripheral Nerve Regeneration

Collagen fibers similar to those used for tendon and ligament repair were employed to evaluate peripheral nerve regeneration in silicone rubber tubes [37]. Approximately 120 collagen fibers (10 mm in length), which had been stabilized by severe dehydration for 3 days, were longitudinally aligned and placed into a silicone tube (1.5-mm inner diameter, 15-mm length). Two groups of prostheses were prepared by varying the cross-link density of the collagen fibers. These differences were characterized by determining the rate of enzymatic degradation [37]. Collagen fiber implants were used to bridge 10-mm gaps in transected sciatic nerves of Sprague–Dawley rats (Fig. 7).

At 4 and 6 weeks postimplantation, the morphologies as well as the axon and blood vessel densities of the newly regenerated tissue in the collagen fiber prostheses were compared with control tissue in which empty silicone tubes bridged the 10-mm gaps. At 4 weeks postimplantation, collagen fibers were still visible in all of the prostheses, indicating incomplete biodegradation. These implants also contained islands of unmyelinated axons and Schwann cells, but few blood vessels and no fascicularization or epineural

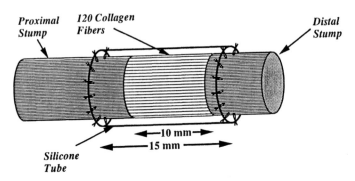

Figure 7 Silicone tube filled with 120 collagen fibers that was used to bridge a 10-mm gap created by excising a rat sciatic nerve.

development. In contrast, the control group exhibited well-developed fascicles around both myelinated and unmyelinated axons and the presence of a prominent epineurium.

At 6 weeks postimplantation, collagen fibers in the prostheses were either visible, suggesting minimal biodegradation, or invisible, suggesting greater biodegradation. Tissue in which the collagen fibers minimally biodegraded contained fewer myelinated axons, lower blood vessel densities, less fascicularization, and decreased epineural development in comparison to the 6-week control group. However, when the collagen fibers biodegraded by 6 weeks, the regenerated tissue exhibited significantly increased unmyelinated and myelinated axon densities and greater epineural development (Fig. 8) when compared with the 6-week control group. These results suggest that collagen fibers that degrade by 6 weeks postimplantation enhance regeneration of sciatic nerve and that the collagen fibers provide a scaffold over which fibroblasts and Schwann cells can migrate and axons can subsequently regenerate.

A follow-up study [47] was undertaken to evaluate the effect of removal of the distal

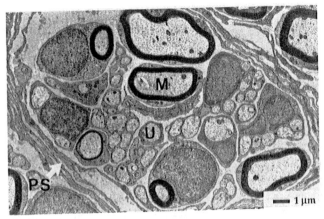

Figure 8 Transmission electron micrograph taken at the midpoint of the 6-week nerve tissue that regenerated through the implant. Large numbers of myelinated (M) and unmyelinated (U) axons appear enclosed in a multilayered perineurial sheath (PS). (From Ref. 37.)

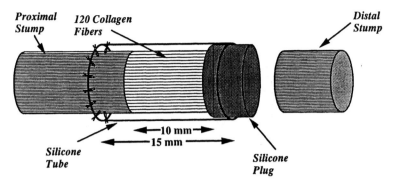

Figure 9 Schematic diagram illustrating a silicone tube 15-mm long with a diameter of 1.5 mm. The proximal stump of the rat sciatic nerve is sutured into one end of the tube filled with collagen fibers, while the other end is sealed with medical-grade silicone adhesive. (From Ref. 47.)

nerve stump from the silicone tube filled with collagen fibers (Fig. 9). Results indicated that, in the absence of the distal nerve stump, an implant containing collagen fibers facilitates Schwann cell migration and axonal elongation only when the implant does not biodegrade prior to nerve regeneration (by 8 weeks postimplantation). These results and results of previous studies suggest that the presence or absence of the distal nerve stump affects the extent of regeneration through a silicone tube filled with collagen fibers. Based on these observations, it is concluded that the results of nerve entubulation studies must be carefully interpreted since the rate of healing is dependent on the extent of nerve injury.

D. Summary

Peripheral nerve repair, in a similar fashion to skin and tendon and ligament repair, is promoted in the presence of a collagen scaffold (collagen fibers). When the collagen fiber implant utilizes both the proximal and distal stumps, sciatic nerve regeneration is enhanced when the collagen fibers degrade by 6 weeks postimplantation. In contrast, implants containing collagen fibers without a distal stump only exhibit enhanced Schwann cell migration and axonal elongation when the fibers do not biodegrade prior to nerve regeneration. These results suggest that when the scaffold degrades too slowly it inhibits axonal elongation, as is the case if it is removed prior to completion of healing. Therefore, facilitation of nerve regeneration requires matching the implant biodegradation rate with the rate of tissue repair. The rate of tissue repair will depend on the distance between the nerve stumps as well as the concentration gradient of neurotrophic factors.

V. CONCLUSION

Type I collagen has been studied extensively as a scaffold material for the repair of soft-tissue injuries not only because it is the primary extracellular matrix component in aligned mammalian connective tissue, but also because it possesses many of the attributes of both synthetic materials and biological grafts. Our laboratory has developed several techniques for engineering insoluble Type I collagen into fibrous scaffolds with hierarchical structures and mechanical properties similar to dense regular and irregular connective tissue. Collagenous matrices for skin replacement, formed by freeze drying and stabilizing insoluble Type I collagen dispersion under optimum conditions, consisted of arrays of fibers that were almost parallel to the surface of the implant. Fibrous collagen scaffolds for tendon and ligament repair were produced by longitudinally aligning high-strength collagen fibers and embedding them in an un-cross-linked collagen matrix. Reconstituted Type I collagen fibers were also used as scaffolds for nerve regeneration by longitudinally aligning them in silicone tubes.

When the tissue-inducing properties of each of these applications were evaluated, results suggested that the rate of tissue ingrowth was optimized when the degradation rate of the collagen scaffold paralleled the rate of wound healing in a particular anatomic site. Initially, the fibrous substructure of the implants provided a scaffold that promoted fibroblast migration and enhanced the aligned deposition of newly synthesized collagen. Upon complete degradation, the implants were replaced by neotissue with fibrillar substructure and mechanical properties similar to the native tissue. In addition, evaluations of cellular and noncellular wound healing responses *in vitro* suggest that these collagen matrices may provide an effective mechanism for predicting implantation responses.

VI. FUTURE CONSIDERATIONS

A. Collagen Fibers with Improved Strength

Results of implantation studies in several different animal models demonstrate that the repair rate of tissue replacements composed of collagen scaffolds is maximized if the scaffold degradation rate parallels the rate of wound healing in a particulate anatomic site [4]. In tendon and ligament replacement, optimum biodegradation occurs within about 20 weeks of implantation [27,28], with loss of about 75% of the initial tensile strength occurring in about 4 weeks [27]. Therefore, in the future design of devices to replace structural tissues, it is necessary to maximize the initial collagen fiber tensile strength to compensate for the rapid loss of implant mechanical properties.

Formation of collagen cross-links is associated with increased load-bearing capacity [16]. Dehydrothermal (DHT) cross-linking induces the physical stabilization of collagen without the introduction of chemical cross-linking agents. It is accomplished by exposing collagen to severe dehydration at elevated temperatures and results in removal of residual water and formation of synthetic peptide bonds [48,49]. After DHT treatment, the helix-to-coil transition temperature is increased, enhancing the thermal stability of collagen without altering its triple-helical structure [49]. Results of pH titration studies indicate that the number of positively and negatively charged amino acids decreases after DHT treatment [49].

A recently completed study suggests that the tensile strength of collagen fibers was optimized by dehydrothermal cross-linking treatment for five days at 110°C [50]. Tensile strength and modulus values of 91.8 and 896 MPa, respectively, were obtained for wet fibers, while dry fibers had values of up to 600 and 8000 MPa, respectively [50]. These values were significantly higher than those previously reported [4,35].

B. *De Novo* Formation of Hydroxyapatite on Collagen Fibers

Bone can be described as a highly ordered composite of Type I collagen integrated with an inorganic mineral phase. *In vitro* models of bone mineralization using collagenous substrates have been conducted in our laboratory using an *in vitro* system of mineralized reconstituted collagen fibers with aligned fibrillar substructure [51]. Collagen fibers were mineralized in a double diffusion chamber saturated with respect to calcium and phosphate. The morphology and ultrastructure of the mineral precipitate were evaluated as a function of the pH of the incubating media. Brushite crystal was observed at acidic pH. Large rectangular crystals formed at pH 5.15 and appear to associate with the collagen fibers. At neutral or alkaline pH, hydroxyapatite crystals were observed in association with the collagen fibers. Spherical aggregates of hydroxyapatite crystals were seen at neutral or alkaline pH, but these structures were reduced in size when formed on collagen at alkaline pH. On close examination, these spherical structures were found to be hollow when viewed in cross-section (Fig. 10).

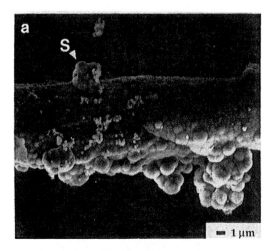

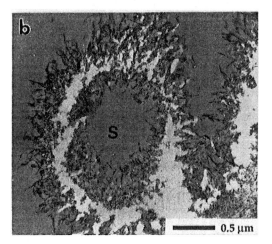

Figure 10 Ultrastructural features of mineralized collagen fibers: (a), scanning electron micrograph indicates a uniform surface coating of mineral precipitate on the collagen fiber formed at pH 7.0, with the addition of spherical crystalline aggregates (S) dispersed over the surface of the fiber; (b), micrograph of a section obtained perpendicular to the axis of the mineralized fiber through spherical structures on the surface of the fiber; these structures appear as hollow aggregates of thin crystal plates.

The crystals precipitated within the interior of the collagen fiber at neutral or alkaline pH were comparable in both size and shape to crystals observed in mineralized turkey tendon and skeletal tissues. These preliminary observations indicate that hydroxyapatite-collagen matrices can be formed *in vitro* that have application for bone repair and replacement. Future studies will evaluate applications in orthopedics, maxillofacial surgery, and dentistry.

REFERENCES

1. Prockop, D. J., Mutations in Collagen Genes as a Cause of Connective Tissue Diseases, *N. Engl. J. Med.*, 26:540 (1992).
2. Birk, D. E., F. H. Silver, and R. L. Trelstad, Matrix Assembly, in *Cell Biology of Extracellular Matrix* (E. D. Hay, Ed.), Plenum Press, New York (1991).
3. Silver, F. H., Y. P. Kato, M. Ohno, and A. J. Wasserman, Analysis of Mammalian Connective Tissue: Relationship Between Hierarchical Structures and Mechanical Properties, *J. Long-Term Eff. Med. Imp.*, 2:165–198 (1992).
4. Christiansen, D. L., G. Pins, M.-C. Wang, M. G. Dunn, and F. H. Silver, Collagenous Biocomposites for the Repair of Soft Tissue Injury, *Mat. Res. Soc. Symp. Proc.*, 252:151 (1992).
5. Silver, F. H., and M.-C. Wang, A Review of the Etiology and Treatment of Skin Ulcers with Wound Dressings: Comparison of the Effects of Occlusive and Nonocclusive Dressings, *J. Long-Term Eff. Med. Imp.*, 2:267–288 (1992).
6. Silver, F. H., and G. Pins, Cell Growth on Collagen: A Review of Tissue Engineering Using Scaffolds Containing Extracellular Matrix, *J. Long-Term Eff. Med. Imp.*, 2:67–80 (1992).
7. Doillon, C. J., F. H. Silver, R. M. Olson, C. Y. Kamath, and R. A. Berg, Fibroblast and Epidermal Cell-Type I Collagen Interactions: Cell Culture and Human Studies, *S.E.M.*, 2: 985–992 (1988).
8. Mian, E., M. Mian, and F. Beghe, Lyophilized Type I Collagen and Chronic Leg Ulcers, *Int. J. Tissue React.*, 13:257–269 (1991).
9. DiMauro, C., A. M. Ossino, M. Trefiletti, P. Polosa, and F. Beghe, Lyophilized Collagen in the Treatment of Diabetic Ulcers, *Drugs Exp. Clin. Res.*, 17:371–373 (1991).
10. Doillon, C. J., and F. H. Silver, Collagen-Based Wound Dressing: Effects of Hyaluronic Acid and Fibronectin on Wound Healing, *Biomats.*, 7:3–8 (1986).
11. Oliver, R. F., R. A. Grant, and M. Kent, The Fate of Cutaneously and Subcutaneously Implanted Trypsin Purified Dermal Collagen in the Pig, *Brit. J. Exp. Path.*, 53:540–549 (1972).
12. Cooke, A., R. F. Oliver, and M. D. Edward, An *In Vitro* Cytotoxicity Study of Aldehyde-Treated Pig Dermal Collagen, *Brit. J. Exp. Path.*, 64:72–76 (1983).
13. Yannas, I. V., J. F. Burke, D. P. Orgill, and E. M. Skrabut, Wound Tissue Can Utilize a Polymeric Template to Synthesize a Functional Extension of Skin, *Science*, 215:174–176 (1982).
14. Doillon, C. J., C. F. Whyne, S. Brandwein, and F. H. Silver, Collagen-Based Wound Dressings: Control of the Pore Structure and Morphology, *J. Biomed. Mat. Res.*, 20:1219–1228 (1986).
15. Doillon, C. J., F. H. Silver, and R. A. Berg, Fibroblast Growth on a Porous Collagen Sponge Containing Hyaluronic Acid and Fibronectin, *Biomats.*, 8:196–200 (1987).
16. Weadock, K. S., R. M. Olson, and F. H. Silver, Evaluation of Collagen Crosslinking Techniques, *Biomat., Med. Dev., Art. Org.*, 11:293 (1984).
17. Doillon, C. J., C. F. Whyne, R. A. Berg, R. M. Olson, and F. H. Silver, Fibroblast-Collagen Sponge and the Spatial Deposition of Newly Synthesized Collagen Fibers *In Vitro* and *In Vivo*, *S.E.M.*, 3:1313–1320 (1984).

18. Doillon, C. J., M. G. Dunn, R. A. Berg, and F. H. Silver, Collagen Deposition During Wound Healing, *S.E.M.*, 2:897–903 (1985).
19. Marks, M. G., C. J. Doillon, and F. H. Silver, Effects of Fibroblasts and Basic Fibroblast Growth Factor on Facilitation of Dermal Wound Healing by Type I Collagen Matrices, *J. Biomed. Mat. Res.*, 25:683–696 (1991).
20. Doillon, C. J., M. G. Dunn, and F. H. Silver, Relationship Between Mechanical Properties and Collagen Structure of Closed and Open Wounds, *J. Biomed. Eng.*, 110:352–356 (1988).
21. Doillon, C. J., A. J. Wasserman, R. A. Berg, and F. H. Silver, Behaviour of Fibroblasts and Epidermal Cells Cultivated on Analogues of Extracellular Matrix, *Biomats.*, 9:91–96 (1988).
22. Doillon, C. J., A. J. Wasserman, R. A. Berg, and F. H. Silver, Epidermal Cells Cultured on a Collagen-Based Material, *Proceedings of the 44th Meeting of the Electron Microscopy Society of America*, San Francisco Press, Inc., 212 (1986).
23. Noyes, F. R., and G. H. McGinniss, Controversy About Treatment of the Knee with Anterior Cruciate Laxity, *Clin. Orthop.*, 198:61–76 (1985).
24. Silver, F. H., A. J. Tria, J. P. Zawadsky, and M. G. Dunn, Anterior Cruciate Ligament Replacement: A Review, *J. Long-Term Eff. Med. Imp.*, 1:135 (1991).
25. Kato, Y. P., Development of a Reconstituted Collagen Fiber Tendon/Ligament Prosthesis: Experimental Studies in Rabbits, Ph.D. thesis, Rutgers University and the University of Medicine and Dentistry of New Jersey, Piscataway, NJ (1991).
26. Goldstein, J. D., A. J. Tria, J. P. Zawadsky, Y. P. Kato, D. Christiansen, and F. H. Silver, Development of a Reconstituted Tendon Prosthesis, *JBJS*, 71-A:1183–1191 (1989).
27. Kato, Y. P., M. G. Dunn, J. P. Zawadsky, A. J. Tria, and F. H. Silver, Regeneration of Achilles Tendon with a Collagen Tendon Prosthesis, *JBJS*, 73-A:561–574 (1991).
28. Dunn, M. G., A. J. Tria, Y. P. Kato, J. R. Bechler, R. S. Ochner, J. P. Zawadsky, and F. H. Silver, Anterior Cruciate Ligament Reconstruction Using a Composite Collagenous Prosthesis, *Am. J. Sports Med.*, 20:507–515 (1992).
29. Viidik, A., Structure and Function of Normal and Healing Tendons and Ligaments, in *Biomechanics of Diarthroidal Joints: Vol. 1.* (V. C. Mow, A. Ratcliffe, and S. L. Y. Woo, Eds.), Springer Verlag, New York (1990).
30. Birk, D. E., and R. L. Trelstad, Extracellular Compartments in Tendon Morphogenesis: Collagen Fibril Bundle and Macroaggregate Formation, *J. Cell Biol.*, 103:231–240 (1986).
31. McBride, D. J., Jr., R. A. Hahn, and F. H. Silver, Morphological Characterization of Tendon Development During Chick Embryogenesis: Measurement of Birefringence Retardation, *Int. J. Biol. Macromol.*, 7:71 (1985).
32. Birk, D. E., J. F. Southern, E. I. Zycband, J. T. Fallon, and R. L. Trelstad, Collagen Fibril Bundles: A Branching Assembly Unit in Tendon Morphogenesis, *Development*, 107:437 (1989).
33. Parry, D. A., A. S. Craig, and G. R. G. Barnes, Tendon and Ligament from the Horse: An Ultrastructural Study of Collagen Fibers as a Function of Age, *Proc. R. Soc. Lond. B*, 203:293–303 (1978).
34. Scott, J. E., and C. R. Orford, Dermatan Sulfate-Rich Proteoglycan Associates with Rat Tail-Tendon Collagen at the d Band in the Gap Region, *Biochem. J.*, 197:213–216 (1981).
35. Kato, Y. P., D. L. Christiansen, R. A. Hahn, S.-J. Shieh, J. D. Goldstein, and F. H. Silver, Mechanical Properties of Collagen Fibres: A Comparison of Reconstituted and Rat Tail Tendon Fibres, *Biomats.*, 10:38 (1989).
36. Williams, B. R., R. A. Gelman, D. C. Poppke, and K. A. Piez, Collagen Fibril Formation: Optimal *In Vitro* Conditions and Preliminary Kinetic Results, *J. Biol. Chem.*, 253:6578–6585 (1978).
37. Rizvi, A. H., A. J. Wasserman, G. Zazanis, and F. H. Silver, Evaluation of Peripheral Nerve Regeneration in the Presence of Longitudinally Aligned Collagen Fibers, *Cells and Materials*, 1:279–289 (1991).
38. Lundborg, G., Nerve Regeneration and Repair, *Acta Orthop. Scand.*, 58:145–169 (1987).

39. Lundborg, G., L. B. Dahlin, N. Danielson, R. Gelberman, F. Longo, H. Powell, and S. Varon, Nerve Regeneration in Silicone Chambers: Influence of Gap Length and of Distal Stump Components, *Exp. Neurol.*, 76:361–375 (1982).

40. Ashur, H., Y. Vilner, A. Finsterbush, M. Rousso, H. Weinberg, and M. Devor, Extent of Fiber Regeneration After Peripheral Nerve Repair: Silicone Splint Versus Suture, Gap Repair Versus Graft, *Exp. Neurol.*, 97:365–374 (1987).

41. Thompson, J. M., and D. J. Pelto, Attachment, Survival and Neurite Extension of Chick Embryo Retinal Neurons on Various Culture Substrates, *Dev. Neurosci.*, 5:447–457 (1982).

42. Johnson, M. I., and V. Argior, Techniques in Tissue Culture of Rat Sympathetic Neurons, *Meth. Enzymol.*, 103:334–347 (1983).

43. Shellswell, G. B., D. J. Restall, V. C. Duance, and A. J. Bailey, Identification of Differential Distribution Collagen Types in the Central and Peripheral Nervous System, *FEBS Lett.*, 106: 305–308 (1979).

44. Roytta, M., V. Salonen, and J. Peltonen, Reversible Endoneurial Changes after Nerve Injury, *Acta. Neuropathol. (Berl.)*, 73:323–332 (1987).

45. McFarlane, K. R., M. Pollock, and D. B. Myers, Collagen Content in Human Ulnar Nerve, *Acta. Neuropathol. (Berl.)*, 50:217–220 (1980).

46. Haftek, J., Stretch Injury of Peripheral Nerve, Acute Effects of Stretching on Rabbit Nerve, *JBJS*, 52B:354–365 (1970).

47. Rizvi, A. H., G. Pins, and F. H. Silver, Peripheral Nerve Regeneration in the Presence of Collagen Fibers: Effect of Removal of the Distal Nerve Stump, *Clin. Mats.* 16:73–80 (1994).

48. Yannas, I. V., and A. V. Tobolsky, Cross-Linking of Gelatine by Dehydration, *Nature*, 215: 509–510 (1967).

49. Silver, F. H., I. V. Yannas, and E. W. Salzman, *In Vitro* Blood Compatibility of Glycosaminoglycan-Precipitated Collagens, *J. Biomed. Mat. Res.*, 13:701–716 (1979).

50. Wang, M.-C., G. D. Pins, and F. H. Silver, Collagen Fibers with Improved Strength for the Repair of Soft Tissue Injuries, *Biomats.*, 15:507–512 (1994).

51. Christiansen, D. L., and F. H. Silver, Mineralization of an Axially Aligned Collagenous Matrix: A Morphology Study, *Cells and Materials*, 3:39177–39188 (1993).

37

Collagen-Based Wound Dressings: How to Control Infection and Immune Response

Subramanian Gunasekaran
Nova Gen Corporation
Newark, California

I. INTRODUCTION

Wound dressings include a wide variety of wound care management devices, drugs, and biological materials. They may vary from simple skin bandages or ointments to full-thickness skin transplants. These dressings could further be classified as temporary wound closures, which include allografts, xenografts, and synthetic or biological materials, and the permanent wound closures, which include autografts, cultured epithelial cells, microskin grafting, and the like [1]. The major challenge in wound care management is choosing an appropriate dressing. Decisions about which dressing to use are often complicated by wound type, the patient's condition or immune response, or by the sudden occurrence of a wound infection.

Much of the confusion over wound care and wound care products can be minimized by using a systematic approach to wound assessment and properly choosing a dressing [2]. Recently, the benefits of the most commonly used wound dressings and their characteristics were discussed to select the appropriate dressing for individual wounds [3]. In that report, it is concluded that there is no "ideal" dressing that can be used on all wounds. This is of course true. However, one may logically select a particular matrix material that can be further modified to suit any wound type. The purpose of this chapter is to describe both the positive and negative features of collagen-based dressings so that such materials may be further developed to meet all challenges in wound care management.

A. Definition

The meaning of *dressing* is a covering applied to a wound to protect it from further damage and to help the healing process. The word *dressing* might have been derived from the old French word *dresser*, meaning to arrange or make things straight, to bring into proper order. The functional expectation of dressing gets changed based on the type of wound being treated. The general wound types are partial-thickness wound, full-thickness wound, infected wound, uninfected wound, moist or dry wound, burn injuries, ulcer wounds, pressure sores, and so on.

B.   Choosing a Dressing

Selection of a dressing may differ based on the wound conditions. There is currently little agreement among surgeons regarding the dressing of choice for different wound types, including wound management for split-thickness skin graft donor sites. A recent review compares the five major groups of dressings: open, semi-open, occlusive, semiocclusive, and biological [4]. The different dressings in each group are described in terms of physiological basis for use, advantages, disadvantages, and practical application. In general, full-skin-thickness wounds larger than 15 square centimeters (cm^2) in area are extremely slow to heal and skin grafting is recommended to achieve a prompt and satisfactory care management of the wound. In this chapter, we consider grafting also as a type of wound dressing method and discuss the relevance and possibility of collagen dressing to replace even a skin graft.

1.   Pressure Sores

More than 70% of pressure sores occur in patients over 70 years of age. Wound dressings for pressure sores and ulcers have to be considered a little differently from other bleeding wounds. The use of special dressings and procedures to facilitate the inflammatory repair response would potentially prevent further tissue trauma. There is considerable doubt about the use of "traditional" wound applications such as gauze or chlorinated lime and boric acid solution (Eusol) [5]. Especially, deeper ulcers (Types 2 and 3) can easily and quickly be treated with collagenlike bioactive dressings that possibly can be made to deliver antiinflammatory drugs to the local wound environment, thus facilitating tissue repair. This may even be true for the sacral (near-anal) sores and cavity ulcers (Type 4), which are more difficult to treat.

2.   Burn Wounds

Burn injuries are another common skin damage for which a dressing is required. Two million people in the United States receive medical treatment each year for burn injuries. Of these patients, 100,000 are hospitalized, and 7800 die as a direct result of their injuries. Of the patients that are hospitalized, 30% to 40% are under 15 years of age and 67% of burn patients are male. Flame burns account for approximately 13% of accidents, scalds account for 85%, and electrical and chemical burns account for approximately 2% [6]. Reports are available on the usage of collagen sponge as a burn dressing [7,8].

C.   Scope

The purpose of this chapter is to critically evaluate and emphasize the significance of collagen-derived resorbable wound dressings. Rather than considering collagen as just a hemostatic agent, if the material is properly developed, the collagen-based product in the near future may function as a permanent wound care management device for varied types of wounds. Therefore, the intent is to focus on the importance of collagen-based wound dressing materials and the current situation in regard to problems in terms of infection and immune response and how to overcome these problems.

II. PRINCIPLES OF SKIN WOUND DRESSINGS AND GRAFTS

A. Role of Hemostasis in Wound Healing

Hemostasis or blood clotting is the prime and foremost event to happen in the process of wound repair. When a wound of the skin or other soft-tissue organ occurs, the body responds with a standard process of repair. The wound healing stages and the associated cells and their important growth factors are given in Fig. 1. As shown in the diagram, the very first step in a normal skin wound repair process is clotting. Therefore, the argument is that an ideal wound dressing in general has to have hemostatic capabilities.

Collagen is the natural hemostatic agent; this protein constitutes a major portion of the extracellular matrix of every tissue. It has specific affinity with different cells and growth factors that play a crucial role in the wound healing process. If a different hemostatic agent is used instead of the extracellular matrix protein collagen, it may be efficient in clotting the blood but may not be further helpful in the normal wound repair process. Again, this is due to the reason that, following clotting, collagen alone could provide proper attachment of the cells and their growth factors responsible for attracting inflammatory cells early to complete the process of normal wound healing. The hemostatic collagen/platelet-induced fibrin clot mesh not only provides structural stability of the wound, but also serves as a guide for where and how the cells and factors would deposit for faster wound repair.

B. The Wound Repair Process

During the first day of the wound repair process, epithelial cells from the skin dislodge from sites away from the injury (from healthy skin) and migrate toward the clot. They secrete enzymes that dislodge the clot and allow it to be reorganized. At this time, the

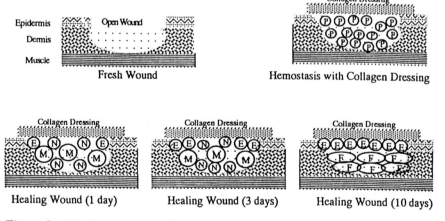

Figure 1 The interface between the dressing and the wound surface is important for proper wound healing. As long as the infection and adverse immune response are under control, the cells appropriately recognize the molecule that constitutes native extracellular matrix protein. Besides cells, other growth factors, cytokines, interleukins, and the like recognize the native protein and bind themselves to exhibit their epitopes in a natural manner. This may not happen with other dressings made up of synthetic materials of nonnative polymers and proteins (P, platelets; M, macrophages; N, neutrophils; E, epithelial cells; F, fibroblasts).

inflammatory cells, neutrophils and macrophages, have invaded the clotted area. The neutrophils are there to digest the bacteria and prevent infection. (In some injuries, dead neutrophil collection can be seen as pus.) Macrophages are there to assist in cleaning up bacteria and dead cells and to secrete more growth factors that attract or form from stem cells, the ultimate repair cells—fibroblasts.

The macrophages and neutrophils are viewed as the key players of the inflammatory response in wound healing. The factors they secrete include PDGF (platelet-derived growth factor), FGF (fibroblast growth factor), TGF β (transforming growth factor-β), ILs (interleukins), EGF (epidermal growth factor), and others. The different growth factors involved including insulinlike growth factor (IGF), in wound healing and the associated cells are given in Table 1.

At Day 3 of wound healing, there is a complete epithelial bridge covering the wound and the site is filled with neutrophils and macrophages. At seven days, the inflammatory process recedes and the repair process begins, with fibroblastic secretion of new extracellular matrix—collagen. Matured skin tissue consists of bricks of fibroblast cells that are mortared by collagen secreted by fibroblast and epithelial cells. A combination of cells and collagen provides a secure bridge over the interrupted skin tissue. Smooth muscle cells and blood capillary cells also grow into the damaged area, while keratinocytes grow on top to seal the wound from the external environment. Much of the secondary cell proliferation occurs in a remodeling process after the initial fibroblast proliferation.

C. Relevance of Using Collagen-based Wound Dressings

The high number of available wound dressings as well as the scientific reports about wound dressings indicate that the problem of an ideal wound dressing material is not yet solved. During the above-mentioned complicated cascade of wound repair events, the primary incident to happen is hemostasis or blood clotting. Collagen is clinically proven to perform as a better hemostatic agent than conventional pressure dressings [9]. Collagen, a major component of blood vessels, mediates primary hemostasis brought about by platelets. The aggregation of platelets induced by collagens is considered an important step in primary hemostasis [10]. The mechanism underlying platelet-collagen interaction is highly complicated; to date various molecules have been suggested to be platelet collagen receptors. It is suggested that P62, a platelet membrane protein with a molecular weight of 62 kilodalton (kDa) in the reducing condition immunoprecipitated by the antibody of a patient with idiopathic thrombocytopenic purpura, could be a platelet collagen receptor. The antibody was also primarily immunoreactive with P62 on immunoblotting. The patient's platelets, which were deficient in P62 and totally lacked collagen-induced platelet aggregation, showed normal platelet adhesion to collagen in the presence of Mg^{2+}. These results indicate that, of the antigens recognized by this antibody,

Table 1 Cells and Growth Factors Involved in Wound Healing

Wound healing stages	Associated cells	Growth factors
Blood clotting	Platelets	PDGF, FGF
Inflammation	Macrophages, neutrophils	PDGF, FGF, TGF-β, IGF, ILs
Repair	Fibroblasts, epithelial cells	EGF, FGF, TFG-β, IGF, ILs

P62 could play an important role in platelet-collagen interaction mediating aggregation, and those multiple membrane molecules, including P62, may participate in platelet adhesion to collagen [11].

In the last few years, many scientific reports about collagen as a wound covering have been published [12–16]. It is definitely essential for the dressings to have hemostatic abilities for bleeding wounds and at the same time it is preferred the dressings are resorbable. The hemostatic ability of a wound dressing is not meant only to stop bleeding; it should also further favor the normal events of wound healing. Among different blood clotting resorbable dressings like collagen, fibrin, cellulose, alginate, and the like, the only natural extracellular matrix material that can bind to relevant repair cells and cytokines (growth factors) is collagen. From this logical viewpoint, it is most relevant to choose collagen as a base material to develop an ideal wound dressing. Because of such a biological role in wound healing, and because of its biocompatibility, bioresorbability, binding ability, and due to recent Food and Drug Administration (FDA) approval of implantable collagen-based devices in the United States, evidently collagenous materials have more possibilities for use as universal wound dressing matrices.

III. COLLAGEN-BASED WOUND DRESSING

A. Collagen Type

Collagen-based wound dressings are biologically derived, native, Type I collagen extracted from well-screened animal tissues. Because of the initial purity of the collagen source and the additional purification process steps involving high-grade chemicals and ultrapure water, the resultant collagenous product can be expected to show consistent *in vivo* behavior. Topical application of absorbable collagen may effectively control bleeding, usually within two to five minutes when applied directly to the bleeding site. It is designed to be totally absorbable if left *in situ* after hemostasis and may act as a good dressing if it can be prevented from getting infected. The positive effect of collagen application on a bleeding wound is well known. In the past few decades, several scientific reports about the significance of collagen as a wound covering have been published [12–17].

B. Effect on Full-Thickness Wounds

Wound dressings are generally tested on experimental full-thickness skin wound animal models. Sedlarik et al. investigated the effect of a collagen sponge on healing of full-thickness wounds in rats [15]. The animals were divided into two controls and two experimental groups. In the control groups, there were air-exposed wounds as well as wounds covered with paraffin gauze. In the experimental groups, the wounds were covered with natural, reconstituted collagen sponge as well as with chemically prepared sponge. All results were compared. The wounds with the collagen sponge covering healed significantly faster. Also, the quality of the wound healing was better in the experimental groups.

Through another full-thickness skin wound repair model in guinea pigs, it was demonstrated that a collagen wound dressing potentially shortens the healing time as well as improves the quality of the wound healing process [16].

C. Effect on Burn Wounds

The porcine collagen membrane was evaluated as a burn wound dressing in deep, partial-skin-thickness burn wounds in rats [18]. Burn wounds of 4 × 4 cm were inflicted by exposure of skin to high temperatures followed by deepithelialization. Wound healing was assessed by planimetry of epithelialization on Day 10 after injury. Open wounds exhibited 24% of wound area reepithelialized. A collagen membrane dressing significantly improved the healing to 69% of wound area ($P < 0.0001$).

In a completely separate experiment, the porcine collagen membrane was applied as a wound dressing to the donor sites of burn patients; its effect on wound healing was compared with that of a petroleum jelly gauze dressing. The donor sites covered with petroleum jelly gauze had reepithelialized by an average of 14.5 days (ranging from 13 to 16 days) after wounding. The wounds dressed with collagen membrane demonstrated a significant increase in the healing rate. Complete reepithelialization was observed by 10.3 days (ranging from 10 to 12 days) after wounding ($P < 0.0001$).

In another human clinical study conducted from 1987 to 1990 [19], it was confirmed in all cases in which they used collagen dressing there was an enhanced metabolic activity in the wound healing process. After 4 to 6 days of treatment, good granulation and vascularization of the wound bed were obtained, so that a skin transplant could be performed, which in all cases healed primarily.

D. Collagen Shields and Role as Drug Delivery Vehicle

The use of collagen shields to enhance comfort and facilitate recovery after eyelid surgery was studied using a clinical experiment [20]; 32 patients undergoing various eyelid procedures were studied. In 10 patients, bilateral surgery was performed and the contralateral eye was used as a control. In patients with unilateral surgery, 20 of 22 patients exhibited comfortable postoperative courses, and in patients with bilateral surgery, 8 of 10 patients had less conjunctival injection, chemosis, corneal staining, or lid edema on the side with the collagen shield.

Another clinical study was conducted to compare the effectiveness of collagen shields in delivering antiinflammatory agents and antibiotics after cataract surgery [21]. A total of 61 patients was randomly assigned to three groups: subconjunctival injection of gentamicin and dexamethasone; the same route plus a collagen shield without any drug; and collagen shield soaked in gentamicin and dexamethasone. All subjects had a manual extracapsular cataract extraction with posterior chamber intraocular lens implantation. Eyes were evaluated at the end of surgery and 24 hours later. The collagen shield achieved a gradual decrease in pain and conjunctival redness after 24 hours. The occurrence of folds in Descemet's membrane was less frequent and aqueous flare less severe than when subconjunctival injections were used. No adverse effect was reported. The studies conclude that the collagen shield could be a safe, better, noninvasive technique because of its double action of bandage and enhancement of drug penetration.

E. Effect on Ulcer Wounds and Pressure Sores

Ulcer and pressure sores are relatively difficult to manage because of their nature and prevalence in up to 75% of the elderly population. These ulcers are painful and debilitating, yet most of the medical treatment is unsuccessful. Surgical treatment consists of repeated skin grafting procedures, which eventually deplete the patient's available donor sites. This initiated *in vitro* culture of human epidermal autograft for potential treatment of leg ulcers in sickle cell anemia. However, because epidermal cells need a supporting matrix in order to be applied to the ulcer, (Reindorf et al.) decided to assess the independent effect of a collagen matrix alone in promoting healing of sickle cell leg ulcers [22]. A collagen matrix dressing, Collistat (Hellistex, Inc.), was applied directly to the ulcers. The ulcers were reviewed every two weeks, and the dressing reapplied every four weeks. Collagen matrix caused the chronic ulcers to heal completely in fewer than three months. This encouraging performance of collagen alone should be greatly enhanced by the *in vitro* epidermal autograft.

As a result of the biological role of collagen in wound healing, and because of its biocompatibility, the use of heterologous collagen-based devices is becoming more widespread. In another study, a lyophilized Type I collagen (Condress) extracted from bovine Achilles tendon through a nondenaturing procedure in the absence of proteolytic enzymes was tested clinically. Electrophoretic analysis and morphological examination by electron microscopy confirmed that the procedure employed to extract collagen did not alter the polypeptidic composition of the molecule and its structure. A gamma-ray dose between 0.5 and 1.5 megarad was found quite adequate to sterilize the final product and certainly was devoid of degradative effect. The finished product had a special absorbing capacity, immersion time, strain resistance, wrinkling temperature, and enzymatic digestion time. The developed product is reported to be nonallergenic and suitable for clinical use. When it has been applied in chronic leg ulcers, pressure sores, or reconstructive surgery, Condress demonstrated the ability to improve wound repair substantially [23].

F. Effect on Cells and Growth Factors

Lyophilized Type I collagen can stimulate wound healing by recruiting a number of different cell types (i.e., platelets and macrophages) and proteins like fibronectin. Platelets and macrophages produce locally acting growth factors that in turn induce fibroblast and epidermal migration, angiogenesis, and increase matrix synthesis. Chronic leg ulcers are the result of microvascular failure owing to ischemia and stasis. When collagen has been used in the treatment of chronic leg ulcers, several observations were made [24]. First, collagen is significantly more effective in stimulating the healing of chronic venous ulcers when compared with hydrocolloids ($p < .05$), with the two products applied upon half of the same ulcer. Second, in the treatment of ulcers due to arterial obstruction, collagen is more effective than hydrocolloids without achieving statistical significance. Third, collagen is very effective in the treatment of ulcers in thalassemic patients. Fourth, telethermographic studies have demonstrated an increase of blood perfusion and histological studies have shown the stimulation of angiogenesis, fibropoiesis, and epidermal growth. Fifth, the application of collagen also determines the maximum obtainable increase under conditions of proven cicatrization difficulty. Finally, enzymatic degradation of collagen has not promoted any bacterial infection and no local or generalized sensibilization phenomena have been observed. The study showed that collagen is a pharmacological approach to wound healing, directly interfering with cellular and noncellular components, and significantly improves the reparative process when delayed [24].

It is predicted that electrical stimulation and growth factors are going to be more popular in the treatment of severe wound conditions like pressure sores and the like [25]. In order to make these healing enhancers function efficiently, collagenlike matrix may seem to play a major role.

G.   Comparing Fibrin and Other Resorbable Dressings

Fibrin glue, another resorbable dressing material, is a hemostatic and adhesive agent that has been used for many years in Europe [26]. It is made by the simultaneous mixing of concentrated fibrinogen complex and bovine thrombin reconstituted in a solution of calcium chloride. Upon mixing, the final stages of the coagulation cascade are mimicked, resulting in formation of a fibrin clot. Fibrin is vital in wound healing because the network formed in the wound acts not only as a hemostatic barrier, but also as a scaffold for migrating cells.

In a recent study [27], fibrin glue (60 milligrams per milliliter [mg/ml]) was applied on 6-mm diameter circular full-thickness wounds in homozygous, genetically diabetic mice. Results showed delayed wound closure as compared with paired control wounds. Dilution of the protein concentration of fibrin glue to 1 mg/ml resulted in higher histological scores compared with those for wounds treated with 60 mg/ml fibrin glue. Finally, the application of an adherent semipermeable dressing (Opsite) over the wound per se resulted in delayed wound closure. Delayed closure may be due to mechanical obstruction to the migration or proliferation of cells that actively participate in the wound healing process, mechanical inhibition of wound contraction, or both. Similar supportive evidence that fibrin glue may prevent the migration or proliferation of wound healing cells resulted from another recent study [28].

H.   Current Market Status of Collagenous Dressings

For a better understanding of the market value of collagenous dressing products, IMS America Limited has provided a survey analysis on the relative market values of different products used as wound dressings. A consolidated account of the report is provided in Table 2. In the given list, collagenous products fall under the category of absorbable hemostats. This category accounts for approximately 75% of the total dressings market value, which is primarily contributed by collagen-based products.

IV.   IMMUNE RESPONSE OF COLLAGENOUS BIOMATERIALS

A.   Collagen Type 1

There are at least 10 different types of collagen that have different amino acid sequences. The most commonly used collagen for medical device formulation is Type 1 collagen. This type of collagen accounts for 90% of the body's connective tissue. It is predominantly found in skin, bone, and tendons. The molecule of collagen has a triple-helical structure that consists of two identical polypeptide chains and a third chain that differs slightly in amino acid sequence. These amino acid chains form a tightly interwoven, triple-helical structure that promotes a unique fibrillar interaction. The coiled configuration of the molecule is maintained throughout the molecule's length except at the short, nonhelical terminal regions. These loose polypeptide projections, called *telopeptides*, are important sites of antigenicity and of cross-link formation within and between molecules.

B.   Method to Minimize the Antigenic Response

The success of a collagen-based device depends on how well the collagen is accepted by the patient's body. However, one should realize that the early inflammation is a normal sign of healing with collagen-based dressings. Dressings such as collagen sponge, polyeth-

Table 2 Dressings, Manufacturers, and Market Size in 1993

Type of dressing	Manufacturer	Product	Market share (millions)
Absorbable hemostats	UpJohn	Gelfoam, Gelfilm	$85.634
	J & J Medical, Inc.	Surgicel, S. Nu-Knit, In-stat, Thrombogen	
	Medchem	Avitene, Hemaflex Sheets	
	Parke-Davis	Thrombostat	
	Jones Medical	Throbinar	
	BD Acute Care Division	Oxycel	
	Astra	Hemopad, Hemotene	
	Calgon Vestal Labs	Helistat	
	Pilling-Weck	Superstat	
	Vitaphore Corporation	Collastat	
Nontransparent occlusive dressings	Convatec	Duoderm	$21.374
	Hollister, Inc.	Restore	
	Coloplast, Inc.	Comfeel	
	Acme United	Lyofoam dressings	
	3M Medical Surgicals	Tegasorb dressing	
	Calgon Vestal Labs	Epi-Lock	
	Sherwood Medical	Ultek dressing, Viasorb	
	Southwest Technologies	Elasto-Gel	
	Synthes	Epigard dressing	
	Beiersdorf	Cutinova	
	NDM	Clearsite	
	J & J Medical Inc.	Ulcer dressing, Nu-Derm	
	Sparta Instrument Co.	Primaderm	
	Bard Patient Care	Intact	
Biological dressings	Bioplasty Inc.	E-Z Derm dressing, Med-iskin, Exoderm	$1.549
	Dow B. Hickam Inc.	Biobrane sheets, gloves	
Wound absorption dressings	Dow B. Hickam Inc.	Granulex, Sorbsan, Pro-derm	$10.341
	Carrington Labs Inc.	Dermal wound gel	
	Calgon Vestal Labs	Kaltostat, Mitraflex	
	Smith & Nephew	Intrasite	
	J & J Medical Inc.	Derbisan	
	Convatec	Duoderm dressing	
	Bard Patient Care	Absorption dressing	
	Hollister Inc.	Dressing wound	
	Pharmaseal Division	Hydra gran	
	Bioderm sciences	Hydron	
	Marion Merrell-Dow	Multipack/Envision	
	Carpace Inc.	Wound dressing	
	Kendall-HPC	Curasorb dressing	

Table 2   Continued

Type of dressing	Manufacturer	Product	Market share (millions)
Nonadherent wet dressings	Sherwood Medicals	Xeroform pet dressing, Xeroflo, Scarlett Red Gauze	$6.355
	Miles Pharmaceuticals	Dome paste bandage	
	Hospital Supplies Div.	Xeroform	
	Purdue Frederick	Betadine	
	Beiersdorf	Gelocast	
	Graham-Field	Medicopaste bandage	
	Sparta Instrument Co.	Gauze dressing, xeroform	
	Convactec	Duoderm dressing	
	Medline	Gauze dressing	
Dressing rolls	Sherwood Pharm	Pharmdine	$1.115
	Kendall HPC/Hospital Supply Division/Hermitage Hospital Products/J & J Medical Inc./Chastron Products Inc./Professional Med. P.	Gauze dressing rolls	
	General Bandage	Gauztex	
	Qualtex/Chaston Products Inc.	Jones dressing	
	Smith & Nephew Rolya	Padding	

Source: Hospital Supply Index™, IMS America, Limited, 1994

ylene glycol, Duoderm, and lanolin ointment induce moderate-to-severe inflammatory changes when placed on the wounds [29]. These wounds reepithelialize significantly faster than gauze-covered wounds used as controls. This contrasts with inert dressings, such as hydrated hydrogel membrane, Carbopol 934P, or Silvadene cream, which did not affect the rate of reepithelialization when compared with the healing of control wounds. Simultaneously, these dressings induced no or minimal inflammatory reaction in the wound tissue. Only when the inflammatory reaction to the wound dressing was excessive (methylcellulose) was the rate of reepithelialization of the wounds significantly inhibited in comparison with control wounds. It is hypothesized that wound dressings, by inducing inflammatory reaction, enhance healing by activating cells such as macrophages or fibroblasts that produce growth factors and other mediators of the repair process.

Therefore, the chronic immune or allergic response is the challenging one for collagenlike medical devices. There are a few tricks to face this challenge. During chemical processing, the antigenic telopeptide region can be broken down by proteolytic enzymes, rendering the collagen less antigenic. Besides enzymatic processing, chemical cross-linking also blocks the antigenic sites on collagen so that the resultant product is less immunogenic. In addition to these processing techniques, principal importance has to be given to right tissue source and maintaining cleanliness during every step of operations.

C. Immunogenicity of Collagenous Devices

Of course, collagenous medical products, as other biological molecules, would tend to cause immunogenic problems if the processing and manufacturing are not done properly. The solution depends on taking care of the purification process of collagen from relevant tissues. This aspect of the problem is better covered elsewhere in this book. There are some recent references that focus on the immunogenicity of bovine collagen implants [30–32].

V. INFECTION IN WOUND HEALING

The incidence of wound infection is remarkably high, especially in developing countries, because of insufficient hygienic and antiseptic measures taken toward wound care management. There are also problems of misdiagnosis and failure to manage wounds according to the appropriate principles. The infection of a wound could have occurred at any period of the wound's duration. Therefore, it is important that the selected dressing is applied following proper adequate asepsis of the wound site, which requires thorough cleansing procedures. In common practice of wound care management, a systemic antibiotic is given to the patient in accordance with the infection type and the patient's gut condition to tolerate the antibiotic. In most cases of serious infection, an additional problem with the systemic approach is that the local concentration of the systemically administered antibiotic may not be sufficient to function efficiently. Considering all of this, it is attractive to develop a wound dressing that is capable of binding a wide spectrum antibiotic that is released into the wound site in a slow and sustained fashion to eradicate the infection and protect the healing wound from secondary infections.

A. Anti-Infective Agents in Wound Treatment

It is out of the scope of this chapter to get into the details of all available anti-infective agents in the treatment of wound infections. However, for better understanding, good reading materials and recent review articles are available that describe common bacterial dermal infections and related bacteriological details of skin and soft-tissue infections [33,34].

Before the treatment and administration of a dressing to the wound, it is essential to diagnose the type of infection properly. Normally, bacterial culture of a sample from the wound site is used to determine the type of microbe. However, there are some recent reports suggesting a possible relationship between the pH and the bacteria in the granulation tissue of the wound [35]. The results of this study showed (1) the optimal pH of the granulation wound is 7.2–7.5 and (2) pH of the granulation tissue of a burn wound is related to quantity and species of bacteria in the granulation tissue. The wound pH is 6.7 or lower when the number of *Escherichia coli* or *Staphylococcus aureus* is over 10^7/gm of granulation tissue. The wound pH is 8.0 when the number of *Pseudomonas aeruginosa* is 10^8/gm of granulation tissue. The authors of the study propose that, because the measurement of wound pH is rapid, simple, and noninvasive, it might be useful in predicting the take rate of a skin graft [35].

B. Current Trends in Infection Treatment

Recent reports are available to indicate no advantage to the early use of systemic antibiotics in patients with burns, but the use of prophylactic antibiotics during excision is still being questioned [36]. Agents like Ofloxacin [37] and zinc oxide [38] are also recommended antimicrobial agents to cure infected wounds. Another effective antibiotic cream often used on infected wounds is silver sulphadiazine (SSD) cream. A clinical trial was devised to determine whether the healing of partial-thickness burns was retarded by the use of SSD. The results implied that SSD did not retard epithelialization of dermal-depth injuries [39].

Potential beneficial effect of postsurgical rinsing with 0.12% to 0.2% chlorhexidine is clinically demonstrated through proper experimental investigations [40]. Medicated aerosol dressings have been prepared with the aim of forming a protective film over the wound after spraying, combining the properties of antiseptics and hemostats. Chlorhexidine acetate, along with three hemostatics (i.e., zinc acetate, methyl cellulose, and Calendula tincture), were used for the formulations. The formulations were found to be satisfactory in their performance and purpose [41].

Another double-blind, randomized clinical trial involving 214 children ages 6 months to 12 years compared the safety and effectiveness of the new carbacephem Loracarbef and cephalosporin Cefaclor for the treatment of skin and skin structure infections. The two agents were given primarily as oral suspensions. Dosages were 15 milligrams per kilogram per day (mg/kg/day) in two divided doses for Loracarbef and 20 mg/kg/day in three divided doses for Cefaclor. Assessment 72 hours after completion of the 7-day course of treatment indicated a favorable clinical response plus eradication of the pretherapy pathogen in 97.3% of the 74 Loracarbef-treated patients eligible for evaluation and 92.3% of 78 evaluable Cefaclor-treated patients. Favorable response rates at a second posttreatment visit 10 to 14 days after the end of therapy were 95.6% in 68 evaluable Loracarbef-treated patients and 86.2% in 65 treated with cefaclor. The incidence of adverse reactions, including gastrointestinal effects, was low in both groups. No statistical difference in clinical or bacteriologic efficacy or safety was detected between patients treated with Loracarbef and Cefaclor [42]. Also, a similar efficacy study on Cefaclor was done on 563 randomized patients in North America [43].

Using an experimental study on deep burn wounds in rabbits, the therapeutic effects of creams of para-chloro-methyl-xylenol (PCMX), a disinfectant phenol compound, and 1% silver sulfadiazine (As-SD) against *Staphylococcus aureus* infections were evaluated. The average bacterial counts per gram of subeschar tissue of cream base, As-SD cream, and PCMX cream groups were 4.69×10^8, 3.05×10^6, and zero, respectively. Gross inspection of the wound surface showed it to be dry and intact in the PCMX cream group, while in the other two groups autolysis of the eschar was seen. Microscopic examination of the pathological sections indicated a generalized lesion with degeneration and necrosis of the epithelium and the dermis, subcutaneous edema, infiltration of inflammatory cells, and degeneration of myofascia. The lesion seen in the PCMX group was mild. Results of the study demonstrated that PCMX is an effective antimicrobial agent against *Staphylococcus aureus*. Its improved therapeutic effect might be due to a higher concentration of PCMX cream than that reported in the literature was used in the present study [44].

All of the above types of anti-infective drugs can potentially be incorporated with collagen dressings to stop wound infection. However, during development of a dressing for an infected wound, care should be taken to maintain an effective concentration of the given antibiotic. Often, the administration of antimicrobial agents is in the form of a hydrophobic cream. This may be just enough to control infection of the wound, but not to enhance the healing process of the wound. The wound repair cascade of events may be affected in a hydrophobic environment and the subsequent speedy recovery may be impeded because of such nonnative conditions. To overcome these fundamental physiological problems in wound healing, properly prepared collagen-based matrix may be an ideal dressing to effectively deliver an antibiotic agent and keep the activity sustained for longer time intervals.

C. Need for Antimicrobial Agent for Collagen Dressing

Collagen is a good culture matrix for various biological cells, including bacterial and fungal strains. The specific cell-binding sites of collagen not only enhance the normal wound healing process, but also may attract microbes that hinder the healing process. Therefore, evidently collagen-containing biomaterials are relatively more susceptible to bacterial infection. Reports are available that a higher incidence of *Staphylococcus aureus* infection occurs among wounds treated with collagen sheet compared with silicone gel treatment [37]. However, in advanced countries in which environmental hygiene normally protects wounds from infection, there may not be a need for an antimicrobial agent added to collagen dressing. The situation is quite different in many other parts of the world. In these developing countries, at least, there is an absolute need for incorporating the proper antimicrobial agent in collagen-based dressings.

Several binding studies of antimicrobial agents to collagen are available in recent literature. For example, mitomycin C and streptomycin A were studied for their interaction with soluble collagens from sepia, fish, and rat skins using ultraviolet (UV) absorption spectroscopy at pH 7.4 and 3.0 and equilibrium dialysis. Both the drugs bound to collagen as shown from association constants and could be quantitatively recovered by prolonged dialysis against water, showing formation of dissociable bonds between drug and collagen. At pH 7.4, mitomycin C showed greater binding affinity than streptomycin A with collagen from all three species. The reverse trend was seen at pH 3.0. Binding capacity of sepia and fish skin collagens for streptomycin A at pH 3.0 was found to be significantly greater than for mitomycin C [45]. The important aspect to be considered here is the proper maintenance of *in vivo* effectiveness of the antimicrobial agent after binding to collagen.

D. Infection Control Methods

As mentioned above in the study that compared silicone gel versus collagen sheet as wound dressings, *Staphylococcus aureus* was detected in 3 wounds out of 10 cases treated with the collagen sheet, whereas no wound treated with the silicone gel developed an infection [37]. This is obviously a major disadvantage in using collagen as a dressing for infective wounds. Only when this problem is addressed can the collagen dressings really be considered much more efficient than the nonbioactive dressings. Therefore, we discuss some of the recent collagen dressing methods that talk about infection control mechanisms.

The clinical effect and the pharmacokinetic profile of gentamycin-containing collagen for local antibiotic treatment was evaluated using 14 patients with localized groin wound graft infection. All patients were treated by surgical revision and the implantation of one collagen sponge containing 130 mg gentamycin besides systemic antibiotics. Of the patients, 13 of the 14 were cured adequately, giving a success rate in this series of 93%. The pharmacokinetic study showed a very high gentamycin concentration in wound fluid, which neatly exceeded the microbial inhibitory concentration (MIC) values for most bacteria normally considered resistant to gentamycin. These high MIC values were sustained for 2 to 3 days. In conclusion, this study demonstrated a good clinical effect of gentamycin-containing collagen with a high cure rate. In the wound fluid, an initial high concentration of gentamycin was achieved that lasted for 2–3 days [46].

Silver-impregnated collagen cuff (Vitacuff) has also been proven to be an effective anti-infective wound care device [47]. Earlier studies suggest the use of resorcinol and resorcinol monoacetate antiseptic agents with collagenous biomaterials as an effective wound infection control method [48].

VI. CONCLUSIONS

A. New Approaches and Proper Cross-Linking Agents

It is important to consider the collagen dressings not simply as a protective layer on wounds, but also as an active pharmacological treatment to manage any type of wound care with sufficient modification to suit the situation [49,50]. Encouragement should be given to new approaches. One recent report talks about the development of a new biomaterial for artificial skin by combining fibrillar collagen with gelatin [51]. The samples cross-linked dehydrothermally are expected to become a useful matrix substance for artificial skin. However, the evaluation of new materials has to be done with an unbiased approach. Some nonprofit organizations like biomaterials societies can come up with a feasible solution for this task. Delivery of new drugs to the wound site is a recommendable approach that is already proven successful in ulcer management [5] and in asepsis management of corneal wounds [52–57].

Also, as discussed in Section IV.C, the allergenic response of collagen has to be minimized by pepsinlike enzymatic treatment and proper cross-linking. Different cross-linking methods are still being evaluated and debated. In one such study [58], an ultraviolet method was preferred over the other two cross-linking methods evaluated. Chemicals like gluteraldehyde and formaldehyde are quite popular in providing such cross-linking to collagen. Still, there may be some deficiencies with these methods.

Dehydrothermal cross-linking is another method investigated in recent years. It is reported through a study that collagen was dehydrothermally treated (heat cured) by heating dry under vacuum at 60°C, 80°C, 100°C, and 120°C [59]. The *in vivo* biodegradation of dehydrothermally treated collagen sponge was investigated using a rat lumbar muscle implantation model for up to 28 days. For all heat-cured collagens, the data strongly indicated that both cross-linking and denaturation/degradation were present in increasing quantities with increasing temperature of treatment.

B. Animal Models

Developing a proper animal model to evaluate a dressing is a major task. When a dressing is developed for more than one type of wound, the evaluation studies have to be conducted in such a way that results of the study must qualify the product to be used on all such wounds. For example, in a study, the investigation of synthetic, adherent, moisture-vapor-permeable dressings was done using mouse and guinea pig systems [60]. Repair tissue was quantified histomorphometrically in full-thickness wounds covered with different dressings for periods up to three weeks. The outcome of the study clearly indicated dressing-induced inhibition of connective tissue could be partially reversed by treatment with transforming growth factor-β Form 1 or 2. The authors suggest such experimental studies in mice and guinea pigs may be useful as models of chronic nonhealing wounds.

C. Increasing Bioactivity by Cell Seeding and Growth Factor Incorporation

Epidermal-cell-seeded matrix seems to be the only possible method to cure some wounds, such as chronic leg ulcers in sickle cell anemia that occur mainly around the ankles. Because epidermal cells need a supporting matrix in order to be applied to the ulcer, properly made collagen may be the future hope for developing epidermal-cell-seeded autografts. In a similar manner, the addition of human basic fibroblast growth factor, a potent angiogenic agent, to the skin-replacement wound dressings can be achieved with collagenous matrices [61].

In another study, healing was evaluated in dermal wounds treated with a collagen sponge seeded with fibroblasts or coated with basic fibroblast growth factor (bFGF). Experimental results indicated that the presence of collagen sponge resulted in increased wound tensile strength, increased collagen fiber diameters, and increased degree of reepithelialization in the upper dermis 15 days postwounding and implantation. These results demonstrated that fibroblast seeding and bFGF coating in conjunction with a Type I collagen sponge matrix facilitates early dermal and epidermal wound healing [62].

A comparative study result showed that seeding of epidermal cells on a more mature dermal equivalent leads to improved differentiation status of the epidermal layer [63]. Another report demonstrated closure of large skin wounds with composite grafts consisting of collagen-glycosaminoglycan (GAG) substrates populated with cultured dermal fibroblasts and epidermal keratinocytes; these were tested in a pilot study on full-thickness burn wounds of three patients as an alternative to split-thickness skin. This cultured skin analog provides an experimental alternative to the split-thickness skin graft that develops histiotypic markers of skin anatomy and antigen expression after wound closure [64].

REFERENCES

1. Kinner, M. A., and Daly, W. L. Skin transplantation. *Critical Care Nursing Clinics of North Am.*, 1992, 4(2):173–78.
2. Cuzzell, J. Z. Choosing a wound dressing: A systematic approach. *AACN Clin. Issues in Critical Care Nursing*, 1990, 1(3):566–77.
3. Fry, M. M. A framework for wound management. *Nursing Standard*, 1993, Mar. 31–Apr. 6, 7(28):29–32.
4. Feldman, D. L. Which dressing for split-thickness skin graft donor sites? *Annals of Plastic Surgery*, 1991, 27(3):288–91.
5. Young, J. B., and Dobrzanski, S. Pressure sores. Epidemiology and current management concepts. *Drugs and Aging*, 1992, 2(1):42–57.
6. Herndon, D. N., Rutan, R. L., and Rutan, T. C. Management of the pediatric patient with burns. *Journal of Burn Care and Rehabilitation*, 1993, 14(1):3–8.
7. Abramo, A. C., and Viola, J. C. Heterologous collagen matrix sponge: Histologic and clinical response to its implantation in third-degree burn injuries. *British Journal of Plastic Surgery*, 1992, 45(2):117–22.
8. Yang, J. Y. Clinical application of a collagen sheet, YCWM, as a burn wound dressing. *Burns*, 1990, 16(6):457–61.
9. Schrader, R., Steinbacher, S., Burger, W., Kadel, C., Vallbracht, C., and Kaltenbach, M.

Collagen application for sealing of arterial puncture sites in comparison to pressure dressing: A randomized trial. *Catheterization and Cardiovascular Diagnosis*, 1992, 27(4):298–302.

10. Kehrel, B., Kronenberg, A., Rauterberg, J., Niesing-Bresch, D., Niehues, U., Kardoeus, J., Schwippert, B., Tschope, D., van de Loo, J., and Clemetson, K. J. Platelets deficient in glycoprotein IIIb aggregate normally to collagens type I and III but not to collagen type V. *Blood*, 1993, 82(11):3364–70.

11. Sugiyama, T., Ishibashi, T., and Okuma, M. Functional role of the antigen recognized by an antiplatelet antibody specific for a putative collagen receptor in platelet-collagen interaction. *International Journal of Hematology*, 1993, 58(1–2):99–104.

12. Mian, E., Martini, P., Beconcini, D., and Mian, M. Healing of open skin surfaces with collagen foils. *International Journal of Tissue Reactions*, 1992, 14(Suppl.):27–34.

13. Micheletti, G., Onorato, I., and Micheletti, L. Heterologous, lyophilized, non-denatured Type-I collagen in dentistry. *International Journal of Tissue Reactions*, 1992, 14(Suppl.):39–42.

14. Mian, M., Aloisi, R., Benetti, D., Rosini, S., and Fantozzi, R. Potential role of heterologous collagen in promoting cutaneous wound repair in rats. *International Journal of Tissue Reactions*, 1992, 14(Suppl.):43–52.

15. Sedlarik, K. M., Schoots, C., Fidler, V., Oosterbaan, J. A., and Klopper, J. P. Rat Model. Comparative animal experiment studies of the effect of exogenous collagen on healing of a deep skin wound. *Unfallchirurgie*, 1991, 17(1):1–13.

16. Sedlarik, K. M., Schoots, C., Oosterbaan, J. A., and Klopper, J. P. Healing of a deep skin wound using a collagen sponge as dressing in the animal experiment. *Aktuelle Traumatologie*, 1992, 22(5):219–28.

17. Palmieri, B. Heterologous collagen in wound healing: A clinical study. *International Journal of Tissue Reactions*, 1992, 14(Suppl.):21–25.

18. Gao, Z. R., Hao, Z. Q., Li, Y., Im, M. J., and Spence, R. J. Rat model porcine dermal collagen as a wound dressing for skin donor sites and deep partial skin thickness burns. *Burns*, 1992, 18(6):492–96.

19. Goudarzi, Y. M., Khodadadyan, C., and Hertel P. Clinical experience with collagenous wound dressing in severe traumatic soft tissue injuries. *Aktuelle Traumatologie*, 1992, 22(5):214–18.

20. Meltzer, M. A., Nassif, J. M., Hyde, K. J., and Arthurs, B. P. Collagen shield contact lens use after eyelid surgery. *Ophthalmic Plastic and Reconstructive Surgery*, 1992, 8(4):290–91.

21. Renard, G., Bennani, N., Lutaj, P., Richard, C., and Trinquand, C. Comparative study of a collagen corneal shield and a subconjunctival injection at the end of cataract surgery. *Journal of Cataract and Refractive Surgery*, 1993, 19(1):48–51.

22. Reindorf, C. A., Walker-Jones, D., Adekile, A. D., Lawal, O., and Oluwole, S. F. Rapid healing of sickle cell leg ulcers treated with collagen dressing. *Journal of the Nat. Med. Assoc.*, 1989, 81(8):866–86.

23. Beghe, F., Menicagli, C., Neggiani, P., Zampieri, A., Trallori, L., Teta, E., and Rosini, S. Lyophilized non-denatured type-I collagen (Condress) extracted from bovine Achilles tendon and suitable for clinical use. *International Journal of Tissue Reactions*, 1992, 14(Suppl.):11–19.

24. Mian, E., Mian, M., and Beghe, F. Lyophilized Type-I collagen and chronic leg ulcers. *International Journal of Tissue Reactions*, 1991, 13(5):257–69.

25. Romanko, K. P. Pressure ulcers. *Clinics in Podiatric Medicine and Surgery*, 1991, 8(4):857–67.

26. Watts, M. T., and Collin, R. The use of fibrin glue in mucous membrane grafting of the fornix. *Ophthalmic Surgery*, 1992, 23(10):689–90.

27. Lasa, C. I., Jr., Kidd, R. R., III, Nunez, H. A., and Drohan, W. N. Effect of fibrin glue and Opsite on open wounds in DB/DB mice. *Journal of Surgical Research*, 1993, 54(3):202–6.

28. Gunasekaran, S., Bathurst, I. C., Constantz, B. R., Quiaoit, J., Barr, P. J., and Gospodarowicz, D. Comparative utility of mineralized collagen as an osteoinductive material. In *HA and Related Materials* (Brown, ed.), CRC Press, Boca Raton, FL, 1994, pp. 171–80.

29. Chvapil, M., Holubec, H., and Chvapil, T. Inert wound dressing is not desirable. *Journal of Surgical Research*, 1991, 51(3):245–52.

30. Soo, C., Rahbar, G., and Moy, R. L. The immunogenecity of bovine collagen implants. *J. Dermatologic Surgery and Oncology*, 1993, 19(5):431–34.

31. Somerville, P., and Wray, R. C. Asymmetrical hypersensitivity to bovine collagen. *Annals of Plastic Surgery*, 1993, 30(5):449–51.

32. Cukier, J., Beauchamp, R. A., Spindler, J. S., Spindler, S., et al. Association between bovine collagen dermal implants and a dermatomyositis or a polymyositis-like syndrome. *Annals of Internal Medicine*, 1993, 118(12):920–28.

33. Ben-Amitai, D., and Ashkenazi, S. Common bacterial skin infections in childhood. *Pediatric Annals*, 1993, 22(4):225–27, 231–33.

34. Yagupsky, P. Bacteriologic aspects of skin and soft tissue infections. *Pediatric Annals*, 1993, 22(4):217–24.

35. Chai, J. K. The pH value of granulating wound and skin graft in burn patients. *Chinese Journal of Plastic Surgery and Burns*, 1992, 8(3):177–78, 246 (in Chinese).

36. Griswold, J. A., Grube, B. J., Engrav, L. H., Marvin, J. A., and Heimbach, D. M. Determinants of donor site infections in small burn grafts. *Journal of Burn Care and Rehabilitation*, 1989, 10(6):531–55.

37. Sawada, Y., Yotsuyanagi, T., and Sone, K. A silicone gel sheet dressing containing an antimicrobial agent for split thickness donor site wounds. *British Journal of Plastic Surgery*, 1990, 43(1):88–93.

38. Agren, M. S. Studies on zinc in wound healing. *Acta Dermato-Venereologica*. Supplementum, 1990, 154:1–36.

39. Stern, H. S. Silver sulphadiazine and the healing of partial thickness burns: A prospective clinical trial. *British Journal of Plastic Surgery*, 1989, 42(5):581–85.

40. Sanz, M., Newman, M. G., Anderson, L., Matoska, W., Otomo-Corgel, J., and Saltini, C. Clinical enhancement of post-periodontal surgical therapy by a 0.12% chlorhexidine gluconate mouthrinse. *Journal of Periodontology*, 1989, 60(10):570–76.

41. Garg, S., and Sharma, S. N. Development of medicated aerosol dressings of chlorhexidine acetate with hemostatics. *Pharmazie*, 1992, 47(12):924–46.

42. Hanfling, M. J., Hausinger, S. A., and Squires, J. Loracarbef vs. Cefaclor in pediatric skin and skin structure infections. *Pediatric Infectious Disease Journal*, 1992, 11(8 Suppl):S27–30.

43. Schupbach, C. W., Olovich, K. G., and Dere, W. H. Efficacy of Cefaclor AF in the treatment of skin and skin-structure infections. *Clinical Therapeutics*, 1992, 14(3):470–79.

44. Ge, S. D. Experimental study of topical uses of para-chloro-methyl-xylenol in burns: Therapeutic effect against *Staphylococcus aureus*. *Chinese Journal of Plastic Surgery and Burns*, 1990, 6(2):115–17, 158–59 (in Chinese).

45. Ramesh, D. V., and Sehgal, P. K. *In-vitro* interaction of mitomycin C and streptomycin A with collagen. *Journal of Pharmacy and Pharmacology*, 1991, 43(11):802–4.

46. Jorgensen, L. G., Sorensen, T. S., and Lorentzen, J. E. Clinical and pharmacokinetic evaluation of gentamycin containing collagen in groin wound infections after vascular reconstruction. *European Journal of Vascular Surgery*, 1991, 5(1):87–91.

47. Babycos, C. R., Barrocas, A., and Webb, W. R. A prospective randomized trial comparing the silver-impregnated collagen cuff with the bedside tunneled subclavian catheter. *Journal of Parenteral and Enteral Nutrition*, 1993, 17(1):61–63.

48. Gunasekaran, S., and Chvapil, M. Collagen based biomaterials: An ideal way of increasing their resistance to infection. *Biomaterials, Artificial Cells and Artificial Organs*, 1988, 16(4):771–84.

49. Murphy, G. F., Orgill, D. P., and Yannas, I. V. Partial dermal regeneration is induced by biodegradable collagen-glycosaminoglycan grafts. *Laboratory Investigation*, 1990, 62(3):305–13.

50. Mian, M., Beghe, F., and Mian, E. Collagen as a pharmacological approach in wound healing. *International Journal of Tissue Reactions*, 1992, 14(Suppl.):1–9.

51. Koide, M., Osaki, K., Konishi, J., Oyamada, K., Katakura, T., Takahashi, A., and Yoshi-

zato, K. A new type of biomaterial for artificial skin: Dehydrothermally cross-linked composites of fibrillar and denatured collagens. *Journal of Biomedical Materials Research*, 1993, 27(1):79-87.

52. Clinch, T. E., Hobden, J. A., Hill, J. M., O'Callaghan, R. J., Engel, L. S., and Kaufmann, H. E. Collagen shields containing tobramycin for sustained therapy (24 hours) of experimental *Pseudomonas* keratitis. *Clao Journal*, 1992, 18(4):245-47.

53. Silbiger, J., and Stern, G. A. Evaluation of corneal collagen shields as a drug delivery device for the treatment of experimental *Pseudomonas* keratitis. *Ophthalmology*, 1992, 99(6):889-92.

54. Assil, K. K., Zarnegar, S. R., Fouraker, B. D., and Schanzlin, D. J. Efficacy of tobramycin-soaked collagen shields versus tobramycin eyedrop loading dose for sustained treatment of experimental *Pseudomonas aeruginosa*-induced keratitis in rabbits. *American Journal of Ophthalmology*, 1992, 113(4):418-23.

55. Herschler, J. Long-term results of trabeculectomy with collagen sponge implant containing low-dose antimetabolite. *Ophthalmology*, 1992, 99(5):666-70.

56. Pleyer, U., Legmann, A., Mondino, B. J., and Lee, D. A. Use of collagen shields containing amphotericin B in the treatment of experimental *Candida albicans*-induced keratomycosis in rabbits. *American Journal of Ophthalmology*, 1992, Mar. 15, 113(3):303-8.

57. Baziuk, N., Gremillion, C. M., Jr., Peyman, G. A., and Cho, H. K. Collagen shields and intraocular drug delivery: Concentration of gentamicin in the aqueous and vitreous of a rabbit eye after lensectomy and vitrectomy. *International Ophthalmology*, 1992, 16(2):101-7.

58. Morykwas, M. J. *In vitro* properties of crosslinked, reconstituted collagen sheets. *Journal of Biomedical Materials Research*, 1990, 24(8):1105-10.

59. Gorham, S. D., Light, N. D., Diamond, A. M., Willins, M. J., Bailey, A. J., Wess, T. J., and Leslie, N. J. Effect of chemical modifications on the susceptibility of collagen to proteolysis. II. Dehydrothermal cross-linking. *International Journal of Biological Macromolecules*, 1992, 14(3):129-38.

60. Ksander, G. A., Pratt, B. M., Desilets-Avis, P., Gerhardt, C. O., and McPherson, J. M. Inhibition of connective tissue formation in dermal wounds covered with synthetic, moisture vapor-permeable dressings and its reversal by transforming growth factor-beta. *Journal of Investigative Dermatology*, 1990, 95(2):195-201.

61. Cooper, M. L., and Hansgrough, J. F. Use of a composite skin graft composed of cultured human keratinocytes and fibroblasts and a collagen-GAG matrix to cover full-thickness wounds on athymic mice. *Surgery*, 1991, 109(2):198-207.

62. Marks, M. G., Doillon, C., and Silver, F. H. Effects of fibroblasts and basic fibroblast growth factor on facilitation of dermal wound healing by type I collagen matrices. *Journal of Biomedical Materials Research*, 1991, 25(5):683-96.

63. Bouvard, V., Germain, L., Rompre, P., Roy, B., and Auger, F. A. Influence of dermal equivalent maturation on the development of a cultured skin equivalent. *Biochemistry and Cell Biology*, 1992, 70(1):34-42.

64. Boyce, S. T., Greenhalgh, D. G., Kagan, R. J., Housinger, T., Sorrell, J. M., Childress, C. P., Rieman, M., and Warden, G. D. Skin anatomy and antigen expression after burn wound closure with composite grafts of cultured skin cells and biopolymers. *Plastic and Reconstructive Surgery*, 1993, 91(4):632-41.

VIII
ADHESIVES, MEMBRANES, COATINGS, AND FILMS

38
Applications of Thin-Film Technology in Biomedical Engineering

Armin Bolz
Friedrich-Alexander-University of Erlangen-Nürnberg
Erlangen-Nürnberg, Germany

I. INTRODUCTION

If alloplastic materials have to be selected for use in biomedical applications, especially in implants, two important requirements must be considered: the compatibility of the solid in the biological environment and the suitability of its mechanical, electrical, or other properties for a given application. With regard to the first requirement, a foreign material placed in contact with a biological fluid must not degrade (unless the degradation is intended) or cause adverse reactions of the body. On the one hand, the material should remain stable with regard to its physical properties; no possibly toxic corrosion products should be released and no diffusion of body fluid components into the material should take place. On the other hand, it must not have harmful effects on the fluid's components; that is, it must not provoke toxic or allergic reactions of the immune system or thrombosis in the case of blood-contacting devices. With regard to the second requirement, the material selection must follow the rules that are well known from implant design. For instance, in the case of mechanically loaded implants, it has to show satisfactory levels of elasticity, rigidity, tensile or bending strength; in the case of electrodes, the electronic properties also have to be taken into consideration, and so on.

A closer examination of these two requirements reveals that they call for two different points of view. On one hand, the biocompatibility requirement is mainly a reflection of the surface characteristics of a solid. On the other hand, the functional properties are largely determined by the bulk. Therefore, an ideal biomaterial meeting both requirements has to have satisfactory surface as well as bulk characteristics. Two drawbacks are related to this *conditio sine qua non*.

First, the needs are as varied as the available implants. Thus, it is necessary to have a variety of good biomaterials available, such as materials with varying degrees of elasticity for tubings, artificial vessels, ocular lenses, or heart valves, and some other materials with

1287

different electrical conductivity for implantable electrical devices. Second, it seems to be almost a paradox of nature that materials with suitable bulk properties fail on account of their poor biocompatibility and vice versa. To illustrate this phenomenon, the artificial heart valve may serve as an example. Although for more than 30 years, research has concentrated on finding a suitable material for this valve, no real success has been achieved. Presently used titanium alloys or LTI (low temperature isotropic) carbon have good mechanical properties for a reliable valve design, but they require anticoagulation therapy, which has well-known disadvantages such as uncontrolled bleeding [1]. This illustrates that, while it is difficult to meet the surface as well as the bulk requirements with a single material, it is almost impossible with regard to all different applications.

The same problem appears in nature, such as in the skeletal bone: Bone must be stable to meet the structural requirements of the body, it must be light not to induce a high body weight, and it must allow good perfusion in order to provide the bone cells with all they need. Figure 1 shows nature's solution, a hybrid design consisting of at least three layers: substantia spongiosa, substantia compacta, and the periosteum.

If we apply this approach of evolution to implant design, we should also create a hybrid design, that is, an implant should be made of a well-known and stable engineering material and a thin, biocompatible and corrosion-resistant surface coating. Thus, on one hand the substrate material can be optimally selected according to the functional needs of the application without regard to its biocompatibility or even toxicity. On the other hand, the coating material will suppress or, in a more exact expression, must not promote the onset of undesirable reactions. Thus, we get a cooperation, or "teamwork," of at least two different materials that facilitates the design of a reliable as well as biocompatible implant.

This approach sounds very simple, and a lot of different coating techniques have already been applied in biomedical engineering. A well-known technique is the plasma spraying process, such as for coating orthopedic devices with hydroxyapatite or calcium phosphate ceramics [3–8] or for producing porous metal surfaces [9] in order to improve the ingrowth of bone. There are also some similar techniques, for example, the flame spraying process [10] and the spraying of powder on top of a hot substrate material [11].

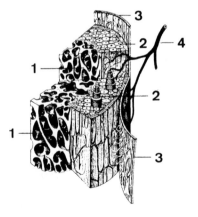

Figure 1   Hybrid design of human femur (1, substantia spongiosa; 2, substantia compacta; 3, periosteum; 4, blood vessels and nerves). (From Ref. 2.)

These powder processes are completed by sintering [12,13] and other high-temperature processes like oxygen diffusion hardening (ODH) [14,15]. Conventional electrodeposition techniques are used to improve electrodes, tools, or orthopedic fixation bars [16–18]. Last, but not least, simple dipping techniques play an important role for catheters and different types of tubings [19–21], especially in order to prepare heparinized, antithrombogenic surfaces [22–26].

Although for some applications very good results have been obtained with these techniques, they show some disadvantages: Above all, there are limitations in the selection of materials because usually it is not possible to achieve ceramic or polymer coatings by electrodeposition or dipping. In addition, these techniques are very coarse and they do not allow an exact tailoring of the surface properties. There are also problems with the adhesion of the coatings. Finally, these coatings contain a relatively high percentage of impurities, which may cause toxic effects or at least affect the desired properties. As a result, thin-film technologies based on high vacuum have become more and more widely used, and scientific as well as industrial users are getting more and more interested.

Therefore, the aim of this chapter is to give a general idea of the different types of thin-film processes, to summarize their advantages and drawbacks, and to compare their usefulness. It is not intended to introduce these technologies in detail in order to make it possible for a potential user to start with the coating the next day. For that purpose, a lot of detailed literature is available, as well as qualified advisors from different thin-film equipment companies who are willing to help.

The following is intended to give an approximate idea of the capabilities of thin-film processes in general. It reviews different applications that have already been realized or are under investigation. Two of them (the development of semiconducting coatings with high hemocompatibility and the coating of pacing leads) are discussed in more detail to demonstrate the tremendous impact of thin-film technology on biomedical engineering.

II. THIN-FILM TECHNOLOGIES AS A TOOL FOR HYBRID DESIGN

The term *thin-film technology* has to be defined. In this context, it is used (as it is understood in the semiconductor industry) for all coating processes that necessitate a high-vacuum chamber and in which the coatings are formed atom by atom. Thus, dipping is not a thin-film technology even if the resulting coatings may be only a few microns in thickness. On the other hand, sintering under high-vacuum conditions is also not considered to belong to this group because it does not meet the atomic requirement, which is absolutely necessary for an exact tailoring of the chemical composition and all properties related to it.

All thin-film processes can be subdivided into three steps. First, the precursor (i.e., the material that contains the elements to be deposited as a thin coating and possibly some other volatile components) must be converted into single atoms or at least molecules. Thus, depending on whether the precursor is chemically pure or bound in a compound, and whether it is solid, liquid, or gaseous, different types of cracking and activating mechanisms have to be employed. Second, these particles have to be transported through the high vacuum inside the reaction chamber to the surface of the substrate, unless the activation process takes place at the deposition site itself. Finally, these particles must be bound and form a layer. As a result, if we want to arrange thin-film techniques according to their underlying physical principles, we have two possible approaches, one from the precursor and another from the particle energy point of view. The other, commonly used differentiation into physical vapor deposition (PVD) and chemical vapor deposition (CVD) techniques is not used in this context as it does not improve the understanding of the specific characteristics of each technology.

Supposing that the precursor is gaseous (e.g., CH_4 for deposition of carbon films), nothing else has to be done aside from controlling the mass flow. Liquid precursors such as metal organic compounds like tetrakis(dimethylamido)-titanium for deposition of TiC_xN_y films have to be heated in a small bottle outside the reactor to provide sufficient vapor pressure; the resulting vapor must then be sent to the process chamber via hot tubing in order to prevent condensation. Solid precursors must be vaporized by using resistance heaters, electron guns, or laser beams, or must be atomized (sputtered) by means of impinging high energetic ions.

The resulting particles are kept in a high-vacuum environment in order to maintain the purity of the precursors; the residual gas pressure should be significantly lower than the process pressure. Then, they are transported by diffusion or electrical fields toward the substrate. On their way to the surface, they may be hit by other particles, which results in a loss of energy. The number of collisions and thus the energy of the arriving particles depend on their mean free path l, that is, the mean way between two collisions (of identical particles), which is calculated according to

$$l \cdot p = const \qquad (1)$$

where p is the pressure and *const* is a constant in the range between $2 \cdot 10^{-1}$ and $20 \cdot 10^{-1}$ Pa (pascal) cm depending on the type of gas (e.g., in air, *const* $= 6.67 \cdot 10^{-1}$ Pa cm) used. Arriving at the substrate, the particles may react depending on their kinetic energy. Figure 2 compares the different mechanisms as a function of energy. Particles with an energy below approximately 0.5 electron volts (eV) in general are too slow to overcome the activation energy barrier for being adsorbed or chemically bound unless their melting point is very high and the substrate temperature low (this exception is valid, e.g., for evaporation techniques). Therefore, they migrate on the surface until they find an active reaction site, or they may even desorb.

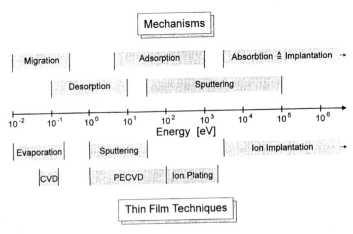

Figure 2   Possible particle reactions and related technologies as a function of energy. (Modified from Ref. 27.)

Table 1 Subdivision of Thin-Film Technologies with Regard to Mean Particle Energy
and Type of Precursor

Mean energy of predominating type of particles	Type of precursor		
	Gaseous	Liquid	Solid
0.1 eV (neutral)	CVD	CVD	Evaporation
1–10 eV (neutral)	PPECVD or plasma polymerization	PECVD or plasma polymerization	Sputtering
100–1000 eV (neutral)	Ion plating	Ion plating	Ion plating
>1000 eV (ions)	Ion implantation	Ion implantation	No established technique

Note: Particle energy depends on whether or not it is an ion (the considered predominating type is indicated in parentheses). In some cases, both types may exist.
CVD, chemical vapor deposition; PECVD, plasma activated CVD.

An easy way for increasing and controlling the energy is the acceleration of ionized particles in electric fields. Particles with energies in the range of 10 eV to 100 eV adsorb or even react at the site of collision. At higher energies, the sputtering mechanism starts, thus releasing substrate atoms. This process may be utilized for cleaning or cross-linking the surface. If the energy is increased above approximately 10 keV, the particles may also penetrate into the substrate.

Table 1 arranges the most important thin-film techniques according to the above-mentioned criteria. Chemical vapor deposition (CVD) and evaporation techniques only use neutral particles with very low particle energies. Acceleration of atoms or molecules can be easily performed in a plasma. Thus, sputtering and plasma-enhanced CVD (PECVD) techniques (or plasma polymerization, which is synonymous) work with partially ionized precursors, resulting in better adhesion. The highest particle energies are used in ion plating and ion implantation processes, which result in the best adhesion properties. However, each technology has its own advantages and drawbacks, and for each application the optimal process should be selected. In the following sections, these techniques are discussed in more detail.

A. Chemical Vapor Deposition

In the chemical vapor deposition process, a chemical vapor phase reaction of the precursor molecules and their subsequent bonding to the surface is induced by simply heating the substrate [27,28]. Its principles are demonstrated in Fig. 3. Three steps determine the quality of CVD-deposited coatings. First, the continuous and well-balanced transport of the precursor to the substrate is essential for the homogeneity of the films. Much effort is necessary (e.g., large sintered plates as porous gas transducers) to achieve an equal and not exhausting flow [29]. Therefore, mainly planar substrates are suited for the CVD process. Otherwise, special fixtures are required or the substrates have to be elevated by the gas stream (fluidized bed process) [30].

Second, the gas must react at the surface of the substrate. In general, the precursor adsorbs and is cracked into its components due to the high substrate temperature. However, in thermal equilibrium, the adsorption rate equals the desorption rate, which would

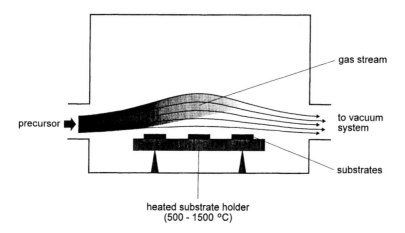

gas stream

precursor

to vacuum
system

substrates

heated substrate holder
(500 - 1500 °C)

Figure 3   Principles of the chemical vapor deposition technique.

result in a vanishing deposition rate. Therefore, the vapor pressure of the precursor must be set above its equilibrium value. After cracking, the volatile residual parts desorb and the main components remain at the surface. Due to the high substrate temperature, the adatoms are allowed to move on top of the surface until they find a suitable binding site. Thus, CVD coatings are generally crystalline. Third, the desorbed particles must be removed from the chamber in such a way that they do not disturb the process; this requires high gas flow and pumping rates.

CVD technology is subdivided into three main groups: the fluidized bed reactor, the cold wall batch reactor, and low-pressure CVD (LPCVD). In the fluidized bed reactor, the reaction takes place in the gas phase as well as on the substrate. Thus, small particles, so-called clusters, reach the substrate's surface and produce an inhomogenous porous film that is similar to a microscopically sintered solid.

A second commonly used reactor type is the cold wall batch reactor. A large carbon- or silicon-carbide-coated graphite susceptor is heated by radio-frequency (RF) coils at the outside of the reactor, inducing eddy currents in the electrically conductive graphite. Thus, the quartz tube used as the reactor chamber remains cool whereas the substrate is heated resulting in a localized deposition.

In order to reduce the substrate temperature, the process has also been modified to the so-called low-pressure CVD (LPCVD). Reducing the temperature to the minimum reaction threshold also necessitates a decrease in process pressure. For instance, whereas the deposition of silicon nitride at atmospheric pressure requires approximately 1000°C for suitable deposition rates, LPCVD deposition of the same material at 800°C requires a reduction in pressure to 0.2 bar. Thus, direct heating with "hot walls" can be employed.

CVD technology is mainly used in the semiconductor industry for depositing all types of silicon-containing layers. In this case, silane (SiH_4) or tetrachlorosilane ($SiCl_4$) is decomposed into silicon and volatile residual components, which are pumped away. Gas mixtures are also used to produce silicon compounds. Another example, which is very important in biomedical engineering, is the manufacturing of LTI carbon [30]. This modification of carbon is used for artificial heart valves due to its biocompatibility and is deposited from hydrocarbons (e.g., propane). However, the high temperatures, in the range of 600°C to 1000°C or more, required by this process may result in a destruction of the substrate material. Polymers will be decomposed, metals will suffer from diffusion processes, and only ceramics may withstand this treatment. For this reason, other technologies with lower substrate temperatures are needed, especially in biomedical engineering.

B. Evaporation

For the evaporation technique, a solid, not thermally degrading (i.e., not polymeric), precursor is required (e.g., a metal or, with some limitations, a ceramic) [27,31]. This material is heated under ultra-high vacuum conditions (usually, the residual gas pressure is well below 10^{-5} Pa) until a sufficient vapor pressure is achieved. If the substrate is kept at a lower temperature than the source, the vapor will condense and form a thin film [32]. The principle is shown in Fig. 4.

In general, two different energy sources are in use: a resistor heater and an electron gun. The cheapest is a resistor heater made of a metal with a high melting point, such as tungsten, molybdenum, or tantalum; sometimes carbon is also used. Small pots containing the coating material are heated simply by applying electric power [33,34]. However, there are a few limitations. If the melting points of the heater and coating material are very similar, the purity of the deposited layers is impaired because of coevaporation of heater material. On the other hand, some coating materials form eutectic alloys with the heater material, which lowers its melting point, and thus the heater is rapidly destroyed.

This problem can be remedied with the use of electron guns. If the coating material is only locally melted, thus forming its own container, negative reactions with the surrounding pot material are impossible. Therefore, all materials can be used and much higher temperatures achieved without destroying the heater. In addition, an electron beam can easily be switched from one pot to another. Thus, the electron beam technique allows fast and easily controllable coevaporation of different materials. In spite of its high costs, it is recommended for high-rate coating with high purity [35,36].

During the past years, a third energy source has become more and more widespread, the laser beam [37,38,39]. Due to its high energy density and independence of electric charges, it is employed for flash evaporation of alloys or ceramics such as high-

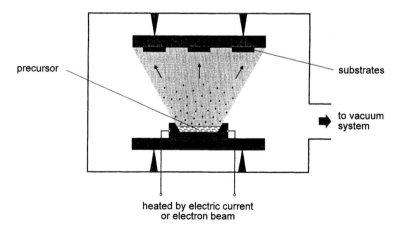

Figure 4 Schematic of an evaporation system.

temperature superconductors. However, a large-scale application is unlikely unless the energy effectiveness of lasers is improved.

In all cases, the resulting vapor pressure P_v is given by the equation of Clausius and Clapeyron. In its integrated form, it can be written as

$$P_v = A \cdot \exp\left(-\frac{B}{T}\right) \tag{2}$$

with A as an integration constant and B as a constant that depends on the heat of vaporization of the coating material. As a result, the vapor pressure is determined in a very sensitive way by the temperature of the source T. If, for example, the temperature of a zinc melt is increased from 390°C to 397°C, P_v is doubled. As the vapor pressure is proportional to the deposition rate, which determines the properties of the coating, high-quality evaporation requires expensive measurement equipment for controlling the source temperature in a closed-loop system.

In first-order approximation, the formation of the layer can be understood as a condensation process with no chemical reactions, which is very similar to the condensation of water on a cold window; the coating material condenses back to its original form [32]. The kinetic energy of the emitted particles E_p is determined by the source temperature T (k is Boltzmann's constant):

$$E_p = \frac{3}{2}kT \tag{3}$$

For usual source temperatures of some thousand kelvin, E_p is in the range of 0.1 eV to 0.5 eV. Due to the fact that the binding energy between two vapor particles is generally much higher than between a particle and the substrate, the arriving particles will move on top of the surface until they find a "colleague." Thus, during the early condensation phase, small islands or nuclei are formed — preferentially at impurities or irregularities of the substrate or at sites with higher binding energy — before they grow into a continuous layer. Due to this low energy of the arriving particles, impurities will remain at the interface between film and substrate, leading to a poor adhesion of evaporated films.

As a rule of thumb, under usual evaporation conditions (substrate is held at room temperature and evaporation temperature of the coating material lies above 1500°C under atmospheric pressure conditions), the condensation coefficient can be assumed to be one (i.e., nearly all particles are bound to the surface). As a result, with modern equipment deposition rates of up to 50 microns per second are achieved [36]. The main applications are optical coatings or coatings on foils.

C. Basic Principles of Plasma-Based Thin-Film Technologies

Chemical vapor deposition has been used in integrated circuit (IC) fabrication almost since the beginning, and it is still used in many cases. Whereas this process is well understood, it requires extremely high temperatures. However, lower temperatures are desired to prevent diffusion of dopants or interactions with the substrate, which may diminish the film quality, and to be free in substrate selection. The simple evaporation technique allows low substrate temperatures but provides poor adhesion. Consequently, plasma techniques have been developed for thin-film technologies that use the physical as well as chemical aspects of the plasma.

A plasma can be thought of as a special state of matter in which the number of positive and negative charges are equal, as they are in a neutral gas, but the individual densities of free charge carriers may be quite high (10^9 to 10^{14} cm^{-3}). The fraction of the original neutral species that has become ionized is called the *degree of ionization*. In plasmas utilized in thin-film technology, only a small percentage of the gas is ionized; usually, their degree of ionization is in the range of 0.5% to 5%.

To form and sustain a plasma requires an energy source to achieve the required ionization. In steady state, the rate of ionization must balance the losses of ions and electrons by recombination or diffusion. Most often, plasmas are initiated and sustained by electric fields produced by direct current (DC) or alternating current (AC) – typically radio-frequency – power supplies. In some cases, the plasma is also induced by a microwave field.

The electric field initially accelerates a few free electrons present in the gas. Although the electric field also acts on the ions, they remain relatively unaffected due to their higher mass. Once these electrons acquire sufficiently high energies, their collisions with gas species result in excitation and ionization processes, the latter generating additional electrons, which are again accelerated by the electric field. In addition, other mechanisms, such as secondary electron emission from the side walls, also produce electrons. This process avalanches quickly, creating the steady state of glow discharge.

Since electrons have a much higher mobility than ions, any floating surface in contact with the plasma will develop a negative potential with respect to the plasma. The resulting electric field reduces the electron current density to the surface until it equals the ion current density and the electrical neutrality of the plasma is thus maintained. The ions are accelerated to the surface by the electric field; they arrive with an energy that is determined by the potential of the surface relative to the plasma. In DC discharges, the potential of floating surfaces normally is a few volts with respect to the plasma, which has insignificant effects. However, if an external voltage is applied, the ion energy easily exceeds some 10 eV. In addition, in RF discharges, powered capacitively coupled electrodes may even develop a so-called DC self-bias voltage (depending on the ratio of surface area of powered electrode and ground) in the range of some 100 volts (V). Detailed explanations of the different processes taking place in a plasma are given in Ref. 27.

There are two aspects of plasmas that are important in thin-film technology: physical and chemical. Many of the physical effects resulting from the plasma state are due to the arrival of photons, electrons, and ions at the substrate. The most important feature of plasma discharges is their ability to generate ions and to accelerate them to energies of up to 1000 eV or more. If the charged particles have sufficient energy, they may have significant effects on the substrate, such as sputtering, cleaning, or stress generation. They may also heat the substrate and influence the film formation. However, it has to be taken into account that, in most plasma systems, the electric field will be perpendicular to the substrate; ions, therefore, reach the sample at normal incidence. Thus, depending on the substrate geometry, some parts of the sample may be affected much less or remain unaffected.

The chemical effect of plasmas results from their ability to generate chemically active species efficiently. This is initiated by the bombardment of molecules and atoms by plasma electrons, which may break the chemical bonds if they have sufficient energy. For example, N_2^+ ions are easily produced in a plasma. When these particles strike a substrate, they may dissociate into N atoms, which are very reactive. While molecular nitrogen is a very stable molecule that is unlikely to react with any substrate material, the addition of nitrogen to a discharge produces highly reactive N atoms. For instance, TiN is easily formed, whereas N_2 does not or only incompletely reacts with titanium. In addition to ions, uncharged radicals may be generated that are extremely reactive. These radicals reach surfaces isotropically from any direction, in contrast to the physically reacting ions mentioned above.

Beside CVD and evaporation, nearly all thin-film processes are based on one or both aspects of the plasma state. Even ion implanters need a plasma to extract the ions that have to be accelerated. However, in a common sense, only PECVD, sputtering, and, to some extent, ion plating, are considered to be "working with a plasma." Their principles are discussed below.

D. Plasma-Enhanced Chemical Vapor Deposition

First, something must be said about the definition of plasma-enhanced chemical vapor deposition (PECVD). PECVD, as described below, is an established commercial technique for the deposition of insulating films such as silicon nitride and silicon oxide in the semiconductor industry [40]. Due to its affinity to thermal CVD, it is called PECVD. However, in some cases, the precursors are monomers and the resulting coatings show polymeric properties, as is the case in most biomedical applications [41,42]. Whereas there is no significant difference in technical details, this type of PECVD is sometimes called a *plasma polymerization process*. Thus, both terms are synonymous. The main application of PECVD is the low-temperature deposition of silicon compounds in IC and solar cell fabrication.

Plasma-enhanced chemical vapor deposition is essentially the extension of simple CVD techniques by using the chemical aspect of a plasma to dissociate and activate the reaction gases. The inelastic collisions between high-energy electrons and gas molecules give rise to highly reactive species. The most important mechanisms are [43]

$$\text{Excitation: } A + e^- \leftrightarrow A^* + e^-$$

$$\text{Ionization: } A + e^- \leftrightarrow A^+ + 2e^-$$

$$\text{Dissociation: } A_2 + e^- \leftrightarrow 2A + e^-$$

$$\text{Electron attachment: } A + e^- \leftrightarrow A^-$$

$$\text{Dissociative attachment: } A_2 + e^- \leftrightarrow A + A^-$$

$$\text{Photoemission: } A^* \leftrightarrow A + h\nu$$

where

A and A_2	= reactants
e^-	= an electron
A^*	= reactant A in excited state
A^- and A^+	= ions of A and B
$h\nu$	= a photon

Thus, the energy of the electrons is used to create reactive and charged species without raising the gas temperature. The reactive species produced in the plasma have lower energy barriers to physical and chemical reactions than the precursor and can react at lower temperatures. Consequently, the plasma process can be accomplished at much lower substrate temperatures than by using thermally driven CVD. For example, deposition temperatures of 700°C to 900°C are required to deposit silicon nitride by thermal CVD, while 250°C to 350°C temperatures are sufficient to deposit similar films by

PECVD. Temperature is still needed to drive the reaction of the reactive species, that is, to provide the energy required to promote surface reactions and desorb by-products, as well as to lower film contamination. Figure 5 shows the basic principles of the PECVD process.

All different types of PECVD reactors commonly use two parallel plates to provide the electric field (DC or capacitively coupled RF) for initiating the plasma; in a few cases, inductively driven plasmas are used. The major difference between the reactor types is the gas flow system; radial flow reactors, inverse radial flow reactors, and hot wall batch reactors are known [27]. In some cases, microwave plasmas are used that are confined by multipolar magnetic fields. In this type of reactor, plasma excitation and plasma-surface interactions are decoupled, and there is no self-bias. Thus, plasma generation and ion bombardment of the substrate can be controlled independently.

A modern and very promising development is the remote PECVD technique that separates the substrate from the plasma [44]. Only the desired species are excited by the glow discharge and then transported to the substrate, where the deposition takes place. Thus, an independent control of the generation of active species and the reaction chemistry is provided. This allows an independent optimization of plasma and substrate parameters.

There is also an inverse PECVD process, plasma etching [27]. It relies on the ability of some reactive gaseous species produced in a discharge not to coat, but to etch a surface. Not only physical sputtering reactions are to be taken into consideration, but also chemical reactions between plasma components and surface molecules. Probably the most important applications of this process are the sputter cleaning of a substrate just prior to deposition and the etching of silicon and silicon oxides in a discharge of CF_4 to form volatile SiF_4 [45].

E. Sputtering

PECVD is a very useful tool for gaseous and liquid precursors, but what has to be done with solid precursors? As discussed in connection with Fig. 2, several different phenomena may occur when an ion approaches the surface of a solid. If its energy is between approximately $5 \cdot 10^1$ and 10^5 eV, the ion impact sets up a series of collisions between

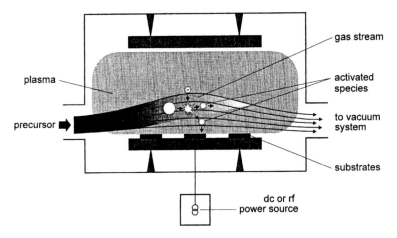

Figure 5 Basic principles of the plasma-enhanced chemical vapor deposition process.

atoms of the solid, which may lead to the ejection of some of these atoms [46]. This ejection process is known as *sputtering* and the solid is called the *target*. Often, it is compared with the break in a game of atomic billiards in which the bombarding ion (the cue ball) strikes a pack of target atoms. Some of them are going back toward the player, that is, they are sputtered out of the target.

In reality, the interatomic potential function is rather different from the hard sphere model. In addition, when sputtering an alloy, the situation is quite different for each element because of the momentum transfer processes responsible for ejecting the atoms and the different binding energies (i.e., the ball radius). However, in practice, it is generally accepted to define a specific sputtering yield (number of ejected particles per impinging ion) for each element and to disregard matrix effects. These specific sputtering yields represent the probabilities of these atoms for being ejected. Figure 6 summarizes the presently known data for one type of primary ion. In consequence, for each element, the fluxes of ejected particles are proportional to the corresponding specific sputtering yield and the number of atoms present within the sputtering depth. It should also be noted that, depending on the ionization coefficient of each element, only a few percent of the ejected particles are ionized.

Consequently, it is very difficult to obtain an ejected flux ratio from an alloy target that represents the original alloy composition. The source composition changes with time. Initially, the more volatile components are preferentially sputtered from the target and the flux is enriched with this component. However, as time goes by, the volatile components of the target gradually exhaust at the surface. In equilibrium (sometimes after several hours of sputtering), the surface composition has changed such that the elemental composition in the flux is the same as in the alloy. Unless the target is very thick, it has to be exchanged before equilibrium is established.

Since the vapor pressure of solids is very low except at elevated temperatures, the sticking coefficient of the different species in the sputtered fluxes effectively equals one and the film composition will be the same as the composition of the flux. Thus, while

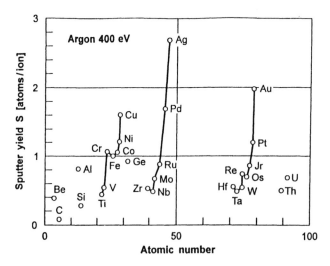

Figure 6   Specific sputtering yield of different elements as a function of atomic number. Primary ion is argon with 400 eV kinetic energy. (From Ref. 27.)

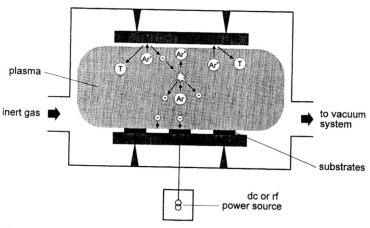

Figure 7 Schematic drawing of a sputtering system (diode system).

sputtering a compound target, film composition will change throughout the coating. Consequently, sputtering of alloys or compounds according to their stoichiometry is only possible if the target is in equilibrium. However, if energetic ions and neutrals reach the substrate and cause sputtering, the film composition may also change. Whereas a prediction of the film composition is very difficult, the structure of sputter-deposited coatings is empirically given by the process pressure and the ratio of substrate temperature and melting point of the coating material [47].

These principles are valid for all types of sputtering processes. A schematic drawing of such a sputtering system is given in Fig. 7. However, some peculiarities have to be mentioned depending on the type of plasma generation.

1.   Direct Current Sputtering

The most simple way to initiate a plasma is to use a high DC voltage. The material we wish to sputter is formed into a target that becomes the cathode of an electrical circuit with a high negative voltage (some thousand volts) applied to it. The substrate we wish to coat is placed on an electrically grounded anode a few centimeters away. Both electrodes are contained in a high-vacuum chamber filled with an inert gas (mostly argon) at approximately 1 to 50 Pa to provide the conditions for a stable glow discharge.

The voltage V required to drive the current I through the system is a function of process pressure. The deposition rate on the substrate depends on the ion flux at the target, that is, the current. However, the amount of sputtering also depends on the ion energy and hence on V. Thus, the choice of sputtering pressure p and the implied choice of V and I are rather important.

However, the requirement for a stable glow discharge sets a lower pressure limit. The discharge is sustained by electrons making ionizing collisions in the gas. The number of ionizing collisions will decrease with decreasing gas pressure according to Eq. 1, so that the discharge current for constant voltage will also decrease; below approximately 0.5 Pa, the sputtering rate in a conventional DC system becomes quite small.

A different problem arises at the other end of the pressure range. Like the collisions of electrons, material sputtered from the target may collide with gas atoms on its way to the substrate. With increasing pressure, the mean free path will be decreased. The result of these collisions is to deflect the sputtered atoms and to thermalize them, that is, to bring their kinetic energy to zero [48,49]. Hence, the deposition rate as well as the adhesion and microhardness of the coatings are decreased. Both problems were overcome by the introduction of magnetron sputtering.

2. Magnetron Sputtering

A magnetron sputtering system equals a conventional DC sputtering system except for the additional permanent magnetic field, which is oriented parallel to the cathode surface [50,51]. The local polarity of the magnetic field is oriented such that the drift paths of the emitted secondary electrons form a closed loop (see Fig. 8). Due to the increased confinement of the secondary electrons in this drift loop compared with a DC or RF diode device, the plasma density is much higher, often by an order of magnitude or more. The result of the high plasma density and its proximity to the cathode is a high current combined with a relatively low discharge voltage. Typical discharge parameters for a magnetron might be a voltage of 500 V and a current of 5 amperes (A), whereas a nonmagnetized diode might operate at 2500 V and 0.5 A [52,53]. In addition, the pressure can be significantly decreased from the 2 to 25 Pa of a typical DC diode, down to 0.3 to 5 Pa for a magnetron arrangement.

Another advantage connected with the high discharge current is the ability to sputter the cathode at a high rate. Deposition rates may be as high as several microns per minute for materials with high sputter yield. Usually, the typical limiting factor in a magnetron device is the ability to cool the cathode.

However, a disadvantage related to the magnetron technique is the inhomogenous erosion of the target [27]. Due to the high plasma density between the two poles of the magnet, the ion flux on the target's surface varies by several orders of magnitude between the outside of the target and the active region. Thus, the target is eroded between the two poles and is sometimes even coated on the outside, resulting in an inefficient use of the target. If noble metals are used, this is really a drawback of the magnetron technique since approximately 80% of the target volume must be recycled. Modern sputtering machines use a movable magnet system.

Another drawback of DC discharges is the substrate and target charging during the

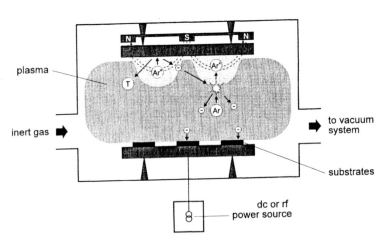

Figure 8   Schematic drawing of a magnetron sputtering system.

deposition of insulators. Insulating layers formed on one of the electrodes would charge up and break down in destructive arcs. These limitations are overcome by the use of RF-driven plasmas.

3. Radio-Frequency Sputtering

So far, only DC discharges have been discussed. In a similar way, the target can be mounted to the powered electrode of an RF discharge utilizing the self-bias voltage for the acceleration of the ions [54,55]. Although radio frequencies above 10 kilohertz (kHz) would be sufficient for most applications, commonly 13.56 megahertz (MHz) are used due to radio communication restrictions. RF sputter deposition is widely used for insulators such as silicon oxide, aluminum oxide, and other oxides, for which the substrate temperature limits preclude other techniques, or when the compositional control is easier to achieve than for alternative methods.

4. Bias Sputtering

Remembering the sensitivity of the nucleation and growth processes to particle bombardment, one would expect to be able to influence the properties of the film by changing the flux and the energy of the incident particles. It is difficult to modify the behavior of the neutral particles directly, but the charged particles can be controlled by changing the local electric field at the substrate. Although the percentage of ions is not more than approximately 1%, this is the basis of the very efficient bias sputtering process. In this technique, the sputtering plasma burns between the cathode and the grounded side walls of the reactor, whereas the substrate is being put on a bias voltage (DC bias for conductive substrates or coatings, RF self-bias in the case of insulating surfaces).

Exact calculations of the effects of an applied bias voltage have to take several factors into consideration: the polarity and magnitude of the bias voltage, the process pressure as it would influence the energy losses of the accelerated ions, the surface area of the electrodes, and so on. All these parameters determine the ion flux and energy density that influence substrate temperature as well as nucleation and crystallization behavior. A good example is the variation of density and resistivity of metal films by means of bias voltage [56]. The microhardness or etch resistance and many other film properties can also be controlled by the bias voltage [57]. A good review of particle bombardment effects on thin-film deposition is given in Ref. 58.

Another effect of the bias voltage is that it may control the incorporation of sputter gas. As mentioned above, the incorporation of sputter gas in the growing films results in higher porosity or even impaired film properties. However, due to their chemical inertness, these sputter gas atoms are loosely bound and can be easily resputtered by a smooth bias-voltage-induced ion bombardment of the substrate during the deposition process [59]. This may also influence other loosely bound particles and improve the crystallinity of the coating.

5. Reactive Sputtering

To complete the discussion of various sputtering techniques, a few words should be said about reactive sputtering. By adding a gas that reacts with a sputtered material in the presence of a plasma, compound films can be deposited using basically the same sputtering system as described above. As an example, which is discussed in detail in Section III.C.2, titanium nitride is often used in biomedical engineering. Its deposition from a titanium nitride target would be quite expensive due to target costs and the difficulties related to the sputtering of compounds mentioned above. However, it is deposited quite easily by adding nitrogen while sputtering a pure titanium target, provided that a sufficient nitrogen supply is maintained. N_2^+ ions are formed in the plasma and bombard both the target and the substrate; the single nitrogen atoms resulting from the impact dissociation react with the sputtered titanium to form titanium nitride. For more details, see Refs. 60–62.

F. Ion Plating

In a common sense, ion plating is a vacuum deposition technique that uses a glow discharge to modify the composition of evaporated films, thus combining evaporation and sputtering. The substrate is fixed to the cathode of a high-voltage glow discharge. Material is then evaporated onto the substrate while maintaining the discharge. As with bias sputtering, there is simultaneous deposition and resputtering, with a balance to ensure net deposition. Unlike those processes, deposition rates in ion plating are high enough that substantial resputtering can be tolerated. Therefore, the acceleration voltages are much higher than compared with bias sputtering systems (usually in the order of some thousand volts).

In addition, these high voltages are used in order to promote forward sputtering and implantation of film atoms into the substrate (see Fig. 2 for the energy-dependent ion-solid interactions). Thus, instead of having a discrete interface, it is spread over a considerable distance, which results in excellent adhesion.

A secondary effect of the ionization is that the ionized material follows electric field lines. Combined with the considerable gas phase scattering that occurs at the pressures normally used and the high mobility of the deposited material due to the high substrate temperatures produced by the ion bombardment, homogenous coverage of three-dimensional objects can be achieved.

Similar results may be obtained if the glow discharge is replaced by an ion gun, which provides the ion flux; an obvious advantage of this technique, called ion-beam-assisted deposition (IBAD), is that it can easily be incorporated into an existing vacuum deposition system. Another advantage is given by the fact that this process allows coating of insulating substrates. As mentioned above, in this case it is necessary to provide a compensating flux of electrons. This is often accomplished by using a thermionic neutralizer filament that emits the appropriate electron current into the beam. Note that the ions are not neutralized in flight but rather the total beam current is zero. However, a drawback is related to this technique. As the energetic ions are uniformly accelerated, it is very difficult to achieve a homogenous coating of three-dimensional substrates.

Furthermore, besides evaporation, any source of material can be used. Thus, *ion plating* in a more general meaning is a genetic term applied to all atomistic film-deposition processes in which the substrate surface and the growing film are subjected to a flux of bombarding particles with energies that are sufficient to cause changes in the film-formation process. Good summaries about the presently used ion plating techniques are given in [63,64].

G. Ion Implantation

Since the discovery of doping effects in semiconductors, it is well known that a few atoms may determine the mechanical, electronic, optical, magnetic, or superconducting properties of a solid. In this regard, the ion implantation technique is of special importance as it allows the introduction of atoms into the surface layer of a solid ($< 1\ \mu m$) with controlled depth distributions.

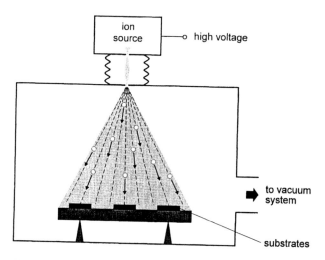

Figure 9   Basic principles of ion implantation.

The implantation system shown in Fig. 9 illustrates the basic elements required for this technique. In the first step, ions must be generated, which is usually achieved by a plasma or a thermionic-driven plasma. These ions are extracted from the ion source and driven into an analyzer magnet. This is almost mandatory for mass separation of the ions in order to eliminate unwanted species that may be present in the beam, depending on the precursor. Behind the magnet system, the ions are accelerated to the required energy and deflected in a beam-sweeping system to obtain uniform implantation.

One of the most important considerations in ion implantation is the depth distribution of the implanted ions. Typical depth profiles in an amorphous substrate are approximately Gaussian in shape and are well described by the theory of Lindhard, Scharff, and Schiott (LSS-theory) [65]. However, in crystalline substrates, the range distribution depends strongly on the orientation of the crystal with respect to implantation direction, that is, on the "channeling effect." If an ion enters the solid almost parallel to a crystal axis or plane, then a correlated series of collisions may steer it through the lattice, thus reducing its rate of energy loss and increasing its penetration depth. The depth profile of such an implantation is difficult to control. Fortunately, in most applications, channeling is negligible.

The effect of the implanted ions on the solid may not only be caused by the ions themselves (e.g., as it is in doping), but also by the lattice disorder and radiation damage that are induced by the incident ions. Up to now, the underlying physical principles are well understood [66], but it is not possible to predict their effects on a material. Therefore, it is still necessary to perform screening tests. The main application of ion implantation is found in IC fabrication and the improvement of wear properties. Its main advantage is that it does not produce an interface, thus providing strongly adhesive coatings.

H.   Comparison of the Different Thin-Film Technologies

In order to provide a basis for making decisions, Table 2 summarizes the most important process parameters and characteristics of the different thin-film techniques. If a thin-film technique must be selected to meet the requirements of a given application, the required film adhesion and maximum substrate temperature need concentrated consideration.

Table 2   Summary of Important Thin-Film Technique Process Parameters

Type of process	Percentage of ions	Energy of neutrals (eV)	Energy of ions (eV)	Noble gas incorporation	Substrate Temperature (°C)	Deposition rate (nm/min)	Adhesion
CVD	0%	0.1	—	—	–1000	100	Good
Evaporation	0%	0.1	—	—	Room temperature	100–1000	Bad
PECVD	0.1–1%	1–10	1–10	—	100–500	10	High internal stress possible
Sputtering	0.1–1%	1–10	1–10	1–20%	100–500	100	Good
Ion plating	5%	0.1–100	300–10000	1–20%	100	100	Very good
Ion implantation	Nearly 100%	—	Higher than 10000	—	Room temperature (if cooled)	Cannot be defined	No interface

III. APPLICATIONS OF THIN-FILM TECHNOLOGY IN BIOMEDICAL ENGINEERING

Although most of these thin-film techniques have been known since the early 1950s, only a few industrial applications have been developed. One reason may be that extensive work was required to develop this technology to the point at which it can be used efficiently. The classic example for thin-film applications is IC fabrication, which has been the pioneer in thin-film technology and has been able to carry the enormous expenditures. Now, almost three decades later, biomedical engineering takes the opportunity of applying these technologies to medical devices. However, we have to admit that biomedical applications of thin-film technology are still in their infancy.

A. General Overview

There are some reviews dealing with special topics [42,67], but, to give an idea of the opportunities and the ideas that have been tried and developed during the last years, Table 3 summarizes some of the already known applications of thin-film technology, which are indicated in the first column. The second and third columns contain the coating material that has been under investigation or is already being used. The last column gives references [68–112]. This list is not complete, but it gives an impression of the large variety of possible applications.

The last section of this chapter draws your attention to two special applications: the development of hemocompatible coatings and the improvement of pacing leads by thin-film technology. The first example is to illustrate that a proper tailoring of the coating properties, and therefore a careful adjustment of the deposition parameters, are necessary in order to obtain optimal results. The second example demonstrates that thin-film technology not only improves a certain feature of an implant, but will sometimes create chances for new developments in the biomedical field that do not seem to be possible without a hybrid design.

B. Amorphous Silicon Carbide Coatings with High Hemocompatibility

Hemocompatibility of alloplastic materials is determined by their surface properties. Above all, smooth surfaces are necessary in order to avoid the activation of the clotting process via trapped corpuscular blood components. However, different materials with the same surface roughness have different effects on the clotting system (i.e., thrombosis is also caused by other interactions).

A well-known hypothesis is based on the hydrophilicity of a material. Materials with an interfacial free energy in the range of 1–3 dyne/cm should provide a hemocompatible surface [113,114]. However, this theory could not be verified in general [115]. Instead, it was theorized that an electrochemical interaction of blood coagulation proteins with the surface is also relevant for hemocompatibility of an alloplastic material.

A first systematic analysis was done by Sawyer, Brattain, and Boddy [116], who found a relationship between electrode polarization and thrombosis during electrolysis of blood. In addition, the work of Sawyer et al. [117] and Baurschmidt and Schaldach [118] proved that thrombosis at artificial surfaces is due to the formation of fibrin, which is induced by an electron transfer from fibrinogen to the solid. As a result, the electronic and electrochemical properties of proteins and solids are key to an understanding of hemocompatibility. Below, a microscopic model for this effect is discussed that allows derivation of the electrochemical requirements for high hemocompatibility [119].

Table 3   Applications of Thin-Film Technology in Biomedical Engineering

Application, suggested use	Coating material or precursor	Technology	Ref. no.
Improving the histocompatibility of Dacron for vascular grafts and percutaneous devices	Titanium	Evaporation	68
Artificial heart valves and other blood-contacting devices, dental implants	LTI carbon	CVD	29, 70, 71
Cementless fixation coating for orthopedic devices	ULTI carbon	CVD	95
Cleaning of metallic implants	No coating, O_2, N_2, Ar plasma	Plasma treatment	72
Changing the hydrolytic degradation rates of absorbable sutures	Methane, trimethylsilane, tetrafluoroethylene	Plasma treatment	73
Improvement of contact lenses	No coating, O_2 plasma	Plasma treatment	74
Avoiding posterior capsule opacification of polydimethylsiloxane contact lenses	No coating, CO_2 plasma	Plasma treatment	75
Increasing the degradation rate of biodegradable polymers (PLA, PBGA)	No coating, N_2 plasma	Plasma treatment	76
Increasing the adhesion of endothelial cells	No coating, N_2, O_2 plasma	Plasma treatment	77
Against high tackiness and hydrophobicity of silicone for contact lenses	Methane	Plasma polymerization	78
Reducing platelet deposition on medical tubings	Hexafluoroethane	Plasma polymerization	79
Providing covalent bonding sites and improving hydrophilicity of blood-contacting devices	Hexane, N-vinyl-2-pyrrolidone	Plasma polymerization	80
Polymer membranes in clinical sensor applications	Various	Plasma polymerization	81, 82
Increasing the binding capability for amine and hydroxyl groups	Glycidyl acrylate	Plasma polymerization	83
Improving the adsorption of albumin; suggested for sensors	Polypropylene	Plasma polymerization	84
Improving the hemocompatibility of blood-contacting devices	Silicone	Plasma polymerization	85, 86
Reducing the adsorption of platelets and leucocytes on Celgard and Silastic membranes	Hexamethylcyclotrisiloxane	Plasma polymerization	87
Improving endothelial cell growth on vascular prostheses	Acetone, methanol, glutaraldehyde, formic acid	Plasma polymerization	88

Table 3 Continued

Application, suggested use	Coating material or precursor	Technology	Ref. no.
Reducing protein attachment and cellular adhesion; improving fouling resistance	Tetraethylene glycol dimethyl ether	Plasma polymerization	89
Tailoring hydrophilicity and hydrophobicity	γ-butyrolactone, hexane, hexamethyl-disilazane	Plasma polymerization	90
Corrosion and wear-resistant coating for general applications	Diamondlike carbon	PECVD	91
Hemocompatible coating for non-flexible blood-contacting devices	Amorphous silicon carbide	PECVD	92
Improving the endothelialization of Dacron for vascular grafts	Carbon	Sputtering	69
Wear-resistant coating for hip joints	Titanium nitride	Sputtering	14, 15
Suggested for improving the in-growth behavior of bone	Titanium	Sputtering	93
Improving pacing and sensing performance of pacing leads	Iridium, titanium nitride	DC sputtering (reactively)	96, 97
Corrosion protection of Co-Cr- and Ni-Cr-based implant alloys	Alumina, silicon carbide	RF sputtering	98
Tailoring the interfacial free energy for high biocompatibility	Fluorocarbon compounds	RF sputtering	99
Hip joint prostheses with improved bond strength of calcium phosphorus	Calcium phosphate	Ion beam sputter deposition	101
Corrosion protection of metal implants, especially orthopedic devices	Titanium nitride	Ion plating	100
Diffusion barrier for harmful metal ions on Co-Cr-Mo-based alloys	Titanium nitride	Ion plating	94
Biocompatibility improvement of 316L implants	Titanium, titanium nitride	Ion plating	102
Improved breakdown corrosion potential for artificial joints	Titanium nitride	Ion plating	103
Texturing of PTFE surfaces for improved soft-tissue response	None	Ion beam etching	104
Corrosion inhibition and improved charge injection capabilities for surgical implants	Noble metals, especially Ir in Ti-6Al-4V	Ion implantation	105, 106
Improving wear resistance of hip joint prostheses	N in titanium	Ion implantation	14, 107–110
Improving wear resistance of hip joint prostheses	N in Ti and 316L	Ion implantation	111
Reduced corrosion in biological media	Au and Rh in Ti-6Al-4V	Ion implantation	112

1. Theory: Hemocompatibility from an Electrochemical Point of View

Several attempts have been made to calculate and to measure the electronic band gap and the density of electronic states of common blood proteins since Szent–Györgyi's first description of the electronic structure of proteins [120]. Quantum mechanical calculations using the linear combination of atomic orbitals (LCAO) theory for each amino acid and combining the resulting basis set of wave functions for different polypeptides demonstrated that most proteins are semiconductors or even insulators with a band gap energy of more than 8 eV and a density of states of approximately 10^{20} cm^{-3} eV^{-1} [121,122]. Temperature-dependent conductivity measurements confirmed these theoretical results [123]. However, increasing the water content of the samples induces a significant decrease of the band gap energy [124,125]. Specifically, the band gap energy of fibrinogen is about 1.8 eV in the hydrated state, and the width of its valence and conduction bands is more than 5 eV [119].

Electrochemical experiments with semiconducting electrodes at different potentials demonstrated that an electron transfer from fibrinogen's occupied valence band states into free states of the solid causes the release of the fibrinopeptides [118]. Of course, this interaction, which is responsible for the cleavage of the peptide bonds, is not an electrostatic one. Instead, an effect of the Franck–Condon principle is proposed as an explanation [126]. The fast electron transfer at the charge transfer point induces an atomic relaxation of the adsorbed terminal groups of the fibrinogen molecule. Figure 10 demonstrates the tremendous changes that are induced by one electron in the atomic structure of glycinamide, the smallest amino acid. The relaxation energy has been calculated by quantum mechanical methods to be 2.08 eV per amino acid, which is large enough for breaking the peptide bonds of the fibrinopeptides with binding energies between 2 and 4 eV [127].

As a result, the most essential aspect in the development of hemocompatible materials is the inhibition of the surface-induced clotting process by inhibiting this charge transfer reaction. Therefore, with regard to hemocompatibility, we have to ask: What properties must a material have so this charge transfer reaction is not induced?

2. Electrochemical Requirements for High Hemocompatibility

The electron transfer reaction can be described as a two-particle tunneling process from occupied states of fibrinogen $D_{Fb}^-(E)$ into free states of the material $D_{Hl}^+(E)$. Disregarding the energy dependence of the tunneling probability $\nu(E)$, the exchange current j_ν between fibrinogen and the material is given by

$$j_\nu = e \cdot \int_{-\infty}^{E_\nu} \nu(E) \cdot D_{Hl}^+(E) \cdot D_{Fb}^-(E) \ dE \qquad (4)$$

$f(E)$ is Fermi's function, E_ν the upper valence band edge of fibrinogen.

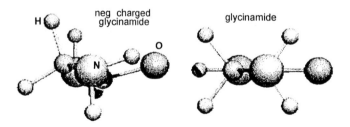

Figure 10   Calculated relaxed structures of neutral and negatively charged glycinamide.

The density of states (DOS) in the valence band of fibrinogen $N_{Fb}(E)$ is approximately 10^{20} cm^{-3} eV^{-1}, and at body temperature almost any state is occupied by electrons, that is, the density of occupied states $D_{Fb}^-(E)$ equals $N_{Fb}(E)$. According to Eq. 4, the number of unoccupied states in the valence band of the solid $D_{Hl}^+(E)$ is the only variable that determines the exact value of the exchange current and thus the exact value of the thrombogenicity. However, $D_{Hl}^+(E)$ depends on the difference E between the energy level under consideration and Fermi's energy, as well as on the density of states $N_{Hl}(E)$ in the material:

$$D_{Hl}^+(E) = N_{Hl}(E) \cdot (1 - f(E)) \qquad (5)$$

An upper limit for $N_{Hl}(E)$ is given by the number of atoms in the solid of approximately 10^{22} cm^{-3}. For example, in common metals, $N_{Hl}(E)$ is 10^{22} cm^{-3} eV^{-1}, resulting in 10^7 unoccupied states 0.9 eV below Fermi's energy and 10^1 unoccupied states 1.3 eV below Fermi's energy. In this case, a high charge transfer current is possible. Thus, in order to decrease the exchange current significantly, $D_{Hl}^+(E)$ has to be lowered.

Since it is very difficult to determine the exact energy dependence of $D_{Hl}^+(E)$, a simplified model should be used to demonstrate the effect. All metals are supposed to have a constant DOS of 10^{22} cm^{-3} eV^{-1}, that is, possible surface variations (e.g., due to oxides) are disregarded. Then, the only way for reducing the density of states of the material is to use semiconductors. Their DOS is regarded to be low and constant (e.g., 10^{16} cm^{-3} eV^{-1}) inside the band gap and high (i.e., 10^{22} cm^{-3} eV^{-1}) outside of it. In first-order approximation, this is really true for crystalline semiconductors; in the case of amorphous semiconductors, exponential tail states with a width of 0.6 eV are assumed (see detail in Fig. 11).

As a result, the energy dependence of $D_{Hl}^+(E)$ is fully determined if the band gap energy and the position of Fermi's energy is defined. Since only the valence band is of interest (Eq. 4), the difference between Fermi's energy and the upper valence band edge of the solid at the surface $E_f - E_{vs}$ is sufficient to calculate the exchange current. The results are demonstrated in Fig. 11. For low values of $E_f - E_{vs}$, the charge transfer is possible between the two valence bands, resulting in a maximum exchange current. However, as soon as $E_f - E_{vs}$ exceeds 0.9 eV (i.e., as soon as the valence band of the material

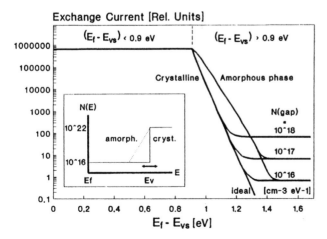

Figure 11   Effect of $E_f - E_{vs}$ on the exchange current, that is, the activation of fibrinogen.

crosses the valence band of fibrinogen) the current decreases by several orders of magnitude. Depending on the band gap DOS, there is a saturation effect for high $E_f - E_{vs}$.

As a result, the most important requirement for high hemocompatibility of a material is a low density of states in the critical energy interval between 0.9 and 1.3 eV below Fermi's energy. (It should be noted that only a small part of the valence band of fibrinogen is responsible for the charge transfer.) In reality, this low density can only be achieved by a semiconducting material with a value of $E_f - E_{vs}$ of more than 1.2 eV (crystalline) or 1.4 eV (amorphous phase). In addition, the conductivity should be more than 10^{-4} (ohm cm)$^{-1}$ in order to stabilize the electrochemical equilibrium at the interface. On one hand, this is an explanation for the empiric finding that passivated metals like titanium show better thromboresistance than noble metals due to the semiconducting properties of most metal oxides. On the other hand, special semiconductors with tailored electronic properties should show superior hemocompatibility.

This is also an outstanding example with regard to the introductory remarks to Section III. Semiconductors are brittle and not suited for mechanically loaded applications in implants. Thus, we have an excellent application of hybrid design; a thin semiconducting film on a stable substrate should result in a reliable and hemocompatible implant.

3. Materials and Methods

For materials and methods, the optimal structure of the coating must be defined first. The deposition of monocrystalline layers, with their good electronic properties with regard to conductivity and low band gap DOS, is technically not feasible on nonplanar implants. Polycrystalline films are unsuited due to their high band gap DOS in the area of the grain boundaries. Therefore, they do not meet the requirement for a low transfer current. For these reasons, amorphous semiconductors are the most promising coating materials.

Judging from the electronic requirements for low band gap DOS, the plasma-enhanced chemical vapor deposition provides the most suitable coating process due to the high inherent hydrogen concentration, which satisfies the electronically active defects in the amorphous layers. At present, amorphous silicon carbide is under evaluation. It is known as a window material for amorphous solar cells because of its excellent electronic properties. The films are deposited from mixtures of silane (10% diluted in hydrogen) and pure methane in a 13.56-MHz capacitively coupled plasma at 10 Pa. In order to achieve a low band gap DOS, the substrate temperature is held at 250°C. The required high conductivity is achieved by n-doping with phosphine (0.1% diluted in hydrogen). The exact details of the sample preparation have been discussed elsewhere [92,119].

The variation of the gas composition gives an opportunity to tailor the band gap $E_g = E_c - E_v$ together with $E_f - E_{vs}$, as well as the conductivity σ. To obtain information about the energy gap, the films were deposited on glass substrates and optical transmission experiments were performed. The spectra were fitted to Eq. 6, given by Mott and Davis for the absorption of amorphous semiconductors, where B is a constant, α the absorption coefficient, and $h\nu$ the photon energy [128]:

$$(\alpha h\nu)^{1/2} = B^{1/2} \cdot (h\nu - E_g) \tag{6}$$

Pure amorphous silicon (a-Si:H) films have a band gap energy of 1.85 eV; an increase in methane concentration leads to an almost linear increase of E_g up to 2.3 eV at 70% methane. With regard to hemocompatibility, $E_f - E_{vs}$ is the interesting parameter, which can be calculated from the band gap energy E_g:

$$E_f - E_{vs} = (E_f - E_v) - \Phi = E_g - E_a - \Phi = (E_c - E_v) - (E_c - E_f) - \Phi \quad (7)$$

where

E_f = Fermi's energy
E_v = upper valence band edge
E_{vs} = upper valence band edge at the interface
E_g = band gap energy
$E_a = E_c - E_f$ = activation energy
E_c = lower conduction band edge
Φ = band bending at the interface

The band bending at the interface Φ has been determined by electrochemical experiments to be approximately 0.1 V. The exact details are summarized in Ref. 119. The position of Fermi's energy in the band gap (i.e., the activation energy) can be derived from the temperature dependence of the conductivity σ according to

$$\sigma = \sigma_0 \cdot \exp\left(\frac{E_c - E_f}{kT}\right) \quad (8)$$

where σ_0 is a constant. For doped a-SiC:H samples, $E_c - E_f$ is in the range between 0.4 and 0.7 eV. As a result, $E_f - E_{vs}$ is between 1.3 and 1.5 eV; the exact data is summarized in Fig. 12. According to the electrochemical model for thrombogenesis, a-SiC:H should display increasing hemocompatibility with increasing carbon content, with a saturation effect above a methane concentration of approximately 50%.

The second important parameter is the conductivity itself. Figure 13 shows the results for the conductivity at body temperature as a function of phosphine doping. At 0.5% phosphine, the required conductivity can be guaranteed. In addition, the band gap DOS was measured with the field effect technique [129]. For methane concentrations below 70%, the field effect DOS of a-SiC:H is about 10^{16} eV^{-1} cm^{-3}.

4. In Vitro *Hemocompatibility*

The hemocompatibility of these a-SiC:H coatings was evaluated *in vitro* with the TIRIF (total internal reflection intrinsic fluorescence) [130] as well as the TEG (thrombelastography) techniques [131].

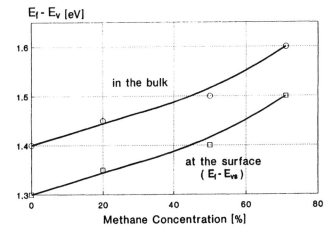

Figure 12 $E_f - E_v$ as a function of methane concentration.

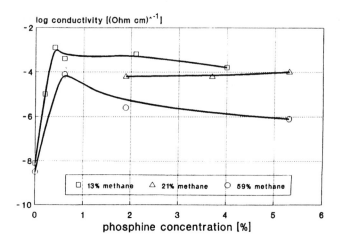

Figure 13   Conductivity of a-SiC:H at body temperature as a function of phosphine concentration.

TIRIF utilizes the intrinsic fluorescence of proteins that contain aromatic amino acids like tryptophan or tyrosin and enables the measurement of protein concentration at interfaces. For this purpose, ultraviolet light (285 nanometers [nm]) is totally reflected at the interface, while the evanescent wave excites only the adsorbed molecules. A photomultiplier detects the emitted fluorescence light as a measure for the amount of protein adsorbed at the interface. After flushing the sample with saline at pH 7.4 and 37°C for one hour, 0.5% bovine fibrinogen solution is added. After different contact times, the interfaces are flushed again with pure saline, and the remaining protein is measured and

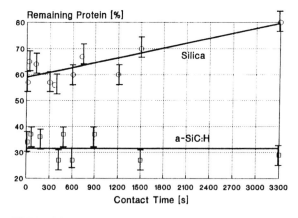

Figure 14   Comparison of the remaining protein concentrations after different contact times for silica and a-SiC:H-coated silica. The a-SiC:H was deposited from a mixture of 50% silane, 49.5% methane, and 0.5% phosphine; the results are normalized to the maximum protein concentration during each adsorption phase.

compared to the maximum adsorbed protein concentration. The results are demonstrated in Fig. 14.

Silica activates fibrinogen due to its low conductivity and leads to an increasing polymerization, which is indicated by the increasing concentration of adsorbed protein. Please note that both materials have the same surface free energy determined by contact angle measurements. On the other hand, a-SiC:H coatings show no time-dependent increase in the remaining protein concentration, thus confirming that no fibrinogen activation and polymerization takes place.

In addition, the hemocompatibility of a-SiC:H was checked with the TEG technique in order to prove the predictions with human blood. The TEG technique measures the elasticity of a growing thrombus as a function of time. For this purpose, fresh blood or platelet-rich plasma fills a slowly rotating pot, and a sensor bar dips into the blood; both pot and bar are made of the material under evaluation. The reaction time is the time between the beginning of the test and the beginning of thrombus formation. The clotting time is indicated by the time between the beginning of thrombus formation and the point at which the thrombus has an elasticity of 20 units. These units are defined by the manufacturer of the test device (Thrombelastograph 2, Hellige, Germany).

The test specimens were milled from pure titanium as a reference material and coated with a-SiC:H of different compositions. In addition, some graphite parts coated with LTI pyrolytic carbon were purchased from Ringsdorff (Germany). All experiments were performed with pooled platelet-rich plasma of the same donors. The maximum time deviation of tests with the same material was 0.5 minutes (min), with the results summarized in Fig. 15.

5. Conclusions

The clotting time of a-SiC:H increases with increasing methane concentration and reaches a plateau above 50% methane. In addition, in comparison to titanium and pyrolytic carbon, the two most frequently used materials for artificial heart valve design, the clotting time is increased by more than 200%. These results prove that a correct tailoring of the electronic properties of the surface by means of thin-film technology results in a device with high hemocompatibility. However, this example also demonstrates the impor-

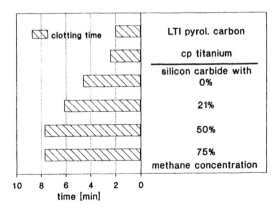

Figure 15 Effect of titanium, pyrolytic carbon, and different a-SiC:H coatings on the thrombelastography (TEG) clotting time.

tance of a very careful optimization of the deposition parameters in order to obtain the best results. The second example below focuses on the chances for new developments in the biomedical field that follow from the application of thin-film technology.

C. Pacing Leads with Fractally Coated Tips

Progress in electrostimulation has to be expected in respect to pacing as well as sensing performance. On one hand, the energy consumption required for the stimulation pulse determines the longevity and size of implantable devices (e.g., pacemakers, implantable defibrillators, neurostimulators, cardiomyoplasty stimulators, etc.). On the other hand, the sensing of electrical signals generated in the body may also be necessary (e.g., pacemakers require a reliable capacity for sensing the intracardiac electrocardiogram in order to synchronize the artificial pacing). In addition, especially in cardiac pacemakers, a third requirement is unsolved. Up to now, all pacemakers are "deaf," that is, they are not able to detect if their pacing pulse resulted in a contraction of the heart. In other words, due to the high polarization artifact voltage of common pacing leads, it is not possible to sense the electrical response of the contracting heart, the so-called ventricular evoked response. Therefore, a third requirement must call for a low polarization artifact. All three requirements, a low energy consumption (i.e., a low pacing threshold), a good signal-to-noise ratio for sensed signals, and a low polarization artifact voltage, are essentially determined by the pacing lead. Especially, the interface between its tip and the body is not a simple ohmic resistor. In the following discussion, these requirements are "translated" into physical terms and a technical solution based on thin-film technology using pacing leads with fractally coated tips is presented.

1. Physical Requirements for Superior Pacing Leads

Physically, an interface between a metal and an electrolyte is characterized by two different charge transport mechanisms, the electronic conduction in the solid and the ionic conduction in the solution. If a voltage is applied between these two phases, ions are driven to the interface and form a layer, which is separated from the metal by a monolayer of adsorbed water molecules, resulting in the so called Helmholtz double layer. This structure is equivalent to a capacitor with water as the dielectric. Thus, the impedance behavior of such an interface (e.g., the impedance of a pacing lead) is equivalent to a high-pass filter. In first-order approximation, its lower corner frequency is determined by the capacity. If we want to minimize energy losses that may arise at this interface during the flow of a pacing current or a signal current (please note that the input impedance of sensing amplifiers in pacemakers is usually relatively low), we have to minimize the lower corner frequency (constant lead resistance is assumed) by maximizing the Helmholtz capacity. This approach has been discussed in more detail in Refs. 96 and 97.

But what about the polarization artifact voltage? By driving a current across this interface (e.g., during a pacing pulse), it is charged according to

$$U = \frac{1}{C_H} \int_0^T I(t)\,dt = \frac{1}{C_H} Q_{St} \qquad (9)$$

where

U = the voltage drop across the Helmholtz double layer
C_H = the capacity of the Helmholtz double layer

$I(t)$ = the displacement current (i.e., the overall current minus losses due to chemical reactions)

T = the pulse width

Q_{St} = the total charge at the interface

At the end of the pacing pulse, the interface remains charged, resulting in a voltage drop U across the Helmholtz double layer according to Eq. 9, which is called the *polarization artifact voltage*.

There are two possibilities to reduce this artifact voltage. On one hand, the charge Q_{St} can be reduced, such as by lowering the pulse voltage, or by the conventional method (i.e., by applying a charge balancing pulse subsequent to the pacing pulse) [132–134]. On the other hand, an increase in capacity of the Helmholtz double layer C_H would have the same effect on the artifact voltage.

At higher electrode potentials, additional effects have to be taken into consideration, such as chemical reactions like the reduction of H^+ ions resulting in a Faraday exchange current and an imbalance in the ionic environment at the interface. The charged reaction products cannot be recharged by a subsequent correcting pulse because they drift away from the interface, obeying a diffusion-limited transport law. Thus, the only way to avoid this contribution to the polarization artifact is to avoid these reactions, that is, to use low-threshold electrodes and to operate them at low output voltages.

As a result, two requirements must be met in order to achieve a pacing lead with a low polarization artifact. First, its threshold voltage should be minimized to avoid chemical reactions at the interface and the resulting diffusion problems, as well as to lower Q_{St}. Second, C_H should be maximized because, according to Eq. 9, a higher C_H results in a lower after potential for the same charge Q_{St}. As mentioned above, the threshold voltage is lowered by increasing C_H. Consequently, all three requirements – a low stimulation threshold, a good sensing performance, and a low polarization artifact voltage – can be combined and be provided by electrodes with high C_H.

2. Technical Realization

In first-order approximation, the Helmholtz double layer is regarded as a parallel plate capacitor having a capacity proportional to the active surface area, which can be varied by changing the surface roughness of the solid. In the technical realization of pacemaker technology, the Helmholtz capacity C_H is usually enhanced by sandblasting, using metal meshes [135], drilling small pores with a laser [136], or sintering small metal spheres [137]. The electrochemically active surface area of these types of tips is approximately five times larger than their geometric surface area. The next significant advance in this direction was the invention of platinum black, resulting in a multiplication factor of about 10 to 50 [137]. With these methods, Helmholtz capacities ranging from 10 to 500 microfarads per square centimeter (μF/cm^2) are achieved. However, these are still too small for a reliable suppression of the polarization artifact. In addition, platinum black is very soft and has only a poor long-term stability.

However, thin-film technology offers a remedy, the fractal coating of the tips [96, 138]. The idea of fractal surface structure is to use its self-repeating properties for enlarging the active surface area. Figure 16 demonstrates the effect in a two-dimensional example. If a given area A0 is covered by a hemisphere, then A0 is doubled. If, on top of this new surface area A1, some new smaller hemispheres are created, then A1 is doubled or A0 is quadrupled. A repetition of this process leads to an increase in active surface area

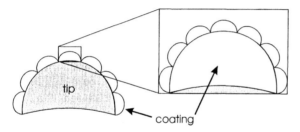

Figure 16 Two-dimensional example of a hemispherical fractal structure (similar to Koch's curve).

by a factor of 2^n concerning the given geometric surface area of the substrate, where n is the number of doubling steps.

This example is valid for all amorphous or metallic coatings with their typical cauliflower surface structure. However, crystalline films may also have fractal properties. Figure 17 shows the example of triangular structures, which are often found in covalently bonded coatings. The properties of a fractal electrode-electrolyte interface are discussed in detail in Ref. 139.

The technical realization of coatings with a fractal surface structure is obtained by bias sputtering. Interestingly, the surface structure is determined mainly by the process pressure and the bias voltage and not, or only to a smaller degree, by the material. For this reason, the surface multiplication factor is not determined by the coating material. In pacemaker technology, titanium nitride and iridium are used due to their high biocompatibility; furthermore, iridium is very stable concerning anodic polarization. Titanium nitride is reactively DC sputtered from a titanium target in a mixed atmosphere of argon and nitrogen; iridium coatings are performed in pure argon. With optimized deposition parameters, values for n of about 10 to 12 are achieved, resulting in specific Helmholtz capacities of up to 50,000 $\mu F/cm^2$. Figure 18 shows the surface morphology of a titanium nitride pacing tip and Fig. 19 shows the same for iridium coatings at different magnifications.

3. Pacing and Sensing Performance of Fractal Pacing Leads

A lot of *in vitro* work has been concentrated on titanium-nitride- or iridium-coated pacing leads with a fractal surface structure. The results prove that the coatings are well adherent, highly biocompatible, and expose an excellent long-term stability with regard to

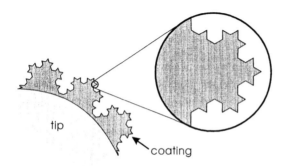

Figure 17 Triangular example of the structure in Fig. 16.

Figure 18   Four different magnifications of a fractal titanium nitride coating deposited by reactive DC sputtering with external DC bias voltage.

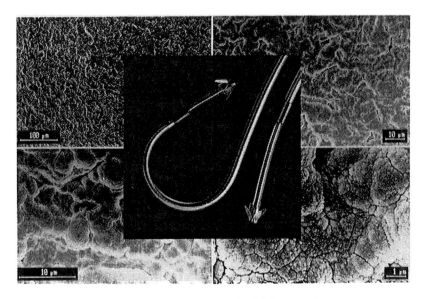

Figure 19   Same treatment as Fig. 18, but for iridium coatings.

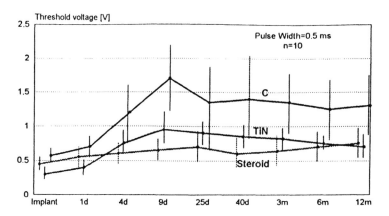

Figure 20   Long-term pacing performance of a carbon lead, a steroid eluting lead, and a fractal titanium-nitride-coated lead. (From Ref. 142.)

corrosion or dissolution. The results have been discussed intensively in Refs. 140 and 141.

For approximately three years, fractally coated pacing leads have been commercially available. The first generation contains leads with titanium-nitride-coated tips (TIR, Biotronik, Inc.), the second one has iridium-coated tips (DNP, IRTI, and SD, Biotronik, Inc.). The indifferent ring electrodes of bipolar leads are always coated with iridium due to the poor stability of titanium nitride under anodic polarization. Currently, even some actively fixated leads are available in a coated modification.

Since the beginning, many clinical research teams investigated the pacing and sensing performance of fractal pacing leads *in vivo* [138,142,143]. Some of their results are summarized in Figs. 20 and 21. The mean pacing threshold voltage is significantly lower if compared with conventional leads. In comparison to steroid eluting leads, which show excellent short-term results due to the suppression of any inflammatory reaction, fractal

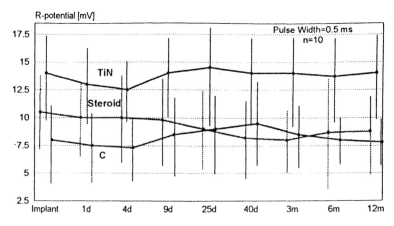

Figure 21   Long-term sensing performance of a carbon lead, a steroid eluting lead, and a fractal titanium-nitride-coated lead. (From Ref. 142.)

pacing leads are advantageous in the long run. With respect to sensing, the results demonstrate the advantage of a low interface impedance, which is very useful, especially in atrial sensing.

However, contradictory results have also been published [144]. However, these results are based on different surface structures. If one compares the scanning electron microscopy (SEM) micrographs, these coatings are much smoother than the coatings presented in Figs. 18 and 19. Thus, these pacing leads have a smaller active surface, a lower Helmholtz capacity, and therefore worse results. This is a good example that, in thin-film technology, it is not sufficient just to define the material and maybe the technology used in order to get reproducible results. Since the properties of the coating are decisively determined by the process parameters and sometimes also by the geometry of the reactor or other specialities of the system, it is crucial to fix all parameters.

4. Polarization Behavior of Fractal Pacing Leads

Two types of fractal unipolar pacing leads have been under investigation in respect to their polarization behavior, the titanium-nitride-coated TIR lead and the iridium-coated IRTI lead, both manufactured by Biotronik. A conventional unipolar lead with smooth elgiloy tip (PE 60/2K10, Biotronik) was used as a control. All three types had the same geometric surface area of 10 mm^2 and were examined *in vitro* as well as *in vivo*.

In vitro, the polarization artifact voltage was determined using an external custom-made pacemaker. All measurements were performed in 0.9% saline against a pacemaker housing. The pacemaker was allowed to apply an autoshort (i.e., to short-circuit the output connectors) for a variable period of time after the pacing pulse. Thus, the artifact voltage could be measured as a function of autoshort duration. The results are shown in Fig. 22. The values for TIR and IRTI leads are equivalent within the accuracy of the measurement. The results prove that the polarization artifact of fractal pacing leads is reduced by more than 90% if compared with conventional smooth types. Conventional leads do not allow a sufficient discharge within the time limit of the evoked response. On the other hand, the polarization artifact voltage of coated electrodes with a fractal surface structure is negligible if an autoshort of more than 50 milliseconds (ms) is applied.

In vivo measurements were performed during pacemaker implantation or exchange

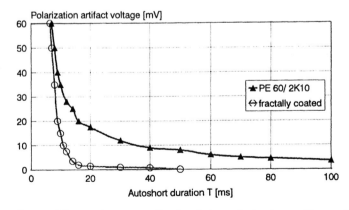

Figure 22 Maximum amplitude of the polarization artifact voltage as a function of autoshort time. Measurements were performed with different unipolar pacing leads powered against a pacemaker housing at 5.0 V, 0.5 ms, and 80 beats per minute (bpm).

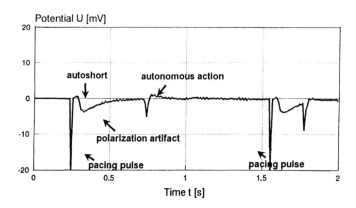

Figure 23 Output signal of an uncoated elgiloy pacing lead six years after implantation (PE 60/2K10, Biotronik) for below-threshold stimulation (90 bpm, 1.1 V) using an autoshort time of 50 ms.

in patients aged between 55 and 75 years. All types of leads were implanted in the apex of the right ventricle. An external pacemaker (EDP 30, Biotronik) was used for stimulation. The voltage drop across the electrode was measured behind the coupling capacitor (10 μF) against a large-area electrode in the body of the patient (i.e., the housing of an implantable pacemaker or the surgical retractor). Due to safety considerations, the signal was recorded via a 4000-V isolation amplifier using a laptop with a 12-bit 4-kHz analog to digital converter.

The illustrations mentioned below show, as a summary of the *in vivo* results, the cathodic (against the housing) pacing pulse as negative values and anodic voltages as positive values. This convention was chosen in order to regard the pacemaker housing as the reference electrode, although the inverse presentation is also used in literature [145,146]. In order to improve the evaluation of the curves, the y-axis was limited to plus and minus 20 millivolts (mV).

The present state of the art in recording the ventricular evoked response is demonstrated in Figs. 23 and 24 using a conventional uncoated elgiloy pacing lead with a smooth

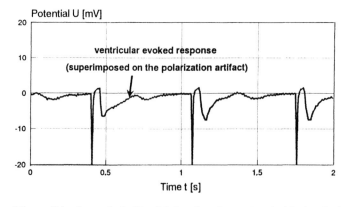

Figure 24 Scenario in Fig. 23, but for above-threshold stimulation (90 bpm, 2.0 V).

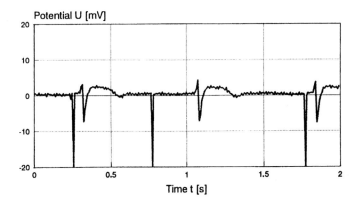

Figure 25 Output signal of a titanium-nitride-coated pacing lead with fractal surface structure (TIR 60-UP, Biotronik) one week after implantation, below-threshold stimulation (115 bpm, 0.3 V).

tip as an example. The threshold voltage was 1.6 V at a pulse width of 0.5 ms. In the example shown in Fig. 23, an output voltage of 1.1 V was used (i.e., the stimulation voltage was below threshold and thus was not effective). It can be seen that the pacing pulses are followed by a short interval of zero voltage due to the autoshort (50 ms) and thereafter the slowly decreasing polarization artifact appears. Some QRS complexes asynchronous to the pacing pulses indicate the autonomous activity of the heart. By contrast, Fig. 24 shows the output signal for an output voltage of 2.0 V. The QRS complexes are replaced by evoked ventricular responses. However, due to the high polarization artifact, the evoked responses are superimposed. Thus, a reliable capture recognition is not possible.

A totally different situation is found if coated pacing leads with a fractal surface structure are used. Figures 25 and 26 show the output signal of a unipolar titanium-nitride-coated lead (TIR 60-UP, Biotronik). Again, the first figure represents ineffective stimulation below threshold, the second illustrates effective stimulation above threshold. Due to the high Helmholtz capacity of these leads, no polarization artifact is observed;

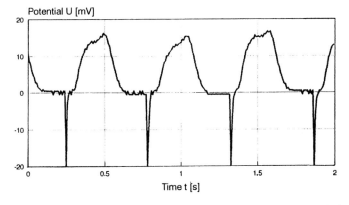

Figure 26 Scenario in Fig. 25, but for above-threshold stimulation (115 bpm, 1.0 V).

after a 50-ms autoshort, the baseline is only disturbed by noise and asynchronous QRS complexes. Consequently, above-threshold stimulation results in undistorted ventricular evoked responses. In order to check the time dependence of these results, especially during the growing-in of the tips, the measurements were repeated during the first two weeks. No change could be observed. Presently, results over more than a half year are available that were recorded with a telemetry system. These results prove the excellent long-term stability of fractally coated pacing leads. Furthermore, the results prove to be independent of the lead type being used (titanium-nitride- or iridium-coated tip). Thus, both new fractal pacing leads guarantee reliable capture recognition even at higher stimulation frequencies when a faster recharge is required.

5. Implications of Thin-Film Technology on Pacemaker Technology

What are the implications these sputter-deposited thin-film coatings with fractal surface structure have on pacemaker technology? Basically, it is the fact that coated pacing leads help to save energy and improve the longevity of implantable devices. In addition, the improved sensing behavior leads to a higher safety margin in demand-controlled devices. All these things can be summarized as a better performance of a device.

However, there are even implications on pacemaker technology that open new opportunities. As mentioned above, the development of a reliable capture recognition algorithm is really a highly important task as it would directly improve the safety of the patient. To date, there are two main methods for detecting the ventricular evoked response, an automatic post pulse compensation technique [132–134] and the use of two different leads for pacing and sensing [147]. The disadvantages of both of these methods, which prevent a capture recognition system, are well known. A recharge pulse requires additional battery power and a second lead suffers from the polarization of the common electrode and requires space in the heart that is not available.

The new method using pacing leads with thin, sputter-deposited coatings with a fractal surface structure avoids these drawbacks: One unipolar (or also bipolar) lead works as an actuator as well as a sensor and its high Helmholtz capacity guarantees a sufficiently low polarization artifact. Thus, in combination with an autoshort circuit, the ventricular evoked response can be reliably detected without additional energy requirements.

However, in some cases, fractal pacing leads do not allow the detection of the depolarization phase of the evoked response as it may lie in the autoshort region. The limitations that result from this disadvantage are specific to the application. Capture recognition for threshold tracking has to give a binary answer: capture or no capture. For this purpose, it is advisable to use the repolarization phase as an indicator since pacing pulse and artifact voltage have an opposite polarity. Capture recognition is, therefore, not limited by the timing restrictions of fractal pacing leads. In addition, research is currently being done to determine if the evoked response can be utilized for measuring the metabolic demand of the human body in order to get an input signal for rate adaptive pacemakers [148,149]. All these new features have been made possible by thin-film technology.

IV. SUMMARY AND CONCLUSIONS

Alloplastic materials in biomedical engineering have to meet functionality as well as biocompatibility requirements. Whereas biocompatibility is mainly a reflection of the surface properties, functionality is determined in first-order approximation by the bulk.

The solution of evolution is a teamwork model, a hybrid design consisting of several different structures and materials. Consequently, an implant should be made of a well-known and stable engineering material and a thin, biocompatible surface coating.

Conventional coating techniques suffer from insufficient material selection and tailoring capabilities. Therefore, the best approach seems to be thin-film technology, which is able to determine the surface composition of a device on an atomic scale. Different techniques are available, ranging from simple thermal evaporation to ion implantation. Due to the semiconductor industry, these techniques are well established and available at a reasonable price.

Although a lot of applications of thin-film technology in the field of biomedical engineering have been investigated, only a few items are already in large-scale production. This fact is not related to a poor feasibility of thin-film technology, but to poor know-how and a high acceptance threshold. Therefore, teamwork between implant designers and thin-film engineers is strongly recommended. The example of the fractal coating of pacing leads is a successful result of such teamwork, and it demonstrates how beneficial this can be. That is, the detection of the ventricular evoked response, which is made possible due to the coating, gives the opportunity to achieve a threshold tracking system or to have a sensor for information about the autonomic nervous system. Thin films not only improve biocompatibility, but also open or stimulate new fields of research and therapy.

REFERENCES

1. Morse, D., R. M. Steiner, and J. Fernandez (Eds.), *Guide to Prosthetic Cardiac Valves*, Springer, New York (1985).
2. Rohen, J. W., and Ch. Yokochi, *Anatomie des Menschen*, F. K. Schattauer, Stuttgart (1982).
3. Zimmerman, M. C., H. Scalzo, and J. R. Parsons, The Attachment of Hydroxyapatite Coated Polysulfone to Bone, *J. Applied Biom.*, 1:295–305 (1990).
4. De Lange, G. L., and K. Donath, Interface Between Bone Tissue and Implants of Solid Hydroxyapatite or Hydroxyapatite-Coated Titanium Implants, *Biomaterials*, 10:121–125 (1989).
5. Ducheyne, P., W. Van Raemdonck, J. C. Heughebaert, and M. Heughebaert, Structural Analysis of Hydroxyapatite Coatings on Titanium, *Biomaterials*, 7:97–103 (1986).
6. Hayashi, K., K. Uenoyama, N. Matsaguchi, and Y. Sugioka, Quantitative Analysis of *In Vivo* Tissue Responses to Titanium-Oxide- and Hydroxyapatite-Coated-Titanium Alloy, *J. Biom. Mat. Res.*, 25:515–523 (1991).
7. Klein, C. P., P. Patka, H. B. M. van der Lubbe, J. G. C. Wolke, and K. de Groot, Plasma Sprayed Coatings of Tetracalciumphosphate, Hydroxyl-Apatite, and α-TCP on Titanium Alloy: An Interface Study, *J. Biom. Mat. Res.*, 25:53–65 (1991).
8. Cook, S. D., K. A. Thomas, J. E. Dalton, T. K. Volkman, T. S. Whitecloud III, and J. F. Cay, Hydroxyapatite Coating of Porous Implants Improves Bone Ingrowth and Interface Attachment Strength, *J. Biom. Mat. Res.*, 26:989–1001 (1992).
9. Luckey, H. A., E. G. Lamprecht, and M. J. Walt, Bone Apposition to Plasma-Sprayed Cobalt-Chromium Alloy, *J. Biom. Mat. Res.*, 26:557–575 (1992).
10. Oguchi, H., K. Ishikawa, S. Ojima, Y. Hirayama, K. Seto, and G. Eguchi, Evaluation of a High-Velocity Flame-Spraying Technique for Hydroxyapatite, *Biomaterials*, 13:471–477 (1992).
11. LaBerge, M., G. Drouin, C. Gelinas, B. Champagne, and J. D. Bobyn, Adherence of HDPE

Powder Coating on Co-Cr Surfaces: Effect of Substrate Preparation and Gas Sterilization, *J. Biom. Mat. Res.*, 24:1427–1438 (1990).

12. Ducheyne, P., S. Radin, M. Heughebaert, and J. C. Heughebaert, Calcium Phosphate Ceramic Coatings on Porous Titanium: Effect of Structure and Composition, Vacuum Sintering and *In Vitro* Dissolution, *Biomaterials*, 11:244–254 (1990).

13. Kim, C. S., and P. Ducheyne, Compositional Variations in the Surface and Interface of Calcium Phosphate Ceramic Coatings on Ti and Ti-6Al-4V due to Sintering and Immersion, *Biomaterials*, 12:461–469 (1991).

14. Streicher, R. M., H. Weber, R. Schön, and M. Semlitsch, New Surface Modification for Ti-6Al-7Nb Alloy: Oxygen Diffusion Hardening (ODH), *Biomaterials*, 12:125–129 (1991).

15. Semlitsch, M. F., H. Weber, R. M. Streicher, and R. Schön, Joint Replacement Components Made of Hot-Forged and Surface-Treated Ti-6Al-7Nb Alloy, *Biomaterials*, 13:781–788 (1992).

16. von Fraunhofer, J. A., and D. Seligson, Cracking of Coated Orthopaedic External Fixation Connecting Bar, *Biomaterials*, 8:228–230 (1987).

17. Troelstra, S. A., Applying Coatings by Electrophoresis, *Philips Technical Review*, 11:293–303 (1951).

18. LPW Taschenbuch für Galvanotechnik, Bd. 1, Verfahrenstechnik, LPW-Chemie GmbH, Neuss, Germany, 13th ed. (1988).

19. Cox, A. J., Effect of a Hydrogel Coating on the Surface Topography of Latex-Based Urinary Catheters: A SEM Study, *Biomaterials*, 8:500–502 (1987).

20. Park, K., F. W. Mao, and H. Park, The Minimum Surface Fibrinogen Concentration Necessary for Platelet Activation on Dimethyldichlorosilane-Coated Glass, *J. Biom. Mat. Res.*, 25:407–420 (1991).

21. Steinburg, J., A. W. Neumann, D. R. Absolom, and W. Zingg, Human Erythrocyte Adhesion and Spreading on Protein-Coated Polymer Surfaces, *J. Biom. Mat. Res.*, 23:591–610 (1989).

22. Ferruti, P., G. Casini, F. Tempesti, R. Barbucci, R. Mastacchi, and M. Sarret, Heparinizable Materials (III). Heparin Retention Power of a Poly(amidoamine) either as Crosslinked Resin, or Surface Grafted on PVC, *Biomaterials*, 5:234–236 (1984).

23. Barbucci, R., M. Benvenuti, G. Casini, P. Ferruti, and F. Tempesti, Heparinizable Materials (IV). Surface-Grafting on Poly(Ethylene Terephtalate) of Heparin-Complexing Poly(Amidoamine) Chains, *Biomaterials*, 6:102–105 (1985).

24. Azzuoli, G., R. Barbucci, M. Benvenuti, P. Ferruti, and M. Nocentini, Chemical and Biological Evaluation of Heparinized Poly(amido-amine) Grafted Polyurethane, *Biomaterials*, 8:61–66 (1987).

25. Arnander, C., D. Bagger-Sjöbäck, S. Frebelius, R. Larsson, and J. Swedenborg, Long-Term Stability *In Vivo* of a Thromboresistant Heparinized Surface, *Biomaterials*, 8:496–499 (1987).

26. Barbucci, R., A. Albanese, A. Magnani, and F. Tempesti, Coating of Commercially Available Materials with a New Heparinizable Material, *J. Biom. Mat. Res.*, 25:1259–1274 (1991).

27. Frey, H., and G. Kienel (Eds.), *Dünnschichttechnologie*, VDI Verlag, Düsseldorf (1987).

28. Van der Brekel, C. H. J., Characterization of the Chemical Vapour Deposition Process, *Philips Research Rep.*, 32:118–125 (1977).

29. Burgers, J. M., *Flow Equations for Composite Gases*, Academic Press, New York (1969).

30. Akins, R. J., and J. C. Bokros, The Deposition of Pure and Alloyed Isotropic Carbons in Steady State Fluidized Beds, *Carbon*, 12:139–152 (1974).

31. Maissel, L. I., and R. Glang, *Vacuum Evaporation: Handbook of Thin Film Technology*, McGraw-Hill, New York (1970).

32. Hirth, J. P., and G. M. Pound, *Condensation and Evaporation—Nucleation and Growth Kinetics*, Pergamon Press, Oxford (1963).

33. Parent, E. D., Power Requirements of Resistance-Heated Intermetallic Evaporation Sources, *J. Vac. Sci. Technol.*, 11:820–822 (1974).

34. Warren, K. A., D. R. Denison, and D. G. Bills: Resistance Heated Sublimator, *Rev. Sci. Instr.*, 38:1019–1022 (1967).

35. Pierce, J. R., *Theory and Design of Electron Beams*, D. van Nostrand, New York (1954).

36. Schiller, S., and G. Jäsch, Deposition by Electron Beam Deposition with Rates of Up to 50 μm s^{-1}, *Thin Solid Films*, 54:9–21 (1978).

37. Fujimori, S., T. Kasai, and T. Inamura, Carbon Film Formation by Laser Evaporation and Ion Beam Sputtering, *Thin Solid Films*, 92:71–80 (1982).

38. Olstad, R. A., and D. R. Olander, Evaporation of Solids by Laser Pulses. I. Iron, *J. Appl. Phys.*, 46:1499–1508 (1975).

39. Olstad, R. A., and D. R. Olander, Evaporation of Solids by Laser Pulses. II. Zirconium Hydride, *J. Appl. Phys.*, 46:1509–1518 (1975).

40. Agajanian, A. K., *Semiconducting Devices: A Bibliography of Fabrication Technology, Properties and Applications*, IFI/Plenum, New York (1976).

41. Inagaki, N., T. Nishio, and K. Katsuura, Some Optical Properties of Polymer Films Prepared by Glow Discharge from Methane, Tetramethylsilane, and Tetramethyltin, *J. Polymer Sci.*, 18:765–770 (1983).

42. Yasuda, H., and M. Gazicki, Biomedical Applications of Plasma Polymerization and Plasma Treatment of Polymer Surfaces, *Biomaterials*, 3:68–77 (1982).

43. Rossnagel, S. M., J. J. Cuomo, and W. D. Westwood (Eds.), *Handbook of Plasma Technology: Fundamentals, Etching, Deposition, and Surface Interactions*, Noyes Publications, Park Ridge, Tennessee (1989).

44. Lucovsky, G., and D. V. Tsu, Plasma Enhanced Chemical Vapor Deposition: Differences between Direct and Remote Plasma Excitation, *J. Vac. Sci. Technol.*, A5:2231–2236 (1987).

45. Mogab, C. J., A. C. Adams, and D. L. Flamm, Plasma Etching of Si and SiO_2 – The Effect of Oxygen Additions to CF_4 Plasmas, *J. Appl. Phys.*, 49:3796–3799 (1978).

46. Behrisch, E. R. (Ed.), *Topics in Applied Physics*, Vol. 47, *Sputtering by Particle Bombardment*, Springer Verlag, Berlin (1981).

47. Thornton, J. A., Influence of Substrate Temperature and Deposition Rate on Structure of Thick Sputtered Cu Coatings, *J. Vac. Sci. Technol.*, 12:830–835 (1975).

48. Somekh, R. E., The Thermalization of Energetic Atoms During the Sputtering Process, *J. Vac. Sci. Technol.*, A3:1285–1290 (1983).

49. Gras-Marti, A., and J. A. Valles-Abarca, Evolution Towards Thermalization and Diffusion of Sputtered Particle Fluxes: Spatial Profiles, *J. Appl. Phys.*, 55:1370–1374 (1984).

50. Varous, T. V., Planar Magnetron Sputtering: A New Industrial Coating Technique, *Solid State Technol.*, 19:62–66 (1976).

51. Vossen, J. L., and W. Kern (Eds.), *Thin Film Processes*, Academic Press, New York (1978).

52. Westwood, W. D., S. Maniv, and P. J. Scanlon, The Current-Voltage Characteristics of Magnetron Sputtering Systems, *J. Appl. Phys.*, 54:6841–6843 (1983).

53. Rossnagel, S. M., Current-Voltage Relations in Magnetrons, *J. Vac. Sci. Technol.*, A6:223–239 (1988).

54. Koenig, H. R., and L. I. Maissel, Application of RF Discharges to Sputtering, *IBM J. Res. Dev.*, 14:168–171 (1970).

55. Suzuki, K., K. Ninomiya, S. Nishimatsu, J. W. Thoman, Jr., and J. I. Steinfeld, Analytical Investigation of Plasma and Electrode Potentials in a Diode Type RF Discharge, *Jap. J. Appl. Physics*, 25:1569–1574 (1986).

56. Maissel, L. I., and P. M. Schaible, Thin Films Deposited by Sputtering, *J. Appl. Phys.*, 36:237–242 (1965).

57. Patten, J. W., and E. D. McClanahan, Effect of Substrate Bias and Deposition Temperature on the Properties of Thick Sputtered Chromium Deposits, *J. Appl. Phys.*, 43:4811–4813 (1972).

58. Mattox, D. M., Particle Bombardment Effects on Thin Film Deposition: A Review, *J. Vac. Sci. Technol.*, A7:1105–1114 (1989).

59. Winters, H. F., and E. Kay, Gas Incorporation into Sputtered Films, *J. Appl. Phys.*, 38: 3928–3934 (1967).

60. Hass, G., and R. E. Thuns, *Physics of Thin Films*, Vol. 2, Academic Press, New York (1964).

61. Rickerby, D. S., and A. Matthews, *Advanced Surface Coatings: A Handbook of Surface Engineering*, Blackie, Chapman and Hall, New York (1991).

62. Thornton, J. A., Recent Advances in Sputter Deposition, *Surface Eng.*, 2:283–292 (1986).

63. Ahmed, N. A. G., *Ion Plating Technologies—Developments and Applications*, John Wiley, New York (1987).

64. Itoh, T., *Ion Beam Assisted Film Growth*, Elsevier Amsterdam (1989).

65. Mayer, J. W., L. Eriksson, and J. A. Davies, *Ion Implantation in Semiconductors*, Academic Press, New York, London (1970).

66. Ryssel, H., and I. Ruge, *Ionenimplantation*, B. G. Teubner, Stuttgart (1978).

67. Gombotz, W. R., and A. S. Hoffman, Gas-Discharge Techniques for Biomaterial Modification, *CRC Crit. Rev. Biocompat.*, 4(1):1–42 (1987).

68. Yan, J. Y. J., F. W. Cooke, P. S. Vaskelis, and A. F. von Recum, Titanium-Coated Dacron Velour: A Study of Interfacial Connective Tissue Formation, *J. Biom. Mat. Res.*, 23:171–189 (1989).

69. Bernex, F., J. P. Mazzucotelli, J. L. Roudiere, N. Benhaiem-Sigaux, J. Leandri, and D. Loisance, *In Vitro* Endothelialization of Carbon-Coated Dacron Vascular Grafts, *Int. J. Artif. Organs*, 15:172–180 (1992).

70. Haubold, A. D., H. S. Shim, and J. C. Bokros, Carbon in Medical Devices, In D. F. Williams (Ed.), *Biocompatibility of Clinical Implant Materials*, Vol. 2, CRC Press, Boca Raton, FL, pp. 3–42 (1981).

71. Kostka, H., S. Birkle, and W. Küsebauch, Pyrographit—Eigenschaften und Anwendungen, *Mat.-Wiss. u. Werkstofftech.*, 20:92–100 (1989).

72. Kummer, F. J., J. L. Ricci, and N. C. Blumenthal, RF Plasma Treatment of Metallic Implant Surfaces, *J. Applied Biom.*, 3:39–44 (1992).

73. Loh, I., H. Lin, and C. C. Chu, Plasma Surface Modification of Synthetic Absorbable Sutures, *J. Applied Biom.*, 3:131–146 (1992).

74. Lukas, J., T. Fenclova, V. Tyrackova, and J. Vacik, The Surface Treatment of Polypropylene Molds and Its Effect on the Quality of Cast Contact Lenses, *J. Applied Biom.*, 3:275–279 (1992).

75. Hettlich, H. J., F. Otterbach, Ch. Mittermayer, R. Kaufmann, and D. Klee, Plasma-Induced Surface Modifications on Silicone Intraocular Lenses: Chemical Analysis and *In Vitro* Characterization, *Biomaterials*, 12:521–524 (1991).

76. Chandy, T., and C. P. Sharma, Effect of Plasma Glow, Glutaraldehyde and Carbodiimide Treatments on the Enzymic Degradation of Poly (L-lactic acid) and Poly (γ-benzyl-L-glutamate) Films, *Biomaterials*, 12:677–682 (1991).

77. Dekker, A., K. Reitsma, T. Beugeling, A. Bantjes, J. Feijen, and W. G., van Aken, Adhesion of Endothelial Cells and Adsorption of Serum Proteins on Gas Plasma-Treated Polytetrafluoroethylene, *Biomaterials*, 12:130–138 (1991).

78. Ho, C.-P., and H. Yasuda, Ultrathin Coating of Plasma Polymer of Methane Applied on the Surface of Silicone Contact Lenses, *J. Biom. Mat. Res.*, 22:919–937 (1988).

79. Yeh, Y.-S., Y. Iriyama, Y. Matsuzawa, S. R. Hanson, and H. Yasuda, Blood Compatibility of Surfaces Modified by Plasma Polymerization, *J. Biom. Mat. Res.*, 22:795–818 (1988).

80. Marchant, R. E., S. D. Johnson, B. H. Schneider, M. P. Agger, and J. M. Anderson, A Hydrophilic Plasma Polymerized Film Composite with Potential Application as an Interface for Biomaterials, *J. Biom. Mat. Res.*, 24:1521–1537 (1990).

81. Davies, M. L., C. J. Hamilton, S. M. Murphy, and B. J. Tighe, Polymer Membranes in Clinical Sensor Applications. I. An Overview of Membrane Function, *Biomaterials*, 13:971–978 (1992).

82. Davies, M. L., C. J. Hamilton, S. M. Murphy, and B. J. Tighe, Polymer Membranes in Clinical Sensor Applications. II. The Design and Fabrication of Permselective Hydrogels for Electrochemical Devices, *Biomaterials*, 13:979–990 (1992).

83. Tanfani, F., A. A. Durrani, M. Kojima, and D. Chapman, Glycidyl Acrylate Plasma Glow Discharged Polymers, *Biomaterials*, 11:585–589 (1990).

84. Sipehia, R., and A. S. Chawla, Characterization of Plasma Polymerized Polypropylene Coatings, *Biomaterials*, 7:155–157 (1986).

85. Chawla, A. S., Use of Plasma Polymerization for Preparing Silicone-Coated Membranes for Possible Use in Blood Oxygenators, *Artif. Organs*, 3:92–96 (1979).

86. Chawla, A. S., and R. Sipehia, Characterization of Plasma Polymerized Silicone Coatings Useful as Biomaterials, *J. Biom. Mat. Res.*, 18:537–545 (1984).

87. Chawla, A. S., Evaluation of Plasma Polymerized Hexamethylcyclotrisiloxane Biomaterials Towards Adhesion of Canine Platelets and Leucocytes, *Biomaterials*, 2:83–88 (1981).

88. Ertel, S. I., B. D. Ratner, and T. A. Horbett, Radiofrequency Plasma Deposition of Oxygen-Containing Films on Polystyrene and Poly(Ethylene Terephtalate) Substrates Improves Endothelial Cell Growth, *J. Biom. Mat. Res.*, 24:1637–1659 (1990).

89. Lopez, G. P., B. D. Ratner, C. D. Tidwell, C. L. Haycox, R. J. Rapoza, and T. A. Horbett, Glow Discharge Plasma Deposition of Tetraethylene Glycol Dimethyl Ether for Fouling-Resistant Biomaterial Surface, *J. Biom. Mat. Res.*, 26:415–439 (1992).

90. Johnson, S. D., J. M. Anderson, and R. E. Marchant, Biocompatibility Studies on Plasma Polymerized Interface Materials Encompassing Both Hydrophobic and Hydrophilic Surfaces, *J. Biom. Mat. Res.*, 26:915–935 (1992).

91. Thomson, L. A., F. C. Law, N. Rushton, and J. Franks, Biocompatibility of Diamondlike Carbon Coating, *Biomaterials*, 12:37–40 (1991).

92. Bolz, A., and M. Schaldach, Haemocompatibility Optimisation of Implants by Hybrid Structuring, *Med. Biol. Eng. Comput.*, 31:S123–S130 (1993).

93. Albrektsson, T., and H.-A. Hansson, An Ultrastructural Characterization of the Interface between Bone and Sputtered Titanium or Stainless Steel Surfaces, *Biomaterials*, 7:201–205 (1986).

94. Wisbey, A., P. J. Gregson, and M. Tuke, Application of PVD TiN Coating to Co-Cr-Mo Based Surgical Implants, *Biomaterials*, 8:477–480 (1987).

95. Cook, S. D., F. S. Georgette, H. B. Skinner, and R. J. Haddad, Jr., Fatigue Properties of Carbon- and Porous-Coated Ti-6Al-4V Alloy, *J. Biom. Mat. Res.*, 18:497–512 (1984).

96. Bolz, A., M. Schaldach, and M. Hubmann, Interface Aspects of Stimulation Electrodes, *Proc. Ann. Conf. IEEE/EMBS*, 13:714–716 (1991).

97. Bolz, A., R. Fröhlich, and M. Schaldach, Elektrochemische Aspekte der Elektrostimulation — Ein Beitrag zur Senkung des Energiebedarfs, In M. Hubmann and R. Hardt (Eds.), *Schrittmachertherapie und Hämodynamik*, MMV-Verlag, Munich, pp. 75–99 (1993).

98. Naji, A., and M. F. Harmand, Cytocompatibility of Two Coating Materials, Amorphous Alumina, and Silicon Carbide Using Human Differentiated Cell Cultures, *Biomaterials*, 12:690–694 (1991).

99. Ruckenstein, E., and S. V. Gourisankar, Preparation and Characterization of Thin Film Surface Coatings for Biological Environments, *Biomaterials*, 7:403–422 (1986).

100. Hayashi, K., N. Matsuguchi, K. Uenoyama, T. Kanemaru, and Y. Sugioka, Evaluation of Metal Implants Coated with Several Types of Ceramics as Biomaterials, *J. Biom. Mat. Res.*, 23:1247–1259 (1989).

101. Ong, J. L., L. C. Lucas, W. R. Lacefield, and E. D. Rigney, Structure, Solubility and Bond Strength of Thin Calcium Phosphate Coatings Produced by Ion Beam Sputter Deposition, *Biomaterials*, 13:249–254 (1992).

102. Gluszek, J., J. Jedrkowiak, J. Markowski, and J. Masalski, Galvanic Couples of 316L Steel with Ti and Ion Plated TiN Coatings in Ringer's Solutions, *Biomaterials*, 11:330–335 (1990).

103. Hayashi, K., I. Noda, K. Uenoyama, and Y. Sugioka, Breakdown Corrosion Potential of Ceramic Coated Metal Implants, *J. Biom. Mat. Res.*, 24:1111–1113 (1990).

104. Taylor, S. R., and D. F. Gibbons, Effect of Surface Texture on the Soft Tissue Response to Polymer Implants, *J. Biom. Mat. Res.*, 17:205–227 (1983).

105. Buchanan, R. A., I.-S. Lee, and J. M. Williams, Surface Modification of Biomaterials through Noble Metal Ion Implantation, *J. Biom. Mat. Res.*, 24:309–318 (1990).

106. Lee, I.-S., R. A. Buchanan, and J. M. Williams, Charge-Injection Densities of Iridium and Iridium-Ion-Implanted Ti-6Al-4V with Relevancy to Neural Stimulation, *J. Biom. Mat. Res.*, 25:1039–1043 (1991).

107. Röstlund, T., P. Thomsen, L. M. Bjursten, and L. E. Ericson, Difference in Tissue Response to Nitrogen-Ion-Implanted Titanium and c.p. Titanium in the Abdominal Wall of the Rat, *J. Biom. Mat. Res.*, 24:847–860 (1990).

108. McKellop, H. A., The Wear Behaviour of Ion-Implanted Ti-6Al-4V Against UHMW Polyethylene, *J. Biom. Mat. Res.*, 24:1413–1425 (1990).

109. Buchanan, R. A., E. D. Rigney, Jr., and J. M. Williams, Ion Implantation of Surgical Ti-6Al-4V for Improved Resistance to Wear-Accelerated Corrosion, *J. Biom. Mat. Res.*, 21: 355–366 (1987).

110. Buchanan, R. A., E. D. Rigney, Jr., and J. M. Williams, Wear Accelerated Corrosion of Ti-6Al-4V and Nitrogen-Ion-Implanted Ti-6Al-4V: Mechanisms and Influence of Fixed Stress Magnitude, *J. Biom. Mat. Res.*, 21:367–377 (1987).

111. Rieu, J., A. Pichat, L. M. Rabbe, A. Rambert, C. Chabrol, and M. Robelet, Ion Implantation Effects on Friction and Wear of Joint Prosthesis Materials, *Biomaterials*, 12:139–143 (1991).

112. Lee, I.-S., R. A. Buchanan, and J. M. Williams, Biocorrosion Studies of Gold and Rhodium Ion Implanted Titanium and Ti-6Al-4V Alloy, Biomedical Mat. and Dev., In J. S. Hanker, and B. L. Giammara (Eds.), *Mat. Res. Soc. Symp. Proc.*, Vol. 110, Materials Research Society, Pittsburgh, 687–695 (1989).

113. Troshin, A. S., *Problems of Cell Permeability*, Permagon Press, London (1966).

114. Weiss, L., *The Cell Periphery: Metastasis and Other Contact Phenomena*, North Holland, Amsterdam (1967).

115. *Guidelines for Blood-Material Interactions*, Report of the National Heart, Lung, and Blood Institute Working Group, NIH Publ. No. 85-2185 (1985).

116. Sawyer, P. N., W. H. Brattain, and P. J. Boddy, Electrochemical Aspects of Thrombogenesis. In P. N. Sawyer, (Ed.) *Biophysical Mechanisms in Vascular Hemostasis and Intravascular Thrombosis*, Appleton-Century-Crofts, New York (1965).

117. Sawyer, P. N., S. Srinivasan, B. Stanczewski, N. Ramasamy, and W. Ramsey, Electrochemical Aspects of Thrombogenesis—Bioelectrochemistry Old and New, *J. Elchem. Soc.*, 121: 221C–234C (1974).

118. Baurschmidt, B., and M. Schaldach, The Electrochemical Aspects of the Thrombogenicity of a Material, *J. Bioeng.*, 1:261–278 (1977).

119. Bolz, A., and M. Schaldach, Physikalische Mechanismen der Festkörper-Protein-Wechselwirkung an der Phasengrenze a-SiC:H—Fibrinogen, *Biomedizinische Technik*, 37: 244–253 (1992).

120. Szent-Györgyi, A., The Study of Energy Levels in Bioelectrochemistry, *Nature*, 148:157–158 (1941).

121. Bakshi, A. K., J. Ladik, M. Seel, and P. Otto, On the Electronic Structure and Conduction Properties of Aperiodic DNA and Proteins: IV. Electronic Structure of Aperiodic Proteins, *Chem. Phys.*, 108:233–241 (1986).

122. Ladik, J., M. Seel, P. Otto, and A. K. Bakshi, On the Electronic Structure and Conduction Properties of Aperiodic DNA and Proteins: I. Strategy and Methods of Investigations, *Chem. Phys.*, 108:203–214 (1986).

123. Eley, D. D., and D. I. Spivey, The Semiconductivity of Organic Substances. Part 6: A Range of Proteins, *Trans. Far. Soc.*, 56:1432–1442 (1960).

124. Rosenberg, B., Electrical Conductivity of Proteins. II. Semiconduction in Crystalline Bovine Hemoglobin, *J. Chem. Phys.*, 36:816–823 (1962).

125. Bone, S., J. Eden, and R. Pethig, Electrical Properties of Proteins as a Function of Hydration and NaCl Content, *Int. J. Quant. Chem.: Quant. Biol. Symp.*, 8:307–316 (1981).

126. Geiger, J., Intensitäten in einem Elektronenbandsystem (Franck-Condon-Prinzip), In H. Gobrecht (Ed.), *Bergmann-Schäfer, Lehrbuch der Experimentalphysik, Band 4, Teil I*, 2d ed., Walter de Gruyter, Berlin, pp. 274–280 (1981).

127. Frisch, M. J., *Program Gaussian 88*, Gaussian, Inc., Pittsburgh (1988).

128. Mott, N. F., and E. A. Davis, *Electronic Processes in Non-Crystalline Materials*, 2d ed., Clarendon Press, Oxford (1979).

129. Cohen, J. D., Density of States from Junction Measurements in Hydrogenated Amorphous Silicon, In J. I. Pankove, *Semiconductors and Semimetals*, Vol. 21, Part C, Academic Press, Orlando (1991).

130. Van Wagenen, R. A., S. Rockhold, and J. D. Andrade, Probing Protein Adsorption by Total Internal Reflection Intrinsic Fluorescence in Biomaterials: Interfacial Phenomena and Applications, *Adv. Chem. Ser.*, 199:351–370 (1981).

131. Hartert, H., Blutgerinnungsstudie mit der Thrombelastographie, einem neuen Untersuchungsverfahren, *Klin. Wochenschr.*, 26:37–40 (1948).

132. Donaldson, R. M., and A. F. Rickards, The Ventricular Endocardial Paced Evoked Response, *PACE*, 5:253–259 (1983).

133. Schaldach, M., G. Boheim, and R. Edelhäuser, Ein Beitrag zur Erkennung evozierter Potentiale, *Biomed. Techn.*, 33(Sonderb. 2):343–344 (1988).

134. Walton, C., A. P. Economides, and S. Gergely, Determination of Myocardial Depolarization and Repolarization Times Using the Ventricular Evoked Potential: Contrasting Effects of Stimulus Interval and Isoprenaline in the Isolated Perfused Rabbit Heart, *PACE*, 12:784–792 (1989).

135. MacCarter, D. J., K. M. Lundberg, and J. P. M. Corstjens, Porous Electrodes: Concept, Technology and Results, *PACE*, 6:427–435 (1983).

136. Hirshorn, M. S., L. K. Holley, M. Skalsky, C. R. Howlett, and E. Musgrove, Characteristics of Advanced Porous and Textured Surface Pacemaker Electrodes, *PACE*, 6:525–536 (1983).

137. Karpawich, P. P., K. B. Stokes, J. R. Helland, C. D. Justice, and J. O. Roskamp, A New Low Threshold Platinized Epicardial Pacing Electrode: Comparative Evaluation in Immature Canines, *PACE*, 11:1139–1148 (1988).

138. Schaldach, M., A. Bolz, J. Breme, M. Hubmann, and R. Hardt, Acute and Long-Term Sensing and Pacing Performance of Pacemaker Leads Having TiN Electrode Tips, In G. E. Antonioli (Ed.), *Pacemaker Leads 1991*, Elsevier, Amsterdam, pp. 441–450 (1991).

139. Liu, S. H., Formation and Anomalous Properties of Fractals, *IEEE Eng. Med. Biol. Mag.*, 11:28–39 (1992).

140. Bolz, A., H. Matlok, M. Still, M. Schaldach, M. Hubmann, and R. Hardt, Langzeitstabilität des Detektions- und Reizschwellenverhaltens von Titannitrid-Herzschrittmacher-Elektroden, *Biomed. Technik*, 35(Ergbd. 3):131–133 (1990).

141. Popp, T., A. Bolz, and M. Schaldach, Untersuchung des Oxidationsverhaltens von Iridiumbeschichtungen für bipolare Stimulationselektroden, *Biomed. Technik*, 37(Ergbd. 2):24–25 (1992).

142. Taubert, G., A. Gebauer, F. Lünninghake, A. Brandes, J. Potratz, and U. Stierle, Long Time Course of Pacing Threshold in Three Bipolar Leads (Siemens 1010T, Medtronic Capsure 5026, Biotronik TIR 60-BP), *PACE*, 16:1179 (1993).

143. Neuzil, P., M. Taborsky, and R. Vopalka, Effect of Fractal Surface Structuring in Coating Electrode Surface—Comparison of Titanium Nitride and Iridium Nitride Materials, *PACE*, 16:1177 (1993).

144. Kreyenhagen, P., and J. Helland, Evaluation of a Titanium Nitride Coated Tip Electrode, In G. E. Antonioli (Ed.), *Pacemaker Leads 1991*, Elsevier, Amsterdam, pp. 451–458 (1991).

145. Auerbach, A. A., and S. Furman, The Autodiagnostic Pacemaker, *PACE*, 2:58–68 (1979).

146. Curtis, A. B., F. Vance, and K. Miller, Automatic Reduction of Stimulus Polarization Artifact for Accurate Evaluation of Ventricular Evoked Responses, *PACE*, 14:526–537 (1991).

147. Callaghan, F., W. Vollmann, A. Livingston, B. Boveja, and D. Abels, The Ventricular Depolarization Gradient: Effects of Exercise, Pacing Rate, Epinephrine, and Intrinsic Heart Rate Control on the Right Ventricular Evoked Response, *PACE*, 12:1115–1130 (1989).

148. Rickards, A. F., Physiologically Adaptive Cardiac Pacemaker, U.S. Patent 4,228,803 (Oct. 21, 1980).

149. Grassi, G., L. Cammilli, L. Alcidi, and P. Maroni, Cardiac Stimulator with Frequency Self-Regulation through the Electrocardiographic T Wave, U. S. Patent 4,556,062 (Dec. 3, 1985).

39
Hydrophilic Lubricity in Medical Applications

Y. L. Fan
Union Carbide Corporation
Bound Brook, New Jersey

I. INTRODUCTION

Hydrophilic surface lubricity is a common phenomenon found on marine life and living tissues in animal and human bodies. This surface lubricity serves to reduce friction in motion or protect tissue from damage or both. It is therefore not surprising that a hydrophilic lubricious surface has become highly desirable for medical devices intended for either transient or permanent implant in the human body [1].

Synthetic and natural-occurring polymeric materials such as polyurethane, acrylic resin, vinyl resin, polyolefin, nylon, and rubber are widely used for making catheters, contact lenses, implant devices, heart valves, intrauterine devices, peristaltic pump chambers, endotracheal tubes, gastroenteric feed tubes, and arteriovenous shunts. It is highly desirable that these materials can be fabricated to provide surfaces that are not only hydrophilic but also have low coefficients of friction when in contact with an aqueous fluid, such as a body fluid. Such lubricious surfaces would facilitate insertion or removal of a medical device into or out of a patient, and would minimize injury or inflammation of mucous membranes as well as aid patient comfort. For ease of handling, it is even more desirable if the surface exhibits a normal plastic feel when dry, but becomes slippery only upon exposure to an aqueous fluid.

Earlier approaches to provide a low-friction surface include the application of a nonpermanent coating such as silicone oil, olive oil, glycerin, or Xylocain jelly, and the use of conventional low-friction materials such as fluoroplastics, silicone, and polyethylene [2–5]. These earlier techniques, while useful, do not satisfy all the needs for a broad spectrum of surgical procedures. In the case of oils being applied, they tend to run off and the lubricity does not last. In the case of the use of low-friction materials such as Teflon® are used, the finished medical devices are difficult to handle because they are slippery at all times.

A variety of approaches have been undertaken to develop hydrophilic lubricious surfaces for medical devices that would provide good handling characteristics when dry but become instantaneously slippery upon exposure to an aqueous body fluid [6–13]. This field was reviewed recently by this author [14,15]. This chapter is an updated version of the earlier review. The author intends to provide the readers the following aspects of hydrophilic lubricity: (1) a systematic discussion and comparison of the various modification methods available currently, (2) a discussion of the commonly used measuring techniques for hydrophilic surface lubricity, and (3) a mechanistic consideration of the hydrophilic surface lubricity. A more general discussion on this subject was provided in a monogram published recently by Technomic Publishing Company [16].

These different techniques for producing lubricious surfaces may be grouped into six categories by virtue of the chemical composition of the finished surfaces: hydrophilic polymers, hydrophilic polymer blends or complexes, interpenetrating polymer networks, chemically reactive hydrophilic polymer coatings, surface grafted hydrophilic monomers, and hydrophilic bulk polymers.

II. LUBRICITY DERIVED FROM HYDROPHILIC POLYMERS

A most commonly used technique for imparting a hydrophilically slippery surface to a medical device such as surgical gloves, catheters, and sutures is the application of an exterior coating containing a suitable hydrophilic polymer. This polymer may be either water soluble or insoluble, but is readily swollen in water. The coating may be applied by any suitable coating process, but a dip-coating process is probably the most commonly used [17].

Poly(vinyl pyrrolidone) or poly(ethylene oxide) are effective hydrophilic polymers for providing a coating surface having a low coefficient of friction when wet while retaining a normal plastic feel when dry [17–21]. (Poly(vinyl pyrrolidone) USP grades are available from GAF and BASF, and poly(ethylene oxide) NF grades are available from Union Carbide.) Coatings made from these polymers have been accepted by the medical communities because of their proven processability, biocompatibility, some anti-thrombogenicity, protein repellent properties, and commercial availability [22,23]. The polymer is dissolved in an organic solvent such as dimethylformamide or tetrahydrofuran and applied by either a dip- or spray-coating process. An acid pretreatment of substrates made from latex rubber may be employed to improve adhesion. A similar effect was described on hydrophilic polyurethane surfaces [24].

Poly(ethylene oxide) has been modified in many ways either to offer an extra niche for the intended application or to be compositionally distinct from the homopolymer. Poly(ethylene oxide) with a free-radical curable unsaturated end group has been made which permitted simultaneous sterilization and curing for medical applications [25,26]. The finished coating is lubricious in water, but otherwise more water resisting than the uncured poly(ethylene oxide). Polyoxyethylene-polyoxypropylene block copolymer coating was found to impart an improved run-down characteristic and reduced tissue drag on some absorbable surgical sutures [27,28]. The lubricity is further improved by formulating with a humectant such as glycerol [29]. Similar lubricating coatings have been prepared using poly(alkylene oxide-glycolic acid ester) block or random copolymers [30,31].

Poly(vinyl alcohol) is used as an interfacial layer to enhance lubrication and shock-absorbing functions of prostheses for damaged bones [32,33]. Maleic anhydride type polymer, more specifically, partially alkylated ester of (vinyl methyl ether-maleic anhydride) copolymer, is used as a coating for catheters and guide wires [34]. It is also used in medical adhesive tapes for wrapping medical instruments such as an optical-fiber scope [35]. A methacrylate-containing pendant methoxy poly(ethylene glycol) moiety is copolymerized with a vinyl comonomer to yield a hydrophilic copolymer suitable for forming a lubricious coating with an improved antithrombogenic property [36]. Latex rubber surgical gloves and condoms can be made more lubricious by incorporating a hydrogel polymer layer such as (2-hydroxyethylmethacrylate-methacrylic acid) copolymer containing a surfactant during the forming operation [37].

Hydrophilic polyurethanes constitute another class of synthetic polymers used for this purpose. Polyurethanes offer a wide range of physical and mechanical properties as well as processabilities and biocompatibilities [38]. Polyurethane homo- and copolymers have been used for friction reduction in pharmaceutical closures [39], as orthopedic casting materials [40], in hydrogel carrier systems [41], and as burn dressings and friction-reduction coatings [42].

Other synthetic polymers useful for affording a friction-reduction coating include poly(alkylene oxalate) [43], poly(vinyl alcohol) [33], water-swellable copolymers [44], ionene (ionic amine) polymers [45,46], and caprolactone homo- and copolymers. The last constitutes a class of biodegradable lubricating polymers [23,47], but the homopolymer is water insoluble, however.

Chitin, poly(N-acetyl-D-glucosamine), and its derivatives have been employed for imparting lubricity on surgical elements (e.g., tubing, catheters, drains, and gloves) [48]. This natural-occurring polymer and its derivatives are thermally stable to endure autoclaving and are biodegradable. Many cellulosic derivatives (e.g., carboxymethyl cellulose, hydroxypropyl cellulose, and methyl cellulose) have been used as a hydrophilic lubricant [49–51]. A slippery coating based on gelatin, alcohol, and formaldehyde was also claimed [52].

Certain hydrophilic lubricious coatings are formulated to be intentionally leachable or absorbable such that the hydrophilic polymer is removed from the device in a controlled fashion [31,47,53]. This type of coating is often used on surgical sutures and shaving devices, among others.

The hydrophilic lubricity of the coating compositions described in this group, while good initially, is often not lasting nor durable. The hydrophilic polymers are readily removed from the substrate surfaces through either leaching or mechanical dislocation due to a lack of adhesion. While resistance to water leaching has been enhanced by cross-linking of the hydrophilic polymer, the finished coating is often nondurable and exhibits inadequate bonding to the substrate in the wet state.

III. HYDROPHILIC POLYMER BLENDS OR COMPLEXES

Hydrophilic coatings that have somewhat improved durability or adhesion or both have been obtained by forming either a polymeric blend or complex with the substrate material. Poly(N-vinyl lactam) was blended with a thermoplastic polyurethane to yield a more durable lubricious coating when wet [19,54]. Slippery polyurethane surfaces having improved abrasion resistance have been obtained by mixing with a dialkyl sulfonate surfactant. The polyurethane may be produced *in situ* from a diol and a diisocyanate in the presence of stannous octoate [55]. A stable hydrophilic polymer blend of poly(vinyl butyral) and poly(N-vinyl lactam) was claimed to retain slippery properties in an aqueous environment [56]. High molecular weight poly(ethylene oxide) was incorporated in a cured polyurethane casting tape to give a controlled slipperiness [57]. A flexible tube for introducing catheters has been made using a poly(vinyl pyrrolidone)-polyurethane blend to impart lubricity and thromboresistance [58].

Such a hydrophilic polymer coating may be rendered more stable by treating with nontoxic organic or inorganic salt such as sodium citrate and sodium chloride, respectively [59]. The substrate surface may be cured with a diisocyanate. Poly(vinyl chloride) (PVC) catheters are rendered lubricious by coating with a two-phase antifriction coating consisting of a poly(vinyl lactam) inner phase and a polyurethane outer phase [60]. A 50% reduction in frictional force was achieved. Another dual-layer coating consisting of an outer layer of poly(vinyl pyrrolidone) and an inner layer of stabilizing polymer such as a cellulose ester was reported to be lubricious and abrasion resistance [61].

Poly(ethylene oxide) and poly(vinyl pyrrolidone) have been used by blending with a structural plastic such as polyurethane using either a coextrusion or coating process [62]. A polymer blend consisting of poly(ethylene oxide) and a mixture of poly(vinyl chloride) and polyurethane is coextruded to produce a medical article having solvent-free lubricious surfaces [63]. The coating was reported to adhere strongly and show no tendency to wash off or separate as solid flakes upon repeated use.

A lubricious coating comprised of a noncross-linked, crystallized polymer or polymer blend is used for modifying the surfaces of diagnostic probes, catheters, and guidewires [64]. A mixture of fatty acid and polymer has been used for developing wet lubricity on nasogastric intubation device upon exposure to stomach fluids. Lubricity is developed when the fatty acid is converted to its salt [65]. A blood vessel catheter is coated with a poly(vinyl pyrrolidone) and a polyurethane mixture to provide antithrombogenic and lubricious surfaces [66]. Another lubricious coating is made from an uniform blend of a hydrophilic polymer and an elastomeric segmented hydrophilic poly(ether urethane) to afford quick lubricity and durability [67].

Polyacrylates such as 2-hydroxyethyl-methacrylate or 2-ethylhexyl acrylate and their copolymers have been used for forming a lubricating layer, bonded to a rubber article, with respect to damp skin [68]. Guidewires are made antithrombotic and slippery by coating with organosiloxane copolymers containing amino-alkyl-siloxane and methyl siloxane [69].

Generally speaking lubricious coatings described in this group exhibit a somewhat better performance than the simple hydrophilic polymer coating mentioned above in terms of longer service time. They are, however, still deficient in abrasion resistance and coating durability.

IV. FORMATION OF INTERPENETRATING POLYMERIC NETWORKS

The need for a greater adhesion between the lubricious coating and the substrate has led to the development of techniques for forming a surface layer of interpenetrating polymeric network (IPN) in which the hydrophilic polymer is anchored permanently with the substrate molecules. In some instances, however, the hydrophilic polymer present in the coating composition is in reality a thermoplastic rather than a thermosetting polymer. The resultant compositions may not be true IPNs, but are mixtures in which the long-chain hydrophilic polymers intertwine physically within the network of a polyurea or polyurea-polyurethane or polyurea-polyacrylate, depending on the nature of the coating composition and substrate used in the device. This approach has received a great deal of attention in recent years, and a variety of IPN-type coatings have been developed using either photo- or radiation-induced polymerization techniques.

The hydrophilic polymer often cited for this application is poly(vinyl pyrrolidone). It is usually used in conjunction with a multifunctional isocyanate with or without the presence of a polyurethane [70–73]. Poly(ethylene oxide) has been applied in pretty much a similar fashion with a polyisocyanate to produce a network structure through mostly physical entanglement [74]. When relatively low molecular weight hydroxyl-terminated poly(alkylene oxide) and/or poly(alkylene glycol) molecules are present in the coating composition, the hydrophilic polymers may become permanently attached to the network structure through the formation of an IPN [75,76]. An IPN structure may be produced in a coating formulation containing an isocyanate, a reactive polyol, and a poly(ethylene oxide) [77].

A similar approach is reported in which poly(vinyl pyrrolidone) instead of poly(ethylene oxide) is used [78]. A slippery coating comprised of poly(vinyl pyrrolidone), a poly(carboxylic acid), or polyanhydride and an N-trimethoxypropylsilyl poly(ethyleneimine) was described as being able to withstand the rigors of sterilization and long-term exposure to human body fluids without substantial loss of lubricity [79–81].

A method to produce an IPN or physical entanglement of a hydrophilic polymer with a polymeric substrate is achieved by presoaking a plastic surface in an aqueous solution of either N-vinyl pyrrolidone or 2-hydroxyl methacrylate or both, and followed by exposure to gamma irradiation [82]. In another method, a non-cross-linked hydrogel such as poly(ethylene oxide) is made a part of a network structure by ultraviolet- (UV) initiated photopolymerization of an acrylic monomer in the presence of a difunctional comonomer such as neopentyl glycol diacrylate. The network may include a drug delivery system in which the therapeutic agent may be either permanently entrapped in the network or can be released from the coating into the living body upon lubrication [83].

Hyaluronic acid and its salts, chrondritic sulfate and agarose, are another class of biocopolymers used for producing a hydrophilic lubricious coating on medical devices. The dimensional stability of such a coating was significantly improved by the incorporation of albumin [84], which was claimed to facilitate the wetting of hyaluronic acid on the more hydrophobic surface of the acrylic polymer anchor coat such as poly(hydroxyethyl methacrylate). This lubricious polysaccharide coating was reported to be particularly useful for providing a long-lasting comfort to wearers of body implants or prostheses. The polysaccharide may be cross-linked to afford a more permanent lubricity. A biocompatible, blood-compatible structure with a low-friction cross-linked hydrophilic polymer coating was reported [85].

Depending on the chemistry involved in the coating processes, a true IPN may or may not be formed in the coating. When the hydrophilic polymer molecules are not permanently bonded in an IPN network, they could leech out, particularly when the hydrophilic polymer molecules exhibit only limited compatibility with the matrix molecules. On the other hand, when true a IPN is produced, good adhesion to the substrate can usually be expected. In reality, however, there may be a distribution of hydrophilic polymer molecules that are bonded in the IPN and those that are only physically entangled in the system. Both are expected to contribute to the observed lubricity. For a tightly cross-linked system in which all hydrophilic polymer molecules are permanently bonded in the IPN network, a reduction in surface lubricity can result because the hydration of the hydrophilic polymer molecules could be retarded by the neighboring hydrophobic molecules. Poly(ethylene oxide) and poly(vinyl pyrrolidone) have been immobilized quite successfully by this technique to provide surface lubricity with a useful degree of durability.

V. CHEMICALLY REACTIVE HYDROPHILIC POLYMER COATINGS

While a good adhesion does not necessarily involve chemical bonding, the presence of chemical bonding in an adhesive joint usually insures a durable bond. Following this thinking, a number of systems have been devised in which the hydrophilic polymer layer is capable of chemically bonding to the substrate, usually through a reactive primer system.

Water-soluble polymers such as cellulosic derivatives, poly(maleic anhydride), poly-acrylamide, or water-soluble nylon have been claimed to bond chemically to a polymeric substrate through a reactive primer such as an aldehyde, epoxy, or isocyanate [86]. Water-swellable copolymers capable of reacting with isocyanates have been reported to produce durable, lubricious coatings [44]. Depending on the chemical nature of the substrate material, however, the reactive primer may or may not bond chemically to the substrate. Polysaccharides such as hyaluronic acid have been bonded to acrylic polymers using a polyisocyanate to afford improved hydrophilicity and lubricity [87,88]. Vinyl pyrrolidone copolymers containing active hydrogen, such as a (vinyl pyrrolidone-2-hydroxyethyl methacrylate) copolymer, have been described as bonding to substrate through a polyisocyanate primer [89].

A hydrophilic lubricious coating having good abrasion resistance is prepared by priming a substrate with a polyisocyanate and followed by contacting with a poly(acrylic acid) solution or dispersion. The coating hydrates instantaneously upon exposure to water and affords a high degree of lubricity. The resultant surfaces exhibit a normal feel when dry, but a fast rate of hydration upon exposure to water [90].

A lubricious coating composition was reported using a water-soluble prepolymer derived from a polyisocyanate and a poly(ethylene oxide) glycol ester of a sulphophthalic acid. Coated sheets are useful as an orthopedic bandage with reduced tack [91]. A hydrophilic coating formulation capable of covalent bonding to substrate was reported that is a reaction product from a mixture of isocyanate, acrylamide, and an acid polysaccharide. The material was reported to exhibit improved low-friction and antithrombogenic properties [92]. A reactive lubricious coating composition is produced by coating a catheter or anesthesia tube consecutively with an isocyanate and poly(N-hydroxypropyl methacrylamide). The coating was reported to afford not only improved wet lubricity, but also better blood, urine, and tissue compatibility, as well as reduction of infection rates [93].

A terpolymer consisting of N-vinyl pyrrolidone, glycidyl acrylate, and vinyl acetate and having a minimum molecular weight of 400,000 has been made to yield a slippery coating composition [94]. An aqueous coating process consists of a first layer of an aqueous acrylic melamine emulsion, followed by a second layer of a water-soluble polymer such as polyacrylamide. The two layers are cured by baking at temperatures above 100°C. This process is useful for coating instruments, including guidewires and catheters [95]. A polymeric complex coating consisting of either poly(ethylene oxide) or poly(vinyl pyrrolidone) with poly(acrylic acid) is used as a lubricious coating composition to provide improved antithrombogenicity and lubricity. The polymeric complexes are immobilized to the substrate surfaces using a reactive polyisocyanate primer [96].

In all these cases, the hydrophilic lubricious coatings have been claimed to be durable and not leachable. A common feature claimed by the patentees of this group is that the hydrated hydrophilic coating is abrasion resistant and would survive either repeated abrasions during a surgical procedure or a long period of implanting in the human body or both. The success of chemical bonding is affected not only by the reactivity of coating chemicals used and the chemical nature of the substrate to be coated, but also by the coating conditions employed. Unless all these required elements are present, this approach may not necessarily lead to a high-quality lubricious coating. Due to the chemical reactivity of the hydrophilic polymers and primers used in these reactive coating systems, a two-step coating process is usually employed. A good process control is therefore also essential for a successful application of the chemically reactive coatings.

VI. SURFACE GRAFTING OF HYDROPHILIC MONOMERS

Instead of solution coating of a hydrophilic polymer, the hydrophilic polymer can be polymerized *in situ* from the corresponding monomer(s) or prepolymer(s) directly onto the surfaces of a medical device. A free-radical polymerization process is usually employed.

A radiation grafting process of hydrophilic monomers onto organic polymeric substrates in the presence of cupric or ferric ions was described by Ratner and Hoffman [97]. Maleic anhydride and other monomeric anhydride have been grafted onto the surface of gloves, sheets, and tubes made from natural rubber by immersing in a polymerization bath at an elevated temperature in the presence of a free-radical initiator. The anhydride-grafted rubber device is subsequently washed with an aqueous alkaline solution to generate carboxylate groups on the surface [98]. According to this technique, the treated rubber articles are provided with a chemically bonded hydrophilic surface that has a lower frictional coefficient to human skin. Acrylamide and other suitable unsaturated monomers have been grafted using gamma radiation to afford self-lubricating fill tubes that would insert easily in retention valves in medical appliances [99].

A similar grafting process was reported recently for forming polyacrylamide on polymeric substrates [100,101]. By this process, an ozone generator is used to generate hydroperoxides on the surface of the polymeric substrate. Polymerization is effected by using a redox initiator in the presence of acrylamide monomer. A high degree of grafting is obtained according to these authors.

A similar technique using either a high-energy radiation source or a free-radical initiator is employed to provide a gastrointestinal catheter having excellent surface slideability and compatibility with living tissues [102]. Good hydrophilic lubricity is also obtained by a plasma-initiated copolymerization of acrylamide and methacrylate. No toxic side effect to humans was reported [103].

A biocompatible, lubricious coating is produced by a peroxide-induced grafting of a hydrophilic vinyl monomer onto an island-shaped, discontinuous undercoat produced from polyethylene, polypentene, and others [104]. A hydrophilic coating is produced by plasma polymerization of N,N-dimethyl acrylamide on the surface of polymeric film that is predeposited with a layer of benzoin ethyl ether photosensitizer. The treated film has a layer of hydrogel coating [105].

Other monomers or monomer-hydrophilic polymer mixtures have been radiation polymerized to produce slippery surfaces [106,107]. The process variables in a photo-induced graft polymerization have been studied by Uyama et al. [108]. They reported that a graft amount of 50–100 $\mu g/cm^2$ of polyacrylamide on a polypropylene substrate was required to afford optimum hydrophilicity.

Poly(ethylene oxide) having an unsaturated end group is cross-linked by radiation to yield an adherent coating through chemical grafting [25,26]. Surface modification that provides both improved hydrophilicity and antithrombogenicity was reported by attaching polysaccharide to functional unsaturated monomers that were grafted onto the substrate through plasma-initiated polymerization [109]. A photo-graftable coating composition consisting of a hydrophilic monomer such as acrylamide, a hydrophobic monomer such as a perfluoroacrylate, a polymerizable poly(ethylene oxide), and a UV initiator was reported [110]. The coating was reported to afford both dry and wet lubricity, and is particularly useful for coating medical devices made of rubber. A hydrophilic lubricious coating is obtained by grafting N-vinyl pyrrolidone onto a hydroperoxidized polymer substrate using a combination of at least two metal salts to provide variable valence metal ions under aqueous alkaline conditions to suppress homopolymerization. This method was reported to afford a poly(N-vinyl pyrrolidone) coating with greater strength and durability than was previously possible [111].

Plastic contact lenses have been treated with electric glow discharge to render the surface hydrophilic without altering the optical characteristics or the physical dimensions of the lenses [112]. A similar glow discharge polymerization technique is employed to impart a hydrophilic, optically clear, impermeable barrier to soft corneal contact lenses [113]. A sulfonation process was reported recently for increasing the wettability and thromboresistance of a wide range of plastic devices [114].

An advantage of the free-radical-initiated grafting process is that a thin and uniform coating, independent of the contour and shape of the medical device to be coated, can be readily obtained. On the other hand, this process may require a higher capital investment for equipment and a more elaborate process control than those of a conventional solution-coating operation. Furthermore, one needs to be concerned with residual monomers in the coated article that may be highly toxic, such as acrylamide. In this instance, an extensive water-washing operation may be necessary to remove the residual monomer. When plasma polymerization is employed, the degree of lubricity of the modified surfaces may be limited due to an excessive cross-linking of the resultant hydrophilic polymer.

VII. HYDROPHILIC BULK POLYMERS

Besides surface modification, lubricious medical devices may be fabricated directly from a hydrophilic polymer, produced by either copolymerization or chemical modification of a base resin. (Ethylene-vinyl acetate) copolymers have been radiation-graft copolymerized with acrylic acid to impart a greater hydrophilicity. The neutralized resin is suitable for making surgical devices having high water-swelling characteristics and slipperiness [115,116]. Low-friction plastic catheters are made from a copolymer of acrylonitrile with either acrylamide or acrylic acid comonomer [108]. The copolymer may be produced by either a controlled hydrolysis of polyacrylonitrile or through copolymerization.

Hydrophilic copolymers derived from N-vinyl pyrrolidone and methyl methacrylate were reported to be useful coating materials for catheters for affording an improved biocompatibility and antithrombogenicity. These non-cross-linked hydrophilic copolymers can be used as an expandable layer on urinary catheters for better securing the catheter in place after insertion [117].

Thermosetting allyl ester copolymers containing maleic anhydride were claimed to form optical parts having excellent abrasion resistance and lubricity [118]. A hydrophilic polyurethane is produced from a long-chain poly(oxyethylene glycol) of molecular weight above 3500, an organic chain extender, water, and an organic diisocyanate. The hydrophilic polyurethane is suitable for making catheters having a good balance of water absorbency, softness, and strength [119].

Organosilicone compounds have been copolymerized with methyl methacrylate to yield an oxygen-permeable contact lens that possesses reduced surface friction and improved wettability [120]. A hydrophilic polymer composition has been rendered bacteriostatic by incorporating a comonomer containing quaternary ammonium salt. For instance, a terpolymer exhibiting such a behavior is made from an ester of an unsaturated carboxylic acid and an alcohol having a terminal quaternary ammonium group, a comonomer, and a cross-linking agent [121–123].

While all the techniques discussed in the above sections are expected to produce lubricious surfaces to different degrees, the rate of lubricity development upon exposure to water may vary considerably. Generally speaking, devices having a hydrophilically modified surface are expected to hydrate at a rapid rate because of a high concentration of hydrophilic moieties on the surface. On the other hand, devices fabricated directly from a hydrophilic resin may hydrate considerably slower because of a lack of sufficient polar moieties on the surface initially. In the latter case, the hydrophilicity of the resin may have to be limited to afford an acceptable water resistance and/or other physical properties. Surface modification with plasma, while highly effective for introducing hydrophilicity, may not necessarily result in lubricity because of the cross-linked structure of the resultant polymeric surfaces. Nevertheless, this technique holds promise for modifying intricate medical devices that might not be easily coated with a conventional solution-coating process.

VIII. MEASUREMENT OF HYDROPHILIC LUBRICITY

The surface hydrophilic lubricity of medical devices such as catheters, drain tubes, guidewires, and others that possess relatively simple shapes is usually measured as a coefficient of friction or frictional force in the presence of an aqueous medium. Ideally, the lubricity of a device should be judged by its rate of hydration, its degree of lubricity, and its resistance to mechanic abrasion or leaching in the hydrated state. A multiple number of tests are often required to fully characterize the lubricity behavior of a device.

The coefficient of friction is usually measured by a sliding-block test in which a rectangular block wrapped in a wet membrane is placed on the surface of a pair of testing specimens in the presence of an aqueous medium such as saline. The platform on which the specimens lie is tilted slowly until the block begins to slide. The angle ϕ is measured. The coefficient of friction is calculated as the tangent of ϕ. This measurement may be repeated after the device is subjected to repeated mechanical abrasion to measure the durability of the hydrated surfaces. A coefficient of friction of 0.2 or less is usually considered to be an acceptable surface lubricity.

The frictional force measurement is carried out by pulling a hydrated device through a curved tubing or a cannula, simulating the contour of the body cavity in which the device is intended to be inserted, or a grommet punctured from an elastic membrane using a force gauge. The lubricity of a device is measured by the pulling force in grams required to overcome the friction between the specimen and the curved tubing or the grommet. The smaller the pulling force required, the higher the lubricity of the device will be. The test may be repeated on the same specimen to measure the degree of surface or coating durability. The effectiveness of a surface-modified device is measured by the reduction in pulling force required in the test in comparison to that of the unmodified device. An 80% or greater reduction in frictional force is not uncommon for devices having a highly lubricious coating or other surface modifications.

The methods for measuring surface lubricity of medical devices were described in more details by Nagaoka and Akashi [124] and Takemura and Tanabe [125], among others [126,127].

IX. A MECHANISTIC CONSIDERATION OF HYDROPHILIC LUBRICITY

According to the modern theory of friction, 90% of the total frictional force is due to adhesion among surface atoms [128]. Thus, the weaker the bond between the surface atoms, the lower the friction and the higher the lubricity will be. Without modification or treatment, most of the medical devices made from either synthetic resins or rubber latex would exhibit a hydrophobic surface. Since water molecules would not wet the surface of such a device during application, direct contact between the medical device and tissues would result in friction when the surfaces are being abraded against each other. If the surface of the medical device is covered by a highly hydrophilic polymer layer, the situation changes dramatically. Such a surface is expected to hydrate rapidly, forming a layer of water-swollen gel (i.e., hydrogel) upon exposure to a body fluid. This water-swollen layer, which is mostly water, forms a fluid barrier between the medical device and tissues. Since fluids form only weak adhesive bonds [121–123], friction between the surfaces is markedly reduced by eliminating any direct solid-to-solid surface contact during motion.

A satisfactory hydrophilic lubricious surface must be able to hydrate rapidly during application, but also exhibit a good adhesion to the substrate in the hydrated state. It is, therefore, not surprising that the top layer of all hydrophilic lubricious coating formulations is composed of a water-soluble or -swellable polymeric material. The molecular weight of the polymer should be sufficiently high such that it would not be leached out of the surface easily.

Even a high molecular weight water-soluble polymer will be leached out or abraded away from the substrate surface unless it is firmly anchored. This may be accomplished by either chemical bonding of the water-soluble polymer to the substrate molecules or the formation of an adherent matrix containing the hydrophilic polymers, the reactive primers, and the substrate molecules. The former is illustrated by the chemical or high-energy radiation grafting process as well as the use of a reactive coating system in which the primer is capable of chemically bonding to both the water-soluble polymer and the substrate. The latter is illustrated by the process of forming a polyurethane-polyurea IPN at the hydrophilic coating-polymeric substrate interface using polyisocyanate and reactive hydrophilic polymers. The IPN layer bridges the hydrophilic polymer layer to the substrate molecules. A similar degree of success has been also demonstrated by physically entrapping a hydrophilic polymer into a cross-linked polymeric matrix.

Thus, the performance of a satisfactory hydrophilic lubricious coating may be visual-

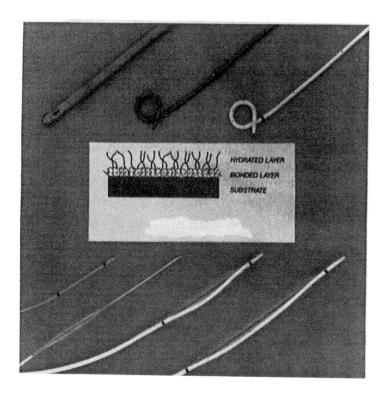

Figure 1 The double-layer model for fast hydration and good bonding. (Courtesy of *Polymer News.*)

ized by the double-layer model shown in Fig. 1. The outer layer of a hydrophilic surface consists of free-flowing polymer chains or segments that are tightly packed on the surface in the dry state, which affords a normal plastic feel to the touch. Upon exposure to water, this exterior layer hydrates instantaneously to produce a lubricious hydrogel layer. The interior layer consists of polymer chains or segments that are either chemically bonded to the substrate or immobilized through an IPN structure or either chemically or physically entangled within a polymeric network generated by the reactive primer molecules. When there is an insufficient population of hydrated molecules in the exterior layer, an inadequate lubricity would result. On the other hand, when the interior layer is insufficiently cured, a condition of poor adhesion or poor wet strength or both could result. Thus, to afford a satisfactory hydrophilic lubricious coating or surface modification for a medical device, a proper choice of coating formulation and processing condition is essential to achieve both a high degree of lubricity and good coating durability. The graph also shows some examples of medical devices that benefit from the use of hydrophilic lubricious coatings.

X. SUMMARY

The different surface-modification techniques for providing a hydrophilic lubricious surface have been grouped into six categories in accordance with the chemical nature of the resultant surfaces. Among these techniques, the formation of lubricious surfaces utilizing either a chemically reactive hydrophilic polymeric coating or a chemical grafting of hydrophilic monomer(s) or the formation of an IPN composition appears to be the most promising. The coated medical device exhibits a normal plastic feel when dry, but becomes instantly lubricious in body fluids. The performance of a satisfactory hydrophilic lubricious coating has been interpreted using a double-layer model in which the outer layer consists of fast-hydrating and mobile polymeric segments, while the inner layer is firmly anchored to the substrate through the coupling action of a reactive primer or direct chemical grafting of hydrophilic monomers or permanent entanglement of hydrophilic molecules with the substrate matrix. Hydrophilic lubricious coatings meeting these criteria have been shown to afford good handling characteristics, quick lubricity in body fluids, and better coating durability in the hydrated state.

REFERENCES

1. Mardis, H. K., and Kroeger, R. M., *Urol. Clin. North Am.*, 15:3, 1988.
2. Heilmain, M. S., Sied, M., and Maddell, S. W., U.S. Patent 4,080,706 (August 23, 1978).
3. Fuji Systems K. K., Japanese Patent 63,164,956 (July 8, 1988).
4. Judd, H. E., et al., U.S. Patent 4,169,163 (September 25, 1979).
5. Wong, E. W. and Ballan, D. G., U.S. Patent 4,838,876 (June 13, 1989).
6. Shook, D. R., et al., *J. Biomech. Eng.*, 108(2):168–74, May 1986.
7. Triolo, P. M., and Andrade, J. D., *J. Biomed. Mater. Res.*, 17(1):149–65, January 1983.
8. Cohen, A., *J. Hosp. Infect.*, 6(Suppl. A.):155–61, March 1985.
9. Pearce, R. S., et al., *Am. J. Surg.*, 148(5):687–91, November 1984.
10. Harrison, L. H., *J. Urol.*, 124(3):347–49, September 1980.
11. Kikuchi, Y., et al., *Cardiovasc. Intervent Radiol. (U.S.)*, 12(2):107–9, March–April 1989.
12. Rodeheaver, G. T., et al., *Surg. Gynecol. Obstet. (U.S.)*, 164(1):17–21, January 1987.
13. Buter, H. K., and Kunin, C. M., *J. Urol. (U.S.)*, 100(4):560–66, October 1965.
14. Fan, Y. L., *Amer. Chem. Soc., Polym. Mater. Sci. Eng.*, 63:709–16, 1990.
15. Fan, Y. L., *Polymernews*, 17(3), March 1992.
16. Ikada, Y., Uyama, Y., *Lubricating Polymer Surfaces*, Technomic Publishing, Lancaster, PA, 1993.
17. Schwartz, A., Graper, J., and Williams, J., U.S. Patent 4,589,873 (June 20, 1986).
18. Althans, W., and Throne, J., European Patent 321,679 (June 28, 1989).
19. Becker, L. F., et al., U.S. Patent 4,835,003 (May 30, 1989).
20. Podell, D. L., U.S. Patent 3,813,695 (June 4, 1974).
21. American Cyanamid Company, European Patent Application 558,965-A2 (March 2, 1992).
22. Goldstein, A., U.S. Patent 4,482,577 (November 13, 1984).
23. Kim, S. W., and Feijen, J., *Critical Reviews in Biocompatibility*, 1(3):229–260, 1985.
24. Gould, F. E., et al., U.S. Patent 4,810,543 (March 7, 1989).
25. Golander, C. G., Jonsson, E. S., and Vladkova, T. G., U.S. Patent 4,840,851 (June 20, 1989).
26. Golander, C. G., Jonsson, E. S., and Vladkova, T. G., European Patent 229,066 (July 22, 1987).
27. Perciaccante, V. A., and Landi, H. P., U.S. Patent 4,047,533 (September 13, 1977).
28. Perciaccante, V. A., and Landi, H. P., U.S. Patent 4,043,344 (August 23, 1977).
29. Hunter, A. W., and Bogdansky, S., European Patent 128,043 (December 12, 1984).
30. Lehmann, L. T., et al., European Patent 258,749 (March 9, 1988).
31. Casey, D. J., et al., U.S. Patent 4,716,203 (December 29, 1987).
32. Gen, S., Ikada, Y., Oka, M., and Okimatsu, H., Japanese Patent 4,303,444 (October 27, 1992).

33. Miller, R. A., British Patent 2,179,258 (March 4, 1987).
34. Murayama, H., and Okajima, N., Japanese Patent 4,144,567 (May 19, 1992).
35. Moriuchi, Y., Japanese Patent 90,306,753 (November 13, 1990).
36. Toray Industries Incorporated, Japanese Patent 2,234,765 (September 17, 1990).
37. Potter, W. D., European Patent Published Application 455,323 (November 6, 1991).
38. Han, D. K., Jeong, S. Y., and Kim, Y. H., *J. Biomed. Mater. Res.*; *Applied Biomaterials*, 23(A2):211–228; 1989.
39. Romberg, V. G., U.S. Patent 4,756,974 (July 12, 1988).
40. Scholz, M. T., et al., European Patent 290,207 (November 11, 1988).
41. Hudgin, D. E., U.S. Patent 3,975,350 (August 17, 1976).
42. Gould, F. E., and Johnston, C. W., German Patent 3,344,001 (June 13, 1985).
43. Shaluby, S. W., and Jamiolkows, D., U.S. Patent 4,105,034 (August 8, 1978).
44. Winn, R. A., U.S. Patent 4,373,009 (February 8, 1983).
45. Schaper, R. J., U.S. Patent 4,166,894 (September 4, 1979).
46. Schaper, R. J., U.S. Patent 4,075,136 (February 21, 1978).
47. Messier, K. A., Rhum, J. D., U.S. Patent 4,624,256 (November 25, 1986).
48. Casey, D. J., U.S. Patent 4,068,757 (January 17, 1978).
49. Smith, F. R., U.S. Patent 4,199,367 (April 22, 1980).
50. Hanke, D. E., U.S. Patent 3,756,238 (September 4, 1973).
51. Trubitsina, N. I., et al., U.S.S.R. Patent 1,265,226 (October 23, 1986).
52. Iwatschenk, P., U.S. Patent 4,798,593 (January 17, 1989).
53. Gillette Co., Netherland Patent 7,904,061 (December 4, 1979).
54. Creasy, W. S., et al., U.S. Paent 4,642,267 (February 10, 1987).
55. Vailancourt, V. L., U.S. Patent 4,705,709 (November 11, 1987).
56. Teffenhart, J. M., U.S. Patent 4,789,720 (December 6, 1988).
57. Yoon, H. K., and Sun, R. L. J., European Patent 266,892 (May 11, 1988).
58. Howard, E. G., World Patent 8,909,246 (October 5, 1989).
59. Takemura, N. and Tanabe, S., European Patent 166,998 (January 1, 1986).
60. Kindt-Larsen, T., Larsen, H. O., Wolff, P., World Patent 9,005,162 (May 17, 1990).
61. Whitbourne, R. J., U.S. Patent 5,001,009 (March 19,1991).
62. Rowland, S. M., and Wright, R. B., U.S. Patent 5,041,100 (August 20, 1991).
63. Karimi, H., Wells, S. C., Spielvogel, D. E., Karakelle, M., and Taller, R. A., U.S. Patent 5,084,315 (January 28, 1992).
64. Ofstead, R. F., U.S. Patent 4,977,901 (December 18, 1990).
65. Etheredge, R. W., and Charkoudian, J. C., U.S. Patent 4,983,170 (January 8, 1991).
66. Tokyo Ai-Tech Kendy, Japanese Patent 8,967,534 (March 22, 1989).
67. Karakelle M., Karimi, H., Lee, M. S., and Taller, R. A., U.S. Patent 4,990,357 (February 5, 1991).
68. Pudell, H. I., U.S. Patent 4,575,476 (March 11, 1986).
69. Gold, P., U.S. Patent 4,534,363 (August 13, 1985).
70. Lambert, H. R., U.S. Patent 4,666,437 (May 19, 1987).
71. Micklus, M. J., and On-Yang, D. T., U.S. Patent 4,100,309 (July 11, 1978).
72. Micklus, M. J., and On-Yang, D. T., U.S. Patent 4,119,094 (October 10, 1978).
73. Hibbs, L., Jang, Y., Liebmann, V. L., Spinks, D., U.S. Patent 5,300,032 (April 5, 1994).
74. Lambert, H. R., U.S. Patent 4,487,808 (December 11, 1984).
75. Creasy, W. S., U.S. Patent 4,847,324 (July 11, 1989).
76. Creasy, W. S., U.S. Patent 4,987,182 (January 22, 1991).
77. Eaton, R. K., U.S. Patent 5,077,352 (December 31, 1991).
78. Eaton, R. K., U.S. Patent 5,160,790 (November 3, 1992).
79. Halpern, G., and Gould, J. U., U.S. Patent 4,657,820 (April 14, 1987).
80. Halpern, G., and Gould, J. U., U.S. Patent 4,810,586 (March 7, 1989).
81. Halpern, G., and Gould, J. U., U.S. Patent 4,722,867 (February 2, 1987).

82.   Goldberg, E. P., Yahiaoni, A., Burns, J. W., Kumar, G., Larson, J. A., Osborn, D. C., Sheets, J. W., Kumar, S. G., and Vahiaoni, A., U.S. Patent 5,094,876 (March 10, 1992).
83.   Buscemi, P. J., and Slaikeu, P. C., U.S. Patent Application 809,889 (December 18, 1991).
84.   Markle, R. A., Brusky, P. L., and Baker, J. H., U.S. Patent 4,943,460 (July 24, 1990).
85.   Kocak, N., U.S. Patent 4,705,511 (November 10, 1987).
86.   Johansson, E. G., and Utassjoeberg, J. M. R., European Patent 217,771 (April 8, 1987).
87.   Beavers, E. M., U.S. Patent 4,663,233 (May 5, 1987).
88.   Halpern, G., et al., U.S. Patent 4,801,475 (January 31, 1989).
89.   Kliment, C. K., and Seems, G. E., U.S. Patent 4,729,914 (March 8, 1988).
90.   Fan, Y. L., U.S. Patent 5,091,205 (February 25, 1992).
91.   Bartizal, D. C., Buckanin, R. S., Ersfeld, D. A., Larson, W. K., Ree, D. K. E., Sandvig, T. C., Scholz, M. T., Bartizal, D., and Buckanin, R., European Patent Published Application 494,083 (July 8, 1992).
92.   Toray Industry Incorporation, Japanese Published Application (September 1, 1990).
93.   Gilding, D. K., Harrison, I. M., Middleton, I. P., and Watson, J. P., World Patent 9,219,290 (November 12, 1992).
94.   Szycher, M. (Ed.), *High Performance Biomaterials*, Technomic Publishing, Lancaster, PA, 1991.
95.   Kamstrup-Larsen, J. V., Larsen, H., Wolff, P., and Larsen, H. O., European Patent Published Application (March 31, 1993).
96.   Fan, Y. L., and Marlin, L., European Patent Published Application (December 16, 1992).
97.   Ratner, R. D., and Hoffman, A. S., U.S. Patent 3,909,049 (February 10, 1976).
98.   Ainpour, P. R., U.S. Patent 4,526,579 (July 2, 1985).
99.   Hyans, T. E., U.S. Patent 4,459,318 (July 10, 1984).
100. DeAntonio, P., Bertrand, W. J., Coulter, S. L., and Mayhan, K. G., Papers presented at 200th Am. Chem. Soc. National Meeting, Washington, D.C., August 26–31, 1990.
101. Janssen, R. A., and Mayhan, K. G., Paper presented at 200th Am. Chem. Soc. National Meeting, Washington, D.C., August 26–31, 1990.
102. Nippon Medical, Japanese Patent 61,209,670 (September 17, 1986).
103. Sumitomo Bakelite K. K., Japanese Patent 63,003,866 (January 8, 1988).
104. Sumitomo Bakelite Company, Japanese Patent 4,215,760 (August 6, 1992).
105. Terumo Corporation, Japanese Patent 4,159,338 (June 2, 1992).
106. Tazuke, S. and Kimura, H., *J. Polym. Sci. Polym. Lett. Ed.*, 36:1087, 1988.
107. Inoue, H. and Kohama, S., *J. Appl. Polym. Sci.*, 29:877, 1984.
108. Uyama, Y. and Ikada, Y., *J. Appl. Polym. Sci.*, 36:1087, 1988.
109. Terumo Corporation, Japanese Patent 4,202,441 (July 23, 1992).
110. Colvin, M., Gupta, A., and Li, C., European Patent application 521,605 (January 7, 1993).
111. Bertrand, W. J., Coulter, S. L., Deantonio, P., and Reich, C. J., World Patent 9,210,533 (June 25, 1992).
112. Gesser, H. D., and Warriner, R. E., U.S. Patent 3,925,178 (December 9, 1975).
113. Peyuman, G. A., Koziol, J. E., and Yasuda, H., U.S. Patent 4,312,575 (January 26, 1982).
114. *Medical Product Manufacturing News*, (5):6, September/October 1990.
115. Fydelor, P., et al., U.S. Patent 4,785,059 (November 15, 1988).
116. Fydelor, P., et al., European Patent 179,839 (May 7, 1986).
117. Brook, M. G., U.S. Patent 4,842,597 (June 27, 1989).
118. ITO Kogaku Kogyo K. K., Japanese Patent 54,110,289 (August 29, 1979).
119. Gould, F. E., Johnson, S. W., U.S. Patent 5,120,816 (June 9, 1992).
120. Neefe, C. W., U.S. Patent 4,280,759 (July 28, 1981).
121. Beede, C. H., Walch, H. L. and Blumig, T., U.S. Patent 4,248,685 (February 3, 1981).
122. Beede, C. H., Walch, H. L., and Blumig, T., U.S. Patent 4,248,685 (February 3, 1981).
123. Beede, C. H., Walch, H. L., and Blumig, T., U.S. Patent 4,111,922 (September 5, 1978).

124. Nagaoka, S., and Akashi, R., *J. Bioactive Compatible Polym.*, 5:212–226, 1990.
125. Takemura, N., and Tanabe, S., U.S. Patent 4,876,126 (October 24, 1989).
126. Takeuchi, M., and Onohara, M., Japanese Patent 633,866 (January 8, 1988).
127. Uyama, Y., Tadokoro, H., and Ikada, Y., *Biomaterials*, 12:71–75, 1991.
128. *McGraw-Hill Encyclopedia of Science and Technology*, 1:429–431, 1987.

40

Tissue Adhesives in Wound Healing

Dale S. Feldman and David H. Sierra
University of Alabama
Birmingham, Alabama

I. INTRODUCTION

The term *tissue adhesive* is a misnomer because the materials used as surgical tissue adhesives can also be used as sealants, drug delivery systems, and as regenerative scaffolds. Fibrin matrices (FMs) appear to have been the first truly successful surgical adhesive technology in medicine. Fibrin has been used in diverse applications, from microsurgical procedures, to cardiovascular surgery, to sustained antibiotic drug delivery. To date, no other adhesive material has been as successful as FM in terms of tissue compatibility, toxicity, and clinical utility. Although characterization of FM in terms of clinical efficacy and biomechanical properties has been reported, studies are limited and often qualitative in nature [1].

Other materials have also been used as surgical tissue adhesives with varying degrees of success. Foremost among these are the cyanoacrylates. Problems with cell toxicity and foreign-body reaction have limited their uses despite superior adhesive and cohesive properties [2,3].

A composite of gelatin, resorcinol, and formaldehyde (GRF) also was used clinically with some success in the 1960s through the early 1980s. The adhesive properties of gelatin are augmented by formaldehyde. Resorcinol, a phenol, is added to decrease the water solubility of the adhesive [3–6].

Another promising adhesive system is mussel adhesive protein (MAP). Originally isolated from mussels [7,8], a composition of several different polyoligopeptides has been produced by means of a peptide synthesizer [9] and by genetic engineering [10]. The MAP, however, is not currently available commercially because of delays in the development of this technology.

The goal of a tissue adhesive is to attach two pieces of tissue together with adequate mechanical strength for the application to last throughout the healing process. Some have suggested that the bond strength should be greater than the tissue [9], although this is probably unnecessary for most applications. In addition, it should perform at least as well as sutures without the problems associated with them: foreign body response, fistulization, dehiscence, time required for application, and uneven stress distribution.

When used as a tissue sealant or fluid barrier, the goal is to prevent fluid or gas leakage within the body. When used as drug delivery systems, the tissue adhesives should protect and serve as a reservoir for the bioactive agent as well as present it to the tissue at the appropriate rate.

In addition, these tissue adhesives should set up quickly in a moist environment, have good handling properties for the surgeon, and lead to a good esthetic result. Further, the adhesive must be able to be approved by governmental agencies for use in the United States.

Typically, additional criteria are added to each of these types of tissue adhesives related and leading to minimal responses (inflammation toxicity, carcinogenicity, viral transmission, etc.) or, in essence, not interfering with the local healing process or having systemic effects. Optimally, however, these tissue adhesives should enhance the local healing process — either stimulating tissue regeneration or speeding up the regenerative process. It is important to note that speeding up healing is normally not sufficient since this could tend to lead to scarring rather than to regeneration. This requires the tissue adhesive to be degradable so it does not interfere in the regeneration or need to be removed later. Therefore, in addition to its other functions, an optimal tissue adhesive should be a degradable regenerative scaffold. For a number of applications, however, this may be the only function of the tissue adhesive.

II. WOUND HEALING

A.   General Wound Healing

In order to understand the use of tissue adhesives as regenerative scaffolds, we have to understand the healing process. Although normal wound healing occurs in specific stages, there is a complex interaction among serum enzymes, locally acting growth factors, platelets, macrophages, fibroblasts, endothelial cells, epidermal cells (if in skin), and the local cellular microenvironment [11].

Initially, disruption of capillaries leads to activation of the Hageman factor, a factor that activates the clotting cascade and the complement cascade [11,12]. The clotting cascade generates thrombin, which stimulates the platelets to release locally acting growth factors [12]. The complement cascade generates C5a, which is a chemoattractant for neutrophils and macrophages [13].

Locally acting growth factors initiate a connective tissue response by causing division and migration of fibroblasts as well as formation of new capillaries [14]. Moreover, a hypoxic wound environment is a potent stimulus for macrophage-derived angiogenic growth factors, which continue the neovascularization initiated by the platelets [13,15]. Other macrophage-derived growth factors, produced because of the wound environment, stimulate fibroblast migration, division, and collagen synthesis [12].

The end result of these complex interactions depends on the location of the wound. In skin, the tissue granulates, forming a neovascularized collagen mesh that contracts, closing the wound space [13]. This granulation is then covered by an epidermal layer, the end result being scar tissue covered by a regenerated epidermis. Attempts at obtaining more dermislike tissue (dermis without appendages) have met with varying degrees of success. Large skin defects (burns, ulcers, full-thickness defects) are treated most successfully clinically by skin grafts. There are clinical investigators, however, examining degradable collagen implants, topical growth factors, and oxygen therapy for creation of a neodermis and epidermal regeneration [13,16,17].

B. Healing Rate

Healing is the rate limiting step in many treatment therapies. Either the options are limited or success is limited by the speed and completeness of healing.

In large or deep defects, such as bone nonunions, extensive burns, and full-thickness skin defects, healing and regeneration occur too slowly on their own and, therefore, the most successful clinical therapy is grafting. This creates additional wounds and, in many cases, such as extensive burns, supply is a problem.

In addition, grafting is not always successful. Free skin grafts require a healthy underlying bed with good vasculature; even then, not all grafts survive [18,19]. Bone graft viability is also dependent on vascular supply. Pedicle skin or bone grafts, grafts with their own blood supply, are not always possible and do not always survive either. A key is the speed of healing across the interface between the bed and graft, particularly the vascular supply reattachment [20]. This requires good attachment to the underlying bed or adjacent tissue so there is no space or fluid collection that can hamper revascularization. Increases in the rate and completeness of healing would increase the options for these types of procedures [21].

Enhancing or speeding regenerative healing is helpful for even small tissue discontinuities (incisional wounds, tears, rupture, surgically created defects). For example, repair of tears of the anterior cruciate ligament is hampered by slow healing, poor vascularity, and difficulty in reattachment. In addition, speeding up of regenerative healing of tissue will reduce rehabilitation time, especially in orthopedic applications because rehabilitation time can be twice as long as the length of immobility [22].

C. Biomaterial Design

Using a tissue adhesive means we are adding an implant or biomaterial to the system. Implants include a wide variety of porous and nonporous, degradable and nondegradable materials. Porous implants have uses that include fixation or stabilization of implants, tissue ingrowth to create a blood-compatible surface for artificial vessel replacement, and scaffolding to help the body repair or regenerate.

Other implants include those for wound dressings for the skin or cornea; for attachments such as sutures or adhesives; for functionality such as contact lenses, intraocular lenses, artificial ligaments or tendons; for facial augmentation; for nerve guides; for fracture fixation; and for dental applications. For the these implants, healing is also a critical component. For wound dressings and attachment systems, the faster the wound healing occurs, the better it is, provided it is regenerative healing. Nerve guides and tissue-augmentation devices also are dependent on the speed of tissue regeneration.

For degradable scaffold systems such as blood vessels or ligaments, healing needs to occur at least as quickly as the degradation. In addition, the type of tissue and mechanical properties are critical for success [23]. For nondegradable systems (e.g., fixation, Dacron® blood vessels, and ligaments) the speed of ingrowth, type of tissue, and mechanical properties are also critical for success [24].

Many investigators have examined and continue to examine ways to enhance and control healing. There still, however, is little known in this area. For porous implants, healing can be altered by changing the implant configuration (pore size, porosity, fiber diameter, etc.), the implant surface (composition, charge, surface energy, etc.), the implant environment (oxygen, magnetic fields, stress, location, etc.), the implant mechanical properties (stiffness, ultimate strength, etc.), or the implant biochemical activity (growth factors and other biochemical agents). Each of these can significantly affect the progression of healing, the rate of healing, and the type of tissue formed.

For nonporous implants, these same techniques, except altering implant configuration, can be used to enhance healing. Similarly, altering the environment or biochemical activity can be used to enhance healing when no implant is present. Orchestration of healing, therefore, should be the goal, controlling both the speed and type of healing. In most cases, regeneration is desired and at the fastest possible rate. In addition, whenever possible, the implant—if there is one—should be degradable. Therefore, the optimal solution is almost always a degradable implant that stimulates regeneration, called *biomaterial enhanced regeneration*.

Tissue adhesives are beneficial in the wound healing process because they cannot only function as an adhesive in the short term, but can be augmented to enhance healing in the long term. They can augment healing by serving as degradable tissue scaffolds and drug delivery systems. Therefore, in most cases, they can and should be designed to be degradable regenerative systems.

III.   TISSUE ADHESIVES

A.   Fibrin

1.   History

Fibrin matrices (FMs) have been used as a biomaterial since the turn of the 20th century. In 1909, Bergel used fibrin fleece as a degradable hemostatic agent [25]. Grey [26] and Harvey [27] produced fibrin sheets for hemostasis in cerebral surgery and as parenchymal tissue dressings. During the 1940s, Cohn and coworkers produced a wide range of fibrinogen and blood protein products as well as evaluated their properties [28–31].

Later, Gerendas produced a fibrin-based material utilizing ground bovine fibrin mixed with glycerol as a plasticizer and under heat and pressure produced a "plastic" that was cross-linked by an aldehyde agent [32]. The mechanical properties could be varied by altering the proportions of the ingredients. This material was formed into suture buttresses, incontinence buttons, and other degradable implants. It is marketed in Europe as Bioplast® by Ethicon.

The first recorded use of fibrinogen as a tissue adhesive was by Young and Medawar in 1940 [33]. In this case, plasma was used for peripheral nerve anastomosis in an animal model. This was followed by use in humans in 1942 by Seddon and Medawar [34]. Tarlov and Benjamin [35] also tried peripheral nerve anastomosis in an animal model using autologous and homologous plasma in conjunction with suture to relieve strain. It was in 1944 that Cronkite et al. recorded the first use of plasma and the then newly produced bovine thrombin to attach skin grafts [36]. Afterward, however, the concept of a fibrin adhesive languished because of its inadequate strength and premature repair failures, mostly due to the low fibrinogen concentration in plasma, normally 2 to 5 milligrams per milliliter (mg/ml) in humans. The material, however, was still in continuous use by researchers such as Medawar [37].

Almost three decades later, in 1972, Matras et al. successfully reintroduced the concept of a "blood glue" [38]. Cryoprecipitate, a plasma product used in factor VIII deficiency therapy, was used in conjunction with a bovine thrombin solution for peripheral nerve anastomosis in research animals. The cryoprecipitate contained elevated concentrations of fibrinogen, factor XIII, fibronectin, and other factors. Success in preclinical studies and appropriate handling properties led to subsequent use in human peripheral nerve anastomosis [39]. Proportions of the cryoprecipitate constituents and production methods were further refined and a commercial version of this technology, Tisseel® or Tissucol®, was launched in Europe by Immuno AG of Vienna, Austria.

The product is marketed as a two-component kit with lyophilized, pooled, human-source fibrinogen/Factor XIII concentrate as Component 1, which is reconstituted with antifibrinolytic solution. Component 2, bovine thrombin, is reconstituted with 40 Mm $CaCl_2$ solution. A competing FM product, Beriplast® (Behringwerke, Marburg, Germany), utilizes a different fibrinogen/factor XIII purification method [40]. A European patent has been issued on this product, described as a single-component FM, which utilizes antithrombin III in the formulation to prevent premature polymerization [41]. A FM product with Cohn fractionation (Centre de Transfusion Sanguine de Lille, France) is marketed as Biocoll® [42]. Also, an adhesive device comprised of alternating layers of collagen and fibrinogen lyophilized together as a composite has been developed for use as a surgical dressing [43]. Thrombin, antibiotics and antiproteases may be incorporated into the collagen layer. All of these products are marketed exclusively in Europe, with the exception of Tissucol, which was recently approved in Japan.

Despite a large body of evidence and a patent for the viral deactivation process, Immuno AG has been unable to gain Food and Drug Administration (FDA) approval for the fibrin product in the United States [44]. This barrier to availability in the United States, accompanied with interest from the clinical success of Tissucol, has led to the development of patient autologous and single-donor-source fibrin. In addition, a few groups in the United States, including the American Red Cross, Baxter-Hyland, Hema Cure, Miles, and the New York Blood Center are actively pursuing clinical products. The majority of fibrin development has centered on fibrinogen/factor XIII isolation and purification, since thrombin is readily available and approved.

In 1983, a cryoprecipitation production method was described that utilizes small amounts (45 ml) of patient autologous blood [45]. Techniques have continued to reduce the amount of blood needed and the time required for collection [46]. A large-scale cryoprecipitation method for use in blood banks was later developed and patented [47,48]. Fibrinogen and factor XIII precipitation by cold ammonium sulfate [49,50], polyethylene glycol [51], cold ethanol (Cohn fractionation) [52], and zinc ion [53] have been developed to produce small amounts of patient autologous fibrin. Each of these methods has advantages and disadvantages that have been reviewed elsewhere [1]. With the exception of cryoprecipitation, none of these methods has been commercialized. This limits the availability of FM to a small but growing number of clinicians who are either aligned with blood bank FM programs or who are themselves both versed in the production methods and willing to produce FM prior to surgery in the operating theater.

2. Chemistry

Fibrinogen is the main structural component in the fibrin matrix. A fibrin clot is cross-linked fibrinogen in a three-dimensional structure in conjunction with platelets and other wound factors. Fibrinogen, a 340-kilodalton (kDa) plasma protein, circulates freely in plasma at a concentration of 2 to 5 mg/ml, with a half-life of 7 days. It is produced primarily by hepatocytes and secondarily by platelets. Fibrinogen separates from blood in the $\beta 2$ globulin fraction and has an isoelectric point (pI) of 5.4. It exists in dimer form; each identical unit consists of three polypeptide chains: α (64 kDa), β (57 kDa), and τ (48 Da), which are bound together by 28 or 29 disulfide bridges, predominantly in the N-terminal region. Fibrinogen possesses a cell surface receptor (integrin) binding sequence (Arg-Gly-Asp-x) on the α-chain. The molecule is primarily rod shaped and measures 45 nanometers (nm) in length at neutral pH with three globular domains: two at each end, D, with a diameter of 60 Å and a smaller one in the middle, E, with a diameter of 40 Å. These regions are linked together by α-helical regions, which make up 30% of the amino acid sequence. There are two random coiled regions of several hundred residues projecting from the D domains that contain amino acids that participate in fibrin cross-linking by factor XIII. The D domains also contain Ca^{2+} binding sites, two with high affinity and four with low affinity [54,55].

Fibrinogen is converted to cross-linked fibrin by a series of reactions. The first step involves proteolytic removal of two fibrinopeptides, A and B, located on the α- and β-chains, respectively. This cleavage alters the charge and conformation of the molecule. The fibrinogen is now called fibrin monomer, and it begins to aggregate with other monomers in a D-E staggered fashion. The fibers then elongate and anastomose, as well as increase in diameter, eventually forming a three-dimensional structure (i.e., a gel) [54,55].

At this point, bonding is primarily hydrogen bonding and electrostatic. The structure is disaggregated by chaotropic agents such as 5M urea, and is subject to rapid dissolution by proteolytic enzymes (e.g., plasmin). It can also be permanently deformed by shearing forces [54,55].

Covalent cross-links are introduced by activated factor XIII via transamination. The cross-links are formed between the τ-chains to form τ-τ dimers, linking the fiber end to end. Two to three sites are available near the C-terminus of the α-chains to form α-polymers. This reaction takes up to 24 hours (h) for completion. The covalent cross-linking of the fibrin network retards proteolytic degradation and alters the mechanical properties [54,55].

3. Properties

Mechanism of Attachment. The adherence of fibrin to various substrates is by three modes: covalent bonds, hydrogen and other electrostatic bonds, and mechanical inter-locking [56–61]. Fibrin has been shown to bond covalently to fibronectin via factor XIIIa catalysis. Fibronectin and fibrin can also covalently cross-link with collagen via factor XIIIa [56]. In skin graft models, fibrin binds preferentially to collagen and not elastin [57,58].

Fibrin binds directly to platelets at the GPIIb/IIIa cell receptor complex, a Ca^{2+} heterodimer (integrin) [59]. Similar cell receptors are found on endothelial cells, primarily megakaryocytes and fibroblasts. Receptor binding uses secondary bonding with RGDx sequences at the N- and C-termini of the α-chain. The fibrin is ultimately linked to the cytoskeleton. Fibrin binding to platelets is of less importance than binding to endothelial cells during wound healing and hemostasis, since the latter binds to collagen to anchor the gel *in situ*. Platelets alter the mechanical properties of fibrin gels by increasing the elastic modulus. This phenomenon is not relevant to gels produced from FM since platelets are absent.

Gel Structure. The structure of the fibrin gel may be altered by changing any one of several factors: fibrinogen, thrombin, or calcium ion concentrations; ionic strength; pH;

and, to a lesser extent, temperature [31,62]. Changes in structure can in turn alter the mechanical properties of the gel, affecting both structural integrity and adhesive qualities. The presence of other moieties with high dipolar moments as well as pressure influences the gelation process [63]. A decrease in gelation time (by high thrombin concentration) or an increase in ionic strength or pH will cause the formation of small-diameter fibrils and pores in the gel structure (<2 micrometers [μm]), which is called a *fine gel*. An increase in gelation time (by low thrombin concentrations) or a decrease in ionic strength or pH will cause the formation of large-diameter fibers and pores (>2 μm), which is called a *coarse gel*. Since the polymerization is initiated by enzymatic action, the reaction rate increases with increasing temperature, reaching a maximum at 37°C. Denaturation begins at higher temperatures.

The presence of fibronectin, however, does not appear to affect pore size or gelation time, although it does increase fibril diameter [64]. Factor XIIIa has no effect on gel porosity [65], as it is responsible for covalently cross-linking adjacent protofibrils, fibrils, and fibrillar structures that are already in intimate contact. Albumin affects gel structure, as measured by turbidimetric means, by controlling the regularity of the structure [66].

Mechanical Characteristics. Since the fibrin matrices act as physical barriers or scaffolds as well as tissue adhesives, both material mechanics and adhesive properties must be considered. Shear loading of fibrin gels has been extensively studied under a wide range of conditions. Fibrin gels have been tested in a Plazek disk apparatus, both dynamically and statically [67–71]. These fibrin gels were made by thrombin, with and without factor XIIIa, as well as with reptilase (ancrod), which causes the cleavage of fibrinopeptide A only. In short, an increase in fibrinogen concentration correlates with an increase in shear modulus (G α [fibrin]$^{2.3}$). Fine gels, regardless of factor XIIIa cross-linking, have a lower shear compliance than coarse gels, although they undergo more permanent deformation (i.e., possess more viscous behavior). The gels consist of rodlike fibers interlocked by steric hindrance; the compliance is dictated by the number of contacts and interactions. In another study, fibronectin was incorporated into fibrin gels under different polymerization conditions [72]. This more closely approximates the physiologic state of a fibrin gel. When compared with gels lacking fibronectin, coarse cross-linked gels demonstrate a large increase in modulus and a decrease in creep compliance, whereas fine gels exhibit a decrease in modulus and a corresponding increase in creep compliance. That is, coarse gels have a decreased susceptibility to permanent deformation, and fine gels show an increase in the ability to be permanently deformed in comparison with fibronectin-free gels.

A decrease in calcium ion concentration below a plateau concentration of 10 mM corresponds to a decrease in shear modulus [73–75]. The increase in modulus corresponds to an increase in α-polymer concentration in fibrin as a function of time, as determined by SDS-polyacrylamide gel electrophoresis. The τ-dimer formation rate is directly proportional to the calcium ion concentration, but the α-polymerization rate is not. These reports indicate that an increase in α-polymerization corresponds to an increase in the number of fiber branches and an increase in fiber diameter, and that Ca^{2+} is required for factor XIII activation and activity. The rate of activation and activity can be controlled to a limited degree by Ca^{2+} concentration.

The tensile properties of fibrin have also been evaluated. Using Tissucol, it was shown that an increase in thrombin concentration up to 500 units/ml (U/ml) corresponds to an increase in ultimate tensile strength and Young's modulus [76]. It was also demonstrated that the increase in ultimate tensile strength and Young's modulus is proportional to the fibrinogen concentration. The mechanical properties increased over time, with no plateau of values up to 100 minutes after polymerization. At 30 minutes, the ultimate tensile strength was reported as 1.11×10^5 Newtons per square meter (N/m^2) and the Young's modulus as 1.48×10^5 N/m^2 at 62-mg/ml fibrinogen concentration. The addition of factor XIII had no effect on the material properties. Similar findings were reported with lower concentrations using a similar model and the same fibrin system [77]. In this study, however, the proportionality of ultimate tensile strength as a function of fibrinogen concentration broke down near the upper limit of the fibrinogen concentration.

The fibrinogen and thrombin volume ratios, in solution, have an impact on mechanical and mixing properties [76]. It was found that sequentially mixed fibrinogen/thrombin material was shown to be unevenly polymerized and nonhomogeneously mixed. Thrombin was released in greater amounts than for simultaneously mixed materials for up to 120 hours after polymerization. The tensile strength of the polymerized sealant was greatest for samples prepared in a 1:1 volume/volume (v/v) ratio (117 grams per square centimeter [g/cm^2]). Samples prepared in a 3:1 v/v fibrinogen/thrombin ratio had a lower tensile strength (86 g/cm^2), whereas a 1:1 v/v preparation with a diluted fibrinogen component (3:1) had the lowest overall tensile strength (19 g/cm^2).

Adhesive Characteristics. An *in vivo* adhesive strength model was developed in which full-thickness incisional wounds were made in a rodent dorsum and repositioned with fibrin sealant (Tissucol) [78]. At time points ranging from 0 to 42 days, the subjects were sacrificed and the repair site fixed in mechanical test grips and strained to failure. The FM was compared with suture closure. The FM repairs had higher bonding strength than suture controls until 4 days postoperatively, when both groups were equivalent. It was noted, however, that the FM groups had delayed development of mechanical strength during the maturation phase (20 and 42 days after surgery).

In another report, investigators used the same method to compare FM formulations and their impact on incisional wound strength 8 days postoperatively [79]. The FM was produced by a previously described method [80] and the fibrinogen, factor XIII, and thrombin concentrations were varied. No differences were detected in wound healing strength by altering the factor XIII/fibrinogen ratio. The authors concluded that a fibrinogen concentration of 39 mg/ml, thrombin concentration of 200–600 U/ml, and the factor XIII concentration present in the FM preparation were ideal for effecting maximal wound healing as determined by Young's modulus, stress, strain, and energy of failure. A higher thrombin and fibrinogen concentration inhibited wound healing and lower concentrations contributed little when compared with suture controls.

A model of skin graft attachment strength that takes into account corporeal stress distribution has been developed [42,44]. Full-thickness rat dorsum is harvested from sacrificed animals and subsequently reattached with the fibrinogen concentrate applied to the skin, and the thrombin/$CaCl_2$ solution applied to the donor site. Force to cause adhesive failure at the skin/donor site interface is measured by applying weights at time points ranging from 1 to 120 minutes postattachment. The cryoprecipitated fibrin sealant had an adhesive strength of 1 gram per square millimeter [g/mm^2] and Tissucol had an adhesive strength of 2 g/mm^2 [44].

In another study, adhesive strength increased with time from about 50 g/cm^2 at 1 minute to 200 g/cm^2 at 120 minutes [42]. In a study involving human split-thickness skin graft adhesive strength [52], a 1-cm^2 piece of graft was attached to a donor site with FM for 3.5 minutes, then pulled off with a dynamometer attached with suture. The breaking strength of glued grafts was 12 to 26 g, with unglued control grafts less than 5 g. No correlation, however, between fibrinogen concentration and breaking strength was observed.

A simple *in vitro* adhesive strength determination model used fresh human dura as the adherend [81]. Overlapping pieces of dura were attached together with two different versions of ammonium sulfate precipitated fibrin as well as Tissucol. The specimens were allowed to set for a period of time and weights were added until adhesive failure occurred. An increase in adhesive strength was reported at 10 and 30 minutes after polymerization for both the commercial and the two ammonium sulfate concentration versions of FM. The commercial FM had a higher bond strength than either of the salt-precipitated fibrin matrices after 10 minutes. At 30 minutes, the high-salt FM had a higher adhesive strength than either Tissucol or low-salt FM. In addition, an ossicular prosthesis was attached to a piece of dura glued to a piece of cardboard and the process was repeated with the same groups, loaded in tension. Under these test conditions, the commercial fibrin sealant had a significantly higher bonding strength at both 10 and 30 minutes after polymerization.

Another test method for adhesive strength using pieces of cotton gauze, fresh rat dermis, or fresh human dura glued to blocks has been described [82]. Mounted adherends were attached together with ammonium sulfate precipitated or polyethylene glycol (PEG) precipitated FMs as well as a commercial preparation (Biocoll). The assemblies were incubated at room temperature at 20% or 100% humidity, and then strained to failure in tensile loading by adding water to a beaker attached to one of the blocks. The commercial material had the highest adhesive strength, followed in decreasing strength by the PEG and then the ammonium sulfate prepared FMs. The adhesive strength was directly proportional to the measured fibrinogen concentration. No variance was attributable to differences in humidity.

A method based on American Society for Testing and Materials (ASTM) adhesive strength tests has been described in which freshly harvested split-thickness rabbit dorsum samples were attached to aluminum jigs, which were subsequently attached together in pairs with designated amounts of fibrinogen and thrombin/$CaCl_2$ solutions [83]. The assemblies were incubated in Ringer's solution at 37°C for 90 minutes, then strained to failure under tensile loading. Both Tissucol and a cryoprecipitated FM were evaluated. The Tissucol had a higher adhesive strength (520 g/cm^2) than the cryoprecipitated material (324 g/cm^2). In a later report, the shear adhesive strength of cryoprecipitated human fibrinogen at six different concentrations was compared with Tissucol [84]. Adhesive strength was evaluated as a function of fibrinogen preparation method, fibrinogen concentration, and polymerization (incubation) time. Split-thickness pigskin was used as the test substrate. Shear adhesive strength was directly proportional to fibrin concentration. No significant increase in adhesive strength over time, however, was noted for the six human cryoprecipitated FM preparations. Tissucol had a significant increase in strength over time; however, its initial strength at 5 minutes was significantly less than a cryoprecipitated fibrin sealant with the same fibrinogen concentration. Adhesive strength for all groups was an order of magnitude less than the tensile cohesive strength at the same concentration. The adhesive failure occurred at the adhesive/adherend interface.

An *in vitro* porcine aorta model has been used to determine the suitability of FM (human cryoprecipitate) in vascular anastomosis [85]. It was found that near maximal strength was achieved in minutes, and no differences were detected in burst strength or tensile strength, of the anastomosis, for variation in fibrinogen concentration. The authors concluded that FM should be used only in smaller-diameter vascular anastomoses because of the relatively low burst and tensile strengths when compared with sutured controls. In addition, even for microvascular anastomosis, the strength is low and probably needs to be augmented by sutures or use of fibrin/collagen composites.

The efficacy of fibronectin ("cold insoluble globulin") as a tissue adhesive was also evaluated [86]. Human fibrinogen/fibronectin mixtures (percent clottable fibrinogen contents ranging from 72–90%) and Tissucol were compared for adhesive strength and SDS-polyacrylamide gel electrophoresis patterns. Rat sciatic nerves were severed and reattached with either one of the fibrinogen/fibronectin mixtures or the commercial FM. Formulations with fibronectin (0% to 70%) showed a decrease in breaking strength, ranging from 1.0 to 5.8 pounds, since fibrinogen concentration decreased with increasing fibronectin content (range of 12 to 50 mg/ml), keeping the total protein concentrations of the formulations constant. The commercial FM with barely detectable fibronectin and a fibrinogen concentration of 80 mg/ml had a breaking strength of 33 pounds. The authors concluded that fibrinogen concentration alone affected breaking strength and that fibronectin demonstrated no advantage for tissue adhesion.

4. Tissue Response

Role in Wound Healing.   The components of FM are part of the normal chain of events in hemostasis and tissue remodeling that occur in wound healing. Fibrin acts as both a hemostatic barrier, adhering to surrounding tissue and cells, and as a scaffold for migrating fibroblasts [87].

Fibrinopeptide A, formed during fibrinogen-to-fibrin conversion, acts as a chemotactic agent for polymorphonuclear leukocytes (PMNs). Other proteases are also involved as mitogens (e.g., collagenase and trypsin). Fibronectin is incorporated covalently into the gel structure to act as an attachment site for migrating fibroblasts. Fibrin degradation products, formed by proteolytic digestion of the fibrin network, stimulate the migration of monocytes, which convert to macrophages. These cells in turn phagocytize fibrin. Stimulated fibroblasts migrate through the fibrin gel structure and deposit collagen (Type III). The infiltrating cells secrete plasminogen activators that further degrade the fibrin network. Neovascularization follows shortly thereafter. This composite of fibrin, macrophages, fibroblasts, collagen, and vascular buds comprises the granulation tissue. The role of deposited fibrin in the formation of normal granulation tissue is well described [88–90]. Since the amount of deposited FM affects the amount of granulation tissue and fibroplasia, care must be exercised in controlling the quantity delivered. Excess amounts of FM can lead to excess fibrous tissue formation [91]. Concentrated fibrin can also inhibit normal cell migration and delay subsequent wound healing [87].

Thrombin also has a role in wound healing. It catalyzes the conversion of fibrinogen to fibrin and activating factor XIII, and is a mitogen for fibroblasts and other endothelial cells. It also affects prostaglandin production and activates protein C, both of which are essential in normal wound healing. Factor XIIIa augments fibroblast proliferation and attachment to the fibrin network by covalent cross-link production. Fibroblast proliferation is further stimulated by factor XIIIa directly. The exact mechanisms by which thrombin and factor XIII contribute to wound healing requires further exploration. Most observations to date are from *in vitro* experiments [92] or in factor-deficient patients [93].

Controversy exists as to whether FMs act as osteogenic or osteostimulatory agents. FM used to fill standardized cortical defects in rabbit tibia has been shown to stimulate neocapillarization and connective tissue cell proliferation at the defect/FM interface [94]. New bone formation was reported to take place by 14 days, whereas unfilled controls developed callus by 31 days. Additional studies have also shown evidence of osteogenesis

[95–101]. But, in contrast, no improvement in osteogenesis in a titanium bone growth chamber filled with FM was found in one study [102], as well as the finding of no significant difference in healing rates in standardized cortical and spongiosa defect repairs in other studies [103–105].

Biocompatibility. The tissue compatibility and safety of FMs have been a concern since the initial development. The safety of both Tissucol and ammonium sulfate prepared FM has been described when used in sensitive anatomic regions such as in the middle ear of chinchillas for the repair of the ossicular chain [106,107]. Tolerance of FM by the cochlea was evaluated by direct application to the oval window, and in some cases entered the vestibule without causing damage to inner ear structures or hearing loss.

Studies with ^{125}I-labeled fibrinogen in FM have shown that the exogenous-labeled fibrinogen is mostly eliminated via urine (>90%) by 10 days, without an antifibrinolytic, in patients undergoing skin grafting [108] and using *in vivo* subcutaneous implantation models [109].

The implantation of species-heterologous fibrin has also been described [110,111]. Porous and solid forms produced from bovine and human plasma were subcutaneously implanted in rodents. Although foreign-body reactions and antibodies to the implants were detected, in this case, by histopathologic and immunologic evaluation, it should be noted, however, that many *in vivo* models use species-heterologous FM with no adverse reactions.

To date, only three serious complications have been reported regarding the clinical use of FM. In two cases, FM in the form of cryoprecipitate was used in conjunction with a balloon catheter to seal a deep hepatic wound track [112]. The first case, in which the wound was caused by a bullet, resulted in fatal hypotension. The second case (knife wound) was salvaged by fluids and vasopressors. The abrupt hypotension was thought to be caused by the large quantities of thrombin/$CaCl_2$ solution (approximately 20 ml). The hypothesis was evaluated with a canine model. Parke-Davis, the thrombin manufacturer (Thrombinar®), reported that thrombin has been associated with 16 cases of hypotension or death. A third complication was reported in a patient with hypogammaglobulinemia. Thrombin was implicated in the hypersensitivity reaction during a procedure to close a bronchopleural fistula [113].

B. Cyanoacrylates

1. History

The cyanoacrylates (CAs) were first synthesized in 1949 (Ardis) [114], but it was not until the adhesive properties were discovered years later that they were considered as a surgical adhesive [115]. The methyl-2-cyanoacrylates (Eastman 910) were the first systems used clinically. Problems with wettability [116–118] and histoxicity lead to the development of longer-chain homopolymers (Table 1) such as ethyl-2-cyanoacrylate (Krazy Glue®), isobutyl-2 cyanoacrylate (Bucrylate®), and butyl-2-cyanoacrylate (Hexacryl®). It was subsequently found that these longer-chain derivatives (isobutyl and butyl) were the least toxic, but still had good adhesive strength [119]. In the 1970s, reports on histoxicity of isobutyl-2-cyanoacrylate began to be published [119–122]. Since then Hexacryl has been the system of choice. Recently, however, Histoacyl® too has shown histoxicity when placed subcutaneously in soft tissue [123].

The histoxicity, as well as reports of carcinogenicity (methyl-2-cyanoacrylate), have prevented FDA approval in the United States, although clinicians in Canada and Europe

Table 1   Structure and Properties of Cyanoacrylates

$$\text{Cyanoacrylate} - \left(-CH_2 - \underset{\underset{OR}{\overset{|}{C}}=O}{\overset{\overset{CN}{\overset{|}{|}}}{\underset{|}{C}}} - \right)_n -$$

Generic name	Proprietary name	R group	Degradation rate, pH = 7 (hour^{-1})	Formaldehyde released, pH = 7 in mg/M polymer
Methyl-2-	Eastman 910	$-CH_3$	3×10^{-3}	797
Ethyl-2-	Krazy Glue	$-CH_2-CH_3$	2×10^{-4}	66
Propyl-2-		$-CH_2-CH_2-CH_3$		
n-Butyl-2-	Histoacryl	$-\underset{\overset{|}{C}H_3}{\overset{\overset{CH_3}{\overset{|}{}}}{C}}-CH_3$	1×10^{-5}	47
Iso-butyl-2-	Bucrylate	$-CH_2-\underset{\overset{|}{C}H_3}{\overset{\overset{CH_3}{\overset{|}{}}}{C}H}$		
n-Heptyl		$-CH_2-CH_2-CH_2-CH_3$		
Fluoroalkyl	MBR 4197R	$-\underset{\overset{|}{C}F_3}{\overset{\overset{CH_3}{\overset{|}{}}}{C}}-H$		

Source: From Refs. 116–119

have used these products for 10 or more years. Recently, however, a U.S. company (Tri-Point Medical) has been trying to get clinical approval for a series of n-butyl-cyanoacrylate compounds [124]. The newer products are claimed to be purer and less histotoxic than previous formulations. Four different formulations (Nexacryl®, Avacryl®, Occyldent®, and TPS®) are being pursued, with different ones for different applications: a tissue adhesive, a tissue sealant for periodontal applications, and a drug delivery system. Some formulations are currently undergoing clinical trials, with Nexacryl the furthest along. There, however, have been a few reports of complications clinically [125,126].

2. Chemistry

The CAs have several advantages that make them useful as a tissue adhesive, including strong, somewhat flexible bond, polymerization in a moist environment, tissue wettability, and ease of application [123,125]. The different CAs are produced by changing the alkyl group in the monomer. As this side chain gets larger, the hardness and tensile strength increases, while the flexibility decreases [126,127]. In addition, the degradation rate also decreases (Table 1).

The CAs are unique in that polymerization occurs at room temperature without the addition of a catalyst, evaporation of a solvent, heat, or pressure [128]. The polymer is made by reacting formaldehyde with alkylcyanoacetate [123]. The structure has a nitrite (CN) and aloxycarbonyl (CO_2R) groups. These groups increase the electronegativity of the ethylene backbone, making one carbon relatively positive and the other relatively negative [128]. This permits a weak base, such as the water from tissue fluid, to complete the anionic polymerization process quickly. The rate of polymerization is inversely proportional to the amount of monomer [128]. The adhesion to tissue is due mostly to strong secondary bonds, although mechanical interlocking is possible as well.

The polymer is not phagocytized *in vivo*, but broken down by hydrolytic chain scission and ester hydrolysis to formaldehyde (Table 1) and alkyl-cyanoacetate as well as CO_2 and water [125–128]. These products are absorbed into the skin and eventually excreted in the urine and feces [129]. The release rate of these compounds is tied to the degradation rate and therefore is lower with larger homologues.

In addition, since branching leads to an increased degradation rate, these products are released faster in isobutyl than n-butyl cyanoacrylates.

3. Properties

As to the properties of CAs, their viscosity can be increased by the addition of other polymers such as polymethacrylate, polyacrylate, polyvinyl acetate, polylactic acid, or cellulose organic esters [130]. In addition, adding energy by heat, ultraviolet (UV) light, x-rays, electron rays, or gamma rays has also been used. Further, to enhance impact strength, various plasticizers have been added: aliphatic monocarboxylic acids, dialkyl esters, and trialkyl phosphates. These enhancements tend to be used only for nonmedical applications [130]. In medical applications, only an FDA-approved dye [0.1% 1-hydroxy-4-[-p-toluidion]-antrachion) is added to give a blue color and allow easy visibility, enabling more precise control of the adhesive film thickness [123,128]. For Histoacryl, each ampule contains 0.3 gm (0.5 cubic centimeters [cc]) of solution. To allow reuse, the plastic tip can be cut off and replaced with a #25 or #27 needle. A new needle needs to be used for each use and the adhesive should be refrigerated between uses. In addition, gloves should be changed before each application to keep the vial free of the patient's blood or body fluids.

For the best results, the tissue should be approximated and as dry as possible. Also, the Histoacryl should be applied in a thin layer (spot welding). Further, although the adhesive bonds almost immediately and with good tensile strength, minimal tension is best. This usually requires adequate deep dermal suture placement. Also, large amounts of the glue should be avoided to prevent thermal damage. The Histoacryl eventually flakes off of the skin in less than 7 days and any excessive glue can be removed with forceps [128].

4. Tissue Response

Histotoxicity is a major concern for this class of tissue adhesive. The methyl cyanoacrylates proved to be unsuitable due to severe tissue necrosis [117,118], the lack of tissue wettability [116], and sarcoma formation (in rats) [131,132]. In addition, the local temperature increase is approximately 4°C [133].

Ethyl-2-cyanoacrylate has also been shown to induce acute inflammatory responses with tissue necrosis, graft ejection, and a foreign-body giant cell response occurring until degradation is complete (typically 6–12 months). The toxicity has been linked to the release of the degradation products (Table 1). When the release is slowed, it is better handled and controlled and thus less toxic. Thus, the concentration has been on the higher homologues [123,134].

The isobutyl cyanoacrylates, although they initially had good results, eventually showed some problems. In the 1970s, for middle ear applications, reports of marked inflammation and damage to bone [120], despite the addition of topical steroids, were published [121]. In addition, when used for intracranial arteriovenous malformations, there have been reports of mural angionecrosis, necrosis of parenchyma, and foreign-body giant cells [122]. Although fluoroalkyl-2-cyanoacrylate has been shown to have less toxicity in the middle ear than isobutyl- or ethyl-2-, it has not been used to a significant degree [125].

The n-butyl cyanoacrylates also showed good results, initially with a mild inflammatory response that resolved within a few weeks [134–145]. After degradation (about 1 year), only fibrous tissue remains. Local temperature rise is only about 1.5°C [133]. Although noncarcinogenic responses have been reported [146–148], the slow degradation and the prolonged tissue exposure is a concern [123]. In the 1980s, however, problems associated with Histoacryl were published, especially when placed subcutaneously, contacting well-vascularized soft tissue [123,149].

It appears that an increase in vascular and lymphatic supply leads to an increase in inflammatory cells at the implant site, leading to an increase in degradation rate. This leads to an increase in toxic product release [123,134].

In general, the butyl-cyanoacrylates show promise as tissue adhesives because of their ease of application, reliability, and good bond strength. They provide better mechanical (Table 2) attachment, both strength and stress distribution, than sutures as well as being less expensive and quicker to apply. There are concerns, however, when applied to well-vascularized tissue. In addition, vascular tissue cannot grow through the material, so true regenerative healing is hampered unless it is made porous or only used to spot weld [123,128]. This causes a reduction in incision strength, compared with sutures, at 14 days even though cyanoacrylates have a higher initial bond strength (Table 2).

C. Gelatin-Resorcinol Formaldehyde

The gelatin, resorcinol, and formaldehyde (GRF) system was developed at the National Heart Institute and Batelle Memorial Institute as a tissue adhesive that would quickly set up and maintain its properties in an *in vivo* water environment [4–6]. The goal was to develop a material that was at least as strong as the tissue it was gluing together, plus have good handling properties, like a high initial tack.

Gelatin was selected because it is denatured collagen, the most common structural protein in the body. Formaldehyde was added to increase cohesive strength and decrease solubility. Although bond strengths were relatively high, the dissolution rate was still too rapid. Resorcinol was then added, since it not only reacted with the formaldehyde to stabilize the system further, but it also reduced the overall viscosity. By reducing the viscosity of the gelatin solution, the concentration of gelatin could be increased without affecting handling. Initially, the formaldehyde was directly applied to the tissue prior to gelatin-resorcinol addition. Eventually, toxicity concerns led to the incorporation of the formaldehyde to create a GRF solution [150].

A 3:1 gelatin-resorcinol mixture with a 40:1 gelatin-resorcinol-to-formaldehyde (37% solution) ratio was selected as having the best overall properties. The setup time at body temperature and pH is about 30 seconds. By lowering the pH, the GRF solution can be

Table 2 Mechanical Properties

Tissue adhesive	UTS (MPa)	Shear strength (MPa)	Modulus (MPa)	Incisional strength (kPa)			Skin graft adhesion (kPa)	
				Post operative	4 days	14 days	Tensile	Shear
Methyl-2-CA	28–55	16–21	210–340	16	9	32		
Propyl-2-CA				30	13	30		
Isobutyl-2-CA	65							
Heptyl-2-CA				32	19	22		
GRF				62				
FM-Tissucol	0.1		0.15				20	52
Sutures				8	16	53		

Source: From Refs. 1, 6, 117

stored indefinitely. Therefore, addition of an inorganic base may be required for activation in some low pH wounds [150].

The toxicity of the GRF system is a concern because of the presence of formaldehyde, although *in vivo* studies have shown minimal toxicity. Contrary to CAs, however, increased toxicity has been found in less vascular tissue like cartilage. The researchers have suggested that increased vascularity more quickly removes the formaldehyde, thus reducing toxicity. Attempts at reducing the formaldehyde concentration, however, have led to reduced strength [150–153].

The product is not presently FDA approved, and thus has had little clinical use in the United States.

D. Mussel Adhesive Protein

Since curing in moist environments is a desired property of a tissue adhesive, researchers have looked at one produced by marine organisms. Marine organisms such as mussels and barnacles produce an adhesive to attach to underwater surfaces in turbulent intertidal zones throughout the world. The blue mussel *Mytilus edulis* synthesizes a polyphenolic protein (MAP, mussel adhesive protein), in a structure called the byssus, which mediates attachment along with other proteins (found in the byssus) such as collagen threads [154,155].

This byssal adhesive is highly cross-linked and thus difficult to analyze, although the non-cross-linked adhesive protein, which is produced and stored in the byssus, has been characterized. It is a 130-kD protein rich in proline, serine, threonine, lysine, tyrosine, and alanine. The protein is hydroxyl-rich, with 60–70% of the amino acid residues containing hydroxyl groups. A high percentage of the proline residues are converted post-translationally to 3- and 4-hydroxyproline, and more than half of the tyrosine residues are also hydroxylated to 3,4-dihydroxyphenylalanine (DOPA). The decapeptide sequence Ala-Lys-Pro/Hyp-Ser-Tyr/DOPA-Hyp-Thr-DOPA-Lys was found to repeat up to 80 times in the mussel adhesive protein (MAP) and thus was considered likely to carry key adhesive sequences. Most of the amino residues form hydrogen bonds, with lysine forming ionic bonds and DOPA forming metal complexes [154–157].

In order to provide sufficient quantities of the MAP for studies, a microbial system to produce the protein has been developed by Genex. The cloning encodes a 24-kD carboxy terminal region of the adhesive protein including 19 decapeptides and 1 hexapeptide. The adhesive protein produced in yeast does not contain the DOPA residue and thus *in vitro* hydroxylation is done using a bacterial tyrosinase. The hydroxylation occurs within 1 h. The stability of the MAP is maintained by lyophilization and storage at $-20°C$ [155].

Although DOPA residues are unusual, they are certainly not unique and have been found in other species. This quinone (oxidized DOPA) tanning process helps to achieve moisture resistance and retard biodegradation in such structures as egg casings in the liver fluke (*Fasciola hepatica*) and protective tubes in the marine worm (*Phragmatopoma californica*) [154].

The adhesive properties of the MAP were tested in an aqueous environment by measuring the amount of protein adherence. It was found that the protein adheres in a moist environment; adhesion is enhanced by hydroxylation of the tyrosine. The amount of protein adhered is directly proportional to the initial protein concentration. The protein is resistant to removal by 0.9% NaCl, 0.5% SDS, Triton X-100, 1% acetic acid, 0.1-N NaOH, and sonication [154,155].

The cohesive strength depends on intermolecular cross-linking and the formation of quinone residues, while the cross-linked gel setting time is dependent on the protein concentration. At 13 mg/ml, the setting time is 2–3 min; at 11 mg/ml, it is 18–20 min; and, at 8 mg/ml, it will not form a gel. The higher concentration systems also seem to have increased cohesive strength, most likely due to the increased cross-link density [155].

The MAP appears to have affinity for both soluble and insoluble collagen. In an *in vitro* study using moist collagen sheets, the MAP consistently showed significantly greater adhesive strength than a fibrin adhesive. In *in vivo* studies, the MAP has shown adequate strength within one minute and increasing strength over several hours [154]. Although minimal toxicity has been reported in numerous cases [158–161], there has been a report of inflammation upon intraocular injection [162].

E.   Other Adhesives

A number of other surgical tissue adhesives have been tried. None of these systems have been used to the same extent or have shown the promise of the ones described above in this chapter, at least in their present forms.

Isocyanates have been used, with the most successful formulation an isocyanate-terminated polyester. When mixed with pyridine, it rapidly bonds to tissue. Moisture, however, greatly weakens the adhesive strength and thus limits its utility. Rubber lattices such as natural rubber/isoprene composites have also been used. Although good bond strengths have been achieved (400 g/cm^2), the time required for *in vivo* setting is relatively long. In addition, the nondegradability of these systems has limited their usefulness. Polyacrylates have shown promise because of the potential reactivity with acid chloride groups present in tissue. Polyacrylchloride does not bond well, while polyacrylamide and glyoxal bond rapidly, but have limited strength [150].

Anhydrides such as copolymers of maleic anhydride and methyl vinyl ether can be cross-linked with polyols, polyfunctional amines, and calcium salts. These materials, however, have low cohesive strength. The dicarboxyanhydride of tartaric acid, which polymerizes with a small amount of base, does not bond well to tissue [150].

Epoxy resins have been examined because of the reactivity of the epoxide group to amines, alcohols, and carboxylic acids. Formaldehyde condensation products of urea, phenol, melamine, casein, and guanidine have also been evaluated. Both of these systems also have a low rate of bond formation [150].

Vinyl-containing polymers such as acrylated gelatin and acrylated polyvinylamine, which are cross-linked by free-radical catalysis, are slow to set up. Other systems such as cross-linked vinyl alcohol, polyvinylamine, polyvinylpyrrolidone, polyethyleneimine, nylon systems, and phosphonyl-chloride-terminated polyols also have been tried with little success [150].

Prolamine, a biodegradable, protein-based viscous gel, is another material that has been tried. It is an amino acid alcohol that polymerizes in the presence of water. It has been used clinically in Europe, but is not presently approved in the United States [163]. In addition, there is a study using oxidized, regenerated cellulose for blood vessel attachment [164].

IV.   CLINICAL APPLICATIONS

The utilization of tissue adhesives in surgery spans more than 25 years. Tissue adhesives have been used for tissue welding, as hemostatic agents, as fluid and gas barriers, as drug delivery systems, and as tissue scaffolds. Although FMs have been used the most extensively, CAs, GRFs, and MAPs could be used in similar clinical applications if fully exploited. In addition, for each application, the matrix should become a more active participant in the healing process than just the passive adhesive and sealant systems presently used. This will more fully utilize the potential of these materials as degradable regenerative scaffolds.

A. Cardiovascular Surgery

Cardiovascular applications for tissue adhesives are numerous [165–168]. They have been used as a hemostatic sealant for vascular graft attachment, cardiovascular patches, heart valve attachment [169], and to preclot porous vascular grafts [170,171]. FM presealed porous alloplastic vascular grafts prevent leakage along the graft better than whole blood. This allows grafts with a larger pore size (coarser weave), which are more flexible than small-pore grafts, to be used more frequently. Both FMs and CAs have been used in microvascular anastomosis [126,172,173]. FMs have been shown to improve early burst strength in rat femoral artery repairs and lead to fewer aneurysms than in the laser-treated arteries [172] and veins [173].

The hemostatic properties of tissue adhesives have been used in a number of applications. FMs have been used to reduce mediastinal drainage after cardiac procedures [174].

While GRF has been used to prevent bleeding in puncture wounds of the heart and aorta [150,175], for blood vessels, collagen fleece has been glued to grafts for hemostasis [169]. In addition, synthetic vascular grafts [Dacron, polytetrafluoroethylene (E.I. du Pont de Nemours)] coated with an FM and growth factors have been shown to improve endothelial cell attachment after presurgical seeding when compared with untreated grafts [176,177].

Both FM and cyanoacrylate have been used for obliteration of arteriovenous (A-V) malformations [178,179] and FM has been used in cutaneous lesions of Kaposi's sarcoma [180]. One of GRF's most effective clinical applications is in the repair of aortic dissections [6,151–153]. The GRF adhesive eliminates the dissection plane as well as strengthens the wall quicker and more effectively than sutures. FM has also been used for this application [181].

B. Orthopedic Surgery

Numerous applications in orthopedics and dentistry can be enhanced by use of tissue adhesives. Orthopedic applications of FM have been reviewed by Schlag and Redl [182]. Uses include filling defects with bone fragments or calcium phosphate ceramic implants, reattachment of osteochondral fragments, bone fragment repositioning in compound or digital fractures, and tendon repair [182,183]. CAs have been used for osteotomy repair [184]. The actual impact of the FM on bone and cartilage healing (improvement in healing rate, wound healing strength development, etc.) remains controversial. Its clinical utility as a procedural adjunct in microsurgery and arthroscopy is clear, however, especially in low mechanical stress conditions. In tendon repair, suturing is necessary for load bearing, but the application of the FM appears to improve the quality of healing scar tissue in both full and partial tears. This, in turn, may reduce the number of sutures needed for stabilization and appositioning, thereby reducing scar formation. Mensical tears have been successfully repaired with arthroscopic application of the FM [185], and remain clinically successful even during six years of clinical follow-up.

In dentistry, the FM has been used to fill extracted tooth sockets and alveolar ridge augmentation with hydroxylapatite [186]. CAs have been used for restoration of broken teeth, as a tooth sealant, and as a periodontal dressing [187].

MAP, because of its adhesiveness for cells, has been used for fixation of osteoblasts and chondrocytes [159,160]. This would have implications for cell seeding in orthopedic applications when using regenerative scaffolds or for better ingrowth for implant fixation.

C. Neurosurgery/Ophthalmic Surgery

Tissue adhesives have been used for many applications in neurosurgery and ophthalmic surgery. FMs and CAs have been employed for peripheral nerve reattachment and dural sealing to prevent cerebrospinal fluid leakage [188,189] with favorable efficacy and clinical utility. Although controlled studies are few and the surgical outcomes usually have not been assessed rigorously, the FM appears to reduce or eliminate the need for sutures in nerve anastomosis, thus reducing the overall procedure time. The FM forms an adhesive "conduit" around the anastomotic site, preventing ingrowth of epineural tissue, giving mechanical stability, and maintaining apposition of the two ends. A major advantage of the FM is that, once the desired appositioning is attained, it may be fixed into place instantly with little or no displacement, which is difficult to achieve with microsuturing. Thus, the major advantage is an increase in the ease of anastomosis and a decrease in procedure time.

All the reports, however, have not been positive. One reviewer noted that the FM should not be placed between the appositioned ends [37]. Another report described the fibronectin component of FM formulations as responsible for fibrosis in nerve anastomosis [190]. Another technical problem is that FM must be confined to the nerve and not be in contact with surrounding tissue in order to prevent adhesion formation. Also, tension must be eliminated to prevent disruption of the repair.

Recent work with growth factor (fibroblast growth factor-1, FGF-1) incorporated porous FM has shown the feasibility of both regenerating a 1-cm length of peripheral nerve tissue, and also regenerating nerve end plates when a severed nerve is glued into muscle. The porosity is critical for preventing the problems seen in nonporous systems when reanastamosing the ends—slowing or preventing of healing—while the growth factor stimulates angiogenesis and nerve regeneration [191].

The FM has been shown to be effective in sealing packing agents in transdural procedures, but there are few reported controlled studies. It was thought by many investigators that the FM increases the ease of performing the procedure and maintains a fluid-tight seal that is equal or superior to suturing or packing techniques. Higher burst pressures were obtained with the FM as compared with sutures in a canine spinal dura puncture model [192]. In a retrospective four-year study, the FM was found to be an effective leak sealant [193]. Also, carbon fiber electrodes have been attached to canine phrenic nerves for diaphragm pacing using the FM [194]. CA has been used in both interventional neuroradiology and for temporary tarsorrhaphy [195].

Both FMs and CAs are used in numerous ophthalmic procedures. There have been reports on usages of FMs and CAs for sealing of lens perforations, eyelid surgery, lacrimal and conjunctival repair, attachment of artificial corneas, retinal reattachment, and emergency procedures [128,197–200]. Both FMs and CAs have been used successfully to close corneal ulcers in humans, although the FM is thought to be better because of its quicker biodegradability [200,201]. In addition, a FM has been used to improve wound healing in lamellar keratoplasty in a rabbit corneal model [202] and MAP has been used to decrease the number of sutures and time in epikeratophakia repair [158].

D. General Surgery/Traumatology

The most common uses of the tissue adhesives have been general surgery and traumatology. Most tissue adhesives have been used successfully for closure of incisional wounds and lacerations [114,128,150]. Both FMs and CAs have been used successfully in controlling bleeding of esophageal varices and ulcers, but the FM leads to less scarring and a better long-term result [202–207]. Repair of ruptured spleens, kidneys, and liver have been done with FMs, for which the tissue adhesive acts as both a sealant and a hemostatic agent [183,184]. Tissue adhesives have been used to repair both pulmonary air leaks and fistulas in animals and humans successfully [153,208–219]. GRF has been able to successfully seal incisions several centimeters long in the lung, although this was only successful in the short term for lung lobectomies [5,153].

For the FM, success with gastrointestinal anastomoses has been equivocal. Colonic anastomoses have shown improved bursting strength in FM-treated animals as compared with those in which sutures were used [220]. Leakage and abscess formation, however, were noted in a rat model [221]. In an irradiated rodent bowel anastomosis model, the FM was shown to increase collagen production at the repair site and have a greater tensile breaking strength than suture controls 7 days after surgery [222]. Good results also are reported for biliary duct repair [223]. The healing rate of duodenal ulcers was improved when Tissucol was applied endoscopically in combination with ranitidine [224].

Packing and hemostasis have been obtained with the tissue adhesives alone and in combination with Gelfoam® (Upjohn, Kalamazoo, MI) or Avitene® collagen fleece (MedChem, Inc., Woburn, MA) in many general surgical applications [128,150,225, 226]. Large liver perforations generated by bullet or knife wounds have been repaired with balloon catheters sealed in the wound track by large amounts (20 cc) of FM [111]. GRF has also been used to seal defects in the liver [5]. Hemostasis was achieved within a few minutes and animals were healthy throughout the 6-month testing period, when all that remained at the incision site was a fibrous scar [5]. Liver biopsy needle tracks have been plugged or sealed with Tissucol in place of Gelfoam [227,228]. The bleeding time was reduced to zero with the use of FM, compared with the 130 to 280 seconds required for Gelfoam controls. Hemostasis in partial nephrectomy has also been described [229].

The FM has been used to seal defects in rodent parietal peritoneum, preventing the formation of adhesions [230,231]. In a detailed rodent histopathological study, the serosa and bowel wall regenerated beneath the fibrin layer [232]. Lactated Ringer's, however, was equivalent to the FM (Tissucol) in preventing abdominal adhesions in rats [233]. Another study showed that the FM antiadhesion activity is dependent on fibrinogen concentration, and that the lesion closure is not required [234]. The efficacy of the FM in preventing peritoneal adhesions is somewhat surprising since fibrinogen is implicated in adhesion formation.

Surprisingly, only a few applications in obstetrics and gynecologic surgery have been reviewed [235,236–240]. An interesting use of the FM has been to create a plug in the uterine cavity after embryo transfer to prevent expulsion or ectopic implantation [237], while CA has been used for episiotomy repair [238] and female sterilization procedures [239].

E. Drug Delivery

Fibrin matrices have been used to deliver drugs and antibiotics in preclinical release and thrombin time studies, and clinically in traumatology. Release appears to follow first-order kinetics. The *in vitro* release rates and antibacterial effects of the FM with carbenicillin, tetracycline, ciprofloxacine hydrochloride, gentamicin, teicoplanin, clindamycin, ampicillin, tobramycin, cefotaxime, and mezlocillin have been evaluated [240-251]. Carbenicillin, gentamicin, and tobramycin were released in clinically effective doses for up to 96 hours, while clindamycin was released for up to 72 hours and ampicillin for up to 48 hours [240]. Clinical efficacious doses of ciprofloxacine hydrochloride and tetracycline were able to be maintained for up to 4 weeks *in vitro*, although *in vivo* times have not been determined [251]. Since the antibiotics appear to be released as the gel degrades, and not by diffusion from the clot, *in vivo* release studies are necessary.

Tissucol and cefotaxime mixtures were utilized to treat osteitis in 46 human subjects [243]. Cefotaxime levels in serum were detected at low concentrations for up to 36 hours. Clinically effective concentrations were measured at the wound site for 84 hours. In another study, clinically effective levels of cefoxitin, gentamicin, ciprofloxacin, and teicoplanin, released from Tissucol, were found *in vitro* [244]. In addition, the FM alone seemed to have antibacterial effects when compared with saline or mouse lung homogenate controls.

The use of fibrin-tobramycin mixtures in a deliberately infected polytetrafluoroethylene aortic graft attachment was studied in dogs [245]. Control and FM (canine cryoprecipitate) control groups developed infection; all subjects in the FM-treated group died. No subjects in the FM-antibiotic group had significant contamination.

In split-thickness skin grafts, however, the FM alone had bacteriostatic effects [246]. Utilizing a rodent model, skin grafts were reattached with saline or FM (Tissucol) in the presence of two different doses of *Pseudomonas aeruginosa* and *Staphylococcus aureus*. The FM was found to improve the percentage of graft take significantly over saline controls at either bacterial dose. The addition of aprotinin did not significantly affect the outcome. The authors hypothesized that the large amount of fibrin overwhelmed bacterial fibrinolytic capacity.

In a murine hepatic contamination model, the application of FM to liver was found to decrease sepsis and adhesion formation significantly when compared with controls (no treatment) [247]. On the other hand, no difference was noted between FM coating and systemic antibiotic treatment in a rodent intraperitoneal contamination model [248]. In this model, a further complication was late abscess formation [249].

In an attempt to increase the release time of dibekacin sulfate from fibrin gels, glutaraldehyde (GTA) was used as a covalent cross-linking agent to attach the antibiotic to the gel structure [250]. Gelation time was increased as a function of increased GTA concentration, although levels below 0.02% did not have a significant effect. The antibiotic was detected at clinically effective levels in rat tibia for up to two weeks. High concentrations of GTA in fibrin gels elicited a strong foreign-body reaction, whereas lower concentrations had no noticeable effect on surrounding tissues.

FM-gelatin-factor XIII and CA mixtures have been used for intraperitoneal cancer therapy [252,253]. For the FM systems with doxorubicin and cisplatin, complete remission was reported in 34 of 37 cases (92%) and partial remission in 3 of 37 (8%). For these cases, gelatin was added as a biological response modifier, effecting macrophage induction to improve T-lymphocyte induction and activity.

Also, 5-fluorouracil drug delivery through an FM has been investigated *in vitro* [254,255]. Up to 120 hours of release have been achieved so far.

In addition, the FM has been used as a drug delivery vehicle for biochemical factors that influence the healing response. Both factors that slow healing (steroids [antikeloid] and hyaluronic acid [adhesion prevention]) as well as those that enhance healing (growth factors) have been used [256].

Further, the FM shows a lot of promise as a degradable regenerative scaffold when incorporated with growth factors. Release rate of the growth factors is closed loop (biofeedback control) because it is tied to degradation, and therefore is tailored to the healing rate of the particular patient and application. In addition, fibrin has innate healing properties; when made porous, it serves as an excellent regenerative scaffold [257].

The use of growth factors (FGF, transforming growth factor [TGF], platelet-derived growth factor [PDGF], bone morphogenetic protein [BMP], etc.) can be used to enhance the regenerative process further in all applications. This can be coupled with any of the other functions of the tissue adhesive or be its only function. For example, neodermis and epidermis can be regenerated in open skin wounds with these systems [258] as well as blood vessels [178] and nerves [191].

F.   Plastic and Reconstructive Surgery

Both FMs and CAs have been used extensively in plastic surgery procedures [1,259,260]. These include blepharoplasty, rhinoplasty, face-lifts, submental liposuction, midbrow-lifts, pina reconstruction, and parotidectomies. The adhesives tend to give more uniform and quicker attachment than sutures, as well as a better cosmetic result [259].

For the FM, grafting of skin has been successful, regardless of the FM production method, in areas of complex anatomical contouring, such as the hands of burn patients [261–263], face-lifts [264], and in rhinophyma repair [265]. Graft attachment strength was improved over staple controls when using a patient-autologous FM preparation [52]. Contraction of full-thickness skin grafts was markedly reduced when an FM (Tissucol) was used instead of staples to attach grafts in a rodent model [266]. The increased adherence of the graft to the recipient tissue bed was considered responsible for this effect. In general, the FM seems to provide close approximation of the graft to the recipient tissue bed, thereby preventing the formation of seromas [267], although seroma formation was not prevented in mastectomy cases [268].

Keratinocytes cultured on sheets of FM (Biocol) have been used as a burn dressing in humans [269]. Full tissue differentiation was noted after one year in the two patients treated.

Both FMs and CAs have been used successfully in a wide variety of craniofacial reconstruction procedures, mainly for bone restructuring [270,271] and in periodontal tissue grafting [271–273]. Cartilage chips may be mixed with an FM for use in rhinoplasty [274].

G.   Otorhinolaryngology

FMs and CAs have been used extensively in otorhinolaryngology [114,123,275,276]. As in other microsurgical procedures, the ease of tissue adhesive delivery under magnification (by a variety of methods into limited spaces) reduces the time to perform the procedure and improves the ease of securing tissue together. Tissue adhesives are well suited for ossicular chain reconstruction (i.e., ossiculoplasty and tympanossiculoplasty) [51, 114,123,128,276,277], as well as for sealing, repair, and replacement of tympanic membranes [276,278,280]. Bone powder and FM are mixed to fill and recontour the bony canal wall and mastoid bone after cholesteatoma resection [280] and mastoidectomy [281,282]. Labyrinthine fistulae and the oval window after stapedectomy have been sealed successfully with tissue adhesives [107,124,283]. FM sealing instead of nasal packing after septal surgery has been found to reduce the length of hospital stay and improve patient comfort [284]. Combined CO_2 laser and FM have been employed in vocal cord lateralization [285]. The laser debulks one or both vocal cords, and the FM glues the cord(s) into the desired position. FMs and CAs have been used for tracheal repair in dogs to reduce the number of sutures to perform anastomosis and form an airtight seal [286].

V. CONCLUSIONS

Presently, there are no ideal tissue adhesives. To complicate this further, each application (hemostatic agent, sealant, drug delivery system, or tissue scaffold) has different requirements. The main characteristics of an ideal tissue adhesive are biocompatibility and adhesive strength as well as the ability to be approved by the FDA. Presently, none of the systems described in the chapter are currently FDA approved, although both FM and CA systems are currently undergoing clinical trials. In the meantime, autologous FM is the only clinically available tissue adhesive.

Compatibility is a concern with both CAs and GRF; both are basically drug delivery systems for formaldehyde and other toxic products. Also, CAs have shown increased irritation when placed subcutaneously in well-vascularized dermal tissue.

Since the optimal compatibility is when the implant degrades and regeneration occurs, adhesives should be designed to enhance regenerative healing. Fibrin not only has inherent healing properties, but also can have its bioactivity enhanced by the addition of growth factors or making the matrix porous.

The GRF and CA systems clearly have the greatest initial strength (Table 2). Therefore, when strength is more important than healing, these systems can be used. The minimum needed mechanical properties for each application, however, are not known, let alone the optimal mechanical properties.

Overall, although the FM appears to be the best choice in most applications, each application has different needs and thus optimal tissue adhesive design. For the most part, it is a trade-off between strength and healing capacity, but other factors, such as ease of use and handling characteristics, are also critical. Presently in the United States, the main concern is getting FDA approval. Without it, the surgeons' options are limited and they are left with autologous FM as the only choice. In Europe, several commercial products are available from regional blood banks and companies involved in the manufacture of human coagulation proteins. The production of FM by these groups permits large-scale manufacture under optimal conditions with the best available methods, quality control, and reproducibility.

While it appears that the FM is gaining greater acceptance and wider application in the United States, a number of advances still need to take place. Primarily, a commercially available, reasonably priced product needs to be produced. Pooled plasma would be the most appropriate source for the large amounts of fibrinogen and factor XIII. These need to be prepared with an effective viral deactivation system. The solvent detergent method appears most promising for viral deactivation since it is used in commercial human antihemophilic factor production [287]. Heat pasteurization processes are not as effective and cause denaturation of the coagulation proteins. Confidence in product safety needs to be established unequivocally. At present, preparation of the FM with patient-autologous plasma is labor intensive and the material properties are too variable, although they do obviate viral transmission concerns. Methods of obtaining fibrin sealant derived from single-donor sources, such as those used in blood banks (e.g., cryoprecipitation for obtaining antihemophilic factor), are most common. Blood centers, however, typically do not provide this service for a variety of reasons (e.g., regulatory, lack of clinical interest, etc.). The use of human thrombin should be further explored to eliminate any possibility of immune reactions triggered by bovine thrombin, as well as infection by potential adventitious agents (i.e., bovine spongiform encephalopathy).

In order for FM technology to gain wider acceptance, the clinical usefulness and efficacy must be rigorously assessed and compared with current repair methods. This requires an understanding of the contribution of the FM constituents in a particular clinical situation since changes in fibrinogen and fibronectin concentrations alter mechanical and wound healing properties. To date, objective assessment has been difficult, although more useful studies have been reported recently. The primary limitation is lack of controls or quantifiable end points. Most reports are "how-to-do-its" (i.e., how the adhesive was used) and subjective evaluations of efficacy. Animal studies have been limited by small numbers, and extrapolation of procedure and wound repair to the human clinical application is difficult. In addition, the choice of repair model is critical: it must be able to differentiate in a quantifiable manner the use of the FM versus the standard means of repair.

Another issue is the contribution of each FM constituent to wound healing, adhesion, and degradation, as well as interaction with therapeutic adjuncts. The production method is important in this regard because different preparative techniques fractionate different constituents with varying yields. This in turn affects the adhesive performance (e.g., mechanical properties, degradation rates, gelation times, uniformity, etc.).

Packaging a convenient preparation for clinicians is crucial. Proper delivery devices (spray, cannula, catheter) must be carefully designed to maximize the ease of application, especially for microsurgical and laproscopic procedures. Since the FM has a variety of functions (hemostasis, tissue welding, etc.) in a wide range of indications, the means of delivery should be designed to maximize its function and increase the ease of delivery to the repair site. For example, spraying the FM in a large, oozing field to obtain hemostasis is much easier than using a dual-syringe applicator. The converse is true when performing an ossicular chain repair through a 7-mm diameter speculum inserted into the mastoid bone.

In comparison to the FM, no other adhesive material is routinely available that provides the same flexibility of use and safety. The FM has many properties desirable in a material involved in wound healing, namely, biodegradability and biocompatibility, as well as the ability to be a regenerative scaffold. It is a material formulated from constituents involved in normal wound healing, it attains adhesion to tissues by physiologic means, and it adheres to wet surfaces. The handling and performance properties may be modulated by altering the formulation. The adhesive may be delivered by variable formats suitable for a given indication or role. Judging by the increased activity in the literature and by the popularity of the commercial products, it is inevitable that FMs soon will be used on a routine basis in the United States.

ACKNOWLEDGMENTS

We wish to thank Lisa Stevens for her help in typing the manuscript. In addition, we would like to thank the graduate students in the Biomedical Polymers Research Group (David Wilson, Abhay Pandit, and Charlene Flahiff), as well as research collaborators at the University of Alabama at Birmingham, the American Red Cross, and Matrix Pharmaceutical, instrumental forces in accomplishing many of the studies presented in this chapter.

REFERENCES

1. Sierra, D., Fibrin sealant adhesive systems: A review of their chemistry, material properties and clinical applications, *J. Biomat. App.*, 7:309–352 (1993).
2. Matsumoto, T., *Tissue Adhesives in Surgery*, Medical Examination Publishing, New York (1972).
3. Sidentop, K.H., Tissue adhesive Histoacryl® (2-cyano-butyl-acrylate) in experimental middle ear surgery, *Am. J. Otol.*, 2:77–87 (1980).
4. Braunwald, N.S., W. Gay, and C.J. Tatooles, The use of a crosslinked gelatin tissue adhesive to control hemorrhaging from liver and kidney, *Surg. Forum*, 16:345–346 (1965).
5. Cooper, C.W., and R.D. Falb, Surgical adhesives, *Ann. N.Y. Acad. Sci.*, 146:214–224 (1968).
6. Bachet, J., F. Gigou, C. Laurian, O. Bical, B. Goudot, and D. Guilmet, Four-year clinical experience with the gelatin-resorcine-formal biological glue in acute aortic dissection, *J. Thorac. Cardiovasc. Surg.*, 83:212–217 (1982).
7. Waite, J.H., Evidence for a repeating 3,4-dihydroxyphenylalanine- and hydroxyproline-containing decapeptide in the adhesive protein of the mussel, *Mytilius edulis L*, *J. Biol. Chem.*, 258:2911–2915 (1983).
8. Waite, J.H., Decapeptides produced from bioadhesive polyphenolic proteins, United States Patent 4,585,585 (1986).
9. Waite, J.H., Nature's underwater adhesive specialist, *Int. J. Adhesion Adhesives*, 7:9–14 (1987).
10. Filpula, D.R., L. Shwu-Maan, R.P. Link, S.L. Strausberg, and R.L. Strausberg, Structural and functional repetition in a marine mussel adhesive protein, *Biotechnol. Prog.*, 6:171–177 (1990).
11. Hunt, T., and W. van Winkle, Fundamentals of wound management in surgery, in *Wound Healing: Disorders of Repair*, Chirugicom, South Plainfield, New Jersey (1976).
12. Davie, E., K. Fugikawa, K. Kurachi, and W. Kisiel, The role of serine proteases in the blood coagulation cascade, *Adv. Enzymol.*, 48:277–318 (1979).
13. Knighton, D., K. Cires, V. Fiegel, L. Ausin, and E. Butler, Classification and treatment of chronic nonhealing wounds: Successful treatment with autologous platelet-derived wound healing factors, *Ann. Surg.*, 204:322–300 (1986).
14. Knighton, D., T. Hunt, K. Thakral, and W. Goodson, Role of platelets and fibrin in the healing sequence: An *in vivo* study of angiogenesis and collagen synthesis, *Ann. Surg.*, 196:379–388 (1982).
15. Knighton, D., I. Silver, and T. Hunt, Regulation of wound healing angiogenesis: Effect of oxygen gradients and inspired oxygen concentration, *Surgery*, 90:262–270 (1981).
16. Lemperle, B., Topical oxygen therapy in the treatment of decubitus ulcers and persistent skin lesions, *Health Technology Assessment Reports*, No. 8, National Center for Health Services Research (1983).
17. Lehman, W., W. James, M. Allo, R. Johnston, and P. MacGanity, Human bite infections

of the hand: Adjunct treatment with hyperbaric oxygen, *Infections in Surgery*, 460–465 (June 1985).

18. Dagher, F.J., *Cutaneous Wounds*, Futura Publishing, New York (1985).

19. Lily, W.R., Re-vascularization of a free skin autograft, *Acta Chirurgica Academiae Sceinti-arum Hungaricae, Tumus*, 12(2):181–192 (1971).

20. Pang, C., and G. Suski, Comparative effects of surgical delay procedure on cutaneous blood flow and skin viability in random and arterial flaps, *Proceedings Plastic Sur. Res Council*, 28:57–59 (1983).

21. Clemmesen, T., *Experimental Studies on the Healing of Free Skin Autografts*, N. Olaf Moller, Copenhagen (1967).

22. Alexander, H., A. Weiss, C. Parsons, E. Straucher, S. Corivran, O. Gona, and C. Mayott, Canine patellar tendon replacement with a polylactic acid polymer-filamentous carbon and tissue degrading scaffold, *Ortho. Rev.*, 10(11):41–51 (1981).

23. Greisler, H., and D. Kim, Aspects of biodegradable vascular prosthesis, in *Vascular Graft Update*, H. Kahalic, A. Kantrowitz, and P. Sung, eds., ASTM, Fairfield, PA, pp. 197–218 (1986).

24. Annis, D., A. Burnat, R. Edwards, A. Highan, K.B. Loveday, and J. Wilson, An elasto-meric vascular prosthesis, *Trans. ASAIO*, 24:209–214 (1978).

25. Bergel, S., Uber wirkugen des fibrin, *Dtsch. Med. Wochenschr.*, 35:633–665 (1909).

26. Grey, E.C., Fibrin as a hemostatic in cerebral surgery, *Surg. Gynecol. Obstet.*, 21:452–454 (1915).

27. Harvey, S.C., The use of fibrin papers and foams in surgery, *Boston Med. Surg. J.*, 174:659–662 (1916).

28. Ferry, J.D., and P.R. Morrison, Chemical, clinical and immunological studies on the prod-ucts of the human plasma fractionation. XVI, Fibrin clots, fibrin films and fibrinogen plastics, *J. Clin. Invest.*, 23:566–572 (1944).

29. Morrison, P.R., and M. Singer, Chemical, clinical and immunological studies on the prod-ucts of human plasma fractionation. XVIII, A note on the absorption rates of fibrin films in tissue, *J. Clin. Invest.*, 23:573–573 (1944).

30. Bering, E.A., Chemical, clinical and immunological studies on the products of human plasma fractionation. XX, The development of fibrin foam as a hemostatic agent and for use in conjunction with human thrombin, *J. Clin. Invest.*, 23:586–589 (1944).

31. Ferry, J.D., The conversion of fibrinogen to fibrin: Events and recollections from 1942 to 1982, *Ann. N.Y. Acad. Sci.*, 408:1–10 (1983).

32. Gerendas, M., Fibrin products as aids in hemostasis and wound healing, in *Fibrinogen*, K. Laki, ed., Marcel Dekker, New York (1968).

33. Young, J., and P.B. Medawar, Fibrin suture of peripheral nerves, *Lancet*, 11:126–132 (1940).

34. Seddon, F.J., and P.B. Medawar, Fibrin suture of human nerves, *Lancet*, 11:87–92 (1942).

35. Tarlov, I.M., and B. Benjamin, Plasma clot and silk suture of nerves, *Surg. Gynecol. Obstet.*, 76:366–374 (1943).

36. Cronkite, E.P., E.L. Lozner, and J. Deaver, Use of thrombin and fibrinogen in skin graft-ing, *J.A.M.A.*, 124:976–978 (1944).

37. Narakas, A., The use of fibrin glue in repair of peripheral nerves, *Orthop. Clin. N. Amer.*, 19:187–199 (1988).

38. Matras, H., H.P. Dinges, B. Manoli, et al., Zur nahtlosen interfaszikularen nerventransplan-tation im tierexperiment, *Wien Med. Woschtr.*, 122:517–523 (1972).

39. Kudema, E., and H. Matras, Die klinische anwendung derklebung von nerveanastomosen bei der rekonstruktion verletzer peripherer nerven, *Wien Klin. Wochenschr.*, 87:495–501 (1975).

40. Fuhge, P., N. Heimburger, H.A. Stohr, and W. Burk, Readily dissolvable lyophilized fibrin-ogen formulation, United States Patent 4,650,678 (1987).

41. Heimburger, N., P. Fuhge, and H. Ronneberge, Single-component tissue adhesives containing fibrinogen, factor XIII, thrombin inhibitor, prothrombin factors and calcium ions in aqueous solution, European Patent 87109374 (1987).

42. Burnouf-Radosevich, M., T. Burnouf, and J.J. Huart, Biochemical and physical properties of a solvent-detergent-treated fibrin glue, *Vox. Sang.*, 58:77–84 (1990).

43. Stemberger, A., Haemostatic tissue-adhesive collagen wound dressing in lint or sponge containing 0.3–2 cm collagen layer coated on a least one side with 0.2–2 mm fibrinogen layer, European Patent 83102773 (1983).

44. Revocation of fibrinogen licenses, *FDA Drug Bull.*, 8:15 (1978).

45. Gestring, G.F., and R. Lerner, Autologous fibrinogen for tissue adhesion, hemostasis and embolization, *Vasc. Surg.*, 17:294–304 (1983).

46. Spodnitz, W., History of Tissue Adhesives, *Proc. of the Sym. Surgical Tissue Adhesives*, 1: 32 (1993).

47. Dresdale, A., B.A. Rose, V. Jeevanandam, et al., Preparation of fibrin glue from single-donor fresh-frozen plasma, *Surgery*, 97:750–754 (1985).

48. Rose, E., and A. Dresdale, Fibrin adhesive prepared as a concentrate from single donor fresh frozen plasma, United States Patent 4,627,879 (1986).

49. Wolf, G., Der Konzentrierte autologe geweb Kleber, *Arch. Ohren Nasen Kehlkopfheil*, 237: 279–283 (1983).

50. Siedentop, K.H., D.M. Harris, and B. Sanchez, Autologous fibrin tissue adhesive, *Laryngoscope*, 95:1074–1076 (1985).

51. Epstein, G.H., R.A. Weisman, S. Zwillenberg, and A.D. Schreiber, A new autologous fibrinogen-based adhesive for otologic surgery, *Ann. Otol. Rhinol. Laryngol.*, 94:40–45 (1986).

52. Dahlstrom, K.K., U.S. Weis-Fogh, S. Medgyesi, J. Rostgaard, and H. Sorenson, The use of autologous fibrin adhesive in skin transplantation, *Plast. Reconstr. Surg.*, 89:968–972 (1992).

53. Bier, M., and P.R. Foster, Purification of antihemophilia factor VIII by precipitation with zinc ions, United States Patent 4,406,886 (1983).

54. Hermans, T., and J. McDonagh, Fibrin: Structure and interactions, *Sem. Thromb. Hemost.*, 8:11–24 (1982).

55. Blomback, B., and L. Hanson, *Plasma Proteins*, John Wiley and Sons, New York (1979).

56. Duckert, F., D. Nyman, and H. Gastpar, Factor XIII, fibrin and collagen, *Thromb. Hemost.*, 63(Suppl.):391–401 (1978).

57. Burleson, R.L., and N. Ennulat, Fibrin adherence to biological tissues, *J. Surg. Res.*, 25: 523–529 (1978).

58. Thornton, J.W., M.J. Tavis, J.H. Harney, H. Pirkle, R.H. Bartlett, and E.A. Woodruff, Graft adherence to wound surfaces: Collagen-fibrin interactions, *Burns*, 3:23–29 (1977).

59. Phillips, D.R., L.K. Jennings, and H.H. Edwards, Identification of membrane proteins mediating the interaction of human platelets, *J. Cell. Biol.*, 86:77086 (1980).

60. Tavis, M.J., J.W. Thorlen, J.H. Harney, E.A. Woodroof, and R.H. Bartlett, Graft adherence to de-epithelized surfaces, *Ann. Surg.*, 194:594–598 (1976).

61. Feldman, M.D., R.T. Sataloff, H.Y. Choi, and S.K. Ballas, Compatibility of autologous fibrin adhesive with implant materials, *Arch. Otolaryngol. Head Neck Surg.*, 114:182–185 (1988).

62. Ferry, J.D., and P.R. Morrison, Preparation and properties of serum and plasma proteins. VIII. The conversion of human fibrinogen to fibrin under various conditions, *J. Amer. Chem. Soc.*, 69:388–400 (1947).

63. Murayama, M., Pressure influences the rate of fibrin polymerization: Decompression accelerates, compression decelerates, *Throm. Res.*, 49:538–539 (1988).

64. Okada, M., B. Blomback, M.-D. Chang, and B. Horowitz, Fibronectin and fibrin gel structure, *J. Biol. Chem.*, 260:1811–1820 (1985).

65. Blomback, B., and M. Okada, Fibrin gels and their possible implication for surface hemor-heology in health and disease, *Ann. N.Y. Acad. Sci.*, 416:397–409 (1983).

66. Carr, M.E., Turbidimetric evaluation of the impact of albumin on the structure of thrombin-mediated fibrin gelation, *Haemostasis*, 17:189–195 (1987).

67. Roberts, W.W., O. Kramer, W. Rosser, F.H.M. Nestler, and J.D. Ferry, Rheology of fibrin clots: Dynamic viscoelastic behavior, *Biophys. Chem.*, 1:152–160 (1974).

68. Gerth, C., W.W. Roberts, and J.D. Ferry, Rheology of fibrin clots. II. Linear viscoelastic behavior in shear creep, *Biophys. Chem.*, 2:208–217 (1974).

69. Nelb, G.W., C. Gerth, and J.D. Ferry, Rheology of fibrin clots. III. Shear creep and creep recovery of fine ligated and coarse unligated clots, *Biophys. Chem.*, 5:377–387 (1976).

70. Rosser, R.W., W.W. Roberts, and J.D. Ferry, Rheology of fibrin clots, IV. Darcy constants and fiber thickness, *Biophys. Chem.*, 7:153–157 (1977).

71. Nelb, G.W., G.W. Kamykowski, and J.D. Ferry, Rheology of fibrin clots. V. Shear modulus, creep and creep recovery of fine unligated clots, *Biophys. Chem.*, 12:15–23 (1981).

72. Kamykowski, G.W., D.F. Mosher, L. Lorand, and J.D. Ferry, Modification of shear modulus and creep compliance of fibrin clots by fibronectin, *Biophys. Chem.*, 13:25–28 (1981).

73. Shen, L.L., R.O. McDonagh, J. McDonagh, and J. Hermans, Fibrin gel structure: Influence of calcium aid covalent cross-linking on the elasticity, *Biochem. Biophy. Res. Comm.*, 56:793–798 (1974).

74. Shen, L.I., J. Hermans, J. McDonagh, R.P. McDonagh, and M. Carr, Effects of calcium ion and covalent crosslinking on formation of elasticity of fibrin gels, *Thromb. Res.*, 6:255–265 (1975).

75. Carr, M.E., L.L. Shen, and J. Hermans, A physical standard of fibrinogen: Measurement of the elastic modulus of dilute fibrin gels with a new elastometer, *Anal. Biochem.*, 72:202–21 (1976).

76. Nowotony, R., A. Chalupka, C. Nowotony, and P. Bosch, Mechanical properties of fibrinogen adhesive material, in *Biomaterials 1980*, G.D. Winter, G.F. Gibbons, and H. Plenk, eds., John Wiley and Sons, London (1982).

77. Welch, M.T., B.R. Smith-Morse, and K.G.M. Brockbank, An assay for assessment of the biomaterial properties of fibrin sealants: Preliminary results, *Trans. Soc. Biomat.*, 14:136 (1991).

78. Jorgensen, P.H., K.H. Jensen, B. Andreassen, and T.T. Andreassen, Mechanical strength in rat skin incisional wounds treated with fibrin sealant, *J. Surg. Res.*, 42:237–241 (1987).

79. Byrne, D.J., J. Hardy, R.A.B. Wood, R. McIntosh, and A. Cuschieri, Effect of fibrin glues on the mechanical properties of healing wounds, *Br. J. Surg.*, 78:841–843 (1991).

80. McIntosh, R.V., N. Docherty, D. Fleming, and P.R. Foster, A high-yield factor VIII concentrate suitable for advanced heat treatment, *Throm. Haem.*, 50:306 (1987).

81. Siedentop, K.H., D.M. Harris, and B. Sanchez, Autologous fibrin tissue adhesive: Factors influencing bonding power, *Laryngoscope*, 98:731–733 (1988).

82. Laitakari, K., and J. Luotnen, Autologous and homologous fibrinogen sealants: Adhesive strength, *Laryngoscope*, 99:974–976 (1989).

83. Sierra, D.H., A.J. Nissen, and J. Welsh, Use of fibrin glue in intracranial procedures: Preliminary results, *Laryngoscope*, 100:360–363 (1990).

84. Sierra, D.H., D. Feldman, R. Saltz, and S. Huang, A method to determine shear adhesive strength of fibrin sealants, *J. Appl. Biomat.*, 3:147–151 (1992).

85. Flahiff, C., D. Feldman, R. Saltz, and S. Huang, Mechanical properties of fibrin adhesives for blood vessel anastomosis, *J. Biomed. Mater. Res.*, 26:481–491 (1992).

86. Steinberger, A., W. Hebeler, W. Duspiva, and G. Glumel, Fibrinogen-cold insoluble globulin mixtures as tissue adhesives, *Thromb. Res.*, 12:907–910 (1978).

87. Schlag, C., H. Redl, M. Turnher, and H.P. Dinges, The importance of fibrin in wound repair, in *Fibrin Sealant in Operative Medicine, Vol. 1: Otolaryngology*, G. Schlag and H. Redl, eds., Springer-Verlag, Heidelberg (1986).

88. Branstedt, S., F. Rank, and P.S. Olson, Wound healing and formation of granulation tissue in normal and defibrinogenated rabbits. An experimental model and histological study, *Eur. Surg. Res.*, 12:12-21 (1980).

89. Holand, B., P. Junker, C. Garbasch, P. Christoffersen, and I. Lorenzen, Formation of granulation tissue subcutaneously implanted sponges in rats, 87(Sect. A):367-374 (1979).

90. Irvin, T.T., *Wound Healing. Principles and Practices*, Chapman and Hall Publishers, London (1981).

91. Immuno AG, Tisseel kit two-component fibrin sealant, steam-treated, Package insert 6208110EZ01/37-4, Immuno AG, Vienna, Austria (1989).

92. Bruhn, H.D., and K.H. Zurborn, Influence of clotting factors (thrombin, factor XIII) and of fibronectin on the growth of tumor cells and leukemic cells *in vitro*, *Blut.*, 46:85-91 (1983).

93. Kitchens, C.S., and T.F. Newcomb, Factor XIII, *Medicine*, 58:413-429 (1979).

94. Bosch, P., F. Braun, J. Eschberger, W. Kovac, and H.P. Spangler, Die Beeinflussung der knochenheilung durch hochkonzentriertes fibrin. Experimenteele untersuchungen am kaninchen, *Arch. Orthop. Unfall-Chir.*, 89:259-264 (1977).

95. Bosch, P., F. Lintner, H. Arbes, and G. Brand, Experimental investigations of the effect of the fibrin adhesive on the kiel heterologous bone graft, *Arch. Orthop. Trauma Surg.*, 96:177-182 (1980).

96. Pfluger, H., P. Bosch, F. Grundschober, H. Kristen, H. Plenk, and S. Schider, Untersuchungen uber das Einwachsen von Knochengewebe in porose metallimplantate, *Wien Klin. Wochensch.*, 91:482-489 (1979).

97. Schumacher, G., A. Braun, and W.D. Heine, Das alloimplant am knochen unter verwendung des fibrinklebesystems tierexperimentelle ergebnisse, in *Fibrinkleber in Orthopaedie und Traumatologie*, H. Cotta and A. Braun, eds., Thieme-Verlag, Stuttgart (1982).

98. Bohler, N., P. Bosch, G. Sandbach, J. Eschberger, and L. Schmid, Experimentelle erfahrungen mit der einkelbung von kortikaliszy-lindern, in *Fibrinkleber in Orthopaedie und Traumatologie*, H. Cotta and A. Braun, eds., Thieme-Verlag, Stuttgart (1982).

99. Bohler, N., P. Bosch, G. Sanbach, G. Schlag, J. Eschberger, and L. Schmid, Der Einfluß von homologem fibrinogen auf die osteomie-heilung beim kaninchen, *Unfallheikunde*, 80:501-511 (1977).

100. Stubinger, B., H.M. Fritsche, W. Erhardt, R. Senekowitsch, et al., Experimentelle anwedung des fibrinklebers zur fixation corticospongioser fragmente und bei der autologen spongiosaplastik, in *Fibrinkleber in Orthopaedie und Traumatologie*, H. Cotta and A. Braun, eds., Thieme-Verlag (1982).

101. Keller, J., T.T. Andreassen, F. Joyce, V.E. Kundsen, P.H. Jorgensen, and Lucht, Fixation of osteochondral fractures. Fibrin sealant tested in dogs, *Arch. Orthop. Scand.*, 56:323-330 (1985).

102. Albrektsson, T., A. Bach, S. Edshange, and A. Jonsson, Fibrin adhesive system (FAS) influence on bone healing rate. A microradiographical evaluation using the bone growth chamber, *Acta Orthop. Scand.*, 53:757-766 (1982).

103. Zilch, H., and F. Noffke, Beeinflußt der fibrinkleber die knochenneubildung? *Unfallheilkunde*, 84:363-367 (1981).

104. Lucht, U., C. Bunger, J.T. Moller, F. Joyce, and H. Plenk, Fibrin sealant in bone transplantation. No effect on blood flow and bone formation in dogs, *Acta Orthop. Scand.*, 57:19-27 (1986).

105. Lambrecht, J.R., and M. Klinger, Resorption of fibrin tissue adhesive sealant by isolated osteoclasts in culture, *Int. J. Oral Maxillofac. Surg.*, 19:177-180 (1990).

106. Siedentop, K.H., D.M. Harris, and A. Loewy, Experimental use of fibrin tissue adhesive in middle ear surgery, *Laryngoscope*, 93:1310-1313 (1983).

107. Harris, D.M., K.H. Siedentop, K.R. Ham, and B. Sanchez, Autologous fibrin tissue adhesive biodegradation and systemic effects, *Laryngoscope*, 97:1141-1144 (1987).

108. Staindl, O., G. Galvan, and M. Macher, The influence of fibrin stabilization and fibrinolysis on the fibrin adhesive system. A clinical study using radioactively marked fibrinogen as a tracer, *Arch. Otorhinolaryngol.*, 233:105–116 (1981).

109. Pfluger, H., Lysis and absorption of fibrin sealant (Tissucol®/Tinsseel®), *in vitro* and *in vivo* experiments, in *Fibrin Sealant in Operative Medicine, Vol. 1: Otolaryngology*, G. Schlag and H. Redl, eds., Springer-Verlag, Heidelberg (1986).

110. Glynn, J.H., and J.H. Richardson, Antigenic properties of fibrin films and foams prepared from human and from bovine blood plasma, *J. Immunol.*, 53:143–152 (1946).

111. Banerjee, S., and L.E. Glynn, Reactions to homologous and heterologous fibrin implants in experimental animals, *Ann. N.Y. Acad. Sci.*, 86:1054–1057 (1960).

112. Berguer, R., R.L. Staerkel, E.E. Moore, et al., Warning: Fatal reaction to the use of fibrin glue in deep hepatic wounds. Case Study, *J. Trauma*, 31:408–411 (1991).

113. Milde, L.N., An anaphylactic reaction to fibrin glue, *Anesth. Analg.*, 69:684–686 (1989).

114. Ardis, A.E., U.S. Patents 2467926 and 2467927 (1949).

115. Coover, H.W., F.B. Joyner, N.H. Shearer, and T.H. Wicker, Chemistry and performance of cyanoacrylate adhesives, *J. Soc. Plast. Eng.*, 15:413–417 (1959).

116. Coover, H., Jr., *Handbook of Adhesives*, Reinhold Publishing, New York (1962).

117. Leonard, F., The N-alkaalpha cyanoacrylate tissue adhesives in surgery, *Ann. N.Y. Acad. Sci.*, 6:203–213 (1968).

118. McKelvie, P., A trial of adhesives in reconstructive middle ear surgery, *J. Laryngol.*, 83: 1102–1109 (1968).

119. Leonard, F., R.K. Kulkarni, G. Brandes, J. Nelson, and J.J. Cameron, Synthesis and degradation of poly(alkyl α-cyanoacrylates), *J. Appl. Polymer Sci.*, 10:259–272 (1966).

120. Kerr, A.G., and G.D.L. Smyth, Bucrylate (isobutyl cyanoacrylate) as an ossicular adhesive, *Arch. Otolaryngol. Head Neck Surg.*, 94:129–131 (1971).

121. Kaufman, R.S., The use of tissue adhesive (isobutyl cyanoacrylate) and topical steroid (0.1 percent dexamethasone) in experimental tympanoplasty, *Laryngoscope*, 84:793–804 (1974).

122. Vinters, H.V., M.J. Lundie, and J.C.E. Kaufmann, Long-term pathological follow-up of cerebral arteriovenous malformations treated by embolization with bucrylate, *N. Engl. J. Med.*, 314:477–483 (1986).

123. Toriumi, D.M., W.F. Raslan, M. Friedman, et al., Histotoxicity of cyanoacrylate tissue adhesives: A comparative study *Arch. Otolaryngol. Head Neck Surg.*, 116:546–550 (1990).

124. Tri-Point Medical's orphan product "Nexacryl" corneal adhesive is approvable, *Medical Devices, Diagnostics, Instrumentation Reports*, 18(5):20–21 (1992).

125. Koltai, P.J., and A.R. Eden, Evaluation of three cyanoacrylate glues for ossicular reconstruction, *Ann. Otol. Rhinol. Laryngol.*, 92:29 (1983).

126. Coover, H.W., and J.M. McIntire, The chemistry of cyanoacrylate adhesives, in *Tissue Adhesives in Surgery*, T. Matsumoto, ed., Medical Examination Publishing, New York (1972).

127. Smith, D.C., Lutes, glues, cements and adhesives in medicine and dentistry, *Biomed. Eng.*, 146:203–213 (1968).

128. Ellis, D.A.F., and A. Shaikh, The ideal tissue adhesive in facial plastic and reconstructive surgery, *J. Otolaryn.*, 19:68–72 (1990).

129. Ousterhout, D.K., G.V. Gladieux, and F. Leonard, Cutaneous absorption of N-alkyl αa-cyanoacrylate, *J. Biomed. Mater. Res.*, 2:157–163 (1968).

130. Billmeyer, F., *Textbook of Polymer Science*, 3rd Edition, John Wiley and Sons, New York (1989).

131. Vinters, H.V., K.A. Galil, M.J. Lundie, and J.C.E. Kaufmann, The histotoxicity of cyanoacrylates, *Neuroradiology*, 27:279–291 (1985).

132. Samson, D., and D. Marshall, Carcinogenic potential of isobutyl-2-cyanoacrylate, *J. Neurosurg.*, 65:571–572 (1986).

133. Hida, T., S.M. Sheta, A.D. Proia, and B.W. McCuen, Retinal toxicity of cyanoacrylate tissue adhesive in the rabbit, *Retina*, 8:148–143 (1988).

134. Toriumi, D.M., W.F. Raslan, M. Friedman, and M. E. Tardy, Variable histotoxicity of histoacryl when used in a subcutaneous site: An experimental study, *Laryngoscope*, 101:339–343 (1991).

135. Matsumoto, T., R.M. Hardaway, K.C. Pani, and P.M. Margetis, Aron Alpha A Sanyo, Japanese tissue adhesive in surgery of internal organs, *Am. Surg.*, 34:263–267 (1968).

136. Kamer, F.M., and J.H. Joseph, Histoacryl: Its use in aesthetic facial plastic surgery, *Arch. Otolaryngol. Head Neck Surg.*, 115:193–197 (1989).

137. Ronis, M.L., J.D. Harwick, R. Fund, et al., Review of cyanoacrylate tissue glues with emphasis on otorhinolaryngological applications, *Laryngoscope*, 94:210–213 (1984).

138. Fung, R.O., M.L. Ronis, and R.M. Mohr, Use of butyl-2-cyanoacrylate in rabbit auricular cartilage, *Arch. Otolaryngol.*, 111:459–464 (1985).

139. Sptinas, M., H. Lossagk, M. Vogel, et al., Retinal surgery using cyanoacrylate as a routine procedure, *Albrecht V. Graefes Arch. Klin. Experimental Ophtalmol.*, 187:89–101 (1973).

140. Aronson, S.D., P.R.B. McMaster, T.E. Moore, et al., Toxicity of the cyanoacrylates, *Arch. Ophthalmol.*, 84:342–349 (1970).

141. Regenbogen, L., A. Romano, M. Zuckerman, et al., Histoacryl tissue adhesive in some types of retinal detachment surgery, *Br. J. Ophthalmol.*, 60:561–572 (1976).

142. Bhaskar, S.N., and J. Frisch, Use of cyanoacrylate adhesives in dentistry, *J. Am. Dent. Assoc.*, 77:831–837 (1968).

143. Kosko, K.I., Upper-lid blepharoplasty: Skin closure achieved with butyl-2-cyanoacrylate, *Ophthalmic Surg.*, 12:424–425 (1981).

144. Siedentop, K.H., Tissue adhesive Histoacryl (2-cyano-butyl-acrylate) in experimental middle ear surgery, *Am. J. Otol.*, 2:77–86 (1980).

145. Orda, R., T. Wiznitzer, and G.M. Goldberg, Repair of hepatic and splenic injuries by autoplastic peritoneal patches and butyl-2-cyanoacrylate monomer, *J. Surg. Res.*, 17:367–374 (1974).

146. Lehman, R.A.W., G.J. Hayes, and F. Leonard, Toxicity of alkyl 2-cyanocrylates, I: Peripheral nerve, *Arch. Surg.*, 93:441–446 (1966).

147. Matsumoto, T., and C.A. HeisterKamp, Long-term study of aerosol cyanoacrylate tissue adhesive spray: Carcinogenicity and other untoward effects, *Ann. Surg.*, 35:825–827 (1969).

148. Soni, N.N., V.E. Whitehurst, R.S. Knight, and J.C. Sinkford, Long-range effects of Ivalon sponge containing isobutyl-2-cyanocrylates on rat tissue: A quantitative planimetric study, *Oral Surg.*, 39:197–202 (1975).

149. Heumann, H., and E. Steinbach, The effects of an adhesive in the middle ear, *Arch. Otolaryngol.*, 106:734–7 (1980).

150. Braunwald, N.S., W. Gay, and C.J. Tatooles, Evaluation of crosslinked gelatin as a tissue adhesive and hemostatic agent. An experimental study, *Surgery*, 59:1024–1030 (1966).

151. Fabiani, J.N., V.A. Jebara, A. Deloche, Y. Stephan, and A. Carpetitier, Use of surgical glue without replacement in the treatment of type a aortic dissection, *Circulation*, 80:264–268 (1989).

152. Bachet, J., B. Goudot, G. Teodore, D. Brodaty, C. Dubois, P.H. De Lentdecker, and D. Guilmet, Surgery of type a acute aortic dissection with gelatin-resorcine-formol biological glue: A 12-year experience, *J. Cardiovasc. Surg.*, 31:263–273 (1990).

153. Guilmet, D., J. Bachet, B. Goudot, C. Laurian, F. Gigou, O. Bical, and M. Barbagelatta, Use of biologic glue in acute aortic dissection. A new surgical technique. Preliminary clinical results, *J. Thorac. Cardiovasc. Surg.*, 77:516–521 (1979).

154. Strausberg, R.L., and R.P. Link, *Protein-Based Medical Adhesives*, Elsevier Science Publishers Ltd., United Kingdom (1990).

155. Filpula, D.R., S.-M. Lee, R.P. Link, S.L. Strausberg, and R.L. Strausberg, Structural and

functional repetition in a marine mussel adhesive protein, *Biotechnol. Prog.*, 6:171-177 (1990).

156.   Waite, J.H., The formation of mussel byssus: Anatomy of a natural manufacturing process, in *Structure, Cellular Synthesis and Assembly of Biopolymers*, S.T. Case, ed., Springer-Verlag, New York, pp. 27-54 (1992).

157.   Laursen, R.A., Reflections on the structure of mussel adhesive proteins, in *Structure, Cellular Synthesis and Assembly of Biopolymers*, S.T. Case, ed., Springer-Verlag, New York, pp. 55-74 (1992).

158.   Robin, J.B., and J. Salazar, The use of mussel adhesive protein in epikeratophakia, *Ann. Spring Meeting, Assoc. for Rsch. in Vision Ophthalmology*, Poster Session (1987).

159.   Pittman, M.I., D. Menche, E.K. Song, A. Ben-Yishay, D. Gilbert, and D.A. Grande, The use of adhesives in chondrocyte transplantation surgery: *In-vivo* studies, *Bulletin of the Hospital for Joint Diseases Orthopedic Institute*, 49:213-221 (1989).

160.   Fulkerson, J.P., L.A. Norton, G. Gronowicz, P. Picciano, J.M. Massicotte, and C.W. Nissen, Attachment of epiphyseal cartilage cells and 17/28 rat osteosarcoma osteoblasts using mussel adhesive protein, *J. Orthop. Res.*, 8:793-798 (1990).

161.   Green, K., R. Berdecia, and L. Cheeks, Mussel adhesive protein: Permeability characteristics when used as a basement membrane. *Short Communication, Current Eye Research*, 3:6 (1987).

162.   Liggett, P.E., M. Cano, J.B. Robin, R.L. Green, and J.S. Lean, Intravitreal biocompatibility of mussel adhesive protein. A preliminary study, *Retina*, 10:144-147 (1990).

163.   Pigott, J.P., D.L. Donovan, J.A. Fink, and W.V. Sharp, Angioscope-assisted occlusion of venous tributaries with prolamine in *in-situ* femoropopliteal bypass: Preliminary results of canine experiments, *J. Vasc. Surg.*, 9:704-709 (1989).

164.   Di Lello, F., D.C. Mullen, and R.J. Flemma, Sutureless fixation of long aortocoronary saphenous vein grafts with oxidized regenerated cellulose, *Ann. Thorac. Surg.*, 47:473-474 (1989).

165.   Baker, J.W., W.D. Spotnitz, and S.P. Nolan, A technique for spray application of fibrin glue during cardiac operations, *Ann. Thorac. Surg.*, 43:564-565 (1987).

166.   Kanchuger, M.S., T.R. Eide, G.R. Manecke, A. Hartman, and P.J. Poppers, The hemodynamic effects of topical fibrin glue during cardiac operations, *J. Cardiothorac. Anesth.*, 3: 745-747 (1989).

167.   Kjaergard, H.K., U.W. Weis-Fogh, H. Sorensen. J. Thiis, and I. Rygg, Autologous fibrin glue – Preparation and clinical use in thoracic surgery, *Eur. J. Cardiothorac. Surg.*, 6:52-54 (1992).

168.   Matthew, T.L., W.D. Spotnitz, I.L. Kron, T.M. Daniel, C.G. Tribble, and S.P. Nolan, Four years' experience with fibrin sealant in thoracic and cardiovascular surgery, *Ann. Thorac.*, 50:40-44 (1990).

169.   Wolner, E., Fibrin gluing in cardiovascular surgery, *Thorac. Cardiovasc. Surg.*, 30:236-237 (1982).

170.   Haverich, A., G. Walterbusch, and H.G. Borst, The use of fibrin glue for sealing vascular prostheses of high porosity, *Thorac. Cardiovasc. Surg.*, 29:252-254 (1981).

171.   Jonas, R.A., F.J. Schoen, R.J. Levy, and A.R. Castaneda, Biological sealants and knitted Dacron: Porosity and histological comparisons of vascular graft materials with and without collagen and fibrin glue pretreatments, *Ann. Thorac. Surg.*, 41:657-663 (1986).

172.   Grubbs, P.E., S. Wang, C. Marini, S. Basu, D.M. Rose, and J.N. Cunningham, Enhancement of CO_2 laser microvascular anastomoses by fibrin glue, *J. Surg. Res.*, 45:112-119 (1988).

173.   Cikrit, D.F., M.C. Dalsing, T.S. Weinstein, K. Palmer, S.G. Lalka, and J.L. Unthank, CO_2 welded venous anastomosis; Enhancement of weld strength with heterologous fibrin glue, *Lasers Surg. Med.*, 10:584-590 (1990).

174.   Spotnitz, W.D., M.S. Dalton, J.W. Baker, and S.P. Nolan, Reduction of perioperative

hemorrhage by anterior mediastinal spray application of fibrin glue during cardiac operations, *Ann. Thorac. Surg.*, 44:529–531 (1987).

175. Seguin, R., J.M. Frapier, P. Colson, and P.A. Chaptal, Fibrin sealant for early repair of acquired ventricular septal defect, *J. Thorac. Cardiovasc. Surg.*, 104:748–751 (1992).

176. Mazzucotelli, J.P., C. Klein-Soyer, A. Beret, C. Brisson, G. Archipoff, and J.P. Cazenave, Endothelial cell seeding: Coating Dacron® and expanded polytetrafluoroethylene vascular grafts with a biological glue allows adhesion and growth of human saphenous vein endothelial cells, *Int. J. Artif. Organs*, 14:482–490 (1991).

177. Griesler, H., D. Cziperie, D. Kim, J. Garfield, D. Petsikas, P. Murchan, E. Applegren, W. Dorhan, and W. Burgess, Enhanced endothelialization of expanded polytetrafluoroethylene grafts by FGF type 1 pretreatment, *Surgery*, 112(2):24–55 (1992).

178. Richling, B., Homologous controlled viscosity fibrin for endovascular embolization, *Acta Neurochirurg.*, 150:132–134 (1982).

179. Berthelsen, B., J. Lofgren, and P. Svendsen, Embolization of cerebral arteriovenous malformations with bucrylate: Experience in a first series of 29 patients, *Acta Radiologics.*, 31:13–21 (1990).

180. Tange, R.A., and F.J. Hadderingh, Utilisation de la Colle de Fibrine dans le Traitment Local du Sarcope de Kaposi au Cours du Sida, Rapport Preliminaire, *Rev. Laryngol.*, 109:187–190 (1988).

181. Dottori, V., S. Spagnolo, G. Passerone, A. Lijoi, L. Barberis, M. Agostini, G. De Gaetano, E. Parodi, M. Maccario, and E.C. Fumagalli, Ten years surgery of aortic dissections and aneurysms: Clinical experience and original contributions, *Minerva Cardioangiol.*, 40:431–436 (1992).

182. Schlag, G., and H. Redl, Fibrin sealant in orthopedic surgery, *Clin. Orthop. Rel. Res.*, 227:269–285 (1987).

183. Ono, K., J. Shikata, K. Shimizu, and T. Yamamuro, Bone-fibrin mixture in spinal surgery, *Clin. Orthop.*, 275:133–139 (1992).

184. Papatheofanis, F.J., Surgical repair of rabbit tibia osteotomy using isobutyl-2-cyanoacrylate, *Arch. Orthop. Trauma Surg.*, 108:236–237 (1989).

185. Ishimura, M., S. Tamai, and Y. Fujisaw, Arthroscopic meniscal repair with fibrin glue, *Arthroscopy*, 7:177–181 (1991).

186. Hotz, G., Alveolar ridge augmentation with hydroxylapatite using fibrin sealant for fixation. Part II: Clinical application, *Int. J. Oral Maxillofac. Surg.*, 20:208–213 (1991).

187. McCabe, M.J., Use of histocryl tissue adhesive to manage an avulsed tooth, *Br. Med. J.*, 301:20–21 (1990).

188. Quinn, J.V., A. Orzowiecki, M.M. Li, I.G. Stiell, T. Sutcliffe, T.J. Elmalie, and W.E. Wood, A randomized, controlled trial comparing a tissue adhesive with suturing in the repair of pediatric facial lacerations, *Ann. of Emerg. Med.*, 22:1130–1135 (1993).

189. Shaffrey, C.I., W.D. Spotnitz, M.E. Shaffrey, and J.A. Jane, Neurosurgical applications of fibrin glue: Augmentation of dural closure in 134 patients, *Neurosurgery*, 26:207–210 (1990).

190. Herter, T., and H. Bennefeld, The influence of fibronectin on the fibrosing of a nerve anastomosis in the rat, *Res. Exp. Med.*, 189:321–329 (1989).

191. Feldman, D., Wound healing application, *Symposium on Surgical Tissue Adhesives*, Atlanta, GA (1993).

192. Cain, J.E., H.G. Rosenthal, M.J. Broom, E.C. Jauch, D.A. Borek, and R.R. Jacobs, Quantification of leakage pressures after durotomy repairs in the canine, *Spine*, 15:969–970 (1990).

193. Matthew, T.L., W.D. Spotnitz, I.L. Kron, T.M. Daniel, D.G. Tribble, and S.P. Nolan, Four years' experience in fibrin sealant in thoracic and cardiovascular surgery, *Ann. Thorac. Surg.*, 50:40–44 (1990).

194. Kimura, M., T. Sugiura, Y. Fukui, M. Togawa, and Y. Harada, Glued carbon fiber electrodes for diaphragm pacing, *Artif. Organs*, 14:390–391 (1990).

195. Brothers, M.F., T.C.E. Kaufmann, A.J. Fox, and J.P. Deveikis, n-Butyl 2-cyanoacrylate-substitute for IBCA in interventional neuroradiology: Histopathologic and polymerization time studies, *AJNR*, 10:777–786 (1989).

196. Diamond, T.P., Temporary tarsorrhaphy with cyanoacrylate adhesive for seventh-nerve palsy, *Lancet*, 335:1039 (1990).

197. Cavanaugh, T.B., and J.D. Gottsch, Infectious keratitis and cyanoacrylate adhesive, *Am. J. Opthal.*, 11:466–472 (1991).

198. Gilbert, C.E., Grierson, and D. McLeod, Retinal patching: A new approach to the management of selected retinal breaks, *Eye*, 3:19–26 (1989).

199. Sheta, S.M., T. Hida, and B.W. McCuen, Cyanoacrylate tissue adhesive in the management of recurrent retinal detachment caused by macular hole, *Am. J. Ophthal.*, 109:28–32 (1990).

200. Lagoutte, F.M., L. Gauthier, and P.R. Comte, A fibrin sealant for perforated and pre-perforated corneal ulcers, *Br. J. Ophthalmol.*, 73:757–761 (1989).

201. Dean, B.S., and E.P. Krenzelok, Cyanoacrylates and corneal abrasion, *Clinical Toxicology*, 27:169–172 (1989).

202. Kim, M.S., and J.H. Kim, Effects of tissue adhesive (Tisseel®) on corneal wound healing in lamellar keratoplasty in rabbits, *Korean J. Ophthalmol.*, 3:14–21 (1989).

203. Antonelli, M., F. Cicconetti, G. Vivino, and A. Casparetto, Closure of a tracheoesophageal fistula by bronchoscopic application of fibrin glue and decontamination of the oral cavity, *Chest*, 100:578–579 (1991).

204. Naga, M., H. Gourbran, M. Said, T. Burnouf, and M. Burnouf Radesevich, Endoscopic injection of biological or non-biological sealants for the initial control of bleeding esophageal varices, *Proc. of the Symposium on Surgical Tissue Adhesives*, 1:27–29 (1993).

205. Scheele, J., H.H. Gentsch, and E. Matteson, Splenic repair by fibrin tissue adhesive and collagen fleece, *Surgery*, 95:6–13 (1984).

206. Kram, H.B., W.C. Shoemaker, S.T. Hino, and D.P. Harley, Splenic salvage using biologic glue, *Arch. Surg.*, 119:1309–1311 (1984).

207. Ishitani, M.B., E.D. McGahren, D.A. Sibley, W.D. Spotnitz, and B.M. Rodgers, Laparoscopically applied fibrin glue in experimental liver trauma, *J. Ped. Surg.*, 24:867–871 (1989).

208. Goldman, C.D., S.H. Blocker, J.L. Ternberg, and E.C. Crouch, Management of experimental pneumothorax in weanling rabbits with the use of fibrin glue sclerosant, *Arch. Surg.*, 121:565–568 (1986).

209. Jessen, C., and P. Sharma, Use of fibrin glue in thoracic surgery, *Ann. Thorac. Surg.*, 39:521–524 (1985).

210. Antonelli, M., F. Cicconetti, G. Vivino, and A. Gasparetto, Closure of a tracheoesophageal fistula by bronchoscopic application of fibrin glue and decontamination of the oral cavity, *Chest*, 100:578–579 (1991).

211. Hanck, H., P.G. Bull, and N. Pridun, Complicated pneumothorax: Short- and long-term results of endoscopic fibrin pleurodesis, *World J. Surg.*, 15:146–149 (1991).

212. Bense, L., Intrabronchial selective coagulative treatment of hemoptysis: Report of three cases, *Chest*, 97:990–996 (1990).

213. Hauck, H., P.G. Bull, and N. Pridun, Complicated pneumothorax: Short- and long-term results of endoscopic fibrin pleurodesis, *World J. Surg.*, 15:146–150 (1991).

214. Mouritzen, C., M. Dromer, and H.O. Keinecke, The effect of fibrin glueing to seal bronchial and alveolar leakages after pulmonary resections and decortications, *Eur. J. Cardiothorac. Surg.*, 7:75–80 (1993).

215. Nicholas, J.M., and S.A. Dulchavsky, Successful use of autologous fibrin gel in traumatic bronchopleural fistula: A case report, *J. Trauma*, 32:87–88 (1992).

216. Salmon, C.J., R.B. Ponn, and J.L. Westcott, Endobronchial vascular occlusion coils for control of a large parenchymal bronchopleural fistula, *Chest*, 98:223–234 (1990).

217. Vincent, T.G., H.J. Van De Wal, J.M. Meijer, C. Van Herwaarden, and L.K. Lacquet, Postponing the limits. Multiple and repeated pulmonary metastasectomy by parenchymal sparing electrocautery excision, *Helv. Chir. Acta*, 57:295–300 (1990).

218. Yasuda, Y., A. Mori, H. Kato, S. Fujino, and S. Asakura, Intrathoracic fibrin glue for postoperative pleuropulmonary fistula, *Ann. Thorac. Surg.*, 51:242–244 (1991).

219. Wood, R.E., S.R. Lacey, and R.G. Azizkhan, Endoscopic management of large, postresection bronchopleural fistulae with methacrylate adhesive (super glue), *J. Ped. Surg.*, 27:201–202 (1992).

220. Kjaergaard, J., P. Nordkild, E. Sjontoft, and A. Hiortrup, Non-sutured fibrin adhesive versus sutured anastomosis. A comparative intra-individual study in dog colon, *Acta Chir. Scand.*, 153:599–601 (1987).

221. van der Ham, A.C., W.J. Kort, I.M. Weijima, H.F. van den Ingh, and J. Jeekel, Effect of fibrin sealant on the healing colonic anastomosis in the rat, *Br. J. Surg.*, 78:49–53 (1991).

222. Saclarides, T.J., D.O. Woodard, M. Bapna, and S.G. Economou, Fibrin glue improves the healing of irradiated bowel anastomosis, *Dis. Conol Rectum*, 35:249–252 (1992).

223. Kram, H.E., M.A. Garces, S.R. Klein, and W.C. Shoemaker, Common bile duct anastomosis using fibrin glue, *Arch. Surg.*, 120:1250–1256 (1985).

224. Ederle, A., C. Scattolini, G. Bulighini, L. Benini, P.G. Orlandi, G. Talamini, and I. Vantinin, Does the combination of a human fibrin sealant with ranitidine accelerate the healing of duodenal ulcer? *Ital. J. Gastroenterol.*, 23:354–346 (1991).

225. Wepner, F., R. Fries, and H. Platz, The use of the fibrin adhesive system for local hemostasis in oral surgery, *J. Oral Maxillofac. Surg.*, 40:555–558 (1982).

226. Haverich, A., W. Maatz, and G. Walterbusch, Evaluation of fibrin seal in animal experiments, *Thorac. Cardiovasc. Surg.*, 30:215–222 (1982).

227. Chisholm, R.A., S.N. Jones, and W.R. Lees, Fibrin sealant as a plug for the post liver biopsy needle track, *Clin. Radiol.*, 40:627–628 (1989).

228. Rodriguez-Fuchs, C.A., and M. Bruno, Plugging liver biopsy sites with coagulation factors, *Lancet*, 11:1087 (1989).

229. Levinson, A.K., F.J. Greskovich, D.A. Swanson, R.A. Stephenson, D.E. Johnson, and B. Lichtiger, Fibrin glue for partial nephrectomy, *Urology*, 38:314–316 (1991).

230. Lindenberg, S., and J.G. Lauritsen, Prevention of peritoneal adhesion formation by fibrin sealant. An experimental study in rats, *Ann. Chir. Gynaecol.*, 73:11–13 (1984).

231. Hjortrup, A., P.J. Nordkild, J. Kiaergaard, E. Sjontroft, and H.P. Oleson, Fibrin adhesives versus sutured anastomosis. A comparative study in the small intestine of pigs, *Br. J. Surg.*, 73:760–761 (1986).

232. Brands, W., I. Jopich, and H. Lochbuhler, Use of highly concentrated human fibrinogen in the pediatric: Surgery. A new therapeutic principle, *Z. Kinderchir.*, 35:159–162 (1982).

233. Caballero, J., and T. Tulandi, Effects of Ringer's lactate and fibrin glue on post-surgical adhesion, *J. Reprod. Med.*, 37:141–143 (1992).

234. de Virgilio, C., T. Dubrow, B.B. Sheppard, W.D. MacDonald, F.J. Nelson, M.A. Lesavoy, and J.M. Robertson, Fibrin glue inhibits intra-abdominal adhesion formation, *Arch. Surg.*, 125:1378–1382 (1990).

235. Schlag, G., and H. Redl, eds., *Fibrin Sealant in Operative Surgery—Plastic Surgery—Maxillofacial and Dental Surgery*, Vol. 4, Springer-Verlag, Berlin (1986).

236. Adamyan, L.V., O.A. Myinbayev, and V.I. Kulakov, Use of fibrin sealant in obstetrics and gynecology: A review of the literature, *Int. J. Fertil.*, 36:76–77, 81–88 (1991).

237. Feichtinger, W., D. Barad, M. Feinman, and P. Barg, The use of two-component fibrin sealant for embryo transfer, *Fertil. Steril.*, 54:733–734 (1990).

238. Adoni, A., and E. Anteby, The use of Histoacryl for episiotomy repair, *Br. J. Obstet. Gyn.*, 98:476–478 (1991).

239. Shuber, F., Transcervical sterilization with use of methyl 2-cyanoacrylate and a new delivery system (the FEMCEPT device), *Am. J. Obstet. Gynecol.*, 160:887–889 (1989).

240. Greco, F., L. dePalma, N. Spagnolo, A. Rossi, N. Speechia, and A. Gigante, Fibrin-antibiotic mixtures: An *in vitro* study assessing the possibility of using a biological carrier for local drug deliver, *J. Biomed. Mater. Res.*, 25:39–51 (1991).

241. Redl, H., G. Schlag, G. Stanek, A. Hirschl, and T. Seelich, *In vitro* properties of mixtures of fibrin seal and antibiotics, *Biomaterials*, 4:29-32 (1983).

242. Lerner, R., and N.S. Binur, Current status of surgical adhesives, *J. Surg. Res.*, 38:165-181 (1985).

243. Zilch, H., and E. Lambris, The sustained release of cefotaxime from a fibrin-cefotaxime compound in treatment of osteitis, *Arch. Orthop. Trauma Surg.*, 106:36-41 (1986).

244. Kram, H.B., M. Bansul, O. Timberlake, et al., Antibacterial effects of fibrin glue and fibrin glue-antibiotic mixtures, *Trans Soc. Biomat.*, 12:164 (1989).

245. Ney, A.L., P.H. Kelly, D.T. Tsukayama, and M.P. Burbrick, Fibrin glue-antibiotic suspension in the prevention of prosthetic graft infection, *J. Trauma*, 30:1000-1006 (1990).

246. Jabs, A.D., T.M. Wider, J. DeBellis, and N.E. Hugo, The effect of fibrin glue on skin grafts in infected sites, *Plast. Reconstr. Surg.*, 89:268-271 (1992).

247. Dulchavsky, S.A., E.R. Geller, J. Maurer, P.R. Kennedy, G.T. Tortora, and S.R. Maitra, Autologous fibrin gel: Bactericidal properties in contaminated hepatic injury, *J. Trauma*, 31: 991-995 (1991).

248. Schwartz, R.J., T.J. Dubrow, R.A. Ival, S.E. Wilson, and R.A. Williams, The effect of fibrin glue on intraperitoneal contamination in rats treated with systemic antibiotics, *J. Surg. Res.*, 52:123-126 (1992).

249. Dubrow, T.J., R.J. Schwartz, J. McKissock, and S.E. Wilson, Effect of aerosolized fibrin solution on intraperitoneal contamination, *Arch. Surg.*, 126:80-83 (1991).

250. Sato, H., K. Ono, and M. Oka, Antibiotic binding fibrinogen as a fibrin glue, *Trans. Third World Biomater. Congr.*, 11:531 (1988).

251. Sugitachi, A., T. Shindo, T. Kido, and T. Kawahara, Bio-adhesiochemo-(BAC)-therapy, *Proc. Amer. Assoc. Cancer Res.*, 33:216 (1992).

252. Nunez, H., R. Hennigh, A. Campagna, S. Harding, W. Drohan, and M. MacPhee, Prolonged release of effective concentrations of antibiotics from fibrin sealant, *Proc. of the Symposium on Surgical Tissue Adhesives*, 1:23 (1993).

253. MacPhee, M., Drug release, *Proc. of the Symposium on Surgical Tissue Adhesives*, 1:97 (1993).

254. MacPhee, M., R. Kidd, A. Campagna, and W. Drohan, Extended delivery of antiproliferative compounds from fibrin sealant, *Proc. of the Symposium on Surgical Tissue Adhesives*, 1:24 (1993).

255. Sierra, D., E. Luck, D. Brown, and N. Yu, Fibrin-collagen adhesive drug delivery system for tumor therapy, *Trans. Society for Biomaterials*, 30:257 (1993).

256. Himmel, H., Plastic and reconstructive surgery, *Proc. of the Symposium on Surgical Tissue Adhesives*, 1:129 (1993).

257. Pandit, A., and D. Feldman, The effect of a porous degradable fibrin scaffold on wound healing, *Trans. Society of Biomaterials*, 30:34 (1994).

258. Wilson, D., D. Feldman, and A. Thompson, Fibrin glue as a matrix for AFGF delivery *in vivo*, *Trans. Society for Biomaterials*, 29:255 (1993).

259. Kamer, F.M., and J.H. Joseph, Histoacryl: Its use in aesthetic facial plastic surgery, *Arch. Otolaryngol. Head and Neck Surg.*, 115:193-197 (1989).

260. Watson, D.P., Use of cyanoacrylate tissue adhesive for closing facial lacerations in children, *Br. Med. J.*, 299:1014 (1989).

261. Saltz, R., A. Dimick, C. Harris, J.C. Grotting, J. Psillakis, and L.O. Vasconez, Application of autologous fibrin glue in burn wounds, *J. Burn Care Rehab.*, 10:504-507 (1989).

262. Stuart, J.D., R.F. Morgan, and J.G. Kenney, Single-donor fibrin glue for hand burns, *Ann. Plast. Surg.*, 24:524-527 (1990).

263. Vibe, P., and J. Pless, A new method of skin graft adhesion, *Scand. J. Plast. Reconstr. Surg.*, 17:263-268 (1983).

264. Marchac, D., E. Pugash, and D. Gault, The use of sprayed fibrin glue for facelifts, *Eur. J. Plast. Surg.*, 10:139-144 (1987).

265. Staindl, O., Surgical management of rhinophyma, *Acta. Otolaryngol.*, 92:137–140 (1981).

266. Brown, D.M., B.R. Barton, V.L. Young, and B.A. Pruitt, Decreased wound contraction with fibrin glue-treated skin grafts, *Arch. Surg.*, 127:404–406 (1992).

267. Lindsey, W.H., T.M. Masterson, M. Llaneras, W.D. Spotnitz, H.J. Wanebo, and R.F. Morgan, Seroma prevention using fibrin glue during modified radical neck dissection in a rat model, *Am. J. Surg.*, 156:310–313 (1988).

268. Jonk, A., J.A. van Dongen, and B.B.R. Kroon, Prevention of seroma following axillary lymph node dissection or radical mastectomy; ineffectiveness of fibrin glue sealing technique (letter), *Neth. J. Surg.*, 39:135 (1987).

269. Ronfard, V., H. Broly, V. Mitchell, et al., Use of human keratinocytes cultured on fibrin glue in the treatment of burn wounds, *Burns*, 17:181–184 (1991).

270. Marchac, D., and D. Rainier, Fibrin glue in craniofacial surgery, *J. Craniofac. Surg.*, 1:32–34 (1990).

271. Vasconez, H., Craniofacial surgery/bone metabolism, *Proc. of the Symposium on Surgical Tissue Adhesives*, 1:132 (1993).

272. Matras, H., The use of fibrin sealant in oral and maxillofacial surgery, *J. Oral Maxillofac. Surg.*, 44:171–176 (1986).

273. Gregory, E.W., and S.J. Schaberg, Experimental use of fibrin sealant for skin graft fixation in mandibular vestibuloplasty, *J. Oral Maxillofac. Surg.*, 44:171–176 (1986).

274. Fontana, A., E. Muti, D. Cicerale, and M. Rizzotti, Cartilage chips synthesized with fibrin glue in rhinoplasty, *Aesthetic Plastic Surg.*, 15:237–240 (1991).

275. Schlag, G., and H. Redl, eds., *Fibrin Sealant in Operative Surgery – Otolaryngology*, Vol. 1., Springer-Verlag, Berlin (1986).

276. Staindl, O., Tissue adhesion with highly concentrated human fibrinogen in otolaryngology, *Ann. Otol. Rhinol. Laryngol.*, 88:413–418 (1979).

277. O'Connor, A.F., and J.J. Shea, A biologic adhesive for otologic practice, *Otolaryngol. Head Neck Surg.*, 90:347–349 (1982).

278. Marquet, J., Fibrin glue in tympanoplasty, *Amer. J. Otol.*, 7:287–289 (1985).

279. Watson, D., and Maguda, T., An experimental study for closure of tympanic membrane perforations with fascia and an adhesive, *South Med. J.*, 58:844–847 (1965).

280. Palva, T., and L.G. Johnsson, Preservation of hearing after removal of the membranous canal with a cholesteatoma, *Arch. Otolaryngol. Head Neck Surg.*, 112:982–985 (1986).

281. Marquet, J., The fibrin seal in otorhinolaryngology, *J. Head Neck Pathol.*, 3:71–72 (1982).

282. Filipo, P., and M. Barbara, Rehabilitation of radical mastoidectomy, *Amer. J. Otol.*, 7:248–251 (1986).

283. Siedentop, K.H., and H. Schobel, Stapedectomy modified by the application of fibrin tissue adhesive, *Amer. J. Otol.*, 12:443–445 (1986).

284. Hayward, P.J., and I.S. Mackay, Fibrin glue in nasal septal surgery, *J. Laryngol. Otol.*, 101:133–136 (1987).

285. Linder, A., and C.E. Lindholm, Vocal cord lateralization using carbon dioxide laser and fibrin glue, *J. Laryngol. Otol.*, 106:226–230 (1992).

286. Kram, H.B., W.C. Shoemaker, S.T. Hino, H.S. Chiang, D.P. Harley, and A.W. Fleming, Tracheal repair with fibrin glue, *J. Thorac. Cardiovasc. Surg.*, 90:771–775 (1985).

287. Horowitz, M.S., C. Rooks, B. Horowitz, and M.W. Hilgartner, Virus safety of solvent/detergent-treated antihemophilic factor concentrate, *Lancet*, 11:186–188 (1988).

41
Thermodynamics of Stimuli-Responsive Polymers for Chemomechanical Membrane Systems

Mario Casolaro and Rolando Barbucci
University of Siena
Siena, Italy

I. INTRODUCTION

A chemomechanical system is a thermodynamic system that undergoes shape change and develops contractile force responding to external stimulus. During the last decade, a number of chemomechanical systems have been investigated [1–8]. These systems include polyelectrolytes in fiber and membrane forms that expand and contract upon changing their degree of ionization or solubility. The reversible dilatation and contraction can be based on the reversible ionization of carboxyl groups, for instance, of a polyacid by alternating addition of alkali or acid, by which the former produces an electrostatic repulsion along the macromolecular chain and causes an expansion of the coiled polymer [9] (Scheme 1). For nonionic and water-soluble polymers, their collapse is due to changes in temperature [10].

The reversible process may be used also to transform information as a signal or receptor. Enzymes, for example, can change their conformation and accumulate strain when they form an enzyme-substrate complex. This mechanical energy of deformation is

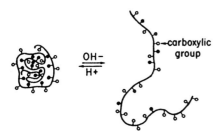

Scheme 1

partially liberated as chemical energy and utilized to promote the enzyme reaction. Thus, the chemomechanical process can be considered as a rather common phenomenon dominating the dynamics and functions of biological systems.

An application of these chemomechanical systems results in membranes having expanding and contracting pores due to the presence of polyelectrolytes grafted on the surface and walls of porous substrates [1–3,11,12]. These polymers change conformation for a process of ionization/deionization of carboxyl groups or protonation/deprotonation of amino groups due to a variation in the environmental pH.

Generally speaking, the principle of the expansion and contraction of the pores is based on a specific interaction between micro- or macromolecular solutes contained in the permeant and the polymer chains grafted onto the membrane, giving rise to a significant conformational change of the grafted polymer. This interaction may occur mainly for two types of processes: (1) physico-chemical, due to changes of pH, temperature, redox, and the like, and (2) biochemical, which occur in the presence of enzymes.

In Type 1, the polymers change conformation according to the different ionization or oxidation state, while temperature-sensitive polymers display volume phase transition properties in water as the temperature rises to a critical value called the *lower critical solution temperature* (LCST) [10]. Some applications in therapeutics and diagnostics of thermally reversible polymers based on poly(N-isopropylacrylamide) (pNIPAAm) and its hydrogels were reviewed by Hoffman [13]. Once the polymer is safely anchored to a porous membrane, it works as well as in the free form and responds to an external stimulus. The grafted polymer can act as a reversible valve, allowing the permeation of substances through the membrane pores under specified conditions [4,14].

In Type 2, the response is obtained with biological components, generally an enzyme such as glucose oxidase (GOD), which acts as signal transducer in glucose-sensitive insulin-releasing systems [5]. These systems are widely investigated for the development of polymeric membranes to control insulin delivery at rates dependent on the external concentration of glucose [3,15–17].

II. THE THERMODYNAMICS OF STIMULI-RESPONSIVE POLYMERS

A. Physicochemical Signals

1. *pH-Sensitive Polymers*

The polarity of a polymer changes greatly with ionization. This polarity change can be reflected in changes of macromolecular size and in changes of water content of a polymer [9,18]. Thus, the thermodynamic study of protonation/ionization is of particular interest for practical utilization, too, to obtain information on conformational changes of polyelectrolytes.

In the case of the protonation of typical polyelectrolyte molecules, the Henderson-Hasselbalch equation is a useful tool that accounts for the dependence of the equilibrium constant K on the charge variation of the molecule during protonation. A generalized form of this equation widely used in the characterization of protonation constants of polyelectrolytes is [19,20]

$$\log K = \log K^\circ + (n - 1) \log [(1 - \alpha)/\alpha] \qquad (1)$$

where α is the degree of protonation of the whole macromolecule and K° and n represent typical parameters of the equation. $K°$ is the value of the intrinsic protonation constant of the repeating monomer unit in the absence of interactions between protonation sites, and n accounts for deviations from ideality of the system at a given ionic strength and polymer concentration. Thus, the n value can be related to electrostatic and hydrophobic influences on the process of protonation [18]. The higher the n value of a polyelectrolyte, the greater is its hydrophilic quality.

This is only one of the several equations used to fit protonation constants of polyelectrolytes. Other approaches lead to similar, but more complicated, analytical forms or contain a series expansion of pH [21]. Both the approximations involved in the derivation of such expressions and varying the effects of protonation on different polymers, these equations often represent merely a convenient manner of fitting the variation of log K with varying charge.

A large volume of experimental work has been carried out on polyelectrolytes carrying acidic groups (i.e., carboxyl) or basic amino groups and the thermodynamic properties for the ionization of these polymers are well established [18]. In some cases, the thermodynamic changes have been calculated also for polymers in different states (i.e., hydrogels) or grafted on the surface of different materials [16,22,23]. Here, we describe a realistic approach by studying the behavior of free and grafted polymers from a thermodynamic standpoint.

We consider here the case of vinyl poly(carboxyl acid)s. They change their chain extension for electrostatic interactions between charged COO $^-$ groups that are sensitive to pH in aqueous media. The uptake of protons leads to a regular decrease in coil dimension because the neutralization of COO $^-$ groups determines a lower electrostatic repulsion between the negatively charged groups on the macromolecular chain. When the polymer is completely protonated, it assumes a compact coil conformation with a lower reduced viscosity (Fig. 1) [24]. Viscometric measurements of dilute solutions are the most convenient experimental tool for the characterization of the dimensions of flexible polymer chains. Besides the short-range and long-range interactions of electrostatic factors affecting polyion expansion, the reduced viscosity (η_{sp}/C) of polyelectrolyte solutions can be related to several other factors (e.g., aggregation, hydration, etc.), all strongly dependent on the ionic strength and polymer concentration [9,18].

The compounds of Table 1 are a series of poly(carboxyl acids) with the amido group in the side chain of a poly(vinyl) structure. These polymers are prepared by radical polymerization of the corresponding monomers, in turn obtained by acylation of acryloyl or methacryloyl chloride with the appropriate amino acid [25,26]. In all cases, the macromolecule exhibits the highest η_{sp}/C in the fully ionized state, reflecting maximum extension of the chain for electrostatic effects. As far as conformational changes are considered, the polymers with chiroptical properties take a random coil conformation, with circular dichroism (CD) spectra, consisting of two dichroic bands.

Figure 2 shows the CD spectra of poly(N-acryloyl-l-valine) at different pH. The first positive band appears near 215 nanometers (nm), when the polymer is mostly in the neutral form. In water, this band is very sensitive to ionization, which induces a red shift and a decrease in intensity [27]. A similar trend is observed for the second negative band at 190 nm. These changes are mainly attributed to the perturbation of all the electronic transitions of the amide and carboxyl chromophores, due to the ionization of the carboxyl group [28]. This behavior is similar to that reported for poly(N-methacryloyl-l-alanine), for which contributions arising from the conformation of the polymer were considered negligible [28]. Charge neutralization always showed lower electrostatic effect, as re-

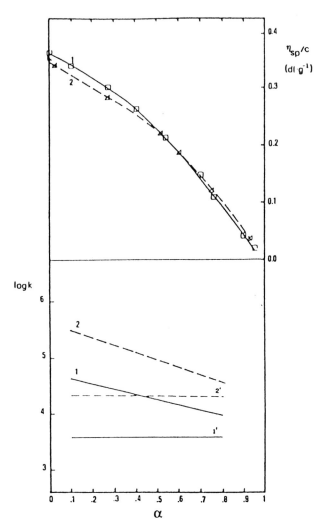

Figure 1   Reduced viscosity η_{sp}/C and basicity constants log K in relation to the degree of protonation α for polymers poly(N-methacryloyl-glycine) (PMAG) (1, ——) and poly(N-methacryloyl-β-alanine) (PMAA) (2, ---) (with corresponding monomers 1′ and 2′) at 25°C in 0.1 M NaCl.

vealed by a sharp decrease in coil size by viscometric data [29]. Hydrophobic groups on polymers may provide further chain contraction and, in some cases, the polymer precipitates because the hydrophobic moiety overcomes the hydrophilic quality of the amido and carboxyl groups at relatively low charge density [18,30].

The coiling process is always accompanied by a decreased value of the apparent basicity constant (log K) of an average ionizable group carried by the polyion. With the exception of some poly(carboxyl acid)s, showing a complex pattern of log K on the degree of protonation α, most of the polymers in Table 1 have been well described by Eq. 1. The log K values decrease linearly with increasing α. Moreover, with the acrylic and

Table 1 Structure of Poly(carboxyl acid)s and Corresponding Henderson–Hasselbalch Parameters for the Protonation Reaction in 0.1 M NaCl at 25°C

Polymer	Structure	α-Range	Log $K^{\circ a}$	n^a
Poly(acrylic acid) (PAA)	$[-CH_2-CH-]_x$ | COOH	0.10–0.80	5.32	1.96
Poly(N-acryloyl-glycine) (PAG)	$[-CH_2-CH-]_x$ | C=O | NH$-CH_2-$COOH	0.11–0.80	4.32	1.50
Poly(N-meth-acryloyl-glycine) (PMAG)	CH_3 | $[-CH_2-C-]_x$ | C=O | NH$-CH_2-$COOH	0.10–0.80	4.26	1.40
Poly(N-meth-acryloyl-β-alanine) (PMAA)	CH_3 | $[-CH_2-C-]_x$ | C=O | NH$-CH_2CH_2-$COOH	0.10–0.80	4.92	1.61
Poly(N-acryloyl-6-aminocaproic acid) (PAC)	$[-CH_2-CH-]_x$ | C=O | NH$-(CH_2)_5-$COOH	0.10–0.64	5.42	1.31
Poly(N-acryloyl-I-valine) (PAV)	$[-CH_2-CH-]_x$ | C=O | NH$-$ C H$-$COOH | C H / \ H_3C CH_3	0.10–0.66	4.59	2.26
Poly(N-acryloyl-I-leucine) (PAL)	$[-CH_2-CH-]_x$ | C=O | NH$-$ C H$-$COOH | C H$_2$ | C H / \ H_3C CH_3	0.11–0.34	4.86	1.96

Table 1 Continued

Polymer		α-Range	Log K°[a]	n[a]
Poly(N-meth-acryloyl-I-leucine) (PMAL)	CH$_3$ $\mid$ $[-CH_2- \underset{\mid}{C}-]_x$ $\mid$ C=O $\mid$ NH$-$C H$-$COOH $\mid$ C H$_2$ $\mid$ C H $/$ $\backslash$ H$_3$C CH$_3$	0.10–0.45	4.84	1.81

[a] $\log K = \log K^{\circ} + (n - 1) \log [(1 - \alpha)/\alpha]$

methacrylic polymers containing the I-leucine and I-valine moieties, the basicity constants decreased to a critical α value, depending on the nature of the amino acid residue (Fig. 3). Poly(N-acryloyl-I-valine) had wide linearity (up to $\alpha = 0.65$), while the polymers with I-leucine residues showed log K/α slope changes around $\alpha = 0.4$ [27]. Thus, the presence of a further methylene group in the side chain affects in a different way the thermodynamic behavior. The trend of log K on α for poly(N-acryloyl-I-valine) suddenly

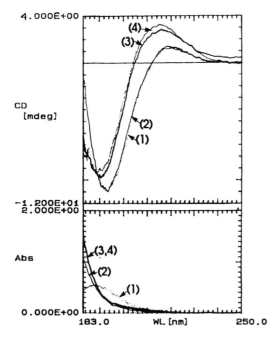

Figure 2 Circular dichroism (CD) and ultraviolet (UV) spectra of poly(N-acryloyl-I-valine) at different pH: (1), 12.5; (2), 7.4; (3), 4.0; (4), 3.5.

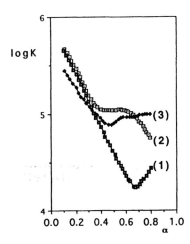

Figure 3 Basicity constants ($-COO^- + H^+ \rightleftharpoons -COOH$) in relation to α in 0.1 M NaCl at 25°C: (1), for poly(N-acryloyl-I-valine) (PAV); (2), for poly(N-acryloyl-I-leucine) (PAL); and (3) poly(N-methacryloyl-I-leucine) (PMAL).

increased at α greater than 0.65. This can be related to the decreased coil size, compelling the isopropyl groups in the side chains into close contact.

A micelle-like behavior is induced by hydrophobic interactions, bringing the polar COO^- groups closer together. Calorimetric data for the protonation of carboxylate groups revealed larger endothermicity as α approached the critical point (Fig. 4). At $\alpha >$ 0.65, a strong endothermic peak superimposes the protonation process. This was ascribed to hydrophobic forces between isopropyl groups that outweigh the repulsive electrostatic interactions of the polymer in the ionized state [9,27]. Upon the collapsing process, the hydrophobic groups release their water molecules, which are structured around the polymer chain, thus increasing the entropy. The size of the isopropyl group is sufficient to allow the macromolecule to take a compact, tightly coiled conformation in which the hydrophobic groups are clustered in microdomains without contact with water. A cloudy solution was observed when cohesive interactions predominated under conditions of almost complete neutralization of the charged groups on the polymer ($\alpha > 0.8$).

When the polymers are attached by one end to a substrate and immersed in solvent, they become systems of practical importance [31]. Equilibrium conformations of, and forces between, such polymer "brushes" have been the subject of both theoretical and experimental study [32–34]. The protonation of the carboxylate groups in polymers grafted on polyurethane (PU) or cellulose (CL) membranes has been studied in heterogeneous phase. The protonation mechanisms revealed a slow kinetics since the thermodynamic equilibrium was longer than for the protonation of carboxylates in the corresponding free polymer analogs.

Large hysteresis loops were displayed between forward and backward titrations with hydrochloric acid and sodium hydroxide titrants at shorter titrant addition times (Fig. 5). The hydrophilicity of the support played an important role in the magnitude of the polymer-substrate interaction [16,22]. It was found that the weaker the interaction was, the shorter the time necessary to reach a condition of thermodynamic equilibrium was, reflected by closer hysteresis curves [22,27]. Equilibrium configuration of the chains

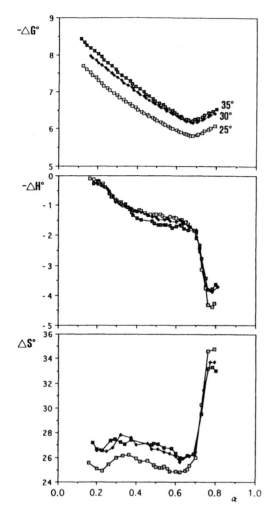

Figure 4 Thermodynamic functions ($-\Delta G°$, kcal mol^{-1}; $-\Delta H°$, kcal mol^{-1}; $\Delta S°$, cal mol^{-1} K^{-1}) for the protonation of the carboxylate groups of poly(N-acryloyl-I-valine) in relation to α in 0.1 M NaCl at different temperatures (°C).

takes longer to attain because intra- and intermolecular interactions hinder the movement of polymer segments. A similar nature (hydrophilic or hydrophobic) of polymer and substrate always leads to a stronger interaction, slowing the kinetics of protonation and thus increasing the time necessary to reach equilibrium pH at each titration point.

A typical potentiometric titration plot, with different hydrophilic components, is showed in Fig. 5. Polyurethane, being a hydrophobic substrate, interacts strongly with the hydrophobic part of the grafted polyelectrolyte. This can be seen for poly(N-methacryloyl-I-leucine) grafted on polyurethane since a larger potentiometric hysteresis loop was observed with respect to the hydrophilic cellulose grafted with poly(N-acryloyl-I-valine). The latter polymer in the free form shows an *n* value of Eq. 1 compatible with a hydrophilic nature.

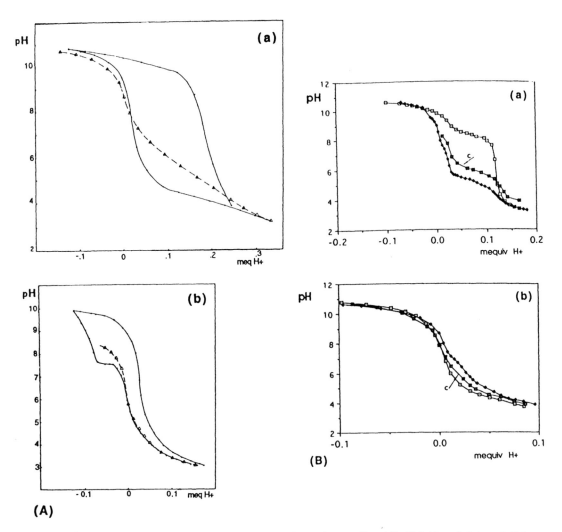

Figure 5 A, Potentiometric titration curves: (a), for poly(acrylic acid) (PAA) and (b), poly(N-acryloyl-glycine) (PAG) grafted on polyurethane film in 0.1 M NaCl at 25°C (stabilization time, 30 min for each titration point; upper curve, titration with 0.1 M NaOH; lower curve, back titration with 0.1 M HCl; the dotted lines are equilibrium curves); B, Potentiometric titration curves: (a), for poly(N-methacryloyl-l-leucine) (PMAL) grafted on polyurethane, and (b), poly(N-acryloyl-l-valine) (PAV) grafted on cellulose membranes (stabilization time, 600 seconds for each titration point) (c, equilibrium curve).

A similar high *n* value was observed for poly(acrylic acid) (PAA) for both substrate grafted and free forms (Table 2) [22,35]. The conditions of thermodynamic equilibrium were reached in different time intervals, depending mainly on the hydrophilic quality of the support. The basicity constants (log $K°$) and the *n* terms of Eq. 1 increased with increasing hydrophilicity, in manner the same as for free basic or acidic polymers [18,22].

Since practically all reactions are accompanied by a heat effect, the course of the

Table 2   Comparison of the Basicity Constants for Free and Cellulose-
or Polyurethane-Grafted Polymers at 25 °C in 0.1 M NaCl

System	$\log K^{\circ\,a}$	$(n - 1)^a$	Time to reach equilibrium pH/h
PAA-free	5.32	0.96	Fast
PAA-g-CL	5.79	0.86	16–20
PAA-g-PU	5.54	0.75	8–12
PAG-free	4.32	0.50	Fast
PAG-g-CL	4.72	0.44	16–20
PAG-g-PU	4.47	0.14	2–10
PAV-free	4.59	1.26	Fast
PAV-g-CL	–	–	3–6
PMAL-free	4.84	0.81	Fast
PMAL-g-PU	–	–	7–14

[a]$\log K = \log K^{\circ} + (n - 1) \log [(1 - \alpha)/\alpha]$

protonation can be followed by observing the heat liberated or absorbed. Such measurements can be made by titration in a calorimeter. The use of calorimetric methods for the determination of analytical species in solutions containing large amounts of insoluble compounds is well documented [36]. A logical extension of this type of application is the study of chemistry occurring at the interface between a solid and a liquid. A solution calorimeter possesses an almost unique property ideally suited to this type of measurement. The response is not significantly affected by the presence of an insoluble substrate as long as the efficient transfer of heat through the solution is not seriously attenuated. The calorimetric technique can quantify and delineate many types of surface interactions, including neutralization reactions.

As evidence, interpretation of curvature obtained in titration enthalpograms plays a key role in the qualitative assignment of the reaction mechanism. Figure 6 shows enthalpograms obtained for the ionization reaction of the poly(acrylic acid) grafted on the cellulose and polyurethane [35,37]. For comparison, the enthalpogram for ionization of the free polymer in aqueous media is also reported. The heat effect associated with the deprotonation reaction of the COOH group in the homogeneous medium is greater with respect to the heterogeneous one, even if the same groups are involved in the proton ionization chemistry. The data are compatible with the likely hypothesis of interaction between the carboxyl groups of the graft chains and the substrate [16,37]. The interaction determines an overall ionization process that is slow and endothermic, probably due to the breaking of hydrogen bonds.

The thermodynamic functions of poly(acrylic acid) grafted on polyurethane is reported in Fig. 7, along with those of the free polymer [35,38]. Unlike the free polymer, the $-\Delta H^{\circ}$ values of the grafted PAA approached zero as the degree of protonation α tended to one. The presence of the PU substrate affected the energetics of the protonation process. This process may be offset for PAA chain segments, elements of which have relative freedom of rotation, on account of changes in chain conformation caused by repulsion of negatively charged carboxylate groups. In the grafted system, the ionization of carboxyl groups of PAA differs from the similar ionization process of statistical coils in solution. The PAA chains fixed at one end to the PU surface ensure a degree of

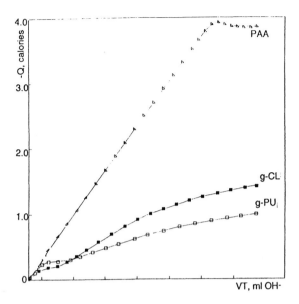

Figure 6 Enthalpograms ($-Q$ in calories versus V_T in milliliters of 0.1 M NaOH) for proton ionization of poly(acrylic acid) (PAA) in the free and grafted (cellulose, CL; polyurethane, PU) states at 25°C in 0.1 M NaCl.

stability toward conformational changes. Similar behavior was found in some porous carboxylic cationites with cross-linked methacrylic acid [39].

An infrared spectroscopic analysis of these systems clearly evidenced the presence of two different COOH groups at low pH, supporting the hypothesis of polymer-substrate interaction [16]. Figure 8 shows the difference infrared (IR) spectra of poly(N-acryloyl-glycine) grafted onto cellulose. By deconvoluting the band relative to the carboxyl C=O stretching in the spectrum of the polymer, two peaks emerged at 1735 cm^{-1} and 1719 cm^{-1}, indicating that there were two "differently behaving" carboxyl groups in the structure of the grafted polymer. The differentiated behavior of the carboxyl groups, reflected in the splitting of the characteristic C=O stretching band, may possibly be attributed to an interaction between these groups and the cellulose substrate due to their similar hydrophilic nature. As a matter of fact, the deconvoluted spectrum of the free polymer did not show the splitting feature, under the same conditions, but there was only a single carboxyl C=O stretching band at 1728 cm^{-1} [30].

On the other hand, when the grafted polymer was in the ionized state, only one band was observed in the deconvoluted spectra relative to the asymmetric or symmetric stretching of the COO$^-$ group, as happens for the free polymer [30]. This indicates that no interaction occurred between the ionized carboxyl and the substrate. Hence, we can say that, when the grafted polymer chains are in extended conformation due to electrostatic repulsion between negatively charged groups, they do not interact with the cellulose moiety, whereas in the unionized state, that is, in compact coil conformation, they do.

2. Temperature-Sensitive Polymers

An important group of water-soluble, nonionic polymers is based on N-alkyl acrylamide homopolymers and copolymers with or without basic or acidic comonomers [40,41].

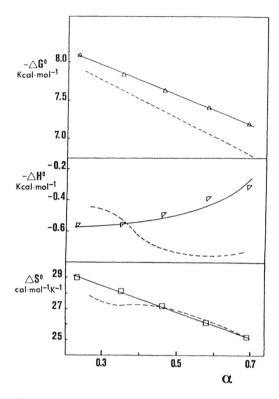

Figure 7 Thermodynamic functions in relation to the degree of protonation α for poly(acrylic acid) (PAA) in the free (---) and polyurethane-grafted (——) state (25°C in 0.1 M NaCl).

These polymers demonstrate good solubility in aqueous media at low temperatures, but separate from solution when the temperature is raised above the lower critical solution temperature (LCST) [10]. In water, the polymer becomes more elongated, presumably as a result of hydrogen bonding between the amide group of the polymer and surrounding water molecules. As the solution temperature rises, H-bonds break and more and more water molecules are released. This release provokes, on a macroscopic scale, the separation of a polymer-rich phase concomitant, on the molecular level, with a collapse of the polymer chains that undergo a change from expanded coils to a more compact conformation [42,43].

The LCST is found in many polymer solutions characterized by hydrogen bonds between solutes and solvents. The formation of these bonds lowers the free energy of solution and the specific molecular orientations required lead to negative entropy changes. This phenomenon is particularly important in aqueous media in which a further negative entropy change is contributed by the hydrophobic effect [44,45]. As hydrogen bonds are thermally labile, a rise in temperature reduces the number of bonds and causes eventual phase separation. A coil-to-globule transition in dilute solutions is observed above the LCST when the enthalpic contribution to the free energy is dominated by the growing entropic component [43,45,46].

Among these kinds of polymers, poly(N-isopropylacrylamide) (pNIPAAm) has received the attention of most investigators [13,45,47–55]. In studying this polymer in

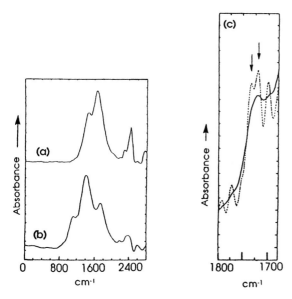

Figure 8   Attenuated total reflection/Fourier transform infrared (FTIR) difference spectra of poly(N-acryloyl-glycine) (PAG) grafted on cellulose: (a), at pH 10; (b), at pH 2; and (c) deconvoluted spectrum at pH 2 in the 1800–1680 cm^{-1} range.

water, Kubota, Fujishjige, and Ando have noted that the pNIPAAm molecule behaves like a flexible coil and the unique thermal behavior of its solution should be attributed mainly to the hydrated structure [42]. A major reason for the great interest in pNIPAAm is having an LCST around 32°C, which allows incorporation of proteins in the aqueous system of this polymer. The reversible phase separation behavior of pNIPAAm has been utilized to make soluble-insoluble enzyme, antibody, and other protein conjugates [56,57]. Having an LCST approaching that of body temperature made pNIPAAm also an attractive system for studies in pharmaceutical applications such as controlled drug delivery and solute separation [58-60]. In solute separation, cross-linked hydrogels have been developed in membrane forms that undergo abrupt, reversible changes in volume in response to minor changes in the gel's environment.

Moreover, the behavior of pNIPAAm grafted on a porous membrane has been recently studied by Iwata et al. in view of constructing temperature-sensitive devices to regulate filtration characteristics [4]. An important and useful feature of thermosensitive polymers is the possibility of controlling their LCST by varying the monomer composition [41,61]. In general, the incorporation of hydrophobic comonomers leads to a lower LCST, while hydrophilic or ionizable comonomers increase the LCST. For the most part, experimental characterization of these LCST phenomena has been limited to simple cloud-point measurements, an approach complicated by variations in precipitated aggregate sizes and settling of precipitates [43,46,62,63].

A more powerful technique that may be applied to the study of the LCST is solution calorimetry [10,43,62-64]. This method provides thermodynamic parameters that lend insight into the forces responsible for the transition. Use of a sensitive scanning microcalorimeter, developed to study structural transitions in proteins, lipids, and nucleic acids, allows the LCST transition parameters to be determined with precision [65–68]. Endo-therms with enthalpies in the order of the strength of hydrogen bonds were observed at temperatures with LCSTs detected by classical cloud-point measurements for an aqueous solution of pNIPAAm. The presence of charge in copolymers also has a major effect [41,61,69–71].

The enthalpy of phase separation decreased with increasing hydrophilicity of comon-omers in copolymers with pNIPAAm. This was found to be a linear function of the LCST [41]. A higher LCST leads to a reduced enthalpy of phase separation due to a smaller amount of structured water molecules at higher temperatures. The amount of structured water is only a function of temperature, while the phase separation temperature is deter-mined by the relative hydrophilicity of the polymer, with charge merely increasing the polymer hydrophilicity [41].

Recently, we studied the thermodynamic behavior of pNIPAAm during the proton-ation of carboxylate groups in copolymers with N-acryloyl-I-valine, which, presenting a structure closer to NIPAAm, revealed copolymers with a random distribution of comon-omers. At a temperature below the LCST, the log K values showed a striking dependence on the number of charged comonomers (Fig. 9). This reflected lower electrostatic effects between neighboring units as the number of charges decreased. With a lower amount of the charged comonomer present in the copolymer, the log K decreased slightly in increas-ing α. In all cases, however, a minimum value of log K was evidenced at a critical α, as described above. As the temperature rose, the critical α shifted to a lower value (Fig. 10). Moreover, the increased log K at temperatures of 30°C and 35°C may be ascribed to the lower hydration of the whole macromolecule, which results in a decreased ionization of the carboxyl groups.

A similar trend was observed for poly(N-acryloyl-I-valine) (Fig. 4). This is consistent with the behavior of polyelectrolytes incorporating hydrophobic comonomers [71–75].

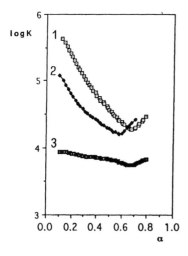

Figure 9   Basicity constants log K in relation to the degree of protonation α for poly(NIPAAm-*co*-AV) with different acryloyl-I-valine (AV) contents at 25°C in 0.1 M NaCl: (1), 100%; (2), 62%; (3), 12%.

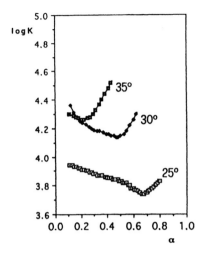

Figure 10   Basicity constants log K in relation to α for poly(NIPAAm-*co*-AV) (AV = acryloyl-I-valine, content 12%) in 0.1 M NaCl at different temperatures (°C).

This phenomenon may be due to the decreased dielectric constant in the polymer environment [72,73] for the lower amount of structured water around the more hydrophobic polymer [71].

Calorimetric titration data for the protonation of carboxylate groups in poly(N-acryloyl-I-valine) and its copolymers with N-isopropylacrylamide revealed the peculiar behavior of hydrophobic interactions (Fig. 11) [76]. Any increase in the content of

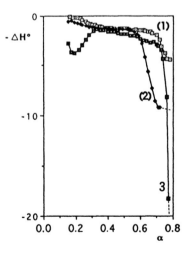

Figure 11   Enthalpy changes ($-\Delta H°$, kcal mol^{-1}) in relation to α for poly(NIPAAm-*co*-AV) with different AV (acryloyl-I-valine) contents at 25°C in 0.1 M NaCl: (1), 100%; (2), 62%; and (3), 12%.

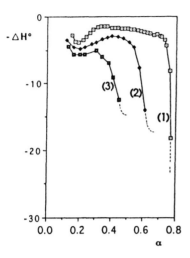

Figure 12   Enthalpy changes ($-\Delta H^\circ$, kcal mol^{-1}) in relation to α for poly(NIPAAm-*co*-AV) (AV = acryloyl-l-valine, content 12%) in 0.1 M NaCl at different temperatures: (1), 25°C; (2), 30°C; and (3), 35°C.

NIPAAm units led to strong endothermicity at a constant temperature of 25°C. As the temperature was increased (Fig. 12), thus approaching the LCST of pNIPAAm, the released water molecules decreased because of the lower hydration of the amide groups.

This is consistent with the LCST phenomenon. As suggested, the LCST behavior is caused by a critical hydrophilic/hydrophobic balance of polymer side groups [40,62,71,77,78]. For pNIPAAm, the amide groups are hydrophilic and the isopropyl groups hydrophobic. The proposed mechanism for the LCST behavior is described next.

At low temperatures, the strong hydrogen-bonding between hydrophilic groups and water outweighs the unfavorable free energy related to the exposure of hydrophobic groups to water, leading to good solubility of the polymer in water. At increasing temperatures, H-bonding weakens, while hydrophobic interactions between hydrophobic side groups increase. Above the LCST, interactions between hydrophobic groups become dominant, leading to entropy-driven polymer collapse and phase separation. The reduced motional freedom of the polymer chain is compensated by a gain in entropy due to the release of structured water around to the hydrophobic groups on the polymer [41].

3.   Redox-Sensitive Polymers

Another kind of stimuli-responsive polymer is compounds with redox properties. They have been investigated for various purposes, such as polymeric reagents or electronic materials [79]. Two classes of such polymers can be distinguished, those characterized by conjugated polymer backbones or by separated redox-active moieties [80]. Two reagents based on lipoamide and nicotinamide derivatives have been considered with a view to application in the biomedical field and as components of drug delivery systems [17,81–83]. The nicotinamide moiety is known to be involved in electron transfer processes of coenzymes in living systems [84]. The reversible redox reaction of the nicotinamide-containing polymer, poly[3-carbamoyl-1-(p-vinylbenzyl)pyridinium chloride] (CVPy), involves the following two species:

Reduced form (hydrophobic)	*Oxidized form* (hydrophilic)

Unlike the reduced form, which is hydrophobic, the oxidized and hydrophilic form of the polymer is freely soluble only in water [85]. It shows a large η_{sp}/C since the presence of positive charges reflects chain extension for electrostatic repulsion. Thus, the polymer behaves as a flexible coil. Ishihara, Kobayashi, and Shinohara reported on the application of this polymer in membranes that act as a regulator of the permeation rate of insulin by means of an oxidation reaction with the hydrogen peroxide formed by the enzymatic reaction of glucose in the presence of glucose oxidase (GOD) [17]. Furthermore, Nishi, Ito, and Imanishi grafted this polymer onto a porous membrane to control the water permeation in different oxidation states [86].

Another redox polymeric system of particular interest in the biomedical field involves compounds containing the five-membered disulfide ring of lipoic acid and its lipoamide derivatives [81–83,87–89]. The five-membered cyclic disulfide is known to act as a coenzyme in oxidative acyl transfer and redox reactions in living systems [90]. These compounds are used as precursors to introduce −SH groups into polymers, improving the powerful reducing character for disulfide bridges in proteins [81,91,92]. Their ability to reduce and maintain thiol in the reduced state arises from the low redox potential for the formation of a sterically favorable dithiolane ring on oxidation [81]. However, the removal of the reducing agent after reduction is usually time consuming and reoxidation may occur during the treatment.

This difficulty could be overcome if the reducing agent was attached to an insoluble carrier. Attachment of lipoic acid to various insoluble polymeric matrices (e.g., cellulose, sephadex or polyacrylamide) results in a potent reducing reagent, the action of which is depicted in Scheme 2. When the disulfide bridge is reduced to a dihydro-type species, it becomes a powerful reducing agent for disulfide compounds similar to dithiothreitol [93,94]. As reported by Gorecki and Patchornik, these insoluble, thiolated polymeric systems were used for the quantitative reduction of oxidized glutathione and cystine as well as for activation of papain. The advantage of this method is an easy removal of the insoluble polymeric reagent [81].

Recently, we reported a synthetic pathway to introduce the −S−S− moiety in polymers by an exchange reaction of the benzotriazole residue in poly(N-alkyl-benzotriazole) with monoamidated lipoyl derivatives [83,85]. These are obtained through an excess of simple diamines reacting with the imidazolide (Scheme 3). The imidazolide, that was not isolated from the alcohol-free chloroform solution, was obtained in the activation of α-lipoic acid with 1,1-carbonyldiimidazole. Insoluble polymers obtained though the exchange reaction were quantitative, operating with only twofold excess of monoamidated lipoyl-derivative. Soluble copolymers with N-vinylpyrrolidinone units were also obtained [83].

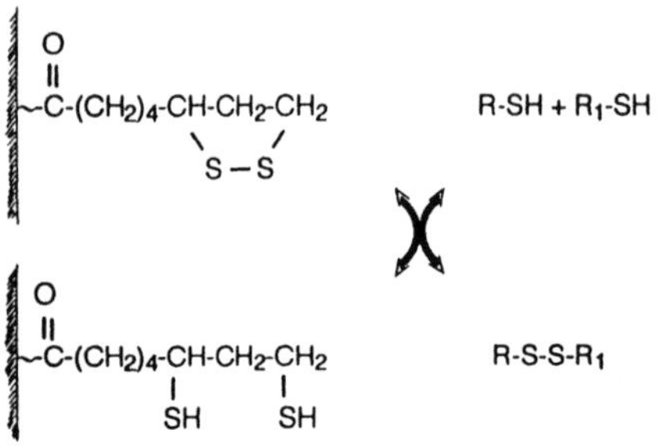

Scheme 2

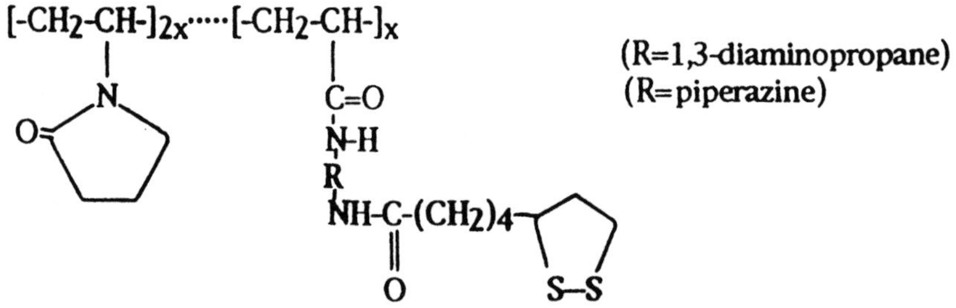

(R=1,3-diaminopropane)
(R=piperazine)

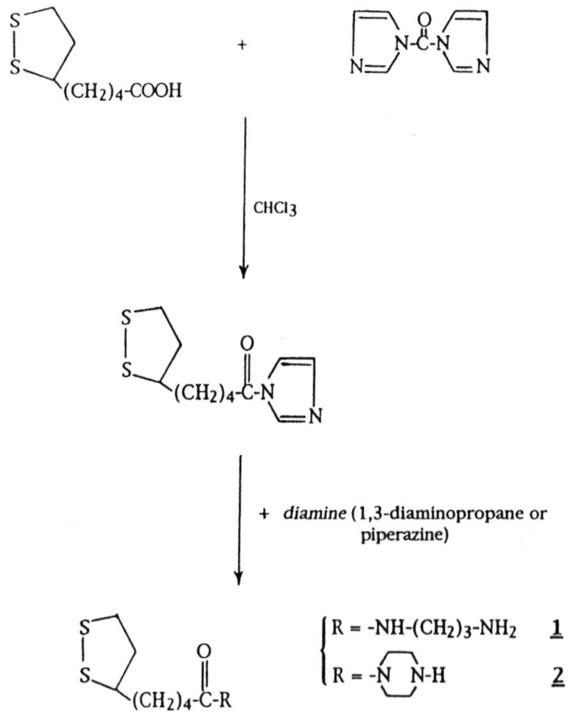

$$R = -NH-(CH_2)_3-NH_2 \quad \underline{1}$$
$$R = -N\underset{}{\diagup\!\!\!\diagdown}N-H \quad \underline{2}$$

Scheme 3

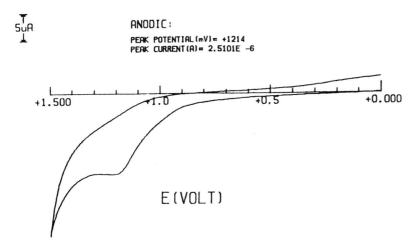

Figure 13 Cyclic voltammogram of the copolymer incorporating N-lipoyl-1,3-diaminopropane residues recorded at a platinum electrode for a dimethyl sulphoxide (DMSO) solution (scan rate = 200 mV sec^{-1}).

The copolymer carrying the N-lipoyl-1,3-diaminopropane residue showed a cyclic voltammetric response at a platinum electrode with only a single oxidation process without a directly associated response in the reverse scan (Fig. 13). The cyclic voltammograms thus consist of a single redox peak having an anode potential that slowly changes with scan rates (Table 3). The cyclic voltammogram is essentially identical to that of the corresponding simple monoamidated compounds except that the oxidation potential is shifted to higher values, increasing with decreasing scan rate [83].

Table 3 Redox Potentials of Low Molecular Weight and Polymeric Compounds (Pt electrode vs ECS)

| Compound | Solvent | Voltage, mV | | Sweep rate, mV sec^{-1} |
		E_a	E_c	
1	DMSO	1070	—	200
	CH$_3$CN	—	−1400	200
Copolymer with 1	DMSO	1214	—	200
		1238	—	100
2	DMSO	1004	—	200

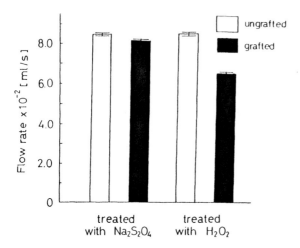

Figure 14   Redox-dependent water permeation through the polyethersulphone porous membrane grafted with poly[3-carbamoyl-1-(p-vinylbenzyl)pyridinium chloride] (CVPy) (pore size, 0.8 μm).

4. *Practical Application of Stimuli-Responsive Polymers*

For practical application of the stimuli-responsive polymers, the redox, temperature, and pH dependence of the water permeability of the graft membranes with the stimuli-sensitive polymers was investigated. It was found that the water, as well as solutes of different molecular weights, permeates the membrane pores at rates strongly dependent on the conformational state of grafted polymers [2–4,11,16,22,27,29,95].

Figure 14 shows the results obtained for the water flow rate through the redox CVPy-grafted polyethersulphone membrane having a pore size of 0.8 micrometers (μm) [85]. The permeation of water containing oxidizer species (H_2O_2, 50 mM) through the membrane grafted with the polymer in the oxidized form was lower with respect to the membrane without graft chains. Moreover, permeation of water containing reducer species ($Na_2S_2O_4$, 20 mM) was found to be higher since the polymer collapsed to the reduced form. The positive charges present on the oxidized polymer chain enhanced chain extension and limited pore size, slowing the water flow, while the reduced state made the polymer chains collapse in a coil.

The pore size of 0.8 μm is probably too large for a suitable control of filtration characteristics. Moreover, the polymer chain length plays a significant role in determining permselectivity. Thus, several studies were carried out on membranes with a reduced pore size of 0.2 μm.

Iwata and Matsuda reported on the ability of ultrafiltration of macromolecular solutes such as dextran and albumin through poly(vinylidene fluoride) (PVdF, pore size 0.22 μm) grafted with poly(acrylic acid) [2]. They found a quick response and a good reversibility and reproducibility on pH changes. This may be related to the different hydrophilic nature of the substrate with respect to the graft polymer, as discussed above for pH-sensitive polymers.

Moreover, Iwata et al. developed temperature-sensitive membranes based on pNIPAAm and its copolymers with hydrophilic (acrylamide) and hydrophobic (n-butyl methacrylate) comonomers grafted on PVdF and demonstrated the water flow rate may be "switched" on and off reversibly as the temperature was cycled [4]. The temperature dependence of permeability changes according to the LCST change of the copolymer.

Similar filtration characteristics were observed on cellulose membranes grafted with pH- and temperature-sensitive poly(N-acryloyl-I-valine). This polymer renders the system suitable for multiple-response stimuli [27,29]. The filtration rate was reversible and had a fast response time when the pH was cycled between acidic and alkaline conditions. In fact, the permeation of water containing hydrochloric acid was higher with respect to water containing sodium hydroxide. In the latter case, the filtration process was instantly blocked.

Figure 15 shows the effect of pH and temperature on the water flow rate through porous cellulose membranes grafted with poly(N-acryloyl-I-valine). Water flow rates could be controlled by changing pH and/or temperature [27]. High pH led to an extended and hydrated macromolecule that obstructed the membrane pores as a barrier to the free water permeation. As the pH fell below the log K value of poly(N-acryloyl-I-valine), the expanded coil of the grafted polymer chain retracted, allowing increased water flow. The water start to flow at a pH less than 4.4, when the carboxylate groups on the chain are mostly protonated. Further protonation led to a sharp increase in flow rate. In the narrow pH range from 4.4 to 4.1, the graft polymer abruptly changed from a hydrated and extended coil to the collapsed state. The above range of pH should be correlated to the range of degree of protonation α for a better understanding.

Figure 16 shows the water flow rate in relation to α as calculated by the log K values of the free polymer at the different temperatures considered [27]. It is interesting to note that the water starts to flow when the polymer is partially charged. This means that it is not necessary to have fully ionized extended or neutral collapsed chains to avoid or enhance water permeation through the pores of the membrane [2,11,95]. Moreover, the increasing water flow is limited to a very narrow range of α (i.e., the range corresponding to the critical value at which the thermodynamic functions changed drastically).

At low pH ($\alpha > 0.65$ for the free polymer), the graft polymer chains shrank and precipitated on the membrane surface, resulting in an overall opening of the membrane pores. Increasing the temperature from 25°C to 30°C led the water flow rate/α plot to shift at higher α values. This is consistent with the higher log K values at 30°C (Fig. 4).

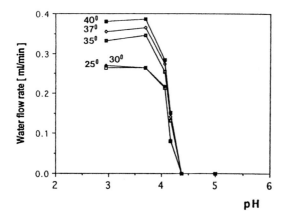

Figure 15 Effect of pH on the water flow rate through poly(N-acryloyl-I-valine) grafted on a porous cellulose membrane (pore size, 0.2 μm) at different temperatures (°C).

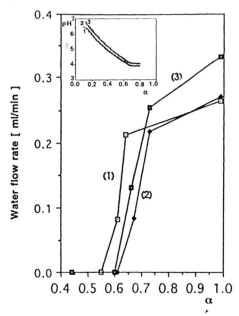

Figure 16 Water flow rate in relation to α for poly(N-acryloyl-I-valine) grafted on a porous cellulose membrane at different temperatures: (1), 25°C; (2), 30°C; and (3), 35°C. The inset shows the relation between pH and α for the polymer at the same temperatures.

Further increase of temperature to 35°C results in a sensitive increase of water flow rate. This may be related to the LCST phenomenon, having the polymer the same charge state for the similarity of log K values at 30°C and 35°C. A plateau in the water flow rate/pH profile occurred when the permeant water at low pH determined a completely uncharged polymer (Fig. 15). The plateau was only shifted to increase the water flow as the temperature was raised above a critical value. This allowed further chain shrinkage in the protonated state of poly(N-acryloyl-I-valine) due to the LCST phenomenon.

The water permeability of this system at a constant pH of 3 in relation to temperature is shown in Fig. 17. At this pH, the neutral form of the graft polymer precipitated and remained stacked on the surface of the membrane, making water permeability easier. As the temperature rose above 30°C, a further increase in water flow rate was observed. This critical temperature is very close to that observed for pNIPAAm. Beyond 31°C, the filtration rate gradually increases with the increase of temperature, reflecting a similar behavior already observed by Iwata et al. for pNIPAAm grafted on porous PVdF membrane [4].

The permselectivity of pH-sensitive systems was investigated for the ability of controlling solute separation of different molecular weight polyoxyethylene (POE) through the hydrophilic porous (0.2-μm) cellulose membrane grafted with poly(acrylic acid) as a function of graft content [3]. Figure 18 shows the results obtained. The higher the graft content is, the lower the permeation will be and more solute will be permeated at low pH. Permeation was also controlled by pore size; the POE with a molecular weight in the range of 6000 to 20,000 was found to be the most suitable solute for permeation studies. Thus, the graft density as well as chain lengths play a role in controlling hydraulic and diffusional characteristics.

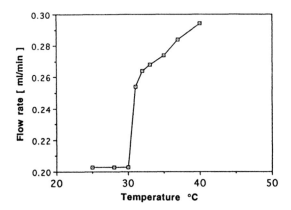

Figure 17   Effect of temperature (°C) on the water flow rate through poly(N-acryloyl-l-valine) (PAV) grafted on cellulose (pH 3).

The problem of chain length, which is the most intriguing problem in this contest, was recently approached by Ito et al. [95]. They revealed the mechanism of permeability control by pore size, as well as graft chain length and concentration, for poly(acrylic acid) grafted onto a straight-pored polycarbonate membrane. The density and the length of the graft chains were controlled by the conditions of graft copolymerization, that is, the time of glow-discharge treatment and the monomer concentration in the polymerization, respectively. Graft chains with too high a density or excessive length restricted the mobility of the poly(acrylic acid) chain, and the pore size became pH independent [5,95].

B.   Biochemical Signals

An enzyme does not affect the thermodynamic equilibrium position of a chemical reaction, merely its rate. Recently, several enzymes were used as components of drug delivery systems for their catalytic specificity [96]. For example, an ingenious system was designed by Heller et al. to release hydrocortisone from a pH-sensitive erodible copolymer combined with urease [97,98]. The enzyme, being sensitive to the concentration of external urea, made the system produce ammonium bicarbonate and ammonium hydroxide, caus-

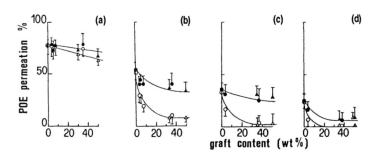

Figure 18   Permeability of polyoxyethylenes (POEs) through poly(acrylic acid) (PAA) grafted on cellulose at pH 7 (lower curve) and pH 2 (upper curve). Molecular weight of POE: (a), 1000; (b), 6000; (c), 20,000; (d), 80,000.

ing the medium pH to increase. Polymer erosion and drug release were accelerated by the product of the enzymatic reaction [99].

While this system has no foreseeable therapeutic application, it established the feasibility of using immobilized enzymes to induce a local pH change, and was the first example of a synthetic self-regulated system. Enormous progress has been made in the development of methods for the immobilization of biological macromolecules, including enzymes, onto solid supports [100,101]. The most obvious benefit to be gained from the use of an immobilized enzyme reagent is the removal of the two problems that have hindered the development of enzymatic methods of analysis. Primarily, the exploitation of the catalytic properties of the enzyme as a reusable reagent considerably reduces the cost of an enzymatic analysis, which in homogeneous solution can be prohibitive. Further, stabilization of an enzyme often occurs in an immobile form, which considerably enhances the resistance of the enzyme to environmental changes.

Enzymes as signal transducers have mostly been applied to the design of glucose-sensitive insulin-releasing systems [15,17,102–105]. In the past, several polymeric membranes based on hydrogels were developed in combination with the enzyme glucose oxidase (GOD) [15,17,106] to control delivery of insulin at rates dependent on the external glucose concentration.

Figure 19 illustrates an alternative system that is based on the "chemical valve" principle. A porous membrane with a specific poly(carboxyl acid) grafted on the surface is prepared as a pH-responsive membrane [1,3]. The enzyme GOD is immobilized on the carboxyl groups of the polymer to make a composite responsive to glucose concentrations. In the absence of glucose and at physiological pH, the carboxyl groups of the polymer are ionized and the repulsion between negative charges should make the polymer chains in an extended conformation, obstructing the membrane pores. The presence of glucose leads the polymer chains to coil since the transformation of glucose to gluconic acid operated by the enzyme GOD lowers the pH of the microenvironment, allowing protonation of the carboxylate groups on the polymer. The pore opens for a prompt permeability of insulin [3,16,107]. The GOD catalyzes the conversion of glucose to gluconic acid and hydrogen peroxide. Both these species can specifically impart external signals to polymers bearing sensitive functional groups. The gluconic acid is a relatively strong acid [108] and can protonate carboxylate or amino groups present on a polymer if

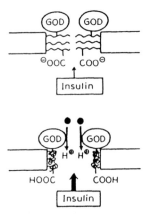

Figure 19   Principle of controlled release system for insulin (GOD = glucose oxidase).

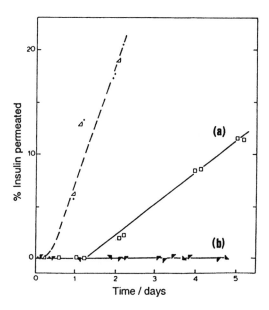

Time / days

Figure 20   Percentage (%) of insulin permeated through poly(N-acryloyl-glycine) (PAG) grafted on porous cellulose membranes: (a), at pH 2.7; and (b), at pH 8.6. Broken line shows the performance of a nongrafted membrane under the same conditions.

their basicity constants are higher [18]. In a similar way, the hydrogen peroxide can impart an oxidation reaction with the redox active moiety of the polymer [17].

In the "chemical valve" system reported above, the insulin molecule was used as a probe to control its permeability through porous cellulose membranes grafted with poly(carboxyl acids) by using the self-regulating system showed in Fig. 19. Preliminary experiments for the insulin permeability in relation to time at two chosen pH's (acid and alkaline regions) always revealed an increasing linearity for both grafted and nongrafted membranes (Fig. 20) [3,16]. Unlike for nongrafted membranes, different slopes were observed at the two pH's for grafted membranes.

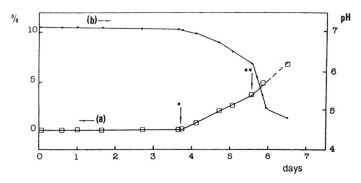

Figure 21   Time response of insulin permeation: (a), percentage (%) of insulin permeated through cellulose membrane grafted with poly(N-acryloyl-glycine) (PAG) and immobilized with glucose oxidase (GOD); (b), pH change profile with glucose addition (* = 50 mg/100 ml; ** = 400 mg/100 ml).

The acidity-enhanced permeation may be related to the conformational change of the graft chains acting as gates. The graft polymer chains are mostly ionized at a pH higher than the log K value of poly(carboxyl acids). This causes repulsion between charged chains, obstructing the membrane pores and preventing insulin permeation. On the other hand, low pH enhanced collapse of the polymer onto the substrate, increasing pore size and enabling faster permeation.

The presence of the enzyme GOD ensures a self-regulating insulin delivery device that resembles an artificial pancreas [107]. Many approaches have been made for this purpose by using hydrogel-based membranes that were sensitive to different external stimuli [5]. In all cases, the control of insulin permeation was associated with the water uptake by the charged groups present on polymers or by the LCST phenomenon for thermosensitive polymers. The glucose-sensitive membrane system of Fig. 19 is based on the chemical valve principle and contains the enzyme GOD immobilized on the graft chains.

Figure 21 shows the time response of insulin permeated with associated pH changes when glucose concentration was at two concentration levels: normal and diabetic. The addition of glucose in the downstream compartment of the reservoir led to an instant increase in the slope of the insulin-permeated/time plot [3,16,107].

REFERENCES

1. Osada, Y. *Adv. Polym. Sci.* 82:1 (1987).
2. Iwata, H., and Matsuda, T. *J. Membr. Sci.* 38:185 (1988).
3. Ito, Y., Casolaro, M., Kono, K., and Imanishi, Y. *J. Controlled Release* 10:195 (1989).
4. Iwata, H., Oodate, H., Uyama, Y., Amemiya, H., and Ikada, Y. *J. Membr. Sci.* 55:119 (1991).
5. Imanishi, Y., ed. *Synthesis of Biocomposite Materials Chemical and Biological Modifications of Natural Polymers*, CRC Press, Boca Raton, FL, 1992.
6. Harland, R.S., and Prud'homme, R.K. *Polyelectrolyte Gels Properties, Preparation, and Applications*, ACS Symp. Series 480, 1992.
7. Takagi, T., Takahashi, K., Aizawa, M., and Miyata, S. *Proceedings of the First International Conference on Intelligent Materials*, Technomic Publ., Lancaster, PA, 1993.
8. Maekawa, S., Gong, J.P., and Osada, Y. *Makromol. Chem., Rapid Comm.* 15:73 (1994).
9. Morawetz, H., ed. *Macromolecules in Solution*, Wiley-Interscience, New York, 1980.
10. Heskins, M., and Guillet, J.E. *J. Macromol. Sci., Chem.* A2(8):1441 (1968).
11. Ito, Y., Inaba, M., Chung, D.J., and Imanishi, Y. *Macromolecules* 25:7313 (1992).
12. Okahata, Y., Noguchi, H., and Seki, T. *Macromolecules* 20:15 (1987).
13. Hoffman, A.S. *J. Controlled Release* 6:297 (1987).
14. Bae, Y.H., Okano, T., and Kim, S.W. *J. Controlled Release* 9:271 (1989).
15. Kost, J., Horbett, T.A., Ratner, B.D., and Singh, M. *J. Biom. Mat. Res.* 19:1117 (1985).
16. Barbucci, R., Casolaro, M., and Magnani, A. *J. Controlled Release* 17:79 (1991).
17. Ishihara, K., Kobayashi, M., and Shinohara, I. *Makromol. Chem., Rapid Comm.* 4:327 (1983).
18. Barbucci, R., Casolaro, M., and Magnani, A. *Coord. Chem. Rev.* 120:29 (1992).
19. Katchalsky, A., and Spitnik, P. *J. Polym. Sci.* 2:432 (1947).
20. Barbucci, R., Casolaro, M., Danzo, N., Barone, V., Ferruti, P., and Angeloni, A. *Macromolecules* 16:456 (1983).
21. Mandel, M. *Europ. Polym. J.* 6:807 (1970).
22. Casolaro, M., and Barbucci, R. *Coll. Surfaces A: Phys. Chem. Eng. Asp.* 77:81 (1993).
23. Khare, A.R., and Peppas, N.A. *J. Biomater. Sci. Polymer Ed.* 4:275 (1993).
24. Barbucci, R., Casolaro, M., Magnani, A., and Roncolini, C. *Macromolecules* 24:1249 (1991).

25. Iwakura, Y., Toda, F., and Suzuki, H. *J. Org. Chem.* 32:440 (1967).
26. Barbucci, R., Casolaro, M., Magnani, A., Roncolini, C., and Ferruti, P. *Polymer* 30:1751 (1989).
27. Casolaro, M. *Reactive Polymers*, 23:71 (1994).
28. Morcellet, S.J., Morcellet, M., and Loucheux, C. *Macromolecules* 17:452 (1984).
29. Casolaro, M., Busi, E., and Landi, F. The Development of Science for the Improvement of Human Life, *Proceedings of the Second Kyoto-Siena Symposium*, Kyoto (Japan), 1994.
30. Barbucci, R., Casolaro, M., and Magnani, A. *Makromol. Chem.* 190:2627 (1989).
31. Milner, S.T., Witten, T.A., and Cates, M.E. *Macromolecules* 21:2610 (1988).
32. de Gennes, P.G. *Macromolecules* 13:1069 (1980).
33. van Lent, B., Leermakers, F., and Schentjens, J. *Macromolecules* 20:1692 (1987).
34. Hadziioannou, G., Patel, S., Granick, S., and Tirrell, M. *J. Amer. Chem. Soc.* 108:2869 (1986).
35. Casolaro, M. In *Chemistry and Properties of Biomolecular Systems*, N. Russo, J. Anastassopoulou, and G. Barone, eds., Kluwer, Dordrecht, 2, p. 127, 1994.
36. Vaughan, G.A., ed. *Thermometric and Enthalpimetric Titrimetry*, Van Nostrand Reinhold, London, 1973.
37. Barbucci, R., Casolaro, M., and Magnani, A. *Clinical Materials* 11:37 (1992).
38. Crescenzi, V., Quadrifoglio, F., and Delben, F. *J. Polym. Sci.: Part A2*, 10:357 (1972).
39. Chernova, A., Yurkenko, V.S., Pisarev, O.A., and Samsonov, G.V. *Polym. Sci., USSR*, 20: 417 (1978).
40. Bae, Y.H., Okano, T., and Kim, S.W. *J. Polym. Sci., Polym. Phys.* 28:923 (1990).
41. Feil, H., Bae, Y.H., Feijen, J., and Kim, S.W. *Macromolecules* 26:2496 (1993).
42. Kubota, K., Fujishige, S., and Ando, I. *Polym. J.* 22:15 (1990).
43. Yamamoto, I., Iwasaki, K., and Hirotsu, S. *J. Phys. Soc. Jpn.* 58:210 (1989).
44. Tanford, C. *The Hydrophobic Effect*, 2nd Ed., Wiley, New York, 1973.
45. Schild, H.G., and Tirrel, D.A. *J. Phys. Chem.* 94:4352 (1990).
46. Fujishige, S., Kubota, K., and Ando, I. *J. Phys. Chem.* 93:3311 (1989).
47. Bae, Y.H., Okano, T., Hsu, R., and Kim, S.W. *Makromol. Chem., Rapid Commun.* 8:481 (1987).
48. Inowata, H., Yagi, Y., Otake, K., Konno, M., and Saito, S. *Macromolecules* 22:3494 (1989).
49. Winnik, F.M. *Macromolecules* 23:233 (1990).
50. Ringsdorf, H., Venzmer, J., and Winnik, F.M. *Macromolecules* 24:1678 (1991).
51. Otake, K., Karaki, R., Ebina, T., Yokoyama, C., and Takahashi, S. *Macromolecules* 26: 2194 (1993).
52. Park, T.G., and Hoffman, A.S. *Macromolecules* 26:5045 (1993).
53. Ringsdorf, H., Simon, J., and Winnik, F.M. *Macromolecules* 25:7306 (1992).
54. Winnik, F.M., Ottaviani, M.F., Bossmann, S.H., Garibay, M.G., and Turro, N.J. *Macromolecules* 25:6007 (1992).
55. Winnik, F.M., Ottaviani, M.F., Bossman, S.H., Pan, W., Garibay, M.G., and Turro, N.J. *J. Phys. Chem.* 97:12998 (1993).
56. Chen, J.P., Yang, H.J., and Hoffman, A.S. *Biomaterials* 11:625 (1990).
57. Takouchi, S., Omodaka, I., Hasegawa, K., Maeda, Y., and Kitano, H. *Makromol. Chem.* 194:1991 (1993).
58. Hoffman, A.S., Afrassiabi, A., and Dong, L.C. *J. Controlled Release* 4:213 (1986).
59. Gutowska, A., Bae, Y.H., Feijen, J., and Kim, S.W. *J. Controlled Release* 22:95 (1992).
60. Palasis, M., and Gehrke, S.H. *J. Controlled Release* 18:1 (1992).
61. Feil, H., Bae, Y.H., Feijen, J., and Kim, S.W. *Macromolecules* 25:5528 (1992).
62. Taylor, L.D., and Cerankowski, L.D. *J. Polym. Sci., Part A: Polym. Chem.* 13:2551 (1975).
63. Wolf, B.A. *Pure Appl. Chem.* 57:323 (1985).
64. Maderek, E., and Wolf, B.A. *Polym. Bull.* 10:458 (1983).
65. Krishnan, K.S., and Brandts, J.F. *Methods Enzymol.* 49:3 (1978).

66. Hinz, H.J. *Methods Enzymol.* 130:59 (1986).
67. Battistel, E., Luisi, P.L., and Rialdi, G. *J. Phys. Chem.* 92:6680 (1988).
68. Jackson, W., and Brandts, J. *Biochemistry* 9:2294 (1970).
69. Ilavski, M., Hrouz, J., and Havlicek, I. *Polymer* 26:1514 (1985).
70. Beltran, S., Hooper, H.H., Blanch, H.W., and Prausnitz, J.M. *J. Chem. Phys.* 92:2061 (1990).
71. Urry, D.W. *Prog. Biophys. Mol. Biol.* 57:23 (1992).
72. Siegel, R.A., and Firestone, B.A. *Macromolecules* 21:3254 (1988).
73. Pradny, M., and Kopecek, J. *Makromol. Chem.* 191:1887 (1990).
74. Wen, S., Xiaonan, Y., and Stevensón, W.T.K. *Biomaterials* 12:479 (1991).
75. Nylund, R.E., and Miller, W.G. *J. Am. Chem. Soc.* 87:3537 (1965).
76. Bekturov, E.A., and Bakauova, Z.Kh., eds. *Synthetic Water-Soluble Polymers in Solution*, Huthig and Wepf Verlag, Basel (Germany), 1986.
77. Otake, K., Inomata, H., Konno, M., and Saito, S. *Macromolecules* 23:283 (1990).
78. Inomato, H., Goto, S., and Saito, S. *Macromolecules* 23:4887 (1990).
79. Manecke, G., and Storck, W. In *Encyclopedia of Polymer Science and Engineering*, Vol. 5, H.F. Mark, N.M. Bikales, C.G. Overberger, G. Menges, and J.I. Kroschwitz, eds., Wiley-Interscience, New York, p. 725 (1986).
80. Laschewsky, A., and Ward, M.D. *Polymer* 32:146 (1991).
81. Goreki, M., and Patchornik, A. *Biochim. Biophys. Acta* 303:36 (1973).
82. Suzuki, T., Nambu, Y., and Endo, T. *Macromolecules* 23:1579 (1990).
83. Casolaro, M., and Busi, E. *Polymer*, 35:360 (1994).
84. Eisner, U., and Kuthan, J. *Chem. Rev.* 72:1 (1972).
85. Casolaro, M., and Ito, Y. *The Development of Science for the Improvement of Human Life: Proceedings of the First Kyoto-Siena Symposium*, F. Casprini, and R. Barbucci, eds., Siena, Italy, 1992.
86. Nishi, S., Ito, Y., and Imanishi, Y. *Proceedings of the 36th Polym. Symp.*, 56 (June 6, 1990).
87. Ise, N., and Tabushi, I., eds. *An Introduction to Speciality Polymers*, Cambridge University Press, Cambridge, England, 1983.
88. Nambu, Y., Kijima, M., and Endo, T. *Macromolecules* 20:962 (1987).
89. Nambu, Y., Acar, M.H., and Endo, T. *Makromol. Chem.* 189:495 (1988).
90. Nambu, Y., Endo, T., and Okawara, M. *Bull. Chem. Soc. Jpn.* 54:3433 (1981).
91. Cobra, J.J., Gival, D., and Porter, P.R. *Biochem. J.* 107:69 (1968).
92. Sperling, R., Burstein, Y., and Steinberg, Y.Z. *Biochemistry* 8:3810 (1968).
93. Ke, B. *Biochim. Biophys. Acta* 25:650 (1957).
94. Fruton, J.S., and Clarke, H.T. *J. Biol. Chem.* 106:667 (1934).
95. Ito, Y., Kotera, S., Inaba, M., Kono, K., and Imanishi, Y. *Polymer* 31:2157 (1990).
96. Lehninger, A.L. *Biochemistry*, Worth Publ., Inc. New York, 1975.
97. Heller, J. *J. Controlled Release* 8:111 (1988).
98. Ishihara, K., Muramoto, N., Fujii, H., and Shinohara, I. *J. Polym. Sci.: Polym. Lett. Ed.* 23:531 (1985).
99. Heller, J., and Trescony, P.V. *J. Pharm. Sci.* 68:919 (1979).
100. Gombotz, W.R., and Hoffman, A.S. *CRC Critical Reviews in Biocompatibility* 4:1 (1987).
101. Srere, P.A., and Uyeda, K. *Methods Enzymol.* 44:11 (1976).
102. Yatvin, M.B., Kreutz, W., Horwitz, B.A., and Shinitzky, M. *Science* 210:1253 (1980).
103. Updike, S.J., Schuls, M., and Ekman, B. *Diabetes Care* 5:207 (1982).
104. Kaetsu, I., Morita, Y., Otori, A., and Naka, Y. *Artif. Organs* 14:237 (1990).
105. Chung, D.J., Ito, Y., and Imanishi, Y. *J. Controlled Release* 18:45 (1992).
106. Ishihara, K., Kobayashi, M., Ishimaru, N., and Shinohara, I. *Polymer J.* 16:625 (1984).
107. Casolaro, M., and Barbucci, R. *Int. J. Art. Organs* 14:732 (1991).
108. Martell, A.E., and Smith, R.M. In *Critical Stability Constants*, Plenum Press, New York, 1974.

IX
CERAMIC MATERIALS

42

A Biologically Active Ceramic Material with Enduring Strength

Thomas D. McGee
Iowa State University, Ames, Iowa
Ann M. Graves and Katherine S. Tweden
St. Jude Medical Center, Minneapolis, Minnesota
Gabriele G. Niederauer
University of Texas Health Center, San Antonio, Texas

I. INTRODUCTION

Ceramic materials are, in general, brittle materials [1]. They fail catastrophically, without ductile deformation prior to failure. This brittle property means that, to be successful as implants, the ceramic must be recognized as a brittle material; the principles of fracture mechanics must be considered in designing both the material and the orthopedic appliance.

Brittle materials contain flaws that depend upon the method of manufacture. New flaws can be introduced during service. The presence of the flaws reduces the fracture strength. For a particular ceramic material with a particular fracture toughness, application of the principles of fracture mechanics can allow a designer to specify the critical tensile stress that must not be exceeded during the lifetime of the implant. That stress depends upon the presence of cracks or flaws, present initially or at subsequent times during service.

Thus, the successful design of an implant requires that the flaw structure be controlled not only at the time of implantation, but also during the life of the implant. Lack of control of flaw structure during service is the reason single-phase, homogeneous materials such as bioglasses and calcium phosphates cannot be used at high stresses. The principles of fracture mechanics that apply are described briefly below.

II. FRACTURE MECHANICS OF BRITTLE MATERIALS

The presence of a Griffith flaw (a tiny crack first proposed by Griffith to explain the weakness of brittle materials) causes stress concentration to occur at the root of the crack tip [2]. Although there are three modes of crack opening, the one that weakens the material most is the Type I mode, for which a tensile stress exists normal to the plane of the crack. The stress concentration factor (SCF) is equal to the square root of the crack length c divided by the square root of the radius of curvature at the root of the crack ρ [3].

$$\text{SCF} = (c/\rho)^{1/2} \tag{1}$$

Because ρ is of ionic dimensions, the stress concentration factor is enormous for expected values of c. This means that an actual stress of the root of the crack σ_r is many times greater than the applied stress σ_a as calculated from ordinary stress analysis. Thus,

$$\sigma_r = (c/\rho)^{1/2}\sigma_a \tag{2}$$

The presence of porosity in a ceramic means that the crack length c will be as large as the largest pore under stress. So, if a 100-micrometer (μm) surface pore is present and the radius of curvature at a crack tip is one nanometer (nm), the actual stress is $10^{2.5}$, or about 300 times, greater than the nominal applied stress. An Haversian system can grow into a 100-μm pore and has often been proposed as a way of bonding bone to a prosthesis. For ceramic materials, this is too risky for almost any expected applied stress. So, large pores are to be avoided.

The strongest ceramics are fine-grained ceramics because the flaws are often about the same length as the grain size. Practical grain sizes often are about 10 μm. From our analysis above, we see that this gives a stress concentration factor of about 30. This means that the material must have a high strength to resist this concentrated stress.

The theoretical strength of a material has been estimated by several methods, and is usually calculated as about 1/10 the Young's modulus E [4]. Actual strengths are much less. This is the reason Griffith proposed the presence of flaws [2]. He calculated the theoretical tensile strength as

$$\sigma_f = (2E\gamma/\pi c)^{1/2} \tag{3}$$

where γ is the fracture surface energy.

We now recognize that γ can include terms for other energy-absorbing mechanisms. Some ceramics have high fracture toughness due to those mechanisms. If c is extracted from Eq. 3, we get

$$\sigma_f c^{1/2} = (2/\pi)^{1/2}(E\gamma)^{1/2} \tag{4}$$

If we let $(2/\pi)^{1/2}$ be a geometric factor Y appropriate to the orientation of the crack and the loading, this can be written

$$\sigma_f c^{1/2} = Y(E\gamma)^{1/2} \tag{5}$$

The product $\sigma c^{1/2}$ is called the stress intensity to distinguish it from the applied stress. For Mode I, this is

$$K_I = \sigma c^{1/2} \tag{6}$$

If the applied stress is sufficient to cause failure, then the critical stress intensity is

$$K_{IC} = \sigma_f c^{1/2} = Y(E\gamma)^{1/2} \tag{7}$$

Note that σ_f is the σ_a, the applied stress, that will cause failure in the presence of a Type I crack. The critical stress intensity cannot be exceeded without failure. Thus, the flaw distribution in a ceramic must be controlled to have a successful implant.

Part of the problem with bioactive ceramic implants is that the flaw distribution has not been controlled. Thus, attempts to use massive tricalcium phosphate and hydroxyapatite in load-bearing applications have not been successful. Even prestressed implants have failed. The failures have often been identified as fatigue failures. However, because of their solubility, such ceramics have changes in the flaw distribution with time. Dissolution at grain boundaries will be preferential, and will introduce larger flaws. Strength can be expected to diminish in a saline environment, whether *in vivo* or *in vitro*.

Some other method is needed, then, to protect the biologically active ceramic from failure by the flaws introduced by its bioactivity. Most of the existing applications, such as the bioglasses and plasma sprayed hydroxyapatite, are doomed to failure because of this requirement. Fracture mechanics cannot be neglected when ceramics are placed in load-bearing applications.

III. PHYSICAL CHEMISTRY CONSIDERATIONS

If a biologically active implant is to be successful, the flaw structure could be controlled if it were made as a composite. This would require a second phase that has enduring strength but would allow the calcium phosphate to control the tissue reaction. A strong, inert, second phase is required, with intimate bonding of one to the other. In principal, polymers, metals, or ceramics could be the second phase. However, the polymers do not bond well to the ceramic and, of themselves, have limited life under load. A polymer/calcium phosphate composite may have some applications, and the elastic properties could be adjusted to be close to those of bone. However, the problem of life under stress makes this a risky approach.

Metal/ceramic composites such as stainless steel/ceramic also look appealing. However, powder metals must be heated in a nonoxidizing atmosphere. At low oxygen partial pressures, the phosphorous in the calcium phosphates will be reduced, be adsorbed by the metals, and cause embrittlement. Deterioration of mechanical properties is highly probably, so this approach is also risky.

A ceramic/ceramic composite has the prospect of bonding the two phases together, and compatible processing conditions of time and temperature. Success would require continuous, interconnected phases. One would provide strength. The other would control tissue response. The inert phase would provide long-term strength. The problem is to choose such a phase.

The calcium phosphates in the $Ca_3(PO_4)_2$ to $Ca_{10}(PO_4)_6(OH)_2$ range of compositions have excellent bioactivity [5]. This is a Ca : P ratio of 1.50 to 1.67. Outside this range, the calcium phosphates are too basic or too acidic, and cause an inflammatory response. The preferred phosphates are salts. When fired at high temperatures, Lewis acids react with Lewis bases to form reaction products. These calcium phosphates result from acid-base reactions. However, when hydroxyapatite is heated, water is removed, producing oxyapatite, $Ca_{10}(PO_4)_6O$, an unstable oxide that disproportionates into the stable oxides $Ca_3(PO_4)_2$ and $Ca_4(PO_4)_2O$, or $[4CaO \cdot P_2O_5]$ [6]. The phase diagram for anhydrous calcium phosphates (Fig. 1) shows that there is a high-temperature form of tricalcium phosphate, α $3CaO \cdot P_2O_5$ that is stable above about 1100°C [7]. It inverts to the β phase on cooling. Because of its refractoriness, high firing temperatures are needed to sinter it so that it is dense.

Choosing a ceramic second phase is difficult because the second phase should not react with the calcium phosphate and alter its bioactivity. In general, complex phase diagrams of ceramics with calcium phosphates are not available. One obvious choice to

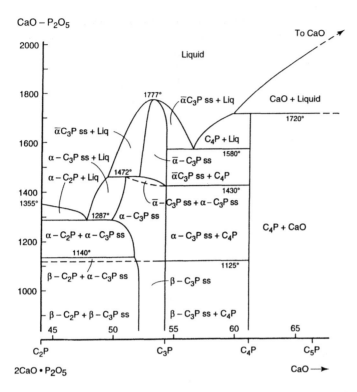

Figure 1 CaO P$_2$O$_5$ phase diagram showing the stable regions of $\hat{\alpha}$, α, and β tricalcium phosphate. (From Ref. 7.)

explore is Al$_2$O$_3$, an acceptable bioinert ceramic for which there is a phase diagram. However, Al$_2$O$_3$, a Lewis acid, reacts with the calcium phosphates to produce low-melting eutectics and intermediate compounds. The ionic radius of Ca^{++} is 0.99 Å. That of P^{+5} is 0.34 Å. In order to prevent solid solution and to find inert, refractory compounds, ions of quite different radii should be chosen.

One possibility is MgAl$_2$O$_4$, a Lewis salt with the divalent magnesium radius of 0.65 Å, and the Al^{+++} ion has a radius of 0.50 Å. Spinel is strong, inert, and biocompatible [8]. Magnesium is too small to proxy for calcium in solid solutions. Aluminum is too large to proxy for phosphorous. The melting temperature of MgAl$_2$O$_4$ is 2105°C, and it also needs a high firing temperature for sintering. It is known to be a strong, bioinert ceramic. This should be a suitable second phase, although no MgAl$_2$O$_4$/Ca$_3$(PO$_4$)$_2$ phase diagram exists. Experimentally, this reasoning was verified. A strong, inert composite was produced that was designated as an osteoceramic.

IV. PROPERTIES OF THE OSTEOCERAMIC

A. Experimental Procedure

1. Osteoceramic Preparation

The osteoceramic was prepared by mixing 50 vol% calcium phosphate tribasic (Mallinckrodt, Inc.) and 50 vol% MgAl$_2$O$_4$ spinel (Baikowski International, Inc.) with a mortar and pestle. The ceramic powder was pressed in a steel die at 4000 pounds per square inch (psi) with a laboratory press and the formed specimens were pressed isostatically at 25,000 psi. The specimens were then sintered to 1500°C for one-half hour (h) with a heating rate of 150°C/h and a cooling rate of approximately 125°C/h.

2. Porosity and Density

The porosity and density of the fired specimens were determined by the Archimedes water displacement method.

3. X-Ray Diffraction

Diffraction patterns were obtained from the fired osteoceramic and from the unfired raw materials using a Siemens D500 x-ray diffractometer with Cu kα radiation. Phases were identified using the American Society for Testing and Materials (ASTM) powder diffraction data file.

4. Mechanical Testing

Bars of the osteoceramic with nominal dimensions of 6.6 × 9.1 × 38.1 millimeters (mm) (0.26 × 0.36 × 1.5 inches [in]) were tested for modulus of rupture (MOR). The MOR tests were performed on an MTS mechanical testing instrument using a test speed of 0.0024 in/min. Osteoceramic bars of the same dimensions were stored in Ringer's solution for seven months before being tested in three-point bending to determine the effects of isotonic saline on the strength of the bars.

Pellets were made from the tricalcium phosphate, the spinel, and the osteoceramic and sintered at 1500°C. The compressive strength of the pellets was determined with an MTS mechanical testing instrument using a centric load applied to a spring-suspended steel bearing block and a test speed of 0.0039 in/min.

Bars of the tricalcium phosphate, the spinel, and the osteoceramic were formed, sintered to 1500°C, and surface ground to insure flat and parallel surfaces. Ultrasonic wave velocities through the samples were measured in order to calculate Young's modulus of elasticity. The average velocities through the samples were determined in water and compared with the velocity of sound in water. The frequency content launched by the transmitting transducer through water was approximately 0–15 megahertz (MHz), with a maximum at about 8.5 MHz. The following equation was used to determine the average ultrasonic wave velocities through the samples.

$$V_{sample} = [1/V_w - t/d]^{-1}$$

where

V_w = the wave velocity through water (centimeters per second, cm/s)
t = time difference between the signals through the sample and the reference (s)
d = the thickness of the sample (cm)

Young's modulus of elasticity E was calculated using the following equation:

$$E = (V_{sample})^2 \times (\rho_b)$$

where ρ_b is the bulk density of the sample (grams per cubic centimeter, g/cm^3).

5. Thermal Expansion

Bars of tricalcium phosphate and of spinel, sintered at 1500°C, were used to determine their mean linear thermal expansion coefficients. A conventional fused-silica dilatometer apparatus was used, heating to 600°C at approximately 200°C/h. The dilatometer utilized a linear variable differential transformer to measure the change in the length of the specimen.

6.   Differential Thermal Analysis and Thermogravimetric Analysis

Differential thermal analysis (DTA) and thermogravimetric analysis (TGA) were conducted on the raw materials, the calcium phosphate, and the magnesium aluminate spinel, using a high-temperature thermal analysis system (Dupont 1090). Aluminum oxide was used as the reference material for the DTA. The tests were run for a temperature range of 25°C to 1150°C with a heating rate of 10°C/min.

7.   Differential Scanning Calorimetry

Thermal analysis of tricalcium phosphate, spinel, and the osteoceramic (all sintered at 1500°C) was performed with a Perkin Elmer Differential Scanning Calorimeter (DSC-4) and a Perkin Elmer System 4 Thermal Analysis Microprocessor. The system utilized ice-bath cooling and a nitrogen purge. The analysis was run for the temperature range of 25°C to 450°C at 10°C/h.

8.   Scanning Electron Microscopy

Etched, polished, and fractured surfaces of the sintered osteoceramic were examined using a JEOL-JSM 840a scanning electron microscope (SEM) at 15 kilovolts (kV). The polished osteoceramic was etched with 1M H_2SO_4 for 30 seconds or with 8% ethylene diamine-tetraacetate (EDTA) for 20 minutes to remove the calcium phosphate partially or completely. Specimens were coated with gold before examination.

B.   Experimental Results

1.   Porosity and Density

The density and porosity data for the fired osteoceramic composite indicate that the pores contained in the composite were predominately closed (Table 1). The fact that the pores are closed prevents transport of body fluids from one location to another. The presence of pores lowers the strength of the composite. The tricalcium phosphate and the spinel also had low porosity.

2.   Chemical Composition and Phase Analysis

X-ray diffraction analysis of the raw materials showed that the crystalline phase of the calcium phosphate tribasic was actually hydroxyapatite $\{Ca_{10}(PO_4)_6(OH)_2\}$. The magnesium aluminate was fully converted spinel ($MgAl_2O_4$). The osteoceramic, sintered to 1500°C, was determined by x-ray diffraction to be α-tricalcium phosphate and $MgAl_2O_4$ spinel. However, amorphous and crystalline phases less than or equal to 5% are not detected by x-ray diffraction.

The Ca/P molar ratio of the osteoceramic was determined to be 1/1.62 using standard wet chemical analysis. This is between the ratios of $Ca_3(PO_4)_2$ and $Ca_{10}(PO_4)_6(OH)_2$ (between 1.50 and 1.67). The calcium phosphates are known to vary in their Ca/P ratio. Solid-solution variations of both $Ca_3(PO_4)_2$ and dehydrated hydroxyapatite [Ca_{10} $(PO_4)_6O$] are known to occur [9,10].

Electron microprobe analysis shows that, in a fired composite with equimolar amounts of $Ca_3(PO_4)_2$ and $MgAl_2O_4$, two continuous phases are present (Whitlockite and spinel), and the calcium and phosphorous do not combine with the magnesium and aluminum [11].

Tweden's results from energy dispersive analysis with the SEM show that the

Table 1   Density and Porosity (1500°C)

	Mean[a]	90% Confidence interval
Ceramic composite		
Bulk density, gm/cc	3.089	3.080 ± 0.007
Apparent density, gm/cc	3.098	3.098 ± 0.006
Apparent porosity, %	0.270	0.270 ± 0.077
Closed pores, %	8.060	8.060 ± 0.182
Bulk/theoretical density ratio, %	91.680	91.680 ± 0.200
Tricalcium phosphate		
Bulk density, gm/cc	2.754	2.754 ± 0.006
Apparent density, gm/cc	2.793	2.793 ± 0.009
Apparent porosity, %	1.402	1.402 ± 0.011
Closed pores, %	10.910	10.910 ± 0.310
Bulk/theoretical density ratio, %	87.690	87.690 ± 0.200
Magnesium aluminate spinel		
Bulk density, gm/cc	3.351	3.351 ± 0.029
Apparent density, gm/cc	3.369	3.369 ± 0.028
Apparent porosity, %	0.536	0.536 ± 0.150
Closed pores, %	5.591	5.591 ± 0.771
Bulk/theoretical density ratio, %	93.870	93.870 ± 0.810

[a]Number of samples = 10

α-tricalcium phosphate and spinel phases in the osteoceramic sintered at 1500°C remain separate [12]. The distribution of the elements magnesium, aluminum, calcium, and phosphorous in the osteoceramic agree with the microprobe results. However, neither method is sensitive enough to detect small amounts of solid solution of one phase in the other phase.

Graves evaluated the osteoceramic sintered at 1500°C with elemental digital mapping using a Kevex Delta X-Ray Microanalyzer with a quantum detector (Fig. 2) [9]. The elemental maps for magnesium, aluminum, calcium, and phosphorous show that the elements magnesium and aluminum were associated, and that the calcium and phosphorous were associated. The magnesium and aluminum did not appear to combine with calcium and phosphorous. This more sensitive technique also did not indicate appreciable solid solution of spinel in calcium phosphate or vice-versa.

3. Mechanical Properties

As for mechanical properties, the fracture strength of the osteoceramic was found to be 70.3 ± 8 megapascal (MPa) for 18 MOR samples. The fracture strength of the bars that had been exposed to Ringer's solution for 7 months was 80.3 ± 9.9 MPa for 18 samples. The strength of the exposed bars was not significantly different from that of the unexposed bars. Therefore, degradation of the osteoceramic strength after 7 months in a physiological environment could not be detected.

Compressive strength tests showed that the osteoceramic was stronger than the tricalcium phosphate and the spinel for a sintering temperature of 1500°C. The mean compressive strengths of the three materials sintered at 1500°C are given in Table 2 in comparison with the strengths of enamel, dentin, compact bone, and cancellous bone [16]. Compres-

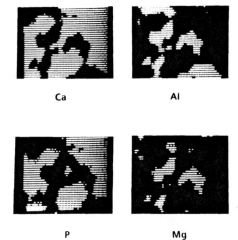

Figure 2   X-ray fluorescent analysis of the distribution of the elements calcium, phosphorous, magnesium, and aluminum using an electron microprobe analyzer (× 200).

sive strengths of α-tricalcium phosphate and the osteoceramic were studied as a function of sintering temperature (Figs. 3, 4). Figure 3 shows that the compressive strength of α-tricalcium phosphate quickly decreases when fired at temperatures above 1400°C, and Figure 4 shows the compressive strength values of the osteoceramic are highest at sintering temperatures of 1350°C and 1400°C. The lower compressive strength of the osteoceramic at sintering temperatures above 1400°C may be due to overfiring and resultant grain growth.

Young's modulus of elasticity for α-tricalcium phosphate, spinel, and the osteoceramic was determined by the ultrasonic velocity method (Table 3). Kingery, Bowen, and Uhlman list the value for the elastic modulus of spinel (5% porosity) to be 238 gigapascal (GPa), which is close to the value listed in Table 3 [13].

4.   Thermal Expansion

The mean linear thermal expansion coefficients of spinel, α-tricalcium phosphate, and the ceramic composite were determined for specific temperature intervals (Table 4, Fig. 5). The spinel and the composite were linear. The thermal expansion data for α-tricalcium

Table 2   Compressive Strength (Room Temperature, Sintered to 1500°C)

Material	Compressive strength (MPa)	90% Confidence interval
Tricalcium phosphate, α-Ca$_3$(PO$_4$)$_2$[a]	46.2	46.2 ± 1.9 MPa
Spinel[a]	186	186 ± 15 MPa
Ceramic composite[a]	299	299 ± 17 MPa
Enamel	241	
Dentin	138	
Compact bone	167	
Cancellous bone	1.86	

[a]Number of samples tested for each material = 10

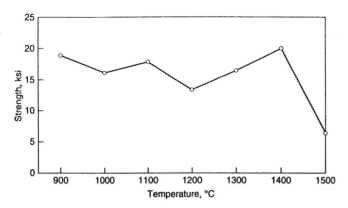

Figure 3 Compressive strength of α-tricalcium phosphate as a function of sintering temperature.

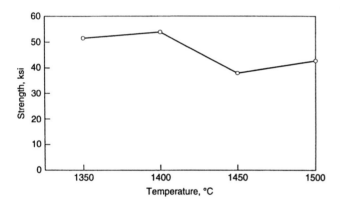

Figure 4 Compressive strength of the osteoceramic as a function of sintering temperature.

Table 3 Young's Modulus of Elasticity (Room Temperature, Sintered to 1500°C)

Material[a]	Young's modulus of elasticity (GPa)	Ref. No.
Tricalcium phosphate, α-$Ca_3(PO_4)_2$	87.1 (12.62×10^6 psi)	
Spinel	224 (32.53×10^6 psi)	
Ceramic composite	114 (16.57×10^6 psi)	
Compact bone	17.2	14
Dentin	13.8	14
Enamel	48	14
Polycrystalline alumina	373	13
Sapphire	392	13
Ti_6Al_4V or Ti	110	14
Pyrolytic carbon	0.14–0.21	16

[a]Number of samples for each material = 10

Table 4   Mean Linear Thermal Expansion Coefficients

Material (Sintered to 1500°C)	Temperature interval	Thermal expansion coefficient (m/m °C)
Ceramic composite	70–435°C	10.72×10^{-6}
Spinel	55–405°C	7.18×10^{-6}

phosphate indicate that two inversions occur in the temperature interval studied (25–450°C). On heating, inversions appeared to occur at 147°C and 236°C. The inversion at 147°C indicates a contraction response, and at 236°C an expansion response. On cooling, similar inversions occurred at approximately 91°C and 203°C. The difference in these inversion temperatures is due to hysteresis caused by the sluggish transformation of tricalcium phosphate.

5. Differential Thermal Analysis and Thermogravimetric Analysis

Differential thermal analysis indicated endothermic reactions occurring in hydroxyapatite at approximately 200°C and 810°C (Fig. 6). There was no indication of thermal reactions in spinel for the temperature range evaluated.

Thermogravimetric analysis of hydroxyapatite indicated a gradual weight loss of about 1.9% from 25–500°C, and a sudden weight loss of about 0.5% at approximately 770°C (Fig. 7). This sudden weight loss corresponds to a sudden partial dehydroxylation of hydroxyapatite and is in agreement with the endothermic reactions measured by DTA.

6. Differential Scanning Calorimetry

A plot of heat flow versus temperature of the α-tricalcium phosphate was obtained from the differential scanning calorimeter (Fig. 8). A reaction was observed at approximately 200°C to 240°C, with hysteresis, which is consistent with thermal expansion results. The differential scanning calorimeter (DSC) plot for spinel shows no indication of thermal changes. The DSC plot of the osteoceramic also shows no thermal changes, indicating that the spinel phase inhibits the phase transformation occurring in the α-tricalcium phosphate phase at 200°C.

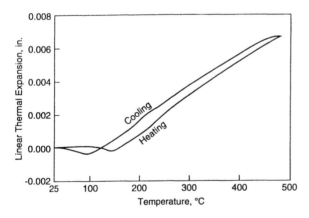

Figure 5   Thermal expansion change of α-tricalcium phosphate as a function of temperature.

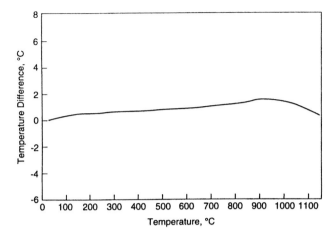

Figure 6   Differential thermal analysis of hydroxyapatite.

7. Microstructure

Electron micrographs of samples sintered at 1500°C are shown in Figures 9–11. Inter-granular-type fracture was observed at the fractured surface of the composite (Fig. 9). The porosity was observed to be evenly dispersed and isolated, and pores ranged in diameter from 1.5 to 2.0 microns.

Phase relief is seen in the composite when polished to emphasize the two phases (Fig. 10). The elevated areas were determined to be spinel, while the depressed areas were α-tricalcium phosphate, based on x-ray fluorescent analysis with the scanning electron microscope. These results confirm that α-tricalcium phosphate is softer than spinel.

When the polished composite was etched to remove the calcium phosphate phase with 8% EDTA for 20 minutes, cubic spinel crystals were revealed (Fig. 11). Cubic spinel crystals and needlelike tricalcium phosphate crystals were revealed in specimens etched with 1M H_2SO_4 for 30 seconds to remove all but the largest calcium phosphate grains.

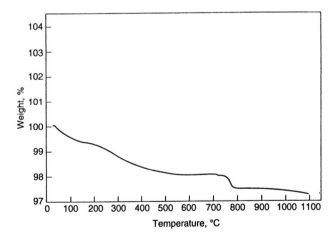

Figure 7   Thermogravimetric analysis of hydroxyapatite.

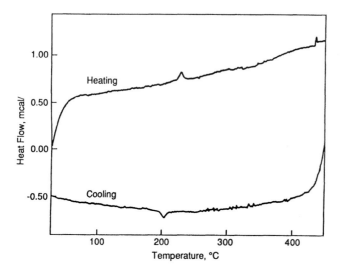

Figure 8   Scanning calorimetric analysis of α-tricalcium phosphate.

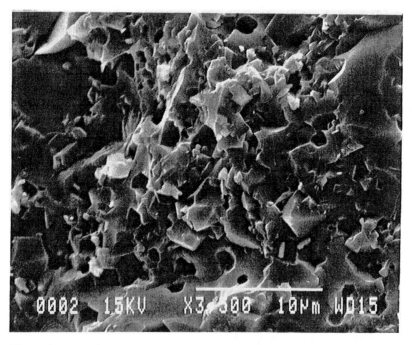

Figure 9   Scanning electron microscopy (SEM) analysis of a fracture surface (×3500).

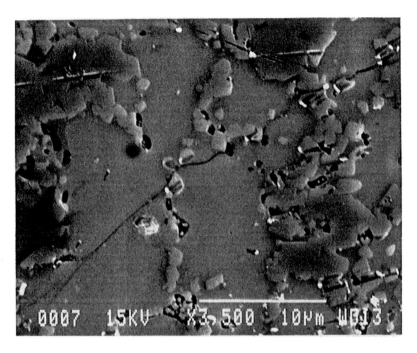

Figure 10 Scanning electron microscopy (SEM) of a surface polished to give relief between the softer (tricalcium phosphate) and the harder (spinel) phases (×3500).

Figure 11 Scanning electron microscopy (SEM) of a surface etched with ethylene diamine-tetraacetate (EDTA) to remove the phosphate phase (×5000).

V. DISCUSSION

An ideal implant would bond to bone and be indistinguishable from bone in all properties. Existing metal and polymer components do not have physical properties that closely match those of bone. The high modulus of elasticity and the high thermal and electrical conductivity of metals are a distinct handicap, giving rise to stress shielding and electrochemical corrosion. Polycrystalline alumina and sapphire have an even higher modulus of elasticity than the metals used in skeletal repairs. A modulus of elasticity similar to that of bone is desirable to minimize stress concentration and stress shielding in the tissue adjacent to an implant. The osteoceramic has strength similar to that of cortical bone and a modulus of elasticity that is about the same as titanium.

The pore structure of the osteoceramic can be controlled [11]. When produced in the dense form, there is no appreciable open porosity; therefore, the osteoceramic does not serve as a wick to transport body fluids. This is especially important in dental applications because oral fluids should not be conducted deep into bone.

Both $Ca_3(PO_4)_2$ and $MgAl_2O_4$ are stable compounds that do not melt at the maximum temperature used in processing. The results from the microprobe, SEM, and EDAX indicate that there is no extensive solid solution or mixing. Therefore, $Ca_3(PO_4)$ and $MgAl_2O_4$ remain as separate stable phases. The microstructure shows that these phases are intermixed to produce a true composite. The structure is similar to a sponge, with the skeleton of the sponge, the spinel, providing strength and insolubility to the implant. The holes in the sponge are filled with α $Ca_3(PO_4)_2$, a ceramic with high biological activity that has variously been characterized as osteophilic or as osteoconductive [14]. The euhedral shape of the spinel crystals and the radiating structure of the $Ca_3(PO_4)_2$ crystals indicates that a small amount of liquid phase was present during firing. Although the extent of solid solution of one phase in another was not determined, it can be predicted from crystal chemical principles that solid solution would be minimal.

When $Ca_{10}(PO_4)_6(OH)_2$ is heated, it loses mechanically held and chemically combined water between 25°C and 800°C, and continues to lose weight at 1100°C. Although the literature suggests several compounds, including oxyapatite, can result when hydroxyapatite is heated, the only crystalline compound obtained in this study was α-$Ca_3(PO_4)_2$ [15]. Therefore, the tricalcium phosphate must be somewhat nonstoichiometric. When the raw material is heated, α-$Ca_3(PO_4)_2$ is obtained, and low-temperature inversions appear at 147°C to 236°C. However, when combined with the spinel, the inversions are inhibited. Therefore, the α-$Ca_3(PO_4)_2$ found in the osteoceramic is in a highly strained condition. This should make it more soluble, hence more biologically active than unconstrained α-$Ca_3(PO_4)_2$.

The compressive strength data also show that the weakness associated with the inversion of the calcium phosphate phase has been eliminated by the additional spinel. The composite is stronger than either of the end components. Thus, a strong, biologically-active ceramic composite has been achieved.

The grain size of the spinel (Fig. 11) is small, about 10 μm maximum. The grain size of the calcium phosphate was also small, but some larger grains existed. This produced the largest flaws in the second-phase spinel. Taking these largest flaws as the critical Griffith flaws allows us to calculate K_{Ic}. Estimating the flaws at 100 μm and using the measured fracture strength of 70 MPa gives

$$K_{Ic} = \sigma c^{1/2} = 70 \text{ MPa} \, (100 \times 10^{-6} \text{m})^{1/2}$$
$$= 70 \times 10^{-2} \text{ MPa} \cdot \text{m}^{1/2}$$

$$= 700 \ KPa \cdot m^{1/2}$$

Taking $E = 114$ GPa and estimating $\alpha = 0.45$ joules per square centimeter (J/m^2),

$$K_{Ic} = 2(114 \ GPa)(0.45 \ J/m^2)^{1/2}/\pi$$
$$= 18 \times 10^4 \ MPa \cdot m^{1/2}$$
$$= 180 \ KPa \cdot m^{1/2}$$

This is reasonable agreement and shows how important flaw structure is.

VI. CONCLUSION

This chapter reports the properties of an osteoceramic, a biologically active ceramic-ceramic composite, that has enduring strength. Application of the osteoceramic as a tissue contact material should be possible if the limitations in fracture of brittle materials are considered in prosthesis design.

REFERENCES

1. Davidge, R.W., *Mechanical Behavior of Ceramics*, Cambridge University Press, New York, 1979.
2. Griffith, A.A., The Phenomenon of Rupture and Flow in Solids, *Phil. Trans. Roy. Soc. Land.* A221:163, 1920.
3. Inglis, C.E., Stresses in a Plate Due to Cracks and Sharp Corners, *Trans. Inst. Nov. Archit.* 55:219, 1913.
4. Gilman, J.J., The Strength of Ceramic Crystals, in *The Physics and Chemistry of Ceramics*, C. Klingsberg, Ed., Gordon and Breach, New York, 1963.
5. Donath, L., Reactions of Tissue to Calcium Phosphate Ceramics, in *Osseo-Integrated Implants*, Vol. 1, G. Heimke, Ed., CRC Press, Boca Raton, FL, 1990.
6. deGroot, K., Bioceramics Consisting of Calcium Phosphate Salts, *Biomaterials* 1:47, 1980.
7. Phase Diagrams for Ceramists, *Am. Cer. Soc.*, 1956.
8. Karagianes, M.T., Westerman, R.E., Rasmussen, J.J., and Lodel, A.M., Development and Evaluation of Porous Dental Implants in Miniature Swine, *J. Dental Res.* 55:85, 1976.
9. Graves, A.M., M.Sc. thesis, Iowa State University, Ames, IA, 1988.
10. Eanes, E.D., *Calc. Tiss. Res.* 5:133, 1970.
11. McGee, T.D., and Wood, J.L., Calcium-Phosphate Magnesium-Aluminate Osteoceramics, *J. Biomed. Mater. Res.* 5:137, 1974.
12. Tweden, K.S., A Comparison of Four Endosseous Dental Implants, Ph.D. thesis, Iowa State University, Ames, IA, 1987.
13. Kingery, W.D., Bowen, H.K., and Uhlmann, D.R., *Introduction to Ceramics*, 2nd Ed., Wiley, New York, 1976.
14. Jarcho, M., Calcium Phosphate Ceramics as Hard Tissue Prostheses, *Clinical Orthop. and Related Res.* 175:259, 1981.
15. deGroot, K., *Bioceramics of Calcium Phosphates*, CRC Press, Boca Raton, FL, 1983.
16. Park, J.D., *Biomaterials Science and Engineering*, Plenum Press, New York, 1984.

43
Calcium Phosphate Biomaterials: Preparation, Properties, and Biodegradation

Racquel Z. LeGeros and John P. LeGeros
New York University College of Dentistry, New York, New York
Guy Daculsi
Faculte de Chirurgie Dentaire, Nantes, France
Regina Kijkowska
Technical University of Krakow, Poland

I. INTRODUCTION AND SCOPE

In the very few papers published between 1920 and 1975, some calcium phosphate (Ca-P) materials were reported to be successful in promoting bone formation when used to repair bone or periodontal defects [1,9,78,87]. The Ca-P materials used in the earlier reports were not characterized; therefore, their success, or lack of it, is difficult to evaluate. For example, the material previously described as "tricalcium phosphate" (TCP) ceramic [45,78] has been shown by x-ray diffraction (XRD) analysis to consist of a mixture of β-tricalcium phosphate (β-TCP) and calcium hydroxyapatite (HA) [58]. Similar materials were subsequently and more appropriately described as biphasic calcium phosphate (BCP) [32]. Calcium phosphate reagents described as "triple calcium phosphate," "tricalcium phosphate," "hydroxyapatite," and "spheroidal hydroxyapatite" were shown to be mislabeled when analyzed using x-ray diffraction [57,59,60]. For example, spheroidal hydroxyapatite was shown to be mostly β-TCP mixed with some HA [57,59].

In the past decade, several commercial Ca-P materials from the United States, Europe, and Japan (Table 1) were introduced, largely through the pioneering efforts of Jarcho [47,48], deGroot [22], and Aoki [3,4]. These materials have gained acceptance for several dental and medical applications for bone repair, augmentation, and substitution. Dental applications include fillers for periodontal bony defects, alveolar ridge augmentation, immediate tooth root replacement, maxillofacial reconstruction, and as adjuncts to uncoated implants or as extenders of autogenous bone or demineralized bone matrix [4,9,17,25,26,32,35,37,40,48,49,58,63,78,79,81,84,85,89,95,97]. Medical applications include ear implants, spinal fusion, and repair of bony defects [1,22,27,37, 40,48,59,63,87,89,92]. Other suggested uses include pulp capping materials [34,50], percutaneous implants [4,22], materials for biocompatibility testing or biomotor applications [13,36,38], and for drug delivery [23,40].

One of the important uses of Ca-P materials, principally, calcium hydroxyapatite

Table 1   Calcium Phosphate Materials

Calcium phosphate ceramics
  Calcium hydroxyapatite (HA), $Ca_{10}(PO_4)_6(OH)_2$
    Commercial products: Calcitite™, Durapatite™, Alveograf™, Osteograf™, Himed™-HA,
    Bioapatite™ (France), Bioroc™ (France), Allotropat™ (Germany), Mitsubishi HA (Japan)
    Noncommercial[a]
  β-Tricalcium phosphate (β-TCP), $Ca_3(PO_4)_2$
    Commercial products: Synthograf™, Augmen™, Himed™-TCP
    Noncommercial[a]
  Biphasic calcium phosphates (BCP) (mixtures of HA and β-TCP)
    Commercial products: Triosit™ (60% HA, 40% β-TCP), Himed™-TCP (85% HA/15%
    TCP; 60% HA/40% TCP)
    Noncommercial[a]
Resorbable calcium phosphate materials
  Unsintered calcium phosphates (calcium-deficient apatites, tricalcium phosphates)
    Commercial products: Osteogen™, Himed™-AP, calcium phosphate reagents
    Noncommercial
Calcium phosphate materials from natural sources
  Freeze-dried/banked human bones, sintered human bone
    Commercial products: From Ultimatics
  Processed/sintered bovine bone
    Commercial product: BioOss™, Osteograf-N™
    Noncommercial[a], BonAp[a]

---

[a]Prepared in laboratories

(HA) ceramic, is to provide bioactive coatings on dental and orthopedic metal implants [15,16,23,31,49,52,54,59,60]. The term "ceramic" used in this text shall refer to sintered materials. Ca-P materials are also used as components in polymeric [11,30,64,86,93] and nonpolymeric [1,30] composites and in cements [12,65].

Ca-P biomaterials, whether used in bulk or as coatings on implants like specially formulated glass ceramics (e.g., Bioglass™) are classified as *bioactive materials* [40,41, 83] because of their ability to participate actively in the formation of an interface with bone considerably stronger than interfaces formed by other nonbioactive materials [42]. These bioactive materials are biocompatible, eliciting responses from bone cells *in vitro* and *in vivo* that are similar to those elicited by bone [6,21,36–38,80–82].

This chapter discusses the following topics relating to Ca-P biomaterials: (1) preparation of the different commercial and noncommercial biomaterials; (2) characterization and properties; (3) composition and properties compared with bone and bone mineral; (4) tissue responses, including biodegradation and the biomaterial-bone interface; and (5) future calcium phosphate materials.

II.   PREPARATION

A.   "Resorbable" Calcium Phosphate Materials

The materials described in earlier reports (1920–1973) as "triple calcium phosphate," "basic calcium phosphate," and "resorbable calcium phosphate" were probably calcium phosphate reagents prepared from synthetic solutions at reaction temperatures below 100°C [1,9,87]. A commercial material recommended for the repair of periodontal de-

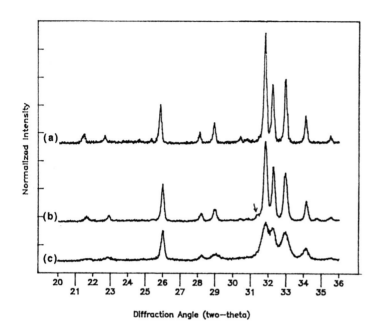

Figure 1 X-ray diffraction patterns: (a), ceramic hydroxyapatite (HA) (Calcitite™); (b), coralline HA (Interpore™ 200); (c), unsintered apatite (Osteogen™). The arrow on (b) indicates presence of Mg-substituted β-TCP.

fects introduced about 10 years ago, described as resorbable HA (Osteogen™) (Fig. 1c), is also a Ca-P material prepared at low temperatures ($\leq 100\,^\circ$C). X-ray diffraction analyses of Ca-P reagents described as tricalcium phosphate, hydroxyapatite or basic calcium phosphate and of the commercially available resorbable HA indicate that these materials are "apatites" (Figs. 1c and 2), that is, calcium phosphate having an apatitic structure but with much poorer crystallinity and variable composition compared to ceramic or mineral hydroxyapatite, $Ca_{10}(PO_4)_6(OH)_2$ (Fig. 1a). These calcium phosphate materials are prepared either by precipitation [39,59] or by hydrolysis of acidic calcium phosphate (e.g., dicalcium phosphate dihydrate, DCPD) or octacalcium phosphate (OCP), $Ca_8H_2(PO_4)_6 \cdot 5H_2O$, or hydrolysis of calcium carbonate, $CaCO_3$, in phosphate solution [59,61,67]:

$$Ca(NO_3)_2 + (NH_4)_2HPO_4 \longrightarrow \text{Apatite} \qquad (1)$$

$$Ca(Ac)_2 + (NH_4)_2HPO_4 \longrightarrow \text{Apatite} \qquad (2)$$

$$CaHPO_4 \text{ or } CaHPO_4 \cdot 2H_2O + NH_4OH \longrightarrow \text{Apatite} \qquad (3)$$

$$CaCO_3 + (NH_4)_2HPO_4 \longrightarrow \text{Apatite} \qquad (4)$$

The apatite compounds can be obtained at low temperatures ($25\,^\circ$C to $60\,^\circ$C) from solutions of basic pH and can be obtained even from acidic solutions (pH 4) when the precipitation or hydrolysis reactions are carried out at higher temperatures, between $80\,^\circ$C and $100\,^\circ$C [59]. Apatites obtained from neutral and acid conditions are usually nonstoichiometric or calcium deficient (Ca/P molar ratio < 1.67) compared to Ca/P = 1.67 for pure stoichiometric calcium hydroxyapatite, $Ca_{10}(PO_4)_6(OH)_2$. Sometimes the

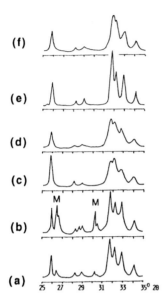

Figure 2 X-ray diffraction patterns of commercial calcium phosphate reagents from different manufacturers (e.g., Fisher, Baker, Mallinkrodt, etc.) labeled as "calcium phosphate tribasic," or "calcium hydroxyapatite" showing apatitic calcium phosphate with poor crystallinity (c, d, e, f), some mixed with another calcium phoshate phase (a, b), e.g., $CaHPO_4$, dicalcium phosphate anhydrous, or monetite (M).

apatites resulting from reactions conducted at low temperatures (25–60°C) are mixed with dicalcium phosphate dihydrate (brushite) or dicalcium phosphate anhydrous (DCPA) (monetite), $CaHPO_4$ (Figs. 1c, 2), probably resulting from incomplete hydrolysis to apatite of these acidic calcium phosphates. Variability in composition and crystallinity in similarly labeled Ca-P reagents was observed from one manufacturer to another and sometimes from one lot number to another in materials obtained from the same manufacturer [57,59,60].

B. Calcium Phosphate Ceramics from Synthetic Sources

1. *Calcium Hydroxyapatite*

Calcium HA ceramic is usually prepared from apatites obtained by precipitation or hydrolysis under basic conditions and subsequently sintered at temperatures between 950°C to 1300°C [10,39,47,59,61,63,68]. Precipitation may be by either Reaction (5) or (6).

$$Ca(NO_3)_2 + (NH_4)_2HPO_4 + NH_4OH \xrightarrow[1100°C]{} Ca_{10}(PO_4)_6(OH)_2 \qquad (5)$$

$$Ca(Ac)_2 + (NH_4)_2HPO_4 + NH_4OH \xrightarrow[950-1100°C]{} Ca_{10}(PO_4(OH)_2 \qquad (6)$$

or by the dropwise addition of phosphoric acid to a saturated solution of calcium hydroxide, $Ca(OH)_2$, as in Reaction (7).

$$Ca(OH)_2 + H_3(PO_4)_2 + NH_4OH \xrightarrow[1100°C]{} Ca_{10}(PO_4)_6(OH)_2 \qquad (7)$$

HA ceramic may also be prepared by sintering the products of hydrolysis of DCPD, DCPA, or OCP in basic solutions [59] or of $CaCO_3$ in phosphate solutions [59], as in Reaction (8) or (9).

$$CaHPO_4 \text{ or } CaHPO_4 \cdot 2H_2O + NH_4OH \xrightarrow[950-1100°C]{} Ca_{10}(PO_4)_6(OH)_2 \quad (8)$$

$$CaCO_3 + (NH_4)_2HPO_4 \xrightarrow[950-1100°C]{} Ca_{10}(PO_4)_6(OH)_2 \qquad (9)$$

To obtain principally HA, critical control of the pH of the reaction and the concentrations of each of the reactants is required [10,39,59].

HA ceramic has also been prepared from commercially available calcium phosphate reagent to which $CaCO_3$ or $Ca(OH)_2$ have been added to minimize the formation of β-TCP upon sintering [22,63]. If this method is employed, it is critical to characterize the Ca-P reagents since variability of β-TCP/HA ratios have been obtained from sintered Ca-P reagents from different manufacturers [57,59,60]. If the unsintered apatite is calcium deficient (Ca/P < 1.67), sintering between 900°C and 1100°C will cause the formation of β-TCP with the HA phase. If the unsintered apatite is calcium rich, sintering will cause the formation of calcium oxide, CaO, with the HA phase. Higher sintering temperatures (> 1300°C) will cause the formation of other calcium phosphates (e.g., α-tricalcium phosphate, α-TCP; tetracalcium phosphate, TTCP, $Ca_4P_2O_9$). As may be expected, sintering at high temperatures causes an increase in crystal size (Fig. 3). Daculsi et al. reported on lattice defects in HA ceramic materials, the number and type depending on the sintering temperature [20]. These changes in physical properties affect the biodegradation of these materials [20,60].

2. β-Tricalcium Phosphate

β-TCP is prepared by sintering calcium-deficient (Ca/P = 1.5) apatite preparations obtained either by precipitation or hydrolysis methods (Reactions 1-9), but under neutral or acid conditions, for example,

$$Ca(NO_3)_2 + (NH_4)_2HPO_4 \xrightarrow[900-1100°C]{} Ca_3(PO_4)_2 \qquad (10)$$

3. Biphasic Calcium Phosphate

BCP, consisting of an intimate mixture of varying ratios of β-TCP and HA, is obtained by sintering apatite preparations of varying Ca/P ratio. The β-TCP/HA ratio of the sintered product depends on the Ca/P molar ratio of the unsintered apatite [57,61,62]:

$$Ca(NO_3)_2 + (NH_4)_2HPO_4 \xrightarrow[900-1100°C]{} Ca_{10}(PO_4)_6(OH)_2 + Ca_3(PO_4)_2 \quad (11)$$

4. Dense or Macroporous Calcium Phosphate Ceramic

HA, β-TCP, or BCP (Figs. 1, 4, 5) can be prepared in dense (microporous) or macroporous forms (Fig. 6). Macroporous ceramic (with pores as large as 500 microns [μ]) is prepared by the addition of volatile components such as hydrogen peroxide or naphthalene before compacting, heating at low temperature to remove volatile agents, and then sintering at high temperatures (950-1100°C) [22,45,47].

(a)

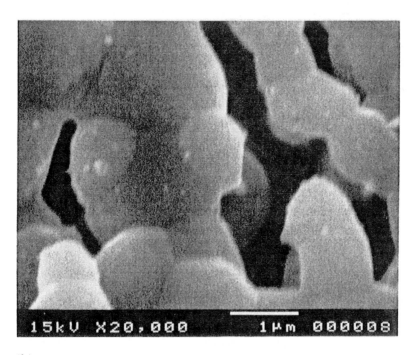

(b)

Figure 3   Scanning electron microscopy (SEM): (a), calcium deficient apatitic material prepared at 60°C; (b), ceramic β-trialcium phosphate (β-TCP) prepared at 100°C by precipitation method and subsequently sintered at >900°C; (c), ceramic hydroxyapatite (HA) precipitated at high PH at

(c)

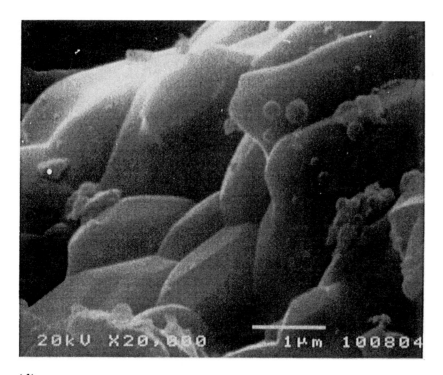

(d)

100°C; (d), HA ceramic obtained by precipitation and subsequent sintering at 1100°C. Note the considerable change in crystal size upon sintering at 1100°C. (SEM courtesy of Prof. T.V. Vijayaraghavan.)

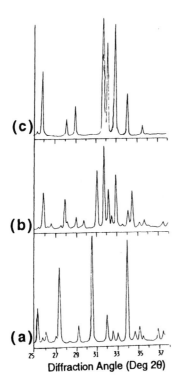

Figure 4 X-ray diffraction (XRD) patterns: (a), ceramic β-tricalcium phosphate (β-TCP); (b), ceramic biphasic calcium phosphate (BCP) consisting of 40% β-TCP and 60% HA; (c), ceramic hydroxyapatite (HA).

Dense or microporous Ca-P ceramic is prepared by compacting or compressing the Ca-P powder under high pressure (20,000 to 30,000 pounds per square inch [psi]) and sintering at high temperatures (1100–1300°C) [22,24,26,47]. Besides the conventional method of uniaxial pressing, other methods of compressing are by hot-pressing techniques or by hot isostatic pressing (HIP) techniques.

The hot-pressing technique employs a programmed simultaneous and continuous application of heat and pressure. This procedure allows densification to take place at much lower temperatures than in the conventional sintering process (e.g., 900°C vs. 1300°C). The lower temperature of densification prevents the formation of other Ca-P phases (e.g., α- and β-TCP, TTCP), which usually form when HA is sintered at temperatures above 1100°C [26].

In the HIP technique, materials are isostatically compressed by gaseous pressures at high temperatures. This process results in greater density, higher compressive strength, and more uniform density in all dimensions than the conventional method of uniaxial pressing. The degree of microporosity depends on the method of compacting and on the sintering temperatures. Dense HA (Figs. 6a and 6b) is described as having a maximum microporosity of 5% by volume with the microporosity of 1 micrometer (μm) in diameter [5,25,26].

Ca-P ceramics (dense or macroporous) are prepared either in particulate or block

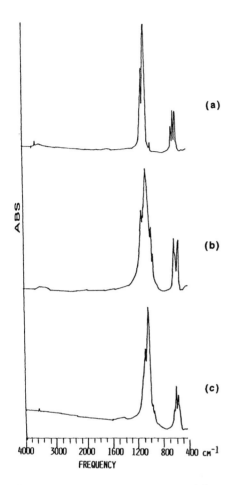

Figure 5 Infrared (IR) spectra: (a), ceramic hydroxyapatite (HA); (b), ceramic β-tricalcium phosphate (β-TCP); and (c), ceramic biphasic calcium phosphate (BCP). These materials give the x-ray diffraction (XRD) patterns shown in Fig. 4.

forms, depending on the intended use. Irregular or spherical particulates are obtained by milling and rolling the compacted sintered blocks. Tooth forms for immediate tooth root replacements are prepared directly from molds; ear implants are carved from block forms [22,25,92]. The dense Ca-P ceramic in block or particulate forms is used in the augmentation of the alveolar ridge for better denture fit [4,17,22,25,48,49] or in orthopedic surgery [4,23,48]. Dense HA blocks are also used as source target materials for coating implants by ion sputtering. Particulate forms of dense or macroporous Ca-P ceramics are used as fillers in bony defects in dental and orthopedic surgery [32,35,37,70,78,95,97], as extenders for autogenous bone or demineralized bone matrix [17,74], as adjuvant fillers with placing of metal implants or for repair of failing dental metal implants [49,74], and for plasma-sprayed coatings on dental and orthopedic implants [23,48,49,52].

(a)   (b)

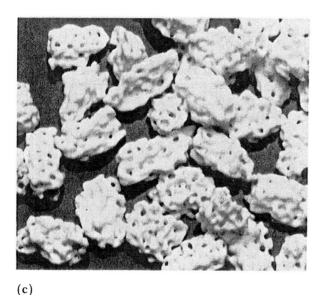

(c)

Figure 6   Scanning electron microscopy (SEM): (a), (b), dense hydroxyapatite (HA) particles; (c), macroporous coralline HA; (d), noncommercial bone apatite.

C.   Calcium Phosphate Materials from Natural Sources

1.   Human Bones: Banked Freeze-Dried Bones or Sintered Bones

Human bones in the banked freeze-dried or sintered forms are commercially available from a few companies. This preparation procedure preserves the microporosity and interconnecting macroporosity of the bone, except when sintered at high temperatures, the size of the macroporosity is reduced.

2.   Processed and/or Sintered Bovine Bone

Commercial and noncommercial processed and/or sintered materials from bovine bone are available (Table 1; Fig. 6c). The macroporosity of the original bovine bone (Fig. 6d) is conserved to a greater or lesser extent, depending on the sintering temperature. The

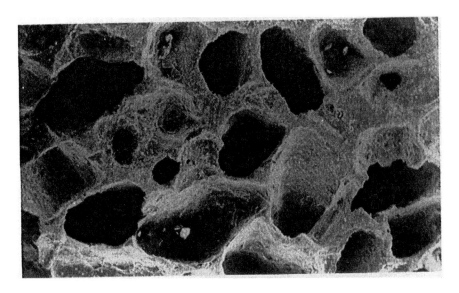

(d)

Figure 6   Continued.

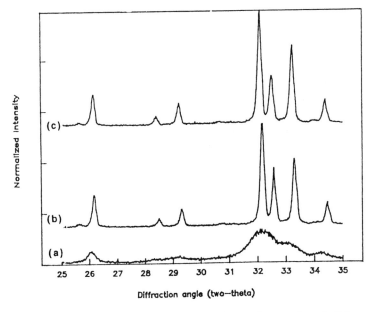

Figure 7   X-ray diffraction patterns: (a), cow bone mineral; (b), noncommercial processed and sintered cow bone; and (c), commercial processed and sintered cowbone.

composition and crystallinity of the bone apatite changes with sintering temperature (Fig. 7).

3. Hydroxyapatite from Corals, Coralline Hydroxyapatite

HA from corals and corraline hydroxyapatite are obtained by hydrothermal conversion of calcium carbonate, $CaCO_3$, in certain coral species (*Porites*) in the presence of ammonium phosphate, $(NH_4)_2HPO_4$, as developed by Roy and Linnehan [88], according to the reaction below:

$$\underset{\text{coral}}{CaCO_3} + (NH_4)_2HPO_4 \xrightarrow[\text{275°C, 12,000 psi}]{} \underset{\text{coralline HA}}{Ca_{10}(PO_4)_6(OH)_2} \qquad (12)$$

The interconnecting macroporosity of the original coral is preserved in this hydrothermal conversion (Fig. 6c). The resulting HA is similar but not identical in composition and crystallinity to the ceramic HA (Fig. 1a vs. Fig. 1b; Fig. 8), resulting in differences in properties (Table 2) discussed in Section IV.2.

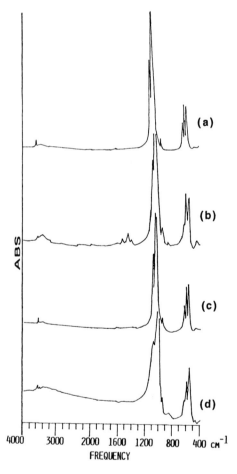

Figure 8   Infrared (IR) spectra: (a), ceramic hydroxyapatite (HA); (b), coralline HA; (c), sintered bovine bone; (d), calcium-deficient apatite (unsintered).

Table 2 Comparative Properties of Ceramic and Coralline Hydroxyapatite

Properties	Ceramic HA[a]	Coralline HA[b]
Composition		
From XRD analysis	HA	HA + β-TCMP[c]
From IR analysis	OH, PO$_4$	OH, PO$_4$, CO$_3$
Chemical analysis	Ca, P	Ca, P, CO$_3$, Sr, Mg
Porosity	Dense	Macroporous
Crystallinity (XRD)		
$\beta_{1/2}$(002) (deg 2θ)	0.12	0.25
Crystal size (TEM)	0.25 × 1.0	0.02 × 0.1 μm
Lattice parameters		
a-axis (+0.003A)	9.419	9.452
c-axis (+0.003A)	6.881	6.893

[a]Ceramic HA, Calcitite
[b]Coralline HA, Interpore
[c]β-TCMP = Mg-substituted β-TCP, based on the d-spacing indicating Mg-for-Ca substitution in the β-TCP or whitlockite structure [12]

D. Calcium Phosphates in Coatings of Implants

Ca-P materials are biocompatible and bioactive, owing perhaps to the similarity of their calcium and phosphate composition to the mineral phase of bone. These materials (calcium and phosphate) are not osteoinductive (do not induce bone formation) as shown by results of implantation in nonbony sites [43]. The Ca-P materials were shown to be osteoconductive, providing a scaffold or template for bone formation (Fig. 9). These attractive features are offset by their lack of strength necessary for load-bearing areas. Therefore, to combine the bioactivity of Ca-P materials and the strength of the metals (e.g., titanium or titanium alloy) used in the manufacture of dental and orthopedic implants, HA ceramic particulates are plasma sprayed onto the metal implants, which may or may not have been previously coated with titanium beads. The HA-coated im-

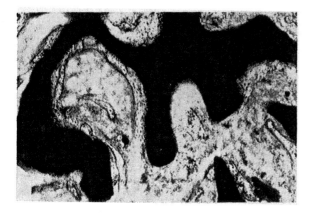

Figure 9 Microradiograph showing bone formation along the edges of BonAp (noncommercial material from processed and sintered bovine bone) showing the osteoconductivity of the material.

plants were reported to exhibit accelerated skeletal fixation and a stronger interface [16,17,22,49].

Coatings can also be deposited by other methods (i.e., electrodeposition or ion sputtering) [23, 31]; however, plasma spraying is the coating method used by manufacturers of coated dental and orthopedic implants.

The composition and crystallinity of the Ca-P coating depend on the following factors: composition of the starting material, methods of coating, and variability in the parameters used in the coating method [23,52–54,59,60,68]. For example, when HA ceramic was ion sputtered, the coating consisted of amorphous calcium phosphate (ACP) containing pyrophosphate (Fig. 10c). When plasma sprayed, the coating consisted of different Ca-P phases (Figs. 10b, 11, 12), including HA (with crystallinity much lower than that of the starting HA ceramic), a substantial amount of ACP, small amounts of α- and β-TCP and TTCP, and sometimes also CaO.

The gradient in composition and crystallinity from the metal/coating interface to the coating surface was demonstrated in an earlier study [59]; the ACP component was greater at the interface than at the surface (Fig. 12). The relative abundance of the Ca-P phases in the coatings varied from different manufacturers or even among those using similar HA ceramic as starting materials [52–54,59] (see Fig. 11). Posttreatments of coatings also affect the final composition of the coating (e.g., heating reduces the ACP component, autoclaving reduces the ACP and TCP components). However, this increase in per crystallinity (crystalline vs. noncrystalline phases) may be at the expense of shear strength between the coating and the metal substrate.

Since different Ca-P phases have different solubilities [52,53,59,68], it is expected that the stability of the coating *in vivo*, and therefore its biodegradation/bioresorption,

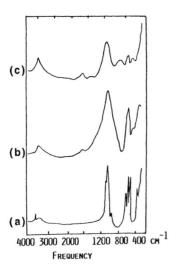

Figure 10   Infrared (IR) spectra: (a), hydroxyapatite (HA) ceramic used as target material for coating; (b), coating obtained by plasma spraying HA ceramic; (c), coating obtained by ion sputtering HA ceramic. The coating obtained by ion sputtering is an amorphous calcium phosphate (ACP) containing some pyrophosphate (peak at about 710 cm^{-1} shown in (c)). Coating obtained by plasma spraying HA ceramic is HA mixed with other Ca-P phases.

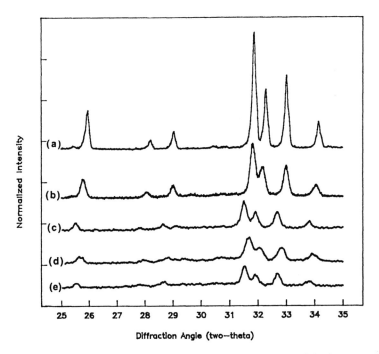

Figure 11 X-ray diffraction (XRD) patterns: (a), of hydroxyapatite (HA) ceramic used in plasma-spraying coatings on commercial dental implants; and (b)-(e), of coatings on different dental implants. Note the differences in crystallinity of the HA component in (b)-(e) compared with that of the starting HA source powder. The ratio of amorphous calcium phosphate (ACP) to HA in the coating increased from (b) to (e). CaO is present in some of the coatings (strongest CaO XRD peak at about 37.45° 2θ, not shown here).

will be affected by the composition of the coating [60]. The stability (or instability) of the coating will affect the stability of the total implant.

III. CALCIUM PHOSPHATE BIOMATERIALS COMPARED WITH BONE AND BONE MINERAL

Bone is an intimately integrated composite of inorganic and organic phases, with an inorganic/organic ratio of approximately 75/25 by weight and 65/35 by volume. The inorganic or mineral phases of calcified tissues (enamel, dentin, bone) were identified to be a calcium phosphate with an apatite structure, idealized as a calcium hydroxyapatite, $Ca_{10}(PO_4)_6(OH)_2$ [24,55,96]. Actually, these mineral phases are carbonate-substituted apatites associated with minor elements (notably, sodium and magnesium) and trace elements [55,59,61]. The Ca/P molar ratios of animal bones are below or above 1.67 (the stoichiometric value for pure HA), depending on the species, age, and type of bone [59]. The bone apatite microcrystals have rodlike or platelike morphology, with average dimensions of 250 × 30 Å, compared with about 2000 Å for HA ceramic (Table 3). The crystallinity (reflecting crystal size and/or strain) of bone apatite is similar to that of dentin; the crystallinities of both bone and dentin apatite are lower than that of enamel apatite (Fig. 13). In terms of carbonate, bone apatite contains much more carbonate than

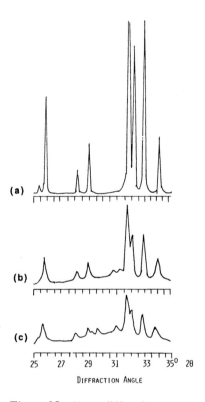

Figure 12 X-ray diffraction (XRD) patterns: (a), of hydroxyapatite (HA) ceramic; (b), coating closest to the metal; and (c), coating at the surface. Note difference in composition and crystallinity; the coating closest to the metal (b) has a higher ratio of amorphous calcium phosphate (ACP) to HA than the coating at the surface, as indicated by higher background in (b) and (c).

enamel apatite (Fig. 14). Compared with HA ceramic, bone apatite differs in crystallinity (Fig. 13c vs. Fig. 1a) and in composition, principally magnesium and CO_3 concentrations (Fig. 14c vs. Fig. 8a) and other properties (Table 3).

IV. CHARACTERIZATION AND PROPERTIES

A. Characterization of Crystal Properties

A combination of analytical techniques are recommended in investigating the crystal properties of Ca-P materials. These techniques include x-ray diffraction (XRD), infrared (IR) spectroscopy, thermogravimetric analysis (TGA); scanning electron microscopy (SEM) and transmission electron microscopy (TEM), and dissolution experiments that give information on solubilities. Each of these techniques gives specific information; however, each technique also has limitations, necessitating the combined use of several techniques for an accurate and comprehensive characterization [19,73].

The XRD profiles give the following information: (1) identification of phases present, if the minor phase is present above 1 wt%; (2) purity of the material, that is, whether there are single or multiphases; (3) crystallinity (reflecting crystal size and/or strain) of

Table 3 Comparative Composition and Crystallographic and Mechanical Properties of Human Enamel, Bone, and Hydroxyapatite Ceramic

	Enamel	Bone	HA
Constituents (wt%)			
Calcium, Ca^{2+}	36.0	24.5	39.6
Phosphorus, P	17.7	11.5	18.5
(Ca/P) molar	1.62	1.65	1.67
Sodium, Na^+	0.5	0.7	Trace
Potassium, K^+	0.08	0.03	Trace
Magnesium, Mg^{2+}	0.44	0.55	Trace
Carbonate CO_{32-}	3.2	5.8	–
Fluoride, F^-	0.01	0.02	–
Chloride, Cl^-	0.30	0.10	–
Ash (total inorganic)	97.0	65.0	100
Total organic	1.0	25.0	–
Adsorbed H_2O^a	1.5	9.7	–
Trace elements: Sr^{2+}, Pb^{2+}, Ba^{2+}, Fe^{3+}, Zn^{2+}, Cu^{2+}, etc.			
Crystallographic properties			
Lattice parameters (± 0.003Å)			
a-axis	9.441	9.419#	9.422
c-axis	6.882	6.880	6.880
Crystallinity index^a	70–75	33–37	100
Crystallite size, Å	1300 × 300	250 × 25–50	2000
Products after sintering (950°C)	HA + TCP	HA + CaO	HA
Mechanical properties			
Elastic modulus (10^6 MPa)	0.014	0.020^b	0.01
Tensile strength (MPa)	70	150^a	100

^{a}Ratio of coherent/incoherent scattering, with mineral HA value as 100
^{b}Values for cortical bone [22, 47], mechanical properties for human enamel [26]

the phases present; (4) percent crystallinity (crystalline vs. noncrystalline or amorphous phase); (5) crystal size and approximate crystal shape, approximated from the dimensions along the *a* and *c* directions [67]; (6) lattice parameters of the HA or TCP phases; (7) structural differences such as α- vs. β-TCP, $Ca_3(PO_4)_2$; and (8) the relative (semiquantitative) abundance of each phase (e.g., β-TCP/HA or ACP/HA ratios). The lattice parameters will indicate substitutions, if any, in the HA or β-TCP phases.

The IR spectra give the following information [33,66,67]: (1) identification of the phase or phases present; (2) functional groups present (e.g., CO_3, PO_4, OH, P_2O_7, from the inorganic phase; $-NH_2$, -CO from the organic phase); and (3) crystallinity (reflecting crystal size and/or strain) as deduced from the resolution of the IR absorption bands. The IR data complement and supplement the XRD data. The difference in composition and crystal properties observed in the XRD diffraction (Figs. 1, 7, 13) is reflected in the IR spectra (Figs. 5, 8, 14). Combined data from XRD and IR provide accurate information on substitutions in the HA or β-TCP structures [55,59,61,63,66,67,69]. Substitution

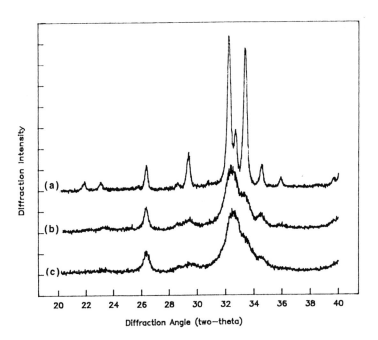

Figure 13 X-ray diffraction (XRD) patterns of apatite obtained from adult human tooth and adult cortical bone after removal of the organic phase: (a), enamel; (b), dentin; and (c), bone.

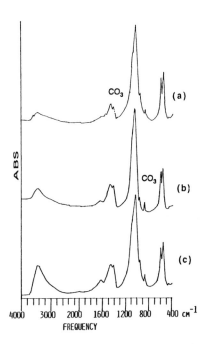

Figure 14 Infrared (IR) spectra: (a), enamel; (b), dentin; and (c), bone apatite. Note the difference in the intensity of the CO_3 absorption bands indicating the higher concentration of CO_3 in bone compared with that in enamel.

of CO_3 in biological apatites of enamel, dentin, and bone (Fig. 14) and in coralline HA (Fig. 8b), but not in ceramic HA (Fig. 8a) or unsintered apatite (Fig. 8d) can be readily shown by IR.

SEM is useful in describing the crystal morphology of the different calcium phosphate biomaterials (Figs. 3, 6). However, SEM is not informative enough when used to characterize surfaces of plasma-sprayed HA coatings (Fig. 15).

Electron microprobe (EDAX) analyses can be used to determine the distribution of ions (e.g., Ca, P) from the implant to the interface and to the bone regions [25,26].

TEM is useful in corroborating the crystal size/shape information from SEM, in demonstrating the presence of lattice defects in Ca-P ceramics [20], and in demonstrating the dissolution process associated with these materials [18,71,72]. TEM was also very helpful in demonstrating the formation of smaller crystallites on the surfaces of retrieved Ca-P ceramic implants [8,51] (see Fig. 16), and characterizing these crystallites as apatite similar to bone apatite [19,62,91]. However, it was IR analyses that demonstrated that these apatite crystallites associated with the Ca-P implant materials were CO_3-apatite similar to bone apatite [43,59,62]. Combined IR and TGA techniques demonstrated the intimate association between the CO_3-apatite and an organic matrix [43,62,71,72], as discussed in Section 5.

Besides the analytical techniques employed, the appropriateness of the sample also has to be considered. For example, information on composition and crystallinity of Ca-P phases in dental implant coatings was obtained from two techniques: (1) XRD analyses of the actual plasma-sprayed coatings on titanium alloy coupons and (2) XRD analyses of coatings scraped from coated coupons or from coated implants. A technical development using a custom-crafted sample holder made possible the direct XRD analyses of coatings on cylindrical dental implants [54]. Results showed that crystallinity and composition of coatings from coupons are different from the coating on the implant in spite of the fact that the coating conditions were presumably similar for both substrates [53]. This difference is probably due to the difference in size and geometries between the coupon and the cylindrical substrates [53,54]. Variations in composition (principally the ACP/HA ratios) and crystallinity of the HA component were observed in coatings of dental implants from different manufacturers [53,68], as shown in Fig. 11. A method employing an algorithm based on real XRD optics which resolve this problem is one that allowed the quantitative characterization of the calcium phosphate phases (including ACP, a significant component in many coatings) present in coatings obtained by plasma spraying HA [54].

The value of characterization of Ca-P materials by combining appropriate analytical techniques cannot be overemphasized. Characterization by Ca/P ratio alone without XRD and IR analyses is insufficient and can even be misleading when dealing with mixed Ca-P phases (e.g., β-TCP + HA, CaO + HA in HA ceramics) or the Ca-P phases in implant coatings. For example, a Ca/P molar ratio of 1.67 can mean pure HA or a mixture of TTCP (Ca/P = 2) plus TCP (Ca/P = 1.5) or CaO plus TCP. A Ca/P ratio of 1.5 does not distinguish between α- or β-TCP since both have the same chemical formula, $Ca_3(PO_4)_2$ but different crystallographic structures, which can only be detected by XRD and IR analyses. Characterization based on solubility product constant is not reliable unless dealing with a material with one phase. Characterization of Ca-P materials consisting of more than one phase (e.g., Ca-P phases in implant coatings) is meaningless unless reinforced by data from other types of analyses. Information and limitations of analytical methods for characterizing the bone-material interface are summarized in Table 4.

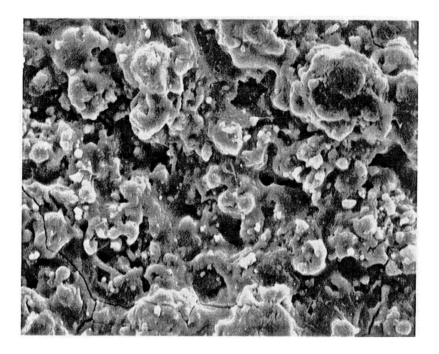

(a)

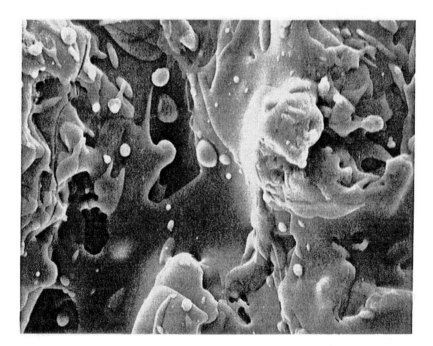

(b)

Figure 15   (a), (b) Scanning electron micrographs (SEM) of coatings on commercial dental implants obtained by plasma spraying hydroxyapatite (HA) ceramic onto titanium alloy substrate showing differences in surface morphology (magnification: ×3000).

B.   Composition

The composition and crystallinity of different Ca-P biomaterials are deduced from their XRD patterns (Figs. 1, 4, 7, 12–13) and IR spectra (Figs. 5, 8, 10, 14).

1.   Calcium Phosphates in Coatings

Characterization of the coatings demonstrated that the composition depends on the coating method: ion sputtering causes the formation of ACP (Fig. 10c), electrodeposition causes formation of ACP or apatite of poor crystallinity, and plasma spraying causes the formation of other calcium phosphates (Fig. 10b) with the HA (ACP, β- and α-TCP and TTCP) [23,31,52–54,59,60,63]. It was also observed that a composition gradient existed from the metal/coating interface to the coating surface (Fig. 12), with the ACP/HA ratio being higher near the interface [53,59]. Since different calcium phosphate phases have different solubilities, it is possible that the stability of the coating, and therefore of the implant, will depend on the composition of the coating. As mentioned above, the crystallinity and composition of the coating depend on methods of coating and other coating parameters. Ca-P phases found in coatings obtained by plasma spraying HA ceramic included HA (with lower crystallinity than that of the HA ceramic used for plasma spraying), ACP, α-TCP, β-TCP, TTCP, and, sometimes, also CaO (Figs. 11, 12).

Table 4   Analytical Methods for Characterizing Biomaterial-Bone Interface

Method	Information	Limitations
X-ray diffraction (XRD)	Phase(s) identification, crystallinity lattice parameters, lattice substitution, preferred orientation	Needs large interface areas, 4 mm^2
Infrared (IR) microprobe	Composition (can detect CO_3 in apatite)	Limited resolution, $>2 \mu m$
Raman microprobe	Composition type of CO_3 substitution (A or B)	Limited resolution, $>2 \mu m$
Light microscopy	Bone contact, organic phase birefringence	Needs large interface areas, $>500 \mu m^2$
X-ray microradiography	New bone formation, nonmineralized or fibrous layer	Limited by resolution
Scanning electron microscopy (SEM)	Crystal morphology, bone contact, surface topography	Limited by resolution, 10 μm
Transmission electron microscopy (TEM)	Crystal size/shape, mineral/organic	Requires preparation of thin sections; introduction of artifacts; qualitative results
Electron diffraction	Identification, lattice patterns Single crystal versus crystal population; preferred orientation	Does not distinguish between HA or CO_3-apatite; qualitative results
Electron microprobe (EDAX)	Elemental composition (Ca, P, Mg, S)	Limited by diameter of electron beam; 1 μm; mostly qualitative results

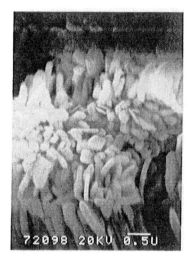

(a)

(b)

Figure 16 Scanning electron microscopy (SEM): (a), coralline hydroxyapatite (HA); (b) and (c), unsintered apatite prepared by hydrolysis of dicalcium phosphate dihydrate (DCPD).

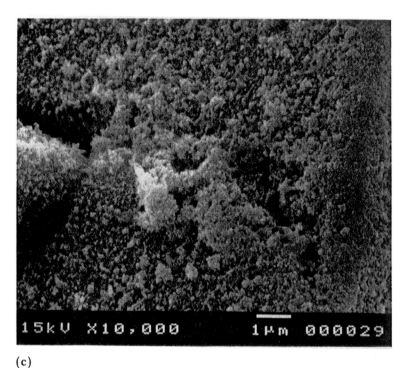

15kV X10,000   1μm 000029

(c)

2. Coralline Hydroxyapatite versus Ceramic Hydroxyapatite

The crystallinity of coralline HA (Interpore™), based on the broadening $\beta_{1/2}$ of the diffraction peaks, is lower than that of ceramic HA (Calcitite™), as shown in Fig. 1, indicating that the apatite crystals in coralline HA are much smaller than those in ceramic HA. This difference in crystal size was verified by SEM (Fig. 16a vs. Fig. 16b) and TEM [70]. The lattice parameters of HA and coralline HA also differ, reflecting the difference in their composition (Table 2). Coralline HA is not as pure as the synthetic ceramic HA. The coralline HA is actually a carbonate-apatite that also contains strontium; the CO_3-apatite phase is mixed with small amounts of magnesium-substituted β-TCP [58,59,70]. These differences in composition and crystal properties probably reflect the difference in their origin and method of preparation [58,59,70].

C. Dissolution Properties

The extent of dissolution of Ca-P materials is commonly investigated in buffers (acetate, lactate, or Tris buffers; pH 3, 6, or 7). *In vitro* dissolution of Ca-P materials depends on the type and concentration of the buffered or unbuffered solutions, pH of the solution, degree of saturation of the solution, solid/solution ratio, the length of suspension in the solutions, and the composition and crystallinity (reflecting crystal size and strain) of the Ca-P material. The degree of micro- and macroporosities, presence of defect structure, and the amount and type of other phases present also have significant influence [20,73].

Powdered coralline HA was shown to have a greater extent of dissolution than ceramic HA (Fig. 17). This is due not only to the difference in their composition (Table 2). Coralline HA contains carbonate, magnesium, and strontium, which have been shown to promote a greater extent of dissolution of synthetic apatites containing these ions [59]. Coralline and ceramic HA also differ in their crystallinity, reflecting crystal size (Figs. 1, 8, 12, 16).

The difference in composition and crystallographic structure (e.g., HA vs. β-TCP; α- vs. β-TCP; ACP vs. HA) of Ca-P materials is reflected in the difference in their stability and solubility. The order of relative solubility of some of the Ca-P compounds are (Fig. 18)

$$ACP \gg TTCP \gg \alpha\text{-TCP} \gg \beta\text{-TCP} \gg AP \gg HA$$

for which AP is an unsintered calcium-deficient apatite and HA is HA ceramic sintered at 1100°C, $Ca_{10}(PO_4)_6(OH_2)$. Substitutions in the TCP or HA structure will affect their extent of dissolution [14,59,60]. The extent of dissolution of BCP materials will depend on their β-TCP/HA ratios (i.e., the higher the ratio is, the greater the extent of dissolution will be) [60,62].

In the case of plasma-sprayed coatings consisting principally of HA, ACP, and much smaller amounts of other Ca-P phases (α- and β-TCP, TTCP) and sometimes CaO, the extent of dissolution will be affected by the type and amount of the non-HA phases.

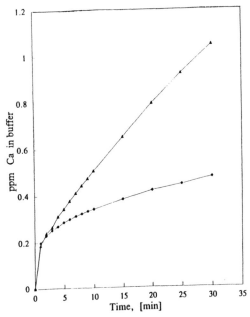

Figure 17 Comparative extent of dissolution of ceramic hydroxyapatite (HA) versus coralline HA in acetate buffer (0.1M KAc, pH 6, 37°C, solid/solution ratio of 25 mg/100 ml). The dissolution is expressed in terms of Ca^{2+} ions released from the material with time [68].

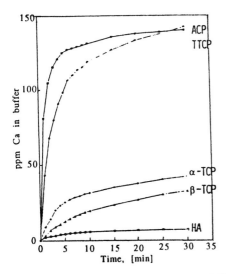

Figure 18   Comparative dissolution of different calcium phosphate phases: amorphous calcium phosphate (ACP) (obtained by plasma spraying) [54]; α- and β-tricalcium phosphate (TCP); ceramic hydroxyapatite (HA); precipitated apatite (calcium deficient) prepared in the laboratory [68].

Coatings with a higher ACP/HA ratios will dissolve or biodegrade to a greater extent than coatings with a low ACP/HA ratio of higher crystallinity [68]. The variability in composition and crystallinity of the coatings (Fig. 11) is reflected in the difference in the extent of dissolution of the coating (Fig. 19) [68].

Preparation conditions such as sintering temperature affect the crystallinity (crystal size and/or strain) of HA (Fig. 3). Sintering temperature also affects crystal perfection. HA sintered at 900°C had a greater number of lattice defects than those prepared at 1200°C [20]. This is reflected in the higher dissolution rate of HA prepared at a lower temperature [60,80].

V.   TISSUE RESPONSE TO CALCIUM PHOSPHATE BIOMATERIALS

The biocompatibility of Ca-P biomaterials is demonstrated *in vitro* and *in vivo* [6,13,36, 38,60,70–72,75,77,80–82,94]. Cellular reactions are responsive to surface properties such as surface roughness [77], surface chemistry [28,29,90], and reactivity of the materials [21,36,80,81,89,95]. Ca-P materials promote cell adhesion and proliferation (Fig. 20) regardless of the composition; cellular action promotes partial dissolution (Fig. 21) of Ca-P materials [18,71,72].

A.   Biodegradation/Bioresorption

The extent of biodegradation/bioresorption of Ca-P materials reflects their solubility or dissolution properties. The factors influencing the dissolution properties were similar to those affecting biodegradation or bioresorption, namely, physical forms (dense vs.

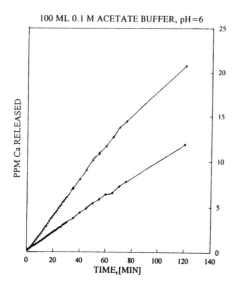

Figure 19 Comparative dissolution of coatings from implants from the same manufacturer. Coated implants were suspended in buffer solutions and release of Ca^{2+} ions monitored using specific ion electrode [68].

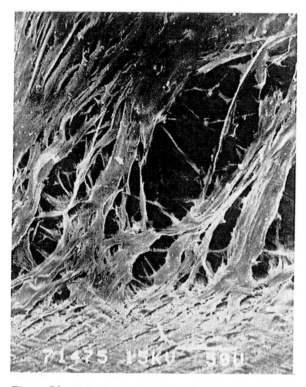

Figure 20 Adhesion and proliferation of cells on surface of ceramic hydroxyapatite (HA).

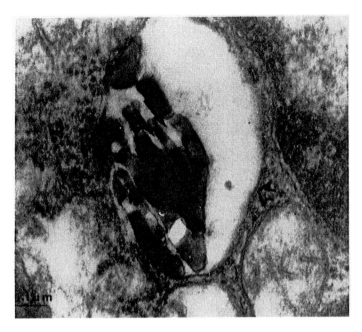

Figure 21 Partial dissolution of biphasic calcium phosphate (BCP) ceramic induced by cellular action [18,60,72].

macroporous), density, composition, and crystallinity (β-TCP vs. HA; coralline HA vs. ceramic HA).

Since biodegradation/bioresorption is related to the dissolution properties of the Ca-P materials, it can be expected that the composition and crystallinity of the Ca-P phases of the coatings will affect the stability of the coating, which will consequently affect the long-term stability of the implant. It is reasonable to assume that Ca-P coatings containing relatively higher concentrations of more soluble (or more biodegradable) Ca-P phases (e.g., ACP, TTCP, α- and β-TCP) will be less stable than those that consist principally of well-crystallized HA.

B. The Calcium Phosphate Biomaterial-Bone Interface

The establishment of a very strong interface between the bioactive material and bone [41] is believed to be influenced by the initial formation of carbonate apatite on the surfaces of these materials [40,41,46,71,72].

The identification of the CO_3-apatite microcrystals associated with much larger Ca-P crystals and their intimate association with an organic phase were determined using TEM, IR, and TGA (Figs. 22 and 23) as reported in Refs. 43 and 62. The presence of these microcrystals is believed to proceed by the dissolution and precipitation processes [18,59,60,62,71,72]. It is proposed that the acid conditions known to be produced by cellular action [7] on the implant surface cause partial dissolution of the crystals of the Ca-P biomaterials (from the bulk or from the coating). This dissolution process increases the levels of calcium and phosphate ions in the biological fluids in the immediate environment, subsequently possibly causing the precipitation of apatite microcrystals that incor-

Figure 22 Transmission electron microscopy (TEM) showing microcrystals associated with much larger crystals of Ca-P biomaterials.

porate some of the ions (Ca^{2+}, Mg^{2+}, Na^+, CO_3^{2-}, HPO_4^{2-}, PO_4^{3-}, etc.) and organic molecules present in the fluid (Fig. 24). The formation of CO_3-apatite can also occur by an indirect process, through the initial formation of nonapatitic Ca-P phases that are more stable under acid conditions (e.g., DCPD, OCP) or under conditions of high Mg^{2+} or CO_3^{2-} (ACP, β-TCMP), which can later hydrolyze to CO_3-apatite [59,60,71,72]. The formation of CO_3-apatite on surfaces of bioactive materials (Fig. 22) *in vitro* and *in vivo* appears to be related to their bioactivity and their ability to form a strong bone-material interface (Table 5) [40–42,60,71,72].

VI. FUTURE CALCIUM PHOSPHATE BIOMATERIALS

At present, there is some disagreement as to the appropriate properties (specifically, crystallinity) and thickness of the implant coatings. Should the coating have high or low crystallinity (high or low HA/ACP)? Should the coating resist or facilitate biodegradation or bioresorption? Should the same Ca-P (dense or macroporous, HA to BCP, ceramic or coralline HA) material be used for every application (repair of bony defects, bone augmentation, bone substitution) or should Ca-P material of different property be used for different applications? Should the material have a high or low level of biodegradation? Is the formation of CO_3-apatite crystallites on the surface of the Ca-P material predictive of the material's *in vivo* performance? If so, to what extent is it predictive?

Exploration and development of calcium phosphate materials for bone repair, augmentation, and substitution continue. Fluoride-substituted apatites (fluoroapatites, FAs)

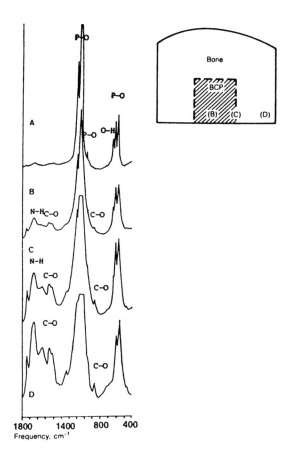

Figure 23 Infrared (IR) spectra: (A), hydroxyapatite (HA) ceramic before implantation; (B), material from the core; (C), material from the interface; and (D), material from the host bone. Note the presence of CO_3-apatite and organic phase in (B), (C), and (D).

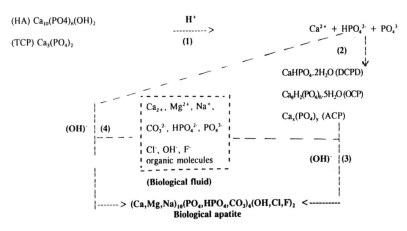

Figure 24 Schematic representation of the processes of dissolution and precipitation involved in the formation of CO_3-apatite on surfaces of Ca-P biomaterials.

Table 5   Effect of Type of Biomaterial on Type of Bonding at the Material-Bone Interface

Material	Biodynamics	Type of bonding
Metals (except titanium) and polymers	Biotolerant	Distance osteogenesis ___| |___ Bone Implant
Ceramic oxides (alumina, zirconia), titanium and titanium alloy	Bioinert	Contact osteogenesis _____	_____ B I
Ca-P biomaterials, special glass ceramics (Bioglass™)	Bioactive	Bonding osteogenesis B_____ _ _ _ _ _ _ _ _____ I

Source: From Ref. 83

are considered for their greater crystallinity and greater stability against dissolution compared with HA [59,69,76,96]. On the other hand, unsintered apatites with properties similar to those of the bone mineral have also been previously recommended [56]. Polymeric and nonpolymeric composites with Ca-P materials [2,11,30,44,64,86,93] should be further investigated since bone, after all, is a composite of a biological polymer (collagen mostly) and bone mineral (CO_3-apatite). Cements derived from apatitic materials or forming apatite as they set [2,12,65] are also being pursued for different dental and medical applications.

ACKNOWLEDGMENTS

The authors gratefully acknowledge the special collaboration of Dr. Guy Daculsi (Nantes), Dr. Orly (Colletica), Dr. M. Gregoire (Nantes), Ms. R. Zhang, Ms. R. Zheng, and Mr. J. Wang for some of work cited in this chapter. The authors also acknowledge the collaboration of Professors T. V. Vijayaraghavan and B. Penugonda (New York University College of Dentistry) for their help in some of the SEM analyses. The valuable technical assistance of M. Retino and the help of Pia Almagro (word processing) are also gratefully acknowledged. This work was supported in part by research grants from NIH/NIDR Nos. DE-04123, DE-07223, and 2 SO7 RR 0762, and by the Calcium Phosphate Research Funds.

REFERENCES

1. Albee, F.H., Studies in bone growth. Triple calcium phosphate as a stimulus to osteogenesis, *Ann. Surg.*, 71:32–36 (1920).
2. Alexander, H., Parsons, J.R., Ricci, J.L., Bajpai, P.K., and Weiss, A.B., Calcium phosphate ceramic based composite as bone graft substitutes. In *Critical Reviews in Biocompatibility*, C. William, ed., CRC Press, Boca Raton, FL, pp. 43–77 (1987).
3. Aoki, H., Kato, K., Ogiso, M., and Tabata, T., Sintered hydroxyapatite as a new dental implant material, *J. Dent. Outlook*, 49:567–575.

4. Aoki, H., *Science and Medical Applications of Hydroxyapatite*. Japan Association of Apatite Science (JAAS), Takayama Press System Center Co., Tokyo (1991).

5. American Society for Testing and Materials, *Annual Book of ASTM Standards*, Section 13, ASTM, Philadelphia, F.1185–1188 (1990).

6. Bagambisa, F.B., Joos, U., and Schilli, W., The interaction of osteogenic cells with hydroxylapatite implant materials *in vitro* and *in vivo*, *Oral Maxillofac. Implants*, 5:217–226 (1990).

7. Baron, R., Neff, L., Louvard, D., and Courtoy, P.J., Cell-mediated extracellular acidification and bone resorption: Evidence for a low pH in resorbing lacunae and localization of 100-kD lysosomal membrane protein at the osteoclast ruffled border, *J. Cell Biol.*, 101:2210–2222 (1985).

8. Beckham, A., Greenlee, T.K., and Crebo, A.R., Jr., Bone formation at a ceramic implant interface, *Calcif. Tiss. Res.*, 8:165–171 (1971).

9. Bhaskar, S.N., Brady, J.M., Getter, L., Grower, M.F., and Driskell, T., Biodegradable ceramic implants in bone, *J. Oral Surg.*, 32:336–346 (1971).

10. Bonel, G., Heughebaert, J.-C., Heughebaert, M., Lacout, J.L., and Lebugle, A., Apatitic calcium orthophosphates and related compounds for biomaterials preparation. In *Bioceramics: Materials Characteristics Versus In Vivo Behavior*, P. Ducheyne and J.E. Lemons, eds., *Ann. N.Y. Acad. Sci.*, 523:115–130 (1988).

11. Bonfield, W., Hydroxyapatite-reinforced polyethylene as an analogous material for bone replacement. In *Bioceramics: Materials Characteristics Versus In Vivo Behavior*, P. Ducheyne and J.E. Lemons, eds., *Ann. N.Y. Acad. Sci.*, 523:173–177 (1988).

12. Brown, W.E., and Chow, L.C., Dental restorative cement pastes. U.S. Patent No. 4,518,430 (May 21, 1985).

13. Cheung, H.S., and Haak, M.H., Growth of osteoblasts on porous calcium phosphate ceramic: An *in vitro* model for biocompatibility study, *Biomat.*, 10:63–67 (1988).

14. Christofferssen, J., and Christoffersen, M.R., Kinetics of dissolution of calcium hydroxyapatite. V. The acidity constant for the hydrogen phosphate surface complex, *J. Cryst. Growth*, 57:21–26 (1982).

15. Cook, S.D., Kay, J.F., Thomas, K.A., and Jarcho, M., Interface mechanics and histology of titanium and hydroxyapatite coated metal implants, *J. Biomed. Mater. Res.*, 23:183–199 (1987).

16. Cook, S.D., Thomas, K.A., Dalton, J.E., Volkman, R.K., Whitecloud, T.S., and Kay, J.E., Hydroxylapatite coating of porous implants improves bone ingrowth and interface attachment strength, *J. Biomed. Mater. Res.*, 26:989–1001 (1992).

17. Cranin, A.N., Tobin, G.P., and Gelbman, J., Applications of hydroxyapatite in oral and maxillofacial surgery, Part II: Ridge augmentation and repair of major oral defects, *Compend. Contin. Educ. Dent.*, 8:334–345 (1987).

18. Daculsi, G., LeGeros, R.Z., and Mitre, D., Crystal dissolution of biological and ceramic apatites, *Calcif. Tiss. Int.*, 45:95–103 (1989).

19. Daculsi, G., LeGeros, R.Z., and Deudon, C., Scanning and transmission electron microscopy and electron probe analysis of the interface between implants and host bone, *Scan. Micros.*, 4:309–314 (1990).

20. Daculsi, G., LeGeros, R.Z., LeGeros, J.P., and Mitre, D., Lattice defects in calcium phosphate ceramics: High resolution TEM ultrastructural study, *J. Appl. Biomat.*, 2:147–152 (1991).

21. Davies, J.E., The use of cell and tissue culture to investigate bone cell reactions to bioactive materials. In *Handbook of Bioactive Ceramics*, Vol. 1, L. Yamamuro T. Hench, and J. Wilson, eds., CRC Press, Boca Raton, FL (1990).

22. deGroot, K., Ceramic of calcium phosphates: Preparation and properties. In *Bioceramics of Calcium Phosphate*, K. deGroot, ed., CRC Press, Boca Raton, FL, pp. 100–114 (1983).

23. deGroot, K., Hydroxylapatite coatings for implants in surgery. In *High Tech Ceramics*, P. Vincenzini, ed., Elsevier Science Publishers, B.V., Amsterdam, pp. 381–386 (1987).

24. DeJong, W.F., La Substance minerals das les os, *Tec. Trav. Chim.*, 45:445–458 (1926).

25. Denissen, H., Dental Root Implants of Apatite Ceramics. Experimental Investigations and Clinical Use of Dental Root Implants Made of Apatite Ceramics. Ph.D. Thesis, Vrije Universiteit te Amsterdam (1979).
26. Denissen, H., Mangano, C., and Cenini, G., *Hydroxylapatite Implants*. Piccin Nuova Libraria, S.P.A., India (1985).
27. Donath, K., Reaction of tissue to calcium phosphate ceramics. In *Osseointegrated Implants, Vol. 1. Basics, Materials and Joint Replacements* G. Heimke, ed., CRC Press, Boca Raton, FL, pp. 99–126 (1990).
28. Doss, S.K., Surface property of hydroxyapatite: The effect on various inorganic ions on the electrophoretic behavior, *J. Dent. Res.*, 55:1067–1075 (1976).
29. Ducheyne, P., Kim, C.S., and Pollack, S.R., The effect of phase differences on the time-dependent variation of the zeta potential of hydroxyapatite, *J. Biomed. Mater. Res.*, 26:147–168 (1992).
30. Ducheyne, P., Marcolongco, M., and Schepers, E., Bioceramic composites. In *An Introduction to Bioceramics*, L.L. Hench and J. Wilson, eds., World Scientific Publishers, London, pp. 281–297 (1993).
31. Ducheyne, P., Van Raemdonck, W., Heughebaert, J.C., and Heughebaert, M., Structural analysis of hydroxyapatite coatings on titanium, *Biomat.*, 7:97–103 (1986).
32. Ellinger, R.F., Nery, E.B., and Lynch, K.L., Histological assessment of periodontal osseous defects following implantation of hydroxyapatite and biphasic calcium phosphate ceramics. A case report, *Int. J. Perio. Restor. Dent.*, 3:223–233 (1986).
33. Fowler, B.O., Moreno, E.C., and Brown, W.E., Infrared spectra of hydroxyapatite octacalcium phosphate and pyrolyzed octacalcium phosphate, *Archs. Oral. Biol.*, 11:477–492.
34. Frank, R.M., Widermann, P., Hemmerle, J., and Freymann, Pulp capping with synthetic hydroxyapatite in human premolars, *J. Appl. Biomat.*, 2:243–250 (1991).
35. Frank, R.M., Gineste, M., Benque, E.P., Hemmerle, J., Duffort, J.F., and Heughebaert, M., Etude ultrastructurale de l'induction osseuse apres implantation de bioapatites chez l'homme, *J. Biol. Bucc.*, 15:125–134 (1987).
36. Frayssinet, P., Primout, I., Rouquet, N., Autefate, A., Guilhem, A., and Bonnevaille, P., Bone cell grafts in bioreactor: A study of feasibility of bone cell autograft in large defects, *J. Mater. Sci. Mater. Med.*, 2:217–221 (1991).
37. Galgut, P.N., Waite, I.M., and Tinkler, S.M.B., Histological investigation of the tissue response to hydroxyapatite used as an implant material in periodontal treatment, *Clin. Mater.*, 6:105–121 (1990).
38. Gregoire, M., Orly, I., and Menanteau, The influence of calcium phosphate biomaterials on human bone cell activities: An *in vitro* approach, *J. Biomed. Mater. Res.*, 24:163–177 (1972).
39. Hayek, E., and Newesely, H., Pentacalcium monohydroxyorthophosphate, *Inorg. Syn.*, 7:63–65 (1963).
40. Hench, L.L., Bioceramics: From concept to clinic, *J. Am. Ceram. Soc.*, 74:1487–1510 (1991).
41. Hench, L.L., and Wilson, J., Surface-active materials, *Biomater. Science*, 226:630–636 (1984).
42. Hench, L.L., Splinter, R.J., Allen, W.C., and Greenlee, T.K., Bonding mechanisms at the interface of ceramic prosthetic materials, *J. Biomed. Mater. Res.*, 2:117–141 (1971).
43. Heughebaert, M., LeGeros, R.Z., Gineste, M., Guilhem, A., and Bone, G., Physico-chemical characterization of deposits associated with HA ceramics implanted in non-osseous sites, *J. Biomed. Mater. Res.*, 22:254–268 (1988).
44. Hong, Y.C., Wang, J.T., and Hong, C.Y., The periapical tissue reactions to a calcium phosphate cement in the teeth of monkeys, *J. Biomed. Res. Mater. Res.*, 25:485–498 (1991).
45. Hubbard, W., Physiological Calcium Phosphate as Orthopedic Implant Material #6. Ph.D. Thesis, Marquette University, Milwaukee, WI (1974).
46. Hyakuna, K., Yamamuro, T., Kotura, Y., Oka, M., Nakamura, T., Kitsugi, T., Kokubo, T.,

and Kushitani, H., Surface reactions of calcium phosphate ceramics to various solutions, *J. Biomed. Mater. Res.*, 24:471–488 (1990).

47. Jarcho, M., Hydroxylapatite synthesis and characterization in dense polycrystalline forms, *J. Mater. Sci.*, 11:2027–2035 (1976).

48. Jarcho, M., Calcium phosphate ceramics as hard tissue prosthetics, *Clin. Orthopaed.*, 157: 259–278 (1981).

49. Jarcho, M., Retrospective analysis of hydroxyapatite development for oral implant applications, *Dent. Clin. North Amer.*, 36:19–26 (1992).

50. Jean, A., Kerebel, B., and LeGeros, R.Z., Effects of various calcium phosphate materials on the reparative dentin bridge, *J. Endo.*, 14:83–87 (1988).

51. Kay, J.F., Physiological acceptance for a ceramic bone implant determined by electron optical analysis. Ph.D. Thesis, Rensselaer Polytechnic Institute, Troy, NY (1977).

52. Klein, C.P.A.T., Wolke, J.G.C., and deGroot, K., Stability of calcium phosphate ceramics and plasma sprayed coating. In *An Introduction to Bioceramics*, L.L. Hench and J. Wilson, eds., World Scientific Publishers, London, pp. 199–221 (1993).

53. LeGeros, J.P., and LeGeros, R.Z., Characterization of calcium phosphate coatings on implants. *The 17th Annual Meeting of the Society for Biomaterials*, May 1–5, 1991, Scottsdale, AZ, Abstr. #192 (1991).

54. LeGeros, J.P., LeGeros, R.Z., Edwards, B., Zitelli, J., and Burgess, A., X-ray diffraction method for the quantitative characacterization of calcium phosphate coatings. In *Characterization and Performance of Calcium Phosphate Coatings for Implants*, ASTM STP 1196, E. Horowitz and J.E. Parr, eds., ASTM, Philadelphia (1994).

55. LeGeros, R.Z., Apatites in biological systems, *Prog. Crystal Growth Charact.*, 4:1–45 (1981).

56. LeGeros, R.Z., Properties of commercial bone grafts compared to human bone and new synthetic bone biomaterials, *Ninth Annual Meeting of the Society for Biomaterials*, Birmingham, AL, Abstr. No. 86 (1983).

57. LeGeros, R.Z., Variability of b-TCP/HAP ratios in sintered apatites, *J. Dent. Res.*, 65:292, Abstr. No. 110 (1986).

58. LeGeros, R.Z., Calcium phosphate materials in restorative dentistry: A review, *Adv. Dent. Res.*, 2:164–183 (1983).

59. LeGeros, R.Z., *Calcium Phosphates in Oral Biology and Medicine. Monographs in Oral Sciences*, Vol. 15, H. Myers, ed., S. Karger, Basel (1991).

60. LeGeros, R.Z., Biodegradation and bioresorption of calcium phosphate ceramics, *Clin. Mat.*, 14:65–88 (1993).

61. LeGeros, R.Z., Biological and synthetic apatites. In *Hydroxyapatite and Related Compounds*, P. Brown, ed., CRC Press, Boca Raton, FL (1994).

62. LeGeros, R.Z., and Daculsi, G., *In vivo* transformation of biphasic calcium phosphate ceramics: Ultrastructural and physicochemical characterizations. In *Handbook of Bioactive Ceramics*, Vol. 2, T. Yamamuro, L. Hench, and J. Wilson, eds., CRC Press, Boca Raton, FL, pp. 17–28 (1990).

63. LeGeros, R.Z., and LeGeros, J.P., Dense hydroxyapatite. In *An Introduction to Bioceramics*, L.L. Hench and J. Wilson, eds., World Scientific Publishing, London, pp. 139–180 (1993).

64. LeGeros, R.Z., and Penugonda, B., Potential use of calcium phosphate as fillers in composite restorative biomaterials. *Second World Congress on Biomaterials*, Washington, DC (1984).

65. LeGeros, R.Z., Chohayeb, A., and Schulman, A., Apatitic calcium phosphates: Possible restorative materials, *J. Dent. Res.*, 61:343, Abstr. No. 1482 (1982).

66. LeGeros, R.Z., LeGeros, J.P., Trautz, O.R., and Klein, E., Spectral properties of carbonate in carbonate-containing apatites, *Dev. Applied Spectroscopy*, 7B:3–12 (1970).

67. LeGeros, R.Z., LeGeros, J.P., Trautz, O.R., and Shirra, W.P., Conversion of monetite, $CaHPO_4$, to apatites: Effect of carbonate on the crystallinity and the morphology of the apatite crystallites, *Adv. in X-Ray Anal.*, 14:57–66 (1971).

68. LeGeros, R.Z., Kijkowska, R., and LeGeros, J.P., Calcium phosphate coatings on implants: Dissolution characteristics. *19th Annual Meeting Society Biomaterials*, Birmingham, AL (April 1993).

69. LeGeros, R.Z., Singer, L., Ophaug, R., and Quirolgico, G., The effect of fluoride on the stability of synthetic and biological (bone mineral) apatites. In *Osteoporosis*, J. Menczel, G.C. Robin, and M. Makin, eds., J. Wiley and Sons, New York, pp. 327–341 (1982).

70. LeGeros, R.Z., Orly, I., Gregoire, M., and Kazimiroff, J., Comparative properties and *in vitro* transformation of HA ceramics in serum, *J. Dent. Res.*, 67:177, Abstr. No. 512 (1988).

71. LeGeros, R.Z., Daculsi, G., Orly, I., and Gregoire, M., Substrate surface dissolution and interfacial biological mineralization. In *The Bone-Biomaterial Interface*, J.E. Davies, ed., University of Toronto Press, pp. 76–88 (1991).

72. LeGeros, R.Z., Daculsi, G., Orly, I., Gregoire, M., Heughebaert, M., Gineste, M., and Kijkowska, R., Formation of carbonate apatite on calcium phosphate materials: Dissolution/ precipitation processes. In *Bone Bonding*, P. Ducheyne, Kokubo, and Van Blitterswijk, eds., Reed Healthcare Communications, pp. 201–212 (1992).

73. LeGeros, R.Z., Daculsi, G., Orly, I., Walters, M., and LeGeros, J.P., Bioceramics-tissue interfaces: Characterization of ultrastructural properties. In *Bioceramics*, Vol. 6, P. Ducheyne and O. Christiansen, eds., Butterworth-Heinemann Ltd., pp. 79–84 (1993).

74. Linkow, L., Bone transplants using the symphysis, the iliac crest and synthetic bone materials, *J. Oral Implantol.*, 11:211–217 (1984).

75. Metsger, D.S., Driskell, T.D., and Paulsrud, J.R., Tricalcium phosphate ceramic – A resorbable bone implant: Review and current status, *J. Am. Dent. Assoc.*, 105:1035–1048 (1982).

76. Moreno, E.C., Kresak, M., and Zahradnik, R.T., Physicochemical aspects of fluoride-apatite systems relevant to the study of dental caries, *Caries Res.*, (Suppl. 1)11:142–177 (1977).

77. Muller-Mai, C.M., Voigt, C., and Gross, U., Incorporation and degradation of hydroxyapatite implants of different surface roughness and surface structure in bone, *Scan. Micros.*, 4: 613–624 (1990).

78. Nery, E.B., Lynch, K.L., Hirthe, W.M., and Mueller, K.H., Bioceramic implants in surgically produced infrabony defects, *J. Periodont.*, 46:328–339 (1975).

79. Nery, E., LeGeros, R.Z., and Lynch, K.L., Tissue response to biphasic calcium phosphate ceramic with different ratios of biphasic calcium phosphate ceramic with different ratios of HA/TCP in periodontal osseous defects, *J. Periodontol.*, 63:729–735 (1992).

80. Nagai, N., and LeGeros, R.Z., Physicochemical properties and tissue reactions to HA materials: Effect of sintering temperature. *19th Annual Meeting Society Biomaterials*, Birmingham, AL (1993).

81. Niwa, S., Sawai, K., Takahashi, S., Tagai, H., Ono, M., and Fukuda, Y., Experimental studies on the implantation of hydroxylapatite in the medullary canal of rabbits, *Biomat.*, 1: 65–71 (1980).

82. Ohgushi, H., Dohi, Y., Tamai, S., and Tabata, S., Osteogenic differentiation of marrow stromal stem cells in porous hydroxyapatite ceramics, *J. Biomed. Mat. Res.*, 27:1401–1407 (1993).

83. Osborn, J.F., and Neweseley, H., The material science of calcium phosphate ceramic, *Biomat.*, 1:108–111 (1980).

84. Passuti, N., Daculsi, G., Rogez, J.M., Martin, S., and Bainvel, J.V., Macroporous calcium phosphate ceramic in dogs, *Clin. Orthop. Rel. Res.*, 248:169–176 (1989).

85. Piecuch, J.J., Augmentation of the atrophic edentulous ridge with porous replaniform hydroxyapatite (Interpore-200), *Dent. Clin. N. Am.*, 291–305 (1992).

86. Rawls, H.R., LeGeros, R.Z., and Zimmerman, B.F., A radiopaque composite restorative using an apatite filler, *J. Dent. Res.*, 64:209, Abstr. No. 307 (1985).

87. Ray, R.D., and Ward, A.A., A preliminary report on studies of basic calcium phosphate in bone replacement, *Surg. Form*, 3:429–439 (1951).

88. Roy, D.M., and Linnehan, S.A., Hydroxyapatite formed from coral skeleton carbonate by hydrothermic exchange, *Nature*, 247:220–227 (1974).

89. Shors, E.C., and Holmes, R.E., Porous hydroxyapatite. In *An Introduction to Bioceramics*, L.L. Hench and J. Wilson, eds., World Scientific Publishers, London, pp. 181–198 (1993).

90. Somasundran, P., Zeta potential of apatite in aqueous solution and its change during equilibrium, *J. Colloid. Interface Sci.*, 27:659–666 (1968).

91. Tracy, B.M., and Doremus, R.H., Direct electron microscopy studies of the bone-hydroxylapatite interface, *J. Biomed. Mater. Res.*, 18:719–726 (1984).

92. van Blitterswijk, C.A., Calcium Phosphate Middle-Ear Implants. Ph.D. Thesis, Rijksuniversitiet te Leiden, Leiden (1985).

93. van Sliedregt, A., Hydroxyapatite/polyactide composites for reconstructive surgery: An *in vitro* and *in vivo* biocompatibility study. Ph.D. Thesis, Leiden University, The Netherlands.

94. Williams, D.F., Review: Tissue-biomaterial interactions, *J. Mat. Sci.*, 22:3421–3445 (1987).

95. Wilson, J., and Merwin, G.E., Biomaterials for facial bone augmentation: Comparative studies, *J. Biomed. Mater. Res. Appl. Biomat.*, 22:159–177 (1988).

96. Young, R.A., and Elliott, J.C., Atomic scale bases for several properties of apatites, *Archs. Oral Biol.*, 11:699–707 (1966).

97. Yukna, R.A., Mayer, E.T., and Brite, D.V., Longitudinal evaluation of durapatite ceramic as an alloplastic implant in periodontal osseous defects after three years, *J. Periodontol.*, 55:633–637 (1984).

44
Mechanical and Biological Characterization of Calcium Phosphates for Use as Biomaterials

Jeffrey M. Toth and Kenneth L. Lynch
The Medical College of Wisconsin, Milwaukee, Wisconsin

Timothy R. Devine
ConvaTec, Racine, Wisconsin

I. MATERIAL CHARACTERIZATION OF CALCIUM PHOSPHATES

Several phases of calcium phosphate ceramics exist that are being used or have been investigated for use as biomaterials for such applications as bone graft substitutes, bioactive coatings for prosthetic devices, dental devices, and controlled drug release. Calcium phosphate phases such as apatite are the principal inorganic constituents of hard tissues found in many biological systems, particularly teeth and bones. Because of their similarity to these naturally occurring calcium phosphate salts, calcium phosphates were synthesized for biomedical applications. Apatites and substituted apatites, in particular calcium hydroxyapatite (HA), as well as tricalcium phosphates (TCPs) and calcium pyrophosphates are the principal phases that have been investigated for use as biomaterials. Mixtures of these phases in various ratios have also been used for implantation purposes. The intent of this chapter is to describe some of the characterization of the material and biological properties of synthetic calcium phosphates for biomedical applications. When known, correlations between the material properties and the biological response to the calcium phosphates are elucidated.

A. Characterization by X-Ray Diffraction

1. *Qualitative Analysis*

X-ray diffraction is a materials engineering tool that can be used to identify the purity and composition of crystalline calcium phosphate materials. Since the biological response to calcium phosphates varies depending on the phase of the material implanted, it is necessary to use x-ray diffraction techniques in order to characterize the phase of the material implanted to ensure quality control of the synthetic ceramic and optimize the biological response to the material. X-ray diffraction will identify the characteristic diffraction lines that correspond to the interplanar spacings of any crystalline material [1,2]. The resultant pattern, which is comprised of plane spacings (located on the *x*-axis) and intensities (located on the *y*-axis), can then be matched to the American Society for Testing and Materials (ASTM) Joint Committee on Powder Diffraction Standards (JCPDS) standards for the various calcium phosphate phases. Hence, the location and intensity of the peaks give the characteristic "thumbprint" that can be used to perform qualitative analysis. One caveat here is that x-ray diffraction will identify the presence of other crystalline phases, but not amorphous or poorly crystalline materials.

Table 1 shows some of the calcium phosphate phases with their corresponding ASTM JCPDS numbers. Our laboratory performs x-ray diffraction analysis on precipitated powders and final calcium phosphate products using copper k-α ($\lambda = 1.54$ angstrom [Å]) radiation at 40 kilovolts (kV) and 20 milliamperes (mA) over a 2-theta range of 5–60° at a scanning speed of 0.05° per 1.0-second (s) photon-counting interval.

X-ray diffraction will also identify other crystalline impurities present in the calcium phosphate. For example, the ASTM standard specification (ASTM F 1185–88) for calcium hydroxyapatite specifies a minimum hydroxyapatite content of 95% as identified by quantitative x-ray diffraction [3]. A similar ASTM specification (ASTM F 1088) exists for β-tricalcium phosphate. ASTM specifications often list the maximum percent of impurity present for a material to be called calcium hydroxyapatite or tricalcium phosphate. In order to identify the relative amount of a crystalline impurity present in the calcium phosphate, it is necessary to perform quantitative analysis. Although the detection limit of impurities by x-ray diffraction depends on several factors (discussed in the quantitative analysis section below), for most crystalline impurities, the detection limit is 1–2 weight percent (wt%). Hence, the detection limit and quantitative analysis are important factors for the identification of impurities in calcium phosphates.

2. Quantitative Analysis

Quantitative analysis of more than one phase is not trivial for calcium phosphate materials. Because quantitative analysis involves measuring the relative intensities of diffraction peaks for the mixture, x-ray diffraction is superior to powder camera methods. The degree to which the two patterns are added depends on the difference between the mass

Table 1   Table of Some Common Calcium Phosphate Phases and Their Corresponding American Society for Testing and Materials Joint Committee on Powder Diffraction Standards Numbers

Crystalline phase	JCPDS number
Calcium hydroxyapatite	9-432
Calcium fluorapatite	15-876
β-Tricalcium phosphate	9-169
α-Tricalcium phosphate	9-348
γ-Calcium pyrophosphate	33-297
β-Calcium pyrophosphate	9-346
α-Calcium pyrophosphate	9-345
Calcium orthophosphate hydrate (brushite)	9-077
Calcium orthophosphate (monetite)	9-080
Tetracalcium phosphate	11-232, 25-1137
Calcium oxide	4-0777

absorption coefficients for both materials, the concentration of the materials, and the relative crystallinity of each. As stated in the section above on qualitative analysis, the ASTM standard specification (ASTM F 1185–88) for calcium hydroxyapatite specifies a minimum hydroxyapatite content of 95% as identified by quantitative x-ray diffraction [3]. It should be noted that some impurities may be easily detectable and others barely detectable at a concentration of 5% due to differences in the mass absorption coefficients and relative intensity of the impurity, as indicated above.

One technique for quantitative analysis is to prepare a calibration curve using the two materials. We have performed such an analysis for mixtures of calcium hydroxyapatite and β-tricalcium phosphate [4]. Using this method, it was first necessary to find pure sintered samples of both hydroxyapatite and tricalcium phosphate to be used as standards in the study. These standards were ground using a mortar and pestle to rule out the possibilities of any preferred orientation effects. Preferred orientation is an undesirable condition in which the distribution of crystal orientations is nonrandom. This may cause errors in absolute intensity measurements.

Next, the pure HA and β-TCP were massed using an analytical balance and further ground together to produce ratios at increments of 10 mass-percent intervals. The mixtures were put into vials, and these vials were then inserted into the padded chamber of a ceramic ball mill and mixed for a half hour. The powdered mixtures were then pressed into specimen holders for the x-ray diffractometer and analyzed using nickel-filtered copper k-α radiation at 40 kV and 20 mA. The I_{100} (highest intensity) peaks of HA and TCP were scanned over the range of 30.5° to 32.5° at a scanning speed of 0.01° per 10-second photon-counting interval (0.06 deg/min).

Five scans were made of each specimen and intensity ratios were calculated from the raw intensity data. For each of the mass ratios, a sample standard deviation and a 95% confidence interval ($p < 0.05$) was calculated. The range of the sample standard deviation was between 0.100 and 0.658, with the average of the sample standard deviations being 0.311. The range of the 95% confidence intervals was between 0.124 and 0.815, with the average being 0.402.

a.   Intensity-Concentration Curve and the Mass Absorption Coefficients.   The relative intensity of the HA peak is the ratio of the intensity of the I_{100} peak for HA of the compound divided by the intensity of the I_{100} peak for the sample of pure HA. A plot of the known mass ratio versus this ratio is known as an *intensity-concentration curve*. This graph yielded a straight line. This result would be expected since the mass absorption coefficients for hydroxyapatite and tricalcium phosphate are within 1% of each other. So, for materials with nearly identical mass absorption coefficients, the mass percentage of a substance in a mixture can be found by dividing the intensity of a particular peak by the intensity of the same peak in the pure substance. In this case, intensity is directly proportional to the concentration.

b.   Calibration Curve for Hydroxyapatite/Tricalcium Phosphate.   Since it is not convenient to fire a pure sample every time a mixture is sintered, it is more convenient to extrapolate the mass ratio from a calibration curve. Therefore, a calibration curve was created for HA/TCP fired at 1100°C for four hours. Figure 1 shows the calibration curve for the HA/TCP ratios. The data for the y-axis were obtained by using the intensities of the I_{100} peaks for HA and TCP in Eq. 1. This is the simple height law equation.

$$\% \text{ HA} = \frac{I_{100}(\text{HA})}{I_{100}(\text{HA}) + I_{100}(\text{TCP})} \times 100\% \tag{1}$$

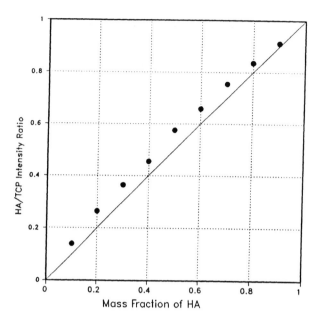

Figure 1 X-ray diffraction calibration curve for hydroxyapatite/tricalcium phosphate (HA/TCP) mixtures. Note that the line does not represent a best-fit line, but represents the 1 : 1 correlation between the mass ratio and the intensity ratio. The difference between the data points and the line is the error in assuming that the observed intensity ratio is equal to the mass ratio.

The calibration curve shows that simply calculating the mass ratios by using the simple height law can be in error by as much as 10% for the ratio of HA/TCP. This points to the need for a calibration curve when doing such an analysis. These errors are caused by differences in the absolute intensity between HA and TCP, not by the slight difference in the mass absorption coefficients as stated above. A calibration curve is an important tool in quantitative x-ray diffractometry since the intensity of an impurity observed is related not only to the concentration of the impurity, but also to the relative crystallinity and mass absorption coefficients as well.

3. Other Characterization

As indicated above, x-ray diffraction can be used to measure factors that have an effect on the crystal structure. In addition to quantitative and qualitative analysis of the calcium phosphates, x-ray diffraction also can be used to make lattice parameter measurements. Particle size broadening (and relative crystallinity) can also be assessed by x-ray diffraction for calcium phosphates. Calcium hydroxyapatite can have many ionic substitutions. It has been known for some time that since these substitutions effect a change in the lattice parameter of hydroxyapatite, x-ray diffraction can be used to measure these changes. Hence, x-ray diffraction analyses along with other characterization techniques are valuable tools for characterizing calcium phosphates.

B. Characterization by Infrared Spectroscopy

The vibrational spectrum has long been used as a technique for evaluating the calcium phosphates, particularly calcium hydroxyapatite (HA). Infrared (IR) spectroscopy complements characterization by x-ray diffraction because it identifies the chemical composition of crystalline as well as noncrystalline calcium phosphates. While x-ray diffraction may identify a calcium phosphate as apatitic, infrared spectroscopy will identify some elements of the composition, particularly group substitution and the presence of carbonate and hydroxide groups.

Some reports of the HA Raman spectrum have been made. More frequently employed is the absorbance spectrum recorded in a transmittance or reflectance mode. Applications for which IR interpretations are employed include distinguishing calcium phosphates, determining degree of crystallinity, and non-calcium phosphate impurity identification. The first report of a hydroxyapatite spectrum with interpretation appears to have been given by Winand and Duyckaerts [5]. Other early reports and assignments were also given and these were in agreement [6–8].

An absorbance spectrum of a well-characterized, stoichiometric, sintered HA from 500 cm^{-1} to 4000 cm^{-1} is presented in Fig. 2. Assignments of the principal bands are given in Table 2 and the frequencies of Fig. 2 are compared with original assignments [7–8]. Orthophosphate ions will give rise to nine vibrations. Normal mode analysis of the tetrahedral symmetry of the free ion predicts a symmetric stretch ν_1, a doubly degenerate O-P-O deformation ν_2, a triply degenerate stretch ν_3, and a triply degenerate bending mode ν_4. Only the triply degenerate modes are infrared active in the free ion. However, the phosphate symmetry is sufficiently perturbed in the apatite lattice to give rise to weak ν_1 and ν_2 absorbance and to lift the degeneracy of the various modes.

For HA, ν_1 is found at 961 cm^{-1}. This assignment is confirmed by the Raman result, in which this band has substantially enhanced intensity relative to the triply degenerate stretch [9]. This would be expected from the fact that, while weakly allowed in the IR, this band is fully allowed in the Raman. The doubly degenerate ν_2 (not seen in Fig. 2) is assigned by analogy to the solution phosphate spectrum to 350 cm^{-1} [8]. Baddiel and Barry assign the low energy component of ν_2 to a transition at 270 cm^{-1} based on Raman data [7]. Two components of the triply degenerate phosphate stretch ν_3 are resolved at

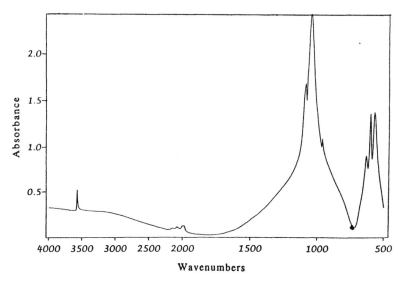

Figure 2 Infrared absorbance spectrum of a well-crystallized, sintered hydroxyapatite, $Ca_5(PO_4)_6(OH)_2$, from 500 cm^{-1} to 4000 cm^{-1}.

Table 2   Assignments of the Principle Infrared Transitions in Calcium
Hydroxyapatite from Figure 2 and as Given in References

Assignment	Ref. 7	Ref. 8	Fig. 2
v_4'	564	561	—
v_4''	574	575	570
v_4'''	603	601	600
$v_L(OH)$	633	631	631
v_1	962	962	961
v_3'	1028	1040	1046
v_3''	1065	—	—
v_3'''	1092	1092	1088
$v_s(OH)$	3566	3570	3571

1046 cm^{-1} and 1088 cm^{-1}. The bands from 550 cm^{-1} to 600 cm^{-1} are assigned to the degenerate phosphate bending mode v_4, the lower energy components of which are not resolved in Fig. 2. Five weak overtone and combination band transitions occur in the vicinity of 2000 cm^{-1} and have been assigned [7].

The absorbance bands of the O-H ion are important characteristic features of HA. Two bands are observed. The 631-cm^{-1} transition v_L is the librational motion of the hydroxide ion in the crystal channel. The hydroxide stretching vibration v_s occurs at 3571 cm^{-1}. These assignments have been confirmed by deuteration experiments [8,10]. Extinction coefficients have been reported [10]. The presence of a new band in the O-H stretching region in spectra taken at elevated temperatures has been reported [10].

Several features regarding O-H motions in the HA spectrum are of note. Published spectra often exhibit broad maxima centered about 3300 cm^{-1} in addition to the sharp band at 3571 cm^{-1} [5,11,12]. This is the result of adsorbed moisture. HA preparations often have a relatively large specific surface area that is hydrophilic. The presence of the adsorbed water will also be seen in a broad bending mode at 1630 cm^{-1}. Hydroxyapatite sintered under a vacuum or in low ambient water vapor conditions shows decreased absorbance intensity in the O-H bands. This is presumably due to the loss of the ion from the lattice. Indeed, the librational band appears more sensitive to this effect than the stretch. Spectra with no discernible intensity at 631 cm^{-1} can still be seen to have a crystalline O-H stretch [13]. Nevertheless, the presence of the two hydroxide transitions is the best infrared evidence for the presence of HA.

Comparisons of HA with other calcium phosphates [5–8] and with so-called nonstoichiometric HA have been done [9,11,14–16]. Most of these have looked at materials with calcium to phosphorus (Ca/P) ratios less than 1.67. Higher Ca/P ratios have also been investigated [17]. A variety of causes may give rise to this nonstoichiometry. These include synthetic design, kinetic factors in precipitation, and decomposition due to sintering or plasma spraying.

Crystallinity has been measured by analysis of infrared spectra. Termine and Posner found a correlation between the measured splitting in the v_4 band and the degree of crystallinity as determined from x-ray diffraction interpretation [18]. It should be noted that the crystal size was held constant in their materials. It is unclear in their discussion how the effect of the O-H libration mode was treated.

Pleshko, Boskey, and Mendelsohn analyzed HA crystallinity using peak analysis of the phosphate stretching region, 900 cm^{-1} to 1200 cm^{-1} [19]. They examined poorly crystalline samples, some of which contained carbonate, and fluoride substitutions. Biological samples were also evaluated. They found that poorly-crystallized apatite spectra could be fit with six band-shape functions, whereas more highly crystalline samples required eight. The poorly crystallized HA crystal size correlates well with the decrease-relative area of one of the component bands. In more crystalline material, it was found that the greater crystallinity correlates with a decrease in the relative area of the resolved ν_1 transition. In a related technique, studies of deconvoluted spectra of the ν_4 bands have been used to evaluate calcium phosphates and maturing bone and enamel [20].

Carbonate substitution occurs readily in the apatite lattice. The CO_3^{2-} can occupy OH$^-$ sites (A type) or PO_4^{3-} sites (B type). In bone, new and mature, there is substantial carbonate content—typically 3–4% [21]. Carbonate inclusion can also occur in synthetic apatites [13,22]. These substituted ions are readily detected in the IR spectrum. IR has been used to monitor CO_3^{2-} deposition in osseous site implants. Analysis of the carbonate transitions has been used to assign relative fractions of carbonate to the A-type and B-type sites [23,24].

C. Methods of Fabrication of Calcium Phosphates

Several methods for the preparation of synthetic calcium phosphate powders for use in the fabrication of bulk ceramics currently exist. The techniques used in each of these methods have advantages and disadvantages in the fabrication of bulk ceramics. First, the properties of these powders, such as impurities, particle size, and particle morphology, will affect the ability of the powders to become bonded in the sintering process. Therefore, the attainment of ideal compressive strengths of the ceramics for many loading applications depends directly on the properties of the starting calcium phosphate powders. Second, the properties of the final ceramic product such as phase present (purity), calcium solubility, and zeta potential are influenced by the properties of the initial calcium phosphate powders. Advantages and disadvantages in terms of both the fabrication and biological effect of the bulk calcium phosphate ceramics prepared by each method are discussed below.

1. Hydrothermal Conversion of Corals (Calcium Carbonate)

One technique that can be used to produce the synthetic calcium phosphate materials is the hydrothermal conversion of the coral skeleton to form predominantly hydroxyapatite. The use of corals from the genera of *Goniopora*, *Porites*, *Favites*, and *Lobophyllia* has been investigated [25–28]. These coral specimens consist of 99% calcium carbonate in the form of aragonite and 1% organic material [27]. After being ultrasonically cleaned to remove organic material, this material is heated at elevated temperature and pressure in the presence of aqueous phosphate solutions [25–28]. This causes a replacement of phosphate ions for the carbonate ions and changes the crystal structure to calcium hydroxyapatite. Typically, the diffraction pattern of this calcium phosphate is predominantly hydroxyapatite.

The porosity of these corals has advantages and disadvantages for implantation purposes. One significant advantage was pointed out early in the investigation of the corals for use as bone graft substitutes. In 1977, Chiroff et al. observed the "remarkable uniformity of pore size and complete interconnection of the pores" [26]. Further studies have shown that certain corals have uniform porosity in the range of 500 to 600 micrometers (μm) in diameter, which is known to be advantageous for osseointegration. In particular, the *Porites* and *Goniopora* corals have an open pore structure (interconnecting pores), which is believed to allow for rapid vascularization and osseointegration of the implant [25–27].

One disadvantage of the corals is that the porosity varies only slightly from 46% to 48% in the genera mentioned above [27]. This may be a disadvantage since the high porosity adversely affects the compressive strength and causes it to be weaker than cancellous bone [25]. The fixed amount of porosity may also be biologically undesirable in certain instances. The highly organized and permeable pore structures may have a biological advantage over the more random and numerous closed-pore structures often produced by naphthalene sublimation or hydrogen peroxide.

For these reasons, the converted and unconverted corals are incorporated into the graft site and are slowly resorbed. The difficulty in obtaining sufficient starting material from commercial sources is a disadvantage for this method. Another disadvantage of the substituted corals is the variability in the final chemical composition of the ceramic. It is known that ions such as magnesium, sodium, chloride, and fluoride, which are present in seawater and become incorporated into the coral, affect sintering parameters of these calcium phosphate implants and may have an effect on the biological response.

2. Preparation by Solid-State Reactions

Calcining is the process of heating a substance to a high temperature, but below its melting point, to cause a loss of moisture, reduction, or oxidation. Some authors have investigated calcining commercially available dibasic calcium phosphate powders such as calcium hydrogen orthophosphate ($CaHPO_4$, known as monetite) or calcium hydrogen orthophosphate dihydrate ($CaHPO_4 \cdot 2H_2O$, known as brushite) to obtain the alpha and beta phases of calcium pyrophosphate [29]. Using these same techniques, commercially available tribasic calcium phosphate, $Ca_3(PO_4)_2$, powders can be calcined to obtain hydroxyapatite (HA), the alpha and beta phases of tricalcium phosphate (TCP), and ratios of these two.

The extreme variability in the original composition of these commercially available powders has frequently led to results that were not reproducible. The purity and true composition of these powders is often in question. The method has a major disadvantage in that the final product depends on the composition of the starting material, which may vary from batch to batch. Thus, the final phase can range from β-tricalcium phosphate to hydroxyapatite, even though the starting powders appear identical by x-ray diffraction. In many instances, the commercially available powder is amorphous initially. Thus, x-ray diffraction is not useful in determining the phase of the dibasic or tribasic calcium phosphate. It is hypothesized that the nonstoichiometry of the commercially available powders, as well as the presence or absence of water in the apatite structure, results in the formation of different phases upon calcining.

Impurities in the starting powders obtained commercially may hinder the ability of these powders to become bonded in the sintering process, thus causing a significant decrease in compressive strength. Hubbard showed evidence of tribasic calcium phosphate powders with nearly identical particle size and morphology, but with radically different compressive strengths after calcining and sintering [29]. Hubbard states that "the impurities effect [*sic*] the bonding and diffusion of the particles" and that "magnesium and sodium probably inhibited strong grain boundary bonding" [29, p. 129] Impurities present in the starting powders may also adversely affect the biological response to the materials. Because of the need for a controlled, predictable, and reproducible biological response to these materials, solid-state reactions are rarely used to fabricate calcium phosphate ceramics for biomedical applications.

3. Precipitation Methods

Calcium orthophosphates, calcium pyrophosphates, tricalcium phosphate, calcium hydroxyapatite, and ratios of hydroxyapatite and β-tricalcium phosphate can be prepared directly or indirectly using precipitation techniques. That is, these phases are precipitated directly, or powders obtained from the precipitation reactions can be calcined to yield the desired phases. HA and ratios of HA/TCP can be precipitated at a basic pH using calcium nitrate, $Ca(NO_3)_2 \cdot 4H_2O$, and ammonium phosphate, $(NH_4)_2HPO_4$, as starting materials.

This technique offers the ability to vary pH of the reaction by adding different amounts of concentrated ammonium hydroxide (NH_4OH), which results in different precipitates. Also, the temperature at which the reaction is carried out can be varied, thus affecting the size of the resultant precipitate particles. A range of the size of these precipitates may be beneficial in sintering since particle size is one of the parameters that affects sintering. In addition, variation of the time at which the precipitate is harvested will greatly affect the resultant phase obtained.

The use of reagent-grade starting materials with minimum amounts of impurities allows the production of pure starting powders. As mentioned in the preceding section, impurities in the starting powders inhibit grain boundary bonding and result in poorly sintered ceramics with low compressive strength values. Impurities may adversely affect the ceramic resorption or other biological response by effecting changes in solubility. It is known that highly substituted apatites elicit a different biological response than calcium hydroxyapatite when implanted [30,31]. Due to the ability to vary reaction parameters and to control the purity of the starting powders, precipitation techniques are invaluable and preferred in preparing calcium phosphate phases for a variety of biomedical applications.

D. Thermal Treatments: Calcining and Sintering and the Production of Macroporosity and Microporosity of Calcium Phosphates

1. Calcining

Once the amorphous or poorly crystalline powders are obtained from the precipitation reactions and dried, they can be calcined to yield a calcium phosphate phase that depends on the composition of the precipitated powder. In the case of fabrication of calcium pyrophosphates, precipitates obtained from the reaction of calcium nitrate tetrahydrate and ammonium phosphate can be calcined to form calcium hydrogen orthophosphate ($CaHPO_4$, monetite by x-ray diffraction). At higher temperatures, this phase can be calcined to obtain alpha and beta phases of calcium pyrophosphate. The transformation of beta calcium pyrophosphate to alpha calcium pyrophosphate is not a chemical reaction. It is a reversible phase transformation that is defined as a change in crystal structure and not a change in composition. Thus, the alpha, beta, and gamma phases of calcium pyrophosphate are allotropes. Chemical reactions and phase transformations upon heating of precipitated calcium phosphate powders and the temperatures at which they occur are summarized below for this system.

$$CaHPO_4 \cdot 2H_2O \xrightarrow{\text{100°C to 300°C}} CaHPO_4 + 2H_2O \qquad (2)$$

$$2CaHPO_4 \xrightarrow{\text{400°C to 500°C}} \gamma\text{-}Ca_2P_2O_7 + H_2O \qquad (3)$$

$$\gamma\text{-}Ca_2P_2O_7 \xrightarrow{\text{750°C}} \beta\text{-}Ca_2P_2O_7 \qquad (4)$$

$$\beta\text{-}Ca_2P_2O_7 \xrightarrow{\text{1170°C}} \alpha\text{-}Ca_2P_2O_7 \qquad (5)$$

2. Isostatic Pressing

Powders can be compacted by placing them in rubber bags made especially for isostatic pressing. The powders are then isostatically compacted in a hydraulic isostatic press at a pressure of 25,000 pounds per square inch (psi). Compaction by isostatic pressure increases the compressive strength of subsequently calcined and sintered ceramics because it removes some porosity from the powders and packs the loose powders closer together, which is beneficial for sintering. Isostatic pressing alone does not appreciably increase the compressive strength of the calcium phosphates.

3. Production of Macroporosity in Calcium Phosphates

To improve osseointegration of calcium phosphate ceramics in the bony tissue, pores are created in the ceramics according to the Hubbard method. This method entails mixing the calcium phosphate powders with sized naphthalene. When the naphthalene is sublimed out, pores are left behind that retain the size of the naphthalene particles. It is known that pores in the range of 150 to 500 μm are optimally beneficial for bony ingrowth. Standard sieves can be used to obtain sized naphthalene particles in the desired porosity range. Shrinkage in porosity size during the second stage of thermal treatment reduces the ultimate size of the pores. This should be taken into account when planning fabrication of porous calcium phosphates.

Sized naphthalene granules can be mixed with a calcium phosphate powder in various amounts to increase or decrease the amount of porosity in the final implant. The mixture should be stirred to create a homogeneous distribution of naphthalene in the calcium phosphate powder before inserting into rubber bags for isostatic pressing. The powder and naphthalene are isostatically compacted as described above. After pressing, the compacted cylinders are removed from the rubber bags and ready for thermal treatments.

The first stage in the thermal treatment is to sublime out the naphthalene. A small tube furnace can be used to perform this operation. Naphthalene can be sublimed from the compacted cylinders by heating at 400°C for four hours. Care should be taken in controlling the maximum temperature in this first thermal treatment since the autoignition temperature of naphthalene is 525°C. In addition, the heating and cooling rate should be controlled to prevent the cylinders from cracking due to thermal shock. Macroporosity is obtained by the large pores created after subliming out naphthalene. This porosity can be varied from 0% to about 70 percent. The calcium phosphate cylinders do not contain any naphthalene after this process.

4. Sintering Parameters and Residual Microporosity

The second stage of the thermal treatment is to effect bonding of the calcium phosphate particles. Cylinders are removed from the rubber bags and fired at different temperatures and times to effect sintering. Sintering is a process by which particles are bonded and become denser (decrease in porosity) when subjected to high temperatures, below the melting temperature of the material. Sintering parameters for the calcium phosphates vary from temperatures of 1000°C to 1300°C and times of 1 to 24 hours. The heating and cooling rate should be less than or equal to 2°C/minute to prevent thermal stresses that could crack the ceramic cylinders. As discussed in the section below, compressive strength is an excellent method to assess the effect of sintering. Microporosity, on the order of a few microns, remaining in the implants is due to gaps left between the sintered particles. While this microporosity is not useful for osseointegration, there is a possibility that it could be useful for cell attachment. Some calcium phosphates that have been subject to hot isostatic pressing (HIP) have almost no microporosity.

E. Characterization of Mechanical Properties

1. Dense Calcium Phosphates

A diamond saw can be used to section sintered cylinders into disks with a height equal to twice the diameter. Since resistance to compressive loading is the most significant property of ceramics, we measure the compressive strength using a servohydraulic material-testing machine (MTS model #809, Minneapolis, MN) in accordance with standard testing methods. Specifically, we use ASTM Standard C773-88 (Standard Test Method for Compressive Strength of Fired Whiteware Materials) for measuring the compressive strength of the calcium phosphates. Compressive strength data can be used as an indicator of optimum sintering conditions for all of the calcium phosphate phases. The cross-sectional area of each of the calcium phosphate disks should be measured with a micrometer and recorded for use in the calculation of the compressive strength. At least 10 specimens should be loaded at a constant rate of 1166 Newtons/second until failure. Using the maximum load before failure, the ultimate compressive strength σ is calculated according to $\sigma = F/A$ where F is the maximum compressive load and A is the cross-sectional area of the specimen.

Sintering of calcium phosphates increases their density and strength. As discussed in the section on thermal treatments, a wide variety of mechanical properties may be obtained due to variability in the shape, size, and purity of the starting powders, as well as the sintering time, temperature, and heating/cooling rate for each of the calcium phosphate phases. As is typical for many ceramics, the compressive strength of the calcium phosphates is quite good. Unfortunately, low values for tensile strength and fracture toughness limit the use of calcium phosphates for low-loading or unloaded conditions.

2. Porous Calcium Phosphates

Although dense calcium phosphates have superior mechanical properties compared with porous calcium phosphates, the lack of porosity does not allow for bony ingrowth or osseointegration of the implant. It is also known that mechanical properties decrease as microporosity and macroporosity of the calcium phosphates increase. Since calcium phosphates are being investigated for use as bone graft substitutes for limited load-bearing situations, it is often important to consider the mechanical properties necessary for each graft site. Due to the ability to tailor-make the calcium phosphates with various sizes and amounts of porosity to foster osseointegration, it follows that the mechanical properties will also vary as the porosity changes.

Although it is not possible to consider the mechanical properties for all applications here, as an example we consider the compressive strength necessary for a synthetic graft for use in cervical spine interbody fusion. We believe that a site-by-site analysis will determine whether calcium phosphates have the needed mechanical properties to be effective as a bone graft substitute in a specific site.

Bone graft substitutes such as calcium phosphates are being investigated for axial load-bearing struts or as interbody grafts for spinal fusion because of the limitations associated with the harvest and quality of autograft as well as the problems associated with the use of allografts. The conundrum in the use of porous calcium phosphates is: What porosity is most effective to promote ingrowth yet strong enough to resist compressive stresses found in the spine? We compared the compressive strengths of tricortical iliac bone grafts versus biphasic 50/50 HA/β-TCP calcium phosphate ceramics of various porosities (15%, 22%, 27%, 32%, 40%, 44%, and 65%) in order to compare the strength of the porous ceramics with standard bone grafts to evaluate the efficacy of these materials to withstand compressive loads found in the spine. Since tricortical iliac bone grafts are used for this surgery, the compressive strength of these grafts was compared directly to the strength of these calcium phosphates.

Iliac Bone Grafts.   A total of six pelvises from fresh-frozen cadavers (ages 58–81) was used in this study. Bone grafts of 2 × 1 cm were obtained from three different regions in the anterior, middle, and posterior superior iliac crests, and a 1 × 1 cm area within the anterior region for the Smith–Robinson graft.

For mechanical testing, each tricortical strut graft was sectioned to a uniform height of 15 mm, and the Smith–Robinson graft was sectioned to a height of 8 mm. The top and bottom of each graft were photographed, and the cross-sectional areas were measured using a computerized digitization technique (Cadkey, Inc., Winder, CN). The prepared grafts were loaded between two parallel plates of a materials testing system (MTS 809). Uniaxial compression load to failure was applied at 1 mm/s under the displacement control. The compressive strength was determined by load to failure divided by the cross-sectional area.

Calcium Phosphate Ceramics.   Precipitation techniques using reagent-grade calcium nitrate tetrahydrate and ammonium phosphate were used to prepare a precipitate that could be calcined to yield a biphasic hydroxyapatite/β-tricalcium phosphate (HA/β-TCP) ceramic. The HA/β-TCP ratio was 50/50 as verified by X-ray diffraction [3]. Sized (range between 355 and 590 μm granular naphthalene was added to the dried powdered precipitate according to the Hubbard method to produce ceramics with various porosities. The naphthalene and precipitate mixture was placed in a rubber bag and isostatically pressed at 170 MegaPascal (MPa). The ceramic cylinder was then heated to remove the naphthalene by sublimation, and was sintered at 1150°C for 4 hours. Next, the ceramic cylinders (1 cm diameter) were sectioned to 1 cm in height, and their porosity was determined by mass and volume measurements. The measured porosities were 15%, 22%, 27%, 32%, 40%, 44%, and 65%. The compressive strength of each specimen was measured according to ASTM test C-773. According to this method, the compressive load was applied at 70 kiloNewtons per minute (kN/min), and the compressive strength was defined by the load to failure divided by the cross-sectional area of the specimen.

Measured compressive strengths of iliac bone grafts and ceramics with the aforementioned porosities were compared using analysis of variance. Mean compressive strengths of the iliac bone grafts and the ceramics with various porosities are shown in Fig. 3. Ceramic specimens demonstrated a fairly uniform compressive strength with small standard deviations as compared to the iliac bone grafts, ranging from 7.1 to 40.9 MPa, with a mean of 16.2 MPa. It was also found that the compressive strength of the ceramics decreased nearly linearly ($R = .957$) as the porosity increased from 15% to 65%. Analysis of variance showed that the mean compressive strength (16.2 MPa) of the iliac bone

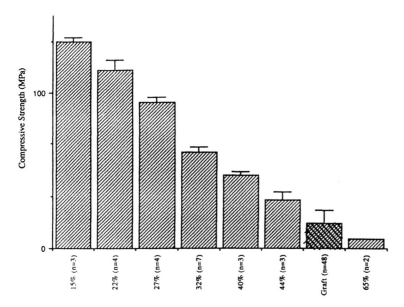

Figure 3   Mean compressive strengths of the 50/50 hydroxyapatite/β-tricalcium phosphate (HA/β-TCP) ceramics with various porosities and the iliac bone grafts. Tested number of specimens are listed in parentheses.

grafts was significantly less ($p \leq 0.002$) than the HA/β-TCP ceramics with porosities less than 44% (mean compressive strengths > 32.2 MPa). The ceramics with 65% porosity were very friable, and hence difficult to prepare according to ASTM Standard C-773. This ceramic was weaker than the bone grafts, although the difference was not statistically significant at $p \leq 0.11$.

Measured compressive strengths of the ceramics would be different if other calcium phosphate phases or other porosity sizes were tested. In addition, if the powdered precipitate had a different morphology or size, or if the sintering conditions were different, the compressive strengths would be affected. For this tested 50/50 HA/β-TCP material, 40% or 44% porosity would provide optimum strength and allow for bony ingrowth to stabilize the implant. 50/50 HA/β-TCP with porosities less than or equal to 44% prepared in this manner are stronger in compression than tricortical iliac crest bone graft. These ceramics would be able to resist compressive stresses found in the spine for use as axial load-bearing struts or as interbody grafts for spinal fusion. In addition, the porosity of the ceramics would allow bony ingrowth to occur to promote fusion. HA/β-TCP 50/50 with 65% porosity was weaker than the bone grafts.

II.   BIOLOGICAL CHARACTERIZATION OF CALCIUM PHOSPHATES

A.   Bone Grafting Materials and the Need for Bone Graft Substitutes

Bone is the second most frequently transplanted tissue in humans [32]. In the United States alone, it is estimated that over 200,000 operations per year are performed that require bone grafts [33]. Surgical uses of bone grafting materials include surgical intervention of osseous nonunions, restoration of the structural integrity of bone after trauma, and filling defects following bone tumor removal. Different grafting materials and graft substitutes are available for use as bone grafts. Each of these has certain advantages and disadvantages for use. An ideal bone graft or bone graft substitute should have certain qualities. It should

1. Be biocompatible, noncarcinogenic, and nonimmunogenic. The material should not initiate an adverse biological response.
2. Exhibit osteogenic properties. The material should actively stimulate the differentiation of mesenchymal stem cells into active osteoblasts.
3. Be uncontaminated. The material should not transmit an infection to the recipient.
4. Be osteoconductive. The material should provide a scaffold for new bone formation.
5. Have structural strength. The material should provide structural strength in a loaded or stressed site.
6. Be available in unlimited quantity. The graft should be large enough to be shaped into the size needed.
7. Be easy to sterilize. The material should not degrade or have altered biological properties as a result of sterilization.
8. Be easy to use. The material should not require undue inconvenience to use.
9. Be inexpensive. The material should be relatively inexpensive.
10. Produce a consistent biological response. The material should have a predictable biological response based on properties determined in quality control of the material.

The intent of the following sections is to characterize and compare several bone graft materials with calcium phosphate bone graft substitutes with regard to the ideal qualities described above.

1. Autogenous Bone Grafts

Autograft bone is bone taken from the patient who will receive it. This graft material may be obtained from the iliac crest, rib, fibula, or tibia. Autogenous cancellous bone graft is currently the preferred graft material for several reasons. First, it is biocompatible and nonimmunogenic since it comes from the same person. Second, it will not transmit a disease to the recipient. Third, it has osteogenic potential since cells on the trabecular surface of the graft become active after transplantation. The majority of the cells inside the graft undergoes necrosis and is ineffective in stimulating osteogenesis [34]. The osteogenesis of the surviving cells of cancellous autografts offers a distinct advantage over cellularly inert allografts and xenografts.

This osteogenic biological response to the autogenous bone graft is best described by Heiple et al., who have conducted research on the subject for over 30 years. They describe the biological response in the following way [34]. At one to two weeks, a fibrous clot is formed around the transplant and necrosis of the marrow inside the graft takes place. At three to four weeks, vascularization of the cancellous autograft takes place, usually in the outer edge of the graft where the hematoma has been resorbed. It has been claimed that this revascularization attracts mesenchymal cells that differentiate into osteoblasts, fibroblasts, and/or chondroblasts, depending on the stimuli present at the site. New bone formation is present on the necrotic trabeculae. Over time, the process of creeping substitution occurs. This term describes the process by which the bone remodels to replace necrotic graft bone with viable host bone [33]. If union occurs between the graft and host bone, a bridging union of bone is present by three months. Somewhere between one and three months, stiffness and then strength are restored to the grafted site. Most investigators have only rarely observed resorption in response to autogenous bone graft, in contrast to allogeneic bone, which is more frequently resorbed [33–35].

Although autogenous bone grafts offer several advantages over other bone grafts, they also have some distinct disadvantages. In the first place, autogenous bone can only be taken in limited amounts from the iliac crests or ribs. Even when a small amount is taken, morbidity at the donor site due to the surgical procedure to remove the graft may be significant. Younger and Chapman found minor complication rates due to the surgery to harvest the graft to be greater than 20% [36]. These minor complications included pain, sepsis, and loss of sensation. In two retrospective studies, major complications such as fracture of the ilia, heterotopic bone formation, and pelvic instability existed in about 5% of the cases [36,37]. Some of these cases require surgical intervention. It is obvious that these problems often outweigh the advantages of using osteogenic autogenous bone as a graft material. Furthermore, patients requiring autogenous bone grafts are often in groups found to be at greatest risk of complications, due to metabolic disease, and offering poor quality bone for transplantation.

2. Allogeneic Bone Grafts

Another graft material, which is the second most frequently used in surgeries requiring a bone graft, is allograft. *Allogeneic bone* is defined as bone taken from a donor of the same species. Just as blood is available from blood banks, allogeneic bone is available from bone banks. Allograft bone has several advantages over autogenous bone. First, the material is available in an unlimited supply. Massive reconstructive surgeries, such as hip arthroplasty revisions, are often performed with allograft bone. Second, there is currently no other graft material or graft substitute that can restore structural integrity to the site as satisfactorily as massive allografts.

Serious problems exist with allogeneic bone. First, the material does not have the osteogenic potential of autogenous bone grafts. Thus, the creeping substitution described above occurs much slower than in response to autogenous bone. Allografts may also elicit an immunological response when implanted. This reaction depends on the degree of mismatch between the transplanted tissue and the host tissue. In cases of significant dissimilarity between the tissues, the graft may become rejected and completely resorbed due to the host response [38].

The problems with immunogenic response and concomitant resorption are suspected as the major cause for fracture of the allografts, which has been reported to occur in roughly 10% of the cases. It has been demonstrated that processing allografts decreases this immunogenic response. In particular, freezing and freeze drying has been shown to decrease the immunogenic response. However, this processing may alter the physical properties of the graft. It is well known that freezing will not kill the human immunodeficiency (HIV) virus that causes acquired immunodeficiency syndrome (AIDS) and several other agents of disease that may be transmitted by the allograft [39]. The effects of irradiation sterilization have been documented by Pelker and Friedlander. They have concluded that sterilization of the allograft with a dose of radiation above three megarads significantly decreases the breaking strength of the graft. In addition, they have determined that this decrease in strength is further compounded by freeze drying [40]. For these reasons, allograft bone may provide initial stability as a bone graft, but inherent problems may cause graft rejection, resorption, and loss of biomechanical stability at the graft site. Further, the possibility of transmission of disease makes the use of allografts precarious.

3. Synthetic Bone Graft Substitutes: Calcium Phosphate Ceramics

Due to the problems with the current bone grafts available, much research has been conducted over the last 25 years to develop synthetic graft substitutes that are biocompatible as well as able to support the necessary loads. Of the materials that have been studied for use as bone graft substitutes, the class of calcium phosphate ceramics has superb biocompatibility over all others.

The calcium phosphates have been shown by many investigators to exhibit the phenomenon of osteoconductivity. *Osteoconductive materials* are defined as materials that are known to support osteogenesis (bone growth) in bony tissue sites. The osteoconductive ceramics have never before been determined to be osteoinductive, that is, able to stimulate osteogenesis actively in bony or nonbony sites. Lane and Sandhu describe osteoinduction as "the differentiation of uncommitted mesenchymal cells into active bone forming cells in the presence of an inductive stimulus" [41]. Similarly, Black has described *osteoinduction* as a "property of a material that produces bone growth associated with *de novo* growth *in vivo*" [42].

The following sections characterize the biological performance, particularly the host response to calcium phosphates, of various phases in osseous and nonosseous tissues. Since various aspects of acute and chronic inflammation are observed as the host response after implantation of calcium phosphates, a description of acute inflammation, chronic inflammation, and the foreign-body response follows. Also, since the host response to calcium phosphates varies with the phase and properties of the phase implanted, the biological response to the various phases is described.

B. Host Response to Calcium Phosphates: The Foreign-Body Reaction

1. Acute Inflammation and the Host Response

When any material is implanted into a tissue, the normal wound healing process associated with that tissue is altered. The material implanted and surgical procedure cause an inflammatory response. The observed inflammatory response is the sum of the acute and chronic inflammatory responses to the surgical procedure and the materials implanted. Acute inflammation following surgical implantation of a sterile calcium phosphate device in bone is usually a short-term process. It can be limited by controlling the trauma when preparing the bony site for implant placement and by appropriate methods of immobilization after device placement. The least traumatic preparation, the best press-fit implantation using sterile technique, and maintenance of apposition at the bone/implant interface to avoid micromotion will mollify the acute inflammatory phase. This also aids in the initiation of healing early in the chronic inflammatory phase.

2. Chronic Inflammation and the Host Response

When discussing the biological characterization of calcium phosphates for use as materials to be implanted in the animal or human body and their relationship to chronic inflammation, it should be clearly understood that such materials must be regarded as foreign bodies in terms of the host-cell/matrix interaction. Thus, a broad range of host-tissue responses could be encountered. These have been grouped under a variety of classifications that tend to oversimplify the complexity of host-organ, -tissue, or -cell responses. The common terminology, for descriptive purposes, ascribed to the host response is generally indicative only of the overall, often short-term, outcome of the result of implantation, and tends to "skip the details" as unimportant.

The following general terms have been used to characterize the host response: bioactive, biocompatible, bioinert, biodegradable, incompatible, and immunogenic [43]. Other terms may be encountered in the literature, but these suffice to cover the broad range of responses. If these terms are unfamiliar, the reader is encouraged to seek a general text in the field of biomaterials for a complete explanation of this nomenclature [44,45].

For biological characterization of materials, it is necessary to approach the biomaterial's host responses with regard to the foreign-body response at the cell/matrix interaction level. In doing so, it will become evident that, while calcium phosphate materials are remarkably "biocompatible" from a clinical viewpoint, host tissues respond in different ways that must be considered when designing devices for useful implantation.

With these introductory remarks in mind, it is important to review what is known about the foreign-body reaction [46–48]. Classical descriptions of the pathological foreign-body reaction have usually been related to penetrating wounds by bullets, shotgun pellets, arrow points, metal fragments, glass, stones, or other traumatic foreign objects. The mode of introduction into the tissues is usually forceful, and often produces associated local tissue damage or destruction. Subsequently, the initial response is acute inflammation, characterized by one or more of the classical clinical signs: (1) *calor*, local tissue temperature elevation; (2) *rubor*, reddening of surrounding tissue; (3) *tumor*, swelling with localized edema; and (4) *dolor*, localized or referred pain. These signs are usually accompanied by hemorrhage and necrosis at the tissue site where the foreign body becomes lodged.

The microscopic acute cellular response is a chemotactic attraction of morphologically identifiable polymorphonuclear neutrophilic leukocytes, so-called pyknotic mononuclear round cells, and more delicately stained mononuclear histiocytes, all of which are phagocytic in their activity. They serve to facilitate healing and repair by removing tissue debris and small foreign particles in advance of the appearance of healing fibroblasts. If damage to tissues is extensive or if infection ensues, the cellular response and acute clinical signs will persist.

In addition, many complex, interrelated physiological reactions are triggered that produce both harmful and beneficial effects that may be somewhat longer lasting, so that the consequences are found in the chronic inflammatory phase as well. Among these is platelet disruption, followed by a cascade of released enzymes and polypeptide activators called kinins that, in turn, may activate other circulating inactive polypeptides. Among the most important of the initial effects is the release of thrombokinase, an enzyme that, along with calcium ions, activates prothrombin to produce thrombin, leading to the conversion of soluble circulating fibrinogen into an insoluble protein called fibrin, which is capable of forming a meshlike matrix that traps erythrocytes and forms a blood clot that aids in hemostasis to diminish hemorrhage at the site. Fibrin also activates factor XII (the Hagemann factor) and bradykinin, known as slow-reacting substance, as well as a group of polypeptide activators called kallekreins. The presence of fibrin in the circulating blood plasma also causes activation of a protective plasminogen, which then becomes a fibrinolysin that can prepare the way for healing as the clot later begins to "organize" and become revascularized to deliver a blood supply to the area, a process known as angiogenesis, necessary for normal healing. This, in turn, may produce more fibrin, and a cyclic effect can occur. The surgical insertion of an implant will initially result in the appearance of most of these same features, both clinical and microscopic.

Other cytokines can activate many functionally inactive circulating polypeptide factors, which in turn trigger a host of other effects. Among these are

The activation of T-lymphocytes to process antigens, which leads to B-lymphocyte and plasma cell proliferation with antibody and gamma globulin production.

An increase of properdin, leading to the alternate pathway of C3 complement activation with subsequent cytolysis.

Prostaglandin E_2 is produced and continues inflammation.

Pyrogens are produced, with subsequent stimulation of the hypothalamus and abnormal irregularity in tissue and body temperature, as well as possible fluid and electrolyte imbalance and loss of organ perfusion associated with shock.

The release of histamine-containing granules from tissue mast cells and histiocytes, which results in increased capillary permeability, edema, and swelling.

Ultimately, mobilization of transforming growth factors (e.g., TGF-β), which are cytokines can direct mesenchymal cells to differentiate into fibroblasts and endothelial cells, leading to angiogenesis, formation of a new blood supply, and healing and repair processes.

This is only a brief overview of events surrounding inflammation.

Three things are important to bear in mind at this point. First, the intense, complex, cellular and biochemical activities surrounding inflammation are all normally physiologically controlled, and they lead beneficially to the healing and repair processes. Second, transforming growth factors are physiologically activated even in the presence of the foreign body. Third, they are necessary to activate mesenchymal cells if healing and repair are to occur. Thus, the apparent disorder of activity surrounding inflammation due to introduction of the foreign body is normally an orderly progression that is necessary to produce healing and restoration of structure and function of damaged tissue, the overall aim of wound healing.

When the foreign body is lodged in tissue for a long period of time (weeks, months, or years) and the phagocytic activity of granulocytic polymorphonuclear (PMN) leukocytes is unable to remove the structural material during the acute phase of inflammation, a condition of chronic inflammation will then exist. The cellular response is marked by disappearance of PMN phagocytes and a shift to an accumulation of mononuclear-type cells occurs. In the early stages, these mononuclear cells are pyknotic and often appear as aggregates in foci. They are then gradually replaced by the proliferation of more delicately stained, pale, epithelioid and histiocyte varieties of phagocytic macrophages. Some arrive through the circulating blood supply and migrate into the tissues, while progressively more proliferate at the site and the absolute numbers of histiocytes increases. Some of these coalesce and form multinucleated giant cells containing from 2 to 20 (or occasionally even up to 200) nuclei. These are also actively phagocytic, attaching themselves to the material structure, and they may be seen with ingested particles in the cytoplasm.

The classical foreign-body reaction is characterized distinctively by the presence of these multinucleated giant cells. The size of the giant cell may have no direct relationship to the size of the residual foreign body since one large giant cell may be found surrounding several small particles, or several smaller giant cells with fewer nuclei may be found strung out along the surface of a much larger particle.

The cardinal clinical signs of acute inflammation may subside, and be only minimal or be totally absent. Usually, if the foreign body is too large to be phagocytized or is especially resistant, granulation tissue consisting of fibroblasts and angioblasts in a fibrous collagen matrix encapsulates the mass inside a dense membrane of fibrous connective tissue. This isolates it from the surrounding tissues, although a blood supply may still exist. The chronic inflammatory process thus isolates the foreign object and facilitates the end result of wound healing. It is especially noteworthy that healing and repair of wounded tissue may proceed simultaneously while accompanying the chronic inflammatory foreign-body reaction.

We now consider the tissue responses to deliberate insertion of calcium phosphate ceramics as foreign-body implant devices.

C. Characterization of Calcium Phosphates in Osseous Tissues

A major benefit of calcium phosphate as a biomaterial for implantation in osseous tissues is the physiological nature of the chemical composition [29,43,49]. Both calcium and phosphate are highly acceptable ions to the normal physiological and cellular functions. Thus, the cell/matrix (foreign-body implant) interactions may be expected to be somewhat more benign than with less physiological materials. Nevertheless, not all calcium phosphate crystalline compositions or geometric fabrications produce the same response; therefore, other factors must be considered when designing a device to fulfill all the requirements for a specific purpose.

Because of the similarity of calcium hydroxyapatite to the apatitic bone mineral, implants of calcium hydroxyapatite have most often been investigated. Pure stoichiometric calcium hydroxyapatite, however, is not the only mineral phase found in bone. Bone mineral is most often a carbonate- and fluoride-substituted apatite structure. The carbonate-substituted form is more soluble than pure apatite, and the fluoride-substituted form is much more resistant to physiological dissolution or phagocytosis than either of these. Other phases of calcium phosphates such as β-tricalcium phosphate and α- and β-calcium pyrophosphate have been investigated for biomedical applications. Of these, the next most frequently investigated has been β-TCP. This composition is readily resorbed, biocompatible, and not so readily replaced by bone. A composite of HA/β-TCP in a 60:40 ratio has also been extensively investigated. We refer to several of these in the following discussions.

As indicated above, due to the relatively poor mechanical properties of the calcium phosphates, the materials lend themselves more to relatively unloaded applications for filling bony defects. In addition, the ceramic should be loaded in a manner that makes use of its greatest property, namely, its compressive strength. Let us consider the actual and potential uses of calcium phosphates as implants in bony sites.

In dentistry, the primary application is as a small, shaped block or as space-filling granules in voids in necrotic bone produced by periodontal disease, and in postextraction sockets in order to maintain the alveolar ridge dimensions as a base for prosthetic devices. Segmental avulsion defects in the mandible can also be replaced with appropriately shaped ceramic devices.

In cranial/maxillofacial reconstruction, calcium phosphate ceramic materials are applied or have potential for application to fill void spaces in mastoid bones after mastoidectomy, as fitted plates and scaffolds bridging segmental bone defects in cleft palate repair and facial reconstruction, and in place of metal plates in reconstruction of the cranial vault. Calcium phosphate ceramics may also be used in reconstruction of the temporomandibular joint as a bone graft supplement.

In orthopedic surgery, the calcium phosphate ceramics have actual or potential uses as a bone graft supplement or substitute in segmental defect repair of long bones following fracture or nonunion, as a cavity filler following tumor excision, in replacing excochleated contents of the femoral head in treatment of ischemic (avascular) necrosis, as a bone-to-bone bridge in arthrodesis or spinal fusions, as a component of collagen/ceramic or polymer/ceramic composites to be used as bone graft supplements, and as a coating for metal prosthetic devices to obtain a better interface for biomechanical bonding to bone. In practically all of these applications, the principal objective is to obtain a well-osseointegrated bone-to-bone bridge with sufficient strength to provide normal structure and function.

It is evident that the ceramic may have many shapes and sizes, but a major requirement is that it should be macroporous at the bone/implant interface to allow bone ingrowth for osseointegration to obtain a secure biomechanical interlocking structure. In a weight-bearing application, the necessity for using a porous structure may place a difficult burden on the designer because of the need for a certain strength of materials to fulfill the required function. This type of loading may necessitate the use of internal fixation.

Ideally, the ceramic, with or without marrow or growth factors, should be capable of enhancing new bone growth to restore healthy structure and function. Again ideally, having served its assigned purpose, it should gradually degrade and be replaced by normal bone tissue. Thus, we see that such a biomaterial needs to be biocompatible, bioactive, and biodegradable. Can calcium phosphates in any composition, shape, or form satisfy all of these requirements?

If all of the requirements are to be dependably fulfilled, it is clear that any device must be tested and monitored accurately to determine whether, or to what degree, it satisfies the four criteria. But, in order to obtain reproducible and predictable results, it is absolutely mandatory that proper quality control should start from the beginning by knowing accurately and precisely the chemical composition and physical characteristics of the raw materials that are used to fabricate the specific device. Further, we must be able to reproduce supplies of the raw material consistently, process it, and fabricate the final structure in the same way every time. Then, it is necessary to test the product with well-defined measurements of performance and biological responses.

By understanding all of the necessary prerequisite qualifications for a true and accurate biological characterization of calcium phosphates used as implants in bone, it will be evident that only some of the pertinent observations can be given here. It should be kept in mind that, no matter how "physiological" or "biocompatible" any of the calcium phosphates may be, the implant is a foreign body to the recipient tissues. Each recipient host subject, whether animal or human, will show some degrees of variation in responding to the foreign body. It is hoped that the host response will be minor and benign so as not to interfere with a successful outcome, that is, the biological performance of the device.

With that in mind, we consider some of the phases of the calcium phosphates and the biological response to each. The sections below summarize the applications and sites in which each of the phases of the calcium phosphates have been used.

1. Apatites

Although almost unlimited substitutions can occur in the apatites, not all of them are physiologically advantageous for implantation. Calcium hydroxyapatite, $Ca_{10}(PO_4)_6(OH)_2$, is the principal calcium phosphate used for biomedical applications. It has been shown by many researchers that bone forms a very close bond with hydroxyapatite. Several investigators have shown that the crystalline planes of hydroxyapatite in the bioceramic matched those of hydroxyapatite in the mineral phase of bone and that biological apatite was deposited directly on the ceramic surface. This strong bonding is particularly evident in calcium-phosphate-coated implants when failure analysis is performed on the implant-bone interface and failure occurs through the bone-ceramic interface as if it were one structure [50]. Adverse biological response to the bulk material has not been documented. Primarily due to the similarity between the mineral phase in bone and these ceramics, the biocompatibility of hydroxyapatite is excellent both as a bulk ceramic and as a coating.

Bulk hydroxyapatite (HA) has been used with success in a variety of applications. HA has been used by some authors with a certain degree of success for the restoration of atrophic mandibular ridges in animal studies as well as human clinical use. Hydroxyapatite has been used as a filler for periodontal and periapical defects, alveolar ridge augmentation, and maxillofacial reconstruction. When used as a porous ceramic, results were very good. Dense, nonporous particles may become encapsulated in fibrous tissue or may migrate from the site. LeGeros summarizes the uses and clinical results of HA in dentistry in a current review article [51]. Ellinger, Nery, and Lynch compared the use of HA with HA/TCP in a periodontal study. They found that hydroxyapatite did not perform as well as a more soluble ratio of HA/TCP that was being tested [52]. In this study, it was found that the hydroxyapatite ceramic was surrounded by fibrous tissue, while HA/TCP was incorporated into adjacent bone.

In orthopedics, hydroxyapatite is primarily used as a coating for orthopedic prostheses. The HA coating is claimed to act as a stimulus for bone growth. These coated prostheses are claimed to be more "forgiving" in the sense that gaps left between the prosthesis and bone will become filled in with bone that is stimulated to grow by the ceramic. Direct ossification on the hydroxyapatite-coated prostheses was observed in human and nonhuman trials [53–56]. This osseointegration occurs despite the fact, or perhaps because, the coating is not pure hydroxyapatite in many cases. Some investigators have detected the presence of other such phases as CaO and tetracalcium phosphate ($Ca_4P_2O_9$) besides hydroxyapatite on coated prostheses [57–58].

Fluorapatite has been shown to be less soluble than hydroxyapatite. The chemical composition of fluorapatite is $Ca_{10}(PO_4)_6(OH)_{2-x}F_x$, where x can vary from zero (hydroxyapatite) to two (fluorapatite). Due to its lower solubility, it has not been used for medical applications. If used in biomedical applications, it would probably be more bioinert than bioactive.

2. Tricalcium Phosphates

Tricalcium phosphates, $Ca_3(PO_4)_2$, are probably the most bioactive of the calcium phosphate ceramics. Their high degree of solubility compared with hydroxyapatite has indicated their use in areas in which a high degree of bioactivity is desired. This may include the coating of dental and orthopedic prostheses as well as in bulk form for use in dental, orthopedic, and plastic reconstructive applications. Unfortunately, this high degree of bioactivity may cause the material to be completely resorbed without a concomitant deposition of bone. This is a problem in certain dental applications, such as atrophic alveolar ridge augmentation, for which a nonresorbable material is desired. Resorption of the material in these sites leads to a clinical failure.

Cutright et al. first investigated the use of tricalcium phosphate ceramics in the tibias of 48 rats [59]. They observed that bone was deposited directly on the ceramics, and that no adverse tissue reaction was observed. The high resorption rate was evident in these early studies. The authors estimated that biodegradation of the ceramic was 95% complete at 48 days after implantation [59]. Metsger, Driskell, and Paulsrud summarized the literature with regard to the various biomedical uses of tricalcium phosphate [60]. Some of these uses are described below.

LeGeros reviews the uses and clinical results of β-tricalcium phosphate in dentistry in a current article [51]. In some cases, such as periodontal osseous defects, TCP was reported to have little benefit in repair of the defect. Metsger, Driskell, and Paulsrud reported that only 50% of human patients had complete bone fill when the material was used to fill bony defects [60]. LeGeros reports work by Osborn and Donath with HA and TCP for use in alveolar ridge augmentation. At four months, it was observed that the pure TCP ceramic disintegrated, while the HA ceramic granules were bonded with mature lamellar bone [51].

Since dental uses of TCP did not show much promise, orthopedic uses were sporadically investigated. In many of the studies in which TCP was used to fill segmental defects in dogs, filling of the defect with bone occurred whether TCP was used or not [61,62]. Investigation of β-tricalcium phosphate granules for filling defects after tumor removal in humans was investigated by Bucholz, Carlton, and Holmes [63]. They showed that these materials were effective in this application and that subsequent fracture did not occur.

β-Tricalcium phosphate granules were also implanted with orthopedic prostheses in hopes that they would stabilize the prostheses and encourage bone formation [64–66]. Berry et al. reported that the interface bond strength between the implant and host bone was not increased by the addition of TCP [64]. Fitzgerald et al. reported that the ceramic granules formed a barrier to bony ingrowth of the prosthesis [65]. Eschenroeder, McLaughlin, and Reger also used β-TCP granules as an interposition material between overreamed femoral intramedullary canals in dogs and a porous-coated intramedullary rod. At six weeks, statistically significant data on the increase in pull-out strength using the granules was not obtained [66]. At six months, the authors claim that the TCP-stabilized specimens exhibited a statistically significant increase in resistance to pull-out force compared with rods implanted without the interposition of TCP granules [66].

Due to problems with resorption of TCP without concomitant deposition of bone, TCP is rarely used clinically. It is useful for applications in which bioactive materials are used as a drug delivery system. Slow resorption of the ceramic allows for slow administration and release of enclosed drugs.

3. Ratios of Hydroxyapatite/Tricalcium Phosphate

Hubbard was the first to investigate ratios of HA/β-TCP for use as a bone graft substitute in 1974 [29]. These materials have often been called biphasic calcium phosphate ceramics since they contain two phases: hydroxyapatite and β-tricalcium phosphate. These composite ceramics offer several distinct advantages over either phase as described above. HA has been characterized as relatively bioinert compared with the tricalcium phosphate ceramic. Although the tricalcium phosphate ceramic phase is more bioactive, it may be quickly resorbed without subsequent bone formation. The biphasic material has the ability to provide a scaffold for new bone growth due to the presence of the HA phase, as well as the ability to foster osteogenesis due to the presence of the bioactive β-tricalcium phosphate phase. The biological response to the biphasic ceramic has been investigated by several authors [63–70]. In addition, some authors have investigated the effectiveness of different ratios in various sites [63–70]. The biphasic ceramic is unique in that various compositions can be fabricated that may be advantageous at a particular site. Other sites, which have different requirements, may necessitate different ratios.

The biphasic ceramic has been investigated in several dental studies, including the repair of periodontal osseous defects and atrophic alveolar ridge augmentation. In these studies, no adverse biological response was detected [69]. The biphasic ceramic was shown to be superior to HA by Ellinger, Nery, and Lynch in the reconstruction of periodontal osseous defects since the HA/TCP was less bioinert than HA [52].

The HA/TCP ceramics have been successfully used by themselves in studies for the filling of defects in long-bone surgery by Daculsi et al. [71]. Passuti et al. have also investigated the use of the HA/TCP ceramics in spinal fusion surgery in the correction of scoliosis [67]. The ceramics were found to provide sufficient fusion mass to stabilize the construct.

4. Hydroxyapatite/Tricalcium Phosphate and Collagen Composite

Some authors have investigated the use of collagen and HA/TCP ceramic materials as bone graft substitutes. Collagen is used as a binder to keep the ceramic granules in the site until fibrous tissue can hold them in place. This is thought to prevent the migration of the ceramic granules. Initial studies have included the investigation of this composite implant in spinal fusion surgery [72,73], in cases of trauma, for the repair of metaphyseal and diaphyseal defects [74–76], for the repair of cranial defects, and for alveolar ridge augmentation [77]. The material has been shown to be biocompatible and not to elicit an adverse biological response. Despite the fact that an immunological response to the collagen is possible, none has been reported in these studies.

5. Calcium Pyrophosphates

In 1974, Hubbard investigated the use of calcium pyrophosphates in the rabbit iliac crest model. After examining the explants, he concluded that "there was not any difference in the physiological response to either the calcium hydroxyapatite material or the calcium pyrophosphate" [29, p. 183]. In addition, active osteoblasts were forming bone appositional to the ceramic. Bone was found in the pores of the alpha and beta calcium pyrophosphate ceramics implanted into the iliac crest. In 1992, we confirmed these results by showing that calcium pyrophosphates elicit a host response in bone that is similar to other calcium phosphate phases, particularly HA and HA/TCP [78]. Others have had reservations about the osteoconductivity of the calcium pyrophosphate ceramics since they contain the pyrophosphate ion, a reported calcification inhibitor.

6. The Influence of Porosity on Biological Characterization

The first generalized biological characterization is related to the porosity of the implant, that is, whether it is nonporous, microporous, or having some measurable degree of macroporosity. The tissue response to very small nonporous or microporous particles is to attempt to remove them by phagocytosis in the acute inflammatory phase. Such small particles are seldom seen in implants except on the surface of larger blocks or spheres. They are to be avoided, because powdered particles (1–10 μm) of any material incite acute inflammation and are chemotactic for PMN leukocytes. When larger blocks or spheres are used, the chronic inflammatory cell response has an early onset. Mononuclear macrophages and giant cells appear and further attempts at phagocytosis are confined to the surface. Hydroxyapatite and fluoroapatites seem impervious to the macrophages. Carbonate-substituted apatites and β-TCP are susceptible to phagocytosis and slow solubilization, thus they are gradually resorbed, but only at the surface.

Simultaneous to the appearance of macrophages, healing fibroblasts proliferate and a fibrous encapsulation develops. Angiogenesis in the very early stages brings a proliferation of thin-walled capillaries, but as the fibrous capsule at the interface increases in depth, the capillaries are obliterated and are found only in the adjacent bone. Solid particles and blocks are effectively "walled off" and isolated from direct contact with bone. Thus, a classical benign foreign-body reaction occurs and the ceramic structure becomes a lasting barrier to bone-to-bone bridging. If surface resorption has caused erosion of the surface, some small areas of point-to-point interface contact with bone may occur by development of osteoblasts at those sites. Such areas may enlarge with time as the resorption and healing proceed, but all too frequently the tenuous foothold of bone-to-bone contact declines and the fibrous capsule replaces it. Histiocytes and giant cells remain in the area near the foreign implant body as long as it remains. Depending upon the location, micromotion may occur, causing the solid particles to be gradually extruded from the site. This is common with all foreign bodies.

Macroporous structures, on the other hand, are responded to in the same way initially. As the macrophages, angioblasts, and fibroblasts proliferate, they are able to invade the pores. When phagocytosis occurs, resorption cavities enlarge, and angioblasts can become endothelial cells forming capillaries and extend the blood supply through interconnecting channels. Healing fibroblasts produce a fibrous collagen matrix. This matrix increases in density and, if the macrophages, histiocytes, and giant cells can successfully resorb the calcium phosphate to increase the areas into which the capillaries can penetrate, the fibrous matrix will extend itself.

Meanwhile, at the ceramic/bone interface, in a favorable environment with an adequate blood supply and various growth factors (which could include TGF-β, bone morphogenetic proteins, osteogenin, and other factors), some of the multipotential mesenchymal cells are induced to differentiate into osteoblasts and an endosteal osteogenic membrane appears. The osteoblasts secrete a fibrous Type I collagen matrix called osteoid, laid down appositionally on the surface of existing bone and also on the scaffold walls of the porous calcium phosphate. Osteoid is the precursor matrix that, when calcified, becomes bone. It is often phenomenal, when viewing microscopic sections, to see the activity of multinucleated giant cells performing their phagocytic activity by ingesting the calcium phosphate foreign-body material, while nearby, with one side of the osteoblast in appositional contact to the surface of that same foreign body, osteoid collagen is being secreted and the matrix becomes calcified. Thus, bone is said to "grow into" the pores.

The foreign-body chronic inflammatory cellular response occurs simultaneously with the advancement of the healing response, which is osteogenesis filling the spaces with normal bone tissue. This phenomenon often occurs in the same microscopic field, only 100 μm apart. In between are seen the healing fibroblasts, histiocyte macrophages, and preosteoblasts, which have all differentiated from the undifferentiated mesenchymal stem cells in response to each specific biochemical messenger. The apparently chaotic cellular activity is remarkably orchestrated.

The rate and distribution of osteogenesis around and within the porous implant will vary considerably, depending upon the size and number of interconnecting channels (i.e., the degree of macroporosity). A pore diameter of less than 150 μm and 30% porosity limits penetration of bone growth to within 1 mm or less of the surface. Larger pores of 400–800 μm appear optimal and are suitable for the fabrication of devices with adequate strength with good bone integration. The percent porosity by weight may be varied selectively to fit the requirements for compressive strength. The more connecting channels that can be provided, the better will be the bone penetration to fill the pores. Pores of large diameter are easily vascularized and bone quickly grows to line the peripheral surface but, for reasons unknown, this causes a transformation to the induction of hematopoiesis and bone marrow then occupies the remaining space.

This same process is seen in marrow cavities that develop following autogenous bone graft transplants. Surfaces once covered by bone are no longer subject to attack by giant cells. However, the giant cells remain in the indigenous cell population as long as the ceramic implant exists at the site—a reminder that this is indeed still a foreign body.

Other major factors that affect the rate and distribution of osteogenesis relate to the physicochemical composition. Bone ingrowth is slow, but usually complete in replamine-form calcium hydroxyapatite blocks because of the interconnecting channels. This apatite structure is made by a hydrothermal exchange of phosphate for carbonate in the coral skeleton. It is possible that the exchange may not be complete, thus leaving an unknown amount of carbonate-substituted apatite. Such a material would be subject to more rapid phagocytosis wherever it exists in the structure. The same is true for ceramics containing β-TCP, which is also readily resorbed. β-TCP alone is often rapidly resorbed in such a manner that the healing does not leave enough time to allow bone growth for replacement as mentioned above. A dense fibrous matrix then develops without proper calcification and true bone is not attained in the healing process. Composites that are made of HA and β-TCP in a 60:40 ratio allow adequate time to provide a scaffold for good appositional bone formation. Granules made from macroporous blocks or cylinders during fabrication and having a diameter of 0.5–1.5 mm are readily osseointegrated into bone of good radiographic density and biomechanical strength when implanted into bone [79]. Other compositions have not been well studied.

The different responses described above are not entirely due to the presence or absence of porosity alone. Healing, with normal bone tissue replacement, requires an adequate blood circulation to be maintained. Revascularization may or may not be due to porosity or composition of the implant. If the angiogenesis is interrupted by micromovement of the implant, only fibroblasts and fibrous tissue will develop. Hence, early and stable immobilization of the implant is a prerequisite for osseointegration. In addition, if a nonporous calcium phosphate ceramic has a crystalline phase that is susceptible to phagocytosis by macrophages, their continued presence, over a long period of time, may be able to alter the surface and allow slow invasion of the solid structure. Angiogenesis can then allow a blood supply to be established and reactivate osteogenesis. The presence of surface scratches, serrations, or fracture of the structure encourages osteogenesis by exposing more surface area to resorption. Porosity, surface area, chemical composition, immobilization, and individual variations in chronic inflammatory response, as well as individual ability to produce and deliver the necessary growth factors, will all affect the range of response and therefore the biological characterization. Accurate and precise quality control and intelligent considerations for standards of use-testing evaluation, indeed the whole process of fabrication from beginning to end, are necessary to predict results accurately.

D.   Characterization of Calcium Phosphates in Nonosseous Tissues

Some of the first investigators of the calcium phosphates did so because they thought that synthetic calcium phosphates (and solutions containing them) had a stimulatory effect on osteogenesis [80–82]. This was suspected because a scientist by the name of H. G. Wells, writing in the *Archives of Internal Medicine* in 1911, proposed a theory that stated that calcium salts exert a stimulatory effect on osteogenic activity [83,84]. Since that time, many authors have investigated the biological response to the calcium phosphate ceramics in nonosseous sites without the addition of an inducing stimulus such as marrow or growth factors and found that the ceramics by themselves did not induce bone in those animal models and at those time periods [85–87].

The host response characteristic of the class of calcium phosphate ceramics has been claimed to be completely that of osteoconduction. Calcium phosphates exhibit the phenomenon of *bioactive osteoconductivity*, that is, they permit bony ingrowth and act as a scaffold in bony tissue sites. They have not been characterized as *osteoinductive*, that is, capable of initiating osteogenesis in nonbony sites. This characterization is based on the work of several authors who have investigated extraskeletal implantation of predominantly the hydroxyapatite phase of the calcium phosphate ceramics.

Piecuch investigated the biological response to coralline replamineform hydroxyapatite implants placed in canine axillary subcutaneous sites. When the explants were harvested at eight weeks, histological observation did not reveal the presence of bone tissue [85]. With these results, the author concluded that "porous hydroxyapatite does not act as a bone-inducing agent".[85] Misiek, Kent, and Carr reported results of chronological explantation of smooth and rough nonporous hydroxyapatites in buccal soft tissue pouches of dogs [86]. At the six-month explantation time, fibrous connective tissue was found in the implants, but osseous tissue was not observed [86].

More recently, Pettis, Kaban, and Glowacki evaluated hydroxyapatite and a composite consisting of demineralized bone matrix and hydroxyapatite in rat subcutaneous tissue. The hydroxyapatite implants did not induce bone formation at the three-week explantation time. Enchondral bone formation was observed in the composite implants in the same time period [87].

Other studies, which are too numerous to be described here, have been conducted in nonosseous tissues with a similar result. Bucholz has summarized the literature with regard to the biological performance of the calcium phosphates in the following way:

> There is no conclusive evidence that any of the porous calcium phosphate biomaterials are osteoinductive. Most are considered osteoconductive, that is, allowing for bone ingrowth from an osseous bed. No bone ingrowth will occur if the implant is inserted into muscle or subcutaneous tissue. [63]

An ideal bone graft substitute would be a material with osteoinductive properties. This material would be able to initiate osteogenesis and promote the healing process in bone. Other uses include reconstructive applications for which a bone-inductive material could be used to restore form and/or function. Several protein or polypeptide materials have been previously reported to be osteoinductive. These include demineralized bone matrix, dentin, osteogenin, and transforming growth factor-β in combination with osteoinductive factor. Some authors have used various phases of the calcium phosphate ceramics as carriers for these growth factors. It has also been shown that, when the calcium phosphate ceramics are loaded with marrow cells in a rat model, bone formation occurs [88–92]. In these same studies, the ceramic phases without marrow or growth factors did not induce bone formation.

Some evidence that the calcium phosphate ceramics may stimulate osteogenesis was reported by Ripamonti. Intramuscular implantation of the porous coralline replamine-form hydroxyapatite in baboons without marrow or growth factors yielded woven and lamellar bone formed by intramembranous osteogenesis at the three-, six-, and nine-month time periods [93,94]. Vargervik has also reported *de novo* bone formation in porous HA blocks without marrow or growth factors. The HA blocks were implanted in subcutaneous tissue, in muscle tissue, and attached to a muscle flap in rhesus monkeys [95]. None of the HA blocks initiated bone formation in any of the sites four weeks after implantation, although enzyme stains indicated activity of osteoblasts [95]. Vargervik consistently found bone in the HA blocks in intramuscular implants and blocks attached to muscle flaps at later time periods. In the subcutaneous sites, bone was found in four out of eight of the HA blocks [95]. Similarly, Shors et al. found bone in 77% of porous HA disks implanted intramuscularly and subcutaneously in dogs at 3, 6, and 12 months [96].

We have reported that a collagen/ceramic composite induced bone formation in a canine subcutaneous model without the addition of growth factors or bone marrow cells [97]. Intramembranous bone formation was first observed in these implants at the seven-to-nine-week harvest times. In a similar study, we attempted to answer questions raised by the earlier study. What caused bone to form in response to the collagen/ceramic composite? Was it the ceramic or the collagen that caused the bone formation? Was the bone formation pathological? Was the osteogenesis in response to the calcium phosphate ceramics limited to hydroxyapatite phases, or would other phases such as the calcium pyrophosphates induce osteogenesis? What was the histogenesis of the bone formation?

Within the past year, we reported the results of the biological response to implantation of the following materials in the subcutaneous tissue of the dog: 60/40 HA/TCP porous granules, α-calcium pyrophosphate porous granules, β-calcium pyrophosphate porous granules, Type I bovine dermal fibrillar collagen, and a collagen and ceramic (60/40 HA/TCP porous granules) composite [78,98–101]. Explants removed at six months containing a ceramic phase induced lamellar bone best described as convoluted, compact, coarse, cancellous primary vascular bone of plexiform structure formed by hypertrophied osteoblasts in intramembranous osteogenesis. Hematopoietic marrow, complete with full red and white cell lines and megakaryocytes, was observed in some of the calcium phosphate explants. The calcium phosphate ceramics were determined to be osteoinductive in that they induced *de novo* bone formation in an ectopic site, namely, canine subcutaneous tissue. The ceramics and composite each were determined to have a unique, consistent, and inherent osteogenic potential in that they consistently produced bone of different quantity and quality, as measured by microradiographic microdensitometry.

Bone was histologically identified in all of the implants containing calcium phosphate ceramics. Bone marrow was found in some of the implants containing the calcium phosphate ceramics. The amount of bone marrow present appeared to correlate with the quantity and quality of bone formation in response to the implants. Bone marrow, complete with full red and white cell lines as well as megakaryocytes, was identified in the decalcified sections. Table 3 summarizes the incidence of bone and marrow formation in response to the implant materials.

Evidence of intramembranous osteogenesis was observed in all of the implants containing a phase of calcium phosphate. Undecalcified sections microscopically examined

Table 3   Incidence of Osteogenesis and Hematopoiesis Following Subcutaneous Implantation of Calcium Phosphates

Material implanted	Bone formation	Marrow formation
Alpha-pyrophosphate (granules)	6/6	0/6
Beta-pyrophosphate (granules)	6/6	1/6
Collagen/ceramic composite	6/6	3/6
60/40 HA/β-TCP (granules)	6/6	4/6
Lyophilized collagen	0/6	0/6

under ultraviolet light revealed no distinct fluorescent calcification front. Rather, diffuse fluorescence of the apatite mineral in the bone was observed. Examination of the decalcified sections with Safranin-O/Fast green stain did not demonstrate the presence of a cartilaginous precursor to bone formation. Hypertrophied osteoblasts of an endosteal membrane were seen forming bone in small trabeculae directly appositional to calcium phosphate granules. Polarized light microscopy revealed that the collagen fibrils in the osteoid were randomly oriented initially. With progressive osteogenesis, the osteoid pattern organized into a lamellar structure that became dense in the form of compact trabeculae. Maturation produced coalescence of lamellar trabeculae around blood vessels, and convoluted, compact, coarse, cancellous bone of plexiform structure developed.

Recently, other authors have reported that they have observed the same phenomenon [102–105]. Several hypotheses for this ceramic-induced osteogenesis have been suggested. At this time, the phenomenon of osteogenesis in response to ectopic implantation of the calcium phosphates remains an enigma. With the exception of pathological bone formation, several hypotheses appear to be likely candidates for the observed phenomenon of osteogenesis in response to ectopic implantation of the calcium phosphates in this canine model. It should also be noted here that this phenomenon is species related. That is, it occurs in canines and primates, but not rodents.

It is postulated that osteogenesis in response to the calcium phosphate ceramics alone has not previously been observed due to three main reasons. First, for whatever reason, the phenomenon is species-related. Second, we have previously reported that, even in the canine subcutaneous model, the first appearance of recognizable bone occurred nearly two months after implantation of the ceramic [97]. Many researchers have removed implants after two-weeks implantation expecting to find bone formation that mimics the swift enchondral-type osteoinduction in response to the bone morphogenic proteins such as osteogenin or transforming growth factor-β (TGF-β) and osteoinductive factor (OIF). Third, the osteogenesis followed an intramembranous endosteal development. Others have looked for the development of cartilage following an enchondral pathway. In those studies, vascularization followed chondrogenesis and bone formation followed in the enchondral pathway.

In our observations, early vascularity leads to primary vascular bone formation in an intramembranous endosteal pathway. With a porous calcium phosphate ceramic and abundant vascular supply, osteogenesis occurs by direct transformation of mesenchymal cells to osteoblasts, followed by endosteal intramembranous osteogenesis. In this manner, the enchondral cartilage phase may be bypassed. Should the *de novo* ossicle of bone persist in the subdermal tissue? It does not have a purpose such as load bearing in that particular site. Secondary osteons (Haversian systems) have not been observed, most likely due to the low state of stress on the bone. If the mechanism for this phenomenon of ossification is due to the natural biological response to primary calcifications (in this case, the implanted ceramics), the bone formation may persist as long as the implant is in place. It is not known whether or not this phenomenon occurs in humans.

As indicated above, the stimulus for the initial application of calcium phosphates to the field of orthopedics was the claim that calcium phosphate salts exert a stimulatory effect on the regulation of bone. Earlier experiments failed to show any osteoinductive effect. Later studies showed that the calcium phosphates were osteoconductive, or acted as a scaffold for the development of bone in bony tissue. When these materials were implanted in nonbony tissue in rodents, osteogenesis was not observed.

Recently, several investigators have shown that calcium phosphates have osteoinductive properties by showing that they induce bone formation in nonbony tissue in dogs, rhesus monkeys, and baboons. Black has defined osteoinductivity as a "property of a material that produces bone growth associated with *de novo* growth *in vivo*" [42, p. 368]. It would appear from these experiments that calcium phosphates indeed have this property in some animal models as described above. Currently, that property or those properties of the calcium phosphates that cause mobilization and cytodifferentiation remain an enigma. Also, whether or not this phenomenon of ceramic-induced osteogenesis occurs in humans is unknown.

III. SUMMARY

Calcium phosphates can be produced that elicit a range of biological response. Nearly all of the calcium phosphate phases have been shown to be biocompatible and bioactive. Much work continues on correlating the material properties measured in preimplant studies with the biological response to the materials. Bioactivity has been demonstrated by active osteoblasts on the surface of the implants for most of the calcium phosphate phases. In addition, many authors have shown convincing evidence of biological apatite in a collagenous matrix on synthetic hydroxyapatite in the same crystalline orientation as the synthetic material.

Certainly, these materials offer a unique interface for osteogenesis. The foreign-body response to these materials can be characterized by the proliferation of mononuclear histiocytes and phagocytic multinucleated giant cells. Some calcium phosphates, particularly calcium phosphate cements, are characterized as having a minimal chronic foreign-body response. Other signs of inflammation are usually minimal or absent.

During the wound healing associated with implantation of the material, angiogenesis takes place and a capillary blood supply develops. Healing fibroblasts attach to the interface and form an appositional fibrous matrix. If vascular penetration can occur into the implant, dependent upon composition and structure, some of the multipotential mesenchymal cells differentiate into osteoblasts and an endosteal membrane develops. This membrane actively produces dense fibrous osteoid that becomes calcified. The orientation of the osteoid is perivascular and appositional to the implant surface, thus primary vascular bone develops from the endosteal membrane activity. Osseointegration thus occurs and the implant becomes a part of the bone structure at the site. If, for any reason, the blood supply is interrupted, the healing fibroblasts predominate and the fibrous matrix becomes avascular. The implant is then enveloped within a fibrous capsule and is isolated from the surrounding bone matrix. This benign outcome of the foreign-body reaction precludes osteogenesis at the interface between bone and implant at whatever sites it occurs. The degree of biocompatibility can be measured in terms of the amount of osseointegration and/or fibrous encapsulation. Restoration of structure and function must be measured by appropriate biomechanical tests and clinical criteria.

It is not only important, but also absolutely necessary, that each fabrication for a specific use must be quality controlled to determine the chemical and physical characteristics from raw material source to end product in order to obtain qualifying data before biological implantation. The designed product must then be biologically characterized by tests that are consistent with intended applications. Animal tests are appropriate and required in most instances for approval to proceed to clinical trials. The animal tests should include implantation of a standardized material in a bony site in order to obtain histological evidence of the cell/matrix interactions and tissue response. Biological characterization involves assessing the histological response at the bone-implant site by a chronological explant study. Preliminary material testing in heterotopic implant sites for a period not less than three to four months in dogs or some mammal other than rodents is advisable. The subcutaneous site is very responsive, reliable, and easily accessible, as well as useful for multiple sample tests. Different calcium phosphate phases, porosities, and geometric configurations can produce greatly different responses that must be characterized. This may help in obtaining regulatory acceptance for clinical trials. The addition of other biomaterials to calcium phosphates to produce composites can greatly alter the responses and the biological characterization should be qualified according to the guidelines suggested here.

REFERENCES

1. Cullity, B.D., *Elements of X-Ray Diffraction*. 2nd ed., Reading, MA: Addison-Wesley, 1978.
2. Klug, H.P., and L.E. Alexander, *X-Ray Diffraction Procedures for Polycrystalline and Amorphous Materials*. 2nd ed., New York: J. Wiley and Sons, 1974.
3. American Society for Testing and Materials (ASTM), Standard Specification for Composition of Ceramic Hydroxylapatite for Surgical Implants, ASTM F 1185-88, December 1988, *Annual Book of ASTM Standards*, 13.01:458–459 (1988).
4. Toth, J.M., W.M. Hirthe, W.G. Hubbard, W.A. Brantley, and K.L. Lynch, Determination of HA/TCP ratios by x-ray diffraction, *J. Applied Biomat.*, 2:37–40 (1991).
5. Winand, L., and G. Duyckaerts, Étude infrarouge de phosphates de calcium de la famille de l'hydroxylapatite, *Bull. Soc. Chim. Belge.*, 71:142–150 (1962).
6. Stutman, J.M., J.D. Termine, and A.S. Posner, Vibrational spectra and structure of the phosphate ion in some calcium phosphates, *Trans. N.Y. Acad. Sci. Ser. II*, 27:669–675 (1965).
7. Baddiel, C.B., and E.E. Berry, Spectra structure correlations in hydroxy and fluorapatite, *Spectrochim. Acta*, 22:1407–1416 (1966).
8. Fowler, B.O., E.C. Moreno, and W.E. Brown, Infra-red spectra of hydroxyapatite, octacalcium phosphate, and pyrolysed octacalcium phosphate, *Arch. Oral Biol.*, 11:477–492 (1966).
9. Whitehead, R.Y., W.R. Lacefield, and L.C. Lucas, Structure and integrity of a plasma sprayed hydroxylapatite coating on titanium, *J. Biomed. Mater. Res.*, 27:1501–1507 (1993).
10. Cant, N.W., J.A.S. Bett, G.R. Wilson, and W.K. Hall, The vibrational spectrum of hydroxyl groups in hydroxyapatites, *Spectrochim. Acta*, 27A:425–439 (1971).
11. Berry, E.E., The structure and composition of some calcium-deficient apatites, *J. Inorg. Nucl. Chem.*, 29:317–327 (1967).
12. Arends, J., J. Christoffersen, M.R. Christoffersen, H. Eckert, B.O. Fowler, J.C. Heughebaert, G.H. Nancollas, J.P. Yesinowski, and S.J. Zawacki, A calcium hydroxyapatite precipitated from an aqueous solution, *J. Cryst. Growth*, 84:515–532 (1987).

13. Ellies, L.G., D.G.A. Nelson, and J.D.B. Featherstone, Crystallographic structure and surface morphology of sintered carbonated apatites, *J. Biomed. Mater. Res.*, 22:541–553 (1988).

14. Ducheyne, P., S. Radin, M. Heughebaert, and J.C. Heughebaert, Calcium phosphate ceramic coatings on porous titanium: Effect of structure and composition on electrophoretic deposition, vacuum sintering and *in vitro* dissolution, *Biomat.*, 11:244–254 (1990).

15. Ducheyne, P., S. Radin, and L. King, The effect of calcium phosphate ceramic composition and structure on *in vitro* behavior. I. Dissolution, *J. Biomed. Mater. Res.*, 27:25–34 (1993).

16. Ishikawa, I., P. Ducheyne, and S. Radin, Determination of the Ca/P ratio in calcium-deficient hydroxyapatite using x-ray diffraction analysis, *J. Mater. Sci.: Materials in Medicine*, 4:165–168 (1993).

17. Bonel, G., J.C. Heughebaert, M. Heughebaert, J.L. LaCout, and Albert LeBugle, Apatitic calcium orthophosphates and related compounds for biomaterials preparation, *Ann. N.Y. Acad. Sci.*, 253:115–130 (1988).

18. Termine, J.D., and A.S. Posner, Infra-red determination of the percentages of crystallinity in apatitic calcium phosphates, *Nature*, 211:268–270 (1966).

19. Pleshko, N., A. Boskey, and R. Mendelsohn, Novel infrared spectroscopic method for the determination of crystallinity of hydroxyapatite minerals, *Biophys. J.*, 60:786–793 (1991).

20. Rey, C., M. Shimizu, B. Collins, and M.J. Glimcher, Resolution-enhanced Fourier transform infrared spectroscopy study of the environment of phosphate ions in the early deposits of a solid phase of calcium-phosphate in bone and enamel, and their evolution with age. I: Investigations in the ν_4 PO_4 domain, *Calcif. Tissue Int.*, 46:384–394 (1990).

21. Heughebaert, M., R.Z. LeGeros, M. Gineste, A. Guilhem, and G. Bonel, Physicochemical characterization of deposits associated with HA ceramics in nonosseous sites, *J. Biomed. Mater. Res.*, 22:257–268 (1988).

22. Radin, S.R., and P. Ducheyne, The effect of calcium phosphate ceramic composition and structure on *in vitro* behavior. II. Precipitation, *J. Biomed. Mater. Res.*, 27:35–45 (1993).

23. Elliot, J.C., D.W. Holcomb, and R.A. Young, Infrared determination of the degree of substitution of hydroxyl by carbonate ions in human dental enamel, *Calcif. Tissue Int.*, 37:372–375 (1985).

24. Rey, C., V. Renugopalakrishnan, M. Shimizu, B. Collins, and M.J. Glimcher, A resolution-enhanced Fourier transform infrared spectroscopic study of the environment of the CO_3^{2-} ion in the mineral phase of enamel during its formation and maturation, *Calcif. Tissue Int.*, 49:259–268 (1991).

25. Bucholz, R.W., A. Carlton, and R.E. Holmes, Hydroxyapatite and tricalcium phosphate bone graft substitutes, *Orthop. Clin. N. Amer.*, 18(2):323–334 (1987).

26. Chiroff, R.T., R.A. White, E.W. White, J.N. Weber, and D.M. Roy, The restoration of articular surfaces overlying replamineform porous biomaterials, *J. Biomed. Mater. Res.*, 11:165–178 (1977).

27. Guillemin, G., J.-L. Patat, J. Fournie, and M. Chetail, The use of coral as a bone graft substitute, *J. Biomed. Mater. Res.*, 21:557–567 (1987).

28. Holmes, R.E., Bone regeneration within a coralline hydroxyapatite implant, *Plastic & Recons. Surg.*, 63(5):626–633 (1979).

29. Hubbard, W.G., Physiological calcium phosphates as orthopaedic biomaterials, Ph. D. dissertation, Marquette University, Milwaukee, WI (1974).

30. LeGeros, R.Z., *Calcium Phosphates in Oral Biology and Medicine*, vol. 15 of *Monographs in Oral Science*, ed. by H.M. Meyers, New York: Karger, 1991.

31. LeGeros, R.Z., Significance of the porosity and physical chemistry of calcium phosphate ceramics: Biodegradation-bioresorption. In *Bioceramics: Material Characteristics Versus In Vivo Behavior*, ed. by P. Ducheyne, and J.E. Lemons, *Annals of the New York Academy of Sciences*, 523:268–271 (1988).

32. Prolo, D.J., and J.J. Rodrigo, Contemporary bone graft physiology and surgery, *Clin. Orthop.*, 200:322–342 (1985).

33. Burchardt, H., Biology of bone transplantation, *Orthop. Clin. N. Amer.*, 18:187–196 (1987).
34. Heiple, K.G., V.M. Goldberg, A.E. Powell, G.D. Bos, and J.M. Zika, Biology of cancellous bone grafts, *Orthop. Clin. N. Amer.*, 18(2):179–185 (1987).
35. Gross, T.P., R.H. Jinnah, H.J. Clarke, and Q.C.N. Cox, The biology of bone grafting, *Orthopedics*, 14:563–568 (1991).
36. Younger, E.M., and M.W. Chapman, Morbidity at bone graft donor sites, *J. Orthop. Trauma*, 3:192–195 (1989).
37. Cockin, J., Autologous bone grafting: Complications at the donor site, *J. Bone Jt. Surg.*, 53B:153 (1971).
38. Bolano, L., and J.A. Kopta, The immunology of bone and cartilage transplantation, *Orthopedics*, 14(9):987–996 (1991).
39. Buck, B.E., and T.I. Malinin, Bone transplantation and human immunodeficiency virus. An estimate of risk of acquired immunodeficiency syndrome (AIDS), *Clin. Orthop.*, 240:129–136 (1989).
40. Pelker, R.R., and G.C. Friedlander, Biomechancal aspects of bone autografts and allografts, *Orthop. Clin. N. Amer.*, 18(2):235–239 (1987).
41. Lane, J.M., and H.S. Sandhu, Current approaches to experimental bone grafting, *Orthop. Clin. N. Amer.*, 18:213–225 (1987).
42. Black, J., *Orthopedic Biomaterials in Research and Practice.* New York: Churchill Livingstone, 1988.
43. Williams, D.F. (editor), *Definitions in Biomaterials. Progress in Biomedical Engineering.* Vol. 4. Amsterdam: Elsevier, 1987.
44. von Recum, A.F. (editor), *Handbook of Biomaterials Evaluation.* New York: Macmillan, 1986.
45. Park, J.B., and R.L. Lakes (editors), *Biomaterials: An Introduction.* 2nd Ed. New York: Plenum Press, 1992.
46. Peacock, E.E., Jr., and W. van Winkle, Inflammation and the cellular response to injury. In *Surgery and Biology of Wound Repair*, Philadelphia: W.B. Saunders, 1970, pp. 1–16.
47. Price, S.A., and L.M. Wilson, *Pathophysiology—Clinical Concepts of Disease Processes.* 2nd ed. New York: McGraw-Hill, 1982.
48. Coleman, D.L., R.N. King, and J.D. Andrade, The foreign body reaction: A chronic inflammatory response, *J. Biomed. Mater. Res.*, 8:199–211 (1974).
49. Damien, C.J., and J.R. Parsons, Bone graft and bone graft substitutes: A review of current technology and applications, *J. Applied Biomat.*, 2:187–208 (1991).
50. Manley, M.T., J.F. Kay, M. Uratsuji, L.S. Stern, and B.N. Stulberg, Hydroxylapatite coatings applied to implants subjected to functional loads, *Trans. 13th Annual Meeting of the Society for Biomaterials*, June 2–6, 1987, New York, p. 210.
51. LeGeros, R.Z., Calcium phosphate materials in restorative dentistry: A review, *Adv. Dent. Res.*, 2(1):164–180 (1988)
52. Ellinger, R.F., E.B. Nery, and K.L. Lynch, Histological assessment of periodontal osseous defects following implantation of hydroxyapatite and biphasic calcium phosphate ceramics, *Int. J. Periodontics & Restorative Dent.*, 6(3):23–34 (1986).
53. deGroot, K., R. Geesink, C.P. Klein, and P. Serekian, Plasma sprayed coatings of hydroxylapatite, *J. Biomed. Mater. Res.*, 21:1375–1381 (1987).
54. Furlong, R.J., and J.F. Osborn, Fixation of hip prostheses by hydroxyapatite ceramic coatings, *J. Bone Jt. Surg. [BR]*, 73B(5):741–745 (1991).
55. Geesink, R.G.T., Experimental and clinical experience with hydroxyapatite-coated hip implants, *Orthopedics*, 12(9):1239–1242 (1989).
56. Geesink, R.G.T., Hydroxyapatite-coated total hip prostheses: Two-year clinical and roentgenographic results of 100 cases, *Clin. Orthop.*, 261:39–58 (1990).
57. Collier, J.P., M.B. Mayor, K.A. Dwyer, D.A. Teske, J. Chae, and V.A. Suprenant, The

unsolved problems associated with the use of plasma-sprayed TCP and HA coatings on orthopedic prostheses, *Trans., 34th Annual Meeting of the Orthopedic Research Society*, February 1–4, 1988, Atlanta, Georgia, p. 51.

58. Ducheyne, P., S. Radin, K. Healy, and J.M. Cuckler, The effect of plasma spraying on the structure and properties of calcium phosphate ceramics, *Trans., 34th Annual Meeting of the Orthopedic Research Society*, February 1–4, 1988, Atlanta, Georgia, p. 50.

59. Cutright, D., and S. Bhaskar, Reaction of bone to tricalcium phosphate ceramic pellets, *Oral Surg.*, 33:850–855 (1972).

60. Metsger, D.S., T.D. Driskell, and J.R. Paulsrud, Tricalcium phosphate ceramic—A resorbable bone implant: Review and current status, *J. Am. Dent. Assoc.*, 105:1035–1038 (1982).

61. Baum, J.R., P.R. Cain, G. Levine, L.W. Gumerman, and M.A. Goodman, Tricalcium phosphate ceramic as a bone graft substitute in the treatment of segmental long bone defects, *Trans., 31st Annual Meeting of the Orthopedic Research Society*, January 21–24, 1985, Las Vegas, Nevada.

62. Lange, T.A., J.E. Zerwekh, R.D. Peek, V. Mooney, and B.H. Harrison, Granular tricalcium phosphate in large cancellous defects, *Annals of Clinical and Laboratory Science*, 16: 467–472 (1986).

63. Bucholz, R.W., A. Carlton, and R.E. Holmes, Hydroxyapatite and tricalcium phosphate bone graft substitutes, *Orthop. Clin. N. Amer.*, 18:323–34 (1987).

64. Berry, J.L., J.M. Geiger, J.M. Moran, J.S. Skraba, and A.S. Greenwald, Use of tricalcium phosphate or electrical stimulation to enhance the bone-porous implant interface, *J. Biomed. Mater. Res.*, 20:65–77 (1986).

65. Fitzgerald, R.H., Jr., E.Y.S. Chao, D.J. McDonald, and G.M. Russoti, A comparison of autografts, allografts, and tricalcium phosphate hydroxyapatite crystals. In *Non-Cemented Total Hip Arthroplasty*, ed. by R. Fitzgerald, Jr., New York: Raven Press, 1988, pp. 159–174.

66. Eschenroeder, H.C., R.E. McLaughlin, and S.I. Reger, Enhanced stabilization of porous-coated metal implants with tricalcium phosphate granules, *Clin. Orthop.*, 216:234–246 (1987).

67. Passuti, N., G. Daculsi, S. Martin, and C. Deudon, Macroporous polycrystalline calcium phosphate implant for spinal fusion in man and dogs. In *CRC Handbook of Bioactive Ceramics*, ed. by T. Yamamuro, L. Hench, and J. Wilson, Boca Raton, FL: CRC Press, 1990, pp. 345–354.

68. Flatley, T.J., K.L. Lynch, and M. Benson, Tissue response to implants of biphasic calcium phosphate ceramic in the rabbit spine, *Clin. Orthop. Rel. Res.*, 179:246–249 (1983).

69. Nery, E.B., and K.L. Lynch, Preliminary clinical studies of bioceramic in periodontal osseous defects, *J. Periodontol.*, 49:523–525 (1978).

70. Frayssinet, P., J.L. Trouillet, N. Rouquet, E. Azimus, and A. Autofage, Osseointegration of macroporous calcium phosphate ceramics having a different chemical composition, *Biomat.*, 14:423–429 (1993).

71. Daculsi, G., N. Passuti, C. Martin, C. Deudon, R.Z. LeGeros, and S. Raher, Macroporous calcium phosphate ceramic for long bone surgery in humans and dogs. Clinical and histological study, *J. Biomed. Mater. Res.*, 24:379–396 (1990).

72. Flatley, T.J., K.L. Lynch, D.A. Ladwig, and D.A. Skrade, Evaluation of Collagen/ ceramic bone graft substitutes with osteoinductive composites in dogs with segmental spinal instrumentation, *Proc. Scoliosis Research Society* Honolulu, HI, September 23–27 (1990).

73. Ladwig, D.A., K.L. Lynch, D.A. Skrade, and T.J. Flatley, Evaluation of collagen and ceramic bone graft substitutes with osteoinductive composites in dogs with spinal fixation, *Trans., 16th Annual Meeting of the Society for Biomaterials*, May 20–23, 1990, Charleston, SC, p. 196.

74. Grundel, R.E., M.W. Chapman, and T. Yee, Evaluation of Type I collagen and a biphasic ceramic in diaphyseal defects of the ulna and metaphyseal defects of the humerus in the canine, *Trans., 33rd Annual Meeting of the Orthopaedic Research Society*, January 19–22, 1987. San Francisco, p. 445.

75. Sandhu, H.S., J.M. Lane, C. Dowling, A. Burstein, K. Piez, and S. Kincaid, Calcium phosphate ceramic supplementation of collagen and marrow composite bone graft, *Trans., 33rd Annual Meeting of the Orthopaedic Research Society*, January 19–22, 1987, San Francisco, p. 93.

76. Uratsuji, M., T.W. Bauer, and M.T. Manley, The evaluation of a new composite of fibrillar collagen and porous calcium phosphate as a bone grafting material, *Trans., 13th Annual Meeting of the Society for Biomaterials*, June 2–6, 1987, New York.

77. Spitzer, W.J., and J. Dumbach, Aufbau des atrophischen kieferkammes mit vorgeformten implanten aus hydroxylapatite-granulat und kollagen — Erste Ergebnisse, *Deutsche Zahnarztliche Zeitschrift*, 43:74–77 (1988).

78. Toth, J.M., Material characterization and biological evaluation of calcium pyrophosphates for use as bone graft substitutes, Ph. D. dissertation, Marquette University, Milwaukee, WI (1992).

79. LeGeros, R.Z., R. Parsons, G. Daculsi, F. Driessens, D. Lee, D., S. Liu, S. Metseger, D Peterson, and M. Walker, Significance of the porosity and physical chemistry of calcium phosphate ceramics, biodegradation and bioresorption, *Ann. N.Y. Acad. Sci.*, 523:268 (1988).

80. Albee, F.H., and H.F. Morrison, Studies in bone growth: Triple calcium phosphate as a stimulus to osteogenesis, *Ann. Surg.*, 71:32–39 (1920).

81. Stewart, W.J., Experimental bone regeneration: Using lime salts and autogenous grafts as sources of available calcium, *Surgery, Gynecology, and Obstetrics*, 59:867–871 (1934).

82. Shands, A.R., Studies in bone formation: The effect of the local presence of calcium salts on osteogenesis, *J. Bone Jt. Surg.*, 19:1065–1076 (1937).

83. Wells, H.G., Calcification and ossification, *Arch. Internal Medicine*, 7:721–753 (1911).

84. Wells, H.G., Pathological calcification, *J. Med. Research*, 14:491–525 (1906).

85. Piecuch, J.F., Extraskeletal implantation of a porous hydroxyapatite ceramic, *J. Dent. Res.*, 61:1458–1460 (1982).

86. Misiek, D.J., J. N. Kent, and R.F. Carr, Soft tissue responses to hydroxylapatite particles of different shapes, *J. Oral Maxillofac. Surg.*, 42:150–160 (1984).

87. Pettis, G.Y., L.B. Kaban, and J. Glowacki, Tissue response to composite ceramic hydroxyapatite/demineralized bone implants, *J. Oral Maxillofac. Surg.*, 48:1068–1074 (1990).

88. Ohgushi, H., V.M. Goldberg, and A.I. Caplan, Heterotopic osteogenesis in porous ceramics induced by marrow cells, *J. Orthop. Res.*, 7:568–578 (1989).

89. Ohgushi, H., M. Okumura, K. Masuhara, V.M. Goldberg, D.T. Davy, and A.I. Caplan, Osteogenic potential of bone marrow sustained by porous calcium phosphate ceramics. In *CRC Handbook of Bioactive Ceramics*, Vol. 2. Calcium Phosphate and Hydroxylapatite Ceramics, ed. by T. Yamamuro, L.L. Hench, and J. Wilson, Boca Raton, FL: CRC Press, 1990, pp. 229–233.

90. Ohgushi, H., M. Okumura, and S. Tamai, Marrow cell induced osteogenesis and chondrogenesis in porous calcium phosphate ceramics. In *Clinical Implant Materials*, ed. by G. Heimke, U. Soltesz, and A.J.C. Lee, Elsevier: Amsterdam, 1990, pp. 225–230.

91. Ohgushi, H., M. Okumura, T. Yoshikawa, S. Tamai, S. Tabata, and Y. Dohi, Regulation of bone development and the relationship to bioactivity. In *Bone Bonding — Reed Healthcare Communications*, ed. by P. Ducheyne, T. Kokubo, and C. Van Blitterswijk, 1992, pp. 47–56.

92. Nade, S., L. Armstrong, E. McCartney, and B. Baggaley, Osteogenesis after bone and bone marrow transplantation, *Clin. Orthop.*, 181:255–263 (1983).

93. Ripamonti, U., Inductive bone matrix and porous hydroxyapatite composites in rodents and

nonhuman primates. In *CRC Handbook of Bioactive Ceramics*, Volume 2. Calcium Phosphate and Hydroxylapatite Ceramics, ed. by T. Yamamuro, L.L. Hench, and J. Wilson, Boca Raton, FL: CRC Press, 1990, pp. 245–253.

94. Ripamonti, U., The morphogenesis of bone in replicas of porous hydroxyapatite obtained from the conversion of calcium carbonate exoskeletons of coral, *J. Bone Jt. Surg.* [*Am.*], 73A:692–703 (1991).

95. Vargervik, K., Critical sites for new bone formation. In *Bone Grafts and Bone Substitutes*, ed. by M. Habal and A. Reddi, Philadelphia, PA: W.B. Saunders, 1992, pp. 112–120.

96. Shors, E.C., R.E. Holmes, and R. Kraut, Bone formation and implant degradation of coralline porous ceramics placed in bone and ectopic sites, *Trans., 39th Annual Meeting of the Orthopaedic Research Society*, February 15–18, 1993, San Francisco, p. 478.

97. Lynch, K.L., J.M. Toth, K.R. Hamson, K.C. Ho, and W.M. Hirthe, Osteoinductivity by subcutaneous implantation of a fibrillar collagen and a calcium phosphate ceramic composite. In *Bioceramics*. Volume 3. Proceedings of the Third International Conference on Ceramics in Medicine, Terre Haute, IN, November 18–20, 1990, Rose-Hulman Institute of Technology Press, Terre Haute, IN, 1992, pp. 295–304.

98. Toth, J.M., K.L. Lynch, and K.R. Hamson, Osteoinductivity by calcium phosphate ceramics, *Trans., Fourth World Biomaterials Congress*, April 24–28, 1992, Berlin, p. 65.

99. Toth, J.M., J.L. Lynch, and D.A. Hackbarth, Jr., Osteogenesis by subcutaneous implantation of calcium phosphates—A histological description, *Trans., 39th Annual Meeting of the Orthopaedic Research Society*, February 15–18, 1993, San Francisco, p. 475.

100. Lynch, K.L., J.M. Toth, D.A. Hackbarth, and K.C. Ho, Heterotopic ossification in calcium phosphate ceramic implants: A foreign body response or natural biological phenomenon? A histological study, *Am. J. Clin. Path.*, 99:348–349 (1993).

101. Toth, J.M., K.L. Lynch, and D.A. Hackbarth, Jr., Ceramic-induced osteogenesis following subcutaneous implantation of calcium phosphates, *Bioceramics*, 6:9–14 (1993).

102. Klein, C., K. de Groot, C. Weiqun, C. Zhang, J. Zhou, L. Yubao, and X. Zhang, Bone formation in non-osseous tissues in dogs induced in porous calcium phosphate ceramics, *Trans., Fourth World Biomaterials Congress*, April 24–28, 1992, Berlin, p. 261.

103. Zhang X., J. Zhou, C. Weiqun, W. Chong, and P. Zhou, A calciumphosphate bioceramics with osteoinduction, *Trans., Fourth World Biomaterials Congress*, April 24–28, 1992, Berlin, p. 332.

104. Yubao, L., X. Zhang, C. Weiqun, L. Yuhua, C. Klein, and K. de Groot, The influence of multiphase calcium phosphate bioceramics on bone formation in nonosseous tissues, *Trans., 19th Annual Meeting of the Society for Biomaterials*, April 28–May 2, 1993, Birmingham, AL, p. 165.

105. Zhang, X., C. Weiqun, J. Weng, L. Yubao, C. Zhang, C. Klein, and K. de Groot, Initiation of osteoinduction in calciumphosphate ceramics without the bone growth factor, *Trans., 19th Annual Meeting of the Society for Biomaterials*, April 28–May 2, 1993, Birmingham, AL, p. 299.

45
Ceramics in Total Hip Arthroplasty

A. Toni, A. Sudanese, S. Terzi, M. Tabarroni, F. Calista,
and A. Giunti
Istituto Ortopedico Rizzoli
Bologna, Italy

I. INTRODUCTION

This chapter is dedicated to presenting the current state of the art in the application of ceramics in total hip arthroplasty (THA). Now, more than 20 years after the first ceramic hip prosthesis was implanted in the human body, the product of the dreams of pioneers concerning the use of these implants must be evaluated in the light of current knowledge to establish the advantages, if any, of the use of ceramics in THA. This contribution is based upon results presented in the literature together with the experimental and clinical experiences we have accumulated since the early 1980s concerning the use of ceramics as bulk and coating materials. The chapter is divided into three main areas: bulk ceramics, plasma spray ceramic coating, and three-dimensional (3-D) porous ceramic coatings.

II. CLINICAL MATERIAL

Our clinical experience is related to the anatomic ceramic hip arthroplasty (An.C.A.), manufactured by Cremascoli, Milan, Italy (Fig. 1). We have implanted 694 implants for index surgery (here we do not consider the 87 cases of revision surgery) from November 1985 to October 1993. The stem, anatomically shaped, is made of cast CoCr alloy and is covered by a madreporic surface in the proximal, anterior, medial, and posterior aspects. The entire stem surface is air-plasma-spray coated with ceramic. The tapered neck can be matched with a dense alumina 32-millimeter (mm) head of three different insertion depths (-4-, 0-, and $+4$-mm modification of neck length). The socket, made of dense alumina, is contoured by a threaded titanium alloy ($TiAl_6Va_4$) ring 18-mm high and is coated by 3-D porous alumina (Poral®) beads on the remaining dome outer surface.

Different types of An.C.A. prostheses were used; the differences between them are reported in Table 1. Three main factors contributed to differentiate the An.C.A. im-

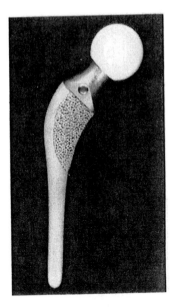

(a)

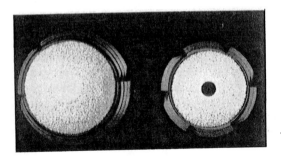

(b)

(c)

Figure 1   Anatomic ceramic hip arthroplasty (An.C.A.) manufactured by Cremascoli, Milan, Italy: (a), stem; (b), socket (top view), with the holed cup made with Ostalox and the cup at the right made with Biolox; (c), same two cups as in (b) (bottom view).

Table 1   An.C.A. Total Hip Arthroplasties Implanted in 1985–1993

Type/N°	Period	Socket alumina	Head alumina	Socket hole	Stem coating	Cement
I/82	November 1985 to August 1987	Ostalox	Ostalox	Yes	Alumina	–
II/22	September 1987 to November 1989	Ostalox	Biolox	Yes	Alumina	–
III/150	September 1987 to June 1990	Biolox	Biolox	–	Alumina	
IV/295	July 1990 to October 1993	Biolox	Biolox	–	Hydroxylapatite	–
V/145	July 1990 to October 1993	Biolox	Biolox	–	–	Yes

Ostalox®: I.M.C., Caravaggio (BG)-Italy
Biolox®: average granule size of 3–4 μ and roughness $\leq 1\mu$

plants: the ceramic material (Ostalox®, IMEC, Caravaggio, Italy; Biolox®, Feldmuhle/ Cerasiv, Plokingen, Germany), the socket design (with or without the dome hole), and the stem coating (alumina and hydroxylapatite [HA]). Ostalox represents a trademark for the IMEC (Caravaggio, Bergamo, Italy) alumina bulk ceramic, which was not responding to the properties defined by the International Organization for Standardization (ISO) 6474 standard in 1979, mainly because of a lack of control of crystal size, which was later found to be larger than 7 microns (μ).

Of the 694 implants, 27 patients (3.9%) underwent revision surgery due to different causes (Table 2). Socket revision rate of Ostalox cups (An.C.A. Types I and II) reaches 9.6%, including one case (No. 4) due to septic loosening. The rate of revision falls to 0.7% with Biolox cups used since 1987. Among stems, the alumina-coated group (An.C.A. Types I–III) presents a revision incidence of 4.3% (11 cases), which drops to 1% (3 cases) with HA-coated stems (used since July 1990), including one acute septic loosening.

It is not the aim of this chapter to describe such clinical data analytically; therefore, it is detailed only when needed to document the topic of ceramics in total hip arthroplasty.

III. BULK CERAMICS

Ceramics are very stable and may well be considered the most chemically and biologically inert of all materials [1]. Ceramics having an aluminum oxide base (Al_2O_3, alumina) have been used since 1970 as a bulk material in the manufacture of components of hip prostheses [2]. Medical applications of alumina followed its use in mechanical engineering and electronics, which started in the 1950s. Besides being inert, ceramics, due to their close atomic structure, are very hard, but also very stiff and brittle, portraying poor flexural, tensile, and impact strength properties. Hence, alumina has only been used in the manufacture of heads and sockets since it could not possibly withstand the stresses applied to the femoral stem of a THA.

Owing to the accepted knowledge that metallic and polymeric debris are implicated in the loosening process that affects THAs [3–10], the main reason for the use of alumina

Table 2 Cases with Revision of An.C.A. Total Hip Arthroplasty

Case no.	Sex	Age	Head	Socket	Stem	An.C.A. type	Follow-up (months)	Cause of revision
1	M	66	O	O		I	2	Socket loosening
2	F	57	**O**			I	9	Head fracture
3	M	54			ALL	I	25	Thigh pain
4 (septic)	F	55	O	O		I	28	Socket loosening
5	F	53	O	O	ALL	I	64	Socket loosening and thigh pain
6	F	59	O			I	75	Socket loosening
7	F	54	O	O		I	78	Socket loosening
8	F	51	O		ALL	I	86	Socket loosening and thigh pain
9	F	54	**O**		ALL	I	87	Head fracture
10	F	63	O	O		I	93	Socket loosening
11	M	61	B	O		II	23	Socket loosening
12	M	75	B	O		II	28	Socket loosening
13	M	61	B		ALL	II	41	Thigh pain
14	F	53	B		ALL	II	52	Thigh pain
15	M	58	B	O		II	65	Socket loosening
16	M	49	B		ALL	III	33	Thigh pain
17	M	58	B	B		III	41	Socket loosening
18	F	62	B		ALL	III	43	Thigh pain
19	F	45	B	B		III	44	Socket loosening
20	F	53	B		ALL	III	44	Thigh pain
21	F	57	B		ALL	III	47	Thigh pain
22	F	68	B		ALL	III	51	Thigh pain
23	M	52	B	B		III	61	Socket loosening
24	F	52	B	B	ALL	III	69	Socket loosening and thigh pain
25 (septic)	M	50	B	B	HA	IV	1	Acute septic loosening
26	F	68	B		HA	IV	29	Thigh pain
27	F	59	B		HA	IV	32	Thigh pain

O, Ostalox; B, Biolox; ALL, alumina; HA, hydroxylapatite. Head revision was due to ball problems (fracture) only in cases in bold. Other head revisions were done routinely when either socket and/or stem were revised.

is to avoid or reduce the wear of articulating surfaces, with respect to the other commonly used materials, metal or polymer (polyethylene). This goal can be achieved in two ways: by substituting the prosthetic epiphyses with ceramic ones and pairing them with polyethylene acetabula (eliminating metal) or by using ceramic material for the head and the socket (ceramic-ceramic), thus eliminating both metal and polyethylene.

In the first case, polyethylene wear is reduced because it is possible to achieve a lower degree of surface roughness with ceramic materials compared to metal alloys [11,12], and thus reduce the abrasive action on the polyethylene. Furthermore, as compared to metal, ceramic material has a surface that is much more resistant to damage by scratching often produced by cement particles [13] or by corrosion in a biological environment. Besides *in vitro* data [14,15], a better tribological performance of ceramic over a metal head when combined with polyethylene has been recently confirmed on surgical and autopsy materials, with a threefold smaller quantity of debris associated with the use of ceramic heads [16]. Among ceramics, zirconia heads (ZrO_2) reduce the amount of polyethylene debris by 40–60% [17] when compared to alumina heads.

The second option is to combine a ceramic head and ceramic socket. In 1970, clinical experience with ceramic-ceramic coupling THA was started by Boutin [2]; work by other German surgeons followed [18–20].

Since then about 2 million ceramic heads have been produced by the five most important manufacturers: Cerasiv/Feldmuhle (Plokingen, Germany), Desmarquest (France), Kyocera (Japan), Metoxit (Switzerland), and Morgan Matroc (United Kingdom).

Cerasiv/Feldmuhle claims to have sold 1.3 million ceramic (Biolox) heads, mainly in European countries. Recording head diameters, 90% of the past production has been of either 32-mm or 28-mm heads. Today, 70% of heads produced by Cerasiv/Feldmuhle have a diameter of 28 mm. In the United States and France, zirconia heads are becoming more popular, with almost a 10% share of the market for ceramic heads. Our clinical experience is limited to ceramic-ceramic THA, which we have implanted since November 1985. This chapter thus is concerned only with this combination, and does not take into consideration the ceramic-polyethylene THA.

Among the causes of failure of ceramic arthroplasties, alumina head fracture and wear of either the head and/or socket should be considered.

A. Alumina Head Fracture

We had two cases with fracture of the ceramic head. Both patients had Ostalox heads (An.C.A. I), which then presented a 2.4% incidence of fracture. None of the remaining 612 Biolox high quality heads presented any fracture. Due to their rarity, a closer description of the two cases is indicated.

Case 2. The patient was a female, 57 years old. Breakage was clinically manifested by crepitation phenomena and mild pain approximately six months after surgery. The patient was referred to us only nine months after implantation, when the head was substituted. We expected the neck to be damaged to a level not compatible with the locking of another ceramic ball. Based on the x-ray evaluation, both the stem and the cup were judged as well osteointegrated. Due to these facts, we decided on a novel solution, using the hole present in the dome of the An.C.A. I cup. A polyethylene insert with 32-mm outer and 22-mm inner diameters, with a self-locking peg for the cup hole, was custom built. At surgery, both the damage of the tapered neck and the stability of the implant were confirmed. The insert peg was then press fitted into the ceramic cup hole and a 22-mm metallic head was coupled with the stem neck. At six-years follow-up to the revision surgery, no radiographic signs of wear or any other local complications were present in this patient.

Case 9. The patient was a female, 54 years old. This patient underwent index surgery for the left hip in April 1986 (An.C.A. I was used). Three years after surgery, she started complaining of a mild hip inguinal pain present when walking and absent at rest. Upon x-ray control, the socket presented a lucent line at the interface with the bone. The pain progressed very slowly. Due to a severely painful opposite arthritic hip, she underwent a second index surgery in March 1992.

Since the left cup moved medially with progressive erosion of the acetabulum, a revision was done on September 1992 with a Harris–Galante–Zimmer cup, fixed by 3 screws, with a polyethylene insert. The stem was stable. The ceramic head was not changed because an Ostalox ball was not available by then and the Biolox ball inner tapered cavity was not matching the An.C.A. I stem. No pathologic signs were reported until June 1993, when, during a forced adduction of the left hip, the patient suddenly heard a noise, which was soon followed by inguinal pain. She reported to the doctor 40 days later and diagnosis of the ceramic ball fracture was made. At surgery, the stem revision also was done due to the wear of the tapered neck caused by ceramic pieces and debris (besides the difficulty in finding the old, matching Ostalox head). An An.C.A. IV stem and ball were then implanted. The polyethylene insert was changed, too.

B. Alumina Wear

For eight patients undergoing revisions of the An.C.A. prosthetic replacement for various reasons, it was possible to ascertain the amount of wear debris in both the soft tissue and the bone tissue around the implant. To make a quantitative evaluation of the wear debris, the tissue of the joint capsule was examined at four sites, each with a 1-square-millimeter (mm^2) area. Measurement of single wear particles and their count per 1000 square microns (μ^2) were carried out by manual image analyzer (Leitz ASM 68 K) and use of an automatic IBM-compatible analyzer (software image measure, Microscience, Inc.). The microscopic images were obtained with a black and white Ikegami ICD 290 camera [21]. Of the patients, six (Cases 1, 3, 4, and 11–13) had An.C.A. I, II, or III implants; the remaining implants, retrieved at 1-month follow-up due to sepsis, was an An.C.A. Type IV (Case 25). One case (Case 2) was complicated by the fracture of the head, and wear of ceramic surfaces could not be defined; only the debris in the surrounding tissues could be evaluated.

The prosthetic surface and the surrounding tissue were studied with a scanning electron microscope (SEM). In cases studied with uncertainty of the composition of the wear material, microanalysis was carried out to identify the elementary composition of the debris. Three cases (Cases 1, 4, 13) with ceramic surfaces did not show any signs of wear, while in another three cases there were signs of gradually increasing wear phenomena, which were minor for two cases (Cases 4, 11) and massive in one case (Case 12). SEM imaging showed the surface of normal ceramic material to be smooth, with small and rare discontinuities among the crystals (Fig. 2). These discontinuities were 2–4 microns, corresponding to crystals that are just a bit deeper than the surface. The initial wear mechanism is manifested precisely by the widening of these gaps.

We may compare the normal prosthetic surface to a cubed porphyry pavement in which the surface is in fact obtained by many small cubes wedged together. Everyone knows how resistant porphyry is even to the heaviest and most repeated loads and how difficult it is to remove a cube from the porphyry "mosaic." Yet, once a small hole, from the loss of even a single cube, has damaged the continuity of the porphyry, it becomes quite easy to remove one cube after the other. Similarly, if one or more crystals are detached from the ceramic surface, a cavity is produced that locally impairs the integrity and the resistance of the surface.

These initial "cavities" may widen in a ceramic surface (Fig. 3) until they coalesce,

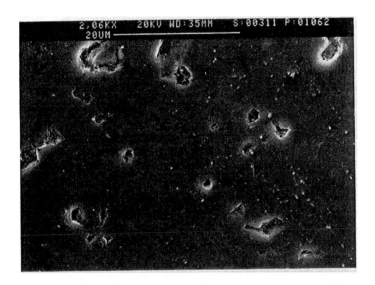

Figure 2 Microphotograph of the normal Ostalox ceramic surface obtained by scanning electron microscopy (SEM). Observe small solutions of continuity 1–5 microns (μ).

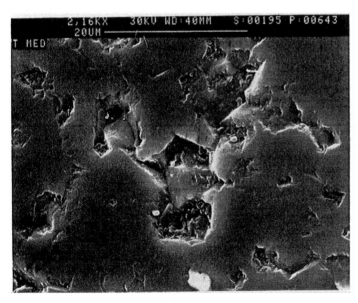

Figure 3 Ostalox ceramic surface obtained by scanning electron microscopy (SEM) with triggering of wear phenomena. In compact surface, observe cavities produced by detachment of adjacent crystals with formation of 10–15-μ cavities.

forming erosive lacunae bordering on islands of surfaces that are still intact (Fig. 4). The process may become further accentuated, causing a breaking up of the ordinate arrangement of the surface crystals, which thus acquires an undulating appearance or, in some cases, a wavy one like that illustrated in Fig. 5. This is the terminal stage of the wear phenomena in the surface layer of a prosthesis. From this moment, alumina can only provide reduced resistance to further wear that may even deform the spherical profile of the prosthesis itself.

Alumina wear was acknowledged to be the cause of loosening in only one An.C.A. I prosthesis. In the other two cases, only minor signs of wear could be seen, such as occasionally found in prostheses that showed loosening caused by other factors. Among these, in a case for which wear was at an initial stage (Case 4), the simultaneous revision of cup and head allowed us to observe the topographical relationship between the worn area in the convex surface of the head and the specular level of the wear phenomena on the border of the acetabulum (Fig. 6a). The edged rim of the prosthetic acetabulum separated the smooth area, produced by lapping and corresponding to the concave hemispherical surface, from that of the plain border of the acetabulum, which, unlike the first, was rough because it had not been lapped. In the ceramic epiphysis, the sickle-shaped eroded surface (Fig. 7) corresponded to the contact area, with the same edge in a rotatory arch of movement corresponding to that produced by articular movement in flexion-extension (approximately 70°). The site was the superoexternal area of the head, corresponding to the upper margin of the acetabulum (Fig. 6b).

The aspect and, in particular, the "alabaster" color of the ceramic debris facilitate easy recognition as compared to metallic debris, which are darker and of a finer-grained structure (Fig. 8). Moreover, microanalysis always resolves cases of doubtful interpreta-

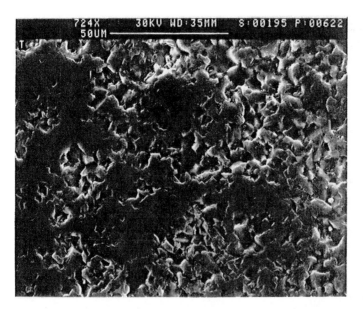

Figure 4 Ostalox ceramic surface obtained by scanning electron microscopy (SEM) with intermediate phase of wear. Next to area of surface that is still compact, observe erosive lacunar areas caused by the confluence of the cavities.

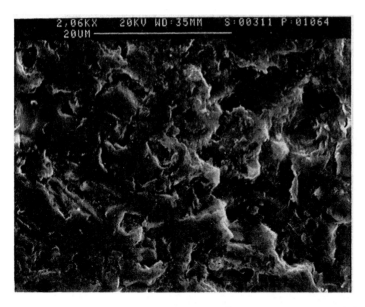

Figure 5 Ostalox ceramic surface obtained by scanning electron microscopy (SEM) with terminal phase of wear. The ceramic surface is irregular because of complete destruction of the superficial crystalline layer.

tion. It is possible to count the debris by subdividing it for each case into intervals in relation to size before representing the quantity per surface unit (μ^2) in each dimensional interval.

The number and distribution measured in three clinical cases (Cases 2, 11, 12) that summarize the tendencies observed in this study are reported in Fig. 9. Case 11 (An.C.A. I) refers to the wear phenomena observed in a prosthesis removed 25 months after surgery for reasons unrelated to wear. Macroscopically, the tissue surrounding the implant was normal, and histological examination of the debris showed that it was minimal. The total quantity of debris was 20,000 debris per μ^2, with the debris being in the minimal dimensional interval ranging 0.1–1 μ.

Case 12 (An.C.A. I) was related to massive ceramic wear at 28 months since index surgery (in 1987). The acetabulum tilted 24 months after implantation, coinciding with the time when the patient began to complain of pain. This symptom gradually increased until the 28-month follow-up, when dislocation of the acetabular component occurred (Fig. 10). Computerized tomography (CT) scans showed osteolysis of the iliac bone, which was filled with radiopaque soft tissue and spread the area of osteolysis to the endopelvic and extrapelvic sites (Fig. 11). The surgical findings showed that the osteolytic area was filled with a white-colored tissue of a cretaceous consistency, similar to tubercular tissue.

However, the presence of macroscopic wear of the acetabulum and ceramic head oriented our diagnosis toward a foreign-body tissue reaction, which was in fact confirmed by histological examination. Surgery was thus completed by substituting the prosthetic acetabulum, associated with homologous bone grafting of the osteolytic cavity. The stem was found to be stable. The dimensional distribution privileged the intervals from 0.2 μ to 2 μ (64% of the debris observed); the overall quantity of debris was 218,000 debris per

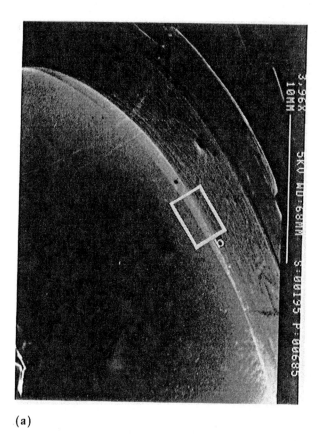

(a)

(b)

Figure 6   Wear at an initial stage: (a), Anatomic ceramic hip arthroplasty (An.C.A.) Ostalox cup microphotograph at lower power (scanning electron microscopy, SEM) in which the flat border is not lapped, and the worn surface is clearly evident at the edge between the border and the joint surface of acetabulum; (b), microphotograph at higher power of worn area corresponding to the framed area in (a) in which the upper portion is the joint surface and the lower portion is the border surface not lapped.

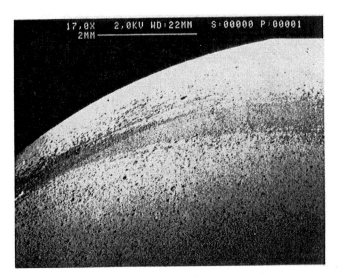

Figure 7 Ostalox ceramic head preparation for scanning electron microscopy (SEM) showing the shape of the wear surface.

μ^2. Thus, if we compare Case 12 with Case 11 at similar follow-up after arthroplasty, wear phenomena was 10 times greater in Case 12. Cellular reaction varied in relation to the quantity and the type of the debris. The smaller debris (1–2 μ) were often found in the intercellular sites, and particles with a microcrystal-like aspect (5–10 μ) were arranged in the extracellular site, without pathological cytological reaction (Fig. 8). Giant cell reaction was a rare occurrence. In the case described above in which there was massive wear of alumina (Case 12) with a high concentration of debris less than or equal to 1 μ, the histiocytic reaction was remarkable, with wide areas featuring necrosis, which in some places was surrounded by septa of fibrous tissue with a lobular shape (Fig. 12). This was

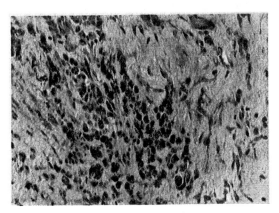

Figure 8 Capsular tissue in which the smaller-size debris (<2 μ) is contained in the intracellular site and is evident as fine cytoplasmatic punctuation or with an opalescent aspect when a bit larger. Other larger-size debris appears in the extracellular site (Hematoxylin Eosin [HE], ×40).

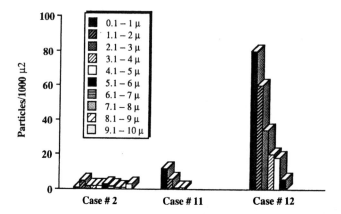

Figure 9 Quantities and size distribution of alumina debris (in the frame legend of dimensional intervals): Case 2 with head fracture; Case 11 with revision due to causes unrelated to wear; Case 12 with loosening due to massive ceramic wear.

the only case that showed evidence of bone tissue resorption due to infiltration of histiocytes filled with alumina powder [21,22].

C.   Zirconia Tribology

The brittleness of ceramic may cause fractures of the prosthetic head when it is coupled with the tapered cone of the metallic stem. Due to its better mechanical features, zirconium oxide (ZrO_2) has been proposed as a substitute for the alumina oxide [17,23–27]. We tested this ceramic compound for its tribologic properties. The zirconium oxide (zirconia) used for testing was formed by 95% zirconium oxide and 5% of yttrium oxide. Such a bioceramic material presents a toughness (10 MPa/m$^{1.5}$) twice as good when

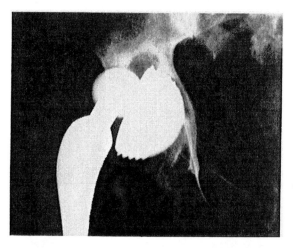

Figure 10 Anatomic ceramic hip arthroplasty (An.C.A.) Type I (Case 12) dislocation of the loosened cup. Osteolysis is present in the iliac bone.

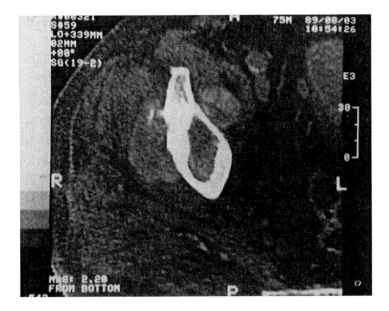

(a)

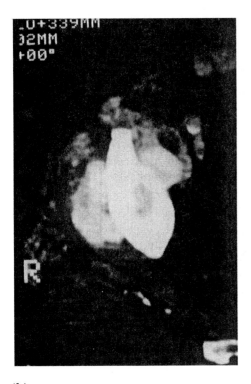

(b)

Figure 11 Case 12: (a), computerized tomography (CT) scan showing the osteolysis of the cancellous bone tissue; (b), same-level CT scan with soft tissue encachement, a pathologic tissue with a whitish aspect due to the density enforcement produced by the alumina debris contained in the foreign-body reactive tissue, presenting around the iliac bone.

1513

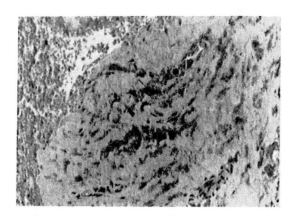

Figure 12   Cases 12, showing massive wear of alumina with a high concentration of debris less than or equal to 1 micron. The hystiocytic reaction is remarkable, with a wide area featuring necrosis (HE, ×40).

compared with alumina. Comparing the other zirconium mechanical data to that for alumina, the zirconium oxide presents a lower compressive strength (3000 vs. 4000 megapascal [MPa]), but has a higher flexural strength (1000 vs. 500 MPa) and lower Young's modulus (206 vs. 320 gigapascal [GPa]). We then made a "disk-on-ring" test according to ISO 6474-1981 (F) to evaluate the zirconia tribological behavior. For both kinds of bioceramics, we kept the same test parameters (see Table 3). The roughness of the disk's surface was Ra = 0.01 μm for the zirconium oxide and Ra = 0.03 μm for the alumina oxide. The alumina oxide sample print depth was 0.002 mm for an average removed material volume of 0.32 cubic millimeters (mm^3) 96 hours after test initiation. The test performed with the zirconium oxide sample ceased after only 5 hours (h) due to the large amount of wear; the average print depth was already 0.5 mm, corresponding to an average removed volume of 80.1 mm^3. The wear rate of zirconium oxide (16 mm^3/h) was 4848 times worse than that for the alumina oxide (0.0033 mm^3/h).

D.   Discussion

The main concern about ceramic-ceramic coupling THA is its brittleness. Ceramic head fracture is the cause of major scepticism of many surgeons. It is not possible to glean from published data the real incidence of this complication since not all the cases reported

Table 3   Test Parameters

Angle of oscillation	±25°
Disk diameter	30.4 mm
Disk height	4 mm
Ring, inside diameter	14 mm
Ring, outside diameter	20 mm
Superficial pressure	20–23 bars
Frequency of oscillation	1.01 mm^2
Surface	160.2 mm^2
Lubricant	Ringer 37°C ± 1

are related to the consistency of the implanted group. Therefore, such case reports cannot contribute to statistics. Nevertheless, some reports may be used, as indicated in Table 4 [28–45]. Data demonstrate the relationship between the evolution of alumina ceramic and its clinical reliability in terms of reduction of fracture incidence. The reliability today of alumina heads is definitively good. Of 612 Biolox implants, we have not experienced any fracture since September 1987. Similarly, Boutin does not report any fracture out of 373 Ceraver ceramic heads since 1979 [28].

In spite of its better mechanical features, zirconium oxide should not be used for ceramic-ceramic coupling prostheses due to its low wear resistance. It may well be that zirconium oxide is the material of choice with which to make ceramic heads in hip prostheses with polyethylene sockets [17].

Another relevant concern about ceramic-ceramic THA could be its effective reduction of wear particles. In fact, despite alumina on alumina being used to avoid wear, ceramic debris were still found on analysis of retrieved THA [28,46–50]. First, we consider the fact that the mechanical and tribological behavior of the ceramic component are strictly related to the material selection. The properties required for alumina ceramic for medical use were defined by the ISO 6474 standard in 1979, and are summarized as an

Table 4   Incidence of Ceramic Head Fracture

Author	Year published	Period implanted	Type THA	Total implants	N° fractures	Fractures, %
Boutin[a]	1981	1970–1972	Ceraver	373	28	7.5
Winter	1992	1974–1979	Friedrichsfeld	100	8	8
Griss	1981	1974–1978	Friedrichsfeld	130	9	6.9
Knahr	1981	1974–1980	Rosenthal	75	8	10.6
Knahr	1987	1976–1979	Rosenthal	67	9	13.4
Mittelmeier	1985	1974–1983	Biolox	877	3	0.4
Mittel-meier[b]	1992	1974–1989	Biolox	3079	3	0.1
Nizard[c]	1992	1977–1979	Ceraver	187	3	1.6
Boutin	1988	1979–1986	Ceraver	373	0	0
Heisel[b]	1987	1974–1986	Biolox	2536	5	0.2
Trepte	1985	1977–	Biolox	41	0	0
Rampoldi	1984	1979–1982	Biolox	59	0	0
Kern[d]	1990	1981–	Biolox	>500	1	0.2
O'Leary	1988	1982–1983	Biolox	69	0	0
Miller	1986	1982–1984	Biolox	231	0	0
Hoffinger	1991	1983–1984	Biolox	119	0	0
Cameron[b]	1991	1980–1990	Biolox	>600	3	0.5
Toni	1990	November 1985 to August 1987	Ostalox	82	2	2.4
Toni	1994	September 1987 to 1992	Biolox	615	0	0

[a]Epoxy linking between head and stem
[b]Mixed coupling surfaces: ceramic on ceramic and ceramic on polyethylene
[c]One of the reported fractures was a 22-mm ceramic head
[d]Ceramic on polyethylene only

alumina content greater than 99.5%, other oxide (SiO_2, CaO, MgO) less than 0.1%, density greater than 3.90 grams per cubic centimeter (g/cm^3), and average grain size less than 7μ. The 1979 standard therefore becomes a threshold between pioneering endeavors and standardized use of alumina ceramic. What happened before 1979 is of historical relevance, with low clinical interest. In fact, the first generation of ceramic-ceramic THA presented high rates of failure, mainly due to the poor properties of the alumina ceramic. Boutin et al. replaced the alumina ceramic of their THA with a more suitable material after 1977 [28].

In 1989, the ISO standard was amended by the draft proposal for a new standard (ISO/DP 6472.2), which, for alumina implant submitted to a high-load application, requires a density greater than 3.94 g/cm^3 and a minimum grain size 4.5 ± 2.6 μ [51]. Due to the fact that CaO impurities favor alumina ceramic fatigue and aging effects [52–54] and SiO_2 promotes grain growth [11], these oxides must be carefully kept at a very low concentration (0.001–0.06%). Since 1989, the main producers of alumina ceramic heads and sockets meet the requirements of the draft proposal.

Other parameters such as design, sphericity, misalignment, and surface roughness may affect the ceramic heads tribology. Walter [51], studying 48 retrievals of ceramic implants, used sphericity deviation to determine the effect of wear (profilometric wear rate). He considered true "wear" to be indicated by those rates exceeding 5μ/year and reported all wear rates exceeding 6.5μ related to cases with either rim contact (possible linear contact with unpolished surface), a central hole (in the dome of the socket), and/or cup malalignment. The cup can be misplaced since the index surgery, or progress to malposition due to loosening.

Another cause of linear contact is repeated subdislocation, which can follow either the vertical placement of the cup or the instability due to a replacement with an excessively short neck length [55,56]. Ceramic components may well be considered as unforgiving for the case of a nonuniform load distribution [57]. Cases with loosened implants maintaining a proper alignment presented a wear rate of 1.8 ± 1.9 μ/year for heads and 2.1 ± 1.6 μ/year for sockets; the rate was lower in cases with septic loosening (average was 0.7 ± 0.3 μ/year). To confirm previous assumptions, all cases with an average grain size less than 4 μ, density greater than 3.94 g/cm^3, porosity less than 1.2%, and an alumina content greater than 99.6% presented a sphericity deviation rate of less than 0.4 μ/year [51].

To define the reliability and evaluate the efficacy of ceramics in improving the clinical outcome of THA, we devote our attention to the clinical experience with the alumina produced accordingly to ISO 6474 or following draft proposal. Sedel et al. reported on 86 cases with a 94% survival at 8 years follow-up with a cemented ceramic-ceramic THA [58]. Other reports for cemented ceramic cups show worse results in elderly patients [28,59] and better results in patients less than 50 years old. This trend is opposite to the tendency usually reported with polyethylene cups. The improved acetabulum trabecular bone mechanical properties of younger patients have been claimed to favor a better adaptation to the stress distribution due to the very stiff ceramic sockets [15].

Mahoney and Dimon refer to 28% stem loosening with Autophor (Mittelmeier) ceramic-ceramic prosthesis, though stating that the metallic stem design was responsible for the unacceptably poor results since the ceramic presented very low wear for retrieved cases [60]. Mittelmeier and Heisel reported a 1.3% incidence of loosening of 3000 cementless cups between 1974 and 1989, with a different incidence for stem loosening: 22% for Mark I type, 8.9% for Mark II type, and 1% for Mark III type [29]. Kummer,

Stuchin, and Frankel, on analysis of removed Autophor ceramic-on-ceramic components, always found a degree of wear, which was catastrophic in a case with vertical malpositioning even with high-purity and high-density alumina ceramic [57]. Dorlot reported on 6 ceramic components implanted after 1977, and retrieved after 9–12 years; in 4 cases, a $0.3-\mu$ wear was detected, which was similar to the value measured on short follow-up retrieval, confirming that usually, after an initial polishing effect, ceramic wear and friction decrease [55]. This has been also proven experimentally by Dorre and Hubner [11]. Nevertheless, Dorlot found 2 cases with tracks on balls articulating with stable sockets, which occurred in patients weighing more than 90 kg; in one case, the track was $44-\mu$ deep, corresponding to a volume of ceramic debris of 18 milligrams (mg) [55].

In exceptional cases, there may have been massive wear phenomena in the alumina of the epiphysis and of the acetabulum. It is not easy to explain the occasional incidence of these cases. In the An.C.A. I prosthesis used between 1985 and 1987, the border was edged (90°) and the plane portion was not polished. With other unfavorable conditions, such as a vertical cup or cup instability, pairing between head and acetabulum could not be stable, with the head tending to subluxate externally. This generates contact on the edge of the acetabulum, with excessive concentration of mechanical stress, enough to favor the triggering of wear phenomena.

This mechanism seems to be confirmed by the semilunate form of wear observed in the head and by the extent and the site of wear on the acetabular border. Both observations correspond to the superoexternal quadrant. The extent corresponds to the circle arch covered by the flexion-extension movement of the hip, with a maximum width of the falciform impression in the epiphysis corresponding to the few degrees of abduction-adduction usually required for walking.

This could explain how wear of the ceramic surface initiates, but it is more difficult to explain the massive wear described in Case 12. The extensive damage of the epiphyseal and acetabular surfaces, with loss of sphericity for both of the prosthetic components, prevents comprehension of the early wear process based on its topographical distribution. What is exceptional, however, is the quantity of debris produced, equal to about 10 times that observed in other patients. Their dimensional range is also characteristically below 1 μ. In the case of breakage of the ceramic head (Case 2), histological evaluation revealed that the quantity of debris was quantitatively comparable to cases of scarce wear of the alumina components (20–22,000 debris per μ^2), though with a wider range of particle dimensions.

It is difficult to state that the poor ceramic properties of Ostalox and the improved ones of Biolox are the cause of the reported different revision rates for the An.C.A. cups, 8.6% for Ostalox and 0.9% for Biolox. In fact, even the surgeon's "learning curve" effect could be advocated due to the fact that the new screwing fixation technique was first experienced with the Ostalox cup (An.C.A. I). Nevertheless, the difference is striking. Cellular reaction to alumina debris has no important pathologic connotations. In most cases, we observed mild histiocytic reactions, while giant cell reaction was practically absent. Thus, it may be interpreted that alumina debris does not cause a particularly intense cellular reaction. However, it is wrong to state that minor cellular reaction is correlated to greater tolerance toward alumina particles; rather, it is a consequence of the scarce quantity of this debris. In fact, in the case in which there was a large amount of debris, the cellular reaction was significant, achieving areas of necrosis (similar to those observed in cellular reactions to metal and plastic materials) with associated periprosthetic osteolytic reaction.

IV. CERAMIC COATING

The ceramic coating of cementless prosthetic hip implants has been advocated so as to achieve a more rapid and extensive secondary stabilization by bony ingrowth [28,61–64]. Among the various ceramic coatings, the air-plasma-spray application of aluminum oxide (Al_2O_3, alumina) has been widely used in clinical applications [65,66]. Such a ceramic coating has been described as being inert and entirely compatible with bone both *in vitro* by means of cell culture techniques [67,68] and *in vivo* with experimental implants [69–73]. The ionic diffusion from Al_2O_3 has been considered negligible and consequently the material has been regarded as extremely safe from a biological perspective [74–76]. Alumina is considered to be a bioinert ceramic leading to "contact osteogenesis" with direct contact between bone and implant surfaces [77]. Other ceramics such as hydroxyl apatite are described as bioactive due to their interaction with bone (bonding osteogenesis) [77]. We experienced consecutively both types of ceramic coatings in the anatomic An.C.A. stems. Some aspects of stability and effectiveness of the two coatings are described below.

A. Alumina-Induced Osteomalacia

Between November 1985 and December 1989, we implanted 254 An.C.A. I, II, and III stems that were alumina coated. Of these, 11 hips subsequently underwent surgical revision (4.3%). Of these patients, 7 (Cases 3, 5, 13, 16, and 20–22 are reported), at an average follow-up period of 23 months following index surgery, complained of thigh pain (absent at rest and progressively worsening with walking) that became functionally limiting on a daily basis. In most cases, the patients realized significant clinical improvement after their index surgery, though this was followed by recrudescent pain.

Radiographic films did not show loosening of the stem. There was evidence of stress shielding of the proximal femur [78]. To eliminate the possibility of occult sepsis, scintigraphy with technetium-labeled leukocytes [79] was performed, with negative results in all cases. Erythrocyte sedimentation rate, white blood cell count, and C-reactive protein titers were noncontributory. Articular needle aspirate for culture was performed in five cases and was negative in all. None of the patients showed signs of allergic reaction, with no anamnestic data on allergic pathology being reported.

The discrepancy between the clinical presentations and the diagnostic workups required caution regarding surgical indications. Since the stem was not loose, caution was required in its removal to avoid major damage to the femoris demineralized bone. In fact, the cortical bone was thin and presented a consistency like wet cardboard due to its severe demineralization.

Macroscopically, the bone tissue appeared to be directly contacting the stem's proximal surface. Using thin osteotomes and curved gouges, the metaphyseal bone was interrupted at the proximal porous stem surface. Then, with some strong hammering, the stem moved upward, allowing it to be removed easily. The smooth surface (placed in the distal two-thirds of the stem) was always extracted without any bone attached, while some spongy bone was to be found in the porous surface. After the stem's removal, a continuous bone lamina was found to reproduce the stem shape in the diaphyseal canal. This bone was sampled accurately for subsequent histological evaluation. Needle biopsies were also taken from the femoral cortex.

Histological sections of the bone obtained adjacent to both the smooth surface and the madreporic regions of the prosthesis were prepared. Only in the latter cases was the bone recovered still attached to the stem's surface. Sections of bone tissue 5 mm thick were derived from samples fixed in buffered formalin at pH 7.2, dehydrated in methanol, then methyl methacrylate embedded and deplasticized before staining. Transverse slices of the stem 5 mm thick, at the level of the macroporous zone, were obtained with the Exakt Cutting system (Kulzer and Co., Germany) with bore nitride saws, taking particular care not to damage the bone spiculae that had grown within the macroporosities.

After fixation in methyl methacrylate, the slices were ground to a thickness of 50 μ for light microscopic evaluation of bone ingrowth. Both the deplasticized sections of the bone samples and those of the stem at the macroporous level were stained with traditional histologic methods, including Paragon (toluidine blue-basic fuchsin), modified Goldner, and Von Kossa stains. A method for detecting aluminum utilizing the ammonium salt of aurine tricarboxylic acid [75], with which aluminum hydroxide stains as a red band, was employed. The details of these techniques were presented elsewhere [80].

Microradiographs were obtained on sections 100 μ thick using Kodak high-resolution radiosensitive slides, Type 1 A (Eastman Kodak Company, Rochester, New York).

Conventional powder x-ray diffraction (XRD) analysis was carried out on the alumina coating of the retrieved stems and on sockets and ceramic heads using a Philips powder Bragg–Brentano goniometer. Wear debris was identified by means of microbeam x-ray diffraction, using the Chesley microcamera, in such a way as to focus the beam on a 50 μ^2 area. Histochemical tests carried out on the bone tissue adjacent to the interface at both the smooth and madreporic surfaces of the prostheses all demonstrated the presence of a nonmineralized layer parallel to the profile of the implant (Fig. 13).

The nonmineralized layer lining the madreporic surface was measured in sections including alloy and bone, measuring 106 μ plus or minus 24 μ standard deviations (SD) thick (range was 68–140 μ). The bone contacting the smooth surface of the stem presented a laminar architecture, which usually remained attached to the femoral bone. The retrieved sample of this lamina was histologically evaluated without the metallic counterpart. The nonmineralized layer thickness was 109.2 μ ± 41.3 μ (range 74–164 μ).

The morphology of the noncalcified zone was lamellar at the light microscopic level with osteon systems typical of mature bone over most of the surface. In several cases,

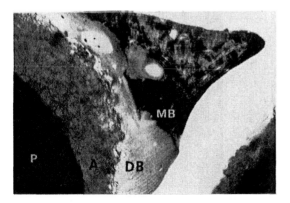

Figure 13   Bone tissue in contact with the porous surface of the prosthesis; observe the pattern of bone tissue that is still visible (P, chrome-cobalt bead; A, layer of alumina coating; DB, demineralized bone tissue; MB, mineralized bone) (Von Kossa, ×16).

individual osteons appeared clearly subdivided into a nonmineralized part immediately adjacent to the prosthesis and a normally mineralized region toward the bone (Fig. 14). The osteoclast-dominated osteolysis of the failed cemented implant was not encountered in this series and the histologic findings were remarkable for the absence of osteolysis per se; likewise, the bone interface could be described as hypocellular with no clearly increased populations of histiocytes or inflammatory cells. The histological evaluation performed on some bone samples retrieved after stem removal and on the small amount of bone still attached to the ceramic surface could not permit any quantitative evaluation of bone ingrowth. Nevertheless, qualitative analysis of the stem-bone interface showed that no fibrous tissue was ever interposed between the porous surface and the bone, while, at the interface between the smooth surface and bone, either a thin fibrous or condroid tissue layer was found in about 20% of the samples.

By means of the specific histochemical technique described above, aluminum was found in all of the examined specimens, both at the level of the smooth stem and at the rough portion of the implant. The metal was found in the highest concentration and identified as a continuous line along the demineralized front (Fig. 15). Only rarely was the linear deposition of aluminum found at the level of the cement lines or throughout the calcified region of the bone. Microradiographs confirmed the absence of calcification in the regions that were in close contact with the prosthesis.

Alumina debris was detected at the stem-bone interface and evaluated quantitatively as number of debris particles per mm^2. Wear debris was seen in 5 of the 6 cases evaluated; particles ranged from 8 to 246 per mm^2, averaging 87.4 $\pm$ 103.1 SD per mm^2. Alumina debris at the stem-bone interface was analyzed by means of microbeam x-ray diffraction and was found to have the same composition as the stem coating (e.g., gamma alumina). In fact, in all observed cases, we found crystallographic transformation of the coating of the stem from a stable phase to gamma phase, a form that is relatively less stable and permits a partial solubility of the alumina coating [81]. On the contrary, ceramic heads and sockets were consistently of the alpha phase. Alumina wear debris was encountered also in the capsular soft tissues in all cases but two. In these cases, the microbeam x-ray diffraction of the particles was not performed.

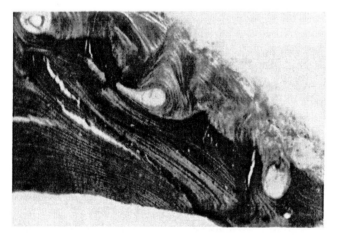

Figure 14   Osteons appeared clearly subdivided into a nonmineralized part immediately adjacent to the prosthesis (upper left, now removed) and a normally mineralized region toward the bone.

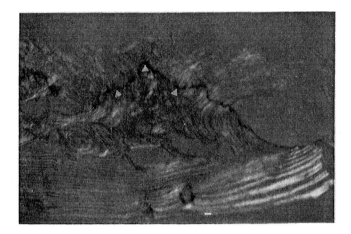

Figure 15   The arrows indicate the rim of concentration of aluminum ions, separating mineralized bone (MB) from demineralized bone (DB) (Aluminon, ×25).

B.   Hydroxylapatite

Bioinert materials such as alumina and titanium alloy permit only a mechanical stability to be achieved with the bone tissue, while the bioactive hydroxylapatite (HA) is supposed to ensure a closer link with bone by achieving a biologic and chemical fixation with the prosthetic surface [82]. We implanted 295 HA-coated cementless An.C.A. IV stems from July 1990 to October 1993. The analysis of the bone-implant interface was completed according to Engh, Bobyn and Glassman's procedures for 65 stems with at least 24 months of follow-up (average 28.2 months) [83]. They were compared with 124 alumina-coated stems of similar length (average follow-up 46.7 months).

We evaluated the implant interface for radiolucent lines (Fig. 16). Since for each stem the seven Gruen [84] regions of interest (ROIs) were examined, we are referring to 1323 ROIs. The incidence of radiolucency in each ROI is detailed in Fig. 17 for both HA- and alumina-coated stems. The ROI with a higher incidence of lucency (21% with alumina) is Location 4, corresponding to the diaphyseal area around the hip of the femoral implant. It sometimes may be troublesome to differentiate between diaphyseal "bone plug" reaction not yet matured contacting the stem surface, and a permanent lucent line, suggesting that fibrous tissue is preventing definitive bone integration. It is a matter of fact that frequently we observed a tip radiolucency evolving toward complete tip bone integration, with total disappearance of the lucent line. This is not the case with other ROIs, in which lucent lines, when present, were steadily maintained or progressively widening, but they never recovered. Data are confirming the lower incidence of radiolucency at the bone-prosthesis interface with HA-coated stems [85–92].

Nevertheless, the result could be biased by the shorter average follow-up of HA with respect to the alumina stem. Considering the radiolucency at the bone-stem interface as a "negative event," we can compare with survival curves [93] the osteointegrated ROIs of all cases with alumina- (151 cases) and HA-coated stems (255 cases) having at least one postoperative x-ray film. Among alumina-coated stems, only shorter ones (of a length similar to HA-coated stems) were studied. Thus, 2765 ROIs were considered (1785 for HA and 980 for alumina), with the survival of osteointegrated ROIs reported in Fig. 18.

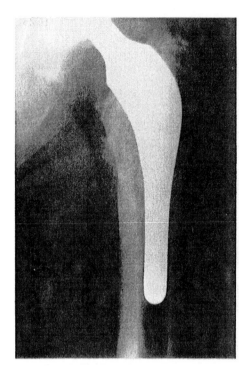

(a)

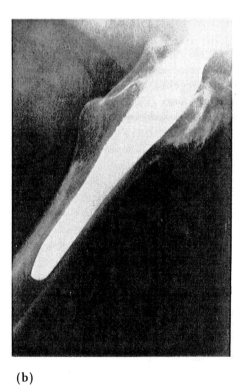

(b)

Figure 16   Anatomic ceramic hip arthroplasty (An.C.A.) IV at 37-month follow-up with a radio-lucent line present distally around the stem tip in both anterior-posterior (A-P) (a) and lateral (b) views.

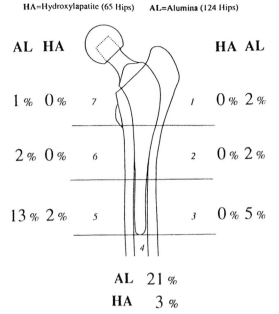

AL HA   HA AL

1 %   0 %   7   1   0 % 2 %

2 %   0 %   6   2   0 % 2 %

13 % 2 %   5   3   0 % 5 %

                    4

AL   21 %

HA   3 %

Figure 17   Incidence of radiolucent lines in the seven regions of interest around the anatomic ceramic hip arthroplasty (An.C.A.) stems (cases with follow-up ≥ 24 months).

Of 295 patients, 2 (0.7%), complaining with persistent thigh pain, underwent revision surgery after index surgery (one at 29 months and one at 32 months). At revision surgery, the stem appeared firmly fixed to the bone. The bone tissue presented the weaker consistency previously described for alumina-coated stem revision. Due to the intention to replace the painful uncemented stem with a cemented one, the stem was removed with the aid of gouges to interrupt the cancellous bone ingrowth into the proximal porous prosthetic surface. The stem was then hammered out of the femoral canal, in which a whitish powder was present at the surface of the bony laminae and was formed as a mold around

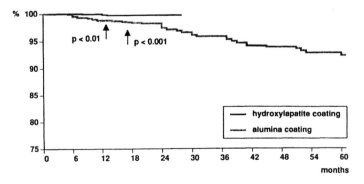

Figure 18   Survival of osteointegrated anatomic ceramic hip arthroplasty (An.C.A.) stem interfaces (considering all seven regions of interest for each stem).

the stem. The HA coating loss was macroscopically evident since the metallic stem surface was largely exposed.

Since the *in vivo* behavior of the HA coating is strongly related to its chemical and physical properties, we must first describe the HA with which we are dealing. These two cases were part of the first lot of HA-coated stems produced by the manufacturer. The characteristics are Ca/P, 1.72; crystallinity, 0.47; amorphous content, 50%; porosity, 5%; average thickness, 130 μ; tensile strength (American Society for Testing and Materials [ASTM] 633C), 3.68 kg/mm^2. Afterward, the production was modified with crystallinity brought up to 0.60 and amorphous content lowered to 40%. These coatings were applied in an air environment. Histological evaluation showed that HA coating is randomly lost. In some regions, it appeared present with a normal thickness (above all, at the beads most outer border), while other parts were detached in the form of gross blocks. Bone trabeculae are normally calcified and contact directly either residual HA or the metallic surface (Fig 19). Only about 10% of the retrieved stem's surface was bone integrated, while only two-thirds of this surface was still HA coated. Sometimes, bone trabeculae were ingrown to the surface of the detached HA blocks, preventing continuity between bone and stem (Fig. 20). The fragmentation of HA coating could impair the long-term secondary stability achievable by bone ingrowth and the "reef" effect of HA coatings' loose fragments may well prevent or slow down the metal alloy's direct osteointegration. Tiny crystals (3–5μ large) were present and tended to cluster (Fig. 21), migrating into the medullary void areas. This debris did not evoke any histiocytic or giant cell reaction. Metallic debris were also detected.

C. Discussion

The application of ceramics in the fabrication of total joint implants continues to receive considerable attention; reports on negative phenomena related to their use are very scarce [94–96].

The histopathological findings of altered mineralization immediately adjacent to the prosthesis prompted us to consider a possible toxic local effect of Al$_2$O$_3$, particularly given the absence of clinical or radiographic stigmata suggesting metabolic bone disease in these patients. These findings appear somewhat analogous to those of osteomalacic

Figure 19   Case 27, showing bone tissue in contact with residual hydroxylapatite coating and with naked metallic surface (fuchsin–light green, ×2.5).

Figure 20 Case 26, in which bone tissue is contacting the fragmented parts of the hydroxylapatite coating, which act as a "reef," preventing contact bone apposition to the metallic surface, visible in the acute space between the beads at right (fuchsin–light green, ×6.3).

bone, in which the newly formed bone matrix fails to mineralize in a normal or timely manner. The presence of mature lamellar bone rather than osteoid bone, however, indicates that the phenomenon is more likely related to demineralization of otherwise normal bone.

Alterations in mineralization induced by the toxic effects of diphosphonates, fluorides, and aluminum have been widely documented [97-100]. Aluminum, as aluminum oxide, constitutes the coating of the prostheses under study. The systemic toxicities of aluminum due to excessive oral or parenteral intake are well known [101]. Aluminum intake in the diet is about 10–100 mg a day and is influenced by the cooking modality and containers used. Absorption in the gastrointestinal tract is negligible in healthy subjects, but becomes significant in patients suffering from renal failure, for which aluminum excretion is impaired. *In vivo*, aluminum ions are largely protein bound via transferin. It is conceivable that this favors aluminum storage in cells with numerous transferin receptors.

The first reports on the possible toxicity of aluminum appeared in the early 1970s

Figure 21 Case 27, in which tiny crystals (3–5 μ large) are present and tend to cluster, migrating into the medullary void areas (fuchsin–light green, ×25).

when Berlyne et al. observed that patients suffering from chronic renal failure, treated with aluminum ion-exchange resins to correct hyperkalemia, showed a significant increase in the plasma concentrations of this element [102]. A similar increase was found in patients who took Al(OH)$_3$ antacids to control hyperphosphatemia or peptic ulcer disease. In 1971, Parsons et al. found, in the bone tissue of nephropathic patients, significant amounts of aluminum that could be related to a previous uremia and to the duration of dialysis [103]. Dent and Winter subsequently reported a case of osteomalacia due to hypophosphatemia induced by excessive intake of aluminum hydroxide [104].

During the following years, these initial and occasional observations of aluminum toxicity were confirmed and it became recognized as the etiologic agent in several dialysis-related complications. Many dialyzed patients, in the long term, suffered from damage of the central nervous system, osteomalacia, and anemia. In support of these observations, Ward et al. in 1978 [105] and Parkinson et al. in 1979 [99] found a correlation between osteomalacia and the aluminum content of water used as the dialysate. As a corollary, patients receiving long-term total parenteral nutrition (TPN) in the presence of a variety of intestinal disorders were found to develop metabolic bone disease. In such individuals with normal kidney function, bone pathology similar to that described for dialyzed patients was noted. In this clinical scenario, the etiologic agent was found to be aluminum that contaminated the casein hydrolysate, the source of protein in the parenteral nutrition formulation [66].

The biochemical mechanisms underlying aluminum toxicity have not been identified with certainty. Numerous *in vitro* experiments have demonstrated interactions of aluminum with essential biochemical processes, but extrapolation to clinical aluminum toxicity is difficult. The documented interaction of aluminum with biologically important substances [106] can be summarized as (1) interactions with proteins (enzymes, calcium-binding proteins, structural proteins, storage proteins), (2) interactions with nucleotides, (3) interactions with phospholipids (cellular membranes), and (4) interactions with minerals (mainly hydroxylapatite).

At the bone-cellular level, Cournot-Witmer et al. in 1981 [107] and Robertson et al. in 1987 [100] advanced the hypothesis that aluminum action on bone cells and matrix mineralization could be modulated by parathyroid hormone (PTH). Other authors have suggested a direct action on osteoblasts, possibly mediated by PTH [108]. Bachra and Van Harskamp hypothesized that aluminum salts could interfere with the deposition of the bone mineral matrix, forming insoluble phosphates and thus creating a local depletion of phosphates able to inhibit regular calcium deposition [109]. Ott et al. confirmed that aluminum deposition at the calcified bone boundary inhibits mineralization of osteoid directly [110]. Maloney et al. further hypothesized that aluminum may bind at the mineralization front and consequently inhibit normal calcification, or that it may bind to existing hydroxylapatite crystals, thus blocking further deposition of calcium phosphate [111].

In other studies, however, it has been demonstrated, by means of double tetracycline labels, that active mineralization may continue despite the presence of substantial aluminum deposits [108]. Talwar et al. highlighted the chemical-physical role of aluminum on the deposition of calcium phosphates in bone [112]. Goodman [113] and Goodman et al. [114], carrying out short-term toxicity tests on rats, confirmed the toxic effect of aluminum on osteoblasts, while in studies on dogs, the role of the altered metabolism of vitamin D in mediating the phenomenon was not able to be defined.

In summary, the specific mechanism(s) for aluminum toxicity in bone remains enigmatic [115]. Although studies of aluminum toxicity are numerous, reports on analogous phenomena related to the implantation of alumina, either as a bulk material or coating, are extremely rare. Lewandowska-Szumiel and Komendar carried out traditional histologic tests and assays of aluminum in the bone tissue of experimental models in which alumina rods were implanted for a period of six–eight months. With normal histological results, they found significant concentrations of aluminum in periprosthetic tissues [95].

Alumina wear debris were found adjacent to the implant in considerable concentration. The crystallographic evidence of the gamma phase detected for this debris proved its origin from the coating, since debris due to wear of head-socket ceramic bearing surfaces always presented in the alpha phase crystallographically. Hydroxylapatite is a biocompatible calcium phosphate ceramic but, unlike tricalcium phosphate, it is not resorbable at neutral pH. Even though dissolution has been observed in an acid environment [116], it was not supposed to dissolve once osteointegrated [82]. The HA stability in biological conditions is related to its chemical composition, degree of crystallinity, modality of plasma spray application, and thickness of the coating [117]. The HA was produced and experimentally tested in animals from 1970 to 1980 [118] and its innovative application in humans has been experienced in oral surgery [119–121]. It was subsequently used as a filler for bone defects to favor bone regeneration [121–123].

The use of dense HA in orthopedic surgery is limited by its low mechanical properties [124]. An optional approach would be to take advantage of the mechanical strength of metal alloy and of the biological activity of coating of metallic surfaces with it and HA [82].

Experimental tests have been performed in animals, mainly dogs, to study the bone reaction to HA-coated implants in both static [82] and dynamic loading conditions [72,82,125,126]. All authors describe an early (3–6 weeks) [82] bone apposition in direct contact with the implant coating surface, without any fibrous tissue interposed, showing a very good osteointegration of the ceramic material.

However, according to Jasty et al. [125], who compared uncoated titanium alloy implants and HA coated ones in dogs, osteointegration undoubtedly occurred earlier with HA-coated surfaces, but there was no significant difference among the two groups referring to the amount of bone ingrown surface at 6 weeks. Moreover, the HA coating did not improve the filling of intraoperative gaps between the bone and the prosthesis, contrary to findings by Geesink, de Groot, and Klein [82]. Nevertheless, the bone ingrown was more intimately associated with the implant surface that was coated with calcium phosphates [125]. A rearrangement of the trabeculae orientation in accordance with Wolff's law at the coated surfaces has also been described [82,126].

Geesink and others started to implant in humans HA-coated total hip prostheses in 1986 [82,85,86]; they were followed by many other surgeons throughout the world [87–92,127,128]. Clinical results reported in the literature are good, with a significant reduction of thigh pain [85–87,89,92,127], which is often present postoperatively in patients with an uncemented stem. This result could be due to the earlier osteointegration, with reduction of the painful micromovements between bone and implant, leading to a quicker secondary stability.

On x-ray films, a lesser incidence of radiolucent lines was observed with HA-coated prostheses [85–92], indicating the absence of fibrous tissue at the bone-implant interface. Definition of stem sinking into the femoral canal by means of plain radiography [90,92] and stereophotogrammetry [92] demonstrated a lower migration with the HA-coated stems, which tend to reach a stable position at three-month follow-up. Our data show an extremely low incidence of radiolucency at the stem-bone interface when the implant was HA coated. The improvement of osteointegration due to HA with respect to alumina (we cannot compare such results with uncoated metallic alloy stems) is statistically ($p < 0.001$) proved by comparison of the survival curves of the osteointegrated ROI at 27-month follow-up, with HA at 99.7% versus alumina at 96.8%, which drops to 92.4% at 60 months.

Bone remodeling was frequently observed (Figs. 22 and 23). The bone trabecular rearrangement bears witness to the mechanical action of HA-coated surfaces in transmitting stresses to the surrounding bone. In Fig. 23, it can be clearly differentiated where bone is massively stress shielded (major trochanter) and where trabecular bone is reinforced to fill the gap present between stem and cortical bone after surgery.

Since 1991, histological studies on autopsy-retrieved HA-coated stems have been reported [77,116,129,130], and other stems could now be studied due to surgical revision [131–133]. Besides observations confirming the expected biocompatibility of HA-coated implants and noting the absence of any flogistic cellular reaction, together with the presence of bonding osteogenesis similar to the fracture-healing process [77,129], some authors reported a significant loss of the HA coating due to partial resorption or to fragmentation into particles that are dispersed into the surrounding tissues [116,130,133]. Bauer at al. describe histological findings suggesting remodeling of cortical and cancellous bone, probably in response to Wolff's law, with associated osteoclast-induced focal loss of HA in areas of resorption of bone, and hyperplastic bridges of bone in areas corresponding to force transmission [130].

Frayssinet et al. confirm this hypothesis, but suggest that only vital bone is resorbed by osteoclasts, while HA debris are found in the macrophages, which also resorb necrotic bone [116]. They state that cells may well distinguish living bone from HA, as they do with the dead bone tissue. Collier et al. [133], studying retrieved HA-coated stems, found a high incidence of coating debonding and loss. This raises some doubts about the use of HA as the sole long-term mechanism of fixation of prostheses due to its limited initial bond to the substrate and the ability of the host to resorb the material over time. Santa-virta et al. [131], examining well-fixed stems, did not find bone ingrowth directly contacting the coating due to a fibrous tissue interposed between the HA and the bone.

Our experience is related to two cases and confirms part of the cited reports. Hydroxylapatite coating, after 29 and 32 months since surgery, is fragmented. The metallic surface is exposed to direct bone opposition. This fact could support the expectation that osteointegration is not impaired by the coating breakdown. Nevertheless, bone trabeculae still bonded to the fragmented HA coating dispersed into the surroundings of the stem ("reef" effect) may jeopardize osteointegration and reduce the likelihood of mechanical stability produced by the earlier osteointegration. Hydroxylapatite debris are present, but they do not cause any major pathologic cellular reaction.

V. THREE-DIMENSIONAL POROUS CERAMIC COATING

A ceramic cup to be fixed to the iliac bone without cement could take great advantage of a 3-D porous surface, similar to the porous surfaces obtained with beads in metal cementless cups. A research project was started in 1983 to construct such a surface for a bulk alumina cup, only using ceramic materials. To reach valid bone-implant interlocking, the porosity should be around 25–45%, with an average pore size ranging between 200 μ and 500 μ. Our aim was to make a porous alumina-glass composite (Poral) to be used as a

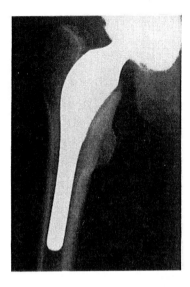

(a)

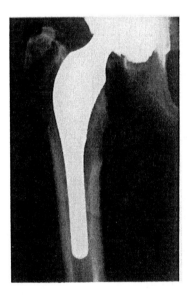

(b)

Figure 22   Anatomic ceramic hip arthroplasty (An.C.A.) with hydroxylapatite coating: (a), post-operative x-ray film showing void space present between stem and cortical bone in the diaphysis; (b), 24-month follow-up x-ray film showing void spaces filled in the distal part of the stem, with trabecular bone remodeling is evident more proximally.

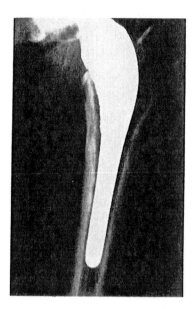

(a)

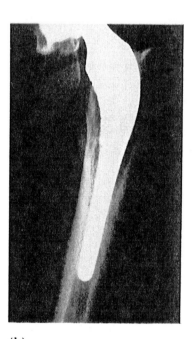

(b)

Figure 23   Anatomic ceramic hip arthroplasty (An.C.A.) with hydroxylapatite coating: (a), post-operative x-ray film showing a gap between stem and cortical bone; (b), the direct bone osteointe-gration of the stem is obtained at 37-month follow-up.

new porous coating for alumina prosthetic components to couple the advantages of the ceramic-ceramic bearing surfaces with the improved biocompatibility produced by the porosity of ceramics instead of a metal alloy. Poral is made by two layers of alumina beads bonded by a high-temperature glass to the outer surface of a hip prosthetic socket. The alumina used was Ostalox in the period November 1985 to November 1989 (An.C.A. I and II), and Biolox from September 1987 to October 1993 (An.C.A. III, IV, and V). The overlapped period is limited to 22 An.C.A. II implants. The bonding glass was obtained with alumina silicates with low alkali and alkaline earth contents. The binding process was performed at around 1400°C.

A. Chemical Stability

Chemical stability of the binding glass has been studied *in vitro* with cups (without the metallic ring) kept in Ringer's solution at 37°C ± 1°C and stirred daily [134]. Since the 99.7% alumina (Ostalox) is totally undissolvable, any weight loss of the ceramic composite may be attributed exclusively to the loss of the glass. The total amount of glass utilized on each composite varied between 697 mg and 723 mg. The glass release was quantified at 500-hour intervals up to 5000 hours. The progressive absolute values are reported in Table 5. The resorption rates calculated at each interval and referred to as the amount in mg per 1000 hours are reported in Fig. 24. The glass presents an initially higher resorption rate (0.00055 mg/h after 1000 hours), which then steadily decreases thereafter, reaching the lowest rate of 0.00038 mg/h at 5000 hours, which represents the actual goal of our experiment. The binding glass, at the rate of 0.00038 mg/h, would require at least 21 years to reach a 10% loss. The composite then may be considered safely stable from a chemical point of view.

B. Shear Strength of Poral Coating

The coating shear strength of Poral was tested with samples loaded to failure using a loading jig attached to an Instron machine (see Fig. 25). The shear strength ranged between 18 and 46 MPa, averaging 32 MPa [64].

Table 5 Cumulative Bioglass Resorption

Hours ($\times 100$)	mg $\times 10^{-1}$
5	0.1
10	0.55
15	0.7
20	0.9
25	1.1
30	1.2
35	1.37
40	1.6
45	1.8
50	1.9

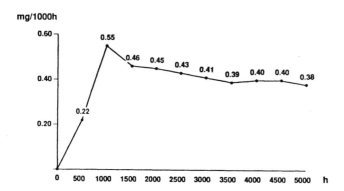

Figure 24 Cumulative resorption rates calculated at each interval and referred to as the amount in milligrams per 1000 hours.

C. Vibrational and Raman Laser Spectroscopy

With a vibrational Raman laser, it was then possible to characterize, at a molecular level, the components of the Poral ceramic composite. Raman spectra were recorded using a Jasco R-300 spectrometer and the 4880-angstrom ($\overset{\circ}{A}$) line of a Lexel Ar^+ laser with a power of -200 milliwatts (mW). The prosthetic socket Raman spectra have been obtained focusing the laser beam in three different positions: beads, socket bulk ceramic, and beads-glass-socket interfaces [64]. They show that external beads and the noncoated socket are made of α-Al_2O_3 (corundum). The beads-glass-socket intermediate position shows the bands of α-Al_2O_3 (corundum) and, moreover, two new bands at -1010 and -360 cm^{-1}. The glass bands are not present (partly masked by the α-Al_2O_3 strong bands) in the Raman spectrum. The presence in the spectrum of the two new bands indicates a structural modification of the interface, denoting more than a simple physical molecular interaction.

In conclusion, the alumina was of an α-Al_2O_3 (corundum) form. The glass was an alumino-silicate with an intermediate structure between vitreous silica and disilicate glasses. The alumina-glass interface has shown the presence of new bands that could be related to a structural interaction between the glass and alumina components.

D. Experimental Bone Ingrowth

Experimental bone ingrowth into porous alumina was studied by implanting 4-mm Poral cylinders in totally anesthetized rabbit distal femoral metaphyses (Fig. 26). The animals were then sacrificed at 2-, 4-, 6-, and 12-week intervals. The specimens retrieved were included in methylmethacrylate, cut with a diamond blade saw, and stained with Paragon. In almost all cases, the ceramic composite was lost during the slice preparation due to its brittleness when thinned to 80 microns. In pictures, the ceramics must then be associated with the void spaces. Two weeks after surgery, connective reparative tissue around the ceramic beads was mainly present (Fig. 27). The bony trabeculae are still highly disorganized. After 4 weeks, bone ingrowth was more evident, and by 6 weeks the bone ingrowth into the porous alumina was impressive. The trabeculae show an intimate contact with the beads (Fig. 28), even at the bottom of the pore, where the bone already

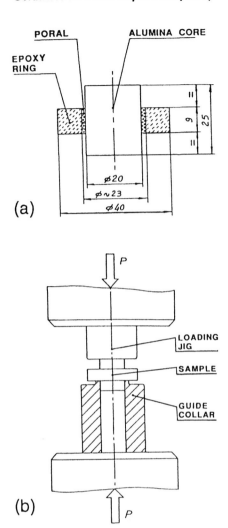

Figure 25 Experimental bone ingrowth into porous alumina: (a), sample used for Poral coating shear strength test; (b), loading jig.

shows an initial adaptive remodeling to the shape of the porosity, though the bony structure was not yet well organized. Almost no connective tissue was interposed between the bone and the ceramic. After 12 weeks, bone remodeling of the new trabeculae can be clearly seen (Fig. 29).

E. Clinical Experience

For the clinical experience, we started implanting the An.C.A. socket with Poral in November 1985. From that date to October 1993, 694 prosthetic cups were implanted, 104 of them constructed with Ostalox and 590 with Biolox. As reported in Table 2, among the 104 An.C.A. Types I and II (Ostalox), socket revision due to loosening was carried

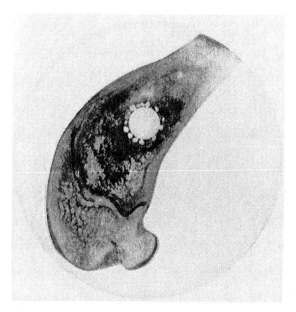

Figure 26   Poral sample implanted in rabbit distal metaphysis (×4).

out in 10 cases (1, 4–12, 15), corresponding to a 9.6% incidence; while only 3 patients of the 590 with the Biolox cup (0.7%) underwent socket revision (Cases 17, 19, 23, 24). No adverse reaction was found at surgery both macroscopically and histologically in the tissue surrounding the Poral surface. The loosened cups presented connective tissue interposed between bone and the porous ceramic. However, x-ray film follow-up of stable cases show quite a different picture, with osteointegration (see Fig. 30), by which bone filled the gap (indicated by arrows in the postoperative film).

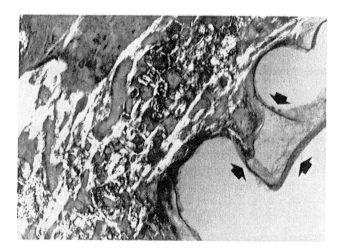

Figure 27   Poral sample retrieved at seven days (void spaces correspond to the missing ceramic); bone tissue is not present in the pore (indicated by arrows), which is filled by connective tissue (Paragon, ×25).

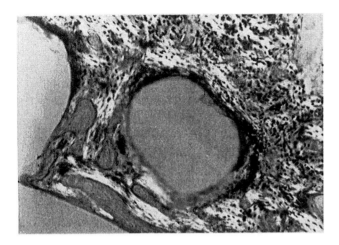

Figure 28 Poral sample retrieved at six weeks; the bone ingrowth into the pore is impressive. The trabeculae reach the bottom of the pore, where the bone shows an initial adaptive remodeling to the pore shape, though the bony structure is not yet well organized (Paragon, ×40).

F. Discussion

The porous alumina showed a proper mechanical strength and a good chemical stability. No cellular adverse reaction was ever observed. The clinical experience confirmed the biocompatibility of such a 3-D porous ceramic coating. The quality of the ceramic is a relevant factor for the clinical results. As with ceramic heads, Ostalox cups proved to be less reliable, as shown by its 9.6% rate of revision. A surgeon's learning curve could also be considered to motivate such a high revision rate, even though it must be underlined that surgeons were already acquainted with the screw cup technique, experienced with the

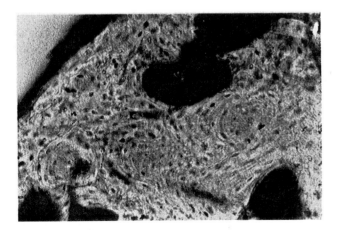

Figure 29   Poral sample retrieved at 12 weeks showing osteonic differentiation of the newly bony trabeculae (Paragon, ×40).

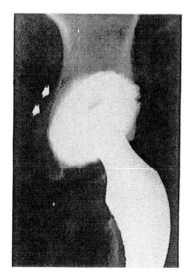

(a)

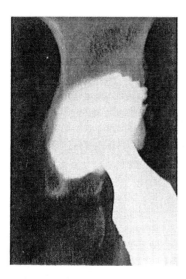

(b)

Figure 30   Biolox cup (anatomic ceramic hip arthroplasty Type IV, An.C.A. IV): (a), postoperative film showing a gap present in the medial area, indicated by arrows; (b), 36-month follow-up with perfect osteointegration of the Poral surface and complete filling of the previous medial gap.

Lord cup for 4 years. Biolox cups, used since 1987, presented a lower revision rate (0.7%). Considering that screwed prosthetic sockets usually have a higher revision rate, we claim that the improved clinical results obtained with An.C.A. Types III, IV, and V are to be attributed to the secondary stability achieved by Poral osteointegration.

VI. CONCLUSION

Bulk alumina is a suitable and reliable material for THA bearing surfaces. The alumina now available is actually reducing the wear so that it is an insignificant problem. Alumina remains the ceramic of choice for ceramic-ceramic coupling, while zirconia may well be the best material to couple with polyethylene.

We are not able to elucidate conclusively the mechanism(s) by which aluminum inhibits or reverses mineralization. Moreover, given the documented chemical stability of alumina, it is difficult to understand how aluminum ions appear to be leached from the coating. We can only suggest that crystallographic transformation occurs from the stable α-phase to the relatively less stable γ-phase, and that the presence of impurities in the form of $Al(OH)_3$ are such that partial solubilization of the alumina coating follows *in vivo*. Given the extensive worldwide use of alumina in clinical applications, the phenomenon we describe deserves careful evaluation as one might hypothesize that periprosthetic decalcification leads to implant loosening. We have described a histological phenomenon that appears to be related to the presence of aluminum at the prosthesis-bone interface. As noted above, we did not find loosening per se, but rather demineralization with residual bone apposition at the interface in the specimens retrieved at revision surgery. The phenomenon described was related to the relatively early onset of pain following the index surgery; 11 of 254 hips (4.3%) have demonstrated progressive, functionally-limiting pain related to the changes described that required revision surgery. These results also underline the need for at least 36 months of clinical validation to elucidate long-term effects of a new coating.

Hydroxylapaptite coating has been used since July 1990 for An.C.A. IV stems. Radiographic results confirm the improved osteointegration reported in the literature, with only very rare lucent lines detectable at the stem-bone interface. Revision surgery was limited to the hips of 2 patients with thigh pain (0.7%), excluding the case of acute septic loosening. These two cases showed that air-plasma-spray HA coating is fragmented and resorbed at the 29- and 32-month follow-up. Since these stems were not loose, such a phenomenon cannot be caused by scratching due to micromovements between bone and implant. At the moment, we do not know if HA coating resorption or fragmentation (reef effect) may jeopardize stem osteointegration. Such a concern remains with regard to the HA coating, which deserves further careful observations.

In conclusion, we may say that, after more than 20 years since the first ceramic THA was implanted in a human, dreams have been followed by facts: ceramic heads have proven to be mechanically reliable and effective in reducing wear, though reported only as episodic events with ceramic-on-ceramic THA. This is thus the only long-term clinically tested choice to replace the widely used polyethylene bearing surfaces, to overcome the wear failure that sooner or later affects this material in THA applications. The absence of interface osteolysis, so often reported for polyethylene THA, is another indirect proof that the ceramic surfaces' wear rate has no pathological relevance. Ceramic coating with HA is promising, even though some concerns about its long-term stability in a biological environment need further clarification. Currently, a 3-D porous coating made of glass-ceramic beads composite is available for the ceramic bulk socket. Poral proved to be biocompatible and clinically advantageous.

REFERENCES

1. Cooke, F.W., Ceramics in orthopedic surgery, *Clin. Orthop.*, 276:135–146 (1990).
2. Boutin, P., Arthroplastie totale de la anche par prothèse en alumine frittée. Etude expérimentale et premiéres applications cliniques, *Rev. de Chir. Orthop.*, 58(3):229–246 (1972).
3. Agins, H.J., E.A. Salvati, C.S. Ranawat, P.D. Wilson, and P.M. Pellicci, The nine to fifteen-year follow-up of one-stage bilateral total hip arthroplasty, *Orthop. Clin. North Am.*, 19:517–530 (1988).
4. Campbell, P., S. Nasser, N. Kossovsky, and H.C. Amstutz, Mechanism of production and histopathological effects of UHMWPE and metal wear debris in porous and cemented surface replacements. In *Symposium on Biocompatibility of Particulate Implant Materials*, K. St. John (Ed.), p. 7, ASTM Committee F-4 on Med. Surg. Mat. Dev., San Antonio, Texas (Oct. 31, 1990).
5. Lombardi, A.V., Jr., T.H. Mallory, B.K. Vaughn, and P. Drouillard, Aseptic loosening in total hip arthroplasty secondary to osteolysis induced by wear debris from titanium alloy modular femoral heads, *J. Bone Joint Surg.*, 71B:1337–1342 (1989).
6. McKellop, H., W. Gogan, E. Ebramzadeh, J. Luck, and A. Sarmiento, Metal-metal wear in metal-plastic prosthetic joints, *16th Annual Meeting of the Society for Biomaterials*, p. 144, Charleston, South Carolina (May 20–23, 1990).
7. Mirra, J.M., H.C. Amstutz, M. Matos, and R. Gold, The pathology of the joint tissues and its clinical relevance in prosthesis failure, *Clin. Orthop.*, 117:221–240 (1976).
8. Nasser, S., P.A. Campbell, D.J. Kilgus, N. Kossovsky, and H.A. Amstutz, Cementless total joint arthroplasty prostheses with titanium alloy articular surfaces. A human retrieval analysis, *Clin. Orthop.*, 261:171–185 (1990).
9. Clarke, I.C., Role of ceramic implants. Design and clinical success with total hip prosthetic ceramic-to-ceramic bearings, *Clin.Orthop.*, 282:19–30 (1992).
10. Pizzoferrato, A., G. Ciapetti, S. Stea, and A. Toni, Cellular events in the mechanisms of prosthesis loosenings, *Clin. Materials*, 7:51–81 (1991).
11. Dorre, E., and H. Hubner, *Alumina*, Springer-Verlag, Berlin (1984).
12. Davidson, J.A., Characteristics of metal and ceramic total hip bearing surfaces and their effect on long-term ultra high molecular weight polyethylene wear, *Clin. Orthop.*, 294:361–378 (1993).
13. Cooper, J.R., D. Dowson, J. Fisher, and B. Jobbins, Ceramic bearing surfaces in total artificial joints: Resistance to third body wear damage from bone cement particles, *J. Med. Eng. Technol.*, 15(2):63–67 (1991).
14. Clarke, I.C., J.M. Dorlot, and J. Graham, Biomechanical stability and design, *Ann. N.Y. Acad. Sci.*, 523:292–296 (1988).
15. Sedel, L., L. Kerboull, P. Christel, A. Meunier, and J. Witvoet, Alumina on alumina hip replacement, *J. Bone Joint Surg. Br.*, 72(4):658–663 (1990).
16. Bos, I., E. Meeuwssen, E.J. Henssge, and U. Lohrs, Differences in polyethylene wear in hip joint prostheses with ceramic and with metal-polyethylene combination of the articulation surfaces — A study of surgical and of autopsy materials, *Z. Orthop. Ihre. Grengzeb.*, 129(6):507–515 (1991).
17. Kumar, P., M. Oka, K. Ikeuchi, K. Shimizu, T. Yamamuro, H. Okumura, and Y. Kotoura, Low wear rate of UHMWPE against zirconia ceramic (Y-PSZ) in comparison to alumina ceramic and SUS 316L alloy, *J. Biomed. Mater. Res.*, 25(7):813–828 (1991).
18. Mittelmeier, H., Zemzntlose Verankerung von Endoprothesen nach dem Tragrippenprinzip, *Z. Orthop.*, 112:27–33 (1974).

19. Griss, P., E. Werner, P. Buchinger, and G. Heimke, Die Mannheimer Oxid-Keramic/Metal-iverbund-Prothesen, *Arch. Orthop. Unfallchir.*, 87:73–84 (1977).

20. Salzer, M., K. Zweymuller, H. Locke, A. Zeibig, N. Stark, H. Plenk, and G. Punzel, Further experimental and clinical experience with aluminum oxide endoprostheses, *J. Biomed. Mater. Res.*, 10(6):847–856 (1976).

21. Pizzoferrato, A., S. Stea, A. Sudanese, A. Toni, M. Nigrisoli, G. Gualtieri, and S. Squarzoni, Morphometric and microanalytical analyses of alumina wear particles in hip prostheses, *Biomaterials*, 14(8):583–587 (1993).

22. Toni, A., S. Stea, S. Squarzoni, A. Sudanese, G. Masetti, N. Maraldi, and A. Pizzoferrato, Considerations on ceramic prosthesis explants, *Chir. Organi Mov.*, 77:359–371 (1992).

23. Shimizu, K., P. Kumar, M. Oka, Y. Kotoura, Y. Nakayama, T. Yamamuro, T. Yanagida, and K. Makinouchi, Time dependent change of the mechanical properties of zirconia *in vivo*. In *Transaction of the 3rd World Biomaterial Congress*, 11, p. 406, Kyoto, Japan, (April 21–25, 1988).

24. Makinouchi, K., T. Yanagida, K. Shimizu, P. Kumar, Y. Kotoura, and M. Oka, Development of zirconia ceramic for implants. In *Orthopaedic Ceramic Implants, Proc. Jap. Soc. Orthop. Ceramic Implants*, H. Oonishi and Y. Ooi (Eds.), p. 4 (1984).

25. Christel, P., A. Meunier, M. Heller, J.P. Torre, and C.N. Peille, Mechanical properties and short term *in vivo* evaluation of Y-PSZ, *J. Biomed. Mater. Res.*, 23:45–61 (1989).

26. Wagner, W., Histologic bone reactions after implantation of alumina and zirconia pins. In *Trans. 7th European Conf. on Biomaterials*, Amsterdam, p. 44 (1987).

27. Kumar, P., K. Shimizu, M. Oka, Y. Kotoura, Y. Nakayama, T. Yamamuro, T. Yanagida, and K. Makinouchi, Biological reaction of zirconia ceramic, *Bioceramic, Proceedings of 1st International Bioceramic Symposium*, 1:341–346 (1988).

28. Boutin, P., P. Christel, J.M. Dorlot, A. Meunier, A. De Roquancourt, D. Blanquaert, S. Herman, L. Sedel, and J. Witvoet, The use of dense alumina-alumina ceramic combination in total hip replacement, *J. Biomed. Mater. Res.*, 22:1203–1232 (1988).

29. Mittelmeier, H., and J. Heisel, Sixteen-years' experience with ceramic hip prostheses, *Clin. Orthop.*, 282:64–72 (1992).

30. Boutin, P.M., THR using alumina-alumina sliding and a metallic stem: 1330 cases and an 11 years follow-up. In *Orthopedic Ceramic Implants, Japanese Society of Orthopedic Ceramic Implants*, H. Oonishi and Y. Ooi (Eds.), Vol. 1, p. 111 (1981).

31. Winter, M., P. Griss, G. Scheller, and T. Moser, Ten- to 14-year results of a ceramic hip prosthesis, *Clin. Orthop.*, 282:73–80 (1992).

32. Griss, P., and G. Heimke, Five years experience with ceramic-metal-composite hip endoprostheses. Clinical evaluation, *Arch. Orthop. Trauma Surg.*, 98(3):157–164 (1981).

33. Knahr, K., M. Salzer, H. Plenk, Jr., F. Grundschober, and W. Ramach, Experience with bioceramic implants in orthopaedic surgery, *Biomaterials*, 2(2):98–104 (1981).

34. Knahr, K., M. Bohler, P. Frank, H. Plenk, and M. Salzer, Survival analysis of an uncemented ceramic acetabular component in total hip replacement, *Arch. Orthop. Trauma Surg.*, 106(5):297–300 (1987).

35. Mittelmeier, H., Report on the first decennium of clinical experience with a cementless, ceramic total hip replacement, *Acta Orthop. Belg.*, 51(2–3):367–376 (1985).

36. Nizard, R.S., L. Sedel, P. Christel, A. Meunier, M. Soudry, and J. Witvoet, Ten-year survivorship of cemented ceramic total hip prosthesis, *Clin. Orthop.*, 282:53–63 (1992).

37. Heisel, J., and E. Schmitt, Implant fractures in ceramic hip endoprostheses, *Z. Orthop.*, 125(5):480–490 (1987).

38. Trepte, C.T., E.F. Gauer, and B.M. Gartner, Results with ceramic-on-ceramic sliding endoprosthesis completely or partially fixed without cement, *Z. Orthop.*, 123(2):239–244 (1985).

39. Rampoldi, A., Mittelmeier's ceramic hip prosthesis, *Ital. J. Orthop. Traumatol.*, 10(3):305–311 (1984).

40. Kern, S., A. Schreiber, and B. Hilfiker, Ceramic head fracture—A rare complication in hip endoprosthesis. A case report and literature survey, *Z. Orthop.*, 128(5):543–548 (1990).
41. O'Leary, J.F., T.H. Mallory, T.J. Kraus, A.V. Lombardi, Jr., and C.L. Lye, Mittelmeier ceramic total hip arthroplasty, *J. Arthroplasty*, 3(1):87–96 (1988).
42. Miller, E.H., R.S. Heidt, M.C. Welch, W.G. Harding, R.S. Heidt, Jr., and S.M. Lawhon, Self-bearing, uncemented, ceramic total hip replacement arthroplasty, *Instr. Course Lect.*, 35:188–202 (1986).
43. Hoffinger, S.A., K.J. Keggi, and L.E. Zatorski, Primary ceramic hip replacement. A prospective study of 119 hips, *Orthopedics*, 14(5):523–531 (1991).
44. Cameron, H.U., Ceramic head implantation failures (letter), *J. Arthroplasty*, 6(2):185–188 (1991).
45. Toni, A., A. Sudanese, D. Ciaroni, D. Dallari, T. Greggi, and A. Giunti, Anatomical ceramic arthroplasty (An.C.A.): Preliminary experience with a new cementless prosthesis, *Chir. Organ. Mov.*, 75(1):81–97 (1990).
46. Plitz, W., and P. Griss, Clinical, histomorphological and materials related observations on removed alumina-ceramic hip joints components. In *Implant Retrieval: Material and Histological Analysis*, A. Weinstein, D. Gibbons, S. Brown, and W. Ruff (Eds.), pp. 131–156, NBS SP-601, U.S. Dept. of Commerce, Washington, DC (1981).
47. Plitz, W., and H.U. Hoss, Wear of alumina-ceramic hip-joints: Some clinical and trobological aspects. In *Biomaterials 1980*, G.D. Winter, D.F. Gibbons, and H. Plenk (Eds.), pp. 187–196, John Wiley, New York (1982).
48. Walter, A., and W. Plitz, Wear of retrieved alumina-ceramic hip-joints in ceramics in surgery. In *Ceramics in Surgery*, P. Vincenzini (Ed.), pp. 253–259, Elsevier, Amsterdam (1983).
49. Walter, A., and W. Plitz, Wear mechanism of alumina-ceramic bearing surfaces of hip-joint prostheses. In *Biomaterials and Biomechanics 1983*, P. Ducheyne, G. van de Perre, and A.E. Aubert (Eds.), pp. 55–60, Elsevier, Amsterdam (1984).
50. Heimke, J., and P. Griss, Five years experience with ceramic-metal-composite hip endoprostheses. II. Mechanical evaluations and improvements, *Arch. Orthop. Trauma Surg.*, 98:165–171 (1981).
51. Walter, A., On the material and the tribology of alumina-alumina couplings for hip joint prostheses, *Clin. Orthop.*, 282:31–46 (1992).
52. Osterholm, H.H., and D.E. Day, Dense alumina aged *in vivo*, *J. Biomed. Mater. Res.*, 15:279–288 (1981).
53. Dalgleish, B.J., and R.D. Rawlings, A comparison of the mechanical behaviour of aluminas in air and simulated body environments, *J. Biomed. Mater. Res.*, 15:527–542 (1981).
54. Krainess, F.E., and W.J. Knapp, Strength of dense alumina ceramic after aging *in vitro*, *J. Biomed. Mater. Res.*, 12:241–246 (1978).
55. Dorlot, J.M., Long-term effects of alumina components in total hip prosthesis, *Clin. Orthop.*, 282:47–52 (1992).
56. Dorlot, J.M., P. Christel, L. Sedel, J. Witvoet, and P. Boutin, Examination of retrieved hip-prosthesis: Wear alumina-alumina components. In *Biological and Biomechanical Performances of Biomaterials*, P. Christel, A. Meunier, and A.J.C. Lee (Eds.), pp. 495–500, Elsevier, Amsterdam (1986).
57. Kummer, F.J., S.A. Stuchin, and V.H. Frankel, Analysis of removed Autophor ceramic-on-ceramic components, *J. Arthroplasty*, 5(1):28–33 (1990).
58. Sedel, L., P. Christel, L. Kerboull, and J. Witvoet, Total hip prosthesis in patient less than 50 years of age. Value of ceramic materials, *Rev. Rhum. Mal. Osteoartic.*, 57(9):605–611 (1990).
59. Witvoet, J., S. Herman, L. Sedel, P. Christel, and D. Blanquaert, Resultats del protheses totales de hance Osteal. A propos de 550 protheses, *Acta Orthop. Belgica.*, 51:288–297 (1985).

60. Mahoney, O.M., and J.H. Dimon, Unsatisfactory results with a ceramic total hip prosthesis, *J. Bone Joint Surg.*, 72/A(5):663–671 (1990).
61. Asada, K., K. Yoshida, A. Shimazu, H. Yunoki, and N. Ishida, Development of alumina ceramic bipolar hip prosthesis and clinical application, *Nippon Seikeigeka Gakkai Zasshi*, 61:155–169 (1987).
62. Boutin, P., and D. Blanquaert, A study of the mechanical properties of alumina-on-alumina total hip prosthesis, *Rev. Chir. Orthop.*, 67:279–287 (1981).
63. Geesink, R.G.T., K. de Groot, and C.P. Klein, Bonding of bone to apatite-coated implants, *J. Bone Joint Surg. Br.*, 70:17–22 (1988).
64. Pizzoferrato, A., A. Toni, A. Sudanese, G. Ciapetti, A. Tinti, and A. Venturini, Multilayered bead ceramic composite coating for hip prostheses: Experimental studies and preliminary clinical results, *J. Biomed. Mater. Res.*, 22:1181–1202 (1988).
65. Huang, G.K., and M. Cao, Experimental study of metallic bone and joint prostheses with plasma-sprayed ceramic coating, *Zentralbl. Chir.*, 117:171–177 (1992).
66. Klein, G.L., A.C. Alfrey, N.L. Miller, D.J. Sherrard, T.K. Hazlet, M.E. Ament, and J.W. Coburn, Aluminum loading during total parenteral nutrition, *Am. J. Clin. Nutr.*, 35:1425–1429 (1982).
67. Malike, M.A., D.A. Puleo, R. Bizios, and R.H. Doremus, Osteoblasts on hydroxyapatite, alumina and bone surfaces *in vitro*: Morphology during the first 2 h of attachment, *Biomaterials*, 13:123–128 (1992).
68. Naji, A., and M.F. Hamrmand, Cytocompatibility of two coating materials, amorphous alumina and silicon carbide, using human differentiated cell cultures, *Biomaterials*, 12:690–694 (1991).
69. Ammar, A., Tissue compatibility of different intracranial implant materials: *In-vivo* and *in-vitro* studies, *Acta Neurochir. Wien*, 72:45–59 (1984).
70. Gatti, A.M., D. Zaffe, G.P. Poli, and R. Galetti, The evaluation of the interface between bone and a bioceramic dental implant, *J. Biomed. Mater. Res.*, 21:1005–1011 (1987).
71. Hayashi, K., N. Matsuguchi, K. Uenoyama, T. Kanemaru, and Y. Sugioka, Evaluation of metal implants coated with several types of ceramics as biomaterials, *J. Biomed. Mater. Res.*, 23:1247–1259 (1989).
72. Hayashi, K., K. Uenoyama, N. Matsuguchi, S. Nakagawa, and Y. Sugioka, The affinity of bone to hydroxyapatite and alumina in experimentally induced osteoporosis, *J. Arthroplasty*, 4(3):257–262 (1989).
73. Uchida, S., A histopathologic comparison of the tissue reaction to prosthetic materials in the knee joint of rats, *Orthopaedics*, 8:1276–1280 (1985).
74. Cini, L., M. Paltrinieri, A. Pizzoferrato, S. Sandrolini, C. Trentani, and C. Zolezzi, Implantation, without cement, of a hip arthroprosthesis coated with alumina, *Chir. Organ. Mov.*, 62:1–27 (1975).
75. Forgon, M., E. Mammel, K. Trombitas, L. Kacsalova, and I. Draveczki, Morphological investigations of a porous aluminium oxide ceramic and the consequences for clinical application, *Arch. Orthop. Trauma Surg.*, 106:385–389 (1987).
76. Plenk, H., Biocompatibility of ceramics in joint prostheses. In *Biocompatibility of Orthopaedic Implants*, D.F. Williams (Ed.), Vol. 1, CRC Press, Boca Raton, FL (1982).
77. Furlong, R.J., and J.F. Osborn, Fixation of hip prostheses by hydroxyapatite ceramic coatings, *J. Bone Joint Surg. Br.*, 73(6):741–745 (1991).
78. Toni, A., C.G. Lewis, A. Sudanese, S. Stea, F. Calista, L. Savarino, A. Pizzoferrato, and A. Giunti, Bone demineralization induced by cementless alumina coated femoral stems, *J. Arthroplasty*, 9:435–444 (1994).
79. Pring, D.J., R.G. Henderson, A.G. Rivett, T. Krausz, R.R.H. Coombs, and J.P. Lavender, Autologous granulocyte scanning of painful prosthetic joint, *J. Bone Joint Surg.*, 68B:647–652 (1986).
80. Stea, S., L. Savarino, A. Toni, A. Sudanese, A. Giunti, and A. Pizzoferrato, Microradio-

graphic and histochemical evaluation of mineralization inhibition at the bone-alumina inter-face, *Biomaterials*, 13:664–667 (1992).

81. Savarino, L., E. Cenni, S. Stea, M.E. Donati, G. Paganetto, A. Moroni, A. Toni, and A. Pizzoferrato, X-ray diffraction of newly formed bone close to alumina- or hydroxylapatite-coated femoral stem, *Biomaterials*, 14, 12:900–906 (1993).

82. Geesink, R.G., K. de Groot, and C.P. Klein, Chemical implant fixation using hydroxylapatite coatings. The development of a human total hip prosthesis for chemical fixation to bone using hydroxylapatite coatings on titanium substrates, *Clin. Orthop.*, 225:147–170 (1987).

83. Engh, C.A., J.D. Bobyn, and A.H. Glassman, Porous coated hip replacement. The factors governing bone ingrowth, stress shielding, and clinical results, *J. Bone Joint Surg.*, 69/B:45–55 (1987).

84. Gruen, T.A., G.M. McNeice, and H.C. Amstutz, "Modes of failure" of cemented stem-type femoral components: A radiographic analysis of loosening, *Clin. Orthop.*, 141:17–27 (1979).

85. Geesink, R.G., Experimental and clinical experience with hydroxyapatite-coated hip im-plants, *Orthopedics*, 12(9):1239–1242 (1989).

86. Geesink, R.G., Hydroxyapatite-coated total hip prostheses. Two-year clinical and roentgeno-graphic results of 100 cases, *Clin. Orthop.*, 261:39–58 (1990).

87. Drucker, D.A., W.N. Capello, J.A. D'Antonio, and L.E. Hile, Works in progress #6. Total hip arthroplasty using a hydroxyapatite-coated acetabular and femoral component, *Orthop. Rev.*, 20(2):179–185 (1991).

88. Manley, M.T., and R. Koch, Clinical results with the hydroxyapatite-coated Omnifit hip stem, *Dent. Clin. North Am.*, 36(1):257–262 (1992).

89. Cook, S.D., J. Enis, G.D. Armstro, and E. Lisecki, Early clinical results with the hydroxyap-atite-coated porous LSF Total Hip System, *Dent. Clin. North Am.*, 36(1):247–255 (1992).

90. Kroon, P.O., and M.A. Freeman, Hydroxyapatite coating og hip prostheses. Effect on migration into the femur, *J. Bone Joint Surg. Br.*, 74(4):518–522 (1992).

91. Abrahams, T.G., and O.D. Crothers, Radiographic analysis of an investigational hydroxy-apatite-coated total hip replacement, *Invest. Radiol.*, 27(10):779–784 (1992).

92. Soballe, K., S. Toksvig-Larsen, J. Gelineck, S. Fruensgaard, E.S. Hansen, L. Ryd, U. Lucht, and C. Bunger, Migration of hydroxyapatite coated femoral prostheses, *J. Bone Joint Surg.*, 75B:681–687 (1993).

93. Mantel, N., and W. Haenszel, Statistical aspects of the analysis of data from retrospective studies of disease, *J. Natl. Cancer Inst.*, 22:719–748 (1959).

94. Donath, K., and A. Kirsch, Welche Bedeutung hat die primare Stabilisation von Implantaten fur die ossare Integration der Eineilphase? *Dtsch. Z. Zaharztl. Implantol.*, 2:11–17 (1986).

95. Lewandowska-Szumiel, L., and J. Komender, Aluminium release as a new factor in the estimation of alumina-bioceramic implants, *Clinical Materials*, 5:167–175 (1990).

96. Ryu, R.K., E.G. Bovill, H.B. Skinner, and W.R. Murray, Soft tissue sarcoma associated with aluminum oxide ceramic total hip arthroplasty, *Clin. Orthop.*, 216:207–212 (1987).

97. Bauer, F.C.H., O.S. Nillson, H. Tornkvist, T.C. Limdholm, and T.S. Limdholm, Effect of a diphosphonate on the inductive activity of rat bone matrix, *Clin. Orthop.*, 185:266–269 (1984).

98. Fleish, H., Experimental basis for the clinical use of diphosphonates in Paget's disease of bone, *Arthritis Reum.*, 23:1162–1170 (1980).

99. Parkinson, I.S., M.K. Ward, T.G. Feest, R.P.W. Fawcett, and D.N.S. Kerr, Fracturing dialysis osteodystrophy and dialysis encephalopathy, *Lancet*, 24:406–409 (1979).

100. Robertson, J.A., A.J. Fenselfeld, C.C. Haygood, P. Wilson, C. Clarke, and F. Llach, Animal model of aluminium-induced osteomalacia: Role of chronic renal failure, *Kidney Int.*, 23:327–335 (1987).

101. Slatopolsky, E., and J. Delmez, Bone desease in chronic renal failure and after renal trans-plantation. In *Disorders of Bone and Mineral Metabolism*, F.L. Coe and M.J. Favus (Eds.), p. 921, Raven Press, New York (1992).

102. Berlyne, G.M., J. Ben Ari, D. Pest, J. Weinberger, M. Stern, G.R. Gilmore, and R. Levine, Hyperaluminaemia from aluminum resins in renal failure, *Lancet*, 5:494–496 (1970).
103. Parsons, V., C. Davies, C. Goode, C. Ogg, and J. Siddiqui, Aluminium in bone from patients with renal failure, *Br. Med. J.*, 4:273–275 (1971).
104. Dent, E., and C. Winter, Osteomalacia due to phosphate depletion from excessive aluminium hydroxyde ingestion, *Br. Med. J.*, 1:551–552 (1974).
105. Ward, M.K., T.G. Feest, H.A. Ellis, I.S. Parkinson, D.N.S. Kerr, J. Herrington, and G.L. Goode, Osteomalacic dialysis osteodistrophy: Evidence for a water-borne aetiological agent, probably aluminium, *Lancet*, 22:841–845 (1978).
106. Van de Vyver, F., and W.J. Visser, Aluminum accumulation in bone, in *Trace of Metals and Fluorides in Bones and Teeth*. N. Priest (Ed.), pp. 41–48, CRC Press, Boca Raton, FL (1990).
107. Cournot-Witmer, G., J. Zingraff, J.J. Plachot, F. Escaig, R. Lefevre, P. Boumati, A. Bourdeau, M. Garabedian, P. Galle, R. Bourdon, T. Drueke, and S. Balsan, Aluminium localization in bone from hemodialyzed patients: Relationship to matrix mineralization, *Kidney Int.*, 20:375–385 (1981).
108. Dunstan, C.R., R.A. Evans, E. Hills, S.Y.P. Wong, and A.C. Alfrey, Effect of aluminum and parathyroid hormone on osteoblasts and bone mineralization in chronic renal failure, *Calcified Tissue International*, 36:133–138 (1984).
109. Bachra, B.N., and G.A. Van Harskamp, The effect of polyvalent metal ions on the stability of a buffer system for calcification *in vitro*, *Calcified Tissue Research*, 4:359–365 (1970).
110. Ott, S., S.A. Maloney, J.W. Coburn, A.C. Alfrey, and D.J. Sherrard, The prevalence of bone aluminium deposition in renal osteodystrophy and its relation to the response to calcitrol therapy, *Engl. J. Med.*, 307:709–716 (1982).
111. Maloney, N.A., S.M. Ott, A.C. Alfrey, N.L. Miller, J.W. Coburn, and D.J. Sherrard, Histological quantitation of aluminium in iliac bone from patients with renal failure, *J. Lab. Clin. Med.*, 99:206–216 (1982).
112. Talwar, H.S., A.H. Reddi, J. Menczel, W.C. Thomas, and J.L. Meyer, Influence of aluminium on mineralization during matrix-induced bone development, *Kidney Int.*, 29:1038–1042 (1986).
113. Goodman, W., Short-term aluminum administration in the rat: Reductions in bone formation without osteomalacia, *J. Lab. Clin. Med.*, 103:749–757 (1984).
114. Goodman, W., D.A. Henry, R. Horst, R.K. Nudelman, A.C. Alfrey, and J.W. Coburn, Parenteral aluminium administration in the dog: Induction of osteomalacia and effect on vitamin D metabolism, *Kidney Int.*, 25:370–375 (1984).
115. Marel, G.M., M.J. McKenna, and B. Frame, Osteomalacia. In *Bone and Mineral Research*, W. Peck (Ed.), p. 335, Elsevier Science Publishers, Amsterdam (1986).
116. Frayssinet, P., D. Hardy, N. Rouquet, B. Giammara, A. Guilhelm, and J. Hanker, New observations on middle term hydroxyapatite-coated titanium alloy hip prostheses, *Biomaterials*, 13(10):668–674 (1992).
117. Serekjan, P., In *Proceedings of Confronto internazionale di esperienze e conoscienze sull' idrossiapatite*, p. 50–51, Rome (1992).
118. Osborn, J.F., and H. Newesely, Dynamic aspects of the implant-bone-interface. In *Dental Implants*, G. Heimke (Ed.), pp. 111–123, Kad Hanser Verlag, Munich (1980).
119. Kangvonkit, P., J.E. Lemons, and V.J. Matukas, Compressive strength measurement and microstructure studies in hydroxylapatite cones, *J. Prosthet. Dent.*, 54:691–696 (1985).
120. Kay, J.F., T.S. Golec, and R.L. Riley, Hydroxylapatite coated subperiosteal dental implants: Design rationale and clinical experience, *J. Prosthet. Dent.*, 58:339–345 (1987).
121. Lemons, J.E., Hydroxyapatite coatings, *Clin. Orthop.*, 235:220–223 (1988).
122. Jarcho, M., Calcium phosphate ceramics as hard tissue prosthetics, *Clin. Orthop.*, 157:259–278 (1981).

123. Oonishi, H., Orthopaedic applications of hydroxyapatite, *Biomaterials*, 12(2):171–178 (1991).
124. De Groot, K., Ceramics of calcium phosphates: Preparation and properties. In *Bioceramics of Calcium Phosphate*, K. de Groot (Ed.), CRC Press, Boca Raton, FL (1982).
125. Jasty, M., H.E. Rubash, G.D. Paiement, C.R. Bragdon, J. Parr, and W.H. Hards, Porous-coated uncemented components in experimental total hip arthroplasty in dogs. Effect of plasma-sprayed calcium phosphate coatings on bone ingrowth, *Clin. Orthop.*, 280:300–309 (1992).
126. Maistrelli, G.L., N. Mahomed, V. Fornasier, L. Antonelli, Y. Li, and A. Binnington, Functional osseointegration of hydroxyapatite-coated implants in a weight-bearing canine model, *J. Arthroplasty*, 8(5):549–554 (1993).
127. Wicki, O., S. Mikic, and P. Gerber, Hydroxylapatite-coated total prosthesis of the hip, *Helv. Chir. Acta*, 57(1):107–115 (1990).
128. Dorr, L.D., and C. Smith, Clinical results from the calcitite-coated press-fit APR hip system, *Dent. Clin. North Am.*, 36(1):239–246 (1992).
129. Soballe, K., K. Gotfredsen, H. Brockstedt-Rasmussen, P.T. Nielsen, and K. Rechnagel, Histologic analysis of a retrieved hydroxyapatite-coated femoral prosthesis, *Clin. Orthop.*, 272:255–258 (1991).
130. Bauer, T.W., R.C. Geesink, R. Zimmermann, and J.T. McMahon, Hydroxyapatite-coated femoral stems. Histological analysis of components retrieved at autopsy, *J. Bone Joint Surg. Am.*, 73(10):1439–1452 (1991).
131. Santavirta, S., D. Nordstrom, P. Ylinen, Y.T. Konttinen, T. Silvennoinen, and P. Rokkanen, Biocompatibility of hydroxyapatite-coated hip prostheses, *Arch. Orthop. Trauma Surg.*, 110(6):288–292 (1991).
132. Bloebaum, R.D., and J.A. Dupont, Osteolysis from a press-fit hydroxyapatite-coated implant. A case study, *J. Arthroplasty*, 8(2):195–202 (1993).
133. Collier, J.P., V.A. Surprenant, M.B. Mayor, M. Wrona, R.E. Jensen, and H.P. Surprenant, Loss of hydroxyapatite coating on retrieved, total hip components, *J. Arthroplasty*, 8(4):389–393 (1993).
134. Toni, A., A. Pizzoferrato, A. Venturini, A. Sudanese, N. Baldini, G. Ciapetti, and S. Stea, Experimental bone ingrowth study of 3-D porous ceramic coating (Poral®) for cementless hip prosthesis. In *Biomaterials and Clinical Applications*, A. Pizzoferrato, P.G. Marchetti, A. Ravaglioli, and A.J.C. Lee (Eds.), pp. 57–62, Elsevier, Amsterdam (1987).

46
Biocompatibility of a New Apatite (HA-SAL1) as a Bone Graft Substitute

Hannah Ben-Bassat and Benjamin Y. Klein
Hadassah University Hospital, Jerusalem, Israel
Isaac Leichter
Jerusalem College of Technology, Jerusalem, Israel

I. INTRODUCTION

Bone is distinguished from other tissues by its hardness and biomechanical strength [1]. It is caused by the deposition of calcium phosphate within the extracellular organic matrix [2–4]. Bone tissue serves not only as a supportive framework of the body, but enters into metabolic interactions with it. For mature bone, 65% of its dry, fat-free weight constitutes its mineral phase [2]. The composition of bone changes with growth and the mineralization process [1–5]. However, the chemical constitution of mature bone in a given species is surprisingly constant [1].

The basic calcium phosphate hydroxyapatite (HA), $Ca_{10}(PO_4)_6(OH)_2$, is the prototype of mature bone mineral. Bone mineral is a microcrystalline, nonstoichiometric, imperfect hydroxyapatite, with high specific and reactive surfaces [1]. This is the basis for the rationale to use HA as a bone graft substitute.

Hydroxyapatite ceramics are used as surgical implant biomaterials as carriers for antibiotics, osteogenic substances, or other drugs and more recently for coating of orthopedic and dental implants [6–9].

The need for bone graft substitutes has become a major orthopedic issue. The composition, microstructure, and molecular surface chemistry of various types of bioceramics have to be tailored to match the specific biological and metabolic requirements of tissues and disease sites [6,7]. Still, not many biomaterials are commercially available or have been approved by the Food and Drug Administration (FDA) as bone graft substitutes [6–8].

This chapter presents baseline studies on structural and biocompatibility aspects of a new hydroxyapatite ceramic HA-SAL1.

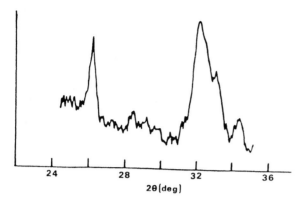

Figure 1   X-ray diffraction of the solid HA-SAL1 precipitated from initially clear solution under microwave irradiation.

II.   HA-SAL1: A NEW HYDROXYAPATITE CERAMIC

A.   Preparation and Characterization of HA-SAL1 Test Disks and Cylinders

A pure hydroxyapatite ceramic (HA-SAL1) was prepared by a new and rapid method developed by Lerner, Sarig, and Azoury: precipitation from aqueous medium by microwave irradiation [10]. Briefly, calcium phosphate precipitation was induced by microwave irradiation for 5 minutes (min). The solid phase was separated by filtration and subjected to elementary x-ray and infrared analysis. The Ca/P ratio was found to be 1.55. The x-ray diffractogram (Fig. 1) and the infrared spectrum (Fig. 2) of the precipitate indicate good reflection from the 202 crystal plane, well-shaped reflection from the 202 plane, and visible shoulders of reflections from the 300 and 112 planes on the pronounced reflection peak of the 211 plane. The infrared spectrum of the precipitate clearly showed the triplet of the (PO_4^{-3}) group peaks at 632, 601, and 565 cm^{-1}. The small peak of 1635 cm^{-1} represents the HPO_4^{-2} group, which indicates that this apatite is not a stoichiometric HA. It contains a dibasic phosphate "impurity" as in natural bone mineral [1]. The broad

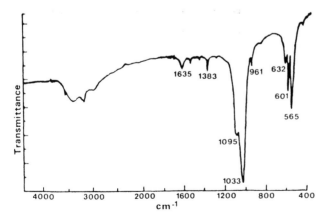

Figure 2   Fourier transform infrared transmittance spectra of the solid HA-SAL1.

peaks (seen in the x-ray diffractogram in Fig. 1) are also consistent with nonstoichiometric HA crystals. All these criteria indicate that the precipitate was composed from hydroxyapatite of satisfactory quality.

Two types of HA-SAL1 disks were prepared and used in the experiments: (1) densely sintered with micropores of approximately 1 micrometer (μm) diameter and (2) porously sintered with coarse pores of 50–100 μm diameter. Their main characteristics are summarized in Table 1. The disks were used either as manufactured (8–10 millimeters [mm] diameter) or after dividing them into 4 roughly equal pieces. For the experiments, the ceramics were washed extensively in distilled water and sterilized by autoclaving at 130°C for 60 min.

HA-SAL1 cylindrical implants of 3 mm diameter and 5 mm length were also shaped out of the HA-SAL1 nonporous ceramic for *in vivo* evaluation in a surgical rat tibia model (see Section IV). They were washed extensively in distilled water and autoclaved.

B.   Scanning Electron Microscopy Appearance of the Apatite Surfaces

Figure 3 shows a scanning electron microscopy (SEM) photomicrograph of HA-SAL1 surfaces. The densely sintered apatite appearance was smooth with a platelike morphology (Fig. 3a). At higher magnification (Fig. 3b), the surface appeared granular, showing fusion of ceramic grains with no apparent microporosity. The porous HA-SAL1 had a pitted surface with micropores of irregular shape and a diameter of 50–100 μm (Fig. 3c).

C.   The Advantages of HA-SAL1

The main advantages of the HA-SAL1 method of preparation are (1) preparation of the powder at room temperature, atmospheric pressure, and short time, which, in comparison with other methods, reduces the cost of production and (2) the possibility to control the degree of crystallinity and the nature and amount of admixtures.

The definition of degrees of hydroxyapatite crystallinity differs markedly when applied by crystallographers as opposed to the definition used by biologists, especially with reference to bone and dental deposits [1,2]. The exact stoichiometry of bone apatite depends upon age, isomorphous substitution, and other factors. Bone mineral is a microcrystalline imperfect hydroxyapatite [1–6], whereas the ceramics available for implantation are usually produced from pure, well-crystallized apatite powders considerably different from bone mineral [6,7,11].

Previous studies showed that apatite of a desired degree of crystallinity with incorporation of accompanying compounds can be produced quickly using simple and inexpen-

Table 1   The Hydroxyapatite HA-SAL1

Material	Porosity	HA%	Composition	
			TCP%	Ca/P ratio
HA-SAL1	Dense	100	–	1.55
HA-SAL-P[a]	Porosity, 40–50% Pore size: 50–100 μm	100	–	1.55

[a]Porous

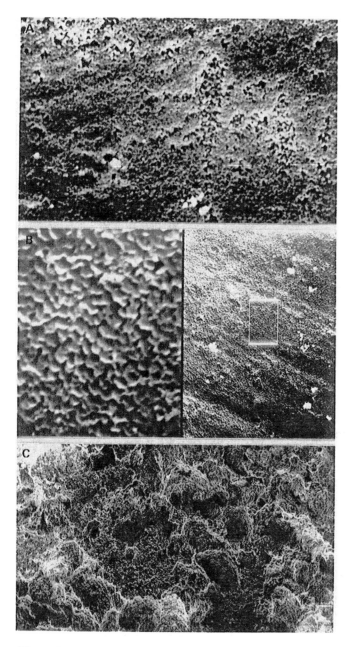

Figure 3   Electron microscopic appearance of HA-SAL1: (A, B) nonporous; (C) porous. Original magnification of (A) was ×250, (B) was ×120, and (C) was ×505.

sive procedures [10]. Microwave energy adsorption during precipitation of calcium phosphate species can change the degree of hydration of calcium in the formed crystals, thus modifying their crystallographic structure [10]. Hydroxyapatite crystals have lattice defects that cause release of Ca^{++} ions in aqueous media [1]. Ca^{++}, in particular, and other bivalent cations such as Mn^{++} and Mg^{++} enhance cell adhesion [12-14], whereas lack of these cations prevents cell spreading, but not cell attachment [15]. Fibronectin has specific binding sites for Ca^{++}, thus increasing adhesion [16].

III. BIOCOMPATIBILITY *IN VITRO* OF HA-SAL1

Hydroxyapatite ceramics seem particularly suitable as bone graft substitutes because their chemical composition is similar to the mineral nature of bone [6-8]. Bone substitute biomaterials may be classified as bonding or nonbonding. "Bioactive" substrata have been shown to establish a direct biological bond with the surface of the implant [11,17,18]. The concept of bioactive materials is intermediate to resorbable and bioinert. The interfacial bond that is formed by these bioactive materials and adjacent tissue is time dependent and its strength, mechanism, and thickness of the bonding zone differ for the various materials [11,12,14,18]. Although the mechanisms of these bonding phenomena are not yet fully understood, they imply cellular, molecular, and ionic activity between tissue and surface [6,12,14].

An important observation in this connection is that many cells fail to proliferate, grow, and differentiate if they cannot adhere to the substrate [19,20]. This anchorage dependence seems to be a prerequisite for bone formation *in vitro* and *in vivo* [6,21-25]. Anchorage, attachment, adhesion, and spreading of cells as well as formation and deposition of bone directly on the implant require a surface that is not only nontoxic, but also allows and favors this cell behavior [12,21,25]. The production of extracellular matrix by the cells and the formation of the interface between cells and material through collagen bonding are also crucial for the development of the bone bond [3,26]. Thus, a satisfactory experimental application of an implant will be largely due to its speed of incorporation into host tissue and its vascularization, followed by ingrowth of new bone [6].

Biocompatibility has been defined as "the ability of a material to perform with an appropriate host response in a specific application'" [27]. Host response is a principal issue in the complex set of phenomena that comprise biocompatibility. The responses of individual cell types are difficult to assess in animal models because of the numerous and complex events that occur at the grafted wound site. *In vitro* models are also useful in this respect. However, in developing *in vitro* models for screening materials, it is recommended to use available human cells and cell lines relevant to the material's application [28,30].

The biocompatibility *in vitro* of HA-SAL1 was evaluated with cells relevant to osteogenesis: human bone cells, cells from a permanent line with osteoblastic properties (Saos2), and vascular endothelial cells participating in angiogenesis. The interaction of these cells with the ceramic was studied using scanning electron microscopy (SEM) and growth assays [29].

A. Morphological Appearance and Growth of Bone and Endothelial Cells on HA-SAL1

1. *Nonporous Disks*

Human-bone-derived cells were seeded on HA-SAL1 disks and examined by SEM after 1, 3, 14, and 30 days of culture. Ultrastructural observation indicated that the cells had anchored, attached, and spread onto the disks already 21 hours after seeding (Figs. 4a, 4b). Most of the cells were well spread, displaying a flat configuration, while cells that were not yet well spread had a rough surface on which blebs could be seen. Still, those cells not yet spread were attached and in contact with neighboring cells (Fig. 4b). Within 3 days after seeding, the cells had colonized large areas of the substrate surface (Fig. 4c). They maintained physical contact with one another through multiple extensions (Fig. 4c). With continued culture, the cells formed a multilayer sheet. They overlapped and superimposed, making it difficult to distinguish the borders of individual cells (Fig. 4d). Cells seeded at a higher density took less time to cover the apatite surface, although there were variations from one culture batch to another. No signs of cytotoxicity or cell degenerations were evident throughout the experiment. The cells retained their typical broad and flattened morphology and exhibited good culture organization.

Growth rate experiments of the human-bone-derived cells on the HA-SAL1 ceramic indicated that the cells multiplied and reached a density of 1500 cells/mm^2 (Fig. 5). Similar experiments with Saos2 cells indicated that these cells multiplied and reached higher cell density (10^4 cells/mm^2), as expected of an osteosarcoma permanent line (Fig. 5).

Concurrent with continuous cultivation of the human-bone-derived cells, formation of ECM-like (extracellular matrix) material (Fig. 6a) and extracellular matrix containing fibrous material secreted by the osteoblasts (Fig. 6b) could be observed. It has not been yet identified or distinguished as collagen. We have also observed globular masses on the cell surface of bone cells in culture (Figs. 6c, 6d). It has been suggested that mineralization begins with matrix vesicles and is followed by additional mineral deposition on collagen fibrils [3,24]. To date, identification of the globular masses (which might also

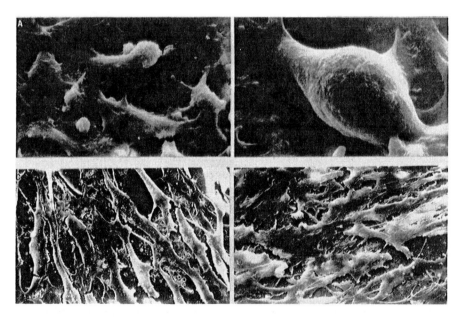

Figure 4 Human-bone-derived cells cultured on HA-SAL1: (A, B), 21 hours after plating; (C), 3 days in culture; (D), 30 days in culture. Original magnification by scanning electron microscopy for (A) was ×1550, (B) was ×3100, (C) was ×710, and (D) was ×600.

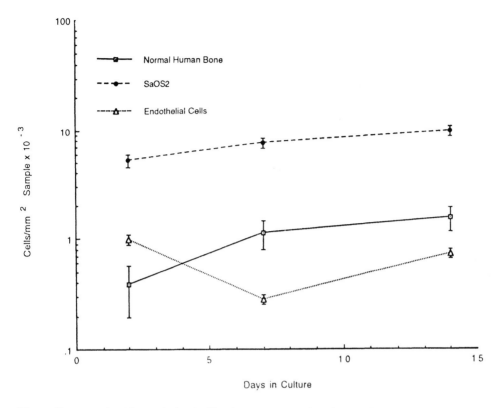

Figure 5 Growth of bone-derived (▣), Saos2 (●), and endothelial (△) cells cultured on HA-SAL1 ceramics (seeding level 4000 cells/mm^2).

be noncollagenous protein mineral globules) [12] using energy dispersive x-ray analysis was unsuccessful since it is too difficult to distinguish the calcium and phosphorus emissions from those of the hydroxyapatite material [17].

The ceramic surfaces of HA-SAL1 did not show microscopic degradation or altered surface morphology resulting from interaction with the cultured cells and/or incubation in protein-containing medium for the duration of the experiments (Fig. 4d).

Light microscopic and SEM examination of the periphery of the specimens and the outgrowth of the cells on the culture dishes showed no signs of toxicity within the test period of 30 days. The cells were well spread, looked "healthy," maintained contact with neighboring cells, and formed a well-organized monolayer (Fig. 7). Mineralized bonelike nodules were detected by alizarin red staining in the cultures growing at the periphery of the apatite disk (generally detectable after 10 days of culturing) (Fig. 8).

The attachment and growth of bovine vascular endothelial cells maintained on HA-SAL1 were also evaluated. SEM experiments showed that, within 2 hours after plating, as many as 60–80% of the cells had attached and spread (Figs. 9a, 9b). Within 2 days after seeding, the cells had formed a well-organized confluent monolayer (Fig. 9c). The cells were homogeneous and closely apposed, with cell borders almost nonapparent. With continuous cultivation, for the duration of the experiment (Fig. 9d).

Growth curve experiments indicated that the vascular endothelial cells had attached

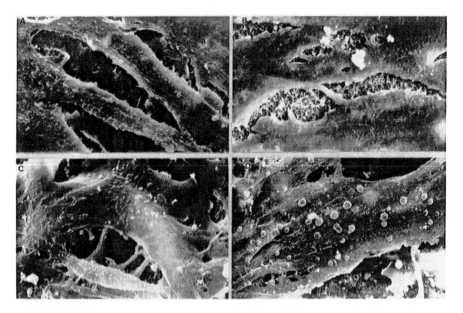

Figure 6   Growth of human-bone-derived cells on HA-SAL1: (A), formation of ECM-like material; (B), secreted fibrous masses; (C, D), secreted globular masses (might be matrix vesicles). Original magnification for (IR) was × 2500, (B) was × 1500, (C) was × 2000, and (D) was × 1000.

on the HA-SAL1 disks and the cell numbers per material sample had increased (Fig. 5), after an initial decrease, probably due to cell overcrowding.

2. Porous Disks

The concept behind porous bioceramics is the ingrowth of tissues into pores, on the surface or throughout the implant, resulting in an increased interfacial area [6,31,32]. The interface is established by the living tissues in the pore, resulting in increased attach-

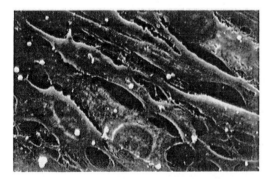

Figure 7   Scanning electron microscopy (SEM) of human-bone-derived cells growing at the periphery of HA-SAL1 specimen looking "healthy" and forming a well organized culture (original magnification × 1200).

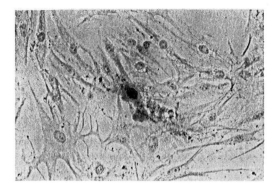

Figure 8   Mineralized bonelike nodules detected by alizarin red staining in the cultures growing at the periphery of the HA-SAL1 disk (original magnification ×400).

ment, termed *biological fixation*. A pore size of 100–150 µm is considered to be necessary for bone ingrowth. Vascular tissue does not seem to appear in pores with a diameter of less than 100 µm. However, there are reports of bone formed within pores not larger than 50–100 µm diameter and connected by holes measuring less than 20 µm at a faster rate than in implants with 200–400 µm pores [6]. One possible explanation might be that smaller pores expose a much larger surface to the invading tissue and thus contribute to mineralization. Bioceramics also contain micropores in the range of several microns

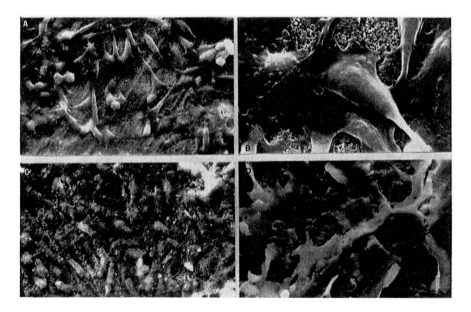

Figure 9   Bovine endothelial cells cultured on HA-SAL1 disks: (A, B), 2 hours after plating; (C), 2 days in culture; (D), 15 days in culture. Original magnification for (A) was ×600, (B) was ×2500, (C) was ×600, and (D) was ×600.

resulting from incomplete fusion of particles during the sintering process [31,32]. It has been postulated that these micropores determine the rate of degradation independent of the chemical composition of the ceramic and are thus ultimately responsible for its biodegradability [6,32,33]. However, it has been shown that hydroxyapatite was not resorbed in spite of the presence of micropores [6].

A preliminary preparation of porous HA-SAL1 was made with a porosity of 40–50% and a pore size of 50–100 μm (Table 1, Fig. 3c). The phenomenological behavior of bone and endothelial cells cultured *in vitro* on this ceramic was evaluated by SEM. The morphology and growth pattern of both cell types was similar to that on the nonporous material, only more elaborate. While the majority of the cells adhered onto the pores surface (Fig. 10a), others extended explorative filopods across the pore, attaching them to the opposite wall (Fig. 10b). With continuous cultivation, cells migrated into the pores, adhered onto the pore surface, and formed a continuous carpet, with the cells covering the densely sintered parts of the material (Figs. 10c, 10d, endothelial and bone cells, respectively). We did not observe sealing of pores with cells in our material, as might be occasionally observed with porous ceramics (Fig. 11).

In experiments of this type, it is especially important to determine the optimal seeding level for individual cell types that allows for adequate cell attachment and multiplication on the ceramic and in the pores. A seeding level that is too high causes saturation and trapping of cells in the pores (Fig. 12a), thus interfering with cell multiplication and resulting in decreased cell density quite soon after plating the cells onto the material. However, depending on the cell type and culture conditions, cell density might increase again. Occasionally, we observed such a growth pattern with bovine endothelial cells (Fig. 5 and Fig. 12a), but not with bone-derived cells, which are less "density" dependent

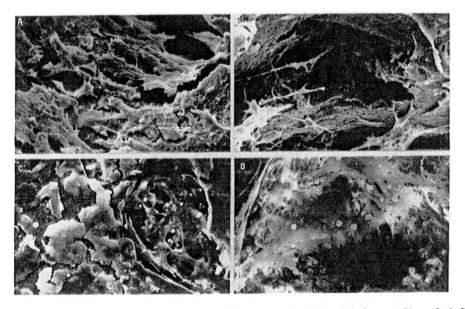

Figure 10   Cells cultured on porous HA-SAL1 ceramic: (A, B, D), bone; (C), endothelial. Cells adhered onto the pore's surface (A, C, D) or extended filopods across the pore (B). Original magnification for (A) was ×600, (B) was ×400, (C) was ×250, (D) was ×300.

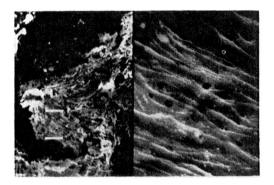

Figure 11 Sealing of pores with cells as seen occasionally with porous ceramics for human-bone-derived cells 14 days in culture (original magnification ×150).

and form multilayers (Fig. 12b), in contrast to the monolayer formed by endothelial cells.

The present experiments demonstrated that cells relevant to osteogenesis, human-bone-derived cells, cells from a line with osteoblastic properties, and endothelial cells attached, spread, proliferated, and formed a well-organized confluent culture on the HA-SAL1 specimens, thus indicating that HA-SAL1 caused no apparent impairment to their cell physiology. The preliminary preparation of porous HA-SAL1 showed similar results with bone-derived and endothelial cells.

The behavior of cells toward substrates is not yet fully understood. It has been suggested that the surface properties of an implant are of central importance for cell attachment and growth [11,12,17,33]. Mechanisms of cells' attachment, adhesion, and spreading have been partly explained by the interaction of physical forces associated with the cell and substrate surfaces [12,13,21,25,34]. It is well known that cells attach and spread better on hydrophilic surfaces [15,21,25,35,36]. Serum proteins can be adsorbed onto the substrate, and cell attachment and spreading is generally better in serum-containing medium [1–5,24,37]. Therefore, mechanical cell-substrate interactions cannot account fully for cell behavior. To this point, an implanted material interacts with pro-

Figure 12 Effect of seeding level: (A), trapping of endothelial cells in pores when plated at too high a seeding level; (B), bone-derived cells were not trapped in the pores, but formed a multilayered culture with many of the cells rounded and not spread. Scanning electron microscopy (SEM) original magnification for (A) was ×75 and for (B) was ×100.

teins in the serum and extracellular fluids before interacting with the cells of the implanted bed [8,21,34]. The role of the immune system within the host response phenomena is also an important issue [27].

Cell growth and differentiation is generally anchorage dependent [20]. Our results showed that human-bone-derived cells were anchored, attached, and spread onto HA-SAL1 within 21 hours after plating and by the third day had colonized and covered large areas of the substrate surface. Endothelial cells had attached and spread within 2 hours after plating on HA-SAL1 and by the second day had formed an almost confluent monolayer [29]. In dynamic conditions, incomplete cell covering of surfaces may be mostly due to lack of rapid and strong adhesion [38]. On HA-SAL1, the attachment and spreading of vascular endothelial cells was remarkably rapid. Since the rate of vascularization and the speed of incorporation of an implant into host tissue determine largely its successful grafting [6,38,39], it is hoped these results will be of benefit to the performance of HA-SAL1 in experimental grafting models.

A high degree of cell adhesion and spreading is regarded as promoting the bonding of an implant with host tissues. During these processes, cells secrete adhesion substances such as fibronectin that are found circulating in plasma or bound to cells and cell matrix [40,41]. Grinell and Marshall have attributed favorable adhesion of cells on hydrophilic surfaces to their ability to bind fibronectin in a more active configuration [42]. Coating substrates with fibronectin increases cells adhesion and growth [42]. It is possible that implanted biomaterials capable of binding adhesion molecules could undergo better bonding with host tissue [8,21,24]. It will be of interest to examine directly the ability of HA-SAL1 to bind adhesion and other proteins, such as transforming growth factor-beta (TGF-β), which has been shown to be growth promoting to osteoblasts [43], to increase fibronectin synthesis [44], and to participate in bone repair and remodeling [45].

IV. FOLLOW-UP OF MINERALIZATION IN SMALL HA-SAL1 IMPLANTS *IN VIVO*

A. Dual Energy X-Ray Absorptiometry: Noninvasive Measurement of Mineral Content at the Site of Implantation

To date, surgeons note no available means of assessing bone ingrowth into porous mineral and find it difficult to compare ingrowth rates into calcium phosphate ceramics and other bone substitutes [45]. Conventional imaging techniques do not determine these properties, yet specialized noninvasive methods are routinely used to assess bone density and mineral content in osteoporosis. Worldwide, this is the most common skeletal disorder; it is characterized by a decrease in bone mass per unit volume (density).

Accurate and quantitative monitoring of minute changes in mineral content at the site of implantation gauges the degree of a bioceramic implant's integration and enhances markedly its clinical evaluation. The mineral content, which can be measured noninvasively, refers to the minerals in both the bioceramic and the host bone. Sequential measurements of bone mineral content at the site of implantation reflect the current rate of remodeling.

In a recent report, we proposed two suitable techniques for such noninvasive sequential analysis of the ceramic implant and host bone: DEXA (dual energy x-ray absorptiometry) and the Compton densitometry [47]. We have used DEXA for measuring the mineral content and assessing bone ingrowth into commercially available bioceramics

[48]. This method relies on the attenuation of a collimated x-ray photon beam as it passes through the examined bone. The degree of beam attentuation is related to the mineral content in bone cross section. The photons so transmitted are detected by a collimated scintillation counter moved back and forth on the other side of the region under investigation.

The device uses an X-ray tube with two filters to provide two photon beams of discrete energies, distinguishable on the basis of their attenuation by bone and soft tissue. One energy serves as a reference for soft-tissue thickness, and the attenuation of the other reflects bone mineral content [49]. Measurements are thus enabled in regions heavily surrounded by soft tissues, such as the hip or spine. Findings are expressed as mass of bone mineral per unit area of the bone scanned (grams per square centimeter, g/cm^2). The scanning of several cross sections at different levels and averaging of results reduces positioning errors of sequential examinations. The degree of mineralization at each measuring point is displayed either in varying shades of gray or colors. Thus, a mineral image of the scan site is depicted (Figs. 13 and 14). The implant can be identified on the image and its mineral content calculated independently of surrounding tissue.

In the present studies, the XR-26 model (Norland Instruments, Inc., Weesp, The Netherlands) was used. This model contains software that enables calculation of mineral content at any desired region within the image. The dimensions of such a region can be predetermined with an accuracy of 0.5 mm. Such resolution is essential for the accurate measurement of changes in the implant's mineral content.

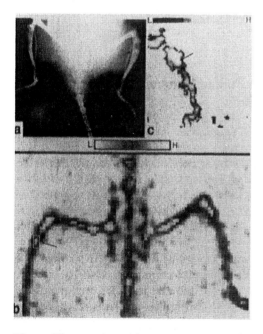

Figure 13 Imaging of hydroxyapatite (HA) implants in rat tibias: (a), radiograph (x-ray) of an implant in the right tibia (arrow); (b), a general view of dual energy x-ray absorptiometry (DEXA) imaging of an HA implant in a rat tibia (arrow), with the intensity scale seen above indicating low (L) and high (H) mineral content; intensity; (c), DEXA imaging of HA implant in rat tibia (arrow), 0.5 mm × 0.5 mm resolution.

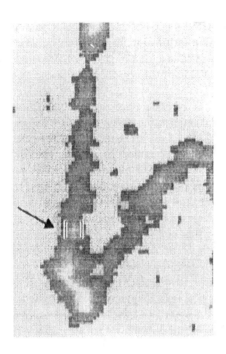

Figure 14   Definition of 2-mm cylindrical hydroxyapatite implant on dual X-ray photon absorptiometry mineral image of proximal rat tibia.

B. Development of a Surgical Model to Follow Mineral Content Changes in Small Ceramic Implants in Rat Tibias

According to sporadic clinical reports, HA ceramics have been used for filling bone defects in tibial plateau fractures and in bone tumor surgery [6–8,46]. However, it was difficult to follow healing progression in these patients. Previously, we demonstrated the use of DEXA *in vivo* for detecting small changes of mineral content (MC) within a ceramic implant in a surgical model we had developed [48]. HA cylinders were implanted in rat proximal tibias and followed for 13 weeks. The precision of the *in vivo* DEXA measurements at the implant site in rats tibia equals 1%, although the ceramic implant presents a high mineral background.

To determine whether the MC changes obtained *in vivo* by the DEXA technique were comparable with changes obtained by invasive techniques commonly used to evaluate bone healing, alkaline phosphatase (ALP) activity within the implants was measured during callus formation. The increase in ALP activity within the implants was found to precede the increase in mineral content as measured noninvasively by the DEXA technique. This was consistent with the timing of ALP activity with respect to mineralization as it occurs during fracture healing. The results showed that DEXA imaging is useful in measuring bone ingrowth in small ceramic HA implants *in vivo* despite the high mineral content background of the implant scaffold [48].

C. Mineralization of HA-SAL1 Cylinders Implanted in Rat Tibias

HA-SAL1 cylindrical implants of 3 mm diameter and 5 mm length were shaped out of HA-SAL1 nonporous ceramic, washed extensively with distilled water, and autoclaved. The surgical model described above was used with modifications [48]. Sabra female rats weighing 250 g were anesthetized and holes of 3 mm diameter were drilled in the right tibial metaphysis, 3 mm below the collateral ligament's insertion. HA-SAL1 cylinders were put in these holes; the contralateral empty holes, as control defects (Fig. 13-*A*), were omitted since they were found to be inappropriate controls [48].

Bone defects implanted with HA-SAL1 cylinders were examined for MC at different time intervals during 8 postimplantation weeks using the DEXA device. Examples of implanted rat tibial DEXA images are shown in Figs. 13b and 13c. In the HA-SAL1 implantation, the region of interest for MC analysis consisted of a 3 mm × 3 mm frame around the implant site. The Day 1 baseline image is shown in Fig. 14. The baseline MC for each implanted defect was determined on Day 1 postimplantation and designated "100%." MC was measured on Days 8, 15, 22, 32, and 53 postimplantation. The MC change (MCC) for each implant consists of the ratio between the measured MC at each

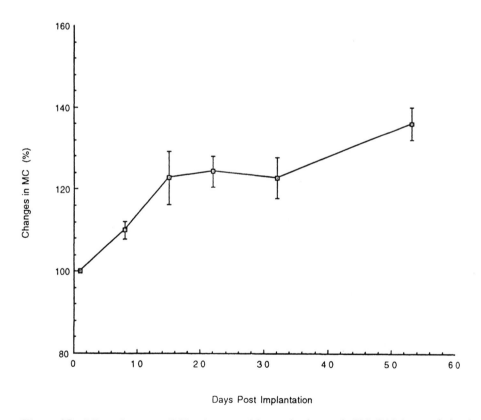

Days Post Implantation

Figure 15 Mineral content (MC) changes with standard error in HA-SAL1 ceramic implants as a function of time after implantation. Ceramic implants were inserted in the right proximal tibias of rats. MC was measured in a region of interest of 3 × 3 mm around the implants. Increment in MC is expressed as the percentage change from the initial postoperative MC values.

time interval and its own Day 1 MC value. In Fig. 15, the mean MCC ± SE (standard error) is expressed in percent deviation from the 100% baseline MC. On Day 53, a maximal mean increase of 36.5% in the MC of the implants was observed. This resulted from apposition of healing fractured-bone callus on the surface of the nonporous HA-SAL1, which showed progressive mineralization. A higher MCC was obtained with porous implants, in which mineralization is extended into the ingrowing bone callus deep into the implants (B. Klein, unpublished observations, 1993). The substantial mineralization onto the HA-SAL1 surface, despite the lack of macropores, is indicative of its biocompatibility *in vivo*.

Previous and present results demonstrate that the noninvasive DEXA technique for measuring bone mass is useful for the follow-up of mineralization status in and around small bioceramic implants [47,48]. The high resolution of the mineralization image obtained with the DEXA device enables the evaluation of even such small implants. For economic and ethical reasons, screening experiments should be performed in small animals, for which repeated measurements are possible without sacrifice of animals at each step during the follow-up. The MC changes determined by the noninvasive DEXA imaging reflect the mineralization process occurring during callus formation in ceramics implanted in bone defects.

The DEXA device could also be used to follow the healing process of bone defects in patients with ceramic implants. The mineral background contributed by the implant itself did not obscure the small changes in MC values caused by callus formation. The curve obtained by sequential DEXA measurements demonstrated a steady increase in MC during the first 8 weeks after implantation. This reflects the massive mineralization of the soft callus in the healing bone.

The ceramic implants used in this study contained 100% HA in the implant composite, with a Ca/P ratio of 1.55. However, the DEXA device cannot discriminate between implant scaffold mineral and newly precipitated mineral. For the duration of the experiments no decrease in MC, which might indicate biodegradation, was observed.

V. SUMMARY

A hydroxyapatite ceramic (HA-SAL1) of desired crystallinity was prepared by Lerner, Sarig, and Azoury from apatite powder using a rapid and inexpensive procedure by precipitation from aqueous medium by microwave irradiation for 5 min [10]. The biocompatibility *in vitro* of two types of HA-SAL1 disks was evaluated: (1) densely sintered with micropores of approximately 1 μm diameter and (2) porously sintered with coarse pores of 50–100 μm diameter. Cells relevant to osteogenesis (human bone cells, cells from a permanent line with osteoblastic properties [Saos2], and endothelial cells participating in angiogenesis) were used in the experiments. The interaction of these cells with the ceramic was studied using scanning electron microscopy and growth assays. On the HA-SAL1 surfaces, the cells attached, spread, and multiplied with no apparent signs of toxicity.

Thereafter, cylindrical implants were shaped out of the HA-SAL1 nonporous ceramic for *in vivo* evaluation in a surgical rat tibia model. Dual energy x-ray absorptiometry (DEXA) imaging was used to follow and quantify peri-implant mineralization. The mineral content change (MCC) for each implant was expressed as an index describing the ratio between the measured MC value and its own baseline value. A maximal mean increase of 36.5% in the implant MC was observed on Day 53. The results demonstrate bone regeneration on the HA-SAL1 ceramic implants *in vivo* and further demonstrate the applicability of DEXA to follow-up mineralization changes in small ceramic implants *in vivo*. This method permits detection of small changes in MC despite the high mineral content background of the implant scaffold.

We presented baseline studies on structural and biocompatibility aspects of a new hydroxyapatite ceramic HA-SAL1. Further studies are in progress to produce porous HA-SAL1 with microscopic architecture approaching that of human cancellous bone and evaluate its performance as a bone substitute in experimental grafting models.

ACKNOWLEDGMENTS

We gratefully acknowledge the support of the Ministry of Industry and Commerce — Chief Scientist's Office, Contract 1944; the Public Committee for the Designation of Estate Funds, Chief Scientist's Office, Ministry of Health; Hadassah Medical Research Services and Development Company Limited; and the Joint Research Fund of the Hebrew University and Hadassah. The excellent assistance of Mr. I. Naveh with the medical photography is gratefully acknowledged.

REFERENCES

1. Posner, A.S., Bone mineral and the mineralization process. *Bone and Mineral Res.*, 5:65–116 (1987).
2. Posner, A.S., The mineral of bone. *Clinical Orthopaedics*, 200:87–99 (1985).
3. Anderson, H.C., Matrix vesicle calcification. Review and update. *Bone and Mineral Res.*, 3: 109–149 (1985).
4. Boskey, A.L., Noncollagenous matrix proteins and their role in mineralization. *Bone and Mineral Res.*, 6:111–123 (1989).
5. Menczel, J., A.S. Posner, and R.A. Harper, Age changes in the crystallinity of rat bone apatite. *Israel J. Med. Sci.*, 1:251–252 (1965).
6. Hench, L.L., Bioceramics: From concept to clinic. *J. Am. Ceram. Soc.*, 74:1487–1510 (1991).
7. De Groot, K., Bioceramics of calcium phosphate: Preparation and properties. In *Bioceramics of Calcium Phosphate*. CRC Press, Boca Raton, Florida, pp. 99–114 (1983).
8. Bucholz, R.W., A. Carlton, and R.E. Holmes, Hydroxyapatite and tricalcium phosphate as bone graft substitutes. *Orthoped. Clinics of North Amer.*, 18:323–334 (1987).
9. Ji, H., and P.M. Marquis, Effect of heat treatment on the microstructure of plasma-sprayed hydroxyapatite coating. *Biomaterials*, 14:64–68 (1993).
10. Lerner, E., S. Sarig, and R. Azoury, Enhanced maturation of hydroxyapatite from aqueous solutions using microwave irradiation. *J. Miner. Sci.*, 2:44–47 (1991).
11. Hench, L.L., and E.C. Ethridge, Biomaterials: An intrafacial approach. *Biophysics and Bioengineering Series*, 4:279–288 (1982).
12. Bruijn, J.D., C.P.A.T. Klein, K. de Groot, and C.A. Van Blitterswijk, The ultrastructure of the bone-hydroxyapatite interface *in-vitro. J. Biomed. Mat. Res.*, 26:1365–1382 (1992).
13. Curtis, A.S.G., Cell adhesion. *Prog. Biophys. Mol. Biol.*, 28:317–386 (1973).
14. Van Blitterswijk, C.A., J.J. Grote, W. Kuypers, C.J.G. Blok-Van Hock, and W.Th. Daems, Bioreactions at the tissue-hydroxyapatite interface. *Biomaterials*, 6:243–251 (1985).
15. Grinell, F., The serum dependence of baby hamster kidney attachment to a substratum. *Exp. Cell Res.*, 97:265–274 (1976).
16. Amphlett, G.W., and M.E. Hrinda, The binding of calcium to human fibronectin. *Biochem. Biophys. Res. Common.*, 111:1045–1053 (1983).
17. Davies, J.E., and T. Matsuda, Extracellular matrix production by osteoblasts on bioactive substrata *in vitro. Scanning Microscopy*, 2:1445–1452 (1988).

18. Ito, G., T. Matsuda, N. Inoue, and T. Kamegan, Biologic comparison of the tissue interface to bioglass. *J. Biomed. Mat. Res.*, 21:485–497 (1987).

19. Folkman, J., and A. Mascona, Role of cell shape in growth control. *Nature*, 273:345–349 (1978).

20. Stoker, M., O. O'Niel, S. Berryman, and V. Waxman, Anchorage and growth regulation in normal and virus transformed cells. *Int. J. Cancer*, 3:683–693 (1968).

21. Bagambisa, F.B., and U. Joos, Preliminary studies on the phenomenological behavior of osteoblasts cultured on hydroxyapatite ceramics. *Biomaterials*, 11:50–56 (1990).

22. Davies, J.E., B. Causton, Y. Bovell, K. Davy, and C.S. Sturt, The migration of osteoblasts over substrata of discrete surface charge. *Biomaterials*, 7:231–233 (1986).

23. Shimazaki, K., and V. Mooney, Comparative study of porous hydroxyapatite and tricalcium phosphate as bone substitute. *J. Orthop. Res.*, 3:301–310 (1985).

24. Weiss, R.E., and A.H. Reddi, Synthesis and localization of fibronectin during collagenous matrix mesenchymal cell interaction and differentiation of cartilage and bone *in vitro*. *Proc. Natl. Acad. Sci. USA*, 77:2074–2078 (1980).

25. Malick, M.A., D.A. Puleo, R. Bizios, and R.H. Doremus, Osteoblasts on hydroxyapatite, alumina and bone surface *in vitro*: Morphology during the first 2 hours of attachment. *Biomaterials*, 13:123–128 (1992).

26. Boskey, A.L., Current concepts of physiology and biochemistry of calcification. *Clin. Orthop. Res.*, 157:225–257 (1981).

27. Remes, A., and D.F. Williams, Immune response in biocompatibility. *Biomaterials*, 13:731–743 (1992).

28. Sgouras, D., and S. Duncan, Methods for the evaluation of biocompatibility of soluble synthetic polymers which have potential biomedical use: 1. Use of the MTT colorimetric assay as a preliminary screen for evaluation of in-vitro cytotoxicity. *J. Materials Science: Materials in Medicine*, 1:61–68 (1990).

29. Ben-Bassat, H., Y.B. Klein, E. Lerner, R. Azoury, E. Rahamim, Z. Shlomai, and S. Sarig, *Cells and Materials*, 4:37–50 (1994).

30. Ali, S.Y., Calcification of cartilage. In *Cartilage: Structure, Function and Biochemistry*, Vol. 1 (Ed. B.K. Hall), Academic Press, New York, pp. 343–378 (1983).

31. de Groot, K., Effect of porosity and physiochemical properties on stability, resorption and strength of calcium phosphate ceramics. In *Bioceramics: Material Characteristics Versus In-Vivo Behavior*, Vol. 523 (Eds. P. Ducheyne and J. Lemons), Ann. N.Y. Acad. Sci., New York (1988).

32. de Groot, K., and R. Le Geros, Significance of porosity and physical chemistry of calcium phosphate ceramics. In *Bioceramics Material Characteristics Versus In Vivo Behavior*, Vol 523 (Eds. P. Ducheyne and J. Lemons), Ann. N.Y. Acad. Sci., New York (1988).

33. Beckman, C.A., T.K. Greenlee, and A.R. Crebo, Bone formation at a ceramic implant interface. *Cal. Tissue Res.*, 18:165–171 (1971).

34. Kitsugi, T. Yamanuro, T. Nakamura, S. Kotani, T. Kokubo, and H. Takenchi, Four calcium phosphate ceramics as bone substitutes for non-weight-bearing. *Biomaterials*, 14:216–224 (1993).

35. De Bono, D.P., M. Pittilo, S. Pringle, and C. Green, Endothelial smooth muscle interactions *in vitro*: Effects of high pH, flow medium and extracellular matrix. *Br. J. Exp. Path.*, 69:209–220 (1988).

36. Sautier, J.M., J.R. Nefussi, and N. Forest, Surface reactive biomaterials in osteoblast culture; An ultrastructural study. *Biomaterials*, 13:400–402 (1993).

37. Tailor, A.C., Attachment and spreading of cells in culture. *Exp. Cell Res. Suppl.*, 8:154–173 (1961).

38. Sirois, E., and C.J. Doillon, Morphology of cells cultivated on collagen-coated vascular prosthesis submitted *in vitro* to a flow under pressure. *Cells and Materials*, 1:291–299 (1991).

39. Grenza, T.E., J.E. Rins, and T.W. Bauer, The rate of vascularization of corralin hydroxyapatite. *Plast. Reconst. Surg.*, 84:245–249 (1986).

40. Yamada, K.M., and K. Olden, Fibronectins—Adhesive glycoproteins of cells surface and blood. *Nature* (*Lond.*), 275:179–184 (1978).

41. Ivarsson, B.L., R.P. Cambria, J. Megerman, and W.M. Abbott, Fibronectin enhances early shear stress resistance of seeded adult venous endothelial cells. *J. Surg. Res.*, 47:203–207 (1989).

42. Grinell, F., and J.L. Marshall, Coating bacterial dishes with fibronectin permits spreading and growth of human diploid fibroblasts. *Cell. Biol. Int. Rep.*, 6:1013–1018 (1982).

43. Robey, P.G., M.F. Young, K.C. Flanders, N.S. Roche, P. Kondaiah, A.H. Reddi, J.D. Termine, M.B. Sporn, and A.B. Roberts, Osteoblasts synthesize and respond to TGF-B *in vitro*. *J. Cell Biol.*, 105:457–463 (1987).

44. Roberts, A.B., M.B. Sporn, R.K. Assoian, J.M. Smith, N.S. Rosh, L.M. Wakefield, U.L. Heine, L.A. Liotta, V. Falanga, J.H. Kerhl, and A.S. Fanci, TGF-B: Rapid induction of fibrosis and angiogenesis *in vivo* and stimulation of collagen formation *in vitro*. *Proc. Natl. Acad. Sci. USA*, 83:4167–4171 (1986).

45. Pleilschrifter, J., and G.R. Mundy, Modulation of TGF-B activity in bone cultures by osteo-tropic hormones. *Proc. Natl. Acad. Sci. USA*, 84:2024–2028 (1987).

46. Uchida, A., N. Araki, Y. Shinto, H. Yoshikawa, and K. Ono, Use of calcium ceramic hydroxyapatite in bone tumor surgery. *J. Bone Joint Surg.*, 72B:298–302 (1990).

47. Leichter, I., and B. Bloch, Evaluation of the calcium phosphate ceramic implant by non-invasive techniques. *Biomaterials*, 13:478–482 (1992).

48. Mosheiff, R., B.Y. Klein, I. Leichter, G. Chaimsky, A. Nyska, A. Peyser, and D. Segal, Use of dual-energy x-ray absorptiometry (DEXA) to follow mineral content changes in small ceramic implants in rats. *Biomaterials*, 13:462–466 (1992).

49. Wilson, C.R., and M. Madsen, Dichromatic absorptiometry of vertebral bone mineral content. *Invest. Radiol.*, 12:180–184 (1977).

47

Characterization of Ion Beam Sputter Deposited Calcium Phosphate Coatings

Joo L. Ong
University of Texas Health Science Center at San Antonio
San Antonio, Texas

Linda C. Lucas and Charles W. Prince
University of Alabama at Birmingham, Birmingham, Alabama

Ganesh N. Raikar
University of Alabama in Huntsville, Huntsville, Alabama

I. INTRODUCTION

Hydroxyapatite (HA) is found naturally in the inorganic portion of bones and teeth. *In vivo*, HA is predominantly impure calcium phosphates (Ca-P) containing many other constituents, such as sodium, magnesium, carbonate, chloride, and fluoride [1–3]. HA is poorly crystallized, with the presence of amorphous phases in at least 30% of mature bone [4–7]. These amorphous phases have been reported to be more metabolically active than the crystalline phases. Furthermore, the incorporation of constituents such as fluoride in natural apatite render chemical stability to the bone, whereas incorporation of carbonate and sodium ions increases the chemical instability of the bone [8–10].

HA and other Ca-P compounds, such as tricalcium phosphate and brushite, can be prepared synthetically. These ceramic biomaterials are known for their bioactive properties, and have been frequently utilized by the dental and orthopedic community. However, bulk HA and Ca-P are brittle and relatively weak when compared with metals and high strength ceramics like alumina and zirconia. Thus, the best use of HA and Ca-P in load-bearing implant applications is as a coating. The ultimate purpose of these bioactive coatings is to obtain better stabilization of metallic implants in the bone and surrounding tissue.

There are numerous methods for producing HA coatings on metallic implants. Foremost is a particulate deposition process known as plasma spraying or arc plasma spraying [11]. This method is commercially employed to produce coatings with a thickness greater than 30 micrometers (μm) [11]. In the plasma spray process, a gas stream carrying HA powders is passed through an electrical plasma produced by a low voltage, high current electrical discharge [11–12]. The powders are then sprayed onto the substrate, on which they solidify. Advantages of plasma spraying include a rapid deposition rate and sufficiently low costs [11]. However, numerous problems with the plasma sprayed coatings have also been cited, including variation in bond strength between the coatings and the metallic substrates [13–16], nonuniformity in coating density [13], and alterations in HA structure as a result of the process [17].

Other experimental deposition techniques include electrophoretic deposition, high velocity oxy-fuel (HVOF) deposition, and radio-frequency sputter deposition. These experimental processes improve the adhesive, composition and structural properties of the coatings. However, these processes also alter the original starting material. For example, structural changes have been observed with the electrophoretic and HVOF deposition processes [18–21]. The radio-frequency sputtering produces amorphous coatings that are reported to have a high dissolution rate [22–23].

II.   ION BEAM SPUTTER DEPOSITION PROCESS

An experimental technique currently used to produce HA and Ca-P coatings is the ion beam sputtering process. This process has been vastly utilized by industry to produce and modify thin films, with the first deposition reported in 1852 [24]. However, only a few studies have been reported on the use of this technique for medical applications [25–27].

Figure 1 shows a schematic of an ion source. The working gas introduced into the discharge chamber is usually argon. Briefly, ions are produced when energetic electrons from the cathode collide with atoms or molecules of the working gas. Most of these ions form beamlets, thereby passing through the holes in the screen grid. The ions in the beamlets are then attracted to the negative accelerator grid, and, after leaving the accelerator grid, form ion beams. The ion beam is neutralized by electrons from the neutralizer; the neutralized beam then bombards the surface of the target.

The growth of the deposited film is caused by the removal of atoms from the target. Figure 2 shows a schematic of a typical ion beam sputtering system. Referring to Fig. 2b, the momentum is transferred from an incident, energetic beam to the atom or molecules of the target material. A large number of atoms are displaced from their normal sites in the crystal lattice, producing a disordered structure. Some of the incident particles also are implanted in the target. Most of the target atoms are displaced from the surface and, if they have enough energy, they escape from the target as sputtered atoms. These sput-

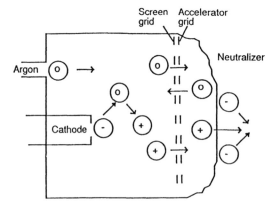

Figure 1   Schematic diagram of ion source used for ion beam sputter deposition. ($\bigcirc$ = working gas; $-$ = electrons; $+$ = ions).

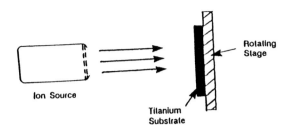

(A)

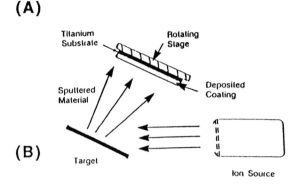

(B)

Figure 2 Schematic diagram of the ion beam sputtering system: (A), ion beam sputter cleaning process; (B), ion beam sputter deposition process.

tered atoms are then deposited onto the substrate that is placed in the path of the sputtered atoms. The substrate stage can be rotated to ensure averaging during deposition. Furthermore, sensitive substrates will not be damaged because the substrates are not immersed in a dense plasma.

Presence of negative ions on the target surface would result in charging problems, especially from compound targets. These charging problems during deposition are eliminated because ion beams of the gridded, Kaufman type and the nongridded Hall effect type are charge neutralized, usually by a hot tungsten or tantalum filament. This low operating pressure allows a significant number of ions to be elastically reflected from the target surface, thereby impacting on the growing film [28]. Films obtained are of better quality and generally denser [29]. However, modification of other properties, such as intrinsic stress, electrical and optical properties, and adhesive strength to the substrate are unavoidable.

Only a few research groups have utilized the ion beam sputtering process to produce Ca-P coatings on a metallic substrate for biomedical applications [27,30–32]. This chapter concentrates on studies conducted by these groups.

III. CALCIUM PHOSPHATE COATINGS

A. Coating Parameters

A schematic of the relative arrangement of the Kaufman-type ion source and the water-cooled rotating substrate holder in the vacuum chamber of the system (Commonwealth Scientific, Alexandria, VA) is shown in Fig. 2. This system is used for studies conducted at the University of Alabama at Birmingham. For the parameters of the coating system, titanium substrates are fixed onto the substrate stage using a conductive adhesive containing silver particles. The chamber is evacuated to a minimum base pressure of 6 × 104 torr. High purity argon (99.999%) is backfilled into the chamber, bringing the pressure to about 4 × 104 torr.

Prior to deposition, the titanium specimens are cleaned using the configuration shown in Fig. 2a. For surface cleaning, the substrate stage is held perpendicularly about 23 centimeters (cm) away from the Kaufman-type ion source for 10 minutes at 40.7 milliamperes (mA) and 500 electron volts (eV). The energetic ion beam, produced from ionizing high purity (99.999%) argon gas, is passed through a 3-cm diameter screen and accelerator grid and is neutralized with electrons from a hot tungsten filament before reaching the substrate.

After cleaning, the stage is rotated so that the substrates are placed in the path of the sputtered atoms (Fig. 2b). The distance between the target and Kaufman-type ion source (3-cm diameter screen and accelerator grid) is approximately 13 cm, while the distance between the target and substrates is approximately 23 cm. The deposition process is accomplished at a chamber pressure of about 4 × 10^{-4} torr, and an ion beam energy of 1000 eV and 40.7 mA. The deposition process is carried out at room temperature at a rate of 0.15 angstroms (Å) per minute, ultimately resulting in a coating thickness of about 0.6 to 1 μm for each sample.

The target used for this system is about 12.5 cm in diameter, and there are numerous ways of producing a target for ion beam sputter deposition. A sintered HA target can be prepared by hot pressing and sintering at 1150°C. This sintered target is then attached to a copper plate of the same diameter using silver paste. The silver paste provides a path for electrical conduction so that charging of the target is eliminated. HA targets can also be produced by coating a copper plate using HVOF or plasma spraying processes. Other methods include melting and pouring a Ca-P glass onto the copper plate [33].

The as-deposited Ca-P coatings are amorphous, and a postdeposition heat treatment is required for crystallization [30]. The thin (0.6- to 1-μm thick) amorphous coatings produced with this process can completely dissolve within 4 hours [31]. For postdeposition heat treatments to form crystalline phases in the coatings, the coatings must be annealed between 500°C and 600°C for 30 to 60 minutes.

B. Analytical Analyses

Since no single analytical technique can truly characterize a material, a variety of analytical techniques must be employed to characterize Ca-P coatings produced by the direct ion beam sputtering process. Surface-sensitive techniques such as atomic force microscopy (AFM), Fourier transform infrared spectroscopy (FTIR), and x-ray photoelectron spectroscopy (XPS) have been used to investigate the topographical, structural, and chemical properties of the coatings.

1. Surface Analyses

Atomic Force Microscopy.   Topographical features with a resolution on the order of 0.1 Å have been reported using AFM [34]. A representative contact mode scan from a Ca-P coating is shown in Fig. 3. The columnar features in Fig. 3a project about 40–80 nanometers (nm) above an otherwise flat surface, with the tallest being about 120 nm high. The mean cross-sectional area of the columns is about 6 ± 3 × 10^4 square nanometers (nm^2). The 1-μm scan in Fig. 3b is taken over a relatively flat region between the large

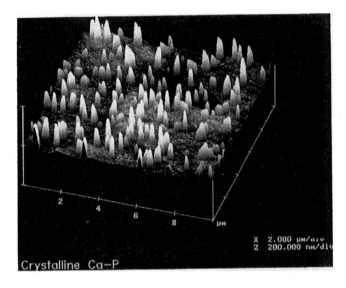

(a)

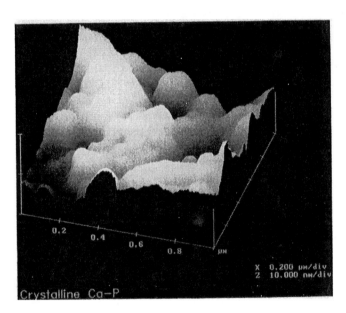

(b)

Figure 3 Atomic force microscopy (AFM) images of calcium phosphate (Ca-P) coatings: (a), 10 μm; (b), 1 μm.

projections in Fig. 3a. Columnarlike growth is also visible on the smaller scale in these areas.

The growth mode may be imposed on the coating by the surface morphology of the substrate so that the cross-sectional areas of the columns represent the underlying microscopic roughness of the surface. This limitation can arise even when diffusion of the coated material is rapid across the surface. Another factor affecting film growth is the angle of the incident beam to the target used during the deposition process [35]. Studies using other materials have suggested that columnar growth is due to sputtered materials arriving from one direction and condensing with little adatom movement [36]. Thus, by altering the direction of the beam relative to the target, a different direction of film growth may be achieved. Changes in surface topography are vital since cellular responses as well as mineralization of the bone may be dependent on the surface topography of the biomaterials [37–38].

Fourier Transform Infrared Spectroscopy.   FTIR analyses have been used to determine the structural and molecular composition of Ca-P coatings [39]. Unlike other techniques such as x-ray diffraction (XRD), the presence of hydroxyl groups after postdeposition heat treatments, as well as the presence of molecular species such as carbonates (CO_3) and phosphates (PO_4) can be evaluated.

The FTIR technique is very useful in determining the chemistry and structure of the ion beam sputter deposited coatings. FTIR spectra of HA target powders and coatings heated to 550°C for 90 minutes are shown in Fig. 4. Using the diffuse reflectance accessory, sharp, distinct PO_4 bands (914 cm^{-1}, 960 cm^{-1}, 1012 cm^{-1}, 1057 cm^{-1}, and 1086 cm^{-1}) can be observed for the HA powders, indicating a crystalline structure (Fig. 4a). Strong hydroxyl (OH) bands are observed at 3572 cm^{-1} and 630 cm^{-1}. Carbonyl (2922 cm^{-1} and 2858 cm^{-1}) and phosphorus-nitrogen (P-N) stretch (1593 cm^{-1}) bands are also observed, indicating some contamination. Nitrogenous species are sometimes found in HA powders since ammonium salt (NH_4^+) is often included in the preparation of HA powders [40–42].

A grazing angle FTIR accessory can be used to evaluate the thin Ca-P sputter depos-

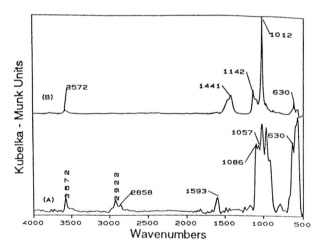

Figure 4   Fourier transform infrared (FTIR) spectra: (A), hydroxyapatite (HA) powders; (B), heat-treated calcium phosphate (Ca-P) coatings.

ited coatings. Figure 4b shows a spectrum for the Ca-P coatings. Distinct PO_4 bands (962 cm^{-1}, 1012 cm^{-1}, 1086 cm^{-1}, and 1142 cm^{-1}) and CO_3 bands at 1400–1500 cm^{-1} are present. A small OH band at 3574 cm^{-1} is shown on the spectrum, indicating that the ion beam deposition process does result in a reduction of the OH groups. The presence of CO_3 is attributed to the ions colliding with the graphite screen and accelerator grids in the ion source, thereby causing the CO_3 ions to be sputtered to the substrates [32]. Investigators have reported that CO_3 absorption bands are characteristic of Type B-CO_3 for all of the heat-treated coatings [43–46]. Furthermore, x-ray diffraction studies have indicated an apatitic structure after heat treatments above 500°C, with contractions in the a- and c-lattice spacings, possibly caused by the presence of CO_3, as well as strain associated with interfacial interactions between the coatings and substrates [47–49].

X-Ray Photoelectron Spectroscopy. Electron spectroscopy for chemical analysis (ESCA) or XPS is a surface-sensitive technique commonly used for elemental characterization. Information within 100 Å of the Ca-P surface can be obtained with this technique [50]. Analyses of HA powders used for the production of the ion beam sputter coater target is critical since HA can be easily contaminated. XPS analyses of HA powders obtained from various sources have indicated the presence of contaminants such as silicon, molybdenum, and fluorine. Concentrations of these elements can range from less than 1 atomic percent to as high as 5 atomic percent. Fluorine is a common contaminant and is known to replace the OH group in hydroxyapatite. Being a lighter element, fluorine is often observed to sputter preferentially to the substrate. This preferential or nonuniform removal of atoms has been observed, especially when composite or multicomponent targets are used [51].

A surface scan of Ca-P deposited coatings is shown in Fig. 5. Surface contaminants such as carbon (C) and fluorine (F) are observed. The atomic concentration of F 1s at a binding energy of 686 eV is 0.3–0.5 atomic percent, whereas the atomic concentration of C 1s is in the range of 22.0 to 25.5 atomic percent. The Ca/P ratio of the surface of the coatings is 2.3 ± 0.1, indicating a calcium-enriched and/or a phosphorus-deficient surface.

2. Analyses After In Vitro Immersion

When an implant is placed in the body, the first 1 nm of the surface interacts with the biofluid and, eventually, the tissue [52]. Thus, it is important to understand the chemical and structural changes that occur in this 1-nm zone [53]. Studies have been conducted to investigate the structural and chemical changes in ion beam sputter deposited Ca-P surfaces after immersion in α-modification of Eagles media (α-MEM) solutions [54].

Fourier Transform Infrared Spectroscopy. Using the grazing angle FTIR accessory, changes in surface structure and molecular composition of Ca-P coatings have been analyzed after immersion in α-MEM solutions for 12 days. Interestingly, coatings after immersion in α-MEM solution exhibit negligible changes, an indication that the surfaces retain their crystalline structure (Figs. 6a and 6b). The PO_4 bands (962 cm^{-1}, 1012 cm^{-1}, and 1086 cm^{-1}), CO_3 bands at 1400–1500 cm^{-1}, and a small OH band at 3574 cm^{-1} are similar to the spectrum for Ca-P coatings prior to immersion in α-MEM.

X-Ray Photoelectron Spectroscopy. While FTIR spectroscopy is capable of detecting molecules exhibiting a change in dipole moment, XPS is capable of detecting all elements and can identify the chemical states of surface species, with the exception of hydrogen and helium [55]. Changes in the calcium and phosphorus concentrations on the surfaces of Ca-P surfaces before and after immersion in α-MEM solutions are provided in Table

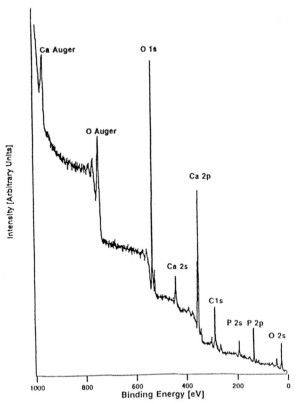

Figure 5 Representative x-ray photoelectron spectroscopy (XPS) surface scan of ion beam sputter deposited calcium phosphate (Ca-P) coatings.

1. Interestingly, no significant change in the phosphorus concentration is observed over time in solution. Table 1 reveals a gradual decrease in the Ca/P ratio from 2.3 ± 0.1 for Ca-P surfaces prior to immersion, to 1.5 ± 0.1 for surfaces of samples after 12 days of immersion in α-MEM solutions. The higher loss of calcium as compared to phosphorus is not surprising for phosphorus-deficient Ca-P coatings. The Ca-P-deposited coatings prior to solution immersion can be said to be phosphorus-deficient and nonstoichiometric since stoichiometric HA is known to have a Ca/P ratio of 1.67.

In addition to a change in phosphorus concentration over time in α-MEM solution, changes in the distribution of phosphate species have also been reported [54]. Figure 7 shows a high-resolution spectrum of P 2p obtained from a Ca-P surface. The Component A peak at 132.8 ± 0.2 eV is assigned to a phosphate (PO_4^{3-}), and the Component B peak at 133.8 ± 0.2 eV is assigned to a hydrogen phosphate (HPO_4^{2-}). Table 2 indicates a significant difference in the distribution of the two phosphate species for the HA standard and Ca-P coatings prior to solution immersion. After the coatings are immersed in the α-MEM solutions, the distribution of PO_4^{3-} and HPO_4^{2-} are more similar to the values shown for the HA standard.

Protein Adsorption on Calcium Phosphate Coatings. The surfaces of implants are not only exposed to inorganic ions, but also to proteins upon implantation (Fig. 8). Studies

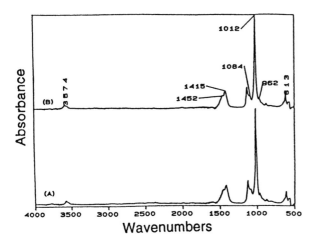

Figure 6 Representative Fourier transform infrared (FTIR) scan of calcium phosphate (Ca-P) coatings: (A), coatings prior to immersion; (B), coatings after immersion in α-modification of Eagle's media (α-MEM) solutions for 12 days.

have targeted the evaluation of protein interactions with the biomaterials. For example, changes in albumin and fibronectin conformation have been observed on the surfaces of biomaterials [56–57]. Such conformational changes were suggested to elicit differences in cellular response among different biomaterials [58].

In a recent study, the conformation of albumin on Ca-P surfaces was examined using Fourier transform infrared spectroscopy with attenuated total reflectance (FTIR-ATR). Using this technique, Ca-P coatings were exposed to α-MEM solutions containing 30 milligrams per milliliter (mg/ml) bovine albumin. A schematic of the FTIR-ATR liquid flow cell accessory is shown in Fig. 9. Two similar ATR germanium crystals are used for assembling the flow cell. However, only one flow channel (with Ca-P coating) is used during individual analysis.

Table 1   Average Calcium Concentration, Phosphorus Concentration, and Calcium/Phosphorus Ratio (±1 Standard Error) of Calcium Phosphate Coatings After Immersion in α-Modification of Eagle's Media Solutions Over Time

Time	Ca concentration	P concentration	Ca/P ratio
HA standard[a]	18.0	11.1	1.62
Control Ca-P coating	18.9 ± 0.1	8.4 ± 0.5	2.3 ± 0.1
3-hour immersion	13.6 ± 0.4	7.6 ± 0.3	1.8 ± 0.1
3-day immersion	12.9 ± 0.3	7.5 ± 0.2	1.7 ± 0.1
6-day immersion	12.7 ± 0.3	7.7 ± 0.4	1.7 ± 0.1
9-day immersion	12.5 ± 1.1	8.1 ± 0.1	1.6 ± 0.1
12-day immersion	12.4 ± 0.5	8.2 ± 0.3	1.5 ± 0.1

[a]One standard analyzed

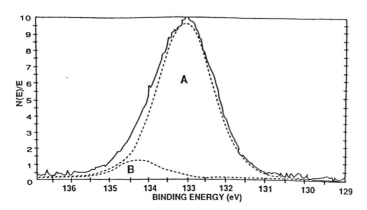

Figure 7   Representative high resolution of P 2p spectrum after immersion in α-modification of Eagle's media (α-MEM) solution (Components A and B correspond to those in Table 2).

Table 2   Relative Distribution (±1 Standard Error) of Different Components for P 2p Over Time in Solution

Sample	Relative distribution of P 2p concentration (%)	
	Component A (PO_4^{3-})	Component B (HPO_4^{2-})
HA standard[a]	91.6	8.4
Control Ca-P coating	65.8 ± 1.8	34.2 ± 1.8
3-hour immersion	84.8 ± 0.5	15.2 ± 0.5
3-day immersion	84.8 ± 4.6	15.2 ± 4.6
6-day immersion	83.8 ± 1.6	16.3 ± 1.6
9-day immersion	86.7 ± 1.0	13.3 ± 1.0
12-day immersion	87.3 ± 0.6	12.7 ± 0.6

[a]One HA and brushite standard analyzed

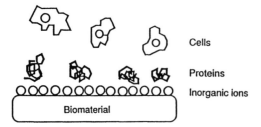

Figure 8   Schematic of biomaterial surface interacting with biological environment.

A representative FTIR-ATR spectrum of bulk albumin (or albumin in solution) and albumin on Ca-P coatings is shown in Figs. 10a and 10b. A total of six components is used to curve fit the spectrum from the bulk albumin (Fig. 10a). The components at positions 1680 cm^{-4} and 1634 cm^{-4} are assigned as β-sheet conformations, whereas the component at position 1655 cm^{-1} is assigned as an α-helix conformation. Figure 10b shows the absorption bands for albumin on Ca-P coatings. A significant difference in the two spectra can be seen. The two adsorption bands for bulk albumin and albumin on Ca-P coatings are deconvoluted and curve fitted with seven components. A shift in the peak position for each component can be observed as compared with bulk albumin. Furthermore, a new component at position 1603 cm^{-1} is also observed.

Other than qualitative differences in the two protein spectra shown in Fig. 10, the α-helix (1655 cm^{-1}) to β-sheet (1684 cm^{-1}) ratio can also be quantified by measuring the intensity of these two specific components. The intensity ratio of the α-helix (1655 cm^{-1}) to β-sheet (1684 cm^{-1}) for bulk albumin is 3.67 $\pm$ 0.16. This ratio is in agreement with other studies [59]. The ratio of the α-helix to the β-sheet for albumin on Ca-P is 2.62 $\pm$ 0.25 and thus is significantly different compared to bulk albumin. This difference in ratio suggests a decrease in the amount of α-helix structure after adsorption on Ca-P surfaces. A decrease in the α-helix structure on other biomaterials has also been reported [60]. Since there is a change in the albumin structure after adsorption on Ca-P coatings, it is not surprising to find the emergence of a new peak at 1603 cm^{-1} for albumin adsorbed on Ca-P coatings. Studies have observed a new peak at 1531 cm^{-1} for adsorbed albumin on other biomaterial surfaces [60–61]. However, the assignment of this new peak at 1603 cm^{-1} is unknown, and further investigation is required for this assignment.

3. Cellular Response to Calcium Phosphate Coatings

In addition to physiological fluid and proteins, rat bone marrow cell cultures have been used to evaluate the cellular response to ion beam sputter coatings [62]. Methodologies for establishing these cultures are provided in several references [63–64]. This chapter summarizes, as examples of the kinds of studies that can be done, the proliferation and synthesis of proteins of rat bone marrow cells.

Hexosaminidase Activity. Most cells contain hexosaminidase, a lysosomal enzyme with activity that is directly proportional to cell number [64]. It cleaves the para-nitrophenol from the β-D glucosamide substrate, thereby producing a color reaction. In a recent study, Ca-P and titanium surfaces were exposed to cultures obtained from rat bone marrow for time periods up to 400 hours. During this study, the hexosaminidase activity was measured as a means of determining cell number. Figure 11 shows no statistical

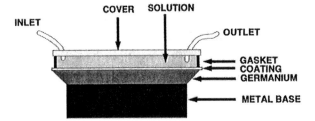

Figure 9 Schematic showing the Fourier transform infrared with attenuated total reflectance (FTIR-ATR) setup used for protein analyses.

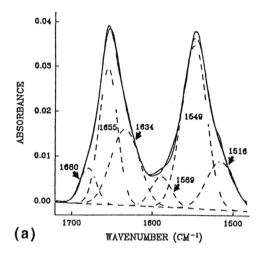

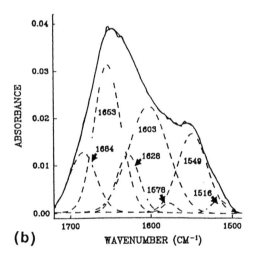

Figure 10   Fourier transform infrared (FTIR) spectra showing the protein absorption bands: (a), bulk albumin; (b), adsorbed albumin on calcium phosphate (Ca-P) coatings.

difference in the hexosaminidase activity of cells on titanium and Ca-P surfaces. Keller et al. [65] and Hestilow et al. [66] similarly observed no statistical difference in cell number between Ca-P and titanium surfaces, and between other metal surfaces and their respective oxides.

Protein Synthesis.   Protein synthesis is an important marker for evaluating cell function. Proteins synthesized and retained by the cells are important for bone formation and calcification. Using ^{35}S-methionine labeling and trichloroacetic acid (TCA) precipitation, the amount of labeled proteins can be measured using a liquid scintillation counter. Figure 12 shows the percent retention of newly synthesized extracellular matrix proteins in the cell layers on titanium and Ca-P surfaces. A statistically lower matrix retention is observed on the Ca-P surfaces during the initial three hours; however, after three hours,

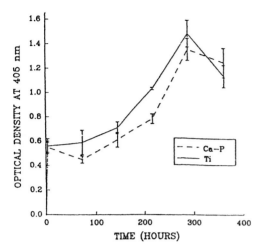

Figure 11   Hexosaminidase activity of cells seeded on calcium phosphate (Ca-P) and titanium surfaces.

no significant difference in the protein synthesis is observed. Although this initial difference is significant, further investigations are needed to study its implication on calcification and cellular functions.

IV. SUMMARY

In summary, all studies on ion beam sputter deposited Ca-P coatings have indicated that chemical and structural changes do occur during the coating process. Although an HA-type coating can be produced, with complete coverage of the substrate, the AFM study

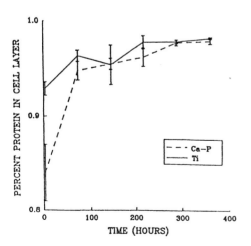

Figure 12   Percent protein retained in the cell layer after seeding cells on calcium phosphate (Ca-P) and titanium surfaces.

indicates a roughened surface with columnar structures having a height in the range of 40 to 80 nm. FTIR analyses indicate the presence of carbonates, and the substitution of carbonates into the coatings is due to the use of graphite screens in the ion source. XPS revealed that the Ca/P ratio of the Ca-P coatings was 2.3, a value significantly higher than 1.62, the Ca/P ratio for HA powders.

After immersion of the Ca-P coatings in α-MEM solutions, no change in the structure is detected using FTIR analyses. However, using XPS analyses, changes in the surface chemistry are observed, with a change in the Ca/P ratio from 2.3 to 1.5 for coatings after 12 days of immersion in solution. Qualitative and quantitative changes in albumin conformation also occur upon adsorption on the Ca-P coatings.

Using rat bone marrow cells for cell culture studies, cell attachment, proliferation, and production of proteins required for bone mineralization occur on the Ca-P surfaces. However, current studies do not suggest a significant difference in cell proliferation and protein synthesis by cells on Ca-P surfaces as compared to cells on titanium surfaces after some incubation period.

The responses of calcifying cells to biomaterials are indeed specific to the material surface. Since cells interact only with the material surface, surface characterization of all materials is vital. The percent crystallinity of HA and Ca-P required for optimum cellular response is currently debated. In addition to crystallinity, there are still missing links required to explain the tissue/biomaterials interface phenomena. Such questions include changes in surface chemistry, the precipitation of organic and inorganic constituents on biomaterial surfaces, and, most importantly, the complex cascade of cellular metabolisms in response to changes at the biomaterial surfaces.

REFERENCES

1. Young, R.A., Biological apatite versus hydroxyapatite at the atomic level, *Clin. Orthop. Rel. Res.*, 113:249–262 (1975).
2. Driessens, F.C.M., and Verbeeck, R.M.H., The mineral in tooth enamel and dental caries, in *Biominerals*, CRC Press, Boca Raton, FL, 1990, pp. 105–161.
3. LeGeros, R.Z., Calcium phosphates in enamel, dentin and bone, in *Calcium Phosphates in Oral Biology and Medicine*, Karger, New York, 1991, pp. 108–129.
4. Bonel, G., Heughebaert, J.C., Heughebaert, M., Lacout, J.L., and Lebugle, A., Apatitic calcium orthophosphates and related compounds for biomaterials, in *Bioceramics: Material Characteristics Versus In Vivo Behavior*, P. Ducheyne and J.E. Lemons (Eds.), New York Academy of Science, New York, 1988, pp. 115–139.
5. Posner, A.S., Crystal chemistry of bone mineral, *Physiol. Rev.*, 49:760–792 (1969).
6. Posner, A.S., Bone mineral on the molecular level, *Fed. Proc.*, 32:1933–1937 (1973).
7. Termine, J.D., and Posner, A.S., Infra-red analyses of rat bone. Age dependency of amorphous and crystalline mineral fraction, *Science*, 153:1523–1525 (1966).
8. Ingram, G.S., Some heteroanionic exchange reactions of hydroxyapatite, *Bull. Soc. Chim. Fr.*, 1841–1844 (1968).
9. Driessens, F.C.M., Van Dijk, J.W.E., and Borggreven, J.M.P.M., Biological calcium phosphates and their role in the physiological of bone and dental tissues. I. Composition and solubility of calcium phosphates, *Cal. Tis. Res.*, 26:127–137 (1978).
10. Van Raemdonck, W., Ducheyne, P., and De Meester, P., Calcium phosphate ceramics, in *Metal and Ceramic Biomaterials*, *Vol. II: Strength and Surface*, P. Ducheyne and G.W. Hastings (Eds.), CRC Press, Boca Raton, FL, 1984, pp. 143–166.
11. Herman, H., Plasma spray deposition processes, *MRS Bulletin*, 12:60–67 (1988).

12. Bunshah, R.F., Deposition technologies: An overview, in *Deposition Technologies for Films and Coatings: Developments and Applications*, R.F. Bunshan (Ed.), Noyes Publications, Park Ridge, NJ, 1982, pp. 1–18.

13. Lacefield, W.R., Hydroxylapatite coatings, in *Bioceramics: Material Characteristics Versus In Vivo Behavior*, P. Ducheyne and J.E. Lemons (Eds.), New York Academy of Science, New York, 1988, pp. 72–80.

14. Filiaggi, M.J., and Pilliar, R.M., Interfacial characterization of a plasma-sprayed hydroxyapatite/Ti-6Al-4V implant system, *Transactions of the 10th Annual Meeting of the Canadian Society for Biomaterials*, 23–25 (1989).

15. Filiaggi, M.J., Coombs, N.A., and Pilliar, R.M., Characterization of the interface in the plasma-sprayed HA coating/Ti-6Al-4V implant system, *J. Biomed. Mater. Res.*, 25:1211–1230 (1991).

16. Whitehead, R.Y., Lucas, L.C., and Lacefield, W.R., The effect of dissolution on plasma sprayed hydroxylapatite coatings on titanium, *Clin. Mater.*, 12:31–39 (1993).

17. Radin, S., and Ducheyne, P., The effect of plasma sprayed induced changes in characteristics on the *in vitro* stability of calcium phosphate ceramics, *Trans. of the 16th Ann. Meeting of the Soc. for Biomater.*, 128 (1990).

18. Ducheyne, P., Van Raemdonck, W., Heughebaert J.C., and Heughebaert, M., Structural analysis of hydroxyapatite coatings on titanium, *Biomaterials*, 7:97–103 (1986).

19. Ducheyne, P., Radin, S., Heughebaert, M., and Heughebaert, J.C., Calcium phosphate ceramic coatings on porous titanium: Effect of structure and composition on electrophoretic deposition, vacuum sintering and *in vitro* dissolution, *Biomaterials*, 11:244–254 (1990).

20. Haman, J., A characterization of hydroxyapatite coatings produced by HVOF (high velocity oxy-fuel) thermal spraying, Master's thesis, University of Alabama at Birmingham, Birmingham, AL, 1993.

21. Boulware, A., Characterization of fluorapatite coatings produced by high velocity oxy-fuel (HVOF) thermal spraying, Master's thesis, University of Alabama at Birmingham, Birmingham, AL, 1993.

22. Coupe, K., Properties of Ca-P thin films produced using RF sputtering techniques, Master's thesis, University of Alabama at Birmingham, Birmingham, AL, 1991.

23. Sioshansi, P., Ion beam processing of biomaterials, in *Surfaces in Biomaterials Symposium*, Algonquin, IL, 1991.

24. Westwood, W.D., Sputter deposition processes, *Mater. Res. Soc. Bull.*, 46–51 (December 1988).

25. Hambleton, J., Schwartz, Z., Gomez, R., Windeler, A.S., Khare, A.G., and Boyan, B.D., Modulation of chondrocyte differentiation by growth on sputter-coated surfaces, *J. Dent. Res.*, 72:390 (1993).

26. Long, L.K., Corrosion evaluations of hydroxylapatite sputter coated implant materials, senior design project, University of Alabama at Birmingham, Birmingham, AL, 1989.

27. Windeler, A.S., Bonewald, L., Khare, A.G., Boyan, B., and Mundy, G.R., The influence of sputtered bone substitutes on cell growth and phenotypic expression, in *The Bone-Biomaterial Interface*, J.E. Davies (Ed.), University of Toronto Press, Toronto, 1991, pp. 205–213.

28. Kay, E., Parmigiani, F., and Parrish, W., Effect of energetic neutralized noble gas ions on the structure of ion beam sputtered thin metal films, *J. Vac. Sci. Technol.*, A5:44–51 (1987).

29. Mattox, D.M., Ion plating technology, in *Deposition Technologies for Films and Coatings*, R.F. Bunshah (Ed.), Noyes Publications, Park Ridge, NJ, 1982, pp. 244–287.

30. Rigney, E.D., Characterization of ion-beam sputter deposited Ca-P films, Ph.D. dissertation, University of Alabama at Birmingham, Birmingham, AL, 1989.

31. Ong, J.L., Lucas, L.C., Lacefield, W.R., and Rigney, E.D., Structure, solubility and bond strength of thin calcium phosphate coatings produced by ion beam sputter deposition, *Biomaterials*, 13:249–254 (1992).

32. Ong, J.L., and Lucas, L.C., Post deposition heat treatments for ion beam sputter deposited calcium phosphate coatings, *Biomaterials* (in press).

33. Burns, J., Optimization of Ca-P ion-sputtered thin films, Master's thesis, University of Alabama at Birmingham, Birmingham, AL, 1990.

34. Rugar, D., and Hansma, P.K., Atomic force microscopy, *Phys. Today*, 43:23–30 (1990).

35. Thornton, J.A., Coating deposition by sputtering, in *Deposition Technologies for Films and Coatings: Developments and Applications*, R.F. Bunshah et al. (Eds.), Noyes Publications, Park Ridge, NJ, 1982, pp. 170–243.

36. Thornton, J.A., High rate thick film growth, *Ann. Rev. Mater. Sci.*, 7:239–260 (1977).

37. Brunette, D.M., Ratkay, J., and Chehroudi, B., Behavior of osteoblasts on micromachined surfaces, in *The Bone-Biomaterial Interface*, J.E. Davies (Ed.), University of Toronto Press, Toronto, 1991, pp. 170–180.

38. Ohara, P.T., and Buck, R.C., Contact guidance *in vitro*: A light, transmission, and scanning electron microscopic study, *Exp. Cell Res.*, 121:235–249 (1979).

39. Bailey, R.T., and Holt, C., Fourier transform infrared spectroscopy and characterisation of biological calcium phosphates, in *Calcified Tissue*, D.W.L. Hukins (Ed.), CRC Press, Boca Raton, FL, 1989, pp. 93–120.

40. Doi, Y., Moriwaki, Y., Okazaki, M., Takahashi, J., and Joshin, K., Carbonate apatites from aqueous and nonaqueous media studied by E.S.R., i.r., and x-ray diffraction: Effect of NH_4^+ ions on crystallographic parameters, *J. Dent. Res.*, 61:429–434 (1982).

41. Kowker, S.E.P., and Elliot, J.C., Infrared absorption bands from NCO^- and NCN^{2-} in heated carbonate containing apatites prepared in the presence of NH_4^+ ions, *Calcif. Tissue Int.*, 29:177–178 (1979).

42. Jarcho, M., Bolen, C.H., Thomas, M.B., Bobick, J., Kay, J.F., and Doremus, R.H., Hydroxylapatite synthesis and characterization in dense polycrystalline form, *J. Mater. Sci.*, 11: 2027–2035 (1976).

43. Harries, J.E., Hasnain, S.S., and Shah, J.S., EXAFS study of structural disorder in carbonate-containing hydroxyapatites, *Calcif. Tissue Int.*, 41:346–350 (1987).

44. Verbeeck, R.M.H., Minerals in human enamel and dentin, in *Tooth Development and Caries*, Vol. 1, F.C.M. Driessens and J.H.M. Woltjens (Eds.), CRC Press, Boca Raton, FL, 1986, pp. 95–152.

45. Nelson, D.G.A., and Featherstone, J.D.B., Preparation analysis and characterization of carbonated apatites, *Calcif. Tissue Int.*, 34:S69–S81 (1982).

46. LeGeros, R.Z., LeGeros, J.P., Trautz, O.R., and Klein, E., Spectral properties of carbonate-containing apatites, *J. Dental Res.*, 43:751–760 (1964).

47. Ducheyne, P., van Raemdonck, W., Heughebaert, J.C., and Heughebaert, M., Structural analysis of hydroxyapatite coatings on titanium, *Biomaterials*, 7:97–103 (1986).

48. LeGeros, R.Z., Trautz, O.R., LeGeros, J.P., and Klein, E., Apatite crystallites: Effects of carbonate on morphology, *Science*, 1409–1411 (1967).◄

49. LeGeros, R.Z., Trautz, O.R., LeGeros, J.P., and Klein, E., Carbonate substitution in the apatite structure (1), *Bull. Soc. Chim. Fr.*, 1712–1718 (1968).◄

50. Paynter, R.W., Introduction to x-ray photoelectron spectroscopy, in *Surface Characterization of Biomaterials*, B.D. Ratner (Ed.), Elsevier Science Publishers B.V., Amsterdam, 1988, pp. 37–52.

51. Zalm, P.C., Quantitative sputtering, in *Handbook of Ion Beam Processing Technology*, J. J. Cuomo, S.M. Rossnagel, and H.R. Kaufman (Eds.), Noyes Publications, Park Ridge, NJ, 1989, pp. 78–111.

52. Kasemo, B., and Lausmaa, J., Biomaterials from a surface science perspective, in *Surface Characterization of Biomaterials*, B.D. Ratner (Ed.), Elsevier Science Publishers B.V., New York, 1988, pp. 1–12.

53. Ong, J.L., Raikar, G.N., Lucas, L.C., Connatser, R., and Gregory, J.C., Spectroscopic characterization of passivated titanium in a physiologic solution, *J. Mater. Sci.: Mater. in Med.* (in press).

54. Ong, J.L., Lucas, L.C., Raikar, G.N., Weimer, J.J., and Gregory, J.C., Surface characterization of ion beam deposited Ca-P coatings after *in vitro* immersion, *Colloids and Surfaces,* 87:151–162 (1994).

55. Seah, M.P., and Briggs, D., A perspective on the analysis of surface and interfaces, in *Practical Surface Analysis,* Vol. 1, D. Briggs and M.P. Seah (Eds.), John Wiley and Sons, New York, 1990, pp. 1–18.

56. Williams, R.L., and Williams, D.F., The spatial resolution of protein adsorption on surfaces of heterogeneous metallic biomaterials, *J. Biomed. Mater. Res.,* 23:339–350 (1989).

57. Hlady, V., and Furedi-Milhofer, H., Adsorption of human serum albumin on precipitated hydroxyapatite, *J. Coll. Interface Sci.,* 69:460–468 (1979).

58. Horbett, T.A., Protein adsorption on biomaterials, in *Biomaterials: Interfacial Phenomena and Applications,* S.L. Cooper and N.A. Peppas (Eds.), American Chemical Society, Washington, DC, 1982.

59. Marx, J., Hudry-Clergeon, G., Capet-Antonini, F., and Bernard, L., Laser Raman spectroscopy study of bovine fibrinogen and fibrin, *Biochim. Biophys. Acta,* 578:107–115 (1979).

60. Lenk, T.J., Ratner, B.D., Gendreau, R.M., and Chittur, K.K., IR spectral changes of bovine serum albumin upon surface adsorption, *J. Biomed. Mater. Res.,* 23:549–569 (1989).

61. Dev, S.B., Rha, C.K., and Walder, R., Secondary structural changes in globular protein induced by a surfactant: Fourier self-deconvolution of FT-IR spectra, *J. Biomol. Struct. Dyn.,* 2:431–442 (1984).

62. Boyan, B.D., Schwartz, Z., Gross, and Sela, J., Biological implications of the use of biomaterials in bone, *J. Dent. Res.,* 72:258 (1993).

63. Boyan, B.D., Schwartz, Z., Dean, D.D., and Hambleton, J.C., Responses of bone and cartilage cells to biomaterials *in vivo* and *in vitro, J. Oral Implant.,* 19:116–122 (1993).

64. Landegren, U., Measurement of cell numbers by means of the endogenous enzyme hexosaminidase. Applications to detection of lymphokines and cell surface antigens, *J. Immunol. Methods,* 67:379–388 (1984).

65. Keller, J.C., Niederauer, G.G., Lacefield, W., Lucas, L., and Zaharias, R., Osteoblast attachment to ceramic implant materials, *J. Dent. Res.,* 71:722 (1992).

66. Hestilow, K.L., Khare, A., Norling, B.K., and Windeler, A.S., Response of cells to metals and their sputter deposited oxides, *J. Dent. Res.,* 71:722 (1992).

X
OTHER MATERIALS

48

Biomedical Properties and Applications of Chitin and Its Derivatives

Mutsuhiro Maeda, Yukio Inoue, and Hideaki Iwase
Juntendo University, Shizuoka, Japan

Koji Kifune
UNITIKA Limited, Kyoto, Japan

I. INTRODUCTION

Chitin, poly-β-(1,4) linked N-acetyl-D-glucosamine is a mucopolysaccharide. It is well known that chitin is also a substrate of lysozyme (muramidase). Chitin, widely distributed in nature, is a substance that sustains and protects the body of crustaceans and microorganisms.

Chitin is insoluble in water and most ordinary solvents. This property of chitin has restricted its use to applications that do not require solubility of the polymer in spite of its abundance in nature and the advantages claimed for the material. Attempts have been made to produce molded chitin, such as fibers and films. The products made by methods using strong acids or dichloroacetic acid were not always of high quality and featured a markedly lower molecular weight due to the use of strong acid and alkali during dissolution. However, since Austin (1975) suggested the possibility of a dissolution method with amide-LiCl (lithium chloride) solvent, Kifune and coworkers (1983) made a great effort and found that mixtures of N-methylpyrrolidone (NMP) or dimethylacetamide (DMA) and LiCl are favorable solvents for chitin.

Chitin is an absorbable and degradable material catabolized by a specific metabolic pathway [1]. It is a natural polymer that is attracting more and more attention as a novel material for medical use because it is not toxic and features low antigenicity, good biocompatibility, and bioactivating effects. Since Prudden et al. reported chitin's ability to promote the healing of wounds, some bioactivating effects based on its immunological adjuvant activity have become apparent [2]. Based on their basic properties, braided fibers, sheets (nonwoven fabrics), and porous sponges have been applied clinically in the form of absorbable suture thread and temporary dressings for skin and soft tissue defects. Furthermore, because of their characteristic features, chitin products and their derivatives are being applied in various fields, especially as a sustained-release vehicle for anticancer drugs, as local hemostatics, as filling for bone defects, as internal fixation devices for bone fractures, and as immunological activation adjuvants for malignancy and/or opportunistic infection.

II. PRODUCTION OF MOLDED CHITINS AND THEIR BIOMEDICAL PROPERTIES

We obtained purified chitin powder (purity 99.9%, diameter 50–100 micrometers [μm], molecular weight 10^6) from the two-stage processing, decalcification with acid treatment and deproteinization with alkali treatment, of crude Japanese pink crab (*Chionecetes opilio*) shells. In the first purification, the shells were treated with 0.1 N (normal) HCl for about 1 hour at room temperature and with 1 N NaOH for 3 hours at 80–90°C. After drying, shells were pulverized to 30-mesh size and again were treated with 0.1 N HCl for 1 or 2 hours at room temperature and with 1 N NaOH for 2 hours at 95°C. The preparation was dissolved in a mixture of dimethylacetamide and lithium chloride (5–9 weight percent [wt%] in saturated amount is favorable) to make dope (molecular weight 10^5).

A. Fiber

The chitin fibers were obtained from the processes of dissolution, filtration, making dope, transport, wet spinning, coagulation, stretching, winding, washing, and drying. Using the solution consisting of DMA and LiCl (94 : 6), highly pure chitin was dissolved to make dope and was then wet molded with a coagulant such as water or middle alcohols (including methyl alcohol, isopropyl alcohol, butyl alcohol, and amyl alcohol). The coagulation velocity and the shape of filter (#1480), size of spinning nozzle (D : 0.06–0.08 mm), reeling speed (10–15 meters per minute [m/min]), and fiber stretch in the wet spinning process are important factors to obtain high strength in extension and stretch of fibers.

The optimum size and physical properties of chitin fibers for practical use are as follows: molecular weight 10^5; fiber denier 0.8–0.9 d; tensile strength 3.5–4.5 g/d; remaining extension 18–15%; and water uptake up to 18.5% in weight. After these fibers were spun, the braider was used to produce the sutures with a variety of thicknesses.

B. Film

The chitin with high purity was dissolved in a mixture of DMA and LiCl. The dope was poured into the frame to make a flat film. As for wet molding of transparent film, choice of coagulants is important because coagulated chitin readily loses its transparency. Butyl alcohol or isopropyl alcohol is suitable for obtaining high transparency. Uniform loading with tension is important to avoid serious contraction in the drying process. We are able to manufacture transparent and high-quality film with a thickness of 30–100 μm, a tensile strength of 60 kilograms per square millimeters kg/mm^2, and an elongation of 14–15%. Film thicker than this level requires gradual coagulation with an additional solvent such as dimethylformide.

C. Rod

The manufacturing process for rods was wet molding and gradual coagulation. The bending strength and elastic modulus were 19.5 kg/mm^2 and 638.9 kg/mm^2 (mean), respectively, in the dry state. In the wet state, the strength decreased to approximately 1/50. The water uptake was more than 100% in weight [3].

Non-woven fabric

Figure 1 Scanning electron microscopic (SEM) finding of chitin sheet.

D. Nonwoven Fabric

Chitin fibers were cut to a length of 5 mm, polyvinyl alcohol (PVA) fibers (type SML) were added as binder at a ratio of 1 : 10 and the fibers were dispersed in water. Then the gel-like product was placed between filter papers and was pressed and treated with a thermal roller at 150°C. Because of its function as a temporary wound covering material, the fabric was manufactured at a thickness of 0.1–0.18 mm. Its scanning electron micrograph is shown in Fig. 1.

There are a number of spaces among the fibers. Chitin fibers absorbed and retained exudate remarkably well in the living body. We designed the sheet to make sufficient room at the maximum swell of fibers. Accordingly, the residual spaces were suitable for transportation of liquids to the outside of the body. The sheet tensile strength of this fabric in the dry state was about 1000 grams (g) per 1-cm width, which was sufficiently tough and resistant for clinical use. The gas permeability is 120–150 cm^3/cm$^2 \cdot$sec.

E. Sponge

An additional treatment to make pores is to coagulate a mixture of chitin dope and a powdered water-soluble substance. PVA powder was used as a water-soluble polymer to make pores so that the pore size and rate could be adjusted by selecting the amount and the size of the powder. After removal of PVA with hot water, lyophilization enabled the manufacture of the high quality sponge. Figure 2 shows the scanning electron microscopy (SEM) findings for chitin sponge. The mechanical strength of chitin sponge (90% porosity) was 6.2 g/mm^2 in the dry state and 3.8 g/mm^2 in the wet state.

III. BIOCOMPATIBILITY AND BIOABSORBABILITY OF CHITIN

Molded chitin tends to be highly hydrophilic. Its high water content of 12–13 times that of body weight (nonwoven fabric) at 25°C has been confirmed. Its ability to absorb serum protein was also found to be high, and increased further by increasing the amount

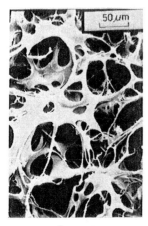

Sponge

Figure 2   SEM finding of chitin sponge.

of amino groups after deacetylation. These facts suggest that the surfaces of molded chitins may readily achieve good bioaffinity when they are applied to the living body.

A.   *In Vitro* Biocompatibility

The ability of animal-cultured cells to adhere to material surfaces is a good method of evaluating their biocompatibility. Chitin films, cellulose films, and polystyrene petri dishes (as a control), were compared using this method. Areolar fibroblasts from a mouse (hereafter referred to as L-cells) on a chitin film showed good cell adhesion and good cell morphology. These findings were identical to that of the untreated control. The growth curve is also similar and the growth rate of L-cells was 70% of that of the untreated control. On the other hand, little cellular adhesion was observed on the cellulose film. While chitin and cellulose have similar chemical structures, chitin seems to be superior to cellulose in terms of biological affinity [4] (Fig. 3). There was no degradation of chitin fibers in Ringer's solution, which is probably due to hydrolysis.

B.   *In Vivo* Biocompatibility and Bioabsorbability

Biodegradation has been confirmed by the fiber, rod, and sponge shapes. The chitin fibers gradually lost their strength when they were implanted into the back muscles of white rabbits. One month after surgery, their tensile strength reached zero, but their initial shape was maintained. Their average molecular weight varied from 10^5 to 10^4 over time.

Nakajima and coworkers measured the generation of N-acetylglucosamine in lysozyme solution by the Morgan–Elson reaction [5]. They showed an increase of N-acetylglucosamine over time, as well as the likelihood of degradation by lysozyme.

The molded chitin was completely absorbed 5 to 6 months later after surgery. However, a large amount of molded chitin, such as rods, was not completely absorbed even one year later. Examination of the chitin rod surface revealed a loss of smoothness *in vivo* in comparison with the samples that were soaked in Ringer's solution.

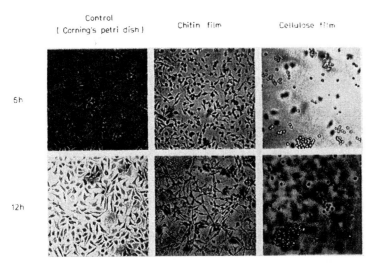

Figure 3 *In vitro* biocompatibility findings of L-cells on chitin film at 5 and 12 hours after incubation.

From histological observation, the inflammatory reaction looks aggressive and is something different from an ordinary foreign-body reaction. The most likely means by which *in vitro* and *in vivo* absorption occurs seems to be that macrophages, histiocytes, and giant cells that surround molded chitin exert phagocytosis during the inflammatory reaction, when lysozyme acts to cut the fiber almost to a hexomer. Moreover, enzymes such as β-N-acetyl-glucosaminase decompose chitin to the final forms of saccharide and N-acetylglucosamine.

The generated N-acetylglucosamine, a common aminoglucose in the body, enters the innate metabolic pathway to be incorporated to form glycoproteins or to be excreted as carbon dioxide gas during respiration [1].

IV. CHEMICAL MODIFICATIONS OF CHITIN

The chemical modifications of chitin are among of the promising means of exploring new, highly sophisticated functions. Chitin derivatives are prepared by such modification reactions as deacetylation; N-derivatives are prepared by such modification reactions as deacetylation, N-acetylation, O-carboxymethylation, sulfation, and complexation with acids, metals, and cations.

A. Deacetylated Chitins

Chitosan is the highly deacetylated product of chitin. Although chitin is insoluble in most solvents, chitosan is readily soluble in acidic solutions, but insoluble in water. However, chitosan, with low molecular weight (less than 2000), is soluble in water. Chitosans are available mostly for industrial and biomedical applications.

Partially deacetylated chitins (DACs) were prepared by deacetylation of chitin or N-acetylation of chitosan. Sannan, Kurita, and Iwakura prepared water-soluble DAC-45 to DAC-55 (45–55% of deacetylation) by homogenous deacetylation of chitin [6]. On the other hand, DAC with heterogenous deacetylation above 60% or below 40% was insoluble in water. The difference in water solubility is presumed to be due to the distribution of the acetamide group in the DAC molecule by Kurita, Sannan, and Iwakura [7]. They reported that the deacetylation in the heterogeneous condition occurred mostly in the amorphous region to form block-type copolymers of GlcNAc and D-glucosamine (GlcN) units. Sashiwa et al. investigated the distribution of the acteamide group in DAC using nitrous acid [8]. The deamination products of various DACs ranged mainly from trisaccharides to monosaccharides, which suggested the random distribution of the acetamide group in DAC molecules prepared by heterogeneous deacetylation. These results were helpful in understanding the mechanism by which the immunoadjuvant activities of these DAC molecules occur.

B. Carboxymethyl Chitins

Carboxymethylation is achieved with monochloroacetic acid and sodium hydroxide. The reaction takes place preferentially in C-6 hydroxy groups. Cross-linked carboxymethyl-chitin or -chitosan show a high capability of separating bovine serum fibrinogen and albumin. The absorption ability of blood protein and calcium and iron ions was studied by Tokura et al. [9]. Carboxymethyl chitin (0.8 substitution degree) showed high immunoadjuvant activity as a result of mouse peritoneal macrophage activation.

C. Hydroxyalkylated Chitins

Hydroxyethyl-chitin (glycol chitin) and -chitosan (glycol chitosan) are water-soluble derivatives. Glycol chitosan was used as a raw material for the membrane matrix.

D. Sulfated Chitins

The sulfation of chitin and chitosan has been one of the most attractive modification fields because of the possibility of preparing in this manner anticoagulant polysaccharides with a structural similarity to heparin. Hirano et al. reported that sulfated derivatives of chitosan such as O-sulfated-N-acetyl chitosan showed the anticoagulant effect with respect to activated partial thromboplastin time, thrombin time, and antithrombin activity [10].

Murata et al. reported on the inhibitory effect of tumor-induced angiogenesis by sulfated chitin derivatives into which the 6-O-sulfate and 6-O-carboxymethyl groups were introduced [11].

V. BIOACTIVATION EFFECTS OF MOLDED CHITIN AND ITS DERIVATIVE

A. Promotion of Wound Healing

The specific chemical agent derived from cartilage that is responsible for the striking biological adjuvant effect of promoting the healing of wounds was reported by Prudden et al. and was identified as chitin [2]. However, few studies concerning this activation effect have been undertaken because of the difficulty of producing pure molded chitin.

Our group conducted a reevaluation of their work from the aspect of adhesive strength found in incised wounds [3]. From the application of the chitin powder to the incised skin of the backs of rats, the resistance to breaking on separation of the full-thickness strip of skin treated with chitin was significantly higher than with absorbable polymer (polyglycolic acid, PGA), nonabsorbable polymer (polyacetal, POM), or skin with no chitin treatment (incised only) on Days 5 and 8. However, no difference was found between areas treated with chitin and areas not treated with chitin on Day 14 after surgery [3].

In addition, Yano et al. performed a similar study to reevaluate the effect of topically administrated molded chitin fiber on tensile strength and the amount of collagen-hydroxyproline [12]. The tensile strength of the incised wounds that were treated with chitin was significantly higher on Days 3, 5, and 7 after surgery. The values of the incised tissues were higher than those of the control group, regardless of whether they had been treated with chitin, except for those with chitin treatment on Day 3 and those without chitin treatment on Day 5. However, the difference in the values between those with and those without chitin treatment was not significant except for Day 7.

Paulette and Prudden documented an increase in the level of collagen in wounds in which "cartilage" was topically injected [13]. In addition, our results indicate that the mechanism of such bioactivation effects might be sought in other events in the process of wound healing rather than simply in an increase in the amount of Type I collagen resulting from collagen synthesis.

Kishimoto and Tamaki immunohistochemically examined the healing process of the experimental third-degree burn wounds treated with a molded chitin dressing and speculated as to its mechanism [14]. Lysozyme-positive cells, which are mainly histiocytes, were found 4 to 14 days after the occurrence of the burn in both the chitin dressing group and the group with undressed wounds. In the group with no dressings, the peak number of these cells occurred on the fourth day, but it lasted from the fourth day to the seventh day in the chitin dressing group. In the early (granulation tissue) stage, more fibroblastic cells and histiocytes were observed in the chitin dressing group than in the group with no dressing. Renewed collagen levels were also found to be different between the two groups. The collagen was found to be thick in the group with no dressing, but it was fine in the chitin dressing group. Even in the late (scar tissue) stage, fine collagen was observed in the chitin dressing group.

It is well known that chitin is a substrate of lysozyme (muramidase), which is found in histiocytes originating from monocytes or macrophages. The biofunction of lysozyme is to decompose chitin to N-acetyl glucosamine-N-acetylemuramic acid. According to Leiborch and Ross, the role of histiocytes in wound healing is biological wound debridement and the promotion of the proliferation of fibroblasts [15]. These two kinds of collagen originate from different fibroblastic cells. Such speculation and the results of these studies suggest that histiocytes might be induced by chitin and that they might promote the proliferation of fibroblastic cells that produced fine collagen in the process of wound healing.

Okamoto et al. demonstrated the increase of interleukin-1 (IL-1) and the fibroblast proliferation factor in the exudates taken from the surrounding areas of molded chitin implantation in canine wounds. They found no difference between chitin and chitosan in terms of the wound healing process in animals. Such wound healing activation effects were also confirmed in the chitin derivatives such as chitosan, partially deacetylated chitin, and N-carboxybutyl chitosan.

According to Muzzarelli [17], lysozome, normally originating in macrophages, hydrolyses susceptible modified chitosan to oligomers that activate macrophages to produce interferon, tumor necrosis factor (TNF), and IL-1. Activated macrophages produce

N-acetyl-β-D-glucosamidase, which catalyzes the production of D-glucosamine and N-acetylglucosamine and substitutes glucosamines from oligomers. These amino sugars are sensitive to fibroblasts, which proliferate under the action of IL-1, for incorporation into hyaluronate and glycosaminoglycans, thus guiding the ordered depositing of collagen, which is also influenced by oligomers.

B. Immunological Adjuvant Activity

Suzuki and coworkers demonstrated the tumor suppression abilities of chitin and chitosan in an experimental system among tumor-bearing mice [18]. They also observed that chitin and chitosan have an excellent ability to activate peritoneal macrophages and their enzymes. In a series of extensive immunological studies regarding chitin and its derivatives, they confirmed that N-acetyl-chitohexose, a water-soluble homologue of chitin, is promising as an immunopotentiator. Nishimura et al. reported that 70% DAC, displayed adjuvant activity for the induction of humoral and cell-mediated immunity, including cytokine production [19,20].

Such DAC was also found to be the most effective chitin derivative for activation of *in vivo* murine peritoneal macrophages for augmentation of host resistance against tumor growth and bacterial infection in mice. A close correlation was observed between the immunological activity and the degree of deacetylation of chitin. The degree of deacetylation was optimal at 70% for the induction of the adjuvant effect.

C. Bacteriostatic Effects

It is well known that chitosan is also used as a preservative in the food processing field. Allan generally documented the growth inhibitory effects of chitosan solutions on bacterial and fungal infections [20]. Suzuki and coworkers reported good protective adjuvant effects of chitin and chitosan on *Staphylococcus aureus* and *Candida albicans* infections in mice [21,22]. They also reported that N-acetyl chitohexaose, a hexamer of chitin, can be used as a biological response modifier for the opportunistic infection of microbes. Ilda et al. indicated that partial (70%) DAC effectively stimulated nonspecific host resistance against Sendai virus and *Escherichia coli* [23].

Seo, Mitsuhashi, and Tanibe showed the good antifungal and antibacterial activity of chitosan (80% deacetylation degree) and developed chitosan blended polynostic rayon fiber for underwear material [24].

Sano et al. demonstrated the good inhibitory effects of water-soluble derivatives of chitin such as low molecular weight chitosan and phosphorylated chitin on the adsorption of oral streptococcal strains, which causes tooth decay and possesses different hydrophobicities, to saliva-treated hydroxyapatite [25].

VI. BIOMEDICAL APPLICATIONS OF MOLDED CHITIN AND ITS DERIVATIVES

Recently, many efforts have been made to apply chitin and its derivatives to utilization for biomedical purposes. Their efficacy has been demonstrated as chelating or absorbing materials of metal and cholesterol [26], cell-separating materials, as membranes for ultrafiltration, as heparinlike substances, as affinity chromatographs, and as carriers for bioreactors. We describe several interesting biomedical applications below.

A. Suture

Chitin is an absorbable suture material with suitable mechanical properties. The chitin suture was prepared from braided fibers (30–50 denier, 30–100 filaments) (Fig. 4). The straight-pull tensile strength of the chitin suture of U.S. Pharmacopeia (USP) 4-0 size in a dry condition was 2.25 pm 0.05 kilograms (kg), which is much the same as Dexon^{TM}, consists of commercially available polyglycolic acid (PGA) and is much stronger than catgut. In a straight pull, the strength of chitin and catgut sutures was less in the wet condition than in the dry condition. However, in a knot pull, the strength of the chitin suture was not less in the wet condition than in the dry condition. The chitin suture had less elongation in a straight pull in a dry condition. In a wet condition, the elongation of the chitin suture was much the same as other suture materials.

Nakajima et al. measured the tensile strength of the chitin suture in the dorsum of rabbits or in incubated body fluids [5]. In the dorsum of rabbits, the tensile strength of the chitin decreased over time and was half of the initial strength on Day 14. Good tensile strength was maintained for a longer period than PGA and catgut sutures in serum, bile, pancreatic juice, and urine. In contrast, the chitin suture was found to be weak against strong acid with a low pH such as gastric juice. These results suggest that the chitin suture has good resistance to hydrolysis and digestive enzymes, and that care must be taken when chitin is exposed to a strong acid in the human body.

Nakajima measured the tensile strength of strips of rabbit skin sutured with chitin and PGA of a standard USP 4-0 suture size. They reported that the chitin sutures demonstrated significantly superior tensile strength after two and three weeks, but that there was no difference in tensile strength between the two sutures by four weeks after surgery. Its clinical application has already been established. The overall clinical results were acceptable; there were no abnormal tissue reactions or infections. There were no allergic reactions, and there were no untoward hepatic, renal, and hematological effects. The wound healing after the application of the chitin suture was reported to be uneventful, but clinically there were no apparent wound healing activation signs in comparison with the conventional sutures.

Suture

Figure 4   SEM finding of chitin suture.

B.   Wound Dressing

Temporary wound dressing materials have been produced by preparing chitin fibers, nonwoven fabrics, and sponges. Chitin artificial skin, in the form of nonwoven fabric, is commercially available in Japan. A number of studies have already reported satisfactory results regarding its wound healing effects, histological reactions, and wound management abilities [27]. The mode of application is similar to that of synthetic and biological dressing materials such as lyophilized dermal porcine skin (LDPS) and collagen fabric and membrane. The chitin artificial skin is superior to the other dressing materials in terms of its analgesic and hemostatic effects, its liquefaction tolerance, and epidermization. It has also been noted that the surface of the healed wound with chitin treatment is smooth and normal. These aspects can be attributed to the bioactivation effects of chitin.

Sponge-type materials were clinically applied for temporary wound coverage of severe soft-tissue damage [28], for nasal and paranasal sinus treatment of intranasal surgery [29], and for postoperative treatment of oral surgery [30]. Such materials were used because of their abilities in the areas of wound protection, pain relief, promotion of wound healing, and hemostatic and bacteriostatic effects. Cotton-type materials were used for pressure sores and good results were obtained. However, we have to mention that it is no use to apply molded chitins once infection is present.

Allan et al. documented the clinical application of chitosan membrane for burn therapy [20]. Biagini et al. recommended N-carboxy butyl chitosan gel for skin tissue repair and speculated that the wound healing mechanism is similar to that of chitin [31].

C.   Sustained Release Drug Delivery

Pangburn, Trescony, and Heller have shown the lysozyme sensitivity of partially deacetylated chitin and its possible use in a self-regulated drug delivery system [32]. We have already reported the sustained release of cisplatin (CDDP) and amikacin (AMK) from capsules made of chitin. By using CDDP-chitin complexes, Suzuki et al. obtained the release of high concentration platinum under an implantation to mouse muscle during more than 8 weeks. Also, many researchers have made great efforts to develop the new drug delivery systems using chitin and its derivatives, such as 5-fluorouracil chitin and chitosan conjugates [34] and 6-0-carboxymethyl chitin gel containing Neocarzinostatin [35].

D.   Topical Hemostatic Agents

Olsen et al. reported the agglutination of blood cells by chitosan acetate added to blood [36]. Malette et al. documented the hemostatic effect of chitosan from implantation of a Dacron vascular prosthesis coated by chitosan solution.

We have developed chitin hemostatic, which is a partially (65%) deacetylated chitin hydrochloride in the form of fibers; it has high water absorption while readily changing to gel form in water [38]. Chitin hemostatic showed a significant reduction of blood loss from a canine iliac bone hole. A significantly high amount of release of β-thromboglobulin and platelet factor IV *in vitro* indicated a platelet activation effect of chitin. Also, some clinical trials are ongoing in Japan [39].

E.   Soft Tissue and Bone Substitute

In the application as a soft tissue and bone substitute, the fibers and sponges of chitin were completely absorbed and disappeared five to six months later. The chitin sponge implanted into the bone of the rabbit femoral condyle was found to be completely absorbed at the fourth month after surgery [4].

Our investigation on the applicability of molded chitin has led to the development of a braided filamentous substitute for extraarticular ligaments and tendons [40]. In an experimental model having a 1-inch (2.5-cm) gap in canine calcaneal tendons, it was shown that grafting with chitin ligaments provided superior reconstruction compared to that with polyester. Mean duration to filling of the gap and to induction of a neotendon composed of bundles of fibrous tissue was 6 weeks, and it took 12 weeks until mechanical properties were recovered to a level of normal tendon (Fig. 5). There is a slight problem with chitin ligaments in terms of initial strength; nevertheless, we obtained satisfactory results with respect to the processes of absorption, replacement, and tendon regeneration using chitin ligaments.

Ohyabu and coworkers implanted chitin sponge to the rabbit meniscal defect and observed the S-100 protein-positive cells (chondrocytic cells) in regenerated tissue and suggested the possibility of meniscal repair by molded chitin [41]. Maeda et al. reported the possibility of osteogenesis induced by chitin in the bone of rabbit femoral condyles [42]. Borrah, Scott, and Wortham documented the osteogenic potential of N-acetyl chitosan in regard to endochondral bone repair and osseous induction in the metacarpal and fibular bone defect model of the rabbit [43].

A composite cement of calcium phosphate and polysaccharide for a bone substitute is currently being developed. This cement is self-hardening when both powder (α-TCP, tetracalcium phosphate) and liquid (chitosan, citric acid, and glucose) components are mixed together and can be molded into a desirable shape.

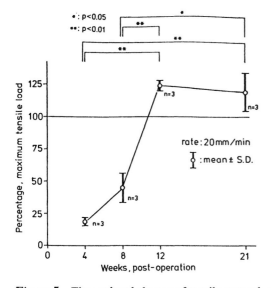

Figure 5   Time-related change of tensile strength of chitin neotendon after surgery.

F. Miscellaneous

1. Ophthalmology

In the field of ophthalmology, Allan et al. conducted a study on chitin n-butyrate and chitosan soft lenses [20]. We also developed chitin soft lenses. Chitin and chitosan oxygen-permeable soft lenses can be considered as a potential eye wound healing dressing.

2. Orthopedic Surgery

In the field of orthopedic surgery, we are developing an intramedullary pin for the internal fixation of fractures of an unloaded site.

3. Cardiovascular Surgery

Chitosan sealant for vascular grafts was reported to have anticoagulant properties in the field of cardiovascular surgery [44].

ACKNOWLEDGMENTS

The authors thank Dr. Seiichi Tokura, professor, Hokkaido University, for his invaluable advice.

REFERENCES

1. Kohn, P., Winzier, R.J., and Hoffmann, R.C., Metabolism of D-glucosamine and N-acetyl-D-glucosamine in the intact rat. *J. Biol. Chem.*, 237:304–308 (1962).
2. Prudden, J.F., Migel, P., Hanson, P., Friedrich, L., and Balassa, L., The discovery of a potent pure chemical wound-healing accelerator. *Am. J. Surg.*, 119:560–564 (1970).
3. Maeda, M., Inoue, Y., Iwase, H., and Kifune, K., Characteristics of chitin for orthopaedic use, In *Chitin, Chitosan and Related Enzymes*, J.P. Zikakis (ed.), Academic Press, New York, pp. 411–415 (1984).
4. Maeda, M., Chitin as a biomaterial. *Bessatsu seikei geka*, 18:196–200 (1990).
5. Nakajima, M., Atsumi, K., Kifune, K., and Kanamaru, H., Chitin is an effective material for sutures. *Jpn. J. Surg.*, 16:418–424 (1986).
6. Sannan, T., Kurita, K., and Iwakura,Y., Studies on chitin 2. Effect of deacetylation on solubility. *Mikromol. Chem.*, 177:3589–3600 (1976).
7. Kurita, K., Sannan, T., and Iwakura,Y., Studies on chitin 4. Evidence for formation of block and random copolymers of N-acetyl-D-glucosamine and D-glucosamine by hetero- and homogeneous hydrolyses. *Makromol. Chem.*, 178:3197–3202 (1977).
8. Sashiwa, H., Saimoto, H., Shigemasa, Y., Ogawa, R., and Tokura, S., Distribution of the acetamide group in partially deacetylated chitins. *Carbohydrate Polymers*, 16:291–296 (1991).
9. Tokura, S., Nishi, N., Nishimura, S., Ikeuchi, Y., Azuma, I., and Nishimura, K., Physico-chemical, biochemical and biological properties of chitin derivatives, In *Chitin, Chitosan and Related Enzymes*, J.P. Zikakis (ed.), Academic Press, New York, pp. 303–325 (1984).
10. Hirano, S., Tanaka, Y., Hasegawa, M., Tobetto, K., and Nisioka, A., Effect of sulfated derivitives of chitosan on some blood coagulant factors. *Carbohydr. Res.*, 137:205–215 (1985).
11. Murata, J., Saiki, I., Makabe, T., Tsuta, Y., Tokura, S., and Azuma, I., Inhibition of tumor-induced angiogenesis by sulfated chitin derivatives. *Cancer Research*, 61:22–26 (1991).
12. Yano, H., Iriyama, K., Nishiwaki, H., and Kifune, K., Effect of N-acetyl-D-glucosamine on wound healing in rats. *Mie Med. J.*, 135:53–56 (1985).
13. Paulette, R.E., and Prudden, J.F., Studies on acceleration of wound healing with cartilage. *Surg. Gynec. Obstet.*, 108:408–415 (1959).
14. Kishimoto, S., and Tamaki, K., Immunohistochemical and histological observations in the

process of burn wound healing in guinea pig skin under chitin membrane dressing. *Acta Dermatol. Kyoto*, 82:471–479 (1987).

15. Leiborch, S.J., and Ross, R., The role of the macrophage in wound repair. A study with hydrocortisone and antimacrophages serum. *Am. J. Path.*, 78:71–100 (1975).

16. Okamoto, Y., Minami, S., Matsuhashi, A., Sashiwa, H., Saimoto, H., Shigemasa, Y., Tanigawa, T., Tanaka, Y., and Tokura, S., Application of chitin and chitosan in small animals, In *Advances in Chitin and Chitosan*, C.J. Brine, P.A. Sandford, and J.P. Zikakis (eds.), Elsevier Applied Science, New York, pp. 70–78 (1992).

17. Muzzarelli, R., Role of lysozyme and N-acetyl-3-D-glucosamidase in the resorption of wound dressing, In *Advances in Chitin and Chitosan*, C.J. Brine, P.A. Sandford, and J.P. Zikakis (eds.), Elsevier Applied Science, New York, pp. 25–33 (1992).

18. Nishimura, K., Nishimura, S., Nishi, N., Saiki, I., Tokura, S., and Azuma, I., Immunological activity of chitin and its derivatives. *Vaccine*, 2:93–92 (1984).

19. Nishimura, K., Ishihara, C., Ukei, S., Tokura, S., and Azuma, I., Stimulation of cytokine production in mice using deacetylated chitin. *Vaccine*, 4:151–156 (1986).

20. Allan, G.G., Altman, L.C., Bensinger, R.E., Ghosh, D.K., Hirabayashi, I., Neogi, A.N., and Neogi, S., Biomedical applications of chitin and chitosan, In *Chitin, Chitosan and Related Enzymes*, J.P. Zikakis (ed.), Academic Press, New York, pp. 119–133 (1984).

21. Suzuki, S., Watanabe, T., Mikami, T., Matsumoto, T., and Suzuki, M., Immunoenhancing effects of N-acetylchitohexose, In *Advances in Chitin and Chitosan*, C.J. Brine, P.A. Sandford, and J.P. Zikakis (eds.), Elsevier Applied Science, New York, pp. 96–105 (1992).

22. Kobayashi, M., Watanabe, T., Suzuki, S., and Suzuku, M., Effect of Nacetylchitohexose against *Candida albicans* infection of tumor bearing mice. *Microbiol. Immunol.*, 34:413–426 (1990).

23. Ilda, J., Uie, T., Ishihara, C., Nishimura, K., Tokura, S., Mizukoshi, N., and Azuma, I., Stimulation of non-specific host resistance against Sendai virus and *Escherchia coli* infections by chitin derivatives in mice. *Vaccine*, 5: 270–274 (1987).

24. Seo, H., Mitsuhashi, K., and Tanibe, H., Antibacterial and antifungal fiber blended by chitosan, In *Advances in Chitin and Chitosan*, C.J. Brine, P.A. Sandford, and J.P. Zikakis (eds.), Elsevier Applied Science, New York, pp. 34–40 (1992).

25. Sano, H., Matsukubo, T., Shibasaki, K., Itoi, H., and Takaesu, Y., Inhibition of adsorption of oral streptococci to saliva treated hydroxyapatite by chitosan derivatives. *Bull. Tokyo Dent. Coll.*, 32:9–17 (1991).

26. Gordon, D.T., and Besch-Williford, C., Action of amino polymers on iron status gut morphology and cholesterol level in the rat, In *Chitin, Chitosan and Related Enzymes*, J.P. Zikakis (ed.), Academic Press, New York, pp. 97–117 (1984).

27. Kifune, K., Clinical application of chitin artificial skin, In *Advances in Chitin and Chitosan*, C.J. Brine, P.A. Sandford, and J.P. Zikakis (eds.), Elsevier Applied Science, New York, pp. 9–15 (1992).

28. Maeda, M., Inoue, Y., Yanagihara, Y., and Iwase, H., Porous chitin (sponge) as a temporary deep wound dressing material. *Seikeigeka*, 43:1441–1446 (1992).

29. Manabe, Y., Saito, H., Takanami, N., and Matsuki, M., Application of chitin and paranasal sinus treatment. *Jibirinsyou Suppl.*, 26:65–69 (1988).

30. Furutani, M., Iida, M., Yamaguchi, Y., Mori, M., Fujita, S., and Sato, T., A new wound dressing of chitin sponge, its clinical application for oral surgery. *Shika Journal*, 24:451–454 (1986).

31. Biagini, G., Muzzarelli, R.A.A., Giardino, R., and Castaldini, C., Biological materials for wound healing, In *Advances in Chitin and Chitosan*, C.J. Brine, P.A. Sandford, and J.P. Zikakis (eds.), Elsevier Applied Science, New York, pp. 16–24 (1992).

32. Pangburn, S.H., Trescony, P.V., and Heller, J., Partially deacetylated chiton: Its use in self-regulated drug delivery systems, In *Chitin, Chitosan and Related Enzymes*, J.P. Zikakis (ed.), Academic Press, New York, pp. 3–19 (1984).

33. Suzuki, K., Yoshimura, H., Matsuura, H., Katoh, T., Nakamura, T., Tsurutani, R., and Kifune, K., A new slow-releasing drug delivery system for chemically combined cisplatin with chitin for intraoperative local application: An experimental study, In *Recent Advance in Diseases of the Esophagus*, Springer-Verlag, Tokyo, pp. 865–870 (1993).

34. Ouchi, T., Inosaka, K., Banba, T., and Ohya, Y., Design of chitin or chitosan/5-fluorouracil conjugate having antitumor activity, In *Advances in Chitin and Chitosan*, C.J. Brine, P.A. Sandford, and J.P. Zikakis (eds.), Elsevier Applied Science, New York, pp. 106–115 (1992).

35. Watanabe, K., Saiki, I., Matsumoto, Y., Azuma, I., Seo, H., Okuyama, H., Uraki, Y., Miura, Y., and Tokura, S., Antimetastatic activity of neocarzinostatin incorporated into controlled release gel of CM-chitin. *Carbohydrate Polymers*, 17:29–37 (1992).

36. Olsen, R., Schwartzmiller, D., Weppner, W., and Winandy, R., Biomedical applications of chitin and its derivatives. In *Chitin and Chitosan*, G. Skjak-Break, T. Anthonsen, and P. Sandford (eds.), Elsevier Applied Science, London, pp. 813–828 (1989).

37. Malette, W.G., Quigier, H.J., Gaines, R.D., Johnson, N.D., and Rainer, W.G., Chitosan: A new hemostatic. *Ann. Thorac. Surg.*, 36:55–58 (1983).

38. Maeda, M., Inoue, Y., Iwase, H., Endo, R., and Kifune, K., Chitin as a topical hemostatic material. *Transactions of the Japanese Society for Biomaterials*, 15:105 (1993)

39. Miyahara, T., Clinical experience of chitin hemostatic. *Sinryou to Shinyaku*, 28:1703–1710 (1991).

40. Maeda, M., Inoue, Y., Iwase, H., and Kifune, K., Experimental study on Achilles tendon reconstruction using molded chitin. *Transactions of the Society of Biomaterials*, 17:96 (1994).

41. Ohyabu, N., Matsui, N., Ohtsuka, T., Taneda, Y., and Kato, T., Experimental study with chitin in the meniscal repair. *Cent. Jpn. Orthop. Traumat.*, 149–150 (1992).

42. Maeda, M., Inoue, Y., Iwase, H., and Kifune, K., Biocompatibility of molded chitin and its applicability as an implant material. *Transactions of the Society for Biomaterials*, 12:21 (1986).

43. Borrah, G.L., Scott, G., and Wortham, K., Bone induction by chitosan in endochondral bones of extremities, In *Advances in Chitin and Chitosan*, C.J. Brine, P.A. Sandford, and J.P. Zikakis (eds.), Elsevier Applied Science, New York, pp. 54–60 (1992).

44. Dutkiewicz, J., Kucharska, M., Papiewski, A., Judkiewicz, L., and Ciszewski, R., Chitosan sealant for vascular grafts with no need of heparinization, In *Advances in Chitin and Chitosan*, C.J. Brine, P.A. Sandford, and J.P. Zikakis (eds.), Elsevier Applied Science, New York, pp. 54–60 (1992).

49

Hydrophilic Polyurethane Coatings: Applications for Hemoperfusion

Ann T. Okkema, James A. Braatz, and Aaron H. Heifetz

W. R. Grace and Co. – Conn., Columbia, Maryland

Raymond M. Hakim and Richard Parker

Vanderbilt University, Nashville, Tennessee

I. INTRODUCTION

One of the primary goals in developing any blood-contacting medical device is to make it blood compatible. The difficulty in achieving this goal is apparent in many devices currently in use, including catheters, hemodialyzers, and blood oxygenators. When blood contacts a foreign surface, proteins immediately adsorb on the surface. Depending on the type of proteins adsorbed and their distribution on the surface, multiple responses may be elicited. These responses include the initiation of the coagulation, contact, and complement cascades as well as cellular adhesion and activation, either by direct contact with the surface or indirectly with the products of these protein pathways [1,2]. The adsorption of plasma proteins such as fibrinogen, fibronectin, and immunoglobulins can promote platelet and leukocyte adhesion and activation, which can lead to thrombus formation. The adsorption of factor XII, the Hageman factor, and its activation by high molecular weight kininogen and prekallikrein initiates the intrinsic coagulation pathway that results in the formation of procoagulant thrombin and subsequent polymerization of fibrinogen to fibrin. The binding of complement protein C3b to a surface may initiate complement activation and subsequent release of inflammatory mediators [3,4]. All of these processes are interrelated and can ultimately culminate in thrombus formation, indicating that the surface is not blood compatible [5,6].

One method to make a surface blood compatible is to reduce protein deposition, which initiates coagulation, complement activation, and cellular adhesion and activation. Since protein binding is dependent on the surface chemistry, morphology, and interfacial energy of the device surface, these responses can be modulated by altering the surface properties. One approach to improving the blood-contacting response is to minimize the interfacial energy between the blood and surface by making the surface more hydrophilic

[7]. A variety of hydrophilic polymer surfaces have been shown to exhibit minimal protein adsorption and, in many cases, reduced platelet adherence [8–12].

A hydrophilic surface can be prepared by (1) manufacturing the device out of a hydrophilic polymer such as a hydrogel or (2) coating the surface of the device with a hydrophilic polymer. Hydrophilic polymers that contain a high percentage of water are mechanically weak and therefore are often used as coatings rather than free-standing gels. In addition, a coating may be preferred when the blood-contacting properties require improvement while the physical and mechanical properties of the device are optimal. When cost is an issue, a thin coating on the order of a monolayer would require a minimal amount of polymer and minimal expense, depending on coating solvent and process.

Cross-linked, water-swollen networks of hydrophilic homopolymers or copolymers are *hydrogels*. Their favorable blood-contacting properties (i.e., reduced protein and platelet binding) can be partially ascribed to their low interfacial energy with water. In addition, the hydrogel's chemical composition, the amount and nature of the water absorbed, the type and number of cross-links, the presence of functional groups, and porosity can affect its blood-contacting properties [13]. Three hydrogels that are commonly used in biomedical applications are poly(2-hydroxyethyl methacrylate) (poly-HEMA) (reviewed in Ref. 10), polyvinyl alcohol (PVA) (reviewed in Ref. 14), and polyethylene oxide (PEO) (reviewed in Refs. 9 and 15).

The imbibed water of a hydrogel can exist in two states: (1) water hydrogen bonded with the polymer network, referred to as *bound water*, and (2) water that is mobile and not associated with the polymer network, referred to as *free water* [10]. The amount and nature of water absorbed by the hydrogel depends primarily on the chemistry of the polymer backbone and affects the mechanical, surface, and transport properties of the hydrogel.

For example, the favorable properties of PEO-based hydrogels are due to the PEO segments' liquid-like properties. PEO tetrahedrally coordinates water molecules through hydrogen bonding with the ether oxygen. This allows it to fit into the voids in the water structure, thereby minimally perturbing the water structure and minimally increasing the interfacial energy [16]. In addition, the great flexibility and mobility of the hydrated PEO chains creates a large excluded volume that minimizes or prevents adsorption of proteins. Hydrophilic PEO surfaces have been prepared by incorporating PEO into a block copolymer, cross-linking it into a hydrogel, covalently binding it to a surface or molecule, or physically adsorbing it to a surface or molecule.

The use of PEO as a nonthrombogenic material in cardiovascular research was initiated by its use in preventing the adsorption of biological macromolecules to chromatographic supports [17]. Subsequently, PEO-based hydrogels were found to have favorable blood-contacting properties (i.e., reduced protein and platelet deposition). These properties are dependent on the chain length of the PEO segment, the degree of phase separation, the mobility of the chain and its ability to migrate to the surface, and the water content of the polymer. While PEO hydrogels have reduced platelet binding, this does not necessarily translate into improved blood compatibility. Llanos and Sefton have shown that PVA hydrogels and PEO-grafted PVA hydrogels have minimal platelet deposition, but are reactive toward platelets, leading to severe thrombocytopenia [9,18]. Therefore, improving the blood-contacting properties of PEO hydrogels remains a challenge.

Uncoated activated charcoal is thrombogenic, resulting in a dramatic decrease in platelet and white blood cell counts, as well as blood coagulation [19,20]. In addition, charcoal fines generated during shipping and clinical handling of brittle-charcoal-containing devices can be released into the bloodstream, in which they can cause microembolisms or become lodged in the lungs, spleen, and kidneys [21]. To improve the blood-contacting properties, the charcoal support is generally coated with a blood-compatible polymer. This coating also acts as a barrier that prevents microparticle release. Ideally, the coating allows adsorption of lower molecular weight toxins while preventing adsorption of higher molecular weight proteins and cells that could result in thrombosis.

One of the first applications of a hydrogel coating on a blood-contacting device used charcoal coated with HEMA in a hemoperfusion device [19]. Charcoal hemoperfusion permits direct and rapid removal of toxins from the blood through contact with an adsorbent material in a fixed-bed column. It is used in the treatment of acute intoxication, liver failure, aluminum removal in chronic hemodialysis patients, and other conditions. A typical hemoperfusion device consists of a cartridge containing between 50 and 200 grams of a nonspecific adsorbent (0.4- to 1.0-millimeter [mm] diameter) with a surface area greater than 800 square meters per gram (m^2/g). Arterial blood flows against gravity at flow rates ranging from 100 milliliters per minute (ml/min) to 400 ml/min, depending on the application. The function and blood-contacting properties of a hemoperfusion device depend on the chemistry and shape of the adsorbent used as well as the casing design and resulting fluid dynamics. Activated charcoal is the most commonly used adsorbent, while silica and polymeric resins have also been used.

Charcoal hemoperfusion adsorbents have been coated with a variety of polymers, including cellulose acetate [22], cellulose nitrate [23], polyHEMA (HEMA) [19,24], polyetherurethane [25], and polyacrylate-polymethacrylate copolymer [26]. In all cases, the targeted toxins were at least partially removed from the circulation. In general, platelet adsorption was reduced for the coated charcoal compared with the uncoated charcoal. The coatings reduced the number of particles released from the support. In addition, the coating reduced packing and clumping of the charcoal, improving flow characteristics. However, the adsorption capacity of the support decreased with increasing polymer coating content. Therefore, the amount of polymer coating must be optimized to improve the blood-contacting properties without compromising the toxin-binding capacity.

A PEO-based polyurethane hydrogel, HYPOL® XP-5, was shown to be protein non-adsorptive, nontoxic, and nonleachable when coated onto silica and polymer tubing [27]. The PEO-based prepolymer was originally named BIOPOL® XP-5, then later named HYPOL XP-5, and was produced by the Organic Chemicals Division of W. R. Grace & Co. – Conn. (now sold as HYPOL® PreMA G-50 by the Hampshire Chemical Corporation, Lexington, MA). This polyurethane hydrogel, which contains approximately 85% water, can be used as a free-standing hydrogel, a conjugate with biological molecules, as well as a coating [27,28].

Based on these favorable properties, this PEO-polyurethane hydrogel was evaluated as a coating on charcoal used in a hemoperfusion device. The primary function of the coated charcoal adsorbent was to remove deferoxamine (DFO) chelated aluminum and iron from patients suffering from aluminum and iron overload. The microparticle release, protein binding, aluminum and iron removal, and *in vitro* and *ex vivo* blood compatibility properties of the hydrogel-coated charcoal were evaluated and are described below.

II. MATERIALS AND METHODS

A. HYPOL XP-5 Polyurethane Charcoal Coating

The HYPOL XP-5 polyurethane prepolymer used to coat the charcoal adsorbent is based on a polyether triol, (molecular weight [MW] $\approx$ 7000), composed of 75% ethylene oxide and 25% propylene oxide units end capped with isophorone diisocyanate [27]. The cross-linked polyurethane hydrogel is formed by reacting the prepolymer with water. The percentages of HYPOL XP-5 polyurethane coatings were calculated as the weight of prepolymer to the weight of the uncoated charcoal and ranged from 0.5% to 60%.

The spherical charcoal adsorbent (BAC-MU, Kureha Chemical Industry Co., Ltd., Tokyo, Japan) has a surface area between 1000 and 1400 m^2/g, and a diameter ranging from 0.4 to 1 mm.

Prior to coating, the charcoal was washed ultrasonically with 2-propanol and water, followed by a 30-minute water rinse. The charcoal was vacuum dried for 1 hour and then heated to 325°C for 12 hours.

For coating levels greater than 10% HYPOL XP-5, 20 grams of washed charcoal was placed in a 500-ml round-bottom flask. Initially, 50 ml of dry acetone were added to the charcoal, followed 30 minutes later with 150 ml of a HYPOL XP-5 acetone solution. The flask was placed on a rotary evaporator (Rotavapor RE120, Büchi) rotated at 90 revolutions per minute (rpm) under vacuum until the acetone was completely evaporated. The coated charcoal was subsequently hydrated in 500 ml of high purity, reverse osmosis/deionized (RO/DI) water (Milli-Qwater, RO10/Milli-Q UF Plus) overnight and then rinsed and filtered with 1000 ml (cc) of Milli-Qwater before vacuum drying at 40°C for 5 days.

For coating levels less than or equal to 10% HYPOL XP-5, 100 grams of washed charcoal were placed in a 500-ml tissue culture roller flask and hydrated at 37°C for 30 minutes using 200 ml Milli-Qwater. Then, 300 ml of a HYPOL XP-5 solution was added to the hydrated charcoal and mixed by rotation for 2 hours at 4 rpm. After coating, the charcoal was filtered using fritted glass filters and rinsed with 1800 ml of Milli-Qwater. The coated charcoal was subsequently vacuum dried at 40°C for 5 days.

B. Minicartridge Preparation

Miniature hemoperfusion cartridges (minicartridges) containing approximately 2.2 grams of charcoal were prepared. The cartridge casing was made from a 10-cc syringe (PLAS-TEPAK, Becton Dickinson) cut at the 5-cc graduation. A nylon filter was placed in the bottom of the syringe and 2.2 g of charcoal was added. A luer lock endcap (Mini-filter, Amicon Ireland, Ltd.) with a nylon filter was glued onto the end of the syringe using a two-part epoxy (Amicon T674A and T674B, Emerson and Cuming, Inc., Woburn, MA).

C. Charcoal Particle Release

For charcoal particle release, 9 ml of water were added to 1 cc of the charcoal support in 15-cc conical tubes and rotated at 30 rpm for 24 hours. Fine particle release of the charcoal supports was determined by measuring light scattering at 550 nanometers (nm) using a UV/VIS Perkin Elmer Lambda II spectrophotometer.

D. Nitrogen Porosimetry

Nitrogen porosimetry characterization of the charcoal supports was performed using a 2400 ASAP Nitrogen Porosimeter (Micromeritics).

E. Solute Binding

Cytochrome C (Type V-A from bovine heart, SIGMA Chemical Co., St. Louis, MO) 2.0 mg/ml in phosphate buffered saline (PBS, pH 7.4) was used to evaluate protein adsorption. The DFO/Fe solution was prepared from equimolar amounts of deferoxamine mesylate (DFO) and ferric nitrate (9 H_2O). The DFO/Fe (ferrioxamine) concentration used was based on (1) a typical clinical 1.5-g dose of DFO administered to aluminum overload patients and (2) the assumption that this amount is completely adsorbed by a full-size hemoperfusion device containing 80 g of charcoal. The starting concentration of DFO/Fe was based on 18 mg of DFO/Fe per gram of charcoal support.

For the cytochrome C and DFO/Fe binding experiments, 10 ml of each solution were added to 1 cc of charcoal support in 15-cc conical tubes. The tubes were rotated on a rotary mixer at 30 rpm at room temperature for 3 hours. Control samples containing no charcoal supports were also rotated. At the following times, 1-ml samples of the solute solutions were removed: $t = 0, 15, 30, 45, 60, 90, 120, 150$, and 180 minutes. The amount of cytochrome C or DFO/Fe remaining in the solution was measured spectrophotometrically at 432 nm using a Lambda II Perkin Elmer UV/VIS spectrophotometer. The amount of solute bound per ml of charcoal was calculated by difference.

F. Minicartridge Perfusion

The perfusion solution was placed in a 500-ml Schott Duran glass bottle fitted with a cap containing inlet and outlet ports. Silicone tubing (size 16, Cole Parmer) was used to connect the bottle to the minicartridge. The test solution was recirculated through the minicartridge at 300 ml/min using a Masterflex pump (Cole-Parmar, Model 7520-25) with a semiocclusive pump head (Easy Load, Model 7518-00, Cole Parmer). The minicartridges were primed with Milli-Qwater for 10 minutes prior to perfusing with the solutions. Samples were taken precartridge using a 3-ml syringe throughout the perfusion.

G. Deferoxamine/Aluminum Perfusion Studies

The minicartridge recirculating perfusion system was used in the following comparison studies: (1)DFO/Fe binding in serum and water, (2) DFO/Fe and DFO/Al binding in water from a single solution, and (3) DFO/Al in water and serum. In the serum experiments, defined bovine calf serum (Hyclone® Laboratories, Inc., Logan, Utah) was used. Equimolar amounts of aluminum atomic adsorption standard (980 micrograms per milliliter [μg/ml] in 1% HCl) and deferoxamine mesylate were prepared for the DFO/Al binding experiments.

The samples from the DFO/Fe in serum studies were analyzed spectrophotometrically at 432 nm. The DFO/Fe and DFO/Al in the other experiments were analyzed using inductively coupled plasma atomic absorption spectroscopy (ICP-AAS Model 3410, Applied Research Laboratory). ICP analyzes the total amount of iron or aluminum in the sample, but cannot determine whether or not the metals are chelated. Additional minicartridge recirculating perfusion studies were performed with unchelated aluminum, which indicated that free aluminum did not adsorb to the charcoal substrate. These results indicate that only the chelated form of aluminum, DFO/Al (aluminoxamine), bound to the support.

A few of the single-component DFO/Al studies were also analyzed using a benzyl alcohol extraction method [29], which separates the aluminoxamine from free aluminum. The extract containing the DFO/Al was then analyzed by Gascoyne Laboratories (Baltimore, MD), using a graphite furnace atomic absorption spectrometer.

H. *In Vitro* Blood Contact

In the single-pass *in vitro* blood-contacting experiments, human blood was perfused through minicartridges at 1.5 ml/min. Blood from 6 volunteers was drawn into 4 60-ml syringes, each containing 1000 units of heparin. Each cartridge was primed with 250 ml of 0.9% physiological saline containing 750 units of heparin, then drained prior to infusion of blood. A minicartridge was placed on the end of a syringe and filled with blood for 1 minute. After 1 minute, blood was infused through the cartridges at 1.5 ml/min using a Harvard infusion pump. Blood was collected into a tube containing ethylene diamine-tetraacetate (EDTA) at the efferent end of the cartridge. A baseline sample of blood (14 ml) was collected from each syringe prior to attaching a minicartridge. Three consecutive samples (14 ml) were collected from the cartridge and then the cartridge was removed, a baseline sample taken, and a new cartridge was attached. From each sample, 2 ml were aliquoted for complete blood counts (CBCs), free hemoglobin, and platelet counts. The remaining blood was immediately centrifuged. The plasma was separated and frozen at $-70°C$ for C3a testing (C3a, Amersham). The order in which the cartridges were tested using each volunteer's blood was randomized.

I. *Ex Vivo* Blood Contact

The *ex vivo* blood-contacting properties of full-size hemoperfusion devices containing 0.5% HYPOL XP-5 coated charcoal were evaluated using a sheep *ex vivo* model. For this study 6 sheep were subjected to a single 3-hour perfusion treatment and then necropsied approximately 48 hours later. Using bloodlines and 0.5% HYPOL XP-5 hemoperfusion devices, 3 sheep were perfused. The other 3 sheep were used as the controls, and perfused using bloodlines only (National Medical Care, Rockleigh, NJ, arterial and venous Bloodlines, 9200 and 9300 series).

Sheep were preselected according to weight (30–40 kg), normal CBC and blood chemistries, absence of parasite infection, and normal respiration. Catheters (Pharmaseal, polyvinyl extension tubes $0.09 \times 0.125 \times 12$ inches) were surgically inserted into the left jugular vein and left carotid artery a minimum of 3 days prior to the perfusion treatments. In addition, a Cordis introducer sheath was positioned in the right jugular vein for the Swan–Ganz thermistor-tipped catheter. Vascular catheters were filled with heparin/antibiotic solution and flushed daily.

Prior to initiating the perfusion treatment, a Swan–Ganz catheter was inserted into a pulmonary artery of the sheep and connected to the cardiac output computer and pressure transducers. The dialysis bloodline tubing set (National Medical Care, arterial and venous bloodlines 9200 and 9300) was connected to the hemoperfusion device and primed with 2 liters of heparinized saline (10,000 units per liter [U/l]). This tubing set was modified with a bypass circuit using two y-joints and a 40-cm length of bloodline. The bypass was positioned approximately 9 cm from the arterial and venous hemoperfusion device connections. For the control experiments, in which only the bloodline was tested, a 25-cm length of dialysis tubing was substituted for the hemoperfusion device.

The sheep was administered a 10,000-U heparin bolus, and then 3 baseline blood samples were taken. The primed system was attached to the carotid artery and jugular vein catheters and connected to the blood pump (Renal Systems, Mini-pump). The circulation was initiated through the bypass at 200 ml/min for 15 minutes. After the bypass recirculation, the blood path was rerouted through the hemoperfusion device for 3 hours. Heparin (6000 U/hour) was continuously infused into the circuit via an infusion line proximal to the perfusion pump.

Arterial and venous blood samples were taken simultaneously at 15, 30, 45, and 60 minutes and arterial blood only at 90, 120, and 180 minutes. Arterial blood was also sampled at 60 minutes and 24 and 48 hours postperfusion. Platelet counts and CBCs were measured on all samples. Pulmonary arterial pressures, body temperature, cardiac output, and device pressures were monitored throughout the perfusion experiments. Blood chemistries were monitored preperfusion and 30 and 60 minutes postperfusion. Necropsies were performed by Pathology Associates, Incorporated, (Frederick, MD), approximately 48 hours posttreatment.

III. RESULTS AND DISCUSSION

Spherical charcoal (BAC-MU) was coated with 0.5% to 60% HYPOL XP-5 polyurethane (weight percent prepolymer to charcoal). The fine particle release, solute adsorption, and *in vitro* blood compatibility properties of these charcoals coated with HYPOL XP-5 were compared with a commercial cellulose-nitrate-coated charcoal (ENCARB®, National Medical Care). Based upon these results, the support coated with HYPOL XP-5 that had properties similar to or better than the cellulose-nitrate-coated charcoal was selected and its *ex vivo* blood-contacting properties were evaluated.

A. Microparticle Release

Since one of the primary functions of a polymer coating is to prevent fine particle generation and release, the protective effect of the HYPOL XP-5 coating was compared with the cellulose nitrate coating. The amount of charcoal particles released for samples coated with 0.5% to 10% HYPOL XP-5 were evaluated. Mixtures of the charcoal samples in water were rotated for 12 hours as a means of mechanically manipulating the charcoal. As shown in Fig. 1, the supports coated with HYPOL XP-5 generated significantly fewer fines than the cellulose-nitrate-coated and uncoated charcoal. Even the 0.5% HYPOL XP-5 coating was an effective barrier.

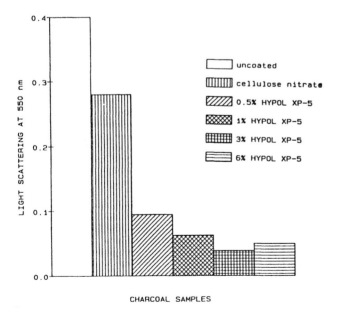

Figure 1 Light scattering at 550 nanometers (nm) resulting from charcoal fines generated by cellulose-nitrate-coated charcoal and charcoal coated with 0.5%, 1%, 3%, and 6% HYPOL XP-5.

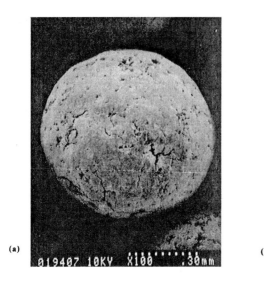

(a)   (b)

Figure 2   Scanning electron micrographs showing the cracks and crevices on the surface of a charcoal bead coated with 10% HYPOL XP-5: (a), 0.3 mm; (b), 100 micrometers.

B.   Polymer Coating Characterization

The HYPOL XP-5 polymer coating was further analyzed using elemental analysis, x-ray photoelectron spectroscopy (XPS), and scanning electron microscopy (SEM) to quantify the coating level and determine the uniformity of the surface coating. SEM analysis showed that the cracks and crevices on the charcoal bead surface were not visibly altered by the coating (see Fig. 2). The polyurethane coating could not be quantified by XPS or elemental analysis because the carbon-based substrate could not be distinguished from the carbon in the polymer coating. In addition, the nitrogen content in the HYPOL XP-5 polyurethane was too low to be used as a quantitative marker.

C.   Nitrogen Porosimetry

Nitrogen porosimetry was used to evaluate the reduction in the surface areas and pore volumes resulting from the HYPOL XP-5 polyurethane and cellulose nitrate coatings. As shown in Table 1, the charcoal coated with 0.5% HYPOL XP-5 and the cellulose-nitrate-

Table 1   Nitrogen Porosimetry Results

Sample	Surface area (m^2/g)	Micropore area (m^2/g)	Pore volume (cc/g)
Uncoated, washed	1470	666	0.49
Cellulose nitrate	1458	606	0.51
0.5% HYPOL XP-5	1444	673	0.48
6% HYPOL XP-5	1295	610	0.43
60% HYPOL XP-5	252	78	0.17

coated charcoal had surface areas and pore volumes similar to the uncoated charcoal. The supports containing greater than or equal to 6% HYPOL XP-5 polyurethane had significantly lower surface areas, suggesting that the polymer was filling the pores. The surface area would not be reduced if increasing the polymer concentration only thickened the coating.

D. Cytochrome C and Deferoxamine/Iron Binding

For the hydrogel-coated charcoal to be effective, protein binding should be reduced, but binding of low molecular weight molecules should be maintained. The solute-binding characteristics for the charcoal coated with 0.5% to 10% HYPOL XP-5 polyurethane were evaluated as a function of time. The effect of the molecular weight of the solute and the type of adsorbing medium (aqueous solutions and serum) on the charcoal supports' binding selectivity were also evaluated.

Two substances of different molecular weights were chosen to evaluate the adsorption properties. Cytochrome C, a protein of approximately MW 12,000, and the chelated product, deferoxamine and iron (DFO/Fe), a clinically important solute of approximately MW 617, were evaluated. Both of these substances are easily assayed spectrophotometrically due to their red color.

As shown in Fig. 3, the HYPOL XP-5 polyurethane coatings reduced protein binding to the charcoal supports. The uncoated charcoal and cellulose-nitrate-coated charcoal adsorbed essentially the same amount of cytochrome C, while the binding capacity and adsorption rate for the charcoals coated with HYPOL XP-5 decreased with increasing coating content.

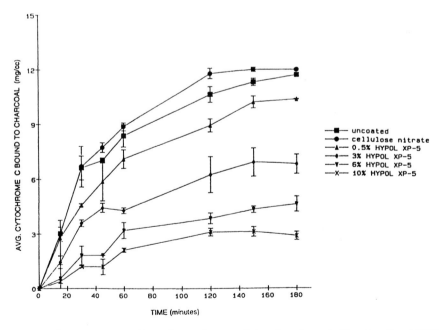

Figure 3 Binding profiles of cytochrome C in phosphate buffered saline (PBS) to uncoated charcoal, charcoal coated with 0.5%, 3%, 6%, and 10% HYPOL XP-5, and cellulose-nitrate-coated charcoal. Error bars (±1 standard deviation) are shown.

As shown in Fig. 4, the HYPOL XP-5 coatings minimally affected the binding capacity for the small solute DFO/Fe. The cellulose-nitrate-coated charcoal, uncoated charcoal, and charcoal coated with 0.5% HYPOL XP-5 had similar DFO/Fe binding profiles in water and adsorbed greater than 50% of the available DFO/Fe in less than 15 minutes.

For a given charcoal support, the differences in the binding profiles of DFO/Fe and cytochrome C are primarily due to the differences in the solutes' molecular weight and the surface area of charcoal available for binding. The low molecular weight DFO/Fe molecule can access the internal pores of the charcoal support, while the much larger cytochrome C molecule is limited to the outer surface of the charcoal bead. These results suggest that, as the polymer concentration increased, more surface area was coated, which reduced the binding capacity of the support for larger molecules. This agrees with the nitrogen porosimetry results.

As shown in Fig. 5, the presence of serum proteins did not affect the binding capacities of the cellulose-nitrate-coated charcoal or charcoal coated with 6% HYPOL XP-5 for DFO/Fe at 36.6 mg/g. This DFO/Fe concentration is well below the maximum binding capacity of the supports, so there was sufficient surface area available for the binding of both DFO/Fe and serum proteins.

E. Deferoxamine/Aluminum Binding

Since the primary application of this hemoperfusion cartridge is the removal of DFO/Al from blood, the binding kinetics of DFO/Al (aluminoxamine) in water and serum were evaluated. The supports' selectivity for DFO/Al and DFO/Fe were compared by evaluat-

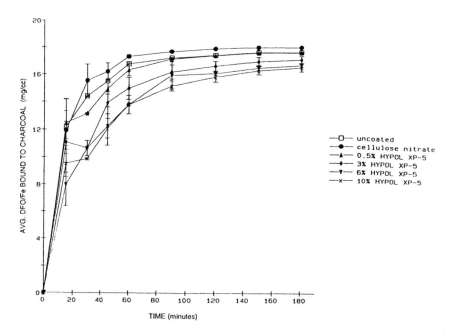

Figure 4 Binding profiles of deferoxamine/iron in water to uncoated charcoal, charcoal coated with 0.5%, 3%, 6%, and 10% HYPOL XP-5, and cellulose-nitrate-coated charcoal. Error bars (±1 standard deviation) are shown.

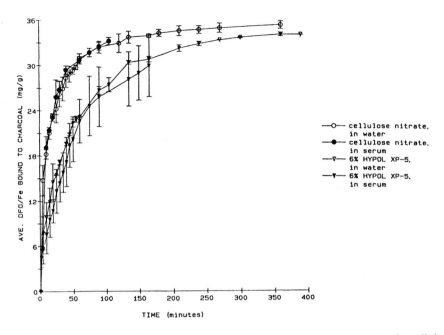

Figure 5 Binding profiles of deferoxamine/iron in water and serum to the cellulose nitrate and charcoals coated with 6% HYPOL XP-5. Error bars (±1 standard deviation) are shown.

ing the adsorption profiles from solutions containing equimolar concentrations of DFO/Al and DFO/Fe in water. As shown in Fig. 6, the DFO/Al and DFO/Fe adsorbed at the same rate for each type of support. In agreement with the DFO/Fe binding results, the charcoal coated with 0.5% HYPOL XP-5 and the cellulose-nitrate-coated charcoal bound essentially the same amount of DFO/Al in water, while the charcoal coated with 6% HYPOL XP-5 adsorbed significantly less.

The charcoal coated with 0.5% HYPOL XP-5 had very similar DFO/Al in serum and DFO/Al in water binding results. However, as shown in Table 2, the maximum amount of solute adsorbed for both cellulose-nitrate-coated charcoal and charcoal coated with 0.5% HYPOL XP-5 was apparently lower in serum than in water. The DFO/Al in serum binding properties of the charcoal coated with 0.5% HYPOL XP-5 was further evaluated using a more sensitive benzyl alcohol extraction technique and graphite-furnace AAS. These results indicated that the charcoal coated with 0.5% HYPOL XP-5 was actually binding 87% of the DFO/Al. Therefore, it appears that, for both charcoal supports, the medium was not affecting the DFO/Al binding capacity at these concentrations.

F. *In Vitro* Blood Contact

In vitro blood-contacting properties of the charcoal supports were evaluated using a single-pass, human blood, minicartridge perfusion experiment. The blood-contacting properties tested included platelet binding, white blood cell (WBC) binding, hemolysis, and complement activation. In all the experiments, an empty "control" cartridge and a cartridge containing cellulose-nitrate-coated charcoal were tested.

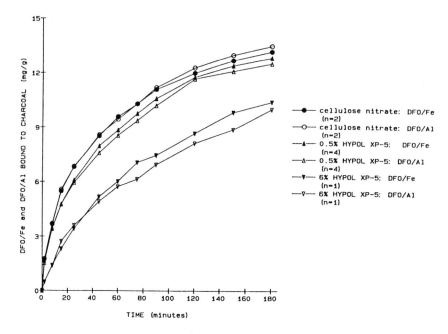

Figure 6 Binding profiles for the cellulose-nitrate-coated charcoal and 0.5% and 6% HYPOL XP-5 coated charcoals, from an aqueous solution containing equimolar concentrations of deferoxamine/iron and deferoxamine/aluminum.

In the first set of experiments, charcoals containing 5%, 10%, 30%, and 60% HYPOL XP-5 polyurethane were evaluated. The platelet and WBC counts were markedly reduced for the supports coated with HYPOL XP-5 and cellulose nitrate, on average 90% and 50% respectively. The HYPOL XP-5 and cellulose nitrate coatings did not significantly affect the cellular binding properties of the charcoal. There were essentially no platelet or WBC binding or complement activation for the empty "control" cartridges, indicating that the casing itself had little or no effect on the cellular binding properties.

Hemolysis or lysing of red blood cells can occur either due to mechanical stress or via an immunological process such as the generation of the complement "membrane attack complex." Hemolysis, as measured by the amount of plasma-free hemoglobin, was minimal and attributed to variations in the charcoal packing of the minicartridge devices.

Table 2 Percent Deferoxamine/Aluminum Bound (Based on Total Available, 18.8 mg/g)

Sample	Water	Serum
Cellulose nitrate	78%	47% (ICP)
0.5% HYPOL XP-5	82%	46% (ICP), 87% (AAS)
6% HYPOL XP-5	62%	30% (ICP)

AAS = graphite furnace atomic adsorption spectroscopy; ICP = inductively coupled plasma atomic adsorption spectroscopy

Complement activation was measured by the amount of C3a generated. There was no significant increase in the C3a levels for any of the coated supports, including cellulose-nitrate-coated charcoal. Typically, a three- to fivefold increase in the C3a plasma concentration is observed within the first 15 minutes of exposure for complement-activating surfaces such as Cuprophan hollow fiber membranes [30].

In another experiment, only WBC and platelet counts for charcoals coated with 0.5% HYPOL XP-5 and cellulose nitrate were evaluated. The charcoal coated with 0.5% HYPOL XP-5 had similar levels of platelet and WBC depletion as the cellulose-nitrate-coated charcoal. However, in this experiment, platelet depletion was considerably less severe, with an average depletion of 60%. In addition, no significant WBC binding occurred for any of the supports.

These *in vitro* results indicate that similar levels of cellular binding were observed for the charcoals coated with HYPOL XP-5 and the cellulose-nitrate-coated charcoal. Varying the amount of the HYPOL XP-5 polyurethane coating did not alter the blood-contacting properties. Although complement activation of the support coated with 0.5% HYPOL XP-5 was not evaluated, the favorable results for the 5%- and 60%-coated charcoal suggests that the charcoal coated with 0.5% HYPOL XP-5 would not be complement activating.

Based on these results, the 0.5%-coated charcoal had the best combination of properties with a high DFO/Al and DFO/Fe adsorption capacity, low protein adsorption, low particulate generations and cellular binding properties similar to ENCARB.

G. Ex Vivo Blood Contact

The *ex vivo* blood-contacting properties of hemoperfusion devices containing charcoal coated with 0.5% HYPOL XP-5 were evaluated using a sheep extracorporeal perfusion system. Sheep are commonly used in the evaluation of dialyzers, blood oxygenators, and hemoperfusion devices due to their size, blood volume, and sensitive cardiopulmonary system [19,31–33].

In this study, three sheep were perfused using a single 0.5% HYPOL-XP-5-coated charcoal hemoperfusion device plus arterial and venous blood lines (National Medical Care, Series 9200 and 9300). In addition, three sheep were treated as controls and perfused using these bloodlines without devices. During each perfusion, the following were evaluated: platelet counts, WBC counts, free hemoglobin, pressure gradient across the device, blood chemistries, and hemodynamic changes. The specific cardiopulmonary responses monitored were the pulmonary arterial pressure (PAP), pulmonary arterial wedge pressure, cardiac output, and blood gases. Notable transient hemodynamic responses can result from anaphylatoxins formed during complement activation, endotoxin contamination, platelet aggregation, and release of vasoconstrictor mediators such as thromboxane A_2.

In general, the 0.5% HYPOL-XP-5-coated charcoal hemoperfusion treatments caused no significant changes in the blood gases or blood chemistries. In addition, the hemoperfusions through the devices did not cause significant hemolysis, as indicated by the relatively constant levels of free hemoglobin in the plasma. The pathological findings indicated there were no significant abnormalities that could be associated with the perfusion experiment. The only abnormalities observed were inflammation and lesions of the catheterized vessels, which were directly related to the surgical procedure.

However, as shown in Fig. 7, there was a dramatic decrease in the platelet counts for

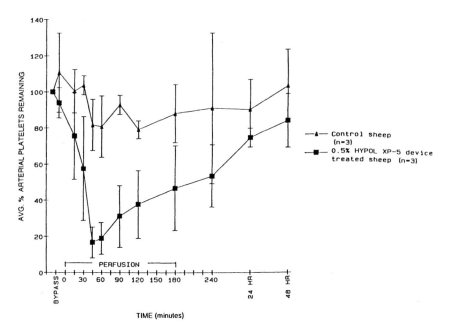

Figure 7   Average percent arterial platelets remaining for three sheep treated with 0.5% HYPOL-XP-5 devices with bloodlines and three control sheep treated with only bloodlines.

the sheep treated with the 0.5% HYPOL-XP-5-coated charcoal devices. The maximum average platelet depletion for the device-treated sheep was 83% of the baseline at 45 minutes, which then rebounded to 50% of the baseline by the end of the 3-hour treatment. The control sheep, treated with bloodlines only, had a maximum average platelet reduction of 20% at 45 minutes, which also returned to 90% by the end of the experiment. Normal platelet counts for both the control and device-treated sheep were essentially restored by 48 hours after treatment. These results suggest that approximately a fifth of the platelet depletion for the device-treated sheep was caused by the tubing surfaces. Despite the transient platelet depletion and relatively high heparin doses used, there were no bleeding problems or related abnormalities observed during the experiment or pathology examination.

Similar transient platelet counts have been reported by several other investigators evaluating hemoperfusion devices [24,25,34]. Weston et al. observed that younger, more active platelets were preferentially removed during perfusion [34]. Gimson et al. showed that platelet depletion, aggregation, and activation were prevented and screen filter pressure reduced by the infusion of prostacyclin, a potent inhibitor of platelet aggregation [35].

In an earlier study, similar platelet depletion levels (85%) were observed for devices containing cellulose-nitrate-coated charcoal (Alukart®, National Medical Care). Since clinical use of Alukart on hemodialysis patients has not been reported to cause any adverse hematological effects related to low platelet counts, this transient platelet decrease, in this sheep model, for the charcoals coated with HYPOL XP-5 and cellulose nitrate is probably a "maximal" response [36].

Similar to the *in vitro* results, WBC depletion was not as severe as the platelet depletion for the charcoal coated with 0.5% HYPOL XP-5. In the first 15 minutes of perfusion, the average WBC depletion was 32% for the charcoal coated with 0.5% HYPOL XP-5; this value returned to baseline levels by the end of the perfusion treatment. Similar WBC depletion was observed for the hemoperfusion device containing cellulose-nitrate-coated charcoal, previously evaluated. No notable WBC binding to the bloodline tubing surfaces occurred with control sheep.

The micrographs in Fig. 8 confirm platelet and WBC binding to the charcoal supports after the 3-hour perfusion treatment. Platelet deposition was nonuniformly scattered across the bead surfaces, with isolated thrombi on some of the charcoal beads. The WBCs appeared to be predominantly located in the surface crevices. These results suggest that the surface roughness and flow dynamics across the surfaces are major determinants of the charcoal's thrombogenicity.

During the 3-hour hemoperfusion treatment, transient pressures developed across the cartridge. As shown in Fig. 9, the magnitude of this pressure gradient varied for each sheep, whereas, the pressure across the bypass circuit for the control sheep did not change during the perfusion (not shown). It was hypothesized that the pressure gradient across the hemoperfusion device was due to platelet and cellular binding to the charcoal, as well as packing and aggregation of the charcoal resulting in reduced flow, leading to increased pressure. However, the time and magnitude of the maximum device pressure gradient did not correlate directly with the platelet and WBC losses since all three device-treated sheep had similar platelet and WBC depletion. These results suggest that cellular binding was not the only cause of this pressure increase.

Interestingly, the pulmonary arterial pressure maxima for the 0.5% HYPOL XP-5 device-treated sheep (see Fig. 10) coincided with the device pressure. In addition, the cardiac output notably decreased for the two sheep eliciting the greatest increase in PAP (VS-75-92 and VS-81-92), while there was essentially no change in cardiac output and

 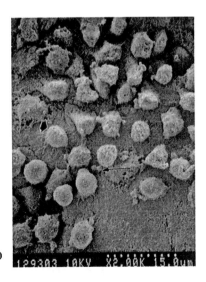

Figure 8   Scanning electron micrograph showing platelets (indicated by small arrow) and white blood cells (indicated by large arrow) adherent to the 0.5% HYPOL XP-5 charcoal surface after a 3-hour perfusion treatment: (a), 150 micrometers; (b), 15 micrometers.

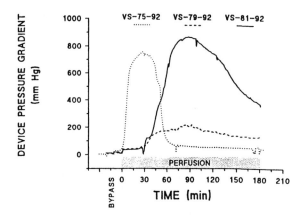

Figure 9   Device pressures for three sheep treated with 0.5% HYPOL XP-5 device and bloodlines.

only a small increase in PAP for sheep (VS-79-93) with the lowest device pressure. The pulmonary wedge pressures and pO_2 did not significantly change for all the device-treated sheep. Since there were no changes in the hemodynamic responses for the control sheep, these responses must be related to the perfusion through the charcoal device.

Cardiopulmonary responses similar to those observed here occur with the generation of anaphylatoxins during complement activation or the release of vasoconstrictors [31,37,38]. These characteristic hemodynamic changes include pulmonary arterial hypertension, systemic arterial hypertension, decreased cardiac output, and hypoxemia [31,38]. In addition to the hemodynamic changes, complement activation is always accompanied by leukopenia, a rapid and severe reduction in the WBC count. Since severe leukopenia and hypoxemia did not occur during the 0.5% HYPOL-XP-5-coated charcoal device perfusions, these *ex vivo* results indicate that the charcoal coated with 0.5% HYPOL XP-5 did not activate complement. These results are in agreement with the *in vitro* results.

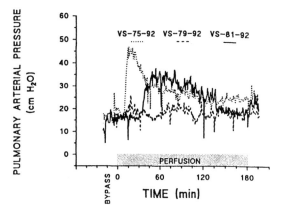

Figure 10   Pulmonary arterial pressures for three sheep treated with 0.5% HYPOL XP-5 device and bloodlines.

On the other hand, vasoactive molecules such as thromboxane can cause similar transient hemodynamic responses [32,34]. Thromboxane A_2 is predominantly synthesized by stimulated platelets as well as polymorphonuclear leukocytes, macrophages, and pulmonary endothelium [38]. Thus, the platelets and other cells adhering to the charcoal support may be releasing a vasoconstricting molecule such as thromboxane A_2 into the blood that is returned to the animal, causing an increase in the PAP.

To determine if thromboxane A_2 release was occurring, thromboxane B_2 levels were evaluated in another *ex vivo* experiment. Thromboxane B_2 is a stable metabolite of thromboxane A_2. Blood samples were taken pre- and postdevice throughout the perfusion treatment. The hematological and cardiopulmonary responses in this experiment were similar to the previous results. For one sheep, thromboxane B_2 levels increased as platelet counts decreased and PAP increased, suggesting that thromboxane release was causing the pulmonary hypertension. However, thromboxane B_2 levels did not significantly change for the other two sheep despite the 80% to 90% decrease in platelet count, 30% to 40% decrease in WBCs, and minor hemodynamic responses.

It can be concluded from these studies that cellular binding in itself is not causing pulmonary hypertension and the increased device pressure gradient. In addition, the differences in the cardiopulmonary responses observed for the sheep are partially dependent on the individual sheep's reactivity.

IV. SUMMARY

The polyurethane hydrogel coating improved the mechanical integrity of the support, reduced protein binding, and maintained its high adsorptive capacity for DFO/Al and DFO/Fe. However, the coating did not significantly affect the blood-contacting properties of the support compared with the cellulose-nitrate-coated charcoal. The blood-contacting properties of a hemoperfusion device are also affected by the irregular surface morphology of the charcoal supports as well as the flow properties resulting from the device design and charcoal packing. It appears that the hydrophilic polyurethane coating applied to these charcoal supports did not completely negate the thrombogenic effect of these other factors.

ACKNOWLEDGMENTS

The assistance of V. Gunther in the coating and characterization of the charcoal supports is gratefully acknowledged. B. Lynn, S. Gibson, and R. Wingard are acknowledged for their extensive assistance with the blood-contacting experiments.

REFERENCES

1. Mason, R.G., H.Y.K. Chuang, S.F. Mohammad, and H.I. Saba, Thrombosis and artificial surfaces, in *The Thromboembolic Disorders*, J. Van der Loo, C.R.M. Prentice, and F.K. Beller (Eds.), Schattauer Verlag, New York, p. 533 (1983).

2. Mohammad, S.F., Extracorporeal thrombogenesis: Mechanisms and prevention, in *Replacement of Renal Function by Dialysis*, J. F. Maher (Ed.), Kluwer Academic, Dordrecht, pp. 229–245 (1898).

3. Kazatchkine, M.D., and M.P. Carreno, Activation of the complement system at the interface between blood and artificial surfaces, *Biomaterials*, 9:30–35 (1988).

4. Chenoweth, D.E., Complement activation in extracorporeal circuits, *Annals New York Academy of Sciences*, 516:306–313 (1987).

5. Szycher, M., Thrombosis, hemostasis, and thrombolysis at prosthetic interfaces, in *Biocompatible Polymers, Metals, and Composites*, M. Szycher (Ed.), Technomic, Lancaster, p. 1 (1983).

6. Hayashi, K., H. Fukumura, and N. Yamamoto, *In vivo* thrombus formation induced by complement activation on polymer surfaces, *J. Biomed. Mater. Res.*, 24:1385–1395 (1990).

7. Andrade, J.D., Interfacial phenomena and biomaterials, *Med. Instrum.*, 7:110–120 (1973).

8. Merrill, E.W., and E.W. Salzman, Polyethylene oxide as a biomaterial, *Amer. Soc. Artif. Int. Organs*, 6:60–64 (1983).

9. Llanos, G.R., and M.V. Seton, Review: Does polyethylene oxide possess a low thrombogenicity? *J. Biomater. Sci. Polymer Edn.*, 4:381–400 (1993).

10. Corkhill, P.H., C.J. Hamilton, and B.J. Tighe, The design of hydrogels for medical applications, *Critical Rev. in Biocompat.*, 5:363–436 (1990).

11. Takahara, A., J. Tashita, T. Kajiyama, M. Takayanagi, and W.J. MacKnight, Microphase separated structure, surface composition and blood compatibility of segmented poly(urethaneureas) with various soft segment components, *Polymer*, 26:987–996 (1985).

12. Okkema, A.Z., T.G. Grasel, R.J. Zdrahala, D.D. Solomon, and S.L. Cooper, Bulk, surface, and blood-contacting properties of polyetherurethanes modified with polyethylene oxide, *J. Biomater. Sci. Polymer Edn.*, 1:43–62 (1989).

13. Bruck, S.D., Aspects of three types of hydrogels for biomedical applications, *J. Biomed. Mater. Res.*, 7:387–404 (1973).

14. Aleyamma, A.J., and C.P. Sharma, Polyvinyl alcohol as a biomaterial, in *Blood Compatible Materials and Devices, Perspective Towards the 21st Century*, C.P. Sharma and M. Szycher (Eds.), Technomic, Lancaster, p. 123 (1991).

15. Graham, N.B., Poly(ethylene oxide) and related hydrogels, in *Hydrogels in Medicine and Pharmacy*, Vol. 2, N. Peppas (Ed.), CRC Press, Boca Raton, FL, p. 95 (1987).

16. Kjellander, R., and E. Florin, Water structure and changes in thermal stability of the system poly(ethylene oxide)-water, *J. Chem. Soc. Faraday Trans.*, 77:2153–2177 (1981).

17. Hiatt, C.W., A. Shelvkov, E.J. Rosenthal, and J.N. Galimore, Treatment of controlled pore glass with poly(ethylene oxide) to prevent adsorption of rabies virus, *J. Chromatogr.*, 56:362–364 (1971).

18. Llanos, G.R., and M.V. Seton, Immobilization of poly(ethylene glycol) onto a poly(vinyl alcohol) hydrogel: 2. Evaluation of thrombogenicity, *J. Biomed. Mater. Res.*, 27:1383–1391 (1993).

19. Andrade, J.D., K. Kunitomo, R. Van Wagenen, B. Kastigir, D. Gough, and W.J. Kolff, Coated adsorbents for direct blood perfusion: HEMA/activated carbon, *Trans. Amer. Soc. Artif. Int. Organs*, 17:222–228 (1971).

20. Chang, T.M.S., Microencapsule artificial kidney: Including updated preparative procedures and properties, *Kid. Intern.*, 10:S2l8–S224 (1976).

21. Hagstam, K.E., L.E. Larsson, and H. Thysell, Experimental studies on charcoal hemoperfusion in phenobarbital intoxication and uremia including histological findings, *Acta Med. Scand.*, 180:593–603 (1966).

22. Yatzidis, H., The use of ion exhange resins and charcoal in acute barbiturate poisoning, in *Acute Barbiturate Poisoning*, H. Matthews (Ed.), Excerpta Medica, Amsterdam, pp. 223–232 (1971).

23. Chang, T.M.S., Removal of endogenous and exogenous toxins by a microencapsulated absorbent, *Canad. J. Physiol. Pharmacol.*, 47:1043–1045 (1969).

24. Gazzard, B.G., M.J. Weston, I.M. Murray-Lyon, H. Flax, C.O. Record, B. Portmann, P.G. Langley, E.H. Dunlop, P.J. Mellon, M.B. Ward, and R. Williams, Charcoal haemoperfusion in the treatment of fulminant hepatic failure, *Lancet*, 1:1301–1307 (1974).

25. Kawanishi, H., M. Nishiki, M. Sugiyama, T. Cho, T. Tsuchiya, and H. Ezaki, Basic study of

polyetherurethane coated bead-type activated charcoal for direct hemoperfusion, *HIJM*, 32: 9–14 (1983).

26. Elkheshen, S., H. Zia, T.E. Needham, A. Badawy, and L.A. Luzzi, Coating charcoal with polyacrylate-polymethacrylate copolymer for haemoperfusion. I: Fabrication and evaluation, *J. Microencap.*, 9:41–51 (1992).

27. Braatz, J.A., A.H. Heifetz, and C.L. Kehr, A new hydrophilic polymer for biomaterial coatings with low protein adsorption, *J. Biomater. Sci. Polymer Edn.*, 3:451–462 (1992).

28. Braatz, J.A., Y. Yasuda, K. Olden, K.M. Yamada, and A.H. Heifetz, Functional peptide-polyurethane conjugates with extended circulatory half-lives, *Bioconjugate Chem.*, 4:262–267 (1993).

29. D'Haese, P.C., L.V. Lamberts, and M.E. DeBroe, Indirect measurement of desferrioxamine and its chelated compounds aluminoxamine and ferrioxamine by zeeman atomic absorption spectrometry, *Clin. Chem.*, 35:884–887 (1989).

30. Chenoweth, D.E., Complement activation during hemodialysis: Clinical observations, proposed mechanisms, and theoretical implications, *Artif. Organs*, 8:281–287 (1984).

31. Walker, J.F., R.M. Lindsay, S.D. Peters, W.J. Sibbald, and A.L. Linton, A sheep model to examine the cardiopulmonary manifestations of blood-dialyzer interactions, *ASAIO J.*, 6: 123–130 (1983).

32. Cheung, A.K., Animal models to study the cardiopulmonary effects of artificial kidney membranes, *Blood Purif.*, 5:155–161 (1987).

33. Dubois, P., T. Egan, J. Duffin, E. Murphy, J. Mates, and J.D. Cooper, Hemocompatibility of the interpulse membrane oxygenator during prolonged venovenous perfusion in sheep, *ASAIO J.*, 7:146–150 (1984).

34. Weston, M.J., P.G. Langley, M.H. Rubin, M.A. Hanid, P. Mellon, and R. Williams, Platelet function in fulminant hepatic failure and effect of charcoal haemoperfusion, *Gut*, 18:897–902 (1977).

35. Gimson, A.E.S., R.D. Hughes, P.J. Mellon, H.F. Woods, P.G. Langley, J. Canalese, R. Williams, and M.J. Weston, Prostacyclin to prevent platelet activation during charcoal haemoperfusion in fulminant hepatic failure, *Lancet*, 26:173–175 (1980).

36. Hakim, R.M., G. Schulman, and J.M. Lazarus, Hemoperfusion in the treatment of aluminum- and iron-induced bone disease, *Amer. Soc. of Nephrol. Abstr.*, 65A (1985).

37. Peterson, M.B., P.C. Huttemeier, W.M. Zapol, E.G. Martin, and W.D. Watkins, Thromboxane mediates acute pulmonary hypertension in sheep extracorporeal perfusion, *Amer. Physiol. Soc.*, 243:H471–H479 (1982).

38. Walker, J.F., The hemodynamic manifestations of blood dialyzer interactions, in *Blood Compatible Materials and Devices, Perspective Towards the 21st Century*, C.P. Sharma and M. Szycher (Eds.), Technomic, Lancaster, p. 271 (1991).

50
Properties, Preparations, and Applications of Bioelastic Materials

Dan W. Urry, Alastair Nicol, and David T. McPherson
University of Alabama at Birmingham, Birmingham, Alabama

Cynthia M. Harris and Timothy M. Parker
Bioelastics Research Limited, Birmingham, Alabama

Jie Xu and D. Channe Gowda
*University of Alabama at Birmingham and
Bioelastics Research Limited, Birmingham, Alabama*

Peter R. Shewry
University of Bristol, Bristol, United Kingdom

I. INTRODUCTION

A. Overview of Properties

1. Bioelastic Materials

Bioelastic materials are based on elastomeric and related polypeptides composed of repeating peptide sequences [1–5]; they may also be called *elastic* and *plastic protein-based polymers* [6–9]. The parent polymers, $(Val^1-Pro^2-Gly^3-Val^4-Gly^5)_n$ or poly(VPGVG), $(Ala^1-Pro^2-Gly^3-Val^4-Gly^5-Val^6)_n$ or poly(APGVGV), $(Val^1-Pro^2-Gly^3-Gly^4)_n$ or poly (VPGG), and $(Val^1-Pro^2-Gly^3-Phe^4-Gly^5-Val^6-Gly^7-Ala^8-Gly^9)_n$ or poly(VPGFGVGAG), derive from sequences that occur in all sequenced mammalian elastin proteins [10–14]. In the most striking example, the sequence $(VPGVG)_n$ occurs in bovine elastin with $n = 11$ without a single substitution [11,15].

2. Inverse Temperature Transition

A fundamental property common to these protein-based polymers is that they are soluble in water at a sufficiently low temperature, but that they hydrophobically fold and associate to form a separate phase as the temperature is elevated through a particular temperature range [16]. The temperature for this phase transition, commonly called *coacervation*, is determined by the hydrophobicity of the amino acids making up the polymer. When more hydrophobic amino acids are included, the temperature is decreased; when less hydrophobic amino acids are included, the temperature is increased. The dependence of the temperature range for the phase transition on the hydrophobicity of the composite amino acids provides the basis for a new hydrophobicity scale dependent for the first time on the hydrophobic folding process of interest [4,17]. Because these protein-based polymers increase order on increasing the temperature through the transition temperature range, the phase transition is referred to as an *inverse temperature transition* and the temperature at which the onset of the transition occurs is designated as T_t [4,16]. Furthermore, the hydrophobicity scale based thereon is called the T_t-based hydrophobicity scale for protein and protein-based polymer engineering.

3. β-Spiral Molecular Structures

A common secondary structural feature, second only in frequency of occurrence to the α-helix and more common than β-sheet structures, is the β-turn [18]; it is a 10-atom, hydrogen-bonded ring in which the C—O of amino acid residue i is hydrogen bonded to the NH of residue $i + 3$. When the β-turn occurs in each repeat of a sequential polypeptide such as poly(VPGVG) or poly(APGVGV) (in this case, the Pro is residue $i + 1$ and Gly is residue $i + 2$) and when the β-turn recurs in a helical array, the resulting structure is called a β-*spiral* [19,20]. The β-spiral may form on optimization of interturn hydrophobic contacts by which it can be a dynamic structure, as is the case for poly(VPGVG) [21], or it may fold with formation of additional hydrogen bonds by which it can become a more rigid structure, as is the case for poly(APGVGV). There are now a number of repeating sequences that are thought to form β-spiral types of structures. Because of the recurring β-turn structure, it is most appropriate to refer to the primary structure as poly(VPGVG) but, for reasons of chemical synthesis, it is the permutation (GVGVP) that is polymerized to give poly(GVGVP). Poly(VPGVG) and poly(GVGVP) should be recognized as equivalent.

4. Elastic β-Spiral Structures

When the series of β-turns fold due to optimization of hydrophobic contacts between the turns of the spiral, when in the process an aqueous cylinder forms inside the β-spiral, and when there are peptide segments suspended between the β-turns, the peptide elements of the suspended segment can undergo large rocking motions, large torsional oscillations, as they are unrestricted by steric and hydrogen-bonded contacts. These rocking or librational motions give a significant degree of entropy to the folded state. When this dynamic β-spiral is extended, the amplitudes of the librational motions become damped, resulting in a marked decrease in entropy. This damping of internal chain dynamics on extension results in an elastic resistance to extension and provides an elastic restoring force as the deforming force is released [22–24].

5. Plastic β-Spiral Structures

The plastic β-spiral structures also may form on increasing the temperature; they may form irreversibly, as is the case for poly(APGVGV), or they may form reversibly, as occurs for poly(VPAVG). Poly(VPAVG) can exhibit elasticity below and through much of the transition temperature range, but becomes a hard plastic once the transition is complete [25]. Because of this, the material may be referred to as an *inverse thermoplastic*.

6. Elastic Matrices

The elastic matrices, whether formed from elastic β-spiral structures or from reversible plastic β-spiral structures, result from cross-linking of the individual polymer chains. While this may be done chemically, photochemically, or enzymatically after introduction of suitably reactive moieties, the elastic matrices can often most easily be obtained by gamma irradiation or by electron beam irradiation of the viscoelastic or coacervate state of the protein-based polymer that is obtained on increasing the temperature above T_t [4]. The elastic moduli of these elastic matrices can vary from 10^4 to 10^5 dynes per square centimeter (dynes/cm^2) for the swollen hydrogel states, to 10^6 to 10^8 dynes/cm^2 for the reversibly elastic states, and to 10^9 dynes/cm^2 and more for the plastic states of the matrices.

7. Thermally Driven Contraction (Thermomechanical Transduction)

The reversible phase transition (i.e., the inverse temperature transitions of hydrophobic folding and assembly exhibited by solutions of these particular protein-based polymers) can be seen in the cross-linked elastic matrices as thermally driven contractions or deswelling. When a weight is hung on a strip of the swollen elastic matrix and the temperature is increased through the range of the inverse temperature transition, the matrix will contract and lift the weight in the performance of mechanical work. Thus, these matrices are capable of performing thermomechanical transduction; they can convert thermal energy into mechanical work [4,16,26,27].

8. The ΔT_t Mechanism of Energy Conversion and First-Order Molecular Machines of the T_t Type

Now, instead of changing the temperature, it becomes possible by different energy inputs to change the value of T_t, that is, to change the temperature range over which the thermally driven contraction occurs. This means, for example, if T_t is just above the operating temperature and an energy input is introduced that lowers the value of T_t to below the operating temperature, that the energy input will also drive contraction and the performance of mechanical work. If the energy input is a change in the concentration of a chemical (i.e., chemical energy) then chemomechanical transduction will have occurred. By means of suitably reactive or responsive groups being a part of the protein-based polymer, it is possible, using the elastic matrices, also to demonstrate baromechanical, electromechanical, and photomechanical transductions. These molecular systems, which can, by hydrophobic folding and assembly, perform useful mechanical work on the introduction of the appropriate energy, are molecular engines. More specifically, they are also called first-order molecular machines of the T_t type as the hydrophobic folding process is directly utilized in the performance of mechanical work [4,16].

9. The ΔT_t Mechanisms of Energy Conversion and Second-Order Molecular Machines of the T_t Type

By including within the elastic or plastic protein-based polymer a suitable pair of functional groups, it becomes possible, by means of the hydrophobic folding transition, to convert, for example, electrical energy into chemical energy or light energy into chemical energy. While these energy conversions are achieved by changing the value of T_t and while they utilize hydrophobic folding or unfolding, they do not involve the performance of mechanical work, which is so readily a part of folding and unfolding. These are called second-order molecular machines of the T_t type, and they include the 10 pairwise energy conversions involving the energy inputs of temperature changes, pressure changes, chemical concentration changes, electrochemical oxidations or reductions, and the absorption or dissipation of electromagnetic radiation [4,16,28].

B. Overview of Preparations

1. *Chemical Syntheses*

Initially and from a several-decade tradition, the protein-based polymers, or sequential polypeptides, have been prepared by chemical means using either classical solution syntheses [29,30], or solid-phase syntheses [31,32]. With appropriate care, both of these chemical approaches have been useful. The particular difficulty with the chemical methods is that the purity of each repeating sequence must be extraordinarily high in order that the properties of the inverse temperature transition of the resulting polymer be adequately maintained. For example, a very small amount of racemization can cause the value of T_t to increase 10°C to 15°C and result in losses in the interesting elastic matrix formation and transductional properties.

Nonetheless, adequate purity has been achieved with specific repeating sequences and, requiring great skill, with the VPGVG sequence in which each of the naturally occurring amino acids has been introduced as a guest residue (X) in Position 4 (i.e., as VPGXG). This was required in order to develop the T_t-based hydrophobicity scale for protein engineering [17]. (Highest molecular weights and better purities were obtained when using the GVGVP and GXGVP permutations [29], which should be recognized as forming equivalent protein-based polymers as the use of VPGVG and VPGXG does.) Exceptionally taxing deblocking of protected functional side chains was required in order that there would be a very limited number of errors in a polymer of 100 to 200 or more repeating units. Furthermore, more complex repeats, such as 30mers, have been successfully synthesized and polymerized [33–35]. As the longer repeats have begun to be designed with more than one functional moiety, however, the syntheses have become more difficult and more taxing such that genetic engineering and microbial biosynthesis becomes an increasingly attractive alternative even for research quantities of materials.

2. *Genetic Engineering and Microbial Biosynthesis*

It is possible to use a molecular biological approach to produce protein-based polymers [36–38]. Recombinant DNA can be used to create synthetic genes encoding multiple repeating units of a given peptide sequence. Using enzymes that "cut and paste" DNA, these synthetic genes may themselves be polymerized to create even longer coding sequences, resulting in protein-based polymers of greater length. These same enzymes can be used to incorporate the polymeric genes into circular plasmid DNA molecules. Plasmids are used as vectors for insertion of foreign genes into bacterial cells for subsequent expression of the gene product [39,40].

Once inside a bacterial cell, a plasmid DNA molecule containing a single protein-based polymer gene can replicate to produce multiple copies per cell. It is then possible to grow a homogeneous bacterial population, arising from a single plasmid-containing cell, capable of producing a single-length, protein-based polymer.

Using this biotechnological approach, we have been able to produce poly(GVGVP) or (GVGVP)$_n$ with values of n ranging from 10 to 250 (unpublished results). Also, it has been possible to exploit the inverse temperature transition properties to purify the elastic protein-based polymer from crude bacterial lysates. At a cooler temperature, the poly (GVGVP) remains in the unfolded soluble state and the insoluble cell debris can be removed by centrifugation; upon heating to above their transition temperatures, for example, to 37°C, the poly(GVGVP) species fold, self-assemble, and settle, forming a new phase that allows selective removal from the remaining solute by centrifugation.

C. Overview of Applications Under Development

1. Medical Applications

Biocompatibility. The primary prerequisite for consideration of medical applications is the determination of the degree of biocompatibility. Extensive studies on three compositions are now completed. These are for poly(VPGVG) and its 20-megarad (Mrad), gamma irradiation, cross-linked matrix X^{20}-poly(VPGVG) [2], poly(APGG) (discussed here), poly(VPAVG) and X^{20}-poly(VPAVG) [unpublished results]. The first two are elastic protein-based polymers and the last is a plastic protein-based polymer. All three compositions have been found to exhibit a remarkable biocompatibility utilizing the following recommended tests (the result is given in parentheses following each test name): (1) the Ames mutagenicity test (nonmutagenic), (2) cytotoxicity-agarose overlay (nontoxic), (3) acute systemic toxicity (nontoxic), (4) intracutaneous toxicity (nontoxic), (5) muscle implantation (favorable), (6) acute intraperitoneal toxicity (nontoxic), (7) systemic antigenicity (nonantigenic), (8) dermal sensitization according to the Magnusson and Kligman maximization method (nonsensitizing), (9) pyrogenicity (nonpyrogenic), (10) Lee–White clotting study (normal clotting time), and (11) *in vitro* hemolysis test (nonhemolytic) [2].

Prevention of Postsurgical and Posttrauma Adhesions. In studies of efficacy as a barrier in the prevention of adhesions involving hundreds of rats and employing a contaminated and bloody peritoneal model, the X^{20}-poly(VPGVG) matrix was found to be effective in 80% of the animals tested when using ethylene oxide gas for sterilization, with the potential for becoming 90% or more effective [41,42]. In a strabismus surgery model in the rabbit eye, preliminary studies showed X^{20}-poly(VPGVG) as well as X^{20}-poly[0.75(VPGVG),0.25(VPGFG)] to prevent adhesion and to be noninflammatory [43]. Work is also under way to determine efficacy in preventing adhesion in models for cardiopulmonary bypass procedures.

Coatings on Catheters, Leads, and Tubings. With the propensity of X^{20}-poly(VPGVG) and X^{20}-poly(APGG) to act innocuously in the body and to be refractory to formation of adhesions, coating these polymers on catheters, leads, and tubings that would reside in the body for days, weeks, and even months could be expected to result in a more ready removal and to do so with minimal damage to tissues that would have been in contact with such temporary devices.

Drug Delivery. There are two special properties of these elastic protein-based polymers that give particular promise for their use in controlled release of drugs. One is that no fibrous capsule forms around X^{20}-poly(VPGVG) even when implanted for months [41,42] such that the bioelastic matrix itself and not the surrounding fibrous capsule would determine release profiles. The second and most enabling aspect is that all of the transductional (energy converting) properties can be employed to control release as these energy inputs can control the degree of swelling or contraction of the elastic matrix.

From the practical point of view, it has been shown that X^{20}-poly(VPGVG) can be heavily doped with drug [44]. Using the model drug biebrich scarlet, it is possible to swell-dope drawing solutions of drug into the matrix and to contract, leaving concentrations of the order of 0.3 M, that is, one drug molecule for each three pentamers, and to observe an interesting diffusional release profile from the contracted state that continues at significant levels for weeks. It is also possible to have within the matrix chemical clocks such as the carboxamide side chain of an asparagine, Asn(N), or a glutamine, Gln(Q), that break down to carboxylates with half-lives that can vary from days to decades, depending on the nearest neighbor residues [3,44,45]. At the rate at which the carboxylates form at the matrix-milieu interface, the surface of the matrix can swell, enhancing the rate of drug release and increasing the expected rate of enzymatic degradation.

Cell Attachment Matrices and Tissue Reconstruction. By adding to the protein-based polymer a cell attachment sequence such as Gly-Arg-Gly-Asp-Ser-Pro (GRGDSP) from fibronectin [46–49] to result in the elastic matrix X^{20}-poly[40(GVGVP),(GRGDSP)] (note that the permutation GVGVP in the polymer is equivalent to VPGVG), a matrix that is refractory to cell adhesion now promotes cell adhesion, cell spreading, and growth to confluence [50–52]. Importantly, this means that cells can migrate into and attach to the matrix and be subjected to and also to sense the tensional forces to which the matrix is subjected in its functional role. It is now appreciated, for example, that the cyclic stretching to which a vascular wall is subjected induces the vascular wall cells to turn on the genes that result in elaboration of the macromolecules required to maintain and to rebuild the tissue required to sustain such tensional forces [53–55]. This has been called *tensegrity* [56], and it provides the basis for the proposal for developing a temporary functional scaffolding into which the natural cells can migrate, attach, and remodel into the natural required tissue [57].

2. *Nonmedical Applications*

Energy-Converting Matrices: Transducers (Sensors/Actuators). With the proper design, elastic protein-based polymers have, to date, been shown to be capable of performing 10 of the 15 pairwise energy conversions involving the 6 energies; motion, pressure, temperature, chemical, and electrochemical and electromagnetic radiation [4,28,58]. Furthermore, the understanding of these energy conversions is such that it is possible to design for enhanced efficiency of energy conversion through an understanding of the subtleties of the inverse temperature transition mechanism. Significantly, it becomes possible to design the transducer for a specific energy input to be sensed and then to design the molecular machine to output (to actuate) energy of the desired form [59]. Thus, a single molecular machine can be designed to be simultaneously the sensor and the actuator for the desired pair of energies. In the context of controlled release, for delivery of pharmaceuticals it could be the integrated diagnostic/therapeutic pair, for the controlled release of crop-enhancement substances, it could mean achieving the desired time of day, the weather condition, and extent of release.

Controllable Superabsorbents. The bioelastic matrices can be essentially dry and then be induced to absorb quantities of water some 10 times the dry weight, that is, to be controllable superabsorbents. A practical example could be a diaper with the particular property that the fluids could be drawn away from the body temperature to cooler, most distant parts. There could also be included in this case the design by which pressure would enhance the swollen state rather than result in the squeezing out of the fluid. Also, as noted below in more detail, the diaper or incontinence apparel would be biodegradable.

Biodegradable Plastics. It can be expected that both the elastic and plastic protein-based polymers would be degradable as protein itself is biodegradable. This is of particular interest for the hard plastic state of the plastic protein-based polymer. As long as the temperature of the inverse temperature transition, that is, the value of T_t, is below the temperature of use, the matrix would remain in the plastic state. The value of T_t could even be well below 0°C, in which case no breakdown would occur even when stored at temperatures below freezing. Importantly, however, there can be included in the plastic carboxamides as the side chains of Gln(Q) and/or Asn(N) residues. Depending on the sequence in which these carboxamide-containing residues occurred, the breakdown to carboxylates at the surface of the plastic would occur with a predetermined half-life with the presence of water at the interface. The carboxylates once formed would raise the value of T_t well above the ambient temperature; the surface layer would swell, and biodegradation could ensue. Thus, there is the potential for biodegradable plastics.

Viscoelastic Bases for Prepared Foods and Related Condiments. The special properties of wheat flour, gained in part due to the storage protein glutenin with its inherent elastic properties, are responsible for the fine, light breads and pastries that can be made from it. Even so, to introduce a truly elastic protein or protein-based polymer could possibly further improve the quality and, when introduced into other grains, could possibly improve the texture of breads, pastries, and other condiments made from these agricultural products. This adds an additional prospect of elastic protein-based proteins to improve the texture and acceptance of other foods.

II. PROPERTIES

A. Inverse Temperature Transitions

A fundamental property of the protein-based polymers that make up bioelastic materials is that, in water, they hydrophobically fold and/or assemble to a state of increased order as the temperature is increased through a transition temperature range [4,16]. This increase in order with increase in temperature is referred to as an *inverse temperature transition*. It occurs as a result of the structure and thermodynamic properties of water that surrounds hydrophobic moieties in amphiphilic polymers such as proteins and protein-based polymers [60,61]. The inverse temperature transition is consistent with the second law of thermodynamics as the total system, water plus polymer, increases in disorder (entropy) as the temperature is increased because the increase in entropy due to the more-ordered water of hydrophobic hydration becoming less-ordered bulk water is greater than the decrease in entropy due to the hydrophobic folding and/or assembling of the polymer [62–64]. This is facilitated in the present case because of the considerable entropy in the particular ordered state in which there is a great amount of internal chain dynamics, giving rise to elasticity [22–24,65].

Differential scanning calorimetry (DSC) curves for three elastic protein-based polymers, poly(IPGVG), poly(LPGVG), and poly(VPGVG), are given in Fig. 1 [66]. The area of the negative extreme gives the magnitude of the endothermic heat for the inverse temperature transition required to destructure the water of hydrophobic hydration. The endothermic heats of the more hydrophobic Ile(I) and Leu(L) containing polypentapeptides are greater than for poly(VPGVG) and, in a similar manner, the temperatures for the transition occur at progressively lower temperatures. The thermodynamic quantities for the transitions are given in Table 1. (Note the T_t value of Table 1 is defined as equal to $\Delta H/\Delta S$ and is not the same value as defined below using temperature profiles of aggregation.)

Another way to follow the occurrence of the inverse temperature transition is to determine the onset of turbidity, as analogs of these polypentapeptides, if not too hydrophobic, are all soluble in water at a lower temperature and aggregate to form cloudy, light-scattering suspensions as the temperature is increased into the transition temperature range. A series of such temperature profiles for aggregation (*TP$_t$*) curves are given in Fig. 2 for the polymer compositions poly[f_V(VPGVG), f_X(VPGXG)] where f_V and f_X are

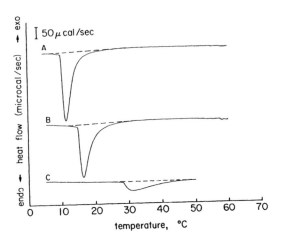

Figure 1 Differential scanning calorimetry data: (A), for poly(IPGVG); (B), for poly(LPGVG); (C), for poly(VPGVG). The relative areas of the curves indicate the relative endothermic heats for the inverse temperature transitions. (Reproduced with permission of Ref. 66.)

mole fractions with $f_V + f_X = 1$, where in these particular examples $f_V = 0.8$ and $f_X = 0.2$ and where X is another naturally occurring amino acid or its modification [16,67]. In particular, Fig. 2a is the concentration dependence for the onset of the aggregation for which the temperature of the inverse temperature transition is defined as T_t (the temperature at half-maximal turbidity) for the high concentration limit that occurs by 40 mg/ml [67]. In Fig. 2b are the set of curves at the high concentration limit for $f_X = 0.2$ for (*a*) tyrosine (Tyr,Y); (*b*) phenylalanine (Phe,F); (*c*) leucine (Leu,L); (*d*) valine (Val,V); (*e*) glutamic acid (Glu COOH,E°); (*f*) lysine (LysNH$_2$,K°); (*g*) glycine (Gly,G); (*h*) tyrosine (TyrPhO$^-$,Y$^-$); (*i*) lysine (LysNH$_3^+$, K$^+$); and (*j*) glutamate (GluCOO$^-$,E$^-$) [16].

When additional values of f_X are included and a plot of f_X versus T_t is given (as in Fig. 3) [17], the plots are seen to be essentially linear to f_X values of 0.5, and it is quite apparent that the values of T_t are proportional to the hydrophobicity of the residue. When extrapolating to $f_X = 1$, that is, to poly(VPGXG), a reference value is obtained; these values are given in Table 2 [4].

Table 1 Thermodynamic Data Derived from Differential Scanning Calorimetry of the Inverse Temperature Transition Exhibited in Water by the Polypentapeptide and Its Analogs

Sample	$T_m{}^a$ (°C)	T_t (°C)	ΔH^b (kcal/mol)	ΔS (cal/mol-K)	ΔQ (cal/g)
Ile^1-PPP	11.6	12.9	3.0	10.4	7.02
Leu^1-PPP	16.4	19.2	2.9	9.9	6.81
PPP	31.1	34.6	1.2	3.9	2.93

$^a T_m$ = temperature of maximum heat absorption. The reproducibility of the temperature is within ±0.1°C.
^{b}The reproducibility of the heats is within ±0.1 kcal/mole.

(A)

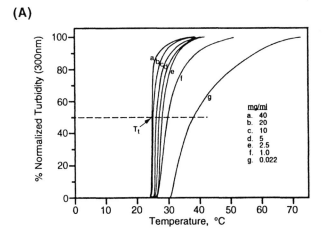

(B)

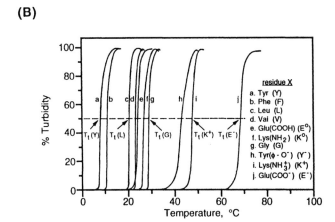

Figure 2 Temperature profiles for aggregation curves: (A), concentration dependence of the temperature of the inverse temperature transition T_t for poly(VPGVF); (B), composition dependence of T_t for poly[f_v(VPGVG),f_X(VPGXG)] where $f_X = 0.2$ for the indicated amino acid residues and states of functional side chains. (Figure 2A reproduced with permission of Ref. 67; Fig. 2B reproduced with permission of Ref. 16.)

Table 2 becomes a T_t-based hydrophobicity scale for protein and protein-based polymer engineering, that is, it becomes possible to introduce any functional group or even a particular sequence of amino acids for a particular functional purpose and then to adjust the composition to place the value of T_t as desired. For example, if it is desirable to have a certain composition of Glu residues at pH 7.4, these will increase the value of T_t such that it will be necessary to add an adequate number of residues more hydrophobic than Val to compensate and to bring the value of T_t back to the desired temperature range. Similarly, if an enzyme site such as a protein kinase site or a cell attachment site is desired and yet the folded state is desired at 37 °C, then the composition can be properly adjusted to have both the added functionalities and the desired folded or unfolded state.

In warm-blooded animals, it matters much less what the heat of the endothermic

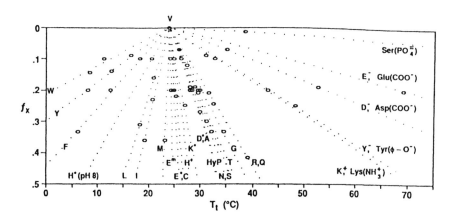

Figure 3 Dependence of the temperature of the inverse temperature transition T_t for poly[f_V (VPGVG), f_X(VPGXG)] as a function of f_X for all of the naturally occurring amino acid residues, in different functional states when relevant, and with phosphorylation of a serine residue, Ser(PO$_4^=$). (Reproduced with permission of Ref. 17.)

hydrophobic folding transition might be and much more whether the temperature of the inverse temperature transition T_t is above the physiological temperature, in which case the protein or protein-based polymer would be hydrophobically unfolded or some 15°C below physiological temperature, in which case the protein or protein-based polymer would be completely hydrophobically folded.

B. Molecular Structure

Based on a very large number of physical studies, including the crystal structure determination of cyclic conformational correlates, the molecular structure of the parent elastic, protein-based polymer poly(VPGVG) is given in Fig. 4. Note in Fig. 4a that the primary structure could also be designated as poly(GVGVP) or as any one of five permutations. The poly(VPGVG) sequence is schematically shown in Fig. 4a, for which the upward deflections for residues P and G are to signify the β-turn structure shown in detail in Fig. 4b [68]. The β-turn is formed by a 10-atom, hydrogen-bonded ring involving the Val^{1}C-O and the Val^{4}NH with Pro^2 and Gly^3 at the corners. The end peptide moiety formed by Pro^2-Gly^3 is directed out of the paper toward the reader and is designated as a Type II β-turn.

At low temperatures, the repeating β-turn is present, but waters of hydrophobic hydration surrounding the Val and, to some extent the Pro, residues prevent folding. On increasing the temperature above T_t, the intramolecular hydrophobic contacts are optimized by the formation of a helical structure schematically shown in Fig. 4c and in more detail in Fig. 4d [69], in which the β-turns are seen to function as spacers between the turns of the helix. This helical recurrence of the β-turn is referred to as a β-spiral, and is shown in detail in Fig. 4e in the spiral axis view above and the side view below. In particular, in Fig. 4e, the contacts between turns of the β-spiral are shown as involving the Val^{1}γCH$_3$ and the Pro^{2}βCH$_2$ moieties [16]. This is one of a family of dynamic helical structures held together by optimizing nonrestricting, interturn, hydrophobic contacts.

Table 2 T_t-Based Hydrophobicity Scale for Protein Engineering

Residue X		T_t linearly extrapolated to $f_X = 1$	Correlation coefficient
Lys(NMeN, reduced)[a]		$-130°C$	1.000
Trp	(W)	$-90°C$	0.993
Tyr	(Y)	$-55°C$	0.999
Phe	(F)	$-30°C$	0.999
His (pH 8)	(H°)	$-10°C$	1.000
Pro	(P)[b]	$(-8°C)$	Calculated
Leu	(L)	$5°C$	0.999
Ile	(I)	$10°C$	0.999
Met	(M)	$20°C$	0.996
Val	(V)	$24°C$	Reference
Glu(COOCH$_3$)	(E^m)	$25°C$	1.000
Glu(COOH)	(E°)	$30°C$	1.000
Cys	(C)	$30°C$	1.000
His (pH 4)	(H$^+$)	$30°C$	1.000
Lys(NH$_2$)	(K°)	$35°C$	0.936
Pro	(P)[c]	$40°C$	0.950
Asp(COOH)	(D°)	$45°C$	0.994
Ala	(A)	$45°C$	0.997
HyP		$50°C$	0.998
Asn	(N)	$50°C$	0.997
Ser	(S)	$50°C$	0.997
Thr	(T)	$50°C$	0.999
Gly	(G)	$55°C$	0.999
Arg	(R)	$60°C$	1.000
Gln	(Q)	$60°C$	0.999
Lys(NH$_3^+$)	(K$^+$)	$120°C$	0.999
Tyr(ϕ-O$^-$)	(Y$^-$)	$120°C$	0.996
Lys(NMeN,oxidized)[a]		$120°C$	1.000
Asp(COO$^-$)	(D$^-$)	$170°C$	0.999
Glu(COO$^-$)	(E$^-$)	$250°C$	1.000
Ser(PO$_4^=$)		$1000°C$	1.000

T_t = temperature of inverse temperature transition for poly[f_V(VPGVG), f_X(VPGXG)]
[a]NMeN is for *N*-methyl nicotinamide pendant on a lysyl side chain, that is, N-methyl nicotinate attached by amide linkage to the ϵ-NH$_2$ of Lys and the reduced state is *N*-methyl-1,6-dihydronicotinamide.
[b]The calculated T_t value for Pro comes from poly(VPGVG) when the experimental values of Val and Gly are used. This hydrophobicity value of $-8°C$ is unique to the β-spiral structure in which there is hydrophobic contact between the Val$^1\gamma$CH$_3$ and Pro^2 βCH$_2$ moieties.
[c]The experimental value determined from poly[f_V(VPGVG), f_P(PPGVG)].

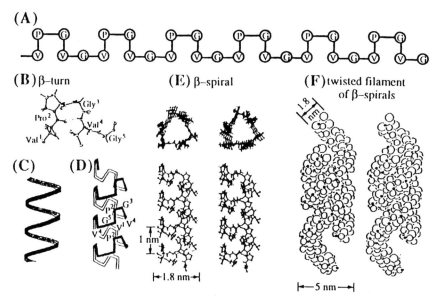

Figure 4   Molecular structure of poly(VPGVG) illustrating: (A), schematic linear representation showing PG insertion of β-turn structures; (B), detailed β-turn structure; (C), schematic ribbon representation of folding to form a helical structure; (D), schematic ribbon represented with β-turn included as spacers between the turns of the helix; (E), β-spiral shown in stereo which is a detailed representation of the helical structure of (D) (upper, spiral axis perspective; lower, side view); (F) representing each of the five residues as spheres of different radii, three β-spirals are shown in stereo pair to have associated to form a twisted filament. (Figure 4A from Ref. 35, 4B from Ref. 68, 4C and 4D from Ref. 69, 4E from Ref. 16, 4F from Ref. 70; all reproduced or adapted with permission.)

As evidenced from electron micrographs of negatively stained incipient aggregates, the β-spirals associate (by means of intermolecular hydrophobic contacts) to form twisted filaments such as those schematically shown in Fig. 4f [70].

Retention of the structure in Fig. 4 requires that the Gly residues be invariant or possibly be replaceable by a D-residue, and as Pro-Gly is the most probable sequence for inserting a β-turn [18], this is a preferred sequence for the $i + 1(2)$ and $i + 2(3)$ residues. Some limited substitution can occur in position $i(1)$, but fortunately position $i + 3(4)$ can tolerate a significant degree of substitution for all of the naturally occurring amino acid residues and many chemical modifications and unnatural analogs. It is the possible variation in position $i + 3(4)$ that makes it possible to develop the T_t-based hydrophobicity scale [16,17]; this position, with some use of position $i(1)$, provides for the great capacity of this structure to be designed for many uses.

C.   Mechanism of Elasticity

On increasing the temperature above T_t of analogs substituted in positions i and $i + 3$, hydrophobic folding and assembly ensues and, on standing, a viscoelastic phase forms in the bottom of the container. This viscoelastic phase, which is yet some 50% water, can be shaped as desired and gamma irradiation cross-linked to form sheets such as seen in Fig.

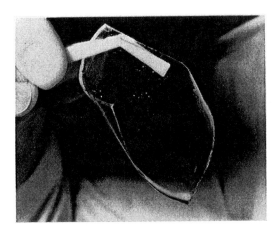

Figure 5 Sheet of X^{20}-poly(VPGVG) obtained on 20-Mrad gamma-irradiation cross-linking. (Courtesy Doug Bryant Photography.)

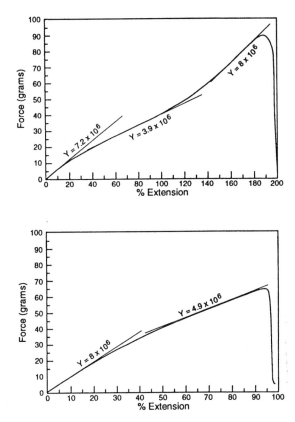

Figure 6 Stress/strain curves showing the Young's (elastic) moduli for various parts of the curves, with the top portion for X^{20}-poly(GVGIP) and the bottom portion for X^{20}-poly[0.75(GVGVP), 0.25(GFGVP)].

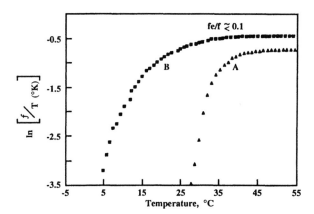

Figure 7 Thermoelasticity data for X^{20}-poly(GVGVP). Curve A is in water and Curve B is in ethylene glycol/water at 30 : 70 ratio by volume (see text for discussion). (Reproduced with permission of Ref. 71.)

5 for poly(VPGVG). As 20 megarads (Mrads) of cobalt-60 gamma irradiation were used, this elastic matrix is designated as X^{20}-poly(VPGVG).

Stress/strain curves for two compositions, X^{20}-poly(IPGVG) and X^{20}-poly[0.75 (VPGVG),0.25(VPGFG)], are given in Fig. 6. These curves are of a shape and of elastic moduli that are appropriate for the vascular wall.

Thermoelasticity data for X^{20}-poly(VPGVG) is given in Fig. 7 as Curve A, for which a strip of X^{20}-poly(VPGVG) was held at fixed extension and then the force was measured as a function of temperature [71]. The data is plotted as $\ln(f/T)$ versus temperature where f is the total force and T is the absolute temperature in °K. Following the analyses of Flory, Ciferi and Hoeir [72], the ratio of the internal energy component of the force f_e to the total force is given by the expression

$$\frac{f_e}{f} = -T\left(\frac{\partial \ln f/T}{\partial T}\right)^{V,L,n} \tag{1}$$

where V, L, and n stand for constant volume, length and composition, respectively. It has been shown for X^{20}-poly(VPGVG) in water that these quantities are essentially constant for the 40°C to 60°C temperature range [67]. From the slope in this temperature range, the f_e/f ratio is calculated to be no more than 0.1 [71]. Since the total force is the sum of the internal energy component f_e, and the entropic component f_s (i.e., $f = f_e + f_s$), this means that X^{20}-poly(VPGVG) is a dominantly entropic elastomer.

The additional curve in Fig. 7, actually Curve B, was carried out in a 30 : 70 by volume mixture of ethylene glycol and water, respectively, where the heat of the endothermic heat (ΔH) of the inverse temperature transition had been reduced to less than 10% of the value in water alone [71]. This means that the entropy change due to solvent, that is, $\Delta S(= \Delta H/T)$, for the transition has been largely removed. Nonetheless, the total force has increased and yet the f_e/f ratio remains less than 0.1. Thus, it appears that the solvent entropy change is not contributing significantly to the elastomeric force. The entropic elastomeric force would necessarily seem to be due to changes in chain configurational entropy.

Since the molecular structure shown in Fig. 4 is not as well described as that of a random chain, the decrease in configurational entropy would need to come from a decrease in entropy on chain extension. This can occur due to a decrease in internal chain dynamics on extension made possible by the large torsional oscillations that can occur in the suspended segment between β-turns in Figs. 4d and 4e. Calculations using Scheraga's ECEPP [22,70,73,74], Karplus' CHARMm [23,75–77], and Kollman's AMBER [24,78] computational programs give similar results. The peptide moieties in the Val^4-Gly^5-Val^1 suspended segment can undergo large-amplitude, low-energy torsional oscillations that become damped on extension. This has been called the *librational entropy mechanism of elasticity* [70]. It is a specific example, with others to come from proteins and protein-based polymers, for which durable elasticity does not require random chain networks.

D. Thermally Driven Contraction (Thermomechanical Transduction)

Shown in Fig. 8 is a space-filling model of a single chain of poly(VPGVG) in an extended state I and a hydrophobically folded (i.e., contracted) state II [79]. Also shown hypothetically is the potential for the thermally driven hydrophobic folding to perform useful work by the lifting of a weight. This is shown schematically in Fig. 9a as a function of temperature [4].

While it has not been possible as yet to see a single chain or a twisted filament do

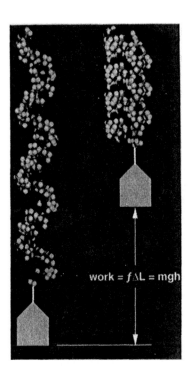

Figure 8   Space-filling model for the molecular structure of poly(VPGVG) in an extended state I and a contracted state II. The hydrophobic folding that occurs on increasing the temperature can be used as shown schematically to lift a weight and thereby perform useful mechanical work. (Reproduced with permission of Ref. 59.)

(A)

(B)

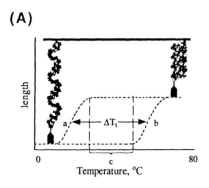

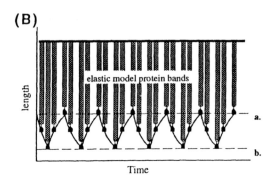

Figure 9 Summary of molecular engines (i.e., first-order molecular machines of the T_t type): (A), thermally driven hydrophobic folding shown to occur at different temperatures depending on different states for the elastic protein-based polymer; (B), means of driving contraction/relaxation using cross-linked elastic model protein bands (i.e., cross-linked matrices of elastic protein-based polymers of the appropriate designs). See text for discussion. (Adapted with permission of Ref. 4.)

mechanical work, it is possible to see a cross-linked matrix such as that in Fig. 5 perform mechanical work. As schematically shown in Fig. 9b, the elastic model protein band, that is, X^{20}-poly(VPGVG), with weight attached reversibly contracts and extends as the temperature is cycled from 40°C to 20°C. This is thermomechanical transduction and it is represented as Postulate 1 in Table 3 [28].

Table 3 Postulates by which Hydrophobic Folding and Assembly Can Achieve All of the Free Energy Transductions of which Living Organisms Are Capable

---

Postulate 1: The input of thermal energy to a protein capable of hydrophobic folding and assembly on increasing the temperature from below to above the temperature T_t of an inverse temperature transition can result in motion and the performance of mechanical work.
    Corollary: Thermomechanical transduction
Postulate 2: Any energy input that changes the temperature T_t at which an inverse temperature transition occurs can be used to produce motion and perform mechanical work.
    Corollary 1   Chemomechanical transduction
    Corollary 2   Electromechanical transduction
    Corollary 3   Baromechanical transduction
    Corollary 4   Photomechanical transduction
Postulate 3: Different energy inputs, each of which can individually drive hydrophobic folding to produce motion and perform mechanical work, can be converted one into the other (transduced) by means of the inverse temperature transition with the correctly designed coupling and T_t value.
    Corollary 1   Electrochemical transduction
    Corollary 2   Electrothermal transduction
    Corollary 3   Baroelectrical transduction
    Corollary 4   Photovoltaic transduction
    Corollary 5   Thermochemical transduction
    Corollary 6   Photothermal transduction
    Corollary 7   Barothermal transduction
    Corollary 8   Barochemical transduction
    Corollary 9   Photobaric transduction
    Corollary 10   Photochemical transduction

E. The ΔT_t Mechanism of Energy Conversion

If the input of a given energy can change the value of T_t, the temperature at which hydrophobic folding and assembly occurs, then that input of energy can be used to perform mechanical work. This is Postulate 2 of Table 3. If two different energy inputs can each change the value of T_t, these two energy forms can be converted, one into the other, by means of the inverse temperature transition of hydrophobic folding and assembly. This is Postulate 3 of Table 3 [28].

1. First-Order Molecular Machines of the T_t Type: Molecular Engines

A construct designed to convert energy from one form to another may be called a *machine*. A machine designed to convert energy into useful mechanical motion is an *engine*. A molecular machine capable of converting energy into useful mechanical motion would be a *molecular engine*. A molecular engine that converts energy into mechanical motion using the inverse temperature transition of hydrophobic folding and assembly is called a *first-order molecular machine of the T_t type*.

Chemically Driven Contraction (Chemomechanical Transduction). As seen in Fig. 2b, protonation of the carboxylate of the Glu residue for a polymer with $f_x = 0.2$ lowers the value of T_t some 45°C, from 69°C to 24°C in 0.15 N NaCl. This means, at neutral pH for the cross-linked elastic matrix X^{20}-poly[0.8(VPGVG),0.2(VPGEG)], that contraction will occur on increasing the temperature above 69°C and, at low pH, the contraction will occur on increasing the temperature above 24°C. Alternatively, it is possible to hold the temperature constant at an intermediate value, for example, at 37°C, and to lower the pH from 7, at which it is in the extended state, to pH 3, at which the contracted state occurs [80]. Thus, by expending the chemical energy of increasing the concentration of proton, the matrix X^{20}-poly[0.8(VPGVG),0.2(VPGEG)] can be chemically driven to perform useful mechanical work. This is *chemomechanical transduction* [80].

Electrically Driven Contraction (Electromechanical Transduction). As may be seen from Table 2, the reference T_t value for N-methyl nicotinate {NMeN} attached by amide linkage to a lysine (Lys,K) side chain is 120°C. On either electrical or chemical reduction to form the 1,6-dihydronicotinamide, the value of T_t is lowered some 250°C to −130°C [81]. Thus, reduction results in a more hydrophobic moiety that lowers the temperature of the hydrophobic folding transition. For the elastic matrix of the composition X^{20}-poly[0.73(VPGVG),0.27(VPGK{NMeN}G)], the matrix will be extended at physiological temperature, but will contract on reduction. This is *electromechanical transduction*.

Pressure-Release-Driven Contraction (Baromechanical Transduction). The volume occupied by a water molecular is smaller in the structure for waters of hydrophobic hydration than in bulk water. The application of pressure, driving the molecular system to the lowest volume, results in an increase of the value of T_t and can cause the contracted matrix to unfold. This is especially the case when there are aromatic side chains present such as phenylalanine (Phe,F), tyrosine (Tyr,Y) and tryptophan (Trp,W). Thus, with an elastic matrix such as X^{20}-poly[0.79(VPGVG),0.21(VPGFG)], the matrix will be extended under 68 atmospheres (atm) of pressure at 13°C, but will contract on releasing the pressure with the lifting of a weight [82]. This is *baromechanical transduction*.

Light-Driven Relaxation (Photomechanical Transduction). When azobenzene {AzB} is attached by amide linkage to the side chain of a Glu(E) residue as in poly[0.68 (VPGVG),0.32(VPGE{AzB}G)] or when cinnamic acid {CA} is attached by amide linkage to the side chain of a Lys(K) residue as in poly[0.8(VPGVG),0.2(VPGK

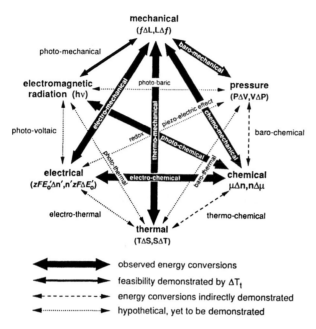

mechanical
($f\Delta L, L\Delta f$)

photo-mechanical

baro-mechanical

electromagnetic
radiation (hν)

photo-baric

pressure
($P\Delta V, V\Delta P$)

electro-mechanical

photo-chemical

chemo-mechanical

thermo-mechanical

piezo-electric effect

photo-voltaic

redox

baro-chemical

photo-thermal

electrical
($zFE_o'\Delta n', n'zF\Delta E_o'$)

electro-chemical

baro-thermal

chemical
$\mu\Delta n, n\Delta\mu$

electro-thermal

thermo-chemical

thermal
($T\Delta S, S\Delta T$)

observed energy conversions

feasibility demonstrated by ΔT_t

energy conversions indirectly demonstrated

hypothetical, yet to be demonstrated

Figure 10   Demonstrated and putative energy conversions using molecular machines of the T_t type (see text for discussion).

{CA}G)], the effect of the absorption of light, which can convert the AzB or CA from a *trans* to a *cis* isomer, is to raise the value of T_t [58,83]. Accordingly, the absorption of light could be expected, under the correct experimental conditions, to lead to unfolding. For reversible conditions, as for azobenzene, in which the light-reacted chromophore can return to the *trans* state in the dark, there can be light-driven relaxation and dark-adapted contraction. This is *photomechanical transduction*.

2.   Second-Order Molecular Machines of the T_t Type

The set of demonstrated and putative energy conversions using molecular machines of the T_t type are represented in Fig. 10 [4]. Those molecular machines that involve the mechanical force apex are the molecular engines as the particular energy input can result in useful mechanical motion. Those pairwise energy conversions not ending at the mechanical force apex involve second-order molecular machines of the T_t type.

This situation is represented in Fig. 11 for electrochemical transaction [28]. A pair of functional groups, the COO⁻ of an Asp(D) or Glu(E) residue and the N-methyl nicotinamide attached by a Lys(K) side chain, are in an elastic protein-based polymer of a composition such that when both functional groups are in their more polar, less hydrophobic state, the value of T_t is just above the operating temperature T_o. With this design of the protein-based polymer, either protonation of the carboxylate due to a chemical energy input or the reduction of the N-methyl nicotinamide due to an electrical energy input will result in hydrophobic folding. The hydrophobic folding will change the property of the second functional group, providing an energy output. For example, reduction of NMeN will cause an increase in the pKa of the carboxylate, resulting in the chemical work of taking up a proton. Or, the chemical energy input of protonation can cause a

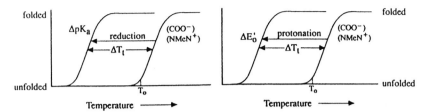

Figure 11 Demonstration of changes in temperature ranges for hydrophobic folding of second-order molecular machines of the T_t type designed to perform electrochemical transduction. (Reproduced with permission of Ref. 28.)

change in the redox potential of the NMeN, resulting in the electrical work of taking up an electron. This design of a protein-based polymer, as well as the capacity to improve the efficiency of energy conversion, is called *poising*. The concept of poising is treated in some detail elsewhere [28].

Each of the other of the 10 pairwise energy conversions not ending at the mechanical apex of Fig. 10 can be considered for specific designs of protein-based polymers, but that is beyond the scope of this chapter. Nonetheless, they could all be relevant, as discussed below in relation to the transducer applications and to controlled-release applications, whether for release of pharmaceuticals or for release of agricultural crop-enhancement substances.

III. PREPARATIONS

A. Chemical Syntheses

As the chemical syntheses have been extensively presented elsewhere, they are only briefly considered here.

1. Polypentapeptide Synthesis

In order to prepare poly(VPGVG) and obtain high molecular weight polymers in good yields, a number of approaches have been undertaken [29]. Great care has to be taken because small levels of impurities can result in termination of the polymerization process and small amounts of racemerization can alter the physical properties. Different pentamer permutations have been prepared and polymerized with different coupling methods. Gly-Val-Gly-Val-Pro was determined to be the best permutation and p-nitrophenyl ester (ONp) was the best polymerization agent.

A 3 + 2 coupling strategy was used [30]; the tripeptide Gly-Val-Gly and the dipeptide Val-Pro were synthesized separately and coupled to obtain the pentamer.

As mentioned above, pentamer purity is essential in obtaining a material with suitable physical properties. Small changes in the preparation of the pentamers can result in a transition temperature that varies as much as 15°C (25°C–40°C). This variance is critical.

A polymer that has a 25°C transition temperature will form a very good cross-linked elastomeric matrix, while a 40°C preparation will not cross-link to form an elastomeric matrix. The solution of this problem is in the purification of the components of the pentapeptide of GVGVP. The dipeptides Boc-VG-OBzl and Boc-VP-OBzl are recrystallized from ethyl acetate/petroleum ether and ethyl ether/petroleum ether, respectively.

Boc-GVG-OBzl is also recrystallized from ethyl acetate/petroleum ether. The polymer obtained from a careful preparation of the pentamer will yield a product with a transition temperature of 25.5°C ± 1°C [67].

2. Polytetrapeptide Synthesis

In order to prepare poly(GGXP), for which X is Ala, Val or Ile, several different coupling strategies were used. In the first approach [27,84] a 2 + 2 strategy was used in which the dipeptides Gly-Gly and X-Pro were coupled together using EDCI/HOBt to give the tetramer, GGXP. The tetramer was converted to a p-nitrophenyl ester and polymerized.

In the second approach [85], a stepwise addition strategy starting at the C terminus was used. Again, the tetramer was converted to a p-nitrophenyl ester and polymerized. The stepwise addition approach gave a better yield and a product with a lower transition temperature as compared to the 2 + 2 approach.

This was expected because the stepwise approach has intermediates that can be purified. In the 2 + 2 approach, only Boc-VP-OBzl and Boc-AP-OBzl can be recrystallized with ethyl ether/petroleum ether. Boc-GG-OBzl and Boc-P-OBzl cannot be recrystallized. In the stepwise approach, not only can the dimers Box-VP-OBzl and Boc-AP-OBzl be purified, but the trimers Boc-GVP-OBzl, Boc-GAP-OBzl, and Boc-GIP-OBzl can be recrystallized with ethyl ether/petroleum ether. This allows that the required high levels of purity can be obtained.

B. Genetic Engineering and Microbial Biosynthesis

An alternative to the organic synthesis of protein-based polymers is a biosynthetic approach using current recombinant deoxyribonucleic acid (DNA) methodologies. Using this approach, a gene encoding the desired peptide sequence is constructed, artificially inserted, and then translated in a microbial host organism. The resulting protein can then be purified, often in large amounts, from cultures grown in fermentation reactors. Molecular biology techniques are used to manipulate the genetic information (i.e., DNA sequences) for their effective expression in the appropriate host organism. The primary tools that make this possible are a refrigerated chest full of enzymes capable of cleaving, joining, copying, and otherwise modifying polynucleotides. In addition, there exist vehicles, or vectors, that make it possible to introduce (i.e., carry) this information into the host organism in a suitable manner for expression.

The most numerous of the DNA-manipulating enzymes are the "restriction endonucleases"; these enzymes are capable of cleaving the double-stranded polynucleotides at sequences specific for, and recognized by, a particular endonuclease species. Depending on the restriction enzyme used, the DNA cleavage will result in "blunt" (flush), double-stranded ends or "sticky" ends with single-stranded extensions of several nucleotides in length. For instance, the enzyme *Sma* 1 recognizes the nucleotide sequence 5'-CCCGGG-3' and cleaves in the middle to leave the double-stranded blunt ends $\binom{-CCC}{-GGG}$ and $\binom{GGG-}{CCC-}$; *Bam*H 1 recognizes the sequence 5'-GGATCC-3' and cuts to leave the sticky ends $\binom{-G}{-CCTAG}$ and $\binom{GATCC-}{G-}$.

Two strands of DNA can be joined, or ligated, using the enzyme ligase. With ligase, the blunt end of one strand of DNA can be ligated to the blunt end of another strand or, the sticky end of one strand can be ligated to another strand having the correct "complementary" sticky end (i.e., proper match of the complementary base pairs, A/T and G/C).

The most common vector employed to insert heterologous genes into a host organism for expression is called a *plasmid*. Plasmids are circular, double-stranded DNA molecules capable of self-replicating inside the cell and existing free from the chromosome. In nature, they typically carry some otherwise nonessential genetic information, that may, in a given context, provide the cell with a definite growth advantage, for instance, resistance to antibiotic. Using the enzymes described above, a plasmid may be opened at a specific site by a given restriction endonuclease, leaving free ends to which a gene sequence with compatible ends can be ligated. Thus, the molecule is recircularized with the gene sequence inserted. Further techniques allow the molecular biologist to then insert the plasmid into a microbial cell in which it is capable of replicating to form many copies per cell.

A number of plasmids have been developed to achieve expression of heterologous genes in a host cell. These expression plasmids carry specific sequences that function as a "promoter" of gene expression. The plasmids have been engineered to contain specific restriction endonuclease sites that allow placement of inserted genes immediately adjacent to a promoter. The promoter is then used to drive expression of the gene, resulting in the production of its respective protein. This biotechnological approach can result in the production of proteins at levels difficult to obtain from natural or organic synthesis sources. In addition, coupled with the ability to perform solid-phase organic synthesis of oligonucleotides, it is possible to custom design genes for proteins that have no natural source.

For a more extensive review of these and other molecular biology and genetic engineering principles and techniques, one may consult the manuals listed in Refs. 39, 40, and 86.

We have been able to use the techniques described above to construct genes for the elastic protein-based polymers $(GVGVP)_n$, for which n ranges from 10 to 250 tandem repeats. These have been "cloned" into plasmids and successfully expressed in the bacterium *Escherichia coli* (*E. coli*). Initially, a gene was constructed using synthetic oligonucleotides to encode $(GVGVP)_{10}$. This gene was made with flanking *Bam*H 1 (G'GATCC) and *Pfl*M 1 (CCANNNN'NTGG) restriction endonuclease recognition sites (see Fig. 12a1). In the *Pfl*M 1 site, *N* stands for any of the four natural bases and the prime indicates the site for chain scission. This gene was inserted into plasmid pUC118 and its sequence was verified by DNA sequence analysis. This 10mer gene was then used as a modular unit for constructing longer genes encoding even higher molecular weights of $(GVGVP)_n$. Large amounts of plasmid containing the 10mer gene were prepared and digested with *Pfl*M 1. The *Pfl*M 1 $(GVGVP)_{10}$ gene fragments were then purified and used in a subsequent ligation reaction to form polymers of $[(GVGVP)_{10}]_n$ (Figs. 12a2 and 12a3). Also, separate adapter oligonucleotides with unique restriction sites were added to this ligation reaction to allow the subsequent cloning of the concatenated gene fragments into various expression plasmids. These adapter oligonucleotides were added at a ratio that would favor the recovery of high molecular weight "concatemers." This process is represented schematically in Fig. 12b.

For recovery and cloning of individual-length concatemer genes, the ligation mixture was digested with *Bam*H 1, then electrophoresed through an agarose slab gel to achieve separation of the various molecular weight sizes. Slices corresponding to different size ranges were then removed from the gel, the DNA was recovered, and then cloned into plasmid pUC118. Gene inserts into this plasmid were analyzed by restriction endonuclease digestion and accurately sized by agarose gel electrophoresis adjacent to a concatemer "ladder" (see Fig. 13). Positive clones in pUC118 were then subcloned into several differ-

(A)

1.

```
                    gly val gly val pro (GVGVP)₈ gly val gly val pro
        CGGGATCCA GGC GTT GGT - - - - - - - - - - - - - - - -CCA GGC GTT GGATCCCG
        GCCCTAGGT CCG CAA CCA - - - - - - - - - - - - - - - -GGT CCG CAA CCTAGGGC

           BamH 1   PflM 1                                    PflM 1   BamH 1
```

2.

```
                    gly val gly val pro (GVGVP)₈ gly val gly val pro
                TT GGT - - - - - - - - - - - - - - - -CCA GGC G
                CG CAA CCA - - - - - - - - - - - - - - - -GGT C
```

3.

```
                    (gly val gly val pro (GVGVP)₈ gly val gly val pro)ₙ
                TT GGT - - - - - - - - - - - - - - - -CCA GGC G
                CG CAA CCA - - - - - - - - - - - - - - - -GGT C
```

4.

```
            met              ((GVGVP)₁₀)ₙ        gly val gly val pro stop
    TCGGATCCAGACC ATG GGC G      TT - - - - - GGC G      TT GGT GTA CCG TAAGCTTGAATTCGGATCCAG
    GACCTCGGTCTGG TAC C          CG CAA - - - - - C      CG CAA CCA CAT GGC ATTCGAACTTAAGCCTAGGTC

       BamH 1   Nco 1                                           Hind 3   EcoR 1  BamH 1
```

(B)

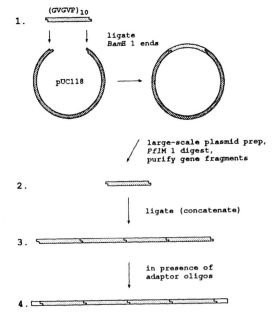

1. (GVGVP)₁₀

   ligate BamH 1 ends

   pUC118

   large-scale plasmid prep, PflM 1 digest, purify gene fragments

2.

   ligate (concatenate)

3.

   in presence of adaptor oligos

4.

ent expression plasmids. Initially, the concatemer genes were cloned into plasmid pQE-32 (Qiagen, Inc.) as a gene fusion behind a sequence encoding six tandem histidines. Expression from this plasmid results in the production of protein with an amino-terminal polyhistidine fusion, specifically, space $MRGSH_6GIQTM-(GVGVP)_n$. This fusion moiety provides the ability to affinity purify the protein by metal-chelate chromotography. Previously, we have reported the construction and expression of a gene encoding $(GVPGV)_{20}$ as a fusion to glutathione S-transferase in plasmid pGEX-3X that could be purified by a glutathione affinity column [38]. Several different size concatemer genes were subcloned into and expressed from the pQE-32 vector in *E. coli*. The poly(GVGVP) polymers that were produced were affinity purified from the bacterial cells and shown to have the requisite glycine, valine, and proline at the expected ratios for poly(GVGVP).

The above procedure established that genes encoding protein-based polymers of high molecular weight, for example, $(GVGVP)_{250}$, can be made and efficiently expressed in an *E. coli* host organism; using a gene fusion and affinity purification approach, it was also possible to highly purify the recombinant protein and show that it was the desired product. However, a primary goal was to show that the protein-based polymers could be effectively expressed as a nonfused "native" protein and be purified by means that exploit the inverse temperature transition properties of poly(GVGVP). Using purified $MRGSH_6$-$GIQTM-(GVGVP)_{120}$, it was shown that the protein-based polymer would reversibly go out of (i.e., coacervate) and into solution by simply increasing and decreasing the temperature; this coming out and into solution could be determined visually by watching the liquid go cloudy and then clear. Subsequently, this property was used to purify histidine fusions with $(GVGVP)_n$ where $n = 40, 140,$ and 250. First, a chilled (on ice) suspension of *E. coli* cells containing the protein-based polymer was lysed by sonic disruption to disperse the cellular components. While still cold, the lysate was centrifuged at high speed to remove the insoluble cell debris. The recovered supernatant was then warmed to 37°C, causing the protein-based polymers to form a visible aggregate, at which point it could be removed from the soluble fraction by centrifugation. Repeating this procedure once again as a wash, resulted in product that was as pure as the affinity-purified material when analyzed by SDS-polyacrylamide gel electrophoresis (see Fig. 14).

To achieve expression of a native, protein-based polymer in *E. coli*, a concatemer gene encoding $(GVGVP)_{120}$ was subcloned from pUC118 into the expression vector pET-11d (Novagen, Inc.) as an *Nco*1 to *Bam*H 1 fragment. The protein is expressed at high levels from this plasmid without the amino-terminal affinity moiety previously used. It was also effectively purified using the temperature-induced aggregation procedure.

Figure 12 (A), 1. The synthetic gene for $(GVGVP)_{10}$ with flanking external *Bam*H 1 and internal *Pfl*M 1 restriction endonuclease recognition sites. The gene and plasmid pUC118 were each cleaved with *Bam*H 1 and mixed together with ligase to recircularize the plasmid with the gene inserted. The plasmid with the cloned gene was amplified in *E. coli*. 2. A large-scale plasmid preparation was cleaved with *Pfl*M 1 and the released $(GVGVP)_{10}$ fragment was purified. 3. The purified *Pfl*M 1 fragment was self-ligated to form the concatemer genes. 4. Addition of adaptor oligonucleotides to the ligation reaction provides terminal sequences containing restriction sites needed for cloning the $(GVGVP)_n$ genes into various expression plasmids. Note that, although the genes are referred to in multiples of 10, the adaptor oligos add one additional pentamer sequence, resulting in $[(10)n] + 1$. (B), schematic illustration of events described in Fig. 12A.

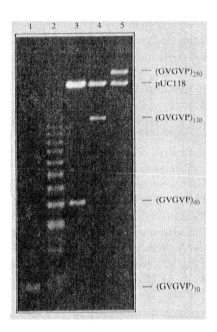

Figure 13 Deoxyribonucleic acid (DNA) samples electrophoresed through an agarose slab gel, stained with ethidium bromide, and visualized with ultraviolet (UV) light: Lane 1, the (GVGVP)$_{10}$ *Pfl*M 1 gene fragment; Lane 2, the concatemer "ladder" formed by ligation/polymerization of the gene in the presence of the adaptor oligonucleotides; Lane 3, the (GVGVP)$_{40}$ gene released from plasmid pUC118 with restriction endonuclease *Bam*H 1; Lanes 4 and 5, the (GVGVP) 120mer and 250mer genes, respectively, released from pUC118 with *Bam*H 1.

IV. APPLICATIONS

With the above knowledge providing the capacity for diverse designs extending beyond that possible with classical organic polymers, with the demonstrated properties providing for wide-ranging functional roles, with the demonstrated microbial biosynthesis and purification providing optimism for production of protein-based polymers at a cost competitive with oil-based polymers, with the environmentally friendly bioproduction requiring no toxic and hazardous chemicals, and with the environmentally clean disposal arising out of controllable biodegradation, the future of bioelastic materials appears both promising and stimulating. The present rate-limiting step for useful products is setting in place a bioproduction, purification, and processing facility. Meanwhile, there are many applications that are under development. A selection of medical and nonmedical applications that are not constrained by confidentiality and licensing agreements is considered below.

A. Medical Applications

Albert Szent-Gyorgyi, after having discovered vitamin C, began working on muscle contraction. During World War II, he was perhaps the first person to see the motion of life outside of living organisms by having fashioned actomyosin into a fiber and, by adding

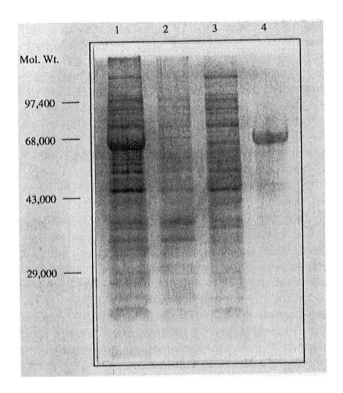

Figure 14 Analysis of microbially expressed $(GVGVP)_{140}$ by electrophoresis through an SDS-polyacrylamide gel followed by staining the gel with $CuCl_2$. Lane 1, crude *E. coli* cell lysate containing $(GVGVP)_{140}$; Lane 2, insoluble cell debris and nonlysed cells removed by centrifugation of cold (4°C) lysate; Lane 3, cell lysate after removal of $(GVGVP)_{140}$; Lane 4, the protein-based polymer removed from the cell lysate by centrifugation following thermally induced aggregation.

the appropriate chemicals, observed fiber contraction. The opening sentence for a series of papers reviewing that work was "Motion has always been looked upon as the index of life" [87].

In the publication that reported the results of the first elastic protein-based polymer designed to demonstrate chemomechanical transduction, it was stated: "To our knowledge, this is the first demonstration of mechanochemical coupling in a synthetic polypeptide and the first system to provide a test of the recent proposal that chemical modulation of an inverse temperature transition can be a mechanism for mechanochemical transduction" [80]. Now we consider elastic protein-based polymers in life-sustaining roles of medical applications.

1. Biocompatibility

For each medical application under consideration, a key first step is determination of biocompatibility of the required composition. This has been accomplished for the elastic protein-based polymer poly(VPGVG) and X^{20}-poly(VPGVG), with the conclusion of extraordinary biocompatibility [3]. The complete set of recommended tests have been completed for the plastic protein-based polymer poly(VPAVG) and X^{20}-poly(VPAVG), with the conclusion of good biocompatibility. Also, nine recommended tests have been completed for the elastic protein-based polymer, poly(GGAP), with again an extraordinary biocompatibility.

The following review of the data for poly(GGAP) is included here as a demonstration of the particular tests used and as a demonstration of the entirely innocuous nature of this protein-based polymer. Furthermore, as noted below, even in the presence of serum, cell adhesion does not occur to X^{20}-poly(GGAP) [85], which implies that this matrix does not support serum protein adsorption to its surface, which could then provide for cell adhesion. The absence of any cell adhesion and the apparent absence of protein adsorption makes this a particularly interesting surface for medical devices. The biocompatibility test results were obtained by North American Sciences Associates (NAmSA®), where the results are kept on file.

The Ames Mutagenicity Test.   The Ames test uses five strains of *Salmonella typhimurium*, each containing a specific mutation in the histidine operon. Without the presence of histidine, these five genetically altered strains (TA98, TA100, TA1535, TA1537, and TA1538) will not grow. To study mutagenicity, these five strains are placed in a histidine-free medium in which only those able to mutate back to the wild, nonhistidine-dependent type will form colonies. The reversion rate is normally constant. Accordingly, if the introduction of a test article causes at least a twofold increase in reversion rate, as compared with spontaneous reversion in the presence of a negative control (in this case, saline), the test article is considered mutagenic and therefore a possible carcinogen.

The test solution was a 4 mg/ml solution in 0.85% saline of poly(GGAP), equivalently poly(APGG). As seen in Table 4, the test article was not found to be mutagenic in this study. In fact, the test article was indistinguishable from the saline negative control. The positive controls were Dexon for TA98, TA100, and TA1537; sodium azide for TA1535; 2-aminofluorene and 2-nitrofluorene for TA1538; and 2-aminofluorene for TA100. Metabolic activation is necessary for 2-aminofluorene to be mutagenic; therefore, the 2-aminofluorene was tested on TA100 and TA1538, both with and without the S-9 mix. The S-9 mix, which is a microsomal fraction from an appropriately induced rat liver homogenate, was used to detect any mutagenic chemicals present that may require metabolic biotransformation to become active mutagenic forms. The innocuous nature of poly(GGAP) becomes apparent when its reversion rates are compared with those of Dexon, which is a sufficiently biocompatible material for regular use as a biodegradable suture material.

Cytotoxicity−Agarose Overlay.   Confluent monolayers of the L-929 mouse connective tissue cells were grown in culture flasks. Minimum essential medium was prepared, then placed in the culture flasks and allowed to solidify over the cells to form the agarose overlay. A 0.1 ml sample of 40-mg/ml poly(GGAP) in 0.90% saline was dosed neat on a 12.7-mm diameter filter disk, which was then placed on the solidified overlay surface. A similar filter disk with 0.1 ml of 0.9% sodium chloride U.S. Pharmacopeia (USP) solution was used as a USP negative control. The positive control was placed in a 25-cm² flask equidistant from the test article on the agarose surface. The flask containing controls and test article was then sealed and incubated for 24 hours at 37°C. After 24 hours, the culture was examined. The test article showed no cell lysis or toxicity (a 0-mm zone of lysis), while the negative controls were nontoxic and the positive control was toxic (with a 9-mm zone of lysis). Thus, poly(GGAP) was found to be nontoxic to L-929 mouse connective tissue cells.

Table 4 Plate Incorporation Assay for the Ames Mutagenicity Test

	Salmonella typhimurium tester strains				
	TA98	TA100	TA1535	TA1537	TA1538
	Number of revertant colonies (average of duplicate plates)				
Saline (−control)	82	180	14	8	6
Saline test article solution (undiluted)	57	172	9	6	6
Saline with S-9 (−control)	92	173	15	11	11
Saline with S-9 test article solution (undiluted)	74	180	12	7	11
Dexon 1 mg/ml (+control)	1048	1408	N/A	247	N/A
Dexon 1 mg/ml with S-9 (+control)	1256	1048	N/A	1048	N/A
Sodium azide 0.1 mg/ml (+control)	N/A	N/A	2752	N/A	N/A
Sodium azide 0.1 mg/ml with S-9 (+control)	N/A	N/A	3176	N/A	N/A
2-Nitrofluorene 1 mg/ml (+control)	N/A	N/A	N/A	N/A	2800
2-Nitrofluorene with S-9 (+control)	N/A	N/A	N/A	N/A	2240
2-Aminofluorene 0.1 mg/ml (+control)	N/A	184	N/A	N/A	11
2-Aminofluorene with S-9 (+control)	N/A	1296	N/A	N/A	2816

Test Article: X^{20}-poly(GGAP), Batch CG65PA
Reports from North American Science Associates (NAmSA). In no case was there a twofold or greater increase in the reversion rate of the tester strains in the presence of the test article solution.
N/A = not applicable

In the cell-attachment studies discussed below, X^{20}-poly(GGAP) was found to be nontoxic to human umbilical vein endothelial cells and to bovine ligamentum nuchae fibroblasts.

Acute Systemic Toxicity.   In tests for acute systemic toxicity, five mice were weighed and then injected by an intravenous route with a 40-mg/ml solution of poly(GGAP) in 0.85% saline at a dose of 50 ml/kg. Five other mice were similarly injected with control saline solution. The animals were then observed at 4, 24, 48, and 72 hours. Significant signs of toxicity are mortality, body weight loss of more than 2 g in three or more mice, convulsions, or prostration. Slight signs of toxicity would be lethargy or hyperactivity. The mice appeared normal; no signs of toxicity were observed for poly(GGAP).

Intracutaneous Toxicity.   A 0.2 ml dose of 40 mg/ml solution of poly(GGAP) in 0.9% saline was injected by the intracutaneous route into five separate sites on the right side of the back of two rabbits; on the left side, 0.9% sodium chloride USP solution was injected as a control. At 24, 48, and 72 hours, the injection sites of the animals were observed for signs of erythema (ER) or edema (ED). In the first study, slight irritation or toxicity was observed. As even a slight irritation was unexpected, the study was rerun at 5 mg/ml. As shown by the results in Table 5, neither irritation nor toxicity was observed. Neither erythema (ER) nor edema (ED) was observed.

Table 5 U.S. Pharmacopeia Intracutaneous Toxicity Observations for Poly(GGAP)

Rabbit no.		24 hours		48 hours		72 hours	
		ER	ED	ER	ED	ER	ED
69235	Test	0	0	0	0	0	0
	Control	0	0	0	0	0	0
69226	Test	0	0	0	0	0	0
	Control	0	0	0	0	0	0

Reports from North American Science Associates (NAmSA). Date prepared, 2-8-93; date injected, 2-8-93; date terminated, 2-11-93. Comments: Not applicable.
Erythema (ER): 0 = none, 1 = barely perceptible, 2 = well defined, 3 = moderate, 4 = severe
Edema (ED): 0 = none, 1 = barely perceptible, 2 = well defined, 3 = raised 1 mm, 4 = raised >1 mm
Rating (test and control): 0.0–0.5, acceptable; 0.6–1.0, slight; <1.0, significant

Systemic Antigenicity Study.   Solutions of poly(GGAP), equivalently poly(APGG), at a concentration of 40 mg/ml volumes in a 0.9% NaCl test solution were injected intraperitoneally at 10 ml/kg body weight three times a week, every other day, until six induction injections were conducted on six guinea pigs. Similarly, three additional guinea pigs were injected intraperitoneally with 0.9% NaCl solution as the control condition. At 10 days after the last intraperitoneal injection, all of the guinea pigs were challenged by an intravenous injection of the poly(GGAP) solution and then observed for any signs of systemic antigenicity. Significant signs of reaction would be face pawing, eye blinking, lethargy, convulsions, and even death. There were no significant reactions to the challenge injection. Therefore, the test article solution would not be considered antigenic in the guinea pig.

Dermal Sensitization Study (A Maximation Method).   In this study, 15 Hartley guinea pigs were used. A 40 mg/ml solution of poly(GGAP) in 0.9% NaCl was the test article. For Induction 1, the 10 animals to be used were clipped, then received three rows of intradermal injections, 2 per row, in an area 2 × 4 cm. The injections were 0.1 ml of Freud's complete adjuvant (FCA), 0.1 ml of the test article, and 0.1 ml of a 1 : 1 suspension of the FCA and test article. The 5 animals to be used as controls at the challenge phase were not induced. For Induction 2, one week later the area was reclipped and a 10% sodium lauryl sulfate (SLS) suspension in petrolatum was massaged into the skin to produce a mild acute inflammation. Any SLS suspension remaining after 24 hours was removed. An area 2 × 4 cm of a Whatman No. 3 MM filter paper, saturated with 0.3 ml of the test article, was topically applied and secured with nonreactive tape and an elastic bandage wrapped around the trunk of each animal. The patch was removed after 48 hours.

Next, the follow-up challenge was performed at 12 days. The area was again clipped and a nonwoven cotton disk in a Hill Top Chamber® was saturated with 0.3 ml of the test article, topically applied, and held in place for a 24-hour period with semiocclusive hypoallergenic adhesive tape and an elastic bandage wrapped around the trunk of the animal. The area was observed 24, 48, 72, and 96 hours after patch removal. The test article, poly(GGAP), did not cause delayed contact sensitization in the guinea pig (see

Table 6). Note that, according to NAmSA, "background or artifactual reactions (0.5 score) were not counted as evidence of a sensitization response. The 0.5 score occurred with the same frequency in the controls as in the test animals.

Rabbit Pyrogen Study.   A 0.9-gram portion of poly(GGAP) was reconstituted in 90 ml of sterile, nonpyrogenic saline. A single dose of 10 ml/kg was intravenously injected into the marginal ear vein of three rabbits. Rectal temperatures were measured and recorded before injection and every hour for 3 hours afterward. A maximum rise of 0.1°C was recorded, with a sum in the three animals of 0.2°C, whereas a summed rise of less than 1.4°C would still be considered nonpyrogenic. Accordingly, the test solution was judged to be nonpyrogenic.

Lee–White Clotting Study.   A 100 mg portion of poly(GGAP) was reconstituted in 10 ml of 0.9% NaCl solution to yield a 10 mg/ml test solution. Next, 0.5 ml of the test solution was added to six siliconized tubes. These tubes were placed in a 37°C heat block along with three siliconized tubes (without material) that served as the control. Finally, 1 ml of fresh canine blood was added to each tube.

At timed intervals, the first tube in each set of three was tilted until nearly horizontal. The tilting procedure was repeated at 30-second intervals until the blood clotted. As seen in Table 7, the replicate mean is greater in each case for the test article, poly(GGAP), than for the control. Clearly, poly(GGAP) does not shorten the clotting time.

Hemolysis Test *In Vitro*.   A clot-free blood sample from a New Zealand white rabbit was collected into an EPTA vacuum tube on the day the test was performed. As usual, four tubes were used in this study: a negative control containing 10 ml of 0.9% sodium chloride (SC) USP solution, a positive control containing 10 ml of USP purified water (PW), and 2 tubes each containing 0.2 ml of the test article, which was a solution of 40 mg/ml poly(GGAP) in 0.9% sodium chloride and 10 ml SC containing 0.2 ml of the test

Table 6   Guinea Pig Sensitization Dermal Reactions, Challenge with X^{20}-Poly(GGAP), Batch CG65PA

Animal No./group	Hours following patch removal (right flank)[a]			
	24	48	72	96
1 Test	0	0.5	0.5	0
2 Test	0	0.5	0.5	0
3 Test	0	0	0	0
4 Test	0	0	0	0
5 Test	0	0	0	0
6 Test	0.5	0.5	0.5	0
7 Test	0	0	0	0
8 Test	0	0	0	0
9 Test	0	0	0	0
10 Test	0	0	0	0
11 Control	0	0	0	0.5
12 Control	0	0	0	0
13 Control	0	0.5	0	0
14 Control	0	0	0	0
15 Control	0.5	0.5	0.5	0

[a]Right flank = test article solution (as received)

Table 7   Lee–White Clotting Study

	Series		
	A	B	Replicate mean
Test Article: Poly(GGAP), Batch CG65PA			
Tube 1	11.5	11.5	11.5
Tube 2	14.5	15.0	14.8
Tube 3	18.0	18.0	18.0[a]
Control			
Tube 1	9.0	7.5	8.3
Tube 2	14.0	12.5	13.3
Tube 3	15.5	15.0	15.3[a]

Note: Time recorded in minutes
[a]Lee-White coagulation time

article, which is again a solution of 40 mg/ml poly(GGAP) in 0.9% sodium chloride. To each tube, 0.2 ml of rabbit blood was added. The tubes were then covered and gently inverted to mix, then placed in a 37°C water bath for 1 hour. After incubation, the tubes were again gently inverted and the solutions were decanted into centrifuge tubes to be centrifuged for 10 minutes at 1000 × g. From absorbance values taken at 545 nm, the percent hemolysis was determined by the equation

$$\frac{\text{Test article} - \text{SC negative control}}{\text{PW positive control}} \times 100 = \%\ \text{hemolysis} \tag{2}$$

The result in each case was 0% hemolysis; poly(GGAP) was found to be nonhemolytic.

The results of the nine biocompatibility tests are summarized in Table 8.

2. Prevention of Adhesions Applications

Results of the use of elastic and plastic protein-based polymers for the prevention of adhesions have been published using two models: a bloodied, contaminated peritoneal model in the rat [41,42] and a strabismus surgery model using the rabbit eye [43]. These results are briefly reviewed below.

A Contaminated Peritoneal Model in the Rat.   The contaminated peritoneal model was developed to test for efficacy of the bioelastic materials in a model of relevance to combat casualty care involving abdominal wounds [41,42]. The model is depicted in Fig. 15. The abdominal wall is scraped with a scalpel until it bleeds. An adjacent loop of bowel is repeatedly punctured with a hypodermic needle until it bleeds and fecal material can be extruded. A loose suture, accessible from outside the animal, is placed around the injured intestine and is used to draw the injured bowel into juxtaposition; and to hold the bowel loosely in contact, with the injured abdominal wall. The abdominal wall is closed. At seven days, the loose suture is cut and withdrawn without opening the peritoneal cavity, and at two weeks the peritoneal cavity is opened and examined for the occurrence of adhesions. As seen in Table 9, adhesions were observed in 100% of the 29 control animals of the study, with significant adhesions occurring 90% of the time. Examples of the adhesions are seen in Figs. 16a and 16b.

Figure 15b schematically shows the placement of the bioelastic matrix between the

Table 8   Summary of Biocompatibility Test Results for Poly(GGAP): Preparation CG65PA

Test	Description	Test system	Results
Ames (mutagenicity), MG019	Determine reversion rate to wild type of histidine-dependent mutants	*Salmonella typhimurium*	Nonmutagenic
Cytotoxicity, agarose overlay, MG030	Agarose overlay to determine cell death and zone of lysis	L-929 mouse fibro-blast	Nontoxic
Acute systemic toxic-ity TU012	Evaluate acute sys-temic toxicity from an IV or IP injection	Mice	Nontoxic
Intracutaneous toxic-ity, TU013	Evaluate local der-mal irritant or toxic effects by injection	Rabbit	Initial test; slight irri-tation; retest, no irritation
Systemic antigenic-ity, TA085	Evaluate general toxicology	Guinea pigs	Nonantigenic
Sensitization (maximi-zation method), TA006	Dermal sensitization potential	Guinea pigs	Nonsensitizing
Pyrogenicity, TU010	Determine febrile reaction	Rabbit	Nonpyrogenic
Clotting study, TA038	Whole-blood clotting times	Dog	Normal clotting time
Hemolysis, CB037	Level of hemolysis in the blood	Rabbit blood	Nonhemolytic

Source: Reports from North American Science Associates (NAmSA), which keeps detailed results in its archives.
IV = intravenous, IP = intraperitoneal

injured wall and the injured bowel with the loose suture passing through the matrix on each side of the bowel loop. This is shown in the animal with the bioelastic matrix X^{20}-poly(GVGVP) at the arrow in the bloodied field of Fig. 16c. As seen in Table 9, in 80% of the 29 test animals in this study, there were no significant adhesions. An example of an animal with no adhesion but with a scarred wall is seen in Fig. 16d. In 10% of the animals, there had occurred a break in the matrix through which the adhesion had grown, as shown in Fig. 16f. If this were prevented, then 90% absence of significant adhesions could result using the bioelastic matrix as compared with 90% occurrence of significant adhesions in the control animal. Even in the cases for which there were judged to be Grade 3 adhesions (i.e., where the bioelastic had become engulfed in the adhesion), on examination the bioelastic matrix was not adherent and could readily be removed from the mass of the adhesion.

In Figure 16e, a small loop of adhesion is seen to have grown around the bioelastic matrix. Significantly, the bioelastic matrix after two weeks in the bloodied and contami-nated site is seen to remain clear and transparent. No fibrous capsule formed around

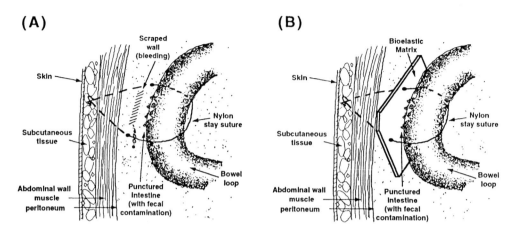

Figure 15 Schematic drawing of the contaminated peritoneal model for testing the efficacy of bioelastic matrices for the prevention of adhesions (see text for discussion): (A), control; (B), with test article. (Reproduced with permission of Ref. 42.)

the matrix. This situation, of relevance to drug delivery, has been observed even after X^{20}-poly(GVGVP) had been in the peritoneal cavity for months. Also, in the original color prints of Fig. 16e, it is apparent that there was no inflammation and no erythema or edema of the abdominal wall against which the matrix had been in contact for two weeks.

Strabismus Surgery Model in the Rabbit Eye.   A schematic representation of the strabismus surgery model in the rabbit eye is given in Fig. 17 [43]. For the positive control (Fig. 17a), a 3 mm × 3 mm lamellar sclerectomy, the scleral defect, was formed underlying

Table 9 Grading of Adhesions on Postmortem Examination Using X^{20}-poly(GVGVP)

Adhesion grade		% Adhesions	
		Control	Gas sterilized
0 } Insignificant	0		59 (17)[a]
1 } Insignificant	10	(3)	21 (6)[a]
2 } Significant	62	(18)	10 (3)[a]
3 } Significant	28	(8)	10 (3)

Scoring criteria for adhesions: for insignificant adhesions; Grade 0 = no adhesions, Grade 1 = single band of adhesion composed of omental fat between omentum and abdominal wall offering no resistance to separation; for significant adhesions, Grade 2 = involving omental fat and intestines with a fibrous band of adhesion tissue between viscera and abdominal wall and moderate force required for separation, Grade 3 = abscessed adhesion involving omental fat, abdominal wall, intestines with fibrous connective tissue proliferation, and sharp dissection needed for separation.
[a]$p < .05$, as compared with control. Amount in parentheses refers to number of rats studied.

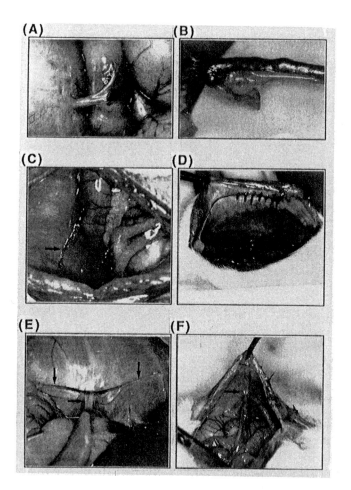

Figure 16   Photographs of the peritoneal cavity of the rat used to test the efficacy of bioelastic materials to prevent adhesions: (A), (B), adhesions formed in control; (C), placement of the bioelastic sheet between injured wall and bowel in a bloodied, contaminated field; (D), absence of adhesions due to placement of the bioelastic sheet; (E), loop of adhesion growing over, but not adhering to, the bioelastic sheet; (F), break in the bioelastic sheet through which adhesions had grown. (Reproduced with permission of Ref. 42.)

the superior rectus muscle and a portion of the muscle capsule was removed directly overlying the scleral defect. This resulted in a dense scar binding the rectus muscle to the sclera, as seen at eight weeks in Fig. 18a.

When the identical procedure is carried out with a bioelastic sleeve having been wrapped around the superior rectus muscle at the position of the muscle capsule and scleral defects (as shown for the test case in Fig. 17b), no scarring is observed and the bioelastic matrix is found to be readily separable from and nonadherent to the sclera and the muscle. At eight weeks (Fig. 18b), in a thin section is the wide region where the layer of X^{20}-poly(GVGVP) (fragmented by the sectioning) had separated muscle from sclera. Note that there is no scar tissue formation and no fibrous encapsulation of this bioelastic

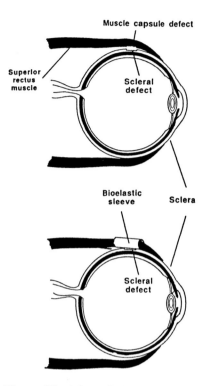

Figure 17   Schematic drawing of the strabismus surgery model used in the rabbit to test bioelastic materials for the prevention of postsurgical adhesions: upper drawing shows positioning of the muscle capsule and scleral defects; lower drawing shows bioelastic sleeve surrounding rectus muscle and thereby presenting a barrier for the prevention of the adhesion of muscle to sclera.

matrix. The X^{20}-poly(GVGVP) had successfully prevented adhesion from occurring. Also, after the few days required for healing from the surgical procedure, there was no inflammation [43].

Also tested was the composition X^{20}-poly[0.75(GVGVP),0.25(GFGVP)]. As seen in the photograph of the eye with the implant in place (Fig. 18c), with this material a glistening fibrous capsule formed, which is seen in the thin section of Fig. 18d. In spite of having elicited fibrous capsule formation with this matrix, there was still no significant scarring or inflammation [43].

What is sought in further developing the material for the strabismus surgery application is a thinner bioelastic matrix that can be handled and implanted and will degrade within a month or two. A key issue in controlling the rate of degradation is discussed below in connection with the drug delivery application.

3.   Drug Delivery

The ultimate objective in controlled release of therapeutic substances was well stated by Pitt and Schindler.

> To develop delivery systems which respond to the biochemical need for the drug comparable to the way the human body exercises control over releases and distribution of its own endogenous biological agents [88].

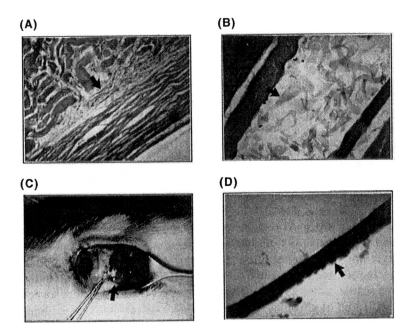

Figure 18 Photograph of histological sections and of the eye when using the strabismus surgery model for the prevention of adhesions in the rabbit eye showing: (A), scar adhering overlaying muscle (upper left) to underlying sclera (middle right); (B), the bioelastic matrix separating muscle (upper left) from sclera (lower right); (C), Phe(F)-containing bioelastic matrix surrounded by a glistening fibrous capsule; (D), sectioned Phe(F)-containing bioelastic matrix with a fibrous capsule noted by the arrow. (Adapted with permission of Ref. 43.)

Of all of the materials currently under development for the delivery of therapeutic substances, none show greater potential than the bioelastic materials. These protein-based polymers appear to be designable to perform all of the energy conversions of which living organisms are capable. Designs for 10 of the 15 pairwise energy conversions of Fig. 10 have been demonstrated and the principles are being developed to carry out the conversions with high efficiency.

In drug delivery or controlled release, there occurs a change in the concentration of the drug, that is, a change in the chemical potential. Thus, the net change is a change in chemical energy. It could be simply passive diffusion from a highly concentrated source, or it could be the change in chemical energy due to a transductional process. For example, the drug-laden matrix could contain a chemical clock that, with a designable half-life, caused the matrix to swell and thereby controlled the rate of release. On the other hand, the drug-laden matrix or vesicle could be under the control of a redox functional group in which enzymatic or other oxidation or reduction could effect swelling or contraction to drive release or the matrix could be under the control of a light-activated chromophore in which light or another part of the electromagnetic spectrum could effect release. All of these and more are possibilities with bioelastic matrices because of their transductional capacities, outlined above in Section II.

Here, the practical elements are simply demonstrated: (1) that the bioelastic matrix can be heavily laden with the model drugs beibrich scarlet and methylene blue; (2) that swell-doped X^{20}-poly(GVGVP) can result in a reasonably sustained release of biebrich scarlet for seven days, from Day 4 to Day 11; (3) that X^{20}-poly[0.9(GVGVP), 0.1(GNGVP)] can similarly be swell doped with biebrich scarlet, but that it totally releases the biebrich by Day 2; and (4) that the matrix X^{20}-poly[0.9(GVGVP), 0.1(GNGVP)], when maintained at 37°C in phosphate buffered saline, swells and begins to fall apart within a week's time.

In Fig. 19a is shown schematically the swell doping process. In Step A of the process, a droplet of 0.1 M biebrich scarlet is added to the bioelastic disk at a sixfold greater volume than the disk. In Step B, the temperature is lowered to 5°C; the disk swells, taking in the droplet, and the drug is allowed to equilibrate throughout the swollen disk. In Step C, the temperature is increased to 37°C; the disk contracts, expelling the excess solution, which is at a decreased drug concentration, and the contracted disk becomes laden with drug. In Step D, the extruded solution is removed as the disk is rinsed twice with sixfold volumes to remove adherent solution; in the case of 0.1 M added solution of biebrich scarlet, the contracted disk has retained biebrich scarlet at a concentration of 0.3 M, which is one biebrich scarlet molecule per turn of poly(GVGVP) β-spiral. The loaded disk is placed in a vial with 1 ml of phosphate buffered saline; every 24 hours, the 1 ml is removed and a new 1 ml is added; the quantity of biebrich scarlet in the removed 1-ml volume is determined spectrophometrically. The data are plotted in Fig. 20; after an

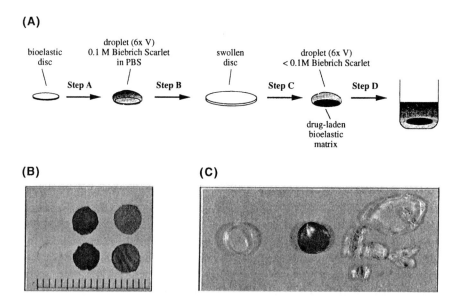

Figure 19 The swell doping process (see text for discussion): (A), swell doping procedure for introducing drug into bioelastic matrices (see text for details of Steps A, B, C, and D); (B), disks of X^{20}-poly(GVGVP) before swell doping (left), after swell doping (middle), and after 11 days of release in phosphate buffered saline for biebrich scarlet (above) and methylene blue (below); (C), disks of X^{20}-poly[0.9(GVGVP),0.1(GNGVP)] before swell doping (left), after swell doping (middle), and after 5 days in phosphate buffered saline in which the matrix has swollen and begun to break up due to the presence of the asparagine (Asn,N) residues.

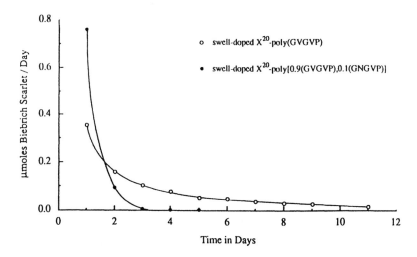

Figure 20   Release profiles of biebrich scarlet from swell-doped X^{20}-poly(GVGVP) (open circles) and from X^{20}-poly[0.9(GVGVP),0.1(GNGVP)] (solid circles). These are the data from the disks of Fig. 19B, upper row, and 19C, respectively (see text for discussion). (Adapted with permission of Ref. 44.)

initial wash out of about four days, there is a reasonably sustained release in the range of 50 to 15 nanomoles (nmoles) per day for a period of one week. Recognizing that the disk volume is only 8.2 microliter (μl), this means that a 1-ml disk could be releasing from 6 millimoles (nmoles) to 2 nmoles per day. Now, it should also be appreciated that the very thin disk, 0.3 mm thick, would provide a limiting poor example of sustained release because there is limited opportunity to establish rate of diffusion from deeper within the disk.

The set of disks used for biebrich scarlet and methylene blue are shown in Fig. 19b before loading after swell doping and after the 11-day release regimen.

The bioelastic matrix X^{20}-poly[0.9(GVGVP),0.1(GNGVP)], in which there is one asparagine (Asn,N) for every 50 residues and that has a carboxamide that can hydrolyze to a carboxylate in phosphate buffered saline at pH 7.4, can similarly be swell doped as discussed above. Its release profile, however, is quite distinct. Due to the carboxamide breakdown to carboxylate, the release is complete by two days (see Fig. 20). Also shown in Fig. 19c is that the disk of X^{20}-poly[0.9(GVGVP),0.1(GNGVP)] has swollen and begun to break up in less than a week in phosphate buffered saline at pH 7.4.

Observed in Fig. 21 is the increase in the value of T_t as a function of time in phosphate buffered saline at pH 7.4 for Polymer I, poly(GVGVP); Polymer II, poly[0.85(GVGVP), 0.15(GVG$_c$VP)]; Polymer III, poly[0.9(VPGVG),0.1(VPG$_c$VG)]; and Polymer IV poly [0.94(VPGVG),0.06(VPG$_c$NG)], where G$_c$ stands for glycolic acid that adds the hydrolyzable ester in the polymer backbone. The ester in the backbone of Polymers II and III causes a breakdown, reflected in the increase in the value of T_t with time, whereas there is no change in the value of T_t with time for poly(GVGVP). The data on Polymer IV shows that the glycolic acid residue and the carboxamide in the side chain add to enhance the rate at which T_t approaches 37°C. At this temperature, the matrix will be totally swollen and possibly even completely dispersed. Thus, matrices containing asparagine and gly-

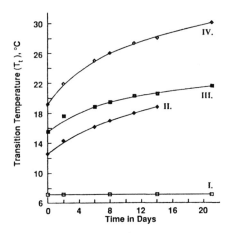

Figure 21   Dependence on time in phosphate buffered saline of the transition temperature T_t for four different compositions of elastic protein-based polymers (see text for discussion and structural identification). (Reproduced with permission of Ref. 44.)

colic acid could be used to control release of drugs that do not exhibit the easy diffusional release of biebrich scarlet and methylene blue.

4.   Cell Adhesion and Growth of Elastomeric Polypeptide Matrices

This section reviews our investigations into the ability of various elastomeric matrices to support cell adhesion and growth *in vitro*. The same simple apparatus was used for both the cell adhesion and the cell growth assays. Glass cell cloning cylinders were placed onto disks of test matrices in the wells of flat-bottomed, 96-well tissue culture plates. An example of a single well is given in Fig. 22. Into the cloning cylinders were placed equal numbers of the test cells in the desired medium. In an adhesion assay, the cells were allowed to settle and adhere for usually 3 or 20 hours before removal of nonadherent cells by gentle rinsing, after which the adherent cells were fixed and subsequently counted and assessed by phase contrast microscopy. In a cell growth assay, the plated cells were

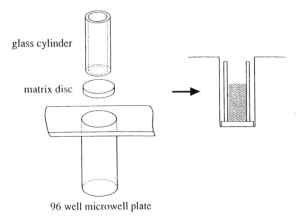

Figure 22   Sketch of a single well used in the adhesion assays.

incubated for 3 days before fixation, observation, and counting. (These methods are described in detail in Ref. 50.) The matrices tested were all prepared by the 20-Mrad gamma irradiation of coacervated, synthetic, elastic protein-based polymers. The cells used were dissociated by trypsinization, then treated with soybean trypsin inhibitor.

Nonadhesive X^{20}-poly(GVGVP).   The elastomeric matrix X^{20}-poly(GVGVP) was found to be a very poor substratum for the adhesion of bovine aortic endothelial cells (BAECs), bovine ligamentum nuchae fibroblasts (LNFs), and human umbilical vein endothelial cells (HUVECs) when tested in serum-free media containing 0.1% bovine serum albumin (BSA) (see Figs. 23 and 24) [50,89].

(A)

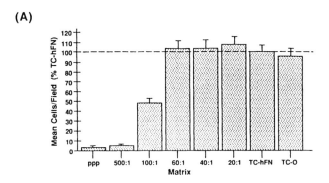

(B)

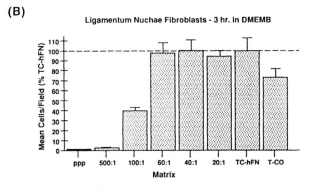

ppp represents X^{20}– poly(GVGVP)
500:1 represents X^{20}– poly[500(GVGVP),(GRGDSP)]
100:1 represents X^{20}– poly[100(GVGVP),(GRGDSP)]
60:1 represents X^{20}– {poly[20(GVGVP),(GRGDSP)] + 2 poly(GVGVP)}
40:1 represents X^{20}– {poly[20(GVGVP),(GRGDSP)] + poly(GVGVP)}
20:1 represents X^{20}– poly[20(GVGVP),(GRGDSP)]
TC-hFN represents tissue culture substratum coated with human fibronectin
TC-O represents untreated culture substratum

Figure 23   (A), Bar graph of bovine aortic endothelial cell adhesion to various matrices or surfaces in a period of three hours in medium 199 containing 25 mM HEPES, 0.1% bovine serum albumin, 100 units/ml penicillin, and 100 μg/ml streptomycin (i.e., M199HB). (B), Bar graph of ligamentum nuchae fibroblast adhesion to various matrices or surfaces in a period of three hours in DMEMB (Dulbecco's modified Eagle medium containing 0.1% bovine serum albumin, 2 mM L-glutamine, 0.1 mM nonessential amino acids, 100 units/ml penicillin, and 100 μg/ml streptomycin). (Reproduced with permission of Ref. 50.)

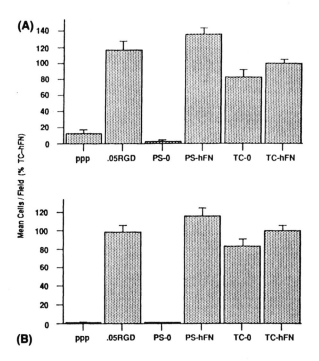

Figure 24   Bar graphs of human umbilical vein endothelial cells (HUVECs) attaching to different substrates: (A), in 4 hours; (B), in 20 hours. The terms ppp, TC-O, and TC-hFN are defined at the bottom of Fig. 23; 0.5 RGD stands for X^{20}-poly[20(GVGVP),(GRGDSP)], PS-O for uncoated plastic cover step, and PH-hFN for human-fibronectin-coated plastic cover slip. (Reproduced with permission of Ref. 89.)

Incorporated RGDS Makes X^{20}-poly(GVGVP) Adhesive.   The synthetic incorporation of GRGDSP, the major integrin binding peptide sequence of fibronectin, into the polypeptide chains before cross-linking resulted in matrices that could support cell adhesion in BSA-containing media. In fact, variation of the ratio of incorporated GRGDSP sequences enabled the modulation of the cell-to-matrix adhesiveness between the extremes of essentially no interaction and equivalence to cell-to-fibronectin-coated substratum adhesion (see Fig. 23). Both BAEC and LNF showed very low levels of attachment to matrices containing 1 GRGDSP to 500 GVGVP sequences, showed intermediate levels of adhesion to matrices containing 1 GRGDSP to 100 GVGVP sequences, and showed maximal adhesion (i.e., levels equivalent to the fibronectin positive controls) to matrices containing from 1 GRGDSP to 60 GVGVP up to 1 GRGDSP to 20 GVGVP sequences. HUVEC were tested for adhesion to matrices containing 1 GRGDSP to 20 GVGVP and were found to have developed high levels of adhesion after both 4 and 20 hours incubation (see Fig. 24).

X^{20}-poly(GVGVP) Matrices Containing RGDS Support Confluent Cell Monolayers. When BAEC or LNF were plated at high density in BSA-containing medium onto matrices containing 1 GRGDSP to 100 GVGVP (or higher proportions of GRGDSP, they formed confluent monolayers (see Fig. 25). This indicates a relatively even distribution of RGDS sites over the matrix surface sufficient to maintain a complete cellular coverage.

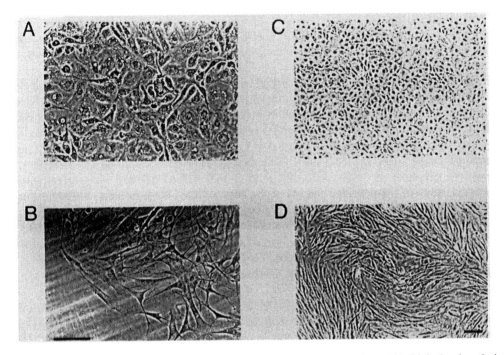

Figure 25   Cell attachment to GRGDSP containing bioelastic matrices: (A), high-density platings of bovine aortic endothelial cells; (B), ligamentum nuchae fibroblast (at edge of cloning cylinder); (C), low-density platings of bovine aortic endothelial cells having grown to confluence in three days' time; and (D), ligamentum nuchae fibroblasts grown to confluence in three days' time. Striations in Fig. 25B are due to the molds in which the bioelastic materials were cross-linked. (Reproduced with permission of Ref. 50.)

Adhesion to X^{20}-poly(GVGVP) in the Presence of Serum.   Although when assayed in the presence of BSA alone, X^{20}-poly(GVGVP) does not support cell adhesion, when assayed in the presence of fetal bovine serum (FBS), a considerable but submaximal level of BAEC and LNF cell adhesion occurs (ref. A). It seems likely that this is the result of adsorption of adhesive components present in serum onto the elastomeric matrix as it was found that X^{20}-poly(GVGVP) pretreated with fibronectin supported significant LNF attachment. It is well appreciated that adsorption of systemic components that generate unwanted effects is one of the great difficulties confronting the development of implantable medical devices. Despite the above finding, implants of X^{20}-poly(GVGVP) have been found to be remarkably biocompatible [2], as also discussed above for poly (GGAP). This is perhaps partly explained by our finding that, even in the presence of FBS, X^{20}-poly(GVGVP) does not support the anchorage-dependent growth of BAEC or LNF to confluency. On the other hand, matrices containing GRGDSP were found to support the growth of BAEC and LNF in a similar manner to the tissue culture plastic positive control (see Fig. 25).

The GRGDSP within X^{20}-poly[(GVGVP),(GRGDSP)] Recognized as a Vitronectin RGD.   The nature of the integrin adhesion receptors responsible for the attachment of the cells to GRGDSP-containing X^{20}-poly(GVGVP) matrices was investigated by means of integrin-specific peptide inhibitors (ref. B). GRGDSP inhibits the binding of both the fibronectin and vitronectin receptors to their ligands, GRGDdSP specifically inhibits only fibronectin-receptor-mediated adhesion, GPenGRGDSPCA (cyclic RGDS) specifically inhibits vitronectin receptor mediated cell adhesion, and GRADSP was used as an inactive control peptide.

Peptides were added at 1 mM to cell adhesion assays to GRGDSP-containing matrices and fibronectin- and vitronectin-coated nontissue culture plastic. The cell types used were LNF, HUVEC, and A375 human malignant melanoma. In each case, the cells plated on the GRGDSP-containing X^{20}-poly(GVGVP) responded in an identical manner to those plated on vitronectin (e.g., see Fig. 26 for A375 human malignant melanoma cells). The presence of 1 mM GRADSP or 1 mM cyclic RGDS completely inhibited cell adhesion, whereas the presence of 1 mM GRGDdSP or 1 mM GRADSP had little effect. On fibronectin, none of the peptides had much, or any, effect. The molecular conformation of the RGD is decisive in terms of receptor specificity as is clearly illustrated by cyclic RGDS being a specific inhibitor for vitronectin while the RGD in vitronectin is actually RGDV. It is probable that the conformation of RGDS within the polypeptide backbone of the elastomeric matrix is such as to be recognized by the vitronectin receptor. This finding could prove useful when fibronectin-receptor-dependent cell attachment was undesirable.

Other methods of incorporation of RGDS into the matrices may need to be developed in order to utilize fibronectin-receptor-mediated cell attachment. Alternative cell attachment peptide sequences may also need another method of incorporation since attempts to promote cell adhesion to matrices using YIGSR, the putative cell attachment sequence of laminin, and REDV, a sequence claimed to be specific for human endothelial cell attachment, did not succeed [52,89].

X^{20}-Poly(GGAP) Does Not Support Cell Adhesion Even in the Presence of Fetal Bovine Serum. A series of interesting polytetrapeptides, poly(GGAP), poly(GGVP), and poly(GGIP), were gamma-irradiation cross-linked to form matrices that were then tested for their cell adhesion properties [85]. In the presence of BSA, neither LNF nor HUVEC showed any attachment to any of the matrices. In the presence of FBS X^{20}-poly(GGIP)-supported cell attachment at near positive control levels, X^{20}-poly(GGVP) supported cell attachment at intermediate levels, but X^{20}-poly(GGAP) did not support any cell attachment, even after 20 hours of incubation (e.g., see Fig. 27 for HUVECs).

A potentially very useful correlation was noted between the mean residue hydrophobicity of the matrices, based on the temperature of the inverse temperature transition hydrophobicity scale and the degree of support for cell adhesion in the presence of FBS. The least hydrophobic, X^{20}-poly(GGAP), did not support the cell attachment. The most hydrophobic, X^{20}-poly(GGIP), supported the highest level of cell adhesion. X^{20}-poly(GGVG) was intermediate in both hydrophobicity and support for cell adhesion.

As appeared to be the case for X^{20}-poly(GVGVP) in the presence of FBS, it is likely that the limited cell adhesion promoted by the FBS is a result of adsorption of adhesive components onto the matrices. In the case of X^{20}-poly(GGAP), it may be that its particular hydrophobicity level is unsuited to adsorption of adhesive components or that other nonadhesive serum components are preferentially adsorbed. The lack of cell adhesive interaction demonstrated in these *in vitro* assays has been supported and extended by the excellent biocompatibility test results for X^{20}-poly(GGAP) (see detailed studies reported above). Thus, X^{20}-poly(GGAP) has good potential for use in applications for which cell attachment is undesirable. It could also provide a nonadhesive background in which the effects of specific adhesion-promoting sequences could be characterized under more physiological conditions.

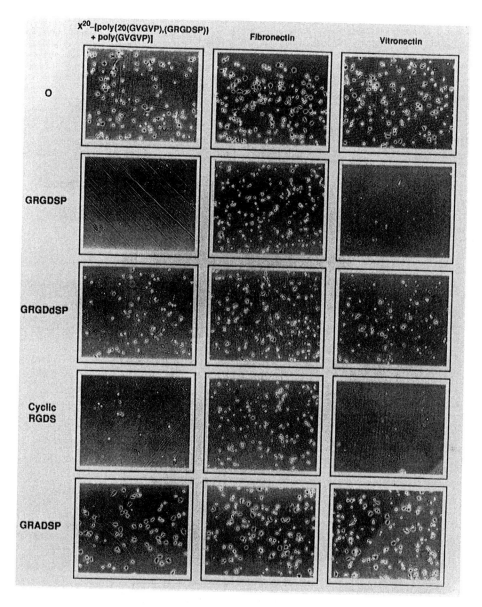

Figure 26 An adhesion of 375 human melanoma cells to GRGDSP-containing bioelastic matrices in the presence of different peptides that differentially compete at the fibronectin and vitronectin cell membrane receptors. It is apparent from comparing the bioelastic matrix with fibronectin- and vitronectin-coated surfaces that GRGDSP in the bioelastic matrix is binding to the vitronectin membrane receptor. The striations are due to the surface of the molds in which the bioelastic materials were cross-linked. (Reproduced with permission of Ref. 89.)

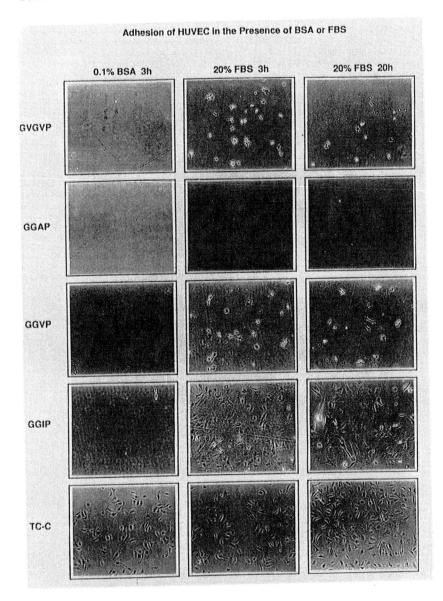

Figure 27 Cell attachment of human umbilical vein endothelial cells (HUVECs) to bioelastic matrices (component repeats noted at left) as compared with TC-C, which is tissue culture plastic coated by fibronectin only in the 0.1% BSA (bovine serum albumin case) Column 1. In Columns 2 and 3, the medium contained 20% FBS (fetal bovine serum) and the incubation period was 3 hours for Column 2 and 20 hours for Column 3. Note that under no circumstances is there cell attachment to GGAP, that is, to X^{20}-poly(GGAP) (see text for discussion). (Reproduced with permission of Ref. 85.)

5. Tissue Engineering/Reconstruction

The properties of bioelastic materials of relevance to tissue engineering/reconstruction are summarized in Table 10. Importantly, basic elastic and plastic protein-based polymers and their cross-linked matrices are biocompatible. Furthermore, their elastic moduli range from 10^4 dynes/cm^2 to 10^9 dynes/cm^2 and, with combinations of the basic units, the full range of values appear to be available. This means that the elastic modulus of virtually any tissue can be matched with a combination of repeating units that, in their homosequential polypeptides, have now been shown to be biocompatible.

Relationship between Elasticity and Tensegrity. It was demonstrated above that the elastic protein-based polymers as cross-linked matrices, that is, X^{20}-poly(GVGVP) and X^{20}-poly(GGAP), provide little or no support for cell attachment. On the addition of the cell-attachment sequence, however, the elastic matrix becomes an excellent substrate for cell attachment, for cell spreading, and for growth to confluence. Thus, there are elastic matrices to which cells can attach using membrane receptors called *integrins* [46–49]. Through the attachments to the integrins in the cell membranes, which are in turn attached to cytoskeletal components, the cells and their inner components can sense the tensional forces to which their extracellular matrix is subjected in its functional role as a body tissue. The concept of tensegrity as stated by Ingber is that the tensional forces sensed by the cell result in intracellular signals by which the appropriate genes are turned on to produce the extracellular matrix that can sustain the tensional forces [56]. This is schematically shown in Fig. 28.

Practicality of a Temporary Functional Scaffolding that Can be Remodeled into a Natural Functional Tissue. Because the bioelastic materials can be prepared with the preferred elastic modulus for any tissue except bone, because the elastic matrix can be prepared with cell attachment sequences to which human cells will attach, and because the bioelastic matrix is essentially biodegradable, the concept of a temporary functional scaffolding that could support cell attachment and normal functioning of natural tissue and could be remodeled into the natural tissue becomes a reasonable consideration. Such matrices could be further prepared to release the appropriate cytokines to enhance such a transformation using any of the possible and most relevant energy conversion processes. The demonstration that functional enzyme sites can also be designed into the basic matrix [90] further extends its versatility of filling a functional, dynamic role. Only bioelastic matrices can provide such versatility and diversity as a biomaterial.

Table 10   Properties of Bioelastic Materials of Relevance to Tissue Engineering

The basic elastomeric polypeptide and its cross-linked matrix are biocompatible and can be designed.
1. To exhibit a range of elastic moduli (covering three orders of magnitude)
2. To exhibit different rates of degradation
3. For various modes of drug and cytokine release
4. To perform free energy transduction involving the intensive variables of mechanical force, temperature, pressure, chemical potential, electrochemical potential, and light or other electromagnetic radiation
5. To contain functional enzyme sites
6. To contain functional cell attachment sites promoting growth to confluence
7. With the proper tissue elastic modulus such that the tensional forces to which the tissue is subjected, can be sensed by the cell, allowing for proper functioning of attached cells

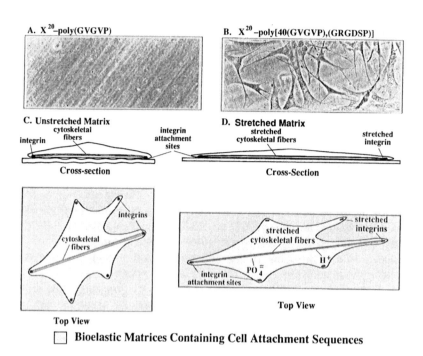

Bioelastic Matrices Containing Cell Attachment Sequences

Figure 28 The concept of tensegrity: (A), the absence of cell attachment to X^{20}-poly(GVGVP); (B), under the same conditions, there is good cell attachment to X^{20}-poly[40(GVGVP),(GRGDSP)]; (C) and (D) combined show the effect of stretching on a cell attached to an elastic substrate shown schematically. The cells attach by means of integrins, which are in turn attached in intracellular cytoskeletal fibers. Just as stretching appropriately designed elastic protein-based polymers results in release or uptake of protons or other chemicals, so too can stretching of cytoskeletal components give rise to similar chemical sequelae (see text for further discussion). (Reproduced with permission of Ref. 4.)

Tubular Constructs of Wide-Ranging Elastic Moduli. Figure 29 demonstrates that the bioelastic materials can be formed into functional shapes such as tubes. The near-transparent small tube at the bottom of the figure is actually a plastic tube with an elastic modulus of some 10^9 dynes/cm^2; it is X^{20}-poly(AVGVP). The next, entirely transparent small tube has an elastic modulus of 10^6 dynes/cm^2 and is X^{20}-poly(GVGVP). The larger, straight, white tube has the stress/strain curve of Fig. 6b and is X^{20}-poly [0.75(GVGVP),0.25(GFGVP)], with an elastic modulus ranging from 5 to 8 $\times$ 10^6 dynes/cm^2. Finally, the tube at the top of Fig. 29 is simply X^{20}-poly(GVGIP) and exhibits an elastic modulus striking similarly to that of vascular wall in the appropriate percent extension range, that is, from 4 to 6 $\times$ 10^6 dynes/cm^2 in the extension range from 30% to 120% (as shown in Fig. 6a).

Figure 29 demonstrates that tubes appropriate to those of the body can be constructed from bioelastic materials. The design variations are virtually infinite. Such tubes could function as temporary functional scaffoldings that could contain appropriate cell-attachment sites, could be remodeled into natural functional tubings by cells that have migrated onto and into the tubes, and could in a natural replacement reconstruction process, result in natural functional vessels.

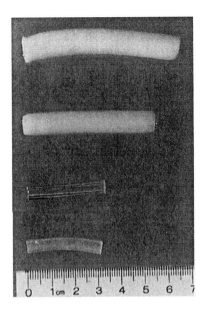

Figure 29   Tubular constructs composed of different gamma-cross-linked elastic and plastic protein-based polymers (see text for discussion).

B.   Nonmedical Applications

The potential nonmedical applications are perhaps even more numerous than the medical applications. The issues related to several are noted below under the headings of transducers, biodegradable plastics, controllable superabsorbents, and food product additives. There are, of course, many others — the controlled release of herbicides, pesticides, and growth factor for agricultural crop enhancement is one; a base for cosmetics is another, and so on.

In general, the nonmedical applications only become possible once low-cost bioproduction is achieved. As this is yet to be realized, the nonmedical applications are discussed in more of a putative manner. The realization of low-cost bioproduction, however, is anticipated within the next year or so, as can be deduced in part from.

1.   *Molecular Transducers (Sensors/Actuators)*

A *transducer* may be defined as a device actuated by (sensing) power from one system and supplying power to (actuating) a second system. Common examples are the telephone receiver sensing an electric power input and supplying an acoustic power output, or a quartz crystal capable of a mechanical power input and responding with an electric power output. The dimensions of power are units of energy (e.g., calories) per unit time (e.g., seconds) such that the treatment above of energy conversion relates directly to transducers with the added consideration of rate.

A protein-based polymer of a particular design is a low-power transducer with a very nonlinear response reasonably well suited to function in a feedback loop to maintain a system in a constant state. The power is limited by the rate of hydrophobic folding and assembly, and the response is nonlinear because it is based on an inverse temperature (phase) transition. The inverse temperature transition indicated in Fig. 11 occurs over a limited temperature range or it can be controlled, for example, by a small change in pH, as when the change in the degree of ionization becomes positively cooperative due to the nature of the competition between polar and apolar species for hydration. Such transducers, which are insensitive to a given power input when outside the transition temperature range, are particularly sensitive when within the responsive range for a particular variable. Accordingly they would be particularly useful when used in a feedback manner to maintain constant a particular variable.

Consideration of the transducer coupling to the input or the output system is quite different, depending on whether it is based on a first-order or second-order molecular machine of the T_t type. If the application utilizes a first-order molecular machine, then there must be a physical attachment to a fiber or a matrix in order to sense or to actuate by changes in length or force. Such fabrications can have limitations, for example, relative to size and strength of a fiber. If the application does not sense or actuate by changes in length or force, as is the case for second-order molecular machines of the T_t type, then such fabrication is not a limitation and a single molecule in solution or in an aggregate can be responsive, for example, to the input of a single photon to actuate an electrode by release of a proton or an electron.

2. *Biodegradable Plastics*

The plastic protein-based polymer poly(VPAVG) exhibits its inverse temperature transition near 30°C; it is soluble in water below 30°C and folds and aggregates to form a more-dense phase when the temperature is increased above 30°C. When the more-dense phase is gamma-irradiation cross-linked, the resulting matrix forms a swollen gel state below 30°C and extrudes all of its water to form a hard plastic state on increasing the temperature above 30°C. An example of the plastic state is seen as the bottom tube of Fig. 29 in tubular form. The elastic modulus of the hard, plasticlike state is of the order of 10^9 dynes/cm^2 or higher.

Plastic protein-based polymers of the general composition, poly[f_v(VPAVG)$_x$$f_x$ (VPAαG)$_x$$f_\beta$(VPA$\beta$G)] can have their value of T_t decreased by introducing more hydrophobic residues as the β-residue, and they can have chemical clocks in the form of asparagine (Asn,N) or glutamine (Gln,Q) residues as the α-residue. This means that the value of T_t can be set such that the swollen state could be reached only at temperatures well above 0°C and therefore would be unattainable. Depending on the half-life of the carboxamide side chain of the α-residue, carboxylates would slowly form at the plastic-water interface, which would raise the temperature of the inverse temperature transition T_t above the ambient temperature such that swelling could begin at the interface with subsequent biodegradation. As long as the value of T_t was greater than the ambient temperature, water at the interface would cause swelling and the surface would become susceptible to biodegradation of the swollen protein-based polymer.

3. *Controllable Superabsorbents*

The increase in volume for X^{20}-poly(VPGVG), which is more than 50% water in the contracted state when T_t is 10°C to 15°C below the ambient temperature, is a factor of 10 when the value of T_t is above ambient temperature. Thus, by controlling T_t, it becomes possible to bring about a superabsorbent state. One need not start with the hydrated state, but rather can begin with a dried or lyophilized, foamlike state of the matrix. In the case of X^{20}-poly(VPAVG), the contracted state, the state some 10°C to 15°C above T_t contains essentially no water such that this material goes from 0% to over 90% water and from a plastic state to a hydrogel state on increasing T_t above ambient temperature or on decreasing ambient temperature below T_t.

There are many ways of changing the value of T_t and they can be categorized in terms of the particular energy input able to do so, such as light, electrical, chemical, and pressure inputs. Depending on the application of interest, any particular energy input could be used to control the extent of absorbency. For a diaper or incontinence apparel, of course, the material would be designed to be in the superabsorbent state just below body temperature such that the fluid would be drawn away from the body surface to more remote, cooler temperatures and both the superabsorbent liner and the plastic outer container could be designed with chemical clocks to achieve a planned degradation.

4. Food Product Additives

Wheat Flours and Doughs. Protein elasticity is also crucial in determining the functional properties of wheat flours and doughs. In this case, the proteins involved are the major storage proteins, which account for about 10% of the grain's dry weight and form a continuous network called *gluten* when flour is wetted and kneaded to give dough. This network entraps carbon dioxide generated by yeasts during proving and becomes expanded to form a light, porous, crumb structure. Leavened bread is, therefore, in essence a protein foam that supports the other flour components, the most important of which is starch. The ability of the wheat gluten proteins to entrap carbon dioxide to form a foam depends on a combination of two physical properties: elasticity and extensibility. A precise balance of these is crucial, as poor quality bread can result when doughs are either insufficiently elastic (weak, too low an elastic modulus) or too elastic (overstrong, too high an elastic modulus). Gluten elasticity is also important for other applications of wheat, including the manufacture of noodles and pasta.

Wheat gluten is a complex mixture of over 50 individual proteins that are classified into two groups present in approximately equal amounts. The gliadins are monomeric proteins that interact by strong, noncovalent forces (chiefly hydrogen bonds and hydrophobic interactions) and contribute to extensibility. In contrast, the glutenins consist of subunits that form high M_r ($\alpha 1$ to 10×10^6) polymers stabilized by interchain disulphide bonds. These polymers appear to be the major determinant of gluten elasticity, although the precise molecular basis for this is not known. However, two features that may be relevant are the number and distribution of disulphide bonds and the presence in one group of glutenin subunits of β-spiral-like structures that may be analogous to those formed by the elastic protein-based polymer poly(VPGVG).

These proteins are called high molecular weight (HMW) subunits of glutenin, and have been studied in some detail because allelic variation in their composition is correlated with differences in the breadmaking quality of wheats [91]. The individual proteins vary from 627 to 827 residues in length (M_rs from about 67,500 to 88,100), and consist of central repetitive domains (481 to 696 residues) flanked by nonrepetitive N-terminal (84 to 104 residues) and C-terminal (42 residue) domains. The repetitive domains consist of tandem and interspersed repeats based on nonapeptide (consensus GYYPTSP or LQQ), hexapeptide (PGQGQQ), and tripeptide (GQQ) motifs, and appears to form β-spiral structures based on repeated β-turns [92]. It has been proposed that these spiral structures are intrinsically elastic (by analogy with elastin) [93,94], although the mechanism is clearly different [95]. The nonrepetitive N- and C-terminal domains appear to be globular and form cross-links via disulphide bonds.

The recent development of routine procedures for transformation of wheat [96,97] will allow the manipulation of wheat proteins by genetic engineering. For example, it should be possible to alter the amounts and properties (including elasticity) of the glutenin polymers by manipulating the amounts, structures, and cross-linking of the individual subunits. However, a more exciting approach would be to add "foreign" proteins in order to introduce specific properties. These could include intrinsically elastic subunits based on the VPGVG pentapeptide motif of elastin or on the appropriate combination of VPGVG and VPAVG, which could result in the preferred elastic modulus. Incorporation of such subunits into gluten via interchain disulphide bonds should result in increased optimal elasticity and possibly improved quality for breadmaking and other applications.

Wheat also provides an opportunity to produce novel protein-based polymers for nonfood use. These could be based on the intrinsic gluten properties, or on foreign proteins such as the $(VPGVG)_n$ polypeptide based on elastin. Wheat starch and gluten are currently separated on an industrial scale, and a similar plant should be suitable for the preparation of novel protein polymers. Such polymers would also have high consumer acceptability, being biocompatible and biodegradable, and have environmental benefits since they are based on a renewable resource rather than on fossil fuels. In addition, they should help to maintain the prosperity of the agricultural sector by providing new end uses for surplus wheat production.

Other Grains.   Concepts similar to those discussed above for wheat flours and doughs for processing into breads and pastries become relevant to grains such as corn, oats, rye, and millet. While varieties of wheats and other grains have been selected for properties favorable for processing into foods, the proteins of grains are principally storage proteins held in readiness for germination; they were not selected in evolution for properties of elasticity, extensibility, and the like that would be optimal for production of breads, noodles, pastries, and so on. Accordingly, the potential of introducing a truly functionally elastic protein-based polymer into grains becomes an interesting prospect.

Food Additives for Improved Textures and Consistency Arising from Viscoelastic Adhesives, Binders, and Bases.   With sufficiently low-cost bioproduction of elastin protein-based polymers, more general use as a food additive or adjunct can be considered to complement desired properties of texture and consistency or to supplant and replace undesired properties of fats, calories, and more. Elastic protein-based polymers are, of course, edible proteins; they can be expressed from aerosol cans as foams with the potential for being a substitute for whipped creams. As gels, foams, and viscoelastic bases, they can be doped with flavors, just as they can be doped with drugs, such that they could be used, for example, in pastry layers and coatings.

As elastic protein-based polymers can be quite adhesive, they could be used as binders for holding together other processed agricultural products. The possibilities of these elastic protein-based polymers, once their properties discussed above have been demonstrated, require little imagination to find myriad applications. With this comes an environmental friendliness in production and disposal and a nutritional aspect of being protein instead of fat.

ACKNOWLEDGMENTS

This work was supported in part by Contract Nos. N00014-89-C-0282 and N00014-90-C-0265 from the Department of the Navy, Office of Naval Research, Naval Medical Research and Development Command; 1R43-HL049705-01 from the National Heart, Lung, and Blood Institute, National Institutes of Health; N00014-89-J-1970 from the Department of the Navy, Office of Naval Research; DAAK60-93-C-0094 from the U.S. Army Natick RD&E Center; 1R43EY09665-01A1 from the National Institutes of Health, National Eye Institute; and P30-AI-277767 from the National Institutes of Health.

REFERENCES

1. Urry, D. W., Bioelastics: "Bioelastics" break new ground in the development of biomaterials, *Res. & Dev.*, 30:57–64 (1989).
2. Urry, D. W., Parker, T. M., Reid, M. C., and Gowda, D. C., Biocompatibility of the bioelastic materials, poly(GVGVP) and its γ-irradiation cross-linked matrix: Summary of generic biological test results, *J. Bioactive Compatible Poly.*, 6:263–282 (1991).
3. Urry, D. W., Gowda, D. C., Harris, C., Harris, R. D., and Cox, B. A., Development of bioelastic materials as biocompatible, transducible and degradable drug delivery matrices, *Poly. Preprints, Div. Polym. Chem., Am. Chem. Soc.*, 33(2):84–85 (1992).
4. Urry, D. W., Molecular Machines: How motion and other functions of living organisms can result from reversible chemical changes, *Angew. Chem.* (German), 105:859–883 (1993); *Angew. Chem. Int. Ed. Engl.*, 32:819–841 (1993).
5. Urry, D. W., Nicol, A., Gowda, D. C., Hoban, L. D., McKee, A., Williams, T., Olsen, D. B., and Cox, B. A., Medical applications of bioelastic materials, In *Biotechnological Polymers: Medical, Pharmaceutical and Industrial Applications* (Charles G. Gebelein, Ed.), Technomic Publishing Atlanta, Georgia, pp. 82–103 (1993).
6. Urry, D. W., Luan, C-H., Harris, R. D., and Prasad, K. U., Aqueous interfacial driving forces in the folding and assembly of protein (elastin)-based polymers: Differential scanning calorimetry studies, *Polym. Preprints, Div. Polym. Chem., Am. Chem. Soc.*, 31:188–189 (1990).
7. Urry, D. W., Peng, S., and Parker, T. M., Hydrophobicity-induced pK shifts in elastin protein-based polymers, *Biopolymers*, 32:373–379 (1992).
8. Urry, D. W., Parker, T. M., Nicol, A., Pattanaik, A., Minehan, D. S., Gowda, D. C., Morrow, C., and McPherson, D. T., The capacity to vary the bioactive role of elastic protein-based polymers, *(Am. Chem. Soc.), Div. Poly. Mat.: Sci. and Eng.*, 66:399–402 (1992).
9. Urry, D. W., Luan, C-H., Peng, S., Parker, T. M., and Gowda, D. C., Hierarchical and modulable hydrophobic folding and self-assembly in elastic protein-based polymers: Implications for signal transduction, *Mat. Res. Soc. Symp. Proc.*, 255:411–422 (1992).
10. Sandberg, L., Leslie, J., Leach, C., Torres, V., Smith, A., and Smith, D., Elastin covalent structure as determined by solid phase amino acid sequence, *Pathol. Biol.*, 33:266–274 (1985).
11. Yeh, H., Ornstein-Goldstein, N., Indik, Z., Sheppard, P., Anderson, N., Rosenbloom, J., Cicila, G., Yoon, K., and Rosenbloom, J., Sequence variation of bovine elastine mRNA due to alternative splicing, *Collagen and Related Research*, 7:235–247 (1987).
12. Indik, Z., Yeh, H., Ornstein-Goldstein, N., Sheppard, P., Anderson, N., Rosenbloom, J., Peltonen, L., and Rosenbloom, J. C., Alternative splicing of human elastin mRNA indicated by sequence analysis of cloned genomic and complementary DNA, *J. Proc. Natl. Acad. Sci., USA*, 84:5680–5684 (1987).
13. Smith, D. W., Sandberg, L. B., Leslie, B. H., Wolt, T. E., Minton, S. T., Myers, B., and Rucker, R. B., Primary structure of a chick tropoelastin peptide: Evidence for a collagen-like amino acid sequence, *Biochem. Biophys. Res. Commun.*, 103:880–885 (1981).
14. Cicila, G., Yoon, K., Ornstein-Goldstein, N., Indik, Z. K., Boyd, C., May, M., Cannizzaro, L. A., Emanual, B. S., and Rosenbloom, J., Elastin gene structure and function, In *Extracellular Matrix: Structure and Function*, Alan R. Liss, New York, pp. 333–350 (1985).
15. Sandberg, L. B., Soskel, N. T., and Leslie, J. B., Elastin structure, biosynthesis, and relation, *N. Engl. J. Med.*, 304:566–579 (1981).
16. Urry, D. W., Free energy transduction in polypeptides and proteins based on inverse temperature transitions, *Prog. Biophys. Molec. Biol.*, 57:23–57 (1992).
17. Urry, D. W., Gowda, D. C., Parker, T. M., Luan, C-H., Reid, M. C., Harris, C. M., Pattanaik, A., and Harris, R. D., Hydrophobicity scale for proteins based on inverse temperature transitions, *Biopolymers*, 32:1243–1250 (1992).
18. Chou, Y. P., and Fasman, G. D., β-turns in proteins, *J. Mol. Biol.*, 115:135–175 (1977).

19. Urry, D. W., Studies on the conformation and interactions of elastin, In *Arterial Mesenchyme and Arteriosclerosis* (W. D. Wagner and T. B. Clarkson, Eds.), Plenum, New York, *Adv. Exp. Med. Bio.*, 43:211–243 (1974).

20. Urry, D. W., A molecular theory of ion conducting channels: A field-dependent transition between conducting and nonconducting conformations, *Proc. Natl. Acad. Sci., USA*, 69: 1610–1614 (1972).

21. Venkatachalam, C. M., and Urry, D. W., Development of a linear helical conformation from its cyclic correlated, β-spiral model of the elastin poly(pentapeptide), (VPGVG)$_n$, *Macromolecules*, 14:1225–1229 (1981).

22. Urry, D. W., and Venkatachalam, C. M., A librational entropy mechanism for elastomers with repeating peptide sequences in helical array, *Int. J. Quant. Chem.: Quant. Biol. Symp.*, 10:81–93 (1983).

23. Chang, D. K., and Urry, D. W., Polypentapeptide of elastin: Damping of internal chain dynamics on extension, *J. Computational Chem.*, 10:850–855 (1989).

24. Wasserman, Z. R., and Salemme, F. R., A molecular dynamics investigation of the elastomeric restoring force in elastin, *Biopolymers*, 29:1613–1631 (1990).

25. Urry, D. W., Jaggard, J., Prasad, K. U., Parker, T., and Harris, R. D., Poly(Val^1-Pro^2-Ala^3-Val^4-Gly^5): A reversible, inverse thermoplastic, In *Biotechnology and Polymers* (C. G. Gebelein, Ed.), Plenum Press, New York, pp. 265–274 (1991).

26. Urry, D. W., Long, M. M., Harris, R. D., and Prasad, K. U., Temperature correlated force and structure development in elastomeric polypeptides: the Ile^1 analog of the polypentapeptide of elastin, *Biopolymers*, 25:1939–1953 (1986).

27. Urry, D. W., Harris, R. D., Long, M. M., and Prasad, K. U., Polytetrapeptide of elastin: Temperature correlated elastomeric force and structure development, *Int. J. Pept. Protein Res.*, 28:649–660 (1986).

28. Urry, D. W., Postulates for protein folding and function, *Intl. J. Quant. Chem.: Quant. Biol. Symp.*, in press (1994).

29. Urry, D. W., and Prasad, K. U., Syntheses, characterizations and medical uses of the polypentapeptide of elastin and its analogs, In *Biocompatibility of Tissue Analogues* (D. F. Williams, Ed.), CRC Press, Boca Raton, Florida, pp. 89–116 (1985).

30. Prasad, K. U., Iqbal, M. A., and Urry, D. W., Utilization of 1-hydroxybenzotriazole in mixed anhydride coupling reactions, *Int. J. Pept. and Protein Res.*, 25:408–413 (1985).

31. Merrifield, R. B., Solid phase peptide synthesis. I. The synthesis of a tetrapeptide, *J. Am. Chem. Soc.*, 85:2149–2154 (1963).

32. Prasad, K. U., Iqbal, M., and Urry, D. W., Synthesis of two component models of elastin, *Proc. Tenth Am. Peptide Symp.*, 1987; *Peptides-Chem. & Biol.* (G. R. Marshall, Ed.), Escom Leiden, pp. 399–403 (1988).

33. Gowda, D. C., Parker, T. M., Harris, C. M., Harris, R. D., and Urry, D. W., Design and synthesis of poly-tricosapeptides to enhance hydrophobic-induced pKa shifts, In *Peptides: Chemistry, Structure and Biology* (R. S. Hodges and J. A. Smith, Eds.), Proceedings of the Thirteenth American Peptide Symposium, Edmonton, Alberta, Canada, in press (1994).

34. Urry, D. W., Gowda, D. C., Peng, S., Parker, T. M., and Harris, R. D., Design at nanometric dimensions to enhance hydrophobicity-induced pKa shifts, *J. Am. Chem. Soc.*, 114:8716–8717 (1992).

35. Urry, D. W., Gowda, D. C., Peng, S., Parker, T. M., Jing, N., and Harris, R. D., Nanometric design of extraordinary hydrophobicity-induced pKa shifts for aspartic acid: Relevance to protein mechanisms, *Biopolymers*, in press (1994).

36. Deguchi, Y., Krejchi, M. T., Borbely, J., Fournier, M. J., Mason, T. L., and Tirrell, D. A., Synthesis and characterization of periodic polypeptides containing repeating-(ALAGLY)XGLUGLY-sequences, In *Mat. Res. Soc. Symp. Proc.*, Periodic Peptide Polymer Prepn Characterization, Conformation Periodic Peptide Polymer. *Escherichia Coli* Periodic Peptide Polymer, 292:205–210 (1993).

37. Capello, J., Protein engineering for biomaterials applications, In *Review Protein Engineering Biomaterial*, Curr. Opin. Struct. Biol., 2:582–586 (1992).

38. McPherson, D. T., Morrow, C., Minehan, D. S., Wu, J., Hunter, E., and Urry, D. W., Production and purification of a recombinant elastomeric polypeptide, G-(VPGVG)$_{19}$-VPGV, from *Escherichia coli, Biotechnol. Prog.*, 8:347–352 (1992).

39. Perbal, Bernard, In *A Practical Guide to Molecular Cloning*, 2nd Edition, Wiley-Interscience, John Wiley & Sons, New York (1988).

40. Ausubel, F. M., In *Current Protocols in Molecular Biology*, Vols. 1 & 2, Greene Publishing Associates and Wiley-Interscience, John Wiley & Sons, New York (1989).

41. Hoban, L. D., Pierce, M., Quance, J., Hayward, I., McKee, A., Gowda, D. C., Urry, D. W., and Williams, T., The use of polypentapeptides of elastin in the prevention of postoperative adhesions, *J. Surgical Res.*, in press (1994).

42. Urry, D. W., Gowda, D. C., Cox, B. A., Hoban, L. D., McKee, A., and Williams, T., Properties and prevention of adhesions applications of bioelastic materials, *Mat. Res. Soc. Symp. Proc.*, 292:253–264 (1993).

43. Elsas, F. J., Gowda, D. C., and Urry, D. W., Synthetic polypeptide sleeve for strabismus surgery. *J. Pediatr. Ophthalmol. Strabismus*, 29:284–286 (1992).

44. Urry, D. W., Gowda, D. C., Harris, C. M., and Harris, R. D., Bioelastic materials and the Δ-T_t-mechanism in drug delivery, In *Polymeric Drugs and Drug Delivery Systems* (R. M. Ottenbrite, Ed.), *Am. Chem. Society Books*, pp. 15–28 (1994).

45. Urry, D. W., Preprogrammed drug delivery systems using chemical triggers for drug release by mechanochemical coupling, *Am. Chem. Soc., Div. Polym. Mat. Sci. and Eng.*, 63:329–336 (1990).

46. Ruoslahti, E., and Pierschbacher, M. D., New perspectives in cell adhesion: RGD, *Science*, 238:491–497 (1987).

47. Ruoslahti, E., Integrins, *J. Clin. Invest.*, 87:1–5 (1991).

48. Pierschbacher, M. D., and Ruoslahti, E., Cell attachment activity of fibronectin can be duplicated by small synthetic fragments of a molecule, *Nature*, 309:30–33 (1984).

49. Pierschbacher, M. D., and Ruoslahti, E., Influence of stereochemistry of the sequence Arg-Gly-Asp-Xaa on binding specificity in cell adhesion, *J. Biol. Chem.*, 262:17294–17298 (1987).

50. Nicol, A., Gowda, D. C., and Urry, D. W., Cell adhesion and growth on synthetic elastomeric matrices containing Arg-Gly-Asp-Ser-[3], *J. Biomed. Mat. Res.*, 26:393–413 (1992).

51. Nicol, A., Gowda, D. C., Prasad, K. U., and Urry, D. W., Promotion of cell adhesion to elastomeric polypeptide matrices by the covalent incorporation of cell attachment sequences, In *Cardiovasc. Sci. & Technol.: Basic & Appl., II*, (J. C. Norman, Ed.), Oxymoron Press, Louisville, Kentucky, pp. 254–257 (1990).

52. Nicol, A., Gowda, D. C., and Urry, D. W., Elastic protein-based polymers as cell attachment matrices, *J. Vasc. Surg.*, 13:746–748 (1991).

53. Leung, D. Y. M., Glagov, S., and Mathews, M. B., A new *in vitro* system for studying cell response to mechanical stimulation, *Exp. Cell. Res.*, 109:285–298 (1977).

54. Leung, D. Y. M., Glagov, S., and Mathews, M. B., Cyclic stretching stimulates synthesis of matrix components by arterial smooth muscle cells *in vitro*, *Science*, 191:475–477 (1976).

55. Wolinsky, H., and Glagov, S., Structural basis for the static mechanical properties of the aortic media, *Circ. Res.*, 14:400–412 (1964).

56. Ingber, D., Integrins as mechanochemical transducers, *Curr. Opin. Cell Biol.*, 3:841–848 (1991).

57. Urry, D. W., Bioelastic materials as matrices for tissue reconstruction, In *Tissue Engineering: Current Perspectives* (Eugene Bell, Ed.), Birkhäuser, Div. Springer-Verlag, New York, pp. 199–206 (1994).

58. Strzegowski, L. A., Martinez, M. B., Gowda, D. C., Urry, D. W., and Tirrell, D. A., Photomodulation of the inverse temperature transition of a modified elastin poly(pentapeptide), *J. Am. Chem. Soc.*, 116:813–814 (1994).

59. Urry, D. W., Biophysics of energy converting model proteins, *Mat. Res. Soc. Symp. Proc.*, pp. 321–332 (1994).

60. Stackelberg, M. V., and Müller, H. R., Zur strwhter der gashydrate, *Naturwissenschaften*, 38:456 (1951).

61. Teeter, M. M., Water structure of a hydrophobic protein at atomic resolution: Pentagon rings of water molecules in crystals of crambin, *Proc. Natl. Acad. Sci., USA*, 81:6014–6018 (1984).

62. Edsall, J. T., and McKenzie, H. A., Water and proteins. II. The location and dynamics of water in protein systems and its relation to their stability and properties, *Adv. Biophys.*, 16: 53–183 (1983).

63. Frank, H. S., and Evans, M. W., Free volume and entrophy in condensed systems. III. Entropy in binary liquid mixtures, partial molal entropy in dilute solutions; structure and thermodynamics in aqueous electrocytes, *J. Chem. Phys.*, 13:507–532 (1945).

64. Kauzmann, W., Some factors in the interpretation of protein denaturation, *Adv. Protein Chem.*, 14:1–63 (1959).

65. Urry, D. W., Thermally driven self-assembly, molecular structuring and entropic mechanisms in elastomeric polypeptides, In *Mol. Conformation and Biol. Interactions* (P. Balaram and S. Ramaseshan, Eds.), Indian Acad. Sci., Bangalore, India, pp. 555–583 (1991).

66. Luan, C-H., Harris, R. D., Prasad, K. U., and Urry, D. W., DSC studies of the inverse temperature transition of the polypentapeptide of elastin and its analogues, *Biopolymers*, 29: 1699–1706 (1990).

67. Urry, D. W., Trapane, T. L., and Prasad, K. U., Phase-structure transitions of the elastin polypentapeptide-water system within the framework of composition-temperature studies, *Biopolymers*, 24:2345–2356 (1985).

68. Cook, W. J., Einspahr, H. M., Trapane, T. L., Urry, D. W., and Bugg, C. E., The crystal structure and conformation of the cyclic trimer of a repeat pentapeptide of elastin, cyclo-(L-Valyl-L-Prolyl-Glycyl-L-Valyl-Glycyl)$_3$, *J. Am. Chem. Soc.*, 102:5502–5505 (1980).

69. Urry, D. W., What is elastin; what is not, *Ultrastruc. Pathol.*, 4:227–251 (1983).

70. Urry, D. W., Venkatachalam, C. M., Long, M. M., and Prasad, K. U., Dynamic β-spirals and a librational entropy mechanism of elasticity, In *Conformation in Biol.* (R. Srinivasan and R. H. Sarma, Eds.), G. N. Ramachandran Festschrift Volume, Adenine Press, New York, pp. 11–27 (1982).

71. Luan, C.-H., Jaggard, J., Harris, R. D., and Urry, D. W., On the source of entropic elastomeric force in polypeptides and proteins: Backbone configurational vs. side chain solvational entropy, *Intl. J. Quant. Chem.: Quant. Biol. Symp.*, 16:235–244 (1989).

72. Flory, P. J., Ciferi, A., and Hoeve, C. A. J., The thermodynamic analysis of thermoelastic measurements on high elastic materials, *J. Polym. Sci.*, 45:235–236 (1960).

73. Momany, F. A., Carruthers, L. M., McGurie, R. F., and Scheraga, H. A., Intermolecular potentials from crystal data. III. Determination of empirical potentials and application to the packing configurations and lattice energies in crystals of hydrocarbonds, carboxylic acids, amines and amides, *J. Phys. Chem.*, 78:1595–1620 (1974).

74. Momany, F. A., McGuire, R. F., Burgess, A. W., and Scheraga, H. A., Energy parameters in polypeptides. VII. Geometric parameters, partial atomic charges, nonbonded interactions, hydrogen bond interactions, and intrinsic torsional potentials for the naturally occurring amino acids, *J. Phys. Chem.*, 79:2361–2381 (1975).

75. Karplus, M., and McCammon, J. A., Dynamics of proteins: Elements and function, *Ann. Rev. Biochem.*, 53:263–300 (1983).

76. Verlet, L., Computer experiments on classical fluids. I. Thermodynamical properties of Lennard-Jones molecules, *Phys. Rev.*, 159:98–103 (1967).

77. Karplus, M., and Kushick, J. N., Method for estimating the configuration entropy of macromolecules, *Macromolecules*, 14:325–332 (1981).

78. Weiner, P. K., and Kollman, P. A., *J. Comput. Chem.*, 2:287–303 (1981).

79. Urry, D. W., Biophysics of energy converting model proteins, *Mat. Res. Soc. Symp. Proc.*, in press (1994).
80. Urry, D. W., Haynes, B., Zhan, H., Harris, R. D., and Prasad, K. U., Mechanochemical coupling in synthetic polypeptides by modulation of an inverse temperature transition, *Proc. Natl. Acad. Sci., USA*, 85:3407–3411 (1988).
81. Urry, D. W., Hayes, L. C., Gowda, D. C., Harris, C. M., and Harris, R. D., Reduction-driven polypeptide folding by the ΔT_t-mechanism, *Biochem. Biophys. Res. Commun.*, 188: 611–617 (1992).
82. Urry, D. W., Hayes, L. C., Parker, T. M., and Harris, R. D., Baromechanical transduction in a model protein by the ΔT_t-mechanism, *Chem. Phys. Ltrs.*, 201:336–340 (1993).
83. Heimbach, C. J., Gowda, D. C., and Urry, D. W., Photochemical transduction using poised elastic polytricosapeptides. In preparation.
84. Luan, C-H., Parker, T. M., Prasad, K. U., and Urry, D. W., DSC studies of NaCl effect on the inverse temperature transition of some elastin-based polytetra-polypenta-, and polynona-peptides, *Biopolymers*, 31:465–475 (1991).
85. Nicol, A., Gowda, D. C., Parker, T. M., and Urry, D. W., Elastomeric polytetrapeptide matrices: Hydrophobicity dependence of cell attachment from adhesive, $(GGIP)_n$, to non-adhesive, $(GGAP)_n$, even in serum, *J. Biomed. Mat. Res.*, 27:801–810 (1993).
86. Sambrook, J., Fritsch, E. F., and Maniatis, T., In *Molecular Cloning: A Laboratory Manual*, Volumes 1–3 (N. Ford, C. Nolan, and M. Ferguson, Eds.), Cold Spring Harbor Laboratory Press, Cold Spring Harbor, New York (1989).
87. Szent-Györgyi, A., Studies on muscle, *Acta Physiol. Scand.*, 9:1–116 (1945).
88. Pitt, C. G., and Schindler, A., In *Progress in Contraceptive Delivery Systems* (E. Hafez and W. Van Os, Eds.), MTP, Kluwer, Lancaster, Great Britain, pp. 17–46 (1980).
89. Nicol, A., Gowda, D. C., Parker, T. M., and Urry, D. W., Cell adhesive properties of bioelastic materials containing cell attachment sequences, In *Biotechnological Bioactive Polymers* (C. G. Gebelein and Ce. E. Carraher, Jr., Eds.), Plenum Press, New York, pp. 95–113 (1994).
90. Pattanaik, A., Gowda, D. C., and Urry, D. W., Phosphorylation and dephosphorylation modulation of an inverse temperature transition, *Biochem. Biophys. Res. Comm.*, 178:539–545 (1991).
91. Shewry, P. R., Halford, N. G., and Tatham, A. S., The high molecular weight subunits of wheat glutenin, (Critical Review Article), *J. Cereal Sci.*, 15:105–120 (1992).
92. Miles, M. J., Carr, H. J., McMaster, T., Belton, P. S., Morris, V. J., Field, J. M., Shewry, P. R., and Tatham, A. S., Scanning tunnelling microscopy of a wheat gluten protein reveals details of a spiral supersecondary structure, *Proc. Natl. Acad. Sci. (USA)*, 88:68–71 (1991).
93. Tatham, A. S., Miflin, B. J., and Shewry, P. R., The β-turn conformation in wheat gluten proteins: Relationship to gluten elasticity, *Cer. Chem.*, 62:405–412 (1985).
94. Tatham, A. S., Shewry, P. R., and Miflin, B. J., Wheat gluten elasticity: A similar molecular basis to elastin? *FEBS Lett.*, 177:205–208 (1984).
95. Belton, P. S., Colquhoun, I. J., Field, J. M., Grant, A., Shewry, P. R., and Tatham, A. S., ^1H and ^2H NMR relaxation studies indicate that the elasticity of the HMW subunits of glutenin is not elastin-like, *J. Cereal Sci.*, in press (1994).
96. Vasil, V., Castillo, A. M., Fromm, M. E., and Vasil, I. K., Herbicide resistant fertile transgenic wheat plants obtained by microprojectile bombardment of regenerable embryogenic callus, *Bio/Technology*, 10:667–675 (1992).
97. Weeks, J. T., Anderson, O. D., and Blechl, A. E., Rapid production of multiple independent lines of fertile transgenic wheat, *Plant Physiology*, 102:1077–1084 (1993).

51
Silicone Biomaterials

John S. Tiffany
INAMED Development Company, Carpinteria, California
Del J. Petraitis
NuSil Technologies, Carpinteria, California

I. INTRODUCTION

Silicone polymers are a relatively recent invention, with commercial production beginning in the 1940s. Shortly thereafter, it was found that glass surfaces treated with silicone fluid delayed the clotting of blood. By the mid-1950s, medical applications of silicones had greatly increased and many studies of the biological properties of these materials were undertaken. Silicones are one of the most widely used and one of the most studied of all artificial materials for medical applications and no discussion of biomaterials is complete without their inclusion.

In this chapter, we give an overview of silicone chemistry, including synthesis, types, and curing mechanisms, followed by a review of some of the more important biomedical applications of these materials. The topic of silicones as biomaterials would easily fill a whole book; we have of necessity left out many significant materials and applications. The reader is referred to the vast medical literature on this topic for information not provided here.

II. GENERAL SILICONE OVERVIEW

A. Synthesis

The starting raw material from which organosilicones are ultimately synthesized is silicon dioxide, one of the most abundant compounds in the crust of the earth. Although silicone dioxide is present in many natural forms, ground quartz is the most commonly used raw material. The initial chemical synthesis step is the reduction of silicon dioxide into virtually pure elemental silicon by using a carbon arc furnace. Scheme 1 illustrates this reaction.

$$SiO_2 + C \longrightarrow Si + CO_2$$

Scheme 1   Reduction of SiO_2.

The next reaction step to produce silicone is the conversion of the elemental silicon into a chlorosilane. This process involves the physical grinding of the silicon and its reaction with methylene chloride in a fluidized bed reactor. This reaction is known as the *direct process* (Scheme 2) and is accomplished at high temperatures in the presence of a copper catalyst.

$$Si + 2CH_3Cl \xrightarrow[\triangle]{Cu} (CH_3)_2Si\,Cl_2 + Byproducts$$

Scheme 2   The direct process.

The next step involves the hydrolysis of the chlorosilane to produce the most basic silicones. Scheme 3 illustrates the most significant hydrolysis product: octamethylcyclotetrasiloxane, more commonly known as D4 or dimethyl cyclic tetramer.

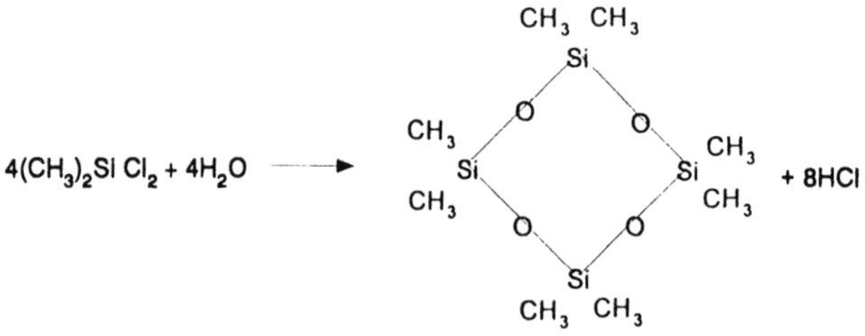

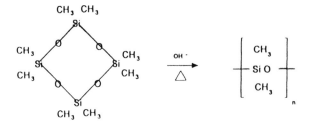

Scheme 3   Hydrolysis production of D4.

At this stage, we have produced the first $-Si-O-Si$ bond linkage and have a true silicone. Although the silicone cyclic species are useful in many applications, including cosmetics, deodorants, skin emollients, and shampoos; additional conversion into polymers is most commonly associated with silicone biomaterials. Scheme 4 is a basic polymerization mechanism.

Scheme 4   Silicone polymerization.

When the silicone polymer is produced with trimethylsiloxy end blocks, it is a nonfunctional fluid and is directly used in some biomaterial applications. This nonfunctional polydimethylsiloxane (PDMS) is used as a lubricant on syringe needles and syringe barrels and as both internal and external catheter lubricants.

The basic characteristics of silicone polymers can be modified by the addition of other substituents onto the silicone backbone. For instance, some of the methyl groups may be replaced with phenyl groups to increase the refractive index of the polymer. Similarly, either the end groups or the backbone methyl groups could be replaced with vinyl or OH functionality to provide moieties to cross-link the polymer into matrices. Such cross-linked polymers can be used directly as gels or they may be modified with fillers and other reinforcing agents prior to cross-linking to produce elastomers with significantly varying physical properties.

Silicone polymers range in viscosities from water consistency, 6.5×10^{-4} Pascal·seconds (Pa·s) (0.65 centipoise) to semisolid, gumlike consistencies with viscosities as high as 5×10^4 Pa·s (50,000,000 centipoise). Lubricious coatings often include blends of low viscosity and medium viscosity silicone polymers to provide multiple-use lubricating coatings. It should be noted, however, that the applications that involve the use of silicones as lubricating agents are only applicable for low shear conditions. Despite the feel of the lower viscosity silicones, they are unacceptable in high shear applications, and will indeed result in seizing if used as metal-to-metal high shear lubricants.

Silicone greases and compounds are generally made by the addition of high surface area, solid powders to medium viscosity, nonfunctional silicone fluid polymers. For instance, the addition of fumed silica with surface areas of 100 square meters per gram (m^2/g) to 200 m^2/g to a 100-Pa·s (100,000 centipoise) silicone polymer results in a smooth, greaselike compound. When fluids are unacceptable because of their tendency to flow off surfaces, these greases are useful as low shear lubricants.

Unlike many organic-based elastomers, cross-linked silicone polymers have limited toughness, with tensile strengths of approximately 0.7 megapascal (MPa) (100 psi, pounds per square inch) and tear strengths of less than 5.25 kiloNewtons per meter (kN/m) (30 ppi, pounds per inch). In order to produce elastomers with useful physical properties, silicone polymers must be reinforced with fillers. To produce elastomers with tensile strengths greater than 10.3 MPa (1500 psi) and tear strengths greater than 17.5 kN/m (100 ppi), fillers such as fumed amorphous silica with surface areas of 150 m^2/g to 400 m^2/g need to be added to the formulations. However, simple addition of these silica materials will not produce a usable base. If the high surface area fumed silica surfaces are not treated to make them more compatible with the silicone polymer, insufficient amounts of filler can be incorporated into a uniform base, and the base will age rapidly to a state of pseudocure known as *creping*, making it unsuitable for additional fabrication. The technology used to treat the filler surface generally involves either a pretreatment on the dry filler or an *in situ* treatment done in the presence of the polymer during the addition of the filler.

Among the filler treatments currently utilized is a treatment to replace the silanol groups on the surface of the silica with trimethyl silyl groups (see Scheme 5).

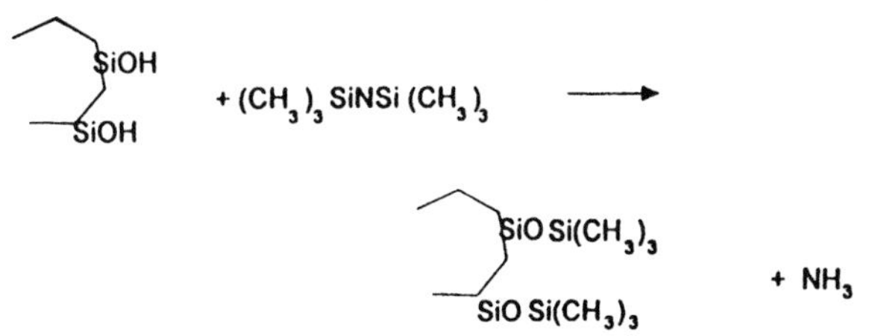

Scheme 5   Filler treatment with hexamethyldisilazane.

Another filler treatment involves the reaction of the surface silanols with short-chain, silanol-terminated polysiloxanes (Scheme 6).

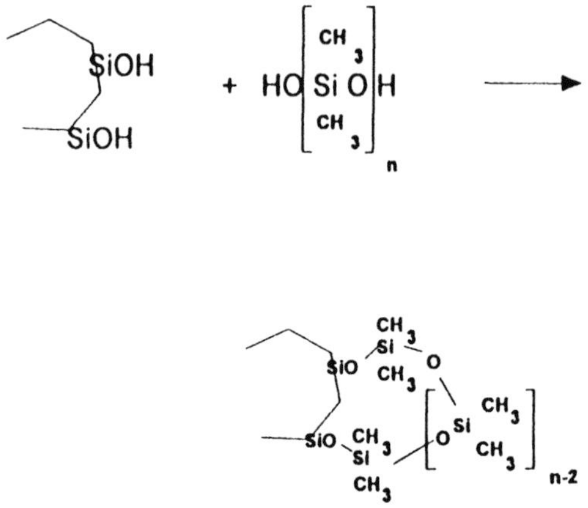

Scheme 6   Filler treatment with silanol condensation.

B.   Silicone Elastomer Cross-Linking Mechanisms

After a base has been properly produced to incorporate adequate filler levels, this base must be cross-linked to form an elastomer. There are currently more than 10 unique silicone cure mechanisms, but only the 5 most significant used in biomaterial applications are discussed. In order to cross-link high consistency elastomers, a peroxide cross-linking agent is utilized. Upon heating, the peroxide forms free radicals that chain transfer to the silicone, initiating a cross-link. Scheme 7 describes an example of this mechanism.

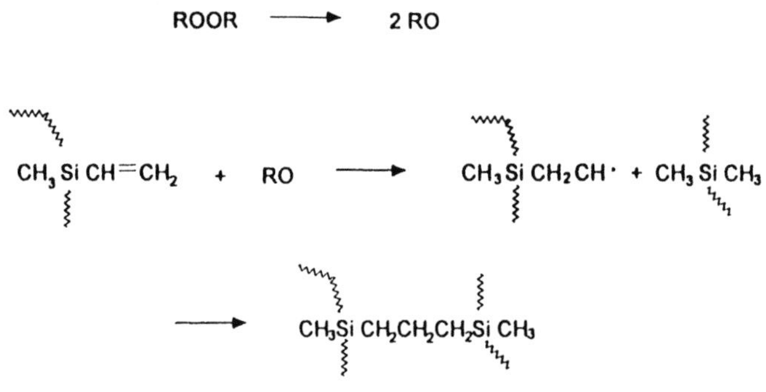

Scheme 7   Peroxide cross-link mechanism.

The choice of peroxide is a function of the fabrication method. For instance, for extruded parts that are cured via hot air vulcanization (HAV), 2,4 dichlorobenzoyl peroxide is used because it has the lowest initiation temperatures. Most peroxides used to cure silicones are not suitable for open-air curing. The oxygen in the air inhibits the cure on the surface, thus requiring closed-mold curing or special ovens designed to eliminate the oxygen from the air during the cure. Peroxide-cured silicones may require a postcure to remove the by-product of the peroxide decomposition. For example, silicone elastomers cured with benzoyl peroxide should be postcured to remove the benzoic acid formed from the benzoyl peroxide. Of course, the actual postcure schedule may depend on the size and thickness of the part or extrusion. Table 1 lists the characteristics of various peroxide choices.

Another method that may be used to remove peroxide residuals from peroxide-cured silicone elastomers is solvent extraction. For instance, a soxhlet extraction using an alcohol/hydrocarbon azeotrope would remove benzoic acid as well as any un-cross-linked silicone species. One of the most significant characteristics of the peroxide-cured elastomers is shrinkage associated with thermal contraction due to the elevated cure temperature and the removal of the peroxide by-product via postcure and/or extraction. Mold and die designs must include factors for shrink compensation to permit the final precise dimensions of the parts.

Table 1 Peroxides

Peroxide	Cure time/ temperature	1/16-inch slab postcure time/temperature	Vinyl specific	Peroxide concentration PPH
50% 2,4 dichloro-benzoyl peroxide in 1000 CS fluid (paste)	5 min/116°C (240°F)	4 hours/150°C (300°F)	No	1.0–2.0, 1.3
50% parachloro-benzoyl peroxide in 1000 CS fluid (paste)	5 min/121°C (250°F)	4 hours/150°C (300°F)	No	1.0–2.0, 1.3
50% benzoyl peroxide in 1000 CS fluid (paste)	5 min/127°C (260°F)	4 hours/150°C (300°F)	No	1.0–2.0, 1.3
100% tertiary butyl pebenzoate (liquid)	10 min/150°C (300°F)	Optional	Partial	0.5–1.5, 0.8
40% dicumyl peroxide (powder)	10 min/150°C (300°F)	Remove odor, 2 hours/150°C (300°F)	Yes	0.8–2.0, 1.0
95% dicumyl peroxide (crystalline solid)	10 min/150°C (300°F)	Remove odor, 2 hours/150°C (300°F)	Yes	0.3–1.0, 0.5
T-Butylperoxide isopropyl carbonate (liquid)	10 min/150°C (300°F)	Optional	Partial	0.3–1.0, 0.5
100% ditertiary butyl peroxide (liquid)	10 min/177°C (340°F)	Optional	Yes	0.3–1.0, 0.5
100% 2,5 bis(tert-butyl peroxy)-2,5 dimethyl hexane (liquid)	10 min/177°C (340°F)	Optional	Yes	0.3–1.0, 0.5
50% 2,5-bis(tert-butyl peroxy)-2,5 dimethyl hexane (powder)	10 min/177°C (340°F)	Optional	Yes	0.5–1.5, 1.0
100% di-tertiary butyl-peroxy-3,3,5-cyclohexane in 40% calcium carbonate	10 min/177°C (340°F)	Optional	Yes	0.5–1.5, 1.0

Another significant cure mechanism used to produce silicone elastomers is the addition-cure mechanism. Scheme 8 describes the chemistry.

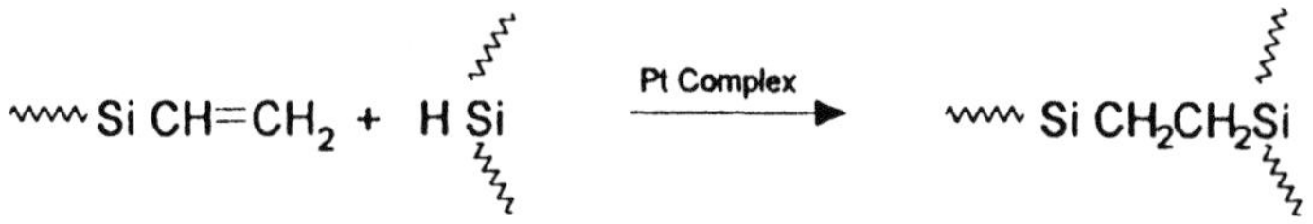

$$\text{\wedge\wedge Si CH=CH}_2 + \text{H Si} \xrightarrow{\text{Pt Complex}} \text{\wedge\wedge Si CH}_2\text{CH}_2\text{Si}$$

Scheme 8 Addition-cured silicone.

Since there is no chemical leaving group, the addition-cured elastomers exhibit no chemical shrinkage and only the thermal contraction factors need to be considered in mold and die design. The cure can be modified to produce variable work times and, in low consistency formulations, can be made to cure at room temperatures. Also, because the properties of the final elastomer are so dependent on the stoichiometry, the modulus can be modified by varying the cross-linker-to-base ratio.

Despite all the desirable properties of the addition cure, there are two characteristics of which the fabricator needs to be cognizant. First, the addition-cure system is susceptible to inhibition. The platinum complex catalyst is used at levels of approximately 10 parts per million (ppm) and as a result can be easily inhibited by relatively small amounts of "poison." Although a variety of materials can act as poisons, the most common inhibitors are sulfur-containing organic elastomers and organometallic compounds. Organotin catalysts used in many silicone room-temperature-cured elastomers are another common source of inhibition.

The second negative characteristic of addition-cured elastomers is their susceptibility to porosity during unconfined heat-accelerated cures due to hydrogen gas. Although this porosity may occur with the low consistency formulations, it is most significant in extrusions that are hot air vulcanized. Care needs to be taken by fabricators to minimize porosity. Minimizing the possible incorporation of moisture during milling and catalysis and extrusion immediately after milling will reduce or minimize the likelihood of porosity forming during the cure.

Another silicone cure mechanism commonly used in biomedical applications is the one-party acetoxy-cured room temperature vulcanizing compound (RTV). This adhesive is supplied as one part and cures into an elastomer upon exposure to moisture (see Scheme 9).

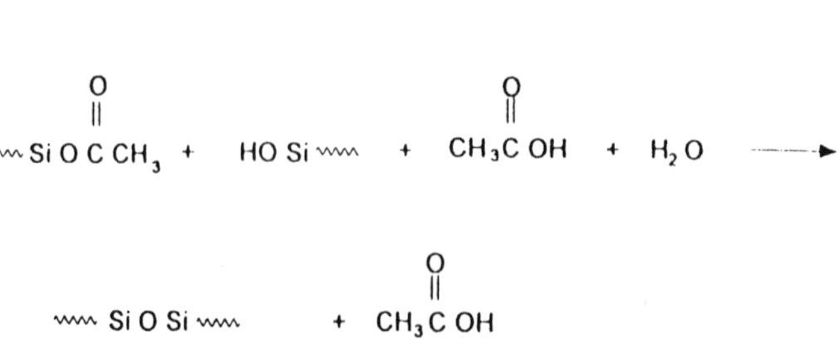

$$\text{\wedge\wedge Si O C CH}_3 + \text{CH}_3\text{C O Si \wedge\wedge} + \text{H}_2\text{O} \xrightarrow{\text{Sn}}$$

$$\text{\wedge\wedge Si O C CH}_3 + \text{HO Si \wedge\wedge} + \text{CH}_3\text{C OH} + \text{H}_2\text{O} \longrightarrow$$

$$\text{\wedge\wedge Si O Si \wedge\wedge} + \text{CH}_3\text{C OH}$$

Scheme 9 Acetoxy cure mechanism.

The acetoxy-cured RTVs require no mixing and are simply dispensed from a tube. Because they require stoichiometric amounts of moisture to cure, only relatively thin sections will cure at a reasonable rate. Acetoxy systems generally provide good adhesion to most substrates without primers and are often used to provide secondary bonding between cured silicone elastomers or between cured elastomers and metal substrates. During cure, these materials do elute acetic acid and they can cause inhibition to addition-cured silicones.

Another room-temperature vulcanizing cure system is the alkoxy two part (see Scheme 10).

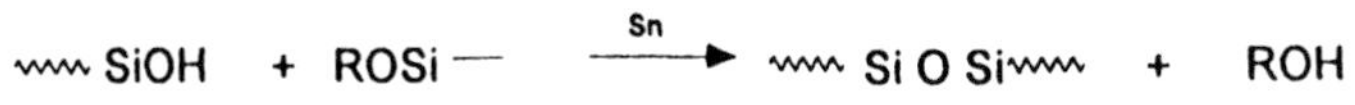

$$\text{\textasciitilde\textasciitilde SiOH} \quad + \quad \text{ROSi} \underline{\quad} \quad \xrightarrow{\text{Sn}} \quad \text{\textasciitilde\textasciitilde Si O Si\textasciitilde\textasciitilde} \quad + \quad \text{ROH}$$

Scheme 10   Alkoxy two-part cure mechanism.

In general, the alkoxy two-part silicone elastomer is formulated with the organotin catalyst as Part B, and the quantity of catalyst added is normally 0.5 weight percent. The most frequently used forms of catalyst are dibutyl tin dilaurate, which gives a 24-hour cure, and stannous octoate, which give a 15–30-minute cure. Generally, the alkoxy two-part elastomers have somewhat lower physical properties and have limited application as silicone biomaterials.

The last silicone cure system that can be used to produce silicone elastomeric biomaterials is the foam system (see Scheme 11).

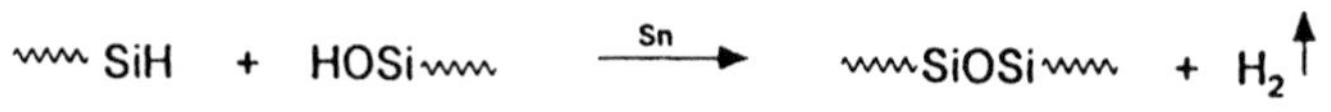

$$\text{\textasciitilde\textasciitilde SiH} \quad + \quad \text{HOSi\textasciitilde\textasciitilde} \quad \xrightarrow{\text{Sn}} \quad \text{\textasciitilde\textasciitilde SiOSi\textasciitilde\textasciitilde} \quad + \quad H_2\uparrow$$

Scheme 11   Foam cure system.

This foam cure system can be catalyzed with either stannous octoate or a platinum complex as the catalyst. Rapid generation of hydrogen as the blowing agent and simultaneous cross-linking make this system result in a cured elastomeric foam in less than 15 minutes. Although most of the silicone foam elastomers have relatively low physical properties, some specialty platinum-catalyzed foam elastomers have been produced with toughness approaching that of flexible urethane foams. It should be noted that virtually all isotactic silicone foams are closed cell and develop a skin on the surface when expanded and cured. It should be also noted that the reaction is very exothermic and, during the cure/blowing process, the foam can be quite hot to the touch.

C. Additives

Other solid fillers are also incorporated into silicone formulations to provide special properties useful in silicone biomaterials. Barium sulfate and tungsten powder are used in some elastomers to provide radiopacity. This radiopacity can then be used as an indicator during x-ray diagnoses to determine position and functionality of various devices, including catheters, drains, and permanent implants. Various solid powder pigments are also incorporated into silicone elastomers for other specific applications.

For instance, various pigments are incorporated into elastomers to act as colorants for visual identification. Other fillers, such as iron oxide, are incorporated to modify the characteristics of the elastomers. Specifically, iron oxide is used in a silicone elastomer formulation for an ultrasound conformal membrane. In this instance, the iron oxide is used to provide appropriate densities for ultrasound applications.

Another solid filler additive that is used to produce elastomer foams is ammonium bicarbonate. The fine NH_4HCO_3 is compounded into high consistency peroxide or addition-cured bases. Then, during HAV cure or heated compression or transfer mold curing, the NH_4HCO_3 decomposes into ammonia, carbon dioxide, and water to form the "blowing agent" gases that create the foam elastomer.

The final filler discussed is perhaps more precisely defined as an additive rather than as a true filler. Such material is a soluble silicone resin that has lower molecular size than the fumed silica and also contains enough monofunction silicone moieties to provide soluble compatibility with the silicone polymers. Diagrammatic representations of these resins are shown in Scheme 12.

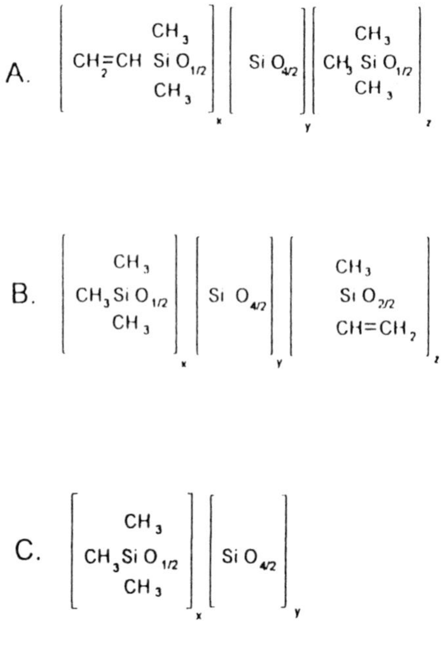

A. $\left[\begin{array}{c} CH_3 \\ CH_2{=}CH\ Si\ O_{1/2} \\ CH_3 \end{array} \right]_x \left[Si\ O_{4/2} \right]_y \left[\begin{array}{c} CH_3 \\ CH_3\ Si\ O_{1/2} \\ CH_3 \end{array} \right]_z$

B. $\left[\begin{array}{c} CH_3 \\ CH_3\ Si\ O_{1/2} \\ CH_3 \end{array} \right]_x \left[Si\ O_{4/2} \right]_y \left[\begin{array}{c} CH_3 \\ Si\ O_{2/2} \\ CH{=}CH_2 \end{array} \right]_z$

C. $\left[\begin{array}{c} CH_3 \\ CH_3 Si\ O_{1/2} \\ CH_3 \end{array} \right]_x \left[Si\ O_{4/2} \right]_y$

Scheme 12   Silicone resin additives.

Examples A and B in Scheme 12 have reactive vinyl functionality in their structure and can be used as reinforcing agents in addition-cured silicones by reacting these vinyl groups during the cross-linking. Although the use of these soluble silicone resins does contribute to tensile strengths, their effects do not significantly increase tear strengths. However, they do produce elastomers that are optically clear and are used to fabricate foldable intraocular lenses and have been evaluated as contact lens materials. A typical resin-reinforced silicone elastomer would have the following properties:

Durometer: 50
Tensile strength: 5.5 MPa (800 psi)
Elongation: 100%
Tear strength: 1.75 kN/m (10 ppi)

III. SPECIFIC MATERIAL PROPERTIES

Silicone elastomer biomaterial physical properties vary from relatively low strength materials to extremely tough materials that are useful in high strength applications. For example, a fast-cure alkoxy two-part elastomer that is based on a very low viscosity polymer and is filled with extending filler may have the following physical properties:

Durometer: 45
Tensile strength: 2.76 MPa (400 psi)
Elongation: 200%
Tear strength: 3.5 kN/m (20 ppi)

Possible applications for these types of materials include external prostheses for facial deformities when ease of use and quick cure properties are desirable and toughness is less significant.

A medium strength elastomer made from an addition-cured, self-leveling base may be filled with high surface area reinforcing silica. An example of this material would have a viscosity of approximately 70,000 centipoise and would have the following properties:

Durometer: 35
Tensile strength: 5.5 MPa (800 psi)
Elongation: 400%
Tear strength: 14 kN/m (80 ppi)

A material such as this could be used for implants, such as for chins and ears, and it could also be used as an encapsulant for the electronic circuitry components of these implants.

Another medium strength silicone elastomer that can be used to fabricate flexible intraocular lenses is reinforced with fumed silica, but maintains optical clarity by designing the base polymer to have a refractive index matching that of the silica.

High strength silicone elastomers used in biomaterial applications would have the following typical properties:

Durometer: 50
Tensile strength: 8.3 MPa (1200 psi)
Elongation: 700%
Tear strength: 26.3 kN/m (150 ppi)

Materials with this range of properties have the most diverse biomaterial applications. These high strength silicone elastomers are used to fabricate tubing, drains, shunts, sheeting, and other parts used throughout the medical industry. Virtually all of the high strength silicone elastomers utilize either the peroxide-cure mechanism or the addition-cure mechanism.

The highest performance silicone elastomers are generally used in applications in which high modulus, high flex life, or extreme abrasion resistance are required. The following properties would be an example of a high performance silicone elastomer:

Durometer: 60
Tensile strength: 11.7 MPa (1700 psi)
Elongation: 700%
Tear strength: 43.8 kN/m (250 ppi)

These highest performance silicones may be used in applications such as orthopedic joint prostheses or for tubing in systolic pumps used for fluid transfer. These highest performance silicone elastomers also have a characteristically high initial modulus with yield points in the upper portion of the stress-strain curve, and a type of tear that is described as "knotty" tear. This knotty tear is particularly significant because it reflects the property by which a nick, cut, or flaw in the silicone would resist propagation. This property in particular enhances the overall apparent toughness of the elastomer, making it suitable for a variety of uses not previously considered.

In addition to these general categories of silicone elastomers, there are a large number of specialty silicone elastomers that were developed to meet the requirements of specialized applications. For example, extremely low durometer, low modulus silicone elastomer was developed to provide cushioning and tear resistance to mate artificial prosthetic legs to amputee stumps. A fast-cure silicone with a cure time of less than five minutes was developed to use as a dental impression material. This fast-cure silicone was further evolved to provide a variety of consistencies, ranging from very flowable to putty, in order to enhance its use as a dental impression material. This fast cure was also modified to provide a cured elastomer with densities less than 1 that could be used to form custom-made earplugs.

Silicone foam elastomers are among the specialty silicones that have found unique medical applications. Preformed foam extrusions are used for scleral buckling to repair detached retinas. Foamed-in-place silicones have been used to fabricate custom prostheses for deformities and have been used to provide shock absorption for artificial limbs. The silicone foams can be fabricated with varying compression densities and toughness characteristics to meet specific needs.

Silicone gels can be formulated to have varying characteristics. The overall cross-link densities can be adjusted by varying the SiH-to-vinyl ratio in the un-cross-linked formulation. By optimizing this ratio, a firm gel will result. Conversely, a very responsive gel can be produced by adjusting this ratio to be either excess in vinyl or excess in SiH. When these formulations are cured, a softer gel will result, with the degree of softness directly proportional to the variance of the ratio from the optimum.

The overall toughness of the gel can be varied by adjusting the degree of polymerization of the base polymer. A relatively tough gel can be formulated by using higher molecular weight polymers in the formulation. Conversely, a very responsive gel can be formulated by utilizing base polymers with molecular weights below 25,000. In a manner similar to elastomers, gels can be formulated to cure at a low temperature or higher temperatures for reducing bioload for subsequent sterilization. Fast-cure gels have been designed to cure in a few minutes at 37°C. Such rapid-curing gels are being investigated for formed-in-place injectable intraocular lenses.

IV. BIOMATERIALS APPLICATIONS OF SILICONES

Biomaterials are generally defined as materials, either of artificial or natural origin, that are used medically to repair, replace, augment, or contact living tissue, and that do not depend on their chemical properties to perform their intended function. Note that this definition does not use the concept of inertness. At the present time, no synthetic materials and few, if any, naturally derived materials have been shown to be totally inert.

Silicones have been used as biomaterials almost since their invention and have a history of producing minimal reaction in contact with living tissue. It is probably safe to say that every living individual in the more developed nations of the world has come in contact with silicones in a medical application.

Living tissue perceives implanted silicone as a foreign body. If the particles are too large to be phagocytized by macrophages, the body attempts to exclude them by forming a fibrous capsule around the material. This is a nonspecific reaction that occurs with any foreign material. As we show below, this reaction may be put to constructive use or it may present potential problems. There is no scientific evidence that there is any specific chemical reaction between a silicone surface and living cells or that silicone can cause any systemic disease. This is because the cross-linked silicone and any unreacted extractables are extremely inert to chemical reactions that are mild enough to take place within the body. The typical silicone polymer is resistant to all but the most severe conditions of temperature, oxidation, reduction, or pH. Below, we list and describe some of the more important applications of silicones in medical devices.

A. Implants

1. Orthopedic

Some of the earliest biomedical applications for silicone rubber were in the field of orthopedic surgery.

Finger and Toe Joints. The strength and flexibility of silicone rubber suggested its potential for use as a joint replacement. Swanson developed a device for this application that appeared to be highly successful and allowed renewed freedom of movement of fingers crippled by arthritis and other degenerative diseases. It later became apparent that the silicone was subject to abrasion, which produced particulate material that activated macrophages or giant cells, as noted above. This in turn led to irritation and release of cytokines that caused bone resorption and loosening of the implants. Redesigned joint implants are still employed, however, under some conditions.

Carpal Tendons. In cases of carpal tunnel syndrome or carpal tendon replacement, silicone rods or stents are placed in the tendon tract and the fibrotic reaction creates a new tunnel. After removal of the stent, a tendon replacement may be implanted. In this case, advantage is taken of the relatively benign reaction of the tissue to form a functional tract.

2. Plastic Surgery

Until 1991, the plastic surgery specialty probably used the largest volume of silicone rubber and gel. Silicones were found to be almost ideal materials for augmentation or replacement of various types of soft tissue.

Mammary Implants. The largest use of silicones in plastic surgery has been in breast augmentation and reconstruction. In the early 1960s, the practice of injecting silicone oil into breast tissue for augmentation was started. It was subsequently found that the fluid had a tendency to migrate through tissue planes in the body to remote areas, where the usual fibrotic reaction might ensue. This caused disfigurement and, in some cases, notably in Japan, adulterants in the fluid caused other reactions.

In 1963, in order to avoid these complications, implants were produced that were composed of silicone gel encased in a silicone rubber bag. This was a great improvement since there was little or no opportunity for migration of the silicone unless the bag was ruptured. The fibrotic reaction still occurred, however, and in some cases the fibrous capsule contracted, leading to a hard, spherical breast. The cause of this contracture has never been elucidated. It only occurs in a fraction of the patients, and frequently is unilateral.

Another finding with gel-filled prostheses was that, because of the high permeability of dimethyl siloxane fluid through dimethyl siloxane rubber, a condition known as "bleed" occurred. This manifested itself in the presence of silicone fluid on the exterior of the implants and in the form of droplets within the fibrous capsule. No correlation has been found between gel bleed and capsular contracture or any other condition. The gel bleed phenomenon was reduced by the inclusion of a barrier layer in the shell of the prosthesis, generally composed of either a diphenyl/dimethyl siloxane copolymer or a fluorosilicone. These are effective in markedly reducing the quantity of bleed. The problem of potential migration of gross amounts of gel in the event of shell rupture still remained.

An alternative approach to the gel problem was to produce a silicone bag implant that contained physiological saline solution instead of gel. These implants still had their problems, including sudden deflation in the event of rupture and a less natural consistency. Capsular contracture still occurred. A benefit was that an uninflated prosthesis could be inserted through a smaller incision with subsequent inflation in place than could a gel-filled device.

Urethane-foam-covered implants were also produced and it appeared that there was a reduced incidence of contracture with their use. It was found that the urethane was not stable to hydrolysis and concerns were voiced regarding the toxicity of the urethanes or their decomposition byproducts. Silicone prostheses with textured silicone outer surfaces are now produced that mimic the surface texture of the foam. These appear to have a reduced contracture rate. This is an additional indication that contracture is not a function of the chemical nature of the silicone.

Facial Implants.   Various implants for facial augmentation are produced, including those for the chin, nose, and malar regions. These are made with either solid silicone elastomer or with a relatively firm gel fill. They may have a patch of Dacron™ velour or molded-in holes to promote fixation by tissue ingrowth.

Custom Implants.   Many implants are produced on a custom basis for reconstruction and augmentation. Manufacturers are able to reproduce a shape in silicone from a plaster moulage provided by a physician. Some common areas for custom plastic surgery implants include ear pinnae, sternum, pectoral and calf muscles, and other body regions. The hardness or consistency may be varied to match the tissue involved. This is usually done by controlling the cross-linker concentration in an addition system.

Tissue Expanders.   Tissue expanders are essentially inflatable silicone bags that are implanted under the skin adjacent to areas where more skin is necessary, such as next to a scar or burn. The expander is inflated with saline solution over a period of weeks, stretching the overlying skin until sufficient new tissue has been created to cover the target defect. The expanders are also used intraoperatively during surgeries such as face lifts to stretch otherwise lax skin. The excess is then excised to leave smooth tissue in place.

3. Ophthalmic Implants

Silicones are frequently used to treat a number of conditions in the eye. Some of these applications were first initiated by using available configurations of material, with specialized devices being developed at a later time.

Intraocular Lenses.   Intraocular lenses (IOLs) are used to replace the natural lens of the eye when it has been clouded by a cataract or trauma. The original IOLs were made of methyl methacrylate and were rigid. It was felt that a lens made with a flexible material might reduce the trauma associated with implantation since it could be folded to be placed through a smaller incision. Silicone elastomer reinforced with a silicone resin having the same refractive index is optically clear, whereas silica-reinforced elastomer is typically hazy. Therefore, resin reinforced PDMS was selected for the first flexible IOLs. The refractive index of PDMS is relatively low, approximately 1.404, so that a high diopter lens needs to be fairly thick. The refractive index may be increased by utilizing a copolymer of dimethyl and diphenyl siloxanes. A mixture cannot be used because they are incompatible and lead to phase separation and clouding. Refractive indices up to 1.46 may be achieved without sacrificing mechanical strength.

Scleral Bands. One of the early applications of silicone in the eye was the use of silicone rubber bands around the sclera of the eyeball to apply pressure. The purpose is to force the exterior of the eye inward to maintain contact with the retina in the case of retinal detachment. Over the years, the bands have become more elaborate.

Ophthalmic Drains. Silicone tubes are used for drainage of both aqueous humor from the eye in the case of glaucoma and tears when the lacrimal duct is blocked.

4. Other Implants

Hydrocephalus Drains. Another application for silicone tubing has been for the relief of hydrocephalus, "water on the brain." This condition may be congenital or may be induced by disease or trauma. In any event, increased pressure due to an accumulation of cerebrospinal fluid in the brain can lead to severe brain damage if not relieved. For this purpose, valves and shunts made of silicone rubber (see Figs. 1 and 2) are inserted that control and drain excess fluid to the cardiovascular system or to the peritoneum, where the fluid is absorbed. These systems may be quite complex and include various valves and ports for external control and injection of drugs.

Sheeting. Silicone rubber sheeting, usually made by calendering gum stock, is used for a large variety of implants, in most cases made by the surgeon by cutting the sheet to the proper size and shape in the operating room. Some of the published applications have been for replacement or augmentation of the dura mater of the brain to protect the

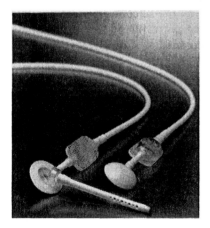

Figure 1   Cerebrospinal fluid ventricular drain, valve, and shunt. (Courtesy Pudenz-Schulte Medical Corp.)

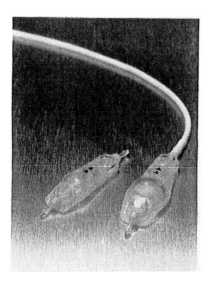

Figure 2   Hydrocephalus shunt valves. (Courtesy Pudenz-Schulte Medical Corp.)

delicate tissue from contact with sharp edges of the cranium, as a support for inguinal and other hernias, to protect the eye from orbital fractures, and to create tissue pockets in which other implants may be placed.

Cardiovascular.

Pacemakers.   Implanted pacemakers and electrode leads are sometimes covered with a layer of silicone that functions both as an electrical insulator and a corrosion barrier. Polyurethanes had also been used for this application, but it was found that the urethanes were not stable and tended to degrade over a period of time. Silicone is both a good insulator and completely stable to attack by body fluids.

Heart valves.   Some of the earliest prosthetic heart valves were of the cage-and-ball type in which the ball was composed of silicone rubber. It was found that a late complication of these implants was the swelling or fracturing of the ball after some years of service. Apparently, this was caused by uptake of blood lipids by the silicone. The problem was corrected by reformulation of the silicone or by replacement of the silicone ball with one composed of metal or glassy carbon. Silicone was also tried as a component of leaflet valves, but presently is not used for this application.

Gastrointestinal Implants.

Antireflux.   Severe gastroesophageal reflux cannot always be treated by medication and in some cases requires surgery. Various surgical remedies have been used, but all require cutting and suturing or stapling of the stomach wall. This can lead to other severe complications. A gel-filled silicone device that is tied around the esophagus below the diaphragm functions to reduce or prevent reflux of the gastric contents with much simpler and less traumatic surgery (see Fig. 3). The device may also be placed laparoscopically, which further reduces the risks of open surgery.

Bariatric.   Severely obese patients that are unable to control their food intake may be helped by at least two different silicone devices. Gastric balloons, while not technically implants, reside in the body for extended periods of time. They are, as the name implies,

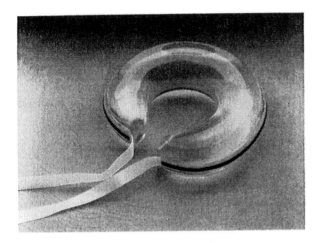

Figure 3   Antireflux prosthesis. (Courtesy McGhan Ltd. Ireland.)

balloons introduced into the stomach through the esophagus and inflated with saline to reduce the effective volume of the stomach and thus limit the amount of food that can be ingested. They also produce the feeling of satiety. These devices are relatively short term, usually being replaced on a regular basis and used for about a year.

For a more permanent effect, a silicone gastric band may be implanted. This is a band placed around the stomach and tightened to create a smaller stomach pouch and again reduce the intake of food. These bands are made with an inflatable balloon for postsurgery adjustment of the stoma size (see Fig. 4). This device may also be implanted laparoscopically. The alternative to these devices is surgery involving the stapling or cutting of the stomach.

Urinary Implants.   There are many silicone devices used to correct problems in the genitourinary tract. Both active and passive devices are available to control incontinence. Active and passive penile implants are also available to correct male impotence. Silicone

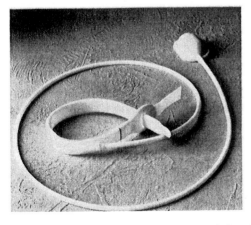

Figure 4   Laparoscopic adjustable gastric band. (Courtesy Bioenterics Corp.)

has been evaluated for an artificial bladder. RTV silicone may be injected into the fallopian tubes as a means of potentially reversible female sterilization.

Drug Delivery.   Because silicone is highly permeable to many substances, it is being used with increasing frequency in drug delivery systems. The rate of delivery is dependent upon the permeability of the particular drug and silicone combination and can be closely controlled. Drugs for female fertility control, cancer chemotherapy, and pain relief are all being delivered to specific sites in the body with reduced systemic toxicity.

B.   Nonimplantables

Silicones have wide usage in applications involving short-term contact with the body that do not involve implantation. Some of these applications are described below.

1.   Tubing

Silicone tubing is used extensively in both drainage and infusion applications. Silicone wound drains, particularly in conjunction with vacuum-activated reservoirs, find very wide use in all fields of surgery. External drainage catheters are also used for urinary, peritoneal, and chest drainage, generally with a gravity system. Feeding tubes and infusion catheters are used to transfer fluids to target sites in the body. Silicone tubes are also used as temporary bypass catheters in operations such as carotid endarterectomies.

2.   Syringe Lubricants

Silicone oil is used as a lubricant for both the needle and barrel of practically all medical syringes. This means that almost all of the population has been inoculated at some time with silicone. In some instances, such as for diabetics who need daily injections of insulin, the quantity of silicone injected in the course of a lifetime can be very significant.

3.   Cosmetics

Silicone fluids are widely used in many cosmetics, including skin creams and cleansers, shampoos, lipstick, deodorants, and hair spray. The silicone imparts a smooth, silky feel to the skin and also may provide water and soil repellency. Volatile cyclic siloxanes are used in some formulations (e.g., hair sprays) for their fugitive properties.

4.   Pressure-Sensitive Adhesives

Pressure-sensitive silicones are used for wound dressings and ostomy bag attachment. These materials are known to be nonirritating, hypoallergenic, and water resistant.

5.   Antiflatulants

Polydimethyl siloxane (simethicone) with fumed silica is used extensively as an antiflatulant, both by itself and in combination with antacids. This combination is also used as an antifoaming agent in blood collection devices and oxygenators.

6.   Stents

Silicone devices are used as stents to create openings and passageways in the body. Because of the benign encapsulation of the device, these passageways may then be used as active channels. The application as a stent in forming a carpal tendon sheath was discussed above. Stents are also used to create artificial vaginas in patients who have a congenital lack of this organ and in cases of sexual change.

7. Wound and Scar Treatments

Silicone sheeting and gel have been found to be effective in reducing the volume and discoloration of keloids and hypertrophic scar tissue. A sheet of either rubber or gel is applied to the scar and left in place. Pressure is not necessary. The dressings are removed and cleaned once or twice a day to prevent infection and irritation. The treatment may take as many as 12 weeks for maximum results. Silicone-based skin creams have a similar effect. The mechanism of action is not known, although there are several hypotheses.

8. Blood Oxygenators

Silicone has the highest oxygen transmission rate of any synthetic polymeric material known. This property and its excellent biocompatibility make it almost ideal for use as a blood oxygenator in bypass applications. During heart or vascular surgery, the blood may be shunted through an external device having a silicone membrane. Exchange of oxygen and carbon dioxide takes place across the membrane. The membrane may be in a variety of forms, the commonest being a simple sheet, either flat or wound concentrically or helically with oxygen channels, or as hollow fibers similar in operation to kidney dialysis units. Silicone surfactants are frequently used in conjunction with these devices to reduce the tendency of the blood to foam. Heparin is also added to reduce clotting. These devices may be operated for several hours at a time without a significant hazard to the patient.

9. Dental Impression Materials

RTV silicones, particularly platinum-catalyzed addition types, are commonly used to prepare impressions of dentition, the alveolar ridge, and other mouth structures for use in preparing orthodontic appliances. These materials set very rapidly at ambient temperature with no exotherm, so that there is no thermal injury to the sensitive mouth tissues. The reproduction of fine detail is excellent and there is very little shrinkage.

10. Contact Lenses

There has been a concerted effort to produce contact lenses using silicone rubber because of its high oxygen permeability, but it is also highly hydrophilic and has a tendency to replace water and adhere to the cornea. Copolymers of silicone and polyethylene glycol have been developed that are sufficiently hydrophilic and do not traumatize the cornea. They are flexible and compliant and allow good oxygenation and nutrition of the corneal tissue.

52
Hyaluronan Biomaterials: Medical Applications

Endre A. Balazs, Edward Leshchiner,
Nancy E. Larsen, and Philip Band
Biomatrix, Incorporated
Ridgefield, New Jersey

I. INTRODUCTION

During the past two decades, hyaluronan (HY, hyaluronic acid) and its derivatives have become important therapeutic agents in medicine. This development was triggered by the recognition in the late 1960s that the highly purified form of hyaluronan (NIF-NaHY, the noninflammatory fraction of Na-hyaluronan), prepared from animal tissues, is extremely biocompatible when applied to such sensitive tissue compartments as the vitreous of the eye and the synovial space in joints [1–3]. A few years later, this discovery led to the medical application of NIF-NaHY in ophthalmic surgery and for the treatment of arthritis in humans and in horses [4,5].

In the mid-1980s, the field of cross-linked hyaluronan derivatives began to develop. Two cross-linked forms of hyaluronan, hylan fluid and hylan gel, were invented and their biological activity and medical usefulness were widely explored [6]. They proved to be just as biocompatible as the native hyaluronan, but had enhanced rheological properties and longer residence time in the tissues than hyaluronan. The therapeutic application of the hylans in a broad spectrum of medical specialties is now in progress. These include viscosurgery, arthritis therapy, adhesion management, topical administration, drug delivery, and soft-tissue augmentation (e.g., dermal, urological, and reconstructive applications). This important development stimulated broad interest in new forms of cross-linked or otherwise modified forms of hyaluronan.

This chapter reviews the chemistry of the various hylans and other cross-linked hyaluronans, focusing on their current usage and proposed medical applications.

A. Hyaluronan

The ubiquitous hyaluronan polysaccharide molecule is located in the intercellular space and fills the space between the collagen and elastin fibers, cell membranes, and basal laminae. It is produced by many cell types in the cell membrane and is, therefore, never stored inside the cell. After synthesis, it is extruded directly into the extracellular space. It is considered as a space-filling, structure-stabilizing, cell-coating and cell-protective polysaccharide. Its primary biological role is to stabilize the intercellular structure of fibrous and membranous proteins. It forms a structurally integrated system with the fibrous proteins of the intercellular space, creating the elastoviscous, protective, lubricating, and stabilizing matrix in which cells are embedded. Hyaluronan solutions are extremely elastoviscous and pseudoplastic. Their exceptionally high rheological properties are present even in highly hydrated polymer systems (water content more than 99%). This combination of high elastoviscosity and low solids content permits unhindered diffusion of metabolites to and from the cells embedded in or separated by the viscoelastic hyaluronan molecular network.

The unusual rheological properties of hyaluronan are based on a relatively simple structure. It is a linear, unbranched (not cross-linked) polyanionic molecular chain consisting of repeating glucuronic acid-N-acetyl glucosamine dimers with a large molecular mass (4–5 million). In an aqueous solution, the hyaluronan molecule behaves as a highly hydrated random coil with very large molecular volume, forcing the molecules to become entangled and interpenetrate at relatively low concentrations (<0.01%), thereby giving rise to hyaluronan's unusual rheological properties [7].

Hyaluronan solutions of low concentration and lower molecule mass than occurring naturally were found to influence various cell activities in *in vitro* studies [8]. These effects could not be demonstrated in *in vivo* systems. The highly elastoviscous hyaluronan solutions also have an effect on cells, especially on the movement of the cell membrane and the cytoskeleton of the cell [9]. The effect of viscoelastic hyaluronan solutions on cells and the intercellular matrix forms the basis of their current medical applications.

The need for enhanced viscoelastic properties and greater solidity than provided by native hyaluronan solutions triggered the interest in producing its cross-linked derivatives. The cross-linking derivation must produce not only rheologically useful, but also biologically compatible, polymeric systems. We review here only those derivatives that have been shown to be useful as therapeutic agents.

B. Hyaluronan Derivatives

The hyaluronan polysaccharide chain contains three types of functional groups that can be used for derivitization, namely, hydroxy-, carboxy-, and acetamido groups. The reducing end groups need not be considered due to the extremely high molecular weight of the polymer. There is an enormous number of derivatives that may be obtained through chemical reactions of these functionalities with various reagents. The practical significance of any hyaluronan derivative is determined, mainly, by its utility in the medical field.

The earliest synthesized derivative of hyaluronan was its sulfate ester [10]. A heterogeneous sulfation process for suspension of hyaluronan in pyridine by a mixture of pyridine and chlorosulfonic acid was developed. A highly substituted product was obtained by running the sulfation reaction repeatedly. The substance was characterized as having strong antienzymatic activity, that is, strong antihyaluronidase and anticoagulant activities, and to be inhibitory of fibroblast growth. A single-phase solution method to obtain hyaluronan sulfate esters was described in Ref. 11, which probably represents the first patented application of hyaluronan technology. Despite the beneficial properties of hyaluronan sulfates, they never have been developed into commercial products, probably due to the availability of other sulfated glycosaminoglycans, especially heparin and chondroitin sulfate.

A substantial number of cross-linked hyaluronan derivatives have been developed. According to the nomenclature suggested by Balazs and Leshchiner [6], cross-linked derivatives of hyaluronan modified only through hydroxy groups (carboxyls and acetamido groups remain unreacted) are named *hylans*. The first cross-linked hyaluronan derivative of the hylan type was obtained using the bifunctional substance 1,2,3,4-diepoxybutane [12]. The diepoxide reacted with hyaluronan in the presence of alkali and sodium borohydride, forming a gel that could swell in water or saline several hundred times. Obviously, the cross-linking was provided by reaction of hydroxy groups with epoxy groups of diepoxybutane.

A large group of hyaluronans cross-linked via their hydroxy group reactivity was developed by Balazs and Leshchiner [13]. Cross-linking agents of various functionalities were used in the reactions, including formaldehyde, dimethylol urea, dimethylolethylene urea, polyisocyanate, and vinyl sulfone. Water-insoluble products were obtained in the shape of a powder, a film, or a coating on various substrates. The products obtained with vinyl sulfone as the cross-linking agent were developed further and the family of hylan gels with properties varying in broad limits was obtained [14]. The following competitive reactions determine the properties of vinyl sulfone cross-linked hyaluronan:

$$|-CH_2OH + O_2S\begin{matrix}CH = CH_2 \\ \diagup \\ \diagdown \\ CH = CH_2\end{matrix} \rightarrow |-CH_2OCH_2CH_2SO_2CH = CH_2 \quad (1)$$

$$\begin{aligned}& |-CH_2OCH_2CH_2SO_2CH = CH_2 \quad (2) \\ + \; & |-CH_2OH \rightarrow |-CH_2OCH_2CH_2SO_2CH_2CH_2OCH_2-|\end{aligned}$$

$$|-CH_2OCH_2CH_2SO_2CH = CH_2 + H_2O \rightarrow |-CH_2OCH_2CH_2SO_2CH_2CH_2OH \quad (3)$$

$$O_2S\begin{matrix}CH = CH_2 \\ \diagup \\ \diagdown \\ CH = CH_2\end{matrix} + H_2O \rightarrow O_2S\begin{matrix}CH_2CH_2OH \\ \diagup \\ \diagdown \\ CH = CH_2\end{matrix} \rightarrow O_2S\begin{matrix}CH_2CH_2 \\ \diagup \quad \diagdown \\ \quad\quad O \\ \diagdown \quad \diagup \\ CH_2CH_2\end{matrix} \quad (4)$$

To simplify the above equations, only primary hydroxy groups of hyaluronan are shown to take part in the reaction. As one can see, the consecutive reactions 1 and 2 lead to formation of bis-(ethyl) sulfone cross-links, whereas Reactions 1 and 3 give 2-hydroxy ethyl-sulfonyl-ethyl pendant groups. Reaction 4 proceeds without polymer and results in formation of 1,4-thioxane dioxide from vinyl sulfone and water. Such important properties of this hylan gel as swelling in various media and rheological properties are determined, mainly, by the ratio of cross-links and pendant groups. Modified hylan gels were also developed by including other polymers containing chemical groups reactive with vinyl sulfone [15] or low molecular weight substances [16]. In general, the hylan gels obtained using this vinyl sulfone process possess an exceptional biocompatibility and other useful properties that make them excellent products for biomedical applications.

A unique soluble hylan polymer (hylan fluid) was developed by Balazs et al. by cross-linking hyaluronan chains to specific protein molecules with formaldehyde during recovery of the polymer from animal tissues such as combs of domestic fowl [17,18]. Some of the important rheological properties of hylan fluid have been described elsewhere [6]. This polymer remains soluble in aqueous media due to the fact that the small amount of formaldehyde combined with the polymer (it may vary in broad limits, depending on the conditions of the process, but usually is around 0.001–0.005%) is not enough for formation of an infinite network. The cross-linking is evident from the higher molecular weight of the hyaluronan polymers obtained in the range of 6 to 24 million. Solutions made from this polymer (hylan fluids) exhibit very high elastoviscous properties [6].

Soluble cross-linked hyaluronan preparations were obtained by Sakurai, Uno, and Okuyama with the use of polyfunctional epoxy compounds in alkaline media and in the presence of an organic solvent [19]. Depending upon reaction conditions, especially the ratio of a polyepoxy compound to hyaluronan, a cross-linked insoluble polymer may be obtained as well. The solubility of the resulting polymer is determined mainly by the cross-linking index (according to the authors' terminology), which is expressed as the number of cross-links per 1000 disaccharide units. The products with a cross-linking index of about 5–10 are soluble and products with a higher cross-linking index (e.g., 40) are insoluble.

Similar cross-linked insoluble gels of hyaluronan produced by reaction with polyfunctional epoxy compounds (e.g., 1,4-butanediol diglycidil ether) in alkaline media were suggested for use as a vitreous substitute and for retinal detachment surgery [20].

The great reactivity of hydroxy groups of hyaluronan toward various reagents was recently used for obtaining new derivatives of hyaluronan which are phosphate esters [21]. These derivatives are obtained by esterification of hyaluronan with phosphoric acid derivatives such as phosphoryl chloride, $POCl_3$, in organic or aqueous solutions and in the presence of acid acceptors. The gel formation occurs very quickly. The consistency of the gel depends greatly on the reaction conditions. A soft gel with the polymer concentration about 1–1.5% and the phosphorus content about 0.1% is described. This gel is said to be autoclavable and it degrades spontaneously in aqueous media at physiological conditions.

All hyaluronan derivatives described above are based on reactions in which hyaluronan plays the role of a polymeric polyhydroxy component. The other group of hyaluronan derivatives consists of substances obtained through reactions in which hyaluronan plays the role of polycarboxylic acid. These derivatives include, mainly, esters and amides and some intermediate products as well. A large group of esters of hyaluronan and various alcohols is described in several patents [22–24]. A convenient method for obtaining hyaluronan esters is described and involves the reaction of tetrabutylammonium salt of hyaluronan in dimethyl sulfoxide with any substance possessing a halogen attached to a carbon. Tetraalkylammonium salts are known to be phase transfer catalysts. The use of tetrabutylammonium salt provides the solubility of hyaluronan in organic solvents and catalyzes its reaction with halogen compounds. The reaction goes according to the following scheme:

$$HY{-}COO^{-}\ \overset{+}{N}(C_4H_9)_4 + RX \rightarrow HY{-}COOR + (C_4H_9)_4\ \overset{+}{N}\overset{-}{X}$$

where HY is the designation for hyaluronan polymer chain and X is a halogen, preferably iodine or bromine. Partial or fully substituted esters of hyaluronan with a variety of alcohols are described. The properties of the esters such as solubility in aqueous or organic media, kinetics of hydrolysis in water, and so on are determined mainly by the nature of the alcohol component of the ester and the degree of substitution of the ester. By using a polyfunctional alkylating agent, a cross-linked ester of hyaluronan can be obtained. Various substances characterized by pharmacological activity may be attached by this method to the hyaluronan macromolecules. One such substance described in the patents is hydrocortisone, which is attached to hyaluronan by reaction of its 21-bromo derivative with tetrabutylammonium hyaluronate. Various medical uses of these hyaluronan esters are suggested, including drug delivery, skin care, and others.

Another method of obtaining hyaluronan esters in which the polymer plays a role of the acid component is described by Malson [25]. It is stated that the reaction of epoxy compounds with hyaluronan in the presence of an acid catalyst proceeds with formation of ester groups rather than ether groups, which form from the reaction of hyaluronan with epoxy compounds in alkaline media. By using polyfunctional epoxy compounds and alternating acidic and alkaline conditions, it is possible to obtain cross-linked polymers containing ether and ester groups simultaneously. Cross-linked gels of hyaluronan obtained by reactions of the polymer with a polyfunctional epoxy compound such as 1,4-butanediol diglycidyl ether are suggested to be used for prevention of adhesion formation in surgery [26].

The other types of hyaluronan derivatives reported in the literature with some practical applications are obtained through a process that involves activation of carboxyl groups with a water-soluble carbodiimide followed by the reaction of the intermediate substance with a nucleophile. When an amino-group-containing substance is used as a nucleophile, the reaction results in formation of an amide bond. The uses of water-soluble carbodiimides for the synthesis of peptides in aqueous solution [27] and for modification of carboxyl groups in proteins [28] were reported. The general mechanism of the reaction was postulated by Khorana and is described below [29].

A carboxyl-containing substance reacts with a water-soluble carbodiimide with formation of an O-acylisourea:

$$R-\overset{\overset{\displaystyle O}{\|}}{C}\diagdown_{OH} \;+\; \underset{\diagdown_{R''}}{\overset{\diagup^{R'}}{\underset{\|}{\overset{N}{\underset{N}{\overset{\|}{C}}}}}} \;+H^+ \to R-\overset{\overset{\displaystyle O}{\|}}{C}-O-\underset{\diagdown_{R''}}{\overset{\diagup^{R'}}{\underset{\|}{\overset{NH}{\underset{NH^+}{C}}}}}$$

The O-acylisourea may rearrange to give N-acylurea:

$$R-\overset{\overset{\displaystyle O}{\|}}{C}-O-\underset{\diagdown_{R''}}{\overset{\diagup^{R'}}{\underset{\|}{\overset{NH}{\underset{NH^+}{C}}}}} \;\to\; \underset{\diagdown_{R''}}{\overset{\diagup^{R'}}{\underset{\|}{\overset{R-\overset{\overset{O}{\|}}{C}-O-N}{\underset{NH}{C=O}}}}} \;+H^+$$

or may react with a nucleophile such as glycine methyl ester to form an amide derivative of carboxyl acid:

$$
\begin{array}{c}
\qquad\qquad\qquad R' \\
\qquad\qquad\qquad / \\
O \qquad NH \\
\parallel \qquad | \\
R-C-O-C \qquad\qquad + \ H_2NCH_2COOCH_3 \rightarrow \\
\parallel \\
NH^+ - R''
\end{array}
$$

$$
\begin{array}{c}
O \qquad\qquad\qquad\qquad\qquad\qquad NHR' \\
\parallel \qquad\qquad\qquad\qquad\qquad\qquad / \\
R-C-NH-CH_2COOCH_3 \ + \ O=C \qquad\qquad +H^+ \\
\qquad\qquad\qquad\qquad\qquad\qquad\qquad \backslash \\
\qquad\qquad\qquad\qquad\qquad\qquad\qquad NHR''
\end{array}
$$

The application of this reaction to various glycosaminoglycans, including hyaluronan, was described by Danishevsky and Siskovic [30]. The yield for the reaction of hyaluronan with glycine methyl ester was about 39%. Water-soluble carbodiimides were used also for attachment to other amino acids as biocompatible spacers for drugs and to various glycosaminoglycans, including hyaluronan, that serve as carriers [31].

The use of carbodiimides for modifying hyaluronan was recently described in a patent publication [32]. The new feature in this work is the use of bis-carbodiimide, which leads to formation of cross-linked derivatives. A slightly modified method for obtaining amide derivatives of hyaluronan and various esters of amino acids is described in a recent patent [33]. These derivatives are described as water-insoluble polymers that form hydrogels that hydrolyze at a rate depending on the nature and the amount of the substituting groups. Polyfunctional reagents are not used in the described method and the resulting products are not cross-linked; the insolubility is provided by introduction of hydrophobic alkyl ester groups. The products are recommended for preventing adhesions after surgery, for drug delivery, and for other medical uses. A similar method is described in another patent publication [34], which uses other polyanionic polysaccharides together with hyaluronan.

II. APPLICATIONS

A. Viscosurgery

The first medical use of NIF-NaHY was in the field of ophthalmic surgery [1]. The new surgical procedure developed during the 1970s was aimed at protecting the tissues during surgery and providing the surgeon with a safe tool to make space and to move tissues [35–37]. This surgical procedure was called viscosurgery [37] and was introduced into medical practice in the early 1980s with the first commercial product of NIF-NaHY under the trademark Healon® (Kabi Pharmacia, Uppsala, Sweden).

The efficacy of viscosurgical devices is based on their elastoviscous properties. They have an important function after surgery as well. When left in the surgical site, they become viscosurgical implants and provide a separation between tissue surfaces and a barrier against postsurgical exudation. The molecular network of hyaluronan forms a barrier to blood, fibrin, and invading cells, thereby preventing adhesion and scar tissue formation and helping to restore normal mechanical function after surgery.

The most important property of viscosurgical devices, besides their rheological properties, is their biocompatibility. NIF-NaHY is not recognized by the body as foreign and does not cause an immunogenic response. Since hyaluronan is present in those tissues in which it is used as a viscosurgical device in the eye (anterior chamber and vitreous), its presence during and after surgery does not elicit a tissue reaction.

NIF-NaHY as a viscosurgical device is primarily used in cataract surgery. It minimizes the surgical trauma during the removal of the cataractous lens and the introduction of its replacement, the intraocular plastic lens. It is also used for replacement of the vitreous after vitreoretinal surgery, mostly in connection with the surgical treatment of retinal detachment. The elastoviscosity of hyaluronan solution has not proved to be sufficient in most cases to retain the retina in its reattached position. Therefore, the newly developed cross-linked hylan gel has been used in several clinical trials as a vitreous substitute. These studies demonstrate that hylan gel can successfully achieve a long-lasting reattachment of the retina.

Healon was the first viscosurgical device marketed worldwide. It is still the most widely used of such devices because it has the greatest elastoviscosity compared with many imitations later developed for the same use. Since hylan fluid is several times more elastoviscous than hyaluronan solutions of comparable polymer concentration, its use as a more effective viscosurgical tool in ophthalmology is being explored.

B. Treatment of Arthritis

Viscosupplementation was introduced to medicine as a new therapeutic modality in the early 1970s (for review, see Ref. 3). This therapeutic concept means that the elastoviscous properties of a tissue compartment are supplemented or augmented in order to influence the pathological process and restore normal physiological function. In various forms of arthritis, especially in osteoarthritis, the elastoviscosity of the intercellular matrix of the soft joint tissues (synovial fluid and tissues, and capsule) is decreased. This decrease is caused by two pathological events.

Due to the inflammatory process in the joint, excess water diffuses from the blood vessels into the tissues. Thus, a joint effusion develops and dilutes the normal matrix, decreasing its elastoviscosity. Alternatively or accompanying this process, the molecular mass of the hyaluronan in the joint fluid and tissues decreases for yet unknown reasons, causing a dramatic drop in the elastoviscous properties of the matrix of these tissues. Both mechanisms lead to a significant drop in the rheological properties of the synovial fluid and the intercellular matrix. This is a pathological event that is characteristic of nearly all arthritic conditions: traumatic arthritis, metabolic arthritis, osteoarthritis, or rheumatoid arthritis.

When this pathological chain of events was recognized [38], the hypothesis was proposed that, by restoring the homeostasis of elastoviscosity of the synovial fluid and matrix of the joint tissue, one can expect improvement of the pathological conditions of the joint. Since the most important symptom of arthritis is pain, one would expect relief of pain after the rheological homeostasis is restored in the joint. This hypothesis was tested first in 1970–1971 in race horses with traumatic arthritis treated with NIF-NaHY (Healon® Biotrics, Inc., Arlington, MA). The elastoviscous 1% solution of about 2 million molecular mass hyaluronan was injected once or repeatedly into the affected joint [39,40].

With this viscosupplementation, the low elastoviscosity of the joint fluid was immediately elevated and maintained at that level for a few days. The pain decreased, the joint mobility increased, and often the rheological properties were restored to a normal level even after the injected hyaluronan was gone from the joint [41]. These studies were then extended to the treatment of human osteoarthritis [42,43]. The results obtained with two or three intraarticular injections one week apart confirmed the same analgesic effect in the human knee joint. Double-blind, controlled studies indicated an effect lasting several months in about 50% of the patients treated [44].

Based on these findings, in the late 1980s two hyaluronan preparations were marketed for viscosupplementation of osteoarthritic joints: one in Japan (Arzt®, Seikagaku and Kaken) and one in Italy (Hyalgan®, Fidia). Both of these preparations consist of a 1% solution of relatively low molecular mass (~700,000) hyaluronan; therefore, the elastoviscosity of these products is relatively low. Nevertheless, investigators from Japan [45], Italy [46], and Great Britain [47] reported successful relief of pain with treatment consisting of 6–12 weekly injections.

Studies in equine arthritis clearly demonstrated that, as the elastoviscosity of the solution used for viscosupplementation increased, the fewer injections that were needed, with the result that greater and more lasting pain relief could be observed [48].

Since the cross-linked derivative of hyaluronan, hylan, is more elastoviscous and has a longer intraarticular residence time, a combination of hylan fluid and hylan gel (hylan gel-fluid 20, Synvisc®, Biomatrix, Inc., United States) was developed and tested in equine and human arthritis. The data showed that three injections, one week apart, were an efficacious treatment of osteoarthritic pain in the knee joint. A nearly complete (>80% improvement) pain relief is achieved in 60–70% of the treated patients for a four-to-six-month period. Longer lasting effects were also observed. Most importantly, no systemic side effects were observed, and local transient discomfort was noted in fewer than 1% of the patients treated [49–52].

The mode of action of viscosupplementation has been the subject of many studies. *In vitro* studies showed that hyaluronan and hylan have a protective effect on cartilage cells against mechanical and chemical damage. This protective effect depends on the elastoviscosity of the solution. Greater rheological properties provide more protection. The protection was tested against such noxious factors as tissue-degrading enzymes and other agents (interleukin-1) released by white blood cells, and oxygen-derived free radicals [53]. Since these enzymes and free radicals are considered as possible causes of osteoarthritis [54], the implication is that such protective mechanisms may play a role. Direct, *in vivo* studies have not yet supported the existence of this mechanism.

Another explanation for the analgesic effect of viscosupplementation could be through its direct effect on pain receptors of the joint capsule. Pain in the joint is triggered by sensory afferent nerve fibers called *nociceptors*. Excitation of nociceptors or afferents of the joint is due to the transmission of mechanical forces to the nerve terminals that are normally activated only by noxious (abnormal, overextended) movement, producing a train of nerve impulses that constitutes a signal of pain for the central nervous system. These signals were produced and measured in normal animal joints that were overextended (noxious movement) and in arthritic animals' joints that were moved within the normal range. In the joints of these animals, hylan gel-fluid 20 (Synvisc) was injected, and within one hour a significant reduction of ongoing or stimulated nerve activity was observed. With nonelastoviscous solutions of hylan or hyaluronan, this effect was not observed [55].

These *in vivo* model studies on arthritic pain strongly suggest that the pain-reducing effect of viscosupplementation is the result of the buffering and screening out of mechanical and chemical stimulatory factors from nociceptors. This effect seems directly related to the elastoviscous properties of the material used for viscosupplementation. Japanese authors using behavioral animal models also reported analgesic effects of hyaluronan solutions. They could demonstrate this effect only if the solution contained high molecular weight (>720,000) hyaluronan molecules [56,57].

Probably the most important factor in the therapeutic process of viscosupplementation is the increase of joint mobility. The analgesic effect triggers a cascade of events, starting with increased joint mobility and resulting in the restoration of the normal flow of hyaluronan in the joint. It flows from the synovial cells in which it is synthesized, to the cartilage surface and through the soft tissues of the joint to the site of outflow, the lymph vessels. It is well known that an immobilized normal joint undergoes pathological changes and the hyaluronan metabolism becomes seriously impaired [58]. With movement of the joint, the homeostasis of the fluid circulation (blood, lymph, synovial fluid) is restored. If the degenerative process in the joint did not progress too far, or if there is no excessive inflammation in the joint, viscosupplementation can restore the homeostasis of the joint.

C. Adhesion Management

Recently, the use of HY and its derivatives in adhesion management has achieved widespread recognition, most notably by commercial ventures, as a potentially unique and effective approach to the prevention or reduction in formation of unwanted collagenous connective tissue following tissue injury. Over 20 years ago, Balazs, Rydell, and Freeman reported on the effect of HY on adhesion formation between tendon and tendon sheaths and between the conjunctiva and sclera [59]. Viscoelastic solutions of HY and water-soluble dry sheets of HY were found to reduce the incidence of adhesion formation without delaying or affecting the healing response to the injury.

Later, in 1980, it was reported that HY was effective in reducing adhesion formation in primary flexor tendon repair [60]. Prior to this investigation, a great variety of materials had been evaluated; none were shown to be clinically useful in preventing or reducing adhesion formation in this wound model.

Since that time, numerous investigations have been carried out using HY and its derivatives in an attempt to identify the most effective form of HY for adhesion management. It is apparent that, in many cases, the optimum form of HY will depend upon the specific indication. For example, in flexor tendon repair the most effective material may be an insoluble gel or film that provides a long-lasting biocompatible (physically and biologically) mechanical barrier to maintain separation of injured tissues and to prevent fibrin network formation between the tissue surfaces. Weiss et al. have shown that the insoluble, viscoelastic, cross-linked HY derivative hylan gel was highly effective in decreasing tendon adhesions in the rabbit extensor hallucis longus tendon following surgical trauma [61,62]. Results with HY solutions have been mixed; it has been reported that solutions of HY such as Viscoat™ (a solution of HY and chondroitin sulfate) and Healon (1% HY) do not significantly reduce adhesion formation in tendon healing [63,64], while others have shown that HY solution has some benefit [65,66]. The observed effect may also be influenced by differences in the severity of tissue trauma.

Management of postsurgical adhesions (abdominal, cardiac, gynecological, neurological) represents another very difficult clinical situation. A family of hyaluronan derivatives known as HAL products: HAL-C™ coating solution, HAL-F™ bioresorbable membrane, and HAL-G™ Gel are being developed for the purpose of reducing the formation of postsurgical adhesions [67,68]. It is interesting to note that two of these products are chemically modified in order to alter the physical properties that influence performance (i.e., viscosity, residence time, and barrier function). HY products derived from hylan have been developed with enhanced rheological properties, extended residence times, and high biocompatibility for use in adhesion management in musculoskeletal, neurological, and abdominal applications [6].

Postoperative peridural adhesion formation following lumbar laminectomy or diskectomy is a major factor in the development of chronic low back pain. Highly viscous solutions of HY (1.9% and 1.0%) [69] and hylan gel [70] have been shown to be effective in reducing postsurgical adhesion formation in animal laminectomy models. The efficacy observed in these models is likely a result of restricted diffusion of the HY from this tissue compartment and, hence, increased residence time.

D. Soft-Tissue Augmentation

Soft-tissue contouring for therapeutic and cosmetic purposes in the fields of dermatology, plastic surgery, and urology has been practiced with a variety of materials for nearly 100 years [71]. Currently, collagen implants are used widely for the correction of soft-tissue defects [72], most notably in the area of dermatology. The problems associated with this implant material relate to its relatively short residence time and temporary beneficial effect and, more importantly, to its biological compatibility and to the overall safety of the material.

Hylan gel, an insoluble, injectable, implantable HY derivative, was developed for use in soft-tissue repair and augmentation [73,74]. Preclinical testing indicates that hylan gel is highly biocompatible (intradermal, subcutaneous, intramuscular) and resistant to resorption and migration (Fig. 1). The lack of significant inflammatory and immune response to hylan gel may contribute to its stability in the tissue, while the cross-linked structure and water insolubility of hylan gel function to restrict diffusion and migration from the site of injection. For these reasons, together with its rheological properties, hylan gel may provide a safe, efficacious alternative to existing products for soft-tissue augmentation in urology (sphincter muscle), dermatology (facial dermis), and breast

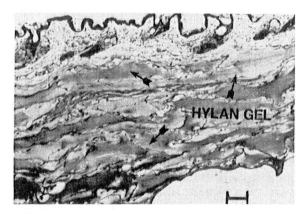

Figure 1   Photomicrograph of hylan gel intradermal implant 6 weeks after implantation in Nu Nu nude mouse with no significant tissue reaction (bar scale represents 100 μm).

augmentation (gel enclosed within an acceptable shell to provide physical compatibility as well as enhanced safety).

E. Drug Delivery

Various polymeric drug delivery systems currently exist for the controlled release of bioactive agents. A primary concern for both topical (transdermal, ophthalmic, wound) and parenteral drug delivery vehicles is their biologic compatibility. Hyaluronan and hylans are well suited for this medical application because of their unusual rheological properties and their extraordinary biological compatibility [6,36,75,76]. In addition, these polysaccharides are physically and chemically versatile, allowing the design and formulation of a variety of systems, each suited for a specific application.

Hyaluronan and hyaluronan derivatives (hylans) have been developed as topical, injectable, and implantable vehicles for the controlled and localized delivery of biologically active molecules (for review, see Ref. 77). Hyaluronan is a natural polysaccharide that exhibits extraordinary biocompatibility and unique rheological properties. Hylans are a family of chemically modified hyaluronans and include hylan fluids, gels, microparticles, and membranes (sheets, coatings). Hylan fluid is a water-soluble, hydrophilic, polyanionic hyaluronan derivative that retains the excellent biocompatibility of the native glycosaminoglycan while providing the basis for various release systems. Hyaluronan and hylan derivatives have been evaluated in *in vitro* and *in vivo* models for their ability to optimize delivery of a variety of pharmacologically active molecules.

In ophthalmology, viscous agents have often been used to increase the corneal residence time of topical therapeutics in an effort to increase the period of drug action substantially [78–81]. Viscous polymers such as cellulose derivatives, poly(vinylalcohol), and polyvinylpyrrolidone are common additives to ophthalmic preparations; however, their effectiveness is limited by their rheological properties and their lack of biocompatibility. The concentrations needed to provide sufficient viscosity are often irritating and incompatible with sensitive ocular issues. Hyaluronan solutions, on the other hand, have tremendous ocular compatibility both internally (when used during ophthalmic surgery) and externally, at concentrations of up to 10 milligrams per milliliter (mg/ml) (1%). Topical hyaluronan solutions (0.1–0.2%) have been shown to be effective therapy for dry eye syndrome [82–85], a condition characterized by pain, itching, burning, and foreign-body sensation. Use of hyaluronan solution by patients with this syndrome resulted in alleviation of symptoms and an increase in tear film breakup time (BUT) [86,87]. Investigators suggest that hyaluronan may interact with the corneal surface and tear film to stabilize the tear film and provide effective wetting, lubrication, and relief from pain caused by exposed and often damaged corneal epithelium [82,86]. The ability to stabilize and interact with the natural tear film is a property unique to hyaluronan, one that apparently results in an increase in the bioavailability of topical ophthalmic medication [86,88].

The most commonly studied hyaluronan delivery system is a pilocarpine-hyaluronan vehicle [86,88]. In one study, a tear sampling technique was used to compare the radioactivity in the tear fluid after instillation of [^{3}H]-pilocarpine solutions in the presence and absence of hyaluronan. It was found that the radioactivity in the tear fluid decreased more slowly from the hyaluronan-[^{3}H]-pilocarpine solution than from the nonviscous [^{3}H]-pilocarpine solutions. In the presence of 0.125% hyaluronan (Healon), the corneal residence time of pilocarpine was doubled and the miotic response was increased proportionately [86,88].

The hylans exhibit biological compatibility that is similar, and often identical, to the parent HY molecule. The physical and rheological properties of hylans, however, are enhanced (hylan fluid) or significantly altered (hylan solids such as gels, membranes, etc.) as compared with native hyaluronan. This physical diversity is a great advantage in the design and manufacture of mono- and bifunctional hylan matrix delivery systems. Bifunctional delivery vehicles are those in which the vehicle has two roles: one as a controlled-release system and another that is a function of its physical properties. A bifunctional hylan delivery system may be a wound dressing material in which the hylan matrix provides a biocompatible, hydrophilic, tissue-contacting surface while delivering a therapeutic agent (i.e., antibiotic or growth factor). The physical diversity of hylan polymers facilitates control of the residence times of the various hylans in specific tissue compartments and also enables control of drug-release kinetics.

Hylan gel and hylan gel-fluid vehicles for gentamicin (a broad spectrum antibiotic) delivery have been developed and evaluated in *in vitro* and *in vivo* models. Controlled-release forms of gentamicin would benefit most applications of gentamicin therapy, including ophthalmic and wound therapy. Preliminary *in vitro* testing indicates that hylan gel has a dramatic effect on the release kinetics of [^{125}I]-gentamicin. In the control (without hylan gel), more than 80% of the gentamicin was released during the first hour, while release of gentamicin from a hylan gel matrix was greatly reduced and characterized by a gradual, sustained elution of gentamicin, requiring approximately 20 hours to achieve 90% release. Ionic interactions between the cationic aminoglycoside and anionic hylan play an important role in determining release kinetics since, in the presence of high salt concentration, release kinetics of gentamicin from hylan gel were identical to those of the control.

In another *in vitro* system, the effectiveness of sustained gentamicin release was demonstrated in a microbiological assay. Hylan gel-gentamicin samples were found to sustain bactericidal activity for over 50 days, while the corresponding gentamicin control (no hylan gel) lost activity after 1–3 days. Hence, hylan gel provides a means of dramatically altering release kinetics without affecting biological activity of the gentamicin molecule.

In a rabbit model, a hylan gel-fluid-gentamicin vehicle was shown to provide superior retention of gentamicin on the corneal surface [K. Green, R. Stone, and T. Reidhammer, 1988, personal communication]. The hylan vehicle was compared with a commercial preparation of gentamicin sulfate. These results demonstrate that hylan gel-fluid greatly increases corneal contact time, thereby maximizing the time available for drug action and possibly reducing the number of dosings that may be required for effective drug therapy. As in the topical ophthalmic hyaluronan systems, it is likely that the pseudoplasticity of hylan gel-fluid contributes greatly to the ability of this material to provide superior corneal contact time; hylan gel-fluids shear "thin" at higher shear rates (i.e., blinking eyelid) and regain original viscosity under low shear (open eye), allowing the material to remain as a thin, stable film on the eye during blinking instead of being pushed off the ocular surface by the movement of the lid. Also, the hydrophilicity of the hylan polymers facilitates good tissue contact, optimizing drug-tissue interaction. The presence of hylan gel particles may act as a drug "reservoir," thereby sustaining higher concentrations of gentamicin on the corneal surface.

Pilocarpine and betaxolol are agents used in the management of the elevated intra-ocular pressure associated with glaucoma. The route of choice for these drugs is direct introduction to the ocular surface or instillation of a drug solution into the conjunctival sac. The site of action of these agents is in the internal eye, which therefore requires drug diffusion across the corneal membrane. Generally, the efficiency of absorption is very low, resulting in a limited or short duration of effect [89]. Corneal absorption is low because precorneal clearance rates are so high and drug is removed from the productively absorbing corneal surface within the first 5 minutes [90–92]. To decrease the rate of fluid removal and turnover from the precorneal space, viscous substances have been included in ophthalmic vehicles that increase corneal contact time and hence intraocular absorption efficiency [78,79]. Though viscous vehicles provided improved results, viscoelastic hyaluronan has been shown to be superior [86].

Several different hylan matrices have been used to improve the delivery of pilocarpine and betaxolol. Hylan gel matrices were loaded with pilocarpine and *in vitro* release kinetics were characterized by a threefold increase in the time required for 50% of the drug to be released, minimization of the undesirable initial burst, and sustained release of the drug for a much greater period of time as compared with the control.

Betaxolol was combined with a chemically modified hylan fluid; sulfate groups were incorporated into the soluble hylan polymer in order to increase the density of negative charge. Increased negative charge produced correspondingly prolonged release rates for betaxalol, indicating that the modification resulted in enhanced interaction between the drug and hylan polymer. The preparation of this vehicle is an example of the use of further chemical modification of the hylan polymer as a means of influencing drug-release kinetics.

Interferon is a protein ($Mr = 18,000$) with antiviral and antiproliferative properties that has been approved for treatment of hairy cell leukemia, genital warts, and Kaposi's sarcoma. *In vitro* assays performed using hylan vehicles loaded with alpha-interferon (IFN, Schering Plough Research, Kenilworth, NJ) indicate that hylan gel-fluid mixtures (hylan G-F 80) and hylan fluid have a substantial effect on the release period of interferon. In the presence of hylan vehicles, there was up to an eightfold reduction in the levels of IFN released over a 24-hour period, as compared with controls not containing hylan. Hylan IFN delivery systems exhibited a reduced "burst" effect in that the amount of IFN released during the initial 1–2 hour period was significantly decreased. The precise nature of the interaction between IFN and hylan is unknown. Associations between proteins and peptides have been described [93,94], and it may be that a specific binding association occurs between hylan and IFN. Release of IFN from the hylan matrix may also be mediated by ion exchange processes and restricted diffusion. Subcutaneous administration of hylan/IFN systems in a rat model resulted in reduced serum levels and presumably higher local levels of IFN as compared with control IFN vehicles. The differences were most notable during the first 24–48 hours. Hylan-IFN therapy may be enhanced by sustaining increased local concentrations of IFN due to controlled or restricted release from a hylan matrix.

Hyaluronan and hylan vehicles are well suited for use in the development of effective delivery systems for pharmacologically active molecules. Hyaluronan-based vehicles offer unique and unmatched biological compatibility, physical and rheological properties, and chemical and physical versatility. This is an area of active research and one with the potential to provide a new approach to administration of a wide variety of pharmaceutics.

F. Percutaneous Embolization

Percutaneous embolization has been used effectively in the treatment of vascular lesions, aneurysms, arteriovenous malformations, and tumors [95–97]. Materials currently available often provide inadequate occlusion, are difficult with which to work, and have poor biocompatibility. Hylan gels for embolization have been developed in which hylan gel forms the matrix for the various components (tantalum, microcrystalline cellulose, hexamethonium chloride [HMC], and thrombin) and provides the physical and biological properties needed for delivery of this material through small-lumen catheters and for tissue compatibility [98].

In this specific application, local intravascular delivery of thrombin is achieved in order to induce rapid, local clot formation by catalyzing the polymerization of fibrinogen to fibrin. Release of HMC is also required to provide local neutralization of heparin, a potent anticoagulant present in the circulation of patients undergoing this medical procedure. Systemic exposure to significant doses of HMC and thrombin is not desirable. A rabbit model was used for the evaluation of this material (auricular artery). Injection of this material intravascularly immediately results in occlusion of the blood vessel; photomicrogaphs of tissue taken 1 week, and 1, 3, and 6 months after injection show an artery occluded initially by injected gel and fibrin and subsequently by increasing fibrotic tissue (Fig. 2). Analysis of blood samples taken pre- and postinjections (1 and 24 hours) revealed no elevation or change in levels of thrombin or HMC, suggesting that systemic exposure to HMC and thrombin was insignificant.

The results from preliminary *in vivo* testing demonstrate that the hylan gel embolization agent forms complete and long-lasting arterial blockage; effective, rapid, local occlusion is due to the combined actions of HMC release (local inhibition of heparin), thrombin release (local catalysis of fibrin formation), and physical obstruction (presence of viscoelastic hylan gel). Hylan gels for embolization may be preloaded with additional active ingredients for the local controlled release of substances with systemic effects that

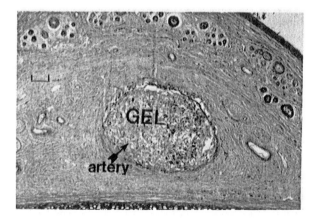

Figure 2 Photomicrograph of occluded artery 1 month after embolization procedure in rabbits. Occlusion is achieved by the gel material and inflammatory cells (bar scale in upper left represents 100 μm).

may be toxic or may interfere with their primary effect on target tissues, for example, immunomodulators or antineoplastic drugs.

G. Topical Applications

Hyaluronan is a critical determinant of the biophysical properties of skin [99,100]. In the dermis, it forms the fluid matrix that embeds collagen and elastin fibers and influences the transport of water, metabolites, and numerous bioactive species controlling cellular function. Hyaluronan has been shown to determine skin turgor (hydration) [101,102], provide viscoelasticity and shock absorption [103,104], and to play an important role in tissue regeneration during healing [105,106]. More recently. its important role in epidermal function has likewise been recognized [107,108].

Early studies on the therapeutic applications of hyaluronan demonstrated that it could control the formation of granulation and scar tissue [40]. Small-scale clinical studies demonstrated its utility for the treatment of nonhealing wounds [109], and a topical hyaluronan ointment for wounds has been widely used in Italy for many years (Connectivina, Fidia SPA).

Widespread use of hyaluronan in topical treatment products was limited by difficulties associated with the production of high molecular weight preparations that maintained the elastoviscosity and matrix-forming properties of the skin's endogenous hyaluronan. Furthermore, it was necessary to develop preparations that were customized for particular topical applications. The first internationally successful topical use of hyaluronan came in the form of a treatment product for normal skin. The hyaluronan preparation developed for this application was based on the association of the hyaluronan with a complex mixture of skin peptides, lipids, and oligosaccharides [110]. This principal of utilizing high molecular weight hyaluronan that has been modified or complexed in order to augment and customize its functionality has become fundamental to all its topical applications.

A wide variety of hyaluronan preparations are now available for topical usage. The technology for its large-scale production from both tissue and bacterial sources has been refined and expanded, and the cost associated with its effective utilization has been significantly reduced. The differences among the available preparations and the principles underlying their utilization in different product categories can be briefly outlined.

1. Importance of High Molecular Weight

The importance of high molecular weight to the matrix-forming properties of hyaluronan, and, consequently to its ability to form a smoothening viscoelastic matrix on the surface of skin, is widely acknowledged [111,112]. Table 1 utilizes profilimetry to compare two identical HY preparations, differing only in their molecular weight. It is clear that the preparation with a molecular weight of 3 million smoothens the cutaneous relief significantly better than the preparation with a molecular weight of 1 million.

More recently, a production technology has been developed that introduces a small number of cross-links into the HY molecule during processing and therefore produces soluble HY molecules with molecular weights in the range of 5 to 25 million [17]. This is especially important because the higher the molecular weight of an HY preparation is, the lower the concentration necessary to form an HY matrix will be, and hence the lower the actual cost for its effective utilization in a formulation will be [113]. Furthermore, the

Table 1 Smoothening of the Cutaneous Relief by Hyaluronan – Effect of Molecular Weight

	Percent improvement	
	MW = 1 million	MW = 3 million
RA (mean surface roughness)	15.6% (p = 0.03)	41.2% (p < 0.01)
RT (depth of roughness)	13.6% (N.S.)[a]	30.6% (p < 0.01)
RP (depth of smoothness)	12.2% (N.S.)[a]	37.2% (p < 0.005)
RZ (mean depth of roughness)	20.0% (p < 0.005)	36.0% (p < 0.005)

Profilimetry was performed on silicone replicas of skin sites to which 0.01 ml of a 1 mg/ml solution of hyaluronan had been applied. The p statistic reported here compares the calculated profilimetric parameters for contralateral treated and untreated sites on the same individual, average over all the subjects used in the study.
[a]N.S. = not significant

matrix formed by lower concentrations of a high molecular weight HY feels better on the skin surface because it minimizes stickiness and the perception of an unnatural film.

2. Synergistic Interactions

A second major principle underlying the topical utilization of hyaluronan is based on its ability to act synergistically with other polymers [114–116]. Synergistic interactions are evident from rheological measurements as well as from evaluation of specific functional parameters. Table 2 shows that, for some systems, the viscosity and elasticity of hyaluronan-polymer complexes can be enhanced as much as 10-fold higher than the additive

Table 2 Rheological Synergy Displayed by Combinations of Hyaluronan with Other Polymers

	Shear viscosity (cps)
2% polyquaternium 24	1,270[a]
0.1% hylan	2,020[a]
Additive	3,290
Measured	33,100[a]
Rheological enhancement	10-fold
0.5% polyox	3,571[b]
0.5% hylan	241,600[b]
Additive	245,171
Measured	897,100[b]
Rheological enhancement	3-fold
20% bovine serum albumin	BDL[c]
0.1% hylan	130[c]
Additive	130
Measured	279[c]
Rheological enhancement	2-fold

[a]Viscosity measured at a shear rate of 0.05 s^{-1}
[b]Viscosity measured at a shear rate of 0.001 s^{-1}
[c]Viscosity measured at a shear rate of 0.02 s^{-1}

rheological properties of the individual polymers. Hylan-polyquaternium 24 complexes have additional functionality due to their substantivity, ability to emulsify lipophilic molecules, and their provision of a supradermal viscoelastic matrix that can improve the uniform delivery of active agents such as sunscreens [117]. Synergistic functionality is also evident in complexes of hylan, albumin, and dextran sulfate. These form a surface matrix with optical properties that improve skin appearance over and above their physical smoothing effects [118].

3. Wound Healing

Hylan matrices have also been used as wound dressings and connective tissue replacement matrices. They have been tested in a wide range of physical forms and compositions: fluids, solids, gels, slurries, absorbent powders, and in combination with other polymers [119]. Several general principles are evident from these studies.

Hylans form an unusually biocompatible interface and provide a physical environment optimized for tissue regeneration. They can serve as a replacement matrix for excisional full-thickness wounds, and have been found to speed healing, reduce scarring, and minimize pain and discomfort [120]. They can also localize and prolong the delivery of wound active agents [77]. By combining and controlling these attributes, hylans can be formulated into materials useful for treating a wide variety of topical wounds.

III. SUMMARY AND CONCLUSIONS

The therapeutic applications of hyaluronan and its derivatives have become firmly established in the medical armamentorium. Elastoviscous NIF-NaHY solutions are widely used as viscosurgical tools and short-acting viscosupplements. Numerous preparations are now available from commercial sources, differing significantly in their physical properties and therapeutic efficacy. The hylan family of cross-linked hyaluronan derivatives expands hyaluronan's applications beyond those of NIF-NaHY. Hylans can be customized to particular medical needs in ways not possible for the nonderivitized hyaluronan molecule. They are available in a wide range of physical forms (fluids, gels, and solids), have enhanced rheological properties and a controllable prolonged residence time. These factors broaden their application and enable therapeutic devices to be customized to particular medical needs. The inherent biocompatibility of hyaluronan and hylan, in combination with their unique physical properties, ensures their continued and expanding therapeutic utilization.

REFERENCES

1. Balazs, E.A. Ultra Pure Hyaluronic Acid and the Use Thereof. U.S. Patent #4,141,973. Feb. 27, 1979.
2. Balazs, E.A., Freeman, M.I., Kloti, R., Meyer-Schwickerath, G., Regnault, F., and Sweeney, D.B. Hyaluronic acid and the replacement of vitreous and aqueous humour. In *Modern Problems in Ophthalmology (Secondary Detachment of the Retina)*, vol. 10, E.B. Streiff (Ed.), Karger, Basel, pp. 3–21 (1972).
3. Balazs, E.A., and Denlinger, J.L. Clinical uses of hyaluronan. In *The Biology of Hyaluronan* (Ciba Foundation Symposium #143), D. Evered and J.J. Whelan (Eds.), Wiley, Chichester, pp. 265–280 (1989).

4. Balazs, E.A. The physical properties of synovial fluid and the special role of hyaluronic acid. In *Disorders of the Knee*, A. Helfel (Ed.), Lippincott, Philadelphia, pp. 63–74 (1982).

5. Balazs, E.A., and Denlinger, J.L. The role of hyaluronic acid in arthritis and its therapeutic uses. In *Osteoarthritis—Current Clinical and Fundamental Problems*, J.G. Peyron (Ed.), Geigy, Rueil-Malmaison, France, pp. 165–174 (1985).

6. Balazs, E.A., and Leshchiner, E.A. Hyaluronan, Its cross-linked derivative—HYLAN—and their medical applications. In *Cellulosics Utilization: Research and Rewards in Cellulosics*. Proceedings of Nisshinbo International Conference on Cellulosics Utilization in the Near Future, H. Inagaki and G.O. Phillips (Eds.), Elsevier Appl. Science, New York, pp. 233–241 (1989).

7. Gibbs, D.A., Merrill, E.W., Smith, K.A., and Balazs, E.A. The rheology of hyaluronic acid. *Biopolymers*, 6:777–791 (1968).

8. Laurent, T.G., and Fraser, J.R.E. Hyaluronan. *FASEB J.*, 6:2397–2404 (1992).

9. Balazs, E.A., and Darzynkiewicz, Z. The effect of hyaluronic acid on fibroblasts, mononuclear phagocytes and lymphocytes. In *Biology of the Fibroblast* (papers of the symposium held in Turku, Finland, 1972), E. Kulonen and J. Pikkarainen (Eds.), Academic Press, London, pp. 237–252 (1973).

10. Balazs, E.A., Hogberg, B., and Laurent, T.C. The biological activity of hyaluron sulfuric acid. *Acta Physiol. Scand.*, 23:168–178 (1951).

11. Hadidian, Z. Sulfuric Acid Esters of Hyaluronic Acid and Process for Producing Therof. U.S. Patent #2,599,172. Jun. 3, 1952.

12. Laurent, T.C., Helsing, K., and Gelotte, B. Cross-linked gels of hyaluronic acid. *Acta Chem. Scand.*, 18(1):274–275 (1964).

13. Balazs, E.A., and Leshchiner, A. Water Insoluble Preparations of Hyaluronic Acid and Processes Thereof. U.K. Patent, GB 2,151,244B. July 17, 1985.

14. Balazs, E.A., and Leshchiner, A. Cross-Linked Gels of Hyaluronic Acid and Products Containing Such Gels. U.S. Patent #4,605,691. Aug. 12, 1986.

15. Balazs, E.A., and Leshchiner, A. Cross-linked Gels of Hyaluronic Acid and Products Containing Such Gels. U.S. Patent #4,582,865. Apr. 15, 1986.

16. Balazs, E.A., and Leshchiner, A. Cross-linked Gels of Hyaluronic Acid and Products Containing Such Gels. U.S. Patent #4,636,524. Jan. 13, 1987.

17. Balazs, E.A., Leshchiner, A., Leshchiner, A., and Band, P. Chemically Modified Hyaluronic Acid Preparation and Method of Recovery Thereof from Animal Tissues. U.S. Patent #4,713,448. Dec. 15, 1987.

18. Balazs, E.A., Leshchiner, A., Leshchiner, A., Larsen, N., and Band, P. Hylan Preparation and Method of Recovery Thereof from Animal Tissues. U.S. Patent #5,099,013. Mar. 24, 1992.

19. Sakurai, K., Uno, Y., and Okuyama, T. Cross-Linked Hyaluronic Acid and Its Use. U.S. Patent #4,716,224. Dec. 29, 1987.

20. Malson, T., and Lindqvist, B.L. Gel of Cross-Linked Hyaluronic Acid for Use as a Vitreous Humor Substitute. U.S. Patent #4,716,154. Dec. 29, 1987.

21. Malson, T., and Lindqvist, B. Cross-Linked Hyaluronate Gels, Their Use and Method for Producing Them. WO #90/09401. Aug. 23, 1990.

22. della Valle, F., and Romeo, A. Esters of Hyaluronic Acid. U.S. Patent #4,851,521. July 25, 1989.

23. della Valle, F., and Romeo, A. Esters of Hyaluronic Acid. U.S. Patent #4,957,744. Sep. 18, 1990.

24. della Valle, F., and Romeo, A. Polysaccharide Esters and Their Salts. U.S. Patent #4,965,353. Oct. 23, 1990.

25. Malson, T. Material of Polysaccharides Containing Carboxyl Groups, and a Process for Producing Such Polysaccharides. U.S. Patent #4,963,666. Oct. 16, 1990.

26. deBelder, A.N., and Malson, T. Method of Preventing Adhesion Between Body Tissues,

Means for Preventing Such Adhesion, and Process for Producing Same Means. U.S. Patent #4,886,787. Dec. 12, 1989.

27. Sheehan, J.C., and Hlavka, J.J. The use of water soluble and basic carbodiimides in peptides synthesis. *J. Org. Chem.*, 21:439 (1956).

28. Hoare, D.G., and Koshland, D.E., Jr. A method for the quantitative modification and estimation of carboxylic acid groups in proteins. *J. Biol. Chem.*, 242:2447–2453 (1967).

29. Khorana, H.G. The chemistry of carbodiimides. *Chem. Rev.*, 53:145 (1953).

30. Danishevsky, I., and Siskovic, E. Conversion of carboxyl groups of mucopolysaccharides into amides of amino acid esters. *Carbohyd. Res.*, 16:199–205 (1971).

31. Sparer, R.V., Ekwuribe, N., and Walton, A.G. Controlled release from glycosaminoglycan complexes. In *Controlled Release Delivery Systems* (Outgowth of the 8th International Symposium on Controlled Release of Bioactive Materials held July 26–29, 1981, in Fort Lauderdale, FL), Marcel Dekker, New York, pp. 107–120 (1981).

32. Prestwich, G.D., and Swann, D.A. N-acylurea and O-acylisourea Derivatives of Hyaluronic Acid. European Patent Application #0416250 A2. Mar. 13, 1991.

33. Hamilton, R., Fox, E.M., Acharya, R.A., and Walts, A.E. Water-Insoluble Derivatives of Hyaluronic Acid. U.S. Patent #4,937,270. June 26, 1990.

34. Burns, J.W., Cox, S., and Waltz, A.E. Water-Insoluble Derivatives of Hyaluronic Acid. U.S. Patent #5,017,229. May 21, 1991.

35. Balazs, E.A. Sodium hyaluronate and viscosurgery. In *Healon (Sodium Hyaluronate): A Guide to Its Use in Ophthalmic Surgery*, D. Miller and R. Stegmann (Eds.), Wiley & Sons, New York, pp. 5–28 (1983).

36. Balazs, E.A. The development of sodium hyaluronate (Healon) as a viscosurgical material in ophthalmic surgery. In *Ophthalmic Viscosurgery—A Review of Standards, Techniques and Applications*, G. Eisner (Ed.), Medicopea, Montreal, pp. 3–19 (1986).

37. Balazs, E.A. The introduction of elastoviscous hyaluronan for viscosurgery. In *Viscoelastic Materials: Basic Science and Clinical Applications*, E. Rosen (Ed.), Pergamon Press, Oxford, pp. 149–165 (1988).

38. Balazs, E.A., and Gibbs, D.A. The rheological properties and biological function of hyaluronic acid. In *Chemistry and Molecular Biology of the Intercellular Matrix*, E.A. Balazs (Ed.), Academic Press, London, pp. 1241–1254 (1970).

39. Rydell, N.W., Butler, J., and Balazs, E.A. Hyaluronic acid in synovial fluid. VI. Effect of intraarticular injection of hyaluronic acid on the clinical symptoms of arthritis in track horses. *Acta Vet. Scand.*, 11:139–155 (1970).

40. Rydell, N.W., and Balazs, E.A. Effect of intra-articular injection of hyaluronic acid on the clinical symptoms of osteoarthritis and on granulation tissue formation. *Clin. Orthop.*, 80:25–32 (1971).

41. Balazs, E.A., and Denlinger, J.L. Sodium hyaluronate and joint function. *J. Equine Vet. Sci.*, 5:217–228 (1985).

42. Helfet, A.J. Management of osteoarthritis of the knee joint. In *Disorders of the Knee*, A.J. Helfet (Ed.), Lippincott, Philadelphia, p. 179 (1974).

43. Peyron, J.G., and Balazs, E.A. Preliminary clinical assessment of Na-hyaluronate injection into human arthritic joint. *Pathol. Biol.*, 23:731–736 (1974).

44. Weiss, C., Balazs, E.A., St. Onge, R., and Denlinger, J.L. Clinical studies of the intraarticular injection of Healon® (sodium hyaluronate) in the treatment of osteoarthiitis of human knees. In *Seminars in Arthritis and Rheumatism*, vol. 11, J.H. Talbot (Ed.), Grune and Stratton, New York, pp. 143–144 (1981).

45. Namiki, O., Toyoshina, H., and Mor, N. Therapeutic effect of intra-articular injection of high molecular weight hyaluronic acid on osteoarthritis of the knee. *Int. J. Clin. Pharm. Ther. Toxicol.*, 20(1):501–507 (1982).

46. Bragantini, A., Cassini, M., De Bastiani, G., and Perbollini, A. Controlled single blind trial

of intra-articularly injected hyaluronic acid (Hyalgan®) in osteoarthritis of the knee. *Clin. Trials J.*, 24:333–340 (1987).

47. Dixen, A.S.J., Jacoby, R.K., Berry, H., and Hamilton, E.B.D. Clinical trials of intra-articular injection of sodium hyaluronate in patients with osteoarthritis of the knee. *Curr. Med. Res. Opin.*, 11:205 (1988).

48. Phillips, M.W. Clinical trial comparison of intra-articular sodium hyaluronate products in the horse. *J. Eq. Vet. Sci.*, 9:39–40 (1989).

49. Adams, M.E. An analysis of clinical studies of the use of crosslinked hyaluronan, hylan, in the treatment of OA. *J. Rheumatol.*, 20(Suppl. 39):16–18 (1993).

50. Balazs, E.A., and Denlinger, J.L. Viscosupplementation: A new concept in the treatment of OA. *J. Rheumatol.*, 20(Suppl. 39):3–9 (1993).

51. Pelletier, J.P., and Martel-Pelletier, J. The pathophysiology of OA and the implication of the use of hyaluronan and hylan as therapeutic agents in viscosupplementation. *J. Rheumatol.*, 20(Suppl. 39):19 (1993).

52. Peyron, J.G. Intra-articular hyaluronan injections in the treatment of OA: State-of-the-art review. *J. Rheumatol.*, 20(Suppl. 39):10–15 (1993).

53. Larsen, N.L., Lombard, K., Parent, E.G., and Balazs, E.A. Effect of hylan on cartilage and chondrocyte cultures. *J. Orthoped. Res.*, 10:23–32 (1992).

54. Greenwald, R.A. Oxygen radicals, inflammation, and arthritis: Pathophysiological considerations and implications for treatment. *Seminars in Arthritis and Rheumatism*, 20:219–240 (1991).

55. Pozo, M.A., Balazs, E.A., and Belmonte, C. Reduction of sensory responses to passive movements of inflamed knee joints by hylan, a hyaluronan derivative. (to be published) (1995).

56. Gotoh, S., Miyazaky, K., Onaya, J., Sakamoto, T., Tokuyasu, K., and Namiko, O. Experimental knee pain model in rats and analgesic effect of sodium hyaluronate (SPH). *Folia pharmacp.*, *Japan*, 92:17–27 (1988).

57. Miyazaki, K., Gotoh, S., Ohkawara, H., and Yamaguchi, T. Studies of analgesic and anti-inflammatory effects of sodium hyaluronate (SPH). *Pharmacometrics*, 28:1123–1135 (1984).

58. Denlinger, J.L. Metabolism of sodium hyaluronate in articular and ocular tissues. Ph.D. thesis, Universite des Sciences et Techniques de Lille, Lille, France (1982).

59. Balazs, E.A., Rydell, N.W., and Freeman, M.I. Effect of hyaluronic acid on adhesion formation. In *Hyaluronic Acid and Matrix Implantation*, 2nd ed., E.A. Balazs (Ed.), Biotrics Inc., Arlington, MA, Appendix 13 (1971).

60. St. Onge, R., Weiss, C., Denlinger, J.L., and Balazs, E.A. A preliminary assessment of Na-hyaluronate injection into "No Man's Land" for primary flexer tendon repair. *Clin. Orthoped. Rel. Res.*, 148:351–357 (1980).

61. Weiss, C., Levy, H., Denlinger, J.L., Suros, J., and Weiss, H. The role of Na-hylan in reducing postsurgical tendon adhesions. *Bull. Hosp. Jt. Dis. Orthop. Inst.*, 46:9–15 (1987).

62. Weiss, C., Suios, J., Michalow, A., Denlinger, J.L., Moore, M., and Tejeiro, W. The role of Na-hylan in reducing postsurgical tendon adhesions. Part 2. *Bull. Hosp. Jt. Dis. Orthop. Inst.*, 47:31–39 (1987).

63. Green, S., Szabo, R., Langa, V., and Klein, M. The inhibition of flexer tendon adhesions. *Bull. Hosp. Jt. Dis. Orthop. Inst.*, 46:16–21 (1986).

64. Meyers, S.A., Seaber, A.V., Glisson, R.R., and Nunley, J.A. The effect of hyaluronic acid on healing of full thickness tendon lacerations in rabbits. Orthop. Res. Soc., 33rd Ann. Mtg., Jan. 19–22, San Francisco, CA (1987).

65. Thomas, S.C., Jones, L.C., and Hungerford, D.S. Hyaluronic acid and its effect of postoperative adhesions in the rabbit flexer tendon. *Clin. Orthop. and Rel. Res.*, 206:281–289 (1986).

66. Amiel, D., Viindeberg, J., Gelberman, F.R., Ishizue, K.K., Sisk, A., and Akeson, W.H. The use of hyaluronic acid in flexer tendon repairs. Orthop. Res. Soc., 34th Ann. Meeting., Feb. 1–4, Atlanta, GA (1988).

67. Burns, J.W. Prevention of post-surgical adhesion formation with sodium hyaluronate-based products. Wound Healing III, January 20–21, Orlando, FL (1992).

68. Skinner, K.C., Colt, M.J., Zerbus, L.A., Kirk, J.F., and Burns, J.W. HAL-F™ Film – A new hyaluronic acid based bioabsorbable barrier for the prevention of abdominal adhesions. Acad. of Surg. Res., 7th Ann. Meeting, September 26–29, Paradise Valley, AZ (1991).

69. Songer, M.N., Rausching, W., Carson, E., Pandit, S.M., and Spencer, D.L. The study of postoperative scar formation following lumbar laminotomy and discetomy in dogs. *Spine, 5:* 550–554 (1990).

70. Weiss, C., Dennis, J., Suros, J.M., Denlinger, J.L., Badia, A., and Montane, I. Sodium hylan for the prevention of postlaminectomy scar formation. Orthop. Res. Soc., 35th Ann. Meeting, February 6–9, (1989).

71. Matton, G., Anseeuw, A., and DeKeyser, F. The history of injectable biomaterials and the biology of collagen. *Aesth. Plast. Surg.*, 9:133–140 (1985).

72. Knapp, J.R., Kaplan, E.N., and Daniel, J.R. Injectable collagen for soft tissue augmentation. *Plast. Reconstr. Surg.*, 60:398–405 (1977).

73. Balazs, E.A., Denlinger, J.L., Leshchiner, E., Band, P., Larsen, N.L., Leshchiner, A., and Morales, B. Hylan: Hyaluronan derivatives for soft tissue repair and augmentation. In *Biotech USA 1988*, Conference Management Corp., Norwalk, CT, pp. 442–451 (1988).

74. Larsen, N.L., Kling, M.B., Balazs, E.A., and Leshchiner, E.A. Hylan gel for soft tissue augmentation. Society for Biomaterials, 16th Ann. Meeting, May 20–23, Charleston, SC, p. 302 (1990).

75. Balazs, E.A. Viscoelastic properties of hyaluronic acid and biological lubrication. Symposium on Prognosis for Arthritis, Rheumatology Research Today and Prospects for Tomorrow, Ann Arbor, MI, 1967, *University of Michigan Medical Center Journal*, (Suppl.):255–259 (1968).

76. Balazs, E.A. The introduction of elastoviscous hyaluronan for viscosurgery. In *Viscoelastic Materials, Basic Science and Clinical Applications*, E.S. Rosen (Ed.), Proceedings of the Second International Symposium of the Northern Eye Institute, Manchester, July 17–19, 1986.

77. Larsen, N.L., and Balazs, E.A. Drug delivery systems using hyaluronan and its derivatives. *Adv. Drug. Del. Rev.*, 71:279–293 (1991).

78. Chrai, S.S., and Robinson, J.R. Ocular evaluation of methyl cellulose vehicles in albino rabbits. *J. Pharm. Sci.*, 63:1218–1221 (1974).

79. Patton, T.P., and Robinson, J.R. Ocular evaluation of polyvinyl alcohol. *J. Pharm. Sci.*, 64:1312–1316 (1975).

80. Saettone, M.P., Giannaccini, B., Teneggi, A., Savigni, P., and Tellini, N. Vehicle effects on ophthalmic bioavailability: The influence of different polymers on the activity of pilocarpine in rabbit and man. *J. Pharm. Pharmacol.*, 34:464–466 (1982).

81. Saettone, M.F., Giannaccini, B., Ravecca, S., LaMarca, F., and Tota, G. Polymer effects on ocular bioavailability – The influence of different liquid vehicles on the mydriatic response of tropocamide in humans and in rabbits. *Int. J. Pharm.*, 20:187–202 (1984).

82. Polack, F.M., and McNiece, M.T. The treatment of dry eyes with Na-hyaluronate (Healon®), *Cornea*, 1:133–136 (1982).

83. DeLuise, V.P., and Peterson, W.S. The use of topical Healon® tears in the management of refractory dry-eye syndrome. *Ann. Ophthalmol.*, 16:823–824 (1984).

84. Stuart, J.C., and Linn, J.G. Dilute sodium hyaluronate (Healon®) in the treatment of ocular surface disorders. *Ann. Ophthalmol.*, 17:190–192 (1985).

85. Sand, B.B., Marner, K., and Norn, M.S. Sodium hyaluronate in the treatment of keratoconjunctivitis sicca. A double masked clinical trial. *Acta Ophthalmol.*, 67:181–183 (1989).

86. Gurny, R., Ibrahim, H., Aebi, A., Burl, P., Wilson, C.G., Washington N., Edman, P., and Camber, O. Design and evaluation of controlled release systems for the eye. *J. Controlled Release*, 6:367–373 (1987).

87. Mengher, L.S., Pandher, K.S., Bron, A.J., and Davey, C.C. Effect of sodium hyaluronate (0.1%) on break-up time (NIBUT) in patients with dry eyes. *Brit. J. Ophthalmol.*, 70:442–447 (1986).

88. Camber, O., and Edman, P. Sodium hyaluronate as an ophthalmic vehicle: Some factors governing the effect of pilocarpine in rabbits. *Current Eye Res.*, 8:563–567 (1989).

89. Mikkelson and Thomas, *J. Ophthalmic Drug Delivery, Pharmaceutical Technology*, pp. 90–94 (1984).

90. Chrai, S.S., Patton, T.P., Mehta, A., and Robinson, J.R. Lacrimal and instilled fluid dynamics in rabbit eyes. *J. Pharm. Sci.*, 62:1112–1121 (1973).

91. Conrad, J.M., Reay, W.A., Polycyn, R.E., and Robinson, J.R. Influence of toxicity and pH on lacrimation and ocular drug bioavailability. *J. Parent. Drug. Ass.*, 32:149–161 (1978).

92. Sieg, J.W., and Robinson, J.R. Mechanistic studies on transcorneal permeation of pilocarpine. *J. Pharm. Sci.*, 65:1816–1822 (1976).

93. Hardingham, T.E., and Muir, H. The specific interaction of hyaluronic acid with cartilage proteoglycans. *Biochim. Biophys. Acta*, 279:402–405 (1972).

94. LeBoeuf, R.D., Raja, R.H., Fuller, G.M., and Weigel, P.H. Human fibrinogen specifically binds hyaluronic acid. *J. Biol. Chem.*, 261:12586–12592 (1986).

95. Luessenhop, A.J., Gibbs, M., and Velasquez, A.C. Cerebrovascular response to emboli. Observations in patient with arteriovenous malformations. *Arch. Neurol. (Chic.)*, 7:264–274 (1962).

96. Djindjian, R., Ciphignon, J., Theron, J., Merland, J.J., and Houdart, R. Embolization by superselective arteriography from the remoral route in neuroradiology, Review of 60 cases. *Neuroradiology*, 6:20–26 (1973).

97. Hilal, S.K., and Michelson, J.W. Therapeutic percutaneous embolization for extra-axial vascular lesions of the head, neck and spine. *J. Neurosurg.*, 43:275–287 (1975).

98. Larsen, N.E., Leshchiner, E.A., Parent, E.G., Hendrickson-Aho, J., Balazs, E.A., and Hilal, S.K. Hylan gel composition for percutaneous embolization. *J. Biomed. Mat. Res.*, 25:699–710 (1991).

99. Comper, W.D., and Laurent, T.C. Physiological function of connective tissue polysaccharides. *Physiol. Rev.*, 58:255–279 (1978).

100. Harkness, R.D. Functional aspects of the connective tissues of skin. In *Chemistry and Molecular Biology of the Intercellular Matrix*, vol. 3, E.A. Balazs (Ed.), Academic Press, New York, pp. 1309–1340 (1970).

101. Yates, J.R. Mechanism of water uptake by skin. In *Biophysical Properties of the Skin*, N.R. Eden (Ed.), Wiley Interscience, New York, pp. 485–502 (1971).

102. Uzuka, M., Nakajima, K., Ohta, S., and Mori, Y. The mechanism of estrogen induced increase in hyaluronate acid biosynthesis. *Biochem. Biophys. Acta*, 627:199–206 (1980).

103. Daly, C.H., and Odland, G.F. Age-related changes in the mechanical properties of human skin. *J. Invest. Derm.*, 73:84–87 (1979).

104. Dikstein, S., and Hartzstark, A. Effect of intradermal hyaluronidase on indentometry. *Bioengineering and Skin*, 4:48–51 (1982).

105. Alexander, S.A., and Donoff, R.B. The glycosaminoglycans of open wounds. *Journal Surgical Research*, 29:422–429 (1980).

106. Longaker, M.T., Chui, E.S., Harrison, M.R., Cromblehome, T.M., Langer, J.C., Duncan, B.W., Adzick, N.S., Veiner, E.D., and Stern, R. Studies in fetal wound healing IV. Hyaluronic acid – Stimulatory activity distinguished fetal wound fluid from adult wound fluid. *Ann. Surg.*, 210:667–672 (1989).

107. Tammi, R., Ripellino, J.A., Margolis, R.U., and Tammi, M. Localization of epidermal hyaluronic acid using the hyaluronate binding region of cartilage proteoglycan as a specific probe. *J. Invest. Derm.*, 90:412–414 (1988).

108. Piepkorn, M., Freeman, P., Carney, H., and Linker, A. Glycosaminoglycan synthesis by

proliferating and differentiated human keratinocytes in culture. *J. Invest. Derm.*, 88:215–219 (1987).

109. Borgonovo, E. Terapia delle sclerodermiti ulcerative con una pomata a base di acido ialuronico. *Gazette Medica Italiana*, 131:84–86 (1972).

110. Balazs, E.A. Hyaluronate Based Compositions and Cosmetic Formulations Containing Same. U.S. Patent #4,487,865 (1984).

111. Balazs, E.A., and Band, P.A. Hyaluronic acid: Its structure and use. *Cosmetics and Toiletries*, 99:65–72 (1984).

112. Hoshizaki, S., and Nakabuta, H. Application of water-holding polymers as a skin moisturizer. Proceedings of XII Congress of the International Federation of the Societies of Cosmetic Chemists. San Francisco, CA (1984).

113. Band, P. Effective use of hyaluronic acid. *Drug and Cosmetic Industry*, 54–57 (Oct. 1985).

114. Balazs, E.A., and Leshchiner, A. Hyaluronate-polyethylene oxide compositions and cosmetic formulations containing same. U.S. Patent #4,303,676 (1986).

115. Brode, G.L., Band, P.A., Goddard, E.D., Barbone, A.G., Leshchiner, A., Pavlichko, J.P., Partain, E.M., and Leung, P.S. Glycosaminoglycan and cationic polymer combinations. U.S. Patent #4,767,463 (1988).

116. Pavlichko, J. Polymer interactions to enhance the function of hyaluronic acid. *Drug and Cosmetic Industry*, 147:26–30 (1990).

117. Band, P.A., Brode, G.L., Goddard, E.D., Barbone, A.G., Leshchiner, E., Harris, W.C., Pavlichko, J.P., Pertain, E.M., and Leung, P.S. Interpolymer complexes between hyaluronan and cationic cellulose polymers. In *Cosmetic and Pharmaceutical Applications of Polymers*, C.G. Gebelein, F. Cheng, and U. Yang (Eds.), Plenum Press, New York, pp. 129–136 (1991).

118. Pavlichko, J., and Band, P.A. The science of minimizing wrinkles. *Soaps, Cosmetics and Chemical Specialties*, 69:33–36 (1992).

119. Balazs, E.A., Band, P.A., Denlinger, J.L., Larsen, N.L., Leshchiner, E.A., and Morales, B. Matrix engineering. *Blood Coagulation and Fibrinolysis*, 2:173–178 (1991).

120. Larsen, N.L., Dursema, H., Miller, K., and Balazs, E.A. Final report: Wound dressing with anti-adhesive antibiotic hyaluronic acid coating. SBIR Contract No. DAMD 17-86-6152, U.S. Army Medical Research Acquisition Activity (1986).

53

Biomaterials Involved in Cartilaginous Implants

Zvi Nevo

Tel Aviv University, Tel Aviv, Israel

Dror Robinson

Tel Aviv University, and Assaf Harofeh Medical Center, Zerifin, Israel

David G. Mendes

Bnai Zion Medical Center, Haifa, Israel

Nachum Halperin

Assaf Harofeh Medical Center, Zerifin, Israel

I. INTRODUCTION

Defects in the articular surfaces of joints are a common cause of disability. Lesions in cartilage might be caused by trauma, osteoarthritis (primary or secondary to metabolic disorders), and inflammatory arthritides. All of these entities share a common denominator of cartilage destruction, which is the end result of different destructive mechanisms. Reconstruction of the cartilage defects by biological implants is an enticing, albeit elusive, prospect. Such a reconstruction could be expected to deteriorate, in the case of the inflammatory arthritides, due to the basic disorder. However, in traumatic lesions, joint resurfacing could, in theory, offer a cure. Similarly, primary osteoarthritis progresses very slowly, and restoration of the articular cartilage could be expected to improve function for prolonged periods.

Current treatment includes either arthrodesis (fusion) of the joint, eliminating pain at the expense of obliterated function, or joint reconstruction using a metal-plastic implant. The latter option, while considered a spectacularly successful treatment, is not adequate for young people due to loosening of the construct. The use of a biological implant capable of healing and incorporating with the neighboring tissues could prevent the problem of loosening and offer a formidable technique for joint reconstruction.

It is beyond the scope of the current article to review in detail the essential needs for a cartilaginous implant. The needs are based on the known failure, beginning very early in life, of cartilage to heal itself upon damage [Buckwalter et al. 1988; Woo and Buckwalter 1988; Robinson, Halperin, and Nevo 1990b; Solursh 1991]. We first state our approach and basic philosophy, proclaiming that the biological resurfacing approach employing cultured cells containing implants is the most promising procedure for cartilage repair. This procedure is already at the clinical experimental stage and believed to be

1717

a replacement for the current procedures of osteochondral grafts [Yamashita et al. 1985; Oakeshott et al. 1988; Meyers, Akeson, and Convery 1989], perforations (drillings), abrasion, or shaving [Milgram 1985].

The procedure of employing skeletal tissue grafts (bone and osteochondral) was meant to restore and maintain the anatomic continuity as well as the mechanical function of the organ. The graft serves basically as a nonvital frame (template-scaffold) for the host cells to invade, populate, and produce matrix. Furthermore, such a graft is believed to release growth factors slowly, recruiting cells to migrate to the area, stimulating the so-called process of ingrowth. Accumulating evidence during the years indicates that such a graft is degraded and replaced by endogenous cells at a very slow pace, demonstrating a defined repair region that remains weak and poorly integrated for many years. In the worst cases, this foreign biological island even creates a lysis center, which is prone to fracture. Prosthetic devices, both metallic and plastic, share the pitfalls of grafts, do not integrate, and have a limited potential for ingrowth. Thus, anchorage is somewhat precarious.

To overcome the above drawbacks of the common orthopedic procedures, the employment of cultured cells embedded in an adhesive substance and impregnated within a scaffold as implants was introduced. Hence, the basic demands from such a structure are that it be elastic, though mechanically firm. Optimally, the scaffold and the adhesive should be biocompatible and biodegradable with a preplanned retention time. The accompanied substances to the cells should create a growth-permissive environment in which the implanted cells can propagate, adhere in the lesion site, and form the typical extracellular matrix components and structures. Finally, the cells' own matrices will replace the artificial implanted scaffolds and integrate properly with the neighboring cells and matrices.

Developments in several different scientific fields have created a new specialty, so-called tissue engineering. A new scientific and technical discipline is emerging for which biocompatible biodegradable adhesives and scaffolds serve as delivery and immobilizing vehicles for cells. These include naturally occurring as well as synthetic biopolymers used as scaffolds. These integrated components in one composite graft are called *in vitro* reconstructed implants.

II. BIOLOGICAL BASIS OF CARTILAGE TRANSPLANTS

Biologically, cartilage is a unique tissue in that it does not require vasculature for nourishment. Instead, it is bathed in synovial fluid, which is a specialized secretion of the synoviocytes lining the internal aspect of the joint capsule. The dependence on synovial fluid requires the cartilage to be thin and also limits the maximal cell density [Stockwell 1979]. In order to perform its lubricating function, cartilage secretes a matrix in which the cells are embedded, each in its own lacuna (chondron). This matrix is semipermeable, limiting access of antibodies and making the cells less antigenic than other somatic cells [Chalmers and Ray 1962; Elves 1974; McKibbin 1971; Malejczyk and Moskalewski 1988]. This defense, while it allows long-term incorporation of biological implants (see below), is not absolute and immune destruction was detected by some authors [Kawabe and Yoshinao 1991]. Embryonal cells are less immunogenic than adult-derived chondrocytes due to the production of anti-Ia (idiopathic antibodies) alloantibodies [Segal et al. 1979]. The limited immune response to cartilage allows the implantation of cartilage allografts without the need for HLA (human leukocytes antigens) matching or immune suppression.

Therefore, the cartilaginous implant represents many unique biological growth features due to the special requirements of the cartilage cells—the chondrocytes. Under the regular growth conditions in monolayers, these cells lose their phenotypic expression easily, undergoing dedifferentiation and attaining typical characteristics of fibroblasts. Chondrocytes emerge as non-anchorage-dependent cells, lacking contact inhibition and forming multilayers, nodulelike structure in monolayers. Furthermore, chondrocytes grow as single cells in suspension as well as in and on hydrated gels [Wakitani et al. 1989; Nevo et al. 1990], resembling the growth characteristics of transformed cells. Chondrocytes have been shown to attain and maintain a unique cellular homeostatic environment by activating a special combination of transporters [Dascalu, Nevo, and Korenstein 1993]. Chondrocytes have unique cytoskeleton elements, transmembranal cell-to-cell and cell-matrix interacting adhesive structures, rich in hyaluronic-acid-binding proteins, and extracellular ligands, differing from the common features of fibroblasts [Nevo et al. 1993c]. All the above information on cartilage cells should be considered and integrated in the design of a cartilaginous implant *in vitro*.

III. REQUIREMENTS FOR CONSTRUCTION OF A BIOLOGICAL IMPLANT

In order to be able to reconstruct articular surfaces with a biological implant, several requirements should be fulfilled. While cell culture is used routinely, reconstruction of tissue *in vivo* is difficult as tissue shape is often dependent on multiple *in vivo* factors. Furthermore, almost all body tissues require blood and nerve supply, making organ reconstruction next to impossible. Articular cartilage, which is devoid of either a blood supply or innervation, is uniquely suited for organ reconstruction. This contrasts with the situation in physis cartilage, which is both innervated and vascularized, making attempts at physis reconstruction futile [Cundy et al. 1991].

In order to reconstitute an articular cartilage implant *in vitro*, a delivery substance is necessary. Such a delivery substance should be biocompatible in such a way that chondrocyte growth is allowed and their phenotype maintained. Cartilage phenotype is notoriously unstable and dedifferentiation into fibroblastlike cells is common [Coon 1966]. Certain substrates are well known to be supportive of cartilaginous differentiation, for example, culture in and on soft gels [Harada et al. 1990; Nevo et al. 1990]. The mechanism of action appears to be prevention of cell attachment and subsequent rounding of the cells. Similar results can be obtained when the cells are grown in a fibrin gel. A more specific mechanism is involved in growing the cells in hyaluronate [Robinson et al. 1993b; Nevo et al. 1993a, 1993b]. Cartilage cells appear to possess hyaluronate receptors and thus the effect of hyaluronate is not only mechanical (prevention of attachment).

IV. SURVEY FOR AN OPTIMAL THREE-DIMENSIONAL GROWTH PERMISSIVE AND ADHESIVE MILIEU

The common monolayer cultures in which a layer of cells grow as a thin film under liquid medium do not match the requirements of a cartilaginous implant. First, this monolayer growth technique does not ensure the preservation of the cells' chondrogenic expression. Chondrogenic expression can be maintained safely only in mass cultures (very high initial cell density) [Nevo et al. 1990]. Second, the monolayer cultures are not in a good format for transportation, nor can such a cell film form the three-dimensional (3-D) tissue needed for an implant device. In the body, cells in cartilage are naturally embedded in massive amounts of extracellular matrix, found in the form of highly hydrated gels. Efforts to mimic the natural cartilage conditions can be obtained by growing chondrocytes in viscous adhesive gel substances. These substances can serve as a growth permissive milieu *in vitro*, as a delivery substance, as well as a biological resorbable immobilizing vehicle to the lesion sites for both cells and bioactive growth factors *in vivo*.

Our first choice for such a desired gel was the plasma clotting proteins, the fibrinogen-fibrin system (like the pioneering milieu-medium for cells in culture). This mixture contained the following constituents per milliliter: 120 mg fibrinogen; 5000 KIU aprotinin (a proteinase inhibitor); 100 millimoles (mM) $CaCl_2$; and 30 units (U) thrombin [Itay, Abramovici, and Nevo 1987; Feldman et al. 1988; Nevo, Robinson, and Halperin 1992]. In general, cells did survive in this mixture (inside CO_2-equilibrated incubators) for hours to a few days. In a detailed study of the effect of the different components on cells' survival and matrix production, it was learned that, with the exception of aprotinin, all the other components had a cytotoxic effect on the two parameters measured. Figure 1 demonstrates the cells in the fibrin gel, as well as reduction in cell numbers and the changes in cell morphology with time *in vitro* (indicating cytotoxicity).

Trials have also been made using an adhesive derived from an oyster for chondrocytes' transplantation [Grande and Pitman 1988]; however, we have no experience with this substance. As it is an alien protein, it can be expected to generate antibody production and immune destruction of the graft.

Hyaluronic acid (HY) emerged as a better noncytotoxic adhesive-viscous gel. It is both biodegradable and biocompatible. HY matches best the requirements of chondrocytes. In a gel made of 2% high molecular weight sodium hyaluronate (above 2×10^6 daltons, a product of Biotechnology General [BTG] Ltd., Kiryat Weizmann, Rehovot, Israel) in (1 : 1 volume to volume [v/v]) phosphate buffered saline at pH 7.2 and complete Ham F-12 medium containing glutamine and antibiotics, chondrocytes survived up to 30 days, either in a CO_2-enriched incubator (5% CO_2 in air at 37°C) or even outside the incubator on the laboratory bench at room temperature (Fig. 2). Tubes containing chondrocytes embedded in hyaluronic acid gel were left on the laboratory shelf for a month. Half of the tubes served as blanks, having been boiled for 5 minutes (min). Both groups, control and experimental, were incubated for 24 hours (h) with 10 μ Ci/ml

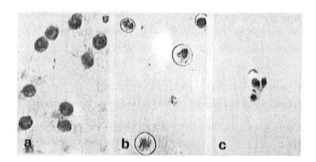

Figure 1   Chondrocytes' survival within fibrin-based adhesive milieu: (a), after 24 hours; (b), after 48 hours; (c), after 7 days (magnification ×630). Note the changes in cell number and cell shape.

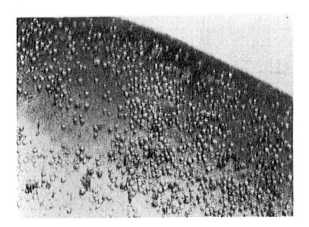

Figure 2   High density of chondrocytes cultured and grown in the hyaluronic-acid-based milieu for 30 days (magnification × 100). The tight density achieved indicates the growth-supportive nature of hyaluronic acid (compare with Fig. 1).

[35]S-carrier-free sulphate. Subsequently, glycosaminoglycans (GAGs) were isolated and radioactivity was measured. The results (cpm [counts per minute] ± SD [standard deviation]) show that the background level in the blanks of ^{35}S-isolated GAG is 494 ± 173 cpm/10^5 cells, while the counts in the experimental group show more than 10 times higher incorporation, 6541 ± 1438 cpm/10^5 cells.

In addition, hyaluronic-acid-based adhesive gel is serving in transportation overseas (Israeli patent #81080, granted Nov. 15th, 1992, to the authors), as well as a delivery substance for installing the cells into defects of articular cartilage. Experiments in animals employing allogeneic embryonal chondrocytes or autogeneic adult bone-marrow-derived chondrocytic-enriched cultures embedded in hyaluronic acid have yielded hyaline cartilage as the reparative tissue developed in the superficial layers of the implant. In the deeper zones of the defect, bony elements have developed three months postoperation (via endochondral ossification). The regenerated tissues were evaluated macroscopically, biochemically, histologically, and histochemically, confirming the development of hyaline cartilage in the superficial zones of the articular layers of the defect and an osseous tissue in the deeper zones (see Figs. 3–7).

In summary, an implant composed of autogeneic or allogeneic embryonal chondrocytes embedded in hyaluronic-acid-based adhesive gel provides a good biological clinical solution to stimulate regeneration and repair of articular cartilage lesions that fail to heal on their own. The average rate of success with this procedure, for which a completely congruent healing was obtained, reached 75% to 85% of the defects in trial. We believe that most failures in our system are due to the migration from the lesion site of the hyaluronic acid gel-cells mixture. Therefore, a study was conducted searching for scaffolds, naturally occurring ones (e.g., cross-linked collagens, proteoglycans, and their copolymers), as well as synthetic biodegradable biopolymers (e.g., polyglycolide, polylactide, and various combinations) for an improved anchorage of the cartilaginous implant in the articular lesion sites (see Fig. 8).

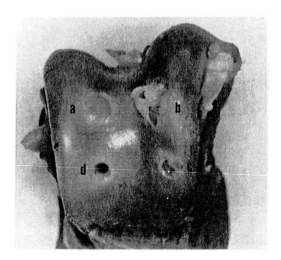

Figure 3   A photomacrograph six weeks after implantation of allogeneic embryonal chick chondrocytes embedded in hyaluronic-acid-based milieu as implant, transplanted into mechanically created defects in the articular surface of the lower tibia of a three-year-old chicken: (a), note a complete healing of the defect with restoration and congruency of the articular surface; (b), note that, in this defect, a slight overgrowth of the reparative tissue occurred, which should be eliminated with time and joint motion; (c), in this defect, the lumen is only partially filled, not reaching the level of the articular surface; (d), the defect is completely empty. The defects in Figs. 3c and 3d should undergo a second implantation treatment (magnification ×8).

V.   COMPOSITE CARTILAGINOUS IMPLANTS: CHONDROCYTE-IMPREGNATED SCAFFOLDS FOR RESURFACING OF ARTICULAR SURFACES

Obviously, neither the cells nor the hydrated polymers used as delivery substances possess the significant mechanical strength necessary for a cartilage-replacing bioprosthesis. The initial mechanical strength should be provided by a scaffold. The scaffold should possess several characteristics: mechanical strength (elasticity and stiffness), porosity (allowing migration of cells), cell attachment sites (either electric charge or specific proteins), biocompatibility (both *in vivo* and *in vitro*), and optimally it should be biodegradable (allowing replacement by cartilage eventually).

A.   Carbon Fiber-Based Scaffold

Carbon fiber pad (Medicarb, Dunlop Medical Products, Leyland Medical International, Ltd., Lancashire, England) is an inert material that can be implanted successfully into joints and bones [Hemmen, Archer, and Bentley 1991; Robinson et al. 1993a; Minns, Muckle, and Betts 1993]. As long as particulate debris is not produced, synovitis is not evoked. However, as the material is brittle, if the mesh is left in a position in which it is exposed to shearing forces, particulate debris will cause foreign-body reaction and synovitis, leading to joint destruction. The material does not allow cartilage cell attachment, and is not particularly supportive of chondrocyte growth. Thus, a common result of such

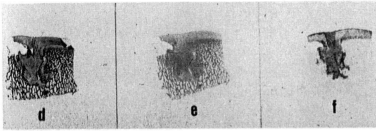

Figure 4   Photomicrographs of histological and histochemical osteochondral sections of a tibial condyle from an adult pig: (a), Masson's trichrome staining for connective tissue constituents and fibrin of a section from a control untreated animal; (b), Mayer's hematoxylin-eosin (H&E) staining of a control animal; (c), an alcian blue staining (pH 1.0) for sulphated glycosaminoglycan detection; (d), Masson's staining of a section through an osteochondral defect, three months postimplantation, of an autogeneic implant composed of bone-marrow-derived chondrocytic-enriched cells embedded in hyaluronic-acid-based glue; note the complete healing of the defect and the sphericity of the articular surface, leveled with the neighboring cartilage; the deep zones of the implanted reparative tissue are still cartilaginous; (e), H&E staining of the same section through the osteochondral defect, three months postoperation; (f), alcian blue staining of the section through the defect; note the more highly stained (bluish) density of the newly formed cartilage in the implanted site (magnification ×4).

implants is a cartilaginous cap covering a loosely fibrotic carbon mesh that is largely empty of chondrocytes (Fig. 9).

B.   Collagen-Based Scaffolds

Collagen is the principal protein responsible for mechanical strength of tissues throughout the body. Therefore, a collagen-based scaffold appears attractive. Indeed, cartilage matrix is made of collagen fibers forming a 3-D network in which proteoglycans and cells (chondrons) are embedded.

Denatured and partially degraded collagen (gelatin) is used clinically for hemostasis and as a temporary space filler. This material possesses a slight mechanical strength when dry and is jellylike when wet. Thus, it cannot act as a scaffold for replacing extensive defects in joint surfaces. However, it is very useful for replacing hole-shaped defects (depth at least twice the diameter of the defect) [Holmes, Volz, and Chvapil 1975].

Collagen can be treated, cross-linked, and made insoluble. The simplest method is

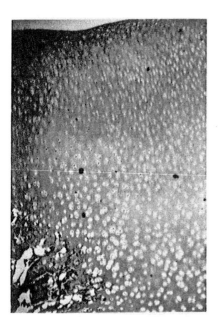

Figure 5 A photomicrograph of the cartilaginous reparative tissue formed in the defect of an adult pig three months after autogeneic bone-marrow-derived chondrocytes embedded in hyaluronic acid were transplanted (as in Fig. 4d) (Masson's trichrome; magnification ×200).

soaking in a cross-linking material such as glutaraldehyde. Unfortunately, the residues of the cross-linking substance are toxic and *in vitro* experiments demonstrated toxicity of the material to embedded cells.

An alternative method is prolonged soaking of collagen in supersaturated sugar solutions. This cross-linking method resembles the glycosylated collagen typical for diabetic

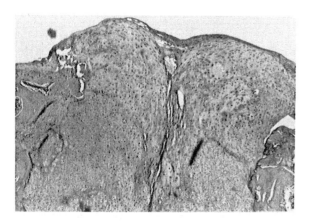

Figure 6 A photomicrograph of the cartilaginous reparative tissue developed in the osteochondral defects of a three-year-old chicken implanted with allogeneic embryonal chick chondrocytes embedded in hyaluronic acid, 2½ months postoperation (hematoxylin-eosin; magnification ×40).

Figure 7 A photomicrograph of the cartilaginous reparative tissue developed three months postoperation in an osteochondral tibial defect of an adult goat implanted with autogeneic bone-marrow-derived chondrocytes embedded in hyaluronic acid. Note the centers of hypertrophic cartilage (HC) and ossifying centers (arrows) (von Kossa, a specific black staining for calcium deposits; magnification ×40).

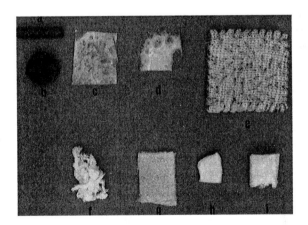

Figure 8 Synthetic and naturally occurring biopolymers used as scaffold or adhesive: (a), carbon fiber rod; (b), a disk of a carbon fiber mesh; (c), a film made of polylactide (85%) and polyglycolide (15%); (d), three-dimensional (3-D) polylactide porous substance; (e), 3-D polyglycolide so-called cartilage pillow (cushion); (f), hyaluronic acid bundles (viscous adhesive gel upon wetting); (g), a piece of a prosthetic collagen meniscal implant; (h), a piece of an artificial skin 91FTS made of collagen-proteoglycan copolymer; (i), absorbable collagen matrix (magnification ×1).

Figure 9   A photomicrograph of the newly formed cartilage above the carbon fiber pad, used as a scaffold impregnated with autogeneic bone-marrow-derived chondrocytes embedded in hyaluronic acid and implanted in an osteochondral defect in a tibia of an adult rabbit (six-weeks postimplantation). Note the intense staining of the cartilage with alcian blue (pH 1.0); no chondrocytes are seen within the pad (alcian blue; magnification ×100).

individuals, the so-called advanced glycosylation end product (AGEs) occurring with proteins. The material does not appear to be cytotoxic *in vitro* to cells.

The next step was conducting studies in which cross-linked collagens and collagen-proteoglycans copolymer were used as scaffolds. Three kinds of such scaffolds (received kindly as gifts) have been involved in experimentation: (1) prosthetic collagen implant (made of collagen and proteoglycan), a product of ReGen Corporation (Franklin Lakes, NJ), provided by Dr. Kevin R. Stone [Stone et al. 1990; Stone et al. 1992], Pacific Presbyterian Medical Center, San Francisco, California; (2) artificial skin 91FTS, a collagen-proteoglycan copolymer of Fibers and Polymers Laboratory Department of Mechanical Engineering (MIT, Cambridge, MA), provided by Dr. I.V. Yannas [Yannas et al. 1989]; (3) Helistat absorbable collagen, a product of American Biomaterials Corporation (Plainsboro, NJ), provided by Dr. Joseph Nichols. Such a material has been shown to be conductive for thin cartilage and fibrocartilage growth *in vivo* [Stone et al. 1990; Stone et al. 1992] and in our studies, both *in vitro* and *in vivo*.

Our preliminary experiments have started using the resorbable collagen enriched with glycosaminoglycans, chemotactic and adherence factors (ReGen Corp. product) as a scaffold, impregnated with allogeneic embryonal chick chondrocytes embedded in hyaluronic acid immediately before the implantation. These cells containing implants were transferred into defects involving the articular cartilage layer and deep into the subchondral bone (osteochondral defect) of three-year-old chickens. Macroscopic examination 2½ months postoperatively looked very promising (Fig. 10). While the untreated defect stayed empty (Fig. 10a), the control (scaffold + HY) was partially filled (Fig. 10b) and the experimental scaffold (scaffold + cells in HY) looked completely congruent (Fig. 10c). The higher the expectations were, the greater was the disappointment. Upon de-

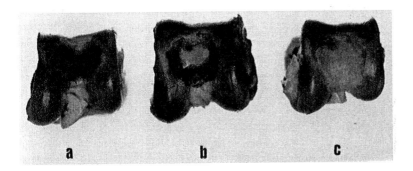

Figure 10   Photomacrographs of damaged tibias of 3-year-old chickens (the defects were created mechanically): (a), an untreated defect 2½-months postoperation; (b), a defect implanted with the collagenous scaffold impregnated with hyaluronic acid as a control; (c), a defect implanted with the collagenous scaffold impregnated with allogeneic embryonal chick chondrocytes embedded in hyaluronic acid; note the completely congruent appearance of the outcome in the experimental articular cartilage surface of the tibia (magnification ×4).

tailed histological examinations (Fig. 11), it was shown that most of the volume occupied originally by the cell-containing implant was empty of cells, only remnants of the scaffold were seen and believed to maintain the form of the implant. Therefore, it appears that this collagenous material is unsupportive for thick cartilage growth *in vivo*. The outcome in this kind of implant resembled, in certain cases, the outcome with the carbon fiber (mentioned above) and synthetic scaffolds described below. A very thin superficial layer of cartilage or fibrocartilage is formed occasionally, overlying the implant, which remains empty.

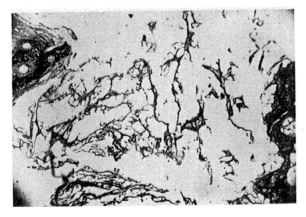

Figure 11   A photomicrograph of a section through the defect implanted with a scaffold impregnated with cells embedded in hyaluronic acid 2½-months postoperation (Fig. 10c). Note the complete lack of cells in the core of the implant (lesion site), with no reparative tissue except remnants of the collagenous scaffold. In certain cases, thin bridges of cartilage-fibrocartilage composed of 2–3 rows of cells were found at the articular surface. Thus, no kind of continuous tissue integrated with its surroundings was formed (Masson's; magnification ×40).

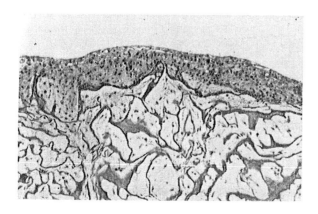

Figure 12   A photomicrograph of a section through the *in vitro* reconstructed composite implant composed of Helistat absorbable collagen of American Biomaterials Corporation impregnated with embryonal chick chondrocytes embedded in hyaluronic acid, 2½ months in culture. Note the thin layer of tissue formed at the scaffold-medium interface (hematoxylin-eosin; magnification × 100).

In an effort to try to understand the reasons for the failure of these naturally occuring scaffolds to support cartilage growth *in vivo*, we performed the following *in vitro* study. Three-dimensional pieces of the three collagenous-based scaffolds (outlined above) were employed in culture, as composite cartilaginous implants, together with embryonal (chick) chondrocytes embedded in hyaluronic acid. The mixture of cells and hyaluronic acid was injected into the scaffolds, as well as smeared over them. The whole implant was placed in a culture dish with a limited amount of medium, not letting the composite implants to dry. After 2½ months in culture, the outcome was examined by histological, histochemical, and immunohistochemical parameters. Figures 12 to 15 demonstrate conclusively, that, even under optimal *in vitro* environmental conditions for growth and

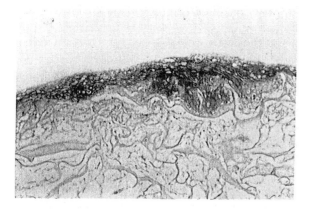

Figure 13   A photomicrograph of the same section as in Fig. 12 stained with alcian blue (pH 1.0). Note the intense color developed at the surface, typical of a young and active cartilaginous tissue (alcian blue; magnification × 100).

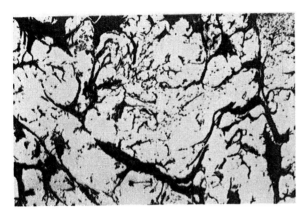

Figure 14 A photomicrograph of the middle core of an *in vitro* reconstructed implant made of ReGen's prosthetic collagen impregnated with embryonal chick chondrocytes embedded in hyaluronic acid. Note the absence of cells impregnated originally in these sites. It is not yet known whether cells lysed or just migrated out (Masson's trichrome; magnification ×400).

chondrogenesis, only a very thin (5 to 10 microns; at most, 10 layers of cells) cartilaginous layer is formed only at the interfaces of the scaffolds with the medium.

Figures 12 to 16 show the resultant tissues developed only at the interfaces with the various collagenous-based scaffolds. Similar results were obtained with all three different collagenous scaffolds. We observed that the less dense the scaffold was, the deeper was the penetration into it by cells (Fig. 16). However, a loose scaffold is problematic for handling. The loose reconstructed implants are more difficult to use, especially to adjust to the 3-D shape and anchor to the lesion margins. The conclusion is that the thin tissue formed in culture at the surface (interface) *in vitro* and *in vivo* resembles the thin layer of hyaline cartilage supported and fed by synovial fluid via diffusion. The limitation appears to be due to the lack of nourishment of more deeply situated cells.

C. Polylactide and Polyglycolide Matrices as Synthetic Scaffolds

The most commonly employed biodegradable synthetic biomaterials are composed of polylactide, polyglycolide, and their copolymers. These materials attracted a lot of attention and many detailed studies have been reported [Vainionpaa, Rokkanen, and Tormala 1989; Majola et al. 1991; Cima, Langer, and Vacanti 1991; Cima et al. 1991; Freed et al. 1993; Pistner et al. 1993]. The polyglcolide, polylactide, and their copolymers in our experimentation were from Bicon, Limited (Tampere, Finland), generously supplied to us in various 3-D formats by Prof. Pertti Tormala, from Tampere University, Technological, Institute of Plastic Technology, Tampere, Finland. Films made of the above combination of synthetic polymers were kindly provided to us by Prof. Donald L. Wise, Center for Biotechnology Engineering, Northeastern University, Boston, Massachusetts.

The retention periods of these substances can be manipulated during the production process and adjusted to the desired time period. These synthetic materials are inert and can be shaped into porous and relatively pliable scaffolds. *In vitro* experiments with the polylactide scaffold looked more promising than with the polyglycolide scaffold, in respect to cell growth, active proliferation, and attachment of chondrocytes to the scaffold's surfaces (Figs. 17 and 18). Further experiments employing these synthetic substances as

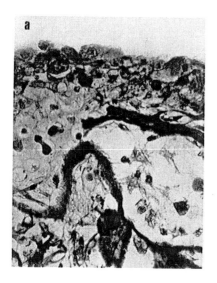

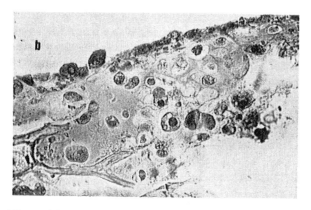

Figure 15 Photomicrographs of immunohistochemical stained sections from an *in vitro* reconstructed implant made of artificial skin collagen from Fibers and Polymer Lab, Massachusetts Institute of Technology, Cambridge, Massachusetts: (a), the double-antibodies peroxidase-antiperoxidase and diaminobenzidine (DAB) staining with polyclonal anticartilage proteoglycan as the primary antibody, staining a thin layer of cells at the interface (magnification ×400); (b), the surface of a ReGen's prosthetic collagen impregnated with cells in hyaluronic acid, stained with a primary monoclonal antibody against cartilaginous keratan sulphate; note the heavy-dense staining of the superficial layer of cells at the interface (magnification ×400).

scaffolds for instantly produced implants (i.e., implants prepared immediately prior to transplantation) resulted in a very poor outcome *in vivo* in spite of the macro appearance of the defect's surfaces (Fig. 19). Hence, these synthetic materials appear to be cytotoxic to cells. This might be related to the degradation products being highly acidic. In conclusion, these substances do not support chondrocyte growth within the scaffolds *in vivo*

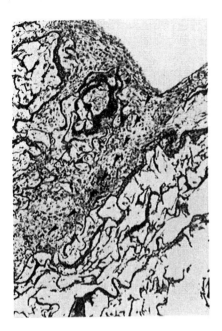

Figure 16 A photomicrograph of a section as in Fig. 12 with a Helistat absorbable collagen scaffold, a product of American Biomaterials Corporation. Note an area with a looser scaffold density that allows cells to penetrate deeper into the scaffold to a depth of 500 microns (arrows) (Masson's; magnification ×100).

Figure 17 A photomicrograph of an *in vitro* reconstructed implant of Bicon's polyglycolide scaffold impregnated with embryonal chick chondrocytes in hyaluronic acid. Cells have been found not to adhere to the polyglycolide surfaces and easily came off and floated (magnification ×400).

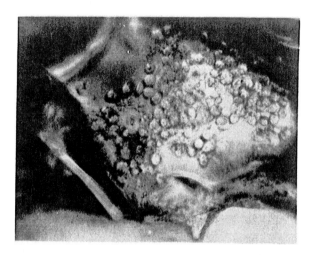

Figure 18   The same implant as in Fig. 17, though the scaffold was made of a polylactide of Bicon. The cells looked "happier" in respect to the rate of cell proliferation and cell adherence (magnification ×400).

(Fig. 20a), but rather tend to induce fibrosis, fibrotic adhesions, and, in the best cases, some thin bridges of cartilage at the articular surfaces of the defects (Figs. 20b and 21).

   Thus, from these *in vivo* studies of both substances embedded with cells and used as cartilaginous implants, it appears that the polymers are not friendly for cells, leaving big, empty spaces without cells, though macroscopically the lesions appeared congruently shaped. Bridges of thin fibrotic tissue were gaping over the defects.

   Employing a synthetic copolymer film of polylactide (85%) and polyglycolide (15%)

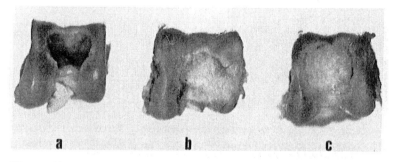

Figure 19   Photomacrographs as in Fig. 10 using polylactide scaffold impregnated with cells just before implantation: (a), an untreated defect 2½-months postoperation; (b), a defect implanted with the polylactide scaffold impregnated with hyaluronic acid (HY) only as a control; (c), a defect implanted with the polylactide scaffold impregnated with allogeneic embryonal chick chondrocytes embedded in HY. Note the completely congruent appearance of both the control scaffold and the experimental (cell-containing) implant (magnification ×4).

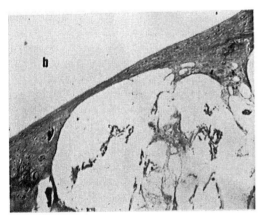

Figure 20   Photomicrographs of sections through defects implanted with a polylactide scaffold impregnated with cells embedded in hyaluronic acid, 2½-months postoperation. Note that, as in the collagen-based scaffolds, no cells survived in the polylactide implant site: (a), note the empty hole in the implanted site (Masson's; ×40); (b), the same as Fig. 20a, with the exception that a thin bridge of a fibrotic tissue remained intact at the articular surface above the implant (hematoxylin-eosin; ×40).

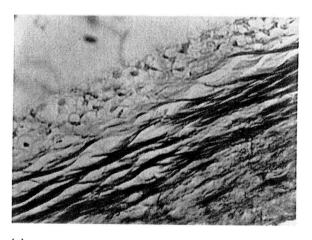

(a)

(b)

Figure 21   Enlargements of the fibrotic tissue seen in Fig. 20b: (a), a surface area showing layers of scaffold remnants and a fibrotic tissue made of 2 to 3 rows of cells (Masson's; ×200); (b), a typical pannuslike granulation tissue developed in response to cartilaginous injury (Masson's; magnification ×400).

just as a coverage for the defect, which was filled with implanted cells embedded in hyaluronic acid, did not live up to the expectations due to the lack of elasticity of the film and the difficulties of tightly suturing it with the neighboring margins of the articular cartilage surface. The film was produced by the Center for Biotechnology and Engineering, Northeastern University.

D. Decalcified Bone Matrix as a Scaffold

The most promising scaffold to date in our laboratory is decalcified bone matrix. The methodology has been detailed by several authors [Nathanson and Hay 1980; Green, Hinton, and Triffitt 1986; Nimni et al. 1988; Katz, Felthousen, and Reddi 1990]. Basically, bones are sequentially washed, defatted, and decalcified. The end material is pliable and naturally porous. This allows easy shaping on one hand and cell migration into the scaffold on the other hand. The material appears to allow cell attachment and is incorporated with the host neighboring tissues. Results of cell-impregnated implants were encouraging. Articular surfaces up to 60% of the entire joint were successfully replaced (Fig. 22). It appears that such an implant prevents, at least in the short run, development of osteoarthritis secondary to the defects in the articulation surfaces, yielding a good hyaline cartilage as the reparative tissue.

VI. MECHANICAL PROPERTIES OF THE REPARATIVE TISSUES IN THE IMPLANTS

The only scaffold used in our laboratory as a cell-containing implant on which detailed mechanical studies were conducted after implantation and regeneration of cartilage is carbon fiber. The results are better than those obtained without a scaffold. However, the implant is not as rigid as the surrounding articular cartilage of a normal joint [Robinson et al. 1993a].

VII. DISCUSSION

Reconstruction of joints by chondrocytes requires a complex implant that is made of three main components: (1) cells, (2) adhesive and growth permissive milieu (delivery substance), and (3) scaffold. Many reports have recently suggested various cell sources

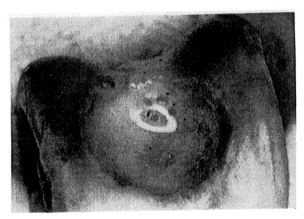

Figure 22   A photomacrograph of a defect implanted with a composite implant made of allogeneic decalcified bone (prepared in our laboratory) impregnated with allogeneic embryonal chick chondrocytes embedded in hyaluronic acid immediately prior to the implantation and transplanted into tibial defect in a three-year-old chicken, three-months postoperation (magnification × 10).

[Cohen and Jonsen 1993; Robinson et al. 1993b], various 3-D bioengineering products as the scaffold [Cima, Langer, and Vacanti 1991; Cima et al. 1991; Edgington 1992; Erickson 1992; Freed et al. 1993; Mikos et al. 1993], and various ways for examination and evaluation of cell response to such substances used as implant materials [Goldring et al. 1990; Wenz et al. 1990; Puleo et al. 1991; Ciapetti et al. 1993].

Cells were shown to be obtainable either from autogeneic mesenchymal stem cells or from allogeneic fetuses [Itay, Abramovici, and Nevo 1987; Harada et al. 1990; Robinson et al. 1990a, 1990b; Robinson et al. 1993a, 1993b; Nevo et al. 1993a, 1993b].

The superior delivery substance to date appears to be hyaluronate as it supports the cartilaginous phenotype. The physical rheological properties of hyaluronic acid, forming a highly hydrated viscous gel, have made it suitable to serve as a lubricant, shock absorber, elastic cushion, and coating agent/separating barrier, evoking minimal immune reaction of cells and matrices, reduced graft-versus-host response, and diminished inflammation [Balazs and Denlinger 1989; Lindblad 1989; Songer, Ghosh, and Spencer 1989]. Hyaluronic acid is a nonantigenic, biocompatible, and biodegradable agent. Other properties of this agent protect tissue from free radicals, clear tissues of cells' debris, decrease postoperative bleeding, and relieve pain [Balazs and Denlinger 1989; Weigel et al. 1989; West and Kumar 1989]. Hyaluronic acid is utilized for viscosupplementation and space filling, as well as a viscosurgical tool in opthalmology, otology, neurosurgery, plastic surgery, dermatology, and cosmetics [Balazs and Band 1984].

Furthermore, hyaluronic acid is used as an antiadhesive and antiulceration agent, controlling granulation tissue and pannus formation, minimizing flexors' deformity and contracture, and stimulating proper regeneration and wound healing processes [Amiel et al. 1985; King et al. 1991; Leardini et al. 1991].

The excellent wound healing in fetal life, characterized by its rapid pace and quality of the reparative tissue (scarless) even in tissuelike cartilage, is commonly related to the richness in the amounts of hyaluronic acid in the fetal tissues [Toole 1982; Adzick and Longaker 1991; Longaker et al. 1991; Mast et al. 1992].

In our study, we have extended the use of hyaluronic acid and employed it successfully as an *in vitro* cell growing milieu, holding cells and growth factors, as well as a delivery substance. Hyaluronic acid controls and modulates in cultures (acts differently on different cells) cell motility, proliferation, differentiation, and attachment to substrates [Robinson et al. 1990a, 1990b].

In the *in vitro* reconstructed composite implants, hyaluronic acid serves as a viscous-adhesive milieu, an anchoring-immobilizing substance, whereas *in vivo*, in the implantation sites, hyaluronic acid helps to integrate and assemble pericellular and extracellular matrices of existing tissues with newly formed reparative tissues [Solursh et al. 1980; Thomas et al. 1992; Turley 1992; Underhill 1992; Yu, Banerjee, and Toole 1992; Knudson 1993; Knudson, Bartnik, and Knudson 1993; Thomas et al. 1993].

The most attractive scaffold appears to be decalcified bone matrix as it is both easily shapable and supportive of cell growth. Such a scaffold should contain many of the growth factors regulating the repair of bone fractures [Nimni et al. 1988; Bolander 1992]. The porous structure and the natural surfaces must be friendly for cells, allowing their proliferation and adherence [Nevo et al. 1993c].

Hence, the biological resurfacing approach is expected to become a leading treatment in the routine clinical procedure in the near future for the repair of articular cartilage defects.

VIII. CONCLUSIONS

A mixture of cultured chondrocytes either of autogeneic or allogeneic origin embedded in a 2% hyaluronic-acid-based adhesive, comprises an optimal cartilaginous implant in respect to

1. Attaining the exact shape of the defect
2. Biological compatibility
3. Integration of the newly formed reparative tissue with its neighbors

Impregnation of such a mixture into a naturally occurring biodegradable, cell-friendly scaffold such as demineralized bone matrix ensures cells' adherence and immobilization, as well as the anchorage of the whole structure within the lesion site, optimizing the chances for hyaline cartilage outcome in the superficial regions of the defect and osseous tissue in the deeper ones.

ACKNOWLEDGMENTS

The authors thank Zoharia Evron, Lidia Lipa, and Ally Sheynin of the Chemical Pathology Department and Asher Pinchasov of the Photography Unit of the Sackler Medical School for their excellent and skillful technical assistance.

The authors thank as well the orthopedic residents Dr. Miriam Efrat-Botnaro, Dr. Daniel Carlin, and Dr. Yiftah Beer for their essential contributions in conducting parts of this cartilage repair research project as their basic science training.

REFERENCES

Adzick, N.S., and Longaker, M.T. Scarless wound healing in the fetus: The role of the extracellular matrix. *Prog. Clin. Bio. Res.*, 365, 177–192, 1991.

Amiel, D., Frey, C., Woo, S.L.Y., Harwood, B.S., and Akeson, W. Value of hyaluronic acid in the prevention of contracture formation. *Clin. Orthop. & Rel. Res.*, 196, 306–311, 1985.

Balazs, E.A., and Band, P. Hyaluronic acid: Its structure and use. *Cosmetics & Toiletries*, 99, 65–72, 1984.

Balazs, E.A., and Denlinger, J.L. Clinical uses of hyaluronan, In *The Biology of Hyaluronan*, CIBA Foundation Symp. No. 143, D. Evered and J. Whelan, eds., Wiley & Sons, Chichester, Sussex, pp. 265–280, 1989.

Bolander, M.E. Regulation of fracture repair by growth factors. *Proc. Soc. Expt. Biol & Med.*, 200, 165–170, 1992.

Buckwalter, J., Rosenberg, L., Coutts, R., Hunziker, E., Reddi, A.H., and Mow, V. Articular cartilage: Injury and repair. In *Injury and Repair of the Musculoskeletal Soft Tissues*, S.L.Y. Woo and J.A. Buckwalter, eds., American Academy of Orthopaedic Surgeons, Rosemont, IL, pp. 465–482, 1988.

Chalmers, J., and Ray, R.D. The growth of transplanted foetal bones in different immunological environments. *J. Bone Joint Surg. Br.*, 44-B, 149–164, 1962.

Ciapetti, G., Cenni, E., Pratelli, L., and Pizzoferrato, A. *In vitro* evaluation of cell/biomaterial interaction by MTT assay. *Biomater.*, 14, 359–364, 1993.

Cima, L.G., Langer, R., and Vacanti, J.P. Polymers for tissue and organ culture. *J. Bioactive & Compatible Polymers*, 6, 232–240, 1991.

Cima, L.G., Vacanti, J.P., Vacanti, C., Ingber, D., Mooney, D., and Langer, R. Tissue engineering by cell transplantation using degradable polymer substrates. *J. Biochem. Eng.*, 113, 143–151, 1991.

Cohen, S.B., and Jonsen, A.R. The future of the fetal tissue bank. *Science*, 262, 1663–1665, 1993.

Coon, H.G. Clonal stability and phenotypic expression of chick cartilage cells *in vitro*. *Proc. Natl. Acad. Sci. U.S.A.*, 55, 66–73, 1966.

Cundy, P.J., Jofe, M., Zaleske, D.J., Ehrlich, M.G., and Mankin, H.J. Physeal reconstruction using tissue donated from early postnatal limbs in a murine model. *J. Orthop. Res.*, 9, 360–366, 1991.

Dascalu, A., Nevo, Z., and Korenstein, R. The control of intracellular pH in cultured avian chondrocytes. *J. Physiol.*, 461, 583–599, 1993.

Edgington, S. 3-D biotech: Tissue engineering. *Biotechnology*, 10, 855–860, 1992.

Elves, M.W. A study of the transplantation antigens on chondrocytes from articular cartilage. *J. Bone Joint Surg. Br.*, 56-B, 178–185, 1974.

Erickson, D. Bioengineering produce versions of body tissues. *Sci. Amer.*, 114–116, Aug. 1992.

Feldman, M.D., Sataloff, R.T., Choi, H.Y., and Ballas, S.K. Compatibility of autologous fibrin adhesive with implant materials. *Arch. Otolaryngol. Head-Neck Surg.*, 114, 182–185, 1988.

Freed, L.E., Marquis, J.C., Nohria, A., Emmanual, J., Mikos, A.G., and Langer, R. Neocartilage formation *in vitro* and *in vivo* using cells cultured on synthetic biodegradable polymers. *J. Biomed. Mat. Res.*, 27, 11–23, 1993.

Goldring, S.R., Flannery, M.S., Petrison, K.K., Evins, A.E., and Jasty, M.J. Evaluation of connective tissue cell responses to orthopaedic implant materials. *Conn. Tiss. Res.*, 24, 77–81, 1990.

Grande, D.A., and Pitman, M.I. The use of adhesives in chondrocyte transplantation surgery preliminary studies. *Bull. Hosp. Joint Dis.*, 48, 140–148, 1988.

Green, E., Hinton, C., and Triffitt, J.T. The effect of decalcified bone matrix on the osteogenic potential of bone marrow. *Clin. Orthop.*, 205, 292–298, 1986.

Harada, K., Oida, S., Sasaki, S., and Enomoto, S. Chondrocyte like colony formation of mesenchymal cells by dentin extracts in agarose gel culture. *J. Dent. Res.*, 69, 1555–1559, 1990.

Hemmen, B., Archer, C.W., and Bentley, G. Repair of articular defects by carbon fibre plugs loaded with chondrocytes. 37th Ann. Meeting Orthop. Res. Soc., March 4–7, Anaheim, CA, p. 278, 1991.

Holmes, M.D., Volz, R.G., and Chvapil, M. Collagen sponge as a matrix for articular cartilage regeneration, *Surg. Forum*, 26, 511–513, 1975.

Itay, S., Abramovici, A., and Nevo, Z. The use of cultured embryonal chick epiphyseal chondrocytes as grafts for defects in chick articular cartilage. *Clin. Orthop.*, 220, 284–303, 1987.

Katz, R.W., Felthousen, G.C., and Reddi, A.H. Radiation sterilized insoluble collagenous bone matrix is a functional carrier of osteogenin for bone induction. *Calcif. Tiss. Intern.*, 47, 183–185, 1990.

Kawabe, N., and Yoshinao, M. The repair of full-thickness articular cartilage defects. Immune responses to reparative tissue formed by allogeneic growth plate chondrocyte implants. *Clin. Orthop.*, 268, 279–293, 1991.

King, S.R., Hickerson, W.L., Proctor, K.G., and Newsome, A.M. Beneficial actions of exogenous hyaluronic acid on wound healing. *Surgery*, 109, 76–84, 1991.

Knudson, C.B. Hyaluronan receptor-directed assembly of chondrocyte pericellular matrix. *J. Cell Biol.*, 120, 825–834, 1993.

Knudson, W., Bartnik, E., and Knudson, C.B. Assembly of pericellular matrices by COS-7 cells, transfered with CD-44 lymphocyte-homing receptor genes. *Proc. Natl. Acad. Sci. U.S.A.*, 90, 4003–4007, 1993.

Leardini, G., Mattara, L., Franceschini, M., and Perbellini, A. Intraarticular treatment of knee osteoarthritis. A comparative study between hyaluronic acid and 6-methyl-prednisolone acetate. *Clin. Expt. Rheum.*, 9, 375–381, 1991.

Lindblad, G.T. Hyaluronic acid preparation to be used for inflammation of skeletal joints. U.S. patent No. 4801, 619, 1989.

Longaker, M.T., Chiu, E.S., Adzick, N.S., Stern, M., Harrison, M.R., and Stern, R. Studies in fetal wound healing. V. A prolonged presence of hyaluronic acid characterizes fetal wound fluid. *Ann. Surg.*, 213, 292–296, 1991.

McKibbin, B. Immature joint cartilage and the homograft reaction. *J. Bone Joint Surg. Br.*, 53-B, 123–135, 1971.

Majola, A., Vainionpaa, S., Vihtonen, K., Mero, M., Vasenius, J., Tormala, P., and Rokkanen, P. Absorption, biocompatibility and fixation properties of polylactic acid in bone tissue: An experimental study in rats. *Clin. Orthop. & Rel. Res.*, 268, 260–269, 1991.

Malejczyk, J., and Moskalewski, S. Effect of immunosuppression on survival and growth of cartilage by transplanted allogeneic epiphyseal chondrocytes. *Clin. Orthop.*, 232, 292–303, 1988.

Mast, B.A., Diegelmann, R.F., Krummel, T.M., and Cohen, I.K. Scarless wound healing in the mammalian fetus. *Surg. Gyn. & Obst.*, 174, 441–451, 1992.

Meyers, M.H., Akeson, W., and Convery, R. Resurfacing of the knee with fresh osteochondral allograft. *J. Bone Joint Surg.*, 71A, 704–713, 1989.

Mikos, A.G., Sarakinos, G., Leite, S.M., Vacanti, J.P., and Langer, R. Laminated three-dimensional biodegradable foams for use in tissue engineering. *Biomater.*, 14, 323–330, 1993.

Milgram, J.W. Injury to articular cartilage joint surfaces. *Clin. Orthop.*, 192, 168–173, 1985.

Minns, R.J., Muckle, D.S., and Betts, J.A. Biological resurfacing using carbon fiber. *Orthop. Intern.*, 1, 414–424, 1993.

Nathanson, M.A., and Hay, E.D. Analysis of cartilage differentiation from skeletal muscle growth in bone matrix. I Ultrastructure aspects. *Dev. Biol.*, 78, 301–333, 1980.

Nevo, Z., Robinson, D., Efrat, M., Halperin, N., and Mendes, D.G. Cartilaginous implants containing cultured cells. *Orthop. Intern.*, 1, 441–446, 1993a.

Nevo, Z., Robinson, D., Efrat, M., Halperin, N., and Mendes, D.G. *In vitro* construction of cells-containing implant for articular cartilage regeneration. *Agents and Actions*, Supplement 39, 231–235, 1993b.

Nevo, Z., Robinson, D., and Halperin, N. The use of grafts composed of cultured cells for repair and regeneration of cartilage and bone. In *Bone*, Vol. 5, B.K. Hall, ed., CRC Press, London, pp. 123–152, 1992.

Nevo, Z., Robinson, D., Halperin, N., and Edelstein, S. Culturing chondrocytes for implantation. In *Methods in Cartilage Research*, A. Maroudas and K. Keuttner, eds., Academic Press, New York, 26, pp. 98–104, 1990.

Nevo, Z., Silver, J., Chorev, Y., Riklis, I., Robinson, D., and Yosipovitch, Z. Adhesion characteristics of chondrocytes cultured separately and in co-cultures with synovial fibroblasts. *Cell Biol. Intern.*, 17, 255–273, 1993c.

Nimni, M.E., Bernick, S., Ertl, D., Nishimoto, S.K., Paule, W., Strates, B.S., and Villaneuva, J. Ectopic bone formation is enhanced in senescent animals implanted with embryonic cells. *Clin. Orthop.*, 234, 255–266, 1988.

Oakeshott, R.D., Farine, I., Pritzker, K.P.H., Langer, F., and Gross, A.E. A clinical and histological analysis of failed fresh osteochondral allografts. *Clin. Orthop. & Rel. Res.*, 233, 283–294, 1988.

Pistner, H., Gutwald, R., Ordung, R., and Reuther, J. Poly(l-lactide): A long-term degradation study *in vivo*. I. Biological results. *Biomater.*, 14, 671–677, 1993.

Puleo, D.A., Holleram, L.A., Doremus, R.H., and Bizios, R. Osteoblast responses to orthopaedic implant materials *in vitro*. *J. Biomed. Mater. Res.*, 25, 711–723, 1991.

Robinson, D., Efrat, M., Halperin, N., Mendes, D.G., and Nevo, Z. Implants composed of carbon fiber mesh and bone-marrow-derived chondrocytes-enriched cultures for joint surface reconstruction. *Bull. Hosp. Joint Dis.*, 53, 75–82, 1993a.

Robinson, D., Efrat, M., Halperin, N., Mendes, D.G., and Nevo, Z. The mesenchymal stem cell in orthopaedics. *Orthop. Intern.*, 1, 448–453, 1993b.

Robinson, D., Halperin, N., and Nevo, Z. Regenerating hyaline cartilage in articular defects of old chickens using implants of embryonal chick chondrocytes embedded in a new natural delivery substance. *Calcif. Tiss. Intern.*, 46, 246–253, 1990a.

Robinson, D., Halperin, N., and Nevo, Z. Use of cultured chondrocytes as implants for repairing cartilage defects. In *Methods in Cartilage Research*, A. Maroudas and K. Kuettner, eds., Academic Press, London, 76, pp. 327-335, 1990b.

Segal, S., Siegal, T., Altaraz, H., Lev-El, A., Nevo, Z., Nebel, L., Katznelson, A., and Feldman, M. Fetal bone grafts do not elicit allograft rejection because of protecting anti-Ia allo-antibodies. *Transplant.*, 28, 88-95, 1979.

Solursh, M. Formation of cartilage tissue *in vitro*. *Cell Biochem.*, 45, 258-260, 1991.

Solursh, M., Hardingham, T.E., Hascall, V.C., and Kimura, J.H. Separate effects of exogenous hyaluronic acid on proteoglycan synthesis and desposition in pericellular matrix by cultured chick embryo limb chondrocytes. *Develop. Biol.*, 75, 121-129, 1980.

Songer, M.N., Ghosh, L., and Spencer, D.L. Effects of sodium hyaluronate in peridural fibrosis after lumbar laminotomy and discectomy. *Spine*, 15, 550-554, 1989.

Stockwell, R.A. *Biology of Cartilage Cells*, Cambridge University Press, Cambridge, 1979.

Stone, K.R., Rodkey, W.G., Webber, R.J., McKinney, L., and Steadman, J.R. Collagen-based prostheses for meniscal regeneration. *Clin. Orthop.*, 252, 129-135, 1990.

Stone, K.R., Rodkey, W.G., Webber, R.J., McKinney, L., and Steadman, J.R. Meniscal regeneration with copolymeric collagen scaffolds. *In vitro* and *in vivo* studies evaluated clinically, histologically, and biomechanically. *Am. J. Sports Med.*, 20, 104-111, 1992.

Thomas, L., Byers, H.R., Vink, J., and Stamenkovic, I. CD-44 regulates tumor cell migration on hyaluronate coated substrate. *J. Cell. Biol.*, 118, 971-977, 1992.

Thomas, L., Etoh, T., Stamenkovic, I., Mihm, M.C., and Byers, H.R. Migration of human melanoma cells on hyaluronate is related to CD-44 expression. *J. Invest. Dermatol.*, 100, 115-120, 1993.

Toole, B.P. Developmental role of hyaluronate. *Conn. Tiss. Res.*, 10, 93-100, 1982.

Turley, E.A. Hyaluronan and cell locomotion. *Cancer Metastasis Rev.* 11, 21-30, 1992.

Underhill, C. CD-44 – The hyaluronan receptor. *J. Cell Sci.*, 103, 293-298, 1992.

Vainionpaa, S., Rokkanen, P., and Tormala, P. Surgical applications of biodegradable polymers in human tissues. *Progr. Polym. Sci.*, 14, 679-716, 1989.

Wakitani, S., Kimura, T., Hirooka, A., Ochi, T., Yoneda, M., Yasui, N., Owaki, H., and Ono, K. Repair of rabbit articular surfaces with allograft chondrocytes embedded in collagen gel. *J. Bone Joint Surg.*, 71B, 74-80, 1989.

Weigel, P.H., Frost, S.J., Leboeuf, R.D., and McGary, C.T. The specific interaction between fibrinogen and hyaluronan: Possible consequence in hematostasis, inflammation and wound healing. *Ciba Found. Symp.*, 143, 261-269, 281-285, 1989.

Wenz, L.M., Merritt, K., Brown, S.A., Moet, A., and Steffee, A.D. *In vitro* biocompatibility of polyetheretherketon and polysulfone composites. *J. Biomed. Mater. Res.*, 24, 207-215, 1990.

West, D.C., and Kumar, S. Hyaluronan and angiogenesis. *Ciba Found. Symp.*, 143, 187-201, 1989.

Woo, S.L.-Y., and Buckwalter, J.A. Injury and repair of musculoskeletal soft tissue. *J. Orthop. Res.*, 6, 907-931, 1988.

Yamashita, F., Sakakida, K., Suzu, F., and Takai, S. The transplantation of an autogeneic osteochondral fragment for osteochondritis dissecans of the knee. *Clin. Orthop.*, 201, 43-50, 1985.

Yannas, I.V., Lee, E., Orgill, D.P., Skrabut, E.M., and Murphy, G.F. Synthesis and characterization of a model extracellular matrix that induces partial regeneration of adult mammalian skin. *Proc. Natl. Acad. Sci. U.S.A.*, 86, 933-937, 1989.

Yu, Q., Banerjee, S.D., and Toole, B.P. The role of hyaluronan-binding protein in assembling of pericellular matrices. *Dev. Dyn.*, 193, 145-151, 1992.

54

Biodurable Polyurethane Elastomers

Michael Szycher and Andrew M. Reed
PolyMedica Industries, Incorporated
Woburn, Massachusetts

I. INTRODUCTION

Polyurethanes (PUs) are a unique class of segmented thermoplastic elastomers composed of alternating rigid and flexible "segments." The short, rigid segments are connected to long, flexible segments, resulting in a complex network in which the hard-to-soft ratio can be stoichiometrically varied and tailored for the intended use.

The generic term *polyurethane* represents one of the most versatile family of synthetic polymers for biomedical applications. A combination of unsurpassed physical and chemical properties coupled with biocompatibility has led to their use in a wide range of biomedical applications, including pacemaker lead insulators, vascular access devices, vascular grafts, and more.

With a range of molecular weights and chemical structures available, a broad range of physical properties can be achieved with polyurethanes, ranging from rigid, structural components to soft, compliant elastomers. The last compositions, in particular, have proven uniquely suitable for implantable devices.

Polyurethanes have been shown to be stable *in vitro* for many years and yet, when implanted, undergo rapid microcracking. Microcracks not only weaken the polymer, but serve as nucleation sites for thrombus formation and often lead to catastrophic failure.

Besides the obvious mechanical requirements, a successful medical-grade polymer designed to interface with the blood environment has to be (1) bioinert, (2) devoid of surface erosions, (3) resistant to molecular chain disruptions, (4) resistant to the uptake of low molecular weight plasma components, and (5) display a low tendency toward surface calcification. In general, polyurethanes meet these requirements. For this reason, biomedical applications of polyurethane elastomers contribute significantly to the quality and effectiveness of the world's health care system. The combination of the properties of polyurethanes coupled with their biocompatibility has led to their use in artificial heart diaphragms, ventricular-assist bladders, arterio-venous (AV) shunts, vascular grafts, pacemaker lead insulators, leaflet valves, and cardiac patches, among other devices [1].

Current activities of device designers, manufacturers, and physicians indicate that devices manufactured from polyurethane elastomers are being increasingly accepted as the biomaterial of choice in most applications requiring compliance with soft tissue and cardiovascular tissue. Initially focused on life-threatening applications, the polyurethanes have gained a commanding position in artificial hearts, pacemaker lead insulators, experimental vascular grafts, over-the-needle catheters, drug delivery systems, and other applications.

However, in the past few years, disturbing reports have questioned the long-term biostability of ether-based polyurethanes. It is now clear that ether-based polyurethanes are susceptible to microcracking due to biological peroxidation of the ether linkage. Because of this failure mode, the classical ether-based polyurethanes such as Biomer (Ethicon, Somerville, NJ), Pellethane (Dow Chemical, LaPorte, TX), and Tecoflex (Thermedics, Woburn, MA) have been withdrawn from the long-term implantable market.

Fortunately, research has uncovered a second generation of polyurethanes based on a polycarbonate diol that appears biostable. The purpose of this chapter is to review critically some of the physiochemical characteristics displayed by these polycarbonate-based elastomers, and to assess their performance when used in the manufacture of permanently implantable medical devices. Many of the most advanced medical devices under development are being fabricated of polycarbonate-based polyurethane elastomers. These polymers offer an unsurpassed combination of biostability, strength, performance, and ease of manufacture.

II. POLYURETHANE CHEMISTRY

The name *polyurethane* is a generic term that refers to polymers containing the —[NH— COO—]— repeating linkage within the molecular chain. This generic name gives no scientific information about the precise chemical composition of the molecular chain. This ambivalence has contributed to a great deal of confusion within the biomaterials community.

In general, polyurethane elastomers are multiphase block copolymers that consist of alternating blocks of "hard" and "soft" segments, and are thus referred to as "segmented polymers" [2]. The inherent chemical incompatibility between the hard and soft segments leads to a phase separation in polyurethane elastomers [3,4]. The extent of the phase separation in polyurethanes determines their physical properties and also their long-term performance in the biological environment.

The properties of segmented polyurethanes are influenced by the hard segment structure, the molecular weight of the soft segment, and the state of microphase separation [5]. A high content of hard segment increases modulus and strength, while decreasing elongation at break. Increasing the molecular weight of the soft segment increases the elongation at break, while reducing the modulus. The chemical nature and molecular weights of the hard and soft segments influence the state of microphase separation in these microdomain-forming elastomers [6].

Polyurethane elastomers are synthesized by the rearrangement polymerization of a diisocyanate and certain active-hydrogen-containing compounds. The main constituents of any polyurethane elastomer are a diisocyanate, a long-chain hydroxy-terminated macroglycol (ethers, polycarbonate esters, etc.), and a chain extender (glycols or diamines). Most of the older polyurethanes are ether-based materials [7]. Medical-grade polyurethanes are listed in Table 1.

A. Biodegradation of Ether-Based Polyurethanes

Segmented poly(urethane etherurea) elastomers were reported to have good *in vitro* stability [8]. However, we are now aware that, under certain *in vivo* conditions, ether-based polyurethanes are subject to significant surface degradation. Investigators have suggested that the major causes of degradation are calcification, hydrolysis, oxidation, and environmental stress cracking (ESC) [9–14].

Degradation by calcification involves the unanticipated surface deposition of calcium phosphate (dystrophic calcification) in a complex interaction between the biomaterial's surface and the biological environment [15].

Degradation by hydrolysis may be due to susceptibility of urethane or urea linkages in the main chain. Hydrolysis can occur through cell-polymer interactions that produce high concentrations of hydrolytic enzymes such as cathepsin B at the polymer-cell interface. Based on this postulated mechanism, Biomer has been reported to be susceptible to degradation after exposure to papain solution, as indicated by the formation of primary aromatic amines [16–17]. Degradation by hydrolysis has also been reported as a possible degradation mechanism in Pellethane [18].

Degradation by oxidation involves the chain scission of the polyether at the $-CH_2-O-$ linkage, as reported by Takahara et al. and Szycher et al. [19,20]. *In vivo* degradation due to environmental stress cracking of an ether-based polyurethane (Pellethane 2363-80A) was first reported in 1981 [21].

Polyurethanes may undergo a variety of degradative mechanisms, depending on their soft segment composition. Takahara et al. reported synthesizing a series of segmented polyurethanes with various soft segments (polyethylene oxide and polydimethylsiloxane) and investigating their respective hydrolytic and lipid uptake stability. The polyethyleoxide-based polymer was susceptible to hydrolytic degradation, while the polydimethylsiloxane-based polymer was degraded by lipid sorption [22].

Table 1 Medical-Grade Polyurethanes

Name	Supplier	Structure	Comments
Biomer	Ethicon, Somerville, NJ	MDI-PTMEG-EDA	Microcracks
Corethane	Corvita, Miami, FL	MDI-PC-1,4 BD	Biodurable
ChronoFlex	PolyMedica, Woburn, MA	HMDI-PC-1,4 BD	Biodurable
Pellethane	Dow Chemical, LaPorte, TX	MDI-PTMEG-1,4 BD	Microcracks
Tecoflex	Thermedics, Woburn, MA	HMDI-PTMEG-1,4 BD	Microcracks
Tecothane	Thermedics	MDI-PTMEG-1,4 BD	Microcracks

MDI = methylene bis phenyl diisocyanate; PTMEG = polytetramethylene ether glycol; 1,4 BD = butane diol; HMDI = hydrogenated MDI; EDA = ethylene diamine; PC = polycarbonate diol

In an effort to enhance the biodurability of polyurethanes and thus reduce the tendency toward environmental stress cracking, Wabers et al. sulfonated the surface of Biomer. They reported that sulfonated Biomer remained mechanically stable, retaining both tensile strength and elasticity after four weeks of subcutaneous implantation, while the unsulfonated Biomer showed marked cracking and a loss of mechanical properties after the same period of implantation [23].

Of all known degradation mechanisms, environmental stress cracking of polyurethanes is arguably potentially the most catastrophic since it may lead to complete physical failure of the implanted device [22]. Because of the recognized importance of environmental stress cracking in ether-based polyurethanes, we decided to develop a polycarbonate-based polyurethane in an effort to overcome the deficiency of current commercially available polyurethanes. In this chapter, we report on the development and *in vivo* testing of a novel, polycarbonate-based polyurethane, trademarked ChronoFlex (PolyMedica, Woburn, MA), which has been shown to be biodurable under the Stokes test [23,24].

B. Environmental Stress Cracking

There are three conditions for environmental stress cracking (ESC) to occur in polyurethanes: (1) the presence of enzymes (formed because of the *in vivo* inflammatory reaction), (2) a susceptible ether-based polyurethane, and (3) the surface of the elastomer must be under mechanical strain [25]. All criteria must be present for environmental stress cracking to occur. If any one of the three conditions is absent, the ether-based polyurethane does not degrade.

Environmental stress cracking is a well-known degradation mechanism in certain polymers. For instance, ESC is commonly observed *in vitro* for plastics such as polyethylene under the action of a detergent and stress [26]. In reference to polyethylene, the polymer needs to be subjected to an offending environment (detergent) and be under mechanical strain (stress) to exhibit the surface cracking phenomenon called ESC. When an offending environment coincides with mechanical stress, the polymer undergoes surface cracking. Notice, however, that ESC may be induced under laboratory conditions on polyethylene. Ether-based polyurethanes have never been shown to undergo ESC under laboratory conditions. ESC has only been reported under *in vivo* conditions. For this reason, Szycher has proposed calling this mechanism biologically induced environmental stress cracking (BI-ESC) [2].

What causes BI-ESC in implanted ether-based polyurethanes? All implanted materials elicit an inflammatory reaction from the host (foreign-body reaction) that is intended to attack, lyse, and eventually remove the invading foreign body. The inflammation is an expected local reaction of vascularized tissue to injury [27]. One of the characteristic features of inflammation in higher animals is exudate formation [28]. The inflammatory reaction is composed of two interrelated responses: (1) a humoral response, composed of a complex mixture of hydrolytic [29] plus oxidative enzymes [30] and the like, and (2) cellular response due to leukocytes and monocytes.

The adhesion of macrophages and formation of foreign-body giant cells (FBGCs) have been observed attached to the surface of biomaterials both *in vitro* and *in vivo* [31–33]. Based upon the cellular response of phagocytosis, polymer biodegradation has been hypothesized as being caused or potentiated by the metabolic activity of hydrogen ions, lysosomal enzymes, and peroxides released from the adherent macrophages at the cell-polymer interface [34].

At present, it is believed that macrophages play a major role in the biodegradation of ether-based polyurethanes. It is known that phagocytic cells such as macrophages and FBGCs have evolved efficient mechanisms for lysing foreign bodies, leading to molecular degradation [35]. While the exact mechanism of foreign-body giant cell formation remains unclear, it is widely accepted that foreign-body giant cells are formed by the fusion of many activated macrophages.

Because of their long life and preferential adhesion to biomaterials, Henson has drawn attention to the macrophages and FBGCs [36]. The presence of phagocytic cells such as macrophages and FBGCs are suspected to be responsible for the *in vivo* degradation of polymers. Upon adherence, macrophages undergo morphologic and surface changes, resulting in membrane perturbation and cell activation. The activated macrophages release metabolic products that are highly concentrated at the interface between cells and the polymer surface. Thus, chemical degradation driven by oxidation or catalyzed by lysosomal enzymes is most likely to occur at the cell-polymer interfacial zones [37].

We thus speculate that macrophages attach themselves to the surface of the polyurethanes and synthesize degradative enzymes that slowly degrade the polyurethanes at the susceptible ether linkages. The oxidative degradation of the ether linkages eventually leads to the phenomenon we call biologically induced environmental stress cracking [38,39].

III. BIOSTABLE POLYURETHANES

Because of a lack of *in vivo* stability to support the use of Pellethane polyurethane in chronic implants, the Dow Chemical Company notified users that, after April 1, 1992, it would no longer supply Pellethane for chronic implantations. This decision was applauded within the biomaterials community as an example of good corporate responsibility. In the summer of 1992, Ethicon informed their customers that Biomer would no longer be offered as a commercially available material.

The removal of Pellethane and Biomer as available biomaterials underscored the need for a new generation of biodurable polyurethane-based elastomers. Coury [40] was among the first to be granted a U.S. Patent for an ether-free polyurethane. The ether-free polyurethane was called Biostable PUR; it consisted of an aliphatic polyurethane without ether linkages in the backbone. It was synthesized from 1,4 cyclohexanediisocyanate, 1,6 hexane diol, 9- or 10-hydroxymethyl octadecanol, and dimer isocyanate. The last, soft-segment-producing monomer was derived from dimerized fatty acids and contained a complex 36-carbon aliphatic hydrocarbon backbone. While the polymer was indeed biostable, it proved much too stiff for most applications (see Table 2, secant modulus).

We have been testing a new polycarbonate-based polyurethane for several years. We theorized that environmental stress cracking was caused by the *in vivo* oxidation of the alpha methyl group in the polyether linkage. Experimental evidence supported this theory since high durometer Pellethane, which contained fewer polyether macroglycol linkages, was less likely to exhibit ESC compared with low durometer Pellethane polymers [41].

Three monomers are used to synthesize ChronoFlex: an aliphatic diisocyanate (hydrogenated methylene diphenyl diisocyanate, HMDI), a polycarbonate-based diol, and 1,4 butane diol. We selected a stoichiometric ratio specifically designed to produce polymers with a low modulus of elasticity. Obtaining a low modulus, ether-free polymer proved very difficult. However, we felt that, in most applications, a low modulus was

Table 2 Physical Properties

	Biostable PUR 88A (Medtronic)	Corethane 80A (Corvita)	ChronoFlex 80A (PolyMedica)
Tensile strength (psi)	3710	5600	5300
Elongation (%)	388	600	585
100% secant modulus	1613	1200	650
300% secant modulus	2500	1600	600

essential to produce catheters with low trauma potential or vascular grafts with proper compliance.

After months of experimentation, we were able to synthesize a polymer with high ultimate tensile strength with a remarkably low secant modulus of elasticity (see Table 2). The new polycarbonate-based polyurethane, tradenamed ChronoFlex, was synthesized from a 2000 molecular weight (MW) polycarbonate macroglycol as shown in Fig. 1. The macroglycol is synthesized by the reaction of 1,6 hexanediol and ethylene carbonate, resulting in poly[1,6-hexyl 1,2-ethyl carbonate] diol. The molecular weight of the macroglycol is controlled by the reaction time and temperature.

The polymer is synthesized by reacting the polycarbonate macroglycol with a stoichiometric excess of HMDI to form an isocyanate-terminated prepolymer, which is then chain extended with 1,4 butane diol to form the elastomer (see Fig. 2).

As the data indicate, we were successful in producing a very strong and yet very compliant polycarbonate-based polymer. We compare Biodurable PUR and Corethane to ChronoFlex since these are two competing polycarbonate-based biodurable polyurethanes.

A. Materials and Methods

Elastomeric tubing was extruded for *in vivo* experiments. Three different polyurethanes were used: the urethane to be studied (ChronoFlex AL-80A) and two positive controls (Tecoflex EG-85A and Pellethane 2363-80A). The samples were formed as short tubes of specified dimensions in order to interface with injection-molded ESC mandrels obtained courtesy of Medtronic, Incorporated, Minneapolis, Minnesota.

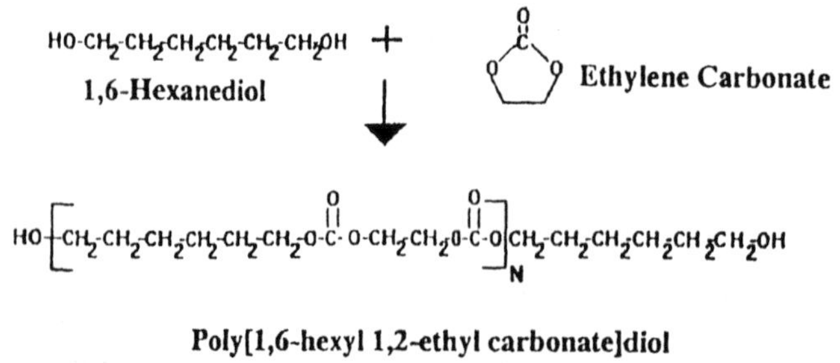

Poly[1,6-hexyl 1,2-ethyl carbonate]diol

Figure 1 Polycarbonate macroglycol synthesis.

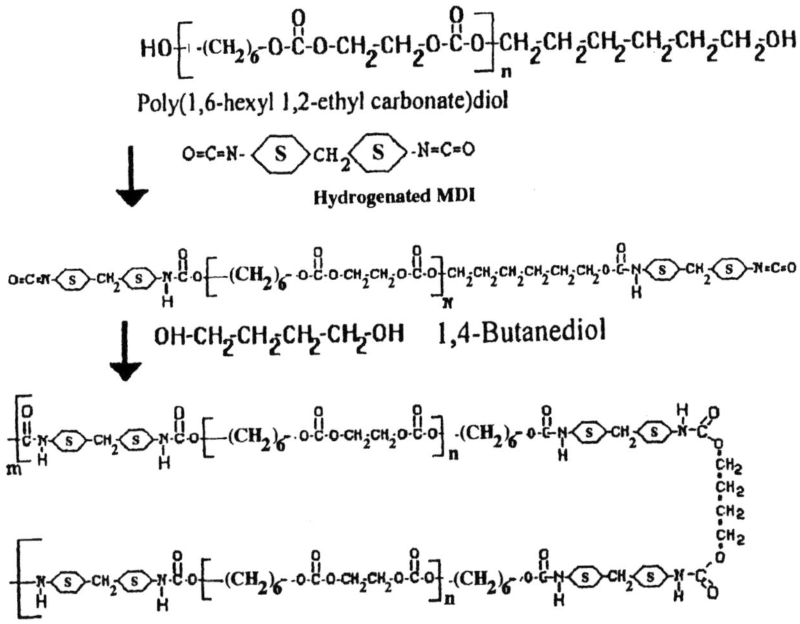

Figure 2   Polycarbonate urethane synthesis.

The ChronoFlex urethane elastomer used for this experiment was formulated to have a durometer of between 80 and 82 on the Shore A hardness scale. Since ChronoFlex urethane is a thermoplastic, it was extruded to make sample tubing. The tubing was extruded by Adam Spence, Incorporated, of Wall, New Jersey.

The extruded ChronoFlex tubing samples had an outside diameter (OD) of 0.085″ (1.93 millimeters [mm]) with an inside diameter (ID) of 0.070″ (1.6 mm). The nominal tubing length was at least 1.5 inches (38 mm), and it had a maximum length of 2.0 inches (51 mm).

Positive control samples were implanted, along with the ChronoFlex samples, as an aid in validating the experimental results. Two materials that have been shown to exhibit stress cracking during prior *in vivo* experiments were used: Tecoflex EG-85A and Pellethane 2363-80A. Tecoflex was chosen as the material for the positive control samples since its aliphatic composition is close to ChronoFlex. Pellethane was chosen to represent a typical aromatic-based PU. Biomer was chosen to test the biostability of a virtually cross-linked, ether-based polyurethane, and Tecothane was tested since it is a polymer that was introduced a short time ago.

The positive control samples were short lengths of extruded Tecoflex EG-80A tubing, with an ID of 0.072 inches and an OD of 0.086 inches (1.82 mm × 2.18 mm). The finished tubing was obtained directly from Thermedics. Pellethane, Tecothane, and ChronoFlex tubing were extruded with an ID of 0.070 and a nominal OD of 0.075. All samples were tested at the highest practical strain levels, namely, 500%.

Injection-molded polysulfone mandrels in the shape of a "dumbbell" were used to maintain the strain level applied to each sample according to the Stokes Method [40]. Illustration 1 depicts the dimensions of the polysulfone mandrel utilized in this study.

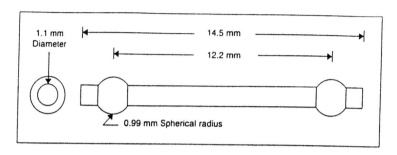

Illustration 1.

Four samples (with identical strain levels) were mounted on mandrels simultaneously. The desired number of cleaned mandrels were placed inside the long piece of tubing. After clamping one end of the tubing, the mandrels were positioned to provide uniform spacing between samples. Then, the free end of the tube was clamped as well. A 2-0 polyester suture was placed around the end of each mandrel. The ligature was as firm as possible without breaking the suture material. The ligature secured the tubing to the short stem projecting from the end of each mandrel.

The distance between the two clamps (which were held at the ends of the tubing) was increased until the tubing was stretched to the desired elongation. The ligature was then placed on the tubing at the end of the mandrel. The clamping force was released, and the individual samples were separated from one another with sharp scissors. Illustration 2 shows the tubing samples mounted on mandrels.

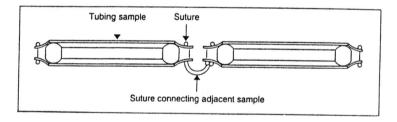

Illustration 2.

Gamma sterilization was used to sterilize the samples. After the sterilization process was completed, the packaged samples were kept in a cool (23–25°C) and well-ventilated area until implantation. A total of eight samples was implanted in each animal. Four samples were placed on each side of the animal. The samples were implanted in two tissue pockets parallel to the animal's spine. Each pocket was about 1.5 inches (37 mm) from the spine, and approximately centered between the neck and tail. This is seen in Illustration 3. The tissue pocket was made between the hypodermis and muscle tissue.

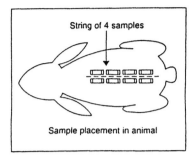

Illustration 3.

B.   Experimental Design

The stressed samples placed in the animals were as follows:

12 Positive control (aliphatic) samples
24 ChronoFlex samples
12 Positive control (aromatic) samples
12 Tecothane samples
12 Biomer samples

The position of the eight samples within any given animal was randomly determined. Adult, female, New Zealand white rabbits (*Orytolagus cuniculus*) were used in the study. Six mature rabbits, each weighing 3 to 5 kilograms (kg), were used. The dorsal sides of the animals were clipped free of fur. Loose hair was removed by a vacuum. Each animal was anesthetized with 30 mg/kg of sodium pentobarbital. The animal's skin at the surgical site was shaven and thoroughly cleaned with a topical antiseptic. Prior to implantation, the area was swabbed with a surgical preparation solution (betadine).

A longitudinal incision (approximately 5-centimeters [cm] long) was made through the skin over the spinal column. The skin was reflected and the attached connective tissue cut to make small subcutaneous pouches on both sides of the spine. The pouches lay between the hypodermis (fat layer) and the muscular tissue. One set of four mandrels was placed in each pouch. The skin wound was then closed with sutures.

At the end of the experimental period, the animals were euthanized by an injection of Somlethol euthanasia solution. The sites containing the mandrels were carefully removed by slicing around the implant site with a scalpel and lifting the tissue and mandrels in a block. The tissues containing the mandrels were placed into 0.9% sodium chloride.

Prior to inspecting each explanted sample for microcracks, all biological tissue was enzyme digested from the sample to provide an unimpeded view of its surface. By comparing the unimplanted samples with the surface-cleaned explanted samples, the presence of biological degradation (such as surface frosting and tubing fracture) could be seen on some of the specimens.

C. Scanning Electron Microscopy

The enzyme surface-cleaned samples were dried in a vacuum for 24 hours at 38°C and placed in a sealed aluminized pouch for storage until testing. Scanning electron microscopy (SEM) analysis was performed by coating the specimen surfaces with gold-palladium to generate secondary electrons, backscattered, electrons and x-ray signals. These signals were collected and displayed on a viewing cathode ray tube and photographed with Polaroid film for use as a permanent record.

The samples were assessed ultrastructurally for topographical characteristics at magnifications ranging from 20 to over 1000 diameters. The surfaces were analyzed for any evidence of microfissures, fragmentation, or signs of chemical degradation. This is shown in Figs. 3 through 5.

Comparing the results of this experiment to other published reports on the biostability of polyurethane served two purposes. It demonstrated that the experimental procedure used in this experiment was similar, or identical, to commonly accepted practice. It also reinforced the hypothesis that ether-free ChronoFlex does not undergo biologic microcracking.

D. Results and Discussion

Once the specimens have been explanted, the Stokes method becomes primarily a visual test. Under scanning electron microscopy, the surface morphology is carefully analyzed for evidence of microcracking or frank polyurethane failure. Table 3 presents the experimental matrix and results.

Figure 6 visually documents our findings that ether-based Tecoflex polyurethane undergoes environmental stress cracking when tested in accordance with this protocol. In the same test and under identical experimental conditions, ChronoFlex polyurethane remained intact, with no evidence of incipient degradation (see Fig. 7).

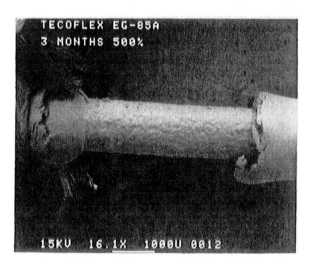

Figure 3 After only a three-month implantation, the Tecoflex EG-85A tubing is clearly seen fractured catastrophically. A small piece of tubing still clings to the left side of the mandrel, maintained in place by the left suture. At the fracture line, the Tecoflex tubing surface is severely degraded by biologically-induced environmental stress cracking (BI-ESC) (three-month implant at 500% elongation).

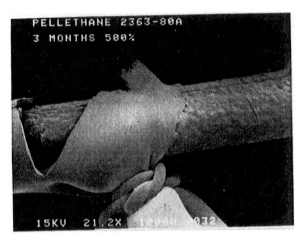

Figure 4 This representative sample of Pellethane 2363-80A tubing shows complete failure of the tubing afer three-month implantation at 500% elongation.

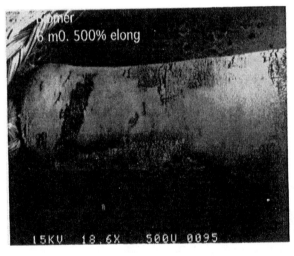

Figure 5 After a six-month implantation, all Biomer samples had undergone severe surface cracking when elongated at 500%.

Table 3 Experimental Matrix Percentage of Specimens Exhibiting Environmental Stress Cracking

Polymer	Strain (%)	Result
Positive control (Tecoflex EG-85A)	500 (3 months)	100% failure, $n = 12$
Positive control (Pellethane 2363 80AE)	500 (6 months)	84% failure, $n = 12$
ChronoFlex 80A	500 (6 months)	0% failure, $n = 24$
Tecothane	500 (6 months)	100% failure, $n = 12$
Biomer	500 (6 months)	95% failure, $n = 12$

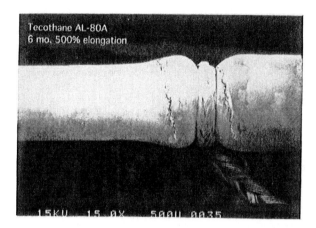

Figure 6   Tecothane is supposedly a Pellethane clone and, indeed, after a six-month implantation, Tecothane 80A does behave like Pellethane and undergoes environmental stress cracking.

We tested these polyurethanes at 500% elongation (to maximize the accelerating effect of strain on ESC). However, it must be emphasized that many polyurethanes have a maximum elongation at break of between 550–625%. By testing polyurethanes so close to their ultimate elongation, we can expect some tensile failures to occur that are unrelated to ESC. However, tensile failures will appear as deep holes, devoid of microcracked edges.

We believe that a more rational test for ESC will consist of testing polymers for a total of six months at an elongation of 400%.

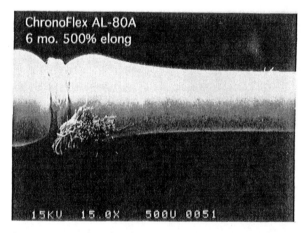

Figure 7   ChronoFlex 80-A, a polycarbonate-based polyurethane elastomer (seen here at the lowest magnification), does not display any evidence of microcracking after a six-month implantation in a rabbit. The tubing can be seen tightly surrounding the environmental surface cracking (ESC) mandrel since the tubing specimen was originally stressed to 500% elongation.

E. Conclusion

The primary purpose of this experiment was to verify that polycarbonate-based polyurethanes are biologically stable. This was ascertained by showing that ChronoFlex tubing
does not exhibit environmental stress cracking when severely stressed for extended periods
of time (six months) under accelerated *in vivo* conditions (500% elongation).

The extruded ChronoFlex tubing displayed no evidence of surface microcracking
after a six-month implantation. It is our hypothesis that polycarbonate-based polyurethane should be more biostable than polymers based on 100% polyether linkages. ChronoFlex is a polycarbonate-based, low modulus polyurethane elastomer.

Based on the results presented in this study, we have demonstrated the viability of a
polycarbonate-based, low modulus segmented polyurethane that is resistant to environmental stress cracking. Thus, we believe we have been successful in developing a biostable
polyurethane elastomer worthy of further evaluation.

REFERENCES

1. H.E. Kambic, S. Murabayashe, and Y. Nose, Biomaterials in Artificial Organs, *C&E News*,
    April 14, pp. 31–44 (1986).
2. M. Szycher, Biostability of Polyurethane Elastomers: A Critical Review, in *Blood Compatible
    Materials and Devices*, C.P. Sharma and M. Szycher (Eds.), Technomic, Lancaster, PA, p.
    42 (1991).
3. S.L. Cooper and A.V. Tobolsky, *J. Appl. Polym Sci.*, 10:1837 (1966).
4. M.D. Lelah and S.L. Cooper, *Polyurethanes in Medicine*, CRC Press, Boca Raton, FL
    (1986).
5. A. Takahara, J. Tashita, T. Kajiyama, and W.J. MacKnight, Microphase Separated Struc-
    ture, Surface Composition, and Blood Compatibility of Segmented poly(urethaneureas) with
    Various Soft Segment Components, *Polymer*, 26:987–996 (1985).
6. A. Takahara, R.W. Hergenrother, A.J. Coury, and S.L. Cooper, Effect of Soft Segment
    Chemistry on the Biostability of Segmented Polyurethanes. II. *In Vitro* Hydrolytic Degrada-
    tion and Lipid Sorption, *J. Biomed. Mater. Res.*, 26:801–818 (1992).
7. M. Szycher, A.A. Siciliano, and A.M. Reed, Polyurethanes in Medical Devices, *Medical
    Design and Material*, 1(2):19 (1991).
8. J.W. Boretos, Long Term *In Vitro* Stability of Segmented Polyurethane, *Proc. 8th Annual
    Soc. Biomater.*, 6:473–476 (1972).
9. K.B. Stokes and K. Cobian, Polyether Polyurethanes for Implantable Pacemaker Leads,
    Biomaterials, 3:225–231 (1982).
10. K. Stokes, P. Urbanski, and J. Upton, The *In Vivo* Autooxidation of Polyether Polyurethanes
    by Metal Ions, *J. Biomat. Sci. Polym.*, 1:207–230 (1990).
11. K.B. Stokes, P. Urbanski, and K. Cobian, New Test Methods for the Evaluation of Stress
    Cracking and Metal Catalyzed Oxidation in Implanted Polymers, in *Polyurethanes in Biomed-
    ical Engineering II*, H. Planck et al. (Eds), Elsevier, Amsterdam, pp. 109–127 (1987).
12. A.J. Coury, K.B. Stokes, P.T. Cahalan, and P.C. Slaikeu, Biostability Considerations for
    Implantable Polyurethanes, *Life Support Syst.*, 5:25–39 (1987).
13. K.B. Stokes, A.J. Coury, and P. Urbanski, Auto-Oxidative Degradation of Implanted Poly-
    ether Polyurethane Devices, *J. Biomater. Appl.*, 1:411–448 (1987).
14. K.B. Stokes, Polyether Polyurethanes: Biostable or Not?, *J. Biomater. Appl.*, 3:228–259
    (1958).
15. F.J. Schoen, H. Harasaki, K.M. Kim, H.C. Anderson, and R.J. Levy, Biomaterials Associ-
    ated Calcification and Strategies for Prevention, *J. Biomed. Mater. Res.*, *Appl. Biomater.*,
    22(A1):11–36 (1988).

16. R.E. Marchant, Q. Zhao, J.M. Anderson, A. Hiltner, and R.S. Ward, Surface Degradation of Biomedical Polyurethanes, in *Surface Characterization of Biomaterials*, B.D. Ratner (Ed.), Elsevier, Amsterdam, pp. 297–311 (1988).

17. R.E. Marchant, Q. Zhao, J.M. Anderson, and A. Hiltner, Degradation of a Poly (Ether Urethane Urea) Elastomer: Infrared and XPS Studies, *Polymer*, 28:2032–2039 (1987).

18. M. Szycher and W.A. McArthur, Surface Fissuring of Polyurethanes Following *In Vivo* Exposure, *Corrosion and Degradation of Implant Materials*, ASTM STP 859, A.C. Fraker and C.D. Griffin (Eds.), Philadelphia, pp. 308–321 (1985).

19. A. Takahara, R.W. Hergenrother, A.J. Coury, and S.L. Cooper, Effect of Soft Segment Chemistry on Biostability of Segmented Polyurethanes I. *In Vitro* Oxidation, *J. Biomed. Mater. Res.*, 25:341–356 (1991).

20. M. Szycher and W.A. McArthur, Surface Fissuring of Polyurethanes Following *In Vivo* Exposure, *Corrosion and Degradation of Implant Materials*, ASTM STP 859, A.C. Fraker and C.D. Griffin (Eds.), Philadelphia, p. 318 (1985).

21. D.J. Parins, K.M. Black, K.D. McCoy, and N.J. Horvath, *In Vivo* Degradation of a Polyurethane, report from Cardiac Pacemakers, St. Paul, MN (1981).

22. A. Takahara, R.W. Hergenrother, A.J. Coury, and S.L. Cooper, Effect of Soft Segment Chemistry on the Biostability of Segmented Polyurethanes, *J. Biomed. Mater. Res.*, 26(6): 801–818 (1992).

23. H.D. Wabers, T.J. McCoy, A.T. Okkema, R.T. Hergenrother, M.F. Wolf, and S.L. Cooper, Biostability, and Blood-Contacting Properties of Sulfonated Grafted Polyurethane and Biomer, *J. Biomater. Sci. Polym.*, 4(2):107–133 (1992).

24. M. Szycher, D. Dempsey, and V.L. Poirier, Surface Fissuring of Polyurethane-Based Pacemaker Leads, *Intl. Biomat. Symp.*, 7:24 (1984).

25. K.B. Stokes, P. Urbanski, and K. Cobian, New Test Methods for the Evaluation of Stress Cracking and Metal Catalyzed Oxidation in Implanted Polymers, in *Polyurethanes in Biomedical Engineering II*, H. Planck et al. (Eds.), Elsevier, Amsterdam, pp. 109–128 (1987).

26. K.B. Stokes, *The Biostability of Various Polyether Polyurethanes Under Stress*, Medtronic, Minneapolis, MN (1983).

27. R.E. Phillips, M. Frey, and R.O. Martin, Long-Term Performance of Polyurethane Pacing Leads: Mechanisms of Design-Related Failures, *PACE*, 9:1166–1167 (1986).

28. R.P. Kambour, A Review of Crazing and Fracture in Thermoplastics, *J. Polym. Sci., Macromol. Rev.*, 7:1–154 (1973).

29. J.B. Walter and M.S. Israel, *General Pathology*, 6th Ed., Livingstone Churchill, New York (1987).

30. J.M. Anderson, Inflammatory Response to Implants, *Trans. Am. Soc. Artif. Inter. Organs*, 34:101–107 (1988).

31. Q. Zhao, R.E. Marchant, J.M. Anderson, and A. Hiltner, Long Term Degradation *In Vitro* of a Poly(Ether Urethane Urea): A Mechanical Property Study, *Polymer*, 28:2040–2046 (1987).

32. B.D. Ratner, K.W. Gladhill, and T. Horbett, Analysis of *In Vitro* Enzymatic and Oxidative Degradation of Polyurethanes, *J. Biomed. Mater. Res.*, 22:509–527 (1988).

33. J.M. Anderson and K.M. Miller, Biomaterial Biocompatibility and the Macrophages, *Biomaterials*, 5:5–10 (1984).

34. R.E. Marchant, K.M. Miller, and J.M. Anderson, *In Vivo* Biocompatibility Studies. V. *In Vivo* Leukocyte Interaction with Biomer, *J. Biomed. Mater. Res.*, 18:1169–1190 (1984).

35. K. Smetana, Multinucleate Foreign Body Cell Formation, *Exp. Mol. Pathol.*, 46:258–265 (1987).

36. P.M. Henson, Mechanism of Exocytosis in Phagocytic Inflammatory Cells, *Am. J. Pathol.*, 101:494–511 (1980) [R. E. Marchant, Cell Adhesion and Interactions with Biomaterials, *J. Adhesion*, 20:211–225 (1986)].

37. Q. Zhao, N. Topham, J.M. Anderson, A. Hiltner, G.A. Lodoen, and C.R. Payet, Foreign

Body Giant Cells and Polyurethane Biostability: *In Vivo* Correlation of Cell Adhesion and Surface Cracking, *J. Biomed. Mater. Res.*, 25:177 (1991).

38. K.L. Spilizeski, R.E. Marchant, J.M. Anderson, and A. Hiltner, *In Vivo* Leukocyte Interactions with the NHLBI-DTB Primary Reference Materials: Polyethylene and Silica-Free Polydimethylsiloxane, *Biomaterials*, 8:12–17 (1987).

39. Q. Zhao, M.P. Agger, M. Fitzpatrick, J.M. Anderson, A. Hiltner, K. Stokes, and P. Urbanski, Cellular Interactions with Biomaterials: *In Vivo* Cracking of Prestressed Pellethane 2363-80A, *J. Biomed. Mater. Res.*, 24:622 (1990).

40. A.J. Coury and C.M. Hobot, Method for Producing Polyurethanes from Poly-(Hydroxyalkyl Urethane), U.S. Patent 5,001,210 (March 19, 1991).

41. A. Takahara, A.J. Coury, R.W. Hergenrother, and S.L. Cooper, Effect of Soft Segment Chemistry on the Biostability of Segmented Polyurethanes. I. *In Vitro* Oxidation, *J. Biomed. Mater. Res.*, 25:314–356 (1991).

XI
Implant Wear and Debris

55
The Wear of Ultra-High Molecular Weight Polyethylene

C. Mauli Agrawal, D. M. Micallef, and J. Mabrey
University of Texas Health Science Center at San Antonio
San Antonio, Texas

I. INTRODUCTION

Total joint replacement has become an effective and popular surgical technique for restoring function to patients whose joints have been afflicted by major trauma or joint degenerative diseases like arthritis. Millions of patients have been fitted with total joint prostheses over the past three decades, and approximately 500,000 total hip and knee replacements are performed every year worldwide [1]. In addition, it is estimated that tens of thousands of shoulder, elbow, and wrist replacements are performed every year. A majority of these prostheses have a metal surface articulating against a second component of ultra-high molecular weight polyethylene (UHMWPE). Over the past 30 years, UHMWPE has evolved as the material of choice in total joint prostheses because of its biocompatibility and relatively superior friction and wear characteristics.

II. HISTORICAL BACKGROUND

In the early 1950s, two English surgeons, McKee and Watson-Farrar, introduced the first clinically successful total hip prosthesis [2]. Both the acetabular and femoral components in this prosthesis were fabricated from a cast cobalt-chrome (CoCr) alloy. It was determined, however, that this design suffered from high frictional torque and severe tissue reactions to wear products.

The great success of total joints today can be traced to the work of Sir John Charnley [12], who realized that the metal-on-metal design was tribologically unsound for *in vivo* applications. In 1959, he introduced poly(tetrafluoroethylene) (PTFE) acetabular cups in conjunction with stainless steel femoral components. Despite its excellent frictional characteristics, PTFE had inadequate wear characteristics, which led to failure of the prostheses after only two to three years in service. In addition, even though PTFE was biocompatible in bulk form, its wear debris unleashed an aggressive foreign-body response. Charnley then successfully turned to UHMWPE for the acetabular component. This material exhibited superior wear characteristics compared with PTFE.

Researchers have also investigated several other polymeric materials [3]. Some materials, like polyamides (Delrin), have been used clinically, but have not proved to be satisfactory [4]. Others have remained restricted to laboratory studies. Ceramic-on-ceramic total joint prostheses were introduced in the 1970s, but the high dimensional accuracy required of components fabricated from these materials and their brittle natures have limited their wide-scale use [5-8]. A summary of some of the different materials clinically evaluated for total joints has been provided by Dumbleton [9].

However, none of these materials have proved to be superior to UHMWPE, which over the past 30 years has emerged as the dominant material used in total joint prostheses [10]. According to Black, "it (UHMWPE) has proven to be the best polymer for load bearing applications in metal-polymer wear pairs whether *in vitro* or *in vivo*" [11, p. 156]. Despite the success of UHMWPE, evidence gradually emerging over the past few years indicates that there are some clinical problems associated with this material.

III. CLINICAL PROBLEMS

Charnley abandoned PTFE as an acetabular surface because of the massive granuloma formation from the resulting wear debris and chose UHMWPE for its superior wear characteristics, concluding that if "wear has to be accepted as inevitable, that the ideal implant will produce very small particles" [12, p. 6]. His choice of UHMWPE was supported by several studies over the next three decades documenting linear wear rates of 0.005 to 0.2 millimeters (mm) per year *in vivo* in successful cases [13-18]. These amounts appear insignificant, but the primary problem with UHMWPE in total hips is the constant generation of submicron particle debris that overwhelms the body's ability to remove the material. This particulate load is of the order of 40 billion per year per hip or greater [19] and significantly contributes to prosthetic loosening and periprosthetic bone loss from osteolysis (see Fig. 1).

While this wear may be continuous, the rate of wear changes over time according to Charnley. Over a 10-year period, the acetabular components in his series had an average total linear wear of 1.5 mm or 0.15 mm per year. However, the rate of wear actually decreased with time, demonstrating a wear rate of 0.18 mm per year for the first five years and 0.10 mm per year during the last four [13]. As has been shown by Livermore, Ilstrup, and Morrey, volumetric wear strongly correlates with the size of the prosthetic femoral head, and components with larger diameters have higher wear rates [20]. The benchmark for failure in total joint arthroplasty is revision of the components. A cumulative acetabular revision rate of 3.5% at 10 years for patients with 32-mm femoral heads compared to a significantly lower rate of 1.5% for 28-mm heads over the same time period has been reported [21].

In addition to the submicron wear debris seen in total hips, certain total knee arthroplasty designs suffer from gross destruction of the UHMWPE components [22-32] secondary to problems with the manufacturing process, the tibiofemoral articular geometry, and the thickness of the implant [25,33]. Heat-treated UHMWPE may delaminate under load, especially in designs based on a cylindrical femoral component articulating with a relatively flat UHMWPE tibial surface. A UHMWPE thickness of less than 4.0 mm is strongly correlated with gross failure of the component [24,25,28,30]. Hood,

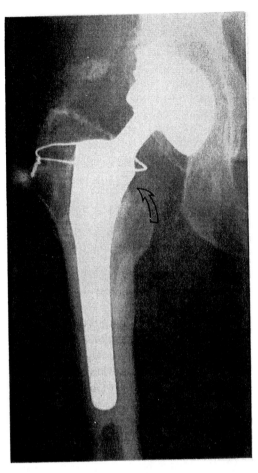

Figure 1 Femoral cortical osteolysis (arrow) of proximal femur after four-year implantation. Cobalt-chrome (32 mm) articulates with ultra-high molecular weight polyethylene (UHMWPE) acetabular cup.

Wright, and Burstein have developed a classification system for the surface damage seen on retrieved UHMWPE total knee components [34] (Table 1) and have demonstrated a significant correlation between patient weight and length of implantation with the surface damage score.

Overloaded with wear debris from the prosthesis, the body's defenses attempt to isolate the material from surrounding tissues. Newman and Scales reported the first adverse clinical reaction to UHMWPE debris in 1951 after removing a failed all-polyethylene resurfacing device from a hip [35]. Retrieved granulation tissue contained an abundance of foreign-body giant cells loaded with irregularly shaped pieces of polyethylene that were "very conspicuous" under polarized light. In one particular study, UHMWPE debris was observed in all specimens from cases with a polyethylene component; this debris stimulated the most significant cellular response when compared to metal or poly(methylmethacrylate) (PMMA) debris [36].

Various authors have studied the interface between the prosthesis and the surround-

Table 1   Modes of Surface Damage of Retrieved Ultra-High Molecular Weight
Polyethylene Tibial Components

Mode	Damage
Pitting	Shallow, irregular surface voids
Embedded acrylic debris	Particles from bone cement
Scratching	Gouging on microscopic scale
Burnishing	Highly polished
Abrasion	Shredded or tufted UHMWPE
Surface deformation	Permanent deformation
Delamination	Surface sheets separated from component

Source: From Ref. 34

ing bone and have identified an intervening membrane with very distinct features [12,37–48], including macrophage infiltrates with intracytoplasmic collections of submicron particles of UHMWPE. One striking feature is the accumulation of very fine particles of UHMWPE in tissue at the distal tip of femoral components that were otherwise well fixed [46] and along the bone cement interface of well-functioning acetabular implants [40]. The technique of visualizing smaller UHMWPE debris, within the range of less than 1 micron (μ) to 5 microns, has been enhanced by the addition of a modified oil red O stain [49,50]. Large fragments stain bright pink, as does the cytoplasm of cells containing smaller-size debris.

Willert and Semlitsch were among the first to propose the concept that UHMWPE wear debris lead to a periimplant giant cell reaction and noted that "the continuous production of foreign material necessitates a constant supply of cells capable of storage" [47, p. 159]. The clinically troublesome sequelae of the body's overreaction to this debris is loosening of the device and/or destruction of the bone supporting the implant. It has been determined that resorption of the proximal femoral neck and lysis of the proximal femur correlate with the extent of both linear and volumetric wear, suggesting an association between UHMWPE wear debris and osteolysis [20]. Osteolysis has been reported both in association with total knee [25–27] and total hip arthroplasties [17,41,44,48,51–55], with the great majority of cases involving total hips. The incidence of osteolysis ranges from 0.13% in one series of 3000 cemented hips [56] to as high as 13% in a series of 154 uncemented hips [54]. In this last series, 75% of the osteolytic cases appeared within three years [54].

Intracellular UHMWPE has been detected in cases of total hips with periprosthetic osteolysis [41]. In these cases, bone resorption has been found to occur in association with macrophages laden with UHMWPE debris and a direct relationship with the number of macrophages present has been determined. Transmission electron microscopy of retrieved tissue has shown intracytoplasmic particles of ovoid-shape UHMWPE particles as small as 0.2 by 0.1 microns [41]. Recent investigations of wear debris utilizing atomic force microscopy have identified similarly shaped UHMWPE particles as small as 0.05 microns in diameter (Fig. 2) [57].

These wear particles unleash a cascade of biologically active molecules called *cytokines* that mediates the surrounding bony destruction. Complicating the picture is the

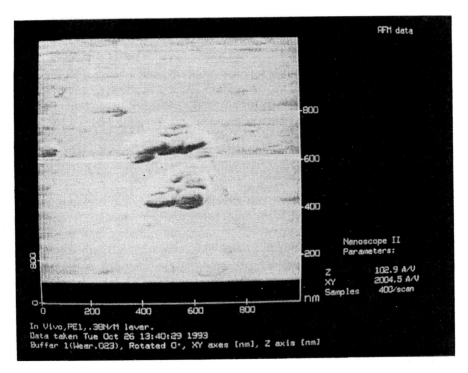

Figure 2 Atomic force micrograph showing ultra-high molecular weight polyethylene (UHMWPE) particles retrieved from human tissue (vertical and horizontal scales in nanometers). (From Ref. 57.)

ability of one cytokine to have multiple biological activities and for multiple cytokines to share common functional properties. These products may be released locally to modulate the activity of other cells or even the cell from which it originated. Other cytokines are produced at distant sites to affect target tissues. Some may even be incorporated into the mineralized bone matrix to be released during the resorptive phase of bone remodeling.

The injection of UHMWPE particles generated *in vitro* into rat's knees containing intraosseous PMMA implants results in active resorption of bone around the implant [58]. A highly cellular layer of connective tissue has been noted some distance from these joints and consists of macrophages and giant cells containing UHMWPE particles. Culturing mouse peritoneal macrophages with 1-micron latex microspheres at a concentration of 108 particles per milliliter (ml) has been shown to increase bone resorption of neonatal mouse calvaria by a factor of 13, while smaller concentrations had no significant effect [59]. This suggests that osteolysis may simply be a function of particle size and number regardless of composition.

The number, size, and shape of UHMWPE particles generated as wear debris will be a function of the properties of the polymer. Thus, any techniques successful in reducing or eliminating the clinical problem will necessarily involve altering either the bulk or surface properties of the UHMWPE components.

IV. PROPERTIES OF ULTRA-HIGH MOLECULAR WEIGHT POLYETHYLENE

UHMWPE is produced by a low pressure polymerization technique using a Ziegler-type catalyst. The resin is mostly linear and, per American Society for Testing and Materials (ASTM) Standard F648-84 for UHMWPE for surgical implants [60], it should have a relative solution viscosity of 2.3 or greater. Commercially available UHMWPE resins have molecular weights ranging from 3×10^6 to 6×10^6 daltons. These extremely high molecular weights impart some unique and desirable properties to the material. For instance, Rose et al. have determined a correlation between higher molecular weights and a reduction in wear [15].

However, high molecular weight also poses several difficulties in processing. Due to its high melt viscosity, UHMWPE is difficult to injection mold and medical devices have to be fabricated using techniques of extrusion, compression molding, and machining. Also, UHMWPE is not easily soluble due to its high molecular weight. Limited solubility can be achieved in xylene, benzene, and decahydronapthalene (Decalin). More detailed information regarding solubility has been provided by Stein [61].

The molecular weight of UHMWPE is an important determinant of its mechanical and physical properties. In general, the molecular weight of polymers can be estimated by dissolving them in a solvent and then using chromatography techniques or measuring the viscosity of the solution. Using these conventional techniques on UHMWPE presents problems because of the low solubility of the very high molecular weight fraction. This difficulty is overcome by dissolving fractions of the polymer in different solvents at elevated temperatures. These fractions are as follows [62]. The lowest fraction, with a molecular weight of about 10^4 daltons or less, is soluble in xylene at 130°C. A middle fraction, with a molecular weight of 6×10^6 daltons or less is soluble in Decalin at 135°C. The higher fraction, above 6×10^6 daltons, is totally insoluble. The molecular weight distributions are very important because crack resistance is extremely sensitive to the amount of low molecular weight components present in the material [62].

ASTM F648-84 also requires that UHMWPE fabricated forms for surgical implants should contain a minimum of extraneous material. A 400-square-centimeter (cm^2) sample should not contain particles greater than 300 micrometers (μm) and in addition there should be no more than 10 particles with dimensions of 300 μm or less. Also, the density of the final form should be in the range 0.93 to 0.94 grams per square centimeter (gm/ cm^2). Other minimum requirements specified by ASTM are listed in Table 2 [60].

UHMWPE exhibits viscoelastic characteristics and creep deformation under an applied load. This property plays a significant role in both the design of total joint prostheses' components and the *in vitro* wear testing of UHMWPE. A limited amount of creep can be beneficial as it promotes better conformity between the bearing surfaces in total hip prostheses. However, large degrees of creep may result in deformation of the UHMWPE components with adverse consequences. The creep behavior of UHMWPE is shown as a function of compressive stress in Fig. 3 [61].

Various grades of UHMWPE have been used for total joint prostheses. The properties of these grades often differ; in addition, these properties may vary within different lots. Since ASTM standards stipulate only the minimum acceptable properties, it is difficult to specify the actual properties of the components unless they are empirically measured or a specification sheet is provided by the supplier.

Table 2   Minimum Requirements for Ultra-High Molecular Weight Polyethylene Fabricated Forms

Property	ASTM test method	Minimum requirement
Tensile strength, 23°C	D 638 (speed C)	
Ultimate	D 638 (speed C)	27 MPa
Yield	D 638 (speed C)	19 MPa
Elongation	D 638 (speed C)	200%
Izod impact strength	D 256 (15° notch A) double	1070 J/M
Deformation under load	D 621 (A) (7 MPa for 24 h)	2% deformation after 90-min recovery
Hardness	D 2240 (Shore D)	60

Source: From Ref. 60

V. WEAR TESTING EQUIPMENT

Since the *in vivo* wear testing of new materials for total joint prostheses is both difficult and expensive, the materials are usually first screened *in vitro*. Most *in vitro* tests simplify and model the *in vivo* conditions and attempt to control closely the parameters of interest. Such tests can determine the friction and wear properties of combinations of materials in a practical and inexpensive way. An extensive review of these wear screening tests has been provided by Clarke [63]. According to Dumbleton [64], absolute predictions of *in vivo* wear rates will not be obtained from laboratory wear testing; only a ranking of the relative wear resistance of combinations of materials can be expected.

Kumar et al. performed a test in which both unidirectional and reciprocating wear

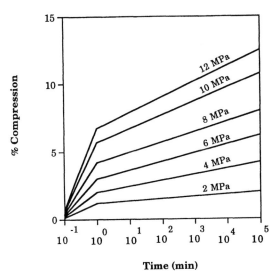

Figure 3   Creep properties of ultra-high molecular weight polyethylene (UHMWPE) under different compressive stresses (measured at 20°C). (From Ref. 61.)

devices were used to evaluate the possible mechanical influences on the wear of UHMWPE against stainless steel, zirconia, and alumina [65]. Using a polymeric pin on a flat of the harder material, the unidirectional device produced wear factors 10–15 times greater than the reciprocating device under similar conditions. Similar findings, though to a lesser degree, were reported by Brown, Atkinson, and Dowson [66], in which the effects of unidirectional and reciprocating motions were investigated during dry sliding of an UHMWPE pin against stainless steel.

A reciprocating machine is recommended by ASTM Standard F732-82 for performing wear tests on materials used for prostheses [67]. Some of the pros and cons of reciprocating motion machines are listed below [68,69]. A reciprocating machine

1. Has a sinusoidal velocity profile
2. May cause the pin to bend (the test apparatus and specimens are not perfectly rigid), resulting in the contact area moving toward the leading edge of the sliding pin, especially during motion reversal, when the pin may rock
3. Is less likely to develop nonphysiologic hydrodynamic lubrication
4. Has more opportunity for wear debris to act as abrasive particles, which occurs *in vivo*
5. Has a ratio of specimen excursion to contact area that can be maintained comparable to that of actual prostheses
6. Has direction reversals that may contribute to wear through a fatiguelike process

Since the motion between the articulating surfaces in total joints is oscillatory, it is more relevant to use a wear test device with reciprocating motion than one providing unidirectional motion.

Primary considerations in the design of a wear test machine are simplicity, reliability, and reproducibility of results [64]. There exist a number of different configurations for these machines. Test results from one configuration will not necessarily be the same as those from another. In general, the *in vitro* test equipment can be divided into four categories, Types I–IV.

A. Type I

Type I machines usually perform simple wear tests with few attempts to simulate physiologic conditions. Tests are often performed at high speeds and loads with lubricants that are not entirely relevant to human joints. Some machines utilize steady speeds with nonreciprocating motion. In recent years, the use of such machines to test materials for total joint prostheses has declined.

B. Type II

Machines belonging to the Type II category usually use pin-on-disk [70], pin-on-plate [71], disk-on-plate [72], or annulus-on-plate [73] configurations (see Section VI.C). These geometries do not accurately simulate total joint prostheses, but they provide a means for evaluating the test materials under closely defined conditions. Type II machines attempt to match the frequency and sliding distance of human motion, and use saline or bovine serum for lubrication. It has been argued that wear tests performed using bovine serum lubrication better approximate *in vivo* conditions because saline lubrication yields a wear mechanism different from that produced in biologic fluids [74].

Machines belonging to the Type II category use a constant load acting between the two articulating surfaces. This load is chosen so that it results in a good approximation of the average contact stresses encountered physiologically. The standards for performing wear tests using Type II machines have been set forth in ASTM F732-82 [67].

Type II machines have two drawbacks. First, they use constant loads instead of the dynamic load cycles experienced by human joints. The effects of dynamic loading on the wear of UHMWPE have not yet been investigated in detail. However, it is possible that such load profiles may contribute to fatigue-related wear of the polymer. Second, pin-on-plate and pin-on-disk configurations of Type II machines usually use an UHMWPE pin on a metal or ceramic counterface. Once again, this configuration may not accurately capture the effects of fatigue loading on the wear process. These effects gain significant relevance in evaluating materials for nonconforming total knee and shoulder prostheses designs in which the motion is a complex combination of rotation and translation. Notwithstanding these limitations, Type II machines have been used extensively for screening materials in the past. They are economical, compact, simple in design, and are easy to operate.

C. Type III

Recent advances in the wear testing of UHMWPE for total joint prostheses have resulted in the development of Type III machines. Machines belonging to this category are similar to Type II machines, but are computer controlled and are capable of producing dynamic load profiles to simulate physiologic loading accurately (Fig. 4). These machines are more economical to operate than total joint simulators and can be used to investigate

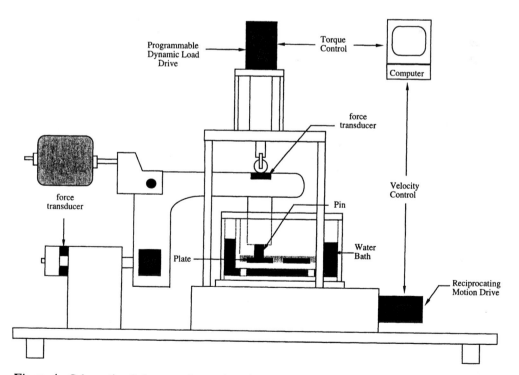

Figure 4 Schematic of pin-on-reciprocating-plate wear test machine.

wear mechanisms and screen new materials using load profiles corresponding to different joints.

D. Type IV

Total joint simulators are classified as Type IV machines. These simulators are very sophisticated wear testing machines that simulate the dynamic load profiles and complex motions of natural joints (Fig. 5) [75]. In state-of-the-art joint simulators, the load profiles and joint movements are variable and computer controlled. The loads are often applied through hydraulic systems. Temperature regulating devices ensure that the tests are performed at physiologically relevant temperatures.

The majority of scientific studies in the public domain that have been performed on joint simulators have used total hip simulators [76]. Knee simulators are more complex to fabricate and run, and relatively few studies using these machines have been reported [77].

Total joint simulators mimic the physiologic conditions relatively closely. However, they are expensive to operate because they require the testing of actual prosthesis components. In addition, the relatively large mass of UHMWPE components compared with the specimens used in Types I to III machines makes it difficult to measure the minute changes in mass due to wear accurately. Despite these limitations, total joint simulators remain the most sophisticated and accurate means of evaluating new prosthesis designs and materials *ex vivo*.

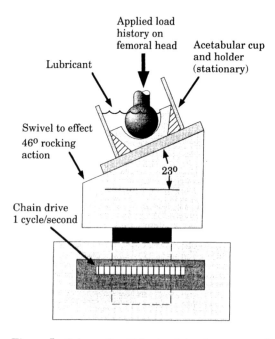

Figure 5   Schematic of hip simulator for wear testing. (From Ref. 75.)

VI. FACTORS INFLUENCING ULTRA-HIGH MOLECULAR WEIGHT POLYETHYLENE WEAR

In the design of a laboratory wear test, there are a number of variables that need to be considered. Even though the test greatly simplifies the physiologic environment in which joint prostheses function, the researcher must decide its complexity and the extent to which it approximates *in vivo* conditions. The following discussion evaluates the variables of importance for both wear testing and data interpretation. It is critical to understand that the wear process is a complex function of many mutually interacting variables. Therefore, isolation of these variables should be attempted with caution.

A. Test Medium

When performing pin-on-flat laboratory wear tests, it is important to simulate normal physiologic conditions as closely as possible in order to produce wear mechanisms similar to those that occur *in vivo*. It has been well established that different fluid media produce different wear rates, frictional coefficients, and wear mechanisms under similar test conditions [65,71,78].

McKellop et al. reported these differences in a study involving a UHMWPE pin loaded (3.45 MPa) against a reciprocating 316 stainless steel disk lubricated with serum, distilled water, or saline solution at 28°C [71]. Wear rates and frictional coefficients were reported for all test cases. Using the same mediums, Kumar et al. compared wear rates and friction of UHMWPE against stainless steel, yttrium oxide partially stabilized zirconia, and alumina at 37°C [65]. They used a similar reciprocating configuration and contact stress as McKellop et al. [71]. Data from these tests for the stainless steel counterfaces are listed in Table 3 and show that the wear rate is highest for saline.

Bovine serum is recommended as the best test medium for laboratory wear testing for prosthetic applications because it is similar to synovial fluid in terms of protein content, and resulting wear rates and wear surface topography resemble those of retrieved total joint replacements [65,71,78].

B. Surface Finish

Polishing the metal or ceramic counterface to a high quality, mirrorlike finish can help reduce UHMWPE wear. Such a polish will decrease the hard asperities that tear and abrade the softer polymer. However, two highly polished mating surfaces may serve to

Table 3   Wear and Friction Dependence on Test Medium

Ref. no.	Bovine serum		Saline		Distilled water	
	Wear factor	Friction coefficient	Wear factor	Friction coefficient	Wear factor	Friction coefficient
71	$0.65 \pm 17\%$ $\mu m/10^6$ cycles	0.07–0.12	$5.20 \pm 17\%$ $\mu m/10^6$ cycles	0.7–0.10	$0.08 \pm 60\%$ $\mu m/10^6$ cycles	0.07–0.13
65	$1.81 \pm 4\%$ mm^3/nm	$0.065 \pm 18\%$	$3.89 \pm 8\%$ mm^3/nm	$0.123 \pm 5\%$	$1.12 \pm 10\%$ mm^3/nm	$0.061 \pm 8\%$

increase adhesive forces between the materials. It has been suggested that textured articulating surfaces may produce an optimal wear resistant system [71]. Such a system may produce low abrasion from asperity contact, low adhesion of the polymer to the smooth metal, and have depressions or channels that act as lubricant reservoirs, allowing for the removal of wear debris. It has been established that the topography of a hard counterface sliding against polymers is a major factor in determining the wear rate in different environments [79].

The effects of counterface roughness on the friction and wear of UHMWPE have been investigated previously [71,80]. Generally, wear rates tend to increase with increasing counterface roughness (Table 4). It has been recommended that surface finish investigations be conducted solely in bovine serum to avoid the formation of a polymeric transfer film that could mask the counterface texture [71]. Furthermore, these tests should continue for a sufficient duration to observe possible changes in the initial surface finish.

It has been reported that, under dry conditions, there is an optimum counterface roughness for minimal wear of UHMWPE [81,82]. However, there is no evidence in the published literature that such an optimal surface roughness exists for fluid environments.

Relatively few studies report any value for the surface finish of the UHMWPE specimens tested. Traditional polishing and lapping techniques usually cannot be used because of the danger of small abrasive particles becoming embedded in the soft polymer [66]. Surface finishes are typically machined, and these reported values have ranged from 0.37–1.5 μm R_a [66,69,78,83,84]. It has been suggested that these machining marks are not relevant because they are removed rapidly during the initial period of the wear test [69,83,84], and this initial wear is not representative of the bulk material [69]. However, it is possible that these original machining marks can be a source of crack initiation and wear debris generation [85].

C. Specimen Geometry

It is usually advisable to keep specimen geometries simple to ensure easy and reproducible fabrication. The pin-on-flat geometry has become the most widely used configuration. Researchers have also reported the use of many other configurations, including block on journal [3], annulus on flat [73], cylinder on flat [86], and disk on flat [72,87]. Figure 6 shows these geometries, and Fig. 7 shows some variations of the pin-on-flat geometry. There is also a strong connection between effective lubrication and specimen geometry. Specimen geometry determines the ease with which lubricants can flow between articulat-

Table 4 Frictional Coefficients of Ultra-High Molecular Weight Polyethylene Against CoCr

Number of specimens	Counterface roughness (μm rms)	Steady-state friction coefficient	Peak friction coefficient
3	Grade A, 0.03–0.05	0.06–0.11	0.15
3	Grade B, 0.07–0.12	0.07–0.12	0.20
3	Grade C, 0.20–0.30	0.11–0.15	0.23
3	Grade D, 0.75–0.76	0.22–0.30	0.30

Source: From Ref. 71

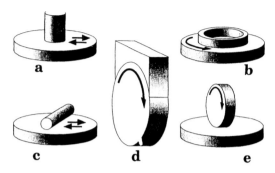

Figure 6   Types of test specimen geometries: (a), pin-on-flat; (b), annulus-on-flat; (c), cylinder-on-flat; (d), block-on-journal; (e), disk-on-disk. (From Ref. 63.)

ing surfaces and remove wear debris. The pin-on-flat geometry is discussed in further detail below.

Many wear test configurations use a flat-on-flat geometry in which a polymeric pin with a flat face is used on a metal or ceramic flat. The flat-on-flat geometry has the advantage that the contact area can be easily calculated and it does not change during testing due to creep and wear. The truncated cone-on-flat geometry can be used to produce smaller contact areas while retaining a larger upper portion to secure the specimen firmly in its holder. This smaller contact area is advantageous because small loads may be used to produce high contact stresses and a more compact wear machine may be designed. However, since the lower portion is conical in shape, the contact area will increase during the wear test in a manner determined by the slope of the taper [83,88].

The hemisphere-on-flat geometry, which is equivalent to a sphere on flat, is most often used when the pin is the harder material, such as a metal or ceramic. The constant radius overcomes problems that arise from misalignments of the pin. An important disadvantage of this specimen geometry is that the contact area continuously increases due to creep and wear during a test. Without accurate knowledge of the contact area, the calculation of the true contact stress becomes impossible [89,90]. In an attempt to overcome this problem, Jones, Hardy, and Crugnola used a hemispherical tipped UHMWPE pin on a metal disk and estimated the contact area by measuring the wear scar on the metal disk [91].

D. Velocity

It has been established that reciprocating motion at a frequency of approximately 1 hertz (Hz) best describes relative movement of a total hip prosthesis. Average surface velocities are in the range of 20 to 50 millimeters per second (mm/s) [64]. It may seem advanta-

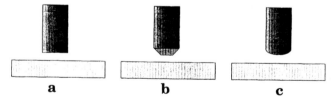

Figure 7   Common pin-on-flat geometries: (a), flat-on-flat; (b), truncated cone-on-flat; (c), hemisphere-on-flat.

geous to conduct tests at high velocities in order to obtain relevant data in a brief period of time. However, this practice is not beneficial because high velocities can increase interfacial temperatures and result in an accelerated wear process [64,75,88,92]. It is recommended by ASTM Standard F732-82 to perform laboratory wear tests at a velocity of 50 mm/s [67].

E. Sliding Distance

It is a common daily occurrence for humans to have a wide variation of gait velocities and stride lengths. Using a pedometer, Seedholm, Dowson, and Wright monitored the daily walking activity as a function of age and occupation (Table 5). It is important to attempt to relate the results of laboratory wear tests to the actual use of a total joint prosthesis. Reporting wear results as a function of the number of test cycles can be misleading without knowledge of the stroke length. Therefore, it is advisable to report wear as a function of sliding distance, which can then provide clinical significance and a comparative test method. It has been estimated that for an elderly person with a 22-mm diameter total hip replacement, the sliding distance under load is approximately 10.7 kilometers (km) per year [94]. Using this estimate and a reciprocating stroke length of 25.4 mm (1.0 inch), 1 million cycles in a laboratory test would be equivalent to 4.71 years of use. However, total joint replacements are becoming increasingly common in younger, more active patients. The sliding distance per year for this group is expected to be much greater. Usually, 1 million test cycles with a stroke length of 25.4 mm are used to simulate 1 year of *in vivo* service [95].

Wear tests must be conducted for a length of time that reasonably predicts the lifetime of a prosthesis. This time is crucial because different wear mechanisms may occur as a function of sliding distance. Tests involving low wear combinations such as cobalt-chrome or stainless steel on UHMWPE are required to last for several million cycles in order to generate enough wear for accurate measurements [74]. Figure 8 show UHMWPE volume lost as a function of test cycles for CoCr and stainless steel counterfaces as reported by McKellop et al. [71].

F. Temperature

UHMWPE is a thermoplastic resin with mechanical properties that are influenced by temperature. It has been shown that increases in temperature result in an increase in creep [75,96], wear [75,82,88,97–99], and oxidation [100] of UHMWPE. It has been

Table 5   Walking Activity According to Occupation

Occupation	Test subjects	Range (steps per day)	Average (steps per day)
Housewives	5	5000–13,000	8,000
Hospital staff	9	7000–30,000	17,000
Elderly people on holiday	9	3000–13,000	6,000
Secretarial staff	7	7500–14,000	11,000
University academic staff	5	6800–13,500	10,200
University technical staff	4	6000–17,000	11,000

Source: From Ref. 93

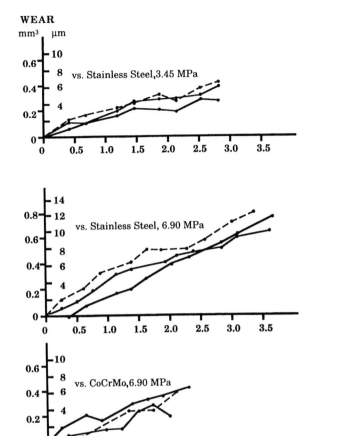

Figure 8 Ultra-high molecular weight polyethylene (UHMWPE) wear against CoCr and 316 stainless steel at 6.9 MPa, 100-cycles/min. (From Ref. 71.)

recommended to conduct laboratory wear tests at a temperature of 37 ± 1°C, or as specified otherwise [67]. Maintaining the test medium at a constant temperature helps to reduce the problem of overheating from periods of high friction [74,101]. However, even if the bulk temperature of the testing media is maintained constant, localized frictional heating occurs and increases the interfacial temperature between the wear specimens [71,92]. This interfacial temperature between sliding contacts is a function of the flash temperature caused by asperity contact and the mean surface temperature that results from the dissipation of frictional heat [82,97,99]. Researchers have developed mathematical models to estimate this interfacial temperature for UHMWPE in dry conditions [99]. Others have studied and measured the resulting interfacial temperature in a dry environment as a function of applied load [83,97,99], sliding speed [82,99], and controlled heating [102].

Davidson, Schwartz, and Lynch investigated the dependence of friction on heating during articulation in a hip joint simulator with distilled water lubrication by using the relationship [92]

$$Q = \mu PV \qquad (1)$$

where

    Q = the frictional heating rate
    μ = the frictional coefficient
    P = the applied normal load
    V = the relative surface velocity

The resulting equilibrium temperature occurs when the rate of heat generation and the rate of heat dissipation of the system are equal [75,92,103]. Figure 9 shows surface heating as a function of time and peak load levels for CoCr on UHMWPE [92]. Ceramic systems produce the lowest levels of heat generation.

G. Molecular Weight

Several investigations have reported lower wear rates of UHMWPE containing higher molecular weight components. Rose et al. investigated wear rates of hip prostheses using a hip simulator machine [15]. Measuring wear by particle retrieval, they reported the prosthesis showing the lowest wear rate (0.3 mg/year) also had the greatest molecular weight on its articular surface. Similarly, the prosthesis showing the greatest wear rates (10.2 mg/year), released the largest particles and had the lowest molecular weight on its surface. A similar trend of increases in wear with reductions in molecular weight (Fig. 10) have been reported elsewhere [104]. Reductions in wear by a factor of 30 or more have been observed when the molecular weight of UHMWPE is decreased from 2×10^{6} to 5×10^{5} daltons [90]. It has been suggested that the greater wear resistance of UHM-WPE with higher molecular weights is due to the extreme ductility and capability for energy absorption of the high molecular weight fractions [90].

H. Load-Contact Stress

Dynamic force profiles for joints have been well documented in the field of biomechanics. Contact stresses specific for each joint or prosthesis are determined when these force profiles are exerted over complex geometries. In order to guarantee identical wear mecha-

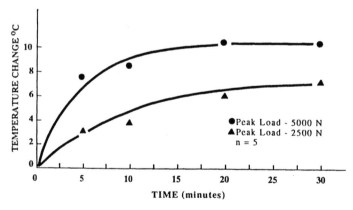

Figure 9 Heat generation for polished CoCr articulating against ultra-high molecular weight polyethylene (UHMWPE) at peak walking loads of 5000 N and 2500 N in 5 ml of water. (From Ref. 92. Reprinted with permission of John Wiley and Sons, Inc.)

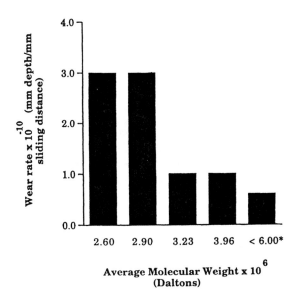

Figure 10 Influence of average molecular weight on wear rates of ultra-high molecular weight polyethylene (UHMWPE) (UHMWPE did not dissolve for molecular weight measurement). (From Ref. 104.)

nisms, it may seem advantageous to reproduce these contact geometries and apply similar loading patterns in laboratory testing. However, reproducing the complexities of joints in simple laboratory tests renders the tests very expensive and time consuming, thus defeating the purpose of screening tests.

For these tests, it is often more practical to use a constant contact stress estimated by calculating the average contact stresses that occur in total joint replacements. Therefore, the actual stress distributions and contact areas in simple screening tests are different from those in total joint prostheses and are determined by test specimen orientation and geometry. However, in more sophisticated systems like total joint simulators, it is possible to emulate complex load profiles.

Typical contact stresses in total hip prostheses have been estimated by various researchers to be in the range of 2.1 to 10.35 MPa [72,88]. Estimates for contact stresses in total knee replacements have been considerably higher [105–107]. McKellop has suggested that testing over a range of estimated contact stresses could help identify possible thresholds at which wear rates dramatically increase [74].

Several investigations have shown a connection between wear rates and load. For an UHMWPE pin on a stainless steel plate configuration, an increase in load leads to an increase in wear. For instance, the average wear rate at 6.9 MPa has been determined to be approximately double the wear rate at 3.45 MPa [71]. Using a disk-on-plate configuration, Rostoker and Galante reported a logarithmic relationship between wear rates and apparent contact stress. The relationship was approximated as

$$W = 0.5 \times 10^{-10}e^{0.21P} \qquad (2)$$

where W is the wear rate in terms of depth of wear per unit sliding distance and P is the contact pressure in MPa after creep has expired [87]. This relationship is graphically depicted in Fig. 11.

A similar empirical relationship developed using a stainless steel hemisphere on an UHMWPE flat wear tester has also been reported [90]:

$$W = 0.834e^{0.0846P} \qquad (3)$$

where W is the weight of wear debris collected in mgf (milligrams-force) and P is the load in newtons. Contact stresses could not be accurately reported due to the spherical nature of the counterface. Both Eqs. 2 and 3 indicate that, for stresses below 5 to 10 MPa, the wear rate is relatively low and marginally dependent on contact stress. However, for stresses above this level, wear rate is extremely sensitive to stress.

For a sphere-on-flat geometry, it has been shown that the maximum contact pressure can be related to contact area according to [108]

$$q_o = \frac{3P}{2\pi a^2} \qquad (4)$$

where

    q_o = the maximum pressure
    P = the compressive force
    a = is the contact radius

However, Rose et al. have brought to attention the difficulty with accurately determining the contact stress used in wear tests with certain specimen geometries and orientations [90]. The typical calculations for Hertzian contact stresses and contact areas are based on linear elastic theory for nonconforming materials. Some typical geometric models and relationships are given in Fig. 12 and Table 6 [109]. However, these relationships do not accurately apply to viscoelastic materials such as UHMWPE. The specimen geometries

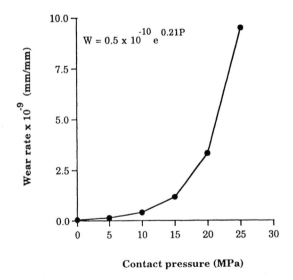

Figure 11   Influence of contact pressure on wear rate. (From Ref. 87.)

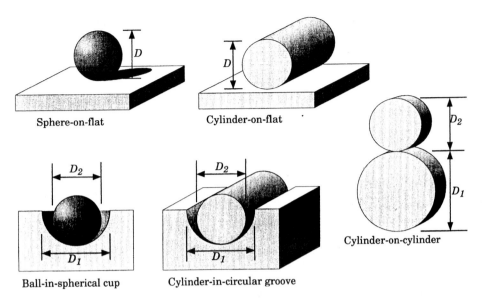

Figure 12   Typical geometric contact models. (From Ref. 109.)

shown in Fig. 12 produce increasing contact areas and decreasing contact stress distributions during a wear test due to the dimensional changes resulting from creep and wear.

In order to demonstrate the magnitude of the differences that arise between calculated and actual values of both contact area and contact stress, a simple test was performed by the authors by applying a load of 92 N on a CoCr pin with a round tip radius of 20 mm against UHMWPE flats. Approximate contact radii were measured at selected time periods using pressure-sensitive film (Itochu International, Roseville, MI) and optical microscopy. Figure 13 shows the relationship between the measured contact radius and time of loading. The final contact radius measured at 72 hours was 1.85 mm, which

Table 6   Contact Stress Relationships for Various Geometries

Geometry of contact	Elastic contact dimension	Maximum compressive stress
Sphere on flat	$a = 0.721 \sqrt[3]{PDC_o}$	$S_c = 0.918 \sqrt[3]{P/(D^2 C_o^2)}$
Cylinder on flat	$b = 1.6 \sqrt{pDC_o}$	$S_c = 0.798 \sqrt{p/(DC_o)}$
Cylinder on cylinder (parallel axes)	$b = 1.6 \sqrt{pC_o/A}$	$S_c = 0.798 \sqrt{(pA)/C_o}$
Ball-in-spherical socket	$a = 0.721 \sqrt[3]{(PC_o)/B}$	$S_c = 0.918 \sqrt[3]{P(B/C_o)^2}$
Cylinder-in-circular groove	$b = 1.6 \sqrt{(pC_o)/B}$	$S_c = 0.798 \sqrt{(pB)/C_o}$

Source: From Ref. 109

where $C_o = \left[\dfrac{(1 - \nu_1^2)}{E_1} \right] + \left[\dfrac{(1 - \nu_2^2)}{E_2} \right]$; $A = \dfrac{(D_1 + D_2)}{D_1 D_2}$; $B = \dfrac{(D_1 - D_2)}{D_1 D_2}$;

P = normal force; p = normal force/length; E_1 = elastic modulus for Material 1; E_2 = elastic modulus for Material 2; $\sqrt{1}$ = Poisson's ratio for Material 1; $\sqrt{2}$ = Poisson's ratio for Material 2; D = diameter when only one body in contact is curved; D_1 = diameter of Body 1 ($D_1 > D_2$); S_c = maximum compressive stress (Hertzian stress); a = radius of the elastic, circular contact; b = width of the contact (used for cylinders)

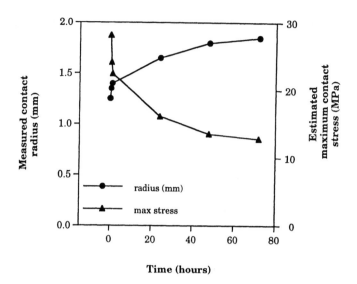

Figure 13   Measured contact radius and estimated contact stress as a function of duration of load (93 N) application using Prescale pressure-sensitive film.

is much larger than the value of 0.998 mm calculated assuming linear elasticity of the material. These measured contact radii can be used in conjunction with Eq. 4 to calculate a "pseudomaximum" compressive stress (Fig. 13). After 72 hours of loading, this stress was estimated at 12.83 MPa. This is far less than the 44.07 MPa calculated analytically assuming linear elastic behavior for UHMWPE. It must be emphasized that these results were produced from a static loading test, whereas the wear tests are dynamic in nature.

I.   Sterilization

In order for successful implantation, total joint prostheses require some form of sterilization to reduce the risk of infection. The typical sterilization procedures of steam autoclaving cannot be used with UHMWPE components because of their tendency to deform during the process [91]. The usual commercial practice is to use a 2.5-megarad (Mrad) dose of radiation from a ^{60}Co gamma ray source.

It has been reported that significant changes occur in the UHMWPE during the sterilization process [110,111]. Radiation sterilization results in the formation of free radicals from the scission of polymeric chains, which occurs primarily in the amorphous regions. These free radicals form cross-linkages, combine with oxygen, or decay by some other mechanism. Structurally, sterilization causes an increase in density and in the degree of crystallinity of UHMWPE. Oxidation results in the formation of hydrophilic carbonyl groups, which causes an increase in fluid absorption. These structural changes have been shown to produce increases in elastic modulus and yield stress, with decreases in tensile strength, ductility, and fatigue properties.

Wear properties of UHMWPE have been studied by various researchers as a function of gamma-irradiation doses. Using solution viscosity techniques, it has been determined that irradiation increases the amount of insoluble gel and also the amount of low molecular weight (<20,000 daltons) material [91]. Using very high doses of irradiation (20–

1000 Mrad) and performing wear tests under dry conditions, Shen and Dumbleton reported increases in both frictional coefficients and wear rates at low loads [97]. Even with lower doses of 0–20 Mrads and a bovine serum environment, it has been reported that wear generally increases with dosage and contact stress [89].

J. Wear Measurements

Upon commencement of sliding under load, materials undergo a "wear-in" period during which the wear rate is high and is a function of specimen geometry and the roughness of the mating surfaces. This period is relatively short and is followed by a period of steady-state wear during which the wear rate significantly decreases. During this last period, the volume of material lost increases linearly with sliding distance. The process of wear causes a release of wear particles, dimensional and topographical changes, and a loss in weight of the specimens. The measurement of these parameters has been attempted by researchers to quantify wear accurately during laboratory testing.

In the past, many investigations estimated wear by measuring dimensional changes in the polymer specimen. This was accomplished with a displacement measuring device [68,71,78,87,91,93,98,112], or with a profilometer to record the geometry of the wear track on a polymeric flat [113–115]. Wear measurements based on dimensional changes are complicated in part due to the viscoelastic properties of UHMWPE. Creep, fluid absorption, and fluctuations in temperature can lead to dimensional changes in the polymer and result in spurious wear estimates.

ASTM F732-82 calls for wear measurements to be made by specimen weight loss. Several researchers have used this technique for wear assessment in both lubricated and dry tests [71,74,83,84,88,116,117]. The potential errors associated with this type of wear measurement are as follows [64]:

1. The specimens may pick up extraneous matter, which can result in inaccuracies.
2. Absorption of fluid from the test medium during the wear test results in a mass gain by the specimens.
3. Some polymeric particles may be still attached to the specimen at the time of weighing. A decision must be made if this polymer is considered wear debris or still part of the specimen. Its removal or inclusion will greatly affect the measurement.

It has been well established that, during a wear test, the UHMWPE absorbs fluid, which results in a gain in mass [71,118]. This potential source of error may be minimized by using two techniques [74]: (1) presoaking the specimens in the test medium for several weeks before initial weighing and (2) using soak control specimens to correct for additional fluid intake during the test. It is recommended that, during the presoaking time, specimens should be periodically cleaned, dried, and massed on an analytical balance with an accuracy of ±10 micrograms (μg) [67]. This procedure should continue until mass gain from absorption reaches the steady state, which is approximately 2 to 3 weeks for UHMWPE [67]. Table 7 outlines one technique for cleaning wear specimens [67].

Wear debris retrieval is a more direct technique to measure wear. These particles may be collected by direct filtration from saline or distilled water test mediums. However, this is not the case when using physiologic test mediums such as bovine serum. Difficulties associated with wear particle retrieval from serum are (1) the small amount of debris is easily lost in other suspended solids in serum, (2) fractions of serum adhere to the

Table 7   Technique for Cleaning UHMWPE Specimens

1. Rinse with tap water to remove bulk contaminants.
2. Wash in an ultrasonic cleaner in a solution of 1% detergent for 15 minutes.
3. Rinse in a stream of distilled water.
4. Rinse in an ultrasonic cleaner in distilled water for 5 minutes.
5. Rinse in a stream of distilled water.
6. Dry with lint-free tissue.
7. Immerse in methyl alcohol for 3 minutes.
8. Dry with lint-free tissue.
9. Air dry in a dust-free environment at room temperature for 30 minutes.

Source: From Ref. 67

polymer surface, and (3) serum is so viscous that direct filtration is not possible [119]. Rose et al. outlined a method to recover wear debris from serum through a stepwise process of chemical digestion, filtration, density column separation, and stereological evaluation of the recovered debris volume [119]. This technique for wear measurement has been used in some studies with success [15,77,90,120]. However, the particle retrieval technique is too laborious and time consuming for widespread use and somewhat impracticable for large-scale laboratory wear testing [63,74].

1. Wear Rate

Several analytical relationships have been developed to calculate wear rates. Ratner et al. derived the following abrasive wear relationship for a metal pressing into a softer polymer [102]:

$$\frac{V}{L} = \frac{\mu W}{Hse} \tag{5}$$

where
V = the volume of polymer removed
L = the sliding distance
μ = the coefficient of friction
W = the applied force
H = the indentation hardness of the polymer
s = the breaking strength of the polymer
e = the elongation to break of the polymer

Archard [121] and Archard and Hurst [122] developed a relationship for adhesive wear showing that the wear rate is independent of the apparent contact area and directly proportional to the applied load. This equation assumes that the contacting asperities plastically deform under load:

$$W = \frac{KsP}{p_m} \tag{6}$$

where
W = the volume removed
K = a constant related to the probability per unit encounter of production of a wear particle

    s = the sliding distance
    P = the applied load
    p_m = the flow pressure of the softer material

These relationships should be used with caution because (1) the breaking strength and elongation to break in Eq. 5 are dependent on strain rate and temperature and (2) the flow pressure in Eq. 6 is difficult to measure because it varies with the method of measurement [69].

Lancaster defined wear rates as [123]

$$\text{Wear rate} = \frac{V}{Ps} \tag{7}$$

where
    V = the volume of material removed
    P = the applied load
    s = the sliding distance

This relationship is simple and useful because it can be used even when more than one wear mechanism is operating and viscoelastic measurement errors associated with polymers are eliminated.

2. Units

Researchers have used a wide variety of units when reporting wear rates. Clarke gives a list of references and the units used in the respective investigation. Wear rates have been reported as volume lost per 10^6 cycles [78,88], mass lost per 10^6 cycles [78,80,117], or linear displacement per 10^6 cycles [71]. Clarke's estimate [94] of 10.7 km of sliding per year of use in a hip prosthesis has been used to report wear rates in linear displacement per year of use [74,88]. However, in recent years, most investigators follow Eq. 7 and report wear rates as cubic millimeters per nanometer (mm^3/nm) [65,66,69,80,83, 84,116,117]. The volume of material removed is calculated by measuring the mass loss and using the density of the polymer as a conversion factor.

In all cases, it is of paramount importance to report clearly all test parameters since changes in any one of them can produce different results. These parameters should include the test materials, fabrication processes, molecular weight, sterilization status and technique, contact stress, frequency, surface velocity, test duration, specimen geometry, test medium, bulk temperature, motion (reciprocating or unidirectional), and any other pertinent information. Furthermore, the initial wear-in region of the test should be included in the data presentation because its extent and nature can provide important information about the dominant wear mechanism, surface topography, and effects of geometrical configuration [64,79].

VII. WEAR MECHANISMS

The wear of UHMWPE occurs predominantly by three different mechanisms: abrasion, adhesion, and fatigue. These mechanisms are usually independent of each other and it is not necessary that all three occur on any one UHMWPE specimen. The occurrence of these mechanisms and the degree of damage caused by them is a function of the ambient environment of the wear couple as well as the material properties and surface morphologies of the two components.

A.   Abrasive Wear

Wear resulting from abrasion is usually manifested as scratches and gouges (Fig. 14), on the UHMWPE surface. A hard metal or ceramic counterface with a rough surface will "plow" through the polymer surface, causing cold flow combined with a mixture of shearing and cutting [86,124]. Rostoker, Chao, and Galante have reported that scratches dominate the wear surface in disk-on-plate tests [124]. These scratches are formed parallel to the direction of relative movement between the counterface and UHMWPE. The abrasive asperities causing the deformation may exist on the counterface (two-body wear) or as particles in the immediate environment (three-body wear). *In vivo* bone cement particles often get trapped between the articulating surface of prostheses and cause three-body wear leading to severe damage [124]. Metal particles that may be generated during the surgical procedure also behave in a similar manner.

The roughness of the counterface plays a critical role in abrasive wear. In an earlier study, Abdallah and Treheux used ceramic counterfaces on UHMWPE and determined that wear increased with an increase in roughness of the counterface [86]. Other studies have also investigated the effects of counterface roughness and arrived at similar conclusions [80]. Lancaster reports that less stiff polymers exhibit a lower degree of dependence of abrasive wear rate on counterface roughness than more rigid ones [125]. As the polymer becomes less stiff (lower elastic modulus), there is less plastic deformation and cutting. This fact may play an important role in determining the abrasive wear of UHMWPE materials with a stiffness that is enhanced by special fabrication techniques.

B.   Adhesive Wear

In the recent past, evidence has been building to show that adhesive wear plays a significant role in the wear of UHMWPE [95]. Wear due to adhesion can proceed by two mechanisms: (1) transfer films and (2) material pull-out. Each of these mechanisms is discussed here.

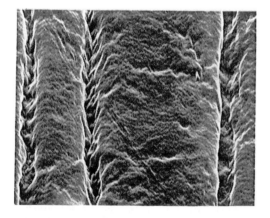

Figure 14   Scanning electron micrograph of ultra-high molecular weight polyethylene (UHMWPE) surface showing scratches. (Micrograph courtesy of Southwest Research Institute, San Antonio, TX.)

1. Transfer Films

Due to the high contact stresses generated at asperities, the UHMWPE can microweld and adhere to the counterface surface. Upon relative movement between the articulating surfaces, this polymeric material transfers to the counterface, where it fills the cavities, coats the asperities, and forms a film. The molecular chains in the films are oriented in the sliding direction [126]. Kumar et al. reported that transfer films were not formed on metal counterfaces when bovine serum was used as the test medium, but were visible when saline or distilled water were used [65]. They also determined that UHMWPE wear against stainless steel was 2 to 3 times higher in saline compared to bovine serum medium. McKellop et al. have suggested that serum proteins act as a boundary lubricant, reducing the probability of transfer film formation [71].

In tests performed on UHMWPE in different mediums, McKellop et al. determined that transfer layers formed on the surfaces of counterfaces lubricated by saline or water were accompanied by an increase in the coefficient of friction [71]. The wear rate was highest for tests run in saline. They also observed that the transfer layers in saline occasionally broke off, resulting in a temporary drop in frictional coefficient. The transfer films in distilled water remained intact once formed, but no films were observed in bovine serum. However, another study has reported the formation of transfer films in serum [78].

The transfer film alters the effective roughness of the counterface and changes the sliding condition to that of polymer on polymer. Blanchet and Kennedy have suggested that transfer film formation, in general, is a two-stage process [68]. The first step is UHMWPE deposition in which strong adhesion takes place at points of real contact between the polymer and the counterface. The nature of this adhesion is unknown, although van der Waals forces, tribochemical reactions, and other mechanisms have been proposed. Once the adhesion takes place, the UHMWPE fails in the subsurface of the bulk under the combination of normal and tangential loads, and transfers to the counterface. During the second step, the deposited polymer is subjected to shear forces by the relative motion. The excess material is removed in the form of small particles and fibrils, leaving behind a smooth, highly oriented film. Once the transfer film forms, it is equally probable for a wear particle to be generated from the polymer surface or from the polymer-coated counterface [127].

Shen and Dumbleton determined that under, dry unlubricated conditions, transfer films of UHMWPE formed on the counterfaces only when the polymer was unirradiated [97]. The authors speculated that irradiated UHMWPE cross-links on the surface, which would prevent the drawing of fibrils for film formation. It is evident that the formation and retention of transfer films are a function of the lubricating media and polymer pretreatment. As transfer films can significantly alter the friction and wear characteristics of UHMWPE, it is important to choose test parameters most relevant to *in vivo* conditions.

2. Material Pull-Out

The material pull-out mechanism for adhesive wear is very similar to transfer film formation. The polymeric material is pulled out from the bulk due to microwelds with the counterface and the polymer surface usually exhibits evidence of fibrils [95]. The material lost from the bulk does not adhere to the counterface to form films, but is lost to the test medium. Recently, it has been speculated that the material pull-out and fibril formation occur predominantly in the lower molecular weight fraction of the polymer on the component surface [128]. The observation that there is a decrease in wear due to an increase in molecular weight lends further credibility to this theory [104].

C.   Fatigue Wear

When a metal or ceramic counterface articulates on UHMWPE, the surface of the polymer is subjected to alternating tensile and compressive stresses. This cyclic loading is schematically depicted in Fig. 15 [75]. Ahead of the counterface and in the direction of sliding, the material experiences a compressive stress. On the other hand, material in the wake of the counterface experiences a tensile force. The stresses on any one point of the polymer surface change in concert with the movement of the counterface. In addition, the polymer within the wear track experiences other cyclic shear and compressive forces associated with the normal force pressing the articulating surfaces together and their relative velocity. This stress state translates along the wear track as the bearing surfaces slide relative to each other and results in cyclic stresses that subject the material to fatigue loading. If the loads exceed the fatigue strength of the material, they may cause the initiation of cracks [117]. Such cracks can be on the surface or subsurface of the polymer. Cracks perpendicular to the surface of UHMWPE have been observed by Dowling et al. on Charnley hip prostheses [129].

On the basis of *in vitro* tests, it has been predicted that fatigue cracks may initiate after approximately 8 years of service for total hip replacements [77]. In general, the contact stresses are higher in knee implants due to less conformity of the articulating components. In metal-backed total knee replacements with thin UHMWPE components, the stresses can increase by a factor of 10 [11]. These high stresses may accelerate the fatigue failure because the rate of fatigue crack propagation in UHMWPE is very sensitive to changes in the range of applied stress intensity [130]. In addition, in most knee prostheses the relative motion between the articulating components is a combination of rotation and translation, which aggravates the fatigue loading.

As the fatigue process progresses, subsurface cracks can initiate that propagate parallel to the surface and then interact with the surface cracks extending into the bulk material to release thin fragments of wear debris.

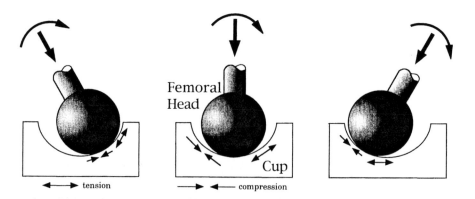

Figure 15   Schematic showing cyclic loading in a joint. (From Ref. 75.)

D. Fold Formation

Another mode of wear that needs to be considered is fold formation. A simplified schematic outlining this mechanism is shown in Fig. 16 [124]. The stress acting on UHMWPE through the counterface results in a pile-up of the polymer adjacent to this counterface. Upon further movement of the counterface, this mound of material is plastically deformed into a fold. After repeated passes of the counterface, the fold is deformed to form a film, which then exfoliates to form a wear particle [124].

VIII. NEW MATERIALS

In an attempt to reduce the wear of UHMWPE, several new materials have been examined for use in total joint prostheses. Ceramic-on-ceramic systems exhibit good friction and wear characteristics [6]. McKellop and Lu have also reported superior performance of zirconia balls on UHMWPE compared with CoCr in wear tests [131]. Similar results have been reported by others [65]. Gradually, ceramic systems are gaining more acceptance and their use is increasing.

Recently, the surface of articulating materials has been modified using techniques of ion implantation [76,132]. These techniques are particularly attractive because they have the potential of modifying the bearing surfaces without altering the bulk properties of the materials. However, the wear data pertaining to these techniques is still inconclusive. McKellop and Röstlund tested nitrogen implanted Ti-6Al-4V counterfaces against UHMWPE in a joint simulator [76]. They reported no change in the wear of UHMWPE, but detected a decrease in the wear of the counterface. On the other hand, Rieu et al. performed pin-on-disk wear tests using nitrogen implanted Ti-6Al-4V and 316L stainless steel counterfaces, as well as ion implanted UHMWPE [132]. They reported improved wear characteristics in all cases. Some preliminary studies on the use of diamondlike carbon (DLC) coatings on metal counterfaces have also been reported [133,134]. These results appear encouraging, but further research is required to ensure the long-term retention of these coatings.

The modification of the bulk properties of UHMWPE is yet another route being currently investigated [135,136]. These modified polymers are known as *enhanced UHMWPE* and usually employ high temperature and pressure to increase the crystallinity of medical-grade UHMWPE [136]. Improvements in the stiffness, strength, creep resistance, and fatigue properties have also been reported [136–138]. Some improvements in the wear characteristics of enhanced UHMWPE compared with UHMWPE have been shown, but so far long-term *in vivo* data are not available [135].

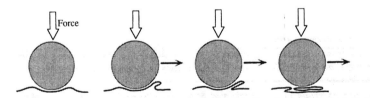

Figure 16   Schematic representation of wear debris generation due to fold formation. (From Ref. 124.)

IX. CONCLUSIONS

The wear of UHMWPE remains a major obstacle to the design of long-lasting total joint prostheses. It is now clear that UHMWPE implants generate submicron wear particles that can initiate a complicated cascade of events resulting in periprosthetic loosening and osteolysis. The sheer number of these particles may be more important than their composition, suggesting that treatments that reduce the number of particles may also reduce the incidence of osteolysis. New techniques are needed to characterize both the wear mechanisms and wear debris associated with UHMWPE. Development of such techniques in conjunction with a concentrated effort to identify the dominant wear mechanisms will eventually lead the scientific community to those variables that control the wear of UHMWPE at the microscopic level.

ACKNOWLEDGMENTS

The authors would like to thank Ms. Cherish Grummon for performing some of the tests included in this chapter. Also, thanks are due to Ms. Joyce Lopez and Ms. Connie Acosta for providing administrative support.

REFERENCES

1. M. Jasty, Clinical reviews: Particulate debris and failure of total hip replacements, *J. Applied Biomat.*, 4(3):273–275 (1993).
2. G.K. McKee and J. Watson-Farrar, Replacement of arthritis hips by the McKee–Farrar prosthesis, *J. Bone Joint Surg.*, 48B(2):245 (1966).
3. H.C. Amstutz, Polymers as bearing materials for total hip replacement: A friction and wear analysis, *J. Biomed. Mater. Res.*, 3:547–568 (1968).
4. E.B. Mathiesen, J.U. Lindgren, F.P. Reinholt, and E. Sudmann, Tissue reactions to wear products from polyacetal (Delrin) and UHMW polyethylene in total hip replacement, *J. Biomed. Mater. Res.*, 21:459–466 (1987).
5. P. Boutin, P. Christel, J.M. Dorlot, A. Meunier, A. Roquancourt, S. Hermen, L. Sedel, and J. Witvoet, The use of dense alumina-alumina ceramic combination in total hip replacement, *J. Biomed. Mater. Res.*, 22:1203 (1988).
6. J.M. Dorlot, P. Christer, and A. Meunier, Wear analysis of retrieved alumina heads and sockets of hip prostheses, *J. Biomed. Mater. Res.*, 23:299–310 (1989).
7. P. Griss, G. Heimke, H. von Andrian-Werburg, B. Krempien, S. Reipa, H.J. Lauterbach, and H.J. Hartung, Morphological and biomechanical aspects of Al_2O_3 ceramic joint replacement: Experimental results and design considerations for human endoprostheses, *J. Biomed. Mater. Res. Symposium*, 6:177 (1975).
8. H. Mittelmeier, Anchoring hip endoprosthesis without bone cement, in *Engineering in Medicine, Advances in Artificial Hip and Knee Joint Technology*, M. Schaldach and D. Hohmann, Editors, Springer-Verlag, Berlin, p. 387 (1976).
9. J.H. Dumbleton, The clinical significance of wear in total hip and knee prostheses, *Journal of Biomaterial Applications*, 3:3–32 (1988).
10. D. Dowson, Friction and wear of medical implants and prosthetic devices, in *ASM Handbook, Friction, Lubrication, and Wear Technology*, S.D. Henry, Editor, ASM International, Metals Park, Ohio, pp. 656–664 (1992).
11. J. Black, *Orthopaedic Biomaterials in Research and Practice*, Churchill Livingstone, New York (1988).
12. J. Charnley, *Low Friction Arthroplasty of the Hip*, Springer Verlag, New York (1979).

13. J. Charnley, Rate of wear in total hip replacement, *Clin. Orthop. Rel. Res.*, 112:170–179 (October 1975).

14. M.J. Griffith, M.K. Seidenstein, D. Williams, and J. Charnley, Socket wear in Charnley low friction arthroplasty of the hip, *Clin. Orthop. Rel. Res.*, 137:37–47 (December 1978).

15. R.M. Rose, H.J. Nusbaum, H. Schneider, M. Ries, I. Paul, A. Crugnola, S.R. Simon, and E.L. Radin, On the true wear rate of ultra high-molecular-weight polyethylene in the total hip prosthesis, *J. Bone Joint Surg.*, 62A:537–549 (June 1980).

16. B. Weightman, S.A.V. Swanson, G.H. Isaac, and B.M. Wroblewski, Polyethylene wear from retrieved acetabular cups, *J. Bone Joint Surg.*, 73B:806–810 (1991).

17. T.H. McCoy, E.A. Salvati, C.S. Ranawat, and P.D. Wilson, A 15 year follow-up study of 100 Charnley low-friction arthroplasties, *Orthopedic Clinics of North America*, 19(3):467–476 (1988).

18. G.H. Isaac, B.M. Wroblewski, J.R. Atkinson, and D. Dowson, A tribological study of retrieved hip prostheses, *Clin. Orthop. Rel. Res.*, 276:115–125 (March 1992).

19. W.J. Maloney and M. Jasty, Wear debris in total hip arthroplasty, *Seminars in Arthroplasty*, 4(3):125–135 (1993).

20. J. Livermore, D. Ilstrup, and B.F. Morrey, Effect of femoral head size on wear of the polyethylene acetabular component, *J. Bone Joint Surg.*, 72A(4):518–528 (1990).

21. B.F. Morrey and D. Ilstrup, Size of the femoral head and acetabular revision in total hip replacement arthroplasty, *J. Bone Joint Surg.*, 71A(1):50–55 (1989).

22. L. Mintz, A.K. Tsao, C.R. McCrae, S.D. Stulberg, and T. Wright, The arthroscopic evaluation and characteristics of severe polyethylene wear in total knee arthroplasty, *Clin. Orthop. Rel. Res.*, 273:215–222 (December 1991).

23. G.A. Engh, Failure of the polyethylene bearing surface of a total knee replacement within four years, *J. Bone Joint Surg.*, 70A(7):1093–1096 (1988).

24. G.A. Engh, K.A. Dwyer, and C.K. Hanes, Polyethylene wear of metal-backed tibial components in total and unicompartmental knee prostheses, *J. Bone Joint Surg.*, 74B(1):9–17 (1992).

25. D.J. Kilgus, J.R. Moreland, G.A.M. Finerman, T.T. Funahashi, and J.S. Tipton, Catastrophic wear of tibial polyethylene inserts, *Clin. Orthop. Rel. Res.*, 273:223–231 (December 1991).

26. D.J. Kilgus, T.T. Funahashi, and P.A. Campbell, Massive femoral osteolysis and early disintegration of a polyethylene-bearing surface of a total knee replacement, *J. Bone Joint Surg.*, 74A(5):770–774 (1992).

27. P.C. Peters, G.A. Engh, K.A. Dwyer, and T.N. Vinh, Osteolysis after total knee arthroplasty without cement, *J. Bone Joint Surg.*, 74A(6):864–876 (1992).

28. J.P. Collier, M.B. Mayor, J.L. McNamara, V.A. Suprenant, and R.E. Jensen, Analysis of the failure of 122 polyethylene inserts from uncemented tibial knee components, *Clin. Orthop. Rel. Res.*, 273:232–242 (December 1991).

29. S.M.G. Jones, I.M. Pinder, and A.J. Malcolm, Polyethylene wear in uncemented knee replacements, *J. Bone Joint Surg.*, 74B(1):18–22 (1992).

30. J.W. Wang and C.C. Lin, A case of massive osteolysis after knee prosthesis polyethylene failure, *Acta Orthopaedica Scandinavica*, 64(4):491–493 (1993).

31. C.G. Moran, I.M. Pinder, T.A. Lees, and M.J. Midwinter, Survivorship analysis of the uncemented porous-coated anatomic knee replacement, *J. Bone Joint Surg.*, 73A(6):848–857 (1991).

32. J.F. Nolan and T.M. Bucknill, Aggressive granulomatosis from polyethylene failure in an uncemented knee replacement, *J. Bone Joint Surg.*, 74B(1):23–24 (1992).

33. T.M. Wright, C.M. Rimnac, P.M. Faris, and M. Bansal, Analysis of surface damage in retireved carbon fiber-reinforced and plain polyethylene tibial components from posterior stabilized total knee replacements, *J. Bone Joint Surg.*, 70A(9):1312–1321 (1988).

34. R.W. Hood, T.M. Wright, and A.H. Burstein, Retrieval analysis of total knee prostheses: A

method and its application to 48 total condylar prostheses, *J. Biomed. Mater. Res.*, 17:829–842 (1983).

35. P.H. Newman and J.T. Scales, The unsuitability of polyethylene for movable weight-bearing prostheses, *J. Bone Joint Surg.*, 33B(3):392–398 (1951).

36. J.M. Mirra, H.C. Amstutz, M. Matos, and R. Gold, The pathology of the joint tissues and its clinical relevance in prosthesis failure, *Clin. Orthop. Rel. Res.*, 117:221–240 (June 1976).

37. J. Charnley, The reaction of bone to self-curing acrylic cement: A long term histological study in man, *J. Bone Joint Surg.*, 52B:340–353 (1970).

38. M. Jasty, W.J. Maloney, C.R. Bragdon, et al., The initiation of failure in cemented femoral components of hip arthroplasties, *J. Bone Joint Surg.*, 73B:551–558 (1991).

39. V.L. Fornasier, J. Wright, and J. Seligman, The histomorphologic and morphometric study of asymptomatic hip arthroplasty — A post-mortem study, *Clin. Orthop. Rel. Res.*, 271:272–282 (1991).

40. T.P. Schmalzried et al., The mechanism of loosening of cemented acetabular components in total hip arthroplasty, *Clin. Orthop. Rel. Res.*, 274:60–78 (January 1992).

41. T.P. Schmalzried, M. Jasty, and W.H. Harris, Periprosthetic bone loss in total hip arthroplasty — Polyethylene wear debris and the concept of the effective joint space, *J. Bone Joint Surg.*, 74A(6):849–863 (1992).

42. S.R. Goldring, A.L. Schiller, M. Roelke, C.M. Rourke, D.A. O'Neil, and W.H. Harris, The synovial-like membrane at the cone-cement interface in loose total hip replacements and its proposed role in bone lysis, *J. Bone Joint Surg.*, 65A(5):575–584 (1983).

43. S.R. Goldring, M. Jasty, M.S. Roelke, C.M. Rourke, F.R. Bringhurst, and W.H. Harris, Formation of a synovial-like membrane at the bone-cement interface. Its role in bone resorption and implant loosening after total hip replacement, *Arthritis & Rheumatism*, 29(7):836–842 (1986).

44. J.T. Wang, Y. Harada, and S.R. Goldring, Biological mechanisms involved in the pathogenesis of aseptic loosening after total joint replacement, *Seminars in Arthroplasty*, 4(4):215–222 (1993).

45. D.W. Howie, B.L. Cornish, and B. Vernon-Roberts, Resurfacing hip arthroplasty — classification of loosening and the role of prosthesis wear particles, *Clin. Orthop. Rel. Res.*, 255:144–159 (June 1990).

46. D.W. Howie, Tissue response in relation to type of wear particles around failed hip arthroplasties, *Journal of Arthroplasty*, 5(4):337–348 (1990).

47. H.G. Willert and M. Semlitsch, Reactions of the articular capsule to wear products of artificial joint prostheses, *J. Biomed. Mater. Res.*, 11:157–164 (1977).

48. S. Santavirta, Y.T. Konttinen, V. Hoikka, and A. Eskola, Immunopathological response to loose cementless acetabular components, *J. Bone Joint Surg.*, 73(1):38–42 (1991).

49. P.C. Peters, T. Vinh, F. Johnson, and G. Engh, The use of oil red O stain to characterize particulate polyethylene and the macrophage respone in cases of osteolysis associated with cementless total hip and total knee arthroplasty, *Transactions of the 38th Annual Meeting of the Orthopaedic Research Society*, 17(2):394 (1992).

50. P.A. Campbell, G. Chun, N. Kossovsky, and H.C. Amstutz, Histological analysis of tissues suggests that "metallosis" may really be "plasticosis," *Transactions of the 38th Annual Meeting of the Orthopaedic Research Society*, 17(2):393 (1992).

51. S. Santavirta, V. Hoikka, A. Eskola, Y.T. Konttinen, T. Paavilainen, and K. Tallroth, Aggressive granulomatous lesions in cementless total hip arthroplasty, *J. Bone Joint Surg.*, 72B(6):980–984 (1990).

52. W.J. Maloney, M. Jasty, A. Rosenberg, and W.H. Harris, Bone lysis in well-fixed cemented femoral components, *J. Bone Joint Surg.*, 72B(6):966–970 (1990).

53. W.J. Maloney, M. Jasty, W.H. Harris, J.O. Galante, and J.J. Callaghan, Endosteal erosion in association with stable uncemented femoral components, *J. Bone Joint Surg.*, 72A:1025–1034 (August 1990).

54. M. Tanzer, W.J. Maloney, M. Jasty, and W.H. Harris, The progression of femoral cortical osteolysis in association with total hip arthroplasty without cement, *J. Bone Joint Surg.*, 74A(3):404–410 (1992).

55. P.P. Anthony, G.A. Gie, C.R. Howie, and R.S.M. Ling, Localised endosteal bone lysis in relation to the femoral components of cemented total hip arthroplasties, *J. Bone Joint Surg.*, 72B(6):971–979 (1990).

56. M. Jasty, W.E. Floyd, A.L. Schiller, S.R. Goldring, and W.H. Harris, Localized osteolysis in stable, non-septic total hip replacement, *J. Bone Joint Surg.*, 68A(6):912–919 (1986).

57. J.D. Mabrey, C.M. Agrawal, J. Lankford, R. Campbell, P.A. Campbell, and T.P. Schmalzried, Atomic force microscopy of *in vivo* and *in vitro* polyethylene wear debris, *Proceedings of the Australian Society of Biomaterials*, Sydney, Australia (Jan. 30–Feb. 1, 1994).

58. D.W. Howie, B. Vernon-Roberts, R. Oakeshott, and B. Manthey, A rat model of resorption of bone at the cement-bone interface in the presence of polyethylene wear particles, *J. Bone Joint Surg.*, 70A(2):257–263 (1988).

59. D.W. Murray and N. Rushton, Macrophages stimulate bone resorption when they phagocytose particles, *J. Bone Joint Surg.*, 72B(6):988–992 (1990).

60. ASTM Standard F648-84, Standard specification for ultra-high-molecular-weight polyethylene powder and fabricated form for surgical implants, in *1991 Annual Book of ASTM Standards, Section 13, Medical Devices and Services*, R. Storer, Editor, ASTM, Philadelphia, p. 201 (1991).

61. H.L. Stein, Ultrahigh molecular weight polyethylenes (UHMWPE), in *Engineered Materials Handbook, Volume 2, Engineering Plastics*, C.A. Dostal, Editor, ASM International, Metals Park, Ohio, p. 170 (1988).

62. A.M. Crugnola, E.L. Radin, R.M. Rose, I.L. Paul, S.R. Simon, and M.B. Berry, UHMWPE as used in articular prostheses (a molecular weight study), *Journal of Applied Polymer Science*, 20:809–812 (1976).

63. I.C. Clarke, Wear-screening and joint simulation studies vs. materials selection and prosthesis design, *CRC Critical Reviews in Biomedical Engineering*, 8(1):29–91 (1982).

64. J.H. Dumbleton, Wear and its measurement for joint prosthesis materials, *Wear*, 49:297–326 (1978).

65. P. Kumir, M. Oka, K. Ikeuchi, K. Shimizu, T. Yamamuro, H. Okumura, and Y. Kotoura, Low wear rate of UHMWPE against zirconia ceramic (Y-PSZ) in comparison to alumina ceramic and SUS 316L alloy, *J. Biomed. Mater. Res.*, 25:813–828 (1991).

66. K.J. Brown, J.R. Atkinson, and D. Dowson, The wear of high molecular weight polyethylene—Part II: The effects of reciprocating motion, orientation in the polyethylene, and a preliminary study of the wear of polyethylene against itself, *Transactions of ASME—Journal of Lubrication Technology*, 104:17–22 (1982).

67. ASTM F732-82, Reciprocating pin-on-flat evaluation of friction and wear properties of polymeric materials for use in total joint prostheses, in *1991 Annual Book of ASTM Standards, Section 13, Medical Devices and Services*, R. Storer, Editor, ASTM, Philadelphia, pp. 227–232 (1991).

68. T.A. Blanchet and F.E. Kennedy, The development of transfer films in ultra-high molecular weight polyethylene/stainless steel oscillatory sliding, *Tribology Transactions*, 32:371–379 (1989).

69. D. Dowson, J.R. Atkinson, and K. Brown, The wear of high molecular weight polyethylene with particular reference to its use in artificial human joints, in *Advances in Polymer Friction and Wear*, L.H. Lee, Editor, Plenum Press, New York, pp. 533–551 (1974).

70. J.R. Cooper, D. Dowson, and J. Fisher, Wear mechanisms of UHMWPE under unidirectional sliding, *Transactions of the Fourth World Biomaterials Congress*, Berlin, p. 619 (1992).

71. H. McKellop, I.C. Clarke, K.L. Markolf, and H.C. Amstutz, Wear characteristics of UHMW polyethylene: A method for accurately measuring extremely low wear rates, *J. Biomed. Mater. Res.*, 12:895–927 (1978).

72. J.O. Galante and W. Rostoker, Wear in total hip prostheses. An experimental evaluation of candidate materials, *Acta Orthopaedica Scandinavica, Supplementum*, 145:5–44 (1973).

73. J.H. Dumbleton, C. Shen, and E.H. Miller, A study of the wear of some materials in connection with total hip replacement, *Wear*, 29:163–171 (1974).

74. H. McKellop, Wear of artificial joint materials II: 12-channel wear-screening device: Correlation of experimental and clinical results, *Engineering in Medicine*, 10(3):123–136 (1981).

75. J.A. Davidson and G. Schwartz, Wear, creep, and frictional heat of femoral implant articulating surfaces and the effect on long-term performance — Part 1, A review, *J. Biomed. Mater. Res., Applied Biomaterials*, 21(A3):261–285 (1987).

76. H.A. McKellop and T.V. Rostlund, The wear behavior of ion-implanted Ti-6Al-4V against UHMW Polyethylene, *J. Biomed. Mater. Res.*, 24:1413–1425 (1990).

77. R.M. Rose, M.D. Ries, I.L. Paul, A.M. Crugnola, and E. Ellis, On the true wear rate of ultrahigh molecular weight polyethylene in the total knee prosthesis, *J. Biomed. Mater. Res.*, 18:207–224 (1984).

78. K.W.J. Wright, H.S. Dobbs, and J.T. Scales, Wear studies on prosthetic materials using the pin-on-disc machine, *J. Biomat.*, 2:41–48 (January 1982).

79. A.E. Hollander and J.K. Lancaster, An application of topographical analysis to the wear of polymers, *Wear*, 25:155–170 (1973).

80. B. Weightman and D. Light, The effect of the surface finish of alumina and stainless steel on the wear rate of UHMWPE, *J. Biomat.*, 7:20–24 (1986).

81. M. Zaki and D. Dowson, The influence of surface roughness and electrical potential upon the friction and wear of polyethylene, in *Biomaterials and Clinical Applications*, A. Pizzoferrato et al., Editors, Elsevier Science Publishers, Amsterdam, pp. 497–502 (1987).

82. T.S. Barrett, G.W. Stachowiak, and A.W. Batchelor, Effect of roughness and sliding speed on the wear and the friction of UHMWPE, *Wear*, 153:331–350 (1992).

83. J.R. Atkinson, K.J. Brown, and D. Dowson, The wear of high molecular weight polyethylene Part I: The wear of isotropic polyethylene against dry stainless steel in unidirectional motion, *Transactions of the ASME — Journal of Lubrication Technology*, 100:208–218 (1978).

84. D. Dowson, J.M. Challen, K. Holmes, and J.R. Atkinson, The influence of counterface roughness on the wear rate of polyethylene, Paper IV, *Proceedings of the 3rd Leeds-Lyon Symposium on Tribology*, University of Leeds, England, pp. 99–102 (1978).

85. H.J. Nusbaum, R.M. Rose, I.L. Paul, A.M. Crugnoola, and E.L. Radin, Wear mechanisms for UHMWPE in the total hip prosthesis, *J. Applied Polymer Science*, 23:777–789 (1979).

86. A.B. Abdallah and D. Treheux, Friction and wear of ultrahigh molecular weight polyethylene against various new ceramics, *Wear*, 142:43–56 (1991).

87. W. Rostoker and J.O. Galante, Contact pressure dependence of wear rates of ultra high molecular weight polyethylene, *J. Biomed. Mater. Res.*, 13:957–964 (1979).

88. H. McKellop, I. Clarke, K. Markolf, and H. Amstutz, Friction and wear properties of polymer, metal, and ceramic prosthetic joint materials evaluated on a multichannel screening device, *J. Biomed. Mater. Res.*, 15:619–653 (1981).

89. R.M. Rose, E.V. Goldfarb, E. Ellis, and A.N. Crugnola, Radiation sterilization and the wear rate of polyethylene, *J. Orthop. Res.*, 2:393–400 (1984).

90. R.M. Rose, W.R. Cimino, E. Ellis, and A.N. Crugnola, Exploratory investigations on the structure dependence of the wear resistance of polyethylene, *Wear*, 77:89–104 (1982).

91. W.R. Jones, W.F. Hady, and A. Crugnola, Effect of gamma irradiation on the friction and wear of UHMWPE, *Wear*, 70:77–92 (1981).

92. J.A. Davidson, G. Schwartz, and G. Lynch, Wear, creep, and frictional heating of femoral implant articulating surfaces and the effect on long-term performance — Part II, friction, heating, and torque, *J. Biomed. Mater. Res.*, 22:69–91 (1988).

93. B.B. Seedholm, D. Dowson, and V. Wright, Wear of solid phase formed high density polyethylene in relation to the life of artificial hips and knees, *Wear*, 24:35–51 (1973).

94. I.C. Clarke, Wear of artificial joint materials I: Friction and wear studies: Validity of wear-screening protocols, *Engineering in Medicine*, 10:115–122 (1981).

95. H.A. McKellop, T. Schmalzried, S.-H. Park, and P. Campbell, Evidence for the generation of sub-micron polyethylene wear particles by micro-adhesive wear in acetabular cups, *Transactions of the 19th Annual Meeting of the Society for Biomaterials*, Birmingham, Alabama, p. 184 (1993).

96. R.G. Voltz and E.L. Gradillas, Thermal deformation of polyethylene in a total knee prosthesis. A laboratory analysis and case report, *J. Bone Joint Surg.*, 60A(5):662–663 (1978).

97. C. Shen and J.H. Dumbleton, The friction and wear behavior of irradiated very high molecular weight polyethylene, *Wear*, 30:349–364 (1974).

98. V.R. Evans and F.E. Kennedy, The effects of temperature on friction and wear in oscillatory motion of polyethylene against stainless steel, *Wear of Materials, ASME*, 1:427–434 (1987).

99. J.M. Challen and D. Dowson, The calculation of interfacial temperatures in a pin-on-disc machine in the wear of non-metallic materials, *Proceedings of the 3rd Leeds-Lyon Symposium on Tribology*, University of Leeds, England, pp. 87–93 (1978).

100. P. Eyerer, M. Kurth, H.A. McKellop, and T. Mittlmeier, Characterization of UHMWPE hip cups run on joint simulators, *J. Biomed. Mater. Res.*, 21:275–291 (1987).

101. R.W. Treharne, R.W. Young, and S.R. Young, Wear of artificial joint materials III: Simulation of the knee joint using a computer-controlled system, *Engineering in Medicine*, 10:137–142 (1981).

102. S.B. Ratner, I.I. Farberova, O.V. Radyukevich, and E.G. Luré, Properties and uses of plastics — Connection between wear-resistance of plastics and other mechanical properties, *Soviet Plastics*, 7:37–41 (1964).

103. J.A. Davidson, S. Gir, and J.P. Paul, Heat transfer analysis of frictional heat dissipation during articulation of femoral implants, *J. Biomed. Mater. Res.*, 22:281–309 (1988).

104. W. Rostoker and J.O. Galante, Some new studies of the wear behavior of ultrahigh molecular weight polyethylene, *J. Biomed. Mater. Res.*, 10:303–310 (1976).

105. K.L. Gunsallus and D.L. Bartel, Stresses and surface damage in PCA and total condylar polyethylene components, *Transactions of the 38th Annual Meeting of the Orthopaedic Research Society*, Washington, D.C., p. 329 (1992).

106. D.L. Bartel, V.L. Bicknell, and T.M. Wright, The effect of conformity, thickness, and material on stresses in ultra-high molecular weight components for total joint replacement, *J. Bone Joint Surg.*, 64A(7):1041–1051 (1986).

107. T.M. Wright and D.L. Bartel, The problem of surface damage in polyethylene total knee components, *Clin. Orthop. Rel. Res.*, 205:67–74 (1986).

108. S.P. Timoshemko and J.N. Goodier, *Theory of Elasticity*, 3rd ed., McGraw-Hill, New York (1970).

109. P.J. Blau, Appendix: Calculations of elastic contact dimensions and stresses, in *ASM Handbook: Friction, Lubrication, and Wear Technology*, S.D. Henry, Editor, ASM International, Metals Park, Ohio, pp. 487–488 (1992).

110. H.J. Nusbaum and R.M. Rose, The effects of radiation sterilization on the properties of ultrahigh molecular weight polyethylene, *J. Biomed. Mater. Res.*, 13:557–567 (1979).

111. R.J. Roe, E.S. Grood, R. Shastri, C.A. Gosselin, and F.R. Noyes, Effect of radiation sterilization and aging on ultrahigh molecular weight polyethylene, *J. Biomed. Mater. Res.*, 15:209–230 (1981).

112. P.S. Wallizer, M. Ben-Dov, M.J. Askew, and J. Pugh, The deformation and wear of plastic components in artificial knee joints — An experimental study, *Engineering in Medicine*, 10:33–38 (1981).

113. P.S. Walker and E. Salvati, The measurement and effects of friction and wear in artificial hip joints, *J. Biomed. Mater. Res., Symposium*, 4:327–342 (1973).

114. J.H. Dumbleton and C. Shen, The wear behavior of ultra high molecular weight polyethylene, *Wear*, 37:279–289 (1976).

115. R.L. Fusaro, Friction, wear, transfer, and wear surface morphology of ultra high molecular weight polyethylene, *Transactions of the ASLE/ASME Lubrication Conference*, Hartford, Connecticut, pp. 1–10 (1983).

116. J.R. Atkinson and R.Z. Cicek, Silane crosslinked by polyethylene for prosthetic applications, *J. Biomat.*, 5:326–335 (1984).

117. B. Weightman and D. Light, A comparison of RCH 1000 and Hi-Fax 1900 ultra-high molecular weight polyethylenes, *J. Biomat.*, 6:177–183 (1985).

118. I.C. Clarke, W. Starkebaum, A. Hosseinian, P. McGuire, R. Okuda, R. Salvey, and R. Young, Fluid-sorption phenomena in sterilized polyethylene acetabular prostheses, *J. Biomat.*, 6:184–188 (1985).

119. R.M. Rose, H. Schneider, M. Ries, I. Paul, A. Crugnola, R.S. Simon, and E.L. Radin, A method for the quantitative recovery of polyethylene wear debris from the simulated service of total joint prostheses, *Wear*, 51:77–84 (1978).

120. R.M. Rose and E.L. Radlin, Wear of polyethylene in the total hip prosthesis, *Clin. Orthop. Rel. Res.*, 170:107–115 (1982).

121. J.F. Archard, Contact and rubbing of flat surfaces, *Journal of Applied Physics*, 24:981–988 (1953).

122. J.F. Archard and W. Hurst, The wear of metals under unlubricated conditions, *Proceedings of the Royal Society*, A236:397–410 (1956).

123. J.K. Lancaster, The influence of substrate hardness on the formation and endurance of molybdenum disulfide films, *Wear*, 10:103–117 (1967).

124. W. Rostoker, E.Y.S. Chao, and J.O. Galante, The appearances of wear on polyethylene – A comparison of *in vivo* and *in vitro* wear surfaces, *J. Biomed. Mater. Res.*, 12:317–335 (1978).

125. J.K. Lancaster, Abrasive wear of polymers, *Wear*, 14:223–239 (1969).

126. C.-L. Su and J.R. Yount, An elastic-plastic stress analysis of a polymeric subsurface with a thin layer under normal and tangential loading, *Wear*, 123:355–367 (1988).

127. M. Vaziri, R.T. Spurr, and F.H. Stott, An investigation of the wear of polymeric materials, *Wear*, 122:329–342 (1988).

128. J. Lankford, G. Dearnaley, C.M. Agrawal, D.M. Micallef, and A.R. McCabe, Enhanced wear resistance of ultra-high molecular weight polyethylene (manuscript in preparation).

129. I.M. Dowling, J.R. Atkinson, D. Dowson, and J. Charnley, The characteristics of acetabular cups worn in the human body, *J. Bone Joint Surg.*, 60:375–382 (1978).

130. G.M. Connelly, C.M. Rimnac, T.M. Wright, R.W. Hertzberg, and J.A. Manson, Fatigue crack propagation behavior of ultrahigh molecular weight polyethylene, *Journal of Orthopaedic Research*, 2(2):119–125 (1984).

131. H. McKellop and B. Lu, Friction, lubrication and wear of polyethylene/metal and polyethylene/ceramic hip prostheses on a joint simulator, *Transactions of the Fourth World Biomaterials Congress*, Berlin, p. 118 (1992).

132. J. Rieu, A. Pichat, L.-M. Rabbe, A. Rambert, C. Chabrol, and M. Robelet, Ion implantation effects on friction and wear of joint prosthesis materials, *J. Biomat.*, 12:139–143 (March 1991).

133. C.M. Agrawal, D.M. Micallef, M.A. Wirth, J. Lankford, G. Dearnaley, and A.R. McCabe, The effects of diamond-like-carbon coatings on the friction and wear of enhanced UHMWPE-metal couples, *Transactions of the 19th Annual Meeting of the Society for Biomaterials*, Birmingham, Alabama, p. 10 (1993).

134. A.K. Mishra and J.A. Davidson, PMMA abrasion resistance of zirconia, TiN and ADLC surfaces, *Transactions of the Fourth World Biomaterials Congress*, Berlin, p. 116 (1992).

135. H. McKellop, B. Lu, and S. Li, Effect of increased density and crystallinity on the wear

of UHMW polyethylene acetabular cups, *Transactions of the Fourth World Biomaterials Congress*, Berlin, p. 117 (1992).

136. S. Li and E.G. Howard, Characterization and description of an enhanced ultra high molecular weight polyethylene for orthopaedic bearing surfaces, *Transactions of the 16th Annual Meeting of the Society for Biomaterials*, Charleston, South Carolina, p. 190 (1990).

137. C.M. Agrawal, K.A. Athanasiou, D.M. Micallef, G. Constantinides, and E. Eifler, Creep indentation characteristics and compressive properties of polyethylene, *Transactions of the 19th Annual Meeting of the Society for Biomaterials*, Birmingham, Alabama, p. 11 (1993).

138. T.M. Wright, K.L. Gunsallus, C.M. Rimnac, D.L. Bartel, and R.W. Klein, Design considerations for an acetabular component made from an enhanced form of ultra high molecular weight polyethylene, *Transactions of the 37th Annual Meeting, Orthopaedic Research Society*, Anaheim, California, p. 248 (1991).

56

The Biological Effects of Wear Debris from Joint Arthroplasties

Stuart B. Goodman
Stanford University Medical Center
Stanford, California

I. INTRODUCTION

Total joint replacement (TJR) is currently one of the most successful operations performed in orthopedic surgery. For the hip and knee, these operations consistently yield excellent clinical results, with 10-year survivals of 90% or more using contemporary prostheses and techniques [Scuderi et al. 1989; Mulroy and Harris 1992]. These results have prompted the use of joint replacement in younger patients, who frequently lead more active life-styles than the sedentary, elderly population for which TJR was originally designed. Furthermore, the general population is aging and the elderly are becoming more physically active. These facts have led to increased expectations by patients undergoing TJR, and increased demands on the prosthesis construct. Recently, new "advanced" prosthetic designs and materials have been introduced, sometimes prior to the completion of detailed experimental and clinical analyses [Bauer 1992; Goodfellow 1992]. Thus, although some technical advances have been associated with an increasing success rate of specific prostheses, the number of cases requiring surgical revision is still increasing. Unfortunately, revision surgery is more complex, the complication rate is higher, and the outcome is poorer compared with primary procedures [Amstutz et al. 1982].

Why do TJRs fail? The most common cause leading to revision is loosening of the components. The etiology of loosening is multifactorial, and includes patient-related variables (e.g., weight, activity level, bone stock, etc.) and mechanical factors, including prosthesis design, material properties, and biological factors. Clinical studies have been useful in delineating the relationship between patient characteristics and the outcome of a specific prosthesis [Linstrand, Stenström, and Lewold 1992; Malachau et al. 1993]. Mechanical factors and material properties can be evaluated in well-designed, laboratory experiments according to published protocols. The biological processes associated with prosthesis implantation are more elusive: the bone and surrounding tissues can sometimes interact with the prosthesis and its by-products in an unexpectedly adverse way, jeopardizing long-term stability of the implant. In this respect, the biological effects of wear debris have played an increasingly important role.

In this chapter, the characteristics of the tissue at the bone-implant interface are reviewed, and the biological effects of wear debris from TJRs are summarized. To increase the longevity of orthopedic implants, new approaches are necessary to mitigate the generation and biological sequelae of particulate debris.

II. THE TISSUE AT THE BONE-IMPLANT INTERFACE IN HUMANS

A. Histology

A good starting point for understanding the characteristics of the bone-implant interface is to review the observations from autopsy studies of retrieved, clinically successful implants. These studies have a distinct advantage over investigations of the tissues surrounding loose implants, which might reflect different stages in the loosening process.

1. *Autopsy Studies of Cemented Joint Prostheses*

Autopsy studies of successful, cemented implants have demonstrated a maturing interface that begins with cell necrosis in the vicinity of the cement [Charnley 1979, 1986]. Over the course of several months, the necrotic tissue is replaced by a fibrous scar that may undergo metaplasia to fibrocartilage or bone over months to years, depending on the local mechanical conditions. The bone-implant interface continues to remodel long after prosthetic insertion [Maloney et al. 1989; Jasty et al. 1990, 1991]. On the femoral side of a cemented hip replacement, a secondary, circumferential trabecular "neocortex" forms adjacent to the cement and is linked to the original, but receded, cortex by a thin, lacelike trabecular network. This gives the appearance of a radiolucent line adjacent to the cement, which may be misinterpreted as radiographic evidence of prosthetic loosening.

In many areas, the original cortex or the neocortex may be intimately associated with the cement mantle, with little intervening fibrous tissue, under the light microscope. Observations of the cement mantle have suggested that prosthetic loosening may begin at the cement-metal interface (rather than at the bone-cement interface), due to the presence of debonding and crack initiation.

On the acetabular side, it has been proposed that wear debris evoke a foreign-body response leading to resorption of bone and progressive undermining of the cement mantle [Schmalzried et al. 1992]. This process begins at the periphery of the cup (at which polyethylene and other wear particles can gain access to the fibrous interface) and progresses toward the apex.

Others have provided evidence suggesting that the femoral and acetabular components of a cemented hip replacement undergo similar mechanisms of loosening. A foreign-body response containing abundant wear debris was observed at the bone-cement interface of cemented components (none of which were regarded as clinically or radiographically loose) harvested at autopsy up to 14 years after implantation [Fornasier, Wright, and Seligman 1991]. The number of histiocytes at the bone-cement interface correlated with the density of polyethylene particles, the thickness of the tissue, and the time after prosthesis implantation.

The previous studies have implicated the ongoing generation of wear debris (especially small, phagocytosable particles of polyethylene) and the associated foreign-body and chronic inflammatory reaction as an important biological mechanism in the evolution of loosening of TJRs.

2. Autopsy and Retrieval Studies of Cementless Prostheses

Few studies have reported histological observations from cementless prostheses retrieved from autopsy or after amputation. In a large series of retrieved, noncemented, porous, coated hip and knee components, none of which were clinically or radiographically loose, Cook et al. [1988 and 1991] observed that fibrous tissue was predominant. The vast majority of the components had ingrowth of bone into less than 10% of the available pore space. Interestingly, a chronic inflammatory infiltrate, composed of lymphocytes and histiocytes, was identified in 14% of cases, suggesting T-cell-mediated delayed hypersensitivity to a complex of metal ions and protein.

Examination of four autopsy-retrieved Zweymuller femoral prostheses showed a preponderance of loose connective tissue abutting the titanium alloy components [Litner, Zweymuller, and Brand 1986]. In some areas, there was direct bone-to-metal contact, without interposed tissue at the light microscopic level. A "secondary bone ring," similar to the neocortex associated with retrieved cemented femoral components was also observed.

Two recent studies have reported the histological results of tissue ingrowth into porous, coated acetabular components retrieved at autopsy. In the first report of 11 components of 1 design harvested up to 75 months postoperatively, bone covered 30% of the metal surface, but composed only 12.1% of the tissue within the pores [Pinhortz et al. 1993]. Organized fibrous tissue was the predominant tissue at the interface. Metallic and polyethylene debris concentrated within the fibrous tissue in the unfilled screw holes. In a similar study of 9 well-fixed, porous, coated acetabular components harvested at autopsy up to 87 months postoperatively, 39% of the prosthetic surface was covered by bone; however, only 13% of the available pore space contained bone [Engh et al. 1993]. Hydroxyapatite coating of implants may facilitate bone apposition, but shed particles of the coating may stimulate a localized, adverse foreign-body reaction [Bauer et al. 1993; Bloebaum and Dupont 1993; Bloebaum et al. 1993; Hofmann, Bachus, and Bloebaum 1993].

Histological studies of clinically successful, well-fixed, cementless implants have demonstrated a predominance of fibrous tissue rather than bone at the implant interface. This tissue may provide a conduit for wear particles from the articulation and the different interfaces of modular components. With time, this debris may migrate to the bone-implant interface at the periphery of the component or through deficiencies such as screw holes.

3. Loose Implants and Osteolysis

When loose implants (whether cemented or cementless) are revised, a characteristic tissue of variable thickness is found at the bone-implant interface. Histologically, this tan-color tissue is composed of a fibrous tissue background stroma containing numerous mono- and multinucleated histiocytes and chronic inflammatory cells [Heilmann et al. 1975; Mirra et al. 1976; Vernon-Roberts and Freeman 1976; Bullough 1973; Willert and Semlitsch 1977; Pizzoferrato 1979; Mirra, Marder, and Amstutz 1982; Bell et al. 1985; Johanson et al. 1987; Maguire, Coscia, and Lynch 1987; Pizzoferrato et al. 1988; Willert, Bertram, and Buchhorn 1990a, 1990b] (Fig. 1). If the prosthesis is composed of metal alone, without polymethylmethacrylate (PMMA) or polyethylene, the interface tissue is primarily fibrous [Kozinn, Johanson, and Bullough 1986]. If the implant employed both

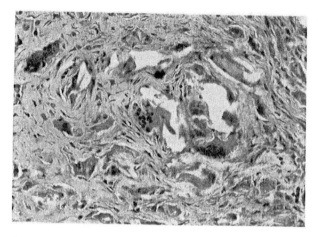

(a)

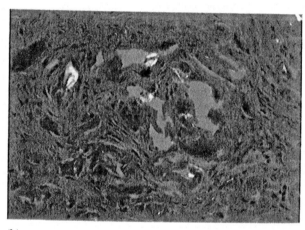

(b)

Figure 1 Tissue harvested from the cement-bone interface of a loose acetabular cup. Histological sections demonstrating macrophages and foreign-body giant cells surrounding and engulfing positively birefringent polyethylene particles. The background stroma is fibrous. Cement debris is dissolved during processing: (a), World Health Organization (WHO) stain; transmitted light, ×250; (b), polarized light, ×250.

metallic and polymeric materials, a foreign-body and chronic inflammatory infiltrate is more prominent. A thin, synovial-like lining sometimes forms immediately adjacent to the cement layer [Goldring et al. 1983, 1986b]. This layer contains large polygonal cells, stacked 2–5 cells thick, and is thought to be due to motion at the bone-cement interface. A similar synovial-like lining surrounds loose rather than tight-fitting metallic screws [Schatzker, Horne, and Sumner-Smith 1975].

Histological observations of the interface surrounding failed, loosened, cemented prostheses first suggested that the generation of wear particles and the process of loosening were related [Willert et al. 1974]. The "loosening membrane," that is, the tissue surrounding a loose prosthesis, contains numerous particles of bone cement, polyethylene, and metallic debris (this, of course, is dependent on the materials used for the prosthesis). Perhaps the easiest (but often the smallest) particles to identify are black metal particles found in the interstitium and within macrophages. These particles range from several microns in diameter to the submicron range, which cannot be seen with the light microscope. In cases in which massive amounts of metallic particles have been generated, the tissues may be gray or black on gross examination [Rae 1975, 1981, 1986; Buchert et al. 1986; Agins et al. 1988; Howie and Vernon-Roberts 1988]. Studies of well-fixed and loose hip implants have noted disproportionately high levels of metal ions in the synovial fluid and local tissues surrounding cemented titanium alloy components compared with cobalt-chrome or stainless steel prostheses [Brien et al. 1992].

Bone cement (polymethylmethacrylate) is dissolved by the organic solvents used in the processing of routine paraffin-embedded sections. The larger polymerized balls of methylmethacrylate form a vacant or negative image [Charnley 1979, 1986]. Particles of polymethylmethacrylate (PMMA) are surrounded by a florid foreign-body response in a fibrous stroma. Smaller particles of PMMA and particles of the radiopaque marker in bone cement (barium sulphate or zirconium) less than approximately 10 micrometers (μm) are phagocytosed by macrophages; larger particles, up to approximately 30 μm, are found within multinucleated foreign-body giant cells. Still larger agglomerates of large and small particles are surrounded by a foreign-body granulomatous reaction [Jones and Hungerford 1987; Goodman, Fornasier, and Kei 1988a].

Polyethylene particles are perhaps the most difficult to identify histologically in retrieved tissues. Using polarized light and special stains such as oil red O, positively birefringent polyethylene debris varies from submicron particles to larger shards up to several millimeters in length [Goodman, Fornasier, and Kei 1988b; Campbell et al. 1992; Peters et al. 1992; Schmalzried et al. 1993] (Fig. 1). The histological reaction to polyethylene wear debris is similar to that seen with bone cement: smaller particles are seen within macrophages and giant cells, whereas larger particles that cannot undergo phagocytosis evoke a granulomatous reaction [Stinson 1964; Goodman, Fornasier, and Kei 1988b; Goodman et al. 1989]. In contrast to bone cement, large numbers of intracellular and interstitial polyethylene particles give the tissue a hazy birefringence under polarized light [Campbell et al. 1992].

Polymers such as PMMA and polyethylene stimulate a very aggressive foreign-body response in particulate form. Charnley [1979] noted this fact when employing another polymer, polytetrafluorethylene, the use of which resulted in rapid clinical and radiographic prosthetic loosening. Similar problems have been noted with Delrin® (Dupont) [Mathiesen et al. 1987; Ahnfelt et al. 1990; Ohlin 1990; Malachau et al. 1993], polyester [Weber 1970], and silicone elastomer prostheses [Gordon and Bullough 1982; Worsing, Engber, and Lange 1982; Rahman and Fagg 1993].

Osteolysis is a term that refers to the destruction of bone seen radiographically as localized or more widespread, large radiolucent areas. Excluding cases associated with infection, histological analysis of osteolytic areas surrounding joint prostheses have demonstrated a florid foreign-body reaction to cement, polyethylene, and/or metallic particles [Jones and Hungerford 1987; Huddleston 1988; Anthony et al. 1990; Maloney et al. 1990a, 1990b, 1993a; Willert et al. 1990a, 1990b; Peters et al. 1994; Schmalzried et al. 1992]. Interestingly, the prostheses do not have to be mechanically loose to demonstrate this reaction. Cemented femoral prostheses made of titanium alloy, especially those with a titanium alloy femoral head, may be associated with higher than expected wear rates, loosening, and osteolysis [Lombardi et al. 1989; Witt and Swann 1991; Buly et al. 1992; Salvati, Betts, and Doty 1993].

Thus, there is substantial histological evidence linking wear debris, prosthetic loosening, and osteolysis. In the next section, we explore the biochemical aspects of this reaction and the biological mechanisms by which the resorption of bone is produced.

B. Biochemistry

As outlined above, the tissue harvested from the bone-implant interface of loose orthopedic implants contains macrophages and chronic inflammatory cells in a fibrovascular stroma. It is not surprising that this tissue is very active metabolically. Most current research efforts have focused on identifying factors in the membrane that regulate inflammatory processes, cellular interactions, and the remodeling of bone. In this respect, cell culture techniques using harvested tissue from the interface have been invaluable.

Prostaglandin E_2 (PGE_2), an arachidonic acid derivative produced by phagocytic cells, has been implicated in the process of aseptic loosening of joint implants [Goldring et al. 1983, 1986b; Jasty, Goldring, and Harris 1984; Goodman et al. 1989; Ayers, Francis, and Duthie 1989; Mather et al. 1989; Ohlin, Johnell, and Lerner 1990]. High levels of this substance have been assayed from cultures of tissue harvested at surgical revision for aseptic loosening. Furthermore, higher levels of PGE_2 were produced in cultures of tissue from clinically and radiologically loose implants compared with well-fixed components [Jasty, Goldring, and Harris 1984; Goodman et al. 1989].

PGE_2 is an extracellular messenger that is produced by inflammatory cells, including polymorphonuclear leukocytes and macrophages [Raisz and Martin 1984; Goetzl and Goldstein 1985; Zurier 1988]. This substance is also known to modulate the intensity and duration of pain to noxious stimuli. In addition, PGE_2 modulates the remodeling of bone and is a powerful inducer of bone resorption *in vivo* and *in vitro* [Klein and Raisz 1970; Raisz and Martin 1984; High 1988]. If an important role were played by PGE_2 in the process of prosthetic loosening, this might explain the pain and periprosthetic lysis of bone commonly seen in the clinical situation. Nonsteroidal antiinflammatory drugs, which are known to inhibit the production of PGE_2, may prove useful in mitigating the symptoms and progression of the loosening process [Goetzl and Goldstein 1985; Goodman et al. 1991]; these drugs have already been used in the treatment of periodontal disease, which in some ways is a process similar to prosthetic loosening [Williams et al. 1985, 1988].

Controversy exists concerning the role of other substances in the etiology of prosthetic loosening. The problem is compounded by the fact that many researchers employ different techniques for harvesting, processing, and assaying the tissues in question. Cytokines, substances produced by one cell that affect the function of other cells, have received much attention. Interleukin-1 (IL-1), interleukin-2 (IL-2), interleukin-6 (IL-6), interleukin-8 (IL-8), tumor necrosis factor (TNF), transforming growth factor-β (TGFβ), and superoxide anions have all been implicated in the biological activities of the bone-implant interface [Appel et al. 1988; Kim et al. 1994; Goodman et al. 1989; Mather et al. 1989; Kossovsky et al. 1991; Chiba et al. 1992; Andrew et al. 1993; Chiba et al. 1993; Nizard et al. 1993; Shanbhag et al. 1993]. These substances mediate many different cellular processes, including the inflammatory response, the immune system, phagocytosis, and the remodeling of bone.

Few studies have directly correlated the presence of specific substances assayed from the tissue harvested from the interface with the degree of bone resorption. Furthermore, these studies are contradictory. In one study, moderate levels of PGE_2 and low levels of collagenase were assayed, but a direct correlation with bone resorption was not reported [Goldring et al. 1983, 1986b]. In another study, only part of the bone resorption (using the mouse calvarial assay) from cultures of loosening membranes harvested from revision cases and the newly formed periprosthetic capsule was correlated with the production of PGE_2 [Ohlin, Johnell, and Lerner 1990]. When PGE_2, IL-1, and collagenase levels from tissues harvested from clinically loose and well-fixed cemented and cementless implants were compared, there was no correlation among the biochemical factors, the degree of bone resorption, and the implant classification [Dorr et al. 1990]. This confusing picture may be due to sampling error, and might be overcome by processing multiple specimens from several different locations.

Biochemical studies of harvested tissue have added another dimension to our understanding of the events at the bone-implant interface of orthopedic prostheses. More comprehensive studies are necessary to correlate the substances that are assayed with loosening and resorption of bone.

C. Immunology

Immunological studies have proven useful in elucidating the properties of the tissue at the bone-implant interface. The use of monoclonal antibodies has allowed precise identification of the different constituent cell types within the harvested tissue using immunohistochemistry; furthermore, using *in situ* hybridization, the location of the messenger ribonucleic acids (RNAs) for specific proteins (cytokines and growth factors) within cells can be determined [Myerson 1988; Nakamura 1990]. Immunological studies can also help determine whether lymphocytes play a role in the tissue response to orthopedic implants and particulate debris.

Immunopathological studies of the tissue surrounding loose, cementless acetabular implants showed an abundance of small particles of polyethylene and titanium [Santavirta et al. 1991]. These particles were thought to cause the migration, adherence, and phagocytosis of CD11b-positive, peroxidase-negative macrophages. There was no apparent histochemical evidence of activation of the immune system (staining for interleukin-2 receptor-positive activated T cells and PCA-1 plasmablasts/plasma cells was negative). In cases of loosening without osteolysis, fibroblastlike mesenchymal cells were predominant; in cases of aggressive granulomatosis, mono- and multinucleated histiocytes were common and activated fibroblasts were more scarce [Tallroth et al. 1989; Santavirta et al. 1991, 1992]. Polymeric and metallic particles were always a prominent finding in cases of aggressive granulomatosis, suggesting an important role for particles in the pathogenesis of osteolysis.

Another group has provided evidence disputing the conclusions that immunological events are not important in prosthetic loosening. Although B-lymphocytes were absent, numerous T-lymphocytes were identified in the tissues harvested from cases with excessive release of titanium alloy (Ti6-Al4-V) particles ("metallosis") [Lalor, Freeman, and Revell 1990, 1991; Lalor and Revell 1992]. Some of the T-lymphocytes were activated (they expressed the HLA-DR antigen). Using the 3.9 monoclonal antibody, it was also shown that the macrophages and lymphocytes interacted in the generation of an immune response to the titanium alloy debris. Two of five patients undergoing revision of titanium-based prostheses had a positive skin reaction to titanium-containing ointment, further suggesting immunologically mediated sensitization [Lalor et al. 1991].

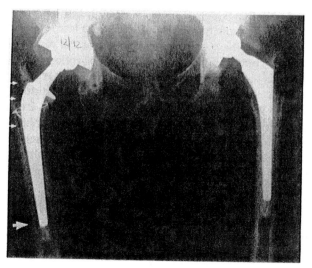

(a)

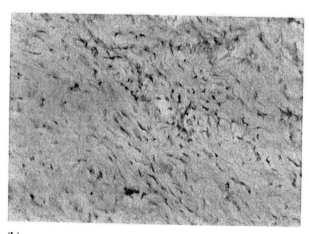

(b)

Figure 2 A loose cemented femoral component that was revised: (a), radiograph showing subsidence of the femoral component of the right hip; note the radiolucent lines surrounding the cement and the presence of cement fractures; (b), immunohistochemical staining of the tissue harvested from the cement-bone interface using monoclonal antibodies directed against human macrophages (EMB11); positive cells stain brown, background cells stain blue ($\times$ 200); (c), immunohistochemical staining using monoclonal antibodies directed against human lymphocytes (CD3); positive cells stain brown ($\times$ 200).

Recently, macrophages, fibroblasts, and T-lymphocytes were identified using immunoperoxidase staining of tissues harvested from clinically and radiographically loose polyethylene cups [Jiranek et al. 1993a]. Interleukin-1β and platelet-derived growth factor-2 messenger RNA was identified within macrophages, and interleukin-1β protein was localized on the surface of macrophages and fibroblasts using *in situ* hybridization.

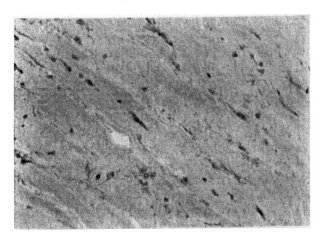

(c)

In a study comparing harvested tissues from clinically loose and well-fixed, cemented and cementless hip and knee prostheses, macrophages and T-lymphocytes but not B-lymphocytes were identified in nearly all specimens [Goodman et al. 1992a]. In an ongoing study, tissue harvested from clinically and radiographically loose cemented prostheses with osteolysis contained greater numbers of macrophages, total T-lymphocytes and cytotoxic T-lymphocytes, and increased staining for interleukin-6 (IL-6) compared with tissues surrounding loose, cemented and cementless prostheses without osteolysis and well-fixed, cemented components [Huie et al. 1993; Goodman et al. 1994] (Fig. 2). These findings suggested a prominent role for T-lymphocytes and specific cytokines in the process of osteolysis.

Are immunological processes involved in aseptic loosening of prostheses? Are particles important to this response? Although these questions can not be conclusively answered at this time, it is well established that lymphocytes modulate many of the functions of macrophages [Unanue and Allen 1987a, 1987b; Johnston 1988; Rentz et al. 1989]. As macrophages make up a large percentage of the cells in tissues from loose prostheses (and especially those with osteolysis), increasing evidence suggests that lymphocytes may play an important role in the body's response to implants and particulate debris.

III. BIOLOGICAL EFFECTS OF PARTICLES *IN VIVO* AND *IN VITRO*

A. Polymers

Polymeric materials are commonly used in joint replacement, as a grout for prosthetic fixation, and as a bearing surface. The generation of polymeric wear debris has been of great concern; Charnley's [1979] use of polytetrafluorethylene proved disastrous because of the unfavorable wear properties of this polymer as a bearing surface, which led to aggressive foreign-body and chronic inflammatory reactions to wear debris. To understand better such reactions and the biochemical and immunological sequelae of implanting polymers in different physical forms, *in vitro* and *in vivo* experiments have been performed.

In general, bulk forms of polymeric materials placed within bone or soft tissue in animals evoke a fibrous tissue encapsulation [Goldring et al. 1986a; Paiement et al. 1986; Goodman, Fornasier, and Kei 1988a, 1988b; Goodman et al. 1990b]. Particles of polymeric materials led to foreign-body and chronic inflammatory reactions, which are dependent on the material used, the particle load, and the characteristics of the particles themselves [Stinson 1964; Goldring et al. 1986a; Paiement et al. 1986; Goodman, Fornasier, and Kei 1988a, 1988b; Goodman et al. 1990b; Perry et al. 1990] (Figs. 3 and 4).

Some of the important particle characteristics include the particle size, shape, surface area, topography, surface charge, and surface energy [Nagura, Asai, and Kojima 1977; Besterman and Low 1983; Kawaguchi et al. 1986; Tabata and Ikada 1988; Barth, Sullivan, and Berg 1991]. Particles that are less than approximately 10 μm are capable of being phagocytosed by macrophages. Larger particles, up to 25–30 μm, can be found within foreign-body giant cells. Still larger particles are often surrounded by numerous mono- and multinucleated histiocytes (macrophages), forming a granuloma. Interestingly, the particulate materials may not have to be phagocytosed to activate macrophages or stimulate the resorption of bone [Howie et al. 1988; Quinn et al. 1992].

Repeated injections of high density polyethylene particles, approximately 20–200 μm

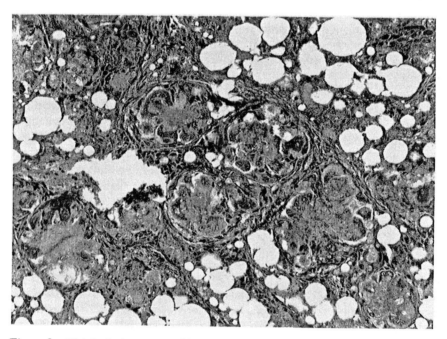

Figure 3   Histological response of bone to the implantation of cement particles in the rabbit tibia. Note the florid foreign-body response composed of mono- and multinucleated histiocytic cells (macrophages) in a fibrous stroma. Several granulomas are evident (World Health Organization stain, × 100).

(a)

(b)

Figure 4   Histological response of bone to the implantation of polyethylene particles in the rabbit tibia. Note the fibrohistiocytic stroma and the foreign-body giant cells surrounding the particles (World Health Organization stain, × 100): (a), transmitted light; (b), polarized light.

in size into the rat knee joint, which contained a non-weight-bearing intraosseous cement plug produced a foreign-body encapsulating membrane that eroded the underlying bone [Howie et al. 1988]. The histological reaction paralleled that seen around loose joint replacements in humans. Biochemical studies have yielded higher levels of PGE$_2$ in culture of the tissue surrounding particulate versus bulk forms of bone cement implanted in bone [Goodman and Chin 1990; Goodman et al. 1991].

Histological and biochemical studies from two animal models of arthroplasty loosening have found high levels of PGE$_2$ in the tissues surrounding loose, cemented prostheses compared with well-fixed implants [Spector et al. 1990; Goodman et al. 1992b, 1993b]. As in the human tissue retrieval studies, the process of implant loosening and the production of PGE$_2$ in the periprosthetic tissues appear to be related. Interestingly, in both of these models of implant loosening, particles of bone cement were added to the interface to simulate the generation of cement debris seen in the human situation [Spector et al. 1990; Goodman et al. 1992b, 1993b]. Thus, it may have been the phagocytosis of the particles themselves rather than the loose prosthesis that led to the high levels of PGE$_2$ production.

The process of phagocytosis of particles results in the release of enzymes, cytokines, and other soluble factors that modulate inflammation and the remodeling of bone [Baggliolini, Schnyder, and Dewald 1982; Williams, Czop, and Austen 1984; Pappatheofanis and Barmada 1991]. Studies in which bone cement (polymethylmethacrylate) particles 50–60 nanometers (nm) in diameter were cultured with polymorphonuclear leukocytes (PMNs) resulted in the release of the lysosomal contents and inhibition of PMN migration in a dose-dependent manner [Pappatheofanis and Barmada 1991]. Increased levels of PGE$_2$, tumor necrosis factor, IL-1-like activity (using the thymocyte proliferation assay), and resorption of bone (using a ^{45}Ca-labeled bone assay) were produced when peripheral blood monocytes were exposed to PMMA particles 1–150 μm in diameter [Herman et al. 1989]. However, when monocytes/macrophages were exposed to Simplex bone cement polymer powder (Howmedica, Rutherford, NJ), increased cellular proliferation and b-glucosaminidase activity (a lysosomal enzyme) were noted, but IL-1-like activity and PGE$_2$ release were not stimulated [Davis et al. 1993].

In another study, phagocytosable particles of PMMA but not polystyrene inhibited deoxyribonucleic acid (DNA) synthesis and impaired the ability of activated macrophages to kill mast cells [Horowitz, Frondoza, and Lennox 1988]. This same group found that phagocytosable cement particles inhibited ^{3}H-thymidine incorporation into macrophages, but led to the release of tumor necrosis factor; larger, nonphagocytosable particles had no effect [Horowitz et al. 1993]. Neither large nor small cement particles affected PGE$_2$ release.

The discordant results from the above experiments are undoubtably due to the different conditions of cell culture and the specific particles chosen for each experiment. It can be concluded, however, that small phagocytosable particles of bone cement or polymethylmethacrylate, under certain conditions, may lead to the release of factors associated with bone resorption.

There is controversy as to whether macrophages (and not osteoclasts alone) can resorb bone. A recent experiment suggested that macrophages do have this capacity. Granulomas were first produced by the subcutaneous implantation of polymethylmethacrylate particles (50–200 μm in diameter) in mice [Quinn et al. 1992]. Scanning electron microscopy of bone explants cultured with the macrophages harvested from the granulomas demonstrated low-grade surface and high-grade lacunar osteolysis. This study suggested that macrophages exposed to polymethylmethacrylate have the capacity to resorb bone. However, it is possible that the stimulated macrophages differentiated into osteoclast progenitors as these cell types have a common cell lineage.

It is extremely difficult to generate/fabricate particles of different polymers with similar physical and chemical properties. Given these restrictions, experiments were carried out to examine the effects of different polymers on peritoneal macrophages *in vitro* [Murray and Rushton 1990]. For latex particles (a negative control) approximately 1 μm in size, a concentration of 10^8 particles/ml (milliliter) (using 10^6 cells/ml) stimulated the release of PGE_2 and ^{45}Ca using the neonatal mouse calvarial assay. Zymogen particles (a positive control) caused similar effects at a concentration of only 10^6 particles/ml. Intermediate concentrations of polymethylmethacrylate and high density polyethylene particles resulted in the same effects. Using this protocol, dose response curves could be generated for different polymers.

In vitro and *in vivo* studies have shown that particles of different polymers elicit foreign-body and chronic inflammatory responses that can result in the release of soluble mediators and the resorption of bone. These mediators regulate many important organ systems, and are important modulators of mesenchymal tissue formation and degradation, including bone [Goldring and Goldring 1990; Nathan 1991; Pfeilschifter, Bonewald, and Mundy 1991; Vitetta and Paul 1991].

B. Metals

Studies in animals and in cell culture have documented that metal particles and their by-products can be deleterious. However, the literature on this subject is confusing and often contradictory.

Phagocytosable particles of cobalt-chrome alloy were more toxic to rat sarcoma cells in tissue culture than stainless steel [Mital and Cohen 1968; Pappas and Cohen 1968]. Due to their high solubility, cobalt from cobalt-chromium alloy, nickel from stainless steel, and vanadium from titanium alloy were found to be toxic to cultured synovial fibroblasts [Rae 1975, 1981]. In a later study by the same author, titanium and titanium alloy particles were apparently well tolerated in tissue culture using mouse peritoneal macrophages or synovial fibroblasts [Rae 1986]. However, others have found that particles of cobalt in low concentrations and titanium, titanium-aluminum, and chromium in higher concentrations were toxic to synovial fibroblasts [Maloney et al. 1993b]. In another study, particles of titanium-aluminum-vanadium induced more release of PGE_2, interleukin-1, interleukin-6, and tumor necrosis factor in tissue cultures of rat peritoneal macrophages compared with cobalt-chromium particles [Haynes et al. 1993]. Supernatants from cultures of macrophages with titanium dioxide particles were associated with increased release of interleukin-1 and resorption of bone using the mouse calvarial assay compared with polystyrene particles of a similar size [Shanbhag et al. 1994]. The macrophage response was dependent on the particle size, composition, and dose (surface area).

Phagocytosable cobalt-chrome particles implanted in the knee joints of rats stimulated inflammatory and foreign-body responses and synovial degeneration after only one week [Howie and Vernon-Roberts 1988]. This reaction was not observed when slightly larger cobalt-chrome alloy particles were implanted in the medullary canal of the rabbit tibia [Goodman et al. 1990a]. Titanium and titanium alloy particles were reported to be well tolerated when injected into the knee joints of mice for up to one year [Rae 1986]. Similar findings were reported using an intraosseous implantation protocol [Goodman et al. 1993a].

The generation of corrosion by-products of implants and their biological consequences have received increasing scrutiny because of the recent use of modular implants
[Merritt and Brown 1985; Collier et al. 1992]. Cell culture studies have shown that some
by-products are compartmentalized within specific cell or tissue types [Merritt, Wenz,
and Brown 1991].

The deleterious effects of metallic particles may be due to the material itself, its
by-products, or third-body wear of an articulating component. The location, particle
load, and the characteristics of the implanted particles appear to be important determinants of their toxicity.

C. Other Particles

Hydroxyapatite and other bioactive ceramic coatings have recently been used to facilitate
bone apposition around cementless implants. As noted above, particles shed from the
coating may stimulate a localized foreign-body reaction [Bauer et al. 1993; Bloebaum
and Dupont 1993; Bloebaum et al. 1993; Hofmann, Bachus, and Bloebaum 1993].

Three types of phagocytosable ceramic particles (hydroxyapatite, tricalcium phosphate, and apatite-wollastonite glass ceramic) produced an inflammatory reaction and
increased release of PGE_2 and tumor necrosis factor in the short term using the air pouch
model in rats [Nagase, Baker, and Schumacher 1988]. Other studies in animals have
demonstrated dissolution or erosion of implanted hydroxyapatite particles in conjunction
with the presence of occasional histiocytes; however, whether encased in bone or exposed
to the open marrow cavity, hydroxyapatite particles promoted bone formation rather
than resorption in the longer term [Jansen et al. 1991; Søballe et al. 1992a, 1992b, 1993;
Goodman et al. 1993a].

IV. CONCLUSION

In vitro and *in vivo* studies, as well as studies of specimens harvested during revision
surgery or retrieved at autopsy, have underlined the general long-term biocompatibility
of bulk implants made of polymers, metals, and other materials. However, in particulate
form, these materials stimulate foreign-body and chronic inflammatory responses that
are capable of inducing the resorption of bone. The degree to which this reaction occurs
is dependent on the location of implantation or the specific model used, the characteristics
of the particles, and the particle load. Because of possible adverse reactions to particles,
it behooves surgeons, material scientists, and biologists to develop new strategies to
prevent the generation of particulate debris from orthopedic implants.

REFERENCES

Agins, H. S., Alcock, N. W., Bansal, M., Salvatti, E. A., Wilson, P. D., Pellici, P. M., and
    Bullough, P. G. (1988). Metallic wear in failed titanium-alloy total hip replacements. *J. Bone
    Joint Surg.*, 70A:347–356.
Ahnfelt, L., Herberts, P., Malchau, H., and Andersson, G. B. (1990). Prognosis of total hip
    replacement. A Swedish multicenter study of 4,664 revisions. *Acta Orthop. Scand.*, (Supple-
    ment 238), 61:1–26.
Amstutz, H. C., Ma, S. M., Jinnah, R. H., and Mai, L. (1982). Revision of aseptic loose total hip
    arthroplasties. *Clin. Orthop.*, 170:21–33.

Andrew, S. M., Andrew, J. G., Hoyland, J. A., Wroblewski, B. M., and Freemont, A. J. (1993). Interleukin-6 in membrane from late loosening of hip arthroplasties. *Transactions of the 39th Annual Meeting of the Orthopaedic Research Society*, 18:243.

Anthony, P. P., Gie, G. A., Howie, C. R., and Ling, R. S. M. (1990). Localized endosteal bone lysis in relation to the femoral components of cemented total hip arthroplasties. *J. Bone Joint Surg.*, 72B:971-979.

Appel, A. M., Sowder, W. G., Hopson, C. N., and Herman, J. H. (1988). Production of mediators of bone resorption by prosthesis associated pseudomembranes. *Transactions of the 34th Annual Meeting of the Orthopaedic Research Society*, 13:362.

Ayers, D. C., Francis, M. J. O., and Duthie, R. B. (1989). Prostaglandin concentrations in the inflammatory membrane around aseptically loose hip prostheses. *Transactions of the 35th Annual Meeting of the Orthopaedic Research Society*, 14:227.

Baggliolini, M., Schnyder, J., and Dewald, B. (1992). Role of phagocytosis in macrophage activation. In *Phagocytosis—Past and Future*. M. L. Karnovsky and L. Bolis (Eds.). London: Academic Press, pp. 339-355.

Barth, E., Sullivan, T., and Berg, E. W. (1991). Particle size versus chemical composition of biomaterials as determining factors in macrophage activation. *Transactions of the 37th Annual Meeting of the Orthopaedic Research Society*, 16:187.

Bauer, G. C. H. (1992). What price progress? Failed innovations of the knee prosthesis. *Acta Orthop. Scand.*, 63(3):245-246.

Bauer, T. W., Stulberg, B. N., Ming, J., and Geesink, R. G. T. (1993). Uncemented acetabular components. Histologic analysis of retrieved hydroxyapatite-coated and porous implants. *J. Arthroplasty*, 8(2):167-177.

Bell, R. S., Schatzker, J., Fornasier, V. L., and Goodman, S. B. (1985). Study of implant failure in the Wagner resurfacing arthroplasty. *J. Bone Joint Surg.*, 67A:1165-1174.

Besterman, J. M., and Low, R. B. (1983). Endocytosis, a review of mechanisms and plasma membrane dynamics. *J. Biochemistry*, 21:1-13.

Bloebaum, R. D., Bachus, K. N., Rubman, M. H., and Dorr, L. D. (1993). Postmortem comparative analysis of titanium and hydroxyapatite porous-coated femoral implants retrieved from the same patient. A case study. *J. Arthroplasty*, 8(2):203-211.

Bloebaum, R. D., and Dupont, J. A. (1993). Osteolysis from a press-fit hydroxyapatite-coated implant. *J. Arthroplasty*, 8(2):195-202.

Brien, W. W., Salvati, E., Betts, F., Bullough, P., Wright, T., Rimnac, C., Buly, R., and Garvin, K. (1992). Metal levels in cemented total hip arthroplasty. A comparison of well-fixed and loose implants. *Clin. Orthop.*, 276:66-74.

Buchert, P. K., Bradley, K. V., Mallory, T. M., Engh, C. A., and Bobyn, J. D. (1986). Excessive metal release due to loosening and fretting of sintered particles on porous-coated hip prostheses. *J. Bone Joint Surg.*, 68A:606-609.

Bullough, P. G. (1973). Tissue reaction to wear debris generated from total hip replacement. *The Hip Society Proceedings*. St. Louis: C. V. Mosby, pp. 80-91.

Buly, R. L., Huo, M. H., Salvati, E., Brien, W., and Bansal, M. (1992). Titanium wear debris in failed cemented total hip arthroplasty. An analysis of 71 cases. *J. Arthroplasty*, 7(3):315-323.

Campbell, P. A., Chun, G., Kossovsky, N., and Amstutz, H. C. (1992). Histological analysis of tissues suggests that "metallosis" may really be "plasticosis." *Transactions of the 38th Annual Meeting of the Orthopaedic Research Society*, 17(2):393.

Charnley, J. (1979). *Low Friction Arthroplasty of the Hip. Theory and Practice*. Berlin: Springer-Verlag.

Charnley, J. (1986). The reaction of bone to self-curing acrylic cement. A long term histological study in man. *J. Bone Joint Surg.*, 52B:340-353.

Chiba, J., Iwaki, Y., Kim, K. J., and Rubash, H. E. (1992). The role of cytokines in femoral osteolysis after cementless total hip arthroplasty. *Transactions of the 38th Annual Meeting of the Orthopaedic Research Society*, 17:350.

Chiba, J., Schwendemen, L., Crofford, T., Booth, R., and Crossett, L. (1993). Comparison of membranes obtained from failed cemented and cementless total knee arthroplasty. *Transactions of the 38th Annual Meeting of the Orthopaedic Research Society*, 18:515.

Collier, J. P., Suprenant, V. A., Jensen, R. E., Mayor, M. B., and Suprenant, H. P. (1992). Corrosion between the components of modular femoral hip prostheses. *J. Bone Joint Surg.*, 74B:511–517.

Cook, S. D., Barrack, R. L., Thomas, K. A., and Haddad Jr., R. J. (1988). Quantitative analysis of tissue growth into human porous total hip components. *J. Arthroplasty*, 3:249–262.

Cook, S. D., McCluskey, L. C., Martin, P. C., and Haddad Jr., R. J. (1991). Inflammatory response in retrieved noncemented porous-coated implants. *Clin. Orthop.*, 264:209–222.

Davis, R. G., Goodman, S. B., Smith, R. L., Lerman, J. A., and Williams, R. J. (1993). The effects of bone cement powder on human adherent monocytes/macrophages *in vitro*. *J. Biomed. Mat. Res.*, 27:1039–1046.

Dorr, L. D., Bloebaum, R., Emmanual, J., and Meldrum, R. (1990). Histologic, biochemical, and ion analysis of tissue and fluids retrieved during total hip arthroplasty. *Clin. Orthop.*, 261:82–95.

Engh, C. A., Zettl-Schaffer, K. F., Kukita, Y., Sweet, D., Jasty, M., and Bragdon, C. E. (1993). Histological and radiographic assessment of well-functioning porous-coated acetabular components. A human post-mortem retrieval study. *J. Bone Joint Surg.*, 75A:814–823.

Fornasier, V. L., Wright, J., and Seligman, J. (1991). The histomorphologic and morphometric study of asymptomatic hip arthroplasty—A post-mortem study. *Clin. Orthop.*, 271:272–282.

Goetzl, E. J., and Goldstein, I. M. (1985). Arachidonic acid metabolites. In *Arthritis and Allied Conditions*. D. J. McCarty (Ed.). Philadelphia: Lea and Febiger, pp. 367–378.

Goldring, M. B., and Goldring, S. R. (1990). Skeletal tissue response to cytokine. *Clin. Orthop.*, 258:245–278.

Goldring, S., Jasty, M., Palement, G., Bragdon, C., Ehrlich, P., and Harris, W. H. (1986a). Tissue response to bulk and particulate biopolymers in a rabbit wound chamber model. *Transactions of the 32nd Annual Meeting Orthopaedic Research Society*, 11:288.

Goldring, S. R., Jasty, M., Roelke, M. S., Rourke, C. M., Bringhurst, F. R., and Harris, W. H. (1986b). Formation of a synovial-like membrane at the bone-cement interface. *Arthritis Rheum.*, 29:836–842.

Goldring, S. R., Schiller, A. L., Roelke, M., Rourke, C. M., O'Neill, D. A., and Harris, W. H. (1983). The synovial-like membrane at the bone cement interface in loose total hip replacements and its proposed role in bone lysis. *J. Bone Joint Surg. (Am).*, 65:575–584.

Goodfellow, J. (1992). Knee prostheses—One step forward, two steps back. *J. Bone Joint Surg.*, 74B:1–2.

Goodman, S. B., Alpert, S., Griffiths, G., Issaz, S., Lee, K., Woolson, S. T., Schurman, D. J., Maloney, W. J., and Weismann, I. (1992a). Immunohistochemical analysis of the membrane surrounding total joint arthroplasties. *Transactions of the Society for Biomaterials Implant Retrieval Symposium*, St. Charles, IL, Sept. 19, 15:33.

Goodman, S. B., and Chin, R. C. (1990). Prostaglandin E2 levels in the membrane surrounding bulk and particulate polymethylmethacrylate in the rabbit tibia. *Clin. Orthop.*, 257:305–309.

Goodman, S. B., Chin, R. C., Chiou, S. S., and Lee, J. (1991). Suppression of prostaglandin E_2 synthesis in the membrane surrounding particulate polymethylmethacrylate in the rabbit tibia. *Clin. Orthop.*, 271:300–304.

Goodman, S. B., Chin, R. C., Chiou, S. S., Schurman, D. J., Woolson, S. T., and Masada, M. T. (1989). A clinical-pathological-biochemical study of the membrane surrounding loosened and nonloosened joint arthroplasty. *Clin. Orthop.*, 244:182–187.

Goodman, S. B., Chin, R., and Magee, F. P. (1992b). Prostaglandin E_2 production by the membrane surrounding loose and fixated cemented tibial hemiarthroplasties in the rabbit tibia. *Clin. Orthop.*, 284:283–287.

Goodman, S. B., Davidson, J. A., and Fornasier, V. L. (1993a). Histological reaction to titanium alloy and hydroxyapatite particles in the rabbit tibia. *Biomaterials*, 14(10):723–728.

Goodman, S. B., Fornasier, V. L., and Kei, J. (1988a). The effects of bulk versus particulate polymethylmethacrylate on bone. *Clin. Orthop.*, 232:255–262.

Goodman, S. B., Fornasier, V. L., and Kei, J. (1988b). The effects of bulk versus particulate ultra-high-molecular-weight polyethylene on bone. *J. Arthroplasty*, 3(Supplement):41–46.

Goodman, S. B., Fornasier, V. L., Lee, J., and Kei, J. (1990a). The effects of bulk versus particulate titanium and cobalt chrome alloy implanted into the rabbit tibia. *J. Biomed. Mat. Res.*, 24:1539–1549.

Goodman, S. B., Fornasier, V. L., Lee, J., and Kei, J. (1990b). The histological effects of the implantation of different sizes of polyethylene particles in the rabbit tibia. *J. Biomed. Mat. Res.*, 24:517–524.

Goodman, S. B., Magee, F., and Fornasier, V. L. (1993b). Radiological and histological assessment of aseptic loosening using a tibial hemiarthroplasty model in the rabbit knee. *Biomaterials*, 14(7):522–528.

Goodman, S. B., Sibley, R., Huie, P., Lee, K., Doshi, A., Roshdieh, B., Woolson, S. T., Maloney, W. J., and Schurman, D. J. (1994). Loosening and osteolysis of cemented joint arthroplasties: a biological spectrum. *Transactions of the 40th Annual Meeting of the Orthopaedic Research Society*, 19:839.

Gordon, M., and Bullough, P. G. (1982). Synovial and osseous inflammation in failed silicone-rubber prostheses. *J. Bone Joint Surg.*, 64A:574–580.

Haynes, D. R., Rogers, S. D., Hay, S., Pearey, M. J., and Howie, D. W. (1993). The differences in toxicity and release of bone-resorbing mediators induced by titanium and cobalt-chromium-alloy wear particles. *J. Bone Joint Surg.*, 75A:825–834.

Heilmann, K., Diezel, P. B., Rossner, J. A., and Brinkman, K. A. (1975). Morphological studies in tissues surrounding alloarthroplastic joints. *Vichows Archiv.*, 366:93–106.

Herman, J. E., Sowder, W. G., Anderson, D., Appel, A. M., and Hopson, C. N. (1989). Poly-methyl-methacrylate-induced release of bone-resorbing factors. *J. Bone Joint Surg.*, 71A:1530–1541.

High, W. B. (1988). Effect of orally administered prostaglandin E_2 on cortical bone turnover in adult dogs: A histomorphometric study. *Bone*, 8:363–373.

Hofmann, A. A., Bachus, K. N., and Bloebaum, R. D. (1993). Comparative study of human cancellous bone remodeling to titanium and hydroxyapatite-coated implants. *J. Arthroplasty*, 8(2):157–166.

Horowitz, S. M., Doty, S. B., Lane, J. M., and Burstein, A. H. (1993). Studies of the mechanism by which the mechanical failure of polymethylmethacrylate leads to bone resorption. *J. Bone Joint Surg.*, 75A:802–813.

Horowitz, S. M., Frondoza, C. G., and Lennox, D. W. (1988). Effects of polymethylmethacrylate exposure upon macrophages. *J. Orthop. Res.*, 6:827–832.

Howie, D. W., and Vernon-Roberts, B. (1988). The synovial response to intraarticular cobalt-chrome wear particles. *Clin. Orthop.*, 232:244–254.

Howie, D. W., Vernon-Roberts, B., Oakeshott, R., and Manthey, B. (1988). A rat model of resorption of bone at the cemented bone interface in the presence of polyethylene wear particles. *J. Bone Joint Surg.*, 70A:257–263.

Huddleston, H. D. (1988). Femoral lysis after cemented hip arthroplasty. *J. Arthroplasty*, 3(4):285–297.

Huie, P., Doshi, A., Lee, K., Woolson, S. T., Maloney, W. J., Schurman, D. J., Goodman, S. B., and Sibley, R. (1993). The mechanism of cell recruitment and osteolysis in total joint arthroplasty: Analysis of the membranous interface using immunohistochemistry and *in situ* hybridization. *Transactions of the 39th Annual Meeting of the Orthopaedic Research Society*, 18:244.

Jansen, J. A., Van De Waerden, J. P. C. M., Wolke, J. G. C., and De Groot, K. (1991). Histologic evaluation of the osseous adaptation to titanium and hydroxyapatite-coated titanium implants. *J. Biomed. Mat. Res.*, 25:973–989.

Jasty, M., Goldring, S. R., and Harris, W. H. (1984). Comparison of bone cement membrane around rigidly fixed versus loose total hip implants. *Transactions of the 30th Annual Meeting of the Orthopaedic Research Society*, 9:125.

Jasty, M. J., Maloney, W. J., Bragdon, C. R., Haire, T., and Harris, W. H. (1990). Histomorphological studies of the long-term skeletal responses to well fixed cemented femoral components. *J. Bone Joint Surg.*, 72A:1220-1229.

Jasty, M., Maloney, W. J., Bragdon, C. R., O'Connor, D. O., Haire, T., and Harris, W. H. (1991). The initiation of failure in cemented femoral components of hip arthroplasties. *J. Bone Joint Surg.*, 73B:551-558.

Jiranek, W. A., Machado, M., Jasty, M., Jevsevar, D., Wolfe, H. J., Goldring, S. R., Goldberg, M. J., and Harris, W. H. (1993a). Production of cytokines around loosened cemented acetabular components. Analysis with immunohistochemical techniques and *in situ* hybridization. *J. Bone Joint Surg.*, 75A:863-879.

Jiranek, W. A., Wolfe, H., Bragdon, C. R., and Goldberg, M. J. (1993b). The role of the immune system in the foreign body response to particulate debris: granuloma formation in immunodeficient mice. *Transactions of the 39th Annual Meeting of the Orthopaedic Research Society*, 18:272.

Johanson, N. A., Bullough, P. G., Wilson, P. D., Salvati, E. A., and Ranawat, C. S. (1987). The microscopic anatomy of the bone cement interface in failed total hip arthroplasties. *Clin. Orthop.*, 218:123-135.

Johnston, R. B. (1988). Monocytes and macrophages. *New England Journal of Medicine*, 381: 747-752.

Jones, L. C., and Hungerford, D. S. (1987). Cement disease. *Clin. Orthop.*, 225:193-206.

Kawaguchi, H., Koiwai, N., Ohtsuka, Y., Miyamoto, M., and Sasakawa, S. (1986). Phagocytosis of latex particles by leukocytes. Dependence of phagocytosis on the size and surface potential of particles. *Biomaterials*, 7:61-66.

Kim, K. J., Chiba, J., and Rubash, H. E. (1994). *In vivo* and *in vitro* analysis of membranes from hip prostheses inserted without cement. *J. Bone Joint Surg.*, 76A:172-180.

Klein, D. C., and Raisz, L. G. (1970). Prostaglandins: Stimulation of bone resorption in tissue culture. *Endocrinology*, 86:1436-1440.

Kossovsky, N., Liao, K., Millet, D., Feng, D., Campbell, P. A., Amstutz, H. A., Finerman, G. M., Thomas, B. J., Kilgus, D. J., Cracchiolo, A., and Allameh, V. (1991). Peri-prosthetic chronic inflammation characterized through the measurement of superoxide anion production by synovial-derived macrophages. *Clin. Orthop.*, 263:263-271.

Kozinn, S., Johanson, A., and Bullough, P. G. (1986). The biologic interface between bone and cementless femoral endoprostheses. *J. Arthroplasty*, 1:249-259.

Lalor, P. A., Freeman, M. A. R., and Revell, P. A. (1990). Immunological studies of the bone-implant interface. *Transactions of the 16th Annual Meeting of the Society for Biomaterials*, 13:203.

Lalor, P. A., and Revell, P. A. (1992). T lymphocytes and titanium aluminum vanadium (TiAlV) alloy. Evidence for immunological events associated with debris deposition. *Transactions of the Fourth World Biomaterials Conference*, p. 438.

Lalor, P. A., Revell, P. A., Gray, A. B., Wright, S., Railton, G. T., and Freeman, M. A. R. (1991). Sensitivity to titanium. A cause of implant failure? *J. Bone Joint Surg.*, 73B:25-28.

Linstrand, A., Stenström, A., and Lewold, S. (1992). Multicenter study of unicompartmental knee revision. PCA, Marmor, and St. George compared in 3777 cases of arthrosis. *Acta Orthop. Scand.*, 63(3):256-259.

Litner, F., Zweymuller, K., and Brand, G. (1986). Tissue reaction to titanium endoprostheses. Autopsy studies in four cases. *J. Arthroplasty*, 1(3):183-195.

Lombardi, A. V., Mallory, T. H., Vaughn, B. K., and Drouillard, P. (1989). Aseptic loosening in total hip arthroplasty secondary to osteolysis induced by wear debris from titanium-alloy modular femoral heads. *J. Bone Joint Surg.*, 71A:1337-1342.

Maguire, S. K., Coscia, M. R., and Lynch, M. H. (1987). Foreign body reaction to polymeric debris following total hip arthroplasty. *Clin. Orthop.*, 21:213-223.

Malachau, H., Herberts, P., Ahnfelt, L., and Johnell, O. (1993). Progress of total hip replacement. Results from the national register of revised failures 1979-1990 in Sweden—A 10 year follow-up of 92,675 THR. Scientific Exhibition presented at the 61st Annual Meeting of the American Academy of Orthopaedic Surgeons, San Francisco, CA, Feb. 18–23.

Maloney, W. J., Jasty, M. J., Burke, D. W., and Harris, W. H. (1989). Biomechanical and histologic investigation of cemented total hip arthroplasties: A study of autopsy retrieved femurs after *in vivo* cycling. *Clin. Orthop.*, 249:129–140.

Maloney, W. J., Jasty, M., Harris, W. H., Galante, J. O., and Callaghan, J. J. (1990a). Endosteal erosion in association with stable uncemented femoral components. *J. Bone Joint Surg.*, 72A: 1026–1034.

Maloney, W. J., Jasty, M., Rosenberg, A., and Harris, W. H. (1990b). Bone lysis in well-fixed cemented femoral components. *J. Bone Joint Surg.*, 72B:966–970.

Maloney, W. J., Peters, P., Engh, C. A., and Chandler, H. (1993a). Severe osteolysis of the pelvis in association with acetabular replacement without cement. *J. Bone Joint Surg.*, 75A:1627–1635.

Maloney, W. J., Smith, R. L., Castro, F., and Schurman, D. J. (1993b). Fibroblast response to metallic debris *in vitro*, enzyme induction, cell proliferation, and toxicity. *J. Bone Joint Surg.*, 75A:835–844.

Mather, S. E., Emmanuael, J., Magee, F. P., Gruen, T. A., and Hedley, A. K. (1989). Interleukin and prostaglandin E_2 in failed total hip arthroplasty. *Transactions of the 35th Annual Meeting of the Orthopaedic Research Society*, 15:498.

Mathiesen, E. B., Lindgren, J. U., Reinholt, F. P., and Sudmann, E. (1987). Tissue reactions to wear products from polyacetal (Delrin) and UHMW polyethylene in total hip replacement. *J. Biomed. Mat. Res.*, 21:459–466.

Merritt, K., and Brown, S. A. (1985). Biological effects of corrosion products from metals. In *Corrosion and Degradation of Implant Materials: Second Symposium*, A. Fraker and C. Griffin (Eds.). Philadelphia: ASTM STP, 859:197–207.

Merritt, K., Wenz, L., and Brown, S. A. (1991). Cell association of fretting corrosion products generated in cell culture. *J. Orthop. Res.*, 9:289–296.

Mirra, J. M., Amstutz, H. C., Matos, R., and Gold, R. (1976). The pathology of the joint tissues and its clinical significance in prosthesis failure. *Clin. Orthop.*, 117:221–240.

Mirra, J. M., Marder, R. A., and Amstutz, H. C. (1982). The pathology of failed total joint arthroplasty. *Clin. Orthop.*, 17:175–183.

Mital, M., and Cohen, J. (1968). Toxicity of metal particles in tissue culture. Part II: A new assay method using cell counts in the lag phase. *J. Bone Joint Surg.*, 50A:547–556.

Mulroy, R. D., and Harris, W. H. (1992). The effect of improved cementing techniques on component loosening in total hip replacement. *J. Bone Joint Surg.*, 72B:757–760.

Murray, D. W., and Rushton, N. (1990). Macrophages stimulate bone resorption when they phagocytose particles. *J. Bone Joint Surg.*, 72B:988–992.

Myerson, D. (1988). *In situ* hybridization. In *Diagnostic Immunopathology*. R. T. Bhan and R. T. McCluskey (Eds.). New York: Raven Press, pp. 475–498.

Nagase, M., Baker, D. G., and Schumacher, H. R. (1988). Prolonged inflammatory reactions induced by artificial ceramics in the rat air pouch model. *J. Rheumatology*, 15:1334–1338.

Nagura, H., Asai, J., and Kojima, K. (1977). Studies on the mechanisms of phagocytosis. I. Effect of electrical surface charge on phagocytic activity of macrophage for fixed red cells. *Cell Structure Function*, 2:21–28.

Nakamura, R. M. (1990). Overview and principles of *in-situ* hybridization. *Clinical Biochemistry*, 23:255–259.

Nathan, C. F. (1991). Coordinate actions of growth factors in monocytes/macrophages. In *Peptide Growth Factors and Their Receptors II*. M. B. Sporn and A. B. Roberts (Eds.). New York: Springer-Verlag, pp. 427–462.

Nizard, R. S., Sedel, L., Simeon, J., Vilette, J. M., and Launay, J. M. (1993). PGE_2 and cytokine levels in failed total hip replacement: Effect of friction combination. *Transactions of the 39th Annual Meeting of the Orthopaedic Research Society*, 18:247.

Ohlin, A. (1990). Failure of the Christiansen hip. Survival analysis of 265 cases. *Acta Orthop. Scand.*, 61(1):7–11.

Ohlin, A., Johnell, O., and Lerner, U. H. (1990). The pathogenesis of loosening of total hip arthroplasties. The production of factors by peri-prosthetic tissues that stimulate *in vitro* bone resorption. *Clin. Orthop.*, 253:287–296.

Paiement, M., Jasty, S., Goldring, C., Bragdon, M., and Roelke Harris, W. H. (1986). Difference in tissue response to particulate biomaterials (metals vs. polymers) in a rabbit wound chamber model. *Transactions of the 32nd Annual Orthopaedic Research Society*, 11:115.

Pappas, A. M., and Cohen, J. (1968). Toxicity of metal particles in tissue culture. Part I: A new assay method using cell counts in the phase of replication. *J. Bone Joint Surg.*, 50A:535–547.

Pappatheofanis, F. J., and Barmada, R. (1991). Polymorphonuclear leukocyte degranulation with the exposure to polymethylmethacrylate nanoparticles. *J. Biomed. Mat. Res.*, 25:761–771.

Perry, M., Frondoza, C., Jones, L., and Hungerford, D. S. (1990). The response of macrophages, fibroblasts and osteoblasts to PMMA and metal particles in tissue culture. *Transactions of the 36th Annual Meeting of the Orthopaedic Research Society*, 15:486.

Peters, P. C., Engh, G. A., Dwyer, K. A., and Vinh, T. (1994). Osteolysis after total knee arthroplasty without cement. *J. Bone Joint Surg.*, 74A:864–876.

Peters, P. C., Vinh, T., Johnso, F., and Engh, G. (1992). The use of oil red O stain to characterize particulate polyethylene and the macrophage response in cases of osteolysis associated with cementless total hip and total knee arthroplasties. *Transactions of the 38th Annual Meeting of the Orthopaedic Research Society*, 17(2):394.

Pfeilschifter, J., Bonewald, L., and Mundy, G. R. (1991). Role of growth factors in cartilage and bone metabolism. In *Peptide Growth Factors and Their Receptors II*. M. B. Sporn and A. B. Roberts (Eds.). New York: Springer-Verlag, pp. 371–400.

Pinhorz, L. E., Urban, R. M., Jacobs, J. J., Sumner, D. R., and Galante, J. O. (1993). A qualitative study of bone and soft tissues in cementless porous-coated acetabular components retrieved at autopsy. *J. Arthroplasty*, 8(2):213–225.

Pizzoferrato, A. (1979). Evaluation of the tissue response to wear products of the hip joint endo-arthroprosthesis. *Biomaterial Medical Review of Artificial Organs*, 7(2):257–262.

Pizzoferrato, A., Savino, L., Stea, S., and Tarbusi, C. (1988). Results of the histological grading on 100 cases of hip prosthesis failure. *Biomaterials*, 9:314–318.

Quinn, J., Joyner, C., Triffit, J. T., and Athanasou, N. A. (1992). Polymethylmethacrylate-induced inflammatory macrophages resorb bone. *J. Bone Joint Surg.*, 74B:652–658.

Rae, T. (1975). A study on the effects of particulate metals of orthopaedic interest on murine macrophages *in vitro*. *J. Bone Joint Surg.*, 57B:444–450.

Rae, T. (1981). The toxicity of metals used in orthopaedic prostheses. An experimental study using cultured humans synovial fibroblasts. *J. Bone Joint Surg.*, 63B:435–440.

Rae, T. (1986). The biological response to titanium and titanium-aluminum-vanadium particles. II. Long term animal studies. *Biomaterials*, 7:37–40.

Rahman, H., and Fagg, P. S. (1993). Silicone granulomatous reactions after first metatarsopha-langeal hemiarthroplasty. *J. Bone Joint Surg.*, 75B:637–639.

Raisz, L. G., and Martin, T. J. (1984). Prostaglandin in bone and mineral metabolism. In *Annual 2, Bone And Mineral Research*. William A. Peck (Ed.). Amsterdam: Elsevier Science Publishers, pp. 286–310.

Rentz, H., Gentz, U., Schmidt, A., Dapper, T., Nain, M., and Gemsa, D. (1989). Activation of macrophages in an experimental rat model of arthritis induced by *Erysipelothrix rhusiopathiae* infection. *Infection and Immunity*, 57(10):3172–3180.

Salvati, E. A., Betts, F., and Doty, S. B. (1993). Particulate metallic debris in cemented total hip arthroplasty. *Clin. Orthop.*, 293:160–173.

Santavirta, S., Gristina, A., and Kontinnen, Y. T. (1992). Cemented versus cementless hip arthroplasty. A review of prosthetic biocompatibility. *Acta Orthop. Scand.*, 63:225–232.

Santavirta, S., Kontinnen, Y. T., Hoikka, V., and Eskola, A. (1991). Immunopathological response to loose cementless acetabular components. *J. Bone Joint Surg.*, 73B:38–42.

Schatzker, J., Horne, J. G., and Sumner-Smith, G. (1975). The effect of movement on the holding power of screws in bone. *Clin. Orthop.*, 11:257–263.

Schmalzried, T. P., Jasty, M., Rosenberg, A., and Harris, W. H. (1993). Histologic identification of polyethylene wear debris using oil red O stain. *J. Applied Biomaterials*, 4:119–125.

Schmalzried, T. P., Kwong, L. M., Jasty, M., Sedlacek, R. C., Haire, T. C., O'Connor, D. O., Bragdon, C. R., Kabo, J. M., Malcolm, A. J., and Harris, W. H. (1992). The mechanism of loosening of cemented acetabular components in total hip arthroplasty. Analysis of specimens retrieved at autopsy. *Clin. Orthop.*, 274:60–78.

Scuderi, G. R., Insall, J. N., Windsor, R. E., and Moran, M. C. (1989). Survivorship of cemented knee replacements. *J. Bone Joint Surg.*, 71B:798–803.

Shanbhag, A. S., Jacobs, J. J., Black, J., and Galant, T. T. (1993). Pro- and antiinflammatory mediators secreted by cells of interfacial membranes from revision total hip replacements. *Transactions of the 39th Annual Meeting of the Orthopaedic Research Society*, 18:517.

Shanbhag, A. S., Jacobs, J. J., Black, J., Galante, J. O., and Galant, T. T. (1994). Macrophage/particle interactions: Effect of size, composition and surface area. *J. Biomed. Mat. Res.*, 28:81–90.

Søballe, K., Brockstedt-Rasmussen, B., Hansen, E. S., and Bünger, C. (1992a). Hydroxyapatite coating modifies implant membrane formation. Controlled micromotion studied in dogs. *Acta Orthop. Scand.*, 63(2):128–140.

Søballe, K., Hansen, E. S., Brockstedt-Rasmussen, H., Jørgensen, P. H., and Bünger, C. (1992b). Tissue ingrowth into titanium and hydroxyapatite-coated implants during stable and unstable mechanical conditions. *J. Orthop. Res.*, 10(2):285–299.

Søballe, K., Hansen, E. S., Brockstedt-Rasmussen, B., and Bünger, C. (1993). Hydroxyapatite coating converts fibrous tissue to bone around loaded implants. *J. Bone Joint Surg.*, 72B:270–278.

Spector, M., Shortkroff, S., Hsu, H. P., Lane, N., Sledge, C., and Thornhill, T. S. (1990). Tissue changes around loose prostheses: A canine model to investigate the effects of an antiinflammatory agent. *Clin. Orthop.*, 261:140–152.

Stinson, N. (1964). Tissue reaction induced in guinea pigs by particulate polymethyl-methacrylate, polythene, and nylon of the same size range. *J. Exp. Pathol.*, 46:135–147.

Tabata, Y., and Ikada, Y. (1988). Effect of size and surface charge of polymer microspheres on their phagocytosis by macrophage. *Biomaterials*, 9:356–362.

Tallroth, K., Eskola, A., Santavirta, S., Konttinen, Y. T., and Lindholm, T. S. (1989). Aggressive granulomatous lesions after hip arthroplasty. *J. Bone Joint Surg.*, 71B:571–575.

Unanue, E. R., and Allen, P. M. (1987a). The basis for the immunoregulatory role of macrophages and other accessory cells. *Science*, 236:551–557.

Unanue, E. R., and Allen, P. M. (1987b). The immunoregulatory role of the macrophage. *Hospital Practice*, 4:87–104.

Vernon-Roberts, B., and Freeman, M. A. R. (1976). Morphological and analytical studies of the tissue adjacent to joint prosthesis investigations into the causes of loosening of prosthesis. In *Advanced Artificial Hip and Knee Joint Technology*. M. Schaldack and D. Hohmann (Eds.). Berlin: Springer-Verlag, p. 148.

Vitetta, E. A., and Paul, W. E. (1991). Role of lymphokines in the immune system. In *Peptide Growth Factors and Their Receptors II*. M. B. Sporn and A. B. Roberts (Eds.). New York: Springer-Verlag, pp. 401–426.

Weber, B. G. (1970). Die Rotations Totalendoprosthese des Huftgelenkes. *Z. Orthop.*, 107(2):304–315.

Willert, H. G., Bertram, H., and Buchhorn, G. H. (1990a). Osteolysis in alloarthroplasty of the hip. The role of bone cement fragmentation. *Clin. Orthop.*, 258:108–121.

Willert, H. G., Bertram, H., and Buchhorn, G. H. (1990b). Osteolysis in alloarthroplasty of the hip. The role of ultra-high molecular weight polyethylene wear particles. *Clin. Orthop.*, 258: 95–107.

Willert, H. G., Ludwig, J., and Semlitsch, M. (1974). Reaction of bone to methacrylate after hip arthroplasty. A long-term gross, light microscopic and scanning electron microscopic study. *J. Bone Joint Surg.*, 56A:1368–1382.

Willert, H. G., and Semlitsch, M. (1977). Reactions of the articular capsule to wear products of artificial joint prostheses. *J. Biomed. Mat. Res.*, 11:157–164.

Williams, J. D., Czop, J. K., and Austen, K. F. (1984). Release of leukotrienes by human monocytes on stimulation of their phagocytic receptor for particulate activators. *J. Immunol.*, 132: 3034–3040.

Williams, R. C., Jeffcoat, M. K., Kaplan, M. L., Goldhaber, P., Johnson, H. G., and Wechter, W. J. (1985). Flurbiprofen: A potent inhibitor of alveolar bone resorption in beagles. *Science*, 227:640–642.

Williams, R. C., Offenbacher, S., Jeffcoat, M. K., Howell, T. H., Johnson, H. G., Hall, C. M., Wechter, W. J., and Goldhaber, P. (1988). Indomethacin or flurbiprofen treatment of periodontitis: Effect on crevicular fluid arachidonic acid metabolites compared with effect on alveolar bone loss. *J. Periodont. Res.*, 23:134–138.

Witt, J. D., and Swann, M. (1991). Metal wear and tissue response in failed titanium alloy total hip replacements. *J. Bone Joint Surg.*, 73B:559–563.

Worsing, R. A., Engber, W. D., and Lange, T. A. (1982). Reactive synovitis from particulate silastic. *J. Bone Joint Surg.*, 64A:581–585.

Zurier, R. B. (1988). Prostaglandins and inflammation. In *Prostaglandins: Biology and Chemistry of Prostaglandins and Related Eicosanoids*. P. B. Curtis-Prior (Ed.). Edinburgh: Churchill Livingstone, pp. 595–607.

57

Materials for Applications That Are Subject to Wear and Friction

Kenneth R. St. John and Audrey K. Tsao
University of Mississippi Medical Center
Jackson, Mississippi

I. INTRODUCTION

The advent of modern technology and novel methods of materials processing has opened a new era of biomaterials applications. These new materials have been derived from existing technology and use in more standard industrial applications. The interaction between human and animal biology has demanded new capabilities from existing materials from the biocompatibility, biomechanical, and biomaterials aspects. These new materials are being utilized in hostile environments that make longevity and durability key factors in their ability to perform successfully. In addition, the relative inaccessibility to replace, exchange, or service these devices makes reliability a primary concern. Nowhere else have materials and the devices that they comprise been asked to perform in such a consistent and predictable fashion.

There are many areas in which biomaterials are utilized for the human body. They cross many fields of expertise, including dentistry and orthodontics, surgical specialties such as orthopedics and plastic surgery, to medical fields such as cardiology and radiology. This wide range of applications demands a wide range of material types and interactions. The biological and material relationships vary depending upon use. Many of these materials and the resultant devices are placed in environments that are subject to wear and friction. These two factors often are intimately related to their ability to survive and function in the long term.

In the selection of materials for use in dynamic load-bearing applications, the amount of relative movement at interfaces must be considered because of the possibility that wear may occur that can result in the liberation of particulate material into the surrounding tissues. The possibilities for relative movement at interfaces include the gross movement that occurs between moving parts, the microscopic movement that can occur at mating surfaces of modular devices, the abrasion that can occur between a moving part and adjacent calcified tissues, and the microscopic adjustments that can occur between an implant and tissues as the implant settles into a final configuration.

In the minimization of wear and the resultant wear debris, it is necessary for material designers, implant designers, instrument manufacturers, surgeons, and other health-care professionals to work as a team, recognizing the part that each must play in this process. Material modifications may enhance the abrasion resistance or reduce the coefficient of friction of a wear couple, thereby reducing the potential for debris generation. The implant designer must consider the consequences of design changes beyond solution of the immediate problem. The ultimate interrelationship of all the system components (including the patients' natural components) in the functioning biomechanical and physiological system must be considered. Manufacturers of instruments, equipment, and supplies for implant devices must work to enhance the repeatability and reliability of the operative procedures. This requires recognition of the consequences that malalignment, surgical errors, and careful design can have on wear problems.

The surgeon and allied health professional form the front line in the careful use of modern technology for precise implant placement to minimize conditions that may be conducive to wear. Continued training and education to understand further the design concepts underlying new technologies become key factors in their proper utilization. In this way, professional decisions in finding solutions to difficult surgical problems can include appropriate use of available device components to enhance the long-term survival of the products. When problems do develop clinically, the early recognition of the symptoms of a wear-related problem can serve to minimize the long-term consequences to the patient and to future patients.

Prosthetic replacement of damaged or malformed biological structures is a common theme in both medicine and dentistry. Congenital malformations can affect the biological performance of many human anatomical structures, leading to early degenerative changes and a decline in the performance of the innate biological structure, thus requiring its replacement for continued function in the human body. In situations in which structures key to life are involved (e.g., the heart valve or human vascular structures), the ultimate survival of the individual may depend on the very materials from which a prosthetic device is manufactured. The repetitive interactions of a beating heart ask that the metal and plastic prosthetic valve continually move in a predictable fashion without excessive wear at the biological and material interface as well as the material-to-material interface between plastic and metal.

II. TYPES AND MECHANISMS OF WEAR

When two surfaces are in contact with each other under load and motion at the interface, there is a resultant frictional force resisting movement. In order for motion to occur, that force must be overcome. This shearing force at the interface has the potential to exceed the intermolecular forces in one or both materials, with resultant wear and generation of debris.

The wear process may result in the transfer of material from the softer to the harder material (transfer film) or in generation of particles of the worn material. These particles, if they are not expelled from the interface, may act as abrasive particles that may accelerate the wear process. The generation of a transfer layer may alter the properties of the wear couple and result in reduced or altered rates of wear. Because of these two phenomena, the wear process and wear models frequently exhibit a "wearing in" period and a "steady-state" period (or an accelerated wear period) after the initial changes in surface properties have occurred. There is also the potential for changes in wear properties as increased conformity of the surfaces is achieved since increased conformity leads to greater surface area of contact and reduced frictional stress.

When the physiological environment is factored into a wear situation, there is the opportunity for lubrication due to the presence of proteins on the material surfaces or in the fluids at the interface. Diffusion of fluids or other physiological products into one or both materials at an interface can change the size, shape, or shear properties of the components and therefore the wear properties of the couple. In addition, the presence of corrosion products on metallic components of the couple can lead to the generation of particulate debris. Nearly all corrosion-resistant implant materials are rendered resistant by the presence of a "passive layer" on the surface that is relatively impervious to ions and inhibits further corrosion. This layer is usually an oxide of at least one component of an alloy and is likely to be less resistant to shear than the underlying metal. If this passive layer is removed as a result of frictional shear forces, it will reform if the right ions are present (usually oxygen), but it can then be further abraded. Metal oxides are essentially ceramic particles and, if they remain in the interface, can be highly abrasive.

Abrasive particles can enter a wear couple from sources other than wear of the interface. Some examples of other sources are residual debris from the surgical procedure, such as bone chips and particles of bone cement, and particles that are generated by failure of other components of the system, such as beads and ceramic coatings. If these particles find their way into the interface, the result can be severely accelerated wear because their presence will not have been included in the design process for the system and their effects may be unpredictable.

III. SITES OF WEAR-SUSCEPTIBLE IMPLANT PLACEMENT

A. Repair or Replacement of Articulating Joints

The function of an articulating joint can be compromised as a result of disease or trauma such as that seen in degenerative arthritis from a previous injury or an inflammatory condition such as rheumatoid arthritis. In these situations, weight-bearing joints such as the knee or hip are replaced with new materials in total joint arthroplasties (Figs. 1 and 2). These constructs of metal and plastic are faced with dual requirements. They must interact between themselves under high stress situations such as walking, jumping, or running, for which the load has been estimated to be four to eight times body weight, with shear and normal stresses seen during the kinematics of gait. They must undergo this loading while being designed to fit and interact within their biological constraints of human bone interfaces, exposure to constant body fluids with immunological responses, and the engineering design constraints of the size of the human knee and hip. Most human biological systems are maintained on at least a cellular level and are not inert structures. These man-made bearing surfaces are being required to substitute for biological bearing surfaces, but are unable to be maintained by biological repair mechanisms. Since substituted prosthetic devices are static and cannot be maintained or serviced once they are implanted, any damage to one of these surfaces is a permanent change in the device (Figs. 3–6).

Maintaining mobility in the joint may also be addressed in another fashion besides the resurfacing of the preexisting biological bearing surface. The entire joint may be

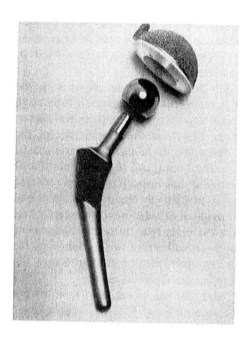

Figure 1   A three-component total hip replacement device consisting of a femoral stem, a femoral head bearing component that is attached with a taper connection, and an acetabular component. (Photo courtesy of Howmedica, Inc., Rutherford, NJ.)

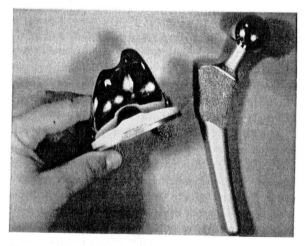

Figure 2   Total hip and total knee prostheses for replacement of the weight-bearing function of the joint. The hip prosthesis on the right would be bearing against a polyethylene acetabular component as seen in Fig. 1.

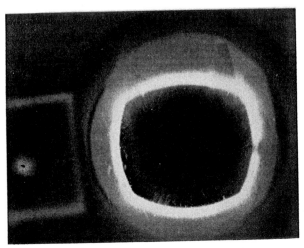

Figure 3 Hip prosthesis showing wear and scratching of the bearing surface. This surface was originally highly polished.

replaced by a completely different mechanism, such as the silastic implants seen in hand metacarpophalangeal joints. In these joints, the resurfacing of a bearing surface is fraught with technical difficulties related to the intrinsic soft-tissue anatomy, its stability (and thus its ability to keep motion occurring in a repetitive, consistent fashion), and the simple constraints of size. These joints require motion as well as stability. Silastic implants are commonly used as a single implant to aid in stability and motion occurs within the implant itself. In this situation, wear and friction are a direct result of the stresses and demands placed upon the materials and designs of the implant.

Prosthetic joints for the temporomandibular joint (TMJ), the shoulder, the elbow, the hip, and many other areas all present unique performance requirements. Each joint exhibits different patterns of load, stability within the prosthesis, and stability within the inherent human tissue envelope, as well as varying degrees of motions in multiple planes. In these situations, the implants and the materials are subject to stress at their biological material interfaces as well as between the device components.

One other area of prosthetic replacement seen today is the interpositional prosthesis. This term generally is applied to a prosthetic replacement that is required to function as a biological spacer to prevent instability with the body due to abnormal biomechanics that would occur if a normal anatomic structure was deficient or absent. In these situations, loading is not always predictable and motion can occur in almost any plane. The implant must resist friction and wear along all its surfaces for a specified time. Although the function of these implants is simply to act as a spacer, biomechanically these implants are asked to be wear, load, and friction resistant. Generally, these implants have not met with a high rate of success due to rapid deterioration under the conditions of use. The silicone rubber radial head prostheses that are used to maintain stability in the elbow are commonly removed once healing has occurred and the inherent biological stability has been reconstituted. If left in place, these implants will ultimately fracture and large amounts of wear debris will cause a biological response worse than the initial problem.

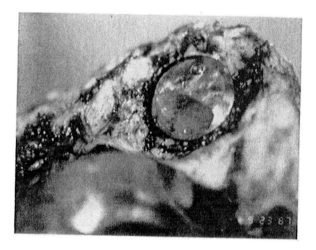

(a)

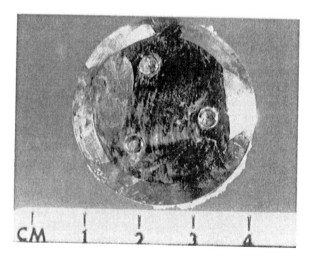

(b)

Figure 4 The metallic backing from a patellar resurfacing knee component. The polyethylene bearing surface has separated from the metal and back and metal-metal wear has occurred: (a), in a patient with metallic wear particles seen in surrounding tissues; (b), a similar component after removal from a patient.

B. Ligament and Tendon Prostheses

The search for a suitable substitute for autogenous tissue has led many investigators to consider synthetic materials for use as ligament replacements in the knee. One of the problems that has been reported for these materials is the generation of wear debris that is released into the joint. The friction in tendon prostheses due to adjacent structures is high as these tissues normally move freely over long distances but are not contained in a

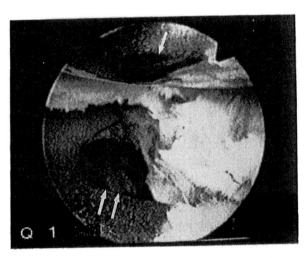

Figure 5   Arthroscopic view of a total knee prosthesis in which the polyethylene liner has failed and wear and burnishing are seen between the femoral condyles (arrow) and the tibial tray (double arrow).

discreet joint. The artificial ligament, while it does not necessarily move, sees varying directions of strain and various conditions of load within the inherent structure as ligaments generally span a mobile joint and lend inherent stability.

The procedure for placing these materials requires the preparation of bony tunnels or grooves through which the prostheses pass. Despite the best efforts of the surgeon, these edges represent raw bone that, in the flexion and extension of the joint function, cause wear of these fibrous materials against the bone and the generation of particulate mate-

Figure 6   Metallic tibial tray showing wear and burnishing from repetitive contact with the polyethylene tibial bearing insert after cracking and partial failure of the tray.

rial. Most of the materials and implant configurations utilized are selected for their tensile mechanical properties and not for their abrasion resistance.

The physical demands upon a material that is placed into the intraarticular environment for the replacement of the cruciate ligament are severe. As the knee passes through its range of motion in the gait cycle, the anterior cruciate ligament prosthesis acts as a check rein to unacceptable anterior translation of the tibia on the femur. In accomplishing this function, the prosthesis is placed in tension at the same time that it is bearing against the anterior edge of the femoral bone tunnel and the posterior edge of the tibial bone tunnel. This situation, with time, can lead to the generation of wear particulates or gross failure of the device. Historically, these devices have been manufactured from polytetrafluoroethylene (PTFE), polyethylene terephthalate, polyethylene, polypropylene, or carbon fibers, all of which have experienced wear against the edges of bony prominences and generation of debris to some extent. One series of studies detailed animal evaluation of possible effects of these particles on the synovium and the anticipated long-term consequences of this wear to the patient [1].

C. Heart Valves

Probably one of the most demanding applications for implant materials is in heart valve replacement. The human heart beats approximately 37 million times a year and, in heart valve replacement, the prosthetic parts may be asked to function for 50 years without failure. The natural heart valves are uniquely suited to this function, in addition to having the capabilities for regeneration and repair. When this repair capability has malfunctioned or some other process leads to the need for replacement, cardiovascular surgeons and device designers are faced with the necessity to use a device that can open quickly, provide minimal impedance to blood flow, and close quickly and completely to prevent regurgitation.

Devices for replacement of heart valve function generally involve some type of occluder that can impede reverse blood flow but move out of the way of flow during heart chamber contraction. The occluder then must be captured by some type of cage, strut, hinge, or other device to prevent it from being carried away by the action of blood flow. Most problems of wear within these devices come from the repetitive relative motion of the occluder against its capturing structure, with progressive erosion of one or both parts of the device. While the actual wear debris generated from this process is most likely not a problem physiologically because of its extremely small size and minimal volume over reasonable time periods, the net result on overall device function can be catastrophic.

In early designs of heart valves, silicone rubber occluders were placed within a metal cage to act as ball valves. These spherical balls had a lower density than blood and floated very easily in the blood flow, therefore opening and closing much as other caged ball check valves work. The problem that developed was twofold. In the first case, silicone rubber materials are susceptible to wear problems, with generation of debris in the absence of other factors. The additional problem that developed with these systems was that the poppets were swelling as lipids diffused into the silicone rubber, leading to increased wear as the balls swelled against the bars of the cage. In some cases, grooves wore into the sides of the poppets and the valves continued to function for a period of time. In other cases, wear did not adequately compensate for swelling and the valves either stuck open or stuck closed. In the most catastrophic cases, the poppets either wore to the point at which they could escape from the cages or fragment, leading to downstream occlusion of vital vessels and death or very serious complications.

The materials that have been used for heart valve applications have included silicone rubber for occluders, pyrolytic carbon, metals, and fabrics for sewing rings. Wear of occluders and the structures that capture the occluders has occurred in clinical application and wear of the fabric sewing rings from contact with occluders has also occurred.

D. Fracture-Fixation Devices

The last group of implants to consider are those used in fracture fixation. These implants are designed to aid in holding bone together until biological healing occurs. Thus, these implants are not necessarily designed to have gross motion. However, fracture healing occurs best when dynamic loading occurs across bone, thus these implants have some inherent flexibility within their materials as well as their constructs to promote biological fracture healing. Patients are encouraged to load their fracture to promote further healing and thus may stress their fracture fixation implant further.

Fracture implants are of varying sizes, shapes, and materials and can be plates with holes for screws to fix them to bone (Fig. 7). They may be cylindrical rods with transfixation screws or bolts placed through the bone and rod to aid in stability (Fig. 8). Thus, there is an interaction between both the biological surfaces and the other metal implant surfaces at which friction can occur in a dynamic loading situation. These areas of friction will promote wear at the interfaces.

IV. MATERIALS OF CONSTRUCTION IN WEAR SUSCEPTIBLE SITUATIONS

A. Metals

The majority of metals are used in one of two types of applications. The first is for temporary use in the human body, such as internal fixation for fractures; the second is as permanent implants functionally replacing a portion of the human anatomy. These functional requirements dictate the type of implant material to be used and the constraints under which it must function. The contact stresses, wettability, frictional constraints, type of loading, and frequency of loading will all have a major impact upon the frictional

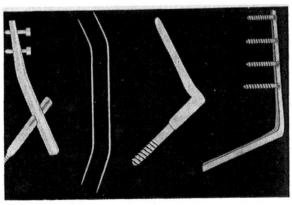

Figure 7 Schematic of several possible configurations of multi-component fracture fixation devices.

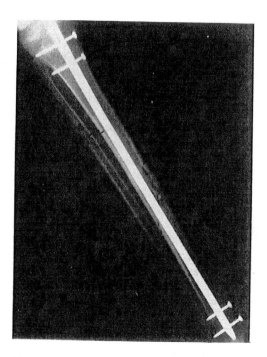

Figure 8   Radiograph of a tibial intramedullary nail with transcortical locking bolts.

and wear characteristics of the individual implants aside from the material properties themselves [2].

1. Stainless Steel

Stainless steel in one of several formulations has been in clinical use for many years. These implants are generally considered to be biocompatible and corrosion resistant and to satisfy the mechanical properties for use as structural bone composites. Fracture-fixation devices of pins, bolts, plates, rods, and some early total joint prostheses use stainless steel as their primary metal material.

In general, the use of stainless steel has been primarily in fracture-fixation devices (Figs. 7–9). Its metallurgical properties and familiarity to the industry make stainless steel an ideal material for these applications. Since these devices, for the most part, need to perform their function for only a given length of time and can be removed from the patient, the friction and wear properties of stainless steel do not adversely affect its normal use. While fretting, friction, and wear are seen at the metal-to-metal interfaces, often the rigidity of the healed composite bone metal structure does not compromise the functional capacity of the implant. The intent of the devices is to allow the bones to be anatomically reduced and held rigidly or semirigidly to allow bone to heal. The function of the implant becomes obsolete after bony healing has taken place. Until bony union is achieved, motion may occur between implant surfaces and, due to the viscoelastic properties of bone, load transfer to the fixation system may cause continued motion at the interfaces after healing.

While movement may be minimal, friction and wear are present between the implant and the host bone as well as between the various components of the implants themselves

[3]. Friction may commonly occur between the screw used to affix a plate or rod into a long bone despite the intent to fix the bone rigidly for union. Other fracture situations may encourage movement between various components of a device. For example, in a sliding hip screw, as the fracture of the femoral neck collapses and union occurs, the screw slides within the metal barrel (Fig. 9) and thus wear and friction occurs [4]. These situations are compounded by the fact that, for the most part, cyclic motion from ambulation or movement inherent in activities of daily living will promote the wear cycle. The fact that most fracture-fixation devices may be surgically removed when indicated may in part alleviate some of the problems with long-term friction and wear [5].

Stainless steel is also commonly used in devices for spinal instrumentation. These applications make use of rods, hooks, sleeves, and other assorted hardware and are potential areas for wear problems. If rigid fusion and/or bony union is obtained, this is usually not a clinical problem. Wear is primarily associated with clinical problems in which gross biological movement continues despite attempts to heal and immobilize the bone.

An additional related use for stainless steel is for mesh, wire, or prostheses, especially for nonroutine surgical applications. While wire and mesh may be used commonly to aid as adjunctive fixation in fracture fixation, there is the potential for interaction between two different devices, such as a wire looped around a butterfly bone fragment in a long bone fracture or a locking pin placed in an intramedullary rod. Mesh and wire as well as custom prostheses are commonly used for reconstruction after tumor resection. Mandibular reconstruction may be performed by contouring a stainless steel wire, rod, or mesh to the desired anatomical shape and fixing to surrounding bone [6,7]. Since this is placed to give structure to an area that is continually moving and subject to repetitive motion grossly, friction and wear may occur along the tissue/metal interface and between metal components, especially if it is anchored within bone. Because of the superficial location of the prosthesis, potential complications of the failure from wear may be detected early in the postoperative period.

Stainless steel is occasionally used in manufacturing total joint prostheses. Charnley used stainless steel one-piece femoral implants in his low friction prosthesis. This was the precursor that formed the basis for the development of modern prosthesis systems. This type of use for this material automatically assumes repetitive cyclic motion from gait. Its use as an articulation and bearing surface requires that wear and friction be minimized for longevity of function.

While the issue of bearing surface wear in the polyethylene counterface material is discussed in another section, it is important to note that the identity and surface finish of the metallic side of the wear couple can have a significant effect on wear of the polyethylene. Wear of both the metal and polyethylene is greatly affected by the potential for three-body wear due to polymethylmethacrylate (PMMA) debris or bone chips left in the joint after insertion of a total hip prosthesis.

The use of forging and casting manufacturing techniques may alter the inherent strength within the steel. In general, it is the high-grade austenitic stainless steels that are required for corrosion and fretting resistance in the wear and friction environment. Increased susceptibility to wear may also occur in those implants that are manipulated at the operating table to conform anatomically. A certain amount of cold working is inherent to these manipulations and can locally change the properties of the device to be used. While fatigue failure is often the catastrophic event that leads to implant failure, the mechanisms of wear and friction are often also present. The debris generated from the wear of

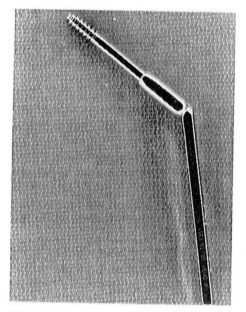

(a)

(b)

Figure 9   A sliding hip screw for the fixation of proximal femoral fractures: (a), the device as it would normally be placed within the patient; (b), the same device showing the amount of sliding that can occur in the process of maintaining compression on the fracture site; (c), a radiograph of a hip screw that has moved to its full extent within the barrel of the plate.

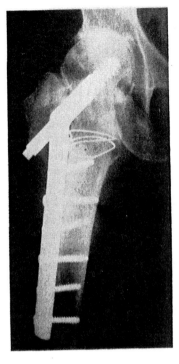

(c)

materials may often lead to a biological response from the surrounding tissues. Isolated cases of tumorigenic associations have been reported [8], but, more commonly, metallic debris and a granulomatous foreign-body response are seen [9].

An additional issue is seen with articulating surfaces composed of stainless steel. Some early hip replacements were composed of stainless steel femoral components [10]. The first total hip replacements were manufactured of 316L stainless steel on the femoral component and Teflon for the acetabular component. Both components were held in place with polymethylmethacrylate as a bone cement [11,12]. In this situation, the wear of the component was directly related to the frictional and material properties of the stainless steel on the other bearing surface as well as the loading configuration of the involved implants.

In general, for all prosthetic joints, the interaction of the bearing surfaces against each other is a primary factor in the occurrence of wear. In the case of stainless steel, the wear rate is also related to the surface finish of the stainless steel. As the roughness of the steel increases, potential wear rates of ultra high molecular weight polyethylene (UHMWPE) also increase, such that a finish of 0.015 microns (μ) will allow an acceptable level of wear [13]. This becomes critical in the clinical scenario in which third-body wear may provoke surface damage of the stainless steel. Polymethylmethacrylate is currently used as an orthopedic bone cement and particles can break off, becoming debris interposed between the articulating surfaces. As such, the initiating frictional surface wear will increase the surface roughness [14]. Once damage has occurred to the surface of stainless steel, accelerated wear linked to the roughness of the surface is seen. Once the wear process has started, it will continue to accelerate and lead to the ultimate failure of a device. As such, stainless steel is no longer commonly used as bearing surfaces for prosthetic joints such as total hip arthroplasties [14].

2. Titanium Alloys

The titanium-containing alloys were chosen for clinical use based upon properties that appeared to solve observed clinical problems, while resistance to wear became an issue only after clinical problems were seen that required solutions. Titanium alloys have been superb biomaterials in many ways. They are extremely biocompatible and inert in the human body, with a modulus of elasticity closer to that of human bone than many other orthopedic implant alloys. Titanium and its alloys were initially chosen for use in total joint replacement because of their lower modulus of elasticity and the potential that the resultant increased load sharing might reduce the proximal femoral atrophy that had been seen around some devices. In addition, in dental implant use, commercially pure titanium had been found to "osseointegrate" or bond to bone without an intervening fibrous tissue capsule, a property that might enhance long-term clinical success if it was seen in joint replacement devices. There are several drawbacks to titanium, including its notch sensitivity, slightly increased difficulty in manufacturing, and its poorer qualities as a bearing surface [15–17].

Titanium has been used as a bearing surface in the past for total joint arthroplasties of the hip and knee, as well as primary endoprostheses for the hip. The first devices that were produced consisted of an integral stem and head that were both composed of the same titanium-aluminum-vanadium alloy. Early clinical results seemed to confirm the acceptability of this alloy for the stem. Over time, tissues adjacent to the device were found to be gray in color; it was assumed that the cause was wear of the head of the device producing particles of titanium and/or titanium oxide (the corrosion product of titanium that comprises the passive layer on the surface of the metal). Over the long term, its use has been uniformly met with problems from wear and its associated particulate metal debris, along with the associated metallosis [18–20].

For titanium bearing surfaces, there have been reports of primary wear problems without associated loosening or third-body wear [18,20]. This wear is only increased when third-body wear is also present such as that seen in fragmentation from PMMA of a loosening prosthesis. Grossly, the titanium wear debris particulate matter presents itself as a black debris that "stains" all living tissue exposed to the implant. Despite the color, the primary problem remains the cellular biological response to the debris since a foreign-body giant cell response often occurs. Radiographically, osteolytic lesions in bone with corresponding soft-tissue inflammation can be detected. Clinically, gross loosening of the implants, with ultimate failure, may occur [21].

Observation of wear-related bone pathology presents a special problem in those implants directly implanted into bone with the expectation of bone ingrowth into the prosthesis to achieve ultimate function and stability. Pads or mesh manufactured from commercially pure titanium may be applied to a titanium alloy device, thus allowing the fixation of these implants directly into bone without the use of a bone cement. This porous coating depends on the gross stability of the implant for its initial success and the ingrowth of bone into these porous coatings for final stability and fixation. Dental tooth implants commonly involve placing a peg directly into bone [22], while total joint arthroplasties commonly use an interference fit with bone or are directly implanted in a bone. These implants, if they do not achieve fixation and stability, will potentially have micromotion occurring between the bone and implant interface. Friction and wear debris produced locally will disturb the bone and implant interface, interrupting the process of bone ingrowth into the surfaces and cause the potential for loosening and failure to magnify.

The result of this early experience and the consequent research is that titanium and its alloys are generally not used in the bearing surfaces of total joint prostheses. More commonly, the bearing surfaces are manufactured from one of the alloys containing cobalt and chromium, which are attached by some mechanism to the titanium alloy components.

The use of combination devices in which components of titanium alloys are attached to bearing components of another material has been generally successful, but the interface between the two components has introduced the opportunity for additional wear and wear-related phenomena to occur. Most of these interfaces feature a mechanical interlock based upon the "morse taper." This mechanism requires carefully machined male and female tapers to be mated with an impacting force, causing the two components to be firmly attached to each other by the interference fit of the matching surfaces. Unfortunately, it would appear that this mating does not prevent microscopic motion with load bearing as the surfaces either move in an oscillatory fashion or settle to a more stable configuration. Particles that appear to have originated from the taper surfaces have been found in periprosthetic tissues and surface degradation that appears to be fretting corrosion has been observed in the taper. Both of these findings suggest the possibility of relative motion at the interface.

Researchers have been studying methods of improving the abrasion resistance of titanium alloys for a number of years, with a goal of returning to a device constructed exclusively of titanium. This need is particularly felt for knee prostheses, in which it is difficult to mate multiple components to have one material for the bearing surface and another for the bone contact surfaces. Some investigators believe it is important to have a bone contact surface of titanium in order to enhance fixation in noncemented applications.

Titanium's ability to be utilized as a bearing surface is so poor that any area of friction from a moving contact has been generally avoided. Since titanium is desirable in many other ways as a biomaterial, an effort has been made to improve its bearing and wear characteristics by surface treatments of it and other various metal implants used today [23–25]. These include the introduction of nitrogen into the surface layers of the metal by conventional nitriding processes [26,27] or ion implantation [24,25,28–30]. Theoretically, this will improve the surface hardness of the titanium by increasing nitrogen interstitial strength or the formation of nitrides in the surface zones. This would decrease the continual removal from abrasion, friction, and wear of the surface protective oxide passivation layer on the surface of the exposed metal. In addition, this may benefit the corrosion resistance of titanium [31,32]. This combination of increased corrosion resistance and wear resistance would enhance titanium's use as a load-bearing surface metal [16] without detrimental effect against other potential complementary bearing surfaces such as ultra high molecular weight polyethylene [16,33].

Titanium is also used for fracture-fixation devices such as rods, screws, and plates, as well as spinal instrumentation. Titanium's modulus of elasticity makes it an ideal implant to allow for bone fixation while still allowing stress transfer to the bone. Similar wear problems and frictional situations will occur in the titanium implant as in the stainless steel implants. Since titanium is extremely notch sensitive, the user must be aware that failure from gross fatigue may be potentiated by the wear phenomenon.

3. Cobalt-Based Alloys

Co-Cr-Mo cast and wrought alloys are among the most wear resistant alloys used today in load-bearing surfaces such as those seen in total joint arthroplasties. The cast Co-Cr-Mo alloys are felt to have superior wear resistance, in part due to the large number of carbides that are formed within the alloy itself due to the manufacturing process [23].

Cobalt/chromium alloys are commonly used as the metal half of the articulating surface in artificial prosthesis due to the superior wear characteristics compared with titanium [21,34]. The micromechanics of Co-Cr-Mo alloy reveal it to be significantly more resistant to surface roughening than titanium alloys [35]. Wear rates of Co-Cr-Mo appear to be on the order of 0.1 micrometer per year (10^6 cycles). Although cobalt/chromium materials may be placed in similar loading situations as titanium, especially in total joint arthroplasties, it appears that the frictional wear problems are fewer [35].

An additional area for friction and wear occurs between two metal surfaces juxtaposed to each other. Although this may not have been initially intended to become an articulating surface, the repetitive cyclic motion in the human body in conjunction with high areas of stress such as seen in the tapered cone connection between a prosthetic femoral head and femoral neck of the femoral component can be an area of concern. The more common failure mode is that of fretting corrosion, especially between two dissimilar metals such as a femoral cobalt/chromium head coupled with a titanium femoral prosthesis. This fretting debris is also a product of relative wear and friction. Abrasive wear may then be initiated into this environment, leading to increased surface damage of the morse taper articulation and failure may then result [36].

Bone ingrowth prostheses manufactured from cobalt/chromium alloys are often made with sintered cobalt-chromium beads applied directly to the implant surface. As such, direct contact between the ingrowth surface and bone will cause high frictional stresses between the implant and bone interface [37]. The potential for wear debris at this interface is theoretically greater.

Other prostheses, such as those used for mandibular reconstruction, are also commonly constructed of cobalt-based alloys. In this situation, friction and wear may also be seen at the junctions of fixation where a screw and nut are used to attach the prosthesis to the bone. A similar loading condition to that for stainless steel fracture-fixation devices is seen here [38].

The foreign-body and inflammatory responses to cobalt-chromium wear particulate debris are similar to those of the other types of metallic debris. The tissue reaction from a biochemical standpoint may depend upon the identity of the particles that are present since iron or chromium salts may be very toxic to the cellular structures surrounding them. In all metal systems, one reason that wear debris is a problem is because it presents to the body a larger surface area per volume of material and may trigger potentially a more toxic biological response [39].

B. Plastics

1. Polyethylene

Various forms of polyethylene have been used in orthopedics and other areas for prosthetic components, either as the entire component or as the bearing surface. In the 1950s, Charnley implanted his first-generation total hip arthroplasty using polytetrafluoroethylene (PTFE) as the material on his acetabular cup component. Loosening with an associated soft-tissue reaction from particles of abraded PTFE were attributed to failure and high molecular weight polyethylene was used instead [2]. Today, the most commonly used polymeric material for bearing surfaces is ultra high molecular weight polyethylene (UHMWPE) [40].

Ultra high molecular weight polyethylene exhibits a range of molecular weights in the material. In addition, cross-linking and structure may be altered by manufacturing processes to change inherent material properties. Thus, density, crystallinity, hardness, and consistency in materials, along with surface roughness, may all contribute to the wear of UHMWPE. Currently, because of its use as a bearing surface, especially in total joint arthroplasty, wear of polyethylene has become a clinical and material issue (Figs. 10 and 11) [12,41–48].

Several issues for polyethylene wear are of concern since it can be placed in several different loading conditions. In total hip arthroplasties in which the bearing surface is a ball-and-socket joint, the components are slightly more constrained and wear tends to be accentuated in the areas in which direct weight bearing occurs [12,49,50]. In total knee arthroplasties in which the motion is significantly less constrained and is a combination of rotational, sliding, and rolling movements, the wear manifests itself in different ways (Fig. 12) [41,45]. Shear forces within the system may also predispose the polyethylene bearing surfaces to wear. Several types of wear have been identified, including pitting, delamination, adhesion, abrasion, and third-body wear (Figs. 13 and 14) [12,14,51,52].

The other component of wear in polyethylene is the contralateral bearing surface. Titanium alloy, stainless steel, cobalt/chrome alloy, and ceramic contralateral bearing surfaces will affect the wear of polyethylene, in part due to their surface finishes and inherent roughness [26,47,49,53]. The relative potential for UHMWPE to cold flow and/ or wear is affected by the amount of stress seen by the polyethylene component. This is affected by the thickness of the polyethylene, any metal supporting surface used to distribute stress within the polyethylene, and the relative hardnesses of the other bearing surfaces [54].

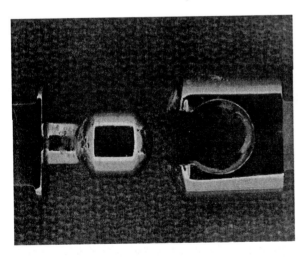

Figure 10 Photograph of a total elbow device that has experienced wear and failure of the polyethylene bearing component.

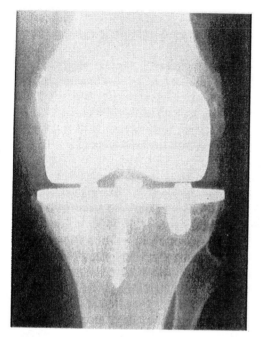

(a)

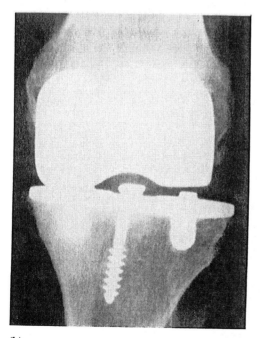

(b)

Figure 11 Radiographs of the knee in a patient who experienced wear of the polyethylene insert: (a), immediately after surgery, the spacing between the two components is correct; (b), with time, the polyethylene on the medial side of the device has worn putting the two metallic components in direct contact with each other.

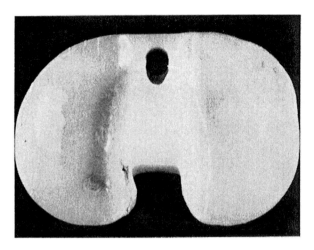

Figure 12 Example of a worn polyethylene tibial insert showing delamination of the polyethylene and cracking.

2. Silicone Rubber

The use of silicone rubber materials in applications subject to wear has included the occluders on heart valve prostheses and as interposition devices in TMJ devices, carpal head prostheses, and tarsal and metatarsal prostheses, as well as the hinge prostheses for metacarpal-phalangeal joint arthroplasty.

When these materials are used in an application in which the rubber is required to articulate against bone or cartilage, a slow erosion of the surface can occur, with liberation of particles, which can then lead to a chronic inflammatory response or bone lysis (Figs. 15 and 16) [55–64]. The term *silicone synovitis* was coined to describe this response to liberated silicone rubber particles [65–70]. In addition, some authors have reported

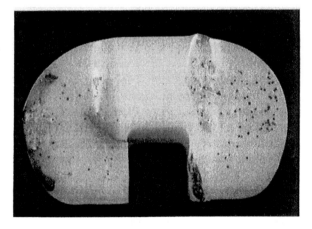

Figure 13 A polyethylene tibial insert that has been subject to third-body wear from exfoliated beads from the porous coating. The beads have become embedded into the polyethylene.

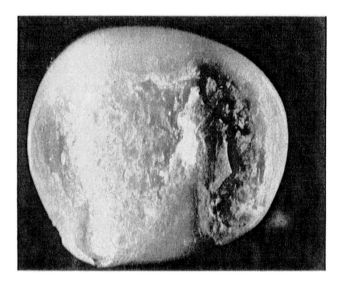

Figure 14   A polyethylene patellar resurfacing device showing gross wear.

malignant lesions in association with these particles [71,72], but with insufficient incidence evidence to suggest a causative relationship. Other studies suggest that the particles are not oncogenic [73].

3. *Polymethylmethacrylate*

Polymethylmethacrylate (PMMA, acrylic) has been used in load-bearing applications subject to wear, but its primary contribution in the consideration of wear-related problems is its potential to act as an abrasive third body in wear couples. The presence of PMMA at the interface can contribute to the wear of the two surfaces intended to be in intimate contact with relative motion. PMMA, in the form of bone cement, is placed or injected into the intramedullary cavity of bone and allowed to cure in place to provide a firm, stable interposition material between the bone and the implant. If the excess cement is not thoroughly cleaned and washed from the adjacent tissues after it cures, it can enter the joint space and provide an additional source of wear. As the patient applies load to the device during the activities of daily living, cracks can form in the cement. Particles may be released that can pass into adjacent tissues or become lodged in the joint space. This fragmentation of the cement and the resultant presence of particles in the joint space and periprosthetic tissues led to the naming of periprosthetic bone atrophy as *cement disease*. This misnomer has since been corrected, with the finding that periprosthetic bone atrophy with the presence of particulate debris occurs in patients who did not receive bone cement during implant placement.

The implant designer, as well as the surgeon and other health-care professionals involved in the performance of total joint arthroplasty, is required to consider the potential for the development of cement debris when designing implants and instruments as well as while using bone cement.

4. *Polytetrafluoroethylene*

PTFE, also referred to as Teflon, has been used in several applications in which wear was a possible issue. Probably one of the earliest uses was in the original Charnley hips [74,75], in which it was found to be less than acceptable for its wear resistance [76]. More recently, it has been used in a ligament replacement device in its expanded and braided form [77]. Reports of the long-term results of the use of the ligament product have suggested that abrasion of the material is associated with effusion and/or rupture. One study reports 34% effusion rates in the operated knees and rupture of the prosthesis in 12% of the patients in a group of patients with greater than two-year followup [78]. The effusions were seen to be associated with foreign-body particles in the tissues and with partial or complete graft rupture.

Another application in which PTFE has been subjected to wear conditions *in situ* is in the treatment of temporomandibular joint problems. Implants were used that consisted of a PTFE bearing surface overlying a porous, carbon or alumina fiber-reinforced PTFE substrate. While initial reports from oral surgeons were positive, with time it was found that the PTFE bearing surface was wearing through and catastrophic wear and debris generation was occurring. A report from 1992 detailing the clinical results from the use of this device in over 680 patients describes failure rates as high as 18% in some series and predicts that the failures in this series may reach 54% [79]. A case report from 1986 tells of a case in which particles of the wear debris migrated to lymph nodes in the submandibular space, leading to pain and necessitating removal [80]. This systemic effect of particles in the lymphatic systems is one of the reasons that the potential for wear and its associated release of particles should be considered in the design of weight-bearing devices. An additional study conducted in the laboratory on a wear simulator yielded data suggesting that the *in vivo* service life of this device might be as short as three years [81]. This report is instructive because it not only describes the testing performed and the results of that testing, but also provides a critique of the preclinical testing performed before bringing the product to market. It also provides recommendations for the process that should be used in the design and conduct of tests for the wear resistance of new materials for this application.

5. *Polyacetal*

Polyacetal has been investigated for use in total joint bearing surfaces in both the hip [82,83] and knee [84,85]. The clinical results were less than satisfactory due to wear when polyacetal was used as the cup material and the counterface was metal, as in the Christiansen hip prosthesis [82,83] (which consisted of cobalt/chromium alloy heads and polyacetal acetabular cups), but results seemed more acceptable when the counterface was polyethylene [85]. This comparison illustrates an important consideration in wear-related design. The specific application with all factors involved must be carefully considered when the materials for wear-susceptible applications are chosen.

C. Ceramic Materials

Wear considerations with the use of ceramics materials fall into two categories. Preparation of one or both components of a wear couple from ceramic materials is usually performed for the reduction of wear. At the same time, the generation of particles of ceramic from wear of a ceramic bearing surface can lead to particles that can participate in three-body wear.

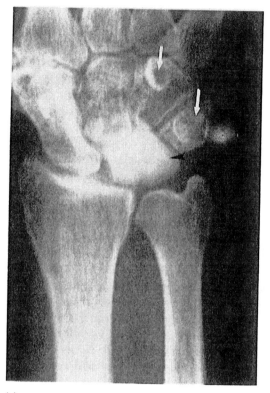

(a)

Figure 15 Radiographs showing lytic areas of bone (arrows) due to wear of a silicone rubber prosthesis (black arrow): (a) anterior-posterior view; (b) lateral view. (Photos courtesy of Alan Freeland, M.D.)

Researchers have found that the amount of frictional heat generated when an alumina material bears against polyethylene or against itself is less than when a cobalt-based alloy bears on polyethylene [86–88]. This has then been translated into the potential for reduced wear of the polyethylene from the combination of reduced damage due to heating and the implied reduced shear forces in the polyethylene. These findings, as well as an analysis of the wear mechanisms involved in the wear of total hip systems [89], have been translated into designs for total hip systems and total knee systems in which the counterface to the polyethylene is a ceramic. Researchers have found clinical evidence of reduced wear of polyethylene when the counter surface is alumina ceramic [90,91]. Additional research on the wear properties of these types of systems has suggested the use of zirconia ceramics in this application as well [53].

Other designs have incorporated ceramic-on-ceramic wear couples for hip prostheses, with results ranging from reports of minimal wear [92] to gross wear [93–95]. The differences in the clinical success with all-ceramic systems have been traced to differences in basic material properties, matching of the components, and surgical parameters such as impingement of the metal implant stem on the ceramic components [96–99].

The above examples of differences in clinical performance for a weight-bearing cou-

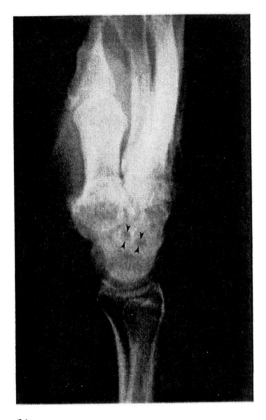

(b)

Figure 15 Continued

ple that is theoretically an improvement illustrate an important consideration in the design of systems subject to wear. The designers of a system that is subject to wear must consider those conditions of use or manufacturing variation that may render a couple less than optimal. In the alumina-on-alumina systems, highly conforming surfaces that were manufactured with a high degree of quality control performed very well unless some outside influence served to upset the ideal situation. Those cases in which impingement occurred against the edge of the cup or in which the degree of conformity was insufficient to prevent local stress fractures of the surface yielded accelerated wear and a situation that was much less than optimal. It is very important to consider the possible modes of surface degradation and wear under both optimal conditions and in those situations in which a subtle change in surgical technique or patient morphology or activity may introduce perturbations into the system.

D. Composite Materials

The use of composite materials for the manufacture of medical implants usually stems from a desire to utilize a material that has properties not available in a homogenous material [100]. The addition of a filler or reinforcing fiber to a polymeric material to alter the polymer properties has, to date, been the primary type of composite evaluated.

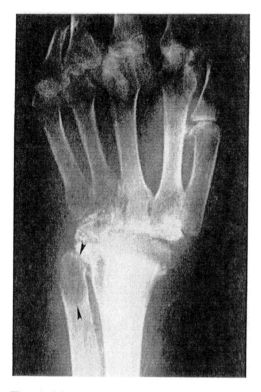

Figure 16 Radiograph showing a lytic lesion (arrow) of the ulna due to silicone rubber wear debris. (Photo courtesy of Alan Freeland, M.D.)

Fillers or reinforcing fibers (which are called *reinforcing material* for the remainder of this discussion) can alter the mechanical properties of a polymeric material by transferring load to the stronger and stiffer reinforcing material, imparting improved resistance to tensile, compressive, or bending loads, while maintaining the light weight and radiolucency of the polymeric material.

Composite materials that have been proposed or used in medical applications include carbon fiber-reinforced polyethylene [40,101], carbon or alumina fiber-reinforced porous Teflon [102], carbon fiber-reinforced polysulfone or polyarylether ketones (polyetherether ketone [PEEK], polyarylether ketone [PAEK]) [103,104], and carbon fiber-reinforced carbon configurations. Other combinations or fiber and polymer candidates may also be under consideration, such as Kevlar® fibers, ceramic fibers, or other polymer matrices. While these materials have shown that they are capable of solving the problems for which they were developed, the question of wear has not been solved for these materials. The addition of reinforcing material to the matrix will improve the bulk properties of the material, but may not have any effect at the microscopic levels seen in wear. In fact, the addition of fibers may have the effect of reducing the resistance of the matrix to wear.

In evaluating fiber-reinforced polyethylene for orthopedic bearing surfaces, some studies found the wear resistance of the fiber-reinforced polyethylene to be poorer than that of the unfilled material [105,106]. This material is no longer available for bearing surface use, primarily because of the appearance of the tissue when wear occurs. Most reports indicate that the tissue response and degree of wear are no worse than for unfilled polyethylene [44,107,108]. One investigator reported tissue response to wear debris from this material in one patient to be, in his opinion, more severe than that seen with the same design of prosthesis with unfilled polyethylene [109]. These observations may be important since they suggest that the issue of wear and wear debris formation may be more important for carbon-reinforced composites than for some other materials since the same level of wear debris formation will provide much more visual evidence for black fibers than for relatively colorless polymers.

The use of reinforced polymers for total joint replacement has grown from the desire to use a material with adequate strength and fatigue resistance but with an elastic modulus lower than that which can be achieved with currently available metal alloys [110,111]. The devices that have been tested experimentally or in clinical trials have incorporated metallic (cobalt/chromium alloy) or ceramic (alumina or titania) bearing surfaces that were attached to the composite stems with a mechanical interference taper connection such as is used to attached the heads to metallic stems. The designers have not yet tackled the problem of the creation of a bearing surface with adequate wear resistance from composite materials.

Other composite systems that have been considered for total joint applications by reason of reduced wear include solid carbon and silicon carbide/carbon composites [112]. These researchers discovered that the tribological properties of these materials exceeded those for alumina ceramic against itself and might represent the next step beyond ceramics as components of total joint systems.

Composites present a new problem for the achievement of wear resistance in medical implants. Frequently, the reinforcing material will be exposed at the surface of the composite, leaving open the possibility that these fibers may pull out of the underlying matrix or fracture during wear. Many high strength fibers have extremely low elongation to failure and are susceptible to bending failure. Since reinforcing fibers would then be released into the wear couple, the potential exists for an increase in wear due to a three-body situation in which the reinforcing fibers will have become an abrasive agent. In addition, the roughening of the surface can cause increased wear as the roughened surface returns to a more polished condition. Finally, the removed fibers may become captured in surrounding tissue or be transported to regional lymph nodes. The ultimate capture of the wear-generated material in adjacent tissues can then lead to problems ranging from observable tissue staining with mild to no tissue response to a chronic granulomatous inflammatory response. The possible final result of sequestration of these particles in tissues may be as severe as the development of neoplastic transformations.

It is noteworthy that Bauer et al. reported that the carbon fibers seen in their patient served as a marker for the tissues to be evaluated for the presence of polyethylene and indicated that similar analyses of lymph nodes in patients with unreinforced polyethylene would likely yield similar results [107]. Groth and Shilling stated that the tissue response seen in their study seemed to be that characteristic of the response to polyethylene and that the response seemed to correspond more closely to the polyethylene particles rather than the carbon fibers [108].

In 1987, Anderson reported on his analysis of lymph node tissue from a patient who had received both a composite ligament replacement device (PTFE and carbon) and a continuous carbon fiber ligament device [113]. The patient's lymph node was found to contain both carbon fibers and a polymeric material with foreign-body giant cells preferentially associated with the polymeric material and benign response to the carbon fibers. He recommended that lymph nodes from patients who had experienced friction failure of polymeric ligament devices be evaluated using polarized light techniques to look for the presence of polymer particles.

As long as composites are not used as either component of a load-bearing wear couple, then the primary concern investigators must consider is the potential for wear at the interface between the implant and adjacent bone, between the implant and bone cement (if used), and, particularly, at the interface that is formed when the components of a multicomponent device are mated.

V. PHYSIOLOGICAL EFFECTS OF CLINICAL DEVICE WEAR

The above discussion of the sites and materials involved in clinical implant wear has alluded to the effects that the generation of wear debris may have upon the surrounding tissues. Among the most critical problems has been the resorption of bone from around total joint prostheses with the attendant loss in bony support for the devices [41,114–116]. Researchers have attempted to identify the mechanisms involved in this bone resorption and the material(s) that may be implicated. In the clinical situation, a number of materials are present simultaneously and more than one type of particle may be present. It is therefore difficult to distinguish the response to individual types of particles when the tissue that has been removed at revision may contain, simultaneously, polyethylene, PMMA, cobalt/chromium alloys, and titanium or its alloys (Fig. 17). Several recent review articles provide some insight into the clinical manifestations of debris-related tissue response and the types of research being performed in the laboratory to elucidate the cellular events that may form a part of this response [117–119].

Most laboratory research to identify the mechanisms of debris-related osteolysis in orthopedic practice have centered on the involvement of interleukins, prostaglandins, and tumor necrosis factor in modifying the bone formation and/or resorption mechanisms such that the net volume of bone present is reduced. Theories of the critical factors for the lytic response to particles have suggested that the critical factor may be material identity, while other researchers suggest that size or shape of the particles may be most important (Fig. 18) [120]. Much of the laboratory research has been based upon the theory that macrophages interact directly with the particles and release or cause the release of the cellular biochemical factors that then act on osteoblasts or osteoclasts. It has also been suggested that there may be direct effects on the bone-forming and resorption cells.

VI. WEAR TESTING

In an effort to characterize the potential for a particular wear couple to generate debris as well as to gauge the possibility for wear-related failures of the bearing surfaces themselves, test methods are used to model the clinical situation. In the area of orthopedic joint replacement, much of the early wear testing was performed using a simple pin-on-disk technique that has been published as a standard test method [121]. While this method has provided some assistance in screening different materials, researchers have found that clinically relevant data are best generated by wear simulators in which actual components are subjected to a combination of cyclic loading and oscillatory motion for a million cycles or more.

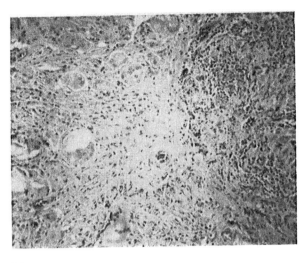

(a)

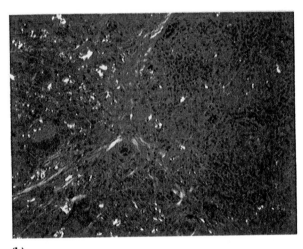

(b)

Figure 17 Microscopic analysis of the periprosthetic tissues adjacent to a worn polyethylene tibial insert: (a), a histiocytic response with foreign-body giant cells; (b), the same tissue viewed with polarized light showing the polyethylene particles in association with the cellular response.

While these methods have not yet been standardized, the results of these studies have been reported in the literature, along with a description of the test setup. Efforts to standardize these tests are currently under way in the American Society for Testing and Materials (ASTM) [122,123], but these efforts are not yet complete. The primary requirements for an appropriate test are

1. The test should be capable of ranking different wear couples for their wear in the same order as is seen clinically.
2. The types of wear seen and the characteristics of the particles generated should mimic, as closely as possible, those seen from clinically retrieved samples.

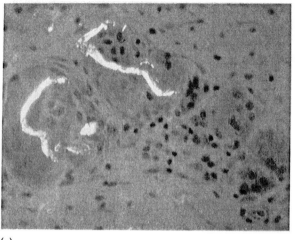

(a)

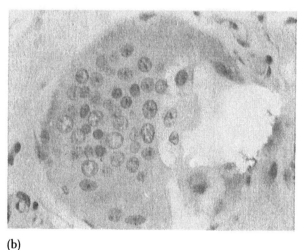

(b)

Figure 18   Microscopic analysis of periprosthetic tissues showing the giant cells in association with polyethylene wear particles: (a), intermediate magnification; (b), high power magnification.

A test that meets these requirements will then allow the material and implant designers to examine wear couples that have yet to be used clinically with reasonable assurance that the results will have meaning as a predictor of clinical performance.

The wear studies performed on the PTFE temporomandibular device exhibit some of the pitfalls that may be encountered in laboratory wear testing [81]. Often, it is easier to validate a particular wear test method after clinical wear failures have occurred. It is important to conduct clinically relevant testing during the preclinical phases of development. Device designers are required to conduct tests in order to prevent unnecessary failures when no clinical data are available for a particular system. In any modeling system, it is not possible to mimic accurately all the variables that might be present in the clinical situation.

It is necessary, therefore, to make compromises and choices in the design of a wear model. Currently, in the field of total joint replacement, it is possible to define, in some ways, the types of device function modeling that may be necessary to compare existing designs with proposed new designs. In many other areas of implant development, the requirements for appropriate modeling of the clinical wear situation may not be as well defined.

VII. CONCLUSIONS

The design process and selection of materials must consider the loading conditions and relative movement that may be required from the device. While the performance requirements for a particular application may dictate the selection of a certain class of materials, differences in composition and processing may have a substantial effect on the wear resistance of the system. Subtle design differences or variations in dimensional tolerances may also contribute to a spectrum of wear performance. It is incumbent upon the designer as well as the medical professional to be aware of the possibility for wear-related problems in service and to attempt to minimize the resultant clinical problems.

Testing protocols for new devices or new materials should consider the potential for the development of wear damage in the planned implementation. If there is a reasonable possibility that wear will occur, a program of testing in a simulated use situation should be a part of the product and process validation procedures. The protocols for clinical trials of devices for wear-susceptible applications should include measurements, when possible, of parameters that might be indicative of changes due to wear. Generally, noninvasive measurement techniques such as measurements from radiographs will not have sufficient resolution for most situations, but can spot gross changes that might be missed if the measurements were not performed. When surgeons remove a device from a patient, whether in the course of normal clinical practice or as a part of a clinical trial, the surfaces should be inspected for signs of changes due to wear. If this practice was to be universally observed, early signs of problems might be recognized and the long-term impact reduced to the patient population at large.

Working as a team, materials scientists, device designers, and medical professionals can develop the capability to recognize and solve wear-related device problems. Early intervention may contribute to longer survival times for devices that experience repetitive loading with motion.

REFERENCES

1. Olson, E. J., J. D. Kang, F. H. Fu, H. I. Georgescu, G. C. Mason, and C. H. Evans, The biochemical and histological effects of artificial wear particles: *In vitro* and *in vivo* studies, *Am. J. Sports Med.*, 16:558–570 (1988).
2. Galante, J. O., and W. Rostoker, Wear in total hip prostheses, *Acta Orthop. Scand. Suppl.*, 145:6–46 (1973).
3. Hayes, W. C., and S. M. Perren, Plate-bone friction in the compression fixation of fractures, *Clin. Orthop. Rel. Res.*, 89:236–240 (1972).
4. Kyle, R. F., T. M. Wright, and A. H. Burstein, Biomechanical analysis of the sliding characteristics of compression hip screws, *J. Bone and Joint Surg.*, 62A:1308–1314 (1980).
5. Weinstein, A., H. Amstutz, G. Pavon, and V. Franceschini, Orthopedic implants — A clinical and metallurgical analysis, *J. Biomed. Mater. Res.*, 4:297–325 (1973).

6. Terz, J. J., E. S. Bear, R. E. King, and W. Lawrence, Jr., Primary reconstruction of the mandible with a wire mesh prosthesis, *Surg. Gyn. & Obst.*, 139:198–200 (1974).

7. Phillips, C. M., Primary and secondary reconstruction of the mandible after ablative surgery, *Amer. J. Surg.*, 114:601–604 (1967).

8. Dube, V. E., and D. E. Fisher, Hemangioendothelioma of the leg following metallic fixation of the tibia, *Cancer*, 30:1260–1266 (1972).

9. Pizzoferrato, A., Evaluation of the tissue response to the wear products of the hip joint endo-arthroprosthesis, *Biomat. Med. Dev. Artif. Org.*, 7:257–262 (1979).

10. Gruen, T. A., and H. C. Amstutz, A failed vitallium/stainless steel total hip replacement: A case report for histological and metallurgical examination, *J. Biomed. Mater. Res.*, 9:465–477 (1975).

11. Gold, B. L., and P. S. Walker, Variables affecting the friction and wear of metal-on-plastic total hip joints, *Clin. Orthop. Rel. Res.*, 100:270–278 (1974).

12. Dowling, J. M., J. R. Atkinson, D. Dowson, and J. Charnley, The characteristics of acetabular cups worn in the human body, *J. Bone and Joint Surg.*, 60B:375–382 (1978).

13. Weightman, B., and D. Light, The effect of the surface finish of alumina and stainless steel on the wear rate of UHMW polyethylene, *Biomaterials*, 7:20–24 (1986).

14. Caravia, L., D. Dowson, J. Fisher, and B. Jobbins, The influence of bone and bone cement debris on counterface roughness in sliding wear tests of ultra-high molecular weight polyethylene on stainless steel, *Proc. Inst. Mech. Eng. Part H: J. of Eng. Med.*, 204:65–70 (1990).

15. McKellop, H. A., A. Sarmiento, C. P. Schwinn, and E. Ebramzadeh, *In vivo* wear of titanium alloy hip prosthesis, *J. Bone and Joint Surg.*, 72A:512–518 (1990).

16. Röstlund, T., B. Albrektsson, T. Albrektsson, and H. McKellop, Wear of ion-implanted pure titanium against UHMWPE, *Biomaterials*, 10:176–181 (1989).

17. Solar, R. J., S. R. Pollack, and E. Korostoff, *In vitro* corrosion testing of titanium surgical implants, *J. Biomed. Mater. Res.*, 13:217–221 (1975).

18. Agins, N. J., N. W. Alcock, M. Bansal, E. A. Salvati, P. D. Wilson, P. M. Pellici, and P. G. Bullough, Metallic wear in failed titanium total hip replacements, *J. Bone and Joint Surg.*, 70A:347–356 (1988).

19. Black, J., H. Sherk, J. Bonini, W. R. Rostoker, F. Schajowicz, and J. O. Galante, Metallosis associated with a stable titanium allow femoral component in total hip replacement, *J. Bone and Joint Surg.*, 72:126–130 (1990).

20. Cameron, H. U., Failure of a titanium endoprosthesis-a case report, *CJS*, 34:625–626 (1991).

21. Salvati, E. A., F. Betts, and S. B. Doty, Particulate metallic debris in cemented total hip arthroplasty, *Clin. Orthop. Rel. Res.*, 293:160–173 (1993).

22. Thull, R., Semiconductive properties of passivated titanium and titanium based hard coatings on metals for implants-an experimental approach, *Med. Progr. Technol.*, 16:225–234 (1990).

23. Pilliar, R. M., Modern metal processing for improved load-bearing surgical implants, *Biomaterials*, 12:95–100 (1991).

24. Rieu, J., A. Pichat, L.-M. Rabbe, A. Rambert, C. Chabrol, and M. Robelet, Ion implantation effects on friction and wear of joint prosthesis materials, *Biomaterials*, 12:139–143 (1991).

25. Rieu, J., A. Pichat, L.-M. Rabbe, C. Chabrol, and M. Robelet, Deterioration mechanisms of joint prosthesis materials: Several solutions by ion implantation surface treatments, *Biomaterials*, 11:51–54 (1990).

26. McKellop, H. A., I. C. Clarke, K. L. Markolf, and H. C Amstutz, Friction and wear properties of polymer, metal, and ceramic prosthetic joint materials evaluated on a multi-channel screening device, *J. Biomed. Mater. Res.*, 15:619–653 (1981).

27. Dumbleton, J. H., and P. Higham, Coating and surface modification, In *Metal and Ceramic*

Biomaterials: Volume 2: Strength and Surface (P. Ducheyne and G. W. Hastings, eds.), pp. 119–141, CRC Press, Boca Raton, FL (1984).

28. Peterson, C. D., B. M. Hillberry, and D. A. Heck, Component wear of total knee prostheses using Ti-6Al-4V, titanium nitride coated Ti-6Al-4V, and cobalt-chromium-molybdenum femoral components, *J. Biomed. Mater. Res.*, 22:887–903 (1988).

29. Williams, J. M., L. M. Beardsley, R. A. Buchanan, and R. K. Bacon, Effect of N-implantation on the corrosive properties of surgical Ti-6Al-4V alloy, *Mater. Res. Soc. Proc.*, 27:735–740 (1984).

30. Sioshansi, P., Ion implantation improves wear resistance of titanium, *Orthop. Today*, 10:24–26 (1990).

31. Buchanan, R. A., E. D. Rigney, Jr., and J. M. Williams, Wear accelerated corrosion of Ti-6Al-4V and nitrogen-ion-implanted Ti-6Al-4V: Mechanisms and influence of fixed-stress magnitude, *J. Biomed. Mater. Res.*, 21:367–377 (1987).

32. Buchanan, R. A., E. D. Rigney, Jr., and J. M. Williams, Ion implantation of surgical Ti-6Al-4V for improved resistance to wear-accelerated corrosion, *J. Biomed. Mater. Res.*, 21:355–366 (1987).

33. McKellop, H. A., and T. Röstlund, The wear behavior of ion-implanted Ti-6Al-4V against UHMW polyethylene, *J. Biomed. Mater. Res.*, 24:1413–1425 (1990).

34. Milliano, M. T., L. A. Whiteside, A. D. Kaiser, and P. A. Zwirkoski, Evaluation of the effect of the femoral articular surface material on the wear of a metal-backed patellar component, *Clin. Orthop. Rel. Res.*, 287:178–186 (1993).

35. Davidson, J. A., Characteristics of metal and ceramic total hip bearing surfaces and their effect on long-term ultra high molecular weight polyethylene wear, *Clin. Orthop. Rel. Res.*, 294:361–378 (1993).

36. Fricker, D. C., and R. Shivanath, Fretting corrosion studies of universal femoral head prostheses and cone taper spigots, *Biomaterials*, 11:495–500 (1990).

37. Shirazi-Adl, A., M. Dammak, and G. Paiement, Experimental determination of friction characteristics at the trabecular bone/porous-coated metal interface in cementless implants, *J. Biomed. Mater. Res.*, 27:167–175 (1993).

38. Cook, H. P., Immediate reconstruction of the mandible by metallic implant following resection for neoplasm, *Ann. Royal Coll. Surg. Eng.*, 42:233–259 (1968).

39. Winter, G. D., Tissue reactions to metallic wear and corrosion products in human patients, *J. Biomed. Mater. Res. Symp.*, 5:11–26 (1974).

40. Tetik, R. D., J. O. Galante, and W. Rostoker, A wear resistant material for total joint replacement—Tissue biocompatibility of an ultra-high molecular weight (UHMW) polyethylene-graphite composite, *J. Biomed. Mater. Res.*, 8:231–250 (1974).

41. Tsao, A., L. Mintz, C. R. McCrae, S. D. Stulberg, and T. Wright, Failure of the porous-coated anatomic prosthesis in total hip arthroplasty due to severe polyethylene wear, *J. Bone and Joint Surg.*, 75A:19–26 (1993).

42. Wright, T. M., and D. L. Bartel, The problem of surface damage in polyethylene total joint components, *Clin. Orthop. Rel. Res.*, 205:67–74 (1986).

43. Wright, T. M., R. W. Hood, and A. H. Burstein, Analysis of material failures, *Orthop. Clin. North Am.*, 13:33–44 (1982).

44. Wright, T. M., C. M. Rimnac, P. M. Faris, and M. Bansal, Analysis of surface damage in retrieved carbon fiber-reinforced and plain polyethylene tibial components from posterior stabilized total knee replacements, *J. Bone and Joint Surg.*, 70A:1312–1319 (1988).

45. Plante-Bordeneuve, P., and M. A. R. Freeman, Tibial high-density polyethylene wear in conforming tibiofemoral prostheses, *J. Bone and Joint Surg.*, 75B:630–639 (1993).

46. Webb, P. J., K. W. J. Wright, and G. D. Winter, The monk "soft top" endoprosthesis: Clinical, biomechanical, and histopathological observations, *J. Bone and Joint Surg.*, 62B:174–179 (1980).

47. Sibley, T. F., and A. Unsworth, Wear of cross-linked polyethylene against itself: A material suitable for surface replacement of the finger joint, *J. Biomed. Eng.*, 13:217–220 (1991).

48. Rose, R. M., A. Crugnola, M. Ries, W. R. Cimino, I. Paul, and E. L. Radin, On origins of high *in vivo* wear rates in polyethylene components of total joint prostheses, *Clin. Orthop. Rel. Res.*, 145:277–286 (1979).

49. Bankston, A. B., P. M. Faris, E. M. Keating, and M. A. Ritter, Polyethylene wear in total hip arthroplasty in patient-matched groups: A comparison of stainless steel, cobalt chrome, and titanium-bearing surfaces, *J. Arthrop.*, 8:315–322 (1993).

50. Rose, R. M., H. J. Nussbaum, H. Schneider, M. Ries, I. Paul, S. R. Simon, and E. L. Radin, On the true wear rate of ultra high molecular weight polyethylene in the total hip prosthesis, *J. Bone and Joint Surg.*, 62A:537–549 (1980).

51. Wright, T. M., C. M. Rimnac, S. D. Stulberg, L. Mintz, A. K. Tsao, R. W. Klein, and C. McCrae, Wear or polyethylene in total joint replacements: Observations from retrieved PCA knee implants, *Clin. Orthop. Rel. Res.*, 276:126–134 (1992).

52. Bloebaum, R. D., K. Nelson, L. D. Dorr, A. A. Hoffmann, and D. J. Lyman, Investigation of early surface delamination observed in retrieved heat-pressed tibial inserts, *Clin. Orthop. Rel. Res.*, 269:120–127 (1991).

53. Kumar, P., M. Oka, K. Ikeuchi, K. Shimizu, T. Yamamuro, H. Okumura, and Y. Kotura, Low wear rate of UHMWPE against zirconia ceramic (Y-PSZ) in comparison to alumina ceramic and SUS 316L alloy, *J. Biomed. Mater. Res.*, 25:813–828 (1991).

54. Bartel, D. L., V. L. Bicknell, and T. M. Wright, The effect of conformity, thickness, and material on stresses in ultra-high molecular weight components for total joint replacement, *J. Bone and Joint Surg.*, 68A:1041–1051 (1986).

55. Verhaar, J., S. Bulstra, and G. Walenkamp, Silicone arthroplasty for hallux rigidus: Implant wear and osteolysis, *Acta Orthop. Scand.*, 60:30–33 (1989).

56. Carter, P. R., L. J. Benton, and P. A. Dysert, Silicone rubber carpal implants: A study of the incidence of late osseous complications, *J. Hand Surg.*, 11A:639–644 (1986).

57. Pellegrini, V. D., and R. I. Burton, Surgical management of basal joint arthritis of the thumb: Part I. Long-term results of silicone implant arthroplasty, *J. Hand Surg.*, 11A:309–324 (1986).

58. Shiel, W. C., and M. Jason, Granulomatous inguinal lymphadenopathy after bilateral metatarsophalangeal joint silicone arthroplasty, *Foot & Ankle*, 6:216–218 (1986).

59. Trepman, E., and F. C. Ewald, Early failure of silicone radial head implants in the rheumatoid elbow, *J. Arthrop.*, 6:59–65 (1991).

60. Rosenthal, D. I., A. E. Rosenberg, A. L. Schiller, and R. J. Smith, Destructive arthritis due to silicone: A foreign-body reaction, *Radiol.*, 149:69–72 (1983).

61. Travis, W. D., K. Balogh, and J. L. Abraham, Silicone granulomas: Report of three cases and review of the literature, *Hum. Pathol.*, 16:19–27 (1985).

62. Dolwick, M. F., and T. B. Aufdemorte, Silicone-induced foreign body reaction and lymphadenopathy after temporomandibular, joint arthroplasty, *Oral Surg. Oral Med. Oral Pathol.*, 59:449–452 (1985).

63. Kircher, T., Silicone lymphadenopathy: A complication of silicone elastomer finger joint prostheses, *Hum Pathol.*, 11:240–244 (1980).

64. Eiken, O., L. Ekerot, C. Lindström, and K. Jonsson, Silicone carpal implants: Risk or benefit? *Scand. J. Plast. Reconstr. Surg.*, 19:295–304 (1985).

65. Peimer, C. A., J. Medige, B. S. Eckert, J. R. Wright, and C. S. Howard, Reactive synovitis after silicone arthroplasty, *J. Hand Surg.*, 11A:624–638 (1986).

66. Smith, R. J., R. E. Atkinson, and J. B. Jupiter, Silicone synovitis of the wrist, *J. Hand Surg.*, 10A:47–60 (1985).

67. Perlman, M. D., A. D. Schor, and M. L. Gold, Implant failure with particulate silicone synovitis (detritic synovitis), *J. Foot Surg.*, 29:584–588 (1990).

68. Peimer, C. A., Long-term complications of trapeziometacarpal silicone arthroplasty, *Clin. Orthop. Rel. Res.*, 220:86–98 (1987).

69. Egloff, D. V., G. Varadi, A. Narakas, C. Simonetta, and C. Canters, Silastic implants of the scaphoid and lunate: A long-term clinical study with a mean follow-up of 13 years, *J. Hand Surg.*, 18B:687–692 (1993).

70. Khoo, C. T. K., Silicone synovitis: The current role of silicone elastomers in joint reconstruction, *J. Hand Surg.*, 18B:679–686 (1993).

71. Digby, J. M., Malignant lymphoma with intranodal silicone rubber particles following metacarpophalangeal joint replacements, *Hand*, 14:326–328 (1982).

72. Benjamin, E., A. Ahmed, A. T. M. F. Rashid, and D. H. Wright, Silicone lymphadenopathy: A report of two cases, one with concomitant malignant lymphoma, *Diag. Histo.*, 5:133–141 (1982).

73. Winson, I. G., J. M. Digby, T. Wickremaratchi, and A. Longstaffe, Malignancy and silastic particles: An animal study, *Clin. Mat.*, 3:177–182 (1988).

74. Charnley, J., Arthroplasty of the total hip in rheumatoid arthritis, *Physiotherapy*, 49:189–191 (1963).

75. Charnley, J., Arthroplasty of the hip, *Lancet*, 1:1129–1132 (1961).

76. Charnley, J., Tissue reactions to polytetrafluoroethylene, *Lancet*, 2:1379 (1963).

77. Bolton, C. W., and W. C. Bruchman, The Gore-Tex™ expanded polytetrafluoroethylene prosthetic ligament: An *in vitro* and *in vivo* evaluation, *Clin. Orthop. Rel. Res.*, 196:202–213 (1985).

78. Paulos, L. E., T. D. Rosenberg, S. R. Grewe, D. S. Tearse, and C. L. Beck, The Gore-Tex anterior cruciate ligament prosthesis: A long-term followup, *Am. J. Sports Med.*, 20:246–252 (1992).

79. Spagnoli, D., and J. N. Kent, Multicenter evaluation of temporomandibular joint Proplast-Teflon disk implant, *Oral Surg. Oral Med. Oral Pathol.*, 74:411–421 (1992).

80. Lagrotteria, L., R. Scapino, A. S. Granston, and D. Felgenhauer, Patient with lymphadenopathy following temporomandibular joint arthroplasty with Proplast, *J. Craniomand. Prac.*, 4:172–178 (1986).

81. Fontenot, M. G., and J. N. Kent, *In vitro* wear performance of Proplast TMJ disc implants, *J. Oral Maxillofac. Surg.*, 50:133–139 (1992).

82. Mathiesen, E. B., U. Lindgren, F. P. Reinholt, and E. Sudmann, Wear of the acetabular socket: Comparison of polyacetal and polyethylene, *Acta Orthop. Scand.*, 57:193–196 (1986).

83. Havelin, L. I., N. R. Gjerdet, O. D. Lunde, M. Rait, and E. Sudmann, Wear of the Christiansen hip prosthesis, *Acta Orthop. Scand.*, 57:419–422 (1986).

84. McKellop, H. A., T. Röstland, and G. W. Bradley, Evaluation of wear in an all-polymer total knee replacement. Part 1: Laboratory testing of polyethylene on polyacetal bearing surfaces, *Clin. Mater.*, 14:117–126 (1993).

85. Bradley, G. W., M. A. R. Freeman, M. A. Tuke, and H. A. McKellop, Evaluation of wear in an all-polymer total knee replacement. Part 2: Clinical evaluation of wear in a polyethylene on polyacetal total knee, *Clin. Mater.*, 14:127–132 (1993).

86. Davidson, J. A., and G. Schwartz, Wear, creep, and frictional heat of femoral implant articulating surfaces and the effect on long term performance—Part I, A review, *J. Biomed. Mater. Res.: Appl. Biomat.*, 21A:261–285 (1987).

87. Davidson, J. A., G. Schwartz, G. Lynch, and S. Gir, Wear, creep, and frictional heat of femoral implant articulating surfaces and the effect on long term performance—Part II, Friction, heating, and torque, *J. Biomed. Mater. Res.: Appl. Biomat.*, 22A:69–91 (1988).

88. Semlitsch, M., M. Lehmann, H. Weber, E. Dorre, and H. G. Willert, New prospects for a prolonged functional life-span of artificial hip joints by using the material combination polyethylene/aluminum oxide ceramic/metal, *J. Biomed. Mater. Res.*, 11:537–552 (1977).

89. Davidson, J. A., Characteristics of metal and ceramic total hip bearing surfaces and their effect on long term ultra high molecular weight polyethylene wear, *Clin. Orthop. Rel. Res.*, 294:361–378 (1993).

90. Zichner, L. P., and H.-G. Willert, Comparison of alumina-polyethylene and metal-polyethylene in clinical trials, *Clin. Orthop. Rel. Res.*, 282:86–93 (1992).

91. Oonishi, H., M. Aono, N. Murata, and S. Kushitani, Alumina versus polyethylene in total knee arthroplasty, *Clin. Orthop. Rel. Res.*, 282:95–104 (1992).

92. Mittlemeier, H., and J. Heisel, Sixteen years experience with ceramic hip prostheses, *Clin. Orthop. Rel. Res.*, 282:64–72 (1992).

93. Winter, M., P. Griss, G. Scheller, and T. Moser, Ten to 14 year results of a ceramic hip prosthesis, *Clin. Orthop. Rel. Res.*, 282:73–80 (1992).

94. Kummer, F. J., S. A. Stuchin, and V. H. Frankel, Analysis of removed Autophor ceramic-on-ceramic components, *J. Arthrop.*, 5:29–33 (1990).

95. Griss, P., and G. Heimke, Five years experience with ceramic-metal-composite endoprostheses: I. Clinical evaluation, *Arch. Orthop. Traumat. Surg.*, 98:157–164 (1981).

96. Walter, A., On the material and the tribology of alumina-alumina couplings for hip joint prostheses, *Clin. Orthop. Rel. Res.*, 282:31–46 (1992).

97. Dorlot, J.-M., Long term effects of alumina components in total hip prostheses, *Clin. Orthop. Rel. Res.*, 282:47–52 (1992).

98. Nizard, R. S., L. Sedel, P. Christel, A. Meunier, M. Soudry, and J. Witvoet, Ten-year survivorship of cemented ceramic-ceramic total hip prosthesis, *Clin. Orthop. Rel. Res.*, 282: 53–63 (1992).

99. Heimke, G., and P. Griss, Five years experience with ceramic-metal-composite endoprostheses: II. Mechanical evaluations and improvements, *Arch. Orthop. Traumat. Surg.*, 98:165–171 (1981).

100. Skinner, H. B., Composite technology for total hip arthroplasty, *Clin. Orthop. Rel. Res.*, 235:224–236 (1988).

101. Sclippa, E., and K. Piekarski, Carbon fiber reinforced polyethylene for possible orthopaedic uses, *J. Biomed. Mater. Res.*, 7:59–70 (1973).

102. Keet, G. G. M., and W. C. Runne, The Anaform endoprosthesis: A Proplast-coated femoral endoprosthesis, *Orthop.*, 12:1185–1190 (1989).

103. St. John, K. R., Applications of advanced composites in orthopaedic implants, In *Biocompatible Polymers, Metals, and Composites* (M. Szycher, ed.), pp. 861–871, Technomic Publishing, Lancaster, PA (1983).

104. Davidson, J. A., The challenge and opportunity for composites in structural orthopaedic applications, *J. Comp. Tech. Res.*, 12:151–161 (1987).

105. Wright, T. M., T. Fukubayashi, and A. H. Burstein, The effect of carbon fiber reinforcement on contact area, contact pressure, and time-dependent deformation in polyethylene tibial components, *J. Biomed. Mater. Res.*, 15:719–730 (1981).

106. Wright, T. M., D. J. Astion, M. Bansal, C. M. Rimnac, T. Green, J. N. Insall, and R. P. Robinson, Failure of carbon fiber-reinforced polyethylene total knee-replacement components. A report of two cases, *J. Bone and Joint Surg.*, 70A:926–932 (1988).

107. Bauer, T. W., M. Saltarelli, J. T. McMahon, and A. H. Wilde, Regional dissemination of wear debris from a total knee prosthesis. A case report, *J. Bone and Joint Surg.*, 75A:106–111 (1993).

108. Groth, H. E., and J. M. Shilling, Tissue response to carbon-reinforced polyethylene, *J. Orthop. Res.*, 1:129–135 (1983).

109. Dannenmaier, W. C., D. W. Haynes, and C. L. Nelson, Granulomatous reaction and cystic bony destruction associated with high wear rate in a total knee prosthesis, *Clin. Orthop. Rel. Res.*, 198:224–230 (1985).

110. Magee, F. P., A. M. Weinstein, J. A. Longo, J. B. Koeneman, and R. A. Yapp, A canine composite femoral stem: An *in vivo* study, *Clin. Orthop. Rel. Res.*, 235:237–252 (1988).

111. Skinner, H. B., Composite technology for total hip arthroplasty, *Clin. Orthop. Rel. Res.*, 235:224–236 (1988).
112. Brückmann, H., G. Keuscher, and K. J. Hüttinger, Carbon, a promising material in endoprosthetics. Part 2: Tribological properties, *Biomaterials*, 1:73–81 (1980).
113. Anderson, J. A., Hexcel Integraft™ stent—Histology review, Prosthetic Ligament Reconstruction of the Knee—Fourth Annual Symposium, Palm Springs, CA (March 26–28, 1987).
114. Peters, P. Jr., G. A. Engh, K. A. Dwyer, and T. N. Vinh, Osteolysis after total knee arthroplasty without cement, *J. Bone and Joint Surg.*, 74A:864–876 (1992).
115. Stulberg, B. N., R. L. Buly, P. L. Howard, S. D. Stulberg, and R. L. Wixon, Porous coated anatomic implant failure: Incidence and modes of failure in uncemented total hip arthroplasty, American Academy of Orthopaedic Surgeons Annual Meeting, Washington, DC (February 20–25, 1992).
116. Lavernia, C., A. K. Tsao, and D. S. Hungerford, Silent osteolysis: Polyethylene debris in cementless total hip replacement, American Academy of Orthopaedic Surgeons Annual Meeting, San Francisco, CA (February 18–23, 1993).
117. Clarke, I. C., P. Campbell, and N. Kossovsky, Debris-mediated osteolysis—A cascade phenomenon involving motion, wear, particulates, macrophage induction, and bone lysis, In *Particulate Debris from Medical Implants: Mechanisms of Formation and Biological Consequences* (K. St. John, ed.), pp. 7–26, American Society for Testing and Materials (STP 1144), Philadelphia (1992).
118. Goodman, S. B., and V. L. Fornasier, Clinical and experimental studies in the biology of aseptic loosening of joint arthroplasties and the role of polymer particles, In *Particulate Debris from Medical Implants: Mechanisms of Formation and Biological Consequences* (K. St. John, ed.), pp. 27–37, American Society for Testing and Materials (STP 1144), Philadelphia (1992).
119. DiCarlo, E. F., and P. G. Bullough, The biologic responses to orthopedic implants and their wear debris, *Clin. Mat.*, 9:235–260 (1992).
120. Tsao, A., L. C. Jones, S. Choung, A. Chen, D. Banes, D. Opishinki, C. Lavernia, and D. S. Hungerford, Quantitative characterization of particulate polyethylene wear debris from failed total joint arthroplasties, Annual Meeting of the Society for Biomaterials, Birmingham, AL (April 28–May 2, 1993).
121. American Society for Testing and Materials, Standard practice for reciprocating pin-on-flat evaluation of friction and wear properties of polymeric materials for use in total joint prostheses, ASTM F732, American Society for Testing and Materials, Philadelphia (1982).
122. American Society for Testing and Materials, Standard practice for gravimetric wear assessment of prosthetic hip designs in simulator devices, Committee F04 Draft, American Society for Testing and Materials, Philadelphia (1993).
123. American Society for Testing and Materials, Standard practice for gravimetric wear assessment of prosthetic knee designs in simulator devices, Committee F04 Draft, American Society for Testing and Materials, Philadelphia (1993).

58

The Low Wear of Cross-Linked Polyethylene Socket in Total Hip Prostheses

Hironobu Oonishi and Yu Takayama
Osaka-Minami National Hospital, Osaka, Japan

Eiji Tsuji
Osaka Prefectural Industrial Engineering Research Institute, Osaka, Japan

I. INTRODUCTION

In 1950, McKee and Wastson-Farran [5] introduced a total hip replacement that used a cobalt alloy metal cup and femoral head. However, progressive loosening inevitably gave poor results. In 1960, Charnley's [2] ideas on joint replacement were colored by studying the natural joint in which extremely low friction was the most striking feature. He chose the material with the lowest coefficient of friction known, polytetrafluoroethylene (PTFE) for the acetabular socket and a metal head 22 millimeters (mm) in diameter for the femur [2,16].

After a little over a year, it became evident that PTFE was wearing much too quickly. A change was made to glass fiber-reinforced PTFE that performed well in the laboratory, but failed because of chemical reactions and scratching by the metal head in the body. At the same time, a dark-horse material was being tested in the laboratory that hardly wore at all compared with PTFE This material was high density polyethylene, which was switched to in November 1962, and which, with various improvements in molecular weight, has been used to the present time.

Polyethylene is a polymer of ethylene CH_2. It can be made in many forms according to the molecular weight and degree of crystallinity, consequently with different mechanical properties. Ultra-high molecular weight polyethylene (UHMWPE) is defined as having an average molecular weight greater than 1.75 million. Margolies (1974) found that all mechanical properties improved with molecular weight up to this value and then remained fairly constant. However, because there is a considerable variation in molecular weight within a given sample, an average molecular weight above 2.25 million is preferred. The RCH 1000 (Ruhr Chemie) versions of UHMWPE are quoted as having a molecular weight of 2 to 5 million.

After the Charnley prosthesis, many surgeons have tried several different prosthesis designs with a metal or alumina head and UHMWPE socket in different combinations. In 1970, to increase the wear resistance of UHMWPE, we made wear tests of UHMWPE irradiated with several high doses of gamma radiation emitted by Co^{60}. As a result, we found that wear, including creep deformation, was smallest at 10^8 rad (100 megarad [Mrad]), although there was an increase in the coefficient of friction. We began to use UHMWPE irradiated with 10^8 rad of gamma radiation clinically in 1971 [6–15].

We first used the Mueller-type total hip prosthesis with a femoral head of 32-mm diameter in 1968. The design and size of the original Mueller prosthesis did not anatomically fit the bone of Japanese women with dysplasia acetabuli. Therefore, together with Shikita [14,15] and in cooperation with the Mizuho Medical Instruments Company, we developed hip prostheses of various sizes to fit the physique of Japanese women with congenital dislocations of the hip and dysplasia acetabuli.

Heads of 28 mm and 32 mm in diameter, several sizes of stems, and three different lengths of necks as in the Mueller prosthesis were prepared for these prostheses using a COP alloy—stainless steel containing 20% cobalt. High-density polyethylene (HDPE) (Million, followed later by UHMWPE) irradiated with 10^8 rad of gamma radiation was used for the socket. The main outer diameters of the sockets were 42, 46, and 50 mm. The prosthesis was named "SOM" after the initials of Shikita, Oonishi, and Mizuho (Fig. 1). We have been using SOM prostheses clinically since 1971. Alumina ceramics were later used by Boutin [1] and Griss et al. [4] in Europe.

We also experimentally confirmed that UHMWPE showed less wear in alumina-to-UHMWPE types than in metal-to-UHMWPE types and, together with Shikita and in cooperation with the Kyocera Corporation, we developed a hip prosthesis with an UHMWPE socket and an alumina head of 28 mm in diameter; the COP alloy was again used for the stem, there were three types of necks, and a dozen or more different stem sizes, thus making the prosthesis applicable to all patients. Nonirradiated UHMWPE was used for the socket; however, irradiated sockets (irradiated Bioceram) were used in nine cases.

The prosthesis was named the Bioceram hip prosthesis. We have used this prosthesis clinically since 1977. We also concurrently used the T-28 (Co-Cr-Mo alloy head with UHMWPE socket) from Zimmer between 1975 and 1981.

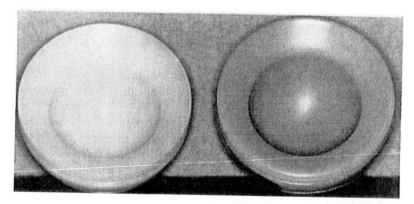

Figure 1   Polyethylene socket of total hip prosthesis: (a), nonirradiated; (b), 100-megarad (Mrad) irradiated.

II. THE EXPERIMENT

A. Materials

For the experiment [14,15], we examined comparatively friction and wear of polyethylene irradiated in several doses sliding against metal and alumina. An Igaki-Shikita (IS) sliding wear test machine cylinder-on-flat was used as a screening test of the friction and wear of irradiated and nonirradiated polyethylene sliding against metal and against alumina. A test strip of polyethylene (50 × 10-mm rectangle, 2-mm thickness, and 1.0-micrometers [μm] Rmax) was rubbed on an end surface of a friction cylinder. The friction, the frictional coefficient, temperature on the faction surface, and reduction in thickness, including creep deformation, were continuously and automatically recorded, mainly by repeated sliding under the same conditions. The reduction in thickness was determined by a drop in the readings of a differential transformer.

The frictional force was measured through a strain gauge. The mean temperature on the friction surface was determined through copper-constanstan thermocouples. This test apparatus can be used under conditions in which lubricants are applied.

Contact pressures of 49, 98, and/or 164 kilograms per square centimeter (kg/cm^2) were applied to the samples. The geometry of the contact area was measured before testing. Contact pressures of 140 to 160 kg/cm^2 are supposed to be representative of the hip joints of Japanese women.

Hizex Million and Million 340M were used as the polyethylene samples. They were irradiated with a gamma ray dose of 3 × 10^5 to 1 × 10^8 rad. Incidentally, a sterilization dose for implant materials is 4 × 10^6 to 6 × 10^6 rad. The gamma radiation was emitted by Co^{60} in air. The transformed surface was cut off about 3 mm below the surface. The friction cylinders were made of stainless steel SUS 304 (Austenite) and alumina. The surface roughness was 0.15-μm Ra on the stainless cylinder and from 0.05 to 0.91 μm Ra on the alumina. Experiments were carried out in some test pieces with physiological saline lubrication and in all test pieces without lubrication.

B. Results

1. Wear Test of Ultra-High Molecular Weight Polyethylene Irradiated by Gamma Radiation Against Stainless Steel

Figure 2 shows the thickness reduction curves for Hizex Million irradiated by gamma radiation without lubrication. The reduction in thickness differed with respect to the different doses of gamma irradiation, the polyethylene received, and the three different contact pressures. It was found that the reduction in thickness was less for this material than for other samples when compared at a contact pressure of 164 kg/cm^2 and a wear time of 100 minutes, irradiated at 1 × 10^8, 5 × 10^7, and 5 × 10^6 rad. Furthermore, from the shape of the curves, it is assumed that the rate of change in the reduction of thickness (reduction in thickness per unit of friction distance) will be less and that, even after longer times, the reduction in thickness will be less than with the other materials.

Under contact pressures of 98 and 49 kg/cm^2, samples exposed to 5 × 10^6 rad showed the lowest reduction in thickness and the lowest change rates. Figure 3 shows the relation among irradiation dose, the frictional coefficient, and the reduction in thickness at a wear time of 100 minutes. Lowered reduction in thickness was shown at doses of 10^5–10^6 rad, indicating improved resistance to wear. However, there was an increase at 1 × 10^7–2 × 10^7 rad and a significant deterioration at 1 × 10^8 rad.

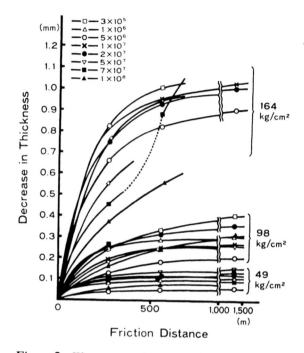

Figure 2   Wear test results. Thickness reduction curves for Hizex Million, irradiated with gamma-radiation, sliding against stainless steel without lubrication.

The coefficient of friction decreased with increased contact pressure in all samples. Under low contact pressure, the coefficient was enhanced slowly from 10^5 to 10^6 rad, and rapidly for values greater than 10^7 rad. The coefficient of friction of all samples tended to decrease 20 to 30 minutes after the onset of testing.

The wear test of gamma-irradiated Million 340M was also performed with and with-

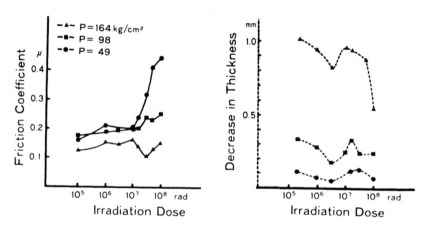

Figure 3   The relation among the coefficient of friction, the decrease in thickness, and irradiation doses to polyethylene.

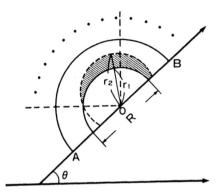

Figure 4   Measurement of the decrease in thickness of the socket on x-ray figure (O = center of wire; R = diameter of head; A, B = outer edge of socket; θ = socket angle; decrease in thickness = $r_2 - r_1$).

out lubrication under a contact pressure of 98 kg/cm^2. Without lubrication, the results were very similar to those above. With lubrication, however, it was difficult to find differences at different doses because the thickness reduction was different even at the same dose. The amount of wear decreased by 1/3 to 9/10. Explanations of these phenomena are very difficult.

2.   Wear Test of Gamma-Irradiated Ultra-High Molecular Weight Polyethylene Against Alumina

The wear test of gamma-irradiated UHMWPE against alumina was performed without lubrication under a contact pressure of 98 kg/cm^2 between an alumina friction cylinder (0.05-μm Ra) and Million 340M, irradiated by gamma rays, in order to observe the effects of various radiation doses. Gamma radiations of 0, 10^3, and 10^6 rad were applied to the polyethylene.

The thickness reduction curves were very similar, with different doses having little effect except for the case of irradiation with 10^3 rad. The coefficient of friction followed the same course, in all cases having a value of about 0.2.

The wear test of stainless steel with almost the same surface roughness as alumina was performed under the same conditions as above. UHMWPE irradiated with 10^6 rad showed almost the same wear, coefficient of friction, and friction surface temperature curves against stainless steel as against alumina.

III.   MEASUREMENT OF THE DECREASE IN THICKNESS OF THE SOCKET

We measured decreases in the thickness of the socket (i.e., wear including creep deformation) over 7 years using a method that we devised [6–15]. Cases for which the socket or stem loosened and the component shifted within 7 years, those not having well-defined radiographic reference points, and those having metal-backed sockets were excluded from these measurements. Measurements were made of 19 SOM, 71 Bioceram, 15 T-28, and 9 irradiated Bioceram joints.

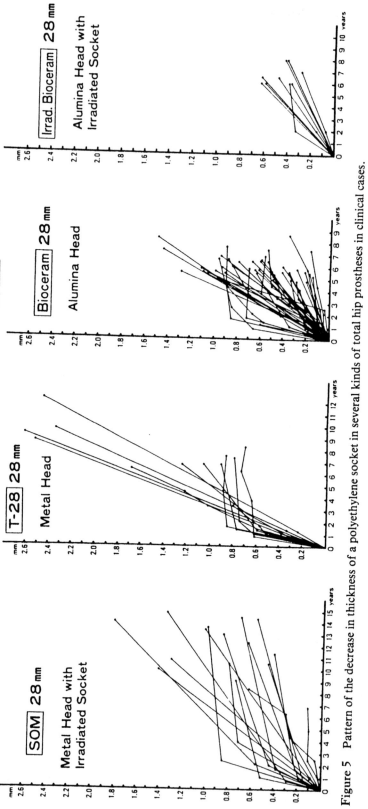

Figure 5   Pattern of the decrease in thickness of a polyethylene socket in several kinds of total hip prostheses in clinical cases.

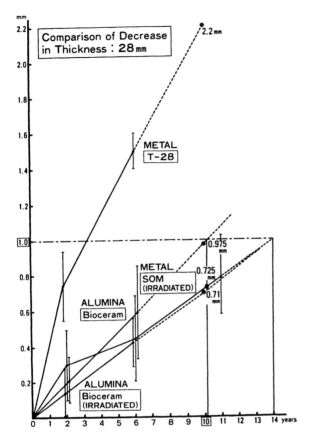

Figure 6   The average decrease in thickness of a polyethylene socket in several kinds of total hip prostheses in clinical cases.

The measurement instruments used were back-light-type digitizers with 1/50-mm resolution, 5 magnifications, a 0.2-mm graduation angle scale that we developed, and small computers. The standard deviation of the values was below 0.05 mm when the measurements were repeated eight times.

The center of the inlet was first determined, and then the wear, measured as a decrease in socket thickness, was obtained by measuring the distance of transfer of the femoral head from the center. The distance r from the center O of A and B at the inlet to the femoral head margin was measured at 10° intervals, the distance r_1 was measured one month after surgery, and the distance r_2 was measured at the time of postoperative observation. The magnification rate due to radiography was corrected, the difference $w = r_2 - r_1$ between the distance at the time of postoperative observation and the distance one month after surgery was taken as the decrease in socket thickness (Fig. 4).

IV.   RESULTS: WEAR PATTERNS

In this section, the word *wear* means the decrease of the socket (i.e., wear including creep deformation). Figure 5 graphically shows the progressive decrease in thickness of SOM, Bioceram, T-28, and irradiated Bioceram sockets.

A. Total Wear

Total wear, that is, the sums of initial and steady-state wear, is clinically important. Figure 6 shows initial wear, the wear during the first two years, total wear for six years, and wear for the longest period measured.

The initial wear in the T-28 was almost double that of the others. Total wear of the SOM, Bioceram, T-28, and irradiated Bioceram were estimated to be 0.73, 0.98, 2.20, and 0.71 mm, respectively, after 10 years.

B. Wear Rate

The wear rate was calculated to determine the decrease in thickness of the sockets each year. The socket thicknesses of the SOM, Bioceram, T-28, and irradiated Bioceram decreased at a steady rate by an average of 0.076, 0.098, 0.247, and 0.072 mm/year, respectively (Fig. 7). A 1.0-mm decrease in socket thickness was estimated to require 13 years for the SOM, 10 years for the Bioceram, 4 years for the T-28, and 14 years for the irradiated Bioceram.

Figure 8 shows x-rays 20 years after surgery. No loosening of the components, no bone necrosis caused by UHMWPE wear particles, and no wear of the socket was recognized on x-ray.

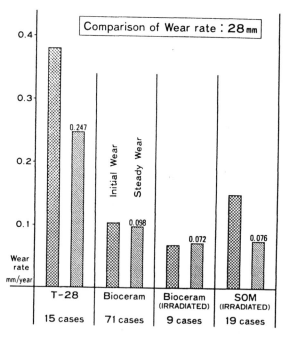

Figure 7   Comparison of the wear rate of a polyethylene socket in several kinds of hip prostheses in clinical cases. In this figure, the word *wear* means the decrease in the thickness of the socket, that is, wear including creep deformation.

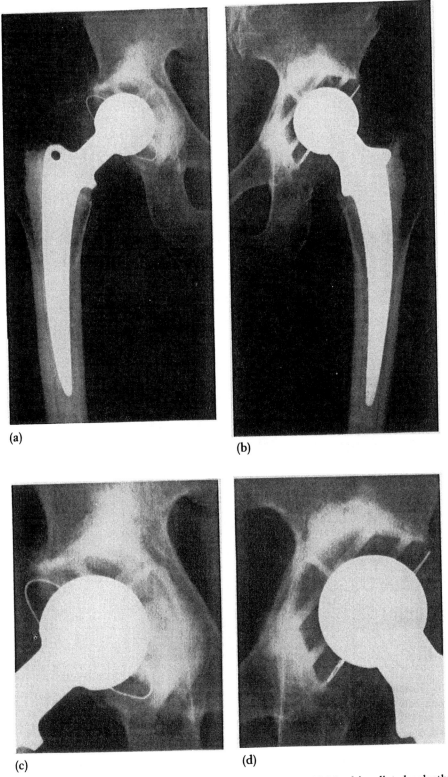

(a)

(b)

(c)

(d)

Figure 8   SOM total hip prostheses combination with a 100-Mrad-irradiated polyethylene socket and a metal head were implanted into the bilateral hip joints. X-ray shows results 20 years after surgery: (a), (c), right hip joint, same patient; (b), (d), left hip joint, same patient.

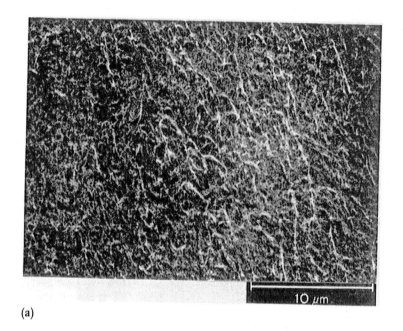

(a)

(b)

Figure 9   A polyethylene socket of a T-28 total hip prosthesis in combination with a metal head and a polyethylene socket without irradiation was observed by scanning electron microscopy (SEM). The socket was removed 10 years after surgery because of the stem sinking: (a), the non-weight-bearing surface; (b), the weight-bearing surface.

V. NAKED EYE OBSERVATION OF THE SURFACE OF THE REMOVED ULTRA-HIGH MOLECULAR WEIGHT POLYETHYLENE SOCKETS

Observations were made by the naked eye of the cases in which loosening and sinking were seen on the stem side only, with no loosening of the socket. In the case of the non-gamma-irradiated polyethylene socket, a smooth and glossy surface was seen on the weight-bearing surface; however, the non-weight-bearing surface was degraded and changed, and was rough with a yellow-brown color.

In the case of the gamma-irradiated polyethylene socket, a smooth and glossy surface was seen on the weight-bearing surface and the non-weight-bearing surface was not degraded or changed as the non-gamma-irradiated polyethylene socket.

VI. SCANNING ELECTRON MICROSCOPY OBSERVATION OF THE WEIGHT-BEARING SURFACE OF REMOVED ULTRA-HIGH MOLECULAR WEIGHT POLYETHYLENE SOCKETS

Scanning electron microscopy (SEM) observations [3,6–13] were made of one T-28 case of six years and one of nine years in which loosening and sinking were seen on the stem side only, and a broad clear zone, with no loosening, was seen covering the socket side. Observations were made of one Bioceram case in which the stem sank and a partial clear zone developed around the socket eight years after replacement.

Observations were made of one case in which a Weber–Stuhmmer-type prosthesis with a 32-mm diameter alumina ball and an UHMWPE socket was used. Creep deformation, loosening of the stem, and a broad clear zone covering the socket developed five years after replacement because the socket was too thin.

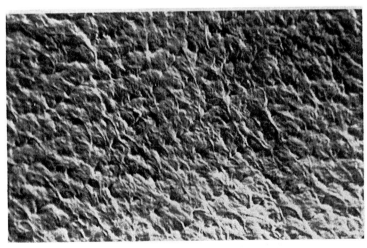

Figure 10   Non-weight-bearing surface of a T-28 polyethylene total hip prosthesis (THP) socket without irradiation; scanning electron microscopy (SEM) observations showed it to be degraded, rough, and a yellow-brown color.

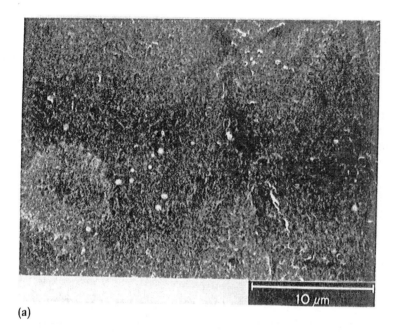

(a)

(b)

Figure 11 The surface of an SOM polyethylene socket with 100-Mrad irradiation was observed by scanning electron microscopy (SEM). The socket was removed 16 years after surgery because of the stem sinking: (a), (b), the non-weight-bearing surface; (c), (d), the weight-bearing surface; (b), higher magnification of (a); (d), higher magnification of (c).

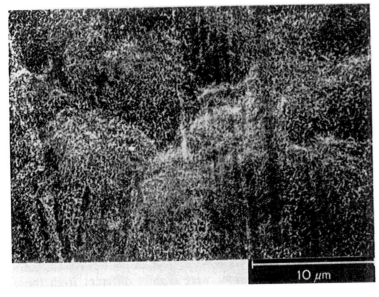

(c)

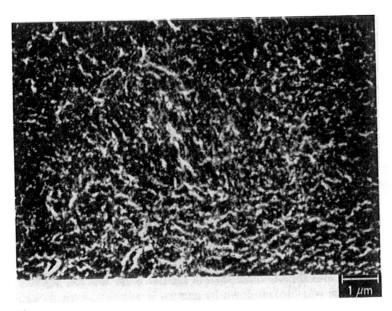

(d)

Figure 11 Continued.

Observations were made of one SOM case of 13 years and one of 14 years in which loosening of the stem and a partial clear zone around the socket were seen, and one case in which the stem broke 14 years after replacement without any clear zone around the socket. As described above, the comparisons were conducted only on cases in which the socket had not loosened.

In the case of the T-28 and Bioceram sockets, the outline of the machine marks on the non-weight-bearing portions was unclear even where they were seen. High magnification observation of the non-weight-bearing portions revealed an irregularly lined pattern of one to several micrometers of smooth ripples.

As Fukubayashi et al. [3] reported, burnishing and small scratches among the burnished areas were seen on the weight-bearing portion in which machine marks had disappeared. However, no pits were observed in our cases. These small scratches were occasionally found together with folding. No scratches were observed on the non-weight-bearing portions. The UHMWPE folding phenomenon was frequently observed (Fig. 9). It was thought to develop from three-body wear resulting from the presence of UHMWPE wear particles between the socket and the femoral head. It is thought that, when the femoral head presses on the tip of the folding UHMWPE, part of the UHMWPE tears away, becoming debris. The authors' observations were slightly different from those reported by Fukubayashi et al. This may be because the authors selected for observation only those cases in which the stem had loosened, but the socket was still firm.

In the case of the Bioceram and the Weber–Stuhmmer-type sockets with alumina head balls, the morphological SEM findings of the sockets were almost the same as those for the T-28. No significant differences were observed.

In the case of non-gamma-irradiated polyethylene, the non-weight-bearing surface degraded and changed, becoming rough with a yellow-brown color observable by the naked eye. Also observed was rough, irregular, scalelike unevenness of about 1 to 2 μm (Fig. 10).

The non-weight-bearing portion in the case of the SOM presented an irregularly lined pattern of smooth and low ripples of about 0.1 μm, which is much smaller than that of the non-gamma-irradiated polyethylene. The weight-bearing portion presented a more clearly outlined pattern, irregularly lined with smooth ripples of about 0.1 μm. Scratches and folding phenomenon, characteristic of non-gamma-irradiated polyethylene, were not observed at all. These findings indicate that there is very little wear for gamma-irradiated polyethylene (Fig. 11).

VII. DISCUSSION

Initial wear in the SOM sockets was slightly more than in the Bioceram sockets, possibly because the polyethylene surface was broken down by gamma irradiation, even though only slightly. A proper dose of gamma radiation produces cross-linking effects under vacuum. However, residual oxygen breaks down the surface slightly since it is impossible to attain a perfect vacuum. It is therefore preferable to remove only the surface layer after gamma irradiation.

Our experimental and clinical studies revealed that the properties and thicknesses of polyethylene used for the socket had considerable influence on the steady static wear of the socket [6–10]. The material, shape (surface roughness), and contour (roundness) on the femoral head side were also important factors in wear. There was considerable difference in wear between the SOM and T-28 prostheses, which differed insignificantly in contour (roundness) and shape (surface roughness) of the femoral head. Thus, we found that not only materials, shape, and contour of the femoral head, but also the properties of the polyethylene socket had a substantial influence on the decrement of the socket thickness. A metal-backed socket and a thick socket of UHMWPE without gamma irradiation can protect against the deformation of the whole shape; however, they cannot protect against creep deformation on a local weight-bearing area. This local creep can weaken the material and increase the likelihood of wear. Around the world, many bioengineers have a misunderstanding about this point.

Wear of the Bioceram socket with the alumina head was lower than the T-28 socket with a metal head. As a result, wear of the socket of the irradiated Bioceram socket (an alumina head and an irradiated socket) was the lowest.

The wear rate of the socket is not always in direct proportion to the rate at which bony tissue surrounding the components or cement becomes necrotic. Experience has indicated that the faster the wear rate of the socket is, the more rapidly the bone necroses. Therefore, socket wear should be as slow as possible.

The alumina head does not cause the deposition of metal elements or corrosion that occurs with metal heads. These results were obtained in our hip simulator tests. Thus, the alumina head was found to be superior to the metal head.

Our cylinder-on-flat study revealed that polyethylene wore the least under gamma irradiation of 1×10^8 rad. The SOM prosthesis socket was irradiated with 10^8 rad. Thus, the experimental and clinical results agreed. From these facts, it was supposed that differences in creep could be more significant than differences in wear.

Our experimental and clinical findings suggest that the best hip prosthesis at present has an alumina head with an UHMWPE socket irradiated with 10^8 rad of gamma radiation. We are continuing to study the optimal dose and the condition of gamma irradiation on sockets. In this study, we will find the relation of a reduction in thickness to the quality and the amount of wear debris generated in test pieces following gamma irradiation of different doses and creep deformation.

REFERENCES

1. Boutin, P., Arthroplastic totale de la hanche par prothése en alumine fritée, *Revue d'Orthopedie*, 58:229–236 (1972).
2. Charnley, J., Surgery of the hip joint, present and future developments, *Br. Med. J.*, 1:821–826 (1960).
3. Fukubayashi, T., and Tomobe, M., Wear of socket observed on the extracted total hip prostheses, *Seikeigeka Mook. Total Hip Prosthesis*, 45:289–298 (1986).
4. Griss, P., Die Aluminiumoxidkeramik-Metall-Verbundsprothese. Eine neue Hüftgelenktotalendo-Prothese zur teilweise zementfreien Implantation, *Archiv für Orthopadische und Unfall-Chirurgie*, 81:259–271 (1975).
5. McKee, G.K., Artificial hip joint, *J. Bone Jt. Surg.*, 33B:465–479 (1951).
6. Oonishi, H., Kotani, T., and Shikita, T., Studies on the shape and contour of the metal prosthetic head in total hip prosthesis, *Proceedings of the 12th SICOT*, pp. 107–123, Excerpta Medica, Amsterdam (1972).
7. Oonishi, H., and Shikita, T., Alumina ceramic total hip prosthesis, *Bessatsu Seikei-Geka No. 3*, pp. 264–279, Nankodo (1983).
8. Oonishi, H., Igaki, H., and Takayama, Y., Comparison of wear of UHMWPE sliding against metal and alumina in total hip prostheses—Wear test and clinical results, *Third World Biomaterials Congress, Transactions*, p. 337, April 21–25, Kyoto, Japan (1988).

9. Oonishi, H., Igaki, H., and Takayama, Y., Wear resistance of gamma-ray irradiated UHM-WPE socket in total hip prostheses — Wear and long term clinical results, *Third World Biomaterials Congress, Transactions*, p. 588, April 21–25, Kyoto, Japan (1988).
10. Oonishi, H., Igaki, H., and Takayama, Y., Wear resistance of gamma-ray irradiated UHMW polyethylene socket in total hip prosthesis — Wear test and long term clinical results, *MRS International Meeting on Advanced Materials*, 1:351–356 (1989).
11. Oonishi, H., Igaki, H., and Takayama, Y., Comparisons of wear of UHMW polyethylene sliding against metal and alumina in total hip prosthesis, in *Bioceramics*, Vol. 1, pp. 272–277, Ishiyaku Euro-American, Tokyo (1989).
12. Oonishi, H., and Tsuji, E., SEM observation on the clinically used gamma-irradiated reinforced HDP socket in total hip replacement, in *Clinical Implant Materials, Advances in Biomaterials*, Vol. 9, pp. 379–384, Elsevier, Amsterdam (1990).
13. Oonishi, H., Takayama, Y., and Tsuji, E., Improvement of polyethylene by irradiation in artificial joints, *Radiat. Phys. Chem.*, 39(6):495–504 (1992).
14. Shikita, T., and Oonishi, H., *Abnutzung und Schmierung der Totalendoprothesen der Huftgelenkes, Der totale Huftgelenkersatz*, pp. 15–28, Georg Thieme Verlog, Stuttgart (1973).
15. Shikita, T., Oonishi, H., Hashimoto, T., Igaki, H., and Sakai, T., Wear resistance of HDP socket in total hip prostheses, *The 1975 Symposium on Biomaterials*, pp. 69–74, Society of Materials Science, Japan, Kyoto University, August 29–30 (1975).
16. Wroblewski, B.M., 15–21 year results of the Charnley low friction arthroplasty, *Clin. Orthop.*, 211:30–35 (1986).

Index

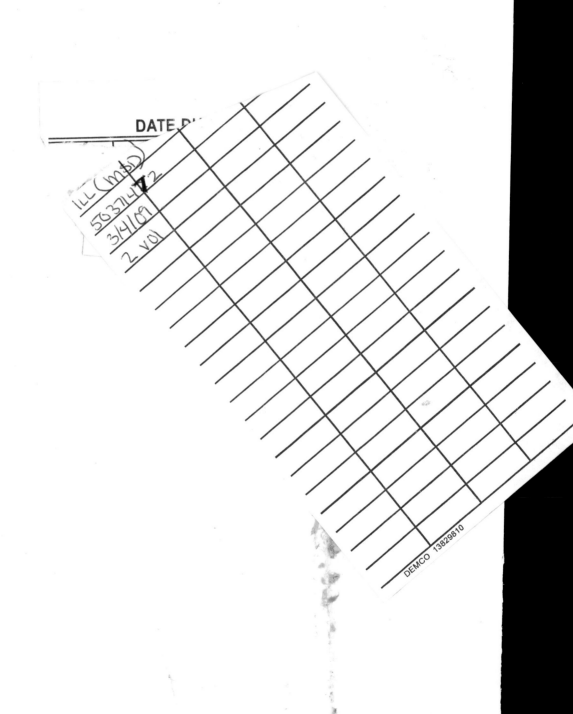

DATE DUE

ILL (IMBD)
56337412
3/4/09
2 vol

DEMCO 13829810